Developing Sustainable Agriculture in Pakistan

Edited by
Iqrar Ahmad Khan
Muhammad Sarwar Khan

CRC Press
Taylor & Francis Group
Boca Raton London New York

CRC Press is an imprint of the
Taylor & Francis Group, an **informa** business

CRC Press
Taylor & Francis Group
6000 Broken Sound Parkway NW, Suite 300
Boca Raton, FL 33487-2742

First issued in paperback 2021

ISBN 13: 978-1-03-209555-4 (pbk)
ISBN 13: 978-0-8153-6653-9 (hbk)

Library of Congress Cataloging-in-Publication Data

Names: Khan, Iqrar A. (Iqrar Ahmad), author. | Khan, Muhammad Sarwar, author.
Title: Developing sustainable agriculture in Pakistan / authors: Iqrar Ahmad Khan and Muhammad Sarwar Khan.
Description: Boca Raton, FL : CRC Press, Taylor & Francis Group, 2018. | Includes bibliographical references and index.
Identifiers: LCCN 2018001401 | ISBN 9780815366539 (hardback : alk. paper)
Subjects: LCSH: Sustainable agriculture--Pakistan.
Classification: LCC S471.P18 K436 2018 | DDC 338.1095491--dc23
LC record available at https://lccn.loc.gov/2018001401

Visit the Taylor & Francis Web site at
http://www.taylorandfrancis.com

and the CRC Press Web site at
http://www.crcpress.com

Publisher's Note
The publisher has gone to great lengths to ensure the quality of this reprint but points out that some imperfections in the original copies may be apparent.

Developing Sustainable Agriculture in Pakistan

Contents

SECTION III Animal Production and Health

SECTION IV Agricultural Incentives for Farmers

Preface

The history of agriculture is the history of humans breeding seeds and animals to produce traits they desire in their crops and livestock. Agriculture plays a pivotal role in the economy and development of Pakistan. It contributes to about 21% of the total GDP and employs nearly 46% of the labor force of Pakistan. However, agriculture production is compromised due to fixed cropping patterns, reliance on a few major crops, narrow genetic pools, and the changing climate. This demands a holistic approach to develop agriculture and to improve the livelihood of the rural populace. This book provides critical analyses of present trends, inadequacies in agriculture, strategic planning, and ways forward to improve programs and policies keeping in view the natural resources, agriculture (crops and animals) production technologies, input supplies, population planning, migration and poverty, and balanced policies on finance, credit, marketing, and trade.

Developing Sustainable Agriculture in Pakistan consists of 38 chapters subdivided into four sections. The flow of chapters in the book is strategically organized to allow for easy reading. It begins with Chapter 1 "Planning for Sustainable Agriculture in Pakistan" in which Drs. Iqrar Ahmad Khan and Muhammad Sarwar Khan comprehensively provide an overview of the latest approaches which could be used to develop sustainable agriculture. Section I (Natural Resources and Input Supplies) begins with Chapter 2, in which Drs. Tahir and Khaliq explain precisely how production factors can be exploited to improve agriculture. In Chapters 3 and 4, Dr. Allah Bukhsh and his team stress upon the necessity of developing water reservoirs to meet ever-increasing requirements for irrigation while exploring the potential of renewable energy sources. In Chapter 5, Drs. Rashid and Nasir discuss environmental degradation and its remedial measures. Dr. Zahir and his colleagues, in Chapter 6, have proposed a number of remedies against factors which cause depletion of soil fertility and decrease productivity. In Chapter 7, Drs. Arshad and Ahmad discuss facts concerning the irrigation system of Pakistan, whereas Dr. Cheema and colleagues, in Chapter 8, describe ICT-based precision agriculture for increased farm productivity and decreased adverse environmental impacts. In Chapters 9–12, Drs. Ali, Khan, Akhtar, Arif, and their colleagues describe the current status and provision of services to farmers for increasing agricultural productivity and improving their livelihood.

Section II (Crop Production Technologies) consists of ten chapters (i.e., Chapters 13 through 22). In Chapter 13, Drs. Ahmad and Husain provide a comprehensive account of climate change and its effects on agriculture. Dr. Murtaza and his colleagues discuss the treatment and management options of low quality waters in Chapter 14, and suggest that changes be brought about in policies for significant improvement and sustainability of crop husbandry. In Chapter 15, Dr. Khaliq and his colleagues discuss production trends, constraints in productivity, and have suggested a way forward for crops of economic significance. In Chapter 16, Dr. Basra and his colleagues stress upon the use of crop rotation and diversification and also suggest introducing new crops in the cropping pattern. In Chapter 17, Dr. Ahmad and his colleagues propose growing high-value horticultural plants including wild-type medicinal plants to improve the pharmaceutical industry. Dr. Siddiqui and his colleagues, in Chapter 18, propose strategies to improve productivity on a sustainable basis while discussing forests and rangeland management issues. In Chapter 19, Drs. Khan and Joyia elaborate on how biotechnology plays a pivotal role in developing GM plants, which are designed to address emerging problems of insects, pests, and diseases under changing climatic conditions. In Chapters 20–22, three teams of authors discuss devastating diseases, insects, and the different uses of methods to combat them.

Section III (Animal Production and Health) is comprised of five chapters (i.e., Chapters 23 through 27). In Chapter 23, Dr. Khan and his colleagues give an insight into problems and solutions related to livestock production, public and private ownership of commodities, and markets. In Chapter 24, Dr. Akhtar and colleagues discuss problems related to the poultry industry. Drs. Javed and Abbas discuss in Chapter 25 how capacity building and integration of new techniques are important for sustainable aquaculture and fisheries in Pakistan. In Chapter 26, Dr. Javed and his colleagues offer a lengthy discussion on the developments and issues related to the livestock and poultry industry and propose suggestions for improvement and value addition to the products of both industries. In Chapter 27, Dr. Sharif and colleagues highlight the problems of malnutrition in children and—considering the severity of the issue—suggest various strategies to alleviate the problem such as school health, nutrition programs, diet diversification, targeted food fortification, nutrition education, and a "one health" approach.

Section IV (Agricultural Incentives for Farmers) consists of eleven chapters. Chapter 28, by Dr. Anjum et al., illustrates the need for gender equality and women's empowerment in different agricultural sectors and classifies the means for improving the economic impact of women's work in agriculture, as well as for enhancing food security and sustenance. In Chapter 29, Dr. Maan and colleagues give a SWOT analysis of the five years population plans and suggest improvements to the quality of schools—through improved curricula and staffing with competent teachers—to reduce the dropout rates of female students. In Chapter 30, Dr. Akhtar and colleagues suggest that reducing the reliance upon foreign debts may result in poverty alleviation in Pakistan. Dr. Farah et al. examine in Chapter 31 the rural–urban migration in Pakistan and suggest effective steps to manage and curb the increasing trend of internal migration. In Chapter 32, Dr. Ali and his team present a critical review of various rural development programs carried out in Pakistan while discussing the highlights of the success story of rural development in South Korea. In Chapter 33, Dr. Ahmad and colleagues describe various challenges confronting the outreach and agricultural extension system in Pakistan. They also refer to social mobilization while emphasizing various opportunities for improvements. In Chapter 34, Dr. Sadaf and her colleagues discuss the future perspectives for Pakistani agricultural price policies in the light of regional and international policies, whereas Drs. Mushtaq and Bashir present the pros and cons of agricultural credit and agricultural cooperatives in Chapter 35 and suggest changes for traditional cooperatives—keeping in view the global economic situation. In Chapter 36, Dr. Ghafoor and colleagues suggest how different initiatives could improve the agricultural marketing system in Pakistan. In Chapter 37, Dr. Ahmad and colleagues, while examining the bilateral trade relations of Pakistan in the region, comment on the fact that value addition in agricultural products is the limiting factor of trade with other countries. In the last chapter, Dr. Khan and colleagues stress upon the need for increasing the production of value-added products through improved supply-chain management, production of innovative nutrient dense foods, and improved storage conditions.

Agriculture is an interdisciplinary endeavor; therefore, it is difficult to cover all aspects of this subject in a single book. The editors of this book are conscious of the fact that there is considerable scope for increasing agricultural productivity by incorporating modern technologies. This is only possible if the farmers have the means necessary and access to credit and free markets. The development of markets where farmers can sell their commodities will directly improve their lives. In this book, we have tried our best to provide a critical overview of the latest trends and future perspectives in agriculture. We hope this book will be a worthwhile resource of up-to-date information for different stakeholders, including policy makers. We also welcome your suggestions, which may help us improve the next edition.

Iqrar Ahmad Khan, PhD
Muhammad Sarwar Khan, PhD

Editors

Iqrar Ahmad Khan has had a long career in education and agriculture and earned his PhD from the University of California, Riverside. He is currently serving as vice chancellor of the University of Agriculture, Faisalabad, Pakistan (since 2008). Dr. Khan has supervised more than 100 graduate students and researchers. Dr. Khan has established a center of agricultural biotechnology and has co-founded a DAAD-sponsored "International Center for Decent Work and Development" (ICDD). He has also helped in establishing a USAID-funded Center of Advanced Studies in Agriculture and Food Security, as well as a French Learning Center and the Chinese Confucius Institute. He has organized numerous international conferences and established academic linkages across continents. Dr. Kahn has also released a potato variety (PARS-70), pioneered research on breeding seedless Kinnow, and discovered new botanical varieties of wheat. Dr. Khan has initiated an internationally acclaimed program to solve the devastating problem of Witches' Broom Disease of lime in Oman. He is currently leading international projects to combat citrus greening disease and mango sudden death. He has published more than 270 articles, five books, and several book chapters.

Dr. Khan has the diplomatic skills to attract international partnerships and establish academic linkages in such countries as Afghanistan, Australia, South Korea, China, Germany, France, Malaysia, Indonesia, Turkey, Iran, India, Oman, Canada, the United Kingdom, and the United States. He has managed collaborative research projects sponsored by national and international agencies. Dr. Khan is a fellow of the Pakistan Academy of Sciences and a member of several professional societies and associations. He has been the recipient of a civil award, *Sitara-e-Imtiaz,* from the government of Pakistan in recognition for his outstanding contributions to the areas of agriculture and food security. Recently, he has also been awarded the *Ordre des Palmes Académiques* (with the grade of Officer) by the French government for his exceptional role as educator.

Muhammad Sarwar Khan has a vibrant career in agriculture, education, and biotechnology and has earned his PhD from the University of Cambridge, UK. The Rockefeller Foundation awarded him a prestigious fellowship under the Rice Biotechnology Program for Developing Countries to carry out research at the Waksman Institute of Microbiology, Rutgers, at the State University of New Jersey. His findings—a research of first-of-its-kind—was published in Nature Biotechnology. Dr. Khan was appointed as national coordinator to train "A" and "FSc" level students by holding training camps across Pakistan to compete for medals in the International Biology Olympiads. He served as the founding head of Biotech Interdisciplinary Division at NIBGE, and is currently serving as the director of the Center of Agricultural Biochemistry and Biotechnology (CABB), University of Agriculture, Faisalabad, Pakistan.

Dr. Khan has supervised more than 100 PhD candidates, MPhil students, and researchers who are now serving at national and international levels in various research institutes and universities. He has vastly published in high impact journals, including *Nature* and *Nature Biotechnology*, and is the author of a number of book chapters and books. Dr. Khan has made significant contributions in the field of agricultural biotechnology. He has developed transgenic sugarcane resistant to top borers and tolerant to herbicides, which was approved by the National Biosafety Committee (NBC) for field trials in 2006–2007. This was the first proposal of endogenously developed GM plants approved by the NBC in Pakistan. Dr. Khan has also pioneered plastid transformation in rice and sugarcane, recalcitrant plant species. He has also knocked out a number of genes from the chloroplast genome of higher plants to assign functions. His current research interests include development of edible-marker-carrying transgenics and cost-effective therapeutics and edible vaccines for animals. Dr. Khan has received prestigious awards, including the President's Medal for Technology, a Gold Medal in Agriculture from the Pakistan Academy of Sciences, a Performance Gold Medal by NIBGE, the Biotechnologist of the Year Award by the National Commission of Biotechnology, and the Best University Teacher Award by the Higher Education Commission of Pakistan. He is also a fellow of the Cambridge Commonwealth society, the Cambridge Philosophical Society, the Rockefeller Foundation, the Pakistan Botanical Society, and the International Association for Plant Biotechnology.

In addition to contributing to innovations in the field of agricultural science, Dr. Khan has served in different senior positions of the Social Safety Net Program of the Government of Pakistan—supported by international donors, including the World Bank—to contribute towards poverty alleviation. During his service there, a number of social protection special initiatives were undertaken to help underprivileged people in Pakistan, especially women for their empowerment.

Contributors

Amjad Abbas
Department of Plant Pathology
University of Agriculture
Faisalabad, Pakistan

Khalid Abbas
Department of Zoology, Wildlife and Fisheries
University of Agriculture
Faisalabad, Pakistan

Qaisar Abbas
Institution of Agricultural and Resource Economics
Faculty of Social Sciences
University of Agriculture
Faisalabad, Pakistan

Arbab Ahmad
Department of Plant Pathology
University of Agriculture
Faisalabad, Pakistan

Ashfaq Ahmad
Department of Agronomy
University of Agriculture
Faisalabad, Pakistan

Burhan Ahmad
Institute of Business Management Sciences
University of Agriculture
Faisalabad, Pakistan

Iftikhar Ahmad
Institute of Horticultural Sciences
University of Agriculture
Faisalabad, Pakistan

Maqshoof Ahmad
Department of Soil Science
University College of Agriculture and Environmental Sciences
Islamia University of Bahawalapur
Bahawalapur, Punjab, Pakistan

Munir Ahmad
Institute of Agriculture Extension and Rural Development
University of Agriculture
Faisalabad, Pakistan

Nazir Ahmad
Faculty of Veterinary Science
University of Agriculture
Faisalabad, Pakistan

Riaz Ahmad
PMIU
Punjab Irrigation Department
Lahore, Pakistan

Saeed Ahmad
Institute of Horticultural Sciences
University of Agriculture
Faisalabad, Pakistan

Shabbir Ahmad
University of Agriculture
Sub Campus Burewala-Vehari
Burewala, Pakistan

Javaid Akhtar
Institute of Soil and Environmental Sciences
University of Agriculture
Faisalabad, Pakistan

Pervez Akhtar
University of Agriculture
Sub Campus Toba Tek Singh
Toba Tek Singh, Pakistan

Saria Akhtar
Department of Rural Sociology
University of Agriculture
Faisalabad, Pakistan

Abid Ali
Department of Entomology
University of Agriculture
Faisalabad, Pakistan

Muhammad Amjad Ali
Department of Plant Pathology
University of Agriculture
Faisalabad, Pakistan

Asghar Ali
Institute of Agricultural and Resource Economics
University of Agriculture
Faisalabad, Pakistan

Safdar Ali
Department of Plant Pathology
University of Agriculture
Faisalabad, Pakistan

Shoukat Ali
Institute of Agricultural Extension and Rural Development
University of Agriculture
Faisalabad, Pakistan

Tanvir Ali
Institute of Agricultural Extension and Rural Development
University of Agriculture
Faisalabad, Pakistan

Waseem Amjad
Department of Energy Systems Engineering
University of Agriculture
Faisalabad, Pakistan

Luqman Amrao
Department of Plant Pathology
University of Agriculture
Faisalabad, Pakistan

Ahmad Din Anjum
Faculty of Veterinary Science
University of Agriculture
Faisalabad, Pakistan

Farkhanda Anjum
Department of Rural Sociology
University of Agriculture
Faisalabad, Pakistan

Raheel Anwar
Infstitute of Horticultural Sciences
University of Agriculture
Faisalabad, Pakistan

Muhammad Jalal Arif
Department of Entomology
University of Agriculture
Faisalabad, Pakistan

Muhammad Arshad
Department of Irrigation and Drainage
University of Agriculture
Faisalabad, Pakistan

Muhammad Imran Arshad
Institute of Microbiology
Faculty of Veterinary Science
University of Agriculture
Faisalabad, Pakistan

Kanwal Asghar
Department of Rural Sociology
University of Agriculture
Faisalabad, Pakistan

Muhammad Ashfaq
Institute of Agricultural and Resource Economics
Faculty of Social Sciences
University of Agriculture
Faisalabad, Pakistan

Ijaz Ashraf
Institute of Agricultural Extension and Rural Development
University of Agriculture
Faisalabad, Pakistan

Bilal Aslam
Faculty of Veterinary Science
University of Agriculture
Faisalabad, Pakistan

Rizwan Aslam
Faculty of Veterinary Science
University of Agriculture
Faisalabad, Pakistan

Muhammad Atiq
Department of Plant Pathology
University of Agriculture
Faisalabad, Pakistan

Hammad Badar
Institute of Business Management Sciences
University of Agriculture
Faisalabad, Pakistan

Allah Bakhsh
Department of Irrigation and Drainage
University of Agriculture
Faisalabad, Pakistan

M. Khalid Bashir
Institute of Agricultural and Resource Economics
Faculty of Social Sciences
University of Agriculture
Faisalabad, Pakistan

Shahzad M. A. Basra
Departments of Agronomy
University of Agriculture
Faisalabad, Pakistan

S. A. Bhatti
Faculty of Animal Husbandry
University of Agriculture
Faisalabad, Pakistan

Masood Sadiq Butt
National Institute of Food Science and Technology
University of Agriculture
Faisalabad, Pakistan

Muhammad Jehanzeb Masud Cheema
USPCAS-AFS
and
Irrigation and Drainage Department
University of Agriculture
Faisalabad, Pakistan

Farah Deeba
Faculty of Veterinary Science
University of Agriculture
Faisalabad, Pakistan

N. Farah
Department of Rural Sociology
University of Agriculture
Faisalabad, Pakistan

Muhammad Farooq
Departments of Agronomy
University of Agriculture
Faisalabad, Pakistan

Umar Farooq
Department of Poultry Science
University of Agriculture
Sub Campus Toba Tek Singh
Toba Tek Singh, Pakistan

Abdul Ghafoor
Institute of Soil and Environmental Sciences
and
Institute of Business Management Sciences
University of Agriculture
Faisalabad, Pakistan

Muhammad Dildar Gogi
Department of Entomology
University of Agriculture
Faisalabad, Pakistan

Amer Habib
Department of Plant Pathology
University of Agriculture
Faisalabad, Pakistan

F. Hassan
Faculty of Animal Husbandry
University of Agriculture
Faisalabad, Pakistan

Sarfraz Hassan
Institute of Agricultural and Resource Economics
University of Agriculture
Faisalabad, Pakistan

Khalid Hussain
Department of Agronomy
University of Agriculture
Faisalabad, Pakistan

Maqsood Hussain
Institute of Agricultural and Resource Economics
University of Agriculture
Faisalabad, Pakistan

Muhammad Iftikhtar
Institute of Agricultural Extension and Rural Development
University of Agriculture
Faisalabad, Pakistan

Muhammad Imran
Faculty of Veterinary Science
University of Agriculture
Faisalabad, Pakistan

Zafar Iqbal
Faculty of Veterinary Science
University of Agriculture
Faisalabad, Pakistan

M. Muzammil Jahangir
Institute of Horticultural Sciences
University of Agriculture
Faisalabad, Pakistan

M. Tariq Javed
Department of Pathology
Faculty of Veterinary Science
University of Agriculture
Faisalabad, Pakistan

Muhammad Javed
Department of Zoology, Wildlife and Fisheries
University of Agriculture
Faisalabad, Pakistan

Nazir Javed
Department of Plant Pathology
University of Agriculture
Faisalabad, Pakistan

Faiz Ahmad Joyia
Centre of Agricultural Biochemistry and Biotechnology (CABB)
University of Agriculture
Faisalabad, Pakistan

Muhammad Kashif
Department of Plant Breeding and Genetics
University of Agriculture
Faisalabad, Pakistan

Abdul Khaliq
Department of Agronomy
University of Agriculture
Faisalabad, Pakistan

Tasneem Khaliq
Department of Agronomy
University of Agriculture
Faisalabad, Pakistan

Izhar A. Khan
Department of Rural Sociology
University of Agriculture
Faisalabad, Pakistan

Iqrar Ahmad Khan
Institute of Horticultural Sciences
Faculty of Agriculture
University of Agriculture
Faisalabad, Pakistan

Ghazanfar Ali Khan
Institute of Agricultural Extension and Rural Development
University of Agriculture
Faisalabad, Pakistan

M. Sajjad Khan
Faculty of Animal Husbandry
University of Agriculture
Faisalabad, Pakistan

Moazzam R. Khan
National Institute of Food Science and Technology
University of Agriculture
Faisalabad, Pakistan

Muhammad Sarwar Khan
Centre of Agricultural Biochemistry and Biotechnology
Faculty of Agriculture
University of Agriculture
Faisalabad, Pakistan

Rashad Rasool Khan
Department of Entomology
University of Agriculture
Faisalabad, Pakistan

Rashid A. Khan
Department of Forestry and Range Management
University of Agriculture
Faisalabad, Pakistan

Zahoor H. Khan
Department of Forestry and Range Management
University of Agriculture
Faisalabad, Pakistan

Aisha Khatoon
Faculty of Veterinary Science
University of Agriculture
Faisalabad, Pakistan

Rakhshanda Kousar
Institution of Agricultural and Resource Economics
Faculty of Social Sciences
University of Agriculture
Faisalabad, Pakistan

M. Ahsan Latif
USPCAS-AFS
and
Computer Sciences Department
University of Agriculture
Faisalabad, Pakistan

Laeeq Akbar Lodhi
Faculty of Veterinary Science
University of Agriculture
Faisalabad, Pakistan

A. A. Maan
Department of Rural Sociology
University of Agriculture
Faisalabad, Pakistan

Hafiz Sultan Mahmood
ABEI
NARC-PARC
Islamabad, Pakistan

Shahid Majeed
Department of Entomology
University of Agriculture
Faisalabad, Pakistan

Asif Maqbool
Institute of Business Management Sciences
University of Agriculture
Faisalabad, Pakistan

Muhammad Aamer Maqsood
Institute of Soil and Environmental Sciences
University of Agriculture
Faisalabad, Pakistan

Muhammad Saleem Mohsin
Institute of Agricultural Extension and Rural Development
University of Agriculture
Faisalabad, Pakistan

Faqir Muhammad
Faculty of Veterinary Science
University of Agriculture
Faisalabad, Pakistan

Ghulam Muhammad
Faculty of Veterinary Science
University of Agriculture
Faisalabad, Pakistan

Anjum Munir
Department of Energy Systems Engineering
University of Agriculture
Faisalabad, Pakistan

Ghulam Murtaza
Institute of Soil and Environmental Sciences
University of Agriculture
Faisalabad, Pakistan

Khalid Mushtaq
Institute of Agricultural and Resource Economics
Faculty of Social Sciences
University of Agriculture
Faisalabad, Pakistan

Ghulam Mustafa
Center of Agricultural Biochemistry and Biotechnology
University of Agriculture
Faisalabad, Pakistan

Abdul Nasir
Department of Structures and Environmental Engineering
University of Agriculture
Faisalabad, Pakistan

Ahmad Kamal Nasir
Electrical Engineering Department
LUMS
Lahore, Pakistan

Khalid Naveed
Department of Plant Pathology
University of Agriculture
Faisalabad, Pakistan

Ahmad Nawaz
Department of Entomology
University of Agriculture
Faisalabad, Pakistan

Muhammad F. Nawaz
Department of Forestry and Range Management
University of Agriculture
Faisalabad, Pakistan

Naima Nawaz
Department of Rural Sociology
University of Agriculture
Faisalabad, Pakistan

Zafar Iqbal Qureshi
Faculty of Veterinary Science
University of Agriculture
Faisalabad, Pakistan

Nasir Ahmad Rajput
Department of Plant Pathology
University of Agriculture
Faisalabad, Pakistan

Mahmood Ahmad Randhawa
Department of Continuing Education
University of Agriculture
Faisalabad, Pakistan

Haroon Rashid
Department of Structures and Environmental Engineering
University of Agriculture
Faisalabad, Pakistan

Muhammad Rashid
Institute of Soil and Environmental Sciences
University of Agriculture
Faisalabad, Pakistan

S. H. Raza
Faculty of Animal Husbandry
University of Agriculture
Faisalabad, Pakistan

Abdul Rehman
Department of Plant Pathology
University of Agriculture
Faisalabad, Pakistan

M. S. Rehman
Faculty of Animal Husbandry
University of Agriculture
Faisalabad, Pakistan

Ayesha Riaz
Department of Rural Sociology
University of Agriculture
Faisalabad, Pakistan

Tahira Sadaf
Institution of Agricultural and Resource Economics
Faculty of Social Sciences
University of Agriculture
Faisalabad, Pakistan

Aqeela Saghir
Institute of Agricultural Extension and Rural Development
University of Agriculture
Faisalabad, Pakistan

Muhammad Sohail Sajid
Faculty of Veterinary Science
University of Agriculture
Faisalabad, Pakistan

Sajjad-ur-Rahman
Faculty of Veterinary Science
University of Agriculture
Faisalabad, Pakistan

Muhammad Kashif Saleemi
Department of Pathology
University of Agriculture
Faisalabad, Pakistan

Aysha Sameen
National Institute of Food Science and Technology
University of Agriculture
Faisalabad, Pakistan

Babar Shahbaz
Institute of Agriculture Extension and Rural Development
University of Agriculture
Faisalabad, Pakistan

Muhammad Adnan Shahid
Water Management Research Center
University of Agriculture
Faisalabad, Pakistan

Amir Shakeel
Department of Plant Breeding and Genetics
University of Agriculture
Faisalabad, Pakistan

Mian Kamran Sharif
National Institute of Food Science and Technology
University of Agriculture
Faisalabad, Pakistan

Aamir Shehzad
National Institute of Food Science and Technology
University of Agriculture
Faisalabad, Pakistan

Muhammad T. Siddiqui
Department of Forestry and Range Management
University of Agriculture
Faisalabad, Pakistan

Muhammad Sufyan
Department of Entomology
University of Agriculture
Faisalabad, Pakistan

Muhammad Tahir
Department of Agronomy
University of Agriculture
Faisalabad, Pakistan

Muhammad Usman
Department of Irrigation and Drainage
University of Agriculture
Faisalabad, Pakistan

Abdul Wahid
Departments of Botany
University of Agriculture
Faisalabad, Pakistan

Waqas Wakil
Department of Entomology
University of Agriculture
Faisalabad, Pakistan

Muhammad Yaseen
Institute of Soil and Environmental Sciences
University of Agriculture
Faisalabad, Pakistan

Muhammad Iqbal Zafar
Department of Rural Sociology
University of Agriculture
Faisalabad, Pakistan

Zahir Ahmad Zahir
Institute of Soil and Environmental Sciences
University of Agriculture
Faisalabad, Pakistan

Muhammad Zia-ur-Rehman
Institute of Soil and Environmental Sciences
University of Agriculture
Faisalabad, Pakistan

Khurram Ziaf
Institute of Horticultural Sciences
University of Agriculture
Faisalabad, Pakistan

1 Planning for Sustainable Agriculture in Pakistan

Iqrar Ahmad Khan and Muhammad Sarwar Khan

CONTENTS

1.1 INTRODUCTION

Sustainable development refers to development which meets the needs of the present without compromising the ability of future generations to meet their own needs (WCED, 1987). The report has left strong imprints on future development policies by considering development and environment inseparable, while focusing on intergenerational equity. The UN Millennium Summit in 2000 passed the following 8 Millennium Development Goals (MDGs), including target dates to achieve these:

1. To eradicate extreme poverty and hunger
2. To achieve universal primary education
3. To promote gender equality and empower women
4. To reduce child mortality
5. To improve maternal health
6. To combat HIV/AIDS, malaria, and other diseases
7. To ensure environmental sustainability
8. To develop a global partnership for development

Although all member countries at that time were committed to achieve the goals by 2015, there was mixed progress where some countries were able to achieve all the goals while others (including Pakistan) remained largely off-track.

In 2015, the United Nations adopted the 2030 agenda for sustainable development. As a responsible nation, we are committed to the UN Sustainable Development Goals (SDGs), once more. Of the 17 indicators/goals, at least 12 are directly dependent on sustainable agriculture and its outcome. The National Assembly of Pakistan has adopted the UN SDGs. The Planning Commission of Pakistan and the Provincial Planning and Development Departments are mandated to ensure that all development spending is targeted towards achieving the SDGs by 2030. Hence, agricultural planning and policy formulation must conform with the SDGs.

Achieving the SDGs means transition to higher productivity and strengthening of rural livelihood. This requires conservation of natural resources (the ecosystem) and building resilience towards climate change. Adoption of emerging technologies (biotechnology and site-specific precision agriculture) and decision-support systems offer new solutions to old problems. The desired transition to sustainable agriculture can only happen under a revamped policy and governance structure which can promote public and private investment in this sector.

Agriculture in Pakistan consists of a vast spread of crops, livestock, fisheries, rangelands, and forestry supported by irrigation network and markets. It ensures food security in the country and contributes 19.8% of the GDP, employs 44% of the workforce, and it provides a livelihood to 66% of the population (>5 million households). The industrial output in the country is dependent on the raw materials and consumption capacity of agriculture-led activities. Exports are also largely dependent on agriculture (65% agro-based).

Characteristically, agriculture in Pakistan is dominated by small farmers growing mainly five crops, and a large population (>70 million each of large and small ruminants) of underperforming livestock. The yields of crops and livestock heads are stagnant. Water scarcity has become evident. Despite stagnation, we have an excess of essential commodities and our farmers are losing money. Furthermore, the surplus of commodities has failed to provide nutritional security for the vulnerable, as a large segment of the population is suffering from nutritional deficiency. We have high costs of production, which makes us uncompetitive in export markets. Our current food imports are worth over $5.3 billion. Sustaining agricultural growth remains an important policy and governance challenge, which would, in turn, determine our ability to comply with the UN's SDGs.

Overcoming stagnation requires continuous development and delivery of technology. Achieving economies of scale and value addition are the other options to make agriculture profitable. Among technologies GMOs, precision focused mechanization, and use of ICTs offer current solutions/

applications. The technology development requires long-term commitment (policy) and investment in agricultural research. We have a large infrastructure and diversified human resource competencies to undertake research challenges, provided we set our priorities right. It also requires international networking, linkages, and a liberal knowledge environment to promote critical thinking and enquiry.

The Federal Government announced a 341 billion Pakistan rupees package for the farmers in 2015 before the onset of Rabi season. This had yielded significant results. The Chief Minister of Punjab addressed the Punjab Agriculture Conference held on the 19th of March, 2016, and announced a 100 billion development package for the farming sector. He also declared the creation of an Agriculture Commission which he planned to chair himself. Lately, the World Bank has stepped in with a new $500 million project called SMART Agriculture (Strengthening Markets for Agriculture and Rural Transformation) in Punjab. There are many other provincial and federal programs for the promotion of agriculture in the country deserving critical analyses to promote sustainability.

There are several pertinent points to ponder including aggressive growth agenda, enhancement of rural economy, global trends in commodity prices, stagnant yields, coupling agricultural growth with research and technology transfer, increasing input use efficiency, market connectivity, backward and forward value chain linkages, land records management, and international linkages (Spielman et al., 2016). It could only happen with a strategic plan for sustainable land and water use.

The Punjab government's agriculture commission created a policy committee, which launched a multipronged review and consultation process. A series of consultative meetings were organized at the farmer's level as a bottom up exercise. This book includes contributions based on the messages gathered during this consultation process. Similar exercises are being carried out in the other provinces of Pakistan, led by the FAO. The FAO is also working with the provincial governments to redefine Agro Ecological Zones (AEZs).

While the world prepares to feed 9 billion people by 2050, we may be expected to feed twice the number of people we have today. The review and consultation exercise has revealed there is no shortage of information but a serious lack of implementation. An analysis of the Agriculture Commission report of Sartaj Aziz (1988) indicated that most of the proposals made then are still valid today but failed to produce results due to inconsistent implementation (GOP, 1988). Hence, it is high time we undertake a SWOT (strength, weakness, opportunity and threat) analysis of our agriculture sector and develop a strategic plan to guide public policy for sustainable agriculture. This book is an attempt to define a framework for sustainable agriculture and food systems. In addition to discussing natural resource and technology aspects, it also delves into the larger human development picture (poverty, gender, and malnutrition) in the country where the rural economy is being drained of skilled manpower and capital. There cannot be a sustainable agriculture without rural development.

1.1.1 Sustainable Agriculture

Agriculture has always evolved through complex interactions between weathering processes of geological material (upper surface of earth or soil, alluvial deposits, mixing organic matter), adaptations to climate, domestication of plants and animals, and anthropological phenomenon. Markets and technological revolutions have become driving forces, which include cultural factors. Intensive agriculture has its toll. The deteriorating soil conditions—as well as the environmental implications of technology and human expectations—demand we examine the challenges and forecast the sustainability of our current agricultural production systems.

The world economic community looks at sustainability as a challenge of global competitiveness which includes productivity of our agricultural and industrial outputs and markets. The global trade of agricultural produce directly affects the income of our farm households, hence their human development index. Our standing on various sustainability indictors is currently dismal (Table 1.1).

High population growth and intensive farming has exerted tremendous pressure on land and other natural resources. Injudicious use of chemical fertilizers and pesticides/herbicides has aggravated the

TABLE 1.1
Important Indices for Pakistan Related to Sustainability

Index	Rank	Source and Data
Global innovation index	113	127 countries data
Global competitiveness index	122	138 countries surveyed (WEF, 2016)
Gender gap index	143	144 countries surveyed (WEF, 2016)
Human capital index	118	130 countries surveyed (WEF, 2016)
Human development index	147	188 countries/territories surveyed (UNDP, 2015)
Enabling trade index	122	136 countries surveyed (WEF, 2016)
Networked readiness index	110	130 countries surveyed (WEF, 2016)

problem and has degraded land and aquifers. The sustainability of natural resources is essential to sustain agricultural activities. Foreseeing similar challenges, Rothamsted (https://www.rothamsted.ac.uk/long-term-experiments) was established in 1843 to investigate the sustainability of agricultural production systems. The long-term experiments conducted there have revealed trends and dangers associated with different agronomic systems of farming. Since then, more than 14 long-term research experiments have been conducted in other parts of the world (Table 1.2), but nothing in this subcontinent. We have been practicing cropping systems and patterns which are bound to decline, that is, wheat-cotton and wheat-rice. Another example is the potato/corn belt, where an extremely exhaustive cropping system is used, lacking a restoration process. Groundwater exploitation has a limit, which is being ignored. Overgrazing in the range lands have created space for intrusive growth

TABLE 1.2
Long-Term Research Experiments (LTRE) in the World

LTRE Name	Year Started	Focus	Site/Location
Russel Ranch	1990	Wheat-tomato	UC Davis
The Morrow Plots	1876	Continuous corn cultivation	Urbana Campaign University of Illinois
Sanborn Field	1888	Rotation Field	Columbia University of Missouri
Magruder Plots	1892	Winter wheat	Stillwater, Oklahoma State University
Callars Rotation	1911	Cotton	Auburn University, Alabama
Permanent Topdressing Experiment	1912	Superphosphate	Rutherglen Center, Victoria, Australia
Crop Residue Management	1931	Crop residue management	Pendleton Oregon State University
LTR-KSU	1961	N, P, and K in irrigated continuous corn	Tribune Kansas State University
No Tillage Plots	1962	No tillage with corn Soybean Oats	Wooster Ohio State University
Belvin Long Term Tillage Trial	1970	No-tillage and moldboard plowing compared	University of Kentucky
INTA Experiment Station	1975	No-till soybean following wheat	Marcos Jaures Argentina
Long Term Ecological Research Plot		Corn-soybean-wheat-rotation with different cropping system	Michigan State University

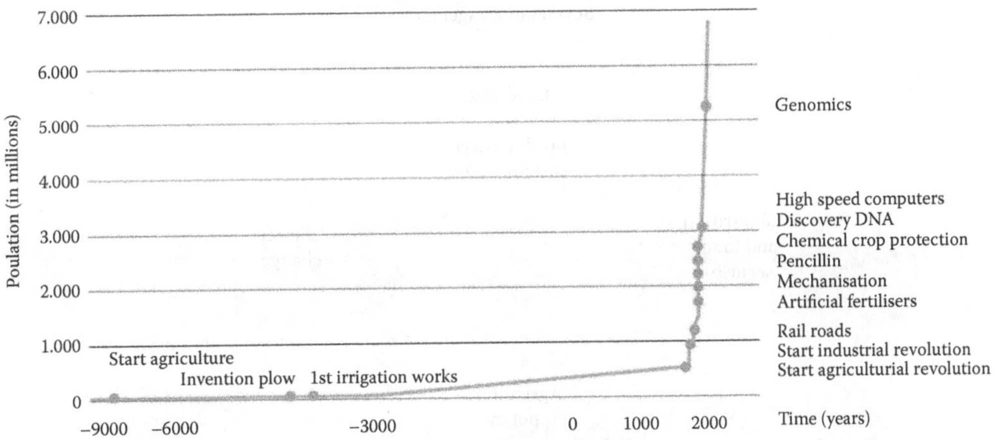

FIGURE 1.1 Innovations and agricultural revolutions in human history.

of weeds, leaving nothing to graze while long-term planning and experimentation are nonexistent. Lack of profitability is also a persistent threat to the sustainability of agriculture systems, forcing migration and brain drain from rural economies. The key to sustainable agriculture lies in focusing on productivity without letting the natural resources decline.

1.1.2 Global Competitiveness and Innovation

It is resolved that innovation is the key to achieving global competitiveness through enhancement of productivity and is a way forward to meeting the SDGs without compromising environmental integrity. At UAF, we wrote Vision 2030 in 2014 where opportunities in agriculture were described as our targets for 2030, a year before the SDGs of 2015 (Khan, 2014). In the year 2016, UAF launched the first innovation catalogue (Khan et al., 2016). Because, we knew that innovations have been the instruments of success during human history.

Innovations are at the heart of sustainable development. Humanity has progressed in overcoming hunger and premature death due to innovations in agriculture and other sectors (advancement in plows, irrigation, fertilizers, green revolution, cotton ginning, vaccines, mechanization, genomics, etc.). Fogel (2004) has developed a history map of innovations in agriculture followed by the industrial revolution (Figure 1.1).

Based upon the review of different indicators and relative positions, sustainable development is only possible in the presence of the right institutions and legal frameworks for incentivizing agricultural innovation through promotion of human capital (required skills, education), agricultural entrepreneurship, infrastructure (research, physical infrastructure), and a mechanism for diffusion of the agricultural innovations along the agricultural value chain. The policies to transform the agriculture sector into an innovation-driven sector (with reliance on smart technologies) can help achieve the goal of sustainable food production and distribution (for an ever-increasing population) and provide foreign exchange (to boost other sectors).

1.2 AGRICULTURAL ISSUES AND ANALYSIS

There can be many ways to look at sustainability challenges. We have undertaken a SWOT (Strength, Weakness, Opportunity and Threat) analysis to narrow down issues facing the sustainability and competitiveness of agriculture in the country. A seven-point agenda has emerged (Figure 1.2). Strategic planning and policy measures would emerge to guide the public decision-making process. At the end of the day, a framework for indigenous solutions through investment in research and development, skill development, and outreach is being envisaged.

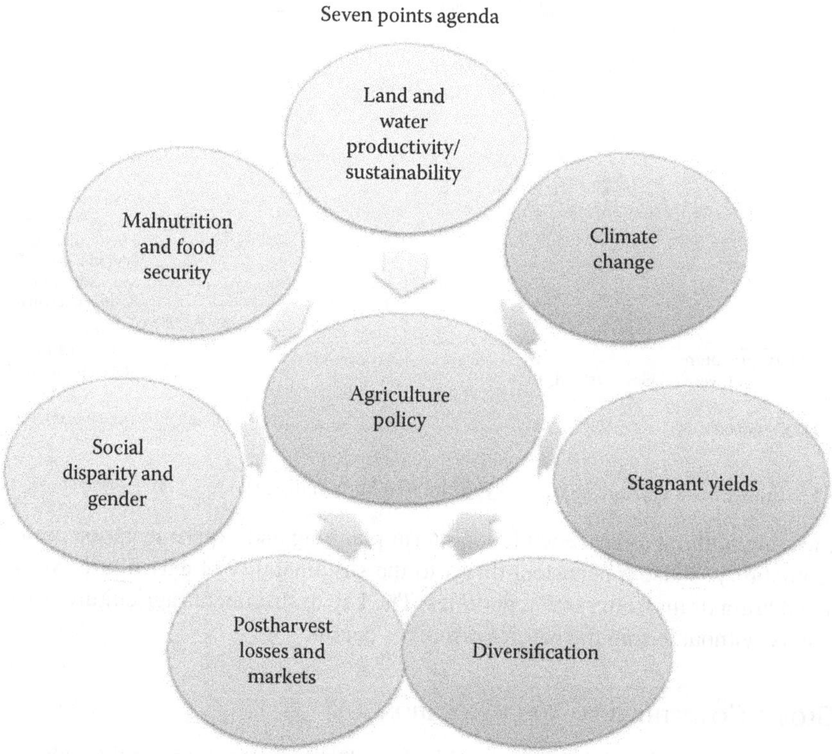

FIGURE 1.2 Schematic layout of important issues in agriculture.

1.2.1 Land and Water Productivity/Sustainability

1.2.1.1 Land and Water Use Policy

It is our major strength that the diversity of topographic features, land textures, climatic variations, technologies, and markets has created a range of agro-ecological divisions/zones. We are also endowed with human resources and indigenous knowledge suited for making agriculture a profitable business, which provides us an opportunity to grow a full range of commodities. The weakness includes subsistence-oriented farming practices, uneven distribution of ownership rights, and fragmentation, which all obstruct optimal land and water use. Land resources are also faced with degradation due to salinity, water logging, soil erosion, soil compaction, desertification, urbanization, and infrastructure projects (the threats) (Khan et al., 2011). High cropping intensity and use of unfit subsoil water are affecting soil health and organic matter content, and therefore sustainability of the system. At present, we have four persons per acre (50 million acres for >200 million people) to feed and clothe, which could increase to 8 per acre by the year 2050. Migration and urbanization have consequences for land use. While we do have more culturable land, we do not have extra water to expand agricultural activity without reducing our current water use.

The productivity of agricultural lands can be enhanced by precise interventions at the micro zones and commodity clusters (at agro-ecological) level by introducing soil and water analysis and plant residue management. The climate change necessitates that we revisit our traditional definition of agro-ecological zones.

There is a need for schemes for wetlands, rainwater harvesting, flood canals, river dredging, river lakes/locks, canal water storage, on-farm storage, and ground water recharge wells. Restriction on groundwater pumping will have to be imposed, sooner or later. Irrigation water should be priced (according to depth of water table) and cultivation of low delta crops should be incentivized for

TABLE 1.3
Current Water Consumption by Five Major Crops

Crop	Water Consumption (MAF)
Wheat	39
Cotton	29
Rice	26
Sugarcane	23
Maize	5

restricting area under rice and sugarcane. It is important to understand the crop share of irrigation water (Table 1.3).

While rice and sugarcane are high delta crops (not to be defended), we use more water for wheat and cotton due to large acreages. Any water savings from wheat and cotton would contribute more in quantities saved. Our average irrigated wheat acreage takes five irrigations per season. This could be reduced to three with better genetics and precision planning, which would translate into saving more water than the entire storage in Tarbela and Mangla dams put together. High Efficiency Irrigation Systems HEIS programs should be reviewed and rewritten after an independent monitoring of the sites developed during the past 5 years. In Punjab, the amount of wastewater disposed after treatment is only 22.11 million cubic feet per day out of 552.23 million cubic feet produced. The lack of watershed management strategies, deforestation, and erosion are sending excessive silt into the rivers, which are silting up our storage dams. The country should have land and water use policies to address these sustainability challenges.

1.2.1.2 Water Governance and Political Challenges

Pakistan's agricultural growth is closely linked with availability of surface water. The surface water supply is stagnant/declining because of the inability to build new water reservoirs and the silting up of existing storage. Moreover, the industrial and municipality water demand is surging, thereby further cutting into the surface water available for the agriculture sector. This leaves the country with the only option to increase water productivity through the use of High Efficiency Irrigation Systems (HEIS), improved irrigation practices and to shift to low delta crops through diversification and promote the development of water efficient crop varieties. Additionally, canals and water courses should be lined, and water should be priced for its rational use. Currently, five crops are major consumers of water and there is an urgent need to rationalize water use during their production.

Due to the concentration of rainfall and glacial melt in river water during the summer months, Pakistan's irrigated agriculture faces water shortages both in time and space. The water shortage is compensated by 50MAF pumped up from the groundwater to meet 104 MAF requirements at the farm gate (Table 1.4). Pakistan's groundwater economy is currently under threat because of discharge rates consistently exceeding recharge rates, and due to climate change related rainfall frequency and intensity uncertainties. Groundwater extraction occurring in the Indian Punjab is also a cause of concern, which was not a factor at the time of signing of Indus Water Treaty.

Table 1.4 shows high water losses at all stages, right from the origin of river flow to the Arabian sea. In order to maintain sustainable use of water resources, losses should be minimized and net groundwater abstraction should be near to neutrality in the long run. One of the reasons for a chaotic water economy is the lack of volumetric pricing mechanism, absence of groundwater rights, and skewed entitlements. In absence of any legal rights, the fuel prices are mistaken as the balancing force for groundwater abstraction. The cheaper energy sources (solar, biogas) and fuel subsidies will further hurt the groundwater economy. Currently, there is lack of proper data about the aquifers and their boundaries. Aquifers should be mapped so that the confined and unconfined aquifers transcending administrative boundaries can be protected through legal rights and water policies.

TABLE 1.4
The Water Budget of the Indus Basin Irrigation System (Values in MAF)

A. Mean annual rivers flow	140
B. Flow to Arabian sea	27
C. River system losses	10
D. Canal losses	26
E. Water course losses	23
F. Water losses till farm gate (B + C + D + E)	49
G. Canal supplies at farm gate (A – F)	*54*
H. Groundwater contribution	*50*
I. Irrigation water at farm gate (G + H)	*104*
J. Field channel losses	10
K. Field application losses	24
L. Total field losses (J + K)	*34*
M. Irrigation water for consumptive use (I – L)	70
N. Rainfall contribution	13
O. Total water availability for crop consumptive use (M + J)	*83*

Source: Adapted from Ahmad, S. and Majeed, R. 2001. *Journal of Engineering and Applied Sciences.*

The irrigation system in the country was an outstanding gravity-driven design at the time of its creation, which was developed as a supply model, that is, "*warabandi*" arrangement. The canal water flows into the farm at fixed time slots per acre on a weekly basis without any need assessment and storage arrangements at the farm level (receiving end). The intensification of cropping systems further increased the demand for water. However, the corresponding investments in water storage (both at source and farm levels) were not made, resulting in groundwater overabstraction. The political realities led to the freezing of water charges and deregulation of groundwater abstraction. Water thefts and distorted allocations are common. There is a clear need to revisit water laws and regulatory mechanisms at the farm level (Cheema, 2012).

Water resource management is further complicated by unresolved transboundary issues with India (Indus Water Treaty) and emerging challenges (no treaty) from Afghanistan. Pakistan, being a lower riparian country, has disadvantages. There also are interprovincial water disputes on the provincial shares, storages and allowable flow in the Indus river down Koteri Barrage, the last diversion before draining (~35 MAF) in to the Arabian Sea. It is alarming to note that we have only a 30 days storage capacity against a 900 and 90 days capacity in the United States and India, respectively. Within provinces, there are significant disparities in irrigation water allocations between and within different canal divisions.

There is a case for developing water stewardship to be socially equitable, environmentally sustainable, and economically beneficial. There could be a value chain approach and/or a community/participatory approach. Water education can play a significant role in conservation and prevention of water pollution. Since the monsoon season is a narrow time bracket, rainwater harvesting is a low hanging fruit to prevent floods and to enhance water availability during droughts. Promotion of on farm storage and separation of rainwater flow from sewerage drains are important challenges. This analysis indicates that treating water is a central issue to sustainability and is a community challenge.

The water-energy-food nexus is strongly linked with climate change. Glacier melt contribute >70% of our freshwater supply. Accelerated melting of snow, due to global warming, may cause more floods and enhanced river flow in the near future as well as drying up of rivers in the coming decades.

1.2.2 Climate Change

Climate change is a continuous process which has created our warm world since the ice ages. The process has been accelerated by man-made interventions. Revolutionary measures are needed to arrest/mitigate these trends and to adapt to the changes. Pakistan is situated in a region regarded as highly vulnerable to adverse impacts of climate change, and this calls for disaster preparedness. Erratic rains, floods, and melting of glaciers in the Himalayan ranges pose threat to our perennial supply of fresh water.

There should be a specific focus on developing monitoring techniques to detect the effects of climate change with an emphasis on productivity (crops, livestock, fishery, and forestry). A decision support mechanism must evolve through prediction models. Education and extension frameworks need to be revamped to disseminate knowledge to communities regarding adapting/mitigating impacts of climate change and preparedness for disaster management. There is a need and opportunity to develop genetics for heat, drought tolerance, and agronomic interventions. Wheat, cotton, and rural poultry should be our near future targets to prepare against the vagaries of climate change.

Small ruminants and camels are uniquely adapted to extreme weather conditions. Our cattle and buffalo breeds are pastoral animals, which are genetically robust and resilient to inclement weather. However, lack of systematic breeding programs has caused genetic deterioration of our milk animals, which deserves urgent attention. We are producing less milk from 70 million heads than 9 million heads in The United States.

The methane produced by livestock and poultry should be converted into valuable options of bioenergy and biofertilizers, while providing an opportunity to claim carbon credits. Puddled rice is a double jeopardy. It is water intensive (inefficient) while producing unwanted methane. The new rice production agronomy is emerging fast and combines direct sowing and AWD (alternate wet and drying) irrigation schemes. It will not only save water but also optimize the plant population, which is the major cause of low productivity of puddled rice.

Breeding efforts to develop heat tolerant germplasm of crops and exploitation of heat adapted animal genetics are making significant progress. Genetic engineering opportunities are also emerging fast. CRISPR/Cas9 is a current technology to edit/tailor the genomes for adaptations and for incorporation of superior characteristics.

An indigenous chicken breed has been developed with 30% less feather load, demonstrating has better heat adaptation as backyard poultry. The naked neck chicken, when crossed with feathered chickens, produces a progeny that has a middle phenotype. The breeding process has been further refined to develop a superior poultry strain for harsh rural environments.

Agro Ecological Zoning (AEZ) was carried out in the 1980s' when most of the present day analytical tools and data were unavailable. Today, we have better access to software and data gathering devices (GIS). It is high time AEZs are redefined to develop decision support systems and to enable precision applications. This will also help the government in policy formulation and long-term strategic planning.

1.2.3 Productivity Gaps and Stagnant Yields

1.2.3.1 Productivity Gap in Crop Sector

Before the advent of settled agriculture, during the times of the hunter-gatherers, the optimum yield was based on the criteria of how much energy is collected per unit of energy consumed. However, with the shift to settled agriculture, the definition of yield became the ratio between number of grains sown and harvested, or as an input/output ratio. As settled agriculture faced land, water, and nutrient scarcity, the yield became defined in terms of spatial boundaries, plant genetic potential, agronomic practices, technology, nutrients, water, and climatic and other agro-ecological parameters.

The agriculture sector of Pakistan is facing severe stagnation in productivity and declining growth. The farmers are leaving their profession. Yield gaps for wheat, rice, maize, cotton, and sugarcane between progressive growers and the national average stand at 43.5%, 45.6%, 58.55%, 30.85% and 61.6%, respectively. Similarly, there are enormous yield gaps when compared with other countries and regions of the world. Major reasons for this difference are unavailability of quality seed, inappropriate sowing (methods and time), weeds, lack of balanced fertilizers, partial mechanization, and excessive use of unfit irrigation water. There are many major pest and disease challenges currently restricting our output (Figure 1.3).

There are different methods to calculate yields depending on which definition of yield is under consideration, namely, theoretical yields, potential yields, water limited yields, attainable yields, and actual yields (FAO, 2015). The yield gaps can be further broken into research and science gaps when the national average yield is compared with research station and global average yields, respectively and extension gaps when actual yields are compared with progressive farmers' yield (Iqbal and Ahmad, 2005).

However, these measures have different sets of data requirement and calculation of yield gaps is always a compromise between the level of analysis and the availability of data for yield and related inputs.

The stagnation is partly due to small farmer's inability to invest and adopt technology. That means >60% of cultivated land is underperforming. Other reasons for stagnation are lack of updated technology and repeat market failures and imperfections. Weed control and plant protection measures provide another opportunity to enhance productivity and bridge the yield gaps. Conventional weed control methods have limited success and research is now moving towards herbicide resistant crop cultivars. An analysis of sugar beet yield in three U.S. states showed higher production with the introduction of Roundup Ready sugar beet cultivars (Figure 1.4). Herbicide tolerant and borer-resistant sugarcane lines have been developed at the Centre of Agricultural Biochemistry and Biotechnology (CABB), University of Agriculture, Faisalabad.

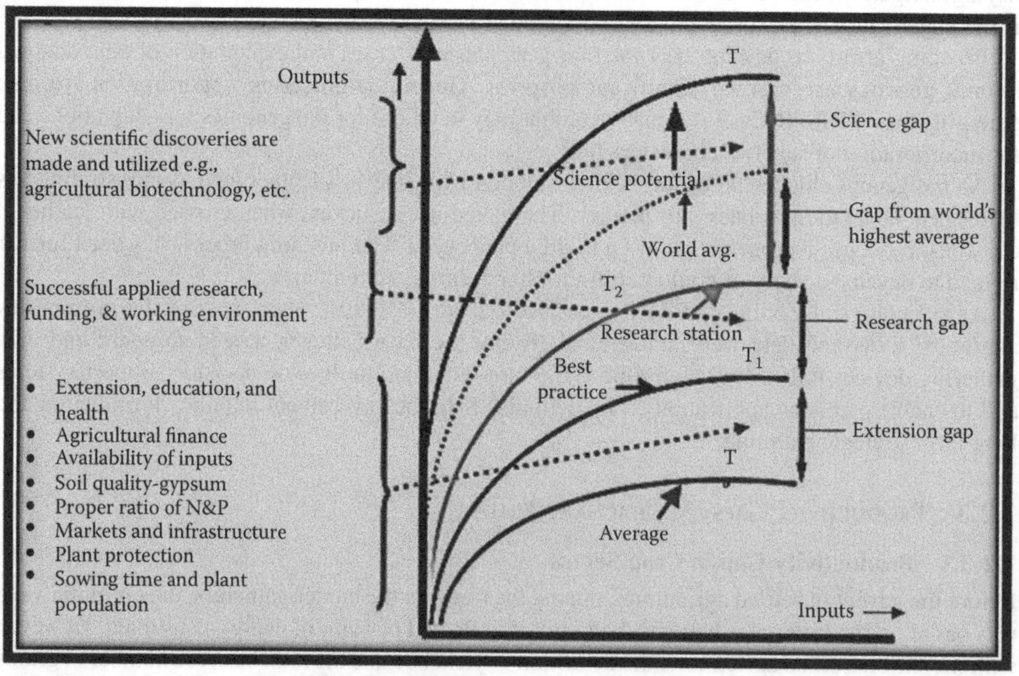

FIGURE 1.3 Productivity gaps in agriculture. (Iqbal, M. and Ahmad, M. 2005. Science & Technology based Agriculture vision of Pakistan and prospects of growth. *Proceedings of the 20th Annual General Meeting Pakistan Society of Development Economics*, Islamabad. Pakistan Institute of Development Economic (PIDE), Islamabad, Pakistan.)

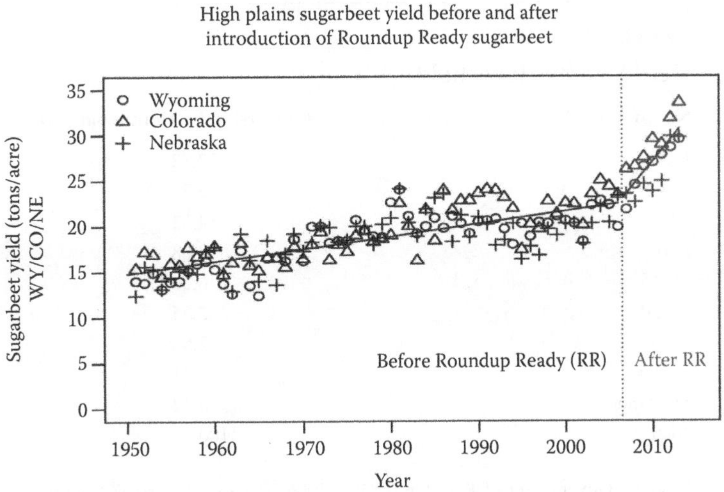

FIGURE 1.4 Roundup Ready crops are genetically engineered to resist the herbicide that kills weeds (http://weedcontrolfreaks.com/2014/02/gmo-failure-to-yield/ accessed on 18-10-2017). (Adapted from USDA–NASS).

Adoption of GMOs can combat biotic and abiotic stresses. Further, in introduction of heat tolerance in cotton allowed late sowing (after wheat) and created a new cropping system/pattern.

1.2.3.2 Productivity Gap in Livestock

Our livestock wealth is massive in terms of numbers and very poor in terms of productivity. While we pride ourselves in our livestock breeds, we must recognize the lack of potential within the pastoral genetics which has evolved to spend energy on fetching water and grass. The milk and meat breeds bred for modern dairy and fattening operations are high converters of energy. A balance must be struck between retaining the native genetics while introducing high performing breeds.

About 51% of the dairying households owned 1–4 animals, 28% households owned 5–10 animals, 14% of dairying households had 11–50 animals, and 7% of the dairying farms in Pakistan had more than 50 animals (Table 1.5). As majority of the dairying households have very small herds and the scope for breed improvement and transfer of technology for productivity enhancements becomes difficult. Only large dairying households (herd sizes more than 50 animals) raising cattle and buffalo can afford to adopt advanced technologies and increase herd productivity. Again, in Pakistan most of such herd sizes are held by the pastoral and nomadic owners in Cholistan and other rangelands, which are highly unlikely to be converted into high yield animals.

The subsistence livestock famers with small herd sizes face low yields and profitability due to poor breeds, lack of proper milk chain, seasonality in production due to heavy dependence on green fodder available only certain months (mainly Jan-April), seasonality in milk consumption (usage is normally low in winter months), lack of cold chain to use surplus milk and market in summer months, high cost of production, and relatively stagnant milk prices (Zia, 2009). An important emerging challenge is how to feed the livestock well. There is a clear case for a scientific management process for rangelands, development of fodder (silage and hey), and animal feed industries.

As such, there are two separate strategic issues to overcome stagnation, that is, to narrow the unachieved gap and to break the barriers in potentials. The narrowing of gap could be achieved by enablement and market incentives. That is a case of optimization of current input/technology uses and practices. The new potentials could be created through research and innovation. That includes investment in human resource, genetics, and precision tools.

TABLE 1.5
Herd Size by Household

No. of Animals	Ownership by Household (%)
1–2	27.32
3–4	23.73
5–6	14.32
7–10	13.68
11–15	6.29
16–20	2.65
21–30	2.58
31–50	2.71
51 or more	6.72
Total	100

Source: Government of Pakistan (GOP). 2006. *Pakistan Livestock Census*. Pakistan Bureau of StatisticsStatistics House, 21-Mauve Area, G-9/1, Islamabad, Pakistan. http://www.pbs.gov.pk/content/pakistan-livestock-census-2006, cited from Zia, U.E. 2009. Pakistan: A dairy sector at a crossroads. In *Smallholder Dairy Development – Lessons Learned in Asia*. Animal production and health commission for Asia and the Pacific, Food and Agriculture Organization of the United Nations Regional Office for Asia and the Pacific, Bangkok. http://www.fao.org/docrep/011/i0588e/I0588E00.htm. Accessed on 18-10-2017.

TABLE 1.6
Impact of Bt Cotton Cultivation in Selected Countries

Country	Insecticide Reduction (%)	Increase in Effective Yield (%)	Increase in Profit (US$/ha)
Argentina	47	33	23
Australia	48	0	66
China	65	24	470
India	41	37	135
Mexico	77	9	295
South Africa	33	22	91
USA	36	10	58

Source: Qaim, M. 2009. *Annual Review of Resource Economics* 1, 665–693.

1.2.3.3 Biotechnology, Environments, and Risk Perception

The adoption of biotechnology applications is very broad. The genetically engineered/modified crops (GM) are a scientifically appropriate tool. The global spread of GM crops is on the rise because of economic (higher yields and lower cost of production) and environmental (less pesticide and less fuel) advantages (Table 1.6). Globally, the biotech crops have shown enormous benefits as the cross-country data shows that there is 33%–77% reduction in use of insecticides in different countries, 37% increase in yield in India, and 470% increase in profit in China due to cultivation of Bt cotton.

The debate on its risk perceptions has been continuing and the world is divided. However, the empirical evidence shows that with the advent of biotechnology the related global environmental effects have been positive, as evidenced in the following 15 years data.

1.3 GLOBAL IMPACTS OF BIOTECH CROPS: ENVIRONMENTAL EFFECTS (1996–2010)

- 443 million kg less pesticide active ingredients used (9.1% reduction)
- 642 liters fuel used
- 17 billion kg reduction in greenhouse gas emission, equivalent to taking 8.6 million cars off the road
- 17.9% reduction in overall environmental impact

(Source: Brookes and Barfoot, 2012)

Moreover, meta-analysis of published studies based on primary data shows that cultivation of GM crops resulted in a 21.6% increase in yield and 68.2% increase in profitability (Figure 1.5). The pesticide cost decreased by 39.2% due to 36.9% reduction in pesticides use (Klümper and Qaim, 2014).

We have failed to fully exploit the opportunities offered by the GM crops. That is partly due to the dysfunctional regulatory framework. It is the combination of GM crops and precision in agriculture that can make our agriculture globally competitive. The future lies in investing in biotechnology development and applications (Malik, 2014).

1.3.1 Five Crops and Diversification

There exist more than 20 cropping patterns in the country as defined by the various agro-ecological zones (Figure 1.6). Yet, our agriculture is characterized by the dominance of five crops, that is, wheat, cotton, rice, maize, and sugarcane. Excepting maize, the other four have an element of promotion by the political economy. The subsistence mentality of the farmer is another big impediment. The small farmer is more concerned about food security and cares less for profitability. There is also a case of lack of technology (seed in particular), skill deficit, and market forces, which keep the farmers hooked to the five crops for which seed and skill are not limiting and which can sell easily. The diversity of our climate and land features is suited to expand the cropping mix through technology adoptions and incentives. We are a net importer of essential commodities, which could be otherwise grown successfully, that is, edible oilseed crops, pulses, high value vegetables, and spices. Soybean is a crop generally rotated with maize, which restores the soil (being a legume) and provides raw material for high value food and feed ingredients. We grow/overproduce maize but not soybean. We import soybean and its products while allowing soil degradation due to lack of crop rotation.

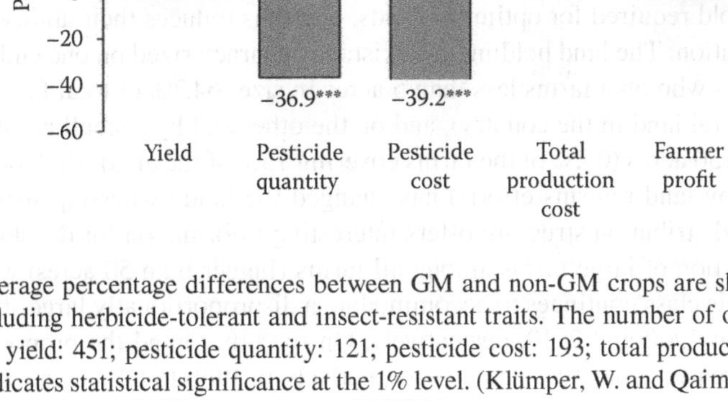

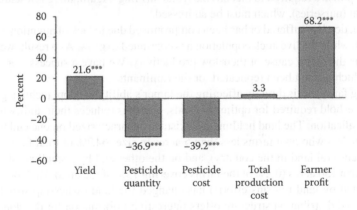

FIGURE 1.5 Average percentage differences between GM and non-GM crops are shown. Results refer to all GM crops, including herbicide-tolerant and insect-resistant traits. The number of observations varies by outcome variable; yield: 451; pesticide quantity: 121; pesticide cost: 193; total production cost: 115; farmer profit: 136. *** indicates statistical significance at the 1% level. (Klümper, W. and Qaim, M. 2014. *PLOS ONE* 9(11), e111629. https://doi.org/10.1371/journal.pone.0111629.)

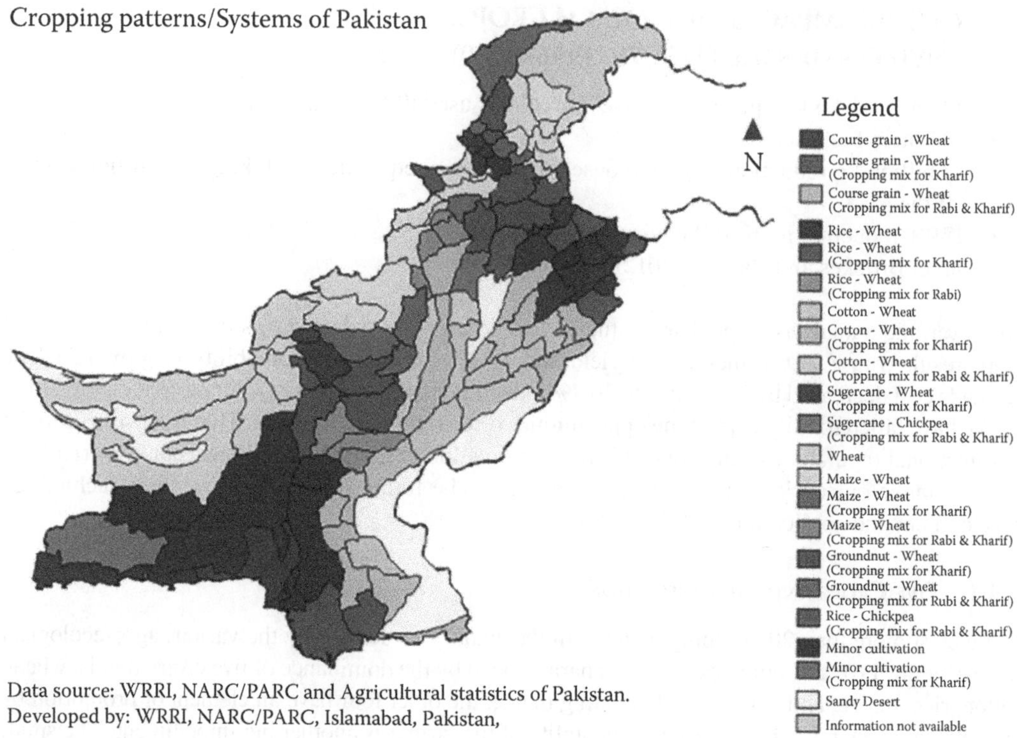

FIGURE 1.6 Topography, soil, climate, and irrigation practices have created more than 20 cropping patterns.

Diversification can be accelerated by incentivizing minor crops with emphasis on fodder, vegetables, fruits, pulses, and oilseed crops. The maize acreage has increased beyond its existing demand. It is high time to promote maize consumption to improve the quality of food and to maintain its demand. Maize farmers should be incentivized to rotate with soybean.

Fruit orchards have remained an isolated activity for the few. A better marketing framework and a value addition pull are required to incentivize fruits farming. Again, there is a scarcity of certified planting material (nurseries), which must be addressed.

Fodder production has suffered or has been compromised due to lack of attention and competition with other crops, while the livestock population has continued to grow. As a result, we have underfed animals which is the major cause of their low productivity. We have a success story in the poultry feed industry which has not been replicated for the ruminants.

The declining farm size is directly affecting the famer's ability to adopt technology or to achieve an economic threshold required for optimum yields, and thus reduces their ability to accept newer options of diversification. The land holding in Pakistan is characterized on one end by a majority of marginal landholders who own farms less than 5 acres in size (64.7% of total farms comprise only 20% of the agricultural land in the country), and on the other end by a small number of very large farms of more than 150 acres (0.2% of the farms covering 13% of the area). High population growth (almost unaffected by land reforms efforts) has changed the land ownership structure over time. The change of land distribution structure offers interesting information for development planning. Although the proportion of large farms among all farms (bigger than 50 acres) was 3.8% in 1960 and 1.2% in 2010, this class continues to accommodate a disproportionally large share of farm area of 27.5% and 24% in 1960 and 2010, respectively. On the other hand the proportion and area of small farms has increased from 19% and 3%, respectively in 1960 to 64.7% and 20%, respectively in 2010. The proportion of farms of small and medium size (i.e., 5–12.5 acres and 12.5–25 acres) has

TABLE 1.7
Agricultural Land Distribution

Category	Operational Holding (acres)	No. of Farms (%)			Farm Area Out of Total Area (%)		
		1960	1990	2010	1960	1990	2010
Marginal	<5	19	47.5	64.7	3.0	11.3	20
Small	5–<12.5	44.3	33.4	24.8	23.6	27.5	26
Medium	12.5–<25	23.8	12.2	6.8	27.0	21.5	18
Large	25–<50	9.0	4.7	2.6	19.0	15.8	13
Very large	>50	3.8	1.8	1.2	27.5	24.0	24

Source: Agriculture Census (1960, 1990, 2010).

also decreased. The average farm size of marginal and small category was 2.2 acres and 7.7 acres in 1990 and 2010, respectively. The fragmentation of land is continuing to drag down farm sizes to uneconomic levels. The disproportionally high number of farms of less than 5 acres poses a serious challenge for sustainability (Table 1.7).

This variation in land ownership structure demands provision of different packages of technological intervention for different clients. There is a need to develop hi-tech machinery models to cater to the needs of different land classes. Similarly, the value chain and marketing system needs to be modeled according to the adaptation capacity of different farming classes. Any policy ignoring this important aspect of farming will not be feasible, both economically and politically.

An important indicator for choosing a crop should be its global/regional competitiveness in terms of cost of production and productivity. Our costs of major crops are high and productivity is low. Obviously, diversification has a future and requires a multipronged strategy of delivering competitive technology packages to ensure fair returns through marketing incentives.

1.3.2 Postharvest Losses and Agricultural Marketing

Globally, food losses and waste accounts for one third of the total food produced for human consumption (FAO, 2011). Food loss is defined as "decrease in mass (dry matter) or nutritional value (quality) of food that was originally intended for human consumption," whereas "food waste" refers to discarding food that is appropriate for human consumption, and "food wastage" as food lost by waste or by deterioration (FAO, 2013). Food wastage has food security and climate change implication and avoiding it can help meet food requirements and reduce the food related carbon footprint. By this definition, if food wastage is considered as a country then it stands at 3rd position in terms of total greenhouse gas (GHG) emissions excluding "land use" change related emissions. Avoidance of food wastage also has significant implications for blue water footprint and biodiversity. In developed countries, there are more losses downstream of the supply chain (i.e., consumption and distribution level) while the developing countries incur more losses during upstream phases (i.e., production, postharvesting). The losses also depend on the commodity under consideration and the cultural and technological context (FAO, 2011). Depending on the crop—and different stages along the value chain, there are variations in losses. Ultimately, the losses at all levels (Table 1.8) drain the competitiveness and profitability of the agriculture sector.

Conservative estimates indicate 16% losses in grains during harvesting and storage due to lack of drying and proper storage structures. The horticultural commodities suffer losses because of faulty harvesting practices and due to lack of cold chains. The bulky nature of fruits and vegetables and

TABLE 1.8
The Food Losses along the Value Chain

	Production	Postproduction	Processing	Distribution	Consumption
Value Chain Stages	Preharvest Harvest Breeding	Handling Storage Transport	Canning Packaging Transformation	Retail Transport	Preparation Table
Causes of Loss	Damage/spillage Left behind in fields Pests/diseases Weather Wrong inputs	Degradation Pests Premature animal death Spillage	Degradation Discard Spillage	Degradation Discard Excess supply Spillage Spoilage	Discard Excess preparation Spoilage

Source: Adapted from Schuster, M. and Torero, M. 2016. In *2016 Global Food Policy Report*. Chapter 3. pp. 22–31. International Food Policy Research Institute (IFPRI), Washington, D.C. http://ebrary.ifpri.org/cdm/ref/collection/p15738coll2/id/130211. Accessed on 20-10-2017.

glut during harvest seasons requires cold storage and processing to avoid losses due to short shelf life. The losses of fresh produce can vary from 25% to 90%, with an average accepted losses of 40%. There are no technological solutions to increase the production of grains by 16% and fresh produce by 40% in a year; however, the prevention/reduction of such losses by investment in postharvest technology, infrastructure, or value addition through processing are possible.

The dairy and meat industries are also victims of market imperfections. The short life of milk and lack of processing facilities are further aggravated due to a very large spread of milk animals across the rural landscape (Riaz, 2008). About 70 million milk animals are owned by small farmers and collection of milk is an insurmountable task. As a result, malpractices are rife to enhance the shelf life of milk. The story of meat marketing is equally complex and its practices obsolete (Jalil et al., 2013). The traditional butchers dominate the meat supply and the need for modern abattoirs and processing plants is evident. The cattle marketing system is full of exploitation. Recently, the Punjab Government has taken a bold step to revamp cattle marketing, the effect of which needs to be watched.

The public procurement of wheat is an important measure, which keeps balance in favor of a plentiful supply of staple food. The economic rational of this intervention is questionable. We offer a support price, which is higher than the international price, but we lack storage facilities resulting in huge losses and quality deterioration of the grain. The market situation is further compounded by the fact that Indian farmers are heavily subsidized and 28 of their commodities are offered support prices. That creates unhealthy competition in the regional trade, which works against the interest of our farmers. The strategic question of support price and public procurement remains a politically sensitive issue. While it is ideal to let the market forces work, it is equally important for the government to ensure that market distortions are not against the farmer and consumer. The markets must be transparent, competitive, and convenient before one expects the market forces to determine the flow of commodities fairly.

The biggest cause of marketing disadvantage to the farmer is his inability to hold the commodity, generally because of lack of storage or because of seasonal debts. Warehousing, trading platforms, and future markets offer some solutions. Other options include investing into value addition along the value chains. This proposition has a unique requirement for every given commodity. The cluster approach can work to support the farmers during the production cycle followed by warehousing/storage and marketing. Heavy losses of produce also occur due to poor transportation, inadequate grading, very heavy spread in price between consumer and farmer, and tough competition with imported goods. The Punjab Government's rural roads program and revamped cattle markets are good examples of corrective measures.

Our lack of competitiveness in the international market is our biggest challenge, due in part to the high costs of production and because of our inability to meet compliance requirements. An increase in our competitiveness would also require narrowing of yield gaps, that is, productivity enhancement. Our regional trade has three major (India, Iran, and China) and several adjoining partners. The current value of trade with India is estimated at $3 billion. The trade with Iran, China and regions along the China-Pakistan Economic Corridor (CPEC) connected by a "one road one belt" has a brighter future. We have to prepare the system to be able to respond to the emerging market access in the near future. This could mean developing new products and skills. Grades, standards, traceability, and SPSS requirements along with international trade barriers (tariff and nontariff) need to be understood as WTO requirements.

Only a small percentage of produce is processed. Food fortification and food safety is also made difficult in part due to the small and informal nature of food processing and distribution. These issues require a deeper understanding of the dynamics of overall food systems, and associated regulatory lapses. Processing and wholesale firms (private or cooperative) could be encouraged to use contracts to directly produce safe and nutritious food supplies. These could support small farmers, kitchen gardening, and small-scale vegetable farming at times.

1.3.3 Social Disparity, Poverty, and Gender Mainstreaming

There are four indicators of global gender gap index: economic participation and opportunity, educational attainment, health and survival, and political empowerment. Pakistan ranks second last in a list of 144 surveyed countries. This relative position shows a dismal situation compared with other countries since 2006, a significant decline over the past 10 years (Table 1.9). In Pakistan, rural women are the largest group at a disadvantage.

Agricultural growth reduces poverty on a much larger scale than growth in other segments of the economy, as studies have shown (Johnston and Mellor, 1961). The growth in agriculture cascades through the economic activity of rural industries and businesses, in turn providing jobs closer to home (suited for women), halting migration from rural areas, and improving livelihoods on a larger scale.

The contribution of rural women to this economy largely goes unacknowledged. Farm household income is usually a mix of on- and off farm engagements. The migration from rural to urban centers has been a continuous process, which has led to the erosion of skills and a transfer of resources from rural to urban areas. The partial migration of family members has benefitted agriculture by allowing some resource transfer back to farming at critical times of the year. Overseas migration from rural areas is an additional debate. The critical issue is how to retain a healthy and skilled workforce in the farming sector while promoting investment available from the income of off farm employment.

Infrastructure and services in rural areas are grossly insufficient and substantial improvements are needed. These include physical infrastructure, education, health facilities, safe drinking water, and sanitation. Above all, the deterioration of social institutions and the disappearance of conflict resolution mechanisms promote out-migration.

The women's contribution to agriculture has been estimated at 43%–80%, a labor force unrecognized and underpaid. In crops like cotton and rice, women contribute directly in the field operations. In family farms, they take part in the whole value chain. The small landless livestock holders involve women in the daily robes. The critical role of gender equality and women empowerment in agriculture is an integral part of ensuring food security and improved nutrition.

In Pakistan, gender roles differ across the provinces and regions, but generally the traditional role of women is that of caretaker, with the major responsibility of tending to the families' domestic needs, including cooking. Along with domestic activities, rural women also play an important role in routine agricultural activities including cleaning seed, cultivating land, harvesting crops, and tending to livestock. Despite their participation in the labor force, women are far less likely to own income-generating assets such as land and livestock or to have a say in household economic decisions.

TABLE 1.9
Global Gender Gap Index for Pakistan for Years 2006/2016 (Rank 0.00 = Imparity & 1 = Parity)

		Year 2016 (145 Surveyed Countries Data)		Year 2006 (115 Surveyed Countries Data)	
Gender Index	**Contextual Data**	**Ranks**	**Score**	**Rank**	**Score**
Global Gender Gap Index		143	0.556	112	0.543
Economic participation and opportunity	Labor force participation; Wage equality for similar work; Estimated earned income (US$, PPP); Legislators, senior officials, and managers; Professional and technical workers	142	0.320	112	0.369
Educational attainment	Literacy rate; Enrolment in primary education; Enrolment in secondary education; Enrolment in tertiary education	135	0.811	110	0.706
Health and survival	Sex ratio at birth; Healthy life expectancy	124	0.967	112	0.951
Political empowerment	Women in parliament; Women in ministerial positions; Years with female head of state (last 50)	90	0.127	37	0.148

Source: World Economic Forum (WEF). 2016. http://reports.weforum.org/global-gender-gap-report-2016/economies/#economy=PAK. Accessed on 22-09-2017.

Studies on Pakistan show that women spend more efficiently on food consumption, families eat more nongrain food items and consume better calories from fruits and vegetables when women have a decision-making power in households.

Despite significant economic growth and potential poverty reduction, many people in Pakistan still do not have economic access to adequate food. The Benazir Income Support Program (BISP) is the largest social protection program with 5.3 million beneficiaries, and it is expanding. BISP has been successful in constructing a National Socioeconomic Registry and a Poverty Scorecard for targeting those in need, and has been a significant step towards achieving the SDGs of eradicating extreme poverty, zero hunger, and the empowerment of women. The BISP in its present framework can work in the shorter term and must evolve into a social protection system.

The youth in rural Pakistan are desperate. Education and skill development opportunities have failed to materialize there. Agriculture there is not envisioned as a career. Rapid urbanization attracts the migratiing youth in a search for better future. This also accentuates the erosion of residual skills from rural regions and produces a constant resource transfer to the urban centers. There is a need to provide them with skills and hope by promoting alternate income generation options and entrepreneurship. This can only happen within the broader framework of rural development as a simultaneous process with agricultural growth.

1.3.4 Malnutrition

Micronutrient deficiency, known as hidden hunger, is widespread in Pakistan and well characterized among rural areas. The National Nutrition Survey (GOP, 2011) revealed that 43.7% of children are stunted while 15.1% are wasted and 31.5% are underweight. The survey data showed that stunting and wasting in 2011 had increased over the past decade. These problems are higher in rural areas and periurban slums compared to urban centers. Suffering is higher amomg women and children, that is, anemia (61.9%), iron deficiency (43.8%), zinc deficiency (39.2%), vitamin A deficiency (54%), and vitamin D insufficiency (40%). This situation demands nutritional interventions to combat the threat of hidden hunger. There is a need to launch a School Nutrition Program. The domestic food

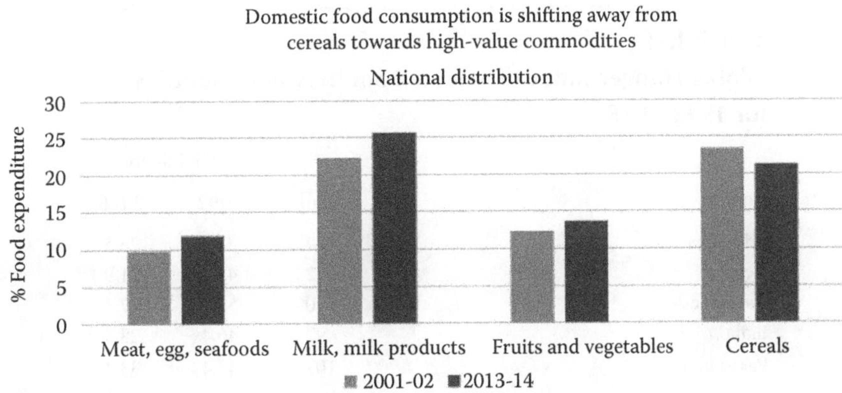

FIGURE 1.7 The shifting pattern of domestic food consumption.

consumption trend is shifting (positively) from cereal consumption to high-value commodities (Figure 1.7). These trends have implications for agricultural planning and policy.

1.3.4.1 Global Food Security

The global food security index covers food affordability, availability, quality, and safety. The index was calculated for 113 countries based on regional diversity, economic importance, population size (to ensure the inclusion of the maximum number of people in the sample). The index results reflect the poor performance of South Asian countries including Pakistan. The results of selected South Asian countries in the calculation of the Global Food Security Index for year 2016 are as in Table 1.10.

1.3.4.2 Global Hunger Index (GHI)

Global Hunger Index (GHI) and Global Food Security index are well known measures related to food security on a global scale. GHI combines undernourishment (proportion of undernourished population), child wasting (proportion of children under 5 who suffer from wasting, i.e., below normal weight to height reflecting undernutrition), child stunting (proportion of children suffering from stunting i.e., low height compared to their age), and child mortality (mortality rate of children below age of 5 reflecting inadequate nutrition and unhealthy conditions). The results of GHI show Pakistan's dismal situation as it is ranked at the 107th position and has made poor progress compared to other selected South Asian countries in the survey. The data of South Asian countries included in the survey of GHI calculations out of the surveyed 118 countries for 1992 and 2016 are as in Table 1.11.

TABLE 1.10
Food Security Index of Selected South Asian Countries

Country	Rank	Overall Score	Affordability	Availability	Quality and Safety
Bangladesh	95	36.8	28.9	46.5	29.7
India	75	49.4	42.0	57.1	46.7
Sri Lanka	65	54.8	51.9	60.1	47.4
Nepal	82	42.9	36.4	47.0	47.9
Pakistan	*78*	*47.8*	*46.3*	*50.4*	*44.5*

Source: The Economist Group- 2017. The Economist Intelligence Unit Ltd. http://foodsecurityindex.eiu.com/ Accessed on 22-09-2017.

TABLE 1.11
Global Hunger Index of Selected South Asian Countries for 1992/2016

Country	Rank	GHI Score 1992	GHI Score 2016
Sri Lanka	84	31.8	25.5
Nepal	72	43.1	21.9
Bangladesh	90	52.4	27.1
India	97	46.4	28.5
Pakistan	107	43.4	33.4

Source: International Food Policy Research Institute (IFPRI). 2016. *Global Hunger Index*. http://www.ifpri.org/topic/global-hunger-index.

The national nutrition survey (GOP, 2011) also revealed poor maternal knowledge about micronutrients. Nutrition education and behavior change initiatives are needed. Families, schools, community leaders, and other stakeholders need to be involved. Avenues for diffusion include social interactive frameworks, such as self-help groups, informal gatherings, and associated activities facilitated by the government and nongovernmental sector. Adolescence is the stage when lifelong nutrition patterns are formed. Viewing this from a life cycle approach to nutrition, it is essential to provide adolescent girls with nutrition education. The horizontal integration of nutrition programs, supported by community mobilization, is suggested to ensure that all marginalized segments of the population are reached. The creation of dedicated nutrition positions in outreach programs and the hiring of qualified nutrition experts in schools are essential. To achieve this, long-term human resource development initiatives are needed, that is, college/university education in nutrition and dietetics as a compliment to public health initiatives.

Our agricultural development scenario of food security needs attention since we are surplus in producing major food commodities (wheat, rice, sugarcane), yet more than half of the population is nutrient deficient. There are two important issues, which are (1) the lack of diversity in diets, and (2) the price/affordability for the consumer. The inclusion of vitamin- and mineral-rich produce and dairy products is needed. The cost of production has to be reduced and market distortions minimized to make these diverse foods affordable. The UN SDG "Zero Hunger" is a complex target which requires a multipronged strategy combining agricultural productivity with access to a full package of nutrition.

1.4 STRATEGIC VISION AND OBJECTIVES

Governments in developing countries prefer policies to appease urban consumers: like overvalued exchange rates discouraging agricultural exports; low domestic prices for agricultural commodities; indirect and implicit taxations; poor financial services and credit for agriculture; and low public investments in rural physical and human capital. This urban biased macroeconomic twist has resulted in an underdeveloped and slow growing agriculture sector. A policy shift is needed to make agriculture competitive, profitable, and sustainable through enablement, efficiency, and value addition for food/nutrition security and socioeconomic development. We have to reform the agriculture sector into a profitable industry by promoting investments in infrastructure, research, outreach, skills, value chains, agroindustry, and rural development. The productivity gap could be narrowed by focusing on land being cultivated by small farmers. The application of ICTs should become integral to the value chain management (Smart Agriculture). The distortion of the terms of trade against agriculture and the rural economy must be stopped. The agriculture-led growth of the economy shall bring prosperity for the masses.

1.4.1 Technology/Precision

- To ensure sustainable use of natural resources (land, water, and air)
- To increase productivity through delivery of quality inputs, credit, and services
- To minimize harvest and postharvest losses from farm to fork
- To promote nontraditional farming segments, practices, and crops/livestock.

1.4.2 Institutional Reforms

- To strengthen the regulatory framework and enabling legislations
- To ensure sustained investment in research/knowledge systems and outreach
- To promote productive employment of rural women and youth through skill development and off farm activities by promoting rural development and alternate incomes.

1.4.3 Infrastructure

- To revamp/invest in marketing systems in order to make them transparent, just, and equitable
- To accelerate interprovincial and regional/CPEC integration of the agriculture sector.

1.5 SHORT-TERM STRATEGIES

Priority 1: The low hanging fruit is to narrow the yield gap between the average and progressive farms. This will require working with the small landholders for the timely provision of inputs, services, and credit along with guaranteed irrigation. The costs of production must be constrained initially by input subsidies and followed by productivity enhancement. The farmer will also respond to the market signals, that is, support price and public procurement initiatives. The current yield gaps and stagnation must be treated separately. The yield gaps can be addressed by the delivery/adoption of available technology while the current stagnation cannot be broken without investment in research to develop new precision tools as well as biological and genetic interventions. Reducing the yield gap by a half is an achievable target for wheat, rice, and cotton by simply optimizing plant populations, enough to accelerate the GDP growth to >4%. This will spare about 2 million Ha of land for crop diversification in Punjab alone. This would require quality seed and precision drilling equipment. The seed industry is currently a victim of incompetent laws and a poor regulatory framework. The breeders have developed a range of varieties which have failed to benefit the farmer due to the faulty seed value chain. It is a distant dream to replace traditional varieties of fruit trees like mango, citrus, and dates due to the lack of a nursery certification system in the country. But, the seed industry would be an easy option to put right. We cannot replace the low yielding livestock in the short term, but we can add high performing cattle to our inventory. Fodder production, silage-making, and markets must be developed to ensure enough animal feed and fodder. We have sufficient mechanical power in the form of tractors but very little as tractor-mounted equipment.

Priority 2: Crop diversification is a challenge. The complication arises from the political economy of food security. We can broaden our choices by focusing on two crops, that is, wheat for food security and cotton for cash. The productivity enhancement of two crops can easily spare land for oil seeds, edible legumes, soybean, fodders, vegetables, coarse grains, and orchards. There are good reasons to deemphasize rice and sugarcane due to water costs to the public. The fifth crop, corn, in its present rotation system is also unsustainable. It must be rotated with a legume, preferably soybean for sustainability. The farmer's uptake of new crops will depend on market signals or a public procurement policy. The government has to offer guaranteed minimum returns for alternate crops. The diversification can also be promoted by crop zoning based on agro-ecological or agro-economic advantages and offering incentives for commodities/products most suited to the zone. A part of the wheat procurement budget should be diverted to minor crops. An alternative to support price/public procurement and subsidies lies in increasing efficiency and precision to reduce the unit cost of production. Livestock breeding and health initiatives must be made into commercially viable propositions to attract the private sector into the business of service providers.

Priority 3: Climate change has provoked new challenges to sustain agricultural productivity. There has to be an elaborate plan to mitigate and adapt to these. The immediate option is to redefine crop zones on the basis of long-term climate trends, soil and water analyses, available technologies, available skills, and current markets and industrial demands. The country could be divided into more than 30 different crop zones and subzones, which would allow a precise decision mechanism for technology transfer and incentive packages.

Priority 4: There should be an emergency plan to curtail postharvest losses by half. This will require an investment in the training programs promoting value addition through product development and for market preparations along the value chain. Home science groups should be incorporated in rural development and extension programs. Investments are also required for transportation and storage infrastructures. The marketing system needs a long-term improvement plan for new markets, legislation, and governance reforms. The Punjab rural roads program must be amplified and the example of cattle markets should be replicated to create a new structure of grain and produce markets. CPEC routes should be marked for the establishment of new agro-processing zones and markets for exports to regional markets. The French government has introduced a new law to mandate the distribution of unsold produce and food at the retail level and in the restaurants for the needy; otherwise, it requires the return of such items to farmers for use as bioenergy or organic matter in soil.

1.6 LONG TERM STRATEGIES

If implemented, the short-term strategies can raise the agriculture sector growth above 4% for the near future. However, for long-term sustainability of the system, as well as agricultural growth and poverty alleviation, the following sections detail the proposed areas of public policy interventions.

1.6.1 Food Security, Nutrition, and Hunger

The food security paradigm must shift from a supply side excess of staple items to an integrated nutritional package where diversified dietary needs are met (zero hunger of SDGs). Food safety issues like pesticide and antibiotic residues in food, mycotoxins, and malpractices associated with food handling must be addressed. All food secure countries in the world have less emphasis on wheat and rice and more on corn, potato, soybean, vegetables, fruits, dates, dairy, and poultry. We need to work on diversification of food supply and consumer habits.

It is pertinent to include food and nutrition subjects in school curricula, coupled with media awareness campaigns and counseling. Legislation for mandatory wheat flour fortification with iron should be introduced/implemented. Breeding programs for genetic fortification of food crops for nutritional enrichment and fertilizer use efficiency are long-term solutions. The greater good could only come from a social and behavioral change towards food through participatory actions. Rural poverty alleviation programs must be focused on the landless and women's enterprises.

1.6.2 Legal Framework and Institutional Reforms

Performance of agriculture is linked with the performance of many public and private sector institutions. This requires legislative and administrative measures, political will, and social movements. There are federal and provincial legal frameworks. With the 18th amendment of Pakistan's constitution, much confusion has arisen which have diminished the role of already underperforming federal institutions. The Irrigation Act, Seed Act, the Plant Breeders Act, the Pesticide Act, the Fertilizer Order, the Cooperatives Act, the Market Act, the Food Act, National Biosafety Committee, and so on are obsolete instruments. The Punjab Government has an Agriculture Commission, which has embarked upon the review of laws and policies impacting agriculture. Similar actions are required to be taken on by the other provincial governments. The Council of Common Interest (CCI) should have an agenda to provide a fresh look at federal laws and policies affecting agriculture and rural development. The CCI could also ask for uniformity among the provinces and all federating units.

1.6.3 Research System and Budgets

The national agricultural research system (NARS) and international agricultural research system (IARS) must compliment each other for a better agriculture (crops, livestock, irrigation, forest, and fisheries). Unfortunately, our NARS is underperforming, full of overlaps, and segmented (research, education, and extension). There are federal institutions, provincial institutions, and universities that have huge investment and strengths. The outcome has been very impressive over a long period of time, however, the current stagnation reflects serious recent malfunctions. There are institutions that have lost their relevance after 18th amendment which include the Pakistan Forest College in Peshawar, the FSC&RD, and to some extent the PARC. Mechanisms are needed for funding research well above the current level of 0.18% of agricultural share in the GDP. Autonomous commodity boards are an option to levy a research tax on value-added agricultural products.

A worldwide recognized system of integrated research, teaching, and extension is that of the United States, called as land grant colleges. It is a tripartite arrangement created through a series of enactments by the U.S. congress (Figure 1.8). However, we have not following this model and have kept research, education, and extension in different domains without any practicable mechanism for integrated effort for agricultural growth. This needs to be aligned to successful land grants college models to improve the efficiency of our public agriculture services.

We also need to create mechanisms to prioritize agricultural research and introduce funding of commissioned research programs. The research should be internationally compatible (scholarship) on the one end, and farmer focused on the other end, with innovations a high priority. At present, investment in commissioned research in the following priority areas is considered essential:

a. Seed production and technology
b. Development of stress-tolerant germplasm for crops and livestock genetics
c. Promotion of new and nontraditional crops
d. Special programs on dates, rangelands, and orchards

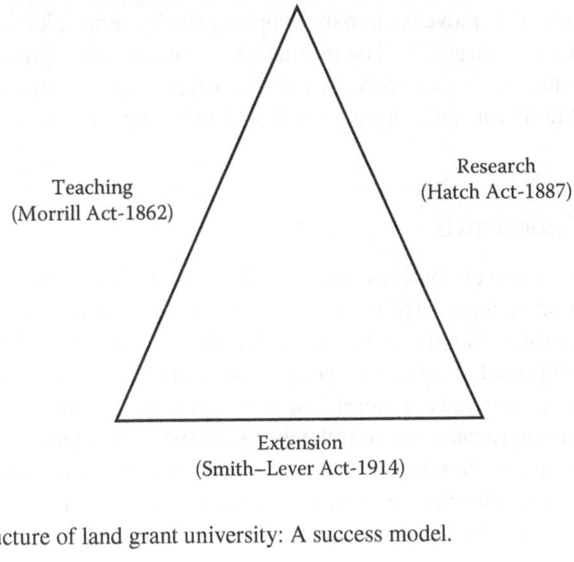

FIGURE 1.8 The structure of land grant university: A success model.

e. Pest management strategies for fruit flies, ticks, pink boll worm, white fly, and vector borne diseases
f. Control of tree dieback (mango and sheeshamin in particular) and citrus greening
g. Ensuring and improving the health and nutritional value of food
h. Mechanization of farm operations (land development to postharvest) including technological adaptation to the Pakistan farm setting and precision farming
i. Increasing viability of horticulture enterprises through zoning and cool chains owned by the marketing cooperatives
j. Use of Information and Communication Technology (ICT) in the transfer of agriculture and food systems knowledge and technology (Smart Agriculture)
k. Policy research and commodity analyses as a regular feature.

1.6.4 Fiscal Policy

The taxation of the agricultural sector and the overall growth rate of the economy are strongly correlated (Krueger et al., 1988; Schiff and Valdés, 1992). The economic growth of a country is strongly linked with growth in its agriculture sector (Johnston and Mellor, 1961) as its growth generates a large multiplier effect (Block and Timmer, 1994). Despite empirical evidence, the agriculture sector has remained subject to heavy taxes (implicit). It is assumed that the agricultural exports earn high profits due to country quota and should be taxed. An explanation for agricultural taxation is given by the fact that the sector is not making proportionate contributions in tax revenue compared to its share in national GDP. Such arguments fail to consider the fact that the agricultural sector provides raw materials and markets which helps the growth of other sectors. Some of the key devices proposed for agriculture taxation include selective commodities taxation, export quota taxation, agricultural income tax, and general sales tax on inputs. It is important to compute all implicit taxes if the goal of the government is to generate tax revenues from agriculture sector, which are commensurate to its share in the GDP. The selective commodity taxation cannot be recommended as it alters resource allocation (the taxed commodity gives incentive to shift to a different commodity). The risk of intersectoral resource allocation can make matters worse for an already dwindling agriculture sector.

Fiscal policies and taxation regimes are important determinants of regional trade and commerce. With the passage of time, support prices of various commodities have been withdrawn except for wheat. Higher general sales tax rates on fertilizer and petroleum products and other taxes on inputs contribute to the escalating costs of production. In order to improve the profitability of various commodities, there is a need to move back to the support price system (selectively) and the provision of inputs at subsidized rates (targeted). The ultimate aim should be progressive liberalization and deregulation to let the market forces work. A rational fiscal policy proposition would be to first provide enabling conditions for agriculture growth and then impose taxes once it progresses at a decent pace.

1.6.5 Credit and Cooperatives

Agriculture is a business and every business requires investment. The farmer is always cash strapped and at the mercy of "rent seekers." He needs credit. Looking at agriculture's share in GDP and corresponding formal credit availability, it is evident that there is a situation of huge underinvestment. The rural areas remain deprived of vast coverage of financial services mainly due to remoteness, high transaction costs, lack of traditional collateral, low literacy to understand procedures, and perverted instruments. The issue is aggravated due to inflexibility on the part of lending agencies. The formal credit services can be made sustainable in rural areas through innovative policies to screen reliable borrowers, monitoring techniques for effective use of loans, and erecting a regulatory environment to encourage sustainable rural financial services (Norton, 2004).

The formal/institutional credit for agriculture is an insignificant component in the entire investment portfolio of agriculture. The vacuum created by the lack of a formal financial stream is being filled by the nonformal sector at exorbitant costs to the farmer. Cooperatives used to play a significant role in the supply of credit and services. In Punjab, there are 136 branches of the Punjab Cooperative Bank, which is dysfunctional. We need to create Marketing and Services Cooperatives (default corporatization) to revive the supply of credit through these branches. This will require market reforms, investment in the improvement of supply chains, promotion of clusters, and enablement for value addition. Revival of cooperatives can boost the productivity of small farmers in many ways. The development of CPEC offers an opportunity of SEZs (Special Economic Zones), which could be agro-focused centers for value addition. This can support credit for entrepreneurship, SMEs, and local employment opportunities. A credit task force of composed of bankers should be constituted to look after these needs.

1.6.6 Input Supply

Seed, fertilizer, agrochemicals, and energy/machinery are the major inputs. The farmers who can better manage these elements are called progressive and can usually harvest optimal yields. The progressive farmers are not necessarily the large owners. They can be landless contractors or small farmers. Generally, large farming operations end up with a better mix of input supplies, and hence produce better yields. The state has a role to play in ensuring timely supply and accessibility of inputs (unadulterated) to the farmers regardless of their ability to pay up front. Now, with ICTs becoming easily accessible, the state must enable the farmer with a "decision support strategy." Use of ICTs for research and development, dissemination of knowledge, and crop/commodity advisories must be fully capitalized. Credit delivery and monitoring should be linked with the ICT services. Precision agriculture has emerged as a tool for efficiency. The ICT strengths in Punjab are enough to take advantage of precision agriculture technologies (Figure 1.9).

1.6.7 Service Provider

Agriculture has now become a high-tech sector in developed countries and the role of agricultural technology is rapidly increasing. In order to remain in the business and be competitive in the world

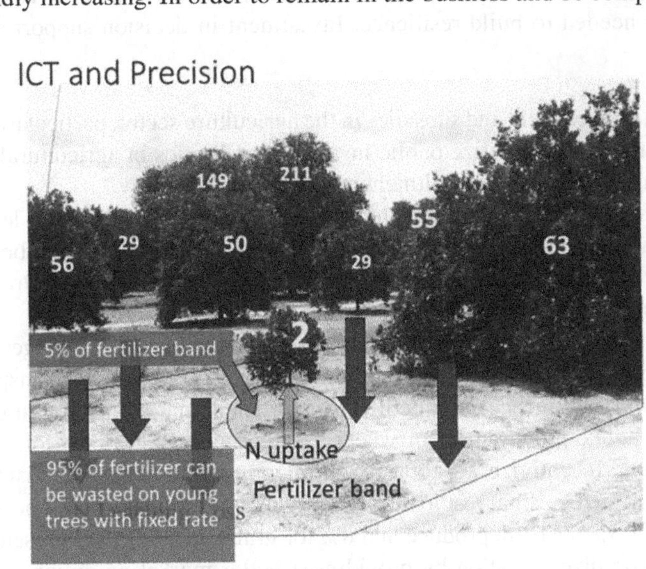

FIGURE 1.9 A definition of precision in terms of site/size specific application. This citrus grove has trees with canopy volumes ranging from 2 cubic meters to 211 cubic meters. A uniform application could be wasteful or insufficient.

at large, technological adoption is an essential condition. There are always risks involved with new technologies but it is proven that the restrictive approach of trade in technology is riskier than the liberal policy of importing technologies (Gisselquist and Grether, 2000). The service providers can be a useful conduct to transform the agriculture sector into a high-tech sector. This is an option to narrow the technology gap and for the small holders to achieve significant savings. This model has been successful in different parts of the world. The idea is to create a range of crop-specific or region-specific entrepreneurial setups with farm machinery and input supplies. These could be matched with credit availability when required. Incentivized farmer's cooperatives could be another option. Entrepreneurs are given loans and incentives to start their businesses (corporatization). The goal should be to elevate the productivity of underperforming small farms and to improve the gains of progressive farmers. Use of agricultural drones is a hot topic of research. Crop monitoring, yield mapping, and agrochemical spraying can be performed by manned and unmanned aerial vehicles.

1.6.8 Rural Development

Agriculture and rural development go together. Rural infrastructure development (roads, school, health) and skill development need massive investments. There is room for social mobilization for collective action and dispute resolutions through community-based organizations. Population welfare, gender mainstreaming, and youth programs must be targeted for rural communities to raise their aspiration and increase their love for agriculture. The agriculture and veterinary universities should be mandated to prefer students' intake from the rural schools. These universities should be mandated to create pre-agriculture programs to promote the attendance of rural youth. Rural poverty needs multidimensional strategies of enablement and job creation along with alternate income generation activities.

1.6.9 Way Forward/Strategic Planning

In the short term, sustainability planning should include education and dissemination of knowledge and skill for enhancing the participation of farming/rural communities. In the medium and long term, the role and effectiveness and relevance of agricultural research and extension, credit services, and marketing are needed to build resilience. Investment in decision support systems would be critical:

1. Analyze public investment and subsidies in the agriculture sector, particularly after the 18th amendment and institutionalize public investment priorities in agricultural infrastructure and marketing with a clear commitment of finances perpetually.
2. Irrigation systems needs improvements from the dam up to the farm level. Real time discharge data should be made public. The tail end farmers must be looked after. Groundwater pumping should be regulated and water should be priced. Promote low delta crops and ban rice cultivation before the onset of rains.
3. Revamp the marketing system on the along the lines of cattle markets to get rid of "Market Administrators" and cartels. That should be aimed at creating a transparent (market information and intelligence), competitive, and efficient marketing structure.
4. Phase out of public procurement of wheat by creating a PPP model for storage (buffer) and distribution. Incentivize warehousing of major commodities and create a commodity exchange. Minimize postharvest losses: Fix targets for reducing overall postharvest losses to the level of 25% for fresh produce and 8% for grains (a half of the present).
5. Incentivize crop diversification by providing a better marketing option or support price for the minor crops/oilseeds/pulses and nontraditional crops. Launch special programs for periurban agriculture to avoid contaminated produce.

6. Create a regional trade policy forum like NAFTA, EU, APAP with a special focus on CPEC. A CPEC think tank on agriculture must work on developing long-term strategies.
7. Seed sector reforms include facilitation for the private sector in seed multiplication and trade. Enforcement of biosafety rules is required to pave the way for the introduction of GM crops. The universities must launch seed science and technology programs. A liberal regime for international partnerships is needed.
8. Balanced use of fertilizer is an utmost need to improve the productivity and to protect our environments. That will also reduce the cost of production. The urea economy of fertilizer industry must be revisited. The attention on P and K has already proved to be a wise step. The crop residue management and micronutrient would be other essential items. Soluble fertilizer formulations are now demanded to promote fertigation.
9. Mechanization and reverse engineering of farm machinery is an opportunity for the small and medium sized enterprises (SME) sector. This includes all tractor mounted equipment from ploughing to postharvest handling and processing. Establishment of service centers/ Rural Business Hubs (RBH) is an option for small rural towns where mechanization and input needs could be met under one roof.
10. Climate change adaptations and mitigation framework needs to be formalized as an essential part of Smart Agriculture; it can be called Climate Smart Agriculture.
11. Invest in skills for value addition and for promotion of SMEs. The rural youth needs to be trained for SME and service delivery options to create alternate income streams for their families (rural nonfarm sector). Value addition training is the low hanging fruit to promote aspirations and create opportunities. Link microfinance and youth loans with skills and entrepreneurship.
12. Gender mainstreaming by extending benefits of women development programs to the rural areas will address inequality by developing women markets and investment in startups. Promotion of bikes for rural girls after matriculation and nutrition awareness of girls are needed. Skill development for women labor forces and awareness about their rights are also important. Future mothers and school lunch programs and curricula are needed. Targeted food fortification and blending options deserve to be tried.
13. Extension and outreach should promote entrepreneurship and aspirations in the agriculture sector. Venture capital and training for future farmers in high-value crops, fruits, and commodities should work. Launch residue management programs and educate farmers on the responsible use of chemicals. Launch crop packages for diversification (alfalfa, oilseeds, pulses, soybean, sorghum, millet, and vegetables) on the basis ofagro-ecological zones. ICT enabled centers with the provision of extension and training of farmers (particularly women and youth) are currently being tried. Special programs for lead farmers and theme leaders (champions of change) should be created.
14. Investments should be made in skill development to reduce postharvest losses and to add value. The quality standards and WTO requirements as well as regional opportunities offered by the CPEC must be addressed to become globally competitive. Comprehensive market reforms program are needed.
15. The investment in research and development should be linked with institutional reforms for the integration of education, research, and extension. Commodity research boards should be institutionalized. Long-term research experiments should be launched to model sustainability of cropping systems.
16. Rural development must include infrastructures for farm to markets at a much larger scale than presently available. Rural life must be made attractive to reduce migration by introducing women and youth development programs along with alternate income propositions (at Markaz level or the small town centers/the new Mandi Towns).

1.7 CONCLUSION

This book is a farmer centric document to be treated as a baseline to establish a continuous review for policies and planning processes. There should be a 1–3 years plan written as a departmental operations manual, which includes investment strategies and implementation targets. Food security must include nutritional security. The political economy of food security must not compromise the profitability of the farmer and transfer resources from rural to urban economies (terms of trade). The immediate targets should be addressing the small farmers' productivity challenges by ensuring quality seed (plant population), machinery, balanced fertilizer, insect pest management (IPM), and weed management. The public procurement of wheat should be phased out and available resources should be used to incentivize crop diversification. The HEIS must be evaluated and redesigned. Medium to long-term plans should be devised for land and water resource management (fragmentation, on-farm water storage, rain water harvesting, water pricing). Grain and produce markets are insufficient and imperfect. Infrastructure and legal frameworks are needed to enhance capacity and to promote the transparency and competitiveness of business systems, which should be free of exploitation by middlemen.

REFERENCES

Ahmad, S. and Majeed, R. 2001. Indus basin irrigation system water budget and associated problems. *Journal of Engineering and Applied Sciences* 20, 67–77.

Block, S. and Timmer, P. 1994. *Agriculture and Economic Growth: Conceptual Issues and the Kenyan Experience.* Harvard Institute for International, Cambridge, MA, USA.

Brookes, G. and Barfoot, P. 2012. Global impact of biotech crops: Environmental Effects 1996–2010. *GM Crops & Food: Biotechnology in Agriculture and Food Chain* 3(2), 129–137.

Cheema, M.J.M. 2012. *Understanding Water Resources Conditions in Data Scarce River Basins using Intelligent Pixel Information Case: Transboundary Indus Basin.* Technical University Delft, The Netherlands, p. 204.

Fogel, R.W. 2004. *The Escape from Hunger and Premature Death, 1700–2100: Europe, American and the Third World.* Cambridge University Press, Cambridge, UK.

Food and Agriculture Organization of the United Nations (FAO). 2011. *Global Food Losses and Waste: Extent, Causes and Prevention.* VialedelleTerme di Caracalla, Rome, Italy.

Food and Agriculture Organization of the United Nations (FAO). 2013. *Food Wastage Footprint: Impacts on Natural Resources.* Available at www.fao.org/docrep/018/i3347e/i3347e.pdf.

Food and Agriculture Organization of the United Nations (FAO). 2015. *Yield Gap Analysis of Field Crops: Methods and Case Studies.* FAO Water Reports 41, FAO Rome-Italy. Available at www.fao.org/3/a-i4695e.pdf.

Gisselquist, D. and Grether, J-M. 2000. An argument for deregulating the transfer of agricultural technologies to developing countries. *World Bank Economic Review* 14(1), 111–127.

Government of Pakistan. 1988. *Report of the National Commission on Agriculture.* Ministry of Food and Agriculture, Islamabad.

Government of Pakistan (GOP). 2006. *Pakistan Livestock Census.* Pakistan Bureau of Statistics, Statistics House, 21-Mauve Area, G-9/1, Islamabad, Pakistan. http://www.pbs.gov.pk/content/pakistan-livestock-census-2006

Government of Pakistan (GOP). 2011. *National Nutrition Survey.* Planning Commission, Planning and Development Division, Islamabad.

International Food Policy Research Institute (IFPRI). 2016. *Global Hunger Index.* http://www.ifpri.org/topic/global-hunger-index

Iqbal, M. and Ahmad, M. 2005. Science & Technology based Agriculture vision of Pakistan and prospects of growth. *Proceedings of the 20th Annual General Meeting Pakistan Society of Development Economics,* Islamabad. Pakistan Institute of Development Economic (PIDE), Islamabad, Pakistan.

Jalil, H., Hussain, S.S. and Saddiqi, A.F. 2013. An empirical study of meat supply chain and prices pattern in Lahore (Pakistan): A case study. *Journal of Supply Chain Management Systems* 2, 44–52.

Johnston, B.F. and Mellor, J.W. 1961. The role of agriculture in economic development. *The American Economic Review* 51(4), 566–593.

Khan, I.A. 2014. *Vision 2030.* University of Agriculture, Faisalabad. http://uaf.edu.pk/downloads/vision2030.pdf. Accessed on 18-10-2017.

Khan, I.A., Zahir, A.Z., Naveed, M. and Rashid, A. (eds.) 2016. *A 101 Innovation Catalogue*. Office of Research, Innovation & Commercialization (ORIC), University of Agriculture, Faisalabad. http://uaf.edu.pk/Catalouge/101/index.html. Accessed on 18-10-2017.

Khan, M.M., Zhang, J. and Hashmi, M.S. 2011. Land distribution, technological changes and productivity in Pakistan's agriculture: Some explanations and policy options. *International Journal of Economics and Management Sciences*. 1, 51–74.

Klümper, W. and Qaim, M. 2014. A meta-analysis of the impacts of genetically modified crops. *PLOS ONE* 9(11), e111629. https://doi.org/10.1371/journal.pone.0111629

Krueger, A.O., Schiff, M. and Valdés, A. 1988. Measuring the impact of sector-specific and economy-wide policies on agricultural incentives in LDCs. *World Bank Economic Review* 2(3), 255–272.

Malik, K.A. 2014. *Biotechnology in Pakistan: Status and Prospects*. Pakistan Academy of Sciences, Islamabad, Pakistan. ISBN: 978-969-8223-14-4

Norton, R.D. 2004. *Agricultural Development Policy: Concepts and Experiences. Food and Agriculture Organization of the United Nations*. John Wiley & Sons, England.

Qaim, M. 2009. The economics of genetically modified crops. *Annual Review of Resource Economics* 1, 665–693.

Riaz, K. 2008. A case study of milk processing: The Idara-e-Kissan Cooperative. *The Lahore Journal of Economics*, 13, 87–128.

Schiff, M. and Valdés, A. 1992. *The Plundering of Agriculture in Developing Countries*. World Bank, Washington, D.C.

Schuster, M. and Torero, M. 2016. Toward a sustainable food system: Reducing food loss and waste. In *2016 Global Food Policy Report*. Chapter 3. pp. 22–31. International Food Policy Research Institute (IFPRI), Washington, D.C. http://ebrary.ifpri.org/cdm/ref/collection/p15738coll2/id/130211. Accessed on 20-10-2017.

Spielman, D.J., Malik, S.J., Dorosh, P., Ahmad, N. 2016. *Agriculture and the Rural Economy in Pakistan: Issues, Outlooks, and Policy Priorities*. University of Pennsylvania Press, Philadelphia, USA. http://www.upenn.edu/pennpress

United Nations Development Program (UNDP). 2015. *Human Development Report*. United Nations Development Program, New York, NY 10017. www.undp.org.

World Commission on Environment and Development (WCED). 1987. Our Common Future. Oxford University Press, Oxford.

World Economic Forum (WEF). 2016. http://reports.weforum.org/global-gender-gap-report-2016/economies/#economy=PAK. Accessed on 22-09-2017.

World Intellectual Property Organization (WIPO). 2016. Global Innovation Index Report: Winning with Global Innovation. Jointly produced by WIPO, Cornell University and INSEAD. http://www.wipo.int/publications/en/details.jsp?id=4064&plang=EN. Accessed on 20-09-2017.

Zia, U.E. 2009. Pakistan: A dairy sector at a crossroads. In *Smallholder Dairy Development – Lessons Learned in Asia*. Animal production and health commission for Asia and the Pacific, Food and Agriculture Organization of the United Nations Regional Office for Asia and the Pacific, Bangkok. http://www.fao.org/docrep/011/i0588e/I0588E00.htm. Accessed on 18-10-2017.

Section I

Natural Resources and Input Supplies

2 Land Use in Pakistan

Muhammad Tahir and Tasneem Khaliq

CONTENTS

2.1 INTRODUCTION

Land is an important nonrenewable resource. Pakistan is located in the northwestern part of the South Asian subcontinent (Table 2.1).It is located at 23–37° North and 61–76° East. The total surface area of Pakistan is 79.61 million hectares excluding the Northern Areas of Pakistan. Out of which 96.9% and 3.1% area is covered by land and water respectively. Pakistan has four provinces, which are Punjab, Baluchistan, Sindh, and KPK province. About 59% of the area of Pakistan is comprised of plateaus and mountains; however, 41% area of Pakistan is covered by deserts and plains (GOP, 2015). The northern region of Pakistan is fenced by the Himalayas, Karakoram, and Hindukush ranges. Pakistan has a 595 km conjoint border with China in the northeast. In the northwest region, the border between Pakistan and Afghanistan is called the Durand line. The boundary of Pakistan with Afghanistan is 2252 km. Iran is in the west of Pakistan, and the length of this border is 909 km. Pakistan has India on the east. The length of border between India and Pakistan is nearly 2912 km. Pakistan has the Arabian Sea on the south. Pakistan has a more than 1046 km long coastline. The main challenge for development efforts in Pakistan lies in its rural sector, which is suffering from widespread poverty, rising unemployment, growing income inequalities, and disproportionally low health and education opportunities. Most of these problems arise mainly, if not solely, from the skewed distribution of land ownership, leading to correspondingly highly unequal distribution of income and social power (Khan and Mahmood, 1997; Hamid and Maliha, 2008). Many economic problems stem primarily from the inability of the agricultural sector, plagued with inequitable land distribution, to provide full employment opportunities and its failure to yield incomes needed for providing a satisfactory living standard to the rural population of the country (Chaudhary, 1994; Khan and Riaz, 2006).

TABLE 2.1
Land Utilization Statistics (Area in Million Hectares)

Year/ Province	Geographical Area	Total Area Reported Col. (4 + 5 + 6 + 7)	Forest Area	Not Available for Cultivation	Culturable Waste	Cultivated Area Col. (8 + 9)	Current Fallow	Net Area Sown	Area Sown More Than Once	Total Cropped Area Col. (9 + 10)	Agriculture Land (13 + 4)	Arable Land (6 + 7)	Uncultivatable Area (4 + 5 + 6)
1	2	3	4	5	6	7	8	9	10	11	12	13	14
Pakistan	79.61	57.99	4.55	23.10	8.27	2.07	6.69	15.38	7.30	22.68	34.89	30.34	35.92
Punjab	20.63	17.50	0.49	2.97	1.52	12.52	1.89	10.63	5.89	16.52	14.53	14.04	4.98

Source: Pakistan Bureau of statistics, Ministry of national food security & Research, 2015.

Note: 3: **Total Area Repored** is the total physical area of the villages/deh, tehsils or districts etc. 4: **Forest Area** is the area of any land administered as forest under any legal enactment dealing with forests. Any cultivated area which may exist within such forest is shown under heading "cultivated area." 5: **Area Not Available for Cultivation** is that uncultivated area of the farm which is under farm home steads, farm roads and other connected purposes and not available for cultivation. 6: **Culturable Waste** is that uncultivated farm area which is fit for cultivation but was not cropped during the year under reference nor in the year before that. 7: **Cultivated Area** is that area which was sown at least during the year under reference or during the previous year. Cultivated Area = Net Area sown + Current Fallow. 8: **Current Fallow** (ploughed but uncropped) is that area which is vacant during the year under reference but was sown at least once during the previous year. 9: **Net Area Sown** is that area which is sown at least once during (Kharif and Rabi) the year under reference. 10: **Area Sown More Than Once** is the difference between the total cropped area and the net area sown. 11: **Total Cropped Area** means the aggregate area of crops raised in a farm during the year under reference including the area under fruit trees.

2.1.1 Physiographic Regions of Pakistan

The Pakistan can be allocated into five physiographic areas:

- The Himalayan Mountain ranges in the northwestern part of the border with China and India. The upper most top, the Godwin-Austin (7610 m) is part of the Trans-Himalayan Range.
- In the north of the border with Afghanistan where the western mountains and the Hindu Kush are located.
- Just to the south of Islamabad, the Pothohar Plateau. Its altitude varies from 300 to 600 m. There is a Salt Range in south of the Pothohar Plateau.
- The Indus Plain extends from the Salt Range to the Arabian Sea. This level plain is mostly made up of alluvium, over 300 m deep, deposited by the Indus river and its tributaries.
- In the southwest, the Balochistan Plateau is situated with an average elevation of about 600 m. From northeast to southwest, the dry hills run across the plateau.

2.1.2 Functions of Land

Land plays important role in agriculture, as a medium for crop production as well as space for livestock. The following are major functions of land;

1. Production function
2. Space function for socioeconomic and infrastructural development
3. Human settlement space function
4. Biotic environmental function
5. Climate regulative function
6. Hydrologic function
7. Waste and pollution control function
8. Storage function
9. Archive or heritage function

2.2 CURRENT LAND USE OF PAKISTAN

The cultivated areas are categorized under the following categories of irrigation :tube well command, canal command, barani, and rod-kohi. The total geographical area of Pakistan is 79.61 million hectares. The total reported area is around 57.99 million hectares, out of which around 8.27 million hectares are classified as culturable waste. This leaves around 22.06 million hectares as cultivated areas. The culturable waste represents 8.20 million hectares, which are appropriate for agriculture if water is made available. Around 4.55 million hectares are under forests. Currently, fallow areas cover 6.68 million hectares, and the net area sown is 15.40 million hectares. Areas sown more than once represent 7.33 million hectares and the total cropped areas cover 22.73 million hectares. The remainder, which is not available and suitable for agriculture and forestry cultivation represents 23.10 million hectares and needs improvement of water resources, one of the leading limitations for the progress of agroforestry.

Most of the farmlands are being underutilized, not reaching their potential, and with modern management, a high level of inputs, the present production of most crops could be quadrupled. However, the yields could be maximized only through a balanced combination of all the production factors. Emphasis on water alone is not enough because irrigation stabilizes production by removing the risk factor, high-yielding crop varieties act as a vehicle for conversion of solar energy into grain and it is soil, water and crop management that enhances the efficiency of this conversion. The gap between the present and the potential crop productions can be attributed mainly to the present low standard of farming, but not to the inadequacy of irrigation water, or the presence of soil problems. The increased emphasis on agricultural and rural development in government policy, during the early 1960s coincided with the advent of the so-called green revolution. This revolution started with the scientific and technological breakthrough in farm inputs, like seed availability of high yielding varieties (HYV), fertilizers and pesticides, and in the rapid expansion of agricultural mechanization, leading to an assured supply of irrigation water and farm power from tube-wells and tractors (Kemper, 2003; Kuriakose et al., 2005; Acumen, 2008).

2.3 LAND USE SYSTEM OF PAKISTAN

The major land use system of Pakistan comprises:

2.3.1 Dry Mountains of the HKH Region

2.3.1.1 Overview

The Hindu Kush-Himalaya (HKH) mountains are rich in natural resources with a wealth of water, extreme elevations, and valleys below mountains capped with snow, landscape, flimsy but diverse ecologies and productive valleys that provide a bundle of biota things and amenities. In Asia, KHK is the source of nine major river systems shared by eight countries. There are four biodiversity hot spots. These countries have established 488 protected areas. It covers 39% (1.6 million sq. km) of the total areas of the HKH. The area offers the necessary supply of fresh water, goods, and services to 1.3 billion downstream mountain people. Upstream it provides this to a populations of 210 million. The river basins are a vital source of biodiversity, water, energy, and food. Conversely, professionals have been worried that rapid global warming is becoming a danger for these ecosystems. Main services and goods delivered by the HKH mountain ecosystems are as follow.

2.3.1.2 Biodiversity: The Source of Fiber, Food, Nutrition, and Medicine

The HKH mountains have high agricultural and biological diversity, including medicinal plants, fiber, and food. There are abundant aromatic and medicinal plants in this region and various types of good quality fiber such as mountain-specific crops (different type of millets, wheat, etc.) and wool, which are in abundant demand in downstream and worldwide markets. Diverse types of forage, timber, fuel wood forests are found at these elevations for supplying to downstream markets. The huge and various amounts of genetic diversity found in these mountains are a significant reserve for the next generations.

2.3.1.3 Water: The Lifeline of People

Aquatic systems are the most important ecosystems the HKH area delivers to downstream mountain people and the world. Commonly denoted as "water tower of Asia" for the plains, 10 main rivers, initiating from the HKH region, supply about 1.3 billion people. Current estimations of the snow level and ice in the HKH Mountains is about 60,000 km^2. Present in the form of ice, snow, and glaciers, an immense quantity of freshwater also exists in groundwater, lakes, and soil.

2.3.1.4 Ecosystem Services

The areas of HKH provide a livelihood and supporting goods to mountain and downstream populations, as well as sustain a safe and healthy environment. For instance, they play a dynamic role in water retention in the form of snow, ice, and groundwater.

Mountains are also responsible for modulating the climate. Mountain soils are pools of carbon, water, and nutrients for soil fertility. Forests in mountains are an important safeguard from natural dangers, ensuring slope constancy, erosion, and avalanches etc. These, a mountain forest has high genetic variation and provides a habitat for wildlife. These forests also play an important role in earth's carbon storage.

The HKH mountains are not only rich in natural resources but also has heights, a delicate biota, steep slopes, uneven basins, distinctive landscape for survey, research and human manipulation.

2.3.2 Balochistan Valley Agriculture and Rangelands

Balochistan is the biggest province (45% of the total area) of Pakistan with 34.5 million hectares. It is situated in the west of the country, sharing borders with Afghanistan to the north and Iran to the west. The rangelands have been broadly classified into poor, medium, and rich potential areas, producing annually less than 50, 60–190 and 200–250 kg dry matter ha^{-1}, respectively. The rangelands classified as poor constitute about 62% of the province and are generally located on the south side of the province. Those classified as medium (25% of the area) and high (13% of the area) potential areas are mostly on the highlands above 1000 m. An other area in these classes is mountain sides and plateau covered by thin soils usually in accessible to grazing.

There are two types of range-ands in Balochistan, known as common and open rangelands. Common rangelands are conventionally owned by tribes, with ordinary institutional measures for their sustainability and effective management. Open rangelands have been increasing on an area as the more exclusive common range-lands have no capability to withstand the animals grazing requirements and area owned by their owners.

Two reasons explain rangeland destruction: an increase in population, and exterior social and economic forces. For example: "In 1980, war in Afghanistan resulted in large migrations to Pakistan. At least 3 million Afghani came into Pakistan with their livestock: 600,000 populations with 4.8 million sheep and goats to Balochistan only. As they have largely been concentrated in many camps along the border areas, their presence has resulted in extreme degradations in the neighborhoods of their camps in northern Balochistan. The effect of a quick increase—equivalent to 14% of the total livestock as well as human population in Balochistan—on the already saturated and fragile environment has been drastic."

2.4 MAJOR CHANGES IN LAND USE SYSTEMS OF PAKISTAN AND IMPACTS ON WATER

2.4.1 Land Use Changes in the Dry Mountains of the HKH Region

- Major changes in the land use system of HKH range started with the construction and completion of HKH highway, which is a unique mountain road in the world. With the construction of the road, farmers from local areas started exporting farm produce to downstream cities, so there was a tremendous change in mountain land use systems which resulted in increased landslides, often damaging the highway.
- Deforestation and overgrazing of watersheds resulted in degradation of rangelands, pastures, and forests in HKH region of Pakistan, where the existing productivity of natural forests is quite low compared to 50 years ago. Clearing of forests and the introduction of field crops and fruit plants have created issues of nonsustainability of watersheds, which are fed into the Indus River.

2.4.1.1 Impacts on Hydrology and Water

- Debris of landslides and sediments in flowing water ultimately enters into Indus river system through nullahs carrying snow and glacier melts. The damming of water in Indus River and change in river morphology are common problems.
- Deforestation and degradation of rangelands resulted in changes in temperature and precipitation regimes affecting occurrence and distribution of precipitation. The ccountry has started facing droughts for the last 5 years, where snow melt was relatively less due to reduced

precipitation. Rangelands are also degraded due to overgrazing by small and large ruminants, which has resulted in a loss of surface cover and an increased runoff in lower valleys.

2.4.2 Land Use Changes in Wet Mountain Regions

- Major change in land use system of the wet mountains occurred due to the reduction of per capita land holding and the division of land under the Muslim heritage. The land holdings are so small that it is not economical to farm the lands. Therefore, farmers started clearing the forest areas to develop terraced lands for cultivation of grain crops for food security.
- Deforestation and overgrazing of wet mountains and watersheds have resulted in degradation of rangelands, pastures, and forests. Clearing of forests and introduction of field crops and fruit plants has created issues of nonsustainability of watersheds, which are fed into the Indus River.

2.4.2.1 Impacts on Hydrology and Water

- Debris of landslides and sediments in flowing water ultimately enters into Indus river system through nullahs carrying runoff water. Changes in river morphology are common problems.
- Deforestation and degradation of rangelands resulted in changes in temperature and precipitation regimes affecting occurrence and distribution of precipitation. The Country has started facing droughts for the last few years, where precipitation and snowmelt was relatively less. Rangelands are also degraded due to overgrazing by livestock, resulting in loss of surface cover, and increased runoff in lower valleys.

2.4.3 Land Use Changes in Indus Basin Irrigated Agriculture

- Irrigated area in the Indus Basin has increased from 9 million hectares in 1950 to about 18 million hectares in 2000. In 50 years, this doubling in irrigated area is a main reason contributing towards changes in land use systems. Cropping intensity was improved from 60% in 1950 to 120% in 2000. This doubling in cropping intensity is a main reason in increasing agricultural production.
- Doubling factors, both in irrigated area and cropping intensity resulted in a four-fold increase in cropped area in Indus Basin irrigated agriculture, which was because of increased water availability due to the construction of reservoirs and an extensive canal network.
- A substantial shift in cropping pattern was observed with improved accessibility of water from the reservoirs (Mangla and Tarbela). Increase in cropped area was observed for grains, cash, and food crops. Therefore, there was a shrinkage in cropped area and conventional oilseeds and coarse grain crops. An increase in area of 52%, 39%, 44% and 36% was attained for sugarcane, rice, cotton, and wheat, respectively. There is 39% increase in the overall cropped area.
- The country's population is mainly concentrated in the cities located within or around irrigated areas.
- Thus fertile lands have been converted into housing colonies, roads, business centers, and industrial towns.

2.4.3.1 Impacts on Hydrology and Water

- Increased diversion of freshwater to irrigated areas coupled with low irrigation efficiency of around 40% in the Indus Basin resulted in waterlogging of around 42% area of the basin. High water table in brackish groundwater zones further resulted in salinization of surface soil due to arid environments.
- Around 20% basin area is affected by salinity. There was a decrease in salinity due to installation of deep tube wells under the Salinity Control and Reclamation Projects (SCARP) and private tube wells in fresh and marginal groundwater zones. However, use

of groundwater of SCARP and private tube wells in brackish groundwater zone resulted in secondary salinization and sodification.

- Increases in irrigated area, cropping intensity, and cropped area resulted into increased canal diversions over the last 30 years, creating a number of storage dams constructed in the Indus basin and leaving very little water in the river system to flow. This has an effect on the river morphology and the ecosystem downstream of the Kotri barrage, where very little water is now entering the Arabian Sea during 9 months of the year (the monsoon season lasts 3 months).
- The construction of the Ghazi Brohta power channels project has affected the river over a length of around 80 km and taken away the rivers right to flow. Indiscriminate use of groundwater and increased number of tube wells have resulted in mining of groundwater, as the water table is lowered in certain basins to the extent of 1–3 meters per annum.
- Lowering of water table makes it not economical—to the extent that pumping without subsidy on electricity is not economical—to raise crops in Balochistan valleys.
- Quality of groundwater is also a concern in areas, especially closer to the coast and in the periphery of the irrigated areas. The intrusion of saltwater into freshwater is now a serious concern due to indiscriminate pumping.
- Natural recharge of groundwater has been reduced tremendously due to the reduced area under water spreading systems like Sailaba and Khushkhaba, which were inbuilt to recharge the groundwater.
- The Karaize system has now been fully abandoned due to the mining of groundwater which has lowered the water table to a level where the Karaizez can't provide water any longer. It is important to mention that the Karaize system was more equitable in distributing water to the poorest compared to any other systems. Because Karaizes are community-based systems and tube wells are on a single ownership basis. Normally, only rich people can afford tube wells.

2.4.4 Farms and Farm Area

According to the estimates of the Agricultural Census (2013–2014), there are 8.26 million farms in Pakistan. These farms are operating over an area of 52.91 million acres of Pakistan. The average size of a farm in the country is 6.4 acres, whereas the cultivated area per farm is 5.2 acres. The total numbers of farms in Punjab province are 5.25 million and cover 64% of the total number of farms in Pakistan. These farms are operating over an area of 29.33 million acres and constitute 55% of total farm area. The distribution of farm area among small and large farms is highly skewed. The average farm size in different regions of Punjab province is about 5.26 acre and the average cultivated area is about 5.1 acres. Farms with less than 5 acres of land constituted 64% of the total private farms but they operated only 23% of the total farm area. Whereas, the farms that are of 5 acres and above in size, comprised only 34% of total private farms in Punjab but they commanded 57% of the total farm area of Punjab. Whereas, the farms that are of 25 acres and above in size, comprised only 2% of total private farms in Punjab but they commanded 20% of the total farm area of Punjab (Table 2.2).

TABLE 2.2
Area Irrigated by Different Sources (Area in Million Hectares)

	Total	Canals	Tube Wells	Wells	Canal Tube Wells	Canal Wells	Others
PAK	18.63	5.99	3.71	0.38	8.11	0.26	0.18
PUNJAB	14.88	3.35	2.82	0.28	8.11	0.26	0.06

Source: Pakistan Bureau of Statistics, Ministry of National Food Security & Research, 2015.

2.4.5 Water Supplies and Sources

The Indus River system in Pakistan serves the world's largest contiguous irrigation network (16 mha) (Tahir and Habib, 2000). Its total length from its sources of origin is about 2880 km. The main source of water in Punjab province is canals that run through its different districts. The length of these canals is 23,184 km and provides about 1.10 lac cusecs water which command a Cultureable Commanded Area (CCA) of about 21 million acres. The second main source of water is private tube-wells which commands about 10.73 millions acres in different districts of Punjab. The most significant change that occurred in the agrarian structure of Pakistan is due to increase in supply of irrigation water, particularly from private tube-wells especially in Punjab. Private tube-wells did not only provide additional water but also provided it on demand. The cultivation of new irrigation-intensive crops is increasing and increased use of fertilizers is resulting in synergistic effects on output and thereby on the profits of farmers (Acumen, 2008). The total area irrigated by all sources of water has increased over the years from 9.3 million hectares in 1965–1966 to 17.9% in 1996–1997 and 26.8% in 2006–2007. More specifically, areas irrigated by canals have increased from 7.5 to 7.8 million hectares whereas those irrigated by tube-wells rose from none to 9.5 million hectares over the same period of time. During the 1990s, more than 100 million acre feet water was diverted annually into the heads; by 2001–2002 only about 79.6 million acre feet were being diverted, then in 2006–2007 availability was restored to about 102 million acre feet. The number of private tube-wells has increased from 506.8 to 964.3 thousand in the last 10 years, nevertheless the level of groundwater abstraction could not increase correspondingly with the number of tube-wells during that same period. About 37.4 million acre feet water for irrigation purposes was provided by private tube wells in 1996–1997; in 2007–2008 this increased to about 40.4 million acre feet. Overmining and attendant saline infringement into fresh groundwater areas damage aquifers permanently (Khan et al., 2011).

2.4.6 Cropping Intensity

Cropping intensity is the number of times a crop is planted per year in a given agricultural area. It is the ratio of effective crop area harvested to the physical area. The overall cropping intensity in the country is 159%. It is highest in the Punjab being 167% and lowest in Balochistan being only 102%. Cropping intensity in Sindh is 162% followed by KPK being 156%. The Barani area covering four districts of Punjab, that is, Attock, Rawalpindi, Jehlum, and Chakwal has very fertile lands and its most popular crops are wheat, jawar, bajra, oilseeds, and maize. The average farm size in this region varies from 4.2 to 9.6 acres and cropping intensity varies from 100% to 117%. The wheat-mixed zone includes six districts with the average farm size ranging from 6.6 to 10.9 acres, while cropping intensity varies from 136% to 161%. The major crops in these regions are wheat, sugarcane, pulses, rice, and cotton. The rice-wheat zone includes six districts. The average farm size ranges from 5.7 to 10.3 acres and cropping intensity varies from 129% to 177%. The major crops in these regions are wheat, rice, millet, and sorghum. The cotton-wheat zone includes twelve districts with the average farm size ranging from 7.7 to 14 acres. The cropping intensity in this zone has wide variability ranging from 113% to 165%. The major crops of these areas are cotton, sugarcane, rice, maize, and vegetables. The mungbean-wheat zone includes four districts of Punjab, that is, Khushab, Mianwali, Bhakar, and Layyah. The average farm size in this zone is fairly large compared to the sizes in other cropping systems of the province and varies from 12.6 to 18.4 acres. The major crops of these areas are mungbean, wheat, guarseed, gram, cotton, and sugarcane. The cropping intensity is relatively low in this region, varying from 103% to 137%, portraying a usual inverse relationship between farm size and cropping intensity.

2.4.7 Motorways

The motorways of Pakistan are a network of multiple-lane, high-speed, limited-access or controlled-access highways, which are owned, maintained, and operated federally by Pakistan's National Highway Authority. The total length of Pakistan's motorways is 872 km as of June 1, 2016. Around

3690 km of motorways are currently under construction in different parts of the country, and most of these motorway projects will be completed in 2020. There are five motorways that are functional in Pakistan and the area under these operational motorways is 22497.6 km^2. Similarly, the total length of Punjab motorways is 654 km. There are three operational motorways in Punjab and they cover an area of 16565.82 km^2.

2.5 LAND USE CLASSIFICATION OF PAKISTAN

The main objectives of land use classification are to measure land and its impact on the ecosystem. To study technical attributes like soil profile, soil texture, agroclimatic conditions etc. Description of classification for land use in Pakistan and for each province is given in appendix (Table 2.1).

2.6 LAND RESOURCES

Since independence, the total cultivated area increased by slightly over 46%. However, there has been virtually no change in total cultivable (cultivated + cultivable) land area since 1960–1961; rather it has gone down as a proportion of reported area from 60% to 52%. Even the net sown area has stabilized around 15–16 million hectares since 1975–1976.

The cropped area has increased from 11.63 million hectares in 1947–1948 to 21.54 million hectares in 1995–1996, a growth of nearly 85% in 49 years. The double cropped area increased steadily from 0.95 million hectares to 6.31 million hectares during the same period and accounts for all the increase in cropped area during the last two decades. Now the current total cropped area is 22.73 million hectares.

The important expansion in the availability of land for agriculture during the first three decades reduced in the eighties. The annual rate of increase of cultivated area from 1980–1981 to 1990–1991 declined to 0.44% compared to 1.1% in the period from 1947–1948 to 1979–1980. Similarly, the rate of increase of cropped area in the 1980s was only 0.9%, in contrast to 1.5% over the previous thirty years. The annual rates of increase of cultivable and cropped areas in 1984–1985 to 1991–1992 were even lower, 0.3% for cultivated area and 0.17% for cropped area, when the rate of population growth was as high as 3.1% a year. Out of the total cropped area of 22.59 million hectares in 1995–1996 nearly 78% (17.58 million hectares) were irrigated, whereas crop production on the remaining 5 million hectares depended upon rainfall.

The irrigated Canal Command Area (CCA) has been grouped into classes on the basis of the severity and nature of its restrictions due to texture, salinity, waterlogging, and sodicity. Out of the total CCA measuring 14 million hectares, approximately 30% (4.2 million hectares) has been classified as class-I land, which has no limitations and is suitable for intensive irrigated agriculture. About 43% (6.0 million hectares) consists of good class-II agricultural land with slight restrictions and is suitable for growing common crops such as wheat and rice. Class-III covers 5.7% (0.8 million hectares) of the CCA, where crop yields are poor, but could be improved by better management. The limitations of class-IV land are severe. This is marginal agricultural land with severe salinity, waterlogging, and sodicity issues where productivity cannot be improved economically.

At this time about one fifth of the cultivated land in the CCA is affected by waterlogging in varying degrees and an even greater proportion suffers from salinity. An additional 1.2 million hectares of land are affected by excessive sodicity. It has been estimated that despite three decades of investments, the water table is still high in 0–1.5 meters in more than 1.58 million hectares of irrigated land. Out of the 1.58 million hectares about 1.46 million hectares have been declared as disaster area.

The uncultivated part of CCA covering 2.3 million hectares comprises of saline soils in the upper Indus and strongly saline gypsiferous soils in the lower Indus plains. The remaining saline, sandy, and eroded soils are not cultivable. An additional 4.7–5.6 million hectares of land are under rain-fed agriculture. Of these rain-fed croplands, 1.62 million hectares receive over 500 mm of rainfall and generally support dependable cropping. Areas receiving 300–500 mm rainfall account for another 0.85 million hectares. The active flood plains (riveraine sailaba) of the Indus river and its major tributaries total about 2.2 million hectares, of which about 0.69 million hectares are under crops. Cultivated torrent sailaba lands constitute an additional 0.97 million hectares within the Indus watershed. Another 0.56 million hectares of torrent sailaba land is cultivated outside the Indus Basin, primarily in Balochistan.

Nearly 60% of Pakistan's land area is cultivated. Much of it is referred to as grazing land, but only about one fifth is covered by usable forage. The remainder has been essentially denuded of useful species by overgrazing and uncontrolled management. More than 60% of all rangeland is in Balochistan. The forestland constitutes slightly over 4% of the total land area of the country excluding Azad Kashmir. Although the area under forests increased from 1.38 million hectares in 1947–1948 to 3.61 million hectares in 1995–1996—an increase of about 162%, it is still considerably short of the required range of 20%–25% of a country's land area in view of environmental and economic considerations.

Our land resource is shrinking due to salinity, waterlogging, sodicity, erosion, and the use some of the best agricultural lands for such nonagricultural uses as urbanization, industrial projects, and highways etc. However, at present land is not the limiting resource as for every net hectare, out of the total reported area sown, there are two uncropped hectares even after excluding the areas not available for cultivation. Thus, at present water not land is the major limiting factor for bringing more areas under cultivation.

2.6.1 Main Problems of Land Resources

It is an irony of fate that on one hand, we are incurring huge expenditure on the reclamation of saline and waterlogged soils, while on the other we are allowing, indiscriminate and irreversible diversion of very good agricultural lands to nonagricultural uses such as urbanization, industrial projects, and other projects etc. Most of the threats to land and soil arise because we expect the soil to perform a range of functions, in some cases many functions at the same site. By steadily increasing the demands on the soil we have often created an unstable system where the soil becomes less resilient and more vulnerable (Lal, 2007, 2009; Pretty, 2008). These threats are increasingly seen as particularly relevant to the biomass production function of soils. The most important causes for reduced land productivity have been water and wind erosion, salinity, sodicity, waterlogging, flooding, and loss of organic matter. According to one finding, 17% of surveyed soils (which include most of the soils usable for agriculture, forestry, or ranching) were affected by water erosion, 7.6% by wind erosion, 8.6% by salinity and sodicity, and 8.65% by flooding and ponding; fully 96% suffered from less than adequate organic matter. These problems often occurred simultaneously and produced synergistic impacts on agricultural productivity. As such, soils in Pakistan have suffered from both water and wind erosion as well as poor organic matter contents, thereby reducing potential productivity of the best soils (Mustafa et al., 2007). The agricultural land in Pakistan has three major issues:

1. Salinity and waterlogging
2. Soil Erosion
3. Floods

2.6.1.1 Salinity and Waterlogging

The concept of salinity in soils denotes the condition in which the total amount of salts in the subsoil is enhanced where ultimately plants become unable to grow root in the soil. Improper irrigation and washing may cause such conditions. Waterlogging describes an elevation of underground water table up to the root zone that causes the conversion of cultivated land into marshy areas.

According to a survey of Pakistan, about 40%, or 6.3 million hectares, of cultivated irrigated land is fertile. About 4.9 million hectares, or nearly one third, has clayey soils needing special tillage implements. About 10%, or 0.6 million hectares of land, have higher sand to silt/ clay ratio. Approximately 40% of the land in the central western parts of the country has been affected by light water and wind erosion. In the southwest and along the southern coastal parts, wind-eroded and salinized soils have predominated (Mustafa et al., 2007). In Punjab and Sindh, 48% of the soils are saline while 18% soils are intensely saline. Of the total salt affected soils of about 6.3 million hectares in Indus plains, only one half has practical significance and its main problem is sodicity, covering about 1.7 million hectares, out of 5.4 million hectares of land, about 1.3 million hectares of cultivated and 0.9 million hectares of uncultivated saline land have drainage problems.

2.6.1.2 Soil Erosion

It is a natural process that some activities of mankind may result in a dramatic increase in erosion rates, especially unsustainable agricultural land use (Lal, 2001). As soil is an essentially nonrenewable resource, when erosion is serious it is generally irreversible and the soil is lost forever. This elimination of nutrient-rich soil makes it barren and causes reduction in per acre yield. For example, arid areas have no foliage because their topsoil is removed by erosion. The wind is also active in dry areas of the Indus Basin as well, and its erosive action is obvious in the region.

About 70% of cultivated barani land has good soils, but has a compact soil layer of 5–8 cm thick that has formed below the plough layer of 7–10 cm. Moreover, soil erosion is a problem in about 1.0 million hectares, especially on mountain lands resulting from extension of cultivation without conservation measures and excessive cutting and grazing. However, there are about 2 million hectares of riverine land of which 0.3 million hectares are cultivated with moisture left by summer floods. This land offers the possibility of irrigation development with tube wells. On other lands, narrow strips of cultivated land bordering the sandy deserts are affected and threatened by wind erosion. Shifting sand is encroaching on cultivated areas and settlements.

2.6.1.3 Floods

Floods are considered one of the major problems of agriculture of Pakistan. They cause major loss of life and property in areas in which they occur. Floods and landslides are mainly natural hazards, intimately related to soil and land management practices, although their impact is often exacerbated by unusual environmental conditions. Landslides have a predominantly local impact on food production although they may temporarily impact food distribution through the disruption of communication networks (Blum, 2013).

The fragmentation of farms into two or more separate holdings is a significant constraint in the agricultural production system in Pakistan, affecting 62% of all farms. Of these, 39% have been fragmented two or three times, 12% four or five times, 7% six to nine times, and 4% more than ten times. Fragmentation is more serious in the KPK and Balochistan, where 77% and 72% of total farms are fragmented. In the Punjab and Sindh, fragmented farms contribute 62% and 52% of total farms, respectively. Not much has been done in this regard to encourage and provide incentives to farmers to consolidate their land holdings for more efficient farming.

Even though land use has improved since independence, the rate of growth of cultivated areas, as well as that of cropped areas, has declined over time. Furthermore, the cultivable waste area has continued to increase, but a large portion of this cultivable wasteland is reclaimable, provided a proper mix of remedies is applied.

2.6.2 Measures to Resolve Problems of Land

Some of the measures suggested by the National Commission on Agriculture, 1988, and in the National Policy of Government of Pakistan, 1991 are given below:

- The saline sodic soils under irrigated agriculture should receive high priority for reclamation with gypsum. Since sodicity, rather than salinity is the main problem in these soils, only moderate leaching is required and soils would improve dramatically by the application of gypsum, which should be made available to the farmers at subsidized rates.
- Special tillage implements, such as chisel ploughs, subsoilers, and disc ploughs etc., with tractors should be promoted and encouraged to break hard pans in the subsoils by subsidizing these operations, especially through the agricultural service companies.
- Low cost fractional tube-well technology, already in widespread use in the Punjab, should be encouraged in order to develop riverine areas.
- Comprehensive soil conservation and gully rehabilitation programs should be implemented in barani and mountain areas in order to check water erosion. Similarly, to control wind erosion, especially in desert areas, sand dune stabilization is worth while if accompanied by controlled grazing, introduction of higher yielding nutritive plant species, increase of vegetation cover, and planting belts of wind break species along the border strips of cultivated area.
- The SCARP program will need thorough reviewing, because the existing high-cost saline groundwater drainage methods have not proved effective, especially in Sindh.
- Reclamation of salt affected land should be treated as an onfarm activity emphasizing good extension work by trained specialists instead of capital intensive engineering projects.
- If surplus water is available after meeting the needs of better land use, the next priority for reclamation should be the permeable saline sodic and strongly saline gypsiferous lands. The latter, occurring under arid climates (rainfall less than 250 mm), naturally contain gypsum and only require leaching coupled with drainage (if needed) for their reclamation.
- The first and foremost requirements in dealing with waterlogged soils is the expeditious removal of all possible excess surface water, especially from the effective root zone of the soil. In sweet water zone vertical draining, the use of tube wells is effective in both lowering the groundwater table as well as supplying additional useable irrigation water. However, in the saline water zones, open or horizontal drainage is the only alternative.
- Programs should be developed to increase the rate of growth of cropped areas through both horizontal and vertical expansion.
- A policy of optimizing land use and cropping systems should be planned and implemented on the basis of soil survey data and agroecological zoning.
- Agra-ecological zones of the country must be redefined by processing the reconnaissance soil survey data along with climatic data, irrigation water resources data, and present land use data.
- Research on soil and water management (especially soil tillage and water scheduling) needs to be strengthened, and take into consideration the agroecological zones, soil types, and farming systems.
- Legislation should be introduced to check nonagricultural uses of good agricultural land.
- To realize the full potential of land, the Department of Soil Survey of Pakistan, Provincial Soil Fertility Organizations, and Department of Land Utilization should be well coordinated and strengthened.
- The farmers must be provided incentives to encourage them to consolidate their fragmented farms. Transfer fees should be abolished for the farmers desiring consolidation of land. Procedures need to be simplified and the law reviewed.
- State lands in possession of landless farmers should be sold out at rates to be fixed by boards of revenue on the basis of the produce index units. Such programs may be implemented by the provinces.

- There are vast cultivable waste lands (about 9 million hectares) which cannot be brought under cultivation due to nonavailability of irrigation water. Over 50% of these culturable wastelands lie in Balochistan alone and receive varying amounts of rainfall. The potential of extending water harvesting and runoff farming techniques to these lands in order to bring more areas under cultivation and to develop rangelands, especially in areas receiving less than 300 mm of annual rainfall. On those cultivable wastelands receiving at least 300 mm or more of rainfall, the feasibility of dry land crop production should be studied until these areas can be brought under regular controlled irrigated agriculture.

2.7 STABILITY OF THE LAND RESOURCES

It is supposed that most of the agricultural land of Pakistan is problem free and fertile; however, it is under the impending danger of deterioration due to salinization and waterlogging.

2.7.1 Hazard of Salinization

Studies of soils all over the Indus plains have revealed that most of the soil salinity in the Indus plains is very old, produced by the process of soil formation much before the introduction of modern canal system. The extent of (secondary) salinity produced as a result of irrigation is limited and does not extend to more than a small proportion of the affected land.

The salinity of soils was well recognized by the British at the time of settlements and colonization in the nineteenth century during the allocation and distribution of canal water in the area. Canal water was not allocated to severely affected soil (Wace, 1934).

There is, however, another process of soil deterioration that should be of great concern to the nation. It is the sodication. There are different symptoms of the sodicity of soil, including reductions in rate of infiltration, hardening of topsoil, and insufficient seed germination, particularly of alkali-sensitive crops. This mode of soil degradation is treacherous, operating insidiously at a slow rate.

2.7.2 Scare of Submergence

The apprehension of submergence of large parts of irrigated plains is based on fantasy, completely detached from the reality. The irrigated Indus plains, although apparently level are far from being flat. There are small but important local variations in relief inherited from depositional patterns of the river alluvia. The rise in water table, due to seepage from the canal system, does not affect all the land uniformly. In depressional areas, representing in-filled channels and basins, the water table comes close to the surface and in some cases, small local areas are submerged and turn into marshes. In most of areas, the water table remains deep enough not to affect crop production and cropping patterns.

The general lay of land and the net input of water in the Indus plains are such that not more than about 1% of the canal commanded area can ever be submerged and the prediction that unless a massive drainage system is provided, certain regions would turn into lakes, has no basis (Choudhri et al., 1978).

2.8 CULTIVATED LAND CLASSIFICATION ON QUALITY BASIS

Canal-irrigated areas are the citadel of agriculture in our country. The canal commands spread over approximately 34 million acres of a variety of land in the Indus plains. Of this, nearly 6 million acres, although within the folds of canal commands are virgin, but the bulk of the land, measuring about 28 million acres is actually under cultivation. The quality of the cultivated land is as discussed below:

2.8.1 Class-I: Very Good Agricultural Land

This land occupies an area of 10.3 million acres. It is well suited for a wide range of crops. Soils in this class have no limitations for crop production. They are medium textured and are easy to work. They are nearly level, deep, and well drained and show the highest response to good management, including the application of fertilizers. They are used for general cropping, vegetables, and orchards.

2.8.2 Class-II: Good Agricultural Land

It occupies 14.4 million acres in the cultivated part of the CCA. Soils in this class have minor limitations for crop production; either the range of suitable crops are somewhat narrow or the management cost is somewhat high. Generally, the net returns from this land are about 25% less than that of Class-I land. The important subclasses within this class are as follows:

1. Subclass irlls-clayey:
   It occurs on an area of about 10.0 million acres. Soils in this subclass are clayey and slowly permeable. Preparation of seedbed is difficult. They become excessively wet during the prolonged rainy season in the areas having annual rainfalls of more than 15 inches. They are used for general cropping in the arid areas. In the subhumid areas in the northeast, the Indus Right Bank and Delta area in Sindh, rice is the main crop. They give high response to good management.
2. Subclass irlls-sandy:
   Soils in this subclass are somewhat sandy (sandy loam). They have a somewhat low capacity for holding water and nutrients. Other soils in this class are underlain by sand at about 30 inches depth. They are used for general cropping and vegetables. They require special water management practices such as frequent but light applications of irrigation water and fertilizer.
3. Subclass irlla-slight salinity and sodicity:
   Soils in this class have minor salinity and sodicity problems generally in the surface to a few inches of depth. Only a small proportion of this land has high water table (at about 5 feet depth).
4. Subclass irllw imperfectly drained:
   Soils in this subclass occupy low-lying areas of land, which accumulate runoff from the surrounding higher land. A part of this area has a water table depth of about three to five feet depth. This land is used for general cropping or rice, but the yields are low to moderate.

2.8.3 Class-III: Moderate Irrigated Land

This land occurs on 1.9 million acres of the cultivated part of the CCA. Soils in this class moderate limitation for crop production. They have a limited range of suitable crops. The net return from this land is generally about 50% of that of class-I land. The important subclasses within this class are:

1. Subclass irllla-porous saline-sodic:
   Soils in this subclass are either saline-sodic to about 36 inches depth or are very strongly saline, containing gypsum with a porous and permeable nature.
2. Subclass irlllw-imperfectly drained:
   Soils in this subclass occupy low areas in the landscape. They have problems of excessive wetness due to either a high water table (at 2–3 feet depth) or an accumulation of rain water. They are mostly clayey and are generally used for growing rice.
3. Subclass irllls-sandy:
   Soils in this subclass are either (sandy loam with loamy sand surface) or have sand at about 20 inches depth. They have a low capacity for holding water and nutrients.

2.8.4 Class-IV: Poor Agricultural Land

This land occurs on about 1.2 million acres in the cultivated part of the CCA. Soils in this class have severe limitations on account of shallow soil depth or strong salinity and sodicity combined with slow permeability. The important subclasses within this class are described below:

1. Subclass irlVs-sandy
   Soils in this subclass are loamy or clayey, but they have sand at about 12 inches depth restricting the root zone. They have a very low capacity for holding water and nutrients. At present, they are used mainly for growing wheat, but the yields are low.
2. Subclass irlVa-dense saline-sodic
   Soils in this subclass are saline-sodic. These soils have been reclaimed to a depth of a few inches, the subsoil being still strongly saline and impervious to water. So these soils are used for growing rice and wheat, but the yields are low.
3. Subclass irlVw-imperfectly drained
   Soils in this subclass have either a high water table or accumulate rainwater during rainy seasons thereby restricting the choice of crops to rice with a low yield.

2.9 UNCULTIVATED AREAS OUT OF CCA

The uncultivated areas out of culturable command areas are used for grazing or forestry. The natural forest areas are confined to northeastern parts of the country where adequate precipitation and the effect of high altitude permit the growing of natural forest. The soils under these forests are generally deep and quite stable in spite of their position on steep slopes. But due to strong biotic pressures, not only are the areas under the natural forests dwindling, but the denuded soils are also subjected to erosion. Far larger areas of the uncultivated land in the country are used for grazing. The grazing potential varies with the availability of moisture influenced by rainfall, temperatures, and soil depth. Soils in these areas show a great range in their physical and chemical characteristics.

2.10 BARANI AREAS

About 11.83 million hectares are under barani farming, which includes rain fed cultivation, riverine sailaba areas (kacha area), and torrent watered sailaba cultivation in Baluchistan and the outwash plains of the Sulaiman and Kirthar ranges.

The main barani areas of the Punjab, Sind, and KPK and part of Baluchistan have been included in soil surveys and the soils have been evaluated in term of their agricultural potential. Under barani cultivation, the class-I and class-II land do not exist because of the limitations imposed by climate, especially inadequate rainfall. The best barani land forms class-III moderate agricultural land. About half of the barani areas have class-III lands. They occur in areas with an annual rainfall of more than 20 inches. A major part of class-III barani lands have no soil limitations, whereas slope, limited soil depth, and erosion are the important soil problems in the remaining part of this class.

The other half of the barani area constitutes class-IV-poor agricultural land. These lands are either located in low rainfall areas, where moisture availability is inadequate and uncertain, or have severe problems of slope, limited soil depth, and erosion. Possibilities for increasing agricultural production on this land are limited. Low rainfall areas and the torrent-watered sailaba areas fall under this class.

2.11 CLIMATE AND PHYSIOGRAPHY OF AGROECOLOGICAL ZONES

2.11.1 Agroecological Zones of Pakistan

Pakistan lies between latitudes 24–37°N and longitude 61–76°E on the globe. The climate of the country is generally characterized as subtropical arid to semi arid. However, small tracts with a temperate subhumid to humid climate are also present in the northern part of country. Overall, the most problematic climatic factor hindering crop productivity in Pakistan is the limited amount of precipitation. The incidence of rainfall in the country is highly variable, with less than 100 mm in some parts of Sindh and more than 500 mm near the foothills of the Himalaya.

Agroecological zones of Pakistan are based on climate, land use and water use. Pakistani land can be divided into the ten following agroecological zones. The following is a general overview of these zones:

1. Indus Delta
   It includes the Thatta, Badin, and Hyderabad. Annual rainfall is between 125 and 250 mm. Main crops are sugarcane, rice, and pulses. Farmers also grow wheat and rapeseed during the winter season.
2. Southern Irrigated Plain
   This includes the districts of Hyderabad, Tharparker, Badin, Sanghar, Dadu, Larkana, Nawabshah, Jaccobabad, Sukkur, Shikarpur, Sibi, and Rahimyar Khan. Annual rainfall around Sukkur is less than 125 mm and increases to 250 mm near the coast. Generally, farmers grow wheat in winter and rice in summer.
3. Sandy Desert
   It includes Tharparker, Nawabshah, Khairpur, Sanghar, Rahimyar Khan, Bawalnagar, Mazaffargarh, Mianwali, and Sargodha. Annual rainfall is between 125 and 250 mm. Summer is very hot and winter is near freezing. Farmers generally grow wheat, millet, caster, and guar.
4. Northern Irrigated Plain
   This includes the regions of Rahimyar Khan, Multan, Mazaffargarh, Vaheri, Sahiwal, Lahore, Kasur, Faisalabad, Jhang, Sheikhupura, Gujranwala, Sargodha, Gujrat, Peshawar, and Murdan. Annual rainfall ranges between 125 and 500 mm in the southern portion of the irrigated plain. Most of the plains are canal irrigated with crops such as cotton, sugarcane, maize, and wheat in the extreme south; while rice, wheat, and berseem are grown in the Northern part of this plain.
5. Barani lands
   These are found in D.I Khan, Bannu, Mianwali, Attock, Abbotabad, Rawalpindi, Chakwal, Jhelum, Gujrat, and Sialkot. Crop production in these regions depends on rainfall, which ranges from 1000 mm per year in the northeast to 200 mm in the southwest. The main crops include maize and wheat in areas above 700 mm of rainfall, groundnut on lighter soils, and millets, sorghum, gram, and lentils in areas with less than 500 mm.
6. Dry Mountains
   Im this region, The districts of Hazara, Mansehra, and Rawalpindi are included in this region. The annual rainfall exceeds 1000 mm over most of the region and the elevation ranges from 1000 to 3000 m. Only 25%–30% of the area is cultivated in terraces. Above 2000 m, only a single crop per year is possible and it is usually maize.
7. Northern Dry Mountains
   This includes the districts of Chitral, Dir, Swat, Malakand, Mohmend agency, Khyber, Bajur, and the tribal areas of Peshawar and Kohat. Annual rainfall ranges from 300 m in tribal Kohat and Chitral to 1000 mm in Swat and Para Chinar. The main crops are maize, wheat, and fodders.
8. Western Dry Mountains
   This consists on Kohat, Bannu, Kurram, North and South Waziristan, Zhob, Loralai, Sibi, Quetta, and Karachi. The annual rainfall in this region varies from 125 mm in the southwest to 500 mm in the north. The main crops in the south of this region (Balochistan) are fruit tree mixed with wheat, vegetables, and fodders. The area above 1000 m further north with more summer rains have the same mix of deciduous fruits and wheat, vegetables, and fodders. At a lower elevation (<1000 m) crops such as maize, sugarcane—with some tropical fruits such as citrus and guava—are common.

9. Dry Western Plateau
   It includes Karachi, Dadu, Mekran, Kharan, Chagai, and Lasbella. The region has hot summers, warm winters, and low variable rainfall ranging from 50 to 200 mm. The cropping systems are based on wheat and summer cereals (i.e., sorghum, millet, and maize) with tropical fruits (mangoes, guavas, chico, citrus, and bananas).
10. Sulaiman Piedmont
    This consists of Karachi, D.I Khan, and D.G Khan. The annual rainfall in this region is low and variable. Most rainfall comes in summer and ranges from 125 to 250 mm per year. The main agriculture system consists of "Sailaba" or "Rod Kohi" where farmers channel flood from a large catchment area into their fields, which are leveled and bunded. Only annual crops are possible consisting of wheat, gram, lentils, oilseeds (Brassica), millet, and sorghum.

2.11.2 Climatic Influences

Pakistan has a diversified climate due to the difference in land topography, altitude, and seasons. Pakistan has four major seasons, including a cold winter from December to February; a moderate spring season from March to May; a hot summer season from May to August and an autumn which starts from late September until December.

Monsoon season in summer has an important impact on the agriculture of Pakistan. Most of the northern areas of Pakistan are situated out of the monsoon rain shadow.

Due to differences in topography, there are significant differences in the temperatures of Pakistan. The regulatory aspects of the climate are:

1. The oceanic influence of the Arabian Sea.
2. The subtropical location of Pakistan.
3. The monsoon winds
4. Higher altitudes in the west and north.
5. A temperature overturn layer at a low altitude of about 1500 m.

2.12 Soils of Pakistan

The upper nutrients containing layers of the earth, cooperative in the growth of vegetation and plants is called soil. Soil contains water, air, organic matter, and minerals etc. It consists of approximately 45% minerals, 5% organic matter, 20%–30% water, and 20%–30% air.

2.12.1 Types of Soil

There are five different soil types, which are described below:

1. Clay soils
2. Sandy soils
3. Peaty soils
4. Silty soils
5. Saline Soil
6. Loamy soils

2.12.2 Classification of Soils According to Regional Basis

The soils of Pakistan can be classified according to their regions:

1. Indus basin soils
2. Bongar soils

3. Khaddar soils
4. Indus Delta soils
5. Mountain soils
6. Sandy Desert soils

2.13 URBANIZATION IN PAKISTAN

Economic theory proposes that the structural change generally occurs in the reallocation of resources from less productivity agriculture to high productivity industrial and service divisions. Cities and towns are put magnets in these divisions to advantage from surpluses and labor migration under employed. Therefore, rapid economic growth and urbanization are interrelated. The pace of urbanization, in turn, to accelerate the migration of a larger reactor.

One difficulty is that the unit cost of investment in infrastructure in rural zones due to the establishment of relatively high deployment over a half long diameter of the population. With a density in the city these fees only for those in rural areas, which is a small part. Therefore, standards for efficient distribution of resources are favorable for investment in urban infrastructure. However, this is the concept of fixed allocated efficiency. In the intermediate and long-term per capita income in urban regions to make higher income elasticity. After returning to these products and farmers are not grains such as wheat or rice is high.

We will continue to promote this cycle of mutual dependence in large supermarkets, hypermarkets, cutting shops, and brokers, who buy their supplies right from farmers. In this process, we have expected that customary goods will also increase the farmers' income. It has been observed, which is the best family in the economy to send their children to school, rather than having them on the farm. This is the case and underdeveloped areas to be capable to catch up with important regional and fusion occurs. It will reduce the income gap between the urban areas and will share the fruits of economic growth on a large scale.

Pakistan is one of the rapidly growing countries in South Asia with regard to urbanization, as the proportion of people in urban areas increased from 17% in 1951 to 38.8% in 2010–2015; predictions indicate that half of the population aged between 10 and 15 years is living in urban areas. The net immigration rate is 2.81% annually towards the urban areas of the country. Metropolitan (100 million people, in 1998) the share of 50% of the total population, so far has been risen to 60% or more. About 80% of GDP and the tax revenues of almost the entire country come from urban areas, which comprise about 60% of the workforce. The poverty rate in urban areas is almost half of that of rural areas. Per capita revenue levels and growth rates are also comparatively high in urban regions, where male and female literacy and admission rates are the best. Despite these positive results, an overall lack of stable, fixed planning, or implementation does not provide a satisfactory picture of the annual growth rate of 3%, which was in the form of an explosion in a large urban setting in Pakistan. Breach of the master plan only in high quality and practical applications that are available for the implementation or rising costs. About 35%–50% of the urban population is living in Katchi Abadis, which are about 300 in numbers for more than fifty years only in Lahore. There are some differences in Karachi now, because there are 500 settlements in accordance with the standards. Lahore, the capital of successive ministers have taken over the past 15 years on the personal interests of the rulers in progress to have a master plan of social development. Sialkot, a typical local citizens and business community, government intervention in infrastructure, productivity and creating relations is taking matters into their own hands. In Karachi, between 2002 and 2008, county town government structure came and some aspects of good governance were seen but since 2008, when city district government system was cancelled without any alternative system of government and local councils and institutional structures are suffering negligence. Land, water, and transport mafias in Karachi are more active involving citizens groups, criminal gangs throughout the city.turning it to the capital city of an invalid argument.

Quetta, Faisalabad, and Rawalpindi have seen mixed results over different time periods, but generally they do not offer a healthy image. Sukkur, Hyderabad, and Gujranwala are usually the poorest example of unplanned urban sprawls. Multan has made some modest progress during the 2008–2013 period, due to the development of infrastructure during this time.

2.13.1 Rural–Urban Integration

About more than half of rural households in Pakistan are destitute, so the head of household usually migrate in search of better jobs, higher wages, and gain professional skills in order to direct a share of their salary to their families back home. Therefore, immigration and rural–urban migration between urban areas and abroad is still an essential strategy. It would be the idealistic to visualize that in the light of this strong push and pull factors and movement to urban areas retreat. On the other hand, the electronic media now available on a large scale, even in the remote villages, has quickened the movement of masses towards urban areas. It would be wise to consider connectivity and economic integration of rural to urban areas and not only hope motivated migration of rural development and disappear. Increases in agricultural productivity through the efficient use of water, land leveling, and high-yielding seeds will further decrease the need for labor. Migration from rural to urban areas is a key element of economic restructuring.

The agricultural sector itself is a big shift in the department. Major and minor crops account for less than the value and proportion of added today than a decade ago in the agricultural sector, livestock and dairy products, fish and share gardening has gradually increased. More and more synergy between the towns and villages of the dairy industry, illustrative examples. Mostof the milk is produced in rural areas, but is consumed in the city.

Milk collection, sharing between rural and urban sectors transport, treatment, disinfection and bottling and retail distribution formation in agriculture, industry and services to benefit the population of the entire value chain. Poultry and meat intake is also a increase in slaughter houses and poultry farms near the city's secular position and will donate to this inter-dependence. More than half of the worth of the product is added here after it leaves the farm.

In agricultural research and development, soil, water and animal husbandry and veterinary services, storage and cold chain, refrigerated transport, what is important to the management of investments to realize their potential supply of agriculture will be invested in roads, logistics, manufacturing, packaging and retail sales organizations to promote the smooth flow of investment, but also increase farmers income, which in turn will drive demand for urban production of consumer goods. Phone, satellite and cable TV channels started the blue differences between urban and rural lifestyles.

Because of land ownership is not clear, complete documentation and record-keeping, to be manipulated and small bureaucratic control hanging over the land market does not work well. 80% of Pakistan's court proceedings belong to a land dispute. If land markets work well and the land becomes mobile there will be more effective dispute resolution among the users.

2.13.2 Managing Urbanization

It must be the starting point for the decentralization of the administrative and financial autonomy to manage the city's point—both big cities and medium-sized cities.

The first 18 amendments to the federal authorities, the authorization of the local financial resources. However, this decentralization will remain incomplete, without the general public until the regime neighborhood is a powerful time requirement in place. The ability of the local government should be enhanced to carry out their duties efficiently so that official improvement should also spread to the land market, labor market, and the provision of basic services.

There should be some large cities in which investment is managed, and the elected mayor should be the head of Municipal Government (MG). Planning and land distribution, its use, as well as housing, infrastructure, park facilities, public transport, water and sanitation, and the development of solid waste disposal must be specifically under the control of MG. The executive authorities and the necessary financial resources should be provided to run the city district government in an appropriate manner.

In the absence of this decision mechanism, Katchi-Abadis and slums, insufficient municipal organization, significant gap between the poor and rich, as well as costly transportation will remain. Corresponding within a city or metropolitan government agencies and jurisdictions to create a parallel fragmentation.

Another task is the deployment of public finance and urban sustainability. Most migrants coming from rural areas were included in the informal economy. This is a critical issue for urban public finance. This inflow of population increases the costs related to the expansion of water sanitation services, sewage treatment, public transportation, and other urban infrastructures and services. While the duties and fees of these services are continuously declining.

Those already in the tax net are requested to pay more of their income, while getting poorer services in return. Similarly, they vote with their feet by moving to the suburbs, where taxes are lower. Another common trend is to induce the official through their relationships for evading taxes. This further reduces the overall tax collection, increasing spendings and affects the financial condition of the city. Municipal pledges to finance massive long-term infrastructure projects cannot be kept because of the poor financial condition of the city government. Therefore, viability of upcoming planning and growth of cities are being limited by shortage of funds.

Preparation and implementation of the overall planning of settlements in the major cities are key factors in urban management. Generating new or existing spaces and rejuvenating old spaces to absorb future migration into cities, we need to invest in infrastructure and basic public services effectively.

There is an other problem that affects overall economic growth, which is avoidance of the law due to local corruption, fraud, nepotism, and favoritism; this prevails in the construction and monitoring mechanism. Introduction of transparency and good governance can bring any change to urban sprawls, shantytowns and slums in the system and improve the water, sewerage and transport facilities in the cities.

2.14 AGRICULTURAL DIVERSIFICATION

In agriculture, modifications can reallocate some of the productive resources in the farm such as land, capital, and agricultural equipment to get higher returns on investment. This diversification also includes risk reduction and responding to altering customer demands, changes in the government's policy, to exterior shocks, and more recently, to the impacts of climate change.

2.14.1 Drivers of Diversification

Diversification can be a response to both prospects and dangers.

2.14.2 Opportunities

Demand of food in developing countries is changing with the improvement in their financial conditions. People are moving away from their traditional food to the one with the larger content of animal products, fruits and vegetables. In order, only progressive farmers are competent to fulfill these needs.

2.14.3 Changing Demographics

Changing demographics influence consumption patterns in the rapidly urbanizing developing countries. In addition, there is a small number of farmers, in percentage terms, at least to provide consumers with a greater quantity.

2.14.4 Export Potential

The Farmers of developing countries can have a considerable success by diversifying into crops that can meet export market demand. While concern about food miles, as well as the cost of complying with supermarket certification requirements such as for Global Good Agricultural Practices (GAP) to meet public requirements may threaten this success in the long run. However, diversification has much potential to meet the export markets.

2.14.5 Adding Value

Pattern in Western witness, now becoming popular in developing countries, the consumer is put in less and less time for food preparation. They are in a growing need for preparing meals at any time and effort from the packaging, such as pre-cut for Salahuddin. This value-added diversification for farmers, in particular can play a key role in the country and supermarket retail opportunities.

2.14.6 Changing Marketing Opportunities

Control can be linked to the farmer's market can open up a variety of possible changes in a new way for government policy. For example, a change in some countries policy to remove the state "market regulation" and a monopoly to hold all dealings it promising for farmers to create new products and buyers contracting directly.

2.14.7 Improving Nutrition

Diversify from the traditional one essential nutrients may be important for farmers in developing countries nutritional value.

2.14.8 Threats

2.14.8.1 Urbanization

This is a chance and a risk, because the pressure on the productivity of land resources in peri-urban areas. If farmers remain on the land, the land that you need to generate more revenue than it is through the cultivation of basic food stuffs. It is in fact close to the market; close to the city explains why farmers diversify their crops and high value.

2.14.8.2 Risk

Farmers face the risk of variable weather and price fluctuations. Diversification is a reasonable response to both. For example, some crops are more drought-resistant than others, but can provide bad economic returns. Similarly, diversification can manage price fluctuations, as not all goods will suffer from lower prices at the same time. In fact, farmers often do the opposite of diversification by planting products that have a high price in one year, while prices collapse in the next year.

2.14.9 External Threats

The farmers depending on exports do not change due to the changing needs of consumers however, most likely due to policy changes and risks. In early 90s Government of Pakistan procured oilseed from the farmers that promotes oilseed cultivation in the area. However, cultivation of oilseeds in Punjab severely decreased with the change of policy.

2.14.10 Domestic Policy Threats

Agricultural production is sometimes undertaken as a consequence of government subsidies, rather than because it is inherently profitable. The reduction or removal of those subsidies, whether direct or indirect, can have a major effect on farmers and provide a significant incentive for diversification or, in some cases, for returning to production of crops grown prior to the introduction of subsidies.

2.14.11 Climate Change

The climate change also leads to changes in the accessibility of water for production. In some countries, including Sri Lanka, India, Canada, Kenya, and Mozambique farmers have begun to diversify in response to climate change. The Kenyan government has implemented policies to encourage crop variation, including the elimination of grants for some crops and the encouragement of land use, zoning, and the introduction of different land tax systems. Climate change is currently threatening agricultural production in different ways: increased soil losses and degradation through increasing extreme weather events causing erosion by water, wind, floods, landslides, desertification, and salinization (Alley et al., 2003; Haron and Dragovich, 2010). It also creates an increasing lack of fresh water for irrigation and competition with water used in biofuel production (Berndes, 2008), industry, and households (Rosegrant and Cai, 2002).

2.14.12 Opportunities for Diversification

In making decisions about diversification, farmers need to look at new income generated by agricultural new farm enterprises in comparison with existing activities, with similar or less risk. In addition, they may lack the market for the product. The United Nations Food and Agriculture Organization (FAO) promoting diversification by small farmers and has issued a booklet identifying mushroom farming, milk production, sheep and goats, fish ponds, and beekeeping among other things as, diversification possibilities.

GENERAL RECOMMENDATIONS

1. A land use policy or a new legislation is needed to restrict unbridled housing schemes around big cities and rural towns, resulting in the pouching of highly fertile lands, and to arrest the land fragmentation process due to laws of inheritance.
2. Reclamation of saline and waterlogged soils must be accelerated. The desertification process due to deforestation, erosion, and overgrazing be addressed. Land degradation is a perpetual threat to agriculture. This continues to occur due to improper land use, intensive cultivation of land without sufficient fallow periods, deforestation, overgrazing, faulty agricultural practices, inappropriate crop rotations, injudicious use of fertilizers, pollution, and over pumping of groundwater. A revision of the present land classification systems is needed.
3. The distribution/allotments and encroachment of state lands must be contained to save the land as a trust for future generations.
4. The productivity of agricultural lands be enhanced by precise interventions at the micro zones- and commodity clusters level by introducing soil and water analysis and market information systems.
5. Punjab should partner with Baluchistan (future food basket) to resolve interprovincial mistrust for water storage and for future land development for agriculture.
6. Develop a scheme for wetlands, rainwater harvesting, flood canals, river dredging, river lakes, river locks, canal water storage, onfarm storage, and groundwater recharge wells.

7. Develop a monitoring and implementation strategy for peri-urban agriculture where contaminated sewerage water and industrial affluent are used for growing vegetables and the rearing of buffalo colonies.
8. Restriction on groundwater pumping must be imposed. Irrigation water should be priced and cultivation of low delta crops be incentivized.
9. Water table rises in waterlogged areas such as Sargodha, Mandi Baha ud din, and areas along the link canals, which needs to be reclaimed. Water productivity in Pakistan, for example, 0.45 kgm^{-3} for wheat is much lower than 0.8 and 1.0 kgm^{-3} in India and the United States, respectively. Water pricing of canal water is negligible and needs rationalization. punjab irrigation & drainage authority (PIDA) needs to be made more functional with the target to enhance recovery for meeting operation and maintenance (O&M) expenditures of the irrigation departments.
10. Develop hydrologic zones to carry out water balance studies for promoting sustainability of water resources in each zone, lining canals, and watercourses.
11. In Punjab, the amount of wastewater disposed of after its treatment is only 22.11 million cubic feet per day out of the 552.23 million cubic feet per day of wastewater produced by the major urban centers.
12. Higher efficiency irrigation systems (HEIS) program be reviewed and rewritten after an independent monitoring of the sites developed during the past 5 years.
13. There is a dire need of development and implementation of a land use policy to save agricultural land from urbanization and industrialization. GIS-based Land Information System (LIS) for land use monitoring and decision-making can be helpful in the land saving process. Legislation to avoid fragmentation, promote land consolidation, and strengthen the role of departments like Soil Survey, Soil Fertility, Soil Reclamation, and Land Utilization in order to realize the full potential of land use.

REFERENCES

Acumen, F. 2008. What It Means to Be Patient: Drip Irrigation in Pakistan's Thar Desert'. http://blog.acumenfund.org/2008/05/12/what-it-means-to-be-patient-drip-irrigation-in-pakistans-thar-desert/. Accessed September 18, 2008.

Agricultural Census. 2013–2014. Government of Pakistan Statistics Division Agricultural Census Organization. 1–52.

Alley, R.B., J. Marotzke, W.D. Nordhaus, J.T. Overpeck, D.M. Peteet, R.A. Pielke, and J.M. Wallace. 2003. Abrupt climate change. *Science*, 299(5615): 2005–2010.

Berndes, G. 2008. Future biomass energy supply: The consumptive water use perspective. *Inter. J. Water Res. Develop.* 24: 235–245.

Blum, W.E.H. 2013. Soil and land resources for agricultural production: General trends and future scenarios A worldwide perspective. *Int. Soil Water Cons. Res.*, 1: 1–14.

Chaudhary, M.A. 1994. Regional Agricultural Underdevelopment, Green Revolution and Future Prospects, (A case study of Pakistan). Paper presented at *10th Annual General Meeting*, Pakistan Institute of Development Economics.

Choudhri M.B., A. Mian, and M. Rafiq. 1978. Nature and magnitude of salinity and drainage problems in relation to agricultural development in Pakistan. Pakistan Soils Bulletin No.8. Soil Conservation. Agricultural Enquiry Committee.

GOP. 2015. *Pakistan Economic Survey 2014–2015*. Ministry of finance, Islamabad, 2: 27–28.

Hamid, H. and Maliha. 2008. Survey of Donor Investments in the Agriculture Sector in Pakistan CFS.

Haron, M. and D. Dragovich. 2010. Climatic influences on dryland salinity in central west New South Wales, Australia. *J. Arid. Environ.*, 74: 1216–1224.

http://www.fao.org/docrep/005/ac484e/ac484e06.htm

Kemper, K.E. 2003. Rethinking groundwater management. Rethinking Water Management: Innovative Approaches to Contemporary Issues. Earthscan, London, p. 120–144.

Khan and H. Mahmood. 1997. Agricultural crisis' in Pakistan same expectation and policy options. *Pak. Dev. R.*, 36(4): 419–466.

Khan, M.M. and A. Riaz. 2006. Powerpoint presentation on_Pakistan—Country Water Highlights' (Adviser to Ministry of Water and Power. Prepared for *ADB Conference on ADB's Water Financing Program 2006–2010*, Islamabad, Pakistan.

Khan, M.M., J. Zhang, M.S. Hashmi, and M.S. HAshmi. 2011. Land distribution, technological changes and productivity in pakistan's agriculture: Some explanations and policy options. *Inter. J. Eco. Manag. Sci.* 1: 51–74.

Kuriakose, A.T., I. Ahluwalia et al. 2005. *Gender Mainstreaming in Water Resources Management.* ARD Internal Paper, World Bank. Washington DC.

Lal, R. 2001. Soil degradation by erosion. *Land. Degrad. Dev.*, 12: 519–539.

Lal, R. 2007. Anthropogenic influences on world soils and implications to global food security. *Adv. Agron.*, 93: 69–93.

Lal, R. 2009. Soils and food sufficiency. A review. *Agron. Sustain. Dev.*, 29: 113–133.

Mustafa, K., M. Shah, N. Khan, R. Khan, and I. Khan. 2007. Resource degradation and environmental concerns in pakistans' agriculture. *Sarhad J. Agric.* 23: 1160–1168.

Pretty, J. 2008. Agricultural sustainability: Concepts, principles and evidence. *Philos. Trans. R. Soc. B.*, 363: 447–465.

Rosegrant, M.W. and X. Cai. 2002. Global water demand and supply projections: Part 2 Results and prospects to 2025. *Water Inter.*, 27: 170–182.

Tahir, Z. and Habib, Z. 2000. *Land and Water Productivity trends: Across Punjab Cananl.* International Water Management Institute (IWMI), Pakistan, 35 p. (IWMI working paper 14).

Wace F.B. 1934. *The Punjab Colony Manual.* Revised Edition. Government Printing Press, Punjab, Lahore.

3 Water
Issues and Remedies

Allah Bakhsh and Muhammad Adnan Shahid

CONTENTS

3.1 INTRODUCTION

Pakistan is an agriculture-based country; the sector having a 21% share of the gross domestic production, employing 43% of the work force, and providing a livelihood to more than 62% of its population [1]. The climate of Pakistan is, however, mostly arid to semiarid with an average annual rainfall of about less than 100 mm in parts of lower Indus plains to more than 750 mm in the northern foothills against average crop water requirements in the range of 1200–1400 mm. Therefore, agriculture in the country heavily depends on the irrigation supplies diverted from the rivers through various dams and barrages. The water storage capacity in the country is very limited in terms of only two major reservoirs, which is further decreasing due to silting up of these reservoirs. Under the changing climate scenario, more variability is expected in the seasonal rainfall and resulting river flows [2], which will create seasonal flooding in monsoon season but water shortage situation for the rest of the year. In this context, there is a dire need to expand water storage capacity to cope with such scenarios, as well as to adopt different management interventions to improve efficiency of the irrigation system.

This chapter presents different aspects of water resources and their usage in Pakistan, which are described with a view for better water management. The first section explains the historical development of the Indus Basin irrigation system, followed by a detailed discussion on transboundary and interprovincial issues and possible options to resolve them. In the subsequent sections, different water resources of Pakistan, such as surface water, groundwater, and rainfall are described along with potential future climate change impacts and the resulting flood management and water harvesting needs. A section on irrigation-water management follows, expanding on the various problems of irrigation systems and their possible solutions. Finally, the suggestions elaborated throughout this chapter have been summarized as a part of water policy framework to promote sustainable irrigation water management in the country.

3.2 HISTORICAL PERSPECTIVE OF INDUS BASIN IRRIGATION SYSTEM

The Indus Basin is one of three major river systems (Indus, Ganges, and Brahmaputra) of Hindu Kush Himalaya (HKH) region, which forms water basins for a major part of South Asia. The lifeline for the Indus Basin Irrigation System (IBIS) is the Indus River, which originates from Tibet, in the northern Himalayas, at an elevation of more than 5000 m. It passes from east to west through Jammu and Kashmir, and then longitudinally south through Pakistan, ultimately flowing into the Arabian Sea. The Indus River receives contributions from eight major tributaries, that is, the Jhelum, Chenab, Ravi, Sutlej, and Beas the on eastern side and the Kabul, Gomal, and Gilgit flowing from west or northwest to the Indus (Figure 3.1). Major portions of the IBIS lies in Pakistan, being ranked

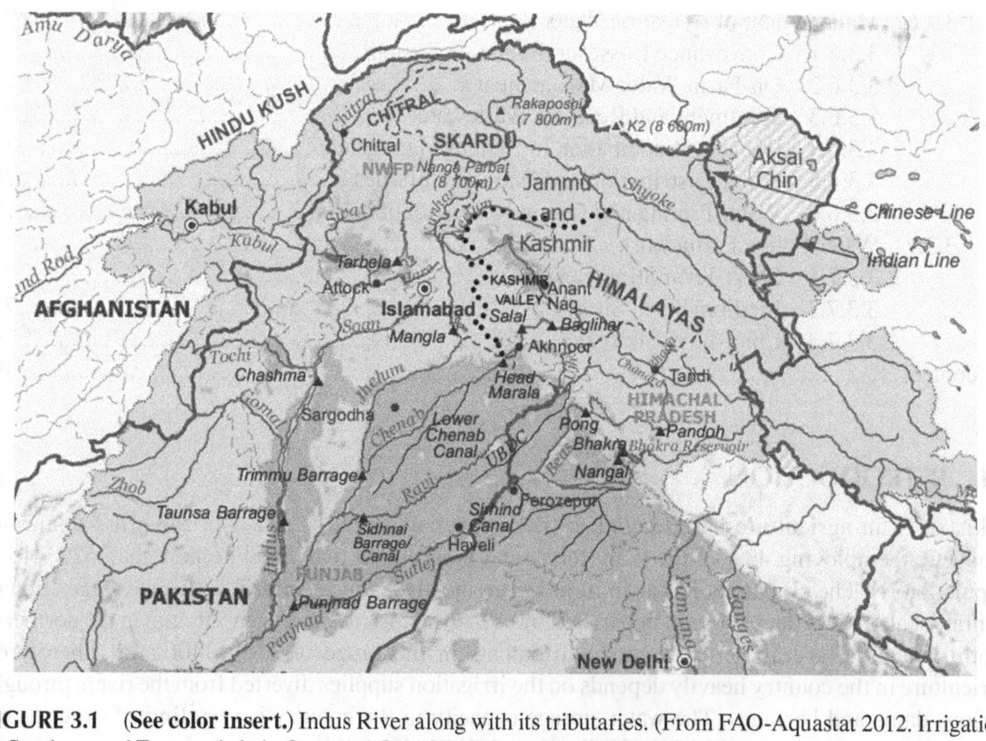

FIGURE 3.1 **(See color insert.)** Indus River along with its tributaries. (From FAO-Aquastat 2012. Irrigation in Southern and Eastern Asia in figures: AQUASTAT Survey—2011. Rome.) [3]

as one of the world's largest contiguous irrigation systems, and a key for the country's agriculture-based economy. The IBIS of Pakistan comprises the world's third largest frozen water reservoirs in the form of glaciers in the north, which along with rainfall contributes on average an amount of 139 MAF to river flows of the Indus River and its tributaries.

Development of irrigation systems in the Indus River Basin dates back to the Harrapan civilization, from 2300 B.C. to 1500 B.C. During the 2nd millennium, various subcontinent emperors constructed limited canal systems to irrigate dry lands along the Ravi, Chenab and Sutlej rivers. The credit for development of the modern IBIS of Pakistan goes to British irrigation engineers who created the original system (1847–1947), which Pakistan inherited in 1947, and to the Pakistani irrigation engineers and institutions like Water and Power Development Authority (WAPDA) who added new dams, barrages, and link and branch canals during the past 60 years. The modern IBIS is also an outcome of the generosity and intellectual input of international experts and institutions like the World Bank, who helped resolve transboundary disputes through agreements such as the Indus Water Treaty (IWT) (1960).

3.3 TRANSBOUNDARY WATER ISSUES

3.3.1 The Indus Water Treaty

After partition of the Indian subcontinent in 1947, the newly constituted states of India and Pakistan were on the verge of war due to water issues—water being lifeline for irrigated agriculture of both countries. The construction of new storage and hydropower facilities during 1947 to 1960 made the situation even worse [4]. Various committees and commissions were formed to solve water issues between the two countries, but couldn't come up with an agreed upon solution.

The World Bank started mediation between the two countries to resolve these issues from 1952 onwards. A continuous effort of 8 years resulted in defining water rights in 1960 with the signing of the IWT to divide river supplies between India and Pakistan with the guaranty of the World Bank. According to the IWT, rights of three main western rivers, that is, Indus, Jhelum, and Chenab, were given to Pakistan, while India was given exclusive rights to the waters of the Ravi, Beas, and Sutlej rivers. Pakistan, however, had agricultural water-use rights to specific tributaries of the Ravi river, as per provisions of annexure–B of the IWT. Similarly, India had also permission to plan run-of-the-river projects on western Rivers for domestic, agriculture, nonconsumptive, and hydropower production purposes (as per IWT annexure–C and annexure–D). There were a number of irrigation canals being fed from eastern rivers in Pakistan before the treaty. As a result of the IWT, a lot of replacement work was done by excavating link canals to feed eastern rivers and their canals by transferring water from western rivers.

3.3.2 Major Reservoirs and Link Canals Constructed by India and Pakistan after IWT

The IWT was successfully implemented and many reservoirs and a network of link canals were constructed by both countries in the first few decades under the Indus Basin Settlement Plan (IBSP). These included construction of large dams such as the Bhakra Dam on the Sutlej river in 1963, the Mangla Dam on the Jhelum river in 1966, the Tarbela Dam on the Indus river in 1976, and the Pong Dam on the Beas river in 1974. However, several dams and hydropower projects were also constructed by India on western rivers, which have always remained controversial. Details of major reservoirs and link canals constructed in the Indus Basin as part of the IWT or afterwards are presented in the Tables 3.1 and 3.2, respectively; and shown in Figure 3.2.

It can be seen that two major reservoirs and a number of link canals were built in Pakistan for diverting water of western rivers to eastern rivers to make up for this deficiency. On the other side, India focused on constructing many reservoirs/storage dams on eastern rivers (Ravi, Sutlej and Beas) to store or divert every drop of water, ignoring the environmental repercussions.

TABLE 3.1
Reservoirs Constructed after IWT

Sr. #	Reservoirs	River with Territory	Construction Year
1	Mangla	Jhelum, Pakistan	1966
2	Chashma	Indus, Pakistan	1971
3	Tarbela	Indus, Pakistan	1976
4	Diamer-Basha	Indus, Pakistan	Under construction
5	Kurramtangi	Kurram, Pakistan	Under construction
6	Munda	Swat, Pakistan	Under construction
7	Bhakra	Sutlej, India	1963
8	Pong	Beas, India	1974
9	Pandoh	Beas, India	1977
10	Salal	Chenab, India	1995
11	Thein	Ravi, India	2001
12	Baglihar	Chenab, India	2004
13	NimooBazgo	Indus, India	Under construction
14	Chutak	Indus, India	Under construction
15	Kishan Ganga	Neelum, India	Under construction

Source: Cheema MJM 2012. Understanding water resources conditions in data scarce river basins using intelligent pixel information, Case: Transboundary Indus Basin, PhD Diss., Technical University, Delft, The Netherlands, 204. [4]

TABLE 3.2
Link Canals Constructed in the Indus Basin

Sr.#	Link Canals	Off-taking Barrage	Linked Rivers	Construction Year	Length(km)
1	Trimmu-Sidhnai	Trimmu	Chenab-Ravi	1965	71
2	Sidhnai-Mailsi	Sidhnai	Ravi-Sutlej	1965	132
3	Mailsi-Bhawal	Sidhnai	Ravi-Sutlej	1965	16
4	Rasul-Qadirabad	Rasul	Jhelum-Chenab	1967	48
5	Qadirabad-Balloki	Qadirabad	Chenab-Ravi	1967	129
6	Balloki-Sulemanki	Balloki	Ravi-Sutlej	1954	63
7	Chashma-Jhelum	Chashma	Indus-Jhelum	1970	101
8	Taunsa-Punjnad	Taunsa	Indus-Chenab	1970	61
9	Beas-Sutlej	Pandoh	Beas-Sutlej	1977	37
10	Sutlej-Yamuna	Nangal	Sutlej-Yamuna	U.C	214
11	Sutlej-Haryana Alwar	Ferozpur	Sutlej-Ganges	P	386

Source: Thatte CD 2008. Indus waters and the 1960 treaty between India and Pakistan. In: Varis, O., Biswas, A.K., and Tortajada, C. (Eds.), *Management of Transboundary Rivers and Lakes*. Springer, Berlin Heidelberg, pp. 165–206; Wilson S (2011). Preparation of sub regional plan for Haryana sub-region of NCR-2021. Interim Report -II. [5,6]
Note: U.C: Under Construction; P: Proposed.

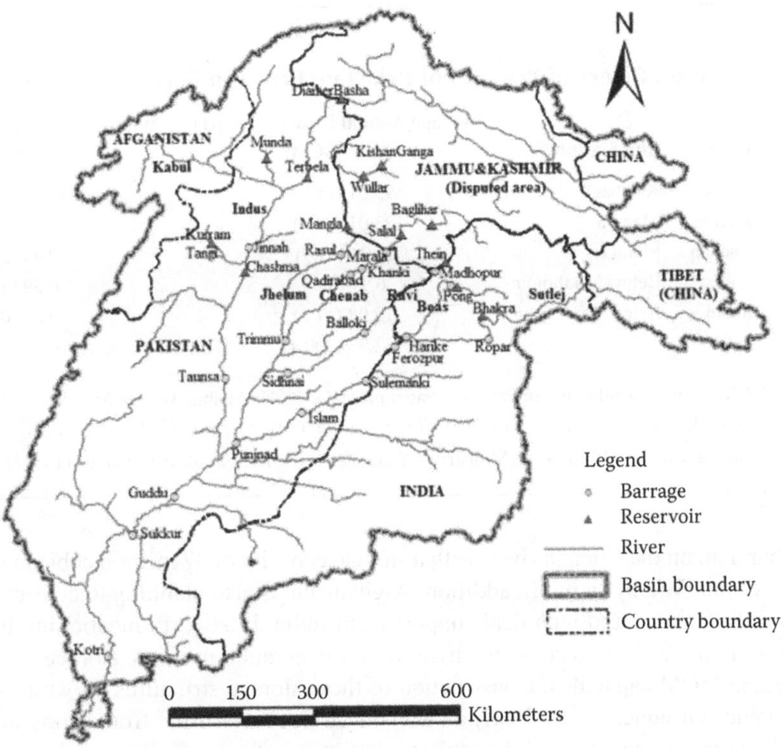

FIGURE 3.2 Hydraulic structures constructed on different rivers in IBIS.

3.3.3 Impact of Transboundary Issues on Pakistan's Water Resources

The transboundary river basins become mostly sources of water conflicts between upper and lower riparian populations. In case of the Indus river basin, the IWT can be considered a successful agreement between Pakistan and India, as it was the best available option for Pakistan to ensure sustainability of river supplies through an international agreement. Moreover, it resulted in the development of a modernized irrigation system in both countries, especially with the construction of two major reservoirs and a network of link canals in Pakistan funded by the World Bank.

The treaty also resulted in an imbalance through a major reduction in surface water supplies to the eastern rivers. However, Pakistan being the lower riparian country had no option but to accept the treaty under the compelling circumstances of water blockage by the upper riparian country, despite several terms of the treaty being against lower riparian rights. The link canals helped to make up the deficiency of the eastern rivers command areas, but groundwater and environmental problems remain in the command areas of these rivers. Table 3.3 presents the pretreaty (1922–1961) and posttreaty recent past situations (2007–2010) regarding average annual surface water supplies in the main rivers of Pakistan.

Table 3.3 shows decreasing flow trends in both eastern and western rivers, with a 92% reduction in average flow of the eastern rivers into Pakistan for the period 2007–2010, due to major diversions of these rivers by India after the IWT. In addition, an approximately 17% reduction in the average flows of western rivers has also been recorded. Reasons for these reductions in western rivers flows may be climate change and its variability [7], as well as upstream interventions by upper riparian populations, which has been a point of dispute between India and Pakistan.

In this regard, several issues have arisen during the last two decades, whose solutions are now part of the Indus Water Commission (IWC) mandate. The Wullar barrage/Tulbul hydropower projects on the Jhelum river, and the Kishan Ganga hydropower project on the Kishan Ganga river, a tributary of the Jhelum river, are a few such examples. Similarly, Pakistan considers the construction

TABLE 3.3
Annual Flows in Main Rivers of the IBIS of Pakistan: Pre- and Posttreaty Scenario

	River	Rim Station	Average Annual Flow (1922–61) MAF (BCM)	Average Annual Flow (2007–10) MAF (BCM)
Western rivers	Indus	Kalabagh	93.01 (114.4)	82.85 (101.9)
	Jhelum	Mangla	23.01 (28.3)	15.69 (19.3)
	Chenab	Marala	25.94 (31.9)	19.43 (23.9)
Eastern rivers	Ravi	Below Madhopur	6.99 (8.6)	0.89 (1.1)
	Sutlej	Below Ferozepur	13.98 (17.2)	0.65 (0.8)
	Total		162.93 (200.4)	119.51 (147.0)

Source: Ahmad S 2009. Water availability in Pakistan: Paper presented by Dr. Shahid Ahmad, Member PARC. In: *National Seminar on "Water Conservation, Present Situation and Future Strategy."* Project Management & Policy Implementation Unit (PMPIU) of the Ministry of Water & Power, Islamabad, Pakistan, p. 114. [7]

of the Baglihar dam on the Chenab river, with a storage capacity of 37 million cubic meters (MCM) as a violation of the IWT by India. In addition, Afghanistan is also planning to control the water of Kabul river with financial and technical support from India. It is worth mentioning that although there is provision in the IWT to construct hydropower generation projects, storage structures must not exceed 12.35 MCM capacities. Construction of these storage structures on western rivers will have catastrophic consequences for Pakistan as reduced flows, resulting from filling of these dams during low flow season, may destroy the rabi season sowing of crops in Pakistan.

There is a dire need to address flaws in different clauses of the treaty causing negative impacts. For instance, construction of unlimited hydropower projects by India under the clause of run-of-the-river projects is not tolerable to Pakistan. International observers may be invited in various disputed projects to inspect whether these are depriving the lower riparian populations of their water rights or not. Moreover, the IWT is almost silent about groundwater aquifers as groundwater development and utilization was almost negligible at that time. However, under the increased groundwater exploitation, it has become a important issue, pushing for the need to consider transboundary aquifer issues while keeping in view the clauses of the IWT which focusing only on surface water supplies. The treaty also does not consider the climate change implications and the ecological changes that have been occurring over the half century since its signing, which may reduce runoff from its mountainous glaciers. Similarly, the treaty doesn't consider today's hot issue of pollution, whose direct victims are always the lower riparian populations [8]. These aspects should be considered while revisiting the clauses of the treaty while maintaining its sanctity.

Another issue is the long-term tension between India and Pakistan. Whenever there is rise of tensions between the two countries, Pakistan is threatened with being deprived of its water rights by India. It has already been mentioned in the IWT, but needs to be emphasized, that no member can withdraw single-handedly from the IWT without the mutual agreement of both countries. Pakistan also needs to sign a similar agreement with Afghanistan regarding the Kabul River supplies.

Moreover, Pakistan needs to engage in serious planning to utilize different options like building small and large dams, check dams, flood canals, and so on to bridge the differences between summer and winter seasonal flows. This is also necessary for checking the extra escape of flows to the sea and making our case stronger, being the lower riparian country. Recently, about 35 MAF of average annual losses of river flow into the sea have been observed, which makes Pakistan's case weaker for stopping India from violating the IWT. Many excellent sites are available in Pakistan for building large water dams to store water and produce cheap electricity with an available hydropower potential of 50,000 MW. Similarly, watershed management practices can be promoted to minimize sediment loadings, and thus help maintain the storage capacity of the existing dams. In this way, high river flow losses into the sea can be minimized by building water storage dams.

3.3.4 Interprovincial Water Issues

Like the transboundary water issues between India and Pakistan, there are also issues among the provinces both in India and Pakistan. Provinces in Pakistan feel concerned over the shared use and the maximum amount of water available for their agricultural, industrial, and domestic needs. Provinces of Sindh, Baluchistan, and Khyber-Pakhtunkhwa (KP) claim that Punjab uses more water to provide benefit to its farmers whereas ground realities reveal that small provinces have higher water allowance compared with that for Punjab. In this context, construction of the Kalabagh dam has not been started despite its most feasible potential and its benefits at the national level—and especially for Sindh—to minimize effects of floods and droughts and the impacts resulting from climate change.

In order to solve water disputes among provinces, the government of Pakistan has constituted various committees. The mandate of all these committees was to negotiate with the relative bodies of the provinces and come up with working solutions. The committees were successful in building a consensus in the form of the signing of the Water Apportionment Accord (1991).

3.3.4.1 Water Apportionment Accord (1991)

The Water Apportionment Accord was approved by the government of Pakistan on 16th March, 1991. This agreement distributed a share of the water resources among the provinces according to requirements and following specific rules. The Indus River System Authority (IRSA), having representation from all four Provinces, was established to monitor water apportionments based on this accord. Province wise water shares are given in Table 3.4 for both the rabi and kharif seasons, based on the agreed percent formula in this agreement. Salient features of this accord include protection of existing canal water uses in each province, recognition of the need for construction of new storage structures at feasible locations, allocation of specific minimum escape to sea below the Kotri Barrage to control sea water intrusion, etc.

3.3.4.2 Reservations of Provinces on Water Resources Development

The water apportionment accord has been very helpful in resolving disputes among provinces regarding judicious distribution of water resources. Despite agreement on the distribution formula, provinces have not been able regarding development of new water reservoirs. There is a need to seriously consider the repercussions of not building any major water reservoir in the country for more than the past 40 years.

TABLE 3.4
Province-Wise Water Shares (1991 Accord) in MAF

Province	Water Shares			Balance Supply Shares (%)[a]
	Kharif	Rabi	Total	
Punjab	37.07	18.87	55.94	37
Sindh	33.94	14.82	48.76	37
KP	3.48	2.30	5.78	14
Civil Canals[b]	1.80	1.20	3.00	
Baluchistan	2.85	1.02	3.87	12
Total	79.14	38.21	117.35	100

Source: Briscoe J and Qamar U 2005. *Pakistan's Water Economy: Running Dry*. The World Bank, Oxford University Press: Oxford, United Kingdom. [9]

[a] Including flood flows & future storage.

[b] Un-gauged civil canals in KP.

TABLE 3.5
Historical and Projected Provincial Allocations of River Water Supplies (MAF)

Province	Pre-Mangla 1960–66	Pre-Tarbela 1967–76	Post-Tarbela 1977–82	Post-Kalabagh (Expected)	Post-Bhasha and Akori (Expected)
Punjab	48.35	49.86	54.51	55.94	61.49
Sindh	36.12	40.67	43.53	48.76	54.31
KPK	4.67	2.43	3.06	5.78	7.88
Baluchistan	–	0.49	1.63	3.87	5.67
Total	89.14	93.45	102.73	114.35	129.35

Source: IRSA (2011). Water Resource Development in Pakistan: Oral presentation by Rao Irshad Ali Khan, Chairman IRSA. In: *Roundtable Discussion on Agriculture & Water in Pakistan*, organized by World Bank, held on March 09, 2011. [10]

Sindh province claims that both upper riparian provinces are overusing the river water than their share, and thus causing reduction in their agreed upon share. This allegation, however, has always been rejected because Sindh canals have much higher water allowance than any other canals in the country and this has caused waterlogging conditions in the Sindh province.

Another argument of the Sindh province is based on a natural climatic and groundwater scenario. It argues that the average annual rainfall in Punjab is about 500–1000 mm as compared to 100–300 mm in Sindh. Similarly, fresh groundwater available annually to Punjab is more in comparison to that of Sindh. It is believed that Sindh is mostly dependent on river water supplies, whereas Punjab may cultivate most of its agricultural lands using alternate water resources. The situation, however, is reverse as heavy pumping of the groundwater in Punjab has deteriorated its quality and lowered the groundwater level there, inducing secondary salinization and rising its costs of exploitation many folds compared with canal water.

Moreover, the apprehension that construction of any new storage structure will reduce the flows to Sindh province is not corroborated by historical record. There has always been increase in the share of water of all provinces after the building of any major water reservoir, such as the Mangla and Tarbela dams (Table 3.5). Severe impacts of the 2010 flood, in the absence of any major reservoir, emphasize the need to build major water storage structures to better control and regulate flood situations. Adding the Kalabagh dam to the IBIS would enable the provinces to irrigate their lands located at higher elevations, such as those in D.I. Khan. Moreover, Kalabagh dam is the only one that can store Kabul river supplies, whereas all other dams built in the upper catchment cannot intercept such flows. Baluchistan has limited water resources and needs to build delay action dams to recharge its fast depleting aquifer as well as store surface water in its area in order to bring new lands under cultivation.

The historic data clearly indicate that on average about 35 MAF of water escapes into sea below the Kotri barrage and that about 14 MAF water is lost between the Sukkur and Kotri barrages. It is clear that construction of any new project, especially the Kalabagh dam, will not disturb the existing shares of the provinces, but will enhance them by checking the flood flows. It is worth mentioning that the need for future water resources development has already been recognized and agreed by all provinces under paragraph VI of the Water Apportionment Accord (1991) [10]. It is also evident that there is a dire need to store surface water by building all sizes of water reservoirs to mitigate the impacts of climate change, which may worsen the frequent occurrence of extreme events such as floods or droughts.

3.3.4.3 Way Forward to Improve Interprovincial Harmony

The IRSA is playing its role to judiciously apportion available water supplies among all the provinces according to their shares. Further development of water resources, however, in terms of new storage reservoirs is imperative to ensure sustainable surface water supplies in future. For this purpose, there is a need to restart a comprehensive dialogue by inviting all stakeholders, especially farmers from all provinces. Technical experts and the media need to play their role by advocating the need to build major storage reservoirs in order to minimize the demand–supply gap and mitigate the climate change consequences on the water and food security of the nation.

Public and private organizations need to take initiatives to address the apprehensions of the provinces, so that a consensus may be reached for development of more water resources. The provinces need to realize that addition of any new dam will enhance the water supply of all provinces. In this regard, historical and future projected provincial allocations reported by IRSA are given in Table 3.5. It can be seen that addition of any reservoir will definitely increase share of all the provinces as per the 1991 Accord. Moreover, under the current energy crisis, the Kalabagh dam will provide 3600 MW electric power and about 35,000 jobs, which will add to the socioeconomic uplift and GDP of the country.

3.3.5 Water Resources: Current Status and Possible Options

3.3.5.1 Surface/Canal Water

About 84% of cultivated areas in Pakistan are exploited through forms of irrigated agriculture, where canal or surface waters are used to produce about 90% of the food products in the country [11]. The IBIS of Pakistan provides annually 139 MAF on average, out of which about 35 MAF is lost to sea unutilized, leaving behind about 104 MAF at the canal heads for diversion into canals. One of the reasons for these losses is the limited storage capacity of three major reservoirs namely the Tarbela, Mangla, and Chashma reservoirs. In this way, Pakistan can only store water for hardly 30 days, as compared to 900 days in the dams of the Colorado and Murray Darling rivers and 120–220 days in India for the Peninsula Rivers [9]. Moreover, existing water reservoirs have lost their storage capacity from 15 MAF in 1975 to 11 MAF in 2011, a 25% decrease mainly due to sediment disposition. Mangla dam raising project has recently helped improving storage capacity to about 14 MAF again (Table 3.6), but diversion supplies seldom meet those anticipated in the Apportionment Accord.

It can be seen that there is a dire need to continuously develop new storage facilities to sustain the depleting storage capacity due to sedimentation, as well as to enhance it for increasing future water requirements. It has been estimated by the Pakistan Water Sector Strategy that there is a need to raise the storage capacity by about 18 MAF by 2025 to meet the projected requirements of the country [12]. In this regard, construction of Diamer-Basha Dam on the Indus is expected to start soon, which is about 315 km upstream of the Tarbela Dam and is a large capacity and multipurpose project.

The importance of storage facilities can be further highlighted by the historic trends of river water supplies and canal water diversions in Pakistan, as presented in Table 3.7. This table shows that river supplies have continuously decreased due to climate change impacts, as well as due to the diversion of eastern rivers' water by India. However, addition of two major storage reservoirs helped Pakistan enhance canal water diversions, which reached a peak in in post-Tarbela period, but have

TABLE 3.6
Storage Capacity of Major Reservoirs in Pakistan

			Capacity During Subsequent Years (MAF)		
Reservoir	Construction Year	Initial Capacity (MAF)	2000	2010	2016
Mangla	1967	5.3	4.54	4.4	7.39 (After Mangla Dam Raising)
Chashma	1971	0.7	0.3	0.2	0.1
Tarbela	1974	9.7	8.8	7.3	6.8
Total		15.7	13.64	11.9	14.29

Source: Briscoe J and Qamar U 2005. *Pakistan's Water Economy: Running Dry*. The World Bank, Oxford University Press: Oxford, United Kingdom. [9]

TABLE 3.7
Historic Trends of River Supplies and Canal Diversions

Year	River Water (MAF)	Canal Water (MAF)	Groundwater Abstraction (MAF)	Remarks
1947	189.00	69.1		Postindependence
1960–1966	189.00	89.14		Before IWT 1960 (Pre Mangla)
1967–1976	176.00	93.45	10.13	After Mangla Dam
1977–1982	160.00	102.73	21.24	After Terbela Dam
1980–1990	148.00	108	35.87	Post-Terbela Peak
1990–1999	142.71	107	48.64	Declining reservoir capacity
2000–2009	138.00	93.5	52.64	Declining reservoir capacity with drought conditions
2010–2015	139.50	103.50	55.5	Reservoir capacity sustained a little due to Mangla Dam raising.

Source: Briscoe J and Qamar U 2005. *Pakistan's Water Economy: Running Dry*. The World Bank, Oxford University Press: Oxford, United Kingdom. [9]

been decreasing continuously since due to siltation of existing storage reservoirs and the lack of addition of any other large reservoir. It can be further observed that groundwater abstraction has increased considerably with time due to rising crop water demands linked to increased cultivated area, cropping intensity, and decreased surface water supplies (Table 3.7).

3.3.5.2 Groundwater

Groundwater is an important source of irrigation water and is used to make up deficiencies in surface water supplies for irrigating lands. The Indus Basin is underlain by an extensive aquifer providing about 50 MAF to augment canal water supplies and covers about 16 Mha of land, of which 6 Mha is freshwater and the remaining 10 Mha is saline [13]. Conjunctive use of groundwater with surface water is in practice on more than 70% of the irrigated areas within the Indus Basin. Figure 3.3 shows that 29% more area came under conjunctive irrigation during the last 20 years.

The groundwater level in most areas of the country was very high in 1980s, posing waterlogging and salinity problems. To cope with these issues, several drainage projects were implemented; tube wells were installed under the Salinity Control and Reclamation Program (SCARP), being the fast way to lower down the groundwater levels.

Gradually, objective of these SCARP tube wells changed from drainage to pumping for irrigation purposes under surface water shortage scenarios. Furthermore, increases in demand due to greater cropping intensity and reduced surface water supplies has urged farmers to install private tube wells. As mentioned in Table 3.7, presently about 50 MAF of groundwater resources are being utilized and the number of public and private tube wells in the country has increased up to more than one million in Pakistan, including 0.8 million in Punjab only. Historical growth of the number of tube wells in Pakistan over the time is shown in Figure 3.4.

Under this increased and unregulated groundwater abstraction, groundwater levels are falling and the quality of groundwater is deteriorating. Estimates indicate that 24.7 metric tons of salts in Punjab and 3.5 metric tons in Sindh are mobilized by the tube wells, thereby posing secondary salinity problems [14]. This indicates that there is not much potential left for further groundwater

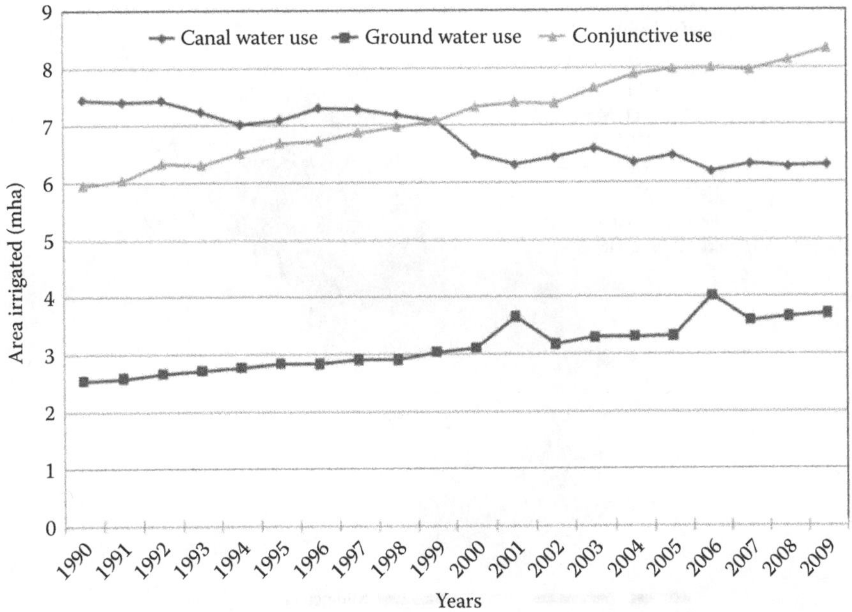

FIGURE 3.3 Surface, ground, and conjunctive water use in IBIS of Pakistan. (From Cheema MJM 2012. Understanding water resources conditions in data scarce river basins using intelligent pixel information, Case: Transboundary Indus Basin, *PhD Diss.*, Technical University, Delft, The Netherlands, 204.) [4]

exploitation. However, there is still no regulation and guidelines regarding tube wells installation nor a regular program regarding groundwater recharge to promote sustainability of groundwater resources.

Another recent problem is the rise of the water table in certain areas, such as for example in Sargodha. This is because of topographic effects along with seepage from the irrigation system in addition to the fact that farmers in these areas are over irrigating their fields. This may result in a waterlogging situation in these areas. There is a dire need to achieve equity in water distribution and create awareness among farmers that the extra application of water is not beneficial at all. Moreover, such excessive water losses in these areas may be minimized or transferred to water shortage areas through some water distribution/water pricing policy mechanism.

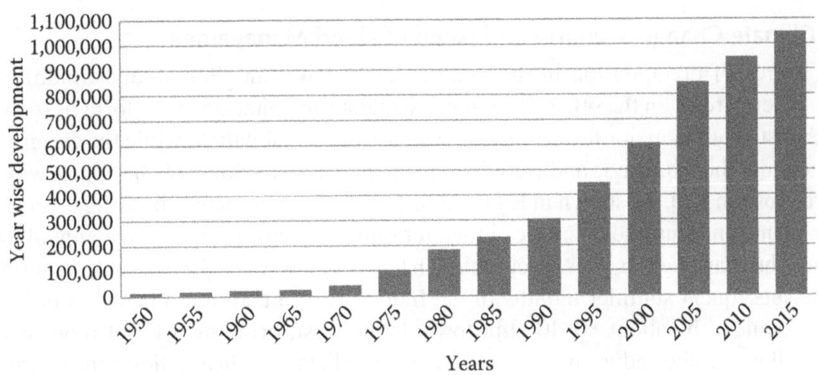

FIGURE 3.4 Historical growth of tubewells in Pakistan.

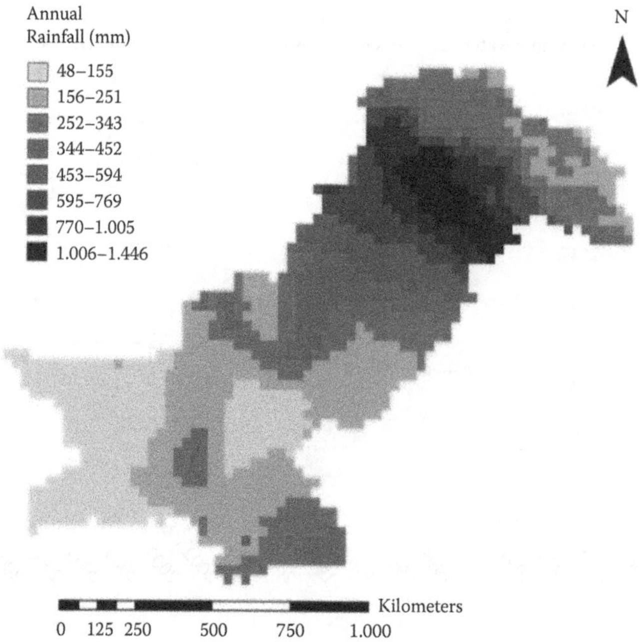

FIGURE 3.5 **(See color insert.)** Regional variations in average annual rainfall based on TRMM rainfall data. (From Shahid MA, Boccardo P, Garcia WC, Albanese A and Cristofori E 2013. Evaluation of TRMM satellite data for mapping monthly precipitation in Pakistan by comparison with locally available data. In: *III CUCS Congress—Imagining Cultures of Cooperation: Universities Working to Face the New Development* Challenges, Turin, Italy.) [15]

3.3.5.3 Rainfall

Rainfall is an important source of water especially in rainfed areas, as well as in irrigated areas as it complements surface and groundwater resources. The climate in most parts of Pakistan is semiarid to arid having an average annual rainfall of 400–600 mm. However, the country has arid/dry zones of Baluchistan and Thar (Sindh), as well as the upper Punjab and northern areas with humid climate. Spatial or regional variation in rainfall in Pakistan is presented in Figure 3.5 by analysis of TRMM satellite precipitation data [15].

In addition to spatial variability, seasonal variability in rainfall touches extremes on both sides, as presented in Figure 3.6, which indicates that rainfall is concentrating in the postsummer rainy months, while winter months are becoming drier.

3.3.5.4 Climate Change Scenario and Need of Flood Management

The main source of precipitation in the plain areas of Pakistan (Punjab and Sindh) is monsoon rainfall. Northern areas, on the other hand, receive major precipitation in the form of snow in winter. Due to such distribution, major flows in Indus river are due to snowmelt, while the eastern tributaries like the Jhelum, Chenab, Ravi, and Sutlej supplement the river flows of the Indus with waters of monsoon rain origin [16]. As shown in Figure 3.6, the changing climate is heading towards increases in summer/monsoon rains and decreases in winter rains. Moreover, the average annual temperature is increasing, thus causing less snow accumulation but more snow and glacier melt, resulting in heavy flows in the subsequent summer months in the Indus river. In addition, the extent of the monsoon region has expanded by about 80–100 km towards the west, bringing a lot of monsoon rains and causing high flows in the Indus river. This scenario predicts very heavy flows in the Indus river in coming decades due to the combined effects of snowmelt and monsoon rains, which would lead to floods [2]. The same has also been reported by Rees and Collins (2005) [17] through long-term

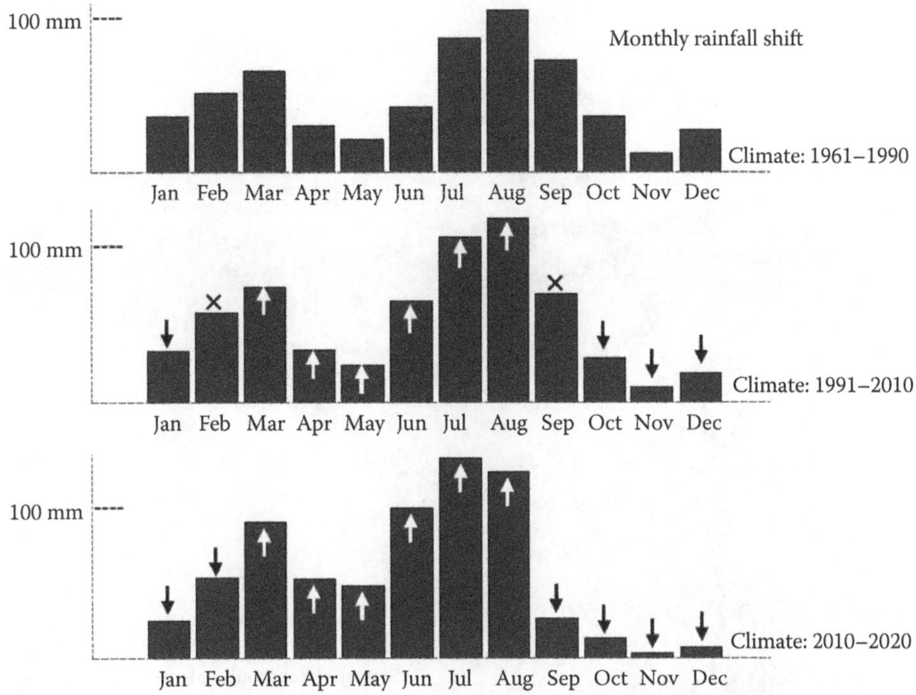

FIGURE 3.6 Patterns of seasonal rainfall variations under changing climate. (From Hanif M 2011. Redistribution of Precipitation (Seasonal Shift) in Pakistan & Super Flood in Pakistan-2010. In: *One Day Colloqium on "Water Crisis and Choices,"* held at University of Agriculture, Faisalabad.) [2]

climate scenarios analysis, indicating about a 50% increase in flows at Besham Qila for the first two decades of the twenty-first century. Afterwards, there will be gradual decrease in flows, up to 40% in the next 80 years, due to the decreasing sizes of glaciers [17].

A complete framework of flood and watershed management is required to mitigate negative flood impacts and harvest their positive impacts, such as rainwater harvesting and groundwater recharge. In this regard, delineation of flood-prone areas may be helpful in identifying potential rainwater harvesting and groundwater recharge sites. Figure 3.7 presents the flood inundation map for Pakistan developed by the historic analysis of 15 years of MODIS daily images, employing a water bodies' classification approach. Such analyses, along with use of modern high resolution remotely sensed data, may be beneficial for flood preparedness, emergency management, as well as investigating suitable options of groundwater recharge and flood-prone agricultural practices.

3.3.5.5 Rainwater Harvesting

As discussed earlier, there exists a high seasonal variability in rainfall in Pakistan with an 80:20% ratio between summer and winter seasons, respectively [2,15]. Rainwater harvesting is an important option to conserve and manage water generated from extreme rainfall events. However, at this point there is no regular program of rainwater harvesting. There is a need to initiate rainwater harvesting not only at the watershed level, but also at the farm level in agriculture and in domestic areas of urban zones.

Rainwater harvesting can be carried out in hilly areas through different watershed management techniques including in situ rainwater harvesting and construction of check dams. Similarly, water storage ponds can be built in the rural/agricultural areas to store excess rainwater or even excess canal water, which can be used later on to irrigate fields. Another important benefit of such water storage ponds is that stored water can be used to recharge groundwater. This is important because

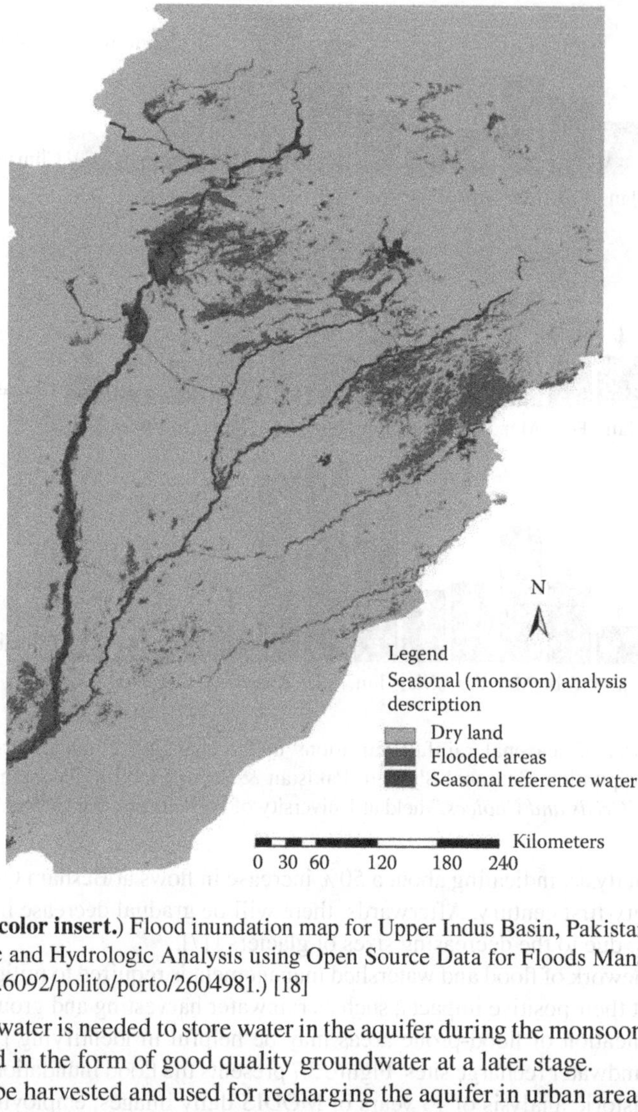

FIGURE 3.7 **(See color insert.)** Flood inundation map for Upper Indus Basin, Pakistan. (From Shahid MA 2015. Geoinformatic and Hydrologic Analysis using Open Source Data for Floods Management in Pakistan. *PhD thesis*, DOI: 10.6092/polito/porto/2604981.) [18]

recharging groundwater is needed to store water in the aquifer during the monsoon and flood seasons, which may be used in the form of good quality groundwater at a later stage.

Rainwater can be harvested and used for recharging the aquifer in urban areas too by employing rooftop rainwater harvesting approaches. Currently, rainwater becomes part of the sewage water and is conveyed to the urban drainage system along with industrial wastewater, causing the loss of good quality water as well as causing urban street flooding situations. The quality of rainwater is usually acceptable after f iltering and can be used to recharge the aquifer.

A model rooftop rainwater harvesting setup has already been established at University of Agriculture, Faisalabad (UAF), which may be replicated at public buildings in urban areas. Figure 3.8 presents the schematic diagram of this setup, where harvested rainwater can be used beneficially for recharging groundwater. A rainwater harvesting and groundwater recharge setup established at UAF is shown in Figure 3.9.

3.3.6 Management of Irrigation Water

3.3.6.1 Conveyance Losses in Irrigation System

It is worth mentioning that out of average canal water diversions of 104 MAF, only 78 MAF are delivered at the canal outlets for the tertiary irrigation system. Considering further conveyance water losses through the irrigation channels, only 60 MAF are available at the farm gate with a canal

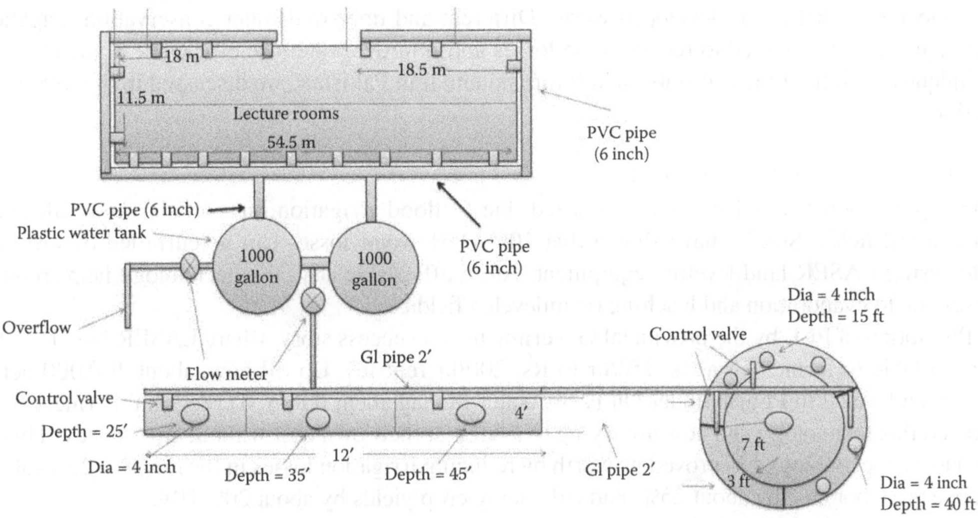

FIGURE 3.8 Schematic diagram of RWH setup and groundwater recharge at UAF.

irrigation system efficiency of about 40%. Through canal lining and rehabilitation of the irrigation system, a major portion of these 44 MAF conveyance losses can be avoided but this will affect groundwater recharge from irrigation systems.

In this regard, different phases of national watercourse improvement projects have helped a lot in improving the tertiary irrigation system. However, still more work is needed to minimize conveyance losses both at canal and watercourse levels. For this purpose, 50% lining of watercourses may be ensured in the first phase. Then, the irrigation canals may be lined, particularly in the saline groundwater zones; while lining of link canals may be carried out in the final phase for saline groundwater zones to check all major conveyance losses and avoid waterlogging conditions.

3.3.6.2 On-Farm Water Management

On-farm water application losses present another grey area contributing to poor overall efficiency of the irrigation system in Pakistan. Out of 60 MAF surface water resources available for farms, only about 50% is available to crops with a 50% irrigation application loss during conventional flood irrigation. This poor irrigation efficiency is also a reason of low water productivity in Pakistan

FIGURE 3.9 RWH and groundwater recharging setup established at FAE&T, UAF.

as compared to that in the developed world. Different and improved water conservation irrigation techniques can be adopted to reduce these losses and improve water use efficiency. Some of these techniques, with their present status and future potential in Pakistan, are discussed in the following sections.

3.3.6.2.1 Precision Land Leveling (PLL)

Huge application losses have been recorded due to flood irrigation on irregular and unleveled agricultural fields. Studies have shown that 20%–25% water losses can be curtailed by leveling fields using LASER land leveling equipment. At an affordable cost, this technology helps reduce losses due to evaporation and leaching on unleveled fields.

Promotion of PLL by the provincial Government is a success story, where LASER land levelers are available to farmers at a Rs. 150/hr to Rs. 200/hr rent fee. Up till now, about 400,000 acres have been leveled in Punjab under On Farm Water Management (OFWM) program. Farmers have adopted this technology and now are trying to avail it at their own cost without any subsidy. This is because the technology has proved its worth by reducing irrigation losses in the range of 25%–30%, decreasing labor cost by about 25%, and enhancing crop yields by about 20% [19].

3.3.6.2.2 Furrow Irrigated Raised Bed Planting

Growing crops on raised beds is one of the improved surface irrigation methods being practiced all over the world with several well documented advantages. In Pakistan, farmers have been found to use ridge-furrow irrigation method for growing certain crops on ridges; however, this may not be considered as a substitute for bed-furrow planting methods on the basis of water savings and increased crop productivity. The faculty of the Agriculture Engineering and Technology and Water Management Research Center (WMRC) at the University of Agriculture, Faisalabad have carried out intensive research and arranged outreach activities over the past 15 years to promote bed-furrow planting for all major crops like wheat, cotton, maize, and rice. This technology has been proved very valuable providing 40%–50% water savings with 10%–20% increases in yields on average [20–23]. In this regard, the bed-planting machine developed at WMRC has been included in the list of equipment being provided by OFWM on subsidy.

3.3.6.2.3 High Efficiency Irrigation Systems

High Efficiency Irrigation Systems (HEIS) are pressurized irrigation systems which use pipes for conveyance of water from source to points of use. Different types of HEIS include drip, trickle, sprinkler, bubbler, rain-gun, center pivot, etc. These methods improve water and nutrient use efficiency to increase crop productivity. Drip irrigation is considered the most efficient system among HEIS, with an application efficiency of up to 90%. The availability of desired soil moisture and nutrients in the root zone helps in enhancing crop yields under drip irrigation in addition to achieving considerable saving of inputs. The technology is very fruitful for orchards, as well as for high value crops like vegetables, cotton, maize, etc.

Although benefits of HEIS are obvious regarding water savings and better crop production. Their sustainable promotion in Pakistan is still a challenge due to many factors including high initial costs, technical complexities, repair and maintenance, and social problems including profitable marketing of produce, etc. Projects for promotion of HEIS by providing subsidy to farmers are already underway, but there is a need to conduct studies regarding reducing cost, irrigation scheduling, fertigation, repair and maintenance, growing profitable high value crops, socioeconomic challenges, and indigenization of HEIS for their successful adoption.

3.3.6.2.4 Perforated Pipe Water Conveyance and Application

Perforated pipe irrigation systems have been developed at UAF, which have potential to increase water application efficiency by 20%–30%. This is because conveyance losses are zero as the pumped water is conveyed through flexible pipes from the source, that is, from tube wells to the fields. This

FIGURE 3.10 Perforated pipe irrigation system installed at farmer's field.

system aims at eliminating traditional watercourses at the farms and delivering irrigation water through pipes. This system is cheap, easy to adopt, and improves irrigation efficiency. The system has been proven useful for irrigation using tube wells, especially under power failure conditions when filled watercourses are immobilized, losing water rather than applying irrigation. Moreover, the perforations made in the pipes help in applying a precise amount of water (Figure 3.10).

3.3.6.3 Treatment and Reuse of Wastewater

Wastewater may be an important alternate source to supplement freshwater supplies; however, its direct use for irrigation purposes poses a serious threat to public health and has led to increased concerns over water quality. In Pakistan, water quality over the years has been deteriorating due to the increased disposal of untreated sewage water. It has been reported that Pakistan produced 962,335 million gallons (4.369 × 109 m^3/year) of wastewater, which included 288,328 (1.309 × 109 m^3/year) and 674,009 (3.060 × 109 m^3/year) million gallons from industrial and municipal use, respectively [24]. Existing Facultative Lagoon (FL) systems in Pakistan are providing only partial treatment to the sewage water and after which it is either discharged into the rivers or used for agriculture purposes. This huge inflow of partially treated sewage to the surface water does not comply with the regulatory requirements of effluents regarding pathogenic microorganism and organic contaminates. Moreover, farmers are mostly using wastewater even without any treatment.

Constructed wetlands (CW) may be adopted as an important modern and energy efficient technology and is gaining lots of attraction. It not only reduces the risk associated with wastewater discharge to the public health and environment but also includes advantages like low maintenance and operation costs, as well as less energy requirements. CW-treated water may be used for irrigation with several advantages, particularly in nonfood crops as compared to the partially treated water, as it also reduces the fertilizer requirements of crops by providing the required nutrients. The technology provides an attractive solution for wastewater treatment while potentially producing a biomass capable of functioning as a renewable energy source, fodder for livestock, and/or for many other beneficial purposes. There is, however, a need to test and evaluate the technique for different agroclimatic conditions and establish demonstration units for its adoption and sustainable dissemination.

3.3.6.4 Farmers Participation in System Management

The participatory irrigation system, which ensures the active participation of stakeholders, always results in better performance of an irrigation system. Establishment of such systems is already in progress in all provinces, for example, the Punjab Irrigation and Drainage Authority (PIDA) and the Sindh Irrigation and Drainage Authority (SIDA). It was observed that Provincial Irrigation

Departments (PIDs) were facing a number of operational and management problems, causing inefficiency in the overall system's operation. These issues, as mentioned by PIDA (2005) [25], included:

- Overall deterioration of system
- General lack of agency responsiveness
- Inequitable distribution of irrigation water
- Inadequate maintenance of irrigation network
- Low irrigation efficiencies
- Escalating gap between the revenues and expenditures
- Environmental degradation
- Lack of farmers' participation in decision-making and management

To overcome these problems, the government launched a participatory irrigation system at the province level (e.g., PIDA in Punjab) comprising structural bodies of Area Water Boards (AWBs), Farmers' Organizations (FOs), and Khal Panchayets (KPs). Salient features of this irrigation management transfer system are as follows:

- Management, operation, and maintenance of irrigation infrastructure by the users
- Obtaining irrigation water and supplying it equitably and efficiently to the farmers
- Assessment and collection of water rates (abiana)
- Remittance of agreed share of collected abiana to the government after retaining its share
- Settlement of water disputes relating to farmers
- Protecting the environment within the command area

These systems, however, are limited to selected areas and their main working is limited to maintenance works of watercourses/distributaries and collection of revenue. Such participatory system can be employed for all management interventions like promotion of Water Conservation Technologies (WCTs) and the formulation and implementation of a groundwater regulation framework. The University of Agriculture, Faisalabad through its Water Management Research Center has successfully implemented a "On Farm Research and Development Component" project funded by JICA to promote different water management interventions through selected FOs in Faisalabad and T.T. Singh. Similar approach may be adopted at the provincial level for achieving the active participation of farmers in all water management activities.

3.3.6.5 Water Distribution and Warabandi Issues

The rotational irrigation system, known as Warabandi, is a supply-based system providing irrigation water to farmers for a fixed amount of time, mostly on a weekly basis, to irrigate their fields based on size of their land holdings. Each stakeholder/farmer has its turn after every 7 days, but the amount of irrigation water depends on the available supplies in the canal. Due to such system, farmers have to adjust their cropping patterns according to water availability instead of demand-based water availability. Sometimes, farmers have to face the issue of surplus water if, for example, their irrigation turn is on a rainy day.

Another demerit of such system is its inability to incorporate water losses from head to tail. Farmers at head get a full supply of water, while the tail farmers have much less water supplies due to conveyance losses. Low canal water allowances, designed on the basis of about a 70% cropping intensity, also become the cause of reduction in water supplies from head to tail. Moreover, canal water allowance is not uniform in all canal command areas, ranging from 3 to 15 cusecs/1000 acres. There is a dire need to rationalize the canal water allowances and employ modern telemetry techniques to monitor the flows for shifting from supply-based to demand-based system. This will help in ensuring availability, adequacy, and reliability of the irrigation water.

3.3.6.6 Water Pricing and Groundwater Regulation

Canal water pricing, or abiana, for irrigation water should be rationalized to recover the cost of water supply as well as promote the efficient use of water. Various policy guidelines were introduced at national and provincial levels on abiana rates (water charges per acre for major crops). The fixed rate of canal water supplies, for example, Rs. 50/acre and Rs. 85/acre for Rabi and Kharif, respectively, does not cover operation and maintenance costs of the system. Especially, it is not justified when compared with groundwater charges. A recent study conducted at UAF under an International Food Policy Research Institute (IFPRI)-funded project revealed that for one acre of wheat in Rabi, Rs. 6000/acre is required for irrigation by groundwater using a diesel-operated tube well. Similarly, for rice crops, it is almost doubled (https://www.ifpri.org/). Moreover, the recovery of abiana is also not satisfactory. According to a report of the planning commission, abiana collection remained less than 50% of the assessed amount.

Another problem is the overexploitation of groundwater, which can actually be linked with water pricing policy. Due to very low rates, or no recovery of abiana, farmers use canal water inefficiently and without any attention to save this precious resource. This results in depriving many farmers, mostly at tail, of their canal water share, and therefore, they have to rely on extensive groundwater pumping for their agriculture usages. This extensive and unregulated groundwater pumping causes problems like groundwater depletion, quality deterioration, and secondary salinization.

There is a need to formulate a complete water pricing and groundwater regulation policy in order to avoid the excessive lowering of the groundwater level. Instead, corporate tube wells can be installed in good groundwater quality zones along the water channels, and may be operated by the registered farmers' bodies like FOs. In addition, increased water rates may help in bridging the differences among surface and groundwater costs and creating awareness among farmers to judiciously apply irrigation water. Integration of such a setup with a demand-based warabandi system may help in reducing groundwater pumping by supplying it through corporate tube wells to the needy farmers. Building of farm water ponds can help in storing irrigation water for transforming supply-based to demand-based systems as efficient irrigation requires its frequent application, which is only possible when irrigation water is available in the form of water ponds.

3.3.7 Water Policy Framework

Water is a key parameter to achieve many of the Sustainable Development Goals (SDGs) formulated by the United Nations (UN) for the overall development and prosperity of humans on the planet. For example, Goal 2 states "End hunger, achieve food security and improved nutrition and promote sustainable agriculture," which may be achieved through sustainable availability and efficient use of water for irrigation. Goal 6 specifically emphasizes on "Ensure availability and sustainable management of water and sanitation for all." Similarly, Goal 13 states "Take urgent action to combat climate change and its impacts." All these SDGs highlight the importance of water at international level for the sustainable development of human beings in rural as well as urban areas, which is only possible through building large water reservoirs to combat flooding and drought conditions.

Water is also one of the absolute priorities of Pakistan Vision 2025, as pillar 4 of this vision emphasizes energy, water, and food security. This pillar of vision 2025 covers all aspects, such as enhancing water storage capacity, investing in proved water conservation techniques to check water losses in all sectors, ensuring safe drinking and sanitation facility for everyone, addressing the burning issues of climate change and resulting disasters through proper watershed management, etc. In addition to measures taken by the federal government, the Punjab government has also launched a comprehensive project entitled "Punjab Irrigated-Agriculture Productivity Improvement Project (PIPIP)" in which different energy and water efficient options like solar-operated tunnel farming units, high efficiency irrigation systems like drip, and so on, are being promoted by providing subsidies to small farmers.

Based on the abovementioned importance of water and the detailed discussion on different aspects of water in previous sections, the following water policy framework is proposed.

There is a need to mitigate the effects of run-of-the-river rights of upper riparian users by keeping in view the decreasing flows in western rivers and the environmental and groundwater aquifer perspective of the lower riparian users, as well as to sign similar agreements with Afghanistan regarding the Kabul river supplies.

Serious planning is needed to implement the water resources development clause of the Water Apportionment Accord, 1991, that is, to build small, medium and large dams, check dams, flood canals, and so on to bridge the differences between summer and winter seasonal flows, and to mitigate both seasonal flooding and water shortage problems.

For building a consensus to develop new water storage reservoirs, or other water resource projects, initiation of a comprehensive mechanism is needed by convening all stakeholders, especially farmers from all provinces, for dialogue and sharing facts. Moreover, the government should take initiatives to address apprehensions against building dams, so that a consensus may be built for the development of more water resources. Historical record proves that building new water reservoirs has always enhanced the share of water for all provinces in addition to mitigating flood and drought effects.

Keeping in view the seasonal rainfall variability and future climate change impacts, a complete framework of flood-risk management strategies need be developed. These include effective flood-alert systems, that is, early warning flood-warning system, preparedness, management, land-use planning, and other structural and nonstructural measures for mitigating floods and their socioeconomic implications.

Watershed management for each creek needs to be promoted to minimize sediment loadings and increase availability of water in hilly areas by adding drop structures, weirs, protecting forests, improving vegetative cover, afforestation, etc.

Different rainwater harvesting options like check dams in hilly areas, water ponds in flat agricultural lands, and rooftop rainwater harvesting for rural and urban areas need to be promoted to recharge groundwater. Need for groundwater recharge has further increased under the scenario of increasing number of tube wells and continuously falling of groundwater levels, as well as decreasing existing recharge through lining of canals and introducing HEIS.

The irrigation system of Pakistan operates at a very low irrigation efficiency due to high conveyance and application losses. Conveyance losses may be minimized by remodeling of the old-designed hydraulic structures (canals, distributaries, outlets, watercourses, etc.), as well as by lining.

Different WCTs like precision land leveling, bed planting, and high efficiency irrigation systems should be further promoted to enhance irrigation application efficiency. In this regard, there is a need to provide training to farmers, promote indigenization of the HEIS, and address the socioeconomic challenges involved in the sustainable adoption of such modern technologies by promoting the interaction with end users/farmers and establishing service centers at the village level.

Low delta crops, that is, crop diversification needs to be promoted to reduce crop water requirements.

Wastewater needs to be treated prior to its discharge to the system and treated wastewater can be used for irrigation while stopping untreated sewage irrigations; wetland concepts may be promoted at the village level to treat wastewater.

Farmers' participation should be encouraged by further expanding the existing participatory irrigation system to all districts and involving them in all aspects, such as watercourse improvement, revenue collection, promotion of modern water conservation technologies, and in devising new policies regarding water pricing and groundwater regulation.

The canal water allowances need to be rationalized in the existing warabandi system. Moreover, on-farm water storage needs to be promoted to shift supply-based irrigation to a demand-based system. Building on-farm water ponds can promote demand-based irrigation supply systems, which is also a prerequisite for HEIS.

There is a need to devise a water-pricing mechanism for canal water charges (abiana), similar to billing patterns of electricity charges, based on an incremental water allowance basis. Moreover, a groundwater regulation policy is needed with the objectives to promote the value and the prudent use of canal water by the farmers to limit uncontrolled overexploitation of groundwater.

This water policy framework, or way forward, may be summarized in different time frames as detailed in the following sections.

3.3.7.1 Short-Term Strategies

Build reasonable numbers of mini and small dams annually and ensure rainwater harvesting at the household level by making it legally binding for citizens to do so. Similarly, lining of watercourses and replacement of field channels with pipes should be encouraged; the use of on farm water storage ponds and HEIS should be promoted to avoid conveyance and application losses.

3.3.7.2 Medium-Term Strategies

Line the irrigation canals in poor groundwater quality zones, ensure wastewater treatment at industrial level and through wetlands, and control untreated sewage irrigation in periurban areas. Promote crop diversification by introducing low delta and high-value crops.

3.3.7.3 Long-Term Strategies

Build multipurpose large dams with at least one mega dam every 10 years. Line link canals to avoid waterlogging, build flood canals for groundwater recharge in floodplains and desert areas, and line river reaches to promote river navigation and groundwater recharge.

REFERENCES

1. Government of Pakistan. 2014. *Pakistan Economic Survey, 2013–14*, Ministry of Finance, Islamabad, Pakistan.
2. Hanif M. 2011. Redistribution of Precipitation (Seasonal Shift) in Pakistan & Super Flood in Pakistan-2010. In: *One Day Colloqium on "Water Crisis and Choices,"* held at University of Agriculture, Faisalabad.
3. FAO-Aquastat. 2012. Irrigation in Southern and Eastern Asia in figures: AQUASTAT Survey—2011. Rome.
4. Cheema MJM. 2012. Understanding water resources conditions in data scarce river basins using intelligent pixel information, Case: Transboundary Indus Basin, *PhD Diss.*, Technical University, Delft, The Netherlands, 204.
5. Thatte CD. 2008. Indus waters and the 1960 treaty between India and Pakistan. In: Varis O., Biswas AK and Tortajada C. (Eds.), *Management of Transboundary Rivers and Lakes*. Springer, Berlin Heidelberg, pp. 165–206.
6. Wilson S. 2011. Preparation of sub regional plan for Haryana sub-region of NCR-2021. Interim Report -II.
7. Ahmad S. 2009. Water availability in Pakistan: Paper presented by Dr. Shahid Ahmad, Member PARC. In: *National Seminar on "Water Conservation, Present Situation and Future Strategy."* Project Management & Policy Implementation Unit (PMPIU) of the Ministry of Water & Power, Islamabad, Pakistan, p. 114.
8. Malik MA, Tahir MA and Bhatti AZ. 2013. *Indus Basin Transboundary Water Issues in Past and Present Perspective*. Technical Journal, University of Engineering and Technology, Taxila, pp. 15–28.
9. Briscoe J and Qamar U. 2005. *Pakistan's Water Economy: Running Dry*. The World Bank, Oxford University Press: Oxford, United Kingdom.
10. IRSA. 2011. Water resource development in Pakistan: Oral presentation by Rao Irshad Ali Khan, Chairman IRSA. In *Roundtable Discussion on Agriculture & Water in Pakistan*, organized by World Bank, held on March 09, 2011.
11. GOP. 2015. *Pakistan Economic Survey, 2014–15*, Ministry of Finance, Islamabad, Pakistan.

12. Qureshi AS. 2011. Water Management in the Indus Basin in Pakistan: Challenges and Opportunities. *Mountain Research and Development*, 31(3), 252–260.
13. Haider G, Prathapar SA, Afzal M and Qureshi AS. 1999. Water for Environment in Pakistan. Paper presented in the *Global Water Partnership Workshop*, 11 April 1999, Islamabad, Pakistan. Available from the author of this article.
14. Kahlown MA and Majeed A. 2004. *Pakistan Water Resources Development and Management*. 1st ed. Pakistan Council of Research in Water Resources, Ministry of Science and Technology, Government of Pakistan, Islamabad, Pakistan.
15. Shahid MA, Boccardo P, Garcia WC, Albanese A and Cristofori E. 2013. Evaluation of TRMM satellite data for mapping monthly precipitation in Pakistan by comparison with locally available data. In: *III CUCS Congress—Imagining Cultures of Cooperation: Universities Working to Face the New Development Challenges*, Turin, Italy, pp. 146–154.
16. Archer DR and Fowler HJ. 2004. Spatial and temporal variations in precipitation in the Upper Indus Basin, global tele-connections and hydrological implications. *Hydrology and Earth System Sciences*, 8(1), 47–61.
17. Rees GH and Collins DN. 2005. Regional differences in response of flow in glacier-fed-Himalayan rivers to climate warming. *Hydrological Processes*, 20(10): 2157–2169.
18. Shahid MA. 2015. Geoinformatic and Hydrologic Analysis using Open Source Data for Floods Management in Pakistan. *PhD thesis*, DOI: 10.6092/polito/porto/2604981.
19. Ahmad B, Khokhar SB and Badar H. 2001. Economics of laser land leveling in district Faisalabad. *Pakistan Journal of Applied Sciences*, 1(3): 409–412.
20. Ahmad N and Mahmood N. 2005. Impact of raised bed technology on water productivity and lodging of wheat. *Pakistan Journal of Water Resources*, 9(2): 7–16.
21. Ahmad N, Arshad M and Shahid MA. 2011. Raised bed technology for crop water productivity of maize and cotton. In: *Proceedings of 21st ICID Congress* held on October 15–23, 2011 at Tehran, Iran. pp: 171–180.
22. Ahmad N, Shahid MA and Anjum L. 2011. Impact of raised bed technology to improve irrigation efficiency and enhance water productivity in rice-wheat cropping system. *Proceedings of International Seminar on the eve of Farm Machinery Week, Golden Jubilee Year Celebrations*, at University of Agriculture, Faisalabad, March 21–26, 2011.
23. Shahid MA, Ahmad N, Saleem M and Akhtar B. 2011. Investigating optimum number of irrigations for wheat under raised bed technology in a semi-arid climate. *International Journal of Agriculture and Applied Sciences*, 3(2): 89–93.
24. Pakistan Water Sector Strategy (PWSS). 2002. National water sector profile, volume 5, October 2002, Ministry of Water and Power, Office of the Chief Engineering Advisor. Available at http://waterinfo.net.pk/cms/pdf/vol5.pdf
25. PIDA. 2005. Scheme for Transfer of Irrigation Management—Farmer Organizations in Punjab, Updated April 2005.

4 Rural Energy Solutions for Community Development

Anjum Munir, Allah Bakhsh, Abdul Ghafoor, Waseem Amjad, and Umar Farooq

CONTENTS

4.1 RURAL ENERGY SOLUTIONS

4.1.1 Biogas Technology

In Pakistan, the history of biogas technology is about 40 years old. Across the country, about 7000 biogas plants have reportedly been installed. Pakistan has the potential of about 5 million biogas digesters. In 1974, the government of Pakistan started a comprehensive biogas scheme and total 4137 floating-drum biogas plants with a capacity of 5–15 m^3 biogas production per day were installed up until 1987. Until 2008, the Pakistan Council for Renewable EnergyTechnologies (PCRET) installed around 2500 biogas plants in different regions of the country. Pakistan has an almost 3000 MW power generation potential in its sugar industry while producing only 700 MW. A private nongovernment organization (NGO) Institute of Research and Social Development (IRSD) installed about 150 biogas plants funded from the United Nations Development Programme (UNDP). In the Sialkot district, NGO 'Koshis' facilitated villagers to construct over 200 biogas plants. In June of 2007, with the assistance of Foundation for Integrated Development Action (FIDA) Punjab Rural Support Program (PRSP) installed 12 dome-type biogas plants in Tehsil Pasrur of Sialkot. From 1982 to 1985, about 1000 biogas digesters were installed in various districts of the Punjab by the Pakistan Council of Appropriate Technology in collaboration with the Agriculture Department (Field Wing). During 2009–2010, The Agriculture Department (Field Wing) launched the program "Adaptation of Biogas technology to mitigate the energy crises" to install 750 family-size biogas digesters in different regions of Punjab. In Faisalabad, the PRSP installed 500 biogas digesters under the "Pakistan Domestic Biogas" Programme (PDBP). Some medium-sized biogas digesters have also been installed in the private sector to meet the energy demands for tube well operations for agriculture purpose, lighting and cooking for domestic purpose, and electricity generation for small industry operation during load shedding hours. The government of Punjab has also installed 50 biogas plants for tube well operation in 2011–2012. UAF installed floating drum, fixed, and bag type biogas plants for tube well operation and farm electrification. A training program titled "Skills for Biogas Plants" for the training of technician and supervisors has also been carried out to build their capacity and to enable them to acquire respectable employments in the field of bioenergy.

4.1.2 Waste to Energy via Biomass Gasification

Biomass is the most economical way of getting energy as an alternate resource. About 81 million ton of biomass is produced in Pakistan annually. Biomass gasification is a thermochemical process of converting biomass into a mixture of combustible and noncombustible gases known as "syngas" or producer gas. The gasifier comprises a reactor, cyclone separator, wet scrubber, water separator, biomass filter, and two heat exchangers. The feeding unit is provided with a bucket elevator for the continuous feeding to the gasifier for smooth operation and power generation. The reactor chamber is divided into four different zones, that is, the drying zone, pyrolysis zone, reduction zone, and combustion zone. During operation, the biomass is fed from the top of the gasifier through the feeding hopper. The combustion takes place in the oxidation zone in which carbon dioxide and water are produced with the limited supply of oxygen, and a significant quantity of heat, produced in this exothermic reaction, is then used for the other reactions, which take place in the reduction zone. The quantity of air is controlled by selecting the number of nozzles and designing their optimum diameter. Gasification of 1 kg biomass has the capacity to produce 2.5 m^3 of producer gas, which can be used to power an internal combustion (IC) engine and generator for electricity generation. The quantity and quality of producer gas depends on the temperature of gasification, size of fuel, and throat inclination angle. Normally, corn cobs are considered to be the most suitable fuel for downdraft gasifiers. It is pertinent to mention here that UAF has already developed a prototype downdraft gasifier for heat applications and power generation. The system comprises a reactor, cyclone separator, wet scrubber, biomass filter, counter current heat exchanger, and refrigeration system for gas purification. The gasifier has been coupled with 15 kVA generator for electricity generation using agricultural waste. The performance evaluation of the system has shown that it is a successful technology for decentralized application in rural community, using agricultural waste. Moreover, UAF has also installed a 100 kW gasification unit in collaboration with the Guangzhou Institute of Energy Conversion (GIEC), China, to address short to medium-size power outages. A briquetting machine has also been installed for making briquettes using agricultural waste, which is an excellent technique to handling raw biomass in compacted form for easier transportation and combustion applications.

4.1.3 Decentralized Solar Technologies for Rural Development

The agroclimatic conditions of Pakistan, ranging from tropical to temperate, allow the growing of 40 different kinds of vegetables and 21 types of fruits. At present, the area under fruit and vegetable cultivation is 0.995 million ha (4.3% of the total cropped area), with a total production of 10.992 million tons. The major factors limiting an increase in area and production are the need for high investments with low returns to the growers. Due to the lack of processing facilities, farmers have to sell their products at very low prices. Moreover, a significant amount of these products are spoiled due to lack of farm gate processing facilities. With the introduction of innovative solar technologies, it is possible to utilize the flux of solar radiant energy in the temperature range of 300°C with an average 50% efficiency. The use of solar energy for high temperature applications is now an economically attractive possibility since the payback period of such a system lies between 1.5 to 2 years (Jayasimha, 2006; Ghafoor and Munir, 2015). The promotion of small-scale agro-based industries for value addition using innovative solar technologies can become a multiplier in rural development. These products can be preserved for a longer period of time for farmers own use who will be able to produce marketable surpluses to increase their income. Currently, the worldwide interest in distillation of herbal or medicinal plants has increased significantly. According to a World Health Organization (WHO) survey, about 70%–80% of the world population relies on nonconventional medicines for their primary health care. This strategy is mostly based on the medicinal plant products known as botanical, herbal, medicines, or phetomedicines (Akerele, 1993; Calixto, 2000; Chan, 2003). In Pakistan, almost 2000 medicinal plant species exist and about 50% of the population in Pakistan is being cured using traditional medicines dispensed by more than 40,000 traditional herbal practitioners. There is a burgeoning need for the promotion of medicinal herb crops in Pakistan. First, because these are reemerging as a health aid due to the mounting costs of prescription drugs in the maintenance of personal health. Second, in the international market, the opportunities are emerging day by day for the trade of medicinal herbs, spices, and essential oils to fetch foreign exchange for the country (Aslam, 2002). Up to now, there has been very little work on the processing of medicinal and aromatic plants in Pakistan. Many stakeholders from all over the world are interested in working on this solar technology. The department of Horticulture, University of Agriculture, Faisalabad, is already conducting research on the distillation of rose oil using energy from fossil fuels. However, by introducing solar-based distillation systems, there will be new possibilities of skill development of academia and stakeholders using training workshops and participatory approaches. Due to functional economical and environmental reasons, there is need to introduce the novelty idea of solar distillery for value addition at farm level. The solar drying of medicinal plants, fruits, vegetables, and so forth plays a vital role in the value addition of agricultural products. The low temperature technology is very useful for the quality drying of different perishable/agricultural products. It will help the farmer dry their farm products without using fossil fuels, thus saving considerable amount of money. This versatile solar tunnel dryer can dry a variety of perishable products effectively.

4.1.4 Way Forward

Keeping in view the above discussion, the following ways forward and policy measures are proposed:

1. Substituting the country's rapidly increasing fossil fuel demand with renewable energy sources and expanding access to affordable energy in remote communities through distributed renewable energy.
2. Mandatory deployment of solar installations in the governmental, public institutional, and agricultural sector.
3. Incentivizing the stand-alone and grid-tied solar PV system, as well as other renewable energy technologies for tube well operation and farm electrification.
4. Implementation of net metering and feed-in-tariff policy to promote solar energy, especially in the rural sector through provision of feed in tariffs can accelerate solar power utilization in rural communities.
5. Formulation of a policy document for energy audit and energy efficiency of conventional buildings and poorly insulated rural buildings.
6. Amendment of building bylaws for rooftop solar installations and promotion of green and zero energy buildings.
7. Maximum use of sunlight to be encouraged during daytime and minimum number of lights, fans, and other electric appliances used during daytime during working hours.
8. Bylaws for installation of occupancy/proximity sensors in public, private offices, agriculture vicinities, and commercial buildings. It is also recommended that motion sensors should be compulsory for switching ON/OFF lights in case of absence of workers.
9. Efficient lighting products should be used and the use of LEDs, energy-efficient fans, motors, and inverter-based air-conditioner should be enforced for all newly constructed buildings and infrastructures, both in residential and rural sector.
10. Government's comprehensive program for replacement of existing outdated/faulty/less efficient/fused appliances by energy-efficient devices through incentivizing to decrease the grid load.

11. Production of efficient appliances to be encouraged through enforcement of MEPS (Minimum Energy Performance Standards) and fiscal incentives for fans, florescent lamps, air conditioners, motors, and refrigerators, UPS, etc.
12. Promoting daylight saving time and enforcement to close markets and shops before 7 or 8 pm to decrease peak load on national grid.
13. Policy needs to be formulated to wisely utilize abundantly available animal dung for biogas production and its utilization for heating, lighting, and rural electrification at farm level.
14. Media involvement for the awareness regarding energy saving, energy efficiency and exploration of renewable energy through documentaries, table discussions, and commercials.
15. Build capacity in research, public awareness, and the private sector.
16. Employment generation through skill development in renewable energy sector.
17. The Government should fix zones to best utilize the wind speed (5–7 m/s) of persistent winds in the coastal regions of Sindh and Baluchistan provinces and in a number of North West frontier valleys to harness the economically viable wind power potential of about 20,000MW in the country.

4.1.5 Short, Medium, and Long-Term Policy

4.1.5.1 Short Term

- Introduction of indigenously produced energy producing units through micro financing, namely solar thermal units (solar-based milk pasteurizer, solar cooker), community-based biogas plants to enhance farm gate processing activities for the processing and value addition, and income generation.
- Capacity building and skill development training programs should be initiated for implementing and promoting appropriate indigenous applicable techniques.

4.1.5.2 Medium Term

- Provision of incentives/subsidized financial support (at least 50%) to attract the farming community to adopt innovative technologies on large scale.
- Mega-projects need to be initiated to explore the available potential at its full extent to address value addition on a larger scale to minimize the 20%–25% existing losses.

4.1.5.3 Long Term

- The Government should fix zones to best utilize the available site-specific energy resources for processing of regional agro-crops, fruits, and vegetables.
- Establishment of cottage industry for the design and developed of on-farm processing technologies on competitive scale to bring cost of technologies to acceptable levels for end-users.

4.1.6 UAF Contribution for Rural Energy Solutions

UAF has already developed demand-based RE technologies for rural communities for farm gate processing that need to be replicated throughout the rural sector with the provision of incentives/subsidies (at least 50%) by the government. In this regard, a solar distillation system was developed and used for the processing of medical and aromatic plants employing a 10m^2 Scheffler reflector; a continuous solar roaster was designed for the roasting of pine nuts, ground nuts, coffee, and so forth with a temperature range varying from 130–180°C to enable it to decrease the product moisture contents 5%–6%; a solar tunnel dryer of 10 m length and 1.33 m width has been developed for the dehydration of perishable agricultural products (apples, mangoes, potatoes, medicinal plants, ripened chilies, apricots, etc.) at an air temperature of up to 60–65°C; a solar multicrop dryer can also be developed for the decentralized drying of paddy, maize, oats, etc. A 2.7 m^2 fixed-focus solar cooker, based on the Scheffler design, has been designed and developed for different applications in rural areas. It will provide a maximum temperature of about 270°C on direct focus, without secondary reflector, giving a power of 478 W which can be successfully used not only for daily household cooking but also for processing perishable stuff to preserve them. To tap the biomass/biogas potential, floating drum type biogas plants (40 and 25 m^3) have been developed and installed near animal sheds on farms in order to use animal dung for biogas production. For this capacity, 1050 kg of dung is required in the feeding chambers to produce 36.75 m^3 per day. Being an agricultural country, livestock is one of the biggest sectors in Pakistan with a total population of 72 million (cows and buffalos) producing approximately 720 million kg dungs. Considering 50% collectability about 360 million kg dung can be used for biogas production. It is estimated that about 18 million m^3 of biogas can be produced (@ 0.035 m^3 per kg) using this collectable dung. This biogas can be used to produce about 2400 MW electricity in the country in rural areas. Also, the agricultural waste/biomass can be addressed by designing a 20 kW biomass gasifier unit using indigenous resources. The biomass consumption rate will be 18–40 kg per hour depending upon type, quality, and moisture contents with a gas production of 102 million m^3 with a heat generation rate of 100,000 kCal/h. The system can be coupled with an 18 kVA generator for power production.

For the handling and processing of raw milk at farm, solar-based milk pasteurization units have been developed and successfully tested for the chilling/pasteurization of 200 liters of milk. The system comprises a pasteurizer (100 liters), a chiller (200 liters), a 2000 W solar PV array, a hybrid inverter, and a 300 Ah battery backup system. The pasteurizer is operated with the help of a vacuum tube collector while a vapor compression refrigeration system has been employed with an inverter technology to chill 200 liters milk in 90 minutes from 30 to 4°C. It is concluded that these decentralized solar-based and bioenergy technologies are sustainable energy solutions using indigenous resources for food processing and value addition, and can play a vital role for income generation and rural community development.

Concluding from the above discussion, three demand-based RE technologies (solar assisted milk chillers, solar cookers, and 25 m^3 biogas plants) have been recommended during the first phase for replication in rural areas where there is no access to electricity/suigas. The cost estimation and revenue generated is summarized in Table 4.1.

TABLE 4.1
Cost Estimations and Revenue Generated by Solar Cookers, Solar-Assisted Milk Chillers, and Biogas Plants

Solar Cookers

No of Villages	No of Cookers	Cost of One Unit (@ Size of 5 m²) PKR	Total Cost (Million PKR)	Subsidy 50% Million PKR	Daily Thermal Unit (kWh)	Annual Unit Used (kWh)	Cost per Unit PKR	Total Saving per Cooker PKR	Total Saving for All Cookers Million PKR
10,000	10,000	50,000	500	250	6	2,190	15	32,850	328.5

Solar-Assisted Milk Chillers

No of Villages	No of Chillers	Cost of One Unit (PKR)	Total Cost (Million PKR)	Subsidy 50% (Million PKR)	Milk Chilled per Day (Liters)	Saving per Liter of Chiller Milk (PKR)	Total Annual Saving (PKR)	Total Saving for All Chillers (Million PKR)	Annual Packing Charges @ PKR 4/Liter for All Chillers (Million PKR)	Remaining Total Saving for All Chillers Million PKR
10,000	10,000	500,000	5,000	2,500	400	10	1,460,000	14,600	5840	8760

Biogas Plants

No of Villages	No of Biogas Plants	Cost of One 25 m^3 Unit	Total Cost (Million PKR)	Subsidy 50% Million PKR	Biogas Prod (m^3/day)/plant @ 70% efficiency	Gas Produced per hr (m^3)	Gas Consumption of Otto Engine (m^3/kW/h))	Total Power Produced (kW/h)	Total Unit Produced per Day (kWh)	Annul Unit Produced (kW)	Cost per Unit Rs/kWh	Total Saving per Year per Plant PKR	Total Saving for All Plants Million PKR
10,000	10,000	50,000	500	250	14	0.583333	0.6	0.972222	23.33333	8516.667	10	85166.67	851.6667

REFERENCES

Akerele, O. 1993. Summary of WHO guidelines for the assessment of herbal medicines. *HerbalGram*, 28:13–19.

Aslam, M. 2002. Conservation, cultivation and trade of medicinal herbs and spices in Pakistan, International workshop on health Challenges of 21st Century and Traditional medicines in SAARC Region.

Calixto, J.B. 2000. Efficacy, safety, quality control, marketing and regulatory guidelines for herbal medicines (phytotherapeutic agents). *Braz. J. Med. Biol. Res.*, 33:179–189.

Chan, K. 2003. Some aspects of toxic contaminants in herbal medicines. *Chemosphere*, 52:1361–1371.

Ghafoor, A., A. Munir. 2015. Design and economic analysis of an off-grid PV system for household electrification. *Renew Sustain Energy Reviews*, 42: 496–502.

Jayasimha, B.K. 2006. Application of Scheffler reflectors for process industry, International Solar Cooker Conference 2006 in Granada, Spain.

5 Environmental Degradation and Remedial Strategies

Haroon Rashid and Abdul Nasir

CONTENTS

5.1 INTRODUCTION

Prevailing environmental conditions are deteriorating day by day in Pakistan. The country is subjected to both natural and anthropogenic sources of pollution. Mainly, environmental issues in Pakistan are directly related to improper management and handling of different environmental scenarios. Determination of the environmental status of Pakistan leads to development of a very complex situation, due to interlinked environmental factors, featuring unknown impacts of different sources of pollution. Intensive research is still required on a priority basis for the proper understanding of the share of the different environmental factors. Nevertheless, based on the available resources and data, it can be clearly observed that various interlinked sources of contamination are posing a severe threat to human health and the environment in Pakistan. Environmental and natural resources are facing a situation of intense stress. Water resources, once available in surplus, are now at an alarming level due to unfair usage and improper management, particularly considering groundwater exploitation in the Province Punjab. Here, the highest rate of deforestation in Pakistan has resulted in an intact forest land cover as low as 4% of the total area. This high level of loss of forests ultimately leads to different other environmental issues such as soil erosion, land sliding, uncontrolled floods, and the most important, changing impacts on climate and altering ecosystems. The associated effects of climate change in the country have increased durations of droughts, thus resulting in less water released to the sea, and an increased salt intrusion into groundwater from the sea [1]. All these environmental factors can be more appropriately termed as "environmental stressors" and do have great socioeconomic impacts in Pakistan. For example drought, water supplies along the Indus, water availability for power generation, agricultural production—particularly in drought affected areas, all are interlinked and once the balance of the system is disturbed, its impact is highlighted on all other associated factors as well. Haphazard industrialization is the major source of environmental degradation of natural resources, leading to severe environmental pollution. Today, one million individual chemicals are manufactured. The number of chemical products is about 90,000 to which 2000 substances are added annually. Half of the chemicals are injurious to health. Consequently, an estimated quantity of 20,000 million tons of industrial waste is generated, 10% of which is hazardous. The growth rate of industrial waste is 10% per annum [2]. Thus, industrialization has a crucial effect on the deterioration of environmental systems in Pakistan.

Experts have reported that environmental degradation may cost Pakistan's economy over Rs.365 billion. Out of this, water-based problems may account for a Rs.112 billion loss, while agricultural pollution may result in an economic damage of Rs.70 billion. Other significant types of environmental degradation such as indoor pollution, urban air pollution, lead exposure, land degradation, and deforestation cause economic losses of approximately Rs.67 billion, Rs.65 billion, Rs.45 billion, and Rs.6 billion, respectively [3].

5.2 INDUSTRIAL POLLUTION AND MUNICIPAL WASTEWATER PERSPECTIVE IN PAKISTAN

Numerous wastes/effluents are produced from industrial activities. Textile, carpets, sports goods, leathers, and surgical instruments are five main types of industries in Pakistan causing environmental pollution. Out of these industries, textile and tanneries industries are the main sources of effluent discharge to the rivers and lakes, which cause environmental problems resulting in serious concerns for human health. Large number of detergents, dyes, acids, and so forth are used in the textile industry, in addition to the huge amounts of chemicals used in the leather industry, both producing high quantities of hazardous effluent. If used without any treatment, effluents from both of these industries may have severe impacts on the environment and human health [4].

The industrial areas of big cities are facing the problems of severe environmental pollution. Due to these conditions, vegetation has been greatly reduced to alarming levels. It has been reported that air pollution in Karachi, Rawalpindi, and Lahore is about 6.4 times higher than the World Health Organization's (WHO) standards and about 3.8 times higher than Japanese standards. In addition to these industries, many other industries, including petrochemical, paper mills, and manufacturing industries release hazardous elements and heavy metals to the environment through air and water. These hazardous elements, after entering the human body, accumulate in various vital organs due to their biologically nondegradable nature and end up causing many serious diseases [4]. Figure 5.1 clearly demonstrates unplanned industrial wastewater and solid waste management practices at different industrial zones in the country.

Untreated municipal wastewater in Pakistan is leading to unhygienic conditions and sanitary problems, which ultimately affect drinking water quality, being supplied though water supply schemes, and water quality leaching into soils and mixing with groundwater aquifers. Sanitary problems are more severe in plain lands due to poor drainage systems plagued by leaky joints and cracked sewers. Such adverse environmental conditions can easily observed in Figure 5.1.

In Pakistan, the untreated wastewater is discharged directly to city sewerage system, natural drains or water bodies, or to nearby fields or internal septic tanks. Mainly, none of the country's cities have any biological treatment process, except Islamabad and Karachi having small-scale treatment setups [5]. According to the 1998 population census, about 1.83×10^7 m^3/hr of wastewater is discharged from the 14 major cities of Pakistan.

Presently, a total quantity of 962,335 million gallons of wastewater are produced all over in Pakistan, including 674,009 million gallons from municipal and 288,326 million gallons from industrial use. It has been reported that the total wastewater discharged to the major rivers is 392,511 million gallons, including 316,740 million gallons from municipal and 75,771 million gallons from industrial effluents. It has also been estimated that around 2000 million gallons of sewage are being discharged to local surface water bodies every day. The paper and board, sugar, textile, cement, polyester yarn, and fertilizer industries produce more than 80% of the total industrial effluents [5].

The problem of industrial water pollution has remained uncontrolled because there have been little or no incentives for industry to treat its effluents, as well as no laws implemented. In KP province, 0.701×10^9 m^3/yr of industrial effluents are discharged into the Kabul river. In Sindh

FIGURE 5.1 **(See color insert.)** Pictorial view of different aspects of environmental pollution in Pakistan: (a) Treated effluent from treatment plant; (b) Chocked effluent-carrying drains; (c) Improper handling of industrial waste; (d) Untreated & mixed industrial effluent.

province, only two out of 34 sugar mills have installed mechanisms for wastewater treatment, mainly because of international pressure as these industries (distilleries) export their products. Similarly in Lahore, a major city of Punjab province, only three out of some 100 industries using hazardous chemicals treat their wastewater. In Faisalabad, there is a wastewater treatment plant, in which wastewater receives primary treatment. In rural areas, wastewater treatment is nonexistent, leading to pollution of surface and groundwater. The fertilizer sector is the only industrial sector in the country that invested significantly in installing wastewater discharge treatment plants.

5.3 SOLID WASTE MANAGEMENT IN PAKISTAN

Solid waste is the unwanted material generated as the result of human and animal activities, which are usually in the form of solids, semi solids or liquids in the container discarded as useless or harmful material. There are different sectors generating solid waste, which concern domestic, industrial, commercial, agricultural, and mineral extraction activities as well as accumulations of refuse in streets and public places.

It has been reported that approximately 56,000 tons per day of solid waste is generated in Pakistan, while the collection rate is only 51%–69%. In most of the cases there are no disposal facilities. The uncollected waste causes clogging of drains, formation of stagnant ponds, and provides breeding places for several kinds of germs causing severe health problems.

In Pakistan, the rate of waste generation on an average varies between 0.23 kg/capita/day and 0.61 kg/capita/day in rural and urban areas, respectively. Growth rate of solid waste generation is about 2.4% per annum. The intense situation of municipal solid waste management, in the form of garbage heaps and the choking of drains, can be observed in Figure 5.2.

FIGURE 5.2 Open dumping of commingled municipal and industrial solid waste.

5.4 HEAVY TRAFFIC LOADS AND ASSOCIATED AIR AND NOISE POLLUTION

Atmospheric pollution is spreading at an immensely high rate in big cities due to exponentially increasing traffic loads on the roads including motorcycles, rickshaws, and trucks (Figure 5.3). Most of the cities in Pakistan are suffering heavily from noise pollution. Karachi city has been found to be one of the noisiest cities in the world [2].

Localities are deprived of peaceful and quiet environments in Pakistan. There is a lack of civic sense that fuels the generation of noise most of the time of the day. Narrow streets in Pakistan, densely populated, have cottage industries installed in their midst, which produce unwanted sounds. The settlements have been developed in an unplanned fashion and there is no provision made for creating acoustic harmony. The play grounds are another source of continuous noise. National events and rituals in Pakistan are conducted in this backdrop of noisy environment. Marriage ceremonies are not held peacefully and people only enjoy them if the bands play at a very high pitch. Music programs are held with disorderly instrumental sounds and haphazard voices. In such a dismal acoustic environment, 20% population of Pakistan has become hard of hearing [2].

In Pakistan, vehicular noises should not exceed the limit of 85 db within the radius of 7.5 meters. Noise levels produced by different vehicles and equipment are presented in Table 5.1.

FIGURE 5.3 **(See color insert.)** Industrial and vehicular air pollution sources: (a) Vehicular pollution on roads; (b) Industrial air pollution.

TABLE 5.1
Noise Levels Produced by Vehicles and Other Equipment

Vehicle/Equipment	Distance (meters)	Noise Level dB (A)
Motorcycle	8	96
Bulldozer	15	94
Heavy truck	15	93
Heavy diesel lorry	7	92
Earth digger	15	80
Tree saw	15	82
Washing machine	15	82
Alarm clock	1	85
Lawn mover	3	102

Source: Mumtaz H. 1998. *Environmental Degradation Realities and Remedies*. Ferozsons (Pvt) Ltd.

5.5 AGROCHEMICALS AND THEIR USAGE IN PAKISTAN

To protect crops from pest attack, use of pesticides by the farmers is increasing. Advancements in scientific chemistry and other biochemical fields have led to the development of various pesticides with different properties and lines of actions. This is helping to control crop losses, but also is creating a lot of pesticide toxicity problems. Excessive use of pesticides is causing air and atmospheric pollution to rural areas along with polluting surface water (Figure 5.4).

Both human and animals are exposed to the uptake of pesticides from air and water and through other sources. Pesticides include insecticides, herbicides, and fungicides. Chemically, pesticides may be categorized into four groups, that is, (i) organochlorines, (ii) organophosphates, (iii) carbonates, and (iv) pyrothroide [6]. When waters contaminated with pesticides are used for drinking purposes, this results in a range of harmful effects from mild headaches and allergies to cancers. The severity of the problem depends on the type of pesticide and the amount of polluted water intake [6]. Agrochemicals also cause nonpoint source of pollution in the form of agricultural runoff (Figure 5.4).

In Pakistan, nearly 50% of the pesticides used are extremely hazardous. Moreover, the use of pesticides has increased manifold in recent years. It has been reported that 80% of the total pesticides

FIGURE 5.4 Nonpoint source of pollution caused by agricultural runoff.

consumed in Pakistan are used in the months of July to October for the protection of cotton crops. The historical analysis of pest occurrence indicates that the current system of pest management, relying mainly on the use of over 6 sprays/season, is responsible for the change in cotton pest complex [5].

There is a need to discourage the indiscriminate use of pesticides and sale of Fouled be strictly banned. Quality control of the chemicals being used against pests and diseases needs to be strictly enforced to avoid their environmental pollution side effects. Moreover, the need of the hour is to encourage use of indigenous products, such as "neem leaves" instead of importing chemicals. It should be the responsibility of the concerned agencies of pesticides to ensure distribution and sale of their products honestly to the users for agricultural crops without hazardous effects [6].

5.6 MAJOR CHALLENGES TO ENVIRONMENTAL ISSUES IN PAKISTAN

The challenges in environmental sector include [3]:

1. Untreated industrial effluent causing severe threat to soil and groundwater resources.
2. Water and air pollution causing health hazards to community.
3. Lack of awareness and severity of the environmental hazards.
4. Land reclamation and deforestation.
5. Management of solid, liquid, hazardous waste, particularly hospital waste.
6. Immense usage of toxic chemical in industries which is causing severe threat to the ozone layer resulting in a global warming effect and climate change.
7. Water resource exploitation.
8. Haphazard industrialization and flood of urbanization.

5.7 REMEDIAL MEASURES AND COPE UP STRATEGIES TO ADDRESS ENVIRONMENTAL ISSUES IN PAKISTAN

5.7.1 Short-Term Strategies

1. Initiation of an urgent country-level movement to spread out tree plantation with a mission to own forests and trees as valuable resource and asset. Government should offer some incentive to those who play significant roles in this movement. Arranging biannual festivals for tree plantation and giving awards to those who contribute the most—in the form of monetary benefits, as well as for example linking this movement with the Benazir Income Support Program or the Laptop Scheme would be positive developments.
2. Union council-level training programs to start "Waste Banks" may prove a very vital component of government policy for poverty alleviation as well as proper solid waste management. In these training sessions. communities should be motivated to take initiatives to segregate solid waste at the source and transform and process it to make usable goods or convert it to energy or fuel. Keeping in mind entrepreneurship projects, such initiatives can prove to be very useful not only for the community itself, but in helping reduce environmental loads in the system along with sharing responsibility with the government.
3. Providing tax subsidies to those industries that have efficient environmental management systems implemented in their industrial units can prove to be an effective strategy to urge industrialists in adopting environmentally friendly techniques.

5.7.2 Long-Term Strategies

1. A culture needs to be developed in which people willingly participate in keeping their environment safe. A lot of work needs to be done to spread awareness to the youth so that they can effectively play their role as custodians of the environment in every respect. In this

regard, children should be the main focus so that it may become a part of their upbringing. Furthermore, the electronic media need to play an important role in promoting the culture in which caring for the environment becomes a priority.

2. Institutions can play an important role in conducting diversified research on different aspects of the environment; furthermore, channelling this research for proper implementation and adoption by government policy-making institutions, as well as commercial organizations, will be necessary. By using the latest technologies and modifications in the conventional industrial processes, environmental pollution sources can be reduced to great extent at the source.
3. Knowledge of National Environmental Quality Standards (NEQS) should be made mandatory to all the industrial office bearers and other commercial organizations' chief executives. Furthermore training workshops should be scheduled across the country in every industrial estate/zone, including every minor and major industrial sector so as to keep the workers well aware of the environmental management systems which must be adopted at all levels of different processing units.

REFERENCES

1. Vaughn B, Carter NT, Sheikh PA, Johnson R. 2010. Security and Environment in Pakistan. *Congressional Research Service, CRS Report for Congress*, http://fpc.state.gov/documents/organization/146411.pdf
2. Mumtaz H. 1998. *Environmental Degradation Realities and Remedies*. Ferozsons (Pvt) Ltd, Lahore Pakistan.
3. PES. 2013. Pakistan Economic Survey Report for Year 2013–2014. Economic Adviser's Wing, Finance Division, Government of Pakistan, Islamabad, http://www.finance.gov.pk/survey/chapters_14/16_Environment.pdf
4. Rizvi SA. 2007. Five Industries Major Polluter in Pakistan. Pakistan Economist, http://www.pakistaneconomist.com/database2/cover/c2007-102.php
5. Murataza G, Zia MH. 2012. Wastewater Production, Treatment and Use in Pakistan. UN Water Activity Information System. Final Country Report, http://www.ais.unwater.org/ais/pluginfile.php/232/mod_page/content/134/pakistan_murtaza_finalcountryreport2012.pdf
6. Alam SM, Khan MA. 1999. Pesticides and Their Effects, Pakistan Economist, http://www.pakistaneconomist.com/database2/cover/c99-9.php

6 Land Degradation
Problems and Remedies

Zahir Ahmad Zahir, Maqshoof Ahmad, and Ghulam Murtaza

CONTENTS

6.1 INTRODUCTION

Land degradation is defined as the depletion of soil fertility and a decrease in the productivity of soil which may be permanent or temporary. Soil fertility is a component of overall soil productivity which deals with its available nutrient status, and its ability to provide nutrients out of its own reserves and through external applications for crop production. The mere presence of plant nutrients for healthy growth in the form available for plant use makes a soil fertile. The soil is, however, only productive when the various nutrients are in appropriate amounts or ratio. There are many factors that affect the availability of nutrients in soils and ultimately their productivity. So, all these contribute to the sustainability of soil or its degradation. It is estimated that about 50% of world's agricultural land is extremely degraded due to natural and anthropic factors (Bai et al. 2008; Agarwal 2008). Every year thousands of hectares of land are being deteriorated in all agrozones, which affect more than 2.5 billion people worldwide (Adams and Eswaran 2000). With the advancement in modern agriculture and urbanization in twentieth century, land degradation has been tremendously accelerated by human activities and thus, in the current twenty-first century, it is the most serious issue on the agenda of food security and the environment. The environmental consequences of land degradation are incongruous terrestrial use, which lead to dreadful conditions of land, aquatic, and vegetation resources resulting in abrupt changes in climate and ecosystem.

Naturally, land degradation is a relatively slow process, but with an unsustainable approach in utilizing land for agriculture, industries, and habitat, the negative results increase in pace, without mitigation. The most important cause of land degradation is considered desertification, which has affected about 40% of agricultural land in the world. Desertification doesn't mean the complete transformation of land into a desert environment, but it includes the degradation of biological ecosystems which become affected to such an extent that their climate becomes too dry and arid for biological diversity (Zamfir

2014). Land degradation results in the deterioration of water resources leading to severe decline in water availability, and causing soil erosion which leads to unfertile soils, deforestation, and reductions in the productiveness of grass, forest, and agricultural lands. Vanishing of biodiversity, due to the development of anthropic climate, also results in the deterioration of productivity of lands.

The causes of land degradation are natural and man-made factors, which directly or indirectly carry out the process. For example, soil erosion—as natural cause of land degradation—directly depends upon the topography of the land, nature, and physical condition of soil, and the intensity and frequency of rainfall in that area. The anthropic or man-made land degradation occurs when lands are utilized without following adequate conservation practices. The intensive use of land for extra production, by using fertilizers without adding organic matter, leads to deterioration of the physical and chemical properties of soil. Man-made land degradation is a major threat to agriculture worldwide and the extent of these man-made activities is increasing day by day leading to irreversible damage to land and water resources. These activities include the overcutting of vegetation, intensive cultivation of land without sufficient fallow periods, deforestation, overgrazing, faulty agricultural practices, inappropriate crop rotations, injudicious use of fertilizers, pollution, and over drilling of groundwater. The combination of all or some of the above activities results in the deterioration of land through salinization, drought, desertification, erosion, decrease in fertility, lowering of the water table, and pollution. The alarming issue of land degradation is emphasized here along with the causes and problems of land degradation. Some strategies to address the issue are also suggested herein, particularly for Pakistan.

6.2 CAUSES OF LAND DEGRADATION

Soil is the only natural medium available for plant growth, thus it is considered the most critical limiting factor in agricultural productivity. Healthy soil is believed to be indispensable for sustainable agricultural production. Any factor, which adversely affects soil health, definitely hampers its production potential to a larger extent and deteriorates its quality. There are number of factors which negatively influence soil health and quality leading to deceases in crop productivity. These factors are briefly discussed as fallows.

6.2.1 Soil Erosion, Nutrient Mining, and Depletion of Soil Fertility

Many factors affect depletion of soil fertility. Mining of soil nutrients from cropped area is real threat to food security and causes environmental degradation. Soil erosion is one of the most serious problems for agricultural lands, forests, and grasslands. Soil erosion is the removal of fertile soil from the surface of land resulting in depletion of soil fertility and productivity. Primarily, it is a natural phenomenon which is responsible for the development of soils from rock parent materials. But increase in the rate of detachment and removal of soil above optimum level results in severe land degradation. Furtermore, accelerated erosion may be attributed to abrupt changes in climate due to human activities resulting in speeding up the process of removal of soil by deforestation, overgrazing, etc.

The agents responsible for erosion include wind and water. Erosion causes considerable damage to the fertile soil layer. In Pakistan, about 11.171×10^6 ha of soils are affected by water erosion, out of which 17.04% lies in Punjab, <1% in Sindh, 38.4% in Khyber Pakhtunkhwa and FATA, 23.6% in Baluchistan, and 20.4% in Northern Areas. About 4.761×10^6 ha of soils are affected by wind erosion, out of which 79.9% lies in Punjab, 13.4% in Sindh, <1% in Khyber Pakhtunkhwa, and 5.9%in Baluchistan province (Table 6.1, Zia et al., 1994). Wind erosion occurs when wind speed exceeds a critical range resulting in removal of fertile soil from land surface. Wind erosion is mainly accelerated by human activities, such as overgrazing of pastures, resulting in removal of surface cover which makes fertile soil prone to abrupt winds causing removal of soil particles through suspension, saltation, and creep. Wind erosion is also accelerated by intensive use of heavy implements for the cultivation of soil. Massive wear and tear of aggregates detaches fertile soil particles, which are blown away by winds, leaving behind barren lands.

TABLE 6.1
Extent of Land Degradation due to Waterlogging, Salinity, and Wind and Water Erosion in Pakistan (Million ha)

	Water Logging				
Province	Pre-monsoon	Post-monsoon	Salinity	Water Erosion	Wind Erosion
Punjab	2.84	3.99	2.643	1.904	3.804
Sindh	4.62	4.75	2.66	0.059	0.639
KP	0.17	0.2	0.522	4.292	0.037
Baluchistan	0.239	0.15	0.394	2.634	0.280
Northern Areas				2.282	
Total	7.869	9.09	6.22	11.171	4.761

Source: Zia et al., 1994. Problem of soil degradation in Pakistan. The collection and analysis of land degradation data. RAPA/FAO Bangkok, Thailand. pp. 179–202.

Removal of soil by water is also an important cause of land degradation. It is most common in mountainous regions and in irrigated areas of the world. Sheet and rill erosion are the typical types of erosion in upland areas, while channel and gully erosion are most common in irrigated areas, and areas around the banks of rivers. Water erosion occurs when large flows of rain water passe through an area and carries with them whole sheets of fertile soil. As a natural process, water erosion is not so much a damaging process because the amount of material transported is about 1 ton ha^{-1} per year, which is negligible amount as compared to man-made water erosion—about 50 ton ha^{-1} per year (Hassan and Rao 2001). According to USDA (1988), the permissible range of load to be removed is 5–12 ton ha^{-1} per year, which can be repaired by the natural process of weathering and soil formation (Young 1998; Tuboly 2000; Hassan and Rao 2001).

The severity of erosion in degrading land can be imagined by assuming that if top soil contains 0.2% nitrogen, the removal of soil 20 ton ha^{-1} per year may remove 80 kg of nitrogen along with other nutrients and organic matter (Natarajan et al. 2010). This loss is equal to several bags of fertilizers of each nutrient removed and cannot be recharged by the addition of fertilizers by humans.

Most of the soils in Pakistan have a poor status of available plant nutrients and cannot support optimum levels of crop productivity. For instance, widespread nutritional deficiencies occur even in relatively fertile soils through water erosion, coupled with continuous nutrient mining by crops in the Pothohar area of Pakistan. Removal of a fertile layer of soil leads to the lowering the productivity of crops and results in increasing food insecurity. Soil fertility status varies with nature of the cropping patterns and management practices. Land degradation, due to erosion encouraged by human activities (mismanagement of natural resources, war, economic policies and natural uncertainties), leaves behind millions of people starved and hundreds and thousands homeless. In Pakistan, almost all available soil is nutrient deficient. Soils are generally deficient in organic matter content—reflecting the severe deficiency of nitrogen (almost 100%), while phosphorus deficiency are common in more than 90% of the soils and potassium in 50% of the soils. Micronutrients such as zinc, boron, and iron are also emerging as deficient (Kobayashi and Nishizawa 2012; Moore et al. 2013). Crop yield increases 30%–50% with balanced fertilization. If 50% soils receive a balanced fertilizer application, agriculture production may increase by 20% compared to current national production levels. Thus fertilizer recommendations/applications on the basis of soil tests are considered prerequisite for higher crop yields.

6.2.2 Soil Salinization

The most devastating effect on crop productivity is caused by the accumulation of high amounts of salts in soils. It is estimated that about 267 million hectares of cultivable land are being degraded by salinization due to poor management of soils around the world (Grewal and Kuhad 2002; Duraiappah and Roy 2007). Water logging is a precursor of soil salinization. In Pakistan, the total geographical area is 79.61 × 10^6 ha, out of which the water logged area is 7.869 and 9.03 × 10^6 ha in premonsoon and postmonsoon season, respectively. Main contributor to water logged area in Sindh is 58.7 and 52.2% in pre-monsoon and post-monsoon season, respectively, Punjab contributes to water logged area by 36.1 and 43.9% in the pre-monsoon and post-monsoon season, respectively, while other provinces, that is, Baluchistan and Khyber Pakhtunkhwa have minor contributions to water logged area. However, the salt-affected soil (saline, saline-sodic and sodic generally termed as salinity) constitutes about 6.22 × 10^6 ha including gypsiferous saline-sodic soils out of which, 42.5% lies in Punjab, 42.8% in Sindh, 8.4% in Khyber Pakhtunkhwa and 6.3% in Baluchistan province (Table 1, Zia et al., 1994). Salinization is a broader term, referring to the building up of free ions in the soil and the development of exchangeable sodium, that is, sodicity. It arises through the unplanned development of agricultural lands and irrigation systems, as seepage of water from the banks of canals results in an increase in water table height, bringing salts above the ground. Opon the evaporation of water, solid salts remain on the surface, increasing the amounts of salts in the surface soil. The problem is also serious in near coastal areas due to the seawater intrusion, which is a natural process, but the development of unplanned irrigation systems can enhance the rate of intrusion rendering millions of hectares of land near the coasts unusable.

There are three main processes of salinization including accumulation of salts in soil from water, salty parent material, and wind deposition. Development of salinity takes place when the physical condition of soil is being altered by human activities—due to the use of heavy machinery, leading to the development of an impermeable layer of soil in the soil profile resulting in the accumulation of salts in the root zone (Dagar and Minhas 2016). Development of salinity through deposits formed by wind depends upon a pertinent source of salt. Salinization of an area may also occur naturally by the weathering of salty parent materials and minerals that releases salts in the soil. These types of salinization are greatly affected by man-made activities and climate change.

Due to the emission of greenhouse gases and pollutants in the atmosphere, the temperature of our planet is rising day by day. In dry and warm areas, high temperatures throughout the summer time result in increase in evapo-transpiration and the development of arid climates leading to the deposition of high amounts of salts in top soil. Leakage of water from unlined water courses and the leaching of extra water from excessive or improperly irrigated fields results in the salinization of low-lying areas.

Soil structure and availability of nutrients are adversely affected by salinity and sodicity. Changes in the distribution of nutrient ions, that is, the proportion of soil solution and exchangeable ions, along with rises in soil pH, osmotic, and specific ions cause deterioration effects in soils. Structural problem in salt-affected soils created by specific conditions (hardsetting and surface crusting) and certain physical processes (swelling, slaking, and dispersion of clay minerals) may affect water holding capacity, water and air movement, seedling emergence, root penetration, erosion, runoff, tillage, and sowing operations. Moreover, serious imbalances in certain plant nutrients usually occur in such soils. These imbalances of nutrients lead to deficiencies of several nutrients and sodium toxicity. All these changes in physicochemical characteristics of soils due to salts also hamper the activity of soil microbes and the development of plant roots, which lead to reduced crop growth and yield.

In Pakistan, the twin problem of waterlogging and salinity has resulted in a lowering of soil fertility and a decline in crop yields. These soils require careful considerations to improve their quality for crop production.

6.2.3 Desertification

According to United Nations Conference on Environment and Development (UNCED) in June 1994 "desertification is the land degradation in arid, semi-arid and dry sub-humid areas resulting from various factors including climatic variations and human activities." Desertification is a condition in which land becomes unproductive and of very less economic importance (Kassas 1995). In extreme cases, it is being converted into a desert landscape, unable to support the populations and societies once dependent upon it, and is considered as the extreme level of land degradation. It was estimated that about 250 million people are affected by desertification and 2 billion are at risk (WIT 2009). Desertification arises in areas having ecological structures which have little rainfall, lengthy arid periods, recurring droughts, and sparse soil cover. The southern areas of Pakistan are highly prone to desertification due to coarse textured soils, less rainfall, and water shortages.

The reason for desertification may be natural or anthropogenic activities. Natural reasons of land degradation due to desertification are sudden changes in climate and prolonged droughts; however, it is difficult to determine whether it is natural or the outcome of human activities. Anthropogenic activities include overstocking, intensive farming, and the removal of vegetation and forests. Studies suggest that deforestation is the major cause of desertification. Overcutting of trees has caused a somewhat irreversible change in climatic conditions of the earth, causing severe disturbance in rainfall distribution resulting in extreme reductions in rainfall in specific areas. A disturbance in hydrological cycle results in the development of desert-like environments in huge areas of the world. It is reported that about one third of earth's land is affected by severe desertification and, due to unsustainable human activities, its extent is increasing every year. The other factors responsible for desertification are socioeconomic and political factors. These are urbanization, unsustainable use and struggle for limited aquatic resources, and administration and political policies.

Land degradation is a serious issue for the farming community in dry areas including southern parts of Pakistan. In these areas, farmers with small landholding are more affected by the consequences of desertification. Overgrazing due to limited supply of resources by livestock leads to removal of soil by runoff and wind in rain fed areas of Asian and African countries. Rain water is the primary source which supports the whole ecosystem. Due to desertification, these zones constantly suffer from the shortage of food stuff and fodder production because of unpredictable and uneven rainfall. About 90% of people affected by desertification are the inhabitants of developing counties, including Pakistan. Unfortunately, due to their poor socioeconomic situation, the problem goes ignored by the authorities and the people, making it more severe (Reddy et al. 2000; Verón and Paruelo 2010).

6.2.4 Pollution

Land pollution is another major source of land degradation. It arises from increasing density of industries, mining, agriculture operations, and urbanization. In Pakistan, pollution is the major cause of land degradation. Industries in large cities are polluting the agricultural land through their effluents. City wastes are not only degrading the land but also deteriorating the environment along with surface and ground water resources. The heavy and unjudicious use of chemicals for agricultural use is another cause of land pollution. The use of these chemicals has drastically increased during the last decades causing severe contamination of cultivated lands.

The continuous release of contaminants in ecosystems results in the degradation of environment and soil, which pollutes and degrades land, making it unfit for use. Land pollution is a major hurdle for achieving the goal of sustainability of present land resources. Soil ecology is endangered by these pollutants and their interference with the environment. The potential sources of pollutants in land include (a) the sewage waste from urbanized areas—the land around major cities of Pakistan is facing this problem, (b) chemicals and hazardous wastes from industries—Kasur, Faisalabad, Karachi, Sialkot are the major cities contaminated, (c) residual waste from agricultural inputs—chemicals, fertilizers, and nonpoint sources which are polluting the soil, and (d) contamination from air pollution.

The accumulation of industrial effluents in soil results in the deterioration of physical, chemical, and biological properties of soil. This change in soil condition causes fatal permanent deterioration of the environment and creating difficult circumstances for human and plants health. Wastes from industries pose very dangerous effects on biodiversity. The most serious issue of industrial waste is the heavy metal toxicity for plants, animals, and human beings. Heavy contamination of soils with industrial wastes not only causes the plants and animals of that habitat to vanish, but it also eliminates the microorganisms inhabiting the soil environment.

The accumulation of toxic heavy metals in soil affects the crop yields in areas receiving industrial, urban, or agricultural wastes. These wastes contain a high concentration of heavy metals like cadmium, chromium, nickel, lead, and others. These metals are also hazardous to plants, severely affecting their productivity. Crops grown in contaminated soils accumulate high amounts of heavy metals in the edible parts, thus resulting in the incorporation of heavy metals into the food chain. These heavy metals are highly injurious for human health causing serious medical problems including cancerous diseases (Francois et al. 2004). For example, mercury is a heavy metal, and leaked into environment through gold mining and metallurgy, accumulates in tissues and causes various muscular disabilities (Gibbs et al. 2006). Cadmium taken in food, substitutes calcium in bones and causes deformation of bones while it also causes disturbances in the function of kidneys, deteriorating its tissues (Hussain et al. 2006).

Land deterioration due to the discharge of pollutants into soil is becoming a serious and neglected problem the world over including Pakistan. The increasing figures of unproductive land are becoming an alarming situation, especially for developing countries. Furthermore, the expanding of municipalities and townships due to growing populations is also encouraging the deterioration of land due to pollution. It is important to address the issue by planned reclamation of land and disposal of urban and industrial wastes, making it possible to limit the land degradation by pollution.

6.2.5 Physical Degradation

The physical condition of soil is a very important property of a land. A good physical state of soil is a prerequisite for the utilization of a piece of land for productive use. The most important physical property of soil regarding its agricultural use is the soil structure. Soil structure refers to the arrangement of chief soil units (sand, silt, and clay) into groups or aggregates. A good structure of soil indicates the capability of a piece of land to support vegetation and other biodiversity. A good structure of soil is identified by the presence of aggregates and organic matter, and its aggregation intensity reflects the productiveness and fertility of land.

The physical condition of the soil (soil structure) dictates all the physical, chemical, and biological processes taking place in the soil. A well-structured soil guarantees the resistance toward degradation and symbolizes the ability of a piece of land to be reclaimed if degraded due to its mismanagement. Soil structure affects the biological functionality and the intensity of soil cover by defining the water permeability, the activity of water, and the amount of water that can be retained by the soil. A good physical condition determines the aeration of the soil, that is, the indication of an ample supply of oxygen to the living fauna and flora of the soil.

The degradation of soil structure results in the compaction of soil, which imposes severe problem for the agricultural use of a land by increasing the labor efforts and energy requirements, and also leads to poor economic production as compared to the efforts spent. The leading reasons of the physical deterioration of land are the mismanagement of soils, intensive use of heavy implements, intensive cropping without resting periods, and lack of organic farming. The soil environments are pretty dynamic and alteration in a single characteristic disturbs the rhythm of other processes taking place resulting in the degradation of whole ecosystem.

In Pakistan, physical degradation of land is increasing at an alarming rate due to intensive cultivation, low organic matter, over grazing, and shortage of water. Pakistan was hit by severe drought from 1997–2003, consequently water shortage for humans, livestock, and agriculture were experienced. Continuing water shortage and drought have taken a heavy toll on Pakistan's economy and increased vulnerability and hardship for many predominantly rural communities. Frequent drought in Pakistan and elsewhere causes widespread damage even to the biological potential of the fertile lands. Drought in many areas in the provinces of Balochistan, Sindh, and southern Punjab has negatively impacted millions of people and livestock heads. Hundreds of people lost their lives, and thousands of livestock and wild ungulates perished due to drought conditions. The drought also severely affected local livelihoods and forced local people to migrate towards the places where they could find jobs and food, and provide feed, fodder, and forage to their livestock, disrupting traditional land use patterns, resulting in the permanent loss of traditional management practices and enhanced the land degradation trends and desertification processes. Similarly, the floods in the monsoon season not only erode the fertile layer of soil but also disturb the microbial population in flood-affected areas, thus degrading soil structure.

In arid and semiarid regions of the country, heavy rains during the monsoon season lead to flooding, which is a regular feature of these areas. According to an estimation, between 1950 and 2001, total losses due to floods were recorded in the order of US$ 10 billion and over 6000 lives (Baig and Shahid 2014). In 2010 and 2014, the country experienced the worst flood of the history of mankind that washed away tons of standing crops, thousands of livestock, human beings, and stored agricultural produce which caused the loss of millions of dollars (Kiani 2014). Excessive flooding ruins the agricultural soils, buries the top soil, leaves the infertile sediments on top, and inhibits cultivation. These floods thus contribute to land degradation and loss of biodiversity leading to an ecosystem imbalance. Deforestation, soil compaction, and soil erosion bring floods and contribute to flooding.

6.2.6 Poor Organic Matter

The decreased levels of organic matter cause a strong reduction in soil fertility as it plays several roles in soil. Decreased organic matter levels result in poor physical, chemical, and biological properties. The microbial activities in soil decrease due to reduced levels of organic matter. These microbial activities play significant roles in nutrient availability and recycling.

In Pakistani soils, the organic matter contents are very low—even below the minimum desirable level. The soil carbon in these soils ranges between 0.52% and 1.38% in different soil series. According to Directorate of Soil Fertility, Lahore, most of our soils have organic matter contents less than 1%. The organic matter contents are controlled by temperature as the mean annual temperature has strong influences on processes of decomposition of organic matter. The high temperature prevailing in Pakistan during the long summer season provides conducive environmental conditions for rapid decomposition of organic matter, leading to loss of soil carbon contents. Aridisol and Entisol are the soil orders that are characterized by lowest soil organic matter among all the soil orders. These two share the largest proportion of soils in Pakistan. So, it can be assumed that our soils naturally have a lesser capacity to hold higher organic matter content. The heavy use of mineral fertilizers by the farmers due to their easy availability, cheaper price, easy-to-handle nature, and higher use efficiencies is the other cause of decrease in incorporation of organic matter in the soil. The increasing prices of mineral fertilizers and soil degradation concerns have forced people to reconsider the organic sources in agriculture.

The stubble management practices used by the farmers leave almost no crop residues in the soil after crop harvest. The farmers use straw and other crop residues as fodder for their farm animals along with the use of animal dung as fuel. These practices are harmful for soil health. According to an estimate, about 50% of the animal droppings are not collected by the farmers, about half of the collected are burnt as fuel, and only one fourth are available for field application. Green manuring is the ignored technology by the farming community because it does not give short-term economic returns to the tenant farming system prevailing in most parts of the country. Intensive soil tillage practice is the other culprit for the lower organic matter content in our soils. Soil tillage improves soil aeration, makes the organic residues bioavailable, and makes them accessible to microbial decomposition, thereby is also the cause of low organic matter contents in the soil. The plow more, earn more slogan, as followed by the farmers, has played a considerable role in the organic matter depletion in Pakistani soils.

6.2.7 Improper Land Use

Improper land use is also the cause of land degradation. In Pakistan, we are not using the lands according to their potential. For example, good fertile land around cities is being converted into colonies and townships, which are also exerting more pressure on city resources. More population in the cities ultimately is the cause of more city effluents and more pollution. During the last decade, cultivated areas have decreased by 0.45% (12568 thousand ha) due to building of new housing estates (urbanization) on good agricultural lands (Economic Survey of Pakistan, 2016). Farm size has decreased to very small unproductive units (Farah et al. 2016) which are not economically viable for agriculture. Land use is not according to the capability classification. Many crops are grown in areas where these are not suitable. For example, farmers in cotton zones are growing rice and sugarcane, which is not feasible, as both the later crops require a high water delta, and the soils in southern Punjab have poor water holding capacity, which increases water consumption. Legumes are rarely included in the cropping system.

Crops are not sown according to the land potential. There is dire need to allocate land for specific crops, for example, light soils can be used for growing pulses and oil seed crops. Over time, we have shifted to intensive agriculture, resulting in little room for leaving the land fallow and for healthy practices such as green manuring. This has weakened the sustainability of productivity in our system.

6.3 FUTURE PROSPECTS AND STRATEGIES TO IMPROVE THE SCENARIO OF LAND DEGRADATION

It is a fact that the land degradation process can only be limited or curtailed, and cannot be stopped completely. It can only be controlled by the "sustainable" practices. Sustainable use of land resources is also important from a social and economic point of view. In order to combat the land degradation, there is a need of developing a set of strategies, involving the stakeholders and nation in conservation activities to control the damage and remedy the lost.

One of the major causes of land degradation is soil erosion, causing the removal of productive portions of land. The utmost need is to follow the conservation practices and implement the policies that can yield fruitful results and improve the conditions. In grazing lands, strict steps should be taken. It is estimated that about 30% reduction in grazing intensity can start the rehabilitation of exposed land by regeneration of vegetation (Pretty et al. 2001; Evans 2005). This issue should be tackled by the understanding of socioeconomic and political barriers to the conservation implementation, although these barriers are too difficult to address. On the other hand, efforts to control erosion and land degradation in cultivated areas can be made by checking the flow of water and developing wind brakes by tree plantation and soil cover. Better land management should be followed by covering the bare soil around the arable lands. The use of minimum or zero tillage techniques should be employed for the minimum disturbance of soils, thus conserving the soil fertility.

Pollution is another environmental issue creating serious problems for many populations through climate change and desertification due to blind cutting of forest trees and burning fossil fuels to meet energy requirements. It is estimated that in the world's most populated country alone, China, every week two new coal plants start emitting hazardous greenhouse gases and wastes that ultimately deteriorate the environment and contaminating large areas of land through disposal of waste and the consequences of mining (FAO 1988; Bertrand 2009). We can harvest the energy of wind and sun by providing inexpensive supplies to reduce poverty. Many countries in the world such as Germany, Denmark, New Zealand, and others are developing renewable energy resources and cutting down the hazardous effluents reducing the speed of land degradation. Everything on earth is linked and the earth is linked to the sun. We can imitate plants and start focusing on capturing sun energy. In one hour, the sun gives the earth the same amount of energy that is equal to the energy consumed by the whole of humanity in a year (Simeon and Ambah 2013). As long as the earth exists, the sun's energy will be inexhaustible. All we have to do is stop drilling the earth and start looking towards the sky for our energy requirement and to learn to cultivate the sun. It is the time to come together and think about what we have remaining. Still, we have the half of the world remaining forests, thousands of water resources, and land. There are the solutions to repair our land and it is up to us what happens next.

6.3.1 Suggestions to Overcome Soil Constraints for Enhancing Agricultural Productivity in Pakistan

In Pakistan, a concerted effort is needed on the part of the public sector with the involvement of all concerned quarters to help raise agricultural productivity through formulation/revision/implementation of policies. The following steps could be useful to achieve this target:

1. Revise the present land classification systems/classification categories to generate a new system which facilitates multipurpose land use (land zoning/land-use maps). With change in climatic patterns and commercialization of the agricultural systems, the farmers have haphazardly shifted towards growing crops without considering the land capability. Although they are getting some benefits, the efficiency of the system as a whole has declined. There is a dire need that land maps be prepared according to their capabilities and clear indications be made how these can be used for a multipurpose farming system with sustained productivity.
2. Crops must be sown according to land productivity/capability. Only those crops should be sown which are suitable for the agroecological conditions of a particular region. Rice and sugarcane should be discouraged in cotton zones.
3. Legislation should be done to keep the farm size under a viable productive unit. Land fragmentation has hampered the adoption of modern production technologies and mechanization of the system. A viable approach and strategy should be developed with the involvement of all stakeholders (farming community, agricultural experts, and civil society) so that a minimum viable agricultural unit can be constituted at each village level where modern production technologies can be used.
4. Housing colonies should be built on lands of B and C categories within a maximum limit of two years. There is increasing trend of bringing agricultural land under civil and industrial use. Such a use needs to be discouraged by restricting these to unproductive lands only. Moreover, the area allocated for estate development should not be kept engaged for unlimited period without any development; rather Government should impose a time limit for such development of towns. Barren lands in the deserts like vast areas surrounding Derawar Fort should be used to establish new cities. The planned cities on such historical points not only attract tourists but also lessen the load on good productive lands around cities.
5. Mineral fertilizers have many negative implications with regard to soil and water health as well as the environment. Moreover, dependence upon sole use of synthetic external inputs has not proven viable for sustainability of the system. This necessitates the integrated use of organic, inorganic, and biofertilizers, which will help a lot towards minimizing reliance on external inputs in addition to increasing the sustainability of the system. The government should encourage the use of organic/biofertilizer supplements, along with mineral fertilizers, by providing some incentives.
6. In the rice-wheat cropping system of Pakistan, a bulk of crop residues (7–10 t ha^{-1}) is produced annually. Wheat straw is being used by farmers for feeding their animals while the rice straw is not suitable for this purpose, due to its high lignin and silica as well as low protein contents. The management of rice straw remains a key issue due to its interference with tillage and other farm operations. So, the farmers burn the residues/stubbles, in the rice-wheat system, with the aim to dispose of crop residues. On one hand, burning helps clear land surface and controls soil borne pest and pathogens but, on the other hand, it kills the beneficial soil insects and microorganisms and contributes to air pollution. The massive

loss of nutrients from the system is another issue linked with such practices. Many cultivars of both rice and wheat release certain secondary metabolites into the rhizosphere during their decomposition that have phytotoxic effects; therefore, the proper management of these residues can be to use them as a natural tool for weed management. A farming system without proper residue management may result in depletion of soil resources due to mining of major nutrients. As a result, a net negative balance and multinutrient deficiencies occur in such a system. Several management options like surface retention, incorporation and mulching, as well as direct seeding in zero tillage technology, can be used as an alternative option to current detrimental practices of stubble management.

7. Balanced use of fertilizer in terms of NPK ratio, keeping in view crop requirements and soil analysis, can significantly contribute in sustaining agricultural productivity. Farmers are using predominantly nitrogenous fertilizers (most commonly urea) and mostly neglecting the use of other nutrients, such as phosphorus and potassium, due to one or the other reason. Fertilizer use should be based on soil testing and crop requirement. Agriculture production can be substantially increased by just focusing on this single factor, that is, balanced fertilizers. Nutrient indexing of all agricultural lands should be carried out in the Punjab, and a cyber-network be developed for fertilizer application as per crop requirement and soil conditions.
8. Salinity problems can be tackled in a better way by adopting the following approaches:
    a. Agronomic options as growing more drought and salt-tolerant crops, on-farm watercourse improvement and precision land levelling, raised bed farming, alternate use of fresh and saline water, and soil amendments such as gypsum and organic matter can help minimize the impact of salts on growing crops.
    b. Engineering approaches consisting of improving/maintenance of irrigation infrastructures, construct drainage facilities, rehabilitation of the existing surface and sub-surface drainage systems, preventing/reducing canal seepage through lining are the long-term solutions for handling the problem of salinity.
    c. Moreover, some policy interventions as incentives for land reclamation, that is, subsidizing gypsum, organic fertilizers/compost/humic acid/biochar, etc., will be helpful. Similarly, conversion of saline wastelands into rangelands and incentives to farmers for to use saline areas for livestock farming will also help to address the issue.
9. Better use of range lands and promotion of agro-forestry will also be helpful in raising the productivity of farmers in the Pothohar region.

Finally, beyond any doubt, the residential colonies on fertile agricultural lands and other human-induced pressures due to commercial activities around the cities are deteriorating precious land and other natural resources; all lead to serious and irreparable damages to our environment. It is, therefore, the need of our time to cope with land degradation problems on a war-footing basis to achieve a sustainable economic future for the country and an ecofriendly environment. The state of affairs needs to overcome the problem of rapidly growing populations, in order to leave this planet healthy and sustainable for the future generations. For sustainable management of land resources, it is no doubt imperative to devise and introduce instruments and frameworks which best match, accommodate, and address the local needs. However, in addition to above-suggested remedial measures, initiation of awareness and capacity building programs, through combined efforts of academetia and extension staff, are equally important. Enabling policies, the creation of suitable conducive environments, and more importantly, their implementation with real spirit to complement the existing efforts are the essential ingredients of any rehabilitation program.

6.4 CONCLUDING REMARKS

It can be concluded that in combating land degradation we are far behind our target. We are still estimating the damage and very little effort is being made to understand the problem. We incur huge damages, in the form of losing approximately 33% of our land, due to different causes of land degradation. It has been estimated that life emerged on earth about 4 billion years ago and that we humans are only 200,000 years of age. Yet, we have succeeded in disrupting the balance that is so essential to life.

In fact, the causes of the land degradation are natural processes, like erosion, salinization, etc., but nature has the tendency to repair them. In last 200 years, we have speeded up all these processes with our materialistic nature. We have over drilled the resources and disrupted the balance so badly that in most of the world there is a radically a new type of urban growth, driven by the urge to survive rather than to prosper. Due to heavy industrialization, the balance of the system is dwindling due to emission of pollutants into the environment. We have also over drilled the resources and deforested our lands. As all the processes are synchronized with each other, disturbance in each results in the disruption of the entire natural environment—which we are facing in the form of land and environmental degradation.

The cost of our actions is high and others pay its price without being actively involved. Land degradation is a global issue and it needs the attention of authorities to think beyond the borders and walls of nations. Degradation is a serious issue for the survival of each and every species on the earth, including *Homo sapiens*. But this issue is not gaining as much importance as it deserves. Though a number of international nongovernment organizations (NGOs) are working on the issue—proving that solidarity between people is stronger than the selfishness of nations, a lot of work and efforts are still needed to solve the issue.

REFERENCES

Adams CR, Eswaran H. 2000. Global land resources in the context of food and environmental security. pp. 35–50. In: Gawande SP (ed.). *Advances in Land Resources Management for the twentieth Century*. Soil Conservation Society of India, New Delhi. 655 pp.

Agarwal KP. 2008. Climate change and its impact on agriculture and food security. *LEISA India* 10:6–7.

Bai ZG, Dent DL, Olsson L, Schaepman ME. 2008. Global assessment of land degradation and improvement identification by remote sensing. Report 2008/01, FAO/ISRIC—Rome/Wageningen.

Baig MB, Shahid SA. 2014. Managing degraded lands for realizing sustainable agriculture through environmental friendly technologies. In: Behnassi M, Shahid SS, Mintz-Habib N (eds.), *Science, Policy and Politics of Modern Agricultural System*, pp. 141–164. Springer, Netherlands.

Bertrand YA. 2009. Home. National Geographic Channel, Europa Corp.

Dagar JC, Minhas P. 2016. Agroforestry for the management of waterlogged saline soils and poor-quality waters, Springer, India.

Duraiappah AK, Roy M. 2007. Poverty and ecosystems: Prototype assessment and reporting method—Kenya case study. *International Institute for Sustainable Development*. Available at: http://www.iisd.org/pdf/2007/poverty_eco.pdf (Accessed on: 09/02/2018).

Economic Survey of Pakistan. 2016. Ministry of Finance, Federal Bureau of Statistics, Islamabad, Pakistan available at: http://www.finance.gov.pk/survey_1516.html. Accessed on: 12/02/2018.

Evans R. 2005. Monitoring water erosion in lowland England and Wales—a personal view of its history and outcomes. *Catena*. 64: 142–161.

Farah N, Khan IA, Manzoor A, Shahbaz B. 2016. Changing land ownership patterns and agricultural activities in the context of urban expansion in Faisalabad, Pakistan. *Pakistan Journal of Social Sciences* 14(3): 183–188.

FAO. 1988. *WCARRD 1979–1989*. Ten years of follow up on impact of development strategies on the rural poor. Rome.

Francois M, Dubourguier HC, Li D, Douay F. 2004. Prediction of heavy metal solubility in agricultural soil around two smelters by the physico-chemical parameters of the soils. *Aquatic Science* 66: 78–85.

Gibbs PA, Chambers BJ, Chaudri AM, McGrath SP, Carlton-Smith CH. 2006. Initial results from long-term studies at three sites on the effects of heavy metal amended liquid sludges on soil microbial activity. *Soil. Use Management*. 22: 180–187.

Grewal MS, Kuhad MS. 2002. Soil desurfacing-impact on productivity and its management. pp. 133–137. In: *Proc. ISCO Conference*, Beijing, China.

Hassan T, Rao S. 2001. Land and water management through watershed technology. Addressing the challenges of land degradation an overview and Indian perspective. pp. 214–233. In: *Proceedings of the National Seminar Land Resource Management for Food and Environmental Security.* Soil conservation society of India.

Hussain M, Ahmed MSA, Kausar A. 2006. Effect of lead and chromium on growth, photosynthetic pigments and yield components in mash beans (*Vigna mungo.* L.). *Pakistan Journal of Botany* 38: 1389–1396.

Kassas M. 1995. Desertification: A general review. *Journal of Arid Environments* 30(2), 115–128.

Kiani K. 2014. Overstated GDP losses. *Dawn News.* September 29, 2014 09:29am

Kobayashi T, Nishizawa NK. 2012. Iron uptake, translocation, and regulation in higher plants. *Annual Review of Plant Biology* 63: 131–152.

Moore A, Hinesb S, Brownc B, Falene C, Martif MH, Chahinea M, Norellg R, Ippolitod J, Parkinsonh S, Satterwhitea M. 2013. Soil plant nutrient interactions on manure-enriched calcareous soils. *Agronomy Journal* 106: 73–80.

Natarajan A, Hegde R, Naidu LGK, Raizada A, Adhikari RN, Patil SL, Rajan K, Sarkar D. 2010. Soil and plant nutrient loss during the recent floods in North Karnataka: Implications and ameliorative measures. *Current Science* 99: 1333–1340.

Pretty J, Brett C, Gee D, Hine R, Mason C, Morison J. 2001. Policy challenges and priorities for internalising the externalities of modern agriculture. *Journal of Environmental Planning and Management* 44: 263–283.

Reddy SG, Singh HP, King C, Dixit S. 2000. Portable rainfall simulator—A participatory action learning tool to understand desertification process. LEISA India supplement. *LEISA Newsletter* 2(1): 16–18.

Simeon PO, Ambah B. 2013. Effect of municipal solid waste on the growth of maize (*Zea mays L.*). *International Letters of Natural Sciences.* 2: 1–10.

Tuboly E. 2000. The United Nations convention to combat desertification. ILEIA, Leusden, Netherlands. *LEISA Newsletter* 2: 14–15.

USDA. 1988. A manual on conservation of soil and water. Oxford and IBH publishing co.pvt. New Delhi.

Verón SR, Paruelo JM. 2010. Desertification alters the response of vegetation to changes in precipitation. *Journal of Applied Ecology* 47: 1233–1241.

World Information Transfer (WIT). 2009. Desertification: Its effects on people and land. *Politics Health Environ. Conf.* 21: 21.

Young A. 1998. Land resources now and for the future. Cambridge University press. Cambridge.

Zamfir HC. 2014. A view on land degradation and desertification issues rares. *Symposium Actual Tasks on Agricultural Engineering Opatija*, Croatia.

Zia MS, Nizami MMI, Salim M. 1994. Problem of soil degradation in Pakistan. *The collection and analysis of land degradation data.* pp. 179–202. RAPA/FAO Bangkok, Thailand.

7 Canal Operation through Management Information System

Muhammad Arshad, Riaz Ahmad, and Muhammad Usman

CONTENTS

7.1 INTRODUCTION

The Indus Basin Irrigation System (IBIS) contains three major reservoirs, 2 head-works, 16 barrages, 2 siphons, 12 inter river link canals, 44 canal arrangements (23 canal system in Punjab, 14 in Sindh, KPK contains 5, while 2 are in Baluchistan) with 107,000 watercourses. The collective length of the canals is about 56,073 km. Moreover, the watercourses, farm channels, and field ditches cover another 1.6 million km.

A distinctive watercourse commands area is between 80.97 and 323.89 hectares. The system uses over 51.168 BCM of groundwater that is pumped through over one million tube wells in Punjab, along with canal supplies [1]. Besides the Indus basin, there are some smaller river basins, which drain straight into the sea. These torrents are flashy in nature and don't have a perennial supply. Approximately 25% of their inflow is utilized for flood irrigation. Presently, the total average annual surface water diversions at the canal heads of the IBIS are 128.781 BCM [2].

The Punjab Irrigation and Power Department is operating and maintaining one of the largest contiguous gravity flow network in the world. The system is serving nearly 8.4 million hectares known as the Culturable Command Area (CCA). Irrigation water is delivered through 58,000 outlets off-taking from canals having a total water conveying capacity of 120,000 cusecs. The irrigation system serves as a lifeline for sustaining agriculture in the province. Irrigated lands supply 90% of the agricultural production, account for about one-fourth of Punjab's GDP, and provide employment to about 45% of the labor force [3].

The Government of Punjab's vision is "to make quantitative and qualitative improvement in the lives of the citizens." To this end, the government is implementing reforms for improving governance, fiscal and financial management, and to improve the efficiency with which public services are delivered. This is in the context of improving irrigation services with an objective to make water entitlements more transparent, enhance farmer confidence, reduce uncertainties, and to increase productivity.

For the implementation of the Monitoring and Implementation Program (MIP) envisaged for well-organized and optimal canal operations, oriented towards equity and transparency, it was planned to launch a Programme Monitoring and Implementation Unit (PMIU) for canal operations and discharge data in the Irrigation and Power Department, controlled by a chief monitor (Team Leader). The PMIU was founded in March, 2006 as a part of organizational set up to implement efficient and optimal canal operations, oriented towards equity and transparency [4].

7.1.1 Mandate of Monitoring Unit

The main mandate of the PMIU is to implement efficient and optimal canal operations oriented towards equity and transparency. In order to meet the PMIU mandate, the following activities are being performed:

- Development of modern tools like the Irrigation Management Information System (IMIS), including Database Management System (DBMS) and Decision Support System (DSS), in order to ensure transparency in getting reliable data available to all the stakeholders, and also to facilitate the high-level management in decision-making;
- Monitoring of channels operation of the entire irrigation network and issuance of a daily channel operation report in order to ensure equitable distribution of canal water;
- Publication of daily discharge data of all the channels of Punjab on the Irrigation Web Site to facilitate rapid monitoring of water distribution in the canal system, especially in between headworks and tail off-takes/outlets, and to reduce grievances of lower riparian users of the canal system;
- Publication of water accounts for main canals regarding entitlements, deliveries, and balance shares are posted on irrigation website in order to ensure transparency in water distribution and easy access to beneficiaries, as well as to enhance Confidence Building Measures (CBM) among the provinces;
- Regular surprise checking of about 10% of the gauges at head and tails of channels to ascertain correctness of data provided by field formations and proper checking for feeding of tails and equitable distribution of water. Accordingly, issuance of biweekly performance evaluation reports (PER), which help in improving quality, efficiency, and accountability of the system;

TABLE 7.1
Criteria for Defining the Status of Channel

Channel Status	Condition
Authorized head	Actual head discharge >= 95% of design discharge or indent
Short from head	Actual head discharge < 95% of design discharge & indent
Dry tail	Actual tail gauge <= 30% of authorized tail gauge
Short tail	30% of auth. tail gauge < actual tail gauge < 90% of auth. tail gauge
Authorized tail	90% of auth. tail gauge <= actual tail gauge <= 115% of auth. tail gauge
Excessive tail	Actual tail gauge > 115% of auth. tail gauge

- Preparation and issuance of discharge tables, on the basis of accurate discharge measurements, and calibration of gauges to facilitate fair and equitable distribution of water among the stakeholders, and also to ensure authorized supply at head and Inter-divisional and Inter-sub-divisional points (Table 7.1);
- Monitoring of *Moghas* (outlets) of tertiary channels (watercourses) to ensure transparent distribution of water supply and to control water theft;
- Calculation of water entitlement of the entire irrigation network to ensure transparency in water distribution; and
- Development of a Complaint Management System (CMS) to improve quality, efficiency, and accountability with which irrigation services are delivered and also the efficient attendance of complaints related to canal operations, shortage of water, water theft and warabandi, etc., lodged by the farming community (received on Toll Free Helpline (0800-11333), Chief Minister Complaint Cell, Chief Minister Open Katchery, and Chief Secretary Petition Cell).

7.2 CURRENT STATUS

At the Irrigation Department, previously, daily data regarding discharges/gauges of rivers, main canals, branch canals, distributaries, and minors was prepared by the field staff in the analog form and retained in the divisional offices; however, gauges/discharges of main/branch canals were transmitted to the Irrigation Secretariat through the canal wire system using telegraph network maintained by the Pakistan Telecommunication Corporation (PTC). Most of the time, daily canal operation data was received with a delay of 2 days. Thus, there was no mechanism for daily transfer of data and daily monitoring of water distribution in a canal system, that is, between the headworks and tail off takes/outlets. There is a dire need of such mechanism/unit to address the above limitations. In this scenario, an independent unit, PMIU, was established in March 2006 as a part of organizational set up in the Irrigation Department to implement efficient and optimal canal operations oriented towards equity and transparency.

7.3 INSTITUTIONAL REFORMS

7.3.1 Provincial Irrigation and Drainage Authority (PIDA)

An institutional reforms process is underway in at the Irrigation Sector in the province. The reforms are basically aimed at better governance, decentralization, farmer's participation in management, efficient water management, and sustainability. An important step in implementation of institutional reforms is the formation of Khal Panchayats and farmers organizations (FO) through social mobilization. During formation of these farmers organizations, representatives of the farmers form a management committee that would hold responsibilities of operation and maintenance of canals and distributaries.

As per PIDA rules, FOs have to be established at distributary level, keeping in view hydrological boundaries of distributaries and minors, technical feasibility, and size of the irrigation system.

1. A "Farmers Organization" established at Distributary level.
2. A "KhalPanchayat" formed at each watercourse level.
3. The KhalPanchayat consists of one chairman and four members, elected out of the farmers/shareholders of the concerned watercourse.
4. The chairmen of KhalPanchayats form the "General Body" of FOs. The general body elects the "Management Committee" of FOs, who is responsible for functioning of the FOs. The Management Committee consists of 4 office bearers and 5 executive members.

7.3.2 Social Mobilization of the Farming Community

Social mobilization staff conducted social mobilization activities at each watercourse level to create awareness amongst farming community of the project area and the process of establishing farmers based entities at watercourse level (KP) and distributary level (FO) started to organize them to take part in Participatory Irrigation Management. It enhanced the interaction with stakeholders and facilitated the formation of KhalPanchayat and FOs. The social mobilization staff conducted the following meetings at each watercourse level.

1. Familiarization meeting.
2. Rapport building meetings.
3. Consultation meeting.

7.3.3 Formation of KhalPanchayats and Farmers Organizations

The formation of FOs has been carried out under the PIDA (FOs) Rules and PIDA FOs (Elections) Regulations 1999 amended to date and PIDA FOs (Registration) Regulations 1999. Under these regulations, the Chief Executive AWB / SE of the Canal Circle is responsible to conduct the elections. The PIDA field staff provided assistance and logistics to an Election Officer and Senior Election Officer appointed by the Chief Executive Area Water Board/S.E. of the circle under the above regulations.

7.4 FARMERS ORGANIZATIONS

Capacity building and training of FOs members is of paramount importance, so that FOs be able to perform all functions independently and efficiently under the IMT scenario. The Training Plan for each AWB has been approved by the Managing Director, PIDA. The trainings to General Body Members, Management Committee Members, FOs Staff are proposed to be imparted in following major areas during the coming years.

1. Water rights of the farmers and their responsibilities.
2. Enhancement of their capabilities of organizational development and managing the system.
3. Technical aspects of O&M of Channels and structures.
4. Procedures of repairs and maintenance of Civil Works on Channels.
5. Administrative and management aspects of irrigation infrastructure at disty level with reference to specific functions/rules/regulations, etc.
6. Maintenance of the records.
7. Regulation of Canal Water supplies and equitable water distribution among the outlets and farmers.

8. Efficient Water Management at farm level.
9. Assessment and collection of water charges.
10. Water dispute resolution process.
11. Legal framework of PIM and its implementation process.
12. Account keeping of funds, expenses, and their Audit.

7.4.1 Revision in FOs/AWBs Rules

A PIDA subcommittee was constituted to review the proposed amendments in FOs/AWB Rules and conducted brain-storming sessions and proposed certain amendments in FOs/AWB Rules, based upon lesson learnt during 3 years performance and FOs in pilot AWB/LCC (East) Faisalabad. Accordingly, revised FOs/AWBs Rules have been approved by the Chief Minister, in Punjab and circulated vide Notification NO. OSD/L&WD/10/2002-II and OSD/L&WD/6/98-III dated December 21, 2010 [5].

7.4.2 Finalization of IMT Agreement

The IMT Agreement as per approved rules is being prepared by PIDA for transfer of Irrigation Management to the FOs of AWB/LCC (East) and AWB/Bahawalnagar Canal Circle, where FOs formation process has been completed. This agreement focuses on the gradual and incremental transfer of powers to FOs for their smooth working and sustainability in the prevalent sociocultural and political circumstances.

7.5 ISSUES OF CANAL OPERATION

Major canal operation challenges and issues are as follows:

- Mainly earthen channels; vulnerable and need continuous maintenance.
- 25%–30% water lost till the farm gate.
- Water availability far short of requirements.
- Distortion due to weak enforcement, gap between demand and supply, fragmentation of holdings.
- Price difference between canal and tube well water.
- Limited inter-river water transfer capacity.
- Water shortages necessitate rotational running, which are compounded during early Kharif and late Rabi period when dams are empty.
- Time lags in conveying water.
- Loss of canal capacities due to aging and deferred maintenance.

7.5.1 Better Enforcement

- Reduction in water theft related losses
- Strict implementation of canal rotation plans
- Employment of canal guards
- External monitoring, shift to smart monitoring
- Farmers feedback system
- Daily review of supply at head and tails
- Canal division wise targets and fortnightly review
- Water theft control teams under Assistant Commissioner
- Nonregistration of water theft cases. Since 2012–2015, 177,500 cases were reported, out of which only 17,000 cases were registered which makes only 10%
- Slow recovery of Abiana / tawan recovery
- Amendments in laws

7.5.2 Impact and Major Tasks and Achievements

The PMIU was established at the Irrigation Department for implementing efficient and optimal canal operations, oriented towards equity and transparency, in pursuance of approved summary which became operational in March 2006. The IMIS and DSM have been developed to provide support to higher management to monitor the key indicators of the system operations, including the discharge of various canals and distributaries, etc. The system is facilitating decision-making regarding water distribution and scheduling it more transparently with the objective of providing canal water to all farmers, as per their entitlements, with emphasis on providing the authorized supply to the tail enders. The live data of all the channels of Punjab is captured daily as well as the statistical data of all canal systems. The live data is updated daily on the website http://irrigation.punjab.gov.pk to make the information available to the end users. The DSM is also facilitating the higher management in monitoring the key indicators of the system, including the discharges at various main and branch canals, and distributaries. It also helps decision-making regarding equitable and transparent water distribution and scheduling.

Following are the major tasks/achievements accomplished so far;

- Establishment of the Data Base Management System (DBMS) for facilitating the decision-makers and monitoring the units to make quick decisions based on correct data. DBMS is serving as a tool for the Best Management Practices (BMP) for the operations of entire canal systems.
- Authenticity of static and live data of water channels available at Irrigation Division, Circle and Zonal levels.
- Development of modern tools like Irrigation Management Information System (IMIS) and Decision Support Model (DSM) for digitization of all the canal data with special emphasis on quick transfer of data from field level to PMIU Office, and development of comprehensive analysis system regarding all parameters of canal operation to ensure equity. Development of (SMS) Data Transmission System (DTS) to have quick transfer of data.
- Digitization of daily gauges and discharges data of all the channels by PMIU was targeted. How far this has been achieved and has been helpful in rapid monitoring of water distribution in canal systems especially in between headworks and tail off-takes/outlets. (It is pertinent to mention that as per previous practice, all the data about discharges/gauges of rivers, main canals, branch canals, distributaries, minor, and the sub-minors was captured by the field staff and retained in the Divisional Offices except some main canals data which was only sent to Irrigation Secretariat).
- Evaluate Transparency in water allocation/scheduling, entitlement, and preparation of operational plans at managerial as well as field operational level and their easy access to beneficiaries. Development of website by PMIU and its proper working with regard to communication of information regarding canal flow data, publication of water accounts for main canal systems, determination of Delivery Performance Ratio (DPR), Capacity Factor (CF), and relevant information to improve transparency and equity.
- Efficient operation of canal based on authentic data with emphasis on supply of canal water at tail ends according to the water entitlement was envisaged. Evaluating this aspect, along usefulness and working of website with respect to posting of distribution programs indicating entitlements and deliveries of channels of main canal systems in Punjab.
- Efficient attendance of complaints with respect to canal operations, shortage of water, water theft, and complaints related to warabandi, etc., by developing a Complaint Management System (CMS). Establishment of a Toll Free helpline to lodge complaints by the farmers and their follow up. In this regard, an auto generated complaint transmission system has been developed for transfer of complaint to concerned XEN, Superintending Engineer, and Chief Engineer for follow up.

- Discharge observations, installation, and calibration of standard enameled iron (EI) gauges at heads and tails of the channels to facilitate fair and equitable distribution of irrigation water among the farmers.
- Monitoring of alteration of outlets (A-Form) process.
- Monitoring of violation of Rotational Program through SMS and triggers generated by the computer application.
- Monthly Collection of data related to Outlet Performance Register Data, Abiana & Tawan cases, Habitual offenders and Cut & Breach Data for development of database in this regard.
- Development of Android applications for monitoring of Flood Fighting Operations.
- Surprise checking of outlets and reporting of water theft incidences through SMS using Android technology. In this regard, Android applications have been designed for location and picture-based evidence for disciplinary action and criminal proceedings against delinquents. In this regard, Android telephones have been provided to the six canal divisions for extension of pilot project from PITB.
- Issuance of Daily Analysis Report regarding channel operations including identification of dry, short and excessive tails.
- Accountability system is in place on the basis of Fortnightly Performance Evaluation Report of Canal Divisions, Circles, and Zones.
- Regular monitoring of supply at critical gauge sites for water accounting in the system to avoid conflict situations among the Canal divisions/Subdivisions.

7.5.3 Water Entitlements and Deliveries

The Indus Water Treaty (IWT) develops a distribution program of 25 main canals (12 main canal systems of Terbela command and 13 main canal systems of Mangla command) for each crop season. In this distribution program, the canal water is distributed in each main canal system according to its authorized share at a 10/5-daily basis. This information was not publically published on any forum. PMIU started displaying this information on the website in a user-friendly way. This module helps users to view the previous, current, and future entitlements of canal water in all main canal systems. PMIU played vital role in displaying this information to all stakeholders including farmers, international donors, and other provinces to ensure the equitable distribution of water according to their share in entire irrigation system, and also to help develop confidence-building measures (CBM) among the provinces.

Key benefits of the water entitlement and deliveries module are given below:

- This is a helping tool for fair distribution of water among the various canal systems.
- At any stage of time, a user can monitor the distribution of water against its entitlement.
- It shows the equitable distribution of canal water.
- It improves the confidence building among the provinces.
- It increases the public awareness regarding water distribution.
- Users can view the water delivered in a selected canal against its entitlement/share at 10 or 5-daily basis. Users can also view the cumulative delivered share and remaining share of a specific canal system; and
- It depicts that PMIU has been playing a vital role in educating not only the irrigation staff but also all the stakeholders regarding their water accounts. At any stage of time, a user can view the water position in the past and also have an idea of the water situation for the future for specific crop seasons.

7.5.4 Relative Delivery Performance Ratio

Delivery performance ratio (DPR) is the ratio of actual discharge to designed discharge. It may also be used as deliverable or scheduled water in order to assess not only the performance of the overall system, but also the contribution of the structural and management components of the system to the performance.

DPR is the most important hydraulic performance indicator. In fact, the DPR ratio is generally considered to be the ratio of actual discharge to designed discharge (DPR = Q_a/Q_d). Relative DPR was used to measure the equity performance indicator. The relative DPR can mathematically be expressed as:

$$\text{DPR at Tail (A)} = \frac{\text{Td}}{\text{ATd}}$$

$$\text{DPR at Head (B)} = \frac{\text{Hd}}{\text{Dd}}$$

where

Td = the actual tail discharge
ATd = the authorized tail discharge
Hd = the actual head discharge
Dd = the designed discharge at head

$$\text{Relative DPR} = \text{A/B}$$

Relative DPR has an ideal value of 1, which means that every changed flow conditions at subsystem head are proportionately distributed among the shareholders, regardless of their location along the distributaries.

7.5.5 Daily Analysis Report Regarding Channel Operations

Analysis of channel operations was not done previously on a regular basis (daily basis) before the establishment of the PMIU. It was only done when required for some reporting purpose. PMIU started analyzing the channel's operations and complaints status on daily basis right after the deployment of databases (DBMS). Once PMIU collected all required data and managed it through DBMS, it became easy and time saving to analyze the channels of entire Punjab. A report is issued, which depicts the status of complaints and analysis report regarding channel operations at zonal level.

The key benefits of daily analysis report include:

- It highlights the weak/neglected aspects of channel operations, especially about dry tails, short tails, and excessive tails.
- It provides a glimpse over the equitable distribution of canal water in a particular zone, circle, and division.
- It acts as a performance indicator of field operators from an administrative point of view.
- It also compares gauges data received from canal division and PMIU staff in order to cross-check and highlight the poor reporting.
- It provides information about the status of complaints.

7.5.6 Performance Analysis Report (PAR)

To ensure equitable water distribution and implementation, PMIU has taken the initiative to compare the canal divisions on the basis of channel operations data received from field and PMIU staff. This evaluation report is compiled from the channel operations data for each 15-day interval, hence producing two reports per month.

Key features of PAR include:

- Healthy competition among the various canal divisions.
- Performance indicator for field operators.
- Highlights the problematic area of the canal division to the XEN so that he can improve lacking areas to ensure the water availability up to the tail irrigators.

Criteria and Parameters: The parameters on which the canal divisions are compared are as follows:

Analysis on the basis of field division data (maximum marks)

- % of channels running with authorized head discharge and tail dry (–15);
- % of channels running short from head and tail dry (–5);
- % of channels running authorized from head and tail short (–12);
- % of channels running short from head and tail short (–4);
- % of channels running with authorized tail (30);
- % of channels running authorized from head and tail excessive (–6);
- % of channels running short from head and tail excessive (–4);
- % of channels of which data was not received (–5); and
- % of water losses/unaccounted water (–25).

Analysis on the basis of PMIU checking (maximum marks)

- % of head gauges checked by PMIU & field staff where difference of 0.2 or more exists on gauge (–19);
- % of tail gauges checked by PMIU & field staff where difference of 0.2 or more exists on gauge (–19);
- % of channels running authorized from head & tail dry (–7);
- % of channels running short from head & tail dry (–3);
- % of channels running authorized from head & tail short (–6);
- % of channels running short from head & tail short (–2);
- % of channels running with authorized tail (15);
- % of channels running authorized from head & tail excessive (–3); and
- % of channels running short from head & tail excessive (–2).

where number of channels against each parameter is calculated by applying the following conditions:

The percentage of channels against the given parameter is calculated as:

$$\%\text{ of Channels} = \frac{\text{Sum of Channels fulfilling the condition for the Parameter}}{\text{Total Analysis Channels in Division}} \times 100$$

For each of the parameters listed above, all canal divisions are relatively marked. Each parameter has the maximum marks. A division with maximum value for a given parameter is marked with maximum marks for that parameter and all the rest of the divisions are marked relatively with the following formula:

$$\text{Divisions Marks} = \left(\frac{\text{Parameter\_Value\_of\_Division}}{\text{Maximum\_Parameter\_Value}}\right) \times 100$$

7.5.7 Development of Software to Calculate Unaccounted Water

Irrigation system in Punjab consists of 56 canal divisions and 25 main canal systems. These main canals pass through one or more divisions to provide water for irrigation. Each division has its authorized share of water form the main canal. The division in which a main canal starts is called "upper division" and it delivers water to the subsequent divisions, which are called "lower division/divisions." It has been a tradition in irrigation systems that lower divisions, most of the time, blame upper divisions for not getting water according to their authorized share.

PMIU developed a DBMS and started monitoring water delivery from upper division to lower divisions through the channels on the critical gauges. These critical points, available on the website for XENs, are where they enter discharge data (gauge, discharge and indent) on a daily basis. SEs can also view the discharges of their respective channels on the critical points of divisions and subdivisions. This information gives a glimpse to the distribution of canal water between divisions and subdivisions at circle level.

PMIU started calculating unaccounted water at division level to ensure the equitable distribution of water and to minimize water theft. This calculation gives an idea of how much water a division receives, utilizes, and delivers to subsequent divisions. This exercise pinpoints the following:

- False reporting of daily gauges data
- Water theft
- Tempered outlets drawing excessive discharge
- Discharges reported through outdated discharge tables

Unaccounted water (water losses) is also the major input in performance evaluation report, which is generated every 15 days. This shows the management skills of field operators regarding canal water distribution and responsibility for transmitting correct information to PMIU office.

Calculation of unaccounted water (water losses) is given below:

Unaccounted water = (total water received – water utilized – total water delivered (if any))

7.5.8 Geographical Information System (GIS) Project

PMIU, in coordination with Agriculture Department and SUPARCO, has developed a GIS, which will be helpful for the spatial analysis of the entire irrigation network of the Punjab.

7.6 STRATEGIES

The PMIU is working in full swing towards achievements of its intended objectives. The website has been made user-friendly and irrigators can watch the data of their channels and also can see the previous 10-daily trends of the flow in the channels on the website. Various analysis tools have been designed to ensure equity and transparency in the system.

The details are as follows:

- Development of irrigation management information system (IMIS) and irrigation database.
- Daily canal flow data collection from field divisions.
- Development of efficient data transmission system through sms (SMS-DTS) using smart phone technology.
- Dissemination of daily canal flow data on irrigation website (http://irrigation.punjab.gov.pk).

7.6.1 Monitoring of Punjab Irrigation System

- Issuance of daily analysis report regarding channel operations including identification of dry, short and excessive tails.
- Monitoring of supply at tails using a GIS web interface.

- Accountability system is in place on the basis of fortnightly performance evaluation report of canal divisions, circles, and zones.
- Daily monitoring and evaluation of canal systems using relative DPR to ensure equitable distribution of canal water.
- Regular monitoring of supply at critical gauge sites for water accounting in the system to avoid conflict situation among the canal divisions/subdivisions.
- Surprise checking of outlets and reporting of water theft incidences through SMS using Android technology.
- Monitoring of alteration of outlets (a-form) process.
- Monitoring of violation of rotational program through SMS and triggers generated by the computer application.
- Developed a complaint management system for efficient redressal of complaints with respect to canal operations, water theft, shortage of water and rotational program issues. A toll free helpline (0800 11 333) has been setup to lodge complaints by the farmers.
- Monitoring of flood fighting operation using smart technology.

7.6.2 Monitoring of Canal Operations Using State-of-the-Art Technology & Equipments

- Development of GIS for spatial analysis of canal operations data.
- Execution of World Bank funded project, "Improvement and modernization of Irrigation and Water Management System" for development of computer-based system to assist in intelligent canal operations by 2016.

7.6.3 Water Resource Management Information System (WRMIS) and Decision Support System (DSS) for Efficient Irrigation Water Management in Punjab

The overall objective of the project is to develop interactive, graphical, web-based Water Resources Management Information System (WRMIS) along with a Decision Support Systemn (DSS).

7.6.4 Water Resource Management Information System (WRMIS)

To make the present irrigation management system more efficient, there is a requirement to develop a DSS incorporating the intelligence of existing canal operators supported by scientific tools such as flow forecasting tools, simulation models for rivers, and canals facilitating the operator in equitable management of precious irrigation flows throughout the system. In this regard, hydrological and hydraulic models are developed to link with the DSS, as shown in Figure 7.1. Further, the DSS is linked with the Irrigation Management Information System (IMIS) for handling large data sets, and provide facility for storage and retrieval of canal and river data, required by the models. The complete application would be Water Resources Management Information System (WRMIS), which will facilitate decision-makers in making informed decisions regarding water availability and system response against various management scenarios.

The overall objective of the WRMIS is to develop interactive, graphical, web-based system along with a DSS. The system will be comprised of following subcomponents.

a. Review/upgrading of WRMIS and DSS.
b. Hydrological modelling for forecast of water availability and determination of Punjab's share.
c. Development of a real-time operations model.
d. Determination and mapping of water table fluctuations.
e. Development of hydraulic models of all main and branch canals of the Punjab irrigation system.

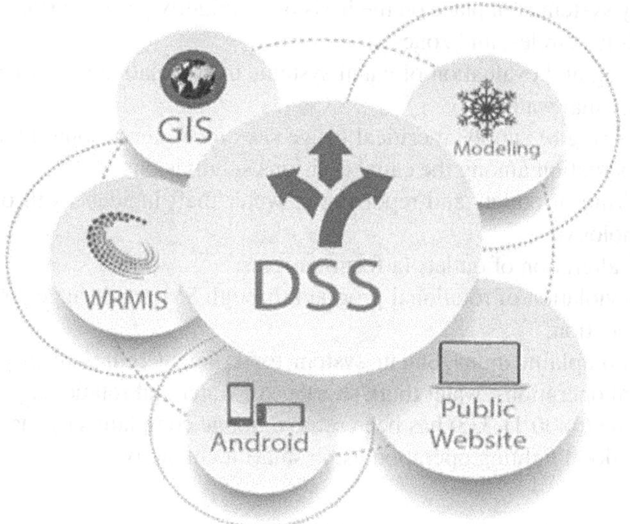

FIGURE 7.1 Water Resources Management Information System (WRMIS) and integration with Decision Support System (DSS).

Task-wise detail of the WRMIS is as under:

A. *Review/upgrading of WRMIS and integration with a DSS:* Requirements analysis is a process of eliciting, analyzing, and documenting requirements. An important step in eliciting requirements is to identify the external and internal stakeholders. Different techniques are used for gathering requirements such as focus group meetings, interviews, questionnaires, workshops, brainstorming, and prototyping. These techniques are used with stakeholders for understanding their existing system, business process, analysing data/information they receive, store, process, and disseminate. In addition to it, problems faced by the stakeholders to perform their daily routine tasks, under their existing environment, are also identified. After eliciting requirements, analysis is performed on it to ensure whether the stated requirements/information are clear, complete, traceable, and consistent.
B. *Hydrological Modelling for Forecast of Water Availability and Determination of Punjab's Share:* Under this task, three snowmelt-runoff models will be used to forecast the water availability at the rim stations on the rivers Indus, Jhelum, and Chenab. Figure 7.2 shows this phenomenon.
C. *Development of a Real-Time Operations Model:* Under this task, a modeling methodology for maximizing water utility during short-term operations of the Punjab Irrigation System will be developed. Moreover, canal operations models will be developed to address day to day irrigation operational management issues and water audit and accounting. The model will necessarily be a network model, capable of handling reservoirs, flow routing, and system operational constraints and will be designed to assist in decision making of Punjab Irrigation system operations, keeping in view the maintaining of an equitable distribution pattern and maximizing water utility.
D. *Determination and Mapping of Water Table Fluctuations:* Under this task of the project, a water table fluctuation maps will be developed. Point data from wells will be converted to groundwater surface maps through appropriate interpolation methods. Rise/fall in groundwater will be computed through subtraction of maps for successive periods.
E. *Development of Hydraulic Models of All Main and Branch Canals of the Punjab Irrigation System:* Under this task, hydraulic modeling of channels in the pilot area will be developed. It will develop, calibrate, and validate hydraulic models of all main canals, branch canals

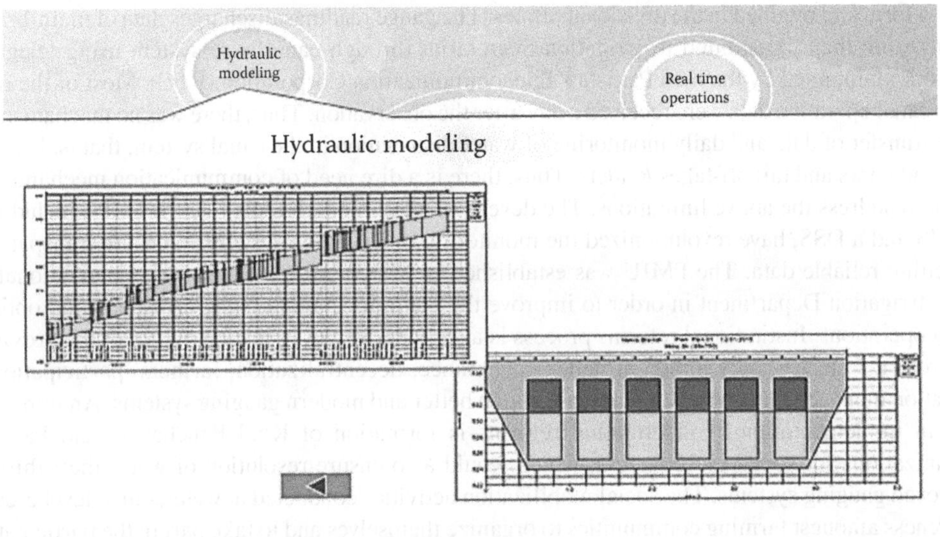

FIGURE 7.2 Model environment for hydraulic modeling.

of the Punjab Irrigation System, and one complete canal system of Eastern Sadiqia Canal (which includes 2 branch canals, 36 distributaries, 57 minors and 11 sub-minors) up to subminor level. The model creates outputs in the form of look-up tables that shall be linked with the DSS-WRMIS to arrive at reliable and equitable water distribution. The model has to detect anomalies in canal operations, such as any unaccounted water usage and system constraints, in order to ensure fair distribution of water in the irrigation canals.

7.7 SUMMARY AND OUTLOOK

Water is important for life as well as for the economic growth of a country; therefore, its sustained supply to meet the demand in domestic, agricultural, and industrial sectors must be maintained. The sources of freshwater in the country include river water, groundwater, and rainfall. The potential of canal water is limited to meeting less than 40% of the demand, while meeting the rest with groundwater and rainfall. The IBIS contains three major reservoirs, 2 head-works, 16 barrages, 2 siphons, 12 inter river link canals, 44 command canals (23 canal system in Punjab, Sindh has 14, and KPK contains 5 while there are 2 in Baluchistan) with 107,000 watercourses. The system uses over 51 BCM of groundwater, pumped through over one million tube wells, along with the canal supplies. Besides the Indus basin, there are some smaller river basins, which drain in to the Arabian Sea. The total capacity of three major reservoirs, at the time of their input to the system, was 15.7 maf and is consistently diminishing at the rate of 0.15 maf per year because of sediment deposition. The water reservoirs play an important role in hydropower production and regulation of irrigation water supplies. Since the construction of the Tarbela Reservoir in 1974, no major reservoir has been added to the irrigation system of Pakistan. If this remains the case and the existing ones continue losing their capacity, the irrigation system may face hard times to regulate the desired water supply system in the country. There is a dire need for the construction of major dams in the country for water conservation and to assure an efficient supply of water for irrigation, domestic, industrial, and other uses.

The Punjab Irrigation and Power Department operates and maintains a major component of the largest contiguous gravity flow network in the world. The system is serving nearly 8.4 million hectors of Culturable Command Area (CCA). Irrigation water is delivered through 58,000 outlets off-taking from canals having total water conveying capacity of 120,000 cusecs.

Previously, the PID was maintaining the daily data regarding the flows through rivers, main canals, branch canals, distributaries, and minors using gauge readings that were prepared by the field staff in analog form and retained in the divisional offices. The gauge readings/discharges data of main/branch canals were then transmitted to Irrigation Secretariat through canal wire system using telegraph network maintained by the then Pakistan Telecommunication Corporation (PTC). Most of the time, daily canal operation data were received a day after the observation. Thus, there was no mechanism for daily transfer of data and daily monitoring of water distribution in the canal system, that is, between the headworks and tail off-takes /outlets. Thus, there is a dire need of communication mechanism or a unit to address the above limitations. The development of modern tools like the IMIS, including a DBMS and a DSS, have revolutionized the monitoring of irrigation systems to ensure transparency in getting reliable data. The PMIU was established in March 2006 as a part of organizational set up in Irrigation Department in order to improve the communication system efficiency and optimal canal operations. Institutional reforms process is also underway in Irrigation Sector of the province. The reforms are basically aimed at better governance, decentralization, farmers' participation in irrigation management, and sustainability through better and modern gauging systems. An important step in implementation of institutional reforms is formation of Khal Panchayats and Farmers Organizations for social mobilization, which would also ensure resolution of water theft through improved gauging systems. The social mobilization activities conducted at watercourse level created awareness amongst farming communities to organize themselves and to take part in the participatory irrigation management. However, the focus on the optimal canal operations, oriented towards equity and transparency, has not yet been achieved and the following relevant issues are not addressed so far:

- Old non-recording (manual type) gauges installed
- Gauge density is low
- Most of the gauges are out of order (about 90,229)
- Real time data is not accessible
- Allocated water allowance is not comparable with actual values
- Disputes among provinces (Water Apportionment Accord, 1991)
- Weak argument on transboundary issues with neighbours.

7.7.1 Outlook

The PMIU has successfully achieved the some of its objectives, including effective monitoring of canal operations of the entire irrigation system of Punjab, suggesting appropriate measures of optimizing canal operations. With the advancement of IT and day-to-day problem faced, there is an urgent need to further develop the tools for converting the existing system towards real-time monitoring with the least human interventions. Consequently, PMIU has planned to carry out the following additional tasks to ongoing activities.

- Installation of remote water level sensing devices in canals
- Live collection of discharge data at barrages and other nodal points
- Data transmission, communication, filtering, and processing
- Online provision of canal flow data for end users
- Better enforcement of rules and strict implementation of canal rotation plans
- Improve the farmers' feedback system
- Daily review of supply at head and tail reaches of canal system
- Amend the canal laws to incorporate farmer participation and efficient recovery system.

7.7.2 Benefits

- Daily review of water supply at head and tail of irrigation system
- Coherence of water provision with sanctioned water allowance
- Flexibility in irrigation possible based on crop water demands in different regions
- High temporal data available for research.

REFERENCES

1. Ahmad D.N. 1993. *Water Resources of Pakistan*. Mirajuddin Press, Lahore.
2. Centre of Excellence in Water Resources Engineering, Lahore. 1997. "Proceedings—Water for the twenty-first Century: Demand, Supply, Development and Socio-Environmental Issues."
3. Engineer Estimates. 2016. Real Time Flow Monitoring System on Punjab Irrigation Network.
4. Federal Bureau of Statistics. 2001. Statistical Division., Govt of Pakistan, "Pakistan Statistical Yearbook 2001."
5. Training Manual. 2008. Prepared for farmers Training Under ACIAR Project (2007–2015).

8 Precision Agriculture and ICT
Future Farming

Muhammad Jehanzeb Masud Cheema,
Hafiz Sultan Mahmood, M. Ahsan Latif,
and Ahmad Kamal Nasir

CONTENTS

Current farming equipment used in the country is not suited to bridge yield gaps between average and progressive farmers due to conventional nature and lack of standardization. The potential crop yield gap can be narrowed by applying site-specific agricultural inputs precisely through decision support mechanisms. The Precision Agriculture is an environment friendly system, which aims at reducing amount of inputs required to grow crops and focuses on increasing yield by increasing the efficiency of inputs. The spatial variability of soils can be mapped and yield monitors can be adopted to map yield. The researchers have developed Precision Seed drills and Variable Rate Applicators of fertilizer and pesticides. Such systems have resulted in 25–50% saving in use of Nitrogen alone. We need to reverse engineer and adapt available technologies and engineering options. The use of agricultural drones and satellites is a hot topic of research. In advanced countries, most of the crop monitoring, yield mapping, and agrochemical spraying is being carried out by manned and unmanned aerial vehicles. It needs to be introduced in the country while developing flight laws. Information and Communication Technologies (ICT) in farm production systems using precision agriculture techniques and mobile agriculture applications (MAgA) for research distribution and real time interaction with farmers could be a game changer. Other ICT applications include dissemination of knowledge and provision of databases for a better planning and land use. Land Record Management Information System (LRMIS) in the Punjab has digitized land records. The province of Sindh is also close to achieving this target. Punjab should share the experience with the Baluchistan province for a better land management in future. Its future uses will include the possibility of better provision of credit facilities and site-specific advisory services.

8.1 INTRODUCTION

Agriculture is considered the backbone of Pakistan's economy. Its contribution in the Gross Domestic Product (GDP) is 20.9% and provides a source of livelihood to 43.5% of the rural population (Anonymous, 2015). These numbers will further decline if we keep on depending on conventional agriculture. The green revolution of 1960s and 1970s boosted agricultural productivity, which was further improved by use of chemicals, fertilizers, and mechanization during 1980s and 90's. However, afterwards the agricultural production became stagnant and horizontal expansion of cultivable land became limited due to increasing population and urbanization. This ever growing population and decreasing agricultural resources (fertile lands) are the major threats to food security and together put great pressure on the country's economy. The shrinkage of land holdings is also a hindrance in achieving higher farm productivity. There are 6.6 million farms households in Pakistan, out of this 86% are small, resource-poor farmers owning less than 50% of the land resources, and the remaining 14% are large resourceful farmers, owning more than 50% the land resources.

The small farmers have less funds available, a poor availability of farm inputs (energy, fertilizer, and pesticide), poor support price mechanism for their products as well as no farm insurance. Moreover, reduction in crop yields, high input prices, inefficient use of inputs, water shortages, and the disliking of Pakistani agricultural products in the international market (Amjad, 2012) makes agriculture a nonremunerative business. They stick to the conventional farm management systems. Such systems consider use of uniform and general recommendations of inputs treating the entire field as a homogeneous entity. This practice not only increases the initial cost due to inefficient use of chemicals, fertilizers, and other inputs but also raises environmental concerns (Corwin et al. 2003). For example, overuse of fertilizers and chemicals for growing crops results in nutrient runoff which pollutes groundwater and thus degrades groundwater quality. Furthermore, agriculture generates greenhouse gases which contribute to the risk of climate change (Schneider and Kumar 2008). The issues can be addressed by changing the farming methods from conventional to conservation, organic farming, and precision agriculture. Introduction of such techniques can help farmers to increase their productivity and thus earn more remuneration from the same farmland.

Here the concept of using agricultural inputs precisely arises; it means doing the right thing at right place and the right time. This type of agriculture is known as Precision Agriculture (P-Ag) or site specific farming (Franzen, 2009). P-Ag is a systematic approach with the involvement of modern instruments (sensors, computers, and satellites) to change the traditional system of farming towards a low input, high efficiency and sustainable farming. The P-Ag system is an environment-friendly system, which aims at reducing amount of inputs required to grow crops and focuses on increasing yield, as well as increasing the efficiency of agrochemicals (Cheema et al. 2016).

The P-Ag concept is well adopted in various developed countries of the world. For example, it started in United States and Europe in early 1980s and early 1990s, respectively. The spatial variability of soils was mapped and yield monitors were developed to map the spatial yield variability. The researchers further developed precision seed drills and variable rate applicators for spot application of seed, fertilizer, and pesticides. The agricultural systems, with on-board sensors, were developed for automated fertilizer spreading and agrochemical spraying. Such systems have resulted in 25%–50% savings in the use of nitrogen only, with an overall saving of 600$ per hectare.

The advent of artificial intelligence and Information and Communication Technologies (ICTs) have opened new horizons in the field of agriculture. Now work is being carried out to make real-time agricultural robots and auto steering technologies to reduce labor requirements. Use of unmanned arial vehicles for agriculture (AgUAV), also a hot topic of research, has begun for crop monitoring, yield mapping, and agrochemical spraying. The Japanese are working extensively on sensor development of digital and hyper spectral image processing on-board UAVs. ICT platforms are now in use for predictions of optimized local farming practices for the farmers. They provide information on crop and soil health, weather conditions, socioeconomic characteristics, labor and inputs availability, and other related variables.

Developing countries like India have initiated its P-Ag programs during the last decade and are working on indigenized solutions of agricultural machinery. However, in Pakistan, P-Ag is a rather newer concept and at the initial stages. On the whole, the farming system of Pakistan is not ready to accept P-Ag technologies. There are few exceptions which can act as incubators for new concepts and high-tech technologies in local conditions (Mahmood et al. 2013). Imported auto steering systems and precision agriculture machinery is being used at a few large farms. On the other hand, small farm machinery, UAVs, and robots can be introduced for the small farms, which will not only restrict soil compaction but also may work on locally available renewable fuels. Sensor-based high-efficiency irrigation systems, variable rate fertilizer, and agrochemical applicators, as well as robots for sowing, weed removal, and harvesting have great potential for small farms.

Though this technology is new for farming communities and for agricultural scientist, a lot of research will be required to develop this technology for small farmers. The objective of this chapter is to describe the P-Ag technologies and their potential in Pakistan in order to increase farm productivity and to decrease adverse environmental impacts.

8.2 PRECISION AGRICULTURE—STATE OF THE ART

P-Ag is an ICT-based system of farm management. The technologies and principles are applied in order to identify, analyze, and manage spatial and temporal variability associated with various characteristics of agricultural production systems. The primary impact of P-Ag is cost reduction and the efficient use of production inputs, the increase in size and scope of farming operations with less labor, to develop a knowledge base through ICT, enhance production processes controls which help in the production of higher value products, and keep track of production for food safety and environmental benefits (Lowenberg-DeBoer and Boehlje, 1996). The key objective of implementing P-Ag is the precise use of agricultural inputs based on soil, climate, and crop needs for improving crop performance and protecting land resources, thus optimizing profitability, agricultural sustainability, and safeguarding the environment (Pierce and Nowak, 1999; Zhang et al. 2002).

The three main elements that constitute P-Ag are information, technology, and decision support. It identifies critical factors which affect crop yield and determine its spatial variability. Thus, sensing variability in crop and soil characteristics is considered as the first step for implementing P-Ag because better information produces better decisions. A large field is divided into a finite number of small management zones to customize field inputs, fulfilling requirements of each zone. The fundamental tools required to map spatial variability and prescribe precise applications include a global positioning system (GPS) based on real-time kinematics, proximal soil sensors, a geographic information system (GIS), a satellite and UAV based hyperspectral imagery, variable-rate systems, and yield monitors (Plant 2001).

Four steps describe the implementation of P-Ag including: (a) characterization of scale and extent in soil and crop variability; (b) interpretation of causes and importance of variability; (c) managing spatial and temporal variability; and (d) monitoring the outcomes of such variability-management practices (Shanwad et al. 2004). Thus, the basic phases of P-Ag concepts include the accurate assessment of variation in soil and crop properties, management of spatial variation, and evaluation in space and time. The flow of information for a cropping season is well described by Gebbers and Adamchuk (2010) and summarized in Figure 8.1.

8.3 PRECISION AGRICULTURE TECHNOLOGIES

The P-Ag technologies include a complete system of software, hardware, and best management practices in order to collect and utilize information properly and efficiently. These technologies are briefly described in the section belows.

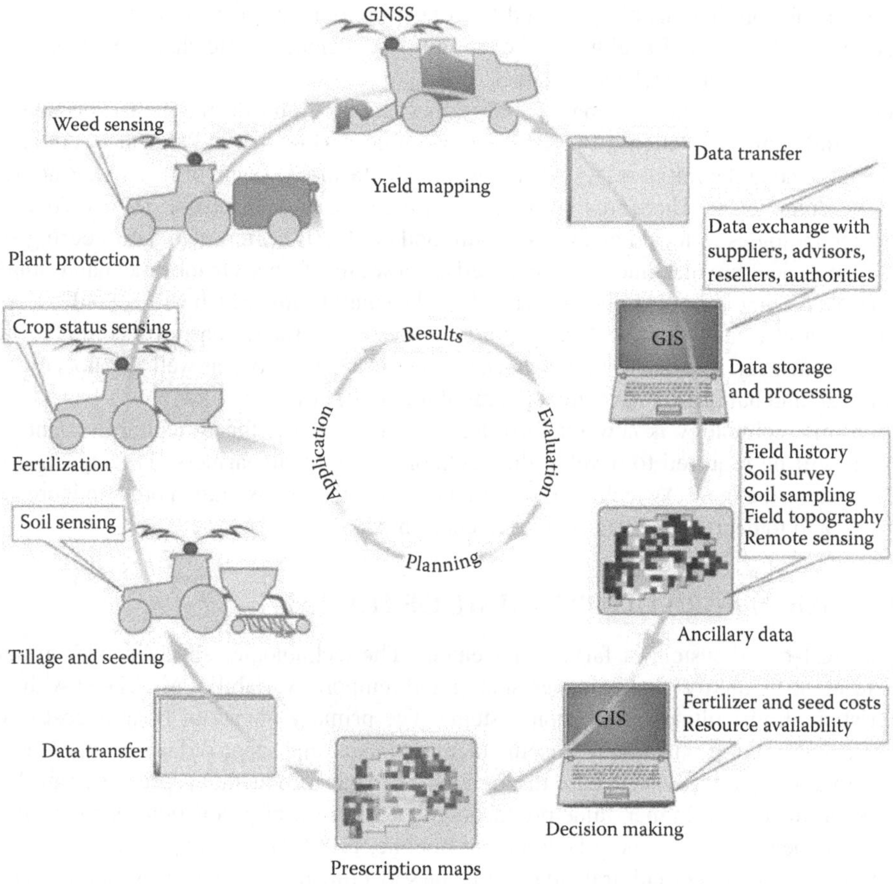

FIGURE 8.1 Precision agriculture information flow for a cropping season. (Adapted from Gebbers, R. and Adamchuk, V.I. 2010. *Science*, 327(5967): 828–31.)

8.3.1 Proximal Soil Sensors for Mapping Soil Properties

Proximal soil sensors observe soils from the ground or within two meters from ground. Remote sensing from an airplane and a satellite can also be applied, but high-resolution data collection is only possible by proximal soil sensors. Therefore, for implementing P-Ag, proximal soil and crop sensors are inevitable. Proximal sensors identify the physical characteristics of soil corresponding to various soil properties. These sensors can be divided into five major groups, based on their sensing workability (Kuang et al. 2012; Mahmood, 2013). The workability of these sensors is based on soil reflectance, soil radiation, resistivity and permittivity, soil conductivity, and strength as well as soil electrochemical properties. Some commonly used proximal soil sensors are vis-NIR reflectance spectrometers, electromagnetic induction sensors (EM38), proximal gamma-ray spectrometer, and ion-selective electrodes. These sensors can either be used manually or mounted on a field vehicle for real-time data collection. Real-time compaction sensors can be used to map compaction data of soil tillage layer by mounting sensors on a shank of a subsoiler.

8.3.2 Unmanned Aerial Vehicle

Use of UAV has increased in most advanced countries associated with agriculture. According to an estimate by AUVSI (2013), the sales of UAVs has doubled during last year, and is expected to be four times that by the year 2025. In agriculture, UAVs are being used for crop area mapping,

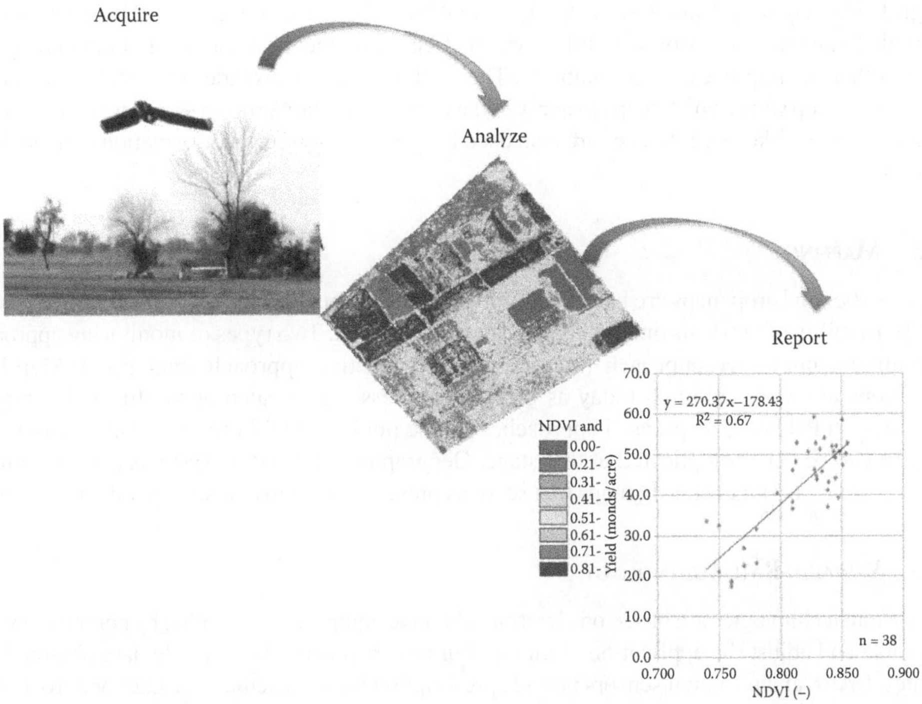

FIGURE 8.2 **(See color insert.)** Information flow from data acquisition to crop health monitoring and yield estimation using UAV system.

crop health monitoring, and plant count at very high resolution (Figure 8.2). These systems can capture visible-infrared (VNIR) and thermal infrared (TIR) images simultaneously to monitor crop growth, water stress, nutrient stress, weeds, precipitation damages, pest infestation, etc. (Misopolinos et al. 2015; López-Granados et al. 2016). Site-specific treatment maps can be produced taking into account different factors, for example flight altitudes, camera types, and weed thresholds. The images obtained can be analyzed and the corresponding reports and prescriptions can then be generated. The Department for Environment, Food and Rural Affairs of the United Kingdom has recently announced 40% grants for research on UAVs, sensors, and analytical software. This will hopefully be a further motivating interest of researchers and farmers into these tools. An integrated package of three components will give farming business tools that are needed to obtain valuable crop information.

The UAVs work in combination with various cameras on board to detect reflectance at various wavelengths. The use of transformed low cost sensors, with very-high spatial resolution systems, can provide important information for site-specific farming (Knoth et al. 2011).

8.3.3 Low Cost Digital Cameras

Cameras on board UAVs for sensing various parameters are costly and increase the overall cost of the system. Researchers have developed new low-cost modified cameras in red and near infrared wavelength bands of the electromagnetic spectrum which are being used to detect plant health and related parameters with reasonable accuracy. More research on various sensors can reduce the costs further, as well as allowing more parameters of soil and moisture to be studied.

8.3.4 Satellite Imagery

Regional scale mapping of crop acreage, crop health, soil moisture, water use, crop yield can be carried out using moderate to high-resolution sensors on board various earth observation satellites.

The grid size may vary from 250 m (freely available) to 30 cm (commercial). The generated grid maps help to identify in-season variability which affects crop yields and assist making management decisions that can improve crop profitability. The satellite imagery provides information on location and extent of crop stress which helps to analyze the causes of certain crop stresses and their remedies. Handheld sensors like GreenSeeker are also used for ground truthing of information collected from satellites.

8.3.5 Mapping

Soil properties and crop maps are basic requirements for site-specific applications of crop inputs. The spatial variability of nutrients and crop yield can be monitored. Two types of monitoring approaches are common: predictive approach (map-based) and reactive approach (real time). Map-based applications are more common today as they require less complicated apparatus and computing algorithms. In Pakistan, map-based approaches can be implemented initially because sensor-based approach can be very complicated at this stage. Geographic information systems (GISs), computer software, are used for generating maps that serve as prescription maps for site-specific management.

8.3.6 Variable-Rate Technology

Variable rate technologies are based on electromechanical equipment controlled by computer systems that enable and adjust the application of an input in near real time. The sample data obtained from satellites, UAVs, and proximal sensors provide prescription for a particular fertilizer or agrochemical to be applied at each unit. A GIS-based soil nutrient map is stored in a computer mounted on a GPS-guided tractor. The decision support system decides the exact amount of fertilizers and other inputs required for each location in the field and sends signals to the variable rate applicator so that the exact amount of input can be applied at each location (Maleki et al. 2007). This operation is done in real time, within seconds. Variable rate fertilizer applicators have been successfully used in the developed world (Zaman et al. 2010). Other variable rate systems include sprayers, granular spreaders, tillage implements, hydrous ammonia spreaders, sensor-based irrigation systems, and agrochemical applicators (Bennett and Brown, 1999; Swisher et al. 2002).

8.3.7 Robotics for Precision Agriculture

With the advancement and availability of low-cost sensor systems and actuators, mobile robotics is also entering into the field of agriculture in advanced countries like United States and Germany. Farm mechanization in Pakistan has still not been adopted throughout the country. UAVs and driverless tractors are examples of aerial mobile robots and ground mobile robots, respectively. The ground mobile robots have advantages because of the accuracy with which they can perform operations during both day and night. These ground mobile robots can perform operations such as—but not limited to—harvesting, weed control, mowing, pruning, seeding, spraying, and thinning. Such mobile robots are used at agricultural farms, nurseries, and orchards. These ground mobile robots use a combination of GPS and IMU (Inertial Measurement Unit) to accurately and autonomously navigate within farm fields. Because such devices are all time ready, therefore, they can be utilized for sowing large areas in a limited time window. Furthermore, in advanced countries' farms, labor is difficult to find at the right time, therefore, such technologies will help farm owners. The only limiting factor at the moment is the cost of such systems. Some examples of ground mobile robots which are being used in practice are BobiRob (Ruckelshausen et al. 2009), ASI Forge Platform (Matese and Di Gennaro, 2015), and Grizzly (Bawden et al. 2014). Fully or partially autonomous ground mobile robots can only be used to implement efficiently variable rate philosophy. With the limited amount of arable land and increasing demand of food for growing populations globally, such automations will be required to increase the throughput of existing grain production techniques.

8.3.8 Data Fusion for Precision Agriculture

Data fusion of inputs of multiple sensors is new in precision agriculture. The interpretation of relationships between a single sensor output and a soil property becomes complex and uncertain (Mahmood et al. 2009), as a single sensor can sense mixed soil properties. Thus, raising questions on the accuracy of the information. The accuracy of a single sensor is often low because all proximal soil sensors respond to more than one soil property of interest. Therefore, conceptually different soil sensors can be combined for providing complementary and robust soil property information (Mahmood et al. 2012). The information acquired from these proximal sensors, in combination with other satellite and UAV mounted sensors, can provide a comprehensive database. A large datasets on agriculture can be generated from this data fusion.

8.4 AGRICULTURAL INPUTS IN PAKISTAN

Pakistan's agricultural system requires set of inputs from sowing to harvesting of crops. The major inputs include seed, water, fertilizer, and agrochemicals. Seed is a basic input while improved seed and planting density is key in raising crop yields. The improved seeds are not only high yielding but also resistant to droughts, salinity, pests, and diseases. In Pakistan, both the public and private sectors are involved in the production, multiplication, and distribution of improved seed. At present, about 20% of the main croplands are sown with improved seed. Recommended number of plants per acre is hardly achieved by conventional sowing methods.

Water is an essential input in agriculture and it is the largest subsector of water use, as it uses around 93% of total water available in the country (Afzal and Ahmad, 2009). The irrigation efficiency is less than 50%, thus causing severe loss of this scarce resource.

Agriculture uses about 14% of the total electricity consumed in the country and 11% of the total diesel consumed in the country. At present, 66,250 metric tons of pesticide are being consumed by the agricultural sector, which is worth of PKR 12.6 billion. Consumption per annum of nitrogen, phosphorus, and potassium is about 3185,881, and 23.7 thousand metric tons in the country, respectively. Fertilizer use for different crops is different. Around 50% fertilizers are used for wheat, whereas rice, cotton, maize, and sugarcane use 14%, 15%, 7%, and 8% of fertilizers, respectively. Fertilizer application is also carried out not according to the requirement of the soils. Per hectare grain yield in the United States is double that of Pakistan, with the application of the same amount of nutrients. Therefore, there is a need to explore potential areas where P-Ag technologies can be introduced keeping in view Pakistan's farm settings.

8.5 POTENTIAL OF PRECISION AGRICULTURE IN THE COUNTRY

The ultimate goal of P-Ag technologies is to gather site-specific information and enable farmers to vary management practices more appropriately within their fields. P-Ag provides understanding of spatial variability of soil properties and corresponding crop status and yields within a field, it identifies the reasons of yield variability and suggests variability-based prescriptions to enhance yields (Aschmann et al. 2003).

It is possible with P-Ag that an integration of a set of appropriate technologies can be made. This will improve traditional cultivation technologies in accordance with the local farming conditions of the country. In many developing countries, intensive horticulture, the animal raising industry, and the value-added processing of primary agroproducts have shown great potential and opportunity (Maohua, 2001). Sophisticated P-Ag technologies, developed and used by advanced countries, created a real challenge to search for P-Ag technologies which are suitable for the farming systems of developing countries.

It is admitted that the only P-Ag technology that has been well adopted in the country is laser land leveling. Precision land leveling is a resource conservation technology, which saves irrigation water and time. It is successful in reducing waterlogging problem in low-lying areas of the farm, where water accumulates during irrigation and rainfall. Another precision technology which has been practiced by small farmers for years is zone management. Small farmers were dividing each acre into 4–8 small management units to customize site-specific application of fertilizers, seed, and irrigation water. However, the accuracy of this manual approach is not so rewarding because it still includes to a large extent uniform application of inputs. Adoption of precision technologies in more areas, such as variable rate application of fertilizers, use of UAVs for mapping tree biomass, yield monitoring, irrigation mapping, crop stress analysis, and mapping soil and crop attributes using proximal and remote sensors, has potential.

Use of UAVs in combination with satellite imagery is practical to monitor crop acreage, health, and yield in Pakistani farm settings. Site-specific fertilizer and agrochemicals can be applied to orchards, as in Malaysia where site-specific fertilizers are being applied to rubber plantations. These variable rate technologies can reduce the amount of fertilizers and pesticides required to achieve good yield and reduce the amount of harmful chemicals, thus being environmentally friendly. The Pothohar area, which is more prone to soil erosion, can be a test case to introduce P-Ag-based tillage and soil erosion control technology which emphasizes mainly on adjusting the tillage depth depending on the soil texture. This system will be linked to a soil database and allows farmers to make site-specific decisions on soil fertility and disease-related problems (Mondal and Basu, 2009).

8.6 CHALLENGES IN ADOPTION OF PRECISION AGRICULTURE

Pakistan is an agricultural country, yet the agricultural sector is not able to meet its potential level of production with its existing resources. Most of modern technologies used are in their infancy in the country. This is attributed to the lack of utilization of modern technologies in agriculture, small landholdings, high input costs, high technology costs, indigenous technology lacks, cropping systems' heterogeneity, and lack of technical knowledge and expertise. Above all, lack of awareness could be the first and foremost reason for this late start of P-Ag in Pakistan (Mahmood et al. 2013).

Small landholdings and high cost of the technology are identified as two major problems in its wider adoption. Only a few farmers can afford to adopt such smart technologies. Other challenges towards the adoption of P-Ag technologies in Pakistan's farm settings are summarized below:

- Heterogeneity of cropping systems is also a challenge in the adoption of precision agriculture in the country.
- Introducing P-Ag technologies in conventional agriculture requires modern technologies; such as smart computers, remote sensors, proximal sensors, yield monitors, actuators, and variable rate applicators together with the competences to use modern GIS software for incorporating GPS information in the outputs of P-Ag technologies. Currently, there is no such facility that provides integrated solutions.
- The lack of legislation from the government towards using environmental friendly production practices in agriculture is a big hurdle in this regard. Laws to reduce environmental contamination, such as groundwater and soil contamination with excessive chemicals and fertilizers, are not available.
- In developed countries, there are skilled organisations and agencies in the private sector that provide consultancy services to farmers in using precision technologies and obtain potential benefits. Unfortunately in Pakistan, there are no such agencies in private sector that can provide such consultancy services to the farmers. If there is any agency/organisation, it does not possess such skill to convince farmers.
- Agricultural machinery manufacturers are producing machinery/implements as a business as usual. They should be encouraged to introduce the modernisation in agricultural sector with the help of research and educational institutions.

- The low export volume of agricultural products is also a reason of not using precision technologies. In the modern world, people are demanding safer foods, which contain very low or minimum amounts of chemicals. The competition in the international market will drive local farmers to use these modern technologies for producing safer foods.

8.7 STRATEGIES FOR ADOPTION OF PRECISION TECHNOLOGIES

The strategies for adoption of P-Ag in Pakistan should take care of problems associated with land fragmentation, shortage of high tech centers for P-Ag, specific software and hardware facilities for P-Ag, and poor economic conditions of farmers. Strict environmental legislation, public concern over excessive use of agrochemicals, economic gains from reduced agricultural inputs, and improved farm management efficiency is required for P-Ag technologies' widespread adoption.

The adoption of these technologies can provide several benefits such as large farms input savings, savings on labor, and equipment use and off course environmental benefits. Sprayers, spreaders, planters, and other application equipment can be connected to a VRT, that works by turning a section of an application equipment on, in areas where application is required, and off in unwanted areas. On an average, it is estimated that around 4.3% savings can be achieved on inputs using VRTs. The payback period of these technologies is less than 2 years (Alabama Extension bulletin, 2010). About 10% input costs can be saved by introducing guidance systems on agricultural tractors which could reduce overlap and input usage. Additional benefits of these systems include reduction in concentration time needed during driving which leads to less fatigue and an increased ability to focus on other tasks. Laser land leveling is one of the techniques used successfully in Pakistan to increase input-use-efficiency. It has to be further promoted as the precise leveling of fields improves significantly irrigation efficiencies. It can increase average application efficiency by 65%, storage efficiency by 70%, and water distribution efficiency by 80%.

These technologies can be used successfully not only in large fields, but also in small fields (Mondal and Basu 2009). Small fields can be consolidated virtually to be considered as big fields for mapping of soil and crop attributes. Small farmers can use low cost and small machine-based VRTs and can start with a single precision application, whereas the progressive farmers should select more than one precision application as a package for growing high-value commercial crops. Some low cost and technology tools have been introduced for small farms in developing countries. This includes the Chlorophyll meter and Leaf Color Chart (LCC), and NDVI measurement instruments that can be used for on-field measurement of the crop nitrogen status. Their usefulness has been proven in various countries like the Philippines, Indonesia, Vietnam, Bangladesh, and India. The use of these gadgets and precise fertilizer applications have resulted in about 40 kg N/ha less fertilizer use, as compared with the previous farmer's practice. GIS is also becoming familiar and is being adopted for use on small farms of Asian countries like Japan, Korea, and Taiwan. These interventions can be introduced among the farming communities of Pakistan.

P-Ag adoption can also be pursued among farmers by developing and popularizing inexpensive electronic gadgets. These gadgets can increase the overall profit of small farmers, which can be instrumental in preparing the platform of P-Ag adoption, by softening the farmer's attitude towards modern technologies. Similarly, UAVs can also be introduced for mapping the farms, identifying diseases, and so on. Most robotic machines and UAVs are compact and thus suitable for small farms.

P-Ag service providers (PASP) have to be present to lease modern equipment, provide technical training/consultancy and backup services to the farmers. They can charge on a per acre basis, which could be an attractive and economical aspect for many small farmers, especially when there is shortage of agricultural labor. Moreover, these PASPs can be used to train progressive farmers and early adopters, expose the neighboring nonparticipating farmers to the new technologies, and show the usefulness of the technology for short and long-term management.

More skilled manpower is required to develop indigenized P-Ag machinery and taking P-Ag forward. Without skilled manpower and research, P-Ag technologies cannot succeed. New study fields, like agricultural mechatronics or robotics, may be introduced in the engineering faculties. The graduates of these fields should be capable of reverse engineering and developing new gadgets for agriculture.

Similarly, mobile apps can be developed which can provide timely information to the farmers about their crop health, nutrients requirements, water applications, and pest management, etc.

By adopting these strategies, it is expected that the current stage of uniform soil and crop management will gradually change to site-specific soil and crop management in the country.

8.8 CONCLUSION

P-Ag deals with the judicious application of farming inputs and is therefore advantageous over traditional agriculture. There is a need to introduce low-cost P-Ag technologies in Pakistan that can be adopted by small farmers. Many precision technologies have a great potential for implementation in the country. With some modifications, partially automatic systems can be used in Pakistan's farm setting. In this perspective, farmers and government authorities should look forward to adopt new and sustainable technologies to increase the efficiency of available resources which would increase crop yields and reduce input costs. A policy road map that can be adopted to familiarize P-Ag in the country is given below:

8.8.1 Way Forward/Action Required

1. Precision
   a. A centralized ICT-based database system with site-specific information on soil, water, crops, etc. should be developed and linked with a land record system.
   b. Virtual clustering of the farms should be carried out.
   c. Variable rate technologies for inputs like fertilizer, pesticides, etc., should be introduced.
   d. Indigenous machinery should be developed using reverse engineering techniques to make advanced machinery affordable for the farmers.
   e. A mechanism of agricultural machinery standardization is required for quality machinery.
   f. The establishment of the Precision Agriculture Service Providers (PASP) should be initiated.
   g. Incentivize local manufacturing of farm equipment, irrigation accessories, and food processing machinery.
   h. Legislation is required to use UAVs and electronic gadgets for site-specific agriculture.
   i. Skill development of agricultural machinery users is immediately required.
   j. ICT services for networking farmer, government agencies, agro banks, academia, and markets has to be developed.
2. Establishment of service centers
   a. Mechanization needs to be met according to the crop clusters and industry needs
   b. Supply of inputs and credit could be combined with the mechanization service that could also act as marketing/trading hubs
   c. The centers can also serve as information centers
   d. Fertilizer sales could be linked with soil and water analysis.

REFERENCES

Alabama Extension bulletin. 2010. Consideration for adopting and implementing precision Ag technologies. *Precision Agriculture series*. http://www.aces.edu/anr/precisionag/Publications. [10.04.2016]

Afzal, N. and Ahmad, S. 2009. Agricultural input use efficiency in Pakistan: key issues and reform areas. *Managing Natural Resources for Sustaining Future Agriculture*, 1 (3), 2009.

Amjad, R. 2012. Key challenges facing Pakistan agriculture: how best can policy makers respond? A note. Planning commission, Government of Pakistan.

Anonymous. 2015. *Pakistan Economic Survey 2014-2015. Ministry of Finance*. Government of Pakistan Islamabad, Pakistan.

Aschmann, S., Caldwell, R., and Cutforth, L. 2003. Affordable opportunities for precision farming. Sustainable Agriculture technical note, United States Department of Agriculture.

AUVSI. 2013. *The economic impact of unmanned aircraft systems integration in the United States*. Association for Unmanned Vehicle Systems International (AUVSI) Economic Report 2013.

Bawden, O., Ball, D., Kulk, J., Perez, T., and Russell, R. 2014. A lightweight, modular robotic vehicle for the sustainable intensification of agriculture.

Bennett, K.A. and Brown, R.B. 1999. Field evaluation of a site specific direct injection herbicide sprayer. In: S. Joseph (Editor), *ASAE Proceedings Paper No. 99-1102*. American Society of Agricultural Engineers, MI, USA.

Cheema, M.J.M., Latif, M.A. and Shafeeque, M. 2016. Proceedings of 2nd National Conference on Precision Agriculture. "Current status and future prospects of adopting precision agriculture in Pakistan's Farm Settings" April 16, 2016. USPCAS-AFS, UAF. ISBN: 978-969-9035-12-8

Corwin, D.L., Kaffka, S.R., Hopmans, J.W., Mori, Y., Van Groenigen, J.W., Van Kessel, C., Lesch, S.M., and Oster, J.D. 2003. Assessment and field-scale mapping of soil quality properties of a saline-sodic soil. *Geoderma*, 114(3–4): 231–59.

Franzen, D. 2009. Site Specific Farming 1: What is Site-specific Farming? SF-1176-1 (Revised).

Gebbers, R. and Adamchuk, V.I. 2010. Precision agriculture and food security. *Science*, 327(5967): 828–31.

Knoth, C., Prinz, T., and Loef, P. 2011: Microcopter-Based Color Infrared (CIR) Close Range Remote Sensing as a Subsidiary Tool for P recision Farming". In: *Proceedings of the ISPRS Workshop on Methods for Change Detection and Process Modelling*, Univ. Press Cologne, Cologne, Germany.

Kuang, B., Mahmood, H.S., Quraishi, M.Z., Hoogmoed, W.B., Mouazen, A.M., and Van Henten, E.J. 2012. Sensing soil properties in the laboratory, *in situ*, and on-line: A review. *Advances in Agronomy*, 114, 155–223.

López-Granados, F., Torres-Sánchez, J., Serrano-Pérez, A., de Castro, A.I., and Peña-Barragán, J.M. 2016. Early weed mapping in sunflower by using UAV technology: Variability of herbicide treatment maps against weed threshold. *Precision Agriculture*, 17(2): 183–199.

Lowenberg-DeBoer, J. and Boehlje, M. 1996. Revolution, evolution or dead-end: economic perspectives on precision agriculture. In: *Proceedings of the Third International Conference on Precision Agriculture*. Robert, P.C., Rust, R.H. and Larson, W.E., eds. American Society of Agronomy, Crops Science Society of America, Soil Science Society of America. Madison, WI.

Mahmood, H.S. 2013. Proximal soil sensors and data fusion for precision agriculture. *PhD thesis*, Wageningen University, Wageningen, Netherlands—ISBN: 978-94-6173-579-9, 205 p. Freely available at: https://library.wur.nl/WebQuery/wda/2032534.

Mahmood, H.S., Hoogmoed, W.B. and van Henten, E.J. 2012. Sensor data fusion to predict multiple soil properties. *Precision Agriculture*, 13(6): 628–645.

Mahmood, H.S., Ahmad, M., Ahmad, T., Saeed, M.A., and Iqbal, M. 2013. Potentials and prospects of precision agriculture in Pakistan—A review. *Pakistan Journal of Agricultural Research*, 26(2), 151–67.

Mahmood, H.S., Hoogmoed, W.B.,and van Henten, E.J. 2009. Combined sensor system for mapping soil properties. In: van Henten, E.J., Goense, D. and Huijsmans, J.F.M., eds. *Precision Agriculture 2009: Proceedings of the 7th European Conference on Precision Agriculture*. Wageningen Academic Publishers, Wageningen, the Netherlands, pp. 423–30.

Maleki, M.R., Mouazen, A.M., Ramon, H., and De Baerdemaeker, J. 2007. Optimisation of soil VIS-NIR sensor-based variable rate application system of soil phosphorus. *Soil and Tillage Research*, 94(1): 239–50.

Maohua, W. 2001. Possible adoption of precision agriculture for developing countries at the threshold of the new millennium. *Computers and Electronics in Agriculture*, 30, 45–50.

Matese, A. and Di Gennaro, S.F. 2015. Technology in precision viticulture: A state of the art review. *Int. J. Wine Res.*, 7, pp. 69–81.

Misopolinos, L., Zalidis, C., Liakopoulos, V., Stavridou, D., Katsigiannis, P., Alexandridis, T.K., and Zalidis, G. 2015, June. Development of a UAV system for VNIR-TIR acquisitions in precision agriculture. In *Third International Conference on Remote Sensing and Geoinformation of the Environment* (pp. 95351H–95351H). International Society for Optics and Photonics.

Mondal, P. and Basu, M. 2009. Adaptation of precision agriculture technologies in India and in some developing countries: Scope, present status and strategies. *J. Progress in Natural Science*, 19, 659–66.

Pierce, F.J. and Nowak, P. 1999. Aspects of Precision Agriculture. *Advances in Agronomy*, 67, 1–85.

Plant, R.E. 2001. Site-specific management: The application of information technology to crop production. *Computers and Electronics in Agriculture*, 30(1-3), 9–29.

Ruckelshausen, A., Biber, P., Dorna, M., Gremmes, H., Klose, R., Linz, A., Rahe, F., Resch, R., Thiel, M., Trautz, D., and Weiss, U. 2009. BoniRob–an autonomous field robot platform for individual plant phenol typing. *Precision Agriculture*, 9(841), p. 1.

Schneider, U.A. and Kumar, P. 2008. Greenhouse gas mitigation through agriculture. *Choices, American Agricultural Economics Association*, 23(1).

Shanwad, U.K., Patil, V.C., and Gowda, H.H. 2004. Precision farming: Dreams and realities for Indian agriculture, Map India Conference. GIS Development Net.

Swisher, D.W., Borgelt, S.C., and Sudduth, K.A. 2002. Optical sensor for granular fertilizer flow rate measurement. *Transactions of the ASAE*, 45(4), 881–888.

Zaman, Q.U., Esau, T.J., Schumann, A.W., Percival, D.C., Chang, Y.K., Read, S.M., and Farooque, A.A. 2010. Prototype Variable Rate Sprayer for Spot- Application of Agrochemicals in Wild Blueberry. *Computers and Electronics in Agriculture*, 2010. 76: pp. 175–82.

Zhang, N., Wang, M., and Wang, N. 2002. Precision agriculture—A worldwide overview. *Computers and Electronics in Agriculture*, 36(2–3), 113–132.

9 Farm Services

Asghar Ali, Sarfraz Hassan, and Abdul Ghafoor

CONTENTS

9.1 INFRASTRUCTURE FACILITIES

9.1.1 Road Infrastructure

The development of local roads is considered necessary because of its necessity for rural development, allowing the farmers' access to the markets for trading agricultural products. It also exposes the undeveloped resources along local roads which are important for diffusion of social services, raising awareness in local areas, and thus in poverty reduction. Provision of road infrastructural facilities, in rural areas of Pakistan, has been shown in Table 9.1. It reveales that in Punjab, farmer's access to paved roads is higher than in other provinces. That is why Punjab's farmers are better off, comparatively. In contrast, average distance of a village from bus stop in Baluchistan is 30.9 km, indicating the underdevelopment in the province.

9.1.1.1 Importance of Paved Road Access

The progress of access from rural areas to markets enables prompt transportation to the markets with proper timing. The same products can be sold with higher prices, which leads to an increase in profitability. At the same time, it becomes easier to procure the inputs necessary for agricultural production. This means, when considering transportation costs, a reduction in the price of inputs leads to an increase in the profitability of farmers. An example of this effect is an increase in the production of sugarcane, resulting from paved roads which reduced transportation costs and ensured its timely delivery and quality to sugar mills. Table 9.2 shows the distance of different wholesale agricultural markets from the Mouza in Pakistan.

TABLE 9.1
Road Infrastructural Facilities to Rural Areas in Pakistan

Province	Proportion of Rural Population with Paved Access (%)	Average Distance to a Bus/Wagon Stop for Rural Population (km)
Punjab	76	3.8
Sindh	63	4.1
NWFP	68	5.7
Baluchistan	27	30.9
Pakistan	68	8.2

Source: Rural Access and Mobility in Pakistan, World Bank, 2010.

TABLE 9.2
Distance of Mouzas from Wholesale Markets

Type of Facility	Mean Distance (km)	Mouzas by Distance from Wholesale Markets (km)				
		Less than 1	1–10	11–25	26–50	51 & above
Livestock market	38	1,056 (2)	13,478 (28)	17,140 (36)	8,656 (18)	7,152 (15)
Grain market	40	1,017 (2)	13,903 (29)	16,164 (34)	8,646 (18)	7,752 (16)
Fruits market	41	692 (1)	13,707 (29)	16,248 (34)	8,490 (18)	8,345 (18)
Vegetable market	41	755 (2)	14,521 (31)	15,984 (34)	8,078 (17)	8,144 (17)
Govt. Procure Center	48	1,330 (3)	14,411 (30)	14,404 (30)	8,103 (17)	9,234 (19)

Source: Pakistan Bureau of Statistics. 2016. Pakistan Mouza Census, 2008. Online available at http://www.pbs.gov.pk/
Note: In parenthesis are percentages.

9.1.2 Health Infrastructure

Better health and healthy living environments contribute to the improvement of family life, human welfare, and ultimately fosters economic growth. In order to achieve a substantial improvement in the health sector, a number of vertical programs are operative in Pakistan. Table 9.3 represents the health infrastructural status across the country, showing the number of Basic Health Units (BHUs) providing health services to rural people, including farmers. These facilities definitely enhance farmers' productivity.

TABLE 9.3
Health Facilities Statistics of Pakistan (NOs.)

Health Facility	2014–2015	2015–2016
No. of hospitals	1113	1142
No. of dispensaries	5413	5499
Basic health units (BHU's)	5571	5438
Maternity & child care centers	671	687

Source: GOP. 2016a. Economic Survey of Pakistan 2015–2016. *Finance Division, Economic Advisor's Wing, Islamabad*; GOP. 2016b. Agricultural Statistics of Pakistan 2015–2016. *Ministry of Food and Agriculture (Economic Wing)*, Islamabad.

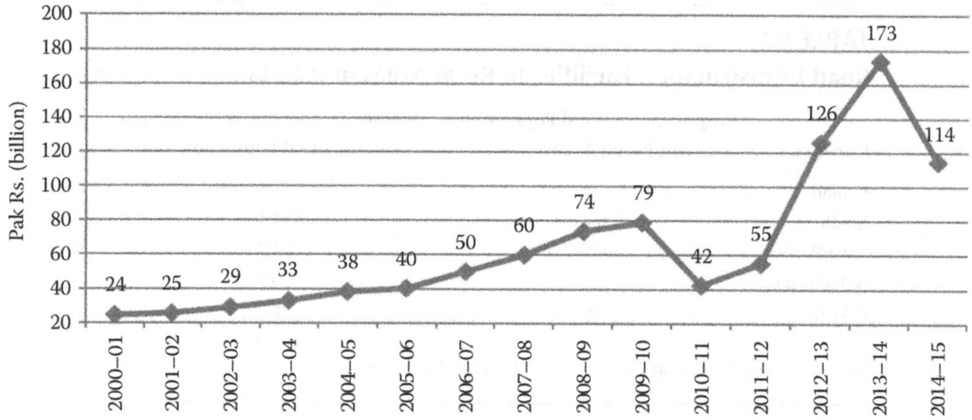

FIGURE 9.1 Total public sector expenditure on health.

TABLE 9.4
Distance of Mouzas from Various Types of Health Facilities

Type of Health Facility	Rural Populated Mouzas	Overall Mean Distance (km)	Mouzas by Distance from Health Facility (km)				
			Less than 1	1–10	11–25	26–50	51 & above
Hospital/dispensary	47482 (100)	17	4838 (10)	25874 (54)	10733 (23)	3877 (8)	2160 (5)
Rural health center	47482 (100)	16	4694 (10)	26390 (56)	10367 (22)	4066 (9)	1965 (4)
Basic health unit	47482 (100)	15	5469 (12)	26678 (56)	9525 (20)	3776 (8)	2034 (4)
Child & mother care center	47482 (100)	22	3002 (6)	23189 (49)	12349 (26)	5320 (11)	3622 (8)
Population welfare center	47482 (100)	19	5466 (12)	23743 (50)	10574 (22)	4611 (10)	3088 (7)
NGO dispensary	47482 (100)	34	1808 (4)	16358 (34)	13674 (29)	9176 (19)	6466 (14)
Private MBBS doctor	47482 (100)	20	5043 (11)	23340 (49)	11468 (24)	4487 (9)	3144 (7)
Midwife facility	47482 (100)	22	11398 (24)	18909 (40)	9444 (20)	4576 (10)	3155 (7)

Source: Pakistan Bureau of Statistics. 2016. Pakistan Mouza Census, 2008. Online available at http://www.pbs.gov.pk/.
Note: In parenthesis are percentages.

The share of health expenditure in the total public sector expenditure is the most significant variable explaining government priorities and the health status of a country. The Figure 9.1 shows the last 15 years' public sector expenditure on the health sector by the government of Pakistan.

Table 9.4 represents the distance that the rural population has to cover in order to reach a health facility. It is clear that a lot of work remains to be done to minimize the distance between the users and the centers providing health facilities. The government should provide the basic health facility within the Mouzas.

9.1.3 Educational Infrastructure

Education plays a pivotal role in the development of a country. A high literacy rate ensures sustainable economic development, labor productivity, and economic prosperity. National equalization of the educational level can eliminate regional inequalities and gender discrimination. It also enables the population to meet the emerging challenges of the modern world and to rapidly adopt to new technologies. It is important for developing countries, where majority of the world's population exists, need to redesign educational policies in line with the advanced nations for enhancing productivity

TABLE 9.5
Number of Mainstream Institutions in Pakistan (Thousands)

Institution Type and Level	2012–2013	2013–2014	2014–2015
Primary	159.7	157.9	158.7
Middle	42.1	42.8	43.2
Matriculation	29.8	30.4	32.6
Higher secondary/intermediate	5.0	5.2	6.0
Degree colleges	1.5	1.1	1.0
Technical & vocational institutes	3.3	3.3	3.4
Universities	0.147	0.161	0.161
Total	241.5	240.9	244.9

Source: GOP. 2015. Ministry of Professional & Technical Training, AEPAM, 2015.

through highly skilled manpower. In Pakistan, the number of educational institutes according to various levels has been shown in Table 9.5 for the years 2012–2013 to 2014–2015.

Table 9.6 depicts the average distance of Mouza from the different levels of educational institutes. It is very interesting to notice that only 15% of the rural population has access to a high school within Mouza or within a radius of 1 km. There is a dire need to minimize this distance so that basic level education can be easily conceivable for the rural masses and farmer's children. Similarly, for skill development programs and training courses, a number of vocational training institutes should also be established at a minimum distance from rural areas.

9.1.4 Telecommunication Facilities

The general importance of telecommunications is well accepted and broadly understood, reflected by its penetration and use. Some of the key areas of impact are:

- Telecommunications provides a technological foundation for societal communications.
- It creates awareness among farmer communities and thus quickens the rate of technology adoption.
- Telecommunications enables participation in the development process.

TABLE 9.6
Distance of Mouzas from Educational Institutes

Type of Educational Institute	Rural Populated Mouzas	Overall Mean Distance (km)	Mouzas by Distance from Educational Facility (km)				
			Less than 1	1–10	11–25	26–50	51 & Above
Primary school	47,482 (100)	10	38,518 (81)	7,251 (15)	994 (2)	464 (1)	255 (1)
Middle school	47,482 (100)	10	14,065 (30)	25,839 (54)	4,912 (10)	1,856 (4)	810 (2)
High/higher secondary school	47,482 (100)	14	7,325 (15)	26,741 (56)	8,243 (17)	3,403 (7)	1,770 (5)
College	47,482 (100)	30	1,924 (4)	12,928 (27)	17,347 (37)	9,785 (21)	5,498 (12)
Vocational center	47,482 (100)	43	756 (2)	10,056 (21)	16,342 (34)	11,632 (24)	8,696 (18)

Source: Pakistan Bureau of Statistics. 2016. Pakistan Mouza Census, 2008. Online available at http://www.pbs.gov.pk/
In parenthesis are percentages.

TABLE 9.7
Distance of Mouzas from Telecommunication Facilities

Type of Facility	Rural Populated Mouzas	Overall Mean Distance (Km)	Mouzas by Distance in from Facility (Km)				
			Less than 1	1–10	11–25	26–50	51 & Above
Fixed line telephone	47,482 (100)	24	9,451 (20)	16,556 (35)	11,722 (25)	5,971 (13)	3,742 (8)
Computer/internet	47,482 (100)	26	5,607 (12)	16,853 (35)	13,611 (29)	7,083 (15)	4,328 (9)
PCO	47,482 (100)	19	16,897 (36)	18,086 (38)	7,009 (15)	3,330 (7)	2,160 (5)
Mobile	47,482 (100)	–	31,707 (66)	–	–	–	–
Wireless phones	47,482 (100)	–	9,715 (20)	–	–	–	–

Source: Pakistan Bureau of Statistics. 2016. Pakistan Mouza Census, 2008. Online available at http://www.pbs.gov.pk/; Punjab Development Statistics 2013. *Bureau of Statistics*. Government of the Punjab, Lahore.

Note: In parenthesis are percentages.

Pakistan Telecommunication Authority (PTA) is the regulatory communications authority, established in 1996. The main service provider is the PTCL with 5 million fixed land lines, 3000 exchanges, and a 2.87 percent tele-density, where rural tele-density is somehow lower than this. There are also five main cellular phone operators. Computer and Internet access is also increasing day by day and plays a very positive role in the rural development. Table 9.7 shows the access of this service to the rural population in the country. The data in the table below is from 2008, but now telecommunication facilities are accessible to each village and most of the rural mass. This is a very quick source of information and farmers can make a lot of farm-level decisions based on the provision of information through these services (Pakistan Telecommunication Authority, 2008).

9.1.5 Electricity

It is the main source of energy in rural areas of Pakistan. Due to limited gas access, the importance of electricity in rural areas is higher than in urban areas. Table 9.8 shows that 35% villages have 100% electricity availability, while others have half or no access. Seeing the energy crises at national level, currently it seems very difficult to obtain 100% availability of electricity in the rural areas of Pakistan. This energy source is crucial for operating electric tube wells, very much important to irrigate crops.

TABLE 9.8
Mouzas Reporting Availability of Electricity

Mouzas Reporting Electricity	Numbers	Percent
All = 100% availability	16,428	35
Mostly = 50% and above	13,789	29
1%–49% availability	8,218	17
No availability	9,047	19
Total Mouzas	47,482	100

Source: Pakistan Bureau of Statistics. 2016. Pakistan Mouza Census, 2008. Online available at http://www.pbs.gov.pk/

9.2 LAND AND LAND RECORD MANAGEMENT SERVICES

The main challenge of development efforts in Pakistan lies in rural areas, which suffer from open poverty accompanied by a number of social, economic, and technological problems. The social problems, for the most part, arise from skewed land distribution patterns which make rural society both inflexible and unjust (Khan, 2011). A number of technological problems also arise from the harshly irregular distribution of land in Pakistan. For example, population pressure on disproportionately scattered lands has forced the continuation of conventional methods of cultivation, and the sovereignty of small cultivation unit sizes and tenancy farming that work in due course to chunk inducements for technological progress. In the same way, many economic problems arise from the failure of the agricultural sector, plagued by unequal land distribution, to supply full employment opportunities and to yield incomes needed for an acceptable living standard to the rural population of the nation (Chaudhary and Herring, 1974).

9.2.1 Land Tenure System

The land tenure system refers to the right of use or right of ownership of land, a system which clearly describes the right of land's holding and its cultivation. Formalization of humankind to land relationship in terms of its use, ownership, distribution, and valuation through land administrative infrastructure is a common way to respond promptly. Experts remained convinced to the wealth potential of land as status symbol both in developed and developing countries (Wallace and Williamson, 2006).

Control and use of land is perhaps the most important factor affecting the allocation of resources and distribution of income. In rural Pakistan, land is the principal form of wealth; class structures and relations are reflected in the land tenure system. Pakistan inherited a system that was characterized by a highly differentiated structure of land interests. A small number of landowners own most of the land that is given to tenants for cultivation on a sharecropping basis. At the time of independence, this landlord–tenant system was dominant in most areas of Sindh and in some parts of Punjab. The other system of cultivation, carried out by the owners themselves, also exists in most parts of Punjab (Sial et al., 2012).

9.2.1.1 Current Structure of the Tenure System in Punjab and Pakistan

In Pakistan, the land tenure system generally is not conducive to progress farming, as cultivators have to give a lion's share of the produce to landlords/owners. The pattern of land tenureship determines the operational status of a farming community. The situations of land occupancy in Pakistan and Punjab during 2010 are shown in Table 9.9.

Table 9.9 reveals that the distribution of land has given rise to massive self-cultivation, as more than 80% of land in Pakistan and also in Punjab is self-cultivated by owners and owner cum tenants, while less than 30% of farmland in Pakistan and Punjab is cultivated by tenants and owner cum tenants.

TABLE 9.9
Numbers of Farms and Farm Area by Tenure in Pakistan and Punjab

Units	Number of Farms (Millions)				Farm Area (Million Acres)			
Classification	Total	Owner Operators	Owner cum Tenants	Tenants	Total	Owner Operators	Owner cum Tenants	Tenants
Pakistan	8.26	6.74	0.6	0.92	52.91	39.43	7.58	5.89
Percentage	100	81.6	7.26	11.14	100	74.5	14.3	11.2
Punjab	5.25	4.32	0.45	0.48	29.32	20.6	5.37	3.35
Percentage	100	82.28	8.57	9.15	100	70.26	18.31	11.43

Source: Pakistan Bureau of Statistics. 2010. Census of Agriculture (2010). Online available at http://www.pbs.gov.pk/.

TABLE 9.10
Numbers of Farms and Farm Area by Size in Pakistan and Punjab

Units	Number of Farms (Millions)				Farm Area (Million Acres)			
Classification	Total	Small	Medium	Large	Total	Small	Medium	Large
Pakistan	8.35	7.52	0.5	0.33	55.59	24.35	8.54	22.7
Percentage	100	90	6	4	100	43.8	15.36	40.84
Punjab	5.45	5.01	0.31	0.13	28.77	16.75	5.07	6.95
Percentage	100	91.9	5.7	2.4	100	58.2	17.6	24.2

Source: Pakistan Bureau of Statistics. 2010. Census of Agriculture (2010). Online available at http://www.pbs.gov.pk/

The distribution of ownership of land is one of the important factors which influence the agricultural productivity and economic well-being of farming communities, and also affects their social and political status in the society. Like all other developing agricultural countries of the world, the distribution of agricultural land is highly skewed in Pakistan.

It is evident from Table 9.10 that land distribution in Pakistan is highly skewed in favor of large farmers; in 2010, the top 4% of farmers owned 40% of farms while the bottom 90% owned less than 44% of farms. Corresponding figures for the Punjab shows that only the top 2.4% of farmers owned more than 24% of the land while the bottom 91.9% owned 58.2% of the land. Widespread poverty and inequitable distribution of farm services and resources are, to a great extent, the results of such a skewed distribution of farmland. The social problems mainly arise from skewed distribution pattern of land ownership which makes the rural society both unyielding and underserved (Raza et al., 2012).

Table 9.11 shows a high rate of increase in the number of farms during 1990–2000 and 2000–2010. This is mainly due to the subdivision of farms as a result of high population growth rate in rural areas.

Table 9.12 shows that the average size of farm in 1960 was 8.78 acres, out of which 6.94 acres were cultivated. In 1972, the farm size increased to 13.06 acres, of which 11.66 acres were cultivated. In, 1980, the farm size decreased to 11.75 acres, of which 10.34 acres were cultivated, and in 1990 farm size again decreased to 9.2 acres, of which 8.3 acres were cultivated. In 2000, the farm size decreased to 7.2 acres, of which 6.6 acres were cultivated, and in 2010 farm size decreased to 5.6 acres, of which 5.1 acres are cultivated.

9.2.2 Land Record Management System

The system of maintenance of land records was initially developed for the purpose of revenue collection. The organized system of written land records was developed after the British occupied Punjab. Land

TABLE 9.11
Total Number of Farms (1960–2010)

Census Year	Total Farms	Change	Index (%)
1960	3,326,217	–	100
1972	2,375,369	−950,848	71.40
1980	2,544,417	169,044	76.50
1990	2,957,453	413,040	88.91
2000	3,864,167	906,714	116.17
2010	5,254,804	1,390,637	157.98

Source: Pakistan Bureau of Statistics. 2010. Census of Agriculture (2010). Online available at http://www.pbs.gov.pk/

TABLE 9.12
Average Farm Size and Cultivated Area, Punjab

Census Year	Number of Farms	Farm Area (Acres)	Ave. Farm Size (Acres)	Ave. Size of Cultivated Area (Acres)
1960	3,326,217	29,313,761	8.78	6.94
1970	2,375,369	31,029,925	13.06	11.66
1980	2,544,413	29,897,882	11.75	10.34
1990	2,957,453	27,204,911	9.2	8.3
2000	3,864,070	27,762,092	7.2	6.6
2010	5,249,804	29,326,443	5.6	5.1
Overtime change 1960–2010	+1,923,587	+12,682	−3.18	−1.84

Source: PBS, 2010 and Author's own calculations.

revenue was a major source of income for the state and maintenance of current land records was a prerequisite for that. Although land revenue has been abolished, the system of maintenance of land records still persists as an essential function of the government. Various attempts have been made to modernize the system, since the 1970s, through introduction of computer technology. The project of Land Records Management and Information Systems is being attempted to build on the successes of the previous attempts, learning from the mistakes and failures of these attempts.

The government of the Punjab has started the computerization of land records with overall objectives to improve service delivery and to enhance the perceived level of tenure security. A Project Management Unit has been set up under the administrative control of the Board of Revenue, Government of the Punjab. The World Bank has agreed to finance the project rollout in 18 Districts of Punjab within the next five years. The project vision is to establish an efficient, accountable, equitable, and secure land record management and information systems. The ultimate vision of the government of the Punjab is to gradually move towards a land titling system (GOP, 2016).

9.2.2.1 Board of Revenue

The Board of Revenue is the successor of the Office of the Financial Commissioner. It was originally constituted under the provisions of West Pakistan Board of Revenue Act, 1957, which on dissolution of One Unit in 1970 became the board of revenue, Punjab. The board is the controlling authority in all matters connected with the administration of land and the collection of government dues including land taxes, land revenue, preparation of land records, and other matters relating thereto. The board is the custodian of the rights of landholders and is the highest revenue court in the province with appellate/provisional jurisdiction against orders of subordinate revenue officers/courts including commissioners and collectors. All revenue officers and revenue courts are subject to the general superintendence and control of the Board of Revenue. The board itself is subject to the administrative control of the provincial government.

9.2.3 Way Forward

- Small farms not only improved land utilization and are more efficient, but also maintained livestock without affecting their cropped acreage. This category of small farms should be supported. They should be facilitated with credit, hired tractors, and subsidized inputs, etc.
- Small farms should be upgraded through redistribution of resumed and undistributed lands, at least up to a subsistence level.

- Although all tenancy categories were equally efficient, the provisions for protection tenants' rights given in the provincial tenancy Acts and Land Reform Regulations should continue to be implemented strictly.

9.3 AGRICULTURAL EDUCATION AND RESEARCH INSTITUTES

9.3.1 Current Scenario of Agricultural Educational Institutes

Agricultural education in Pakistan has been continuously upgrading. Being an agricultural economy, major portions of the GDP comes from agriculture and it is considered a supporting hand for the economy of Pakistan. In such an intensive agricultural setup, agricultural education in Pakistan has not been given very much importance. Compared to new technological breakthrough, inventions, increase in crop and livestock intensity, and diversification in agriculture, rate of growth in skilled persons and agricultural graduates is not compatible. Modern agricultural practices are increasingly turning out to be knowledge-based and hence gaining expertise in them is not an easy task for many of our farmers. Keeping in view that increased productivity must be the sole objective of agricultural research, our scientific community is leaving no stone unturned for bringing about paradigm changes in agricultural education in the country. There are major agricultural universities in Faisalabad, Lahore, Multan, Rawalpindi, Dera Ghazi Khan, Peshawar, Dera Ismail Khan, Tandojam, and Lasbella. All these universities have been continuously providing trained agricultural graduates. These graduates are working in research organizations, NGO's, universities, field jobs, and in banks. As stated earlier, the research and development in agriculture is not as quick as it was supposed to be, particularly considering the scope and need of agriculture in Pakistan.

At present, there is a constant rise in enrollment among agricultural universities. This trend is basically due to greater awareness among the masses regarding the role of agriculture and its future perspectives. A decade before, only students with low scores and from rural backgrounds chose agriculture as a profession. Nowadays the picture is totally different. Agriculture has become a subject opted for by students from urban settings as well. The funding in agricultural research is increasing. More and more foreign scholarships and grants are being awarded to agricultural graduates. The outlook of agriculture has changed by the integration of many other modern fields of learning in agriculture. Jobs are also on a gradual rise. There are two agricultural universities in top seven universities of Pakistan according to latest HEC ranking report. These are: University of Agriculture Faisalabad (UAF) and Peer Mehr Ali Shah Arid Agricultural University Rawalpindi (PMAS AAUR). University of Agriculture, Faisalabad is by far the largest agricultural university in Pakistan; it operates the Division of Education and Extension and the Water Management Research Centre along with six faculties. University of Agriculture, Faisalabad is the only Pakistani institution to be ranked among the top 100 global universities in the field of agricultural science. As of 2015, the university employed 450 PhD-qualified scientists (in headcounts), representing a substantial increase over the 2010 level of just 150. Over the next 10 years, UAF is expected to graduate an additional 100 PhD-qualified scientists per year, most of them will only be in their thirties. Significant numbers of PhD-qualified scientists will also graduate from other universities in Pakistan.

Table 9.13 shows the number of agricultural institutions in provinces, at the federal level, and Azaad Jammu Kashmir (AJK) from the 60s until to day. The total number of agricultural institutions has increased over time. Punjab shows somewhat additional growth in the establishment of agricultural institutions. Table 9.14 highlights the top agricultural institutions present in Pakistan.

9.3.1.1 Main Services Provided by Agricultural Educational Institutes in Pakistan

These agricultural educational institutes provide a platform for the most effective interaction between the farming communities and are considered as an affective operative source of human development. More than 85,000 graduates have been qualified in Pakistan up to 2015. Well-established extension and advisory services are available to farmers for assistance at their

TABLE 9.13
Agricultural Educational Institutes in Pakistan

Year	1960	1970	1980	1990	2000	2010	2015
Punjab	1	2	3	3	6	10	12
Sindh	–	1	1	1	2	2	3
KPK	–	1	1	2	2	2	2
Baluchistan	–	1	1	1	1	2	2
Federal	1	1	–	–	1	2	2
AJK	–	–	–	–	–	–	1

Source: HEC 2015. *5th Ranking of. Pakistani Higher Education Institutions*. HEIs.

TABLE 9.14
Top Agricultural Educational Institutes in Pakistan

University of Agriculture, Faisalabad, Punjab
Pir Mehr Ali Shah Arid Agriculture University, Rawalpindi, Punjab
University of Veterinary and Animal Sciences, Lahore, Punjab
Agricultural University, Peshawar
Sindh Agriculture University, Tandojam
Lasbela University of Agriculture, Water and Marine Sciences
Muhammad Nawaz Shareef University of Agriculture, Multan

Source: HEC 2015. *5th Ranking of. Pakistani Higher Education Institutions*. HEIs.

doorstep and acquaint the farming community with modern technology for higher production on a sustainable basis. These institutes also conduct farmers training and field days periodically. They also arrange on regular basis farmers' festivals, trainings, workshops, and seminars, and so on to bridge the gaps between educational institutes and farming communities. Research and development, diffusion and adoption of modern technologies, and developing linkages among farming communities are also the priorities of these institutes.

These institutes also play an important role in crop maximization, agriculture sustainability through fellow farmers, empowerment through capacity building, demonstration trials at field level, development of farmer groups, good communication skills, agricultural information services, and gender development through women extension services. Some of these institutes develop most effective interactions with farmers and other stakeholders with its alumni through the Farmers Syndicate Hall. They organize different training courses for in-service governmental and nongovernmental employees, that is, nominees of the Pakistan Army (both at commissioned and noncommissioned levels), banks, NGOs, and other agencies responsible for agricultural and rural development.

These institutes have various national and foreign-funded research projects on food security, mobile apps, gender empowerment, communication, rural development, and cyber extension, which ultimately facilitate the farmers. They are operating Business Incubation Centers (BICs) in order to increase the role of the private sector in agriculture. BICs accelerate the successful development of startup and fledgling companies by providing entrepreneurs with an array of targeted resources and services.

9.3.2 Current Scenario of Agricultural Research Institutes in Pakistan

In Pakistan, 209 agricultural research institutes are conducting research, excluding the private sector. Pakistan Agricultural Research Council (PARC) is the country's principal agricultural research and development agency. It has a wide-ranging mandate to coordinate research among all agencies (federal, provincial, and higher education). This institute has 12 research subinstitutes; National Agricultural Research Center (NARC) is the largest one among these. PARC accounts for 13% of the country's total agricultural research capacity. Punjab, the largest province, accounts for 1145 Full Time Equivalents (FTEs) researchers at the Punjab Agricultural Research Board (PARB)—a provincial body that guides research planning and resource allocation, and the Ayub Agricultural Research Institute (AARI), which manages 28 crop-related research institutes and units employing half the province's agricultural researchers. Similarly, provincial agencies in Baluchistan, KPK, and Sindh are employing 242, 327, and 377 FTEs researchers, respectively. The higher education sector accounted for 15% of the nation's agricultural research capacity in 2012, representing a substantial increase over levels in the early 1990s. A number of private companies are also engaged in the research process and in providing services to farmers. As active breeding programs focused in areas of Bt cotton, hybrid maize, vegetables, and several other crops have been operating in the country (Table 9.15). Different diversified information regarding these research institutes have been provided in Table 9.16.

9.3.2.1 Main Services Provided by Agricultural Research Institutes in Pakistan

These research institutes provide services to farmers relating to issues of crop production technologies, fertilizer usage, biofertilizer, soil, water, pesticide, and weedicide use. The institutes provide training at research centers and at the farms relating to kitchen gardening, tunnel farming, plant clinics, optimal use of fertilizers, and monitoring of inputs of preservice and inservice for farming communities.

- The institutes are continuously working on the development of production technologies of different crops as well as research on water resources.
- Different research institutes have established Farmer Field Schools (FFS) in which farmers are being trained regarding various technologies/interventions.
- Services related to supply of seed kits, seed quality testing, seed production, procurement, processing, export, sales, and marketing have also been provided by these institutes.
- These institutes also provide different services to farming community in the form of water user's associations, improvement of watercourses, different irrigation schemes, precision land leveling, High Efficiency Irrigation System (HEIS), soil and water conservation, and renewable energy technology.
- Research organizations are providing training to farmers, extension workers, and pesticides dealers for pest scouting, quality control of pesticides, registration of pesticides distributors, and standardization of pesticides.

TABLE 9.15
Research Focus of Agricultural Research Institutes (Percent)

Major Crops	50	Livestock	18
Wheat	19	Natural Resources	3
Fruits	12	Fisheries	4
Cotton	10	Forestry	1
Rice	9	Other	15
Vegetables	9		

Source: ASTI-IFPRI 2015. Agricultural research and development indicator factsheet of Pakistan. *Agricultural Science and Technology Indicators*. November 2015.

TABLE 9.16
Key Indicators of Pakistan Agriculture Research and Development 2000–2015

Key Indicators			2000	2009	2015
			Total Agricultural Research Spending		
PKR (million constant 2011 prices)			5,737.2	7534.6	8120.2
PPP dollars (million constant 2011 prices)			236.2	309.5	333.6
Growth (%)				31	8
			Total Number of Agricultural Researchers		
Full Time Equivalents (FTEs)			3453.7	3555.3	3678.3
Growth (%)				3	3
			Agricultural Research Intensity		
Spending as a Share of Agricultural GDP			0.20%	0.19%	0.18%
FTEs/100,000 farmers			18.47	14.79	14.43
Financial Resources (based on federal and provincial agencies only)					
Spending Allocation			**Salaries**	**Operating Costs**	**Capital Investments**
Percentage			78	18	3
			Allocation of Govt. agencies by Cost (Percent)		
PARC			87	13	0
NARC			81	16	3
Other Federal			72	25	3
Baluchistan			92	8	0
KPK			86	11	2
Punjab			72	22	5
Sindh			72	20	8
Autonomous Territories			98	2	0
Research Capacity (%)	**NARC**	**PARC**	**Other Federal**	**Province**	**Higher Education**
	7	6	14	58	15
Total Government Agencies (No.)			**Federal**	**Provincial**	**Higher Education**
			64	114	31
Researcher Profile (%)				**Male**	**Female**
				88	12
Number by Qualification (FTEs)			**PhD**	**MSc**	**Bsc**
			761.6	2492.3	424.3

Source: ASTI-IFPRI 2015. Agricultural research and development indicator factsheet of Pakistan. *Agricultural Science and Technology Indicators*. November 2015.

Note: Research conducted by private sector is not included due to unavailability of data.

- Farm advisory services and services related to funds, commercialization, agricultural marketing information, price control mechanisms, crop estimates, modern agricultural machinery, good governance initiatives, installation of digital rate board, optimal utilization of land, cost of production, and harvest prices are being given to farmers and other stakeholders.
- Nowadays, these research institutes use print and electronic media for technology transfer. Free advisory services are available to flower growers on production technologies of commercial floriculture crops, training on floriculture, landscaping, and ornamental plants of all types cultivated for sale on economical rates.

- Free distribution of seeds of major crops and vegetables for increasing production to promote kitchen gardening for farming communities.

9.3.3 Limitation/Constraints

- Despite various policy reforms introduced over the past decade, Pakistan remains vulnerable to food insecurity; agricultural research and development investment remains too low to address this challenge—in the last five years Pakistan has invested just 0.18% of its agricultural GDP in agricultural research and development—considerably less than its South Asian neighbors and only a fraction of the internationally proclaimed target of at least 1%.
- Pakistan's capacity to deliver effective agricultural research outputs is hindered by critical human resource challenges. Long-term recruitment restrictions have left many federal and provincial research agencies with aging pools of researchers. Given the official retirement age of 60 years, large-scale capacity losses are imminent in the coming years.
- Moreover, limited opportunities for promotion and training, as well as a lack of performance-based merit systems, constitute key impediments to staff motivation.
- Average researcher qualifications at the provincial research and development agencies are considerably lower than those of researchers working for the federal government and higher education agencies because the average salary levels in the provinces are considerably lower. This, in addition to restricted recruitment and training, as well as lack of performance-based incentives makes the provincial agencies less attractive as employers.
- The subject of agriculture at school level is lacking.
- There is limited access of educational and research institutes to small farmers.

9.3.4 Way Forward

- To address the country's agricultural productivity challenges, Pakistan needs to increase its investments in strategic agricultural research areas.
- The government needs to clearly define its long-term research and development priorities and secure sustained funding, not only in support of salary-related expenditures, but also to cover day-to-day costs of operating research programs.
- More creative mechanisms also need to be explored to stimulate private funding for agricultural research and development.
- It remains critical, however, that investments be made in staff training in order to counteract the imminent loss of senior researchers to retirement.
- In addition, the government should create and monitor performance standards for federal and provincial researchers and introduce a system of performance-based promotions and salary levels.
- Government should increase the number of laboratories in the country, farm machinery per acre basis—too low compared to neighboring countries—and trained extension workers, FAs, and researchers with up to date knowledge.

9.4 EXTENSION SERVICES

Agricultural extension is believed to be an indispensable pillar both for the progress of rural communities and as a part of a strategy of agricultural development to improve the sustainability of farming systems, promote agricultural diversification, and integrate farmers into dynamic markets. Agricultural extensions system play multipurpose role with the provision of need-based and demand-based knowledge. They provide agronomic techniques in a systematic fashion to improve production, income, the rural populations' welfare, and mitigate their problems.

9.4.1 Extension Infrastructure in Pakistan

In a global context, the agricultural services are facing new challenges regarding increasing demands for food, declining cultivated areas, and fiscal constraints in the public sector. International organizations and donor agencies have suggested the governments of developing countries reform and modify their existing public sector structures with purpose-specific and need-specific approaches (Umali and Schwartz, 1994). This shifting of the public agricultural extension philosophy reflects an all-inclusive tendency towards privatization (Dancey, 1993).

Agriculture extension in Pakistan was a solely public funded service for almost 40 years, from the period since independence in 1947 until 1988. During this period, successive governments experimented with several different models and policies of extension services ,with the view to increase their efficiency, achieving only limited success.

In Pakistan, numbers of public extension approaches and models have been tried and discarded based on the traditional linear approach. Mostly multisectoral extension programs focused on rural and community development. Examples of these were: The Village Cooperative Movement (early 50's); Village Agricultural and Industrial Development Program (Village-AID Program, started in1952); Basic Democracies System (BDS, 1959–1970); Rural Works Program (RWP, 1963–1972); People's Works Program (PWP, 1972–1975); Integrated Rural Development Program (IRDP, 1970–1978); Inputs at Farmers' Doorsteps Approach (1970–1978); and the Training and Visit system (TandV, 1980–1994).

The ratio between field-assistants and farmers under the traditional extension system was 1:1000. In the 1980s, the training and visit program was introduced in Punjab and Sindh with the assistance of the World Bank. Under this program, the field-assistant to farmer ratio was reduced to 1:650. The agriculture extension service established under the TandV system has gradually weakened during the1990's. There were no proper facilities for regular backup training of the staff; funds for traveling and daily allowances were drastically reduced and adaptive research farms discontinued. Despite such poor developments, the extension staff kept on maintaining limited contacts with farmers, organize field days and field seminars. Some other programs such as the Agriculture Development Corporation, Barani Area Development Program, Crop Maximization Program, Integrated Rural Development Program, Integrated Pest Management Program, Farmers' Field School Approach, Productivity Enhancement Model Sargodha, and Electronic media in agricultural knowledge transfer remained operative

In 1988, a new experiment was tried to allow the private sector and especially the input supply companies to initiate their own advisory services along with their products delivery system. The idea was to increase the impetus for commercialization of agriculture in order to attain food security. The government of Pakistan appointed a commission to look into its poor agricultural performance. The commission suggested the inclusion of the private sector in reshaping agricultural extensions: "The most important shift needed is to encourage the private sector to provide the total package of services (specialized cultivation operations, spraying, and harvesting) and not just the sale of specific inputs." Since this suggestion, these advisory service have become a regular provision for most leading companies and a rough estimate suggests that almost 70%–80% of advisory services are now being provided by private companies, which have established their own independent structures to provide advisory services.

The private sector concentrates more on the needs of large farmers and is enthusiastic to participate in practices which encourage profit maximization. By contrast, public extensions focus on small farmers in order to increase their socioeconomic condition with a more educationally oriented role, along with providing support to strengthen their capacity building in order to improve their livelihood (Saravanan, 2001).

Public and private Extension Field Staff (EFS) identified difficulties they faced in transfer of technology. Majority of EFS faced lack of transport facility in large areas of their jurisdiction. Private EFS also faced active geographical mobility in their respective areas. Large areas of jurisdiction and poor transportation are always the cry of public and private extension field staff, which is confirmed by various studies. As a result, EFS do not pay regular visits to farmer's field and thus are unable

TABLE 9.17
Agricultural Extension Infrastructure in Pakistan

Province	Agriculture Officer	Field Assistant
Punjab	763	3264
Sindh	573	1026
Balochistan	586	1016
Khyber Pakhtunkhwa	222	539
AJK, GB, FATA, ICT	180	673
Total	2324	6518

to diffuse new technologies (Mirani and Memon, 2011). Table 9.17 shows the infrastructure of agricultural extension in four provinces and in AJK, GB, and FATA area.

9.4.2 Problems of Agricultural Extension in Pakistan

Agricultural extension faces two foremost challenges. First, it is assumed that information and organization in the agriculture sector do not receive the due importance they deserve, whereas they clearly have higher significance and greater importance.

Extension staff are not well equipped with the appropriate skills to fulfill their functions efficiently. Therefore, it is important that the skills of extension agents be improved, their working knowledge updated, and that they are able to generate innovative ideas in order to develop agricultural practices which can meet complex demand patterns, reduce poverty, and enhance ecological resources.

Agricultural extension departments are faced with a number of challenges and constraints which are common to many extension institutions in developing countries. These include: inadequate finances and funding, lack of qualified and trained extension staff; poor, weak, and deteriorated infrastructure; lack or weak coordination mechanisms and functional linkages with the other institutions in both the public and private sectors; absence of quality control and impact assessment mechanisms; unclear extension mandates and lack of job descriptions by staff (FAO, 2005).

9.4.3 Way Forward

Pakistan needs to formulate a vibrant national agricultural extension policy that could gain and ensure political support and commitments for funding. Both farming communities and the civil society undervalue extension professionals. It is very important to value and respect the extension professionals to enable them to work with the missionary spirit and dedication they require. To make preservice education more capable to cater to the needs of the farmers, radical changes are required in the curriculum to accommodate the changes happening in extension reforms. Extension organizations have too many superiors and bosses; however, such a problem can be checked by introducing decentralization, and launching extensive capacity building programs for the decentralized units, without politicizing the process. Diversity in extension services needs to be promoted, which can be achieved by involving both public and private institutions. Farmers could be empowered through the formation of farmers' groups. By doing so, they can help themselves and watch after their interests and benefits; therefore, they can create an effective lobby for getting effective extension services. Extension should decide on a privatization option only where and when it is socially and economically practicable, viable, and feasible. The extension services certainly can have a constructive role for all the communities, particularly in postwar, postdisaster, and epidemic situations. The mandate of extension services needs to be broadened to include the development of rural human resources. The development and application of information technology tools can certainly facilitate the extension work; however, this development must not be viewed and taken as the replacement of extension services (Baig and Aldosari, 2013).

9.5 FARM MECHANIZATION SERVICES

9.5.1 History of Mechanization in Pakistan

Mechanization refers to the replacement of animal and human power with machines. The use of mechanical power in the agriculture sector of Pakistan was initially used in early 1950s for groundwater pumping to irrigate agricultural fields, using private tube wells. The progress of tube well installation in the 1950s, however, was slow despite of the full decade (1959–1969) of development; their number did not exceed 4200 by 1960. After 1959–1960, private tube wells gained momentum and the recorded number of tube wells reached 25,000 by 1964. Furthermore, the seed-fertilizer revolution and increasing number of tube wells promoted the use of tractors and tractor-driven tillage tools in mid-1960s. Unfortunately, the mechanization received bitter opposition of Planning and development (P&D) bureaucrats, agricultural economists, and policy makers. They were of the view that agricultural machinery would replace labor and thus create unemployment in the country. However, after the mid-1960s, this proved to be a mere apprehension and the use of tractors made headway due to:

a. The tube wells and seed-fertilizer revolution doubled labor requirement in the agriculture sector, resulting in unprecedented labor shortages during peak seasons.
b. Increased cropping intensities, with the use of tractors and tillage implements, compared with animal and human power sources.
c. As a consequence of these developments, bullock's prices, wages, and opportunity cost of feeding bullocks rose tremendously.

Therefore, the migration of rural people from villages to urban areas in search of jobs (public or private) and establishment of their business created labor shortages in the agriculture sector. These factors were sufficient to convince the farming community to adopt the use of tractors and related equipment in order to alleviate power constraints and to keep costs low. The historical progression of tractors in the agriculture sector of Pakistan is shown in Figure 9.2. The figure clearly shows that tractor numbers increased rapidly after 1975 (Chaudhary et al., 1998).

The number of tractors (privately owned and government owned) by type of ownership in different administrative units of the country are shown in Table 9.18. The table clearly depicts that the number of privately owned tractors is much higher than government owned tractors. The maximum number of tractors is reported in Punjab Province. Similarly, the number of tractors, by horsepower, found in the different administrative units of the country is shown in Table 9.19. The table shows that the maximum number of tractors, in all administrative units of the country, are in the range of 45–55 hp tractors. It is worth mentioning here that since most of the farms are less than 10 hectares, therefore,

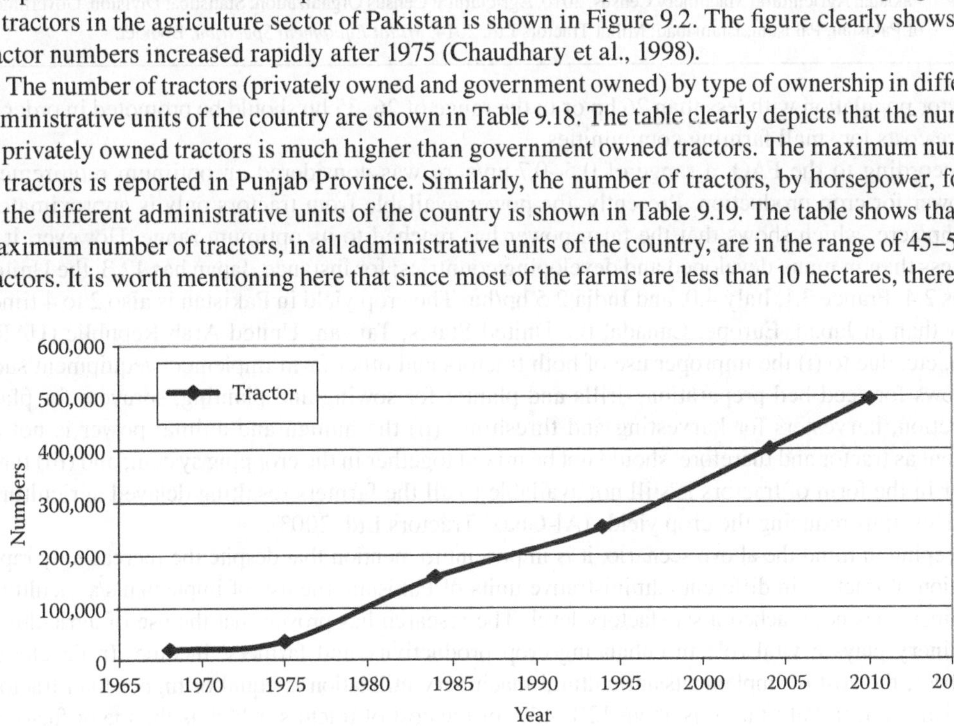

FIGURE 9.2 Historical perspective of tractors in agriculture sector of Pakistan.

TABLE 9.18
Number of Tractors and Tube Wells by Type of Ownership

	No. of Tractors			No. of Tube Wells		
Administrative Unit	**Privately Owned**	**Government Owned**	**Total**	**Privately Owned**	**Government Owned**	**Total**
Punjab	331,273	632	331,905	788,117	40,666	828,805
Sindh	36,082	163	36,245	37,213	1011	38,232
KPK	23,967	302	34,269	16,988	1426	18,418
Balochistan	9124	120	9244	17,625	1606	19,233
Pakistan	400,446	1217	401,663	859,943	44,709	904,688

Source: Pakistan Agricultural Machinery Census. 2010. Agricultural Census Organization, Statistical Division, Government of Pakistan, Pakistan, Islamabad.

TABLE 9.19
Number of Private Tractors by Horsepower

Administrative Unit	Total Tractors	<26 hp	26–35	36–45	45–55	56–65	66 and Above
Punjab	331,273	85	405	16,442	255,158	18,434	18,434
Sindh	36,082	4	19	255	28,325	6,428	1,047
KPK	23,967	282	13	388	18,780	3,980	472
Balochistan	9,124	–	11	300	8261	348	196
Pakistan	400,446	371	448	17,385	310,524	50,971	20,149

Source: Pakistan Agricultural Machinery Census. 2010. Agricultural Census Organization, Statistical Division, Government of Pakistan, Pakistan, Islamabad; Millat Tractors Ltd. 2014. *Millat Equipment Spectrum*. Booklet.

a tractor population with less than 26 hp or in the range of 26–35 hp should be promoted in order to reduce costs for small farming communities.

According to the FAO, a range of 0.5–0.7 hp/acre was considered an optimum requirement of power for crop production. Presently, the power available from tractors only is approximately 0.58 hp/acre, which shows that the farm power has reached to its optimum range. However, it is still less than in many developed and developing countries; for instance, Japan has 12.3, the United States 2.4, France 3.1, Italy 4.0, and India 2.5 hp/ha. The crop yield in Pakistan is also 2 to 4 times lower than in Japan, Europe, Canada, the United States, Taiwan, United Arab Republic (UAR), India, etc. due to (i) the improper use of both tractors and other farm implements/equipment such as plows for seed bed preparation, drills and planter for sowing and planting, sprayers for plant protection, harvesters for harvesting and threshing, (ii) the human and animal power is not as efficient as tractor and therefore, should not be mixed together in the cropping system, and (iii) farm power in the form of tractors is still not available to all the farmers resulting delayed agricultural practices, thus reducing the crop yields (Al-Ghazi Tractors Ltd, 2003).

Keeping in mind the above scenario, it is important to mention that despite the increasingly rapid adoption of tractors in different administrative units of Pakistan, the use of implements/agricultural machinery has not reached a satisfactory level. The research has proven that the use of agricultural machinery plays a vital role in enhancing crop productivity and farmers' income. In developed countries, the cost of implements/agricultural machinery utilization is equal or higher than tractors cost. However, in Pakistan, it is about 12%–15% of the cost of tractors, which is the major factor in reduction of crop yield. Another important factor is the lack of knowledge and information available to the farming community about the selection and use of agricultural machinery (Hunt and Wilson, 2016).

TABLE 9.20
Status and Trend of Selected Implements and Farm Machinery of Pakistan

Implements/ Machinery	Year 1984 Total in Pak.	Year 1984 Punjab	Year 1994 Total in Pak.	Year 1994 Punjab	Year 2004 Total in Pak.	Year 2004 Punjab	Year 2010 Total in Pak.	Year 2010 Punjab
Cultivator	146,863	123,755	236,272	203,444	369,866	317,506	512,067	445,276
MB plow	7,319	2,780	28,413	17,980	40,050	27,093	42,477	36,937
Disc Harrow	8,140	2,734	13,233	8,302	23,764	16,032	25,663	221,24
Ridger	4,711	4,030	10,984	10,872	71,338	66,806	192,167	173121
Drill	11,251	10,669	64,126	60835	708,10	66,700	295,184	251,112
Trolley	98,787	81,668	176,412	145,557	242,655	195,332	28,6041	248,732
Thresher	78,377	71,195	112,707	96,655	137,270	122,737	353,768	265,546
Reaper	–	–	8,073	–	13,600	12,528	66,958	58,099
Combine harvester	–	–	359	–	3,355	2,899	29,344	21,369
Chisel plow	712	–	6,535	–	8,514	6,719	–	–
Rotavator	2,101	–	5,594	–	47,919	44,192	210,394	196,234
Blades	69,004	–	164,489	–	233,126	189,965	315,423	285,645
Sprayers	–	–	20,778	–	21,756	20,976	1,438,991	1,121,110
Laser land leveler	–	–	–	–	2,785	1,692	7,756	4,843

Source: Pakistan Agricultural Machinery Census. 2010. Agricultural Census Organization, Statistical Division, Government of Pakistan, Pakistan, Islamabad;

9.5.2 Farm Machinery Statistics

The status and trends of selected farm machinery in Pakistan, during last few decades, is shown in Table 9.20. The table shows that the rate of agricultural machinery adoption is slower compared to the adoptability of tractors. It is therefore, extremely important to enrich the agriculture at par with tractors population especially by promoting the use of latest available agricultural machinery (Culpin, 2000; Dixit et al., 2014).

9.5.3 Way Forward

The following strategies are proposed for adaption of full mechanization in the country.

- Agricultural extension departments of different provinces can help in introducing the different agricultural implements/machines, offering alternatives from the commonly used manual and conventional agricultural tools. This can be accomplished by appointing assistant engineers/scientists at Tehsil level like agricultural engineers, agronomists, soil scientists, plant protection officers, and so on in order to popularize the recommended machinery to perform different agricultural practices.
- Provincial agricultural engineering departments should cooperate to demonstrate the recommended agricultural implements by setting up demonstration plots at Union Council level. This would help the farmers in appreciating their use and benefits and consequently adopt them without any hesitation.
- Standardized local design and development of different low-cost agricultural machinery should be made available and encouraged.
- Machinery pools should be established at Tehsil level, fully equipped with all kind of power sources and agricultural implements/machines at low rental rates for farmers.

- Farmers may be advised to select and purchase selective agricultural machinery by alternate farmers and promote mutual exchanges to mechanize their farms economically.
- The media should play a positive role to promote mechanization through popular talks, talk shows with agricultural specialists, and programs to demonstration different types of machines under actual field conditions.
- The government should announce subsidies to attract the farming community to adopt and promote mechanization.
- The private sector should be encouraged to establish agricultural machinery workshops at each Tehsil and provide the necessary machines to the farmers at low rents.
- Pakistan Agricultural Machinery and Implements Manufacturers Association (PAMIMA) needs to be encouraged to play its due role of upgrading manufacturers premise facilities, creating their own research and development, and producing quality products at competitive prices to meet WTO challenges.
- Use of renewable energy sources, for example, biogas and biomass should be promoted at farms to decrease the costs of operation. For this purpose, biogas-operated tube wells using 25 and 40 m^3 floating drum or fixed dome biogas plants can be installed to operate 0.75 cusec tube wells. Similarly, solar energy can be used for drying fruits, vegetables, medicinal plants, and cereal crops using hybrid solar dryers. Biomass gasification units can be installed for engine powered farm electrification.

9.6 CREDIT SERVICES

In Pakistan, a majority of farmers (i.e., 90%) are small farmers having less than 5 hectares of land. These farmers are caught up in the vicious circle of low productivity and poverty. To break this vicious circle we have to modernize our agriculture which needs additional investment. Additional investment either comes through savings or by borrowing. The savings rate in Pakistan is very low, so the only option left to the farming community is borrowing. Farmers borrow funds to meet their social, consumption, and production needs. They borrow both from institutional and noninstitutional sources. In Pakistan, institutional lending is done against collateral, which is difficult to manage, especially by small farmers. That is why a majority of small farmers took loans from noninstitutional sources where the cost of credit was very high. According to a Pakistan rural credit survey in 1985, 63% of small farmer loans were from noninstitutional sources.

9.6.1 Credit Disbursed by Source

There are two major sources of agricultural credit in Pakistan; formal/institutional and informal/noninstitutional. Formal sources are comprised of financial institutions like cooperative banks, privatized, and nationalized commercial banks such as United Bank Limited (UBL), Muslim Commercial Bank (MCB), Allied Bank Limited (ABL), Habib Bank Limited (HBL), National Bank of Pakistan (NBP), and the Taccavi credits and Zarai Taraqiti Bank Limited (ZTBL). Informal sources include friends, neighbors, relatives, commission agents, professional moneylenders, and traders. Nowadays, a number of NGOs are also providing agricultural credit to rural communities (Iqbal et al., 2003).

9.6.1.1 Government/Formal

Institutional credit disbursed by various sources in Pakistan is presented in Table 9.21. As indicated in the table, the average annual amount of credit disbursed was Rs. 103 and 51,5875 million for the year 1960–1961 and 2014–1915, respectively, showing an increase of 4970 times. But if we look at the credit availability per hectare for the same time period these figures are Rs. 7 and 22,700, respectively, that are for less the credit needs of the farmers. The share of commercial banks in agricultural credit disbursement was zero until 1970–1971, but now this share has increased to 50%. Islamic banks and microfinance banks are positive additions to the credit business. Numbers of

TABLE 9.21
Credit Disbursed by Agencies (Rs. Million), Total Cropped Area (Million Hectares), and Loan Amount (Rs/Hectare)

Years	ZTBL	Taccavi Banks	Cooperatives	Commercial Banks	Islamic, MFBs[a], DPB[b]	Total	Total Cropped Area	Loan Amount	Commercial Bank
1960–1961	30.9	15	58	0	0	104	15	7	0
1970–1971	92.7	10	55	0	0	158	17	10	0
1980–1981	1,066	9	1,128	1,816	0	4,020	19	208	0.45
1990–1991	8,323	56	3,017	3,518	0	14,221	22	652	0.25
2000–2001	27,610	0	5,124	12,056	0	44,790	22	2,032	0.27
2010–2011	65,361	0	7,162	140,312	50,187	263,022	23	11,577	0.53
2014–2015	95,826	0	10,486	262,912	146,650	515,875	23	22,696	0.51

Source: State bank of Pakistan, 2015; Pakistan Bureau of Statistics. 2010. Census of Agriculture (2010). Online available at http://www.pbs.gov.pk/

[a] Micro Finance Banks.

[b] Domestic Private Banks.

NGOs are also providing loans to the farming community and their role must be appreciated and highlighted. Newly entered organization should be facilitated so that small farming community can easily reach that are located in far long areas of Pakistan.

Unfortunately, this sector is not getting its due share in input use; for example, the total credit disbursed to the private sector in 2015 and 2016 were Rs. 3608 and 3915 billion, respectively. Out of this, agricultural credit was Rs. 267 and 284 billion, respectively. In percentage term it is only about 7.5% of the total, which is far less than its due share of 20%.

One major problem with institutional credit is that the funds, which are being allocated for agriculture, are not being disbursed. As indicated in Table 9.22, the total funds allocated for agriculture in 2014–2015 and 2015–2016 were Rs. 500 and 600 billion, respectively, while the released amount was only 326 and 385 billion, that is, about 64% of the total allocation.

TABLE 9.22
Indicative Targets and Actual Disbursement of Agriculture Loans (Rs. Billion)

	2014–2015				2015–2016			
Banks	Target 2014–2015	Disbursed	Target Achieved (%)	% Share in Total Disbursed	Target 2015–2016	Disbursed	Target Achieved (%)	% Share in Total Disbursed
Major commercial banks	252.5	167.4	66.3	51.4	305.7	198.8	65	51.6
ZTBL	90	56.2	62.4	17.2	102	55.3	54.2	14.3
DPBs	115.5	72.1	62.4	22.1	131.8	84.8	64.4	22
PPCBL	11.5	5.8	50.9	1.8	12.5	6.1	48.8	1.6
MFBs (9)	28.2	20.7	73.6	6.4	40.1	34.5	86	8.9
Islamic banks (5)	2.3	3.8	134.7	1.2	7.9	6	75.9	1.6
Total	500	326	65.2	100	600	385.5	64.3	100

Source: State bank of Pakistan, 2015.

ZTBL: Zarai Taraqiati Bank Limited: DPBs: Domestic Priavte Banks.

PPCBL: Punjab Provincial Corporative Bank Limited: Commercial Banks: Include ABL, HBL, MCB, NBP, and UBL.

MFBs: 9 Micro Finance Banks: 5 Islamic Banks.

9.6.1.2 Private/Informal

Informal sources are insufficient and nonreliable. No comprehensive data are accessible on the quantity of credit provided by informal credit sources. As such it is difficult to find out solid proof regarding the relative share of these sources as part of the total credit supply. However, some rough estimates and few reports illustrate that the formal credit sources have been able to meet only 50% of the total credit supplies of the farm sector and it may be assumed that the remaining share is met by informal sources of credit (Irfan et al., 1999).

9.6.2 Credit Utilization by Purpose

The composition of loans advanced by ZTBL/ADBP in terms of short-term (production) and medium-term loan (machinery) is presented in Table 9.23. As seen from the table, a major change has taken place in the composition of loan as only 18% of loans were allocated for seed and fertilizer purposes until 1980, while this figure increased to 54% in the year 2001. On the other hand, loans for tube wells and tractors, over the same time period, were 65% and 27%, respectively.

9.6.3 Credit Utilization by Farm Size

According to the census of agriculture of 2010, there were 8.34 million farms in Pakistan, out of them 7.45 million, or 90%, were small farmers. Small farmers did not get their due share of total credits disbursed in Pakistan. As mentioned earlier, institutional credit is linked with guarantees, which are difficult to manage by small farmers. In Pakistan pass book issued to the farmer (being a proof of collateral), difficult to get. According to the credit finance survey of 1985, only 9% of the farming community had a pass book. The pass book percentage for small farmers having less than 5 hectares of land varied from 0.5% to 12% only while this percentage for large farmers, having lands of 20 and 60 hectares, were 37 and 49%, respectively. That is why the major share of credit goes to the large farmers. In 2010, 1.08 million farms obtained credits from different institutional sources, which constituted 12.9% of the total farming community. The small farmers who utilized credit facilities were 11.8% of the total small farming community, whereas the large farmers who utilized the credit facility were 23% of the large farming community (Table 9.24).

9.6.4 Problems/Constraints

- Funds allocated for agricultural credit are few in comparison to their due share. The share of agriculture in GDP is over 20% and the share of agricultural credit amongst total credit disbursed to private sector is only 7.5%.

TABLE 9.23
Purpose-Wise Distribution of Institutional Credit Disbursed by ZTBL/ADBP and Commercial Banks in Pakistan (Percent Shares)

Year	Seed	Fertilizer	Sub-Total	Tube Well	Tractor	Sub-Total	Other	Grand Total
1975–1980	2.8	15.2	18.0	4.3	60.4	64.7	17.3	100
1981–1985	9.6	37.7	47.3	2.2	28.7	30.9	21.8	100
1986–1990	7.3	29.6	36.9	4.1	24.3	28.4	34.7	100
1991–1995	9.3	29.9	39.2	2.5	25.2	27.7	33.1	100
1996–1901	12.9	41.3	54.2	3.3	15.4	18.6	27.2	100

ZTBL: Zarai Taraqiati Bank Limited.

TABLE 9.24
Agricultural Credit by Source and Size of Farm

Size of Farm (Acres)	Total HH under Agri. Debt	Source of Credit					
		Z.T.B.L	Commercial Banks	Financial Institutions	N.G.Os	Commission Agents	Friends and Relatives
Total Farm HH[a]	1,081,907	382,320	50,885	29,157	15,703	230,502	493,382
Under 0.5	154,205	23,504	3,345	3,846	2,145	22,710	106,270
0.5 to under 1	158,631	31,904	5,237	5,234	2,649	36,192	93,218
1 to under 2	225,783	70,362	9,373	8,050	3,182	51,161	107,712
2 to under 3	174,891	69,002	7,989	3,706	3,162	39,936	69,799
3 to under 5	164,640	73,192	8,528	4,706	2,142	35,250	61,270
5 to under 10	119,641	63,417	8,181	2,260	1,688	26,944	34,777
10 to under 20	58,226	34,231	4,603	951	618	13,039	15,079
20 to under 40	19,492	12,900	2,379	316	86	3,878	3,821
40 to under 60	3,255	2,113	626	34	4	671	609
60 and Above	3,154	1,699	624	49	35	700	828

Note: Agriculture census of Pakistan, 2010.
[a] Household.

- The meager amount allocated for agriculture is not actually utilized; for example during 2015 and 2016, only 65% of total allocation was used.
- Farming communities in general, and small farmers in particular, need credit. The small farmers are the most deprived component of the farming community. According to a census of agriculture in 2010, only 13% of the farming community had used credit facilities. Among small farmers (90% of the farming community) only 11.8% used this facility. On the other hand, among large farmers (10% of the farming community), 23% used this facility.
- To obtain credit from institutional sources collateral is needed. A pass book is the proof of this collateral, which is very difficult to get, and according to one estimate different categories of small farmers having less than 5 hectares of land possessed the pass book to the range of less than 5 to 12%. Whereas 49% of large farms had a passbook.
- Formal credit facilities are not reaching 7.45 million small farmers, located in 45,000 villages of Pakistan. This is why, according to one estimate, 63% of farmers'credit requirements were met by noninstitutional sources.
- Farmers need credit to meet their social, consumption, and production requirements. The loans provided by institutional sources do not even meet production needs; the funds provided are not properly utilized by farmers.

9.6.5 Way Forward

- A comprehensive rural finance survey—like that of 1985—should be carried out to identify the farmers' credit needs, problems, shares of institutional and noninstitutional credit, cost/interest rates on these loans, purposes of obtaining credit, and its utilization and problems in the repayment process.
- Amounts allocated for agricultural credits should be increased and its share in total credit disbursed should be increased to 20%, that is, the share of agriculture in the GDP.
- Small farmers must be given top priority in credit disbursement according to their contribution to the agricultural sector.

- Credit should be given to both farming and nonfarming communities living in rural areas for social consumption, commercial, agribusiness/cottage industry with the purpose to generate farming income in the rural area. The proper utilization of agricultural credit should slow the rural–urban migration.
- As can be seen from the recent past the probability and intensity of extreme climatic event has increased; therefore, to safeguard small farmers from these events, interest free loans should be provided according to their needs.
- As agriculture is a risky business, the chance of default is high. The private commercial banks hesitate in providing loans to small farmers who are the most vulnerable to climate and economic shocks. To minimize such risks insurance mechanisms, such as ZTBL provides on small scale, must be attached with the loans.
- As land records have been computerized, the revenue department must be directed to issue a pass book to every farmer, especially smaller ones on a priority basis.
- It must be completed in shortest possible time (maybe less than a week) and the process be made more farmer friendly. Recovery procedures in case of defaults are very harsh and humiliating and can in some cases lead to the sale of farmland/death of the farmers. It must be improved and made farmers friendly. For this purpose, the revenue department must be engaged in it, along with the insurance cover on the loan.
- With the law of inheritance farms are divided and farm sizes are falling drastically below the economically feasible land holding size. To counter this, one requires the use of cooperative farming in agriculture. The farmer's cooperatives must be revisited/revived and the credit disbursed through them should be increased. In case of tenants who do not possess collateral, loans must be granted on personal guarantees.

9.7 MARKET SERVICES

Marketing holds a crucial role in any business and assists both producers and consumers. The producers equipped with market knowledge are capable of formulating innovative and greater strategies to provide consumers with goods and services in a well-organized and most effective way. Agricultural marketing has two aspects—the product side and the farm input supply side. It is simply the application of rules and principles of marketing in the agriculture sector. Agricultural marketing includes all the business activities engaged with planning, production, transformation, sorting, grading, transportation, and distribution of goods and services associated with agriculture as preferred by agricultural producers (farmers) and ultimate consumers (Mohy-Ud-Din and Badar, 2011). The major stakeholders in Pakistan's agricultural marketing system include producers, traders, trade supporters, decision-makers, and consumers.

Agricultural commodities are collected from widely scattered villages by government and semigovernment procurement organizations for particular items (chiefly food and export items), and by brokers, commission agents, and traders for other products. The products head to their final destinations in urban centers or for export, via the large networks of town markets (*mandis*) and through a many market intermediaries (Etzel et al. 2005). There are 87 market committees and 246 agricultural markets in Punjab, out of which, 149 are grain markets and 97 are fruit and vegetables markets (AMIS, 2016). Figure 9.3 explains the organizational chart of the Department of Economics and Marketing, Government of Punjab.

9.7.1 PROCUREMENT OF CROPS

Provincial government agencies provide services to farmers for the procurement of their produce. In Pakistan, main institutions providing the procurement services are the Pakistan Agricultural Storage and Services Corporation (PASSCO) and Provincial Food Departments (PFDs). PASSCO procures agricultural commodities such as wheat, paddy, pulses, and potatoes on behalf of government in

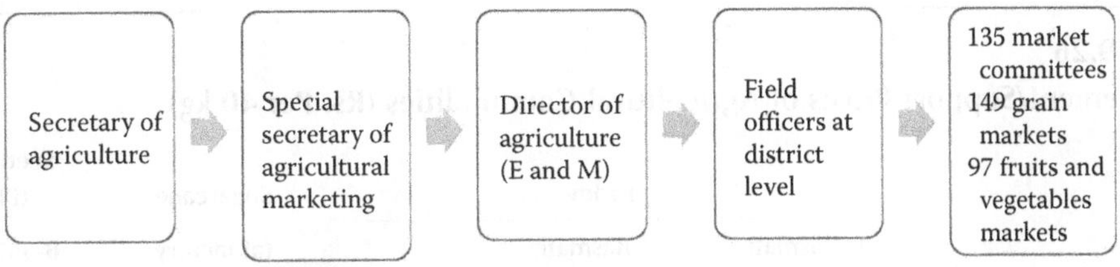

FIGURE 9.3 Organizational chart of agricultural marketing wing.

order to stabilize the prices while also constructing cold storage facilities (Hussain, 2004). Similarly, the vision of provincial food departments is to provide quality flour at reasonable prices and safeguarding the interests of wheat growers by providing guaranteed price support. During 2016, wheat procurement targets in Pakistan were set to about 7 million tons. This task was assigned to PASSCO and PFD. The historical data of procurement, release and stock of wheat in Pakistan is presented in Table 9.25.

9.7.2 Support Prices

Prices support for different agricultural products are listed by the Agricultural Policy Institute (API). the government acts as a buyer in the market to protect the interest of producers with weak holding power, especially in glut periods when there are dangers of price crashes. Thus, the government usually fixes the intervention prices of the main agricultural products, that is, seed cotton, lint, cotton, wheat, gram, sugarcane, cleaned rice, paddy, onions, potatoes, tobacco, and oilseeds (Altaf, 2006). The purpose of price support is to provide a floor (minimum level) in the market rather than replacing the open market function (Mohy-Ud-Din and Badar, 2011). Table 9.26 presents the historical trends of price support for wheat, paddy, sugarcane, and seed cotton from 2007 to 2015 in Pakistan.

9.7.3 Government Market Services (Input Output Prices)

Several procurement and export agencies came into being in the 1070s, such as the Cotton Export Corporation of Pakistan (CEP), Rice Export Corporation of Pakistan (RECP), Ghee Corporation of Pakistan (GCP), and Agricultural Marketing and Storage Limited (AMSL) (GOP, 1988). An

TABLE 9.25
Procurement, Releases, and Stocks of Wheat (000 Tons)

Year	Procurement	Release	Stocks in 1st May
2007–2008	3918	6320	136
2008–2009	9200	5784	822
2009–2010	6715	5985	4223
2010–2011	6150	6404	3186
2011–2012	5792	5820	3506
2012–2013	7910	6363	1681
2013–2014	5948	6452	1177
2014–2015	6131	3957	3351
2015–2016	7050	4468	5016

Source: GOP. 2016a. *Economic Survey of Pakistan 2015–2016*. Finance Division, Economic Advisor's Wing, Islamabad; GOP. 2016b. *Agricultural Statistics of Pakistan 2015–2016*. Ministry of Food and Agriculture (Economic Wing), Islamabad.

TABLE 9.26
Procurement/Support Prices of Agricultural Commodities (Rs. Per 40 kg)

		Paddy			Sugarcane	Seed Cotton (Phutti)
Fiscal Year	Wheat	Basmati 385	Basmati Super/2000	Irri-6	(at factory gate)	B-557, F-149 Niab-78
2007–2008	625	–	–	–	64	1,025
2008–2009	950	1,250	1,500	700	80	1,465
2009–2010	950	1,000	1,250	600	101	–
2010–2011	950	–	–	–	125	–
2011–2012	1,050	–	–	–	151	–
2012–2013	1,200	–	–	–	171	–
2013–2014	1,200	–	–	–	171	–
2014–2015	1,300	–	–	–	181	3000
2015–2016	1,300	–	–	–	177	3000

Source: GOP. 2015. Ministry of Professional and Technical Training, AEPAM, Islamabad; GOP. 2016a. *Economic Survey of Pakistan 2015–2016.* Finance Division, Economic Advisor's Wing, Islamabad; GOP. 2016b. *Agricultural Statistics of Pakistan 2015–2016.* Ministry of Food and Agriculture (Economic Wing), Islamabad.

institution, namely the Agricultural and Livestock Products Marketing and Grading Department (ALPMGD), is working at the federal level to advise and guide the government about marketing aspects. The API directs the government of Pakistan in setting adequate domestic and international agricultural price policies. The Trading Corporation of Pakistan (TCP) regulates cotton markets. The Pakistan Tobacco Board fixes the prices and grades of tobacco. The provincial agricultural departments, for example, the PFDs manage the agricultural marketing at the provincial level. In Punjab the marketing of agricultural products is administered by the Directorate of Agriculture (Economics and Marketing). Moreover, the Punjab Agricultural Marketing Company (PAMCO) is working to improve transportation, processing, and storage services. The Punjab Institute of Agricultural Marketing (PIAM) was established to provide the stakeholders of agricultural products with sufficient skills and training (GOP, 2011). All these institutes/agencies are providing services to farmers in one way or another to protect their interest.

9.7.4 Local/Private Market Services

The private sector is involved in a number of market activities, for example, farm input supply, assembly, grading, sorting, packaging, storing, and transportation of agricultural produce from producers to consumers. Numerous processing entities—cotton ginning factories, flour mills, oil extraction mills, rice mills, and sugar mills—have been operating over time in Pakistan. These entities purchase commodities directly or indirectly from farmers and utilize form changing activities (processing and value addition) to produce goods needed by the customers. Table 9.27 shows the number of different private units providing services, along with their worth, during the year 2001 and 2006.

The cold and other storage facilities are privately owned to a large extent at market level—not at farm level. The prices of all the agricultural commodities are automatically set by market forces exerted by demand and supply, except for a few goods like wheat, for which support price mechanisms are working. With the forward and backward liaison, agro-based industries have high multiplier impacts in terms of job creation and value addition (FAO and UNIDO, 2010). These local/private market services play an important role in uplifting the socioeconomic status of farmers by improving their livelihoods.

TABLE 9.27
Major Agro-Based Industries in Pakistan in 2000–2001 and 2005–2006

	2000–2001		2005–2006	
Industries	**No.**	**Value (M Rs.)**	**No.**	**Value (M Rs.)**
Food manufacturing	880	188,610	1,197	509,980
Dairy products except ice cream	10	10,877	47	57,675
Canning of fruits and vegetables	10	1,516	24	7,335
Vegetable ghee	56	43,962	92	89,238
Cotton seed oils	39	2,019	6	596
Rice milling	228	4,255	315	9,089
Wheat and grain milling	300	27,475	440	59,276
Refined sugar	62	55,585	72	115,264
Leather & leather products	82	18,655	142	26,742
Tobacco Manufacturing	12	23,959	13	52,401
Manufacturing of textile	1063	320,932	1,646	844,841
Cotton spinning	214	157,030	504	379,690
Cotton weaving	108	51,945	577	297,943
Finishing of textiles	304	30,388	311	79,237
Ginning and baling of fiber	334	55,573	540	130,680

Source: GOP. 2011. *Economic Survey of Pakistan 2010–2011*. Finance Division, Ministry of Finance, Government of Pakistan.

9.7.5 Storage Services

Storage services greatly facilitate stabilizing highly volatile agricultural prices, which can badly affect the farmer. It also ensures an uninterrupted supply of agricultural goods throughout the year—even during off-seasons. Pakistan currently experiences from extreme lacks of adequate and state-of-the-art storage facilities at both farm and market level. Holding power of the farmers depends heavily on storage facilities. Insufficient storage, especially in case of perishable commodities, leads to the loss of produce during various handling operations. Public storage facilities are not adequate to accumulate large surpluses of agricultural produce in Pakistan (GOP, 2016). There is a need for increases in storage facilities to safeguard the interest of farmers.

Table 9.28 shows the stagnant situation of storage facilities for wheat, rice, and cotton, both at provincial and federal level. So, these facilities should be increased and services be enhanced over the time, keeping in view the production statistics.

9.7.6 Problems in Agricultural Marketing

An efficient marketing system is indispensable for agriculture of Pakistan. It keeps farm prices at sound levels and that food items remain affordable for consumers. The marketing system of Pakistan has improved with time but still there are several pitfalls in the existing marketing system. The main problems faced by agricultural marketing are described below.

- Less marketable surplus due to small land holdings
- High marketing margins
- Poor condition of farm-to-market roads, leading to high transformational costs and postharvest losses
- Poor storage facilities and less staying power of farmers

TABLE 9.28
Government Storage Capacity in Pakistan (000 Tons)

Agency	2006	2007	2008	2009	2010	2011
Wheat	4339	4339	4339	4339	4339	4339
Rice	826	826	826	826	826	826
Cotton	77	77	77	77	77	77
Total Capacity	5242	5242	5242	5242	5242	5242

Source: GOP. 2011. *Economic Survey of Pakistan 2010–2011*. Finance Division, Ministry of Finance, Government of Pakistan.

- Poor postharvest management practices
- Exploitive role of middlemen
- Highly fluctuating and unpredictable produce prices, particularly affecting small farmers.

9.7.7 Way Forward

- Crate awareness among the farmers and provide them with marketing training
- Improvement in proper farm-to-market roads
- Enhance advanced storage facilities
- Establish an appropriate number of new *mandi*s in different towns
- Sustain the efficient flow of market information
- Ensure the provision of farmer friendly credit facilities
- Fixing the precise procurement targets of the government
- Attempt to rationalize the role of middlemen to reduce marketing margins.

9.8 SOIL AND WATER TESTING SERVICES/DIAGNOSTICS

9.8.1 Laboratory Testing Services/Diagnostics

In this era of modern technology, soil and water testing is becoming an essential component of modern farming in developed and developing countries. Laboratory tests basically focus on essential nutrients for crop growth, mostly supplied by using organic or inorganic fertilizers, such as nitrogen (N), phosphorus (P), and potassium (K). Based on the type of soil, tests are also conducted for secondary nutrients such as calcium (Ca), magnesium (Mg), and sulfur (S). In dry regions, other micronutrients such as iron (Fe), zinc (Zn), manganese (Mn), copper (Cu), and boron (B) are also measured, because the deficiencies of these essential elements are more frequently associated with calcareous soils. These areas may also have excessive/toxic levels of elements like boron and higher levels of sodium (Na) and magnesium (Mg), which adversely affect soil physical properties. Since nutrient behavior is governed by soil properties and environmental conditions, testing and measurements of such soil properties is required. These include pH, salinity, organic matter (OM), calcium carbonate ($CaCO_3$), texture, and aggregate stability. In dry areas, the existence of gypsum ($CaSO_4.2H_2O$) is also of concern.

Soil testing is an important method to analyze the status of soil fertility and optimize fertilizer requirements for good crop productivity. These tests also help to identify the specific soil-associated problems, which can then be eliminated by using agronomic/cultural practices or by the addition of soil nutrients/amendments.

Keeping in view the importance of soil testing, the government of Punjab established soil and water testing laboratories at the district level in 1981 to respond to the requirements of the farming community. These laboratories are providing various types of services to farmers. Soil samples are collected from the farmers' field at their demand, or supplied by the farmers, and then analyzed for their physical and chemical soil characteristics at an affordable cost to the farmers. The farmers are then given advise/recommendations to improve soil fertility based on the analysis reports.

9.8.2 Objectives of Soil and Water Testing Laboratory

The soil and water testing laboratories are providing various kinds of services to the farmers. The major objectives of soil and water testing laboratories include:

- Advisory services through soil testing to enhance soil fertility levels using suitable dose/ amount of recommended fertilizers thus reducing the cost of input fertilizer.
- Analysis of soil fertility status and associated physiochemical characteristics of different type of soils.
- Control of soil salinity.
- Analysis of surface and groundwater quality for its suitability for irrigation.
- Providing facilities for each farmer to analyze soil and water samples at nominal costs.
- Diagnose the soil related problems and recommendations for farmers for its reclamation through application of suitable nutrients/amendments.
- Analysis of soil fertility status for balanced use of fertilizers for various crops.
- Diagnose salinity and sodicity hazards of soils and their reclamation.
- Field experiments at farmer's fields to formulate the fertilizer recommendations for a variety of agricultural crops.

9.8.3 Current Situation

The government of Pakistan has established soil and water testing laboratories in each district. The functions of these laboratories are to provide services for soil and water quality testing. Parameters tested include electrical conductivity (EC), sodium adsorption ratio (SAR), residual sodium carbonate (RSC), and pH, etc. Some of the renowned private organization as the Fauji Fertilizer Company Limited and SGS Pakistan (Private) Limited are also rendering laboratory services to the farmers.

9.8.4 Type of Tests

The water quality analyses are performed in laboratories to evaluate water for irrigation purposes. The important indicators of water analysis determined through these tests are given in Table 9.29. Similarly, the important indicators of soil analysis which are determined during soil testing are given in the Table 9.30.

TABLE 9.29
Various Indicators Identified through Water Testing

EC (dS m^{-1})	CO_3^{2-} (me/L)	HCO_3^- (me/L)	Cl^- (me/L)	$Ca^{2+}+Mg^{2+}$ (me/L)	Na^+ (me/L)	RSC (me/L)	SAR

TABLE 9.30
Various Indicators Identified through Soil Testing

pH_s	EC (dS m^{-1})	CO_3^{2-} (me/L)	HCO_3^- (me/L)	Cl^- (me/L)	$Ca^{2+}+Mg^{2+}$ (me/L)	Na^+ (me/L)	SAR

It is important to mention that soil and water testing laboratories are too few at the district level and are not sufficient; most farmers have to travel long distances to conduct tests of their soil and water.

9.8.5 Way Forward

- Establishment of soil and water testing laboratories at Tehsil and Union Council levels in order to facilitate the analysis of soil and water samples.
- Provision of government or privately owned mobile testing laboratories equipped with the latest equipment and trained manpower to render soil and water testing services.
- Provision of latest infrastructure/equipment and trained manpower to perform soil and water quality test in order to ensure the authenticity of analysis reports and the recommended doses of fertilizers/nutrients.
- Awareness programs should be launched through extension workers and the media among the farming communities in order to introduce the importance of soil and water quality analyses, which play a vital role in enhancing crop productivity.

9.9 CROP INPUTS (SEED, FERTILIZER AND PESTICIDES) SERVICES

9.9.1 Seed Services

The agricultural production process is very complex. On one hand, it is subjected to nature and carried out in open environment, and on other hand, a large number of inputs of different quality are used in it. These inputs are land, seed, fertilizer, pesticides/weedicides, water, machinery, and human labor and management. After land, seed is the most important input in agriculture as the response/deficiency of other inputs like fertilizer, pesticides, and so on depends on it. In late 1960 s and early 1970 s, a green revolution came into agriculture with the introduction of high yielding varieties of wheat and rice. These varieties were more responsive to fertilizer and irrigation. In the recent past, major breakthroughs in maize and poultry production came with the introduction of hybrid verities of maize and poultry breeds. Table 9.31 explains the yield gap of major crops at different levels.

As indicated from Table 9.31, large yield gaps in major crops exist between the national average and progressive farmers. Per hectare yield gaps between research stations and traditional farmers varied from 52% to 44% during 1988–2015. The rice (basmati) yield gap between research stations and average farmers increased from 10% to 29% during the same time period. Thus, there is an immense difference in resource-based cropping intensity and management of research station,

TABLE 9.31
Yield Gaps in Major Crops (kg/hectare)

Years		1988				2015		
Major Crops	Experiential Station	Progressive Farmer	Average Farmer	Traditional Farmer	Yield Gap (%)	Progressive Farmer	Average Farmer	Yield Gap (%)
Wheat	6425 (100%)	2909 (45%)	1408 (21%)	1975 (30%)	52	5070	2858	44
Rice (Basmati)	6850 (100%)	1495 (21%)	1357 (19%)	1186 (17%)	10	4148	2953	29
Cotton	2527 (100%)	1903 (75%)	1269 (50%)	991 (39%)	44	4148	1705	59
Sugarcane (Tonnes)	183 (100%)	60 (32%)	38 (20%)	45 (24%)	36	108	62	43
Maize	6873 (100%)	4546 (66%)	1600 (23%)	–	64	7098	5724	19

Source: Agriculture Commission Report 1988 and Punjab Growth Strategy 2015.
Note: Figures in parentheses are yield gap in percentages.

progressive farmer and average farmer. Consequently, it is better to estimate the yield gap between progressive farmers and average farmers. These yield gaps vary from 20% to 60% in all major crops. From 1988 to 2015, the yield gap became wider and wider, an important reason being the nonavailability of certified seeds.

No one can disagree with the importance of good quality seed in agriculture; however, the farmland area planted with good quality seeds is small. Table 9.32 shows the distribution of certified seeds of main crops during different time periods.

A significant increase in the distribution of improved seed can be observed in case of wheat, rice, and maize from 1978 to 2015. Improved seeds came either by selection of good planst from the field, as lots of variability exist in each plant variety in nature, or through the outcomes of the plant breeder's effort. To produce good quality seed, skilled manpower and well organized and fully equipped lab facilities are required. Skilled manpower and lab facilities are available to some extent in Pakistan, that is why some good varieties of wheat, cotton, maize, and rice are available here to farmers. Prebasic seeds, basic seeds, and certified seeds are all produced in a well-organized manner by following quality controls measures. After the approval of varieties, a seed is distributed among farmers in different steps, which involve seed production, seed processing and multiplication, and seed storage and distribution. Comprehensive monitoring and evaluation is needed at all stages, from seed production to final distribution to farmers but unfortunately the actual state of affairs regarding quality control is very poor.

To provide certified seeds to 8.5 million farms, improved seeds must be produced or multiplied on a larger scale. According to a national commission report on agriculture in 1988, 850 hectares in Punjab and 175 hectares in Sindh were allocated for this purpose. To meet the basic seed requirements, another 850 hectares in Punjab and 700 hectares in Sindh were needed. The cropping intensity and cropped area has normally increased over time so the area required to produce certified seed should have gone up rapidly. Seed processing in the past were mostly done by public sector seed companies/organization one in Punjab and other in Sindh and that two were confined to the wheat, cotton and rice seed. Now, large numbers of public and private companies are involved in the production, processing, and distribution of seeds. The relevant information is given in Table 9.33.

Throughout the world, seed certification is a well-managed quality control process. However, the situation in Pakistan is very disappointing. There are number of cases where new varieties reaches farmers or seed companies before their approval. Recently, a study was carried out by the University of Agriculture Faisalabad to check the quality of BT cotton seed. A number of samples were taken which showed that BT genes were either absent or germination levels were very low. As a result, during that year, massive attacks of insects destroyed 30% crop in Pakistan. Presently, the seed act of 1970 is in force, which has many shortcomings hile not properly implemented. A new seed act is pending since 2009.

Seed are produced during one cropping year/season and distributed to farmers during the next season. Meanwhile, it must be stored in proper warehouses having good ventilation and temperature

TABLE 9.32
Distribution of Improved Seed in Pakistan (000 Tonnes)

Pakistan	1978	1981	1985	2010	2014	Area (MH)	2015
Wheat	32.21	50.14	57.06	266.35	271.24	9.068	315.03
Paddy	0.69	0.75	1.87	22.2	49.62	1.81	58
Maize	0.59	0.6	1.33	8.74	15.59	2.72	26.02
Cotton	15.28	19.35	20.26	12.46	20.68	4.53	24.4
Gram	0.58	1.06	1.25	0.2	0.02	1.31	0.06
Grand Total	49.35	71.9	81.77	312.63	359.18	19.438	423.51

Source: Agriculture Commission 1988 and Agricultural Statistics 2014–2015.

TABLE 9.33
Total Number of Registered Seed Companies in Pakistan

Category	Punjab	Sindh	KPK	G.B	Balochistan	Total
Public Sector	1	1	1	–	1	4
Private Sector	600	91	20	2	7	720
Private-Multinational	4	1	–	–	–	5
Total Active	605	93	21	2	8	729
Cancelled	129	14	7	–	0	150
Total	734	110	28	2	8	879

Source: Federal seed certification, 2012.

control facilities. Seed must be packed in proper bags, but unfortunately in Pakistan it is being kept in plastic bags, mostly in ginning/rice factories, where appropriate temperature and air circulation are not maintained, thus seeds heat up possibly resulting in poor germination and lower yields.

9.9.1.1 Way Forward

New seed comes from prebasic seed, having some genetic needs. To preserve the prebasic seed in the country, seed bank must be established and the necessary funds/manpower/equipment must be provided by the federal government. The university of Agricultural Faisalabad is most ideal place for this purpose. The long awaited seed act must be implemented with letter and sprit so that plant breeders can be incentivized for their efforts and the availability of quality seeds is ensured for farmers. A strict quality control system must be followed at all stages of seed development, that is, from approval of a variety to the final distribution to farmers. New varieties of seed should not reach farmers and seed companies before their approval. Good quality/true to type seeds must be sold in the market along with germination rate/expiry date/ISO number and the name of companies labeled on the bags. Company warehouses and packaging material should be regularly monitored by the agriculture department (Afzal, 2010; Afzal et al., 2009).

New pests and diseases are emerging in Pakistan indicating the weakness of our quarantine laws. Existing laws are very old and must be reenacted to meet the challenges of the times, with strict enforcements to monitor the import of any kind of seed, plant, or animal in Pakistan.

9.9.2 Fertilizer Services

Fertilizer is an important and costly input in the agriculture production process. Per hectare fertilizer application has increased from 10 kg in 1970 to 170 kg in 2016. Although it is comparable with our neighboring country India, it is less than in Egypt, Germany, and the Netherlands where fertilizer use per hectare are 446, 196, and 291 kg, respectively.

Overall fertilizer use has been increasing in Pakistan but it is not being applied using a balanced ratio, which adversely affects the efficiency of fertilizer use. In 1988, NPK was used with a ratio of 2.1:1:0.5. Using fertilizers with the recommended ratio increases the efficiency of fertilizer use and crop yields but also reduces the production costs and increases profitability.

Fertilizer requirements vary between crops and regions and are heavily dependent on soil fertility. Soil fertility analyses must be done before fertilizer applications. Over time, cropping and land use intensity have increasing to meet food and fiber demands of rapidly increasing populations and to earn precious foreign exchange. Crops are extracting both micro- and macronutrients from the soil; however, we are only applying macronutrients such as nitrogen and phosphorus. Thus, deficiency of micronutrients takes place in soils. Deficiency of zinc is common in rice zones. Similarly, Pakistani

TABLE 9.34
Projected Fertilizer and Insecticides Use

	Fertilizer Offtake (000 N/Tonnes)				Import of Fertilizer		Use of Insecticides		
Years	N	P	K	Total	000N/Tonnes	Years	Quantity Tonnes	Value (M. Rs)	Growth (%)
1955	6.6	–	–	6.6	–	2001	21,255	3,471	–
1960	31	0.4	–	32.4	–	2002	31,783	5,320	50
1965	69.8	1.2	–	71.1	–	2003	22,242	3,441	5
1970	251.5	30.5	1.2	283.2	–	2004	41,406	7,157	95
1975	445.3	102.5	2.8	550.6	–	2005	41,561	8,281	96
1980	842	226.9	9.6	1079.5	–	2006	33,954	6,804	60
1985	934.8	293.8	24.7	1253.3	–	2007	29,089	5,848	37
1990	1332.4	408.9	42.6	1783	–	2008	27,841	6,330	31
2001	2,264	677	23	2966	580	2009	28,839	8,981	36
2005	2,796	864	33	3694	784	2010	38,227	13,473	80
2010	3,275	860	24	4360	1444	2014	23,147	14,058	9
2015	3,309	995	33	4316	750	2015	23,157	13,520	9

Source: Agricultural Statistics, 2015.

soils are also deficient in organic matter. Organic matter not only provides micro- and macronutrients to the plant but also improve the soil structure and thus the water holding capacity of soil. It also increases the efficiency of fertilizer. In fertile soil, organic matter is normally greater than 2%, but unfortunately most Pakistani soils have organic matter contents which are less than 0.5%. Various sources of organic matter are farmyard manure, green maturing, crop residuals, and composts which can be generated from city-based waste.

Fertilizer prices are increasing faster than the prices of crops. For example, urea, DAP, and wheat prices between 2009 and 2010 were Rs. 850, 1700 per bag, and Rs. 950/40 kg, respectively. Whereas during 2015, these prices increased to Rs. 1850, 3500, and 1300, respectively. Fertilizer prices increased by almost 100% when wheat prices increased by only 30%. This shows the adverse terms of trade against agriculture and thus affect badly the profitability of farming. There are nine major fertilizer-producing companies out of which three have more than a 75% market production share. The production capacity and actual potential is given in Table 9.34 (Ali et al., 2015).

Phosphorus fertilizer is mostly imported to meet crop demands. There are a number of issues in the importation of fertilizer, such as issuing of tender and delivery time and the prices of fertilizer. Most of the time, prices of imported fertilizers are normally higher than the prices of locally produced fertilizers. Farmers have some serious and genuine concerns regarding the quality of fertilizers, their availability during peak crop season, and their prices. Substandard fertilizers, particularly DAP, are available on the market at the time of peak demand and thus artificial shortages of fertilizers are created to charge higher prices to the farmer.

9.9.2.1 Solutions

Fertilizer is a very costly input, which must be used efficiently. Soil analysis must be carried out before the application of fertilizer so that the proper dose and balanced fertilizer ratio can be determined. The existing soil and fertilizer testing labs are either missing or deficient in manpower/material. Such labs should be established at Tehsil level to insure efficient use of fertilizer and its quality. The University of Agriculture Faisalabad has taken a lead in this regard, providing online E-Guidance to farmers regarding proper balanced use of fertilizers for each major crop, in light of soil analyses performed for different districts of Punjab.

Fertilizer markets must be properly monitored to ensure the availability of quality fertilizer to farmers at reasonable prices. As international fertilizer prices tend to be higher than the local fertilizer prices, local production must be encouraged. Local fertilizer producing companies are producing noticeably less than their capacity. Proper raw materials, for example gas, must be supplied to them so that they can produce cheap fertilizers to meet local demand and reduce reliance on imported fertilizer. To resolve this issue related to imported fertilizers, import tenders must be issued in a timely and transparent fashion to ensure a timely supply to the farmers' doorstep. Fertilizer prices are high and could be significantly reduced by reducing the sales taxes imposed on them.

9.9.3 Pesticides Services

Healthy plants and crops can lead to good crop yields. Crop health and yields are adversely affected by disease and pest attacks. Over time, crop experts are trying to control disease and pest attacks by applying cultural practices, using insecticides and pesticides, and by developing crop varieties which are resistant to disease and pest attacks. Significant improvements in disease control have been made by evolving disease/pest resistant varieties of wheat (to control rust and smut), sugarcane (to control smut), and cotton (to control whitefly and pink bollworm). The process of evolving new varieties will continue as new diseases/pests emerge—when old ones are eliminated, or as they develop resistance against these varieties. To control diseases/pests, the short-term measure, though costly and environmentally unfriendly, is the use of pesticides and insecticides. According to an estimate, crop loses in Asia as whole is more than 20% due to these menaces. According to another estimate, by the International Rice Research Institute, crop yields could be increased about 2.5 times if proper management of crop, including pest management operations were followed.

Pesticides/insecticides used in Pakistan are mostly imported. The amount of pesticides imports, along with percentage growth rates, are listed in Table 9.34.

The information regarding the pesticides used on different crops, along with farm sizes is given in Table 9.35. It is evident from that on an average on all farm sizes, only 24% of the cropped area is covered by the pesticides/insecticides. Insecticides and pesticides are mostly sprayed on cotton and sugarcane; the respective percentage of cotton and sugarcane area are about 59% and 29%,

TABLE 9.35
Use of Plant Protection Measures on Important Crops by Size of Farm-2010

Size of Farm (Hectares)	Area Covered by Plant Protection: Area	Area Covered by Plant Protection: % of Total Crop	Percent of Crop Area Covered by Plant Protection Measures: Wheat	Cotton	Sugarcane	Rice	Maize
Total	6,711,306	24	20	59	29	39	11
Under 0.5	168,759	19	18	55	13	35	4
0.5 to under 1	433,723	23	20	59	21	38	6
1.0 to under 2	931,644	24	20	59	28	38	9
2 to under 3	1,017,258	25	22	59	32	40	12
3 to under 5	136,8138	26	22	61	29	41	15
5 to under 10	1,315,139	25	21	61	30	41	12
10 to under 20	791,881	24	19	56	29	35	16
20 to under 40	398,720	23	16	56	27	26	23
40 to under 60	109,365	21	14	54	28	26	26
60 and above	176,666	23	17	49	30	38	14

Source: Agricultural Statistics, 2015.

respectively. Farmer complaints regarding the poor quality and high prices of pesticides are very common. Lack of timely availability, and the poor quality of pesticides, destroyed 30% of the cotton production in 2015–2016.

9.9.3.1 Way Forward

Pesticide use cannot be avoided and is costly as well as environmentally unfriendly. It should be used only when it becomes necessary. Pest scouting should be regularly carried out by farmers, crop must be treated by quality pesticides/ insecticides when it becomes necessary. Proper training of extension workers regarding pest scouting and identification of pest/diseases must be regularly conducted. These extension workers should guide the farmers in this respect. The important elements of integrated pest management are first, the development of disease and pest resistant varieties, second, the biological control of pest by identifying, preserving, and multiplying the predator insects, third the use of cultural operations, and the fourth the use of pesticides to control pest. Although good progress has been made in developing disease and pest resistant varieties, this process should be continued as new pests and diseases emerge and resistances develop. Information technology should be used to control pest and diseases in an effective way.

9.10 LIVESTOCK SERVICES

Being a major player in the national economy, the livestock sector is accepted as an economic engine for poverty alleviation in Pakistan. Livestock is the primary activity, along with crop husbandry, in rural areas of Pakistan. This sector contributes 56.3% to agriculture value addition, and 11.8% to the GDP in Pakistan. The sector grew by 4.1% in 2014–2015. In Pakistan, the cattle population was the largest in 2014–2015, at 41.2 million heads, comprising buffaloes (35.6 million), sheep (29.4 million), and goats (68.4 million).

The income, food, and employment generated from this sector are very important for rural households. According to the latest estimates, more than 8 million families in Pakistan are engaged in raising livestock. The population of buffaloes, cattle, and goats has been increasing over time in Pakistan. This significant increase in livestock population is encouraging, given the importance of this sector for ensuring food security in the country.

Veterinary services are considered a national and global public good and thus have become the responsibility of national governments and the international community. The basic purpose of veterinary public service is to support the access of animals and animal products to regional, national, and international markets. The official veterinary services are aimed at: (1) control of animal diseases with the ability to rapidly detect and diagnose them; (2) minimize and control along the food chain (veterinary public health) the risks of zoonosis and food borne diseases; and (3) ensure the welfare of animals. The veterinary services in the country are evaluated on the basis of four criteria, including (1) human, physical and financial resources; (2) technical authority and capability; (3) interaction with stakeholders; and (4) access to markets.

A department of veterinary services was started by the British in the subcontinent. This department was renamed the Animal Husbandry Department in the newly created Pakistan. Over the years, the provincial departments dealing with veterinary services used different nomenclature including animal husbandry, livestock and dairy development, etc. A national veterinary laboratory, Islamabad was set up by the federal government to act as regulatory body to check drugs and vaccines—not part of the mandate of the Ministry of Livestock and Dairy Development but regulated by the Ministry of Health, investigate drug residues in imported and export materials, as well as to provide diagnostic and training services.

Provincial governments are manage the majority of institutions which carry out activities relating to veterinary services. Table 9.36 shows that there are 963 veterinary hospitals, 2869 veterinary dispensaries, and 2875 veterinary centers in the public sector, which are mandated to control livestock diseases and provide prophylactic vaccination against major infectious diseases of animals.

TABLE 9.36
Number of Veterinary Service Institutions in Pakistan

Region	Research/Vaccine Production Institutes	Veterinary Hospitals	Veterinary Dispensaries	Veterinary Centers	Diagnostic Labs
Total	6	963	2869	2875	72

TABLE 9.37
Livestock Extension Infrastructure in Pakistan

Province	Veterinary Officers	Veterinary Assistant
Punjab	1403	2271
Sindh	358	748
Balochistan	631	1153
Khyber Pakhtunkhwa	338	1600
AJK, GB, FATA, ICT	323	1346
Total	3053	7118

Source: Agricultural Census, 2010.

There are 72 diagnostic labs meant to facilitate the diagnoses of farm animal diseases. As for as the veterinary extension staff is concerned, there are 3053 veterinary officers and 7118 veterinary assistants working in Pakistan, providing extension services to the livestock farmers (Table 9.37).

These institutions submit monthly progress reports, including infectious diseases reports. These institutions are also supposed to provide flash reporting in case of disease outbreak. A network of district diagnostic laboratories exists, which is supported by regional and provincial diagnostic laboratories and national reference laboratories.

Recent surveys have shown that most of the cattle markets across the country are devoid of any infrastructural facility like shade, shelter, sanitation, drinking water, etc. The Punjab livestock Breeding Act 2014 was notified to regulate the livestock breeding services, improve the genetic potential of breeds, and build cattle markets along modern lines. It has yet to come into effect. However, Punjab is moving ahead to restructure the cattle markets and has decided to establish at least 13 model markets in the province. The existing markets of Lahore, Sheikhupura, Nankana sahib, and Kasur districts are being reorganized and upgraded under the umbrella of a model of private–public enterprise, called the Cattle Market Management Company (CMMC).

In Punjab, cattle markets are traditionally held in 14 towns on days fixed by the local administration. These towns include Arifwala, Sheikhupura, Pasrur, Chiniot, Khushab, and Bhalwal. Similar markets are also arranged in eight towns of Sindh including Thatta, Mirpur Sakro, Talhar, and Golarchi where different kinds of buffaloes, goats, sheep, and so on are auctioned or traded. The other veterinary services are provided publically and privately through different types of livestock markets in Pakistan. The following are the five types of markets which are providing different livestock services:

1. Live animals markets
2. Meat markets
3. Milk markets
4. Hides and skins markets
5. Wool markets.

The animal quarantine department looks after the import and export of live animals and animal products. This department has its offices at all exit and entry points (international ports and airports) of the country. The department is understaffed, underbudgeted, has minimum laboratory expertise, and quarantine houses for 200 animals only at each location.

The government has control over these type of markets and generating fund by levying charges on the entry and exit of large and small ruminants especially at Eid ul Izha occasion. These markets are not efficiently working and have problems of inadequate basic facilities, no weighing machines, and no market committee control.

On the other side of abattoir services, there are some issues which need to be resolved quickly such as insufficient numbers of these markets, outdated and unhygienic conditions, lack of basic slaughtering facilities, rudimentary disposal systems of by-products, and conventional flaying methods adopted.

Marketing services of meat include the government control over retail prices as well as the certification of meats by ensuring the slaughtering of healthy and young animals. Sanitary measures are implemented for meat transportation and sale to ensure hygienic meat.

Milk marketing services include milk inspection systems which have rules on proper milking, hygienic milk handling, and prevention from adulteration. The number of milk process plants is limited. Marketing of hide and skin services are working poorly in Pakistan. The major issue with these services is the unsatisfactory treatment of hides and skins, due to lack of proper training of butcher and traders, which result in the production of defective hide and skins of animals. Marketing of wool has very important incentive for the proper development of animal body. Proper wool marketing services are absent, which include animal shearing systems and the poor sorting and grading system at the farm level.

9.10.1 Poultry Industries in Pakistan

In Pakistan, per capita consumption of meat is only 7 kg and 65–70 eggs annually. Whereas the developed world is consuming about 40 kg of meat and over 300 eggs per capita per year. As per standards of the World Health Organization, daily requirements of animal protein for a person is 27 g, whereas the public is consuming 17 g only. We are already consuming less animal protein as per the required standards. Thus, there is need to look into this industry more deeply.

The Pakistan Poultry Association was formed in 1978, which is the sole representative body of poultry farmers and allied sectors, and is registered with the government of Pakistan. Here, we present current status of the Pakistan poultry industry and its future expected growth to achieve national requirements and export targets, which would greatly depend upon the government's support to the poultry industry.

Current turnover of Pakistan' s poultry industry is about Rs. 750 billion, which generates employment and income for about 1.5 million people. This sector is one of the most organized agro-based sector of Pakistan. Its growth rate is 10%–12% per annum. At present over Rs. 190 billion worth of agricultural produce and by-products are being used in poultry feeds. There are over 25,000 poultry farms spread deep into rural areas across the country, from Karachi to Peshawar. Capacity of farms ranges from 5000 to 500,000 broilers. Marketing channels of broilers and eggs are predominantly in the unorganized sector. About 40%–45% of total meat consumption is being fulfilled from poultry sector. Poultry slaughterhouses, processing/value addition in organized way is only 5%–6% of the total sector. Annually, we are producing 18,000 million eggs and 2250 million kg of chicken meat.

9.10.2 Problems in the Livestock Industry in Pakistan

9.10.2.1 Health

Livestock health is a limiting factor to productivity. A major problem is the lack of knowledge and awareness about the productive benefits of disease control. Those farmers, who are aware of the benefits, have limited access to appropriate vaccines and therapeutic drugs. Animal production systems with varying capacities are affected by different types of diseases. In short, diseases can seriously affect productivity and profitability. Vaccination and treatment of animals is generally ignored by livestock farmers, which results in huge losses regarding productivity and numbers of heads. Farmers used to inject oxytocin to lactating animals for milk let down, which is hazardous for reproductive organs and productivity as well. Farmers use the traditional methods for animal treatments which exacerbate the problems for animal health.

9.10.2.2 Diagnosis of Diseases

Lack of diagnostic facilities for diseases is a major factor causing low productivity. Due to insufficient diagnostic laboratories, veterinarians use hit and trial methods for diagnoses and cures, which results in inefficiency of the treatment. Mastitis (inflammation of udder) is a major problem in lactating animals. It significantly decreases milk production. Farmers have no awareness about the diagnosis and cure of this disease. Farmers usually care for their diseased animals for 2 to 3 days, but this disease needs care during 10 days. External and internal parasites of animals also cause low productivity. It is important to check fecal samples every month to diagnose and deworm animals, which is usually not performed by farmers. In Pakistan, quacks (neem hakeem) are very active in curing animals in the villages. These nontechnically proficient persons often treat animals with hit or miss methods, which sometimes even cause animal deaths.

9.10.2.3 Animal Productivity/Genetic Potential

Pakistan has breeds with low genetic potential Sire (bull). The female breeds with the best potential, such as Sahiwal cow and Nili-Ravi buffalo, are rarely found at the farms of small and medium farmers. There is need to save and exploit the genetic potential of these high yielding breeds. It is a common observation that there is a trend among farmers to cross the animals with imported semen. This practice is a great threat to our local and potential breeds. It is interesting to remark that Australia had demanded 100% pure Sahiwal breed which Pakistan could not provide. It shows that the country is losing the breed.

9.10.2.4 Lack of Livestock Credit

Huge investment are needed to establish modern livestock farms,. Unlike the crop sector, the livestock sector requires more capital. The absence of credit disbursement to small and medium-scale farmers, the involvement of poor in the commercialization of livestock production is restricted.

9.10.2.5 Poor Livestock Extension Activities

The livestock extension wing in the country is performing poorly and is biased toward large farmers, neglecting poor rural livestock-keepers. The public sector follows a top-down transfer of technology approach. It is now universally accepted that it is not result-oriented. Instead, a bottom-up approach should be adopted in which the participation of the livestock farmers is ensured. Extension programs focus only on large ruminants and other species are almost excluded—a problem which needs to be addressed. The extension's messages are not frequently disseminated through print and electronic media.

9.10.2.6 Poor Marketing System

Proper marketing systems encourage the animal sector productivity. The private sector has organized the farmers' associations for their own interest. These associations collect milk for the organizations. Regarding marketing, farmers are at the mercy of beoparies and dodhies. These market players exploit poor farmers. There should be a systematic marketing system that could ensure the farmers' profit share.

9.10.3 Way Forward

The livestock policy covers every aspect of the development of the sector, clearly fixing the responsibilities of implementation for the federal, provincial, and district governments. Under the policy, the federal government should be responsible for national policies, planning and economic coordination, import/export of animals and animal products, quarantine, research, and international coordination while the provisional governments should work for livestock development, veterinary vaccine production, research, etc. The district governments should look after veterinary health services, breeding services, and livestock markets. Substantial support should be made available to the private sector for the modernization of techniques, better marketing, training of butcheries and slaughterhouses. Livestock insurance schemes should be started for all the farmers in order to safeguard them from natural calamities.

REFERENCES

Afzal, M. 2010. Improving veterinary service in Pakistan. *Pakistan Vet. J.*, 2009, 29(4): 206–210.

Afzal, N. and Ahmad, D. S. 2009. Agricultural input use efficiency in Pakistan. *Manag. Nat. Resour. Sustain. Future Agri.* 1: 1–12.

Al-Ghazi Tractors Ltd. 2003. Technical Features of FIAT Ghazi vs. MF-260 Turbo. Leaflets.

Ali, M., Ahmed, F. and Channa, H. D. 2015. Pakistan's Fertilizer Sector: Structure, Policies, Performance and Impacts. Working Paper, Pakistan Strategy Support Program, International Food Policy Research Institute, Washington, DC.

Altaf, Z. 2006. *Agricultural Support Prices in Pakistan: Dogma and Doctrinaire.* Directorate of Agricultural Information, Punjab, Lahore.

AMIS. 2016. Agricultural Marketing Information System. Online available at http://www.amis.pk.

ASTI-IFPRI. 2015. *Agricultural Research and Development Indicator Factsheet of Pakistan.* Agricultural Science and Technology Indicators, Washington, DC. November 2015.

Baig, M. B. and Aldosari, F. 2013. Agricultural extension in Asia: Constraints and options for improvement. *J. Anim. Plant Sci.* 23(2): 619–63.

Chaudhary, M. G. and Herring, R. 1974. The 1972 land reforms in Pakistan and their economic implication: A preliminary analysis, Islamabad, Pakistan. *Pak. Dev. Review.* 13(3): 245–279.

Chaudhary, F. M., Saqib, G. S. and Rizvi, M. A. 1998. *Fundamental of Tractor Mechanics and Energy Conservation.* National Book Foundation, Islamabad, Pakistan.

Culpin, C. 2000. *Farm Machinery.* 10th Edition, John Wiley & Sons, Inc., USA.

Dancey, R. J. 1993. The evolution of agricultural extension in England and Wales- Presidential address. *Journal of Agri. Economics*, (44): 375–393.

Dixit, J., Sharma, S. and Ali, M. 2014. Present status, potential and future needs for mechanization of agricultural operations in Jammu and Kashmir State of India. *Agric Eng Int: CIGR Journal*, 16(3): 87–96.

Etzel, M. J., Walker, B. J. and Stanton, W. J. 2005. *Marketing.* McGraw-Hill Inc., USA.

FAO 2005. Food and Agriculture Organization of the United Nations Regional Office for the Near East Cairo, Egypt.

FAO and UNIDO. 2010. Agro-Industries for Development.

GOP. 1988. *Agricultural Statistics of Pakistan 1987–1988,* Ministry of Finance, Islamabad.

GOP. 1990. *Food and Marketing Margins.* Ministry of Food, Agriculture & Cooperatives, Islamabad.

GOP. 2010. *Economic Survey of Pakistan 2009–2010.* Finance Division, Ministry of Finance, Government of Pakistan.

GOP. 2011. *Economic Survey of Pakistan 2010–2011.* Finance Division, Ministry of Finance, Government of Pakistan.

GOP. 2015. Ministry of Professional and Technical Training, AEPAM, Islamabad.

GOP. 2016a. *Economic Survey of Pakistan 2015–2016.* Finance Division, Economic Advisor's Wing, Islamabad.

GOP. 2016b. *Agricultural Statistics of Pakistan 2015–2016.* Ministry of Food and Agriculture (Economic Wing), Islamabad.

GOP. 2016. Project Management Unit, Board of Revenue, Government of Punjab. Online available at http://www.punjab-zameen.gov.pk/.

HEC. 2015. *5th Ranking of Pakistani Higher Education Institutions.* HEIs, Islamabad, Pakistan.

Hunt, D. and Wilson, D. 2016. *Farm Power and Machinery Management.* 10th Edition, Waveland Press, Inc., Illinois, USA.

Hussain. 2004. *Economic Management in Pakistan 1999–2002.* OUP Catalogue. Oxford University Press, Karachi, Pakistan.

Iqbal, M., Ahmad, M., and Abbas, K. 2003. The Impact of Institutional Credit on Agricultural Production in Pakistan. *The Pakistan Development Review*, pp. 469–485.

Irfan, M., Arif, G., Ali, S. M., and Nazil, H. 1999. The Structure of Informal Credit Market in Pakistan. PIDE Research Report No. 168.

Khan, M. A. J., Lodhi, T. E., Idrees, M., Mahmood, Z. and Munir, S. 2011. Training needs of agricultural officers regarding mechanized farming in Punjab, Pakistan. *Sarhad J. Agric.* 27(4):633–636.

Millat Tractors Ltd. 2014. *Millat Equipment Spectrum*. Booklet.

Mirani, Z. and Memon, A. 2011. Farmers' assessment of the farm advisory services of public and private agricultural extension in Hyderabad district, Sindh. *Pak. J. of. Agri. Res.* 24(1–4): 56–64.

Mohy-Ud-Din, Q and Badar, H. 2011. *Marketing of Agricultural Products in Pakistan: Theory and Practices*. Higher Education Commission, Faisalabad, Pakistan.

Pakistan Agricultural Machinery Census. 2010. *Agricultural Census Organization, Statistical Division*, Government of Pakistan, Pakistan, Islamabad.

Pakistan Bureau of Statistics. 2010. Census of Agriculture (2010). Online available at http://www.pbs.gov.pk/

Pakistan Bureau of Statistics. 2016. Pakistan Mouza Census, 2008. Online available at http://www.pbs.gov.pk/

Pakistan Telecommunication Authority. 2008. Annual Report of PTA, Government of Pakistan. Online available at http://www.pta.gov.pk/

Punjab Development Statistics. 2013. Bureau of Statistics. Government of the Punjab, Lahore.

Raza, S. A., Ali, Y. and Mehboob, F. 2012. Role of agriculture in economic growth of Pakistan. *Munich Personal RePEc Archive* 1–8.

Saravanan, R. 2001. Privatization of agricultural extension: In: C. P. Shekara. (Ed.) *Private Extension in India: Myths, Realities, Apprehensions and Approaches*. National Institute of Agricultural Extension Management (MANAGE), Hyderabad, India. pp: 60–71.

Sial, M. H., Iqbal, S. and Sheikh, A. D. 2012. Farm size productivity relationship: Recent evidence from central Punjab. *Pak. Econ. Soc. Rev.* 50(2): 139–162.

Umali, D. L. and Schwartz, L. 1994. Public and Private Agricultural Extension: Beyond Traditional Frontiers.

Wallace, J. and Williamson, I. P. 2006. Building land markets. *Land Policy Journal* 23(2): 123–135.

10 Input Supplies
Production of Quality Seeds

Muhammad Sarwar Khan and Faiz Ahmad Joyia

CONTENTS

10.1 INTRODUCTION

Seed is the backbone of the agricultural sector and the seed value chain in Pakistan is a missing link in enhancing farm productivity. Though annual seed demand in Pakistan is huge—reaching 170 million tons in 2014–2015, yet it is informal as farmers rely on their own seed for most of the crops, which are not of high quality. Without a steady supply of high quality seed, crop productivity as well as crop quality would greatly decline (Vellve, 2013). Genetic improvement techniques, based on plant biotechnology and plant breeding, bestow an increase in crop yield without relying on an increase in land area. It may also contribute to global food security as it is a cost-effective tool for increasing the nutritional value of crops (Xu et al., 2006). Molecular breeding of varieties specifically adapted to the unique conditions require traits that help plants to grow well under unfavorable environmental conditions. It can be achieved by introducing genes for important agronomic traits like increase in water and nutrient use efficiency, weed competitiveness, pest and disease resistance, early maturity, abiotic stress tolerance, that is, resistance to drought, salinity, waterlogging, and extreme temperatures (Brookes and Barfoot, 2006; Seo et al., 2011). Nevertheless, all such efforts would be useless without an effective seed production and distribution system.

Pure and healthy seed of high yielding varieties is the most indispensible input for sustainable crop production. Genetically modified seed possess various measurable attributes, which help to determine its value for farmers with high production potentials while reducing other costs like pesticides, fertilizers, etc. With the increased access to information technology, the farmers are becoming more aware of the importance of high quality seed (Whitford et al., 2013; Zhang et al., 2014). Progressive growers always demand high quality seed including GMOs and hybrids owing to their higher germination potential, homogeneity, and enhanced yields. Such high quality seeds are selected with improved characteristics resulting from heterosis or combining the abilities of parent plants (Stangland and Lynch, 2014) and genetic modification through modern biotechnological approaches. All categories including hybrid, nonhybrid, and GM seeds have their pros and cons. Hybrid seed is more costly than nonhybrid. Fresh hybrid seed needs to be bought every planting season and farmers cannot replant their grain as seed without major reductions in yield due to poor and patchy germination and reduced yield potential. Hybrid seed guarantees a uniform crop with better yield in cross-pollinated crops. It is the first filial generation of a cross between two inbred lines. If its seed is harvested and kept for next sowing season, it turns up as a second filial generation with higher genetic segregation rate resulting inpoor yield potential. Subsequent segregating generations keep on losing uniformity causing a decline in crop productivity. High prices of hybrid seed have become an extra burden for low-income farmers—in addition to fertilizers and pesticides. However, nonhybrid seeds exhibit lower yield potential and higher susceptibility to disease and pests. Due to the high cost of hybrid seed, poor farmers are preferably using nonhybrid seeds available in local markets or their own stored seed. The only benefit associated with nonhybrid seed is the production of reproducible seed (Bekele et al., 2008). Alternatively, genetically modified seeds may possess both qualities. Genetic modification may bestow upon it a superior production ability while farmers can use their own seeds for multiple generations before the seed loses its vigour or the expression of engineered trait(s) may decline.

During the year 2014–2015, Pakistan imported 752,384 M. tons of seed (Table 10.1) of different crops worth more than US$ 798 million (Pakistan Economic Survey, 2014–2015). This includes the major crops like maize, rice, sorghum, berseem, lucerne, potato, and vegetables.

Local seed production of major crops like wheat and cotton is more or less meeting the domestic needs. Most of the farmers produce their own wheat seed irrespective of land holding size. Seed companies of the government and private sector are contributing to the supply of wheat seed for farmers. However, wheat seed is neither certified nor graded or processed in most cases.

Cotton seed production in Pakistan is a major issue. The seed cotton is ginned on ginning machines that can never serve the purpose of a seed processing unit. Cotton of different varieties is brought by different farmers at ginneries and, after removing the lint, whatever is left is declared as seed. Studies have shown that the germination percentage of such seeds is always below 30%. Cotton seed is purchased/produced and sold by more than a thousand registered and unregistered seed companies. Sometimes, cottonseed is sold before it is approved by the competent statutory and regulatory bodies. Moreover, a single variety/line is sometimes sold with different names by

TABLE 10.1
Pakistan Seed Market Size 2014–2015

Crop	Area Planted (Acres)	Volume (M. Tons)	Value (US $)
Cereal crops	31,044,295	258,712	330,687,290
Oil seeds	1,870,383	8,009	21,593,118
Pulses	4,547,270	72,845	70,994,737
Fodder crops	3,853,200	96,590	51,332,894
Pasture grasses	2,568,800	19,110	25,409,868
Vegetables	1,210,300	9,198	110,643,717
Spices	14,820	120	505,263
Industrial (cotton)	7,410,000	45,000	26,644,737
Tubers	296,400	242,800	161,052,632
Total	52,815,468	752,384	798,864,256

Source: Pakistan Economic Survey. 2014–2015. Ministry of Food, Agriculture and Livestock, Islamabad different companies. The cotton farmers do have the desire to cultivate new varieties of cotton and are always requesting the breeders and seed companies for even minute quantities of seed of newly approved varieties or elite lines. Sometimes, breeders provide seed to farmers for trial purposes and the farmers themselves start multiplying and selling promising lines locally.

The total seed demand of fodder crops is almost entirely being met with imported seed since crops are not sown for seed purpose. As the fodder is based on the most nutritive and palatable foliar plant parts, fodder crops are generally harvested before the commencement of the reproductive phase because during this phase photosynthates are routed towards flowers and seed, thus decreasing the nutritive value of foliage. In Pakistan, there is no research organization working on the production of genetically modified (GM) fodder crops. Hence, in Pakistan no GM fodder seeds are being produced in-spite of increasing fodder demands for increasing the herbivore's population.

The exotic seeds produced in different environments are not well adapted for local environmental conditions and there is always a threat of yield decline. The absence of apposite technologies and regulatory frameworks related to varietal development, seed production, harvesting, threshing, processing, labeling, storage, marketing, IPRs, breeders rights, and other activities in this chain is a big constraint for seed development and production in the value chain of the country. Due to these factors, locally produced seed is neither of higher quality nor in sufficient quantity to meet domestic needs. Consequently, it is imported, costing huge amounts of foreign exchange. Furthermore, the import of seed is often poorly handled regarding the rules of quarantine, thus there is a high potential for introducing new diseases, insects, and pests.

10.2 CONSTRAINTS AND WAY FORWARD

A number of critical factors—listed below—can help comprehend numerous issues in this value chain, which needs to be addressed appropriately to uphold the agriculture, and ultimately, the economy of the country.

10.2.1 Variety Development

The varieties are developed through various selection methods in accordance with the breeding objectives. The breeders set their objectives and develop the varieties taking proper time; however, sometimes the seed goes in the hands of some private seed companies before proper evaluation, which results in a quicker runoff of the varieties. In Pakistan, the number of varieties or hybrids developed is very large for some crops like wheat, rice, cotton, and sugarcane but not sufficient for other crops. This depends upon the available trained manpower, which for some crops is adequate while inadequate for others. Moreover, the number of varieties approved each year should be kept to appropriate levels, which otherwise results in rapid variety extinction and greater variability in the final produce at national level.

Huge efforts are needed to develop advanced lines which are distinct, uniform, and stable (DUS) across the agroecological zones. Such lines possess valuable characters of agronomic and economic importance. The federal seed certification and registration department (FSC&RD) in Pakistan do not register such advanced lines as germplasm lines, which is discouraging for breeders. It results in the loss of such precious material that otherwise would become the basis of variety development.

Variety development is long and laborious work. Biologically, it depends upon the crop reproduction system and the objectives of variety development. A crop plant may be self-pollinated, cross-pollinated or often cross-pollinated. Breeders achieve high level homozygosity in cultivars of self-pollinated crop varieties (like wheat and rice), a workable homozygosity in cross-pollinated crops (like cotton and sorghum), and complete heterozygosity in case of hybrid cultivars. Therefore different breeding methods are adopted in different crops. Hybridization is done when the genes from different plants are to be brought together. On the other hand, biotechnological techniques are a good alternative if single or few genes are to be transferred to an otherwise good variety.

Biotechnology has an added advantage of transferring genes from unrelated plants or genes from any organism to a specific crop plant. Varietal development through conventional plant breeding, as well as modern biotechnological methods, have specific objectives like seed yield, biomass yield, disease resistance, insect resistance, abiotic stress tolerance, early maturity, enhanced nutritional quality, and ease of harvesting.

10.2.2 Seed Multiplication

Seed farms established in the public sector are used to produce seeds of different crops and meet almost 30% of the seed demand in Pakistan, while another 30% is met by private companies through either seed multiplication or import. The seed imported by private companies is sold to farmers according to their own business rules. These activities discouraged the local seed commerce and encouraged the import of seed, which is still at this time a dominant activity. Seed multiplication from the breeder's foundation seed should be based on standard international procedures and protocols. It should be free of diseases, insect pests, and other such impurities. It should be grown on plenty of land with ample inputs to guarantee its health and vigor. The crop for seed purpose should be free of any biotic, abiotic, edaphic, or environmental stresses.

10.2.3 Harvesting and Threshing

The crops sown for seed purpose should be harvested and threshed with the most appropriate machines at the proper physiological stages of maturity. Harvesting immature or overmature crops will deteriorate seed quality. Moreover, inapposite threshing practices produce damaged seeds. The best time to harvest would be during rain free, clear, and dry season at optimum crop maturity. Sophisticated instrumentation is available to measure seed moisture content, which will enhance the quality of the harvested and threshed seeds. Higher seed water content at the time of threshing would result in seed breakage. Moreover, it enhances postharvest losses, fungal, and bacterial infection rates, thus deteriorating the storage life as well as the germination percentage of seed.

10.2.4 Drying and Cleaning

For desiccation, the threshed seed should be placed on tidy daises in a moisture-free environment having appropriately high temperature. It is much better if it is made of concrete or at least with mud plaster. The place should have good ventilation. The seed should be cleaned using all possible ways to remove contaminations. Mechanized processes of seed cleaning and drying are also available, which should be adopted to attain the highest international standards.

10.2.5 Packaging and Storage

Temperature and relative humidity are the most important factors affecting seed quality and storage longevity. Both seed quality and storage longevity are reciprocally proportional to the temperature and humidity. Lower are the temperature and relative humidity, higher will the quality and longevity of storage be, and vice versa. For storage purposes, seed should be packaged in bags of appropriate materials, which can guarantee humidity protection and repel stored-grain pests. Fungicide and insecticide treatment are necessary to avoid pest attacks during storage as well as germination. The seed lots should be periodically inspected for seed quality and scouted for pests. Characteristically, good quality seed is healthy, physiologically mature, and pure to type—having uniform size, shape, and color. It should be free of insect-pests, their propagules (such as eggs, pupae, hyphae, or spores), inert matter, other variety/crop/weed seeds, and damaged seeds. It is necessary that storage facilities be equipped to maintain these characteristics of quality seed.

10.2.6 Marketing

A major portion of the seed in the country is furnished by the farmers. It is relatively inexpensive and easily accessible. In such a situation, it is desirable to train farmers in the maintenance of their own seed, as a part of their main crop production. Seed marketing in Pakistan is mainly carried out by the provincial seed corporations of the public sector, while commission agents and seed companies market seeds for the private sector. However, in the case of private commission agents, the seed lots are not appropriately labeled, which should otherwise clearly show the crop and variety name, germination percentage, production and sowing timing, origin and net weight, certification class of seed, seed type, whether it is open pollinated or hybrid, and certifying agency. It should be clearly mentioned that farmers should not keep seed they produce themselves after sowing hybrid seeds. The label should have a proper color clearly indicating the type of seed. The government should exercise its authority to enforce the truthfulness of labels. A third key player are the private seed companies, which sometimes produce own seed, but mostly purchase it from different farmers after they clean and pack the seeds with their brand names, and sell it out to the farmers. Seed acts, seed rules, and breeders rights in this domain must be observed and protected by the law enforcement agencies of the government at federal as well as provincial levels. Furthermore, an advisory group consisting of government representatives, scientists, academia, farmers, and seed companies should be set up at national and provincial levels to regulate these laws and take care of the rights of all stakeholders including farmers, breeders, and sellers. The farmers lack the ability to observe all these sale requirements, hence promoting their capacity building is also urged. Farmers should be encouraged to buy seeds from professionally trained seed producers or reputable seed companies.

The increase in production of agricultural crops depends not only on the development of higher yielding seed varieties but also on the efficiency of the systems available to ensure that these seeds reach farmers on time. Effective seed marketing is thus an essential component of activities to improve food security. Seed marketing should aim to satisfy the farmer's demand for reliable supply of a range of improved seed varieties at an acceptable price. However, the difficulties of organizing effective seed delivery systems, especially to small-scale farmers, have often been neglected in comparison with the supply of progressive growers and large land holders (Ashfaque, 2013). There is a need for strong coordination between public and private seed sectors to undertake efficient seed distribution and minimize the monopoly of any sector (Spielman et al., 2014).

10.2.7 Maintaining Genetic Purity

Maintenance of genetic purity is inevitable to sustain seed quality, farmer's reliability, and viable crop production. Numerous factors including breeding methods, natural crossing, mechanical seed mixing, mutations, and genetic drift play an active part in seed quality deterioration. The presence of phenotypic variation in a varietal stand not only diminishes the crop productivity but also deteriorate crop quality. Hence, it is obligatory to renew or reconstitute the varietal seeds periodically for its longevity in the field. This is especially the case with open pollinated varieties as they are naturally cross-pollinated.

10.3 AGRONOMIC PRACTICES OF SEED PRODUCTION

Best agronomic practices, resulting high crop productivity, will also yield good quality seed. Hence, to attain good quality seed essential agronomic activities include selection of the best crop varieties, most fertile pieces of land, precise sowing methodologies, ideal sowing dates, optimum plant populations, and appropriate supply of inputs. It also entails the effective control of weeds, insects, diseases and other pests, as well as roughing, availability of pollinators if required, timely harvesting, and proper threshing. Then, seeds should be dried, cleaned, graded, treated, and stored in ventilated and moisture free places. All these activities must be performed according to the crop requirements precisely and timely.

10.4 SEED TESTING

Purity testing, health testing, and GMO detection at genetic and molecular levels has become robust with the advent of modern biotechnological diagnostic tools. Currently, the Federal Seed Certification and Registration Department (FSC&RD) performs seed regulatory functions under the Seed Act of 1976, through its 16 seed testing laboratories/offices located in various ecological zones of the country. However, the department lacks modern biotechnological tools for diagnostic purposes at large. After the 18th constitutional amendment, the management of this department is in a confused state, which is hampering the seed sector and especially the establishment of modern seed testing infrastructures and the hiring of skilled personnel. Hence, there is an urgent need of establishing DNA barcoding and forensic laboratories, at least at division levels, not only for certification of new varieties but also for testing of seeds provided by the public and private sectors on an annual basis.

10.5 GM SEED AS PREFERABLE PLATFORM OF BIOPHARMING

Seeds can be employed as innovative and scalable platforms of "biopharming" and can be used to manufacture pharmaceutical proteins, other therapeutics, and expensive molecules. Hence, GM seeds may be regarded as small packets of biopharmaceuticals. This system is much more advantageous being cost effective, involving transport and storage friendly products, presents minimum human pathogenic contamination, and is easily scaled up (Chebolu and Daniell, 2009). It is anticipated that the existing systems will be augmented soon for seed-based biopharmaceuticals, which will enhance the worth of agriculture (Raskin et al., 2002). Seeds of cereals like rice and dicots like soybean seem to be even more advantageous for the production of recombinant molecules. Because endosperm specific expression systems can yield greater amounts of recombinant proteins with added benefits of long term storage without the need of any cold chain. Significant advancements have been reported for the exploitation of GM seeds as "recombinant protein bioreactors." Significant exmples are human lactoferrin (Nandi et al., 2002; Suzuki et al., 2003), human lysozyme (Yang et al., 2001; Huang et al., 2002a,b), human insulin-like growth factor-1 (hIGF-1) (Panahi et al., 2004), recombinant pollen allergen (Okada et al., 2003), recombinant human α1-antitrypsin (rAAT) (Terashima et al., 2002; Trexler et al., 2002), and industrial enzyme like transglutaminase by over-expressing transglutaminase (rTGP) gene (Claparols et al., 2004).

10.6 FUTURE PROSPECTS

1. The private sector should be encouraged to establish systems of seed development, processing, grading, packaging, and marketing. The government may provide a liberal credit to these companies to keep consumers prices at lower levels.
2. Promising varieties with one or two specific weaknesses should be transformed for producing GM crops and their seeds should be provided to farmers on subsidized rates.
3. The variety development process would be enhanced by enacting seed rules and breeders' rights acts. These rules have long been put on the table but have not been properly executed and implemented.
4. Legislation in the seed sector is direly in need to frame appropriate regulations for the seed sector. The germplasm is registered after 2 years of DUS (Distinctive, Uniform, and Stable) testing and a variety/cultivar is registered after its approval at international level. The seed rules in Pakistan should have the provision to register the germplasm lines.

5. Trained personnel for farming communities, science, and academia is urgently needed in the seed sector. Universities should play vital role in this regard. University of Agriculture Faisalabad has accepted the challenge of capacity building through the initiation of new degree programs in seed science and technology.
6. Seed testing labs should be established in each district, which should have state-of-the-art infrastructures and skilled human resource.
7. The government should encourage research and development activities in variety development, seed multiplication, and the use of certified seed of the approved crop varieties.
8. Groups of progressive growers may be formed at village level to monitor the supply of quality seeds to the farming community.

10.7 STRATEGIC ACTION PLAN

Short Term

- Seed treatment with fungicides should be made compulsory for seed sale
- Effective quarantine measures during seed import

Medium Term

- Trained manpower for improving the informal seed sector through on-farm hands on training
- National private sector should be strengthened through provision of soft loans through the commercial banking sector
- Punjab Seed Corporation (PSC) should focus on crops where private sector is not providing seed

Long Term

- Immediate implementation of regulations like the Amended Seed Act and Plant Breeders Rights
- Policy research needs to be reoriented to the informal sector
- Variety release procedures should be simplified and made more transparent
- The role of FSC&RD should be redefined

REFERENCES

Ashfaque, A. 2013. Rural marketing strategies for selling products & services: Issues & challenges. *J Bus Manage Soc Sci Res.* 2(1):2319–5614.

Bekele, A. S., A. K. Tewodros and Y. Liang. 2008. Technology adoption under seed access constraints and the economic impacts of improved pigeon pea varieties in Tanzania. *J Int Assoc Agri Econ.* 39:309–323.

Brookes, G. and P. Barfoot. 2006. Global impact of biotech crops: Socio-economic and environmental effects in the first ten years of commercial use. *Agri Bio Forum.* 9:139–151.

Chebolu, S. and H. Daniell. 2009. *Plant Produced Microbial Vaccines.* Springer, Berlin Heidelberg, 332:33–54.

Claparols, M. I., L. Bassie, B. Miro, S. Del-Duca, J. Rodriguez-Montesinos, P. Christou, D. Serafini-Fracassini and T. Capell. 2004. Transgenic rice as a vehicle for the production of the industrial enzyme transglutaminase. *Transgenic Res.* 13:195–199.

Huang, J. M., S. Nandi, L. Y. Wu, D. Yalda, G. Bartley, R. Rodriguez, B. Lonnerdal and N. Huang. 2002a. Expression of natural antimicrobial human lysozyme in rice grains. *Mol Breed.* 10:83–94.

Huang, J. M., L. Y. Wu, D. Yalda, Y. Adkins, S. L. Kelleher, M. Crane, B. Lonnerdal, R. L. Rodriguez and N. Huang. 2002b. Expression of functional recombinant human lysozyme in transgenic rice cell culture. *Transgenic Res.* 11:229–239.

Nandi, S., Y. A. Suzuki, J. M. Huang, D. Yalda, P. Pham, L. Y. Wu, G. Bartley, N. Huang and B. Lonnerdal. 2002. Expression of human lactoferrin in transgenic rice grains for the application in infant formula. *Plant Sci.* 163:713–722.

Okada, A., T. Okada, T. Ide, M. Itoh, K. Tanaka, F. Takaiwa and K. Toriyama. 2003. Accumulation of Japanese cedar pollen allergen, Cry j 1, in the protein body I of transgenic rice seeds using the promoter and signal sequence of glutelin *GluB-1* gene. *Mol Breed.* 12:61–70.

Panahi, M., Z. Alli, X. Y. Cheng, L. Belbaraka, J. Belgoudi, R. Sardana, J. Phipps and I. Altosaar. 2004. Recombinant protein expression plasmids optimized for industrial *E. coli* fermentation and plant systems produce biologically active human insulin-like growth factor-1 in transgenic rice and tobacco plants. *Transgenic Res.* 13:245–259.

Pakistan Economic Survey. 2014–2015. Ministry of Food, Agriculture and Livestock, Islamabad.

Raskin, I., D. M. Ribnicky, S. Komarnytsky, N. Ilic, A. Poulev, N. Borisjuk, A. Brinker et al. 2002. Plants and human health in the twentyfirst century. *Trends Biotechnol.* 20:522–531.

Seo, Y. S., M. S. Chern, L. E. Bartley, T. E. Richter and M. Han. 2011. Towards a rice stress response interactome. *PloS Genetics.* 7(4):e1002020.

Spielman, D. J., D. E. Kolady, A. Cavalieri and N. C. Rao. 2014. The seed and agricultural biotechnology industries in India: An analysis of industry structure, competition, and policy options. *Food Policy.* 45:88–100.

Stangland, G. R. and P. J. Lynch. 2014. Plants and seeds of hybrid corn variety CH923684. Monsanto Technology Llc. U.S. Patent 8,816,174.

Suzuki, Y. A., S. L. Kelleher, D. Yalda, L. Y. Wu, J. M. Huang, N. Huang and B. Lonnerdal. 2003. Expression, characterization, and biologic activity of recombinant human lactoferrin in rice. *J Pediatr Gastr Nutr.* 36:190–199.

Terashima, M., N. Hashikawa, M. Hattori and H. Yoshida. 2002. Growth characteristics of rice cell genetically modified for recombinant human α1- antitrypsin production. *Biochem Eng J.* 12:155–160.

Trexler, M. M., K. A. McDonaldand and A. P. Jackman. 2002. Bioreactor production of human α1-antitrypsin using metabolically regulated plant cell cultures. *Biotechnol Prog.* 18:501–508.

Vellve, R. 2013. *Saving the Seed: Genetic Diversity and European Agriculture.* London.

Whitford, R., D. Fleury, J. Reif, M. Garcia, T. Okada, V. Korzun and P. Langridge. 2013. Hybrid breeding in wheat: Technologies to improve hybrid wheat seed production. *J Exp Bot.* 64:5411–5428.

Xu, K., X. Xu, T. Fukao, P. Canlas and R. M. Rodriguez. 2006. Sub1A encodes an ethylene responsive-like factor that confers submergence tolerance to rice. *Nature.* 442:705–708.

Yang, D. C., L. Y. Wu, Y. S. Hwang, L. F. Chen and N. Huang. 2001. Expression of the REB transcriptional activator in rice grains improves the yield of recombinant proteins whose genes are controlled by a *Reb*-responsive promoter. *Proc Natl Acad Sci USA.* 98:11438–11443.

Zhang, H., X. Chenxi, Y. He, J. Zong, X. Yang, H. Si, Z. Sun, J. Hu, W. Liang and D. Zhang. 2014. Mutation in CSA creates a new photoperiod-sensitive genie male sterile line applicable for hybrid rice seed production. *Proc Nat Acad Sci USA.* 110:76–81.

11 Fertilizers and Gypsum

Javaid Akhtar, Muhammad Yaseen, Muhammad Rashid, Ghulam Murtaza, and Muhammad Aamer Maqsood

CONTENTS

11.1 FERTILIZERS

11.1.1 Role of Fertilizer in the Crop Production

In Pakistan, fertilizer is most important and expensive input for agricultural production. It is a great challenge to increase crop productivity with the aim to fulfill the food demand over the next 50 years due to increasing populations, decreasing arable land, rapid global climate change, increasing water scarcity, and rising prices of agricultural inputs. The role of fertilizers in this scenario has gained increasing importance to obtain a high agricultural production from the same piece of land, while soil fertility decline. To meet the expected food demand without substantial increases in food price an increase of 70%–100% in production, compared to present productivity, will be required (UN, 2011). Furthermore, high cereal production is only possible through high inputs of fertilizers. Currently, 40%–60% of the cereal production depends upon fertilizer use and by 2050, almost 110% grain production will have to depend on fertilizers (Tilman et al., 2011).

At present, fertilizer use in the country is imbalanced. The data on fertilizer use reflect the extreme predominance of nitrogen in nutrient consumption over many years. This has created a serious imbalance between nitrogen, phosphorous, and potassium while the use of micronutrients is still at small scale. This scenario is observed in developing countries particularly, where populations are increasing rapidly but access to agricultural inputs is limited (Herder et al., 2010). Therefore, optimum utilization of chemical fertilizers in intensive cropping systems is necessary for adequate nutrient supply and optimum crop yield. The overall use of nitrogen, phosphorus, and potassium fertilizer is currently about 45 million tons (Heffer and Prud'home, 2012) and by 2030, the demand will rise by 20 million tons. The demand has increased at the rate 1.5%, 2.3%, and 3.7% annually for nitrogen, phosphorous, and potasium, respectively (Vance et al., 2003).

This chapter highlights the present status and role of fertilizers in crop production and the constraints associated with fertilizer supply, as well as second generation approaches used to increase fertilizer efficiency, in order to develop guidelines for the balanced use of fertilizers in the country.

11.1.2 Issues Related with Supply of Fertilizer at Critical Time

Optimum crop production depends upon timely application of fertilizers to crops, which in turn depends upon the timely supply of fertilizers. The last 5 years' data concerning fertilizer use in Pakistan provides an understanding of issues related to fertilizer supply and use. The domestic production of nitrogen fertilizers, from July to March 2011–2012 decreased by 1.4% as compared to the previous year's production, because the fertilizer industry experienced a curtailment of natural gas, which is the raw material for urea. However, a timely import of urea covered up shortages in supply and thus increased total availability of fertilizer by 16.3%.

The prices of urea rose by 81.4% in July to March 2011–2012 as compared to the same period of the previous year. The prices of Diammonium phosphate (DAP), Calcium Ammonium Nitrate (CAN) and Nitrophos (NP) also rose by 38.8%, 75.5%, and 45.7%, respectively, compared with the same period in 2010–2011. In 2010–2011, domestic production of fertilizer was 2,287,000 tons, while in 2012 it decreased to 2,255,000 tons. Fertilizer availability at critical crop stage is reduced due to following factors:

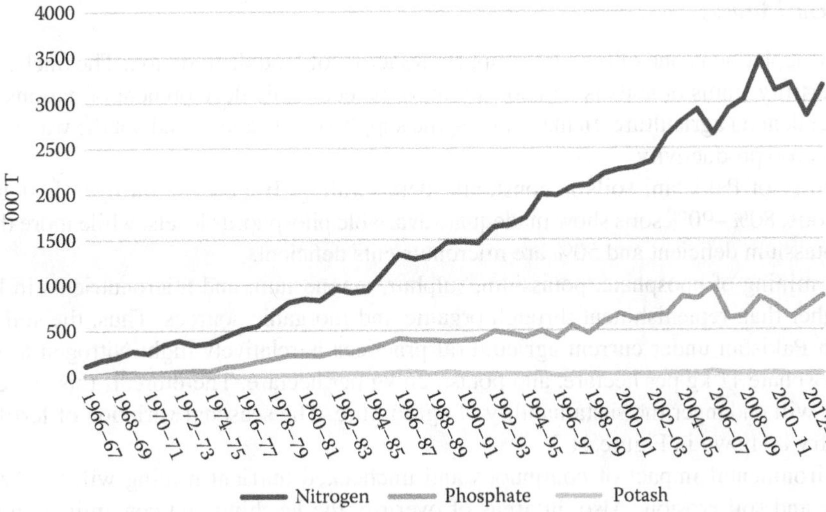

FIGURE 11.1 Changing trend in the use of NPK fertilizers over time.

- Power crisis
- Gas shortage
- Transportation/handling issues
- Increased energy cost
- High fertilizer cost.

Overall, the fertilizer off take in 2011–2012 was 4.1% less against achieved and 7.5% less against target as compared to 2010–2011. The Figure 11.1 and Table 11.1 show the use of NPK fertilizers through time and the ratio of N:P:K applied by the farmers.

TABLE 11.1
Trend in the Ratio of N:P:K Applied by Farmers Over Time

NPK Ratio through Time

Year	N:P:K Ratio
1999–2000	3.71:1:0.03
2000–2001	3.33:1:0.03
2001–2002	3.64:1:0.03
2002–2003	3.62:1:0.03
2003–2004	3.74:1:0.03
2004–2005	3.24:1:0.04
2005–2006	3.45:1:0.03
2006–2007	2.72:1:0.04
2007–2008	4.65:1:0.04
2008–2009	4.67:1:0.04
2009–2010	4.05:1:0.03
2010–2011	4.08:1:0.04
2011–2012	5.18:1:0.03
2012–2013	3.83:1:0.03
2013–2014	3.47:1:0.02
Desired ratio	2.0:1:0.4

11.1.3 Soil Mining

Soil fertility depletion is one of the most important factors of land degradation. The unchecked and declining fertility status of soils is a major threat to the economic development of nations who are mostly dependent on agriculture. In many areas, the supply of nutrients—and not the water supply—often limits crop productivity.

The fertility of Pakistani soils is constantly deteriorating. In general, nitrogen is deficient in almost all soils, 80%–90% soils show inadequate available phosphorus levels, while more than 40% soils are potassium deficient and 50% are micronutrients deficient.

Nutrient mining of phosphate, potassium, sulphur, magnesium, and micronutrients in Pakistan is quite higher than replenishment through organic and inorganic sources. Thus, the soil nutrient depletion in Pakistan under current agricultural practices is relatively high: Nitrogen at 9 kg per hectare, phosphate 11 kg per hectare, and potash 26 kg per hectare. Therefore, if this current trend continues it will endanger the sustainability of agriculture. The seasonal variation of fertilizer use in the country is shown in Figure 11.2.

The environmental impact of continuous and unchecked nutrient mining will be the loss of biodiversity and soil erosion. Also, in areas of overuse, the leaching and contamination of soils, ground water, and surface water may take place. Soil mining is comprised of two components, as explained in the following sections.

11.1.3.1 Soil Nutrient Mining by Natural Processes

Soil fertility decreases when nutrients present in soil are depleted by crop uptake while addition of nutrients through fertilizers and organic manures does not take place. This poor soil fertility leads to poor crop yields.

Depletion of nutrients from soil also takes place when the nutrient-rich topsoil is eroded. Different other factors are also involved in nutrient depletion—including overtillage which damages soil structure, and low amount of nutrient addition. Soil salinity may also decrease nutrient availability in some cases.

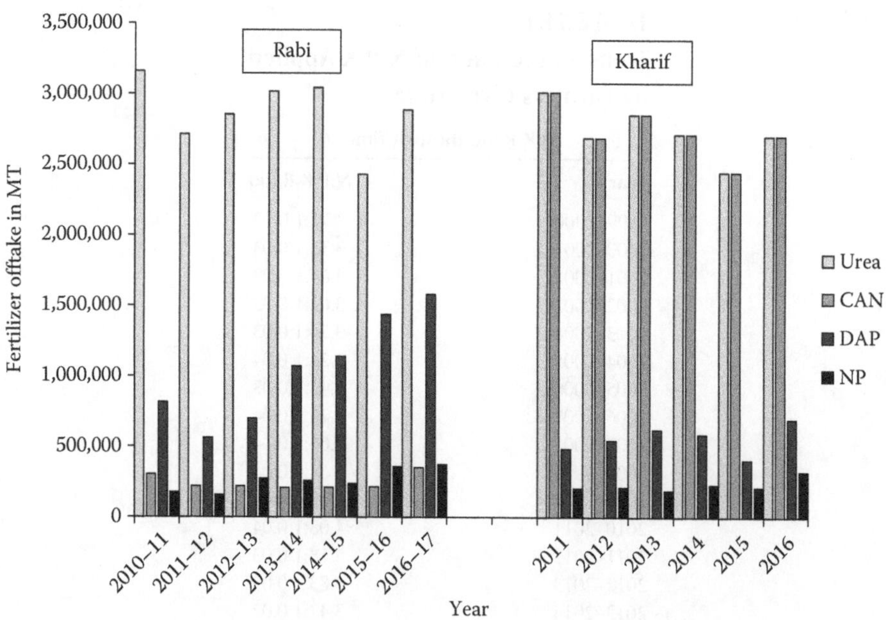

FIGURE 11.2 Trend of fertilizers off take during rabi and kharif cropping seasons.

Mutert (1996) reported that in Asia a negative nutrient balance for rice and other major crops occurred, particularly in developing countries with large size populations. In Pakistan, there are number of factors involved which affect the sustainable growth of agriculture. Among these factors, soil fertility depletion through mining of soil nutrients from cropped area is a real threat to food security and causes environmental degradation. Most of the Pakistani soils are poor in plant available nutrients and cannot support optimum level of crop productivity (Ahmad and Rashid, 2003). Widespread nutritional loss from fertile soils occurred through water erosion, coupled with continuous nutrient mining by crops in the Pothwar area of Pakistan (Rashid et al., 1997).

11.1.3.2 Soil Nutrient Mining by Human Induced Processes

Human induced activities include nutrient loss by harvesting of crops and other anthropogenic activities. Nutrient depletion by human activities can be due to overcultivation, insufficient supply of nutrients, accelerated soil erosion caused by inappropriate land use, poor soil management practices, and unbalanced fertilization, etc.

11.1.4 Current Status of Fertilizer Use and Crop Removal in Pakistan

In Pakistan, the total area under cultivation is about 22 m ha. Total share of grain crops is about 55%, followed by cotton and sugarcane 18%; pulses 5%; oilseed crops 3%, and fruit/vegetables 4%. Wheat is the main food crop of Pakistan and is used as a staple food. Wheat occupies of 36.3% of the total cropped areas.

Five major crops (wheat, rice, maize, cotton, and sugarcane) consume about 87% of total fertilizers in Pakistan. Wheat consumes about 45%, and cotton 23%. Sugarcane has the third highest nutrient use per ha. Fruit and vegetables share is 5.6%.

Fertilizer use at farm level is very low for potash and phosphorus. Only 2% of farmers use potash while 92% and 83% of the farmers apply nitrogen and phosphate, respectively. Consequently, there has been a decline in nutrient offtake in the recent years (Table 11.2).

11.1.5 Impact of Variable Fertilizer Application Rates on Soil and Crop

Fertilizer application rates in Pakistan vary with farm size. As farm area increases, fertilizer application rates decrease. In Pakistan, fertilizer usage on various crops is not according to the crop demand. Imbalance use of fertilizer, that is, the overfocus on nitrogen and minimal use of potassium and micronutrients is an important aspect to address. This nutrient mismanagement has resulted in potassium-deficient soils and affects crop yields. Different farm-level surveys should be conducted to assess fertilizer use by crop and its impact on productivity. The main source of information is fertilizer deliveries in a specific ecological zone from which per hectare fertilizer use can be calculated.

It is crucial to improve fertilizer use efficiency (FUE) to achieve optimum crop yields. Improvement in FUE and agricultural production with reduced inorganic fertilizer use, for economic and environmental reasons, will require a new generation of technologies (Dobermann, 2005) such as the controlled and slow release of fertilizers (Trenkel, 2010) or nutrient solubilizing bacteria and fungi impregnated fertilizers (Adesemoye et al., 2009).

11.1.6 Fertilizer Supply in the Context of Future Scenario

In Pakistan, the total availability of urea—including 122 thousand tones of imported supplies and 2451 thousand tons of domestic production—was about 2959 thousand tones against the off take of 2716 thousand tons, leaving an inventory of 184 thousand tons for Rabi 2014–2015 (NFDC, 2016).

TABLE 11.2
Month-Wise Comparative Nutrient Offtake (Tonnes) during Year 2015–2016 Compared with Previous Years

Year	2011–2012	2012–2013	2013–2014	2014–2015	2015–2016	Percent Change
			Total Nutrients ($N + P_2O_5 + K_2O$)			
July	353,248	271,491	291,051	345,145	283,164	−18.0
August	347,920	206,677	318,589	401,924	240,713	−40.1
September	362,517	273,135	385,177	291,383	105,478	−63.8
October	396,068	269,343	446,627	424,219	493,489	16.3
November	348,223	360,153	558,071	522,389	447,209	−14.4
December	451,127	566,392	461,323	491,670	492,628	0.2
January	309,366	292,560	323,695	361,072	321,673	−10.9
February	92,632	228,812	263,968	280,752	216,541	−22.9
March	180,061	259,048	256,002	258,609	238,430	−7.8
April	207,073	193,628	170,893	237,889	202,371	−14.9
May	220,600	310,032	300,904	319,420	287,739	−9.9
June	592,116	389,940	312,787	381,947	419,198	9.8
Total	3,860,952	3,621,210	4,089,087	4,316,419	3,748,633	−13.2
Percent change		−6.2	12.9	5.6	−13.2	
			Nitrogen			
July	272,117	212,118	244,245	266,940	235,805	−11.7
August	296,246	173,686	269,155	332,093	219,282	−34.0
September	289,765	167,955	290,354	227,674	97,096	−57.4
October	288,485	160,017	292,746	258,181	214,152	−17.1
November	259,616	222,006	350,046	311,692	285,840	−8.3
December	378,733	493,432	368,772	373,471	403,602	8.1
January	292,339	260,477	302,229	316,688	292,933	−7.5
February	80,272	210,581	207,331	240,601	184,696	−23.2
March	148,616	215,844	207,766	221,057	198,321	−10.3
April	169,778	165,859	141,856	195,658	168,288	−14.0
May	197,289	262,701	234,762	248,943	235,924	−5.2
June	533,262	308,777	275,268	315,650	358,239	13.5
Total	3,206,519	2,853,453	3,184,530	3,308,648	2,894,178	−12.5
Percent change		−11.0	11.6	3.9	−12.5	
			Phosphate			
July	78,067	58,377	45,302	76,295	46,471	−39.1
August	49,935	32,192	48,096	67,987	20,916	−69.2
September	70,462	102,024	92,300	59,925	7,287	−87.8
October	104,783	105,790	150,439	162,756	275,242	69.1
November	86,467	136,445	206,522	207,620	159,263	−23.3
December	70,248	70,785	91,462	114,794	86,822	−24.4
January	16,526	31,557	20,697	42,886	27,917	−34.9
February	11,403	16,626	53,952	37,996	29,994	−21.1
March	30,070	41,084	45,676	33,858	37,672	11.3
April	35,846	26,940	26,755	40,542	32,521	−19.8

(Continued)

TABLE 11.2 (*Continued*)
Month-Wise Comparative Nutrient Offtake (Tonnes) during Year 2015–2016 Compared with Previous Years

Year	2011–2012	2012–2013	2013–2014	2014–2015	2015–2016	Percent Change
May	21,682	46,170	63,476	67,757	49,771	−26.5
June	57,701	78,954	36,215	62,570	58,860	−5.9
Total	633,190	746,944	880,892	974,986	832,737	−14.6
Percent change		18.0	17.9	10.7	−14.6	
			Potash			
July	3,064	994	1,504	1,890	887	−53.1
August	1,739	799	1,338	1,844	515	−72.1
September	2,289	3,157	2,523	3,784	1,096	−71.0
October	2,800	3,536	3,443	3,282	4,096	24.8
November	2,140	1,702	1,503	3,077	2,106	−31.6
December	2,146	2,175	1,089	3,406	2,204	−35.3
January	500	525	769	1,598	848	−46.9
February	958	1,607	2,684	2,155	1,851	−14.1
March	1,375	2,119	2,560	3,694	2,437	−34.0
April	1449	829	2,282	1,689	1,562	−7.5
May	1630	1,161	2,667	2,720	2,044	−24.8
June	1153	2,210	1,304	3,726	2,098	−43.7
Total	21,243	20,814	23,666	32,865	21,744	−33.8
Percent change	−2.0	13.7	38.9	−33.8		

Source: NFDC. 2016. Fertilizer Review 2015-16, National Fertilizer Development Centre (NFDC), Planning and Development Division, Islamabad.

The future growth rate of fertilizers is expected to be approximately 4%–5%. The demand for DAP will probably remain higher than for urea because of greater advertising efforts by the local manufacturer-Fauji Jordan Fertilizer Company (FJFC). DAP remains that most important among the phosphoric fertilizers because of different benefits including its easy handling.

If present circumstances are addressed, the demand for various types of fertilizers will progressively increase. The government is expected to continue the subsidy on fertilizers. If it continues to do so in the coming years, fertilizers (mainly urea) will be sold at discount rates, which will increase the fertilizer use.

11.1.7 Challenges for Plant Nutrient Management

Plant nutrition management faces many challenges which include preservation of sustainable crop productivity and improving the quality of land and quality of water resources to fulfill food demands. Environmental hazards can be reduced by conforming fertilizer application to crop requirements and adopting sensible soil and water conservation methods. The loss of soil fertility from frequent nutrient mining, along with unbalanced plant nutrition practices, poses a serious threat to agricultural production in many countries.

The sources of plant nutrients must be described and information should be disseminated among farmers on their efficient use for agricultural intensification and their potential impact on the environment.

11.1.7.1 Reasons for Nonadoption of Suitable Plant Nutrient Management Practices in Pakistan

Maximum fertilizer use efficiency is obtained when all other factors of production such as seed, water, and cultivation practices are also optimum.

- Our farmers do not have access to pure and certified seeds, resulting in nutrient losses due to low nutrient efficient varieties.
- Either extremely low or high irrigation is a cause low FUE.
- Despite the large-scale mechanization of agriculture in different parts of Pakistan, most of the agricultural operations are carried out manually using simple and conventional tools that cannot render high nitrogen use efficiency (NUE).
- Farmers have small and fragmented land holding due to which they are unable to adopt and use modern techniques of improving nutrient use efficiency.
- Poor agricultural marketing has resulted in the farmer's distrust of the extension services and government policies.
- Inadequate storage facilities both on the farmer and government sides, as well as misguided government policies cause shortages of fertilizer supply.
- Lack of capital to purchase fertilizers, land implements, and approved crop seed result in low nutrient use efficiency.
- Automation of agriculture is growing but it is still at very low level in most areas of Pakistan and, consequently, the old and conventional techniques of production cannot increase the production and the nutrient use efficiency.

11.1.7.2 Enhancing Fertilizer Use Efficiencies (FUE): Conventional to Second Generation Approaches

Fertilizer use efficiency can be defined as the extent of fertilizer use recovery in agriculture per crop unit. Among various inputs, fertilizer is a costly input; therefore, this input must be utilized efficiently. Fertilizers are considered efficient when the maximum yield is obtained with the minimum possible application. It is indeed difficult to quantify the efficiency of a particular fertilizer since it depends on:

- Losses due to leaching
- Losses in gaseous forms
- Immobilization by chemical precipitation, adsorption on exchange complex and microbial cells
- Chemical reactions between various components in fertilizers during mixing, before application to soil
- Physical properties of soil
- Chemical properties of soil
- Fertilizer characteristics

Fertilizer recommendations are based on the nutrient supplying power of soils as determined by soil tests. For efficient use of fertilizer, 4Rs (right source, right rate, right place, and right time) is the best practice (Table 11.3).

11.1.7.3 Practical Approaches to Increase Efficiency of Fertilizers

In principle, maximum efficiency of agricultural input will be possible when maximum productivity is obtained per unit area per unit input applied. This will be possible when all inputs are used

TABLE 11.3
4R Principles of Nutrient Stewardship

Right Source	Right Rate	Right Time	Right Place
1. Supply in plant available forms	1. Appropriately assess soil nutrient supply	1. Assess timing of crop uptake	1. Recognize root-soil dynamics
2. Suit soil properties	2. Assess all available indigenous nutrient sources	2. Assess dynamics of soil nutrient supply	2. Manage spatial variability
3. Recognize synergisms among elements	3. Assess plant demand	3. Recognize timing of weather factors	3. Fit needs of tillage system
4. Blend compatibility	4. Predict fertilizer use efficiency	4. Evaluate logistics of operations	4. Limit potential off-field transport

simultaneously in appropriate amounts and manners over a given area. The following approaches must be considered to get high FUE as well as crop yield:

- Deviation from normal planting or sowing time suited for a particular crop variety in a particular locality will affect the efficiency of fertilizers.
- Optimum plant spacing and maintenance of optimum plant population are essential to get maximum benefits from the applied fertilizers.
- Effective organic matter recycling is essential to maintain fertility and productivity. Response for a nutrient is generally higher in soils supplied with adequate amounts of organic matter.
- While applying organic manures having high C:N ratio, ensure that adequate amounts of nitrogenous fertilizers are applied to soils to compensate biological locking up of nitrogen in microorganisms.
- Include a legume either in rotational sequence or as an intercrop. Legumes, besides fixing atmospheric nitrogen, transform nonavailable native phosphates and precipitated or fixed fertilizer phosphates into available forms.
- Excessive irrigation should be avoided, as it results in the loss of nitrogen and potassium fertilizers. Further, there is also gaseous loss of nitrogen under waterlogged condition. There should not be excess water in the soil, particularly at the time of fertilizer application.
- Fertilizer application should be made after draining excess water.
- Climate also affects crop response towards FUE. Crop response to phosphatic fertilizers is generally greater in dry conditions.

11.1.7.4 Fertilizer Application Practices

Balanced use of fertilizers must be according to soil requirements, which can be determined by soil tests. Efficiency of a straight fertilizer containing single nutrients depends on the sufficiency of other nutrients in soil. Phosphates, in general, are more efficient when the entire dose is applied as basal dressing, potash entirely as basal dressing or partly as basal and rest in split doses depending on the soil texture. Nitrogen should be applied in 2–3 (or 4) split doses. Water-soluble phosphatic fertilizers should be placed 4–6 cm below the soil surface and 4–6 cm away from the seeds to ensure maximum availability to plants. Insoluble fertilizers should be thoroughly mixed in soil. Sometimes it is better to cure the urea by mixing the urea fertilizer with 5–10 parts of soil thoroughly and keeping the mixture overnight. This enhances the conversion rate of urea into ammonium carbonate. Whenever nitrogen loss in drainage water is suspected due to leaching, efficiency of the nitrogen fertilizer can be increased by mixing the fertilizer with crushed neem seed (5:1 parts) (Ali et al., 2017; Yaseen et al., 2017). Use of controlled release fertilizer to reduce leaching can be very effective.

Zinc deficiency is becoming more and more widespread. In such cases, application of zinc sulphate, at the rate of 10–25 kg of zinc sulphate as basal dressing, not only corrects the zinc deficiency but also enhances the efficiency of the other applied fertilizers. Foliar application of fertilizers should be resorted to under certain soil and climatic conditions, which can improve FUE. This is beneficial, particularly in calcareous soils with high pH, where soil-applied nutrient mobility is a problem (Yaseen et al., 2013). Moreover, foliar application of micronutrient is a relatively economical and efficient method. For more economical and efficient use of major nutrients, it is better to integrate soil application of nutrient with fertigation and foliar application. This practice will not only help in saving fertilizers but will also improve crop yield.

11.1.7.5 Controlled and Stabilized Fertilizer Technologies and Their Scope in Pakistan

Pakistan, being energy deficient, has to avoid unnecessary pressure on imported raw materials and gas. It is therefore necessary to find ways and means to reduce the use of fertilizers and other energy intensive inputs. Advances have been made in finding alternate ways of improving availability of nitrogen. Controlled release fertilizers (CRFs) and the use of stabilizers for nutrients is the best way for fertilizers to release nutrients in a controlled manner. For example, in the case of N stabilizers (nitrapyrin, dicyandiamide, n-butyl-thiophosphoric triamide) which cause inhibition of nitrification processes or activity of urease enzymes in soil. CRFs will improve nitrogen use efficiency (NUE) by reducing leaching losses of nitrogen as nitrate. CRFs have chemical compositions of low solubility compounds, which are in a coated form. These CRFs are more expensive than immediately water-soluble fertilizers and are used for high-value crops. CRFs are widely used in the agriculture sector and polymer-coated fertilizers is one type of CRF.

11.1.8 Nutrient Application: Recommended versus Applied

The deficiency of primary macronutrients and some micronutrients in Pakistani soils is alarming and so is the status of current fertilizer use in the country. Consumption of N, P, and K has not shown a balanced growth over the past many years. Nitrogen is used in largest quantity as compared to other nutrients and is showing consistent growth. It reached 3.5 million nutrient tons by the end of first decade of the twenty-first century. The average use of nitrogen in Pakistan was less than 15 kg/ha during 1969–1970 (Table 11.4), which increased to 43 kg/ha during the next 10 years and became 70 kg/ha

TABLE 11.4
Total Fertilizer Nutrients Offtake and Use of N, P, and K (kg/ha) in Pakistan

	Total Nutrients Offtake (000 Nutrients Tons)			Use of Nutrient (kg/ha)			
Year	Nitrogen	Phosphorus	Potassium	Nitrogen	Phosphorus	Potassium	NP Ratio
69–70	274	37	1.34	14.5	1.6	0.1	7.47
79–80	806	228	9.6	43.2	11.3	0.5	3.53
89–90	1468	382	40	70.5	18.5	1.2	3.84
99–00	2218	597	19	96.5	26.0	1.0	3.71
05–06	2927	850	27	130.0	37.8	1.8	3.44
06–07	2649	979	43	114.5	42.3	1.3	2.71
07–08	2925	630	27	125.0	26.9	1.1	4.64
09–10	3476	860	24	146.1	36.2	1.0	4.04
11–12	3207	633	21	137	27	0.9	5.06
12–13	2853	747	21	125	32.8	0.9	3.82
14–15	3309	975	33	145	43	1.4	3.39

Source: NFDC. 2016. Fertilizer Review 2015-16, National Fertilizer Development Centre (NFDC), Planning and Development Division, Islamabad.

during 1989–1990. By the end of twentieth century, the average use of nitrogen was about 100 kg/ha, which increased to 146 kg/ha over the next 10 years, which is the highest figure for the country.

The growth rate of phosphorous use has remained very slow over the same period. This fertilizer was introduced in the country in 1959–1960 but its use in 1969–1970 was only 1.6 kg/ha, which increased to 11.3 kg/ha over the next 10 years. Its average use in the country was 26 kg/ha at the end of twentieth century, that is, one fourth of the use of nitrogen. The average use of phosphates peaked to 42 kg/ha in 2006–2007 but it fell back to 27 kg/ha during the next year. The phosphate use remained variable in the range of 27–37 during the next 7 years and during 2014–2015, it again achieved the status of year 2006–2007, that is, 42 kg/ha.

The use of potassium has barely improved in Pakistan, reaching its highest point at only 45 thousands nutrient tons in 1987–1988. Since then its use has been variable but lower than this figure. In 1996–1997, its use was 8.4 thousand nutrient tons. In 2009–2010, the use of this nutrient was 24 thousand nutrient tons.

The above statistics reflect the extreme predominance of nitrogen in nutrient consumption in the country, which has created a serious imbalance in the use of macronutrients. However, during last few years the trend of NP balance appears to have improved.

11.1.9 Constraints in Balance Use of Fertilizer

11.1.9.1 Price of Fertilizers Created Imbalance in the Use of N and P

The main reason for the increasing imbalance in fertilizer use have been varying price structures over time. In the beginning, the prices of DAP and urea were almost same till the middle of 1986. At this point, the price of DAP started moving up, in comparison to urea, until at the end of 1990s, the price of DAP became twice that of urea. During 2006–2007, phosphorous was subsidized, which led to the improvement of use of phosphorous in relation to nitrogen. This was the year when use of phosphorous became highest but again the situation was reverted because of the withdrawal of subsidies for phosphorous fertilizers. Cost of phosphorous fertilizer application was Rs. 4450 per acre in 2008–2009 but in 2012 it increased by 83% to Rs. 8125 per acre. Overall, Pakistani farmers are paying very high prices for all the inputs including water, seeds, fuel, and pesticides—in addition to fertilizers.

During the last few years, again the reduction in the price of phosphorous fertilizer has shown positive impact on the use of phosphorous and ratio of NP improved from 5.06 in 2011–2012 to 3.39 in the year 2014–2015.

11.1.9.2 Flawed Subsidy Transmission Mechanism

A subsidy diffusion mechanism is based on the support price fixed by the government. Urea, imported by the Trading Corporation of Pakistan is sold to the National Fertilizer Corporation (NFC) at the international market price, which is set by the Economic Coordination Committee. NFC then dispenses the fertilizer across the country, through its marketing wing, the National Fertilizer Marketing Limited (NFML). Unfortunately, this creates plenty of chances of corruption by various methods. The subsidy is untargeted. Press reports declared that the mechanism for the circulation of imports is substandard.

11.1.9.3 Gas Shortage

According to the National Gas Allocation and Management Policy of 2005, after domestic consumers, the fertilizer industry is given the highest importance but the current policy decisions do not reflect the prioritie laid out in the policy. To divert gas into the fertilizer sector is not economical. The government allocation policies are inconsistent, and should be unpromising for future investment decisions. Nine fertilizer manufacturers have contracts with the government to assure gas supplies for 9 months of the year, and at least one of them has formally gone to court to force the government to supply gas to its plant. The consequences of this will be perhaps seen in the future, whether such a course of action will be repeated again by the fertilizer manufacturers.

11.1.9.4 Adulteration and Black Marketing

Fertilizers, particularly those imported and sold by the private sector are repeatedly reported to be contaminated. Unauthorized dealers and substandard fertilizer products are major cause of adulteration, especially in the DAP. Even with the availability of fertilizer testing labs, the unawareness of the farmers is still a major constraint in this situation.

11.1.9.5 Fertilizer Recommendations

The fertilizer recommendations given by various departments are general in nature. The fertilizers have become a very expensive input and such recommendations are not commensurate with the investment made. Moreover, the general recommendations remain the same over the years. An example for wheat crop can be seen in Table 11.5.

11.1.10 Nutrients Use Based on Soil Test and Target Yield

The nutrients are to be applied in amounts and proportions needed by a specific crop in a specific soil. It is a proven fact now that by balancing the use of nitrogen and phosphorous, the yields of wheat, rice, cotton, and maize can be increased by 25%–45%. The fertility of every field is variable and nutrients are to be applied based on the nutrient status of the soil and the target yield of the specific crop. No system in the country was available to provide the guidance to the farmers for the nutrients to be applied on the basis of soil tests and target yields.

11.1.10.1 Soil Test-Based Fertilizer Prediction Models

The University of Agriculture, Faisalabad has developed soil test-based fertilizer use prediction models to predict the nitrogen and phosphorous requirements for getting the desired yield of

TABLE 11.5
Fertilizer Recommendations for Wheat over the Years (1973–2016)

	Irrigated				Rain-Fed		
Soil Type	**Poor**	**Medium**	**Fertile**	**Progressive Farmers**	**<350 mm**	**350–500 mm**	**>500 mm**
Year	**N:P:K**	**N:P:K**	**N:P:K**	**N:P:K**	**N:P:K**	**N:P:K**	**N:P:K**
1973		114:85:0					
1985–1986	136:111:57	114:67:57	84:57:57		57:57:57	84:57:57	114:57:57
1993–1994	136:114:62	114:84:62	84:57:62	158:114:62	57:57:62	84:57:64	114:84:62
2004	130:114:62	104:84:62	79:57:62	158:114:62			
2006–2007	130:114:62	104:84:62	79:57:62	158:114:62	57:57:62	84:57:62	114:57:62
2008–2009	130:114:62	104:84:62	79:57:62	158:114:62			
2009–2010	130:114:62	104:84:62	79:57:62	158:114:62			
2010–2011	130:114:62	104:84:62	79:57:62	158:114:62	57:57:30	114:114:30	114:114:30
2011–2012	130:114:62	104:84:62	79:57:62	158:114:62	57:57:30	114:114:30	114:114:30
2012–2013	130:114:62	104:84:62	79:57:62	158:114:42	57:57:30	114:114:30	114:114:30
2015–2016	130:114:62	114:84:62	79:57:62		57:57:30	114:114:30	114:114:30

Source: NFDC. 2016. Fertilizer Review 2015-16, National Fertilizer Development Centre (NFDC), Planning and Development Division, Islamabad.

major crops. With these models, fertilizer requirements of every field could be determined for desired yields of the crop to be sown. It will lead to the balanced use of nitrogen and phosphorous. Potassium has to be used only when benefit exceeds the cost involved. Zinc and boron should be used according to the recommendation of the department only in deficient soils. Moreover, the availability of the nutrients from the soils is also a issue because of the low organic matter content of our soils. To improve the availability of nutrients, use of chemical fertilizers is to be integrated with organic sources, that is, farm yard manure, green manures, compost, or pressed mud, etc.

11.1.10.2 Use of ICT for Site Specific Fertilizer Application

The dream of balanced nutrition could only be achieved if guidance is available to farmers at their doorstep. At present, the staff of the agriculture department cannot reach to more than 10% farmers of the province. The information to every farmer could only be provided at his doorstep using information and communication technologies (ICT). The use of ICT for dissemination of information to the farmers is not a new concept now. Some countries are already disseminating such information to farming community this way.

University of Agriculture, Faisalabad developed a website: www.fertilizeruaf.pk to provide information regarding soil test-based fertilizer prediction models to farmers at their doorstep. The addresses of all soil testing laboratories in Punjab have also been uploaded on this website. The production plans for all major crops are available on this website. Awareness of this website has been promoted among the farming community through a United States Agency for International Development (USAID) project being implemented through International Center for Agricultural Research in the Dry Areas (ICARDA). There were 50,000 hits in one year on this website, and it has created a notion for the balanced use of fertilizers among the farming community.

11.1.11 Integrated Nutrient Management (INM) System for Improving Soil Fertility Status and Crop Yield Quality

Mineral fertilizers have a significant importance in crop production and are an indispensable component of today's agriculture, but recovery of nitrogen in the soil plant system seldom exceeds 50%, whereas the remaining is lost through different means like leaching, volatilization, denitrification, etc. (Abbasi et al., 2003). The same is the case with other nutrients as phosphorous recovery efficiency is 10%–25% and potassium recovery is up to 50% while micronutrient use efficiency is 5%–10%. Thus, strategies for increasing and sustaining agricultural productivity will have to focus on using available nutrient resources more efficiently, effectively, and sustainably than in the past. In this scenario, integrated nutrient management (INM) using organic manures with mineral fertilizers is advocated as a viable approach not only in maintaining and sustaining proper plant growth and productivity, but also for providing stability to crop production (Ahmad et al., 2008; Aulakh and Grant, 2008). Thus, neither the organic manure alone nor the chemical fertilizers can achieve the yield sustainability under any cropping system where nutrient depletion and turnover in soil–plant systems is so great.

11.1.12 Future Outlook

Fertilizer use in Pakistan is imbalanced and far below that of other countries in the world. The information given in Table 11.6 depicts the fertilizer use in various countries.

Improvement in agricultural production on a sustainable basis needs special attention of the government who needs to make policy decisions so that farmers (particularly the small farmers) could be encouraged to engage themselves devotedly in the profession of agriculture. The new policy must address following topics.

TABLE 11.6
Fertilizer Use in Different Countries (kg ha^{-1})

Country	Nitrogen	Phosphorus	Potash	Total
China	205	77	52	333
Japan	131	158	109	398
Holland	408	69	79	556
Korea	230	109	128	467
Egypt	259	37	7	304
Indian Punjab	133	35	2	170
Pakistan	96	25	1	121

11.1.12.1 Promotion of Balanced Use of Fertilizer

The contribution of the balanced use of fertilizers in increasing yields varies from 30% to 60% in different crop production areas of the country. One kilogram of fertilizer nutrient produces about 8 kg of cereals, 2.5 kg of cotton seed, and 114 kg of stripped sugarcane. The land used under the single cropping systems is more prone to depleting soil fertility because only certain essential plant nutrients are intensively removed by the same crop, year after year.

Balanced use of fertilizers is indispensable for sustaining optimum yields and profits. It ensures high productivity without polluting the environment. This dynamic approach responds to the needs of higher productivity and the emergence of any new nutrient deficiencies. Balanced and efficient uses of fertilizer are the two aspects of any sound soil fertility management program.

Aulakh and Malhi (2004) stressed on the significance of nutrient interactions as agriculture becomes more intensive because interaction between two or more nutrients can be positive or negative. A deficiency of any one of the other 17 essential plant nutrients can affect the absorption and function of nitrogen, thus causing reductions in nitrogen use efficiency. The impact of balanced use of fertilizers on crop yields can be viewed from the information given in Table 11.7.

11.1.12.1.1 Use of Potassium Fertilizer

The department of agriculture is recommending potassium fertilizer for the last 30 years at least @ 60 kg/ha but its use has not crossed the limit of 2 kg/ha (Table 11.3). The maximum off take of this

TABLE 11.7
Effect of Balanced Fertilization on Yield and Nitrogen Agronomic Efficiency

	Yield (t ha^{-1})			Agronomic Efficiency, kg Grain kg N^{-1}		
Crop	Control	N alone	+PK	N alone	+PK	Increase
Rice (wet season)	2.7	3.2	3.8	13.5	27.0	13.5
Rice (summer)	3.0	3.4	6.2	10.5	81.0	69.5
Wheat	1.4	1.8	2.2	10.8	20.0	9.2
Pearl millet	1.0	1.2	1.6	4.7	15.0	10.3
Maize	1.6	2.4	3.2	19.5	39.0	19.5
Sorghum	1.2	1.4	1.7	5.3	12.0	6.7
Sugarcane	47.2	59.0	81.4	78.7	227.7	150.0

Source: NFDC. 2016. Fertilizer Review 2015-16, National Fertilizer Development Centre (NFDC), Planning and Development Division, Islamabad.
Note: Assumes a typical N harvest index of 56%.

nutrient in any year was 45 thousand nutrient tons. The annual removal of potassium by different crops is in the range of 30–80 kg/ha. It is obvious that there is continuous depletion of this nutrient from the soils. The research trial data showed that use of potassium increased the yield of various crops (Ahmad and Chaudhry, 2000).

11.1.12.1.2 Micronutrients and Other Fertilizer Products

It is a documented fact that among the micronutrients, zinc deficiency is widespread mainly in rice and citrus (Ahmad and Rashid, 2003). Deficiency of zinc has been reported in wheat crop also but not to the level of rice and citrus. The second deficient micronutrient is boron as established in cotton. The iron deficiency in deciduous fruit plants has also been reported (Ahmad and Rashid, 2003). No doubt, the micronutrients are equally important but it is quite unfortunate that their standard sources are difficult to decide due to mushroom growth of products in the market. The farmer is totally confused about their use and crops are suffering from their deficiency. For zinc, hundreds of solid and liquid products are being sold and farmers are being cheated. Like the major nutrients, three or four standard products need to be approved by the government, publicized, and marketed so that farmers could decide about the use of the deficient micronutrients.

11.1.12.2 Policy Decisions

11.1.12.2.1 Assessment

Efficient assessment of fertilizer requirements is very important to make decisions and projections about fertilizer requirements, import of fertilizer raw materials and products, and budget financing in order to fulfill the nutrients requirement of farmers for optimum yields.

11.1.12.2.2 Marketing

Effective marketing structures must be planned through the partnership of the government and private sectors to achieve efficient fertilizer distribution and usage according to farmer requirements. Such marketing structures would also be helpful to make comprehensive policies which create the right balance in private and government as partners for production, distribution and import of fertilizers, which at present is highly dependent on political and economic conditions of the country.

11.1.12.2.3 Transport and Storage

Efficient transportation and storage infrastructure will ensure timely, efficient delivery and use of fertilizers. Moreover, investment in transportation and storage facilities will help to find out alternate means of transportation in connection with location and size of storage facilities to make an efficient supply chain.

11.1.12.2.4 Training

To educate farmers, farm support policies should include comprehensive training programs on new crop production technologies. Efficient use of fertilizers and other inputs, approaches and dissemination of technologies through extension services to farming community should be a part of the training. The training should be "requirement-based" learning programs with integration of experiences of the farming community. A comprehensive package of technology including balanced use of fertilizers as a part of the integrated plant nutrient system (IPNS) should be prepared and introduced.

11.1.12.2.5 Pricing

Price of input is most important factor which affects revenue of farmers by enhancing crop production, whether or not subsidy is given on agriculture inputs. Most of the developing countries give subsidies on inputs in agriculture. Whereas, developed countries have introduced farm support policies rather than subsidies to sustain the crop production. In developing countries, subsidy on agricultural inputs re-adjusts pricing that forces an increase in the demands of plant nutrients. In any case, subsidy,

TABLE 11.8
Unit Price of Crop in Relation to Price of Urea and DAP

	Price (Rs/40 kg)		Urea Bag/40 kg		DAP Bag/40 kg	
Crop	**2008**	**2015**	**2008**	**2015**	**2008**	**2015**
Wheat	940	1300	1.61	0.68	0.75	0.38
Cotton	3200	2200	5.47	1.16	2.56	0.65
Coarse rice	750	550	1.28	0.29	0.60	0.16
Maize	900	800	1.54	0.42	0.72	0.24
Sugar cane	81	180	0.14	0.09	0.06	0.05
Urea	585	1900				
DAP	1250	3400				

whether properly used or misused, is a burden on country budget and needs to be shifted to develop sustainable farm support policies.

Other incentives for fertilizer use may include guaranteed support prices for agricultural produce, duty-free imports of fertilizer, and tax exemptions for credit and investment in fertilizers and crop production. All these measures affect the profitability of fertilizer use and provide an economic motivation for increasing crop production.

The pricing and subsidy mechanism should be designed to keep pace between the income of farmer from output and expenditure incurred on input (fertilizer). The information given in Table 11.8 highlights the widening gap between commodity prices and the unit cost per bag of fertilizer (urea and DAP) in 2008 as compared to 2015. Due to the manifold increase in price gap and the drastic reduction in exchange rate, fertilizer cost pressure on farmers has been significantly increased and has resulted in a reduction of the famers' ability to invest in crop.

11.1.12.2.6 Legislation

New amendments in the Fertilizer Act should be introduced for advanced fertilizer technologies including controlled and slow release, polymer-coated and impregnated/bio-stimulant coated fertilizers.

11.1.12.2.7 Packaging

Fertilizer packaging, other than reducing losses, should not be allowed to add extra cost to fertilizers. It must provide accurate and complete information on the bag.

11.1.12.2.8 Advice and Planning

Well-integrated fertilizer polices should be framed by the consultation of two vital stakeholders, including the government and private sectors, with a priority to coordinate on country agricultural and food security policies. This platform will be responsible for all aspects regarding fertilizer pricing and marketing, forecasting demand, and identifying the research and extension priorities.

11.1.12.2.9 Financing

Fertilizer demand is highly dependent on cropping season, which leads to variable traders' cash flow and will determine the need for credit in domestic and foreign currency.

11.2 GYPSUM

11.2.1 Role of Gypsum in Saline-Sodic/Sodic Soils

Land degradation and scarcity of good quality water are the biggest issues worldwide. In Pakistan, 10 million ha soils are salt-affected and about 80% of the salt-affected soils of Punjab are saline-sodic and need external calcium sources for their reclamation (Ghafoor et al., 2012). Kim et al. (2017) documented the positive effects of gypsum application for the reclamation of salt-affected soils of costal tidelands. Similarly, many scientists also documented the supportive role of gypsum application for the reclamation of salt-affected soils by improving the soil physical properties such as flocculation, aggregate stability, and infiltration rate in Egypt (Helalia et al., 1992), California (Lebron et al., 2002), India (Ekwue and Harrilal, 2010), and Iran (Yazdanpanah et al., 2011; Mazaheri and Mahmoodabadi, 2012). Gypsum is considered a cheaper and better source of calcium. Different amendments are in use for the reclamation of saline-sodic/sodic soils and gypsum has long-term effects when compared with acids (Zia et al., 2006). It has been reported through experiments on a variety of soils, that the most effective Ca^{2+} concentration for Na–Ca exchange was 6–10 $mmol_c\ L^{-1}$ in irrigation water/soil solution (Ghafoor et al., 2004). For saline-sodic and sodic soils, gypsum is considered more economical, while acids can reclaim such soils at relatively faster rates but at a higher costs and also affect soil health (Mace et al., 1999; Ghafoor et al., 2004). Frenkel et al. (1989) advocated that mixed application of acids and gypsum is better and faster to reclaim saline-sodic soils than either applied alone.

11.2.2 Gypsum Application

In Pakistan, leaching prior to gypsum application into plough layer of saline-sodic soils does not seem to be of any practical significance. Gypsum ($CaSO_4 \cdot 2H_2O$) is a neutral salt of Ca^{2+} and has a purity of ≥65%. It is mined in the salt range area of Punjab and powdered for agricultural use. It has a low solubility of 28 $mmol_c\ L^{-1}$ at 25°C, and it seldom exceeds 15 $mmol_c\ L^{-1}$ in soil (Rhoades 1982). However, because of its low dissolution rate it effectively sustains the electrolyte concentration for longer, which in turn is very useful for improving the downward movement of water in soil (Ghafoor et al., 2004; Kahlon et al., 2012). It also reduces the leaching of unreacted calcium, especially in coarse textured soils. However, it takes relatively more time compared to acids to reclaim saline-sodic/sodic soils. Its local accessibility at cheaper rates and its harmless nature are the main reasons of its popularity among farmers. Particle size has monetary concerns since grinding to finer size-grades becomes expensive. It has been found that gypsum of 16 mesh size can reclaim soils very efficiently if brackish water is used for irrigation (Ghafoor et al., 2004). Generally, gypsum passed through a 30 mesh sieve is considered better (Malik et al., 1992) and the same is supplied in bags and in bulk to farmers in Pakistan. Gypsum stones can be used for the reclamation of saline-sodic soils and water by placing gypsum stones in water channels.

In Pakistan, Malik et al. (1992) conducted 55 experiments on different soils, five each in 11 districts of the Punjab province of Pakistan. They have reported value to cost ratio of 1.8–4.6 for crops like wheat, rice, cotton, and berseem. It has been observed that reclamation of sodic/saline-sodic soils with acid is 5–10 times more expensive than gypsum (Qadir et al., 2001). Similarly, the quantity and related cost of a calcium source, such as gypsum, increases with the concentration of Na^+ in saline-sodic/sodic soils (Vyshpolsky et al., 2008). Even the physical presence of gypsum in soils reduces crusting, hard-setting, and soil strength (Ghafoor et al., 2004; Ahmad et al., 2016) to favor seed germination, which is one of the greatest problems in salt-affected soils. Some multilocation studies carried out by the authors in the Indus Plain of Pakistan indicated that gypsum is the most easily available, economical, effective, and easy to use amendment for reclamation of saline-sodic soils (Murtaza et al., 2009, 2015; Ghafoor et al., 2012). The economics of treatments for the reclamation of saline-sodic soil under the International Waterlogging and Salinity Research Project is given in Table 11.9.

11.2.3 Availability of Gypsum

Pakistan has gypsum reserves in the form of mountains with an estimated amount of 5–6 billion tons. It is mainly present in the salt range of Pakistan. Major reserves of gypsum are found in Mianwali, Quaidabad, Dera Ismail Khan, and Dera Ghazi Khan Districts. Gypsum crushed mostly in Dera Ismail Khan and Dera Ghazi Khan has better quality. Generally, gypsum mined from deeper

TABLE 11.9
Economics (Rs/ha) of Organic and Inorganic Amendments to Reclaim Saline-Sodic Soil Using Drainage Water after 3 Rices + 3 Wheats

Treatment	pH_s	EC_e (dS m^{-1})	SAR	Net Income
Original soil (Khurrianwala series)	7.9–8.4	8.5–32.3	21.0–77.5	–
S1B9 sump water alone, FDPA	8.4	9.8	22.9	28,427
Soil-applied gypsum @ 50% SGR	8.4	8.4	21.8	28,380
Water-applied H_2SO_4 @50% @ WRSC	8.4	10.3	23.9	−11,719
Soil-applied gypsum @ 100% SGR	8.3	8.5	20.9	35,714
FYM @ 25 Mg/ha/annum	8.4	10.1	16.4	35,713

Note: EC_{iw} 2.93–3.21 dS m^{-1}, SAR_{iw} 12.0–18.2, RSC_{iw} 3.7–10.0 $mmol_c$ L^{-1} (Ghafoor et al., 1997).

layers has better quality compared to that from the upper surfaces of mountains. According to an estimate, the requirement for reclamation/amelioration of various categories of gypsum for different types of soil is 50×10^6 tons per annum. Despite its immense current demand, gypsum production is less than a million ton (0.5–0.7 million tons) per year leaving a very large gap which needs to be filled. Gypsum is also desirable for the amelioration of low quality tube well water used for irrigation.

11.2.4 Gypsum Subsidy

Considering the extent and severity of the problem under discussion, the government of Pakistan subsidized gypsum and allocated reasonable amounts to provinces in 1983 through the Punjab Agricultural Development and Supplies Corporation (PAD & SC) as major agencies. However, this facility was taken up by some farmers; however, most of the small farmers, representing nearly 93% of the total farming community, were deprived of this opportunity. The quality of gypsum supplied at very cheap rates was of poor quality due to the presence of impurities like calcite, dolomite, and clay contents affecting the soil health negatively. Furthermore, awareness about beneficial aspects of gypsum, credit, storage, and transportation facilities remained main hurdles for gypsum marketing (Rafiq, 2001). The Directorate of Soil Fertility Research, Punjab issued soil analysis reports for gypsum requirement and Punjab government sold 5,59,800 million tons of gypsum from 1983 to 1997–1998. Distribution of gypsum in Sindh was channeled through main and small stores located all over the province. In 1972, the price of gypsum was 60 rupees per ton, which rose to Rs. 80 to 160 per ton during the 1980s. Finally, in 1992 the price went up to Rs. 300 per ton. Fauji Fertilizer Company (FFC) also distributed 1200 tons at the rate of Rs. 500 per ton in the early 1980s, which was stopped due to low profitability and competition with government agencies.

During 2006–2012, av 50% subsidy on gypsum purchase was provided through a project of the Punjab Government entitled "Pakistan Community Development Project for Rehabilitation of Saline and Waterlogged Lands" in PindiBhattian, Shorkot, Hafizabad (Sahiwal tehsils), Jhang, and Sargodha Districts.

Now, as a part of "Kisan Package" announced recently, the government of Punjab is interested in reintroducing subsidies on gypsum to the farmers. For this purpose, significant amounts of funds have been already been allocated to supply gypsum on subsidized rates to the farmers and land owners of degraded lands.

11.2.5 Size and Quality

The solubility of gypsum depends upon the rate of Ca^{2+} diffusion from solution to exchange sites and the rate of Na^+ diffusion away from the exchange sites into the bulk soil solution. It has been

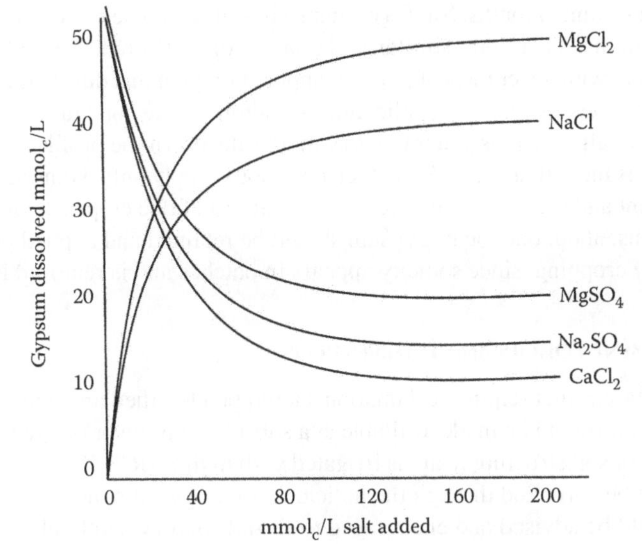

FIGURE 11.3 Effect of salts on gypsum solubility (relative presentation).

observed that 16–30 mesh size gypsum is sufficient to meet the dissolution requirement for sodic/saline-sodic soils. Actually, the dissolution of gypsum depends more on the type of salts present in saline-sodic soils (Figure 11.3). This fact is of practical significance that the type of electrolytes present in soil solution has very high effect on gypsum dissolution while the particle size of gypsum has less effect on its dissolution. Hence, in sodic soils (with NaCl as dominant salt), dissolution of the added gypsum could be reasonably high enough to induce desorption of Na^+ by Ca^{2+} even with coarser particles of gypsum, which allows for a reduction in grinding costs. Recently, Ahmad et al. (2016) concluded that gypsum applied at 25% soil GR (SGR) was the most favorable and economical for optimal increase in saturated hydraulic conductivity (Ks) and effective removal of Na^+ and salts from loamy sand soils, while gypsum applied at 50% and 75% SGR stayed optimal for sandy loam and clay loam soils.

Higher solubility of gypsum in $MgCl_2$ solution is due to the formation of $MgSO_4$ ion pairs, which increases its solubility by consuming SO_4, which is one of the products of gypsum dissolution. The lower solubility of gypsum in $CaCl_2$ solution is due to common ion effects. The solubility of gypsum will be further decreased in the presence of $NaHCO_3$ and Na_2CO_3 solutions due to the formation of $CaCO_3$ and possible coating of lime on gypsum with time, respectively.

11.2.6 Tips for Economical Reclamation of Saline-Sodic/Sodic Soils

The success of any technology is dependent upon its cost/benefit ratio and is considered a key factor for adoption by farmers. In most of the studies, economic evaluation of treatments is overlooked. However, to economize the reclamation of sodic/saline-sodic soils the following considerations are important.

1. Application of organic manures (decomposition of which releases organic acids and helps maintain infiltration rates) reduce gypsum application rates, for example, if 10 tons/acre of FYM are added, then one can safely decrease the gypsum application to about 50% of the soil GR compared to gypsum alone at application of 100% soil GR.
2. Prevailing atmospheric temperatures should be considered because adsorption–desorption are the physical-chemical reactions, and high temperatures generally increase their rate.

Hence during summer months, Na–Ca exchange in soils could be more efficient than during the winter months. Therefore, similar reclamation of soils could be realized during the summer season, with lower rates of gypsum application than that during the winter months, with the same rates of gypsum application. In addition, during summer, monsoon rains provide good quality surplus water to carry more salts down the profile.

3. If the soil GR is more than 5 tons/acre, then one should apply only 5 tons/acre of gypsum in one installment and the remaining after at least harvest of one crop each of rice and wheat. In all situations, about one ton of gypsum should be retained and applied on sodic patches after a year of cropping, since sodicity appears in patches and is removed in patches.

11.2.7 Government Policies and Interventions

- High priority areas that require reclamation should be identified and notified.
- Quality gypsum should be made available at a subsidized price or on credit.
- Deterioration of soil structure in areas irrigated with high SAR/RSC government tube well waters should be corrected through the judicious use of gypsum, at government expense.
- Farmers should be advised and encouraged to install shallow depth tube wells because in general shallow depth water is of better quality.
- Recycle city wastewater locally to reclaim the salt-affected soils using amendments.
- Land leveling should be promoted to increase water use efficiency and decrease salinity and sodicity of soils.
- Gypsum subsidy and free supply in high priority areas identified earlier may be provided to farmers, particularly to small farmers. It should be made obligatory by law for farmers to get their tube well water analyzed for which a mobile teams of experts should be organized to collect, analyze, and provide advice on reports.
- Local requirements should be fulfilled before exporting of gypsum. There are reports that high quality gypsum is being exported to India through the Wahga boarder near Lahore, which should be immediately banned—while there is still need and demand of high quality gypsum in Pakistan.

11.2.8 Future Plan/Recommendations

There is dire need to start tube well water analyses for site-specific recommendations by the Agriculture Department and it must be made obligatory for the tube well owners to get the water analyzed and act upon advice.

- Doorstep provision of quality gypsum may be ensured to farmers on credit or free of cost for the sustainability of tube well irrigated agriculture since small farmers may not be in a position to buy gypsum on cash payment.
- Government should take strict action to stop adulteration in gypsum.
- Gypsum manufacturing plants should be installed in those areas where raw materials have at least 70% purity.

APPENDIX 11.A: SWOT ANALYSIS

Fertilizer Issues and Way Forward for Sustainable Crop Production

Vision:

How to make agriculture profitable, sustainable, and internationally compatible.

- Sustainable crop production with the application of the right source at the right rate, time, and place. Improving current fertilizer use efficiency and mitigate the effect of climate change.

Objectives:

- To increase productivity per unit area on a sustainable basis with improved quality
- To ensure sustainable use of natural resources
- Integrated nutrient management

Strategies/Action Plans:

Productivity could be increased by:

- Ensuring timely availability of unadulterated and quality fertilizers.
- Exploring hi-tech fertilizer technologies.
- Maintaining soil health and water quality for good crop production.
- Integrated use of crop residues and other biological sources with mineral fertilizers.

Regulatory arrangements could be strengthened by:

- Implementation of existing fertilizer act/laws in letter and spirit.
- Capacity building of students to transfer technology package developed at UAF to farmers of their native region.

STRENGTHS

- Four seasons are conducive to good agriculture.
- Pakistan possesses vast fertile and irrigated agricultural lands for cultivation.
- Soils are suitable for growing diversified agronomic crops, vegetables, fruit, and ornamental plants.
- Fertilizer production particularly nitrogen (urea & CAN) and phosphorus (NP & SSP) sources.
- Soils have K-containing minerals.
- Well-established canal system; more than 80% of all areas fall under irrigated agriculture.
- Pakistan has topographic advantages for the cultivation of diversified crops and fruit orchards.
- Farmers are receptive and are willing to accept new technologies, developments, and interventions.

WEAKNESSES

- Punjab has skewed distribution of land and water.
- It has suboptimal/insufficient/limited farm mechanization.
- Outreaching is weak. As a result, there is ineffective transfer of technology, which is a major hurdle in the development of agriculture.
- Water and fertilizer-use efficiencies are very low. More than 70% of installed tube wells pump poor quality water. There is no regulatory mechanism to regulate the pumping of water.
- Most soils are alkaline and calcareous, low in organic matter, naturally low to medium in soil fertility.
- Burning of crop residues wastes natural resources and enhances pollution.

- Wrong method of fertilizer application.
- At present, fertilizer use in the country is imbalanced. The recommended fertilizer application rates for N:P:K is 2:1.5:0.5 but farmer application rates are 3:1:0.02, indicating more use of nitrogen fertilizers.
- Lack of up-to-date and extensive soil fertility analyses. Existence of knowledge gaps is also harming the agriculture sector badly. Lack of adoption of new technology and knowledge is hurting the agriculture industry.
- Farmers are reluctant to adopt higher and balanced use of fertilizers, due to lack of guaranteed support prices for agricultural production.
- The situation has worsened because of duty on imports of fertilizers and the absence of tax exemptions for credit and investment in fertilizer and crop production.
- Lack of proper coordination between government and private sectors to achieve efficient fertilizer distribution and usage.

Opportunities

- Agriculture production can be substantially increased just by:
  - Improve N:P_2O_5 balance
  - Promoting regular use of potash and micronutrients (Zn, Fe, B), where required
  - Proper integration of FYM/biological practices with fertilizers (Integrated Plant Nutrition)
  - Proper method and time of application with improved techniques like foliar fertilization and fertigation.
- Nutrient indexing of all agricultural land should be carried out in Punjab and a cyber-network should be developed for fertilizer application as per crop requirement and soil conditions.
- Following the Pesticide Act, the Fertilizer Act should be enforced strictly in letter and spirit.
- National plant nutrition focal points should be set up to advise on policy and to regulate the availability, quality, production, and trade of plant nutrients (in particular fertilizers) in a way that complements the overall plant nutrition strategy of the government. Then this policy should be implemented.
- Management of crop residues—instead of burning—can boost SOM level.

Threats

- Loss of soil fertility due to continual mining by crop removal without adequate replenishment, combined with imbalanced plant nutrition practices, poses a serious threat to agricultural production.
- Burning of crop residues (7–10 t ha^{-1}) in rice-wheat. cotton, and sugarcane cropping systems.
- Almost 100% soils are deficient in nitrogen, 90% in phosphorus, 47% in potassium. Micronutrients (Zn, B, and Fe) are also deficient in all cropping systems.
- Majority of the soils being low in organic matter, alkaline, and calcareous in nature with light to medium in texture interfere with availability of nutrients according to plant needs.
- Widening ratio in N:P.
- Shortage of canal water and use of brackish water to fulfill water requirements.

REFERENCES

Abbasi, M.K., Z. Shah and W.A. Adams. 2003. Effect of the nitrification inhibitor nitrapyrin on the fate of nitrogen applied to a soil under laboratory conditions. *J Plant Nutr Soil Sci.* 166: 1–6.

Adesemoye, A.O., H.A. Torbert and J.W. Kloepper. 2009. Plant growth-promoting rhizobacteria allow reduced application rates of chemical fertilizers. *Microb Ecol.* 58: 921–929.

Ahmad, N. and A.G. Chaudhry. 2000. *Fertilizer Use at Farm Level.* NFDC Publication 4/2000, Islamabad.

Ahmad, N. and M. Rashid. 2003. *Fertilizers and Their Use in Pakistan.* Extension Guide, Islamabad.

Ahmad, R., M. Naveed, M. Aslam, Z.A. Zahir, M. Arshad and G. Jilani. 2008. Economizing the use of nitrogen fertilizer in wheat production through enriched compost. *Renew Agric Food Syst.* 23:1–7.

Ahmad, S., A. Ghafoor, M.E. Akhtar and M.Z. Khan. 2016. Implication of gypsum rates to optimize hydraulic conductivity for variable-texture saline-sodic soils reclamation. *Land Degrad Develop.* 27: 550–560.

Ali, I., A. Mustafa and M. Yaseen. 2017. Polymer coated DAP helps in enhancing growth, yield and phosphorus use efficiency of wheat (triticum aestivum L.). *Journal of Plant Nutrition*, 40(18) https://doi.org/10.1080/01904167.2017.1381118.

Aulakh, M.S. and C.A. Grant. 2008. *Integrated Nutrient Management for Sustainable Crop Production.* The Haworth Press, Taylor and Francis Group: New York, USA.

Aulakh, M.S. and S.S. Malhi. 2004. Fertilizer nitrogen use efficiency as influenced by interactions with other nutrients. In: *Agriculture and the Nitrogen Cycle. Assessing the Impacts of Fertilizer Use on Food Production and the Environment.* Mosier AR, Syers JK, Freney JR (eds) SCOPE 65, pp. 181–191. Island Press. Washington, DC.

Dobermann, A. 2005. Nitrogen use efficiency state of the art. *IFA International Workshop on Enhanced-Efficiency Fertilizers*, Frankfurt, Germany.

Ekwue, E.I. and A. Harrilal. 2010. Effect of soil type, peat, slope, compaction effort and their interactions on infiltration, runoff and raindrop erosion of some Trinidadian soils. *Biosyst Eng.* 105: 112–118.

Frenkel, H., Z. Gerstl and N. Alperovitch. 1989. Exchange induced dissolution of gypsum and reclamation of sodic soils. *Eur J Soil Sci.* 40: 599–611.

Ghafoor, A., M.R. Chaudhry, M. Qadir, G. Murtaza and H.R. Ahmad. 1997. *Use of Agricultural Drainage Water for Crops on Normal and Salt-Affected Soils without Disturbing Biosphere Equilibrium.* Pub no. 176, IWASRI, Lahore, Pakistan, p. 135.

Ghafoor, A., G. Murtaza, M.Z. Rehman and M. Sabir. 2012. Reclamation and salt leaching efficiency for tile drained saline sodic soil using marginal quality water for irrigating rice and wheat crops. *Land Degrad Develop.* 23: 1–9.

Ghafoor, A., M.M. Qadir and G. Murtaza. 2004. *Salt-Affected Soils: Principles of Management.* Allied Book Centre, Lahore, Pakistan.

Heffer, P. and M. Prud'homme. 2012. *Fertilizer Outlook 2012–2016.* International Fertilizer Industry Association, Paris.

Helalia, A.M., S. El-Amir, S.T. Abou-Zeid and K.F. Zaghloul. 1992. Bio-reclamation of saline-sodic soil by Amshot grass in northern Egypt. *Soil Till Res.* 22(1–2): 109–115.

Herder, G.D., G.V. Isterdael, T. Beeckman and I.D. Smet. 2010. The root of new generation revolution. *Trend Plant Sci.* 15: 600–607.

Kahlon, U.Z., G. Murtaza and A. Ghafoor. 2012. Amelioration of saline-sodic soil with amendments using brackish water, canal water and their combination. *Int J Agric Biol.* 14: 38–46.

Kim, Y.J., B.K. Choo and J.Y. Cho. 2017. Effect of gypsum and rice straw compost application on improvements of soil quality during desalination of reclaimed coastal tideland soils: Ten years of long-term experiments. *CATENA* 156: 131–138.

Lebron, I., D.L. Suarez and T. Yoshida. 2002. Gypsum effect on the aggregate size and geometry of three sodic soils under reclamation. *Soil Sci Soc Am J.* 66: 92–98.

Mace, J.E., C. Amrhein and J.D. Oster. 1999. Comparison of gypsum and sulfuric acid for sodic soil reclamation. *Arid Soil Res Rehabil.* 13: 171–188.

Malik, D., G. Hussain and S.J.A. Sherazi. 1992. On-farm evaluation of gypsum application and its economics. *Soil Health for Sustainable Agri. Proc. 3rd Natl. Congr. Soil Sci.*, pp. 407–420. March 20–22, 1990, Lahore, Pakistan.

Mazaheri, M.R. and M. Mahmoodabadi. 2012. Study on infiltration rate based on primary particle size distribution data in arid and semi arid region soils. *Arab J Geosci.* 5: 1039–1046.

Murtaza, G., A. Ghafoor, G. Owens, M. Qadir and U. Z. Kahlon. 2009. Environmental and economic benefits of saline-sodic soil reclamation using low-quality water and soil amendments in conjunction with a rice-wheat cropping system. *J Agron Crop Sci.* 195: 124–136.

Murtaza, G., B. Murtaza and A. Hassan. 2015. Management of low-quality water on marginal salt-affected soils with wheat and sesbania crops. *Commun Soil Sci Plant Anal.* 46: 2379–2394.

Mutert, E. 1996. Plant nutrient balance in Asia and the Pacific region: Facts and consequences for agricultural production. *APO-FFTC Seminar on Appropriate Use of Fertilizers* on November 1, 1995, pp. 73–112, Singapore.

NFDC. 2016. Fertilizer Review 2015-16, National Fertilizer Development Centre (NFDC), Planning and Development Division, Islamabad.

Qadir, M., A. Ghafoor and G. Murtaza. 2001. Use of saline-sodic waters through phytoremediation of calcareous saline-sodic soils. *Agric Water Manage.* 50: 197–210.

Rafiq, M. 2001. Gypsum marketing in Pakistan-its problems and prospectus. *Int J Agric Biol.* 3: 336–338.

Rashid, A., E. Rafique and N. Ali. 1997. Micronutrient deficiencies in rain fed calcareous soils of Pakistan. II. Boron Nutrition of the peanut plant. *Commun Soil Sci Plant Anal.* 28: 149–159.

Rhoades, J.D. 1982. Reclamation and management of salt-affected soils after drainage. *Proc. Soil and Water Management Seminar*, pp. 1–123. November 29–December 2, Canada.

Tilman, D., C. Balzer, J. Hill and B.L. Befort. 2011. Global food demand and the sustainable intensification of agriculture. *Proc Nat Acad Sci.* 108: 20260–20264.

Trenkel, M.E. 2010. *Slow- and Controlled-Release and Stabilized Fertilizers: An Option for Enhancing Nutrient Use Efficiency in Agriculture*, 2nd ed. Intl Fert Ind Assn, Paris.

United Nations. 2011. World Population Prospects: The 2010 Revision, Standard variants. Updated: 28 June 2011. http://esa.un.org/wpp/Excel-Data/population.htm (accessed May 15, 2013).

Vance, C.P., C. Uhde-Stone and D.L. Allan. 2003. Phosphorus acquisition and use: Critical adaptations by plants for securing a nonrenewable resource. *New Phytol.* 157(3): 423–447.

Vyshpolsky, F., M. Qadir, A. Karimov, K. Mukhamedjanov, U. Bekbaev, R. Paroda, A. Aw-Hassan and F. Karajeh. 2008. Enhancing the productivity of high-magnesium soil and water resources through the application of phosphogypsum in Central Asia. *Land Degrad Develop.* 19: 45–56.

Yaseen, M., W. Ahmad and M. Shahbaz. 2013. Role of foliar feeding of micronutrients in yield maximization of cotton in Punjab. *Turk J Agric For.* 37(4): 420–426.

Yaseen, M., M.Z. Aziz, A. Manzoor, M. Naveed, Y. Hamid, S. Noor and M.A. Khalid. 2017. Promoting growth, yield and phosphorus use efficiency of crops in maize-wheat cropping system by using polymer coated diammonium phosphate. *Commun Soil Sci Plant Anal.* 48(6): 646–655. DOI: 10.1080/00103624.2017.1282510.

Yazdanpanah, N., E. Pazira, A. Neshat, H. Naghavi, A.A. Moezi and M. Mahmoodabadi. 2011. Effect of some amendments on leachate properties of a calcareous saline-sodic soil. *Int Agrophys.* 25(3): 307–310.

Zia, M.H., A. Ghafoor and T.M. Boers. 2006. Comparison of sulfurous acid generator and alternate amendments to improve the quality of saline-sodic water for sustainable rice yields. *Paddy Water Environ.* 4: 153–162.

12 Input Supplies
The Starring Role of Pesticide Inputs in Agricultural Productivity and Food Security

Muhammad Jalal Arif, Muhammad Dildar Gogi, Ahmad Nawaz, Muhammad Sufyan, Rashad Rasool Khan, and Muhammad Arshad

CONTENTS

12.1 INTRODUCTION

In 1950, the population of the world was 2.5 billion, which grew to 7.4 billion in 2016. It is estimated that the world population will reach the alarming figure of 9.1 billion in 2050. The population of the world is currently increasing at 1.2% per annum, which means the addition of 77 million individuals annually. About half of this annual increase is contributed to by India, China, Pakistan, Nigeria, Bangladesh, and Indonesia. According to an estimate, about 95% of the world population growth will occur in developing countries (Cohen, 2005; Carvalho 2006). In developing regions of the world, 800–840 million people are undernourished and face food insecurity (Carvalho 2006). In these developing countries, the main sources of food supplies are livestock products and cereal grains; however, 80% of the population of these countries lives in villages, deriving their livelihood and foods directly from the agriculture sector. In these countries, a huge gap exists between food production and feeding populations. This gap is expected to increase by the year 2050 and should be minimized by enhancing the agricultural productivity and food security (Carvalho 2006).

Genetically engineered organisms and synthetic chemical-free organic agriculture are the modern agricultural developments of the last decade. Use of biotechnological approaches and promotion of genetically modified crops including colza, soybean, tomatoes, and maize addressed the food security requirements and nutritional crises providing promising solutions to these issues. But, genomic development and GMOs patency are in the grip of private research companies, which surprisingly are surpassing and beating public research organizations (Pingali and Traxler, 2002). Monsanto, Zeneca, Aventis, Novartis, and so on are considered the leading and prominent private companies revolutionizing agriculture with modern biotechnological approaches. These private companies have declared that GMOs may prove resistant to pests, drought conditions, frost, and so forth and would bring a revolution in the agriculture sector (Pingali and Traxler, 2002). Glyphosate was the most widely and commonly used weedicide, which guarantees higher yields of crops. However, soybean and many other crops have been genetically modified for resistance to glyphosate-based weedicides; while weeds have become resistant to glyphosate-based insecticides. This changing scenario in GM-crop tolerance and weed resistance to glyphosate has diverted the dependency of growers from glyphosate herbicides to genetically modified glyphosate or herbicides resistance/tolerance plants. But the surprising and interesting fact is that both the glyphosate herbicide and herbicide tolerant GM-crops are the monopoly products of the same company (Sharpe, 1999). Most of countries are now showing disinclination in licensing GM crops although the company pioneering these products has confidently declared these crops economical, socially acceptable, and ecologically best-fit products (Falcon and Fowler, 2002).

Organic agriculture has been developed and initiated as recent paradigm in agriculture ecosystem during twenties and grown on millions of hectares. The area-wide implementation of organic agriculture guarantees chemical free healthy food and environment, avoiding application of agrochemicals, and ensuring the normal functioning of biotic components of agro-ecosystems. Nonetheless, it has been reported that the organic agriculture system, in spite of guaranteeing healthy and quality agricultural products, is lacking the potential to produce quality products in a quantity needed to meet the food requirements of the community and humanity. It is therefore a fact that organic agriculture can only improve the food safety but contributes little in coping with the issue of food security. However, ever-increasing populations of the globe demand for increases in agricultural production, consistant with the rate of population increase (Carvalho 2006).

The current system of agriculture production depends mostly on agrochemicals, which play an integral and key role in crop production and protection. Presently, there are two geographically opposite schools of thought on the use of agrochemicals in agro-ecosystems. The technologically advanced countries like Canada, the EU, the United States, etc., have restrained the use of agrochemicals through approval and enforcement of EPA legislations (Carvalho 2006). The main objective of such legislation is to protect community and environment from the toxic residues of agrochemicals. In these countries, this objective is achieved by testing toxicological aspects of agrochemicals and enforcing maximum residual limits (MRLs) of agrochemicals in our food and water. These countries are encouraging and enforcing to rely less on toxic and persistent chemicals, and more on less persistent, soft, and ecofriendly green products (Carvalho 2006). However, these green agrochemicals are not affordable by the stakeholders of developing countries because of their higher prices than conventional pesticides (Carvalho 2006). On the other hand, the concept about the use of agrochemicals in developing countries is unlikely different. In these countries, the growers rely on only those agrochemicals which provide quick solutions to their production and protection issues, as well as prove efficient in securing better crop yield and more income. They prefer to use cheap agrochemicals regardless of their expiry issues and/or their higher environmental impacts, heath issues, and MRLs issues in food commodities. Such agrochemicals are either very easy to synthesize or donated by developed countries (Carvalho 2006). Many tropical countries having the capacity to produce and formulate agrochemicals (like India), are investing in the import/production of such cheap agrochemicals, especially pesticides, which are then marketed to different developing countries. For example, the most prominent sales route in the tropics for enormous use of organochlorine compounds had been India, Bangladesh, Philippines, and Latin America. Chemical volatilization is the most important process, which accounts for the transboundary spread of residues of highly persistent agrochemicals to higher latitudes and countries of temperate and polar regions (Carvalho 2005). Most of the highly persistent organic pollutants (POPs) have been banned and still a majority of such compounds are on the hit list of the Environmental Protection Agency (EPA)—because POPs have bio-accumulative properties, a toxic activity, and a high persistence potential over years (Pelley 2006).

The hampering of agricultural productivity in Pakistan is associated with different limiting factors, which can be divided into four major and twenty-eight subcategories (Figure 12.1). Agricultural productivity can be enhanced by using different means (Figure 12.2) as solutions used either alone or grouped in an integrated way (Carvalho 2006). Among these means, increasing agricultural land area appears as a difficult chore as areas of agricultural land are drastically decreasing in almost all the regions of the world. In Pakistan, agricultural land is being converted into residential colonies, which has resulted into a drastic reduction of agricultural land. The means addressable for enhancing productivity and food security areb the efficient use of various types of inputs (seeds, irrigation, fertilizers, and pesticides) and using GMOs crops and varieties resistant to pests (Carvalho 2006). Along with other factors, agrochemicals (fertilizers and pesticides) were largely responsible for bringing the green revolution around the globe (Borlaugh and Dowswell 1993). Dependence on fertilizers and pesticides, as external inputs, is a short-term and immediate solution for large-scale commercial and intensive production of crops, animals, and for crop pest and disease vector management in agricultural systems (2015).

Pesticides have not only improved the agronomic practices, crop efficiencies, and yields but also ensure better pest management in agro-ecosystems. Smart and safe application of pesticides guarantees better sustainable agriculture. More than 800 million people are in the grip of food insecurity and more than 30% yield losses (preharveting and postharvesting) are caused annually by crop pests. The world will be in a position to provide sustenance for 9 billion humans on earth in 2050, when pesticides will be applied effectively and in a comprehensive integrative way (Welle et al. 2014).

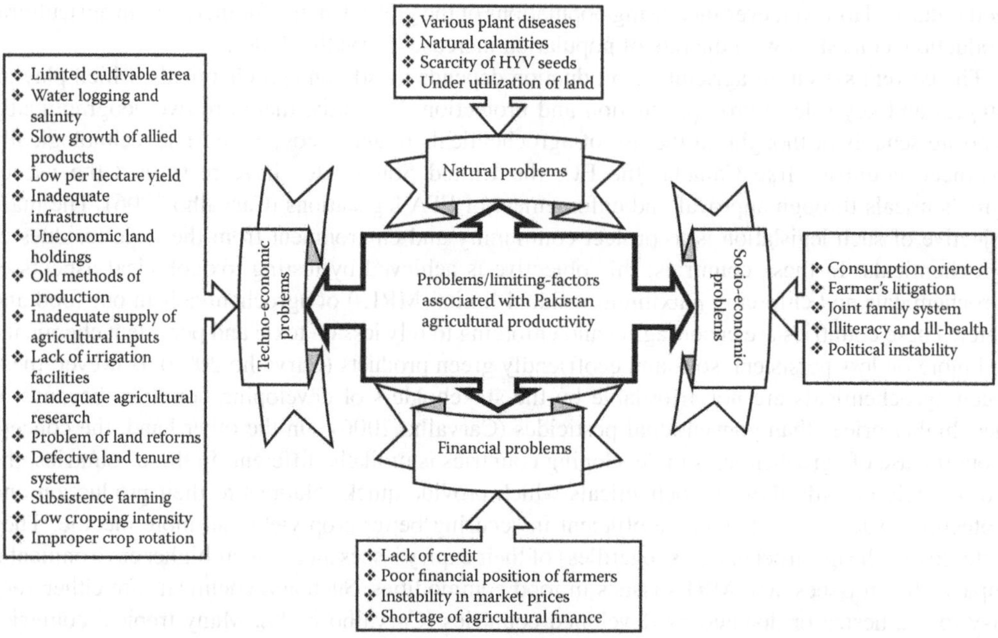

FIGURE 12.1 Pictorial depiction of the problems/limiting-factors associated with agricultural productivity in Pakistan.

Most economists and analysts consider that changes in the present economic monarchy, global trade system, and supply-chain system of food resources and agriculture development are issues that need to be resolved to bring about the sustainable eradication of hunger and poverty (Sachs, 2005). Irrespective of the different other development plans being launched in the world to cope with hunger, malnutrition, and poverty, a drastic development change in the agricultural sector and increase in agriculture production are the most important issues to be addressed in regions of

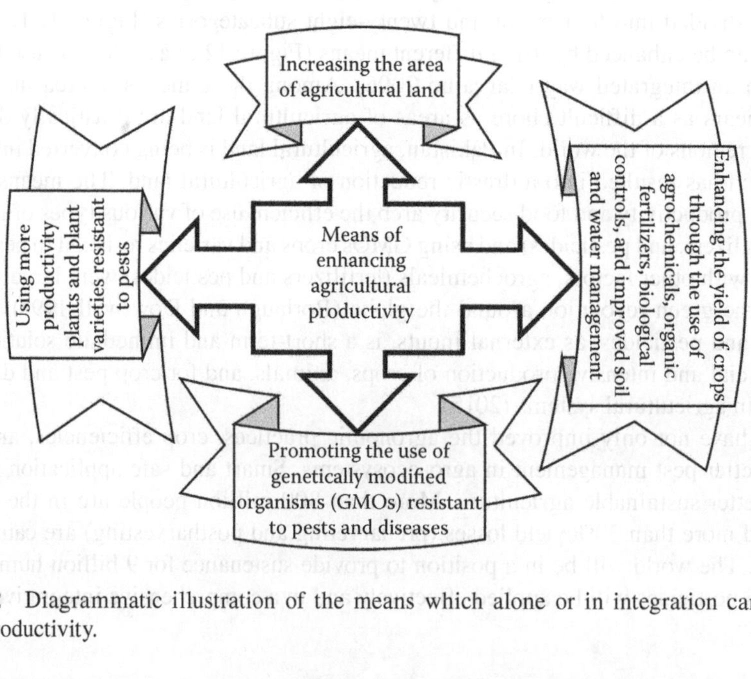

FIGURE 12.2 Diagrammatic illustration of the means which alone or in integration can enhance the agricultural productivity.

the world where there is poverty, hunger, and malnutrition. Agriculture development plans should consider the following tools for enhancing agriculture production:

- Development of pest resistant and climate resilient cultivars with high yield potential
- Establishment of irrigation system in fields which ensure the rational use of irrigation water
- Introduce and implement measures which prove effective in preventing pesticide contamination in water
- Deployment of such intensified agronomy system/reforms that ensure the development of new cultivars, prevention of soil erosion, amelioration of salt-related soil problems, etc.
- Selection and recommendation of highly nutritive plant species in the agriculture system
- Judicious, sensible, and moderate usage of various inputs including fertilizers, pesticides etc.
- Arrangement of awareness campaigns and conduct mass-scale training and education programs for the stakeholders about the precise and accurate use of agrochemicals and ways and means for soil protection
- Ensure the enforcement of all possible precautionary principles—mandatory for developing, testing/evaluating—before mass-scale introduction of genetically modified crops in the fields
- Encourage the research community to develop—and stakeholders to adopt—the organic farming system and integrated pest management (IPM) approaches to diminish usage of synthetic chemicals in order to cope with issues of environmental contamination, human health, and food safety and security

This chapter focuses on the critical analysis of use of pesticides as inputs to enhance agricultural productivity and food security. In this chapter we examine the forms and practical uses of pesticides, their history, case studies of the role of pesticides in pest suppression and crop productivity, the necessity and concerns surrounding the use of pesticides, the challenges of pesticides, and the future of pesticides in Pakistan. SWOT analysis highlighting the strengths, weaknesses, opportunities, and threats related to pesticides usage is presented in this chapter (Figure 12.3). Furthermore, the issues of pesticides—from import to consumption—are examined and the potential strategies/action-plans for addressing these issues are comprehensively discussed.

12.2 PESTICIDE INPUTS: FORM AND PRACTICES

12.2.1 Types of Pesticide Formulations

Pesticide can be classified in different groups on the basis of target pest, mode of action, and mode of entry to the target site (Table 12.1). A formulated material consists of active ingredients and other inert materials in specific proportions for the effective control of pests. The actual poison, also called active ingredient (AI), is seldom used in its purest and actual form for use against pests; rather the AI is admixed with different adjuvants for making the formulation more efficient and easier to calibrate, tank-mix, and apply. The marketed pesticides consist of different types of formulations based on (i) the chemistry of the AI, and (ii) different formulation offer different advantages (Waxman 1998). A detailed view of available pesticide formulations is given below.

- *Emulsifiable Concentrates (EC or E).* These are the formulations in which the AIs are not soluble in water. So, to make such AIs soluble in water, emulsifiers are mixed in these formulations which create suspensions when dissolved and agitated slightly in water. The equipment suitable for the application of ECs formulation include sprayers and mist-blowers operated at low-pressure, low-volume. The solvents used may cause rapid deterioration of plastic, rubber, and iron parts of the sprayers unless these parts are prepared from some resistant materials like neoprene rubber.

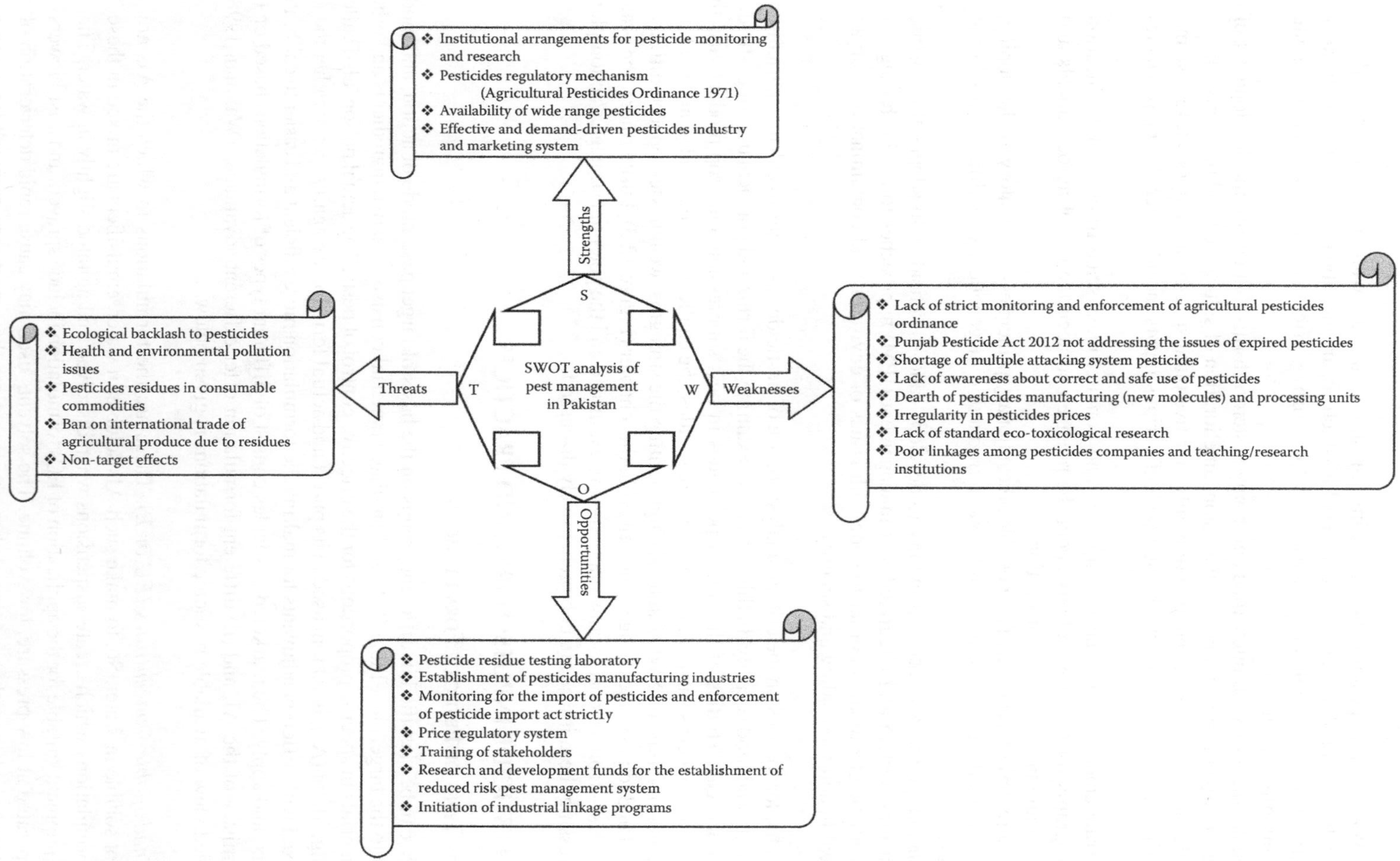

FIGURE 12.3 Diagrammatic illustration of SWOT analysis of the pesticides situation in Punjab, Pakistan.

TABLE 12.1
Pesticides Classification on the Basis of Target Pests, Mode of Entry, and Mode of Action

Sr. No.	Categories of Pesticide	Target Pests
1	Acaricides.Miticides	Mites, ticks, and spiders
2	Algicides	Algae in aquatic habitat
3	Antifouling agents	Organisms active on underwater surfaces
4	Antimicrobials	Microorganisms including bacteria and viruses
5	Attractants/Lures	Attract a variety of pests
6	Avicides	Birds
8	Biopesticides	Derivatives from natural sources like animals, plants, bacteria, and certain minerals. They are lethal for a variety of pests
9	Biocides	These chemicals are lethal for microorganisms
10	Defoliants	Cause foliage to drop from a tree or growing plant, generally are sprayed for timely harvesting
11	Desiccants	These toxicants result in drying and death of living tissues
12	Disinfectants and sanitizers	Pathogens
13	Fungicides	Pathogenic fungi
14	Fumigants	Pests harboring in hidden and closed places like buildings or soil
15	Herbicides	Weeds, grasses and other unwanted plants
16	Insect growth regulators	Insect pests
17	Insecticides	Insects and arthropods
18	Miticides	Mites pests
19	Microbial pesticides	Pathogenic microorganisms
20	Molluscicides	These chemicals are lethal for members of phylum Molusca
21	Nematicides	Pathogenic nematodes
22	Ovicides	These toxicants are lethal the egg development and kill the embryo pests inside the eggs. The target pests include orthropod pests
23	Pheromones	These chemicals modify the behavior of pests. They include sex, aggregating, alarming, trail, and spacing pheromones etc. These are species specific and target the relevant pest species
24	Piscicides	These are lethal for fish species
25	Plant growth regulators	Substances (except Macro- and miro-nutrients based fertilizers.) used for changing the growth behavior of plants
26	Predacides	Such toxicants are lethal for vertebrate and invertebrate predators
27	Repellents	These chemicals repel the pests from settling, oviposition and feeding. Almost all kinds of pest-like insects, orthropods, birds, etc., are targets of such insecticides
28	Rodenticides	These toxins are lethal for rodents like mouse, porcupines etc.
29	Synergists	Chemical that are nontoxic but have the potential to increase the performance of another chemicals
Based on Mode of Entry		
1	Stomach Poisons	These poisons enter body of pests through digestive tracts and kill. These poisons target chewing pests
2	Contact Poisons	The chemicals kill the pests through contact action. All crawling pests are the targeted by these poisons
3	Systemic Poisons	These poison are applied to plant foliage or on soil. From there, these poisons penetrate through foliage or absorbed through roots and then translocate through plant system. These poisons kill the sucking pests
4	Translaminar Poisons	These are poisons which translocate downward to lower surface of the leave when applied on upper side of leaf. Sucking arthropods feeding on the lower side of leaf are target pests

(Continued)

TABLE 12.1 (*Continued*)
Pesticides Classification on the Basis of Target Pests, Mode of Entry, and Mode of Action

Sr. No.	Categories of Pesticide	Target Pests
5	Fumigants/respiratory poisons	These chemicals get volatilized and their vapor enter into the body of organisms through respiration and kill them Almost all kinds orthropods are target of such insecticides
Based on Mode of Action		
1	Physical poison	These poisons kill the organisms especially insects by physical action like spiracular blockage and disruption of cuticle. These include dusts and oils. Almost all kinds insect are target of such insecticides Almost all kinds insects are target of such insecticides
2	Protoplasmic poison	These toxic chemicals precipitate proteins. Almost all arthropods are target
3	Respiratory poison	These chemicals disrupt the release and functioning of respiratory enzymes. Almost all kinds terrestrial insects are target of such insecticides
4	Axonic and synaptic poisons	Chemicals disrupt the transmission and induction of nerve impulses through axons and synapse across two neurons. Almost all kinds insects are target of such insecticides
5	Chitin synthesis inhibitors	These insecticides inhibit the synthesis of chitin and disrupt the molting process. Almost all kinds insects are target of such insecticides

- *Wettable Powder (WP or P).* These types of formulations are prepared by mixing the AIs in different types of adjuvants including surfactants, dispensers, and inert carriers like silica, clay, talc, etc. The formulated product usually contains 25% AI or even more depending upon the needs. For making sprayable tank-mixtures of such formulations, these are agitated constantly and vigorously in the tank because WPs don't make solutions but rather suspensions in water. In contrast to ECs, even very high concentrations of WP formulation have the least burning effects on foliage. However, these formulations have abrasive properties and may wear machine nozzle and spray pumps more quickly.
- *Water-Soluble Powders or Soluble Powders (WSP or SP).* In these formulations, the AIs are in the form of finely ground solid powders, which are soluble in water and may consist of small quantities of wetting-agents. These formulations do not require vigorous and constant agitation, as required by wettable powders and flowables, because WSPs always give solutions without any suspension or precipitation when tank-mixed in water. However, they can present an inhalation hazard to applicators during mixing. The proportionate quantity of the AI in SPs/WSP formulations ranges between 15% and 95%.
- *Water-Soluble Bags/Packs (WSB or WSP).* Some solid formulations are prepared in inert water-soluble bags/packs that can be directly placed and dissolved directly in the tanks of sprayers. The main advantages associated with such formulations include elimination of dose-measurement requirements and minimized threat of exposure to sprayers during tank-mixing procedures. These bags have unique physical properties and biodegradabilities.
- *Water-Soluble Concentrates, Liquids or Solutions (S, WS, WSC or WSL).* The AI used to prepare such formulations is water-soluble. Therefore, water or any other solvent is used for mixing with the AI to prepare the formulation. WSC or WSL types of formulations do not need constant agitation for mixing because such formulations always produce a real solution in water. Water-soluble concentrates are often liquid, salt, or amine solutions.
- *Oil Solutions (OS).* The oil solutions formulations are prepared by formulating the AI in oil or in any organic solvent with the addition of any wetting-agent or emulsifier. These formulations are in almost ready-to-use form. Mostly pesticides, made for household and garden applications, are prepared in the form of oil solutions in a variety of bottles, cans, and plastic containers. These pesticides are mostly available on the market in the form of

oil-based concentrates; that is why they are diluted with kerosene-oil, diesel, or any other oil before use. A disadvantage associated with OS formulation is its phytotoxic impacts on vegetation or plantation.

- *Flowables or Sprayable Suspensions (F, FL).* Thee formulations are prepared for those AIs which are nonsoluble in water and organic solvents and sometimes are called water-dispersible suspensions. For preparation of such formulations, the AI is saturated on a diluent (clay, etc.) which is subsequently pulverized into an extremely fine textured powder. This powder then is then mixed with small quantities of liquid to transform it into a thick paste, which can be measured by volume like WPs. For preparing their sprayable tank-mixed solution, FL formulations are admixed in water with moderate and adequate agitation before application on crops. Such formulations bear the advantages of both WP and EC formulations. These formulations are becoming increasingly popular. However, FL formulations impose a little phytotoxic impacts on plants/crops.
- *Water-dispersible granules or Dry Flowables (WDG or DF).* Water-dispersible granules are finely divided powders formulated into concentrated, dustless granules. They resemble wettable powder formulations. To maintain a uniform spray mixture, DFs form a suspension in water and require little agitation. These formulations contain a high percentage (75%–90%) of the AIs.
- *Ultralow-Volume Concentrates (ULV).* The formulation may contain only the AI or the AI in a small amount of solvent. They are used as a spray application of undiluted formulation at a rate less than or equal to 1/2 of a gallon per acre and often require specialized application equipment. A large area can be sprayed with a small volume of liquid. These formulations now are applied on few sites and are often prohibited unless it is specifically designated on the label, or authorized based on an official written or published recommendation.
- *Encapsulated or Microencapsulated Formulations.* These formulations are prepared by incorporating dry or liquid forms of AI, in permeable polymer or plastic spheres of diameter 15–50 μm, and then mixing these spheres with wetting agents, thickeners, and water. This is done to rationalize the required concentration of an AI in a slow release and flowable form. These formulations usually contain 2 lbs per gallon of AI. In contrast, they may pose significant hazards for bees, which may take the capsules with pollen back to the hive.
- *Dusts (D).* Dusts are consist of low concentrations of AI (1%–10% by weight), combined with an inert carrier (talc, chalk, clay, nut hulls, or volcanic ash). Dusts are formulated for application in the dry form. Therefore, they are known as the simplest formulations of pesticides to manufacture, ready to use, and easy to apply. Dust particles may drift long distances away from target sites even when wind velocities are low; due to their drift potential, herbicides are not formulated as dusts. Most portions of dust formulations drift (upward and downward); in agriculture, only 10%–40% of the material reaches the target site.
- *Granules (G).* Granular and dust formulations are quite similar, having small differences of size of their inert materials. The granular formulations are prepared by coating active ingredient in the range of 2%–40% by weight on ant absorptive material. These formulations can be applied at any time of the day without the risk of drift in winds up to 20 mph. They can also be applied at planting time to protect roots from pests or to introduce a systemic effect for the above-ground parts of the plants.
- *Poisonous Baits (PB).* They can be defined as a pesticide mixed with an edible material (attractive to a particular pest). The pests are killed by consuming poison in a single feeding or over a period of time. Baits do not need to be applied to whole areas and are mostly used to control rodents and fruit flies. They must be placed where it is likely they will be consumed. The percentage of AI in bait formulations is quite low (<5%). In bait formulation, a minute quantity of AI is used for application on large area.

- *Fumigants*. Fumigants are defined as a chemical or mixture of chemicals which volatilizes into gas, vapor, fume, or smoke at normal room temperature and prove lethal to different pests. They are a rather loosely defined group of formulations. They may be in form of gases and volatile liquids and solid materials. Fumigants are used for disinfection of buildings, enclosures, or other materials/objects kept in any enclosure. Fumigants are applied to soil too for the treatment of high-value crops at nurseries and greenhouses to control nematodes, diseases, and insects (larvae and adults) as well as to deteriorate the germeability of weed seeds. The treated soil may require covering with plastic sheets for several days to retain the volatile chemical, allowing it to exert its maximum effect on the target pests.

12.2.2 Procedures for Application of Pesticides

In the present scenario of plant protection systems in agriculture, modern machinery and equipments are required for precise and accurate application of pesticides and the economical control of pests. The selection of equipment should be made as per type and size of the job for which the equipment is being selected. For safe and efficient application of agrochemicals and the proper functioning of associated equipments, careful reading of the label of agrochemicals and appropriate selection, operation, calibration and maintenance of the equipments are pre-requisites. Application methods for agrochemicals depend on the type of crop and pest; however, directions written on label regarding application procedure should strictly be followed.

Methods of pesticides application

Band application: In this method, pesticides are applied on parallel strips or bands rather than applyied uniformly on the whole crop or field.

Basal application: This method is used to apply herbicides for control of weeds. In this method, the chemical is directed to towards the basal portion of small trees, brush, hedges, and so forth for the suppression of unwanted vegetation.

Broadcast application: It is the application of pesticides uniformly on an entire area or field.

Crack and crevice application: It involves the placement or plunging of pesticides in small quantity into the pest's hidding places like cracks and crevices in building, cabinets, baseboards, etc.

Directed-spray application: This method of pesticides application targets the pest only to minimize the chances of exposure of nontarget organisms.

Foliar application: In this method, the pesticides are directed sprayed on the leaves of a plant.

Soil application: This method involves the direct placement of pesticides on or in the soil rather than application on the plant's parts.

Soil incorporation: This method involves the utilization of tillage, rainfall, or irrigation equipment to incorporate the pesticides deep into soil around the root zone of plants.

Soil injection: This method is used to apply pesticides deep down into the soil's surface using pressure.

Space treatment: This method is used for treatment of enclosed areas with pesticides.

Spot treatment: It involves the application of any pesticide to a specific small area or surface.

Tree injection: Used to inject pesticides with pressure in the tree stem under the bark.

12.2.3 Pesticide Registration

The Department of Plant Protection (DPP), with its head office at Karachi, is responsible for registration and other regulatory aspects of pesticides in Pakistan. The regulation of pesticides (import, manufacture, formulation, refilling/repacking, sale, etc.) is controlled under the Agricultural Pesticides Ordinance of 1971, through the Agricultural Pesticides Rules, 1973 (Annex III). Generally, once a pesticide is registered, its registration is renewed periodically. Currently, almost 26 pesticides have been deregistered and their import banned in Pakistan (DPP 2105). In Pakistan, three types of pesticide registration practices are performed using the form-1, form-16, and form-17. Application for registration of pesticides has to be made to the Federal Government using Form-1 (DPP 2015). The information such as proof of manufacturing, registration, use, and data of pesticides is required. The Pakistan Embassy in the country of manufacture has to endorse that information. Upon receiving the application, the Federal Government sends the application together with a sample of pesticide to the pesticides laboratory for test or analysis. On receipt of the result of the test, the Federal Government may forward the same to the Provincial Government and Federal Agencies to conduct biological tests under field conditions. The tests should cover two crop seasons. Form-2 is used to assign a certificate with a registration number and the certificate of registration is valid for 3 years. The registration may be renewed for a further period of 3 years if an application is made on this behalf in Form 3, and a certificate of renewal of registration is filed in Form-4 (http://plantprotection.gov.pk/wp-content/uploads/Act%20and%20Rules%20PDF/AGRICULTURAL%20PESTICIDE%20RULES%20(APR).pdf). The application for registration may be rejected if the results of the analysis tests do not match with the information supplied. For attaining a registration certificate, the applicant must deposit a nonrefundable fee of 25,000/Rs. for each formulation of pesticide. However, certificate bearers can renew their registration certificates by depositing a nonrefundable fee of 5000/Rs. Similarly, for the registration of a manufacturing or formulation plant, an application is made in Form–18 and Form–20 and by depositing a 25,000/Rs. registration fee (DPP 2015).

Similarly, before pesticides can be imported to Pakistan—except for experimental purposes—they must be be registered. Import of pesticide cannot be allowed in the country without strict compliance to the details mentioned in the application submitted for registration. Pesticides should be imported only by persons who have a Certificate of Registration is issued by the government. For the import of pesticides not bearing any trade names, it is mandatory for the importers to submit an application to government for seeking approval to imports such pesticides using Form-16. The importer should also present guarantees that he has the necessary storage facilities, which are open for inspection by any official of the government, as well as the required documents, including proofs of manufacturing, registration, use, and data of pesticides which should be endorsed by the Pakistan Embassy in the country of manufacture. If the pesticides are registered in the country of manufacture, but not registered under Form–1 or Form-16, then the importers should comply with the conditions specified in Form-17.

The provision of the Pesticide Ordinance included the appointment of inspectors and the provisions of procedures for taking samples for quality analysis; however, this proved unsatisfactory due to different constraints (Jabbar and Mallick 1994). The main focus of the pesticide business in Pakistan is the import and sale of pesticides as local production is only a small proportion of the total agrochemical business in Pakistan. The members of APTAC (Agricultural Pesticide Technical Advisory Committee) advise the government on technical matters, but there is not a single independent institute capable of conducting pesticide research. However, the general scenario vis-à-vis over-use, strict, rigorous, inflexible implementation and enforcement of regulations appears austere because agricultural pesticides policies are concerned largely on quality and less on quantity of pesticide usage (DPP 2015).

12.3 HISTORICAL PERSPECTIVES OF PESTICIDES AS INPUTS IN PAKISTAN

12.3.1 Global History of Pesticide Use

Pesticides have been used to control insects and diseases as early as 2500–1500 BC by the Sumerians and Chinese (Schumann 1991). Pliny, in 79 AD, advocated the use of arsenic as an insecticide. In 900 AD, the Chinese were using arsenic and other inorganic chemicals in their gardens to kill insect pests. The naturally occurring pesticide nicotine was extracted from tobacco leaves in the seventieth century and used against the *Plum curculio* (true weevil) and the lace bug. Approximately 100 years later, Prevost described the use of copper sulfate for the inhibition of smut spores (Waxman 1998).

Important pesticides such as rotenone were developed in 1850 from the roots of derris plants and pyrethrum from the flower heads of chrysanthemum. These are still widely used for the effective control of different pests. During the same time, new inorganic materials were introduced for controlling insect pests of different crops. For example, investigations lead to the introduction of impure copper arsenate (Paris green) in 1867 for control of the Colorado beetle in the United States. In 1882, a substance for the control of the potato blight and vine mildew was accidently discovered by Millardet when they treated roadside vines with a mixture of copper sulfate and lime in order to discourage pilfering of the crops. Lead arsenate was used for control of the gypsy moth in 1892. Later on, the established effectiveness of copper sulfate, lime, and water mixtures were established against vine mildew and the mixture was named "Bordeaux mixture." In 1897, formaldehyde was introduced for fumigation. Copper arsenate was replaced by calcium arsenate in 1912. In the United States, calcium arsenate was effectively used to control cotton boll weevil. Similarly, organomercurials were used for the first time for seed dressings in 1913 against cereal smut and bunt diseases.

The modern era of synthetic pesticides really starts in 1930s. Important examples include alkyl thiocyanate insecticides and the first organic fungicides (dithiocarbamate fungicides). In 1939, a Swiss chemist, Paul Hermann Müller discovered the powerful insecticidal properties of dichlorodiphenyltrichloroethane (DDT). In 1943, within 4 years of its discovery, DDT became the most widely used single insecticide around the globe. Organophophates were developed as warfare agents during World War II. Unfortunately, both organochlorine and organophosphate pesticides are highly poisonous to nontarget organisms. This type of property forced scientists to develop more selective and less poisonous pesticides. From the development of DDT to the present-day pyrethroids and new chemistry pesticides, marvelous developments in biopesticides have occurred (Lamberth et al. 2013). In Pakistan, as well as in other developing countries, agrochemicals—especially pesticides—are the most reliable tool and weapon against prevailing pest problems, particularly since there exists no satisfactory and acceptable alternative control methods. The risks associated with the use of toxic compounds for crop production are highlighted in the 1962 book *Silent Spring* written by Rachel Carson. In recent decades, humankind has made excellent progress in the manipulation of genes. The need of the era is to develop such seeds which possess the genetic ability to kill or inhibit disease-causing pests.

12.3.2 Global Market of Pesticides

User-level estimates of conventional pesticide sales placed the value of the world pesticide market at about $25,280 million in 1993 with about 4500 million lbs of AI. The share of herbicides was 46% followed by insecticides (31%), fungicides (16%), and miscellaneous pesticides (6%). Among pesticide-consuming countries, the U.S. market was the largest, with a share of 34% of the total world market. The worth of the pesticide market was about $8484 million and 24% of total AI (Waxman 1998). The world market of pestcides increased to $32 billion in 2001. The worldwide use of pesticides increased to 5.2 billion pounds in 2006 and 2007, and its worth increased to more than $35.8 billion in 2006, reaching more than $39.4 billion in 2007. Again, herbicides accounted for the largest portion of total use, followed by other pesticides, insecticides, and fungicides. The U.S. pesticide market accounted for 22% of the total world pesticide amount used and sales amounted to approximately $12.5 billion. Herbicides such as glyphosate, atrazine, metolachlor-S, acetochlor, 2,4-D, pendimethalin, as well as the fumigants metam-sodium, dichloropropene, methyl bromide, and chloropicrin were among the top ten pesticides used in agriculture by weight (EPA 2014).

12.3.3 Pesticide History and Market in Pakistan

The history of synthetic pesticides use in Pakistan starts in 1947, with only 508 hand sprayers and 16 vehicles. The chemical pesticides were used for the first time to combat locust attacks in 1951, using aircrafts for aerial sprayings. These aircrafts were used as well to combat the *Pyrilla perpusilla* infestations on sugarcane in KPK (Khyber Pakhtunkhwa). After initial successes, pesticide use was extended to cotton, rice, and orchards throughout the country.

The era of pesticide use in Pakistan can be divided into distinct six periods depending upon the mode of distribution, pricing, and time period. First, (i) 1947–1965 (the pesticides were used free of cost by the government), (ii) 1966–1974 (the government provided the pesticides for a flat rate of Rs 0.25/L and subsidized 75% of the price), (iii) 1975–1979 (25% pesticide distribution was through the public sector while 75% was through the private sector and the government provided a 50% subsidy on ECs/WPs and a 75% subsidy on granules), (iv) 1980–1985 (complete withdrawal of pesticide subsidies in Punjab, Sindh, and KPK except in Baluchistan, FATA, and AK), (v) 1986–1991 (complete withdrawal of subsidies in all provinces except Baluchistan while 100% of the distribution was channeled through the private sector), and (vi) 1992–1993 (duty and surcharge exempted on imports of weedicides and only duty exemption on pesticides) (Jabbar and Mallick 1994). In 1954, the pesticide business started with the import of 254 metric tons of formulated products and increased to 20,648 metric tons in 1986–1987 (Jabbar and Mallick 1994). The AI consumption of pesticides increased at a rate of 25% per annum, from 906 metric tons in 1980 to 5519 metric tons in 1992. The share of the pesticide market on a provincial basis was 90% for the Punjab, 8% for Sindh, and 2% for KPK and Balochistan (Khooharo et al., 2008). Most pesticides used in Pakistan were insecticides (74%), followed by herbicides (14%), fungicides (9%), acaricides (2%), and fumigants (1%) (Khan 1998).

The sale of pesticides in 1995 was worth 9 billion rupees (US$ 222 million). Pyrethroids account for the greatest share with 45% of the market by value, followed by organophosphates with 39%, chlorinated hydrocarbons 9%, and carbamates 4%. Currently, more than 600 pesticides have been registered in Pakistan with a total business worth of about Rs. 32.75 billion representing a total import of 43,590 metric tons of formulated products. In contrast to the world market, insecticides are about 52% of the total pesticides with a business worth of about Rs. 17.03 billion. The weedicides business is now increasing in Pakistan. The current share of weedicides among total imported pesticides in Pakistan is about 31% (Rs. 10.15 billion), followed by fungicides 16% (Rs. 5.24 billion) and others 1% (Rs. 0.33 billion). The 46% (Rs. 7.83 billion) of total insecticides are being used to control the sucking insect pests. The rest of the pesticides are mostly used to control bollworms and other insects. A total of 933 pesticide formulations have been registered by 391 national and multinational pesticide companies; of these 26 pesticides have been banned. Most imported pesticides are used in Punjab (88.3%) followed by Sindh (8.2%), KPK (2.8%), and Baluchistan (0.7%). In cotton, 60% of the pesticides are sprayed against various pests while 07% in Peddy, 04% in cereals, 02% in sugarcane, and 27% in some other crops. (DPP 2015).

12.4 PESTICIDES AND PEST MANAGEMENT: CASE STUDIES FROM PAKISTAN AND THE GLOBE

The usage of pesticides as insecticides, fungicides, herbicides, rodenticides, and so forth protects crops from different types of pests, contributes to significant reductions in preharvesting and postharvesting crop losses, improves the yield of economical crops including maize, vegetables, potatoes, cotton, fruits, and stored commodities, safeguards cattle from diseases and ticks, and protects humans from different types of vectors responsible for the transmission of various diseases (Taylor et al. 2003; Carvalho 2006; Tagar et al. 2015).

Globally, pesticides have contributed enormously to crop productivity by suppressing the intensity of pests, reducing the density of crop competitors, and ameliorating the condition of the crop plants. The historical record clearly indicates that significant improvements in yield occurred for several crops already after the mere suppression of weeds by effective herbicides. For example, weeds in cucumber, dry bean, sorghum, peach, potato, and rice fields were eradicated reducing nutrient competition for the actual crop (Comes et al. 1962; Burnside and Wicks 1964; Mueller and Oelke 1965; Daniell and Hardcastle 1972; Glaze 1975; Nelson and Giles 1989). A 4-year study showed that wheat weeds were eradicated using 2,4-D (Alley 1981). An analysis of the 1961–1975 data from the University of Minnesota and the University of Illinois indicated that weeds were suppressed drastically in corn and soybean with herbicides (Dexter 1982; Schroder et al. 1984). In the United Kingdom, the application of MCPA (2-methyl-4-chlorophenoxyacetic acid) to cereals reduced weeds (Lever 1991). Herbicides have been credited with being the main factor in suppressing wheat weeds in Canada (Freyman et al. 1981). In Australia, herbicides—instead of tillage—resulted in an extraordinary suppression of weeds, better water conservation in the soil profile, and an increase in grain yields (Wylie 2008). The use of herbicides is cited as a primary factor in suppressing peanut weeds (Grichar and Colburn 1993), rice weeds (Smith et al. 1977), blueberry and sugarcane weeds (Yarborough and Ismail 1985; Yarborough et al. 1986), and cranberry weeds (Dana 1989; Eck 1990).

In the United States, data collected since the early 1960s indicate that weed suppression and yield increases are attributable to herbicide application in *Oryza sativa* (rice), *Phaseolus vulgaris* (kidney bean), *Sorghum bicolor* (sorghum), *Solanum tuberosum* (potato) (Comes et al. 1962; Burnside and Wicks 1964; Mueller and Oelke 1965; Nelson and Giles 1989), *Triticum aestivum* (wheat) (Alley 1981), and *Zea mays* (corn) (Dexter 1982). Similarly, weed suppression and associated increases in yield have been documented in cereal crops in United Kingdom (Lever 1991), in grain crops in Australia (Wylie 2008), in wheat in Canada (Freyman et al. 1981), and in soybean and maize in Argentina (Penna and Lema 2003). Application of atrazine resulted in weed suppression and a doubling of maize yields in Nigeria (Benson 1982). Application of herbicides increased yields in Zimbabwe by 50% for maize (Chivinge 1990) and 33% in Kenya for maize as well (Muthamia et al. 2002).

In Africa, a drastic decrease in pest pressure due to pesticide application has been reported for various cropping systems, schemes, and situations . The major insect pests of cowpea in sub-Saharan Africa have been controlled by spraying insecticides (Dugje et al. 2009; Kamara et al. 2010). In western and southern Africa, attacks by fungal diseases and insect pests have been the main cause of low groundnut yields, which were improved by 80% with the application of fungicides and insecticides (Naab et al. 2005). In India, sprays of insecticides resulted in a 90% reduction of pod borer infestations; in addition, application of fungicides caused a 60% reduction in the incidence of the ascochyta blight (Ameta et al. 2010; Maheshwari et al. 2012). Application of systemic fungicides by seed-treatment or foliar spray produced an excellent suppression of downy mildew diseases, which cause a 20%–90% yield losse of maize crops in different Asian countries like Indonesia, the Philippines, and India (Mikoshiba 1983; Putnam 2007). Ward et al. (1999) documented an excellent suppression of grey leaf spot disease in Africa. *Striga* weed has been well controlled with a systemic herbicide—imazapyr—when it is coated on the seeds of IR-maize varieties and maize plants have been also protected from the side effects of *Striga* weed (De Groote et al. 2008).

Agrochemicals and agricultural technologies invented for improving agricultural productivity, food safety, and food security were later implemented for the promotion and modernization of agricultural farms. Crop dusting techniques were first practiced in the Mississippi Delta on cotton crops in early 1922. In Pakistan, pesticides were used for the first time in 1950 against locust attacks. This pest was successfully controlled with insecticides and now outbreaks are very rare. Only 254 tons of pesticides were imported; while the import and distribution of pesticides was completely controlled and subsidized by the government of Pakistan in 1980.

About 80%–90% of imported pesticides are used for the control of cotton pests (Khan et al. 2002; Qamar-ul-Haq et al. 2008). The gigantic cost intricate in agrochemical based agriculture production system involve production losses, resistance development in pests, damages to animals and human health, loss of biodiversity and residues monitoring costs in food crops, fruits and other consumable commodities. According to health hazard data of women cotton pickers from the cotton belt, approximately 2.23 million women cotton pickers suffered from lethal chronic diseases due to the indirect exposure to pesticides sprayed on cotton. A huge amount of 105 million rupees (as big share of their income) are spent by the women for treatment (NARC 2000; Khan et al. 2002; PCCC 2006). Before 1980, the pesticide import and distribution was controlled by the Plant Protection Department, Government of Pakistan and used for the locust control program, malaria control program, and pest control programs in cotton, sugarcane, rice, tobacco, etc. (Ahmad 1988).

In 1991–1992, after the transfer of the pesticide sale and distribution business to the private sector in Pakistan, pesticides were sprayed on about 3.8 million ha (18% of total cropped area) against epidemic outbreaks of locusts, sugarcane pyrilla, white-backed plant-hoppers, crickets, etc. (Jabbar and Mallick 1994). Experience shows that high yielding varieties perform better when external inputs like fertilizers and pesticides are applied. Generally, the losses due to uncontrolled pests are between 15%–25%. Looking at the yield figures of 1991–1992, the prevention of 8%–10% losses due to pests added another million bales to that year's cotton production. The discussions with experts in Pakistan Agricultural Research Council (PARC) have suggested that during the 1991–1992 cotton season the cotton leaf curl virus disease was responsible for an estimated loss of 3 million bales (25%), which in the following year was reduced by management of its vector whitefly through the use of insecticides (Jabbar and Mallick 1994). During the era after the 1980s, successful management of diseases by the use of pesticides (Khan 1990a; Jabbar and Mallick 1994) reduced the effects of the following: the cotton leaf curl virus in cotton belt of Sindh and Punjab, the banana bunchy top in Sindh (Badin, Mirpur Khas, Hyderabad, Sanghar, Nawabshah and Thatta districts), sugarcane pyrilla in the sugarcane belts, rice insect pest complexes (yellow rice stem borer, white backed plant hopper, and leaf roller are major pests of rice), wild boar and rodents [two major problems causing substantial losses (up to 100%) in sugarcane, maize, groundnut, rice, wheat, potatoes, etc.], and fruit pests in the northern areas (codling moth in apples, pears and walnuts in Gilgit, Skardu and Hunza valleys; red spider mites in apples, plums, almonds, peaches, grapes, and walnuts throughout the northern areas, etc.). For example, the Agriculture Extension Department of the Government of Punjab, sought advice from the University of Faisalabad. Racumin, warfarin, T10, and ZnPH2 were used for the control of vertebrate pests. (Khan 1990b). Temik 10% was concentrated to 100% AI and filled into capsules which were then placed in dough baits to kill wild boars. But by this highly concentrated poison threatens other animals that eat this bait (Jabbar and Mallick 1994). Since the advent of pesticides in Pakistan (1950), pesticides consumption (5519 metric tons in 1991) has shown a consistent rise, reaching in recent years 43,560 metric tons of formulated products (insecticides 52%, weedicides 31%, fungicides 16%, and others 1%) (Department of Plant Protection 2015). 1 One billion U.S. dollars worth of biopesticides are sold in the global pesticides market. But this market value is far less than global pesticides market of value 40 billion U.S. dollars (Popp 2011).

12.5 PESTICIDES RELATED AGRICULTURAL PRODUCTIVITY: GLOBAL AND NATIONAL PERSPECTIVES

Globally, pesticides have contributed enormously to crop productivity. The historical record clearly indicates that significant improvements in yield occurred for several crops only after the introduction of effective herbicides. For example, cucumber yields increased by 24%, dry bean yields by 38%, sorghum yields by 34%, peach yields by 167%, potato yields by 29%, and rice yields by 160% (Comes et al. 1962; Burnside and Wicks 1964; Mueller and Oelke 1965; Daniell and Hardcastle 1972; Glaze 1975; Nelson and Giles 1989). A report of four year study revealed that application of 2,4-D contributed 255 kg/ha increase in the wheat yield through suppressing weeds competition for wheat crop (Alley 1981). Dexter (1982) documented 15% and 19% more yield in cotton and soybean crops, respectively just due to weed suppression by application of weedicides in USA. In the UK, approximately 20% increase in yield of cereals was attributed to the application of MCPA (2-methyl-4-chlorophenoxyacetic acid) (Lever 1991).Herbicides have been documented as major factor that contribute in increasing wheat yields twice in Canada (Freyman et al. 1981). According to Wylie (2008), application of herbicides in Australia resulted in 15%–25% increase in grain yield with an addition of 27 mm more water in the soil profile. Weed control by the use of herbicides is reported as main factor in increasing 20% corn yields from 1964 through 1979, 62% soybean yield from 1965 through 1979 (Schroder et al.1984), 50% peanut yields (Grichar and Colburn 1993) and reasonable rice yield (Smith et al. 1977). The production of blueberry increased more than tripled since the introduction and usage of herbicides in Maine (1980) (Yarborough and Ismail 1985; Yarborough et al. 1986).

The introduction and use of dichlobenil, norflurazon, and glyphosate as weedicides for the suppression of weeds in cranberry fields doubled cranberry yields from 1960–1980 (Dana 1989; Eck 1990). Significant increases in the yield of sugarcane crops in Louisiana, during the 1950s, as well as in rice, dry bean, sorghum, and potato yields in the United States during the 1960s were mostly due to the introduction of weedicides in cropping systems (Comes et al. 1962; Burnside and Wicks 1964; Mueller and Oelke 1965; Nelson and Giles 1989). Similar yield responses have been documented for cereals in the United Kingdom (Lever 1991), for grains crops in Australian (Wylie 2008), and for wheat in Canada (Freyman et al. 1981). Better yield responses of crops are attributed to the fact that weed suppression results not only in less competition but also in better moisture conservation of the seedbeds due to less cultivation required for weed removal (Nalewaja 1975). An area-wide adoption and implementation of weedicides, especially glyphosate, in Argentina has been the main factor behind the economic sustainability and production expansion of two important crops, maize and soybean (Penna and Lema 2003). During early 1970s, less than 1 million ha of field crop areas received herbicide applications, which gradually increased, until 2005, to more than 70 million ha (Zhang 2003). Weeds caused more than 40% rice yield losses during 1973 in China in spite of hand weeding. However, in 1988, with most rice growers using herbicides on large areas, yield losses in rice crops due to weeds were estimated at only 6%–8% (Moody 1991). Dong et al. (2010) documented 2.8% and 1.5% losses in rice yields due to above-canopy and below-canopy uncontrolled weeds in the Yunnan Plateau, respectively. The research results indicated that areas receiving herbicides in Russia covered 25 million ha in 1968, which increased to 47 million ha in 1973 (Keiserukhshy and Kashirsky 1975). This increase in herbicide usage resulted in a 50% increase in yield of cereal crops on state farms (Chenkin 1975). However, during the 1990s, wheat production decreased drastically because of the severe decline in the use of herbicides due to the USSR's dissolution in 1990–1991 (Zakharenko 2000). The value of the herbicide market in Russia was 2.8 times higher in 2010 than in 2003 because the Russian government amended and introduced agricultural policies that were favoring the supply, availability, and use of herbicides on a large scale for crop production systems (McDougall 2013). Research results have confirmed that application of imazapyr as a systemic herbicide—by the seed coating method—resulted in season-long control of *Striga* weeds in maize fields resulting in three- to fourfold higher maize yields (Kanampiu et al. 2003).

Research reviews confirm the potentials of pesticides in suppressing pest pressure and enhancing yield recovery of crops under different cropping schemes and situations in Asia, Africa, and other continents. In sub-Saharan Africa, cowpea is a major dietary protein source for rural communities; however, its production is hampered due to the attacks of different insect pests. Fortunately, two to three applications of insecticides on this crop help the growers in reducing the pest pressure and enhancing yields tenfold (Dugje et al. 2009; Kamara et al. 2010). Very low yields of groundnut in Africa were attributed to the lack of application of fungicides on this crop. Application of fungicides on this crop improved groundnut yield by up to 80% due to suppression of the leaf spot disease (Naab et al. 2005). Annual yield losses of 30% in pulse crops in India—due to severe infestation of insects, diseases, and weed pests—can be significantly decreased by the use insecticides, fungicides, and herbicides by large-scale growers (Dhar and Ahmad 2004). Ameta et al. (2010) and Maheshwari et al. (2012) concluded that a 90% suppression of pod borer infestations and a 60% decrease in blight incidences could be achieved with insecticides and fungicides application, respectively. Sekhon et al. (2004) reported two times higher yields of bean crops in India when herbicides were used as compared to conventional weed removal practices. In Indonesia, the Philippines, and India downy mildew diseases cause 20%–90% yield losses in maize (Mikoshiba 1983; Putnam 2007). Seed-treatment with systemic fungicides alone and application of foliar spray in combination with seed-treatment enhanced maize yields up to 8%–10% and 34%, respectively. Approximately 50%–60% yield losses, caused by grey leaf spot disease in Africa under different production and protection scenarios, were minimized with the application of fungicides (Ward et al. 1999). In Zambia, approximately 27%–54% differences in grain yield (due to variation in genotype susceptibility) were recorded between treated and untreated fields (Verma 2001).

12.6 PESTICIDES USE: NECESSITIES AND CONCERNS

12.6.1 Pesticides Benefits

The discrepancies between the increasing food demands of the world population and the global agricultural output have resulted in alarming situation at both national and international levels (Smil 2000; Ingram 2011). Increasing food demands towards high quality foods, especially in the developing countries, are projected to increase by 70% (FAO 2009). Worldwide, more than 800 million people are subjected to food insecurity, and by 2050 this may reach up to one billion, while at present 130 million children are malnourished (FAO 2012). In addition to preharvest crop losses, food chain losses are also comparatively high (IWMI 2007). Globally, on average 35% of potential crop yields is lost to preharvest pests and without pesticide use, crop yields losses to pests could reach 70% depending on area and environmental conditions (Oerke 2005). Increasing food requirements can only be met either by cultivating new lands or intensifying the existing agricultural areas. Due to the scarcity of additional agricultural land, sustainable production on existing land is by far the better choice.

Given these limitations, agricultural intensification (Merrington et al. 2004) along with the use of chemicals such as pesticides play an important role in increasing crop productivity. The valuable outcomes of pesticides use indicate that pesticides enhance the living standards of people around the globe by increasing agriculture outputs. Different segments of activity of society such as forestry, agriculture, community health, and domestic activities attain significant benefits by using pesticides at different scales (Aktar et al. 2009). The better yield (and worth of yield) nurtures many indirect economic effects, which accrue from the additional crop production moving through the economy and creates new jobs and earnings of the workers. However, the agricultural producers are the major consumer with about USD 40 billion worth of pesticides per annum and a total consumption of more than 2 million tons/year. Europe and the United States consume 69%, leaving 31% for the remaining world (Abhilash and Singh 2009). In Pakistan, pesticides consumption was 7000 metric tons per annum in 1960, which has grown to 16,226 metric tons in 1976–1977, about 14,848 tons in 1987, and had reached 78,132 tons in 2003 (Syed and Malik 2011). Pesticide consumption data indicate that about 80% of total pesticides are being used on cotton crops, while the rest is used on other crops like paddy, sugarcane, maize, fruits, vegetables, and tobacco, etc. (Economic Survey of Pakistan 2006). Pesticide is a substance or mixture of substances which are used to preserve crops by managing, killing, or repelling pests. Farmers in advanced countries suppose a four- or fivefold return of worth spent on pesticides (Gianessi and Reigner 2005, 2006; Gianessi 2009). According to some studies, an additional one dollar spent on pesticides increased the value of outputs by about USD 4, indicating a high level of productivity of crops (Headly 1968). The pesticides costs are relatively low compared to total production costs recorded in different studies for pest management. For example, in most of the E.U. countries pesticides account for approximately 7%–8% of total farm production costs, increasing to 11% in France and Ireland; while only 4% in Slovenia (Popp 2011). In the United States, pesticides account for about 5%–6% of the total farm inputs for the management of pests (USDA

2010). According to the national pesticide benefit studies in the United States, about were spent on pesticides and their application for crop use every year. In the United States, pesticide use on crops saves around USD 60 billion by spending only USD 9.2 billion every year (Gianessi and Reigner 2005; Gianessi and Reigner 2006; Gianessi 2009). This indicates a net return of about USD 6.5 for every dollar spent on pesticides and their application.

In addition to enhancing crop yields and saving livestock, pesticides have also had a direct impact on human health safety. The use of pesticides has saved millions of lives by killing pests that carry or transmit severe diseases. For example, the prevalence of the deadly disease malaria (transmitted by infected mosquitoes) has been significantly decreased due to the use of pesticides. Similarly, other diseases such as the bubonic plague and typhus, carried by rat fleas and transmitted by lice, respectively, were minimized due to pesticide practices.

12.6.2 Pesticides Concerns

At the same time, while pesticides play a significant role regarding pest management, their release into the environment may have numerous adverse toxic short and long-term effects on ecosystems. Pesticides hold a distinctive position among environmental hazards due to their acute and chronic toxicity. It is also well recognized that acute poisons cause health disturbances such as seizures, rashes, and gastrointestinal illness (Guidotti and Gosselin 1999). Similarly, according to various long-term studies, chronic effects such as cancer and adverse reproductive outcomes have been seriously associated with the extensive use of pesticides (Sanborn et al. 2007).

12.6.3 Loss of Species Diversity

In addition to target pests, pesticides generally harm all forms of life in the ecosystem. Though there are some pesticides that are considered to be selective in their mode of action, their range is only limited to test specimens. Use of pesticides may eliminate key species and their impacts on the biodiversity of plants and animals constitute a major adverse effect, vital for the functioning of the entire ecosystem. This may disturb the dynamics of the food web in communities by intruding the existing dietary associations between different species. The loss of key species due to pesticides results in cascading effects which alter trophic interactions and other food linkages, ultimately resulting in the extermination of other species in the ecosystem. For example, *Enhydra lutris* as a keystone marine species confines the population density of sea urchins (Mills et al. 1993). Consequently, the foremost direct and/or indirect adverse effects of pesticides on the environmental appear in the form of the loss of plant's and animal's biodiversity in agricultural landscapes.

12.6.4 Effect on Pollinators and Predators

The loss of key natural pollinators like butterflies and honeybees due to the pesticide application is a major setback for seed and fruit production (Hackenberg 2007). According to one estimate, U.S. farmers lose up to $200 million a year due to reduced crop pollination as pesticides have eliminated about a fifth of honeybee colonies in the country (Miller 2004). Similarly, the broad spectrum use of pesticides disrupt the natural balance between predators and pests. This disruption in balance prompts the escalated use of pesticides to enhance crop productivity. Due to this imbalance outbreaks of secondary pests become a damaging problem and predictably about one-third of the 300 most damaging insects in the United States went from secondary to major problem after pesticide use (Miller 2004). In addition to pest outbreaks, some pests species become resistant to pesticides as a result of frequent pesticide application in specific environmental conditions (Brown 1971). According to some estimates about 520 insect and mite species, nearly 150 plant pathogen species, and about 273 weed species are now resistant to pesticides (Stuart 2003). Despite some efforts in this regard, this problem continues to increase and spreads to other species as well.

12.6.5 Effects on Soil and Water

Pesticide application contaminates soil ecosystems and persists there for several years, adversely affecting arthropods, earthworms, fungi, bacteria, and protozoa—small organisms vital to soil ecosystems where they fix nitrogen or break down organic matter to enable chemical reactions (Pimentel et al. 1997). Pesticides reduce species diversity in soils, soil fertility, as well as the total biomass of these biota. For instance herbicides reduce the vegetative cover of soils resulting in soil erosion through runoff which leads to disturbed soil structures and therefore creates an imbalance in soil fertility. Similarly, pesticides applied in the ecosystem find their way into water bodies, through runoff, resulting in the loss of clean water sources for communities. For example, in the United States, according to the U.S. Geological Survey, pesticides pollute every stream and about 90% of the wells sampled (Gilliom et al. 2007). Pesticide residues were present in rain and groundwater bodies. Similarly, in another study pesticide concentrations exceeded the limit in drinking water sampled from river water and groundwater (Bingham 2007).

12.6.6 Pesticides Residues in Food

The presence of toxic compounds in the environment and food chains due to the indiscriminate use of pesticides (Khan et al. 2010; Ahmed et al. 2011) causes a serious concern regarding the MRLs (maximum residue levels). An MRL is the maximum concentration of a pesticide residue that is legally permitted on a particular food. There are some agrochemicals—comprising organochlorines (OCs) and organophosphates (OPs)—which might pose serious threats for the health of both wildlife and humans, being persistent and toxic (Eqani et al. 2012a). For example, in Pakistan one third of the total imported pesticides used for fruits and vegetables production are such agrochemicals. These pesticides enter in our food chain in the form of residues in food. They cause toxic effects directly or through the food chain such as water, fruits, vegetables, or any other way. They also cause chronic effects over long periods of exposure, which include carcinogenic effects, mutagenic, and teratogenic neurotoxic effects, etc. One of the main reasons for the lower export values of fruits and vegetables in Pakistan is the higher amounts of toxic compounds present in the form of residues. There are almost 144 consignments confiscated from Europe in only 2010–2011 because of pesticide residues and other compounds. Many studies in Pakistan have reported pesticide residues in blood serum and fat samples of residents in the Balochistan, Sindh, and Punjab provinces (Naqvi and Jahan 1996; Parveen et al. 2004). Globally, food safety is a main public concern and most of the governments and private and international organizations have set MRLs for food commodities to control the unacceptable risks of human toxicity (Latif et al. 2011). According to some studies, the concentrations of OCPs, Ops, and other types of agrochemicals appeared at high levels in different fruits and vegetables (Masud and Farhat 1985; Cheema and Shah 1987; Masud and Hassan 1995). Similarly, the occurrence and residue levels of frequently used pesticides such as cypermethrin, methamedophos, monocrotophos, cyfluthrin, dieldrin, and methyl parathion were noticed among different mango varieties (Hussain et al. 2002). One of the main causes of higher amounts of pesticide residues in crops is the transfer of ownership of pesticides to the private sector in 1989, after which private companies motivated users to apply more than recommended doses to increase sales (Tariq 2005). In another study in the United States, several thousand food samples were analyzed—8 fruits and 12 vegetables, and about 73% had pesticide residues. Similarly, among five crops (apples, peaches, pears, strawberries, and celery) pesticide residues were detected in 90% of the samples. About 37 different pesticides were detected in apples (Groth et al. 1999).

12.7 CHALLENGES OF PESTICIDES MARKET IN PAKISTAN

Suppression of pest through synthetic chemicals is the fastest and most certain way of pest management. It has the benefits of speed of control in circumstances of immense pest outbreaks against biological and cultural control tactics, which work over a longer time period. In Pakistan, the use of pesticide started in 1950s for the control of locust and the Government imported 254 tons of formulated pesticides in 1954 (Habib 1996) and then pesticide business began in the country. Till 1980, for the import and distribution of pesticides in Pakistan, the Plant Protection Department was responsible through the national agricultural extension system. Most of the imported pesticides were used for aerial spraying against locust, sugarcane pests, rice, cotton, fruit crops and tobacco. The aerial spraying was free of charge and pesticide cost was subsidized. While there are serious problems of environmental and ecological nature due to the excessive dependence on pesticides, pesticides perseverance in the food chain and the development of pesticide resistance in pests (Brown 1971) are the two major problems yet. During the handling, transfer to smaller bottles/containers, and storage of surplus toxicant, a number of accidents happened (Ahmad 1988). Studies reveal that in cotton growing zones, the natural enemies' populations have decreased as much as 90% (Husnain 1999). The misuse and overuse of pesticides in agriculture cause health and environmental problems. Strict monitoring and enforcement of the Agricultural Pesticides Ordinance is direly needed to control the proper handling, marketing, and use of pesticides. Shortage of multiple attacking system pesticides is another challenge for effective chemical control of certain crop pests.

Postharvest losses are very high due to the scarcity of proper harvesting equipment and a dearth of postharvest technology. Technically enriched, economically sound, and highly efficient crop production, protection, and harvesting services are badly needed to address the issues of low crop productivity. By considering the role of agriculture in the economy of a country, role of synthetic chemicals in the crop protection to meet the world's increasing food demand, the concerned departments (government, private sector or traders) involved in this sector must need to review their policies. To protect crop plants by using pesticides is needed but the negative externalities related to pesticide use have also increased. These externalities, including damaging impacts on marine life and beneficial flora and fauna increase the mortality rates of humans in developing countries and, most importantly, excessive exposure to pesticides creats severe health issues for farmers.

No doubt pesticides have contributed enormously to crop productivity of various crops nationally as well as globally. In addition to enhancing crop yield and saving livestock, pesticides have also had a direct impact on human health safety. In Pakistan, pesticide usage is facing many challenges including the lack of pesticide manufacturing and processing units at any scale in Pakistan, the risk of pollution, the causalities and ecological backlashes due to their misuse. Other challenges include health issues, pesticides residues in food (food safety), week and flexible pesticide registration and monitoring systems, monopoly-based pesticide marketing systems, irregularity in pesticide pricing, import of cheap or donated pesticides, poor quality pesticide formulation adjuvants, and lack of pesticide-based multiple-attacking-systems for pest, as well as lack of awareness for efficient application strategies. The following are the major challenges/issues associated with use of pesticides inputs.

- Punjab Agricultural Pesticides Amendment Act 2012 does not address the issues of expired pesticides
- Shortage of multiple-attacking-system pesticides
- Health and environmental pollution issues
- Pesticides manufacturing (new molecules) and processing units
- Lack of policy regarding pesticides prices
- Lack of standard ecotoxicological research
- Ecological backlash to pesticides
- High pesticides residues in consumable commodities
- Poor linkages among pesticides companies and teaching/research institutions.

12.7.1 Misuse of Pesticides

The regular underdose and overdose application of pesticides results in the development of resistance in pests against these chemical. Nearly all crops are damaged by insect pests, but cotton is very much vulnerable to attack of pests and the effective production of cotton in Pakistan is mainly dependent on pesticide use. In current years, the attacks of cotton whitefly and cotton bollworm in Bt and non-Bt cotton increased the reliance on pesticides, and usage has increased more than 65%. On the other hand, adulterated and impure insecticides are also marketed, which produce ineffective pest control and hence result in pesticide resistance development in the target pests. The government of Pakistan has established the regulatory mechanism for monitoring pesticide usage in the country, but strict monitoring and implementation is yet required.

12.7.2 Health Issues and Challenges

Pesticide poisoning in humans is the main concern with the use of pesticides. Farmers do not realize how much the pesticides are toxic to human health and environment. The farmers, importers, and distributors of pesticides are not fully aware of the health risks. There are the different pathways or routes through which pesticides enter the body but the most sensitive organs are the skin, lungs, eyes, and digestive systems. The pesticides are classified into acute and chronic poison categories on the basis of exposure time and concentration. The acute effects include nausea, headache, vomiting, fatigue, weakness, and sometime unconsciousness. While chronic effects are carcinogenic, tumors, teratogenic or birth defects, behavioral changes, disturbances in the immune system, reproductive changes, and many allergic reactions—particularly of the skin—can sometimes be mistaken for signs of acute exposure (Repetto and Baliga 1996). Chronic pesticide poisoning can be measured through pesticide residues and the cholinesterase levels in the blood, mother's milk, and fatty tissues.

In Pakistan, cotton picking is mostly carried out by women; therefore, the extensive use of chemicals on cotton crops results in a greater exposure to poisonous or toxic chemicals by women pickers. During recent years cotton crops have been under severe attack by some bollworms—American bollworm and pink bollworm—and some sucking pests such as the whitefly, resulting in the heavy spraying of pesticides during the season—even in the late season. In most of cases pesticides are applied in the late season and even during the cotton-picking season (Orphal 2001), which is a serious health concern issue for women pickers who feed their infants. Most of the women cotton pickers are illiterate and not well informed about the hazardous effects of pesticide poisoning and exposure. A survey study conducted in the cotton growing areas of the Punjab, Pakistan has shown that 68% of the female farm workers (cotton-pickers) are not well aware of the pesticide's hazardous effects and 95% of the female workers did not take any protective measures against pesticides. However, male farm workers were conscious of the health issues caused by pesticide poisoning and 86% of the surveyed men were aware of pesticide exposure and adopted precautionary measures or used personal protective equipment.

12.7.3 Pesticides Applicators

The farm workers or farmers who apply most of the pesticides do not have sufficient knowledge about the formulation of pesticides, especially the concentration of AIs in the product material. Mostly, they mix the different chemicals together and face severe problems in case of pesticide accidents or poisonings. Some farmers have knowledge about the hazardous effects of pesticide but don't care during handling and spraying of pesticide materials; their behavior towards personal protective equipment does not reflect their knowledge of pesticide safety information. Their clothing is not adequate during spraying applications in the field. Thus, farmers are highly at risk for adulterated pesticides and the mishandling of pesticides. There is no regulatory system in Pakistan for monitoring the health of workers or pesticide handlers involved in the formulation, packaging, spraying, and disposal of pesticides (Inayatullah and Haseeb 1996).

12.7.4 Pesticides Residues in Food and the Environment

There are many reports which indicate the presence of pesticide residues in food materials in Pakistan. The majority of the farmers apply pesticides near harvesting time, or even dip food commodities—mostly fruits and vegetables—in pesticides just before selling them in commercial markets. The presence of pesticides residues in consumable food is a major concern for the usage of pesticides on large scale. Most of studies have shown pesticide residues in vegetables and fruits. Ahad et al. (2001) analyzed the samples of four vegetables, that is, brinjal (*Solanum melongena*), okra (*Hibiscus esculentus*), bitter gourd (*Momordica charantia*), and gourd (*Citrullus vulgaris*) and found that 60% of the samples were above the maximum residual level of pesticides, mainly dichlorvos, fenitrothion, carbofuran, methyl-parathion, and azinphos-methyl. Most of farmers dispose of pesticides into canals, streams, and rivers, which is major threat to the environment. From 2007 to 2010; 134 food export consignments were rejected by European countries due to the presence of pesticides residues, heavy metals, and afla-toxins found in chilies, spices, dry fruits, pickles, and brown rice. Pesticides are mainly blamed in mango export failures to several destinations like Europe, America, the Middle East, and Australia. A basket survey from Lahore, Sheikhupura, Mureidke, and Gujranwala showed that 3% of the fruits and 7% of the vegetables sampled had pesticide residues greater than the MRLs. No official data is actually available for this situation in Pakistan since 2013. Almost all major insect pests of field crops and stored commodities have developed moderate to high levels of resistance against widely used insecticides. Pesticide use has resulted in a 10%–50% increase in productivity by reducing 30%–70% of the pest infestations. In contrast, overuse has resulted in higher residue limits in over 50% of fruits, vegetables, and cereals.

12.7.5 Pesticide Registration and Monitoring System

There should be a strict regulatory system for the registration of pesticides and importation and distribution of pesticides should be in compliance with the protocol established by the plant protection department. The weak monitoring system for the import, storage, sale, transport, and distribution of pesticides is a main threat for the effective implementation of pesticides.

12.7.6 Pesticide Marketing System

Pesticide companies have attractive packages and marketing systems to sell their products to farmers, which rely only on chemical pesticides for the control of pests.

12.7.7 Pesticide Resistance

The continuous dependence on pesticides and their injudicious use results in the development of resistance in insect pests and hence increases the demands for alternative pesticides. The manipulation of alternative pest control tactics (Integrated Pest Management Programs) can result in the delaying of resistance related issues.

12.7.8 Use of Illegal and Banned Pesticides

The use of illegal and banned pesticides should be strictly prohibited in the country.

12.7.9 Infrastructural Challenges

There should be adequate laboratories facilities to analyze pesticide residues and their toxicological aspects.

The issues mentioned above can be addressed by implementing the following strategies:

- Provision of regulation in the Punjab Agricultural Pesticides Amendment Act of 2012 addressing the issue of expired pesticides
- Import of multiple-attacking-system pesticides
- Create awareness about the correct and safe use of pesticides through training and media campaigns
- Establishment of pesticides manufacturing and processing units
- Enforcement and monitoring of the pesticides prices regulatory system
- Development of ecotoxicological research laboratories
- Establishment of pesticide resistance monitoring and mitigation laboratories
- Establishment of pesticide residue testing laboratories to address the issue of residues in food commodities as well as the development of reduced-risk pesticides and GMC-based pest management strategies
- Development of strong industry-academia/research linkages.

12.8 FUTURE OF PESTICIDES: A PAKISTAN PERSPECTIVE

Pakistan has chiefly an agricultural heritage where agriculture is the anchor of our economy. About 70% of its population depends on agriculture directly or indirectly and the country has around 68% agro-based industries. Pesticide application increases day after day. In Pakistan, mostly insecticides (74%) comprise the major part of pesticides used, which is followed by herbicides (14%), fungicides (9%), acaricides (2%), and fumigants (1%) (Khan 1998). In 1980, synthetic pyrethroids became familiar when deltamethrin, fenvalerate, and permethrin were commercially introduced in Pakistan. During 1980–1985, many brands of pyrethroids were provided to farmers for application in crops. It was estimated at the time that synthetic pyrethroids accounted for more than 70% of the entire pesticide market (Malik 1986). The share of the phoshphate group in the market was 39% while chlorinated hydrocarbons took 9% of the market. During 1984, the share of carbamate pesticides was 4% (Memon 1986). As the use of pesticide in controlling pests to prevent crop loss rose, pesticide toxicity cases also increased so that there is huge requirement to produce other ways for sufficient plant defence.

In Pakistan, the agrochemicals industry is largely partitioned without any processing or basic manufacturing units which can deliver an opening for the emerging local capability in our state. The government has methodically brought a high urgency to the enhancement of farm production, naturally prioritizing more on approaches for protecting crops. In Pakistan, crops of chief importance are wheat, rice, maize, sugarcane, cotton, vegetables, and fruits. Accounting for about 70% of total pesticide use, cotton ranks as the most important crop for the agrochemical industry. Pesticides are widely considered for managing the targeted pest species and for that the key types of pesticides are fungicides, herbicides and insecticides. Main part in pesticide market is of the insecticides. Maximum part in the group of insecticides is taken by pyrethroids and organophosphates (Ops). Herbicides usage is more and more significant nowadays because of the increasing expense and scarcity of labor.

In the 1980's, development of generic pesticides strategies brought about the introduction and selling of pesticides by several local establishments. In the agrochemical industry, the market might have looked at as three large groups including multinational firms, well systematized local businesses by nation-wide advertising systems, and native dealers. Because of several market entrance limitations—including a process of registration that require up to 3 years, multinational companies were capable of efficiently marketing their products and cognate the benefits of remarkable limitations for above a decade. Since the incorporation of the general plan, as well as advanced procedures of registration, the share of these companies has declined in the market. But Multi-National Companies in early 80's developed fast national distribution networks than local corporations thus providing ease for general product markets. Native dealers include corporations that market the general products as well as help in extremely good environment but they can perform their processes in rural marketplaces or small townships. Yearly, the world-wide market of pesticides is estimated to be about 30 billion USD and few manufacturers control it by supplying unpackaged active components. In the generic market of pesticides, India and China have dominated, generating significant export income from it. There is no manufacturing of pesticides or their processing at any scale in Pakistan, although many units for the formulation of pesticides have been introduced which are restricted to formulate liquid and granular insecticides. The technical grade chemicals and emulsifiers are mostly imported From China, whereas key solvents used are locally available. The formulation industry is in a good situation for increasing the value-added chain, disposing of the necessary information base, management capabilities, and money investment. Additionally, with the increasing attention focused on nonchemical tactics, the manufacturing of a biopesticide prodcuts by global companies should be considered, as such products might propose long-standing possibilities of export.

From the start, the use of pesticides increased tremendously agricultural productivity through controlling pest attacks. With initial successes at home and in other countries, the government introduced intervention plans to support the utilization of chemical pesticides. These programs, such as subsidies on pesticides and provision of easy availability of these chemicals, modified the competition towards approval of pesticide-based pest control technologies compared with other alternative pest control technologies. The pesticide subsidies and propesticide elongations encourage farmers to apply more chemical pesticide instead of employing other pest control methods. This is how almost all agricultural measures (direct or indirect) strengthen the dependence on chemical-based control. On the other hand, alternative pest management schemes originate from the market. Instead of being thought of as able to protect crops, pesticides are now associated to negative factors like resistance of pests, threats to human health and biological ecosystems. New information reports that the rate of insect resistance development to chemical insecticides is increasing. Despite enormous increases in pesticide use, cotton crops cannot be correctly protected from pest damage. Pesticide utilization has negative effects and has produced inefficient production, nevertheless the use of chemical pesticides is still the only crop protection technology and there is no other option. To solve pest resistance or the efficiency of accessible pesticides, farmers can either use more pesticides or change to a more damaging one. This chemical pesticide-based control of pests closes a vicious cycle, which has self-strengthened pesticide use in the country.

Marketplace investigations have exposed three potent product lines for the production of pesticides. These are the manufacturing of naphtha-based "Petkolin" from industrial alcohol produced in sugar distilleries, copper oxychloride (50wp), and sulphur (80wp) pesticides. As mentioned earlier, the past 10 years have witnessed important unions amongst the large worldwide manufacturing companies, shortening the scope of industry momentously. As limitations have increased, multinational companies are either merging with other competing companies or quitting the market.

12.8.1 Production

There are approximately 40 companies involved in the formulation of pesticides. The locally formulated products comprise liquid granules pesticides and powders, accounting for 67% of the local market. Most raw materials are AIs and pesticides in the product form are imported.

12.8.2 Demand

The pesticide application for better crop production increased from 906 metric tons in 1980 to 5519 metric tons in 1992. It was stated by Tariq (2002) that consumption of pesticides has increased enormously over the last 2 decades. Consumption of pesticides grew approximately 70 times, out of which around 80% is used for cotton, whereas increase in cotton production was twofold. The worth of pesticides has surpassed 12–14 billion Rs. including the production price. The present market size of Pakistan for pesticides is approximately 100,000 of AI with a turnover of Rs. 30 billion annually. Pesticide import has increased to 23,033 metric tons in July–May 2013–2014. Consumption is equal to imports due to limited local formulations. In spite of heavy use of pesticides, pest infestations are becoming more severe and efficacy of the products is diminished because of wrong timing and quantity of use, failure of pesticide ingredients due to resistance development in pests and contamination.

12.8.3 Supply

In the early 1980s, the private sector entered the pesticide industry in which MNCs (multinational companies) played key role over a decade. Farmers used Chinese generic products and quit using imported products because of the fast attacks of pests. Up to now, in Pakistan, there are around 3000 registered products and about 100 active suppliers in the market. Most products are not efficient due to their lengthy procedure of registration, application and effect on insect pests. Currently, there are approximately 70% Chinese generic products and 30% branded products.

Recent developments in science and technology have resulted in the development of novel products, which are popular as "New Chemistry Insecticides." These new insecticide molecules are more efficient and hence are cost effective due to their smaller application doses. The European pesticide vendors have invested more in research and development budgets to explore new molecules. However, the government's policies are more inclined to support generic pesticide products to combat the high prices of monopoly-based, patented new chemistry molecules. The government strategy is to improve the business environment of pesticide businesses.

12.8.4 Environmental Aspects

The injudicious application and overuse of pesticides cause severe health hazards to consumers, users, wildlife, and the environment. Use of pesticide chemicals on crops and the increasing demand for their use has made it indispensable to set up strict administrative controls on the production, storage, and usage of pesticides. Before their large scale production and commercial marketing, new products will need to pass through regulatory standards which should include testing for fitness and evaluation of efficiency.

12.8.5 The Ecotoxicology Research

A number of advisor reports for improving environmentally associated research in Pakistan have appeared over the years (Sadar 1981; Schwass 1986a,b) but their endorsement and implementation is still distant largely due to lack of funds. The need of an institution appropriately set up for monitoring the environmental effects of pesticides is felt even more these days. For the registration of any pesticide, information on the fate of pesticides in the environment is a prerequisite. It should comprise fate in soil, air, constancy, water, ill effects on beneficial insects and nontargeted species, aquatic and soil organisms, and mammalian toxicity describing the potential hazards to the health of human beings. Unfortunately, there is no requirement for such information. At the moment all pesticides are either imported as finished products or locally formulated from imported AIs, hence the data provided from the original country of manufacture is accepted by default. Since the behavior of a pesticide may change in different environments, accepting the unknown data poses a risk of overlooking possible threats to the local ecology. A report (Calderbank 1988) has recommended the establishment of an ecotoxicological center. The objective is to develop an organization with laboratories and the necessary expertise to measure and monitor environmental pesticide-related pollution in Pakistan. Once established, such facilities and expertise could be extended to monitor the impacts of other potential environmental pollutants in Pakistan.

12.8.6 Recommendations

- Recognize the situations under which pesticide chemicals may be direly needed for application in future pest control/management.
- Define the different kinds of chemicals that are most suitable for pest control tactics based on their environmental impact.
- Discover the most encouraging prospects to diminish environmental and health hazards.
- Endorse a suitable part of the pesticide sector in developing and researching pesticide products, to endorse their testing, registration, usage methods, and to promote community consciousness about the safe use of pesticides.
- Formulate policy changes and research investment plans emphasizing the development of pesticides and application technologies that are compatible with pest management approaches and impose the least hazardous health impacts.
- Encourage and endorse social as well as scientific initiatives to develop and use substitutes to pesticides.
- Promote the capability and inspiration of agrarian labors to minimize their exposure to possibly injurious substances as well as to improve and implement strong worker-safety rules.
- Lessen the adverse effects to nontarget organisms by prudent selection of chemical pesticides, employment of precise application technology, and identifying the environmental and financial influences for usage of pesticide in agricultural systems.
- Establishment of pesticide residue testing laboratories at divisional level.
- Monitoring and enforcement of APO 1971 and Rules 1973.
- Formulate price regulatory systems and their enforcement at district level.
- Extensive training of stakeholders.
- Involvement of research organizations for the development of reduced risk pest management system in Pakistan by providing research and development funds.
- Introduce policies that encourage huge investments in the industrialization of pesticide manufacturing units to encourage innovative pest management technologies in Pakistan.
- Establishment of strong linkages among pesticides companies and research institutions for the proper management of insecticides problems.

12.8.7 Action-Plan/Way-Forward

Short Term

- Import of multiple-attacking-system pesticides.
- To address the issue of pesticides residues in food commodities, it is imperative to establish practical pesticides detection laboratories at divisional level for testing the toxicological aspects of agrochemicals and fix and enforce MRLs for registering pesticides. The government should initiate projects to establish such laboratories and should upgrade existing pesticides testing laboratories for this purpose. The government should also take strict action on MRLs issues by passing an act similar to the seed act.
- Many countries having the capacity to produce and formulate agrochemicals (India, China), are investing in the import/production of cheap agrochemicals, especially pesticides, and then are marketing them to different developing countries. Most of the Pakistan-based pesticides companies are importing AIs/formulated-materials of such insecticides from these countries either on low/subsidized rates or on a donated basis. The government should strictly monitor the import of such pesticides from these countries and enforce the pesticide import act to check the formulation and registration of such pesticides.
- To address the issue of irregularities in pesticide prices, the government should pass and enforce a pesticide price act to establish a uniform pricing system for markets throughout

Pakistan. These should be under the strict vigilance of price monitoring teams at district level.
- Arrangement of awareness campaigns and the conduct of mass-scale training and education programs for stakeholders about the precise and accurate use of agrochemicals.
- Ensure the enforcement of all possible precautionary principles, which are mandatory while developing, testing/evaluating, and introducing GM-crops on a mass-scale in the fields.
- Encourage the research community to develop and identify stakeholders to adopt the organic farming system and integrated pest management (IPM) approaches to diminish the usage of synthetic chemicals. This will be necessary to ultimately cope with issues of environmental contamination, human health, food safety, and security.
- State and research organizations should approve and enforce an agenda which ensures the stabilization of the functioning of agricultural activity and society; concentrate more on the programmatic investments for the development of sustainable agriculture; focus on the training and education of the agriculturists and other stakeholders in the technologies of implementation, agrochemical application, conservation of soil and water resources, protection of genetic diversity, and application of precautionary principles recommended for GM-crops cultivation. Endowment and implementation of all of these efforts should be a part of the national agenda and these efforts must be implemented to least at the regional level, specifically in regions that have been declared most impoverished.
- It is very necessary to establish a new corporation in the public sector that may provide three important inputs to growers, that is, seed, fertilizers, and pesticides at reasonable prices.
- The provincial/federal government must develop an organic farming agriculture policy, modeled on the success stories of developed nations. This government-enforced policy must focus on tax enticements, incentives, and encouragement, a short-term financial support program during the transitory phase period, technical training, education and support for farmers, the development and implementation of organic-farming programs, and the provision of research and development funds for organic agriculture in Pakistan.
- The government should introduce policies which encourage investments in the industrialization of pesticide manufacturing and processing units, and which foster innovative pest management technologies development units in Pakistan.
- Involvement of the pesticide industry in research planning meetings/seminars/workshops, etc.

Medium Term

- Amendment of regulation in APO 2012 addressing the issue of expired pesticides.
- Enforcement and monitoring of a pesticide price regulatory system.

Medium-Long Term

- Setting up pesticides resistance monitoring and mitigation laboratories.
- Instituting hi-tech pesticide residue testing labs.
- Testing and evaluation of reduced-risk pesticides and GMC-based pest management tactics.
- Development of ecotoxicological research laboratories.

Long Term

- Establishment of pesticide manufacturing and processing units (Figure 12.4).

12.9 CONCLUSION

Pesticide application in agriculture resulted in improvements in food production, and these have been discussed in this chapter. In this chapter, strong evidence has been shown in favor

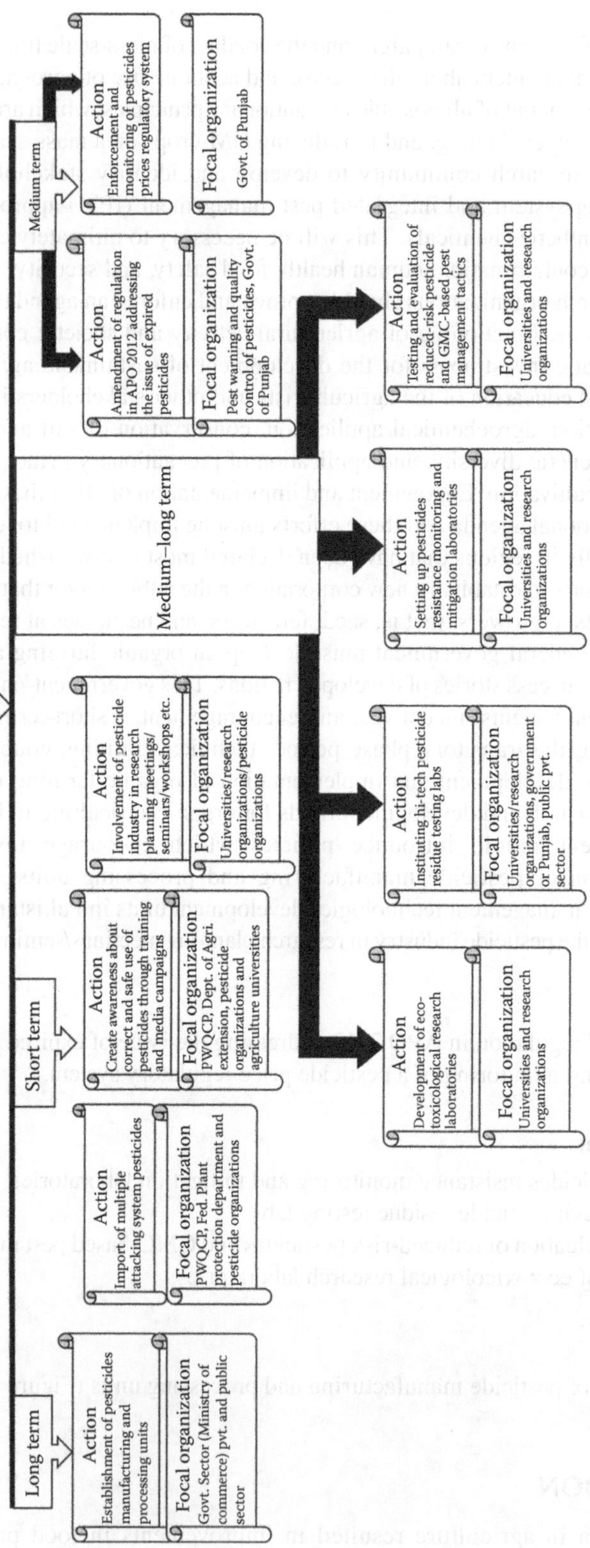

FIGURE 12.4 Diagrammatic illustration of action-plan/way-forward for coping with the issues of pesticides in Punjab, Pakistan.

of the direct role of pesticides to obtain efficient, economical, and effective control of weeds, insects, fungi, bacteria, and rodent pests. We have shown their indirect role in achieving drastic increases in crops yields and for the economical production of higher quantities of food. During last 30 years, large numbers of new chemical pesticides have been explored and introduced for the production and protection of agriculture systems. These new pesticides are comparatively safer for environmental and nontarget species being more pest-specific while having extremely low application rates (grams or ounces per hectare) and better application through improved application techniques such as site-specific and precise application technologies. The prominent examples of such pesticides include herbicides of the sulfonylurea group, fungicides of the piperidinylthiazole group, and insecticides or acaricides of mectin group (Lamberth et al. 2013). However, despite the opportunities, challenges remain with such new chemistry pesticides. These hurdles are developmental costs, market acceptability, effectiveness and bioactive life, as well as their impact and safety for the environment and nontarget organisms. However, making such new pest management technologies acceptable to society will require focused, rigorous, and strenuous efforts as well as coordinated and synchronized approaches on the part chemical companies, academics, the government, and citizen groups alike.

12.10 POLICY POINTS

Agricultural productivity and food security needs to adapt to the unremitting population growth and address the issues of undernourishment, health, and welfare of the population. Presently, agriculture needs another green-revolution, which appears questionable. But use of agrochemicals, high-yielding varieties, and intensive irrigation can play a sound role in enhancing agricultural production. Further improvements in production and food security can be achieved by the utilization of modern technologies, while insuring biodiversity conservation, environmental protection, and food security. Otherwise, utilization of modern technologies may become a great threat to the quality of human health, the environment, and the biodiversity of our ecosystem. State and research organizations alike should approve and enforce an agenda which ensures the stabilization of a functional agricultural sector in society and concentrate more on investments for the development of sustainable agriculture. They should focus on the training and education of agriculturists and other stakeholders in technology implementation and the following areas: Agrochemical application, conservation of soil and water resources, protection of genetic diversity, and the application of precautionary principles that are recommended for GM-crops cultivation. These efforts should be a part of the national agenda—implemented at least at regional level, specifically in impoverished regions. The issues of agricultural productivity can be addressed if the state is successful in preparing and implementing such national agriculture plans which critically address such issues as education and poverty of farmers, tenant and rural community problems, high market prices of agrochemicals, insignificant financial support/subsidies from the government, access to dubious companies and dearth of credible companies, flood of substandard and low quality agrochemicals in the markets, and propaganda of NGOs against pesticide application to conserve biodiversity.

It is hard to estimate the losses easily, but roughly growers incur net losses of more than Rs. 500 hundred million annually and these figures vary with region. In 1998, this loss was estimated at Rs. 30 billion for cotton alone in the country. Pesticides protect only 33% of the cropped area, while 67% remains at the mercy of nature. Insufficient awareness about the correct usage of pesticides, their rapid increases in prices, and the sale of substandard pesticides in the market also contribute to losses. According to a field study, it was concluded that crop yields might have been increased from 2% to 5% if growers followed appropriate pest management practices. Postharvest losses are very high due to the scarcity of proper harvesting equipment and the dearth of postharvest technologies. Technically enriched, economically sound, and highly efficient crop production, protection, and harvesting services are badly needed to address the issues of low crop productivity. It is truly necessary to establish a new corporation in the public sector that may provide seed, fertilizer, and pesticide inputs to growers at a reasonable price. Poor quality pesticides currently available should be discouraged by strict legal policies. The local pesticide industry should be encouraged to use specific local herbs. The government should aerial spray cropping areas during periods of heavy pest attack. The prices of pesticide companies should be monitored strictly.

Keeping in view the difficulties in the use of pesticides and their adverse effects, there is a dire need to switch to alternative strategies like the ones adopted successfully in developed nations. Our growers may also follow suit to save their money, environment, and grow certified produce, but policy makers continue to neglect the issue due to some mala fide intents or flawed policies. Illiteracy, unawareness, and traditional farming continue to be obstinate obstacles in providing innovative agricultural methods and alternatives for growers. Integrated pest management is good alternative to create organic farms and use methods to minimize pest populations, socioeconomical losses, and ecological damage. Agricultural systems should also reduce their dependence on pesticides. Instead, they should switch to natural and organic fertilizers. Pest management systems should be established on the pillars of biological, cultural, and mechanical methods with the aim of enhancing the biodiversity of beneficial fauna. These tactics are practicable in horticultural and urban environments and have been tested and recognized in technologically advanced countries. The latter have seen a growth in the demand for their products in the world, and holds great prospects for the future of developing countries, such as Pakistan, to produce less environmentally hazardous certified products in an age of global competition.

Pakistan has several kinds of herbs and botanics which can be used for local environmentally friendly pesticide manufacturing at low costs. In this regard, The Agriculture University, Faisalabad and Tando Jam and its affiliated research centers can play an important role. It is also proposed that the provincial/federal governments must develop organic farming agriculture policies inspired by the developed nations' success stories in this regard. The government-enforced policies must also focus on tax enticements, incentives, and encouragements, as well as short-term financial support programs for: Transitory phase periods, technical training, education, support for farmers, the development and implementation of organic-farming programs, and research and development for organic agriculture. The federal and provincial government should cooperate—they should initiate and establish a national action programs on pesticides alternative resources. The agricultural policy should be easily available to the concerned stakeholders, agricultural communities, and the related public through electronic/print media in local/regional languages with large-scale incentives. The government should slash general sales tax on input purchases like pesticides seeds, fertilizers, and other mechanical tools in order to meet global challenges.

REFERENCES

Abhilash PC and Singh N. 2009. Pesticide use and application: An Indian scenario. *J. Hazard Mat.* 165(1):1–12.

Ahad K, Hayat Y and Ahmad I. 2001. Capillary chromatographic determination of pesticide residues in multan division. *Nucleus*, 38(2):145–149.

Ahmad Z. 1988. Privatization of pesticides sales and distribution—An overview. *PAPA Seminar* June 2. Faisalabad.

Ahmed A, Randhawa MA, Yusuf MJ and Khalid N. 2011. Effect of processing on pesticide residues in food crops: A review. *J. Agric. Res.* 49: 379–390.

Aktar W, Sengupta D and Chowdhury A. 2009. Impact of pesticides use in agriculture: Their benefits and hazards. *Interdisp. Toxicol.* 2: 1–12.

Alley HP. 1981. Weed control. In: Kolp BJ, Alley HP, Fornstrom KJ and Hough HW (eds.), *1981 Winter Wheat Production in Wyoming.* Bulletin 603, Revised, Agricultural Experiment Station, University of Wyoming, Laramie.

Ameta OP, Sharma US and Jain HK. 2010. Relative efficacy of Flubendiamide 480SC against *Helicoverpa armigera* (Hubner) in chick pea. *Pestol.* 34(11): 31.

Benson JM. 1982. *Weeds in Tropical Crops: Review of Abstracts on Constraints in Production Caused by Weeds in Maize, Rice, Sorghum, Millet, Groundnuts and Cassava. FAO Plant Production and Protection Paper.* Food and Agriculture Organization of the United Nations, Rome, 32(1):63 pp.

Bingham S. 2007. *Pesticides in Rivers and Groundwater.* Envir. Agency, UK.

Borlaug NE and Dowswell CR. 1993. *Fertilizer: To nourish infertile soil that feeds a fertile population that crowds a fragile world. Keynote address at the 61st Annual Conference.* International Fertilizer Industry Association (IFA), May 24–27, New Orleans, Louisiana, USA, 18 p.

Brown AWA. (1971) Pest resistance to pesticides. *Pesticides in the Environment Sterens, R.W.*, 1: 457–552. New York.

Burnside OC and Wicks GA. 1964. Cultivation and herbicide treatments of dryland sorghum. *Weeds* 12: 307–310.

Calderbank A. 1988. Book review: *Occupational Hazards of Pesticide Use.* Turnbull GJ (ed.), Taylor & Francis, London, 1985. pp. xi+ 184, ISBN 0-85066-325-3. *J. Chem. Technol. and Biotechnol.* 42:242.

Carvalho FP. 2005. Residues of persistent organic pollutants in coastal environments—a review. In: Gomes FV, Pinto FT, Neves L, Sena O, Ferreira O (eds.), *Proceedings of the First International Conference on Coastal Conservation and Management in the Atlantic and Mediterranean (ICCCM'05)*, Tavira, Portugal, April 17–20. FEUP, Universidade do Porto, ISBN: 972-752-083-9 pp. 423–431.

Carvalho FP. 2006. Agriculture, pesticides, food security and food safety. *Int. Sci. Policy.* 9: 685–692.

Cheema AA and Shah FH. 1987. Pesticides residues in fruits and vegetables. *Pak. J. Sci. Ind. Res.* 30: 511–12.

Chenkin AF. 1975. Economic effect of Plant Protection in the Russian Federated Republic. Pp. 27–31. *VIII International Congress of Plant Protection.* Moscow, USSR.

Chivinge OA. (1990) Weed science technological needs for the communal areas of Zimbabwe. *Zambezia*, 17(2): 133–143.

Cohen JE. 2005. Human population grows up. *Sci. Amerc.* 293:48–55.

Comes RD, Timmons FL and Weldon LW. 1962. Chemical control of annual weeds in Pinto and Great Northern field beans. *Agric. Experi. Station, Uni. Wyoming, Bull.* 393:15.

Dana MN. 1989. The American cranberry industry. *Acta Hortic.* 241:287–294.

Daniell JW and Hardcastle WS. 1972. Response of peach trees to herbicide and mechanical weed control. *Weed Sci.* 20:133–136.

De Groote H, Wangare L, Kanampiu F, Odendo M, Diallo A, Karaya H and Friesen D. 2008. The potential of a herbicide resistant maize technology for *Striga* control in Africa. *Agric. Syst.* 97:83–94.

Department of Plant Protection (DPP). 2015. Govt. of Pakistan. Ministry of national food security and research. http://plantprotection.gov.pk/ (visited on 15-02-2016).

Dexter AG. 1982. Weedonomics. *Proceedings, north central weed control conference*, Des Moines, Iowa, December, 8–10.

Dhar V and Ahmad R. 2004. Integrated pest management in chickpea and pigeonpea. In: Birthal PS and Sharma OP (eds.) *Integrated Pest Management in Indian Agriculture*, 109–117. Chandu. Press, Delhi, India.

Dong K, Chen B, Li Z, Dong Y and Wang H. 2010. A characterization of rice pests and quantification of yield losses in the japonica rice zone of Yunnan, China. *Crop Prot.* 29:603–611.

Dugje IY, Omoigui LO, Ekeleme F, Kamara Y and Ajeigbe H. 2009. *Farmers Guide to Cowpea Production in West Africa.* Int. Inst. Tropic. Agric. (IITA), Ibadan, Nigeria, http://www.iita.org/c/document_library/get_file?uuid=dd0fe400-eb90-470c-9dc1-f679c5d66a81&groupId=25357 (Accessed on 10 June 2016).

Eck P. 1990. *The American Cranberry.* Rutgers University Press. New Brunswick, New Jersey, 420 pp.

Economic Survey of Pakistan. 2006. Finance division, government of Pakistan, Islamabad. http://www.finance.gov.pk/survey/sur_chap_05-06/02-Agriculture.

Environmental Protection Agency (EPA) of United States, About Pesticides. 2014. http://www.epa.gov/pesticides/about/index.htm

Eqani SAMAS, Malik RN, Katsoyiannis A, Zhang G, Chakraborty P, Mohammad A and Jones KC. 2012. Distribution and risk assessment of organochlorine contaminants in surface water from River Chenab, Pakistan. *J. Environ. Monit.* 14: 1645–1654.

Falcon WP and Fowler C. 2002. Carving up the commons—emergence of a new international regime for germplasm development and transfer. *Food Policy* 27:197–222.

FAO. 2009. *Feeding the world in 2050. World agricultural Summit on food security 16–18 November 2009.* Food and Agriculture Organization of the United Nations, Rome.

Food and Agriculture Organization of the United Nations. 2012. *The State of Food Insecurity in the World.* The Food and Agriculture Organization of the United Nations, Rome, Italy.

Freyman S, Palmer CJ, Hobbs EH, Dormaar JF, Schaalje GB and Moyer JR. 1981. Yield trends on long-term dryland wheat rotations at Lethbridge. *Can. J. Plant Sci.* 61:609–619.

Gianessi LP. 2009. *The Value of Insecticides in U.S. Crop Production. Crop Life Foundation.* Crop Prot. Res. Inst. (CPRI), Washington, DC.

Gianessi LP and Reigner N. 2005. *The Value of Fungicides in U.S. Crop Production. Crop Life Foundation.* Crop Prot. Res. Inst. (CPRI), Washington, DC.

Gianessi LP and Reigner N. 2006. *The Value of Herbicides in U.S. Crop Production. 2005 Update. Crop life Foundation.* Crop Prot. Res. Inst. (CPRI), Washington DC.

Gilliom RJ, Barbash JE, Crawford CG, Hamilton PA, Martin JD, Nakagaki N, Nowell NH et al. 2007. *Pesticides in the Nation's Streams and Ground Water, 1992–2001.* U.S. Geological Survey circular 1291, Virginia, USA. Revised February 15, 2007.

Glaze NC. 1975. Weed control in cucumber and watermelon. *J. Am. Soc. Hortic. Sci.* 100:207–209.

Grichar WJ and Colburn AE. 1993. Effect of dinitroaniline herbicides upon yield and grade of five runner cultivars. *Peanut. Sci.* 20:126–128.

Groth E, Benbrook CM and Lutx K. 1999. Do You Know What You're Eating? An Analysis of US Government Data on Pesticide Residues in Foods, http://www/consumers union.org/food/do_you_know2.htm (January 19, 2003).

Guidotti TL and Gosselin P. 1999. *The Canadian Guide to Health and the Environment.* University of Alberta Press, UK.

Habib N. 1996. *Invisible Farmers: A Study on the Role of Woman in Agriculture and Impact of Pesticides on Them.* Khoj Research and Publication Centre, Lahore, Pakistan, pp. 129.

Hackenberg D. 2007. *Letter from David Hackenberg to American Growers.* Plattform Imkerinnen, Austria.

Headly JC. 1968. Estimating the productivity of agricultural pesticides. *Am. J. Ag. Ec.* 50: 13–23.

Husnain T. 1999. Pesticide Use and its Impact on Crop Ecologies: Issues and Options. SDPI Working Paper Series, SDPI, Islamabad, pp. 73.

Hussain S, Masud T and Ahad K. 2002. Determination of pesticides residues in selected varieties of Mango. *Pak. J. Nutr.* 1: 41–42.

Inayatullah C and Haseeb M. 1996. Poisoning by pesticides. *Pak. J. Medic. Res.* 35(20): 57–58.

Ingram J. 2011. A food systems approach to researching food security and its interactions with global environmental change. *Food Secur.* 3: 417–431.

IWMI. 2007. *Water for Food, Water for Life: A Comprehensive Assessment of Water Management in Agriculture.* Earth scan and Colombo. International Water Management Institute, London

Jabbar A and Mallick S. 1994. Pesticides and environment situation in Pakistan. Working Paper Series # 19, Sustainable Development Policy Institute (SDPI), Islamabad, Pakistan.

Kamara AY, Ekeleme F, Omoigui LO, Abdoulaye T, Amaza P, Chikoye D and Dugje IY. 2010. Integrating planting date with insecticide spraying regimes to manage insect pests of cowpea in north-eastern Nigeria. *Int. J. Pest. Manage.* 56(3): 243–253.

Kanampiu FK, Kabambe V, Massawe C, Jasi L, Friesen D, Ransom JK and Gressel J. 2003. Multi-site, multi-season field tests demonstrate that herbicide seed-coating herbicide-resistance maize controls *Striga* spp. and increases yields in several African countries. *Crop Prot.* 22:697–706.

Keiserukhshy MG and Kashirsky OP. 1975 Economics of plant protection in the USSR. *VIII International Congress of Plant Protection*, Vol. II, Moscow, USSR.

Khan MA, Muhammad I, Iftikhar A and Manzoor HS. 2002. Economic evaluation of pesticide use externalities in the cotton zones of Punjab, The Pakistan Development Review 41:4 Part II:683–693.

Khan MH. 1990a. Wild boar: Identification, biology and behaviour. In: Brooks, J.E., E. Ahmad, I. Hussain, S. Munir, A.A. Khan, (eds.) *A Training Manual on Vertebrate Pest Management*, 149-151. GOP/USAID/DWRC/Vertebrate Pest Control Project. NARC. PARC, Islamabad.

Khan MH. 1990b. Chemical control of wild boar. In: Brooks JE, Ahmad E, Hussain I, Munir S, Khan AA, (eds.) *A Training Manual on Vertebrate Pest Management*, 161-163. GOP/USAID/DWRC/Vertebrate Pest Control Project. NARC. PARC, Islamabad.

Khan MJ, Zia MS and Qasim M. 2010. Use of pesticides and their role in environmental pollution. *World Acad. Sci. Eng. Technol.* 48:122–128.

Khan MS. 1998. Pakistan crop protection market. *PAPA Bullt.* 9: 7–9.

Khooharo AA, Memon RA and Mallah MU. 2008. An empirical analysis of pesticide marketing in Pakistan. *Pak. Econ. Soc. Rev.* 57–74.

Lamberth C, Jeanmart S, Luksch T and Plant A. 2013. Current challenges and trends in the discovery of agrochemicals. *Science* 341:742–745.

Latif Y, Sherazi STH and Bhanger MI. 2011. Monitoring of pesticide residues in commonly used fruits in Hyderabad region, Pakistan. *Amerc. J. Anal. Chem.* 2(8A): 46–52.

Lever BG. 1991. *Crop Protection Chemicals (Ellis Horwood Series in Applied Science and Industrial Technology).* Ellis Horwood Ltd. UK. 500 pp.

Maheshwari SK, Bhat NA, Shah TA, Shukla AK and Hare K. 2012. Effect of fungicidal seed and foliar applications on *Ascochyta* blight of pea. *Ann Plant Prot Sci.* 20(1): 240–241.
Malik TM. 1986. Pyrethroids – the harder weapon. *Pak. Agric.* 23.
Masud SZ and Farhat S. 1985. Pesticides residues in foodstuffs in Pakistan organochlorine pesticide in fruits and vegetables. *Pak. J. Sci. Ind. Res.* 28: 417–422.
Masud SZ and Hassan N. 1995. Pesticides residues in fruits in food stuffs in Pakistan Organochlorine, organophosphates and pyrethroids insecticides in fruits and vegetables. *Pak. J. Sci. Ind. Res.* 35: 499–504.
McDougall P. 2013. http://phillipsmcdougall.com (30 July 2013).
Memon NA. 1986. Why not promote indigenous pesticides? *Pak. Agric.* 31.
Merrington G Nfa LW, Parkinson R, Redman M and Winder L. 2004. *Agricultural Pollution: Environmental Problems and Practical Solutions.* CRC Press, USA.
Mikoshiba H. 1983. Studies on the control of Downy Mildew disease of Maize in tropical Countries of Asia. *Technical Bulletin of the Tropical Agricultural Research Center No.* 16: 62.
Miller GT. 2004. *Sustaining the Earth.* 6th edition. Thompson Learning, Inc. Pacific Grove, California, USA.
Mills LS, Michael E, Soulé D and Doak F. 1993. The Keystone-species concept in ecology and conservation: Management and policy must explicitly consider the complexity of interactions in natural systems. *Bio. Sci.* 43(4): 219–224.
Moody K. 1991. Weed management in rice. In: D. Pimentel (ed.). *Handbook of Pest Management in Agriculture.* 301–328. Vol. 3. CRCN Press, Boca Raton, Florida.
Mueller KE and Oelke EA. 1965. Water grass control in rice fields with Propanil and Ordram. *Calif Agr* 19(7): 10–12.
Muthamia JGN, Musembi F, Maina JM, Ouma JO, Amboga S, Murithi F, Micheni AN et al. 2002. Participatory on-farm trials on weed control in smallholder farms in maizebased cropping systems. Pp. 468–473. Friesen DK and Palmer AFE (eds.) *Proceedings of Seventh Eastern and South Africa Regional Maize Conference*, Nairobi, Kenya, February 5–11.
Naab JB, Tsigbey FK, Prasad PVV, Boote KJ, Bailey JE and Brandenburg RL. 2005. Effects of sowing date and fungicide application on yield of early and late maturing peanut cultivars grown under rainfed conditions in Ghana. *Crop Prot.* 24:325–332.
Nalewaja JD. 1975. Herbicidal weed control uses energy efficiently. *Weeds Today*, (Fall):10–12.
Naqvi SNH and Jahan M. 1996. Pesticide residues in random blood samples in human population in Karachi. *J. Coll. Phys. Surg. Pak.* 6: 151–153.
NARC. 2000. *Pesticide Health Hazards to Women Cotton Pickers In Pakistan, IPM News, National IPM Programme, Islamabad.* Volume 1, Issue 4. National Agriculture Research Centre, Islamabad, Pakistan.
Nelson DC and Giles JF. 1989. Weed management in two potato cultivars using tillage and pendimethalin. *Weed Sci.* 37:228–232.
Oerke EC. 2005. Crop losses to pests. *J. Agr. Sci.* 144: 31–43.
Orphal J. 2001. Economics of Pesticide Use in Cotton Production in Pakistan, *Diploma Thesis*, University of Hannover, Germany.
Parveen Z, Khuhro MI, Rafiq N and Kausar N. 2004. Evaluation of multiple pesticide residues in apple and citrus fruits. 1999–2001. *Bull. Environ. Contam. Toxicol.* 73: 312–318.
PCCC. 2006. *Cotton Production Plan, 1989–1990.* Pakistan Central Cotton Committee, Karachi.
Pelley J. 2006. DDT's legacy lasts for many decades. *Environ. Sci. Technol.* 40:4533–4534.
Penna JA and Lema D. 2003. Adoption of herbicide resistant soybeans in Argentina: An economic analysis. In: *Economic and Environmental Impacts of Agrotechnology*, 203–220. Kluwer-Plenum, New York.
Pimentel D, Wilson C, McCullum C, Huang R, Dwen P, Flack J, Tran Q, Saltman T and Cliff B. 1997. Economic and environmental benefits of biodiversity. *Biosci.* 47(11): 747–757.
Pingali PL and Traxler G. 2002. Changing locus of agricultural research: Will the poor benefit from biotechnology and privatization trends? *Food Policy* 27:223–238.
Popp J. 2011. Cost-benefit analysis of crop protection measures. *J. Consumer Prot. Food Safety* 6(Supplement 1):105–112.
Putnam ML. 2007. Brown stripe downy mildew (*Sclerophthora rayssiae* var. *zeae*) of maize. Plant Health Prog, Downloaded from website: https://www.plantmanagementnetwork.org/pub/php/diagnosticguide/2007/stripe/ (Accessed on 14 June 2016).
Qamar-ul-Haq AT, Ahmad M and Nosheen F. 2008. An analysis of pesticide usage by cotton growers: a case study of district Multan, Punjab-Pakistan. *Pak. J. Agri. Sci.* 45(1): 133–137.
Repetto R and Baliga SS. 1996. *Pesticides and the Immune System: The public Health Risks.* New York, World Resources Institute, USA.

Sachs J. 2005. *The End of Poverty: Economic possibilities for our time.* New York: The Penguin Press, USA.
Sadar MH. 1981. *Organizational Improvement and Research Needs for More Effective and Safer use of Pesticides.* Report No. TOKTEN, CSO, PARC, Islamabad.
Sanborn M, Kerr KJ, Sanin LH, Cole DC, Bassil KL and Vakil C. 2007. Noncancer health effects of pesticides. Systematic review and implications for family doctors. *Can Family Phys.* 53(10): 1712–1720.
Schroder D, Headley JC and Finley RM. 1984. The contribution of herbicides and other technologies to corn production in the Corn Belt region, 1964 to 1979. *Central J. Agric. Econ.* 6(1): 95–104.
Schumann GL. 1991. Plant Diseases: Their biology and social impact. *Amerc. Phytopathol. Soc. Press*, St. Paul, Minnesota, USA.
Schwass R. 1986a. *Sustainable Development in Pakistan - Some Key Issue.* Pakistan National Conservation Strategy, Phase-1 Report.
Schwass R. 1986b. *Summary Report of the NCS for Pakistan Workshop*, Islamabad. World Conservation Strategy of the International Union for the Conservation of Nature and Natural Resources (IUCN). August 1: 25–28.
Sekhon, HS, Singh G, Sharma P and Sharma P. 2004. Agronomic management of mungbean grown under different environments. Pp. 82–103. In *Proceedings of the Final Workshop and Planning Meeting, Improving Income and Nutrition by Incorporating Mungbean in Cereal Fallows in the Indo-Gangetic Plains of South Asia DFID Mungbean Project for 2002–2004*, Punjab Agricultural University, Ludhiana, Punjab, India, May 27–31.
Sharpe M. 1999. Focus. Towards sustainable pesticides. *J. Environ. Monit.* 1:33N–36N.
Smil V. 2000. *In Feeding the World. A Challenge for the Twenty-First Century.* Cambridge, MA: MIT Press.
Smith RJ, Flinchum WT and Seaman DE. 1977. Weed Control in U.S. Rice Production. Agriculture Handbook No. 497, U.S. Department of Agriculture, 78 pp.
Stuart S. 2003. Development of Resistance in Pest Populations, http://www.nd.edu/_chem191/e2.html (January 20, 2003).
Syed JH and Malik RN. 2011. Occurrence and source identification of organochlorine pesticides in the surrounding surface soils of the Ittehad Chemical Industries Kalashah Kaku, Pakistan. *Environ. Earth Sci.* 62: 1311–1321.
Tagar HK, Bullo A, Shah SR and Shah SMM. 2015. Sustainable development goals: End poverty, food security and healthy lives through human resources and managing agriculture productivity (A case of Pakistan). *Int. J. Innovative Res. Develop.* 4(10):283–289.
Tariq MA. 2002. Need to tap agriculture sector. Daily Dawn, Economic and Business Review. Downloaded from website: http://www.dawn.com/news/14790/need-to-tap-agriculture-sector (Accessed on 9 June, 2016)
Tariq MI. 2005. Leaching and degradation of cotton pesticides on different soil series of cotton growing areas of Punjab, *Pakistan in Lysimeters. Ph.D thesis*, University of the Punjab, Lahore, Pakistan.
Taylor M, Klaine S, Carvalho FP, Barcelo D, Everaarts J. (eds.) 2003. *Pesticide Residues in Coastal Tropical Ecosystems. Distribution, Fate and Effects.* Taylor and Francis, London.
USDA. 2010. Commodity costs and returns: U.S. and regional cost and return data. http://www.ers.usda.gov./data/costsandreturns (Accessed on 20 September 2010).
Verma BN. 2001. Grey leaf spot disease of maize Loss assessment, genetic studies and breeding for resistance in Zambia. *Seventh Eastern and Southern Africa Regional Maize Conference Proceedings*, Nairobi, Kenya, February 11–15.
Ward JMJ, Stromberg EL, Nowell DC and Nutter FW Jr. 1999. Gray leaf spot, a disease of global importance in maize production. *Plant Dis.* 83(10): 884–895.
Waxman MF. 1998. *The Agrochemical and Pesticides Safety Handbook.* CRC Press, USA.
Welle SC, Culbreath AK, Gianessi L and Godfrey LD. 2014. The Contributions of pesticides to pest management in meeting the global need for food production by 2050. Issue Paper # 55, *Council for Agricultural Science and Technology (CAST)*, Ames, Iowa.
Wylie P. 2008. High Profit Farming in Northern Australia—A New Era in Grain Farming.
Yarborough DE and Ismail AA. 1985. Hexazinone on weeds and on low bush blueberry growth and yield. *Hortic. Sci.* 20:406–407.
Yarborough DE, Hanchar JJ, Skinner SP and Ismail AI. 1986. Weed response, yield, and economics of hexazinone and nitrogen use in low bush blueberry production. *Weed Sci.* 34:723–729.
Zakharenko V. 2000. Bioeconomic methods and decision-making models for herbicide use in Russian agriculture. *3rd International Weed Science Congress*, Foz do Iguassu, Brazil, June 6–11.
Zhang ZP. 2003. Development of chemical weed control and integrated weed management in China. *Weed Biol. Manag.* 3: 197–203.

Section II

Crop Production and Health

13 Climate Change and Agriculture

Ashfaq Ahmad and Khalid Hussain

CONTENTS

13.1 BASIC CONCEPTS OF CLIMATE CHANGE

13.1.1 Rationale

Agriculture industry is most vulnerable to future climate change (CC) and variability. The accelerating pace of change in climate is making situation more severe due to the occurrence of unprecedented events of weather. For the last 10 years, Asia has been facing serious threats because of CC and variability such as tsunamis, floods, and droughts. Extreme events are expected to have large negative impacts on crop productivity (Challinor et al. 2007; Wheeler et al. 2000). The most intense impacts of CC over the next few decades will be on food and agriculture systems. Some Asian countries, for example Pakistan and India in semiarid environments are at risk of rising temperatures and rainfall variability, which could result in increasing demands of water for agricultural production (Anwar et al. 2013). The rise in temperature and reduction in rainfall under semiarid environments are likely to reduce production of maize, wheat, rice, and minor crops in the coming 20 years (Lobell et al. 2008). Food security will substantially be affected by these adverse changes. Since the 1990s, increasing prices of different commodities along with decreasing cropped area per capita have caused reductions in food productivity. All these factors are eroding food security in many countries (Brown and Funk 2008).

The term *Climate* refers to long-term weather conditions prevailing over a particular area and time, this time period may be 30 years. However, some scientists define climate as "... average weather conditions in particular region/place and time period." However, *weather* is fundamentally the way in which the atmosphere is behaving. It mainly consists of the short-term atmospheric conditions prevailing over a particular area in a given time. Climate mainly controls the distribution of crops across the globe while weather primarily decides the productivity of crops.

According to the fifth assessment report of the Intergovernmental Panel on Climate Change (IPCC), *CC* is a change in the mean state of weather statistics over three decades of time period. It is mainly due to change in natural internal processes or external forcing like disparities in the solar cycles and volcanic eruptions as well as persistent changes in atmospheric composition due to anthropogenic activities. However, the United Nations Framework Convention on Climate Change (UNFCCC) defines CC as "change of climate which is associated directly or indirectly with anthropogenic activities that alter the composition of the global atmosphere." But climate variability is dealt with natural internal causes. It may be of two types (i) internal variability which comes in response to natural internal processes within the climate system, and (ii) external variability which are variations within a climate system due to anthropogenic external forcing.

13.1.1.1 Future Trends

Future climate will be determined by committed warming caused by past anthropogenic emissions as well as future anthropogenic emissions and natural climate variability. The change in global average surface temperature for 2016–2035 relative to 1986–2005 is similar under four Representative Concentration Pathways (RCPs). It may likely be in the range 0.3–0.7°C. The assumptions are (i) no volcanic eruptions in future, (ii) no more variations in some natural sources (CH_4 and N_2O), and (iii) little change in total solar irradiance. Nevertheless, the extent of the projected CC is significantly affected by the choice of emissions scenario in mid-twenty-first century. It has been projected under scenario RCP8.5 that there will be temperature extremes over most of land areas. Similarly, the increase in average surface temperature of the globe by the end of this current century (2081–2100) in relation to 1986–2005 is probably going to be 0.3–1.7°C for RCP2.6, 1.1–2.6°C for RCP4.5, 1.4–3.1°C for RCP6.0, and 2.6–4.8°C for RCP8.5. Similarly, variations in precipitation/rainfall will not be identical. Meanwhile, regions of high latitudes are likely to face an escalation in annual average precipitation for RCP8.5 scenario (IPCC 2013).

13.1.2 Sectoral and Regional Trends in Greenhouse Gas Emission

At the global scale, anthropogenic activities are major emitters of greenhouse gases (GHGs) emission. Carbon dioxide, methane, nitrous oxide, and fluorinated gases are major GHGs being emitted due to anthropogenic activities.

Carbon dioxide (CO_2): The primary source of CO_2 is use of fossil fuels. Similarly, land use and land use change is also other major source of CO_2 emission. Carbon dioxide can also be emitted due to direct human impacts on forestry and other land use like deforestation, soil degradation, and land clearing. Likewise, emission of CO_2 is greater from a disturbed (tilled) soil than an undisturbed soil (no till) (Figure 13.1).

Methane (CH_4): Methane (CH_4) has about 23-times more global warming potential than CO_2 (IPCC 2014a,b,c). The major sources of methane emission are (i) organic decay, (ii) wetlands, (iii) rice cultivation, (iv) biomass burning, (v) natural gas and oil extraction, (vi) enteric fermentation of cattle, and (vii) refuse landfills. Methane production in ruminant animal is under anaerobic conditions and this emission is called enteric fermentation. Methane is also produced in soil due to metabolic activities of a highly specific bacterial group called "methanogens." Their activity rises under submerged and anaerobic conditions which develop in flooded rice. This flooded condition limits the supply of oxygen in the soil resulting in an increase in microbial activities, leading to emissions of methane.

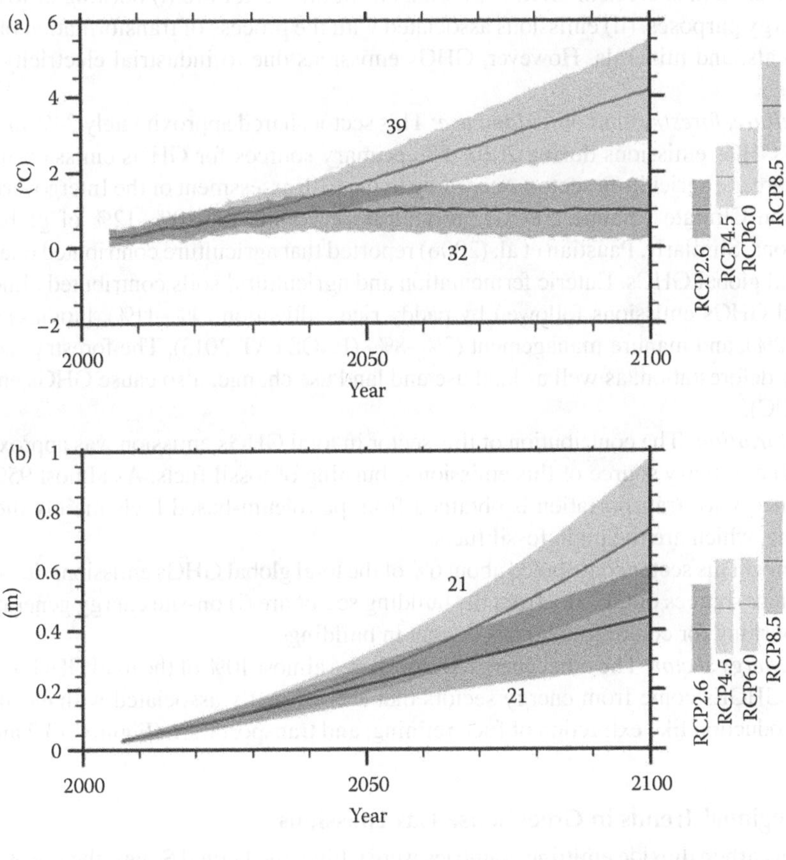

FIGURE 13.1 **(See color insert.)** Change in global mean surface temperature (a) and variations in global average sea level rise (b) from 2006 to 2100 as projected by multi-model simulations.

Nitrous oxide (N_2O): Nitrous oxide has 298-times more global warming potential than carbon dioxide. Major sources of nitrous oxide are (i) nitrogenous fertilizers, (ii) burning of biomass and fossil fuels, (iii) forests, and (iv) grasslands. Soils are major contributors of nitrous oxide emissions. Nitrous oxide emissions from agricultural soil denotes the loss of soil nitrogen and the reduction of nitrogen use efficiency.

Fluorinated gases (F-gases): The primary sources of emission of fluorinated gases are (i) industrial processes, (ii) refrigeration, and (iii) consumption of a different variety of consumer products. Fluorinated gases include hydrofluorocarbons (HFCs), sulfur hexafluoride (SF6), and perfluorocarbons (PFCs).

Black carbon: The primary sources of this solid particle or aerosol are incomplete combustion of fossil fuels or the burning of biomass. However, it is not a gas but its contribution to warming of the atmosphere is nevertheless considerable.

13.1.2.1 Sector-Wise Contribution in Global Greenhouse Gas Emission

Emission of global GHGs has been categorized into the following sectors;

- *Electricity and heat production:* This sector contributed almost 25% of global GHGEs during 2010. The primary sources for GHGs emissions are (i) burning of gas and coal, and (ii) oil consumption in generation of electricity and heat.
- *Industry:* The contribution of this sector was about 21% of total GHGs emission during 2010. The main sources of GHGs emissions from this sector are (i) burning of fossil fuels for energy purposes, (ii) emissions associated with the process of transformation of metals, chemicals, and minerals. However, GHGs emissions due to industrial electricity use are excluded.
- *Agriculture, forestry, and other land use:* This sector shared approximately 24% of the total global GHGs emissions during 2010. The primary sources for GHGs emission from this sector are the agriculture sector. According to the fifth assessment of the Intergovernmental Panel on Climate Change (IPCC), agriculture accounts for 10%–12% of global GHG emissions. Similarly, Paustian et al. (2006) reported that agriculture contributed one third of the total global GHCs. Enteric fermentation and agricultural soils contributed almost 70% of total GHGs emissions followed by paddy rice cultivation (9%–11%), biomass burning (6%–12%), and manure management (7%–8%) (FAOSTAT 2013). The forestry sector and mainly deforestation, as well as land use and land use change, also cause GHGs emissions (LULUC).
- *Transportation:* The contribution of this sector in total GHGs emission was approximately 14%. The primary source of this emission is burning of fossil fuels. As almost 95% of the total energy for transportation is obtained from petroleum-based fuels mainly diesel and gasoline, which are the main fossil fuels.
- *Buildings:* This sector contributed about 6% of the total global GHGs emissions during 2010. The major sources of GHGEs from the building sector are (i) on-site energy generation, (ii) fuels burning for cooking in homes or heat in buildings.
- *Other energy sector:* The other energy sector shares almost 10% of the total GHGs emission. These GHGEs come from energy sectors that are indirectly associated with electricity or heat production like extraction of fuel, refining, and transportation (Figures 13.2 and 13.3).

13.1.2.2 Regional Trends in Greenhouse Gas Emissions

The top seven carbon dioxide emitting countries were China, the United States, the European Union, India, Russia, Japan, and Canada during the year 2011. These data on greenhouse emissions include emissions of carbon dioxide from combustion of fossil fuels, cement manufacturing, and gas flaring. However, sources and sinks of emissions related land use changes are excluded in these estimates.

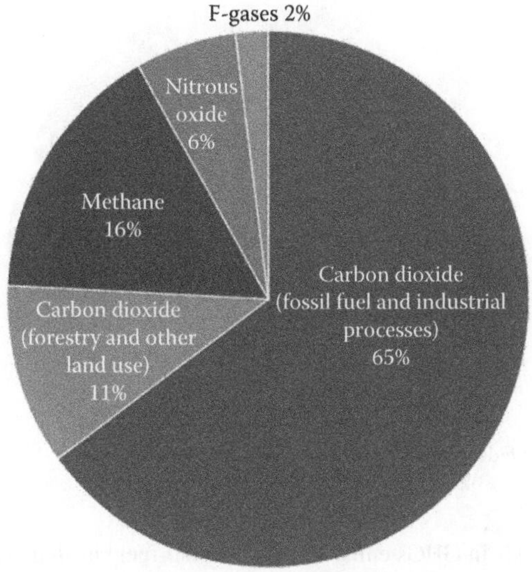

FIGURE 13.2 **(See color insert.)** Global GHGs emission during year 2010. (From IPCC, 2014a)

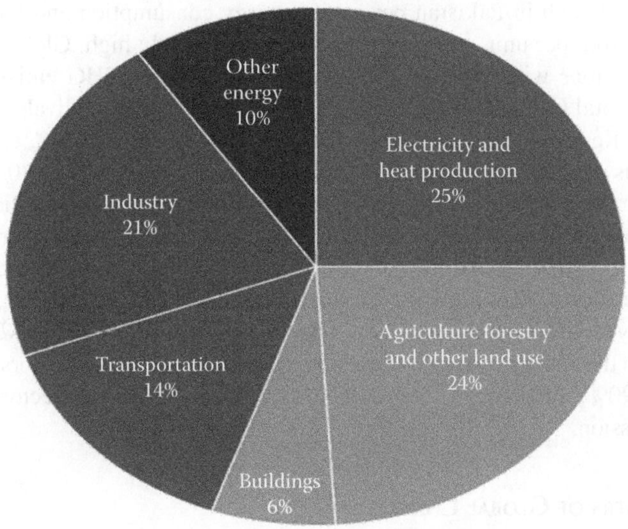

FIGURE 13.3 Sector-wise GHGs emission during year 2010. (From IPCC, 2014a)

These estimates indicate that agriculture, forestry, and other land use contributed almost 8 billion metric tons of CO_2 equivalent in net global GHGs emissions (FAO 2014), or approximately 24% of total GHGs emissions (IPCC 2014a,b,c) (Figure 13.4).

13.1.3 Sectoral Contribution in Greenhouse Gas Emission in Pakistan

It has now become overwhelmingly obvious that CC is going to affect primarily poor nations. These regions are also identified as CC "hotspots" with low food production/security and are prone to high climatic stresses. In Pakistan, although the agriculture and livestock sectors grew rapidly at 2.9% and 4.1% during the year 2013–2014, respectively (Pakistan's Bureau of Statistics, Government of Pakistan), these sectors are most certain to be impacted by CC due to their heavy dependence on natural resources.

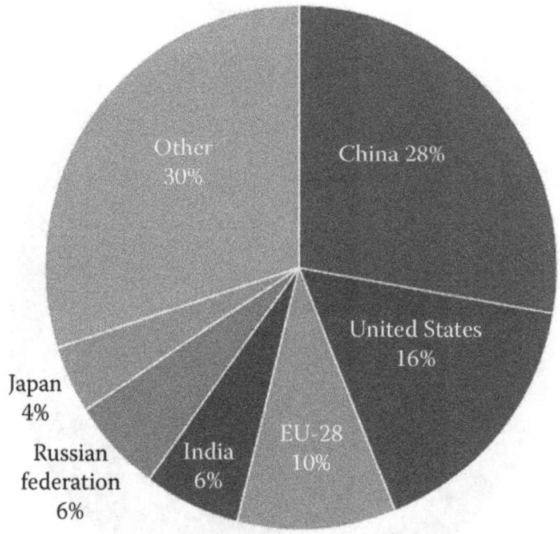

FIGURE 13.4 Regional trends in GHGs emission due to fossil-fuel burning, manufacturing of cement, and gas flaring.

Moreover, Pakistan is one of the lowest contributors of global GHGs, nevertheless receiving great threats from CC. Although in Pakistan per capita energy consumption and total CO_2 emission is low, CO_2 emission from per unit energy consumption is relatively high. Globally, Pakistan ranked 135th in GHG emissions with only a 0.8% contribution in total GHG emission (PAEC-ASAD 2009). In Pakistan, total GHG emission was 341 million tons of CO_2 equivalents ($MtCO_2e$) during 1990–2012 (World Resource Institute 2016). These emissions include 54% carbon dioxide, 36% methane, 9% nitrous oxide, 1% carbon monoxide, and minute concentrations (0.3%) of nonmethane volatile organic compounds. The energy sector in Pakistan is the largest significant sector of GHG emissions. With one estimate, 157 million tons of CO_2 were emitted by the energy sector including transport during 2007–2008, around 51% of the country's total emission. Other contributors include agriculture and livestock (39% emission), the industrial sector with around 6%, land use, land use change, and forestry (LULUCF) is about 3% while 1% from waste products (NEEDS 2011).

In Pakistan, both the energy and agriculture sectors are the major contributors of GHG emissions, which account for 90% of Pakistan's total GHG emissions. Both of these sectors need focus on the issues of GHG emission.

13.1.4 Key Drivers of Global Change

It has been projected that concentration of CO_2-eq will cross from 750 to 1300 ppm by the end of twenty-first century and therefore, mean surface temperature of the globe will increase from 3.7 to 4.8°C as compared to preindustrial levels during the current century. The various drivers possibly responsible for this projected increase in GHGs are;

1. Economic growth such as Gross Domestic Product (GDP)/capita
2. Population growth (capita)
3. The energy intensity required for per unit economic output (energy/GDP)

Globally, population and economic growth will be the significant drivers of CO_2 emissions due to fossil fuel combustion. Over the last decade, economic growth was an important driver of global carbon dioxide emissions while population growth remained mainly steady. However, population of the globe has increased from 3.7 to 6.9 billion during 1970–2010. Nevertheless, changes in technology, economic structure as well as changes in other inputs like labor and capital will cause a steady decline in energy intensity of economic output worldwide. The decline in energy requirement will have an offsetting impact on global carbon dioxide emissions. Similarly, innovations in technology and infrastructural choices are affecting greenhouse gases emissions due to productive growth, energy and carbon intensity as well as consumption patterns. Innovations in technology will improve labor and resource productivity. However, the directions and speed of change in technology mainly depends on existing policies in the country. Technology is also known as a central element for infrastructural choices and spatial organization in cities, which have long-term impacts on GHG emissions.

13.2 CHANGING SCENARIO IN AGRICULTURE

13.2.1 CC and Crop Productivity

Agriculture is highly dependent on weather, and therefore changes in global climate could have major effects on crop yields, and thus food supply. Climate change is threatening crop production and food security. Climate change may result in a decrease in cereals production by 7–10% by the end of this century. Projections of CC and its impacts on crops production are inherently uncertain. But uncertainty can often be quantified by projecting future GHGEs, temperature, precipitation, and their influence on climate overall. There are at least two ways that agriculture may be influenced by the prospective CC. First, agriculture is sensitive to climate as changes in climate attributes affects plant growth. Second, agriculture may have an important role to play in retracting global warming through mitigation of GHGEs. The impact of global CC can now be seen at the local levels in most parts of Pakistan. Most Pakistani farmers have been dealing with climate vulnerability, climate shifts, climate shocks, water stresses, and unsustainable production over a decade. In some regions, there is an urgent need to develop and use climate resilient crop varieties that are resistant to biotic and abiotic stresses such as drought, erratic climate, and flooding, as well as pests and diseases resulting from changing climate. A case study was conducted by Ahmad et al. (2015) to assess the vulnerability of rice-wheat cropping systems to CC in Punjab-Pakistan. Mid-century climatic projections for the period 2040 to 2069, using Representative Concentration Pathway (RCP) 8.5 were generated with the help of five general circulation models (GCMs). These projections showed increasing trends in maximum and minimum temperatures of about 2°C. Two crop models, the Decision Support System for Agro-technology Transfer (DSSAT) and the Agricultural Production System sIMulator (APSIM) were calibrated, evaluated, and validated with experimental data for both rice and wheat crops. Further, a weather baseline of years (1981–2010) was developed to make a baseline of yield trend for both crops in order to compare with projected weather scenarios. Then, both crop models were run with weather baseline along with projected delta scenarios of five GCMs (mid-century) to compare the yield trend of each GCM with the baseline. Furthermore, survey data of 155 farmers from five districts (Sialkot, Gujranwala, Sheikhupura, Nankansahib, and Hafizabad) were collected to validate both crop models (DSSAT and APSIM) at the farmer's field scale. Both calibrated models were run with baseline and projected scenarios in order to evaluate the impact of CC on rice-wheat cropping systems. The results of DSSAT and APSIM showed that mean reductions in paddy yield were 15.2% and 17.2%, respectively. However in the case of wheat, simulated mean grain yield reduction was 14.0% with DSSAT and 13.76% with APSIM model.

13.2.2 Impacts of CC on Livestock, Poultry, Fisheries, and Forest Production

The livestock sector, contributing 11.8% in the national GDP and 56.1% in the agricultural value addition (Anonymous 2013–2014), is also among the forefront of industries which are expected to be affected by CC. A variety of factors are responsible for the rapid change in livestock systems of developing countries. Globally, it is expected that human populations will increase rapidly and will reach to 9.2 billion by 2050; hence, the demand for livestock products will continue to increase significantly during the coming decades (Delgado et al. 1999). So, to cope with the augmenting demands for livestock and their products new institutional innovations, technologies, and policies are required. Impacts of CC on livestock are outlined and organized under various headings: quantity and quality of feeds; heat stress; water; livestock diseases and disease vectors; and biodiversity.

1. *Feeds:* Several impacts have been expected on feed crops and grazing systems by CC, especially through changes in atmospheric CO_2 concentrations and temperature. Increment in the atmospheric level of CO_2 reduces water loss by plants through partial closure of stomata and hence improves water-use efficiency. The exact phenomenon behind higher temperature role in lowering plant growth is not known but it may be due to more water utilization and higher radiation levels (Rötter and van de Geijn 1999).
2. *Heat Stress:* Scientists have not yet focused on the effect of climatic change on heat stress of animals, especially in subtropics and tropical areas. But it can be assumed that heat exchange between environment and animals will be greatly influenced by warming and hence growth, feed intake (SCA 1990), reproduction, mortality, maintenance, and production will be adversely affected.
3. *Water:* Freshwater resources around the globe are fairly scarce as they contribute only 2.5% of total water resources. Animal water requirements have been shown to increase with temperature. Water utilization by Zebu cattle (*Bos indicus*) are 3 kg per kg utilization of per dry matter (DM) at 10°C, which increases to 5 kg per DM kg at 30°C, and 10 kg per kg utilization of DM at 35°C (NRC 1981).
4. *Diseases and their vectors:* Climate change impacts distribution of infectious diseases through changing various disease determinants including pathogens, hosts, vectors, and epidemiology, and so on as described by Baylis and Githeko (2006). Higher temperatures can be a source for the rapid propagation of various pathogens as it may uplift their development rates, and parasites or pathogens may reside a part of their life cycle outside their host (Harvell et al. 2002).
5. *Biodiversity:* Climate change has already affected biodiversity badly with substantial impacts. Amongst nearly 4000 breeds belonging to the ass, water buffalo, goat, cattle, horse, sheep, and pig species of livestock recorded during the twentieth century, 16% of them have become extinct while 12% of them have became scarce (Ehrenfeld 2005). According to the FAO (2007), 20% of the remaining animal breeds are at risk of extinction. Similarly, 20%–8% of mammalian species in developed countries are at risk while this affects 7%–10% in developing countries (CGRFA 2007). Using the data of ten countries, an association of the climate change (CC) has recently been developed with the parameters like species of the animals, per animal income, and number of animals per farm (Seo and Mendelsohn 2006).

Poultry includes birds reared for economic purposes. These include domestic fowl, broiler, layer, turkey, duck, goose, ostrich, etc. The poultry sector is important as it contributes 6.1% and 10.8% to the agricultural and livestock sectors, respectively while its share in total GDP is 1.3% and its share in total meat production of the country is 28.1% (Anonymous 2013–2014). It has a significant contribution in terms of fulfilling the animal protein requirement of humans, while being used in making various poultry by-products like blood meal, meat meal, and bone meal to fulfill protein and other essential nutrient requirements. They are raised with relatively low capital investment and readily available household labor. Scavenging village chickens have cultural, social, nutritional, economic, and sanitary functions in human life. The inherited capacity of the poultry has a major role in the production and growth of the poultry sector; however, it does not only depend upon inherited capacity but also to a great extent upon the environment. Environmental factors which affect the productivity and growth of poultry include temperature, relative humidity, intensity of light, sunshine or day length, housing system, and ventilation

(Alade and Ademola 2013; Olanrewaju et al. 2010). Regarding productivity, there is need for good housing systems to maintain optimum environmental temperatures and reduce heat stress with cooling and ventilation. Studies by the Department for the Environment, Food and Rural Affairs (DEFRA) on broiler hens have reported that when a poultry houses are experimentally exposed to potential future changes in climate they exceed critical temperatures on 30% more occasions, as compared to 10% increases in ventilation. Moreover, other dramatic environmental conditions, such as storms, rain, and rate of air flow might increase stress and adversely affect production and growth of the flock.

Climate change is affecting fisheries by affecting the distribution and productivity of marine and freshwater fish stocks. Other impacts include habitat damage, ocean acidification, water scarcity, etc. Fisheries are also being exposed to some other direct and indirect CC impacts such as human migration and displacement, coastal communities' disturbance, and infrastructure change due to sea level rise, changes in distribution, intensity, and frequency of tropical storms. Climate change impacts on fisheries are uneven for different geographical areas, regions, and countries but its vulnerability completely depends on weather-related hazards associated with climate variability. Pakistan is among countries potentially subjected to drastic CC impacts on its fisheries industry (Allison et al. 2005).

Forest covers around 4 billion hectare of land surface globally (FAO 2005). Pakistan is included in the countries where 20%–100% of growing stocks of wood was removed from the forests. Impact of climate changing indicators like increase in temperatures and elevated CO_2 is not very prominent but CC induced modifications such as forest fires, insects, pest pathogens outbreak and extreme events like flooding and droughts (Kirilenko and Sedjo 2007).

13.2.3 Impacts of CC and Water Resources

Water resource stress will increase globally due to CC; these effects are obvious in parts of Europe, Central and South America, southern Africa, and south and eastern Asia. In these countries CC will increase runoff. The countries having the capacity to tackle and store this runoff could benefit to some extent; however, the benefits may be offset when runoff increases in the wet season while limited amounts of runoff water would be available during the dry season (Arnell 2004). There are several indicators of water resource stress such as per capita availability of water and the ratio of water requirement to available water. Whenever water requirement increases to the level of available water, stress will appear with its potential impacts. Falkenmark and Lindh (1976) figured out some standard of water stress. They revealed that if water requirement is greater than 20% of total water availability, water stress will often be a limiting factor on development; however, if this difference is 40% or more it will create severe stress. Water stress will be problematic when any region has less than 1700 m^3 water per capita per year (Falkenmark and Lindh 1976). These numerical indicators can be indicators of water resource depletion under changing climate in any country, nevertheless the overall consequences of water stress would depend on water resources management.

Pakistan has 40 million acre feet (MAF) of surface water resources and around 56 MAF of underground reserves. Currently, annual availability of water has been reduced up to 1000 m^3 per capita, which was fivefold higher in the 1950s with 5140 m^3 per capita (Hussain and Mumtaz 2014). Pakistan is moving at quite a fast pace toward water limitations and approaching towards water scarcity. Pakistan water resources are being affected by CC in the face of fast glacier melts, changing rainfall patterns, extreme weather events, such as droughts and floods. Pakistan experienced severe floods during 1950, 1956, 1957, 1973, 1976, 1978, 1988, 1992, 2010, 2011, 2012, and 2014. In contrast to flooding, Pakistan has also faced its worst droughts during 1998–2004. The water crisis either in the form of flooding or droughts is drastically affecting the economy and development of the country.

The adverse impacts of CC on water resources can only be minimized with conservation of water, which is being wasted into the sea. Ground water depletion is also very high in most parts of the country. This depletion is due to unnecessary pumping of water for high water demanding crops and a shift in rainfall patterns and amounts. Most of Pakistan's freshwater is being wasted into the sea due the few number of water reservoirs and the low storage capacity of available water reservoirs. No doubt, Pakistan gets huge amounts of freshwater from glaciers melting in the Himalayas, but CC is triggering the fast melting of the snow resources, which increase the rate of flooding in the country; this phenomenon will accelerate and become more frequent in near future due to CC. Water availability will be high during the wet season, while it will worsen the situation during the dry season with low water availability to all sectors of activity. In both scenarios, the country's growth and development will fully be reduced and disturbed. Furthermore, impacts of changing climate on water scarcity issues in Pakistan can be minimized with the integration of local, provincial, and national plans, knowledge and research by addressing administrative, financial, technical, social and environmental reservations/issues.

Climate change implications can be minimized with two basic approaches: Adaptation and mitigation including watersheds management, improvement of existing catchments and water bodies, judicious use of present sources, development and exploration of new water sources, campaigning for water conservation techniques, using adequate drainage methods, efficient design of water storage facilities, reducing water conveyance losses (distribution and supply), improving irrigation systems, and waste water utilization.

13.2.4 Water, Energy, and Food Nexus under Changing Climate

Water, energy, and food are quite essential for poverty alleviation, human well being, and sustainability. Global projections indicated that freshwater, energy, and food demand will increase significantly over the next decades due to growing population pressure and mobility, economic development, international trade, urbanization, diversifying diets, cultural, and technological changes, and CC (Hoff 2011). Energy, food and water are closely linked with each other. Water is the main input for agricultural goods production while energy is directly or indirectly required for food and water production and distribution. The agriculture sector is the largest user of water of both surface and subsurface origins. Globally, agriculture is using 70% of the total water consumption while around 30% of the total energy is being consumed in food production and distribution globally (World Water Development Report 2014). Total global water consumption for irrigation is projected to increase by 10% in 2050 (FAO 2011). High population pressure is increasing water, energy, and food consumption over time. This situation is expected to worsen in the near future as 60% more food will be needed in order to feed the world population in 2050, while global energy consumption is projected to grow by 50% by 2035 (IEA 2010). The water, energy, and food nexus exhibit synergies and tradeoffs simultaneously. Their interaction is positive in one way but negative in another. High availability and water use is promoting the food production while at the same time water use for food production is reducing river flows, which directly reduce hydropower potential. Similarly, the use of surface irrigation in the form of high-pressure irrigation system may conserve water but results in higher energy consumption. Understanding these synergies and tradeoffs and maintain a balance among them is vital to ensure a water, energy, and food nexus beneficial under changing climate conditions. Most of the community is well aware of water, energy, and food challenges and also addresse them individually.

Individual approaches for single source management will not be helpful for achieving the desired goal of sustainability and food production. Therefore, a nexus approach of management through enhanced discussion, coordination, and collaboration is needed to ensure the benefits and appropriate safeguards may be put in place against nexus tradeoffs. Nexus interactions are quite dynamic, complex, and cannot be looked at individually. Nevertheless, a nexus approach will help us in better understanding dynamic and complex interrelationships between water, energy, and food while sustaining limited resources (Figures 13.5 and 13.6).

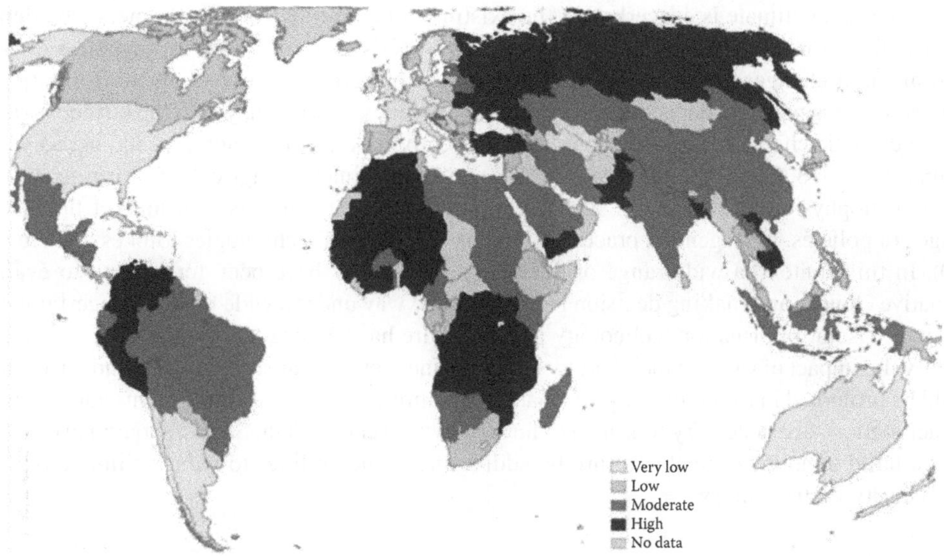

FIGURE 13.5 Comparative vulnerability of national economies to climate change impacts on fisheries. (From Allison EH et al. 2005. Effects of climate change on the sustainability of capture and enhancement fisheries important to the poor: Analysis of the vulnerability and adaptability of fisher folk living in poverty. Fisheries Management Science Programme MRAG/DFID, Project no. R4778J, Final Technical Report, London, p. 164.)

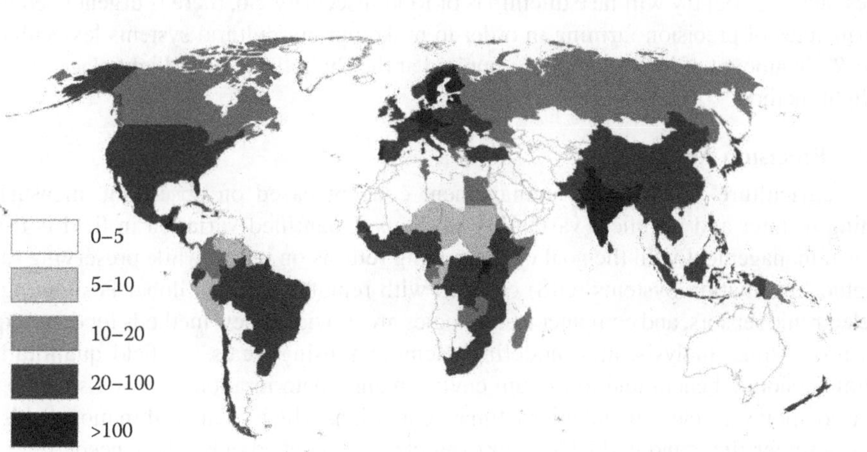

FIGURE 13.6 Global wood harvest computed on a per country base, $m^3\ km^{-2}$ and white areas correspond to low harvest or no data. (From Kirilenko, AP and RA Sedjo 2007. *PNAS*. 104:19697–19702.)

13.3 RISK AND UNCERTAINTY ASSESSMENT OF CLIMATE CHANGE (CC)

13.3.1 Rationale

In agricultural systems of Asia, farmers mostly have small land holdings and are severely vulnerable to the changing climate. However, Pakistan has been declared among the top ten most vulnerable countries to changing climate. The geographic location of this country makes it more prone to CC as it lies in a semiarid to arid part of the world. Changes in climatic conditions are affecting agricultural produce and this will become more catastrophic in future. The increasing pace of changing climate is exacerbating the existing vulnerabilities of poor farmers who depend on agriculture for their survival (Slingo et al. 2005; Nelson et al. 2009). In response of these environmental changes, consistent efforts in the field of agronomy, use of agro-informatics for future risk assessment and decision-making as well as policy making may help the farmers to adapt to these climatic changes. These changes are inherently linked with water, soil and agrodiversity (Adams et al. 1998, Cline 2007). Moreover, these are determinants of agricultural farming systems and their biophysiological changes such as productivity, pests and disease, market fluctuation, changes in policies, management practices, land use, and use of technologies (Stokes and Howden 2010). In this context, a wide range of decision options/tools have been formulated to evaluate alternative choices and making decision in a systematic way under a wide range of uncertainties in outcomes. Use of information technology in agriculture has become indispensable as it is used to quantify the impact of CC. Financial protection of farmers by insurance companies and government should be promoted in the country against natural calamities. Early warning systems and automatic weather stations are necessary to sensitize farmers with weather data. So, we should have to limit CC and build a more sustainable future by addressing vulnerabilities to reduce climatic risks for food security in the country.

13.3.2 Agriculture Sector Monitoring

Nowadays agro-ecosystems are facing serious threats and risks associated with crop production. These risks and uncertainties in crop production will be greater in the future as the global climate is changing day by day. Climate change will put huge pressure on natural resources and livelihood management. The productivity of major crops will undergo shuffling phases due to scarcity of resources, and the country will face dilemmas of food insecurity. So, there is urgent need to use the "4 R" principles of precision farming in order to make our agricultural systems less vulnerable to CC. The *R*ight amount of inputs should be applied at the *R*ight time in the *R*ight place with the help of the *R*ight method.

13.3.2.1 Precision Agriculture

Precision agriculture is a farming management concept based on observing, measuring, and responding to inter and intrafield variability in crops. Quantified variation in field is treated for whole farm management with the goal of optimizing returns on inputs while preserving resources. Geographic Information Systems (GIS) coupled with remote sensing, Global Positioning System (GPS), electronic sensors, and computer technologies are providing new methods for data acquisition, storage, processing, analysis, and modeling. Remote sensing treats the field quantitatively for maximum economic benefit and minimum environmental deterioration. The site-specific farming (SSE) based upon variable soil and microclimate conditions which occur within most fields reduces waste, because fertilizer and herbicide—for example—are applied only where needed.

i. *Remote Sensing*: It is used to gather information about a distant object without making physical contact with it. Comprehensive and reliable information on land use cover, forest area, soils geological information, extent of wetlands, agricultural crops, water resources and hazards, or natural calamities like droughts and floods can be obtained through remote sensing. Remote sensing techniques, especially for crops, are being used to monitor and sense the crop stages, crop stresses (biotic and abiotic), soil salinity, lack of appropriate agricultural inputs, improper crop management practices, and greenhouse gases mapping.
ii. *Global Positioning System*: It is a satellite-based tool which can be used to locate the position of an object anywhere on the earth. Global positioning system are useful to record data on slope of field, nutrients status for crop, and final yield. The outputs of GPS are used in GIS for further analysis. It is an efficient tool to monitor crop yields accurately if it has been coupled with GPS receivers. The information gathered from various satellites and

receivers of GPS can be combined to generate field crops management approaches for the application of herbicides, irrigation, and fertilizers as well as harvesting of crops (Liaghat and Balasundram 2010).

iii. *Geographic Information System (GIS)*: Burrough and McDonnell (1998) has defined GIS as a potent set of tools to collect, store, retrieve, transform, and display spatial data from the real world for a particular set of purposes. This technology is being integrated with remote sensing to give accurate and reliable crop yield estimates, which help to minimize uncertainty in the grain industry. Similarly, this technology is recognized as a vital tool for integrating different maps and satellite information in crop growth models, which in turn helps in understanding the interactions among complex natural systems. GIS is not being used to make maps but to produce images, animations, drawings, etc.

13.3.2.2 Climate and Weather Forecasting

Recent advances in climate modeling have resulted in an increased ability to predict rainfall in many parts of the world with a lead time ranging from a few days to a few months, by using dynamic forecasts or statistical methods (Njau 2010). Climate forecasts can help farmers reduce their vulnerability to drought and climate extremes, while also allowing them to maximize opportunities when favorable rainfall conditions are predicted (Patt et al. 2005; Phillips et al. 2001; Roncoli et al. 2009). Technically, climate forecast information is now much improved (downscaled and a good degree of confidence in predicting the frequency of dry spells) (Hansen et al. 2011). Roncoli et al. (2009) found that when farmers in Burkina Faso were engaged in pilot projects and intensive interactions with researchers, they were likely to use climate forecast information to change management practices. Receiving training and education is seen as critical to the successful uptake of forecast information (Hansen et al. 2011).

13.3.2.3 Early Warning System

Climate change will enhance temperature and drought regimes across various parts of the world. This changing climate will cause food insecurity in the country, so development of early warning systems and other programs will prove effective tools (Brown and Funk 2008). In future, various parts of the globe will have greater intra- and interannual variations in climatic elements (Ganor et al. 2010). Hence, it is necessary to make our farmers climate ready with early dissemination of weather information so they can adjust their time of field operations.

13.3.2.4 Agricultural Information Management and Dissemination System

Marginalized groups of farmers often do not have equal access to these technologies (Roncoli et al. 2009; Vogel 2000). Agricultural Information Management and Dissemination System (AIMDS) should be introduced at the farms for climatic, weather, technical, and market information as well as complaints of farmers and crop management adaptive options. Also, increase awareness among the public to help defend against climatic variability and risks. Concepts of natural disasters should be changed into manageable options and should be taken as an opportunity.

13.3.3 Quantification of CC Impacts

The rise in temperature and rainfall are causing adverse effects on agricultural productivity. However, there are mainly two approaches to quantify CC impacts on crops. These are (i) fields with controlled environment experimentation, and (ii) the use of simulation tools.

13.3.3.1 Field with Controlled Environment Experimentation

Quantification of the impact of weather elements on crop growth can be assessed with the growing of crop in Free Atmospheric Carbon dioxide Enrichment (FACE) or Free-Air Temperature Enrichment (FATE).

Free Atmospheric Carbon dioxide Enrichment (FACE): This system provides opportunities to check the direct impact of elevated carbon dioxide on growth, phenology, and yield of plant without altering the microclimate. The main aims of this system are:

i. To maintain a steady flow of carbon dioxide within environment so that external fluctuation in CO_2 will not have an effect on plant growth
ii. To provide the reliable field data on plant response in order to develop and evaluate mechanistic growth models so that the impact of future climates may be assessed
iii. To evaluate global CO_2 fluxes and the carbon balance between atmospheric CO_2 concentration and vegetation

Free-Air Temperature Enrichment (FATE): This system has been designed to evaluate the impact of rising temperature on plant growth, development, and yield. Its aims are similar as those of FACE. However, the interactive effect of elevated CO_2 and temperature on plant growth and yield should be studied in field sets. So, combining of the FACE system with the FATE needs to be fully explored in order to expose natural agro-ecosystem to that of elevated carbon dioxide concentration continuously along with increased temperature.

13.3.4 Application of System Modeling in Decision Support

The models have been used with success for simulating efficient production, crop adaptation, improved risk management, and sustainable production (Mohanty et al. 2011). The purpose of crop simulation models application in field is to control the seasonal management and particularly, to provide the farmers with an opportunity to minimize input (Booltink et al. 2001). However, these tools are very effective in the formulation of policy interventions. Hence, modern research in agriculture should be disseminated towards farmers, policy makers, and other organizations involved in the decision-making process so that sustainable agriculture can be accomplished over the wide range of variations in climate. In this context, Decision Support System for Agro-technology Transfer (DSSAT) and Agricultural Production Systems sIMulator (APSIM) are the cropping system models used in decision-making processes. The formulation of effective policies is the need-of-the-hour to make farmers more resilient against any natural disaster.

13.3.4.1 Application of Simulation Models to Assess CC Impact

To assess the impact of future CC on crop production, in order to develop adaptation packages ,is necessary to save our natural resources from degradation and to feed our nation. There are different steps to assess CC impacts, these are following;

1. *Weather Baseline:* To develop a baseline of weather, a weather data set of the past 30 years is used. This weather data may be either observed or model outputs. The main purpose of baseline development is to compare the change in crop productivity due to future climate scenarios with current climatic conditions. There is a series of GCMs being used to generate weather baseline and on the basis of weather elements trend in baseline, future climatic scenarios can be developed.
2. *Climate Scenarios:* Climate scenarios are generated on the base of global GHGs emission rates. These are compulsory to assess the impact of future CC on crop productivity. These scenarios depict the climatic conditions that will prevail in a particular part of a century. There are several steps to carry out CC analysis. First of all, socioeconomic models are being used to develop future scenarios for deriving the GHG emissions. So, future scenarios have been classified as A1, A2, B1, and B2 (IPCC 2001). However, A1 and B1 mainly deal with socioeconomic changes at global scale, while A2 and B2 show changes at regional level. Likewise, A1 and A2 scenarios are mainly based on economic sustainability and do not deal with environmental issues, whereas B1 and B2 scenarios are based on environmental sustainability.

13.3.4.2 Integrated Assessment of Crop, Climate, and Economic Modeling

The agriculture sector is the most vulnerable sector, which ultimately has impacts on its dependants and associates. To study all the complications of agriculture under changing climate, biophysical and economic aspects should be considered and integrated (Hillel and Rosenzweig 2010). The Agricultural Model Inter comparison and Improvement Project (AgMIP) is an initiative to integrate the crop, climate, and economic modeling for a comprehensive analysis. The AgMIP research activities are being performed by four teams for climate, crop, economic, and information technology with guidance provided by a leadership team (Rosenzweig et al. 2013). The biggest advantage of integrated assessment and linking models of various disciplines is the provision of a common platform for specialists and stakeholders to develop Representative Agricultural Pathways (RAPs) in order to enhance food security (Webber et al. 2014). For the integrated improvements, climate modeling and impact assessment communities have developed new concepts of Representative Concentration Pathways (RCPs) and Shared Socio-economic Pathways (SSPs). Representative Agricultural Pathways (RAPs) have been developed to link the RCPs and SSPs as a way to extend to agricultural models (Antle 2011). Modeling approaches are being used to assess the impact of CC on crop and water productivity, soil health, and livestock adoptability.

13.4 CLIMATE SMART APPROACHES

Climate Smart Agriculture aims to increase productivity and income, build resilience to CC, reduce GHGEs and enhance achievement of national food security and development goals in a sustainable way. Many agricultural technologies and practices are considered climate smart approaches including minimum tillage, residue incorporation, nutrient and irrigation management, methods of crop establishment, and soil amendments. These can improve crop water and nutrient use efficiency and enhance crop yields, while reducing Greenhouse Gas (GHG) emissions (Sapkota et al. 2015). Similarly, use of improved seeds, rainwater harvesting, crop/livestock insurances, information and communication technology (ICT)-based agro-advisories can help to reduce the impact of CC and variability on the agriculture sector (Altieri and Nicholls 2013).

13.4.1 CC Mitigation Practices

Mitigation highlights the *causes* of the problem, which involve reducing greenhouse gases concentrations in the atmosphere. According to the United Nations Framework Convention on CC (UNFCCC), mitigations are human intervention to reduce GHGs emission: Either reduction of its sources or enhancing the sinks. Nowadays, mitigation is more cost effective than adaptation to CC in developing countries. There are various mitigation options to CC in agro-ecosystems. Some of them are discussed below.

The capturing and storage of atmospheric carbon in agro-ecosystem is known as carbon sequestration. Carbon in the soil acts as Soil Organic Carbon (SOC) or Soil Inorganic Carbon (SIC), SOC is known as one of the CC mitigating options (Zhang et al. 2014). Nevertheless, any small change in SOC may affect atmospheric CO_2 concentration (Gu et al. 2009). However, management of SOC will play a pivotal role to enhance its amount in soil to provide food security, as well as to mitigation the future climate. However, SOC is mainly related with input of organic carbon to the agro-ecosystems. Xu et al. (2014) reported a linear and positive relationship of SOC with crop productivity as the higher amount of SOC will result in higher crop productivity. Resultantly, production of higher biomass would result in more organic carbon inputs into soils and therefore, rise in soil carbon sequestration in the form of SOC (Xu et al. 2011). There are some agricultural practices which may lead to enhanced SOC like organic amendments such as green manuring, straw/stubble management, and farmyard manure. They are mainly recommended as an alternative to mineral fertilizers to provide essential plant nutrients (Gu et al. 2009). These amendments are deemed good practices to improve soil fertility and sustain crop yields as well as to enhance carbon sequestration (Gu et al. 2009; Zhang et al. 2009). Any activity which will disturb stabilization of SOM is responsible for enrichment of atmospheric CO_2 through degradation of SOC. Many investigations have proved that no-tillage or reduced tillage provides more protection to SOM as compared to conventional practices (Pinheiro et al. 2004; Six et al. 2000). Adoption of these strategies will make our agricultural systems more sustainable and resilient against future climatic changes trends.

In the case of methane emissions from rice cultivation, the mitigation practices are (i) alternating wetting and drying, (ii) aerobic/upland rice cultivation, (iii) system of rice intensification, and (iv) integrated nutrient management.

Similarly, methane emission from ruminants (enteric fermentation) can also be minimized by changing the composition of feed.

Nitrous oxide emission can be reduced by (i) site-specific, efficient nutrient management (Pathak 2010), (ii) application of synthetic nitrification inhibitors like dicyandiamide (DCD) and nitrapyrin, and (iii) use of organic/natural nitrification inhibitors like neem (*Azadirachta indica*) cake, neem oil, and karanja (*Millettia pinnata*) seed extract.

13.4.2 CC Adaptation Strategies

Adaptation strategies deal with the negative impacts of CC and have the objective of reducing the vulnerability of human and natural agro-ecosystems. Their aims are to make agro-ecosystem more resilient against natural catastrophes. The adaptation strategies in response to develop resilience in our agro-ecosystems are discussed in Sections 13.4.2.1 to 13.4.2.8.

13.4.2.1 Developing Climate-Ready Crops

There is a need to develop crop varieties having greater potential for yield and resistance to abiotic stresses (water logging/flooding, drought, salinity) in order to maintain yield stability. The best approach is improvement in germ plasm of major filed crops. This approach will ensure tolerance against abiotic stresses. The plants having these desired traits will be enabled to perform well under a wide range of stresses. The manipulation of gene pyramiding in genetic engineering is necessary to pool up the desired traits in a single plant to develop true to type plant. The other breeding approach is transformation of C_3 rice to C_4. This will help in boosting the productivity of rice as well as making it resource efficient. Similarly more steps in breeding programs should be adopted to enhance radiation-use efficiency of major field crops.

13.4.2.2 Crop Diversification

It is replacement of conventional cultivars and livestock breeds with newly advanced cultivars and animal breeds in order to maintain high production in spite of changing climate. Crop diversification is known as defense tool against all abiotic stresses as it exploits the efficient utilization of natural resources. Crop diversification also covers the growing of alternate crops in existing intensive cropping systems. This will develop resilience in existing cropping systems against future changes in climate. The significant constraint of crop diversification is that it deals with switching of crop, which may be costly and possibly result in less profitability.

13.4.2.3 Adjusting Cropping Season

It is growing of crops in such a manner that coincides with prevailing weather conditions of the growth season. This will help crops in completing their life cycle in a preferable environment. Adjustment of planting times has a vital role in minimizing the devastating impacts of rising temperature on sterility of reproductive parts of crops. It is also necessary to minimize yield instability by avoiding grain filling stage from the warmest period. Adaptation strategies to minimize the impacts of extreme climatic events, which are now seen in semiarid and arid regions of Pakistan, include altering the cropping calendar to benefit from wet periods and escape from extreme weather events during growing season. The existing cropping systems should be subtituted to include suitable advanced cultivars, which would enhance the crop intensities; the sowing of various types of crops should be considered as well. Farmers can adapt to varying moisture regimes by altering crops.

13.4.2.4 Resource Smart Technologies

The resource-conserving technologies (RCTs) are those actions which increase resource or efficiencies of inputs and give instant and discernible economic benefits like reductions in the cost of production; savings on fuel and labor requirements, and timely crop establishment resulting in higher yields. Similarly, wheat productivity under heat and water shortage environments can be boosted significantly with the adoption of RCTs. Resource conservation technologies, such as no till or zero tillage, can permit farmers to grow wheat in time after harvesting of rice so that wheat may head and fill grains prior to onset of hot conditions of the premonsoon season. The resource-conserving technologies in rice-wheat cropping system also have a prominent impact on reduction of GHGEs and adaptation to future changing climate (Pathak et al. 2011). Hence, soil and water management is very much vital for adaptation to CC. In future, due to increases in temperatures and changing patterns of precipitation, water shortages will become more frequent as water becomes a rare resource. So, it is necessary to consider those actions that have positive impacts on water conservation and use in order to boost crop productivity. In this context, farmers should be trained and inspired to adopt on-farm water saving techniques and select advanced genotypes for field crops.

13.4.2.5 Allocation of Crops in Alternative Areas

There is a need to recognize crops and regions that are highly prone to changing climate/variability, and then relocate those crops to more suitable areas. For instance, increased temperatures would have adverse impacts on the quality of aromatic crops such as basmati rice and tea. So, alternative regions that are becoming more favorable for these crops need to be recognized and evaluated for their suitability.

13.4.2.6 Land Use and Land-Use Change Management Practices

Changing land-use practices like rotating or shifting crops, changing the intensity of inorganic fertilizers and application of pesticides, proper utilization of labor and capital inputs may help in reducing the risks from changing climate. Similarly, adjustment in cropping sequences, such as changing the timing of sowing, fertilization, irrigation scheduling, spraying, and harvesting will make existing cropping systems less prone to changes in climate. Likewise, changing planting time can also assist the farmers to regulate growing season length, which would better fit with changing environments.

13.4.2.7 Pest Management Strategies

The rising temperature and variations in precipitation have a strong correlation with incidence of insects/pests and diseases of major crops. CC will have potential effects on pest–host relationships as it affects the pest and host population as well as pest–host interactions. A few potential adaptation packages against pests could be:

i. The development of pest and disease resistant cultivars
ii. The prefered adoption of integrated pest management (IPM)
iii. Pest forecasting with advanced tools like simulation models
iv. Pests and diseases management with natural pesticides, bacterial, and viral pesticides as well as pheromones to disturb pest reproduction
v. Bio-agents that also have a key role in pest management

13.4.2.8 Weather Forecasting and Insurance Schemes for Crops

Climate influences the distribution of crops over different regions of the world, while weather influences crop production and productivity. Among the various weather elements affecting crops, temperature and rainfall are perhaps the important ones. It is the reason for the development of special flora and fauna in different climatic zones of the world seen in tropical, temperate, and cold regions. Weather forecasting will help farmers to decide on farm operations to carry out; for example, pesticides and micronutrient spraying, irrigation scheduling, fertilizing, and harvesting. Moreover, weather forecasting can play a vital role in the management of crop diseases and insect pest attacks. Also, farmers get alerts allowing them to manage different abiotic stresses such as heat waves, frost, storms, and heavy rainfall.

Crop insurance is a contractual arrangement that compensates when yields or revenues falls below threshold (Schnitkey and Sherrick 2014). Simulated insurance is an innovative idea to compensate farmers' losses on modeled stresses, such as floods or heavy rainfalls, rather than on observed ones. This approach will save time, costs of verifying observed losses, build confidence of consumers, and will be especially viable for small-scale farmers.

13.4.3 Mitigation Policies and Institutions

Strong relationship among researchers, academia, social mobilizers, and farming communities is the key step towards knowledge-smart approaches. These types of relationships paved the way towards actual solutions of field problems of the farming community. Farmer-to-farmer learning, farmer's networks on adaptation technologies, seed and fodder banks, crowdsourcing seed, market information, and kitchen gardening are important aspects towards solutions of technical issues at the farmer's doorstep.

Those institutions are essential to create and transfer useful information and guide people by interpreting the new technologies into understanding and working on it. The institutional approaches, such as farmer field schools, can guide and facilitate farmers for the implementation of new techniques. Other approaches might include shows on radio stations popular among farming communities, which share agricultural information that is easily available, useful, useable; promoting weather-related knowledge to local rural people; foster agricultural plots exhibitions for the community and push for the exchange of ideas between farmers. The profits achieved by using sustainable management techniques for the land usually take time to evolve. Since poor farmers have no resources to access credit and markets, and they are unable to adapt to these new techniques to adopt to climate resilient agriculture, the strong institutional relationships will have to maintain agricultural markets and financing mechanisms, which are very important. The most efficient technique allows researchers, private sector investors, community members and policy makers to collectively describe issues that are planned to resolve.

13.5 WAY FORWARD

The world population is growing at a high rate and feeding this growing population is the challenge of the day, which will become even more of a test in the near future. Worldwide, 870 million people live currently in hunger, mostly in Asia and Africa. Scientists have predicted that the CC, chiefly based on GHGEs produced by human activities, will have more detrimental global effects in the future. It has now become overwhelmingly clear that the CC is going to affect primarily poor nations. These regions are also identified as CC "hotspots" with low food production/security and are prone to high climatic stresses. The effects of CC on the agriculture sector can be minimized by CC impact assessment (capacity building on CC impact assessment), CC monitoring (data pool and baseline development on food crops, agroforestry, livestock and fisheries for all stakeholders, the development of weather forecasting and early warning systems like the "Pakistan Automated Environment Monitoring Network (PAEMN)," and a cyber infrastructure for data handling and archiving). Other methods to prepare for CC include field experimentation using FACE, FATE, Controlled temperature, and CO_2 chambers, CC mitigation (natural resource management improving soil health, soil conservation, water conservation, reforestation, biochar application to improve carbon sequestration, quality seed etc. to minimize CC impacts on short to long term basis), CC adaptation (development of adaptation strategies for wheat, rice, maize, legume crops and livestock, agroforestry and fisheries management against CC in vulnerable environment, application of model evolved mitigation strategies under field condition for field testing, further screening of cropping systems, livestock, forests and fisheries production under redefined agro ecological zones).

REFERENCES

Adams RM, BH Hurd, S Lenhart and N Leary. 1998. Effects of global climate change on agriculture: An interpretative review. *Clim. Res.* 11:19–30.

Ahmad A, M Ashfaq, G Rasul, SA Wajid, T Khaliq, F Rasul, U Saeed et al. 2015. Impact of climate change on the Rice–Wheat cropping system of Pakistan (Chapter No. 7). The agricultural model inter-comparison and improvement project (AgMIP) integrated crop and economic assessments, Part 2. ICP series on climate change impact, adaptation, and mitigation—Vol. 3. In *Joint Publication with American Society of Agronomy, Crop Science Society of America, and Soil Science Society of America.* [Rosenzweig C and D Hillel (Eds.)]. Imperial College Press, London, Chapter No. 7, Page No. 219–258.

Alade OA and AO Ademola. 2013. Perceived effect of climate variation on the poultry production in OkeOgun area of Oyo state. *J. Agric. Sci.* 5:176–182.

Allison EH, WN Adger, MC Badjeck, K Brown, D Conway, NK Dulvy, A Halls, A Perry and JD Reynolds. 2005. *Effects of climate change on the sustainability of capture and enhancement fisheries important to the poor: Analysis of the vulnerability and adaptability of fisher folk living in poverty.* Fisheries Management Science Programme MRAG/DFID, Project no. R4778J, Final Technical Report, London, p. 164.

Altieri MA and CI Nicholls. 2013. The adaptation and mitigation potential of traditional agriculture in a changing climate. *Clim. Chang.* 140:1–13. DOI: 10.1007/s10584-013-0909-y

Anonymous. 2014. *Economic Survey of Pakistan.* Pakistan Bureau of Statistics, Economic Division, Islamabad, Pakistan.

Antle JM. 2011. Parsimonious technology impact assessment. *Am. J. Agric. Econ.* 93:1292–1311.

Anwar MR, DL Liu, I Macadam and G Kelly. 2013. Adapting agriculture to climate change: A review. *Theor. Appl. Climatol.* 113:225–245.

Arnell NW. 2004. Climate change and global water resources: SRES emissions and socioeconomic scenarios. *Global Environ. Change* 14:31–52.

Baylis M and AK Githeko. 2006. *The effects of climate change on infectious diseases of animals.* Report for the Foresight Project on Detection of Infectious Diseases, Department of Trade and Industry, UK Government, p. 35

Booltink HWG, BJ Van Alphen, WD Batchelor, JO Paz, JJ Stoorvogel and R Vargas. 2001. Tools for optimizing management of spatially-variable fields. *Agric. Syst.* 70:445–476.

Brown ME and CC Funk. 2008. Food security under climate change. *Science* 319:580–581.

Burrough PA and RA McDonnell. 1998. *Principles of Geographical Information Systems.* Oxford University Press, New York, p. 333.

CGRFA (Commission on Genetic Resources for Food and Agriculture). 2007. *The State of the World's Animal Genetic Resources for Food and Agriculture.* FAO, Rome, p. 523.

Challinor A, T Wheeler, C Garforth, P Crauford and A Kassam. 2007. Assessing the Vulnerability of food crop systems in Africa to climate change. *Clim. Change* 83:381–399.

Cline W. 2007. *Global Warming and Agriculture: Impact Estimates by Country.* Center for Global Development and Peterson Institute for International Economics, Washington.

Delgado C, M Rosegrant, H Steinfeld, S Ehui and C Courbois. 1999. Livestock to 2020: The next food revolution. Food, Agriculture and the Environment Discussion Paper 28.IFPRI/FAO/ILRI, Washington, DC, USA.

Ehrenfeld D. 2005. The environmental limits to globalization. *Conserv. Biol.* 19:318–326.

Falkenmark M and G Lindh. 1976. *Water for a Starving World.* West- view Press, Boulder, CO.

FAO. 2005. Global Forest Resources Assessment: Progress towards sustainable forest management. FAO Viale delle Terme di Caracalla, Rome, Italy.

FAO. 2007. Safety of fishermen. FAO Viale delle Terme di Caracalla, Rome, Italy.

FAO. 2011. FAOSTAT online database. http://faostat.fao.org/. Accessed on December 2011.

FAO. 2014. *Agriculture, Forestry and Other Land Use Emissions by Sources and Removals by Sinks.* (PDF, 89 pp., 3.5MB) Exit EPA Disclaimer Climate, Energy and Tenure Division, FAO, Rome.

FAOSTAT. 2013. The statistical database of the Food and Agriculture Organization (FAO), Rome.

Ganor E, I Osetinksy and A Stupp. 2010. Increasing trend of African dust, over 49 years, in the eastern mediterranean. *J. Geophys. Res.* 115:201–215.

Gu YF, XP Zhang, SH Tu and K Lindstr Ömc. 2009. Soil microbial biomass, crop yields, and bacterial community structure as affected by long-term fertilizer treatments under wheat-rice cropping. *Europ. J. Soil Biol.* 45:239–246.

Hansen, JW, SJ Mason, L Sun and A Tall. 2011. Review of seasonal climate forecasting for agriculture in sub-Saharan Africa. *Exp. Agric.* 47:205–217.

Harvell, CD, CE Mitchell, JR Ward, S Altizer, AP Dobson, RS Ostfeld and MD Samuel. 2002. Ecology—Climate warming and disease risks for terrestrial and marine biota. *Science* 296:2158–2162.

Hillel D and C Rosenzweig. 2010. Conclusion: Climate change and agro-ecosystems: Main findings and future research directions. In *Handbook of Climate Change and Agro-Ecosystems: Impacts, Adaptation, and Mitigation.* [Hillel D and C Rosenzweig (Eds.)]. ICP Series on Climate Change Impacts, Adaptation, and Mitigation Vol. 1. Imperial College Press, London, pp. 429–435.

Hoff H. 2011. Understanding the Nexus. *Background Paper for the Bonn 2011 Conference: The Water, Energy and Food Security Nexus.* Stockholm, Sweden: Stockholm Environment Institute (SEI).

Hussain M and S Mumtaz. 2014. Climate change and managing water crisis: Pakistan's perspective. *Rev. Environ. Health* 29:71–79.

IEA. 2010. Sustainable Production of Second-Generation Biofuels (Report). http://www.iea.org/papers/2010/second generation biofuels.pdf.

IPCC. 2001. *Climate Change*, The Scientific Basis, Cambridge University Press, Cambridge, 881pp.

IPCC. 2013. Climate change 2013: The physical science basis. In *Contribution of Working Group I to the Fifth Assessment Report of the Intergovernmental Panel on Climate Change* [Stocker, TF, D Qin, G-K Plattner, M Tignor, SK Allen, J Boschung, A Nauels, Y Xia, V Bex and PM Midgley (Eds.)]. Cambridge University Press, Cambridge, United Kingdom and New York, NY, USA, 1535 pp., doi: 10.1017/CBO9781107415324.

IPCC. 2014a. Climate change 2014: Mitigation of climate change. In *Exit EPA Disclaimer Contribution of Working Group III to the Fifth Assessment Report of the Intergovernmental Panel on Climate Change* [Edenhofer O, R Pichs-Madruga, Y Sokona, E Farahani, S Kadner, K Seyboth, A Adler, I Baum, S Brunner, P Eickemeier, B Kriemann, J Savolainen, S Schlömer, C von Stechow, T Zwickel and JC Minx (Eds.)]. Cambridge University Press, Cambridge, United Kingdom and New York, NY, USA, pp. 1–1435.

IPCC. 2014b. Agriculture, forestry and other land use (AFOLU). Chapter 11. In climate change 2014: Mitigation of climate change. In *Contribution of Working Group III to the Fifth Assessment Report of the Intergovernmental Panel on Climate Change* [Edenhofer, O, R Pichs-Madruga, Y Sokona, E Farahani, S Kadner, K Seyboth, A Adler et al. (Eds.)]. Cambridge University Press, Cambridge, United Kingdom and New York, NY, USA, 811–921.

IPCC. 2014c. Climate change 2014: Synthesis report. In *Contribution of Working Groups I, II and III to the Fifth Assessment Report of the Intergovernmental Panel on Climate Change.* (PDF, 80 pp., 4.2MB) Exit EPA Disclaimer [Core Writing Team, RK Pachauri and LA Meyer (Eds.)]. IPCC, Geneva, Switzerland, p. 151.

Kirilenko, AP and RA Sedjo. 2007. Climate change impacts on forestry. *PNAS.* 104:19697–19702.

Liaghat S and S Balasundram. 2010. A review: The role of remote sensing in precision agriculture. *Am. J. Agric. Biol. Sci.* 5:50–63.

Lobell, DB, MB Burke, C Tebaldi, MD Mastrandrea, WP Falcon and RL Naylo. 2008. Prioritizing climate change adaptation needs for food security in 2030. *Science* 319:607–610.

Mohanty M, KS Reddy, ME Probert, RC Dalal, AS Rao and NW Menzies. 2011. Modelling N mineralization from green manure and farmyard manure from a laboratory incubation study. *Ecol. Mod.* 222:719–726.

NEEDS. 2011. https://unfccc.int/files/adaptation/application/pdf/pakistanneeds.pdf.

Nelson GC, MW Rosegrant, J Koo, R Robertson, T Sulser, T Zhu, C Ringler et al. 2009. *Climate change: Impact on agriculture and costs of adaptation.* Food Policy Report 21, International Food Policy Research Institute (IFPRI), Washington DC, USA.

Njau LN. 2010. Seasonal-to-interannual climate variability in the context of development and delivery of science-based climate prediction and information services worldwide for the benefit of society. *Procedia Environ. Sci.* 1:411–420.

NRC. 1981. *Effect of Environment on Nutrient Requirements of Domestic Animals.* Subcommittee on Environmental Stress, National Research Council. National Academy Press, Washington, DC.

Olanrewaju HA, JL Purswell, SD Collier and SL Branton. 2010. Effect of ambient temperature and light intensity on growth performance and carcass characteristics of heavy broiler chickens at 56 days of age. *Int. J. Poultry Sci.* 9:720–725.

Pathak H. 2010. Mitigating greenhouse gas and nitrogen loss with improved fertilizer management in rice: Quantification and economic assessment. *Nutr. Cycling Agroecosys* 87:443–454.

Pathak H, YS Saharawat, M Gathala and JK Ladha. 2011. Impact of resource-conserving technologies on productivity and greenhouse gas emission in rice-wheat system. *Greenhouse Gas Sci. Technol.* 1:261–277.

Patt A, P Suarez and C Gwata. 2005. Effects of seasonal climate forecasts and participatory workshops among subsistence farmers in Zimbabwe. *PNAS* 102:12623–12628.

Paustian K, JM Antle, J Sheehan and EA Paul. 2006. *Agriculture's Role in Greenhouse Gas Mitigation.* Pew Center on Global Climate Change, Arlington.

Phillips J, E Makaudze and L Unganai. 2001. Current and potential use of climate forecasts for resource-poor farmers in Zimbabwe. In *Impacts of El Nino and Climate Variability in Agriculture* [Rosenzweig, C (Ed.)]. American Society of Agronomy Special Publication, Madison, Wisconsin, pp. 87–100.

Pinheiro EFM, MG Pereira, LHC Anjos and PLOA Machado. 2004. Densimetric fractionation of organic matter in soil under different tillage and vegetation cover in Paty do Alferes, State of Rio de Janeiro (Brazil). *Rev. Bras. Cienc. do Solo.* 28:731–737.

Roncoli C, C Jost, P Kirshen, M Sanon, KT Ingram, M Woodin, L Somé, F Ouattara, BF Sanfo and C Sia. 2009. From accessing to assessing forecasts: An end-to-end study of participatory climate forecast dissemination in Burkina Faso (West Africa). *Clim. Change* 92:433–460.

Rosenzweig C, JW Jones, JL Hatfield, AC Ruane, KJ Boote, P Thorburn, JM Antle et al. 2013. The agricultural model inter-comparison and improvement project (AgMIP): Protocols and pilot studies. *Agric. For. Meteorol.* 170:166–182.

Rötter R and SC van de Geijn. 1999. Climate change effects on plant growth, crop yield and livestock. *Clim. Change* 43:651–681.

Sapkota TB, ML Jat, JP Aryal, RK Jat and A Khatri-Chhetri. 2015. Climate change adaptation, greenhouse gas mitigation and economic profitability of conservation agriculture: Some examples from cereal systems of Indo-Gangetic Plains. *J. Integr. Agric.* 14:1524–1533.

SCA (Standing Committee on Agriculture). 1990. *Feeding Standards for Australian Livestock: Ruminants.* CSIRO Publications, East Melbourne, Australia.

Schnitkey G and B Sherrick. 2014. *Crop Insurance Encyclopedia of Agriculture and Food Systems*, Academic press, CA, USA, pp. 399–407.

Seo S and R Mendelsohn. 2006. The impact of climate change on livestock management in Africa: A structural Ricardian analysis. CEEPA Discussion Paper No. 23, Centre for Environmental Economics and Policy in Africa, University of Pretoria.

Six J, ET Elliott and K Paustian. 2000. Soil macro-aggregate turnover and micro-aggregate formation: A mechanism for C sequestration under no-tillage agriculture. *Soil Biol. Biochem.* 32:2099–2103.

Slingo JM, AJ Challinor, BJ Hoskins and TR Wheeler. 2005. Food crops in a changing climate. *Philos. Trans. Roy. Soc.* 360:1983–1989.

Stokes, CJ and SM Howden. 2010. *Adapting Agriculture to Climate Change: Preparing Australian Agriculture, Forestry and Fisheries for the Future.* CSIRO Publishing, Collingwood, p. 296.

Vogel C. 2000. Useable science: An assessment of long-term seasonal forecasts amongst farmers in rural areas of South Africa. *S. Afr. Geogr. J.* 82:107–116.

Webber H, T Gaiser and F Ewert. 2014. What role can crop models play in supporting climate change adaptation decisions to enhance food security in Sub-Saharan Africa? *Agr. Syst.* 127:161–177.

Wheeler TR, PQ Craufurd, RH Ellis, JR Porter and PV Prasad. 2000. Temperature variability and the annual yield of crops. *Agric. Ecosyst. Environ.* 82:159–167.

World Water Assessment Programme. 2014. The United Nations World Water Development Report 2014: Water and Energy. Paris, UNESCO.

Xu MG, YL Lou, XL Sun, W Wang, M Baniyamuddin and K Zhao. 2011. Soil organic carbon active fractions as early indicators for total carbon change under straw incorporation. *Biol. Fertil. Soils.* 47:745–752.

Xu YM, H Liu, XH Wang, MG Xu, WJ Zhang and GY Jiang. 2014. Changes in organic carbon index of grey desert soil in Northwest China after long-term fertilization. *J. Integrative Agri.* 13:554–561.

Zhang LM, MG Xu, YL Lou, XL Wang, S Qin, TM Jiang and ZF Li. 2014. Changes in yellow paddy soil organic carbon fractions under long-term fertilization. *Scientia Agricultura Sinica* 47:3817–3825.

Zhang WJ, MG Xu, BR Wang and XJ Wang. 2009. Soil organic carbon, total nitrogen and grain yields under long-term fertilizations in the upland red soil of southern China. *Nutr. Cycl. Agroecosys.* 84:59–69.

14 Treatment and Management of Low Quality Water for Irrigation

Ghulam Murtaza, Muhammad Zia-ur-Rehman, Muhammad Aamer Maqsood, and Abdul Ghafoor

CONTENTS

14.1 RATIONALE

The Indus Plains are located in arid and semiarid climatic zones of Pakistan where monsoon rains (July to September) are erratic. These rains frequently lead to heavy flooding which is now considered as a major impact of climate change and global warming. Extreme examples of such climate changes include flood events during the years of 2010, 2011, and 2014, which claimed huge number of human and animals lives, besides the billion-dollar loss of infrastructure, soil quality, and crops. The major source of water for agriculture, industry, and domestic usage is the Indus Basin Irrigation System (IBIS) which is the world's single largest irrigation system covering a total area of 17.2 Mha. Farmers in arid and semiarid regions of the Indus Plains have also been facing challenges in achieving sustainable irrigation for many years. The major factor responsible for the demise of ancient civilizations is crop failures due to salinization of agricultural soils (Qadir et al. 2009, 2014; Zhao et al. 2013).

About 70%–80% of the total river flow from snow melt and monsoonal precipitation happens in summer. The climate change not only limits the water availability but also damages crops and hence proves a financial constraint for the resource-poor communities. The soil salinity is increasing both in extent and intensity, which is an alarming threat for the future. Such losses could be further aggravated due to low precipitation and high temperatures, even if continue at the current rate. Salinity and drainage problems are persisting in many key production areas of the world due to lack of comprehensive regional efforts. The most affected areas include the southwestern United States, Australia, Iran, and major portions of the Indo-Gangetic Plain (Ritzema et al. 2008; Singh et al. 2010; Emadodin et al. 2012; Wichelns and Oster 2014). Insufficient leaching of salts, due to low annual precipitation along with higher temperatures, aggravates the salt stress in already salt-affected regions (Sommer et al. 2013). A drastic decrease in water availability has been observed in Pakistan, which was 5300 m^3/year/person in the 1950s and has been decreased since to 1100 m^3/person/year in 2014. More worrisome is that predictions are less than 1000 m^3 per capita by the year 2025. An amount of 1000 m^3/person/year is the threshold level. However, at global level it was 16,800 and 6,000 m^3/person/year in 1950 and 2014, respectively (IWMI 2015). The World Bank report included Pakistan in the list of 17 countries predicted to suffer from severe shortages of water by the year 2025. The development of additional surface water storage for drought periods is immensely needed since about 41938.3 to 45638.8 million $m^3$of surface water is being discharged into the Arabian Sea. The political and personal interests of all legislators, policy makers, and technocrats should take a back seat over national interests in order to constitute strict policies to limit harm to the biosphere equilibrium.

The increased cropping intensity and preferential usage of freshwater for nonagricultural sectors compelled the farmers to use brackish groundwater as a supplement to freshwater (Shah et al. 2007; Grattan et al. 2015). Presently, in Pakistan, about 60%–70% undergroundwater is hazardous (Ghafoor et al. 2004) being pumped by more than 1.07×10^6 tube wells (pumping 9.05×10^6 ha–m groundwater) (Anonymous 2014). The area irrigated by different sources (canal, tube well, wells, and wastewater) in Punjab Province of Pakistan is given in Table 14.1. The irrigation with poor quality water may have several detrimental effects, for example, soil salinization and/or sodication, high bulk density, low infiltration rate, nutrient imbalance, and specific ion toxicity in plants resulting in poor growth and ultimately loss in economic yields of crops. Lowering down of water table causes subsoil drying which is another future concern (Ghafoor et al. 2004; Wichelns and Qadir 2015). This issue must be addressed wisely by enforcing strict prohibiting laws against overpumping of groundwater and recharging groundwater by rainwater harvesting.

TABLE 14.1
Area Irrigated (Mha) by Different Sources in Punjab, Pakistan

Territory/ Year	Canal	Wells	Tube Wells	Others	Total
Pakistan 1959–60	8.73	0.78	0.15	0.67	10.33
Punjab 1980–1981	4.58	0.60	3.04	2.18	10.40
Punjab 1990–1991	4.31	0.51	4.82	2.99	12.63
Punjab 2000–01	4.13	0.51	5.66	3.22	13.85
Punjab 2013–14	3.56	0.51	11.12	4.00	19.19

Source: Anonymous 2014. *Agricultural Statistics of Pakistan 2013–14*. Food, Agri. and Livestock Div., Ministry of Food, Agri. and Livestock, Govt. of Pakistan, Islamabad, Pakistan.

Due to rapid migration to urban areas, large volumes of freshwater are being diverted to industrial, commercial, and domestic sectors where in turn wastewater is generated (Lazarova and Bahri 2005; Tchobanoglous et al. 2007). This enormous amount of wastewater generated is used for irrigation by farmers of urban and periurban areas due to shortages of canal water. Also farmers are using raw sewage water intentionally because of its nutrient enrichment and cheaper source of water (Keraita and Drechsel 2004; Scott et al. 2004; Murtaza et al. 2010) as it is also considered a viable option for disposal.

The groundwater, floodwater, and mixed raw city effluents may pose a serious threat owing to high electrical conductivity (EC), sodium adsorption ratio (SAR), and/or residual sodium carbonate (RSC), heavy metals (Hussain et al. 2006; Murtaza et al. 2010; Qadir et al. 2010), high magnesium to calcium ratio (Qadir et al. 2009), diseases, pathogens (Trang et al. 2007), detergents, azo-dyes and pesticide residues (Liang et al. 2013), etc. However, management and use of poor quality waters depend upon a number of factors including crop type, plant growth stage, physicochemical properties of soils, reactions at the soil–water interface, climate, farmer's motivation, quality of drainage water, and socioeconomic requirements and conditions.

This chapter addresses solution-based efforts to treat and/or manage low quality water which have been critically evaluated to reshape the future policies for safe and sustainable use of poor quality water and to save the biosphere equilibrium. Moreover, it will highlight the economic feasibility of using different amendments including gypsum, acids and/or acid formers, and organic materials for brackish water treatment while causing the least disturbance to the biosphere.

14.2 BRACKISH WATER

Brackish water contains high soluble salts (high EC) and undesirable ionic ratios (high SAR and/or RSC). Low supply of canal water under the threats of climate change, shrinking capacity of water reservoirs, and competition from nonagricultural demands along with intensified cropping (Murtaza et al. 2009) have increased pumping of groundwater over the years. The number of tube wells has increased in Pakistan from 3000 to a million during the last 60 years (Anonymous 2015). In Punjab alone, private tube wells increased from nearly 10^4 in 1960 to about 5×10^5 in 2000 and around 8×10^5 in 2015. Consequently, soils are deteriorating due to buildup of different types of salts in groundwater (Ashfaq et al. 2009). Unfortunately, 60%–70% of the pumped groundwater is brackish due to salinity and/or sodicity. These groundwaters are supplemented with different types of salts, which change the quality of soil accordingly (Masood and Gohar 2000). Long-term use of brackish waters for agriculture will increase the soil salinity and sodicity, decline the crop yield and its quality/shelf life, and create environment issues (Ghafoor et al. 2001a; Oster and Grattan 2002).

The Salinity Control and Reclamation Project (SCARP) was launched by Water and Power Development Authority (WAPDA) in the early 1960s to control waterlogging with the installation of tube wells and the construction of surface and tile drains. One of the major reasons for the failure of this project was the pumping of brackish water by deep tube wells (of 3–5 cusecs capacity) and irrigation of crops without testing its quality along with the inappropriate mixing ratio of saline/sodic tube well waters with canal waters. Huge expenditures were incurred for the execution of this project. About 20,000 large capacity tube wells were installed, tile drainage systems were provided over areas covering 14,000 ha, and more than 10,845 km of surface drains were constructed. As a result of these drainage efforts, farmers pumped more and more groundwater to increase cropping intensity. Consequently, waterlogging in Punjab was alleviated but this created salinity problems, particularly in those areas where tube well waters were brackish. Therefore, a strong coordination is required among irrigation, agriculture, and social/extension departments to improve the sustainability of agriculture in the future.

Subsoil drying is another future concern, which must be addressed wisely, caused by drawdown of the water table (Konikow and Kendy 2005). Massive pumping of groundwater is a key factor responsible for low quality of groundwater and net negative water balance may be due to insufficient water recharge. As per report of the Chairman of the WAPDA, subsoil water has lowered to alarming extent in most of cities of Pakistan. Presently, WASA is pumping water from Lahore city at a depth of about 229 meters and subsoil water is present only at a depth of 305 meters (Mahmood, 2016).

The scarcity of quality irrigation water in Pakistan as well as in other parts of the world has attracted international attention. Additionally, some private organizations are marketing their products to treat the brackish water without proper testing and validation under local agro-ecological conditions, as discussed in a later section. According to Murtaza et al. (2009), with proper management practices, poor quality water could be used for irrigation without a significant loss to soil and plant health.

14.3 WATER QUALITY CRITERIA FOR IRRIGATION

Water quality means suitability of irrigation water to meet the needs of farmers for irrigation purposes under a set of agroclimatic and socioeconomic conditions. The quality of irrigation water is a function of type, nature, and total quantity of dissolved salts. Salts in irrigation water are present in variable concentrations and are mainly derived from the dissolution or weathering of rocks and soil minerals. Irrigation water quality has a marked effect on soil physical and chemical properties, irrigation system performance, and crop yields. Thus, it is necessary to understand the criteria for irrigation water quality. Bauder et al. (2011) from Colorado State University used five different categories in describing quality of irrigation water with reference to crop production and soil health: (i) Salinization (Total solute contents), (ii) sodication (relative proportion of sodium), (iii) pH (acidic or basic), (iv) alkalinity (carbonate and bicarbonate), and (v) specific ion interaction. Another guideline was proposed by University of California Committee of Consultant's (1974), namely the Water Quality Guidelines, which were prepared in collaboration with the U.S. Salinity Laboratory staff. The guidelines are used in irrigated agriculture for evaluation of surface water, groundwater, drainage water, sewage water, and wastewater.

In Pakistan and India, numerous scientists proposed various water quality guidelines (Yunus 1977; Hussain 1978; Sheikh 1989; Gupta 1990) but proper support of experimental data is lacking. Generally, the guidelines developed by the US Salinity Lab. Staff (1954); Ayers and Westcot (1985) are the ones being followed to predict the effects of water quality on soils and crops' health more accurately. Extensive experimentation is needed to develop the water quality guidelines giving due importance to physicochemical properties of soils for specific agroclimatic regions. The comprehensive classification of low quality water is presented in Table 14.2, which needs specific management practices. This classification can serve the purpose more precisely and provide a better understanding at regional levels.

14.4 TREATMENT OPTIONS

14.4.1 Salinity of Waters

14.4.1.1 Dilution

Salinity of irrigation water (EC_{iw}) is an index of total soluble salts in a substrate. The primary effect of salinity is the unavailability of water due to increased competition of the plant with ions for water in the soil solution. The only treatment option to bring a decrease in total soluble salts is dilution with low salt water (Ghafoor et al. 2001a). Use of any chemical amendment such as gypsum, acid, or acid formers will elevate the salt contents in irrigation water making it even less suitable for agriculture. However, the use of high EC_{iw} water showed better results during initial reclamation phase of saline-sodic/sodic soils because higher electrolytes concentration improves the hydraulic properties of soil (Shainberg and Letey 1984; Murtaza et al. 2009; Ghafoor et al. 2012; Cucci et al. 2015). Addition of organic matter is also a good option to counter or delay the appearance of adverse effects of high EC_{iw} on soils and crops (Ghafoor et al. 2012; Diacono and Montemurro 2015).

TABLE 14.2
Grouping of Groundwater for Quality Parameters (Proposed by the Authors)

	Water Quality Parameter		
Water Quality Class	EC_{iw} (dS m^{-1})	SAR_{iw}	RSC ($mmol_c$ L^{-1})
A. Suitable	<1.5	<10	<2.5
B. Saline			
Marginally saline	1.5–3.0	<10	<2.5
Saline	≥3.0	<10	<2.5
High-SAR saline	≥3.0	≥10	<2.5
C. Saline-sodic			
Marginally saline-sodic	1.5–3.0	10–15	2.5–4.0
Moderately saline-sodic	1.5–4.0	10–15	≥4.0
Highly saline-sodic	≥4.0	≥15	≥4.0
D. Sodic			
Marginally sodic	0.5–3.0	10–15	2.5–4.0
Moderately sodic	1.5–3.0	≥15	≥4.0
Highly sodic	<3.0	≥15	≥4.0

Source: Ghafoor et al. 2004. *Salt-affected Soils: Principles of Management.* Allied Book Centre, Lahore, Pakistan.

14.4.1.2 Desalination

Desalination is a procedure to remove excessive salts from water to make it suitable for irrigation. Installation of water desalting units working on the distillation principle can play a vital role for the sustainability of irrigated agriculture. This technology is limited to circumstances where low-cost energy is available because a large amount of input energy is required during the desalination process.

Another effective technique used to extract most of the soluble salts from water is reverse osmosis. The brackish inflow water is allowed to pass through a semipermeable membrane applying pressure. The concentrated brine is then collected downstream for either safe disposal or treatment to extract commercial-grade salts. This technology is well established and fabricated at economical costs in the United States, Saudi Arabia, China, the ex-USSR, and Morocco under government management (Burn et al. 2015). But this technology appears difficult and expensive to implement in Pakistan presently due to energy crises.

Nanofiltration is another technique being used for water desalination which may become a promising technique due to the high quality water it produces (Al-Amoudi 2016). It removes regulated and unregulated organic compounds, and organic and inorganic ions. Waters containing low total dissolved solids (TDS) are used for softening of polyvalent cations and disinfection removal by producing precursors such as synthetic organic matter. Such unit is working in Saudi Arabia which treats 8.6 ML/day; it decreases hardness from 7500 to 220 mg/L, and treats 80%–95% of divalent cations and reduces TDS from 45,460 to 28,260 mg L^{-1} (Tanninen et al. 2005). This technique is very effective but very costly for irrigation of crops at field level.

Another membrane separation technique is known as electrodialysis in which ion-exchange membranes are used for the removal of ions. Waters containing dissolved salts are passed through electric fields through ion-selective membranes in multicompartment cells (Strathmann 2004; Taylor and Bolto 2011). Selective monovalent membranes are now available to facilitate the monovalent elimination of such ions as K^+ and Na^+, which results in lowering SAR—an important parameter for irrigation water (Burn et al. 2015). However, this technique may be uneconomical for use on a large scale, especially in energy-deficient countries. Therefore, this desalination method has never been practiced in Pakistan and is as such premature to comment on.

14.4.2 Sodicity

14.4.2.1 Sodium Adsorption Ratio (SAR)

This parameter can be calculated using ionic concentrations of soluble cations (relative proportion of Na^+ to Ca^{2+} and Mg^{2+}) in water or soil solution (US Salinity Lab. Staff 1954) to evaluate the sodicity hazard of waters and soils.

14.4.2.2 Dilution

Like EC_{iw}, mixing with low SAR or high $Ca^{2+} + Mg^{2+}$ water will decrease SAR of water and it will be reduced by the square root times of the dilution factor (DF) when rainwater is used for the dilution purposes, that is,

$$\text{Blended water SAR} = \text{Original SAR}/(\text{DF})^{1/2}$$

14.4.2.3 Addition of Calcium Source

Decrease in Na^+ concentration in irrigation water is not economically viable for agriculture except when diluting with low Na^+ water. Calcium concentration in irrigation water can be increased by lining the water courses with gypsum or lime stones as baffled structures/geometry to maximize the water–stone contact as well as to enhance the flow velocity of water. Some success stories of improving the water quality with gypsum lining in water courses are presented in Table 14.3. This strategy is economical and ecofriendly besides some problems associated with rodents, weeding and continuous cleaning of water courses. Also, gypsum stone dissolution decreases with time due to lime ($CaCO_3$) coating the gypsum stone surfaces, especially with high RSC. It may become cost-intensive (Malik et al. 1992; Oster 1994) when flushing with common commercial acids becomes required to maintain the dissolution rate of the gypsum stones to sustain the effectiveness of water treatment.

14.4.3 Residual Sodium Carbonate (RSC)

This parameter also indicates the sodicity hazard to be created by irrigation water. The calculation of RSC depends on the theory of quantitative precipitation of Ca^{2+}, CO_3^{2-}, and HCO_3^- (Eaton 1950)

TABLE 14.3
Water Amelioration through Gypsum Stone Lining in Water Courses

Unamended Water			Amended Water			
EC (dS m^{-1})	SAR	RSC (mmol$_c$ L^{-1})	EC (dS m^{-1})	SAR	RSC (mmol$_c$ L^{-1})	Source
3.6	21.0	11.5	4.0	15.8	6.0	Qureshi et al. (1975)
3.5	19.5	12.9	3.9	12.0	6.5	Qureshi et al. (1977)
1.7	9.3	7.1	2.1	6.0	4.0	Hussain et al. (1986)
1.2	14.4	5.0	1.6	6.8	0	Chaudhry et al. (1984)
1.8	9.8	7.1	2.1	8.7	4.6	Ghafoor et al. (1987)
2.9	31.2	11.4	3.4	18.2	6.3	Choudhary et al. (2004)
2.01	8.88	8.5	2.9	6.8	6.2	Yaduvanshi and Swarup (2005)
2.7	22.03	7.0	3.1	16.5	3.4	Izhar-ul-Haq et al. (2007)

in soils upon irrigation and this is not necessarily true all the time. Rate and quantity of precipitation of salt is governed by those ions present in lowest quantity. Upon dilution, the freshly precipitated $CaCO_3$ is dissolved and the reverse process (reprecipitation) may take place upon increase in soil solution concentration by evapotranspiration.

14.4.3.1 Addition of Calcium Source

This process may be obtained via lining of water channels using gypsum stones as irregular structures or in any geometric structure to promote water flow as well as to enhance the dissolution by increasing water–stone contact. The calculated quantity of gypsum powder required to decrease hazards of RSC (or to mitigate the negative impacts produced by high RSC of water) may possibly be applied in soil before plantation of the planned crops.

14.4.3.2 Neutralization of CO_3^{2-} and HCO_3^-

This treatment may be achieved by mineral acids (H_2SO_4, HCl, HNO_3) or acid precursors such as sulfur (S), calcium polysulfide (CaS_5), and iron sulfide (FeS) but the choice is highly site-specific because this process will increase EC_{iw} at the same time.

14.4.3.3 Dilution

The practice of mixing low RSC water (good quality water) with water having high RSC, or water containing relatively high $Ca^{2+} + Mg^{2+}$, will effectively decrease the negative impacts of RSC and the expected benefits will depend upon the degree of dilution.

For lowering the RSC, application of gypsum is the most reliable and cost-effective. Similarly, acids could serve the same purpose but are about 5–8 times more costly than gypsum (Ghafoor et al. 2004). The superlative economical and feasible solution for lowering the high SAR and/or RSC of waters is soil-application of gypsum before planting each crop, while knowing the water requirements of crops and the intensity of the SAR and/or RSC of water and soil. The combined usage of brackish water and canal water is not a conceivable approach in our conditions as supplies of canal water are not controlled by the farmers.

The effectiveness of sulfurous acid generator (SAG) and alternate amendments were compared during rice growth by Zia et al. (2006). Only small reductions in RSC (5.4 to 3.6 $mmol_c\ L^{-1}$) and pH (7.5 to 7.2) occurred in SAG treated saline-sodic tube well water. The SAG treatment had little effects on EC or SAR of water (Table 14.4). Gypsum application for the treatment of saline-sodic water was about six times cheaper compared to SAG and sulfuric acid treatments. It was found that soil-applied gypsum remained economical to ameliorate the sodic hazard of tube well water and rice production (Ghafoor et al. 2004).

Some multilocation studies conducted by the authors in the Indus Plains of Pakistan for the safe use of brackish water and for reclamation of salt-affected soils are presented in Table 14.5.

14.4.4 High Concentration of Mg^{2+}

Irrigation water in some areas may have high Mg^{2+} with Mg^{2+} to Ca^{2+} ratios > 1 (Qadir et al. 2009). Application of this water for reclamation and irrigation purposes may give rise to exchangeable Mg^{2+} in soils (Ghafoor et al. 2004; Karajeh et al. 2004). At high concentration, exchangeable Mg^{2+} may cause degradation of soils due to its adverse effects on the physical properties of soil. Zhang and Norton (2002) leached two soils in packed soil columns with successive concentrations of solutions containing either Ca^{2+} or Mg^{2+}. They concluded that Mg^{2+} reduced saturated hydraulic conductivity (K_s) through increased clay swelling, disaggregation, and clay dispersion compared to that with Ca^{2+}. A recent information (Smith et al. 2015) supports the general conclusion that the harmful effects of four common cations on the hydraulic properties of soil were in the relative order of $Na^+ > K^+ > Mg^{2+} > Ca^{2+}$. In improving the soil structure and water permeability, Mg^{2+} was found to be less effective than

TABLE 14.4
Effects of Sulfurous Acid Generator (SAG) on Tube Well Water Quality

Treatment	pH	EC ($dS\ m^{-1}$)	SAR	RSC ($mmol_c\ L^{-1}$)	Source
Untreated tube well water	8.5	2.7	13.2	2.7	Kahlown et al. (2000)
SAG treated water	2.8–4.0	2.3	12.1	Nil	
Untreated tube well water	8.4	1.9	14.2	2.9	Kahlown and Gill (2008)
SAG treated water	3.5	2.4	10.1	Nil	
Mixed water (one part treated and four parts untreated water)	7	2	11.5	0.6	
Untreated tube well water	7.5	3.3	16.6	5.4	Zia et al. (2006)
All SAG treated water	2.5	7.4	17.0	0.1	
SAG treated water (mixed)	7.2	3.7	16.8	3.6	

Ca^{2+} due to its greater hydration energy and larger hydration shell. It is a general observation of authors that with increasing EC of groundwater of Punjab in particular, the concentration of Mg^{2+} relative to Ca^{2+} increases. Mostly, Ca^{2+} and Mg^{2+} are determined collectively instead of individually. It is highly recommended that Ca^{2+} and Mg^{2+} should be determined separately from groundwater to quantify land degradation due to Mg^{2+}.

14.4.5 EC_{iw} to SAR_{iw} Ratio

At a given SAR_{iw} or SAR_{ss}, low EC_{iw} or EC_e decreases the infiltration rate of soil by increasing the diffused double layer thickness, DDL (Ayers and Westcott 1985; Ghafoor et al. 2004). Water with high EC_{iw} and low SAR_{iw} was found better for salt-affected and normal soils because of the positive effects of electrolytes on water infiltration and soil hydraulic conductivity, while the reverse was true for low EC_{iw} with high SAR_{iw} (Chinchmalatpure et al. 2014). Therefore, at the beginning of

TABLE 14.5
Irrigation and Soil Management Strategies for Using Saline-Sodic Tube Well Waters

Plant/Crop Rotation and Amendment	Soil Texture[a]	Duration (Year)	Water (EC; SAR; RSC)[b]	Decrease/Increase over Initial (%) EC_e[b]	SAR[b]	Source
Rice-wheat, cyclic use of water	SCL	3	3.32; 16.29; 5.25	−30	−50	(1)
Cotton-wheat, cyclic use + Gypsum	SL	3	2.9–3.4; 12.0–19.4; 4.6–10	+15	+7	(2)
Rice-Wheat, gypsum	SCL	3	1.78; 11.80; 1.90	−20	−95	(3)
Gyp + green manuring every year	SCL	3	3.94; 20.10; 8.98	−57	−52	(4)
Gyp (100% SGR) + green manuring	SiL	3	2.7; 8.0; 1.3	−70	−66	(5)
Sesbania-wheat (gypsum @ water GR to lower SAR at 10)	SL	2	3.94; 18.20	−35	−49	(6)

Source: (1) Qadir et al. 2001; (2) Murtaza et al. 2006; (3) Murtaza et al. 2009; (4) Ghafoor et al. 2011; (5) Ghafoor et al. 2012; (6) Murtaza et al. 2015.

[a] SCL, Sandy clay loam; SiL, Silty loam; SL, Sandy loam

[b] EC ($dS\ m^{-1}$); SAR ($(mmol\ L^{-1})^{1/2}$); RSC ($mmol_c\ L^{-1}$).

saline-sodic and sodic soils reclamation, waters containing high EC:SAR ratio were found equally effective (Ghafoor et al. 2001b); however, as soil reclamation progresses, one has to switch to better quality water having low EC_{iw} and low SAR_{iw} (Khandewal and Lal 1991; Mandal et al. 2008; Murtaza et al. 2009). It has been found that considerably greater concentrations of the electrolyte are required to cause flocculation of soil clays than what is needed for the dispersion of soil clays (Quirk 2001). It is important to consider this quality parameter for sustainable management of brackish water in order to reclaim the salt-affected soil as well as to maintain a leaching fraction. From different studies conducted in the Fourth Drainage Project Area (FDPA), Faisalabad, it was concluded that brackish water could be used safely at the beginning of reclamation programs (Ghafoor et al. 2011, 2012; Murtaza et al. 2009). However, this parameter is not given a great deal of importance at present and must be considered for future guidelines on water quality for irrigation purposes.

14.4.6 Infiltration Problems

Infiltration rate refers to the rate at which water enters into a soil from the surface. An infiltration rate (IR) greater than 12 mm h^{-1} is considered high while at 3 mm h^{-1} is considered low (US. Salinity Lab. Staff 1954). Apart from water quality, many other factors like soil physical characteristics (soil texture, type of clay mineral, pore-size, soil structure, and its water contents) and chemical characteristics (exchangeable cations) also affect infiltration rates. The IR generally decreases with either decreasing salinity (EC) or increasing SAR and increases with increasing EC (Oster 1994; Murtaza et al. 2006). Therefore, EC and SAR must be considered for appropriate assessment of the ultimate effect of water on its IR.

14.4.7 Specific Ion Effects

When concentration of an ion relative to the total ion concentration increases beyond a certain limit, then it could cause toxicity or deficiency of other ions and is referred as a specific ion effect. Ions like Cl^-, B (H_3BO_3, $H_2BO_3^-$) and Na^+ can cause specific toxicities to plants when these are higher in concentrations in plant root zones (Grattan et al. 2015). Sodium concentrations above 60% of the total cations in water or soil solution induce Ca^{2+} and/or K^+ deficiency and scorching of leaf tips in plants in addition to deteriorating the soil chemical and physical properties (Nishanthiny et al. 2010). The concentrations of Cl^- and SO_4^{2-} also show a relationship with each other. The water is considered more hazardous if the ratio of $Cl^-:SO_4^{2-}$ becomes greater than 1:3. The SO_4^{2-} ions are considered more dangerous for the plant's roots as they perturb internal metabolic processes of the plants. About 5–10 $mmol_c\ L^{-1}$ Cl^- becomes detrimental to sensitive plants. However, there is a relatively wide range of sensitivities among different plants to Na^+ and Cl^- effects, which are more prominent under sprinkler irrigation. As stated above, generally waters having high EC contain more Mg^{2+} than Ca^{2+}. If Mg^{2+} is more than Ca^{2+}, it adversely affects plants as well as soil health. Similarly, increasing Mg^{2+} over Ca^{2+} in irrigation water affects plant nutrient availability and uptake. For most of the soils and crops, a $Ca^{2+}:Mg^{2+}$ ratio $\geq$ 1.0 is considered safe. As compared to other plants, grasses are more sensitive to Mg^{2+} since they acquire high Mg^{2+} in their living tissues. From previous studies, it was also noted that high Mg^{2+} water is comparatively more detrimental to rice yields as compared to cotton and wheat crops (Ghafoor et al. 1997; Huber and Jones 2013). This aspect requires further research since some groundwaters have more Mg^{2+} than Ca^{2+} in arid regions (Ghafoor et al. 2004). However, the productivity of Mg^{2+} dominated soils can be improved by applying an adequate amount of Ca^{2+} (in the form of lime or gypsum) to such soils, which will increase the levels of Ca^{2+} on the cation exchange complex (CEC) to overcome the unfavorable impacts of Mg^{2+} (Vyshpolsky et al. 2008).

Besides some of the discussed technologies for improvement of brackish water, a number of other untested technologies are also marketed to farmers. These technologies include EM/BM technology from Japan, SAG by Sweet Water Solution from the United States, Electro-Magnetic Membrane technology from Germany, and RISTECH Technology from UK. These technologies claim their novelty in desalination of brackish water, changing the soil texture (a static property for more than 3 years) over a period of 2–3 years, converting sodium into nitrogen (a nonscientific claim), and claiming bacteria in these recipes eat salts as well as sodium present in soil or water. Unfortunately, none of these technologies have been proven by any research institutes or any university farms. In spite of this fact, these are just making money by fooling farmers on a large scale. Immediate attention is required at government level to stop their illegal or unauthorized business and save hard earned monies of farmers from being wasted on these ineffective technologies.

14.5 WASTE AND GREY WATER

Wastewater is being increasingly used in the agricultural sector to cope with the shortage of freshwater resources. A growing demand for water to produce food, supply industries, and support human populations has led to competition for freshwater supplies.

Use of wastewater has been seen as an attractive and viable option to meet the growing needs for water. The agriculture sector is the largest user of surface water, groundwater, and particularly wastewater globally accounting for about 70% of water use. Irrigation with wastewater can help alleviate strain on water resources by providing a reliable and year-round availability of water with many nutrients for crop growth (Dickin et al. 2016).

This is very important in regions where climate change is expected to exacerbate water stress and increase precipitation variability (Hanjra et al. 2012). It is a common practice in developing countries such as Pakistan to discharge domestic and industrial sewage directly into a sewer system and have no separate arrangement for their disposal (Qadir et al. 2010). Unfortunatcly, this wastewater is not properly treated because of nonfunctioning and mainly nonavailability of treatment plants.

Generally, sewage of these drains have high EC, SAR, RSC, and high concentrations of numerous heavy metals, organics and associated pathogens which need site-specific management and remediation strategies. The use and most importantly the recycling of wastewater for irrigation is an attractive option for farmers because of its continuous supply and nutrient content (Ensink et al. 2004; Murtaza et al. 2010). It can have positive effects on agriculture and income of communities, although may adversely affect the soil physical and chemical properties, besides contaminating the human food chain (Qadir et al. 2010; Hanjra et al. 2012; Vincent 2014; Rehman et al. 2015).

Recent estimates depict that the total volume of wastewater generated in Pakistan is about 962,335 million gallons (4.369×10^9 m^3/y), of which 674,009 million gallons (3.060×10^9 m^3/y; and 5.54×10^9 m^3/y) is generated by municipalities and 288,326 million gallons (1.309×10^9 m^3/y) from industries (Table 14.6). Wastewater is not subjected to any treatment in Pakistan and there is no biological treatment process in cities except for Karachi and Islamabad that treat only a small proportion of wastewater before its disposal. It has been estimated that the amount of treated wastewater is not more than 1% (Murtaza and Zia 2012). The wastewater after different treatment processes usually flows into open drains and there are no provisions or restrictions for its reuse for agriculture.

In 2008, the agriculture sector used about 94% of the total water withdrawal. While municipal and industrial water withdrawal shares were 5% and 1%, respectively (Government of Pakistan 2008).

In Punjab, the amount of wastewater disposed of after its treatment is only 0.63 million cubic meters per day (out of 15.64 million cubic meters per day) of wastewater produced by an urban population of 27.476 million. The annual wastewater production from some major cities of Pakistan is presented in Table 14.7. It is interesting that only 20% of sewage water is being treated at the primary level in Faisalabad by the domestic wastewater treatment plant Uchkera.

TABLE 14.6
Sector-Wise Wastewater Production in Pakistan during the Year 2010

Sr. No.	Source	Volume $10^6\ m^3\ y^{-1}$	Volume Percent (%)
1	Industry	395	6
2	Commercial	266	5
3	Urban residential	1,628	25
4	Rural residential	3,059	48
5	Agriculture	1,036	16
Total		6,414	100

Source: FAO Report—UNW-AIS. Individual's capacity development on the safe use of wastewater in agriculture in Pakistan, Available at http://www.ais.unwater.org/ais/pluginfile.php/232/mod_page/content/124/pakistan_murtaza_finalcountryreport2012.pdf

The treatment of wastewater at the source is highly recommended. Different technologies like chemical and plant-based are in use for the treatment of wastewater (Murtaza et al. 2014; Rehman et al. 2015).

Use of the low cost agrowaste derived biosorbents is an efficient and emerging solution for wastewater treatment which reuses it for the irrigation of crops, trees, forests, and greenbelts (Abdolali et al. 2014; Niazi et al. 2016).

TABLE 14.7
Wastewater Produced Annually from Major Cities of Pakistan

City	Urban Population (million)	Wastewater Generation Population Equivalent per Day (1 ped)	Wastewater Discharge (cumecs)	Wastewater Treatment (%)
Lahore	6.748	231	3304	1
Faisalabad	2.830	180	1278	20
Gujranwala	2.148	180	312	Nil
Multan	1.623	180	235	Nil
Rawalpindi	2.318	180	171	Nil
Sargodha	0.876	145	99	Nil
Sialkot	0.855	145	92	Nil
Gujarat	0.676	145	42	Nil
Sheikhupura	0.827	145	416	Nil
Jhang	0.782	145	21	Nil
Peshawar	1.439	180	228	N/A
Karachi	13.205	411	5400	2
Quetta	0.896	145	95	N/A
Hyderabad	1.578	180	218	N/A
Islamabad	0.689	135	77	20

Source: Japan International Cooperation Agency 2010. Preparatory study on Lahore water supply, sewerage and drainage improvement project in Islamic Republic of Pakistan; Pakistan Water Supply and Sanitation 2013. World Bank. Urban Development. Annual Plan 2013–14.

Unfortunately, in developing countries—including Pakistan—there are weak national policies and poor implementation of these concerning the sustainable use of wastewater. Awareness should be raised among different sections of the society including the policy makers, general public, organizations, industrialists, and farmers through a well-coordinated program aimed at preserving the future of the nation.

14.6 CONCLUSIONS AND RECOMMENDATIONS

In order to minimize the impact of poor quality tube well water, immediate preventive measures are required to avoid soil salination as well as sodication. The nonjudicious use of poor quality water can deteriorate soil health, ultimately affecting all living organisms, including human beings. Thus, it is important to reclaim the saline and/or sodic soils and to adopt the appropriate measures that can minimize the different hazards associated with use of low quality irrigation waters. Use of gypsum along with green manure or farm manure before each crop is the most economical solution for high SAR and/or RSC waters, considering the delta of crop water and the severity of the SAR and/or RSC problem. The combined application of brackish and canal waters is not possible under the present conditions, as canal water is not at the disposal of each farmer. While cyclic (alternate irrigation with brackish and canal water) irrigation is already in practice but it is recommended that farmers not use brackish water for irrigation at critical stages of crop growth. It is wise to use poor quality water on poor soils and good water on good soils.

It must be made compulsory and legal for the owners of tube wells to get site-specific recommendations after testing of tube well water by soil and water experts. For the sustainability of irrigated agriculture and for safety of the environment, a doorstep supply of gypsum must be ensured on credit.

The judicious use of water at regional and national levels should be encouraged. The recently developed idea of virtual water should be popularized and brought into action by including virtual water information on manufacturing products and agricultural goods before marketing to consumers, along with raising awareness through various campaigns aimed at consumers as well.

The training of government officials—especially from the field of agriculture extension—and farmers should be implemented on a regular basis. The sedimentation of dams, which cause a gradual decrease in the storage capacity of water reservoirs, is predicted to cause a loss of 33% in overall storage capacity until 2020. Therefore, developing and extending the water storage capacity and desilting the existing dams through physical removal, as well as forestation in catchments areas, will be important steps to take.

City wastewater poses serious threats due to toxic heavy metals. Application of soil-applied gypsum, other economical calcium salts, mono-ammonium phosphate, and DAP decreases the availability of metals. Soil application of acids, acid formers, and EDTA salts for those crops which are not directly consumed by human beings enhances the availability of metals in crops and increases their removal from the soil. However, wastewater should not be used for growing leafy vegetables as these vegetables can accumulate high concentrations of metals in their leaves.

Another possible approach for the usage of brackish water is the cultivation of salt-tolerant crops having less irrigation requirements. This will also help to control waterlogging. It is proposed that the breeders should come forward and develop salt-tolerant and drought-resistant varieties of different crops.

Uniform irrigation water allocations in all the provinces of Pakistan is recommended to achieve sustainable irrigation and to avoid waterlogging and salinity. A high surface irrigation water allowance (8 to 17 cusecs per 1000 acres) has been allotted in several canal commands in Sindh, while in Punjab it varied from 2.73 to 4.2 cusecs per 1000 acres only. In order to efficiently meet current water requirements of the agricultural, domestic, and industrial sectors an *ab initio* (from the beginning) level of water reallocation and efficient water management, with consideration to groundwater quality and its safe yield, is recommended (van Steenburgen et al. 2015).

14.7 POLICIES

In order to control water quality problems, the following policy points must be considered: (1) Enhancement of irrigation efficiencies and maintenance of proper leaching fraction; (2) construction of water storage facilities to implement demand-specific irrigation water supply; (3) promotion of water use associations; (4) maintenance of the existing drainage infrastructure with site-specific emphasis on new projects; and (5) strengthening the advisory services to farmers and training to the extension staff for site-specific solutions of problems.

Brackish water can be used safely on a sustainable basis with the following measures: (1) Volumetric consumption as the ultimate basis for water charges; (2) demonstration studies in community management of selected distributaries on a pilot scale; (3) groundwater recharge through holistic approaches, encompassing both the demand management and methods to increase water recharge; (4) educating farmers and promoting extension activities, including in-service staff training; (5) review funding, with the goal to use a 60:40 percent ratio for research and administration, respectively; (6) In time and space credit facility for gypsum, especially for small and resource-poor farmers, (7) Compulsory analysis and advice solutions of tube well waters through mobile teams of experts.

REFERENCES

Abdolali A, Ngo HH, Guo WS, Lee DJ, Tung KL and Wang XC. 2014. Development and evaluation of a new multi-metal binding biosorbent. *Bioresour Technol.* 160: 98–106.

Al-Amoudi A. 2016. Nanofiltration membrane cleaning characterization. *Desalin Water Treat.* 57: 323–334.

Anonymous. 2014. *Agricultural Statistics of Pakistan 2013–14.* Food, Agri. and Livestock Div., Ministry of Food, Agri. and Livestock, Govt. of Pakistan, Islamabad, Pakistan.

Anonymous. 2015. *Agricultural Statistics of Pakistan 2014–15.* Food, Agri. and Livestock Div., Ministry of Food, Agri. and Livestock, Govt. of Pakistan, Islamabad, Pakistan.

Ashfaq M, Griffith G and Hussain I. 2009. *Economics of Water Resources in Pakistan*, Pak TM Printers, Lahore, Pakistan. 230 p.

Ayers RS and Westcott DW. 1985. Water quality for agriculture. *Irrigation and Drainage Paper* 29, FAO, Rome, Italy.

Bauder TA, Waskom RM, Sutherland PL and Davis JG. 2011. *Irrigation Water Quality Criteria.* Colorado State University Extension, Fort Collins, CO, USA.

Burn S, Hoang M, Zarzo D, Olewniak F, Campos E, Bolto B and Barron O. 2015. Desalination techniques—A review of the opportunities for desalination in agriculture. *Desalin.* 364: 2–6.

Chaudhry MR, Hamid A and Javid MA. 1984. *Use of Gypsum in Amending Sodic Water for Crop Production.* Mona Recl. Expt. Proj. Publ. No. 136. WAPDA Colony, Bhalwal, Sargodha, Pakistan. 23 p.

Choudhary OP, Josan AS, Bajwa MS and Kapur M.L. 2004. Effect of sustained sodic and saline-sodic irrigation and application of gypsum and farmyard manure on yield and quality of sugarcane under semi-arid conditions. *Field Crop Res.* 87: 103–116.

Chinchmalatpure AR, Bardhan G, Nayak AK, Gururaja RG, Chaudhari SK and Sharma DK. 2014. Effect of sodium adsorption ratio with different electrolyte concentrations on saturated hydraulic conductivity of selected salt affected soils of Gujarat. *J Indian Soc Soil Sci.* 62: 1–8.

Cucci G, Lacolla G, Pagliai M and Vignozzi N. 2015. Effect of reclamation on the structure of silty-clay soils irrigated with saline-sodic waters. *IntAgrophys.* 29: 23–30.

Diacono M and Montemurro F. 2015. Effectiveness of organic wastes as fertilizers and amendments in salt-affected soils. *Agric.* 5: 221–230.

Dickin SK, Schuster-Wallace CJ, Qadir M and Pizzacalla K. 2016. A review of health risks and pathways for exposure to wastewater use in agriculture. *Environ Health Persp.* 124: 900–909.

Eaton FM. 1950. Significance of carbonate in irrigation waters. *Soil Sci.* 69: 123–133.

Emadodin I, Narita D and Bork HR. 2012. Soil degradation and agricultural sustainability: an overview from Iran. *Environ Develop Sustain.* 14: 611–625.

Ensink JHJ, Simmons RW and van der Hoek W. 2004. Wastewater use in Pakistan: the cases of Haroonabad and Faisalabad. In: Scott CA, Faruqui NI and Raschid-sally L (eds.), *Wastewater use in Irrigated Agriculture*, pp. 91–99, CAB International, Wallingford, UK.

Government of Pakistan. 2008. Agriculture statistics of Pakistan. Economic Wing of the Ministry of Food, Agriculture and Livestock, Government of Pakistan.

Ghafoor A, Gill MA, Hassan A, Murtaza G and Qadir M. (2001a). Gypsum: an economical amendment for amelioration of saline-sodic waters and soils, and for improving crop yields. *Int J Agric Biol.* 3: 266–275.

Ghafoor A, Muhammed S and Yaqub M. 1987. Use of saline-sodic water for reclamation of salt-affected soil and for crop production. *Pakistan J Soil Sci.* 2(1–4): 17–21.

Ghafoor A, Murtaza G, Maann AA, Qadir M and Ahmad B. 2011. Treatments and economic aspects of growing rice and wheat crops during reclamation of tile drained saline-sodic soils using brackish waters. *Irrig Drain.* 60: 418–426.

Ghafoor A, Murtaza G, Rehman MZ, Saifullah and Sabir M. 2012. Reclamation and salt leaching efficiency of treatments for tile drained saline-sodic soil using marginal quality water for irrigating rice and wheat crops. *Land Degrad Develop.* 23: 1–9.

Ghafoor A, Nadeem SM, Hassan A and Sadiq M. 2001b. Reclamation response of two different textured saline-sodic soils to EC_{iw} to SAR_{iw} ratios. *Pakistan J Soil Sci.* 19: 92–99.

Ghafoor A, Qadir M and Murtaza G. 1997. Potential for reusing low quality drainage water for soil amelioration and crop production. In: *Proc. Int. Symp. Water for the twenty-first Century: demand, Supply, Development and Socio-Environmental Issues*, pp. 411–420. June 17–19, Lahore, Pakistan.

Ghafoor A, Qadir M and Murtaza G. 2004. *Salt-affected Soils: Principles of Management.* Allied Book Centre, Lahore, Pakistan.

Grattan SR, Díaz FJ, Pedrero F and Vivaldi GA. 2015. Assessing the suitability of saline wastewaters for irrigation of Citrus spp.: Emphasis on boron and specific-ion interactions. *Agric Water Manage.* 157: 48–58.

Gupta IC. 1990. *Use of Saline Water in Agriculture.* Oxford and IBH Publ. House Co. Pvt. Ltd., Bombay, India.

Hanjra MA, Blackwell J, Carr G, Zhang F and Jackson TM. 2012. Wastewater irrigation and environmental health: implications for water governance and public policy. *Int J Hyg Environ Health* 215: 255–269.

Huber DM and Jones JB. 2013. The role of magnesium in plant disease. *Plant Soil* 368: 73–85.

Hussain G. 1978. Determination of irrigation water quality standards. *Ph.D. Diss.*, Colorado State University, Fort Collins, USA.

Hussain SI, Ghafoor A, Ahmad S, Murtaza G and Sabir M. 2006. Irrigation of crops with raw sewage: hazard and assessment in effluent, soil and vegetables. *Pakistan J Agric Sci.* 43: 97–101.

Hussain T, Muhammed S and Nabi G. 1986. Potential for using brackish groundwater for crop production. In: Ahmad R and Pietro AS (eds.), *Proc. Prospects for Biosaline Research. Workshop US-Pak. Bio Saline Res* pp. 469–476. September 22–26, 1985, Karachi, Pakistan.

IWMI. 2015. *International Water Management Institute, IWMI Annual Report 2015*, Colombo, Sri Lanka, p. 28.

Izhar-ul-Haq, Muhammad B and Iqbal F. 2007. Effect of gypsum and farmyard manure on soil properties and wheat crop irrigated with brackish water. *Soil Environ.* 26: 164–171.

Japan International Cooperation Agency. 2010. Preparatory study on Lahore water supply, sewerage and drainage improvement project in Islamic Republic of Pakistan.

Kahlown MA, Bajwa ZI, Abaidullah M and Hanif M. 2000. *Sulphurous Acid Generator for the Treatment of Brackish Groundwater and Reclamation of Salt-Affected Soils.* Mona Reclamation Expt. Project Publ. No. 247. Bhalwal, Sargodha, Pakistan. p. 23.

Kahlown MA and Gill MA. 2008. Managing saline-sodic groundwater in the Indus Basin. *Quart Sci Vis.* 9: 1–10.

Karajeh F, Suleimenov M, Karimov A, Vyshpolsky F, Mukhamedjanov Kh and Bekbaev U. 2004. *Technology of Irrigation, Water Saving and Improving Soil Fertility in Arys Turkestan Canal Command Zone.* Kazakh Research Institute of Water Management: Taraz; 18 (in Russian).

Keraita BN and Drechsel P. 2004. Agricultural use of untreated urban wastewater in Ghana. In: Scott CA, Faruqui NI, Raschid-Sally L (eds.), *Wastewater use in Irrigated Agriculture.* CABI Publishing, Wallingford, UK, pp. 101–112.

Khandewal and Lal P. 1991. Effect of salinity, sodicity and boron irrigation water on properties of different soils and yield of wheat. *J Indian Soc Soil Sci.* 39: 537–541.

Konikow LF and Kendy E. 2005. Groundwater depletion: a global problem. *Hydrogeol J.* 13: 317–320.

Lazarova V and Bahri A. 2005. *Water Reuse for Irrigation: agriculture, Landscapes, and Turf Grass.* CRC Press, Boca Raton, FL, USA.

Liang X, Ning X, Chen G, Lin M, Liu J and Wang Y. 2013. Concentrations and speciation of heavy metals in sludge from nine textile dyeing plants. *Ecotox Environ Safe.* 98: 128–134.

Mahmood Z. 2016. Searching the solution for water constraints: opportunities and expectations. The Daily Jang. June 24, 2016 http://www.jang.com.pk.

Malik D, Hussain G and Sherazi SJA. 1992. On-farm evaluation of gypsum application and its economics. In *Soil Health for Sustainable Agriculture.* In: *Proc 3rd Nat.Congr Soil Sci.* March 20–22, 1990, Lahore, Pakistan. pp. 407–420.

Mandal UK, Bhardwaj AK, Warrington DN, Goldstein D, Tal AB and Levy GJ. 2008. Changes in soil hydraulic conductivity, runoff, and soil loss due to irrigation with different types of saline–sodic water. *Geoderma* 144: 509–516.

Masood S and Gohar MS. 2000. *Participatory Drainage and Groundwater Management under Punjab Private Sector Groundwater Development Project*. National seminar on drainage in Pakistan, Mehran University, Jamshoro, Sindh, Pakistan.

Murtaza G, Ghafoor A, Owens G, Qadir M and Kahlon UZ. 2009. Environmental and economic benefits of saline-sodic soil reclamation using low-quality water and soil amendments in conjunction with a rice-wheat cropping system. *J Agron Crop Sci.* 195: 124–136.

Murtaza G, Ghafoor A and Qadir M. 2006. Irrigation and soil management strategies for using saline-sodic water in a cotton-wheat rotation. *Agric Water Manage.* 81: 98–114.

Murtaza G, Ghafoor A, Qadir M, Owens G, Aziz MA, Zia MH and Saifullah. 2010. Disposal and use of sewage on agricultural lands in Pakistan: a review. *Pedosphere* 20: 23–34.

Murtaza G, Murtaza B and Hassan A. 2015. Management of low-quality water on marginal salt-affected soils with wheat and Sesbania crops. *Commun Soil Sci Plant Anal.* 46: 2379–2394.

Murtaza G, Murtaza B, Niazi NK and Sabir M. 2014. Soil contaminants: sources, effects and approaches for remediation. In: Ahmad et al. (eds.), *Improvement of Crops in the Era of Climatic Changes*. Springer Science+Business Media, New York, USA, pp. 171–196.

Murtaza G and Zia MH. 2012. FAO Report—UNW-AIS. Individual's capacity development on the safe use of wastewater in agriculture in Pakistan, Available at http://www.ais.unwater.org/ais/pluginfile.php/232/mod_page/content/124/pakistan_murtaza_finalcountryreport2012.pdf

Niazi NK, Murtaza B, Bibi I, Shahid M, White JC, Nawaz MF, Bashir S, Shakoor MB, Choppala G, Murtaza G and Wang H. 2016. Removal and recovery of metals by biosorbents and biochars derived from biowastes. In: Prasad MNV and Shih K (eds.), *Environmental Materials and Waste Resource Recovery and Pollution Prevention*. San Diego, CA, USA.

Nishanthiny SC, Thushyanthy M, Barathithasan T and Saravanan S. 2010. Irrigation water quality based on hydro chemical analysis, Jaffna, Sri Lanka. *Am Eurasian J Agric Environ Sci.* 7: 100–102.

Oster JD. 1994. Irrigation with poor quality water. *Agric Water Manage.* 25: 271–291.

Oster JD and Grattan SR. 2002. Drainage water reuse. *Irrig Drain Syst.* 16(4): 297–310.

Pakistan Water Supply and Sanitation. 2013. World Bank. Urban Development. Annual Plan 2013–2014.

Qadir M, Ghafoor A and Murtaza G. 2001. Use of saline-sodic waters through phytoremediation of calcareous saline-sodic soils. *Agric Water Manage.* 50(3): 197–210.

Qadir M, Noble ad, Qureshi AS, Gupta RK, Yuldashev T and Karimov A. 2009. Salt-induced land and water degradation in the Aral Sea basin: a challenge to sustainable agriculture in Central Asia. *Nat Resour Forum* 33: 134–149.

Qadir M, Quillérou E, Nangia V, Murtaza G, Singh M, Thomas RJ, Drechsel P and Noble ad. 2014. Economics of salt-induced land degradation and restoration. *Nat Resour Forum* 38: 282–295.

Qadir M, Wichelns D, Raschid-Sally L, McCornick PG, Drechsel P, Bahri A and Minhas PS. 2010. The challenges of wastewater irrigation in developing countries. *Agric Water Manage.* 97: 561–568.

Quirk JP. 2001. The significance of the threshold and turbidity concentrations in relation to sodicity and microstructure. *Aust J Soil Res.* 39: 1185–1217.

Qureshi RH, Aslam Z, Salim M and Sandhu GR. 1977. Use of saline-sodic water for wheat production. In: *Proc. Water Management for Agri. Seminar*, pp. 329–336. Nov. 15–17, Lahore.

Qureshi RH, Hanif M, Rajoka MI and Sandhu GR. 1975. Use of saline-sodic water for crop production. In: *Proc. The Optimum Use of Water in Agriculture*. Cent-Scientific Programme Report No. 17, pp. 63–69, Ankara, Turkey.

Rehman MZ, Rizwan M, Ghafoor A, Naeem A, Ali S, Sabir M and Qayyum MF. 2015. Effect of inorganic amendments for *in situ* stabilization of cadmium in contaminated soil and its phyto-availability to wheat and rice under rotation. *Environ SciPollut Res.* 22: 16897–16906.

Ritzema HP, Satyanarayana TV, Raman S and Boonstra J. 2008. Subsurface drainage to combat waterlogging and salinity in irrigated lands in India: lessons learned in farmers' fields. *Agric Water Manage.* 95: 179–189.

Scott CA, Faruqui NI and Raschid-Sally L. 2004. Wastewater use in irrigated agriculture: management challenges in developing countries. In: Scott CA, Faruqui NI, Raschid-Sally L (eds.), *Wastewater use in Irrigated Agriculture*. CABI Publishing, UK.

Shah T, Burke J and Villholth K. 2007. Groundwater: a global assessment of scale and significance. In: Molden D (eds.), *Water for Food, Water for Life*, London, UK and IWMI, Colombo, Sri Lanka. pp. 395–423.

Shainberg I and Letey J. 1984. Response of sodic soils to saline conditions. *Hilgardia* 52: 1–57.

Sheikh IA. 1989 Country report on problem of waterlogging and salinity in Pakistan. In *Proceeding Information Seminar on Waterlogging and Salinity Research in Some Major Problem Countries*. May 27–28, 1989, Lahore, Pakistan.

Singh A, Krause P, Panda SN and Flugel WA. 2010. Rising water table: a threat to sustainable agriculture in an irrigated semi-arid region of Haryana, India. *Agric Water Manage*. 97: 1443–1451.

Smith CJ, Oster JD and Sposito G. 2015. Potassium and magnesium in irrigation water quality assessment. *Agric Water Manage*. 157: 59–64.

Sommer R, Glazirina M, Yuldashev T, Otarov A, Ibraeva M, Martynova L, Bekenov M et al. 2013. Impact of climate change on wheat productivity in Central Asia. *AgricEcosyst Environ*. 178: 78–99.

Strathmann H. 2004. Operating principle of electrodialysis and related processes. In: Strathmann H. (ed), *Ion-Exchange Membrane Separation Processes*, Elsevier, Amsterdam, The Netherland. pp. 147–233.

Tanninen J, Kamppinen L and Nyström M. 2005. Pretreatment and hybrid processes. In: Schäfer AI, Fane AG and Waite TD (eds.), *Nanofiltration—Principles and Applications*. Elsevier, Oxford, UK. pp. 253–254.

Taylor R and Bolto B. 2011. Electrodialysis—a mature membrane desalting process with a bright future? *Water J*. 38: 96–99.

Tchobanoglous G, Asano T, Burton F, Leverenz H and Tsuchihashi R. 2007. *Water Reuse: issues, Technologies, and Applications*. McGraw-Hill, New York, pp 674685. Baecker AAW, Roux KI, Av H 1989 Microbiological contaminants of metal-working fluids in service. S Afr J Sci. 85: 293295.

Trang DT, van Der Hoek W, Tuan ND, Cam PD, Viet VH, Luu DD, Konradsen F and Dalsgaard A. 2007. Skin disease among farmers using wastewater in rice cultivation in Nam Dinh, Vietnam. *Trop Med Int Health*. 12: 51–58.

University of California Committee of Consultants. 1974. Guidelines for interpretation of water quality for agriculture. University of California, Davis, USA. 13 p.

US Salinity Lab. Staff. 1954. *Diagnosis and Improvement of Saline and Alkali Soils*. USDA Handbook 60, Washington DC., USA.

van Steenburgen F, Basharat M and Lashari BK. 2015. Key challenges and opportunities for conjunctive management of surface and groundwater in mega-irrigation systems: lower Indus Pakistan. *Resources* 4(4): 831–856.

Vincent S. 2014. Environmental health monitoring: a pragmatic approach. *Int J Waste Res*. 4172: 2252–5211.

Vyshpolsky F, Qadir M, Karimov A, Mukhamedjanov K, Bekbaev U, Paroda R, Aw-Hassan A and Karajeh F. 2008. Enhancing the productivity of high-magnesium soil and water resources in central Asia through the application of phosphogypsum. *Land Degrad Develop*. 19: 45–56.

Wichelns D and Oster JD. 2014. Beyond California: an international perspective on the sustainability of irrigated agriculture. In: Chang AC, Silva DB (eds.), *Salinity and Drainage in the San Joaquin Valley*, California: Science, Technology, and Policy. Global Issues in Water Policy 5. Springer, New York, USA.

Wichelns D and Qadir M. 2015. Achieving sustainable irrigation requires effective management of salts, soil salinity, and shallow groundwater. *Agric Water Manage*. 157: 31–38.

Yaduvanshi NPS and Swarup A. 2005. Effect of continuous use of sodic irrigation water with and without gypsum, farmyard manure, pressmud and fertilizer on soil properties and yields of rice and wheat in a long term experiment. *NutrCyclnAgroecosys*. 73: 111–118.

Yunus MM. 1977. Water quality in the Indus Plains. In *Proceeding Water Management for Agriculture (EXXON Seminar)*, Nov. 15–17, Lahore, Pakistan.

Zhang XC and Norton LD. 2002. Effect of exchangeable Mg on saturated hydraulic conductivity, disaggregation and clay dispersion of disturbed soils. *J Hydrol*. 260: 194–205.

Zhao K, Li X, Zhou X, Dodson J and Ji M. 2013. Impact of agriculture on an oasis landscape during the late Holocene: palynological evidence from the Xintala site in Xinjiang. *NW Chin. Quatern Int*. 311: 81–86.

Zia MH, Ghafoor A, Saifullah and Boers ThM. 2006. Comparison of sulfurous acid generator and alternate amendments to improve the quality of saline-sodic water for sustainable rice yields. *Paddy Water Environ*. 4: 153–162.

15 Stagnant Yields

Abdul Khaliq, Amir Shakeel, Muhammad Kashif, and Ghulam Mustafa

CONTENTS

15.1 INTRODUCTION

15.1.1 Growth in Agriculture—Historical Perspective

Agriculture is the mainstay of the economy of Pakistan. There has been a simultaneous rise in national economy and agriculture; therefore, any serious fluctuation in agricultural growth has had a concomitant influence on the national economy (Figure 15.1). During the era of the "Green Revolution" in the 1960s, short statured varieties of wheat and rice were introduced, which were fertilizer responsive, efficient in light harvest, and had higher turnover of photosynthates towards grains. In fact, the productivity of these cereals marked an unprecedented quantum leap. Owing to such productivity, this decade recorded an average growth rate of 5.1%.

The growth in agriculture, however, declined during the 1970s to 2.4%. The colossal policy of nationalizing private entrepreneurs had overall negative implications for the economy. Moreover, the process of varietal development and their release was too slow and their potential achieved a plateau. However, there was enormous public sector investment in agriculture during the 1970s. Important institutions like Tarbela Dam, the Pakistan Agricultural Research Council, the Training and Visit Program of Agricultural Extension, the Seed Certification and Registration Departments/Seed Corporations, the On Farm Water Management, and the Barani Area Development Programs were commissioned during this period. The Cotton Export Corporation and Rice Export Corporation were established during that decade for facilitating export of indigenous produce of the cotton and rice respectively (Hanif et al., 2004).

Such high investments in the agriculture sector during the seventies produced results in eighties and thus the average growth rate reached 5.4% during this period. The Agricultural Prices Commission (APCom) was established to regulate support prices of major crops. Procurements were done by the public sector which minimized the role of the private sector. Problems of crashing prices during periods of surplus were addressed to some extent. Besides, it created security and protection for the farming community in generating farm incomes. Before this era, the emphasis was on new technologies, inputs, and investments without taking care of the issue of distortions on the produce. In fact, many corrective measures were taken during the eighties to nullify the distortions and addresse stability of produce prices.

The government experienced a heavy drain of exchequer on the procurement of agricultural produce. Having realized this, the government reconsidered its public sector intervention policies.

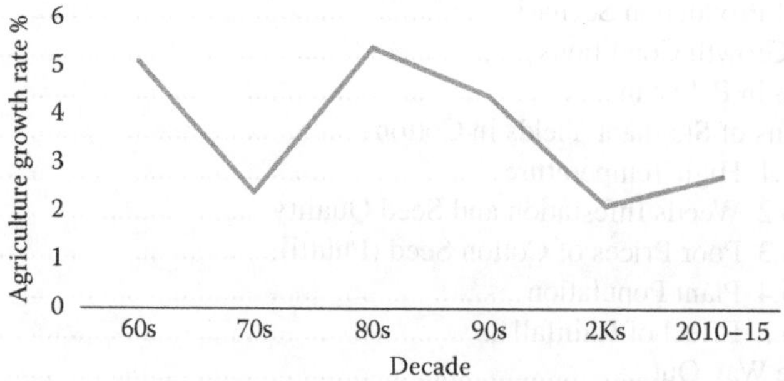

FIGURE 15.1 Growth rate of agriculture during last 5 decades. (From various issues of Economic Survey of Pakistan.)

Deregulation, privatization, and restructuring the size of the government and refixing priorities were carried out. Many of the public sector institutions involved in the procurement of agricultural commodities were closed at the start of the 1990s. The Cotton Export Corporation, Rice Export Corporation, and Agricultural Marketing and Storage Limited ceased their activities. The procurement activity was limited to a number of restricted commodities so that both the frequency of interventions and size of procurements were curtailed. This again resulted in price stability and farm economy started dwindling. Huge attacks of cotton leaf curl viruses in the start of 1990s threatened the future of the textile sector in Pakistan. Moreover, the epidemic of the American bollworm and army worm inflicted heavy losses in cotton crops and put the economy at risks during middle of that decade. The late the 1990s also witnessed severe droughts incurring massive losses in both the crop and livestock sectors. Wide fluctuations occurred in the growth of the agriculture sector during that time. Nonetheless, despite these challenges, Pakistan started exporting wheat during this era, and the average growth rate during the decade stood at 4.4%.

The agriculture growth witnessed mixed trends during the first half of the new millennium. An unprecedented drought hit the country during 2000–2001 and 2001–2002, which lowered agricultural added value and hence produced negative growth during these years. During 2002–2003 and 2004–2005, the overall agricultural growth showed a modest to strong recovery due to relatively better availability of irrigation water. The following year, 2005–2006, again recorded a decline in the agriculture sector owing primarily to poor performance of major crops. However, in 2006–2007 the agriculture sector grew by 5.0% as against the preceding year's growth of 1.6%. In fact, major crops recovered from a negative 4.1% the preceding year to a positive 7.6%. Wheat production was 10.5% higher than the preceding year while sugarcane production improved by 22.6%.

The agriculture sector exhibited a growth of 3.3% during 2012–2013. Agriculture subsectors improved so that a growth of 3.2, 3.7, 0.1, and 0.7% was recorded in crops, livestock, forestry, and fisheries, respectively. A rise of 2.3 and 6.7% was recorded in important crops and other crops, respectively. Cotton ginning declined by 2.9%. Weather conditions and the water situation had a negative impact of Kharif crops, and productivity of rice and cotton crops dropped in 2012–2013. A positive growth (2.9%) was recorded for the agriculture sector during 2014–2015, as compared with the previous year growth of 2.7% as all related agriculture subsectors performed relatively better. However, the performance of important crops remained weak. Only cotton and rice production could register a positive growth (9.5 and 3.0%, respectively) while production of sugarcane, maize, and wheat was lower by 7.1, 5.0, and 1.9%, respectively as against the previous year's estimates.

15.2 GAPS IN PRODUCTIVITY

Pakistan faces low levels of productivity of crops. Nonetheless, there is large variation in the productivity of crops between progressive growers and smaller landholder subsistence farmers. Yield gaps for wheat, rice, maize, and cotton between progressive growers and the national average stand at 43.5, 45.6, 58.55, and 30.85%, respectively (Figure 15.2). On the extreme end, gaps for sugarcane yields stand at 72.8% in Sindh and 61.6% in Punjab.

This chapter discusses production trends, constraints in productivity, and suggests ways forward for the crops of economic significance and important for food security.

15.3 FOOD GRAINS

15.3.1 Wheat

15.3.1.1 Brief Overview

Its center of origin is in southeast Asia, belonging to the central Asiatic center. Wild relatives such as durum, enikorn, and so on still grow in Syria, Iraq, Lebanon, Israel, and Turkey. Archeologists have found carbonized wheat grains in Egypt, Turkey, Iraq, and other countries. Ancient farmers

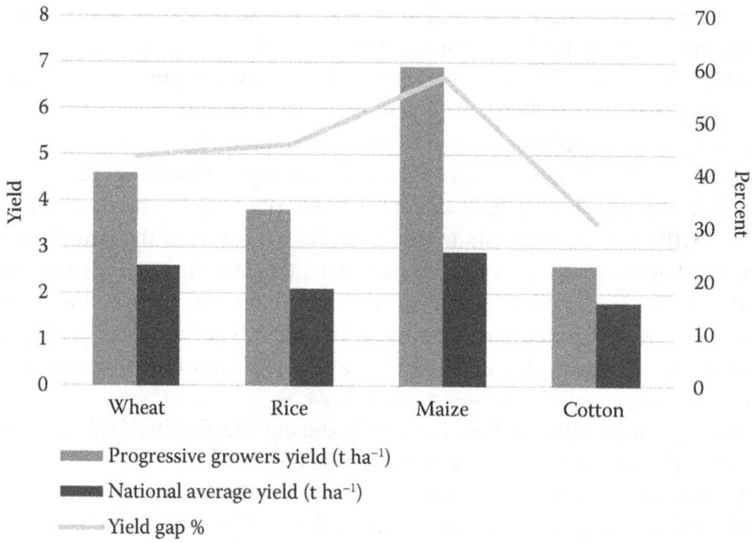

FIGURE 15.2 Yield gap of major crops in Pakistan.

selected genotypes based upon good seed yield and used the harvest for sowing the next year. Many centuries of cultivation, along with selections, resulted in improved grain size, grain number per plant, nonshattering, hulled-free threshing, and high yields. Nowadays wheat is grown all around the world. It is highly self-pollinated, photosynthetically a C_3 plant, and grows well in cooler environments.

15.3.1.2 Types of Wheat

Mainly two types of wheat are grown globally, first is *Triticum aestivum* commonly known as bread or spring/winter or hexaploid wheat. It covers 90% of all wheat cultivation areas. The second type is *T. turgidum*, commonly known as durum wheat, and accounts for the remaining 10% wheat crops in the world. Based on the growth habit, *Triticum aestivum* is further divided into two types.

15.3.1.2.1 Winter Wheat

Wheat with winter growth habits may be distinguished from spring wheat habit because an exposure in the seedling stage to near freezing temperatures is required before flowering will occur (vernalization). Further, it has the ability to withstand freezing temperatures.

15.3.1.2.2 Spring Wheat

This group of wheat is photoperiod sensitive and flowers during periods of declining day lengths. These are grown in such areas where winter temperatures are too high to meet vernalization requirements, and are cool and most favorable for spring wheat, but not too low to kill spring wheat by frost.

All types of cultivated wheats are exceptionally diverse in that the physiological characteristics of wheat plants adapt them to a wide range of climatic environments and the chemical and physical characteristics of gluten contribute to a wide use of wheat grain for many food products.

15.3.1.3 Ideal Growth Conditions

Today, hexaploid bread wheat (winter/spring) is cultivated on 95% of the fields and tetraploid durum wheat on 5% of the fields (Shewry, 2009). Mostly, wheat is grown in semiarid areas and accounts for 37% of the areas of developing countries. It is cultivated at latitudes 57° N to 47° S, having annual precipitations ranging from 520 to 1150 mm, and optimum temperatures between 20°C and 24°C. It requires temperatures between 16°C and 22°C during the vegetative stage, and 25°C for 4–5 weeks during the reproductive stage (Shewry, 2009).

15.3.1.4 Global Scenario

Wheat is one of the world's leading cereals, fulfilling 50% of the world's food requirements, along with seven other crops including three cereals—maize, rice, and barley (Sleper and Poehlman, 2006). Wheat feeds one third of the world's population and is declared the staple food of 43 countries. Globally, total wheat production stood at 729.5 million tons contributing 30% to grain production, of which 18% was traded internationally. In the total global production of Asia it contributed 320.3 million tons (FAO, 2013–2014).

15.3.1.5 Wheat in Pakistan

15.3.1.5.1 Pre "Green Revolution" Era

During the first decade after independence, wheat breeding in the Punjab focused on grain quality and rust resistance. In 1957, two tall bread wheat varieties, C271 and C273, were released for commercial use. These had hard vitreous grain with good gluten strength, excellent chapattis, and good grain storage characteristics (ISA-2, 2015). But these varieties had low yield potential to meet the production requirements of the country (Figure 15.1). Due to higher plant height lodging was one of the major problems, which led to no grain to poor grain formation during the stages of anthesis to grain filling and maturity. Moreover, these varieties were not responsive to fertilizers. The growth period, from sowing to grain maturity, extended beyond 170 days. It allowed only mono cropping of wheat leaving little space for cultivation of other crops between consecutive wheat crops.

15.3.1.5.2 Post "Green Revolution" Era

During early 1960s, there was a lot of work on developing wheat varieties that were short in stature, photosynthetically efficient, and which translocated a higher proportion of the assimilates into grain. Introduction of such a Mexican semidwarf variety, Maxi-Pak, in Pakistan during 1965 revolutionized wheat production in the country (Borojevic and Borojevic, 2005). This Green Revolution increased wheat production by three folds over the next 50 years. Wheat yields jumped from 2525 to 6252 kg/ha, almost a threefold increase. Today, wheat varieties trace their lineage to several parents, each contributing to the final variety (Figure 15.3). With the induction of reduced height genes (Rht) plant height not only reduced but stems became stronger and shorter. The plants became responsive

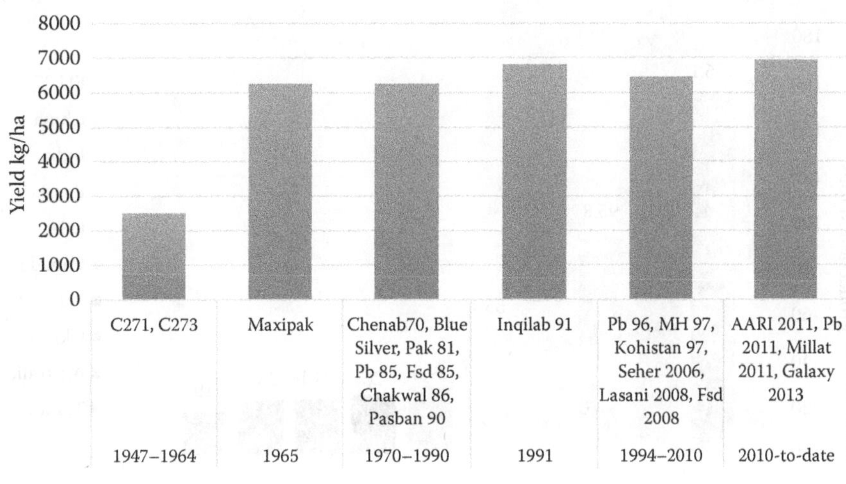

FIGURE 15.3 Average yield potential (kg ha^{-1}) of elite wheat varieties under cultivation since 1947.

to fertilizers resulting in high levels of yield within same field. Due to stiffer stems and a reducued intermodal distance the plants became resistant to lodging as well.

15.3.1.5.3 Present Scenario

Wheat, being the staple food of Pakistan, occupies about a 68% area of food crops annually. Pakistan ranked 7th among the top 10 wheat producing countries during 2013–2014 (Figure 15.4). Wheat is cultivated in canal irrigated and northern rainfed areas. The area sown during 2014–2015 was 9180 thousand hectares with a production of 25.5 million tons. It contributed 10% to the value added in agriculture and 2.1% to GDP (Economic Survey of Pakistan, 2015–2016). The total area and production of wheat showed a decrease of 0.2% and 1.9% over the last year 2014–2015, respectively. The average yield was estimated 2775 kg/ha marking a decrease of 1.7% against the previous year's yield of 2824 kg/ha. Many factors are thought to be involved in this decline in area, production, and yield of wheat, which are discussed further in the following sections.

15.3.1.5.4 Trends in Wheat Production

The "Green Revolution" that started in mid 1960 s in Pakistan was characterized by a remarkable increase in yield potential and by developing a plant architecture which had an increased fertilizer response, reduced height and lodging, maximum grain setting, and an increased 1000 grain weight. An overview of three components related to wheat, such as area, production, and yield, highlights that there has been a steady increase in production—from 7 to 29 million tons—from the early 1960s to the present (Figure 15.5). The area under wheat cultivation did not increase as expected, and is still under 9 million hectares. No doubt the yield trend line passes linearly through the last 5 decades, nevertheless the national average yield is quite low compared to other wheat producing countries in the world (Figure 15.4). Wheat yields in Pakistan stand around 2.5 tons per hectare, which is quite low compared to the potential available with the released varieties. The experimental trials indicate that the wheat yield potential is more than 8 tons per hectare given optimal provision of all agronomic requirements of the crop. However, practically even the most progressive farmers are not surpassing 5 tons per hectare (Figure 15.6). It is because though newly released varieties are showing good yield potential but they are not displaying the discrete level of increase over the previously released varieties. There is about a 60% yield gap in wheat, which needs to be narrowed down for its cultivation to become a success story in the country. Punjab is the major wheat-producing province of the country, producing 19.7 million tons from 6.9 million hectares with an average of 2.8 tons per hectare (Anonymous, 2013).

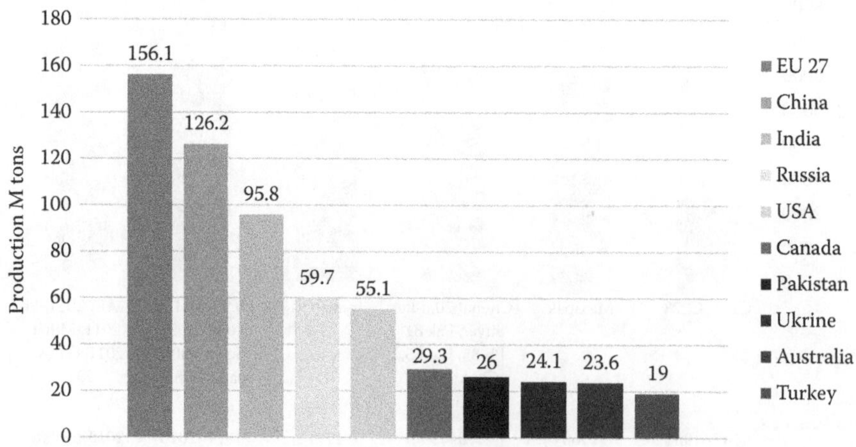

FIGURE 15.4 Top 10 wheat producing countries during 2013–2014.

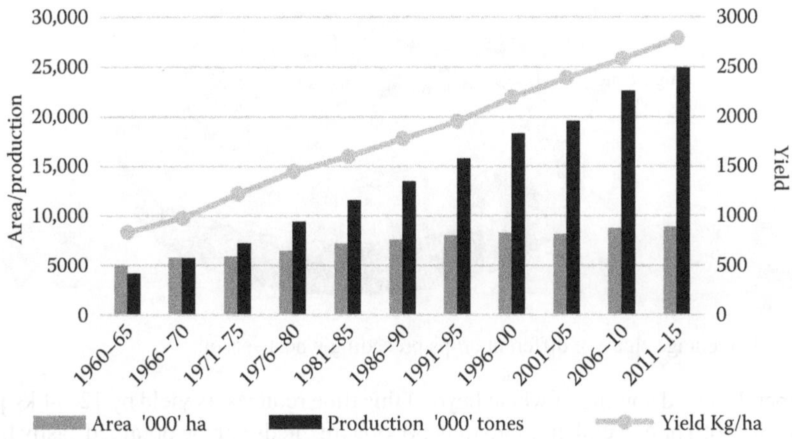

FIGURE 15.5 Area, production and yield trends of wheat over past 55 years.

Yield statistics show that an enormous increase in average yield was observed during 1965–1970 with the release of Maxipak—as compared to varieties before 1965, but then between 1971–1990 varieties released for general cultivation in Punjab did not show further significant increases in yield (Figure 15.1). Then again, with the release of Inqilab-91, a boost in average yield potential close to 30% was registered; however, after that no variety was able to match the potential of this variety (Figure 15.3). In 2014–2015, Galaxy-13 was released as a promising variety to replace Inqilab-91 but could not better it. Nonetheless, it is going be rejected as it is susceptible to rusts and also has issues of shattering. In fact, the era of quantum leaps in wheat yields, as realized during the mid-sixties and in the nineties, seem to be over as yields have experienced a plateau during the recent past. The varietal performances, in terms of improved wheat yields, are not satisfactory.

15.3.1.6 Reasons for Stagnant Yields in Wheat

15.3.1.6.1 Delayed Harvesting of Kharif Crops

Delayed harvesting of kharif crops is the foremost reason for low productivity and instability in wheat yield. About 78% of wheat by area is cultivated after the harvesting of different kharif crops among which cotton is the main crop (Figure 15.7). For the past 5 years, cotton growers have extended the growth duration up to December in a quest to catch as many pickings as possible, and hence, delaying the sowing of wheat. Sowing of wheat is in fact pushed back to the last week of December, and even in some cases up to early January, instead of its recommended sowing time of

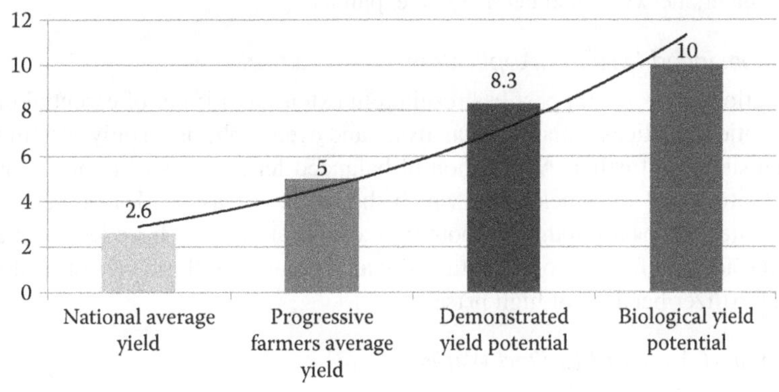

FIGURE 15.6 Difference of average yield potential (t ha^{-1}).

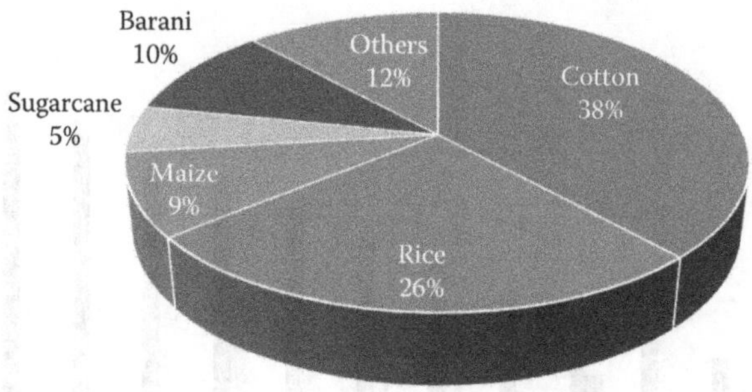

FIGURE 15.7 Percentage share of different crops preceding wheat season.

15th November. Delayed sowing of wheat beyond this time reduces its yield by 12–14 kg per acre per day. Sown by the 20th of November, more than 50 kg/acre yields can be obtained easily by following the prpoer agronomic practices; after the 20th yield, drops to 40 kg/acre and after December it dips below 30 kg per acre. Nonavailability of soil moisture in rainfed areas also delays wheat sowing in these areas.

15.3.1.6.2 Improper Seed Rate

The Agriculture Department, Government of Punjab, advises farmers to use 50 kg per acre seed rates for normal sowing and 60 kg per acre for late sowing of wheat. There are other opinions as well. A seed rate of 40–45 kg per acre for timely sown crops is likely to achieve 8 million plants per acre. The plants need space and soil to grow the proper number of tillers and canopy for maximum growth and development. If the number of plants per acre is increased they will be congested and every individual plant will suffer for lack of nutrients, moisture, and space, eventually leading to poor growth and development, fewer number of tillers, weak tillers, small spike length, small and thin seeds, poor thousand grain weight, and ultimately low grain yield. The new recommendations of seed rate for wheat are not suitable. To obtain a uniform compensation effect among plants, the ideal plant density for wheat can be achieved by applying 45 kg per acre for timely sowing, and 50 kg per acre for late sowing.

15.3.1.6.3 Use of Old Varietal Seeds

Breeders working in different research stations/institutes have developed new high yielding and disease resistant varieties, which have improved crop yield. Presently only 20% of the total seed requirements are met through certified seed, leaving rest of the areas sown by farmer's own seed or seeds from local merchants, which have no quality assurance. The farmers should be trained for producing their own quality seed that would improve wheat productivity on small farms. The authentic seed dealers network also needs to be expanded.

15.3.1.6.4 Unbalanced Fertilizer Application

Intensive cultivation in the recent past has resulted in extensive mining of essential plant nutrients. Faulty of fertilization practices without soil analyses and overemphasis on only certain nutrients have deteriorated the situation further. Application of balanced fertilizers still remains an elusive goal. At present, NP ratio is 4:1 for wheat crop, which should be at least 1.5:1, or otherwise plants gain height and ultimately crops are lodged. About 93% of wheat growers have less than a 10 hectares' farm size, but the account for 65% of the total cultivated areas. Small growers often do not apply the recommended fertilizer because of high prices.

15.3.1.6.5 Rainfed Area and Limited Water

In Punjab, the canals are closed during winter for annual cleaning. Both early and late sown crops of wheat suffer severe shortages for the first irrigation during the critical growth stage of crown root development and tillering. Farmers have limited access to tube well water as well. Such shortages of water during the critical growth stage of wheat have severe impacts on yield. A shift in rainfall patterns has also become a major problem over last 5 years. In barani areas, the main source of moisture is rain, which does not occur at proper time. Sometimes, wheat sowing is delayed too much in the hope of another rainfall to replenish soil moisture in barani areas.

15.3.1.6.6 Stressful Environment

Abiotic and biotic stresses have adversely affected crops. When frost occurs during the heading it results in damage of spikelets, especially for crops sown early; then, sudden increases in temperature result in poor seed setting. Aphid attacks during the past few years are also affecting wheat production. More recently, army worm attacks have been reported in some wheat fields.

15.3.1.6.7 Faulty Marketing System

A faulty marketing system is also one of the major problems. When wheat production is sufficient, farmers face difficulties in obtaining adequate prices from private wheat buyers.

15.3.2 Rice

15.3.2.1 Brief Overview

Rice (*Oryza sativa* L.) is an important cereal crop in the world consumed by millions of people as a staple food. In South Asia, it accounts for more than 50% of crops. In these cropping systems, rice is grown in sequence with wheat, maize, and legumes, etc. China is the largest producer of rice in the world (FAO, 2013). It is a versatile crop grown from the Himalayan foothills to the irrigated Indo-Gangetic plains. Most of the world's rice comes from India, China, Japan, Indonesia, Thailand, Burma, Bangladesh, and Pakistan. Asia leads the world's rice production with a contribution of about 90% (Matloob et al., 2015). People in Asian countries consume 75% of the world's rice production, making it an important commodity for food security in Asia (Seck et al., 2012). Rice also holds a distinctive position in Pakistan's agriculture-based economy. It is the second most important staple cereal after wheat, and the third largest crop in Pakistan after wheat and cotton. High quality rice is grown in Pakistan for both domestic consumption and export. Pakistan's share of the world rice trade is around 11% (Ali et al., 2014). It is an important cash crop accounting for 3.1% of value added in agriculture and 0.6% in GDP. In Pakistan, rice fulfills about 8% of daily calorie intake and per capita consumption is 20 kg annually. During 2015–2016, rice was sown on an area of 2.75 million hectares with a total production of 6.81 million tones and an average yield of 2346 kg ha^{-1} (Anonymous, 2016). About 12% of the cropped area fall under rice cultivation, 67% of which lies in Punjab. In the country, two types of rice cultivars (coarse and fine) are grown for food. Of the total area, 62% is under basmati, 27% under International Rice Research Institute (IRRI) coarse varieties, and 11% under other varieties (Ali et al., 2014). "Basmati" (fine grain aromatic rice) is an important brand of Pakistan that has a key role in earning foreign exchange for the country due to its significant value in the international market (Anonymous, 2013). The United Arab Emirates, Malaysia, Bangladesh, Iran, Indonesia, and Saudi Arabia import good quality rice from Pakistan (Akbar et al., 2011).

15.3.2.2 Conventional Rice Production System: Challenges and Constraints

The most common method of rice crop establishment in many Asian countries is still transplanting in puddled soils (wet-tillage) with continuous flooding. Although this method furnishes adequate weed control through water ponding, yet, it is a gigantic water consumer (Kumar and Ladha, 2011; Matloob et al., 2015). A large amount of water and labor is required for transplanted rice. The soil health is also adversely affected by puddling due to the dispersion of soil particles, soil compaction, and drudgery of tillage operations requiring more energy in succeeding crops such as wheat (Singh et al., 2002). In the rice-wheat system, soil management practices for rice has deleterious effects on the soil environment for the succeeding wheat and other upland crops, also require more energy for tillage operations in succeeding crops. Fujisaka et al. (1994), on the basis of a diagnostic survey conducted in several rice-wheat areas in South Asia, observed low wheat yields in a rice–wheat system, mainly due to deterioration in soil structure and the development of subsurface hardpans. Surface evaporation and percolation through the puddling process causes high losses of water. Rice consumes 21% of available fresh water in Pakistan where acute water shortage has already hampered agricultural production. Both surface underground water resources are already shrinking and water has become a limiting factor for agricultural productivity (Mann et al., 2007).

Nowadays, food security is the most important incentive to feed the increasing population of the world. Rice is an important commodity to fulfill food demands of this rapidly growing population. Generally, rice is cultivated by transplanting 30–40 days old seedlings in nurseries raised on beds and which are puddled and flooded (Ehsanullah et al., 2007; Chauhan, 2012). In addition, through nursery transplantation, the required plant population (250,000 plants ha^{-1}) is seldom maintained, primarily due to shortage of skilled labor and high costs of this operation (Baloch et al., 2000). Suboptimum plant population is of significant importance among the various factors limiting the paddy yield in transplanted rice cultures (Mann and Ashraf, 2001). Besides suppressing weeds, this method also provides a better growing environment to rice plants; however, this method requires a large amount of water during its growth period (Rao et al., 2007; Chauhan and Johnson, 2009). Rice is grown in continuously flooded conditions, with ponding depths of 50–75 mm for most of the growing season, maintained by 15–25 irrigations (Ahmad et al., 2007). All over the world water is becoming scarce. Both quantity and quality of water is seriously threated by melting of glaciers, high demand by urban and industrial users, salinization, and chemical pollution. Out of 90% water being used for agriculture, about 45% is consumed by irrigated rice. Therefore, about 22 million ha of Asian rice fields will suffer from economic water scarcity, and 17 million ha may experience physical water scarcity by 2025 (Tuong and Bouman, 2003). The problem is further aggravated by increasing costs of fuel and electricity, declining water tables, and climatic change (Vorosmarty et al., 2000; Rosegrant et al., 2002). Scarcity of water, shortage and high cost of labor, and reduced profit are the main problems associated with conventionally flooded rice systems, which threaten its economic viability (Bhushan et al., 2007). Labor is required to transplant the nurseries to the main field but laborers are reluctant to do such a tedious job, therefore, labor costs have increased. Transplanted rice in flooded fields also suffer from transplanting shock, tissue damage, late maturity, and less water use efficiency (Parthasarathi et al., 2012). Another problem associated with flooded rice is methane emission. During last 150 years, its concentration has been increased from 0.74 to 1.73 μmol mol^{-1}. It is a greenhouse gas and is 120 times more potent than CO_2 (Lelieved et al., 1998). The rice yields of the country are either stagnant or declining day by day.

Although the yield of basmati rice is much lower than IRRI varieties the demand is high, in both national and international markets. Therefore, in recent past, the majority of Pakistani farmers preferred growing transplanted basmati rice despite lower yields, high production costs, and intensive water requirements. Nevertheless, with the introduction of herbicides and the evolution of high yielding, short stature, and lodging resistant varieties, transplanting techniques will not be necessarily practiced by farmers in the future. Furthermore, lower benefit cost ratio with reduced productivity per unit area ever increasing costs of inputs, compaction of soil structure due to puddling, and failure of nursery because of unfavorable weather conditions, nutrient(s) deficiency, toxicity, and plant protection problems have agitated the conviction of farmers in this system.

15.3.2.3 Rice Cultivation Zones

In Pakistan, rice is grown under 4 distinct agro-ecological zones with diverse climatic, hydrological, and edaphic conditions (Figure 15.8). Each zone has its own critical problems associated with rice production. Packages of agronomic practices have been developed for the specific agro-ecological conditions of each zone.

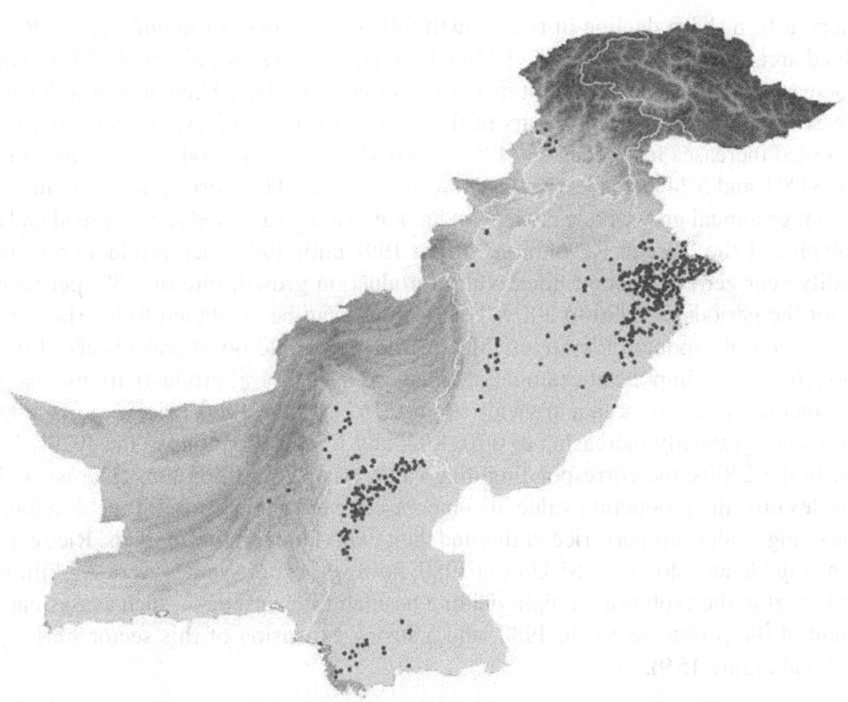

FIGURE 15.8 **(See color insert.)** Rice growing areas of Pakistan. Each green dot denotes 5000 ha of rice. (From http://ricepedia.org/index.php/pakistan.)

Zone I covers the northern high mountainous areas. Here, the climate is subhumid, with Short monsoonal summers (750–1000 mm average annual rainfall). Cold tolerant rice varieties are recommended for this zone to cope with the cold. Zone II is located in the "Rechna Doab" that is, the land between two rivers—Ravi and Chenab. The climate is semiarid, subtropical with 400–700 mm rainfall. The rice-growing season is fairly long and suitable for cultivating coarse (nonbasmati) as well as fine-grained aromatic (basmati) rice varieties. The premium quality basmati rice grows along the "Kalar tract," which lies in this zone. The Kalar tract is characterized by heavy clay soil with good water holding capacity. Zone III is a large tract of land on the west bank of the Indus River. The climate is arid, subtropical with 100 mm average annual rainfall. Heat tolerant coarse-grained varieties are grown here as temperatures may exceed 50°C in this zone. Zone IV is the Indus delta where the climate is arid tropical marine with no marked seasons, suitable for growing coarse grain varieties (Salim et al., 2003; Bashir et al., 2007). Apart from these major rice-growing areas, rice patches are also found in other cropping patterns based on cotton, sugarcane, and maize.

15.3.2.4 Production Trends-Historic Perspective

After independence in 1947, territories with enormous agricultural potential collectively known "as the bread-basket of India" comprised Pakistan. About 31% of irrigated areas of undivided India, containing 25% of the areas under cereal cultivation, were inherited by Pakistan. Of this total area under cereals, 35% was devoted to wheat production, while rice was cultivated on 32% of these areas. Since then, things have changed dramatically and the country's food production has increased fivefold and hunger has never been an issue except in some rare extreme cases. National average rice yields have almost doubled, compared to 1947, with a concurrent sevenfold increase in total production (Anonymous, 2001). This increase can be attributed to increase in yield growth that has positively contributed towards production growth. During 1958–1967, increase in rice growth was observed to be 1.8%. This was followed by a period of pronounced growth of 6% from 1967 to

1973. Afterwards, a sharp decline in rice growth followed (0.56% per annum from 1967 to 1997) as more land area was allocated to low yielding basmati varieties (Kataki et al., 2001). Rice yields are still increasing; nevertheless, current increases in rice yields have been quite low during the last 2 decades as compared to the early years of the "Green Revolution." Area, yield, and production of rice recorded increases for 3 decades (1961–1980). During this period, growth rates in area and yield were 3.68% and 5.24%, respectively (Wang et al., 2012). This corresponded to an increase of 8.91% in average annual growth rate of rice production, which was owed to the spread and adoption of technologies of the "Green Revolution." From 1981 until 1995, rice production growth rates were steadily near zero. A second hike, with a production growth rate of 3.3% per annum, was observed for the periods of 1996 to 2009. This increase can be attributed to hybrid varieties and the intensification of production systems. Meanwhile, a sizeable boost was observed in terms of rice exports. Rice is an important example of enhanced agricultural productivity in Pakistan, with concurrent increases in both area and yields (Figure 15.2). In the 1960s, rice exports amounted to 0.14 million tons, gradually increasing to 0.70, 1.02, and 1.59 million tons in the 1970s, 1980s, and the 1990s. In the 2000s, the corresponding ding value was 2.76 million tons (Dorosh and Wailes, 2010). Besides this, the production value also increased over time because of the development and cultivation of high value aromatic rice grains and their export to lucrative markets. Rice export value was very negligible and stood at 385 US$ in 1980; however, export values were 2.2 billion US$ in 2010, due in part to the evolution of high-yielding basmati rice varieties—such as Basmati-385, the development of the private sector in 1988, and a strong expansion of this sector starting in 1992 (Table 15.1 and Figure 15.9).

15.3.2.4.1 Current Scenario

Despite gains, rice production has been set back by natural disasters, socioeconomic constraints, and government policy in recent years. During the year 2010, harvested areas and yields of rice declined by 18% and 30%, respectively compared to 2009 (Anonymous, 2011). Destruction caused by the floods of July 2010 was the prime cause of this loss in productivity. During 2014–2015, areas under rice cultivation increased by 3.6% over the previous year, and the highest ever production of 7005 million tons was obtained—3% higher over the previous year's production during the same period (Anonymous, 2015). This boost in rice production was attributed to an increase of irrigation water and its timely availability, as well as the fact that more areas were under rice cultivation, especially of hybrid rice. This increased growth rate was followed by a decline of 4.9% and 2.7% in rice area and production, respectively during 2015–2016 (Anonymous, 2016). This time a reduction of the area under rice cultivation was because of reduced net returns owing to the low prices of

TABLE 15.1
Growth Rate of Rice (Percent per Annum) Production in Pakistan

Period	Area	Production	Yield
1961–1980	8.91	3.68	5.24
1981–1995	0.66	0.69	−0.03
1996–2009	3.26	1.60	1.60
2010–2015	4.45	9.04	3.76

Source: Wang H, Velarde O and Abedullah A 2012. In: Wang H, Pandey S, Velarde O and Hardy B (eds) *Patterns of Varietal Adoption and Economics of Rice Production in Asia*, International Rice Research Institute, Los Baños (Philippines). pp. 67–71

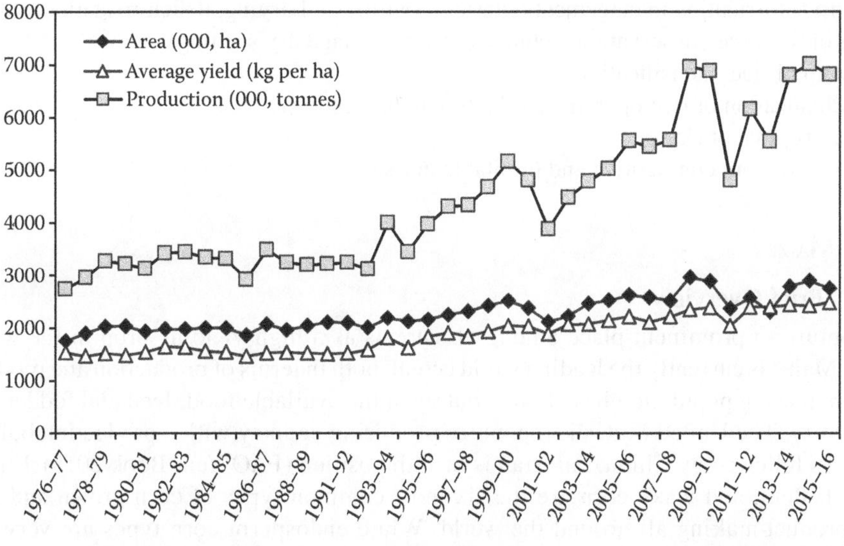

FIGURE 15.9 Production, area, and yield of rice in Pakistan (1976–2016). (From FAO-Food and Agriculture Organization 2013. *FAOSTAT Database "Countries by Commodity (Rice, Paddy)"*. FAO, Rome, (Accessed on May 2014).)

rice in both the national and international markets. Decreased economic returns and ever rising costs of production force rice growers to substitute rice with more economical crops like maize and fodder. Heavy rains in July 2015 also adversely affected rice crop. Depressed prices were due to the abundant supply and sluggish exports of basmati rice and large carryover stock from previous year's bumper crop.

15.3.2.5 Causes of Low Yields

There are numerous bottlenecks in the production and yield improvement of rice. Some of these are listed as here.

- Predominant cultivation of low yielding high-quality basmati rice
- Uncertainty of climatic optima
- Dwindling natural resource base
- Suboptimal plant population due to manual transplanting
- Nonavailability of labor during peak periods and rising wage rates
- Acute water shortage
- Supplemental irrigation with brackish underground water
- Biotic (Rice blast, leaf folder, weeds) and abiotic stresses (drought, salinity, waterlogging)
- Inefficient harvesting and postharvest management
- Cold damage to rice crop in the northern mountain rice growing areas
- Nutrient deficiencies and disorders (zinc)
- Reduced market prices leading to switching to more profitable crops

15.3.2.6 Way Forward

The following interventions are suggested as a way forward to improve rice cultivation in the country.

- Aerobic rice
- Direct seeded rice (dry and wet seeding)
- Hybrid rice

- Improved irrigation management (alternate wetting and drying, deficit irrigation)
- Use of resource conservation technologies (zero tillage dry seeding)
- System of rice intensification
- Mechanization of rice operations (planting to harvesting)
- Support price of rice
- Assess to more competitive and profitable markets

15.3.3 Maize

15.3.3.1 Brief Overview

Maize captures a prominent place among all cereals as a high yielding crop in the world and Pakistan. Maize is currently the leading world cereal, both in terms of production and productivity. Rapidly increasing population has already outraged the available food, feed and fodder supplies in the country. The United States is top maize producing country with a production bulk of 363 million tons followed by China. Pakistan is in 14th position (FAO Year Book 2015). Flint, dent, pop, sweet, flour, and waxy corn are the six most common types of corn grown and used for specific product-making all around the world. White endosperm corn types are very popular for human food and fetch premium prices in the market. Yellow endosperm corns are high yielders compared to white corn but are not accepted by the dry-milling industry for its yellowish flour color.

15.3.3.2 Maize in Pakistan

Maize is an important crop and is in third position after wheat and rice in Pakistan. Maize is cultivated as a multipurpose crop (food grain for humans, feed for poultry, and forage for livestock) by average to poor farmers that have below 10 livestock heads or bear marginal lands. It is cultivated in various cropping systems in rotation with a number of crops. Some growers cultivate maize as a sole cash crop as it is a commercial/industrial crop. Nonetheless, a significant amount of maize grain is consumed as feed (Figure 15.10). Maize is grown in almost all provinces of Pakistan, but Punjab and KP are the main contributors with about 97% of the total grain production. Only 2%–3% of total maize is produced in Sindh and Balochistan. Maize is also an important crop of AJK with about 0.122 million hectares of area during autumn.

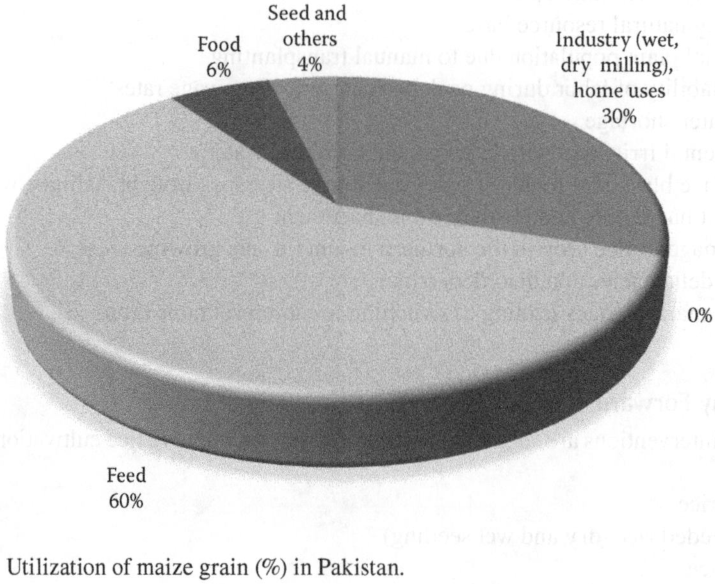

FIGURE 15.10 Utilization of maize grain (%) in Pakistan.

15.3.3.3 Types of Maize

Maize is grown twice a year in Pakistan, especially in Punjab, and is divided into two groups on the basis of sowing season.

15.3.3.3.1 Spring Maize

Spring maize is planted from the first week of February to the first week of March in Punjab. Its cultivation has increased due to the active involvement of multinational seed companies in Pakistan. Merely a few hybrids are dominantly grown by progressive farmers. Although the soil and climatic conditions of Pakistan are favorable for maize production its per hectare yield is very low in comparison to other maize growing countries of the world. In Punjab, the highest yields are obtained from the spring maize due to single hybrid sowing (Figure 15.11). Since the introduction of spring maize cultivation in Pakistan, there has been a gradual increase in cultivated areas of maize in the irrigated lowland areas of Punjab.

15.3.3.3.2 Autumn/Kharif Maize

Autumn/kharif maize is sown in Punjab from July to August, but there are exceptions with farmers sowing as early as May or as late as September. These farmers are more concerned with the production of fodder. Approximately 60%–70% of maize in Punjab is grown for green fodder by small farmers, while 30%–40% of maize is sown for specialty fodder crops by farmers with relatively large holdings. Maize in central Punjab is grown primarily during the summer season within a mixed cropping zone situated between the predominantly rice-wheat system to the north, and the cotton-wheat system to the south. Generally, maize is grown in small, scattered plots; extensive monoculture is not common. Although the area of autumn maize is almost double than that of spring maize, the yield is low because of large share of open pollinated varieties and fodder crop cultivated during autumn season. It is only recently that multinational companies have introduced hybrids for autumn crops and that grain yields have significantly increased (Figure 15.12).

15.3.3.4 Trends in Maize Production

Early maturing varieties of wheat and maize, developed during the past two decades, have led to greater crop intensification. The number of farmers in KP and northern Punjab that follow a maize-wheat rotation has steadily increased (Figure 15.13). From the 1960s to 1990s, the combined production of two season crops was no more than 1.5 ton per hectare, but with the advent of multinational companies after 2000, overall production and yield per acre significantly increased

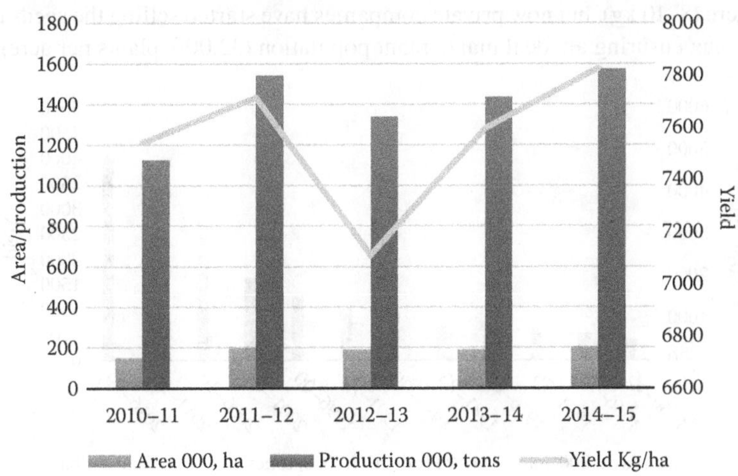

FIGURE 15.11 Area, production, and yield trend of spring maize in Punjab.

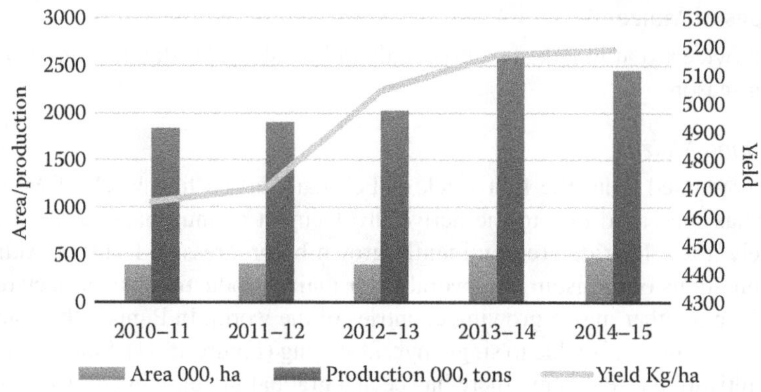

FIGURE 15.12 Area, production and yield trend of Autumn maize in Punjab.

during the past decade to more than 4 ton per hectare. In Punjab, progressive farmers planting only maize hybrids are getting around 10 tons grain per hectare from the spring crop. They are also adopting improved production technologies delivered by multinational companies. Spring maize cultivation is now gaining importance in KP as well, with about 30%–35% of the maize areas under cultivation of hybrids almost exclusively supplied by multinational seed companies.

15.3.3.5 Silage Production

Maize is called the king of crops for silage. Nowadays, maize as a silage crop is recognized by local farmers, with about a 4% annual increase in silage. It is reported that cotton and rice farmers are shifting towards maize cultivation due to high yields of hybrids grains and quality silage. Recently, increase in silage export has further uplifted adoption of maize crop.

15.3.3.6 Reasons for Fluctuating Yields

15.3.3.6.1 High Prices of Commercial Hybrids

The hybrids available on the market mainly belong to multinational seed companies, which are imported and are very costly (Rs. 600–750 per kg). Progressive farmers are able to buy the latest hybrids but resource poor farmers are not able to buy these seeds and the other essential inputs of production technology. That is why areas under spring crops have diminished; resource poor farmers cannot afford its cultivation during spring. Moreover, the seed was previously sold on the basis of seed rate per acre (8–10 kg), but now private companies have started selling the seeds on the basis of seed count per bag ensuring an ideal maize plant population (32,000, plants per acre) and claiming

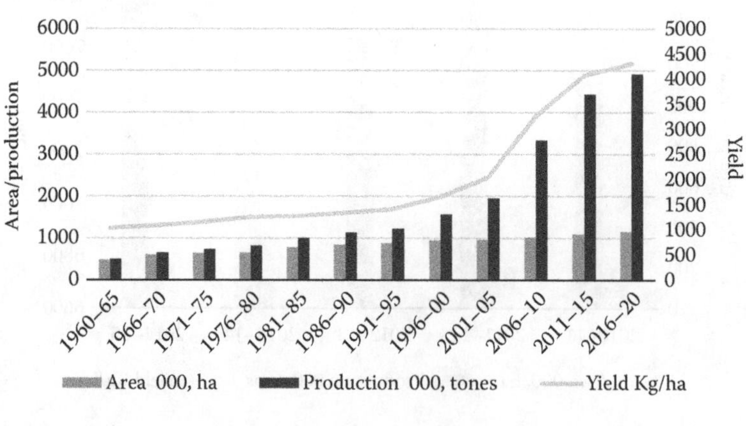

FIGURE 15.13 Trend in maize area, production, and yield for the past 50 years.

a 90% germination rate. Almost all the hybrid seeds planted in the country are imported and multinational companies sell them at a very high price, which is a significant burden on the farmers. Local varieties cannot compete with imported hybrids. In 2011, Pakistan Agricultural Research Council (PARC) reported that only three institutes in Punjab (MMRI, CCRI, and NARC) were involved in local hybrid development. They have so far released only 10 hybrids—quite insufficient compared to multinational and private seed companies which develop dozens of hybrids annually. Along with hybrids, locally 30 OPV's have also been released but have low turnover in terms of grain production despite their production for silage purposes.

Multinational seed companies should start seed production locally, which will reduce the cost of hybrid seeds. Local seed companies should be encouraged by the public sector to develop more local hybrids.

15.3.3.6.2 High Cost of Inputs and Lack of Improved Production Technology

Inputs like fertilizers, herbicides, and insecticides are costly and are beyond the reach of resource-poor farmers. Sometimes, progressive farmers limit some of the inputs to minimize expenses per unit area. Lowering the prices with subsidies may encourage farmers to use better inputs, which will help yield good crop harvests. Local seed companies should promote some mutual sharing programs; for example, cooperative farming schemes for lowland or resource-limited farmers to help them benefits from modern production technologies using cost sharing mechanisms.

15.3.3.6.3 Nonavailability of Quality Seed

A majority of the farmers have claimed that they were interested in buying high-grade seeds; however, they were not available locally and often when they did manage to get some these were not of pure quality and their input resources were wasted. They were reluctant to plant their own local or mixture of seeds, which were poorer in performance than high quality seeds. There should be proper infrastructures for seed production where farmers can buy them. There is a need to establish cooperatives among the farmers, which will manage the seed and production technologies and disseminate them to interested farmers.

15.3.3.6.4 Climate Change

Maize is a sensitive crop and is affected by various types of biotic and abiotic stresses. The spring season crop is severally affected by heat stress which causes plant mortality at the early vegetative stages, and flowering and grain-filling during the later stages. In 2012–2013 a decline in grain yield was observed (Figure 15.11). The month of May is very important when the emergence of reproductive organs is in progress and pollination is taking place. Any stress during this phase severely hampers yields. High temperatures dehydrate the released pollens very quickly due to dry winds, as maize is a cross-pollinated plant and wind is the primary source of pollen dissemination. May of 2016 was the hottest month on record during the past 5 years and definitely affected pollination and seed setting. Preliminary estimates reveal severe yield declines with observed bareness of the cobs. In Punjab and Sindh hot season temperatures rising beyond 45°C cause tassel blast in maize and loss of viability of pollen and even silk. There is no suitable variety or hybrid that resists high temperatures. The only solution suggested or practice is to change planting dates to avoid temperature stresses during the reproductive stages of the crop. Imported hybrids are developed in cooler environments and are often unable to withstand high temperatures when planted in our local environments. It is therefore suggested that tolerant hybrids should be developed locally under high temperature stresses.

Autumn crops are affected by either excessive water—due to monsoons, or water shortages, nevertheless the growers during this season usually have plenty of groundwater and are able to avoid loss of pollen viability due to heat stress. Farmers face scarcity of tube well water for irrigation due to power failures. On the other hand, untimely and intensive rains cause excessive standing water in fields which results in root lodging in maize. Soil salinity due to high water tables and waterlogging are also common.

15.3.3.6.5 Instability in Prices

Unstable and ever-fluctuating market prices cause large reductions in planting areas. Increases in price support for wheat (being a food security item) gives incentives for wheat cultivation and thus reduces areas growing spring maize resulting in enormous production reductions. The autumn crop is a chance grain crop, otherwise farmer completely cut maize for fodder purposes and prepare for wheat sowing. In recent years, the failure of cotton and rice crops has caused a shift towards maize cultivation as an alternative cash crop. Disease epidemics in cotton, shortages of water supply, and volatility in prices of paddy have been the reasons for such shifts.

15.3.3.6.6 Postharvest Losses

The ideal moisture content for storing maize is below 13%, which is hard to achieve in reasonable time for autumn crops. If the crop was sown in August, it will reach physiological maturity in November, during which time the air temperature is hardly sufficient for seed drying, with rather damp and foggy weather conditions present in most of the Punjab. The farmers usually leave cobs piled up on the soil near or around field areas. It is exposed to rodents, insects, domestic animals, rainfall, and is also vulnerable to high levels of aflatoxin from soil. Due to this practice, even though high yields are expected, grain appearance and quality are deteriorated lowering market prices. These farmers should be aware of and be able to use proper drying and storage techniques.

15.3.3.6.7 Less Frequent On-Farm Activities and Demonstration

On-farm research and demonstration activities are not sufficient to train farmers for cultivation of maize as a commercial cash crop. These activities are very important for the assessment of new genotypes, evaluation of elite lines, selection of varieties, development of hybrids, and the dissemination of production technologies. Distortion of the market by middle men, at the time of maize harvest, needs to be checked by having certain institutional developments minimize producer risks and exploitation.

15.4 COTTON

15.4.1 Brief Overview

Cotton, commonly known as White Gold, is the main cash crop of Pakistan. The area under cultivation of cotton crops has increased significantly during last 55 years from around 1293 thousand hectares (1960–1961) to 2917 thousand hectares (2015–2016). Punjab is the major producer of cotton in the country with about 80% of the total cultivation area. It accounts for 5.1% of the value added in agriculture and about 1.0% to GDP. Besides providing raw material to hundreds of ginning factories and textile mills in the country, it contributes by two-third to the country's export earnings in the form of cotton commodities including textile. The livelihood of a large number of rural farmers is directly dependent on this crop and it employees millions of people along the entire cotton value chain.

15.4.2 Species of Cotton

There are 50 species of *Gossypium* but only four species are cultivated. Globally, *Gossypium hirsutum* is grown on more than 90% of the cultivated areas, followed by *G. barbadense* (8%–9%), *G. arborium* and *G. herbacium* (1%). In Pakistan, *G. hirsutum* is grown on almost 99% of the total cotton areas and is of economical and industrial significance in the country.

15.4.3 Global Production Scenario

Pakistan is the fourth largest cotton producer and the third largest exporter of raw cotton in the world. It is a leading exporter of yarn in the world. However, the average cotton yield of Pakistan ranks 13th in the world. Pakistan annually imports around 1.5–2 million bales of cotton to meet growing demand of local textile mills. It has now become vital for Pakistan to increase its yield per acre.

15.4.4 Ideal Growth Conditions

Cotton is grown between latitudes 37° north and 30° south in temperate, subtropical, and tropical regions of every continent.

Cotton requires long vegetation periods (150–180 days) without frost, constant temperatures between 18 and 30° with ample sunshine and fairly dry conditions. A minimum of 500 mm of water between germination and boll formation is required. Deep, well-drained soils with fairly good fertility are also required for its successful cultivation.

15.4.5 Cotton in Pakistan

Since the 1980s, the production of cotton entered a high-demand era in Pakistan. This was the period when cotton varieties, with high quality seed and high yield potential, paved the way for the production of cotton in the country. Production strategies changed from low-input to high-input, and self-sustaining to market oriented, legally protected and considerably mechanized system. The yields of 159 kg per hectare, during 1947–1948, improved to 277 kg per hectare, during 1975–1976, and grew further to 769 kg per hectare in 1991–1992. Cotton is one of the major crops of Punjab and has a long cultivation history there. Presently, there have been 7-fold increase in yields, 10-fold increases in production since 1947. Since, the highest production years were 1991, 2004, and 2011, whereas lowest production years have been 1976, 1983, 1993, and 2015.

Since 1960, there has been significant increases in cotton yields as the areas and production have increased. The environmental conditions were favorable for cotton growth and there was no attack of insects, pests and diseases, all this leading to good harvests. In 1988, cotton variety S12 was developed by the Cotton Research Station, Multan. It was a versatile variety and gave high yields until 1991–1992. In 1993, S12 became susceptible to cotton leaf curl virus and production dropped drastically. As this variety was grown on all major cotton producing areas, the production declined steadily. Nonetheless, the production improved during 1995–1996 with the introduction of NIAB-Krishma, FH-634, and CIM-448 which were resistant to the cotton leaf curl virus (CLCuV). The year 1998 again witnessed a sharp failure of cotton crops as a new Burewala strain of CLCuV struck cotton fields of the major cotton growing areas. After 2000, Bt varieties were introduced, which offered safety against bollworms. In 2004–2005, cotton was a bumper crop with the highest yield yet as the environmental conditions were quite favorable for cotton growth and did not experience any epidemic of disease or virus (Figure 15.14).

During 2009–2010, there was an increasing trend for the cultivation of Bt. Cotton varieties. Nonetheless, the production during 2010–2013 dropped by 11.3% as compared with previous year owing primarily to a decrease in cultivated areas and production due to floods, widespread attacks of CLCuV and sucking pest/insects in core and noncore areas. Moreover, the shortage of water due to canal closures during floods caused fruit shedding in certain areas. During 2013–2014, cotton production decreased again due to decline in the area sown as the farmers did not get proper rates of cotton (phutti), discouraging the growers to cultivate cotton. Instead the farmers shifted to cultivation of maize and rice in some districts of Punjab due to their better market returns. The year 2014–2015 also witnessed a huge decline in cotton production to 10.07 million bales, compared with 13.96 million in 2014–2015, and the initial estimate of 15.5 m bales. Unfavorable climatic conditions, disease epidemics, and a mixture of varieties under the name of Bt cotton were the major reasons for such unprecedented declines in cotton production.

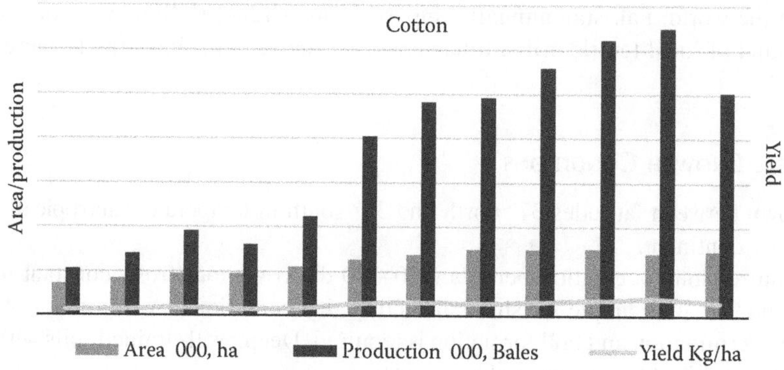

FIGURE 15.14 Area, production, and yield of cotton over 50 years.

15.4.6 Reasons of Stagnant Yields in Cotton

Cotton yields have been stagnant for the last several years due to factors discussed in the following subsections.

15.4.6.1 High Temperature

High temperatures reduced yields due to flowers shedding and less boll setting. In addition, high temperature coupled with high humidity boosted the whitefly population, which remained one of the major causes of yield reductions during 2015.

15.4.6.2 Weeds Infestation and Seed Quality

The extent of the weed population was enormous causing multiple negative effects such as nutrient depletion, competition for nutrients, sunlight, hibernating sites for insects/pests, and posing problems in picking. Moreover, high rains and humid environments led to the robust spread of weeds, and many farmers failed to control weeds successfully. Among biotic factors, whitefly, pink bollworm, armyworm, and jassid were the most damaging pests. Many farmers cultivated so-called Bt cotton varieties. However, there was large variation in crop stand throughout the cotton areas indicating that the varieties were not true to type. Although, availability of certified seeds has increased from 27% (2014) to 35% (2015), it clearly needs to be improved further (Table 15.2).

15.4.6.3 Poor Prices of Cotton Seed (Phutti)

Farmers could not fetch good prices for their produce, especially during July to September last year, which changed their mindsets about the cultivation of cotton. Although inputs were available

TABLE 15.2
Availability of Certified Cotton Seed From 2010–2016

Year	Total Seed Requirement (MT)	Availability of Certified Seed (%)
2010–2011	32,000	18
2011–2012	32,000	1.4
2012–2013	32,000	1.3
2013–2014	32,000	2.4
2014–2015	32,000	27
2015–2016	32,000	35

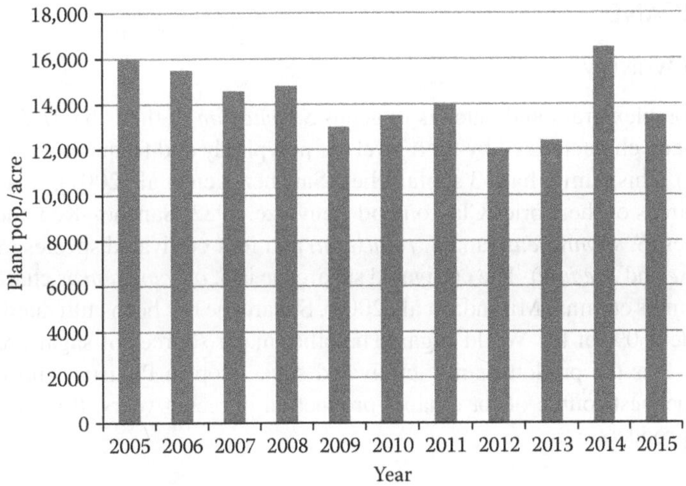

FIGURE 15.15 Plant population trend of cotton in Punjab 2005–2015.

in the market, farmers used these inputs quite reluctantly as they were not getting good price of their produce. Fertilizer and pesticide use was lower compared to 2014. Delayed planting, low plant population, poor plant growth/canopy further lowered the production.

15.4.6.4 Plant Population

Plant population was much lower during 2015 as compared to 2014. Apparently the germination of the seed was poor due to erratic rains in some cotton areas during the planting season. Scarce cotton populations led to the significant spread of weeds, which were not controlled (Figure 15.15).

15.4.6.5 Effect of Rainfall

Amongst the abiotic factors, climatic change, particularly early monsoons in the months of June and July, had an adverse effect on the growth and development of cotton crops. Unusually high rains (>300 mm) during the months of June, July, August, and September brought crops under severe stress at an early growth stage. High rains induced shedding of flowers, squares, and bolls, induced high plant mortality and leaching of nutrients from soil as well as high pest spread and stunted plant growth. Poor sunshine and high humidity, due to extended wet and cloudy weather, adversely affected nitrogen application, pesticide spraying, weed control, and other agricultural practices. All these adversities resulted in unprecedented declines in yields of cotton.

15.4.7 Future Way Out

Nonetheless, there are intrinsic effects of fluctuations in temperature, precipitation, floods, and other climatic conditions on agricultural productivity, the challenges faced by the cotton sector are more complicated than mere agronomic ones. At present, medium-staple cotton varieties are grown in the country and long-staple cotton is imported for blending. The productivity of cotton in Pakistan remains consistently lower than that of other major producers in the world. The overproduction of sugarcane, induced by an inflated price supports, has been cannibalizing land and water resources used for cotton, especially in the prime cotton zones. We need a systemic approach with some long-term policy interventions such as the development of indigenous varieties with double Bt genes, shifts in production from medium to long and extra-long staple cottons, innovations in harvesting and postharvest handling for quality enhancement, and securing the cotton belt from encroachment by other crops.

15.5 SUGARCANE

15.5.1 Brief Overview

Sugarcane is a complex grass and belongs to genus *Saccharum* of the tribe *Andropogonoaeae* and family *poaceae*, and characterized by high levels of polyploidy and frequently aneuploidy (Daniels and Roach, 1987). This family has 13 subfamilies (Sanchez-Ken et al., 2007). It is found in tropical and subtropical areas of the world (Clayton and Renvoize, 1982; Sanchez-Ken and Clark, 2010). It has two wild types (*S. spontaneum* and *S. robustam*) and four cultivated species (*S. officinarum*, *S. barberi*, *S. sinense* and *S. edule*). The cultivated sugarcane is *S. officinarum* as characterized by high levels of sucrose in its culmns (Miranda et al., 2008). Sugarcane has been cultivated since prehistoric times and provides 60% of the World sugar. The other major sources of sugar are sugar beet, and synthetic sugars. It is the predominantly cultivated sugar crop in Pakistan and is also becoming popular as the cheapest source of for ethanol production in recent years. It is the raw material for many industrial products.

15.5.2 Sugarcane Production in Pakistan

Sugarcane is the prime source of sweetener in our country and is second most important cash crop in Pakistan after cotton. It contributes 3.2% in agriculture value added and 0.6% in GDP (Anonymous, 2016). Pakistan is the 4th largest grower of sugarcane but average production has been very low (56.7 tons per hectare) as compared to developed countries (64–70 tons per hectare). Sugarcane was cultivated on an area of 1141 thousand hectares during 2014–2015, with a total production of 62.7 million tons. Pakistan ranks 6th in yield and 9th in sugar production in the world. During the past 40 years, there has been a continuous trend of increase in area as well as production; however, yields per hectare have been far less than for other advanced sugarcane producing countries.

During 2015–2016, the area under sugarcane cultivation decreased to 1132 thousand hectares as compared with the previous year's area of 1141 thousand hectares, a decline of 0.8%. On the other hand, sugarcane production increased from 62.8 million tons to 65.5 million tons, a 4.2% increase in production with an overall better yield (57.8 tons per hectare) compared the previous year (55.0 tons per hectare). Sugarcane is predominantly a tropical crop but its cultivation has extended rapidly over the subtropics between the latitude 37° North and 31° South and is cultivated mostly in Carabbean and Florida but also in Central and South America, India. The world's sugar production is centered around and on the equator. This belt extends northward to and somewhat beyond the tropic of Cancer and southward slightly past the tropic of Capricorn. This belt can be broadly divided into (1) a subtropical zone, irrigated with moderate temperature, between 26° N latitude and 30° N latitude; (2) an irrigated arid, subtropical zone between 24° N latitude and 26° N latitude; and (3) a temperate zone of north Punjab and N.W.F.P. between 32° N latitude to 34° N latitude. There are three major agroecological zones of sugarcane cultivation in Pakistan: (a) extreme temperate conditions in KP; (b) subtropical arid zones of the Indus valley; and (c) coastal areas of southern Sindh (Leff et al., 2004).

During the last decade, the year 2013–2014 has been the best for sugarcane with the highest ever production (67.4 million tons). The production of sugarcane decreased during 2014–2015 by 7.1% and was recorded as 62.8 million tons. Inspite of decrease in area under cultivation during 2015–2016, total production increased to 65.4 million tons (Figure 15.16). The decline in area was attributed to reduced purchases of cane by the mills and delayed payments to the farmers. The increase in yields was owed to favorable weather conditions in 2015–2016.

15.5.3 Global Scenario of Sugarcane Production

Sugarcane is the main sugar crop, which contributes to more than 60% of world sugar supply. Brazil is the largest sugarcane growing country followed by India, China, Pakistan, Thailand, Mexico,

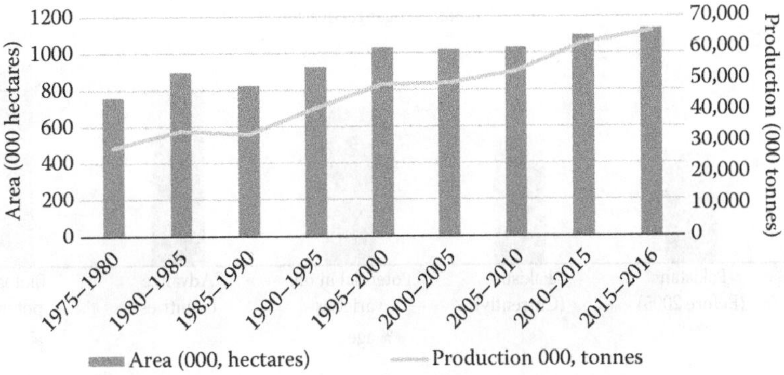

FIGURE 15.16 Trends in area and production of sugarcane in Pakistan over the past forty years.

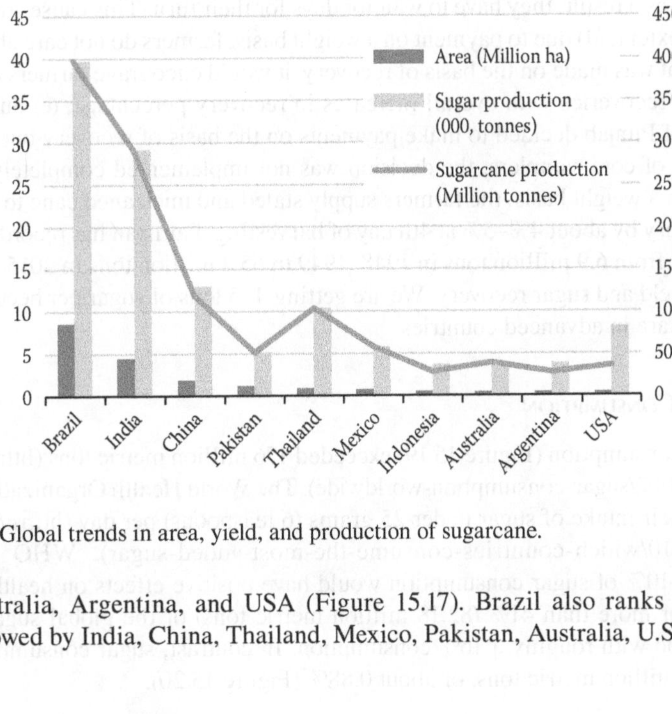

FIGURE 15.17 Global trends in area, yield, and production of sugarcane.

Indonesia, Australia, Argentina, and USA (Figure 15.17). Brazil also ranks first in sugarcane production followed by India, China, Thailand, Mexico, Pakistan, Australia, U.S.A, Indonesia, and Guatemala.

15.5.4 Trends in Sugar Recovery

Brazil is the largest sugar producing country (166.3 million metric tons) and has the highest sugar recovery of 14.50%; it is followed by Australia (13.80%) whereas average sugar recovery in our country merely stands at 10%. Some progressive growers and efficient sugar mills are getting 11.60% sugar recovery (Figure 15.18). Major impediments causing this low recovery are (a) the percentage of sucrose in cane, which varies from 8% to 16%, and depends to a great extent on the variety of cane. Most of varieties grown in Pakistan have lower sugar contents with low recovery. Recently, efforts have been made to promote high sugar varieties, which has increased sugar recoveries from 8.74%, during 2002–2003, to 10.00% in 2012–2013; (b) most sugarcane is planted after wheat, in the months of April–May, or in berseem during the month of April, instead of February–March. This results in delayed maturity of the crop leading to reduced sugar recovery; (c) unscheduled supply of cane to the mills is also one of the reasons for low sugar recovery. Sugarcane mill (cane collection department) mostly does not give the supply schedule to the growers. The farmers bring their cane to mill without

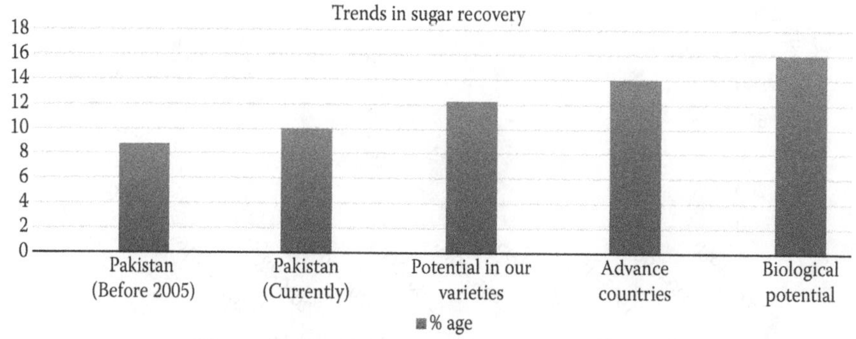

FIGURE 15.18 Comparative trends in sugar recovery in Pakistan in comparison with advanced sugar producing countries.

prior appointment as a result, they have to wait for days for their turn. This causes reduction in sugar recovery to great extent; (d) due to payment on a weight basis, farmers do not care about the maturity of cane. If payment was made on the basis of recovery, it would encourage farmers to grow varieties with higher sugar recoveries with overall increases in recovery percentage; (e) during 2012–2013, the Government of Punjab decided to make payments on the basis of recovery but due to problems in the availability of core samplers, the decision was not implemented completely; and (f) due to payments made on a weight basis, the farmers supply staled and uncleaned cane to the mills, which reduced the recovery by about 4%–5% at 4th day of harvesting. Pakistan has recorded a tremendous rise in production, from 6.9 million tons in 1948–1949 to 65.4 million tons in 2015–2016, with little improvement in yield and sugar recovery. We are getting 4–5 tons of sugar per hectare compared to 6–13 tons per hectare in advanced countries.

15.5.5 Sugar Consumption

The global sugar consumption (Figure 15.19) exceeded 176 million metric tons (http://www.statista.com/statistics/496002/sugar-consumption-worldwide). The World Health Organization recommends that adults keep their intake of sugar under 25 grams (6 teaspoons) per day (https://www.weforum.org/agenda/2015/10/which-countries-consume-the-most-added-sugar). WHO estimated that reductions in 5%–10% of sugar consumption would have positive effects on health. In 2014/2015, Asia accounted for more than 41% (82.18 million metric tons) of the global sugar consumption, followed by Europe with roughly a 16% consumption. In contrast, sugar consumption in Oceania amounted to 1.72 million metric tons, or about 0.88% (Figure 15.20).

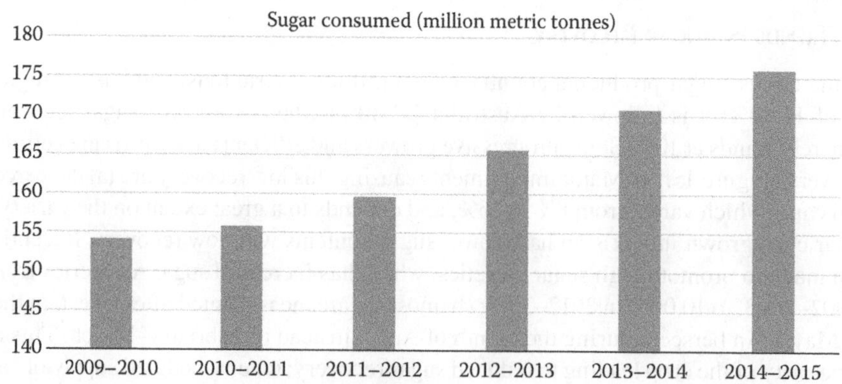

FIGURE 15.19 Global trends of sugar consumption.

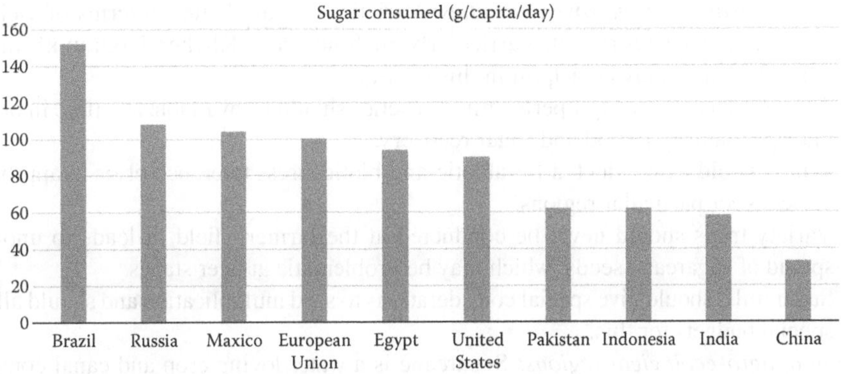

FIGURE 15.20 Difference in sugar consumption among different nations.

15.5.6 Sugar Prices and Self Sufficiency

Pakistan is among the highest sugar-consuming countries. Limited sugar supplies and the steady increase in prices have negatively affected household sugar consumption. While total consumption of sugar is increasing, it is growing at a slower rate than Pakistan's population growth rate of 2.5%. Consequently, per capita consumption of sugar is on the decline in Pakistan. Bulk consumers such as bakeries, makers of candy and local sweets, and soft drink manufacturers account for about 60% of the total sugar demand. Pakistan was a sugar importer in 2010–2011 but has been self-sufficient since 2012.

15.5.7 How to Promote Sugarcane Crop

The areas under sugarcane cultivation have been stagnant over the years but the production varied over the years. It is because most of the cultivars are not resistant to biotic and abiotic stresses. Yield per hectare is very low in our country, which indicates that we have been lagging behind the potential production compared to the rest of the world. The possible reasons are lack of a proper sugarcane breeding program, conventional methods of cultivation with little or no mechanization, and marketing distortions.

We not only get lower farm yields, but also have not developed the allied industries which can use its by-products and contribute to the national economy. We only export ethanol and bagasse (80% of total) for a worth of 1.0 billion US$ annually. The only outcome of the 20% bagasse we process is ethanol. If sugarcane by-products, including bagasse/molasses, are used wisely they could create a multibillion dollar industry linked to sugarcane. In countries like Brazil, sugar is the by-product and other things like bagasse (this is by-product for us) are used to produce several valuable products. They also get 50% of their energy from sugarcane biofuels.

Even as a neglected industry, it still contributes to value added in agriculture (3.5%) and GDP (0.7%). Given proper consideration by the government and the industry, this crop can do wonders for the economy of the country. The public sector needs to formulate such policies for investment in the expansion of sugarcane through research innovations and setting up industries for value addition. The overall recommendations to improve sugarcane crop are as follows.

Opting for efficient production technologies: Usually the farmers' practice of fertilization is not based on soil analysis. There is dire need for nutrient indexing of the whole sugarcane growing area. Farmers should be educated about the concept of balanced fertilizer application with proper methods and at the proper time so that higher fertilizer use efficiency is realized. Proper row spacing, method of sowing, sowing time, seed rate, weed control, and insect pest management are other components of a successful package of production.

Promoting cultivation of improved varieties: Varieties that are better in terms of yield and recovery should be promoted, particularly in Sindh and Khyber Pakhtunkhwa. The following measures may be helpful in this regard.

- At least two to three high-performing varieties should be available at a time in order to manage maturity period and sugar recovery.
- Trials should be conducted for abiotic and biotic stress tolerance prior to approval of varieties for particular regions.
- Variety trials should never be conducted at the farmer's field. It leads to unofficial spread of sugarcane seeds, which may be problematic at later stages.
- Sugar mills should give special considerations to seed multiplication and should allocate special budgets for this.

Rezoning of agro-ecological regions: Sugarcane is a water loving crop and canal command areas with well-drained soils are most suitable for its cultivation. There is dire need of rezoning of the agro-ecological regions for sugarcane cultivation so that the highest yields are realized. This would also help to classify varieties adaptable to specific ecological conditions with sustained yield.

Encouraging sowing of high sugar cultivars: Payment should be made to the farmers on the basis of sugar recovery. This would promote sowing of high sugar varieties and hence better incentives to the industry as well.

Proper management of crop harvest: Sugar recovery is strictly dependent on the time period from harvesting to crushing. Maximum recovery is reported to be attained with 24–48 hr of harvesting so, harvesting should be properly planned keeping in view this aspect of recovery.

Payment of proper price to the famers: Most of the farmers in our country have small landholdings and are completely dependent on the income generated from their agricultural production. Proper price of the crop should be paid to farmers in time so that they are better able to manage their fields for better productivity.

Mechanization of sugarcane cultivation: Sugarcane is a labor-intensive crop. With the increasing trend in shift of farm labor to urban areas, the growers find it difficult to manage the crops properly. Farm labor also is not working properly. Every 1% increase in trash (extra dry leaves) to cane reduces sugar recovery by 0.12%. So, opting for mechanized farming may be helpful to reduce recovery losses as well as to improve crop yields. Governments should encourage mechanization of sugarcane cultivation by importing the proper machinery.

Proper use K fertilizer: Potassium is one of the key nutrients playing an important role in enhanced sugar recovery. The farmers use urea as a first choice and then DAP, whereas use of potassium fertilizers is very limited. This not only increases chances of crop lodging but also result sin poor sugar recovery. Farmers need to be made aware of using potash fertilizers.

Opting technological innovations: Technological innovations are the need of the time. Employing advanced techniques, sugarcane genotypes with improved sugar recovery and with improved biotic and abiotic stress tolerance can be developed. Similarly, use of slow release or controlled release fertilizers will be of great impact for enhancing nutrient use efficiency, and ultimately leading to improved yields.

Promoting research and development: A sound varietal development program run as a joint public-industry venture should be in place. Currently, variety developmental program is dependent on imported fuzz from Brazil, the United States, Australia, and Sri Lanka, which is not properly managed and planned (http://www.psmacentre.com/misc.php?miscid=11&type=sugarcane_varities&status=1). In fact, the sugarcane breeding program in the country is obsolete and needs innovative interventions. Establishment of a Sugarcane Research and Development Board (SRDB) is a good initiative by the Government of Punjab and is likely to contribute towards sugarcane research and development.

REFERENCES

Ahmad MD, Turral H, Masih I, Giordano M and Masood Z. 2007. *Water Saving Technologies: Myths and Realities Revealed in Pakistan's Rice-Wheat Systems.* International Water Management Institute, Colombo, Sri Lanka, 44p.

Akbar N, Jabran EK and Ali MA. 2011. Weed management improves yield and quality of direct seeded rice. *Aust. J. Crops Sci.* 5: 688–694.

Ali A, Erenstein O and Rahut DB. 2014. Impact of direct rice-sowing technology on rice producers' earnings: Empirical evidence from Pakistan. *Dev. Stud. Res.: An Open Access J.* 1: 244–254.

Anonymous. 2001. *Social Development in Pakistan.* Annual Review 2001. Growth, inequality and poverty. Social Policy and Development Center, Karachi. Oxford University Press, Karachi, p. 4.

Anonymous. 2011. *Pakistan Economic Survey 2010–2011.* Economic Advisor's Wing, Ministry of Finance, Islamabad, Pakistan, p. 19.

Anonymous. 2013. *Pakistan Economic Survey 2012–2013.* Economic Advisor's Wing, Ministry of Finance, Islamabad, Pakistan, p. 21.

Anonymous. 2015. *Pakistan Economic Survey 2014–2015.* Economic Advisor's Wing, Ministry of Finance, Islamabad, p. 28.

Anonymous. 2016. *Pakistan Economic Survey 2015–2016.* Economic Advisor's Wing, Ministry of Finance, Islamabad, p. 28.

Baloch MS, Awan IU, Jatoi SA, Hussain I and Khan BU. 2000. Evaluation of seeding densities in broadcast wet seeded rice. *J. Pure Appl/. Sci.* 19: 63–65.

Bashir K, Khan NM, Rasheed S and Salim M. 2007. Indica rice varietal development in Pakistan: An overview. *Paddy Water Environ* 5: 73–81.

Bhushan L, Ladha JK, Gupta RK, Singh S, Tirol-Padre A, Saharawat YS, Gathala M and Pathak H. 2007. Saving of water and labor in a rice-wheat system with no-tillage and direct seeding technologies. *Agron. J.* 99: 1288–1296.

Borojevic K and Borojevic K. 2005. The transfer and history of reduced height genes (Rht) in wheat from Japan to Europe. *J. Hered.* 96: 455–459.

Chauhan B and Johnson D. 2009. Ecological studies on *Cyperus difformis, Cyperus iria* and *Fimbristylis miliacea*: Three troublesome annual sedge weeds of rice. *Ann. Appl. Biol.* 155: 103–112.

Chauhan BS. 2012. Weed ecology and weed management strategies for dry-seeded rice in Asia. *Weed Technol.* 26: 1–13.

Clayton WD and Renvoize SA. 1982. Gramineae. In: Polhill RM (ed) *Flora of Tropical East Africa*, Balkema, Rotterdam. Part 3: 700–767.

Daniels J and Roach BT. 1987. Taxonomy and evolution, In Heinz DJ (ed) *Sugarcane Improvement through Breeding.* Elsevier, Amsterdam. pp. 7–84.

Dorosh PA and Wailes E. 2010. The international rice trade: Structure, conduct, and performance. In: Pandey S, Byerlee D, Dawe D, Dobermann A, Mohanty S, Rozelle S and Hardy B (eds) *Rice in the Global Economy: Strategic Research and Policy Issues for Food Security*, International Rice Research Institute, Los Banos (Philippines). p. 362.

Ehsanullah N, Jabran AK and Habib T. 2007. Comparison of different planting methods for optimization of plant population of rice (*Oryza sativa* L.) in Punjab (Pakistan). *Pak. J. Agric. Sci.* 44: 597–599.

FAO-Food and Agriculture Organization. 2013. *FAOSTAT Database "Countries by Commodity (Rice, Paddy)".* FAO, Rome, (Accessed on May 2014).

Fujisaka S, Harrington L and Hobbs PR. 1994. Rice-wheat in South Asia: Systems and long-term priorities established through diagnostic research. *Agric. Syst.* 46: 169–187.

Hanif M, Khan SA and Nauman FA. 2004. *Agricultural Perspective and Policy.* Ministry of Food, Agriculture and Livestock, Islamabad, Pakistan.

http://www.psmacentre.com/misc.php?miscid=11&type=sugarcane_varities&status=1

http://www.statista.com/statistics/496002/sugar-consumption-worldwide

Kataki PK, Hobbs P and Adhikary B. 2001. The rice-wheat cropping system of South Asia: Trends, constraints and productivity—A Prologue. In: The rice-wheat cropping system of South Asia: Trends, constraints, productivity and policy. *J. Crop Prod.* 3: 1–26.

Kumar V and Ladha JK. 2011. Direct seeding of rice: Recent developments and future needs. *Adv. Agron.* 111: 297–413.

Leff B, Ramankutty N and Foley JA. 2004. Geographic distribution of major crops across the world. *Global Biogeochemical Cycles*, 18: GB1009.

Lelieved J, Crutzen PJ and Dentener FJ. 1998. Changing concentration, lifetime and climate forcing of atmosphere methane. *Tellus Ser. B.* 50B: 128–150.

Mann RA, Ahmad S, Hassan G and Baloch MS. 2007. Weed management in direct seeded rice crop. *Pak. J. Weed Sci. Res.* 13: 219–226.

Mann RA and Ashraf M. 2001. Improvement of Basmati and its production practices in Pakistan. In: Chaudhary RC, Tran DV and Duffy R (eds) *Specialty Rice of the World: Breeding, Production and Marketing*, Food and Agricultural Organization of the United Nations, Rome. pp. 129–148.

Matloob A, Khaliq A and Chauhan BS. 2015. Weeds of rice in Asia: Problems and opportunities. *Advance. Agron.* 130: 291–336.

Miranda LLD, Vasconcelos ACM, Landell MG and Xavier MA. 2008. Viveiro de mudas. em: Leila luci dinardo-miranda; antõnio carlos machado de vasconcelos; marcos guimarães de Andrade landell. (org.). cana-de-açúcar. led. *Campinas* 1: 535–546.

Parthasarathi T, Vanitha K, Lakshamanakumar P and Kalaiyarasi D. 2012. Aerobic rice-mitigating water stress for the future climate change. *Int. J. Agron. Plant Prod.* 3: 241–254.

Rao AN, Johnson DE, Sivaprasad B, Ladha JK and Mortimer AM. 2007. Weed management in direct-seeded rice. *Adv. Agron.* 93: 153–255.

Rosegrant MW, Cai X, Cline SA. 2002. *Gloxbal Water Outlook to 2025: Averting an Impending Crisis.* International Food Policy Research Institute, Washington USA, pp. 2–24.

Salim M, Akram M, Ehsan M, Ashraf AM. 2003. Rice, a production handbook. Pakistan Agricultural Research Council, Islamabad, Pakistan.

Sanchez-Ken JG and Clark LG. 2010. Phylogeny and a new tribal classification of the Panicoideaes. 1. (*Poaceae*) based on plastid and nuclear sequence data and structural data. *Am. J. Bot.* 97: 1732–1748. doi:10.3732/ajb.1000024.

Sanchez-Ken JG, Clark LG, Kellogg EA and Kay EE. 2007. Reinstatement and emendation of subfamily Micrairoideae (*Poaceae*). *Syst. Bot.* 32:71–80. doi:10.1600/036364407780360102.

Seck PA, Diagne A, Mohanty S and Wopereis MCS. 2012. Crops that feed the world 7: Rice. *Food Sec.* 4: 7–24.

Shewry PR 2009. Darwin review—Wheat. *J. Expt. Bot.* 60: 1537–1553.

Singh Y, Singh G, Singh VP, Singh RK, Singh P, Srivastava RSI, Saxena A, Mortimer M, Johnson DE and White JL. 2002. Effect of different establishment methods on rice-wheat and the implication of weed management in Indo-Gangetic plains. In: *Proceedings of the International Workshop on Herbicide Resistance Management & Zero Tillage in Rice-Wheat Cropping System*, Department of Agronomy, CCS Haryana Agricultural University, Hisar, India. pp. 188–192.

Sleper DA and Poehlman JM. 2006. *Breeding Field Crops.* Blackwell Publishing, New York, USA.

Tuong TP and Bouman BAM. 2003. Rice production in water-scarce environments. In: *Proc. Water Productivity Workshop*, November 12–14, Colombo, Sri Lanka. International Water Management Institute, Colombo, Sri Lanka.

Vorosmarty CJ, Green P, Salisbury J and Lammers RB. 2000. Global water resources: Vulnerability from climate change and population growth. *Science* 289: 284–288.

Wang H, Velarde O and Abedullah A. 2012. Pattern of varietal adoption and economics of rice production in Pakistan. In: Wang H, Pandey S, Velarde O and Hardy B (eds) *Patterns of Varietal Adoption and Economics of Rice Production in Asia*, International Rice Research Institute, Los Baños (Philippines). pp. 67–71.

16 Crop Diversification

Shahzad M. A. Basra, Maqsood Hussain, Abdul Wahid, and Muhammad Farooq

CONTENTS

16.1 INTRODUCTION

A national policy to maximize the utilization of agricultural resources, for the overall agricultural development in the country, is termed "crop diversification." It offers many viable options to the farmers to grow a variety of different crops on their farms and it decreases the negative effects of monoculture and provides a better system of crop specialization for efficient resource utilization. Crop diversification maximizes profit for the farmers from whole farm operations.

Two types of crop diversification exist at farm level: One is vertical diversification and second is horizontal diversification. In vertical diversification, farmers engage in different value addition or other businesses such as poultry, livestock, or fish farming along with crop cultivation. In horizontal diversification, farmers grow many different new kinds of crops such as vegetables, minor crops, and fruits along with major traditional crops.

In Pakistan, the root of economic development is agriculture, which contributes by ~60% towards employment, food security, and income. There is a need over time to enhance the agricultural productivity to meet the increasing demand for food. In addition to the perpetual roles of agriculture in terms of high yielding varieties of wheat and rice, the growth of agriculture is also acknowledged for diversification in the context of high quality commodities. It is hoped that the progress of Pakistan's agricultural exports will become smoothed by the implementation of the provisions of World Trade Organization's (WTO) Agreement on Agriculture (AoA) through restricting the agricultural subsidies of industrialized countries. Hence, to facilitate the agricultural exports of nontraditional commodities, there is need for agricultural diversification. Nature has endowed Pakistan with diverse types of soils, climates, and agro-ecological resources. Farmers can grow a mixture of crops, off-season vegetables, cut flowers, and can rear special breeds of livestock due to presence of wide diversities across the country (Naseer, 2016).

TABLE 16.1
The Share of Total Cropped Area of Different Crops (1990, 2000, and 2010)

Area under Crops (000 ha)	1990	2000	2010
Total cropped	21,820	22,040	23,670
Wheat	36	37	39
Cotton	12	13	13
Pulses	7	6	6
Sugarcane	4	4	4
Vegetables	1	1	2
Oilseed	2	3	3
Fruits	2	3	4
Fodder	13	11	11

Source: Abro, A.A. 2012. *Int. J. Buss, Mgt. Eco. Res.*, 3: 536–545.

During the year 2010, 76.37% of the total cropped land was shared by six major crops (wheat, cotton, rice, maize sugarcane, and gram), while 23.63% was cultivated with other minor crops, which shows that there is a plenty of demand for crop diversification to increase the earnings of farmers, especially the small farmers. However, the availability of infrastructures and facilities such as transportation, market, storage house, irrigation, and electricity are the key elements determining the income from diversification (Abro, 2012). Diversification is an essential part of the process of structural transformation of an economy. In Pakistan, during the period from 1990 to 2010, the crop diversification process has begun because of the introduction of modern agricultural technologies. The share of total cropped areas for different crops (1990–2010) is presented in Table 16.1.

The adoption and implementation of crop diversification in Pakistan has been influenced significantly by economic factors. Moreover, the development of drought-resistant crops, penetration of rural markets, and agricultural infrastructure have contributed in this regard. Thus, in the next stage of crop diversification, more and more growth is possible from the production of value-added agricultural commodities. The value added commodities are providing higher yields as well as earlier returns compared to traditional food grains. However, the adoption of value added commodities is comparatively slow in Pakistan because of the lack of appropriate markets, institutions, and unenthusiastic policy response.

Crop diversification is a vital tool to balance and to improve the utilization of agricultural resources in agro-ecosystems. It offers a wider approach to grow several crops and reduce the negative impacts of monoculture and its associated problems. As mentinned before, there are two approaches for crop diversification: Vertical (value addition or business farming) and horizontal (different crops like vegetables, minor crops). In this approach, farmers can grow different crops having commercial value due to the wide diversity of ecosystems. Crop diversification strengthens the economic infrastructure of an ecosystem. In this chapter, significance of crop diversification in national scenario will be discussed, and potential alternative crops will be proposed for inclusion in the existing cropping system of Pakistan.

16.2 CROP PRODUCTION AND ECONOMIC SCENARIO

Price volatility in inputs and product prices in agriculture is very common. Climatic factors such as floods and droughts, as well as outbreaks of diseases can significantly affect the market. Crop diversification also provides protection to the farmers from climatic and biological risks and uncertainty. Farmers are now more exposed to market forces because domestic markets are linked with international markets through exchange markets. Crop diversification is a risk management plan and an important step for the shift from subsistence to commercial agriculture, most of the time on economic considerations.

In agriculture, the integral part of management is to avoid or overcome the shocks of risks. Government policy to provide training and information to farmers can be very helpful in assessing and managing these risks. The crop diversification itself is a very useful risk management tool and is always a good policy to minimize risk. At farm level, the income variability can be reduced through the diversification of the enterprise by combining different production packages. Different crops and livestock can be included in the diversification. When low income from one enterprise is simultaneously counterbalanced by a high or satisfactory income from other enterprises, it is termed as an efficient diversification.

In farming, business risk is an important feature. The factors that impact and cause large fluctuations in farm income are uncertainties in weather, prices, yields, global market, and are government policies. Selections among alternatives that reduce financial effects result from such uncertainties that come under risk management (USDA, 2016). There are several types of risks; however, production risk and price or market risk are very important, as explained below.

16.2.1 Production Risk

Risk in production systems from the natural indeterminate growth processes of crops and livestock. The risk factors, which affect both the quality and quantity of produce are diseases, pests, and weather change. Production risk rises with the introduction of new technologies and several strategies can be used to reduce them.

16.2.1.1 Risk Reducing Inputs

These are the inputs involved in production that improve the quality and quantity of the produce. To reduce the risk of low yield, composts and fertilizers are used. To reduce the risk of crop damage, Integrated Pest Management (IPM) and pesticides are used. To reduce the risk of low rainfall, irrigation is used.

All inputs do not necessarily reduce risk. Crops will still depend on rainfall, which may or may not be favorable along with fertilizer application. In low soil moisture levels, the application of fertilizers results in low yield. Farmers do not experience one type of production risk at a time only. They also face the risk of weeds, pests, and unfavorable weather at the same time. Using one type of input, which reduces one type of risk, will not prevent low yields caused through insect and pest damage. As an example, the use of hybrid seeds may increase yields when rainfall is good but produces poorer yields than traditional varieties of seeds when rainfall is poor. Thus, farmers should calculate the benefit cost ratio of the input before using an input as a risk-reducing strategy.

16.2.1.2 Risk Reducing Technologies

By using new practices and technologies farmers can reduce specific risks common to their area of production. For example, new varieties of seeds are being developed and livestock are being bred for their unique characteristics.

16.2.1.2.1 Selection of Low-Risk Activities

By selecting a farm enterprise that has low risk is a way to reduce production risks. Farmers choose reliability over probability in these situations. An activity is abandoned by farmers when it has a high potential for income but also carries a high risk for loss, and another less risky activity is chosen instead. For example: a high yielding variety of millet or sorghum is not preferable for a farmer, which fails in drought compared with low yielding but drought resistant varieties.

16.2.1.2.2 System Flexibility

Flexibility is an important strategy for the management of risk in farming systems. A flexible farming system provides the farmer with a wide range of options for short-term and quick changes in sales and production. Examples of flexibility are: (*i*) adjustments of land area under cultivation or number of farm animals kept, in response of market changes; (*ii*) leave land unplanted (fallow) when rainfall is low to avoid risk expenditure on inputs; (*iii*) intensified farming if future prices are likely to rise; and (*iv*) labor utilization instead of hiring or purchasing farm machinery.

16.2.1.2.3 Production Diversification

Diversification of the production system is a successful risk management and low risk spreading strategy, owning to no change in all farm operations and enterprises, which are affected by changing the conditions. Some ways include:

1. Management of many farm enterprises together at a time (or in the same season); farmers may be apprehensive that diseases and pests can cause failure of their normal crop stand. In such conditions, farmers may grow more than one crop (i.e., multiple enterprises) in the same season. They grow more diseases and pest resistant crops. As mostly maize and sorghum are often grown together, maize is likely to fail in drought but is more resistant to bird damage, whereas sorghum is drought resistant but at risk to bird damage.
2. Engaging the same farm enterprise in different physical locations; farmers diversify to avoid risk by cultivating crops on different soil types. Crops on sandy upland soil may fail in a dry year; crop on wet, river-valley land may fail in a wet year. By cultivating crops on different soil type locations, farmers divide risks.
3. Engaging the same farm enterprise over successive periods of time (or seasons); farmers can diversify it over time. To avoid the risk of water stress, spread out (staggered) planting can be used. If an early planted crop does not receive sufficient rain, then a late planted crop may not be affected in the same way. In a similar way farmers may increase the production to meet the increased demand for vegetables and poultry on festival days.
4. Off-farm activities also produce income. Many farmers get income from the off-farm activities. Production diversification sum up can be used to manage yield, income and price risk. In the diversification, a farmer can divide the farm risks over several enterprises. By observing a farm as a whole, one can observe the effects of production diversification (Kahan, 2013).

16.2.2 Price or Market Risks

Market risk is the uncertainty in price producers pay for inputs or receive for commodities they produce. From one commodity to another, the nature of price risk differs. Marketing risk or price risk is due to the uncertainty of future market prices and the variability of product prices that the farmer faces when they are making a decision about the produce. Many ways are used to overcome the variability in price, as discussed in the following sections.

16.2.2.1 Spreading Sales

Sale spreading is a process in which the crop harvest is stored and sold at different times according to price variability. For example, as vegetables are processed by solar drying they can be sold at times of short supply, thus a farmer can store his produce and wait for better prices.

16.2.2.2 Direct Sales

Direct selling includes selling the produce directly to final consumers to avoid the risk and to enhance profitability. In direct sales, small farmers can never benefit by direct selling to final consumers.

16.2.2.3 Market Price Information

In managing price risk, the tracking of relevant price information is a key element. The short-term market price information provides the farmers information about decision on selling the produce. This involves date to date price information and up-to-date information on demand and supply. To make planting decisions and to plan marketing strategies, long-term market information can be used which covers annual or quarterly price reports from market information services, as well as contracts with companies providing services such as storage, transport, and inputs (fertilizers and seeds) (Kahan, 2013).

16.3 PATTERNS OF CROP DIVERSIFICATION

Crop diversification intensity is the number of crops produced along with their combined level of spread or concentration variation in the density of crop(s) in a region at certain period. Less diverse provinces are those which produce less numbers of crops, whereas those provinces which produce relatively greater numbers of crops are more diverse. The pricing policies of the government also negatively impinge on the farmers' land allocation decision to different crops, due to the relative prices of inputs and outputs.

The factors that influence area allocation patterns among small farmers are; farm size, investment constraints, and food self-sufficiency. Among larger farmers these factors usually are more related to economic constraints based on relative crop prices rather than resource constraints.

Crop diversification is basically a shift from the agricultural to the industrial setting. The diversification of commercial crops in agriculture has become a very important policy to increase income and minimize risks due to crop failures. The adoption of crop diversification policy can lead to both individual and social gains (Haque, 1996). Crop diversification can also be utilized to reduce poverty, create employment, and to protect the environment from hazardous effects (Hayami and Otsuka, 1995).

16.4 CROP DIVERSIFICATION AS A STRATEGY FOR VARIOUS NATIONAL COMMITMENTS

Crop diversification may help meet the national commitments for food and nutritional security. Crop diversification, as a strategy for various national commitments, is discussed below.

16.4.1 Food and Nutritional Security

Crop diversification is a vital tool for achieving sustainable agriculture growth and ensuring food and nutritional security (Behera et al., 2007). Changing lifestyles and enhanced urbanization have altered the need for food consumption (food grains to nonfood grains). Food consumption may be influenced by food prices and household incomes. Yield stability is an important component targeted by smallholder farmers. For economic returns and food, smallholder farmers rely on seasonal yields. Therefore, the consequences of yield fluctuation can important since they impact availability of food and income for other basic necessities for the family. That's why focus on crop diversification can help farmers to cope with food insecurity due to possible enhancements in yields (Cowger and Weisz, 2008) and it will result in stability of yields and an insurance effect (Yachi and Loreau, 1999), because if one crop fails to perform, they can rely on the other crops. Enhancing crops diversity by intercropping of sugarcane, maize, tobacco, wheat, broad bean, and potato were reported to improve yields for some combinations between 33.2% and 84.7% (Li et al., 2009). Likewise, it was reported that cereals grain yield increased in field trials of cereals variety mixtures (Kiær et al., 2009).

Crop diversification can cause a reduction in risks related to climatic variability and biotic stresses, especially in fragile ecosystems, and reduce fluctuations in commodity by introducing unique varieties or increasing locally adapted varieties. This will help to improve food security and generate income for resource-poor farmers (Mahapatra and Behera, 2004). In an agro-ecosystem, crop diversification not only helps in the better exploitation of agro-ecological mechanisms, but also provides household income, permitting buying of alternative foods and also offers diversity for human diets. Thus, diversification in production and consumption by inclusion of various plant species can play a role in improving food security, health, and nutrition. Nutritional balance of the diet can be promoted by the diversified production systems, which enhance dietary diversity (Ali and Farooq, 2003).

16.4.2 Natural Resource Management for Sustainable Agricultural Development

Crop diversification is an approach to increase the use of several natural resources, including land and water, for the development of agriculture and gives the farmers different options to produce various crops in different agroclimatic conditions (Acharya et al., 2011). Sustainable agriculture can be viewed as a production system which contributes in improving ecological resilience, livelihoods of people, and economic security. For achieving sustainable agriculture, the potential for enhanced crop diversification relies on local scenario, involving critical factors such as farm assets, human capital, off-farm income, migration and input and output market infrastructure (Smale and King, 2005). Reduction in nutrient losses, better resource use efficiency, enhanced stability, and improved productivity of production systems are the ecological benefits obtained from crop diversity (Hooper and Vitousek, 1997; Tilman et al., 1997; Reich et al., 2001). The possibility of total crop failure is reduced due to the higher level of crop diversity which increases system economic stability under a wider range of ecological conditions (Naeem et al., 1994; Chapin et al., 1997).

Diversification is an important requirement to restore and increase the value of degraded natural resources. New cropping pattern and cropping systems have been introduced on various occasions to enhance or retain the importance of natural resources, especially land and water. For example, in India, especially in West Bengal, wheat has been introduced in rice systems for better utilization of residual moisture and therefore, wheat irrigation requirements have been diminished (Behera et al., 2007). On the other hand, in Punjab, the problems of salinity and waterlogging have been increased due to unjudicious crop-mixing (such as wheat-rice). This challenges sustainable agriculture production and farm income (Vyas, 1996). Crop diversification can achieve the objective of optimum use of water, protecting land from salinity and water logging.

16.5 POTENTIAL ALTERNATE/NEW CROPS

Food security around the world is threatened by several factors such as growing populations and the resultant increased food demands. The production of higher quantities of grain crops is directly linked to optimal input use in crop production. Our consistent food habits and limited food diversity are also among the other factors which may pose as a significant threat to the availability of food for people. Human beings are currently dependant upon four cereal crops (rice, wheat, barley, and maize), the production of which may not be compatible with the existing population pressure. Thus, there is a dire need to bring in a radical shift in existing cropping patterns and a possible change in food preferences. To achieve this objective, it is imperative that new crops be introduced among the farmers, marketers, and consumers. However, this may not be possible unless these become known to relevant stakeholders in terms of potential yields, production technologies, and health benefits.

Among the important crops considered as potential alternate/new crops for this are quinoa (*Chenopodium quinoa*), moringa (many specifies but we shall discuss the most important *Moringa oleifera*, Amaranth (*Amaranthus* spp.), teff (*Eragrostis tef*), stevia (*Stevia rebaudiana*), chia (*Salvia hispanica*), and soybean. However, there may be questions and reservations about their place in the existing cropping systems of major crops in Pakistan, as well as farmer's preferences to adapt to them as new or alternate crops.

16.5.1 Quinoa

Quinoa is listed as one of the world's healthiest foods. The United Nations, Food and Agricultural Organization (FAO) officially celebrated the year 2013 as "The International Year of the Quinoa" due to its important role in the achieving world food security, impressive biodiversity, and high nutritive value (UN News Centre, 2013). Its grains are rich in proteins (De Bruin, 1963), essential amino acids, minerals (Zn, Mg, Cu, Fe, Ca), and vitamins (B2, E, A) (Repo-Carrasco et al., 2003). Quinoa is not a new crop. It has been cultivated for ages in Bolivia and the Peruvian Andean regions (Garcia et al. 2007). Since it is originally from a very harsh climateso it has a climate-proof behavior and can be grown on a variety of soils, ranging from clayey to sandy soils having pH 4.5–9. It is well adapted to frost (Jacobsen et al., 2005, 2007), salinity (Jacobsen et al., 2003; Ruffino et al., 2010; Hariadi et al., 2011), and can tolerate water-deficit conditions (Jensen et al., 2000; Garcia et al., 2003, 2007). Due to these features, quinoa is gaining worldwide popularity and its demand is continuously increasing (Jacobsen, 2011).

16.5.1.1 Botanical Description

Quinoa (*Chenopedium quinoa* willd.) belongs to family Amaranthaceae and subfamily Chenopodioideae. Quinoa was domesticated within the interior basin of Lake Titicaca at ≥3500 m above sea level. Subsequently, the crop experienced prolonged selection in an environment that is extremely adverse with respect to abiotic stresses (except for heat). The crop is grown mainly for its grains. Since, quinoa does not belong to Poaceae, so it is a pseudocereal (not a true grain). It is also C_3 halophyte crop plant. Quinoa plant can grow up to a height of 1–3 m depending upon sowing density. Its seed germinate after few hours of imbibition. The roots penetrate almost 30 cm deep in rhizosphere. Stem is cylindrical, 2–3.5 cm in diameter. Seeds are round to flattened, 1.5–4 mm in diameter; 1 g of lot contains 350 seeds (Ruales and Nair, 1994). Seeds also vary in color and size (Mujica, 1994), the color may vary from black to white and grey.

16.5.1.2 Nutritional Value

Grains of quinoa are mainly used in food items. Both sprout and grain are good example of functional food, which lowers the risk of many diseases or exerts positive effects on health other than its nutritional value. Public interest is increasing worldwide to replace common grains of common cereals with high nutritional value grains like quinoa (Pasko et al., 2009). Quinoa can be used as green fodder. The residues of quinoa crops are used as feed for sheep, pigs, cattle, poultry, and horses. There are several quinoa products used in food, pharmaceuticals, and cosmetics.

Quinoa grains contain all essential amino acids including lysine, threonine, tryptophan, which are deficient in most cereals. Protein quality of quinoa resembles that of milk casein protein (Repo-Carrasco et al., 2003). The grain protein contents ranges from 13% to 19% (Wright et al., 2002). Almost 58.1%–64.2% of the dry matter of its grain is starch, out of which 11% is amylase (Lorenz and Coulter, 1991; Jian and Kuhn, 1999). Quinoa starch granules have a polygonal form with a diameter of 2 μm. The starch granule of quinoa can beneficially use as biodegradable fillers in polymer packaging, which contain an excellent freezing ability (Ahamed et al., 1998). Similarly, quinoa flour has low percentage of glucose and fructose, and a high concentration of D-xylose and maltose, which are used in malted drink formulations. In quinoa, Ca, Mg, Fe, Cu, and Zn are present in high concentrations compared to rice, wheat, barley, and maize (De Bruin, 1963; Repo-Carrasco et al., 2003; Dini et al., 2005). All the main vitamins, especially α-carotene, niacin, thiamine, riboflavin, and α-tocopherol (vitamin E) are present in quinoa grains. Quinoa grain also contains fatty acids beneficial to health like linoleic and oleic acids (Ruales and Nair, 1994).

16.5.1.3 Quinoa Success Story of Pakistan

Quinoa was introduced in Pakistan during 2009 by Alternate Crops Lab, University of Agriculture, Faisalabad (Basra et al., 2014). Quinoa accessions showed variation in their adaptation to local climate. Stable accessions were further tested on salt-affected and polluted soils in different agro-ecological zones. The basic production technology has been developed and disseminated to local farming communities. Yield performance and nutritional profile so far explored represent that local produce has comparable yield and nutritional characteristics with the grain produced in the native lands. Its local consumption is increasing and exporters have been interested in increasing its production to make it an exportable commodity.

16.5.1.4 Production Technology

Quinoa can be grown in Pakistan as a rabi crop. It can be sown between mid-November to mid-December throughout Pakistan, except in the northern areas. Two to three ploughings, followed by planking, are necessary to prepare soils for sowing of quinoa. Optimum seed rate for quinoa is 5–8 kg ha^{-1} on flat or ridge sowing in lines manually or by seed drills on well-drained soils in *watter* condition. Quinoa is very responsive to nitrogen (N), so a fertilizer at the rate of N:P:K of 75:60:50 kg ha^{-1} is recommended. Phosphorous (P) and potassium (K) should be applied as basal dose N should be applied in two splits; half at the time of sowing and half at 70 days after sowing. Weed infestation could cause severe damage during the early phenological stage up to bud formation stage. Weeding is mostly done manually. However, experimentation is under progress for chemical weed control. Quinoa plants resemble its wild relatives *Chenopodium album, C. murale,* so care must be taken for proper weed identification and their removal. At maturity, quinoa plants change color from green to brown, orange, or red depending upon the accession. The crop is harvested manually or by combined harvest at the nail dented stage. Normally, threshing is done after 7 days of cutting when seed hulls are removed easily by wheat thresher with slight modification.

16.5.1.5 Local Acreage and Yields

Farmers have started growing quinoa in Pakistan. According to information gathered from farmers, the estimated acreage under quinoa cultivation was above 800 acres during 2015–2016 and the average yield was 22 monds per acre, which is comparable to yields in the Andes region. Its local consumption is increasing, however, most of the production was exported. Farmers earned 4–6 times more than with wheat. So, quinoa has already been recognized as a cash crop and an exportable commodity.

16.5.1.6 Quinoa World Trade and Prospects in Pakistan

The world trade of quinoa has grown in recent years. There is a sharp increase in quinoa exports of from Latin America in the region of the three Andean countries that account for over 80% of global exports since 2006. In 2012, quinoa exports accounted to approximately 131 million USD, with high concentrations in both destination and origin. For instance, 10% of global exports originated from the United States, 6% from the European Union, and 84.2% from Bolivia, Ecuador, and Peru (Bazile et al., 2015).

During 1992 and 2007, the international price (FOB) of quinoa remained at USD 1–2/kg. In 2009, the price rose sharply to around USD 2.9/kg, and then stabelized at around USD3/kg. This increase in price proves its high demand on the international market (Bazile et al., 2015).

Stagnant yields and poor marketing systems had very poor impacts on agriculture in Pakistan. Although wheat is a crop which has relative crop insurance due to government procurement, its profit margins are meager due to high input costs. Quinoa is a natural alternative to wheat, presently ensuring 4–6 times more profits if exports are patronized by the government. Also, quinoa is such a perfect food that it can provide a choice for local people concerned with their health. In developing quinoa cultivation in the coming years, the main driving force is the growing demand of quinoa worldwide. There is a trend for replacing or fortifiying common foods like wheat, rice, and corn with crops having health promoting properties and quinoa is one of these. Since Pakistan has already taken the initiative of screening germ plasm for local conditions and jumped into world trade, perhaps Pakistan can secure a handsome share of the international market, which would result in a revolution for the local economy.

16.5.2 Moringa

Moringa, due to its very high nutritional, medicinal, pharmaceutical, and industrial usage is known as a *miracle tree*. It is native to Pakistan. It is an evergreen to deciduous, medium-sized tree, which may grow up to 12 m in height with spreading and brittle branches, feathery foliage of tri pinnate leaves, open crown of drooping, and with whitish bark of corky and thickly fissured. It has been renowned for various uses for many centuries and was used as an ornamental plant during first 2 decades of the nineteenth century (Morton, 1991). Later on, in the twentieth century, it was introduced as a food plant (Muluvi et al., 1999). It is well known for its highly nutritional leaves, flowers, edible fruits, roots, and seed oil. Many parts of the plant are widely used in traditional medicines (Foidl et al., 2001).

16.5.2.1 Taxonomy and Genetic Diversity

Moringa belongs to family Moringaceae, single genus *Moringa* with 13 known species (Nadkarni, 1976). The most popular and widely cultivated specie is *M. oleifera* also known as the 'drumstick' tree (Ramachandran et al., 1980) native to sub-Himalayan tracts of Bangladesh, Pakistan, Afghanistan, and India, now distributed to the Arabian Peninsula, South Asia, tropical Africa, the Caribbean, and tropical South America and Central America (Somali et al., 1984;Mughal et al., 1999; Fahey, 2005; Palada et al., 2007; Nouman et al., 2013). Other moringa spp. are also useful but are rare, these include *M. drouhardii*, *M. concanensis*, *M. longituba*, *M. peregrina*, *M. stenopetala*and *M. ovalifolia* (Jahn, 1986; Morton, 1991). In Pakistan, only *M. oleifera* and *M. concanensis* are reported. *M. peregrina* is rare and only reported in remote areas of Tharparker (Qaiser, 1973; Manzoor et al., 2007), while *M. oleifera*, most commonly and locally known as "Sohanjna," is mostly raised in the plains of Punjab, KPK, Sindh, and Baluchistan (Anwar et al., 2005).

16.5.2.2 Nutritional Composition of Moringa

Moringa is gaining popularity as a highly nutritious plant which nourishes deprived people and can be a good source of nutrition for people of all age groups (Fuglie, 2001). Moringa plants contain rich amount of protein, relatively higher than the protein of eggs and milk, comparable to World Health Organization (WHO) recommended intakes (Freiberger et al., 1998). Likewise, moringa also comprises important quantities of minerals such as Ca, P, Fe, Cu, Zn, I, S, selenium (Se), and Mn, as well as vitamins, especially vitamin A and C (Leonard and Rweyemamu, 2006). It is a natural plant, which is simply accessible when green vegetables are not when food is scarce. A 100 g of moringa leaves contain substantial amounts of nutrients, which can fulfill 1/3 of the daily requirement of women. Utilization of different parts of the moringa plant can be a cheap and sustainable approach to prevent malnutrition in the developing countries (Fuglie, 2005).

16.5.2.3 Moringa Health Claims

Various parts of the moringa (leaves, seeds, root, pods, flowers, bark, gum, and seed oil) have been traditionally used for the treatment of conditions such as cardiovascular diseases, renal function disorders, hepatic and gastrointestinal diseases, infections, swelling, and pain (Siddhuraju and Becker, 2003). Moringa leaves and seeds are often used for arthritis, possessing anti-inflammatory properties (Sutar et al., 2008).

Moringa also exhibits antispasmodic activities making it useful in gastrointestinal motility disorders and diarrhea (Gilani et al., 1992), as well it demonstrates anti-ulcer properties (Pal et al., 1995). Several parts of the plant have antitumor (Makonnen et al., 1997), antipyretic, anti-inflammatory, and anti-epileptic properties (Pal et al., 1995). Other major therapeutic effects of the moringa are diuretic (Morton, 1991), hypolipidemic (Mehta et al., 2003), antihypertensive

(Dahot, 1988) antidiabetic, hepatoprotective (Ruckmani et al., 1998), antifungal, and antibacterial (Nikkon et al., 2003). The leaves are a natural treatment for insomnia (Fuglie, 1999; Fahey, 2005). These uses emphasized the utilization of moringa leaves in the food industry as a synergistic natural product, applied to cultural foods and the pharmaceutical industry and as a preventative for many diseases (Middleton et al., 2000; Miean and Mohamed, 2001; Fahey, 2005). Moringa is also rich in antioxidants like polyphenols, flavonoids, and carotenoids (Dillard and German, 2000). The most important compounds present in moringa from a medicinal point of view are kaempferol, quercetin, zeatin (Makonnen et al., 1997) caffeoylquinic acids, and rutin (Fuglie, 2005).

16.5.2.4 Water Purification

An interesting use of moringa is for the purification of drinking water (Berger et al., 1984; Olsen, 1987; Gassenschmidt et al., 1995). In the seeds, polyelectrolytes are present which act as coagulant for reducing viral and bacterial contamination from drinking water and for removing turbidity. It is being practiced in rural societies in Malawi, Sudan, Myanmar, India, and Indonesia (Jahn, 1986; Sutherland et al., 1989; Nyein and Aye, 1997; Mandloi et al., 2004). Seed powder in cloth is suspended in water overnight or for a few hours to coagulate impurities. The purified water is poured leaving behind the coagulated particles on the bottom. Up to 99% of colloids can be removed. This method of water purification is easy and cheap, only two seeds required in treatment of one liter of water. A majority of Pakistan's population has no access to clean drinking water, therefore water clarity by moringa seed can be a very cheap solution.

16.5.2.5 Crop Growth Enhancer

Moringa leaf extract (MLE) obtained from fresh leaves of moringa possesses very effective crop growth enhancing capabilities being rich in some plant hormone such as cytokinin in the form of zeatin (Basra and Lovatt, 2016) and many growth promoting compounds like ascorbates, phenols, osmoprotectants, and antioxidants (Makkar et al., 2007; Rady et al., 2013). MLE applied as either a seed priming agent or plant foliage has been known to increase the plant tolerance to abiotic stresses (Yasmeen et al., 2012) and improves crop production by several folds (Foidl et al., 2001; Table 16.2). This practice has been introduced and has gathered very good feedback from farmers for wheat,

TABLE 16.2
Yield Improvement in Various Crops Due to Foliage Applied Moringa Water

Moringa Extract Type	Crop	Stage of Application	Yield Improvement (%)	Reference
Leaf extract	Tomato	2 week after emergence until physiological maturity	141% fruit weight	Culver *et al.* (2012)
12.5% aqueous leaf water extract	Pea	Flower initiation till maturity with one week interval	40.22 dry weight of pea pod/kg	Singh *et al.* (2013)
MLWE (1:30)	Wheat	Booting/Heading	18.5%	Yasmeen *et al.* (2013)
MLWE (1:30)	Wheat	Tillering + Jointing + Booting + Heading	10.7	Yasmeen *et al.* (2012)
2% MLWE	Wheat	Tillering + Booting	40%	Afzal and Iqbal (2015)
MLE (1:30)	Common Bean	25 and 40 DAS	14.2%–15.87%	Rady *et al.* (2015)
MLE (1:32)	Maize	2 weeks after emergence and at every 2 weeks thereafter	31.52	Biswas *et al.* (2016)
MLE (3%)	Maize	30 and 50 DAS	24.30	Kamran *et al.* (2016)
MLE 3.0	Maize	30 and 45 DAS	52.0	Jahangeer (2011)
MLE 2.0	Canola	30 and 60 DAS	15.7	Iqbal (2011)

maize ,and cotton. This aspect of moringa can be exploited for industrial formulation and direct use by the farmers to increase farm income by harvesting higher yields.

16.5.2.6 Moringa Seeds Oil (Ben Oil)

Moringa oil is commonly known as "ben oil." It ranges 25%–45% (Anwar et al., 2005; Ayerza, 2011), and is like olive oil and is rich in stearic, palmetic, oleic, and behmic acids. This oil is also used in cosmetics, soaps, and in cooking (Delaveau and Boiteau, 1980; Ramachandran et al., 1980; Szolnokim, 1985). Due to its power for absorbing and retaining odors, the oil of *M. oleifera*is is highly valued by perfumers and by watchmakers as a lubricant (Ramachandran et al., 1980).

16.5.2.7 Animal Fodder and Poultry Feed

During dry season, moringa provides sufficient fodder for the livestock as it is one of the most nutritious trees. Moringa has high sprouting capacity after pruning and it produces good quality higher leaf biomass per unit area (Foidl et al., 2001; Nouman et al., 2013). Moringa can be grown on wide variety of soils and can easily tolerate drought spells up to 6 months, growing well where annual rainfall occurs between 250 and 1500 mm year^{-1} (Abdulkarim et al., 2005). Moringa supplement is used for both fattening and increase in milk production. Feeding with Moringa fresh leaves with tender stems can increase milk production and weight by 43%–65% and 32%, respectively. Animal health can be improved with moringa as it improve the digestibility of other foods which animals eat (Fuglie, 2000). Although moringa is a tree, it can also be grown intensively as a field crop for fodder purposes. Moringa crop once planted lasts for several years with less than 10% annual mortality (Foidl et al., 2001). The crop can be harvested nine times a year and annually produces 650–700 metric tons of green mass which is equivalent to 100 to 110 metric tons of dry mass, 7000 kg of lipids, with 65% being omega-3 fatty acids, 17.5 metric tons of pure protein, 10 metric tons of fermentable sugars, 45 metric tons of hemicellulose and cellulose and approximately 8 metric tons of starch (Fuglie, 2000; Foidl et al., 2001; Reyes, 2006). Spacing also affects the crop. To harvest the maximum biomass with better nutritional composition, moringa crops are grown with a narrow row spacing of 15 cm × 30 cm with cutting intervals of 30 days (Basra et al., 2015). Recently, moringa has been introduced as fodder and farmers are using it with very good results. Since it is a native plant it has a very good potential both fattening and milk production of the local livestock.

16.5.2.8 Production Technology

Moringa can be cultivated through seeds (sexual propagation) or stem cuttings (asexual propagation). In southern Punjab, propagation through stem cutting is most common. Two sowing seasons in Pakistan are best for moringa plantation: Spring (end of Feb to March) and monsoon (July–August). Except winter, Moringa can be grown year round. The best time for stem cutting plantation is the month of March and early April. Moringa trees, propagated through stem cutting, can bear fruit within one year. This method is most commonly adopted in southern Punjab where 4 feet long mature stem cutting are planted in pit in such a way that a 1/3 of the stem is buried in soil. For moringa tree propagation through seed, a 1 foot pit is dug and filled it with mixture of sand, silt, and well decomposed organic matter. Two seed are sown in one hole and water is applied. After a few weeks seedlings will emerge. When seedlings reach a height of 6 inches, maintain one seedling in each hill by thinning (Palada and Chang, 2003). When growing moringa for fodder or biomass production, it is cultivated by seed with narrow spacing (plant to plant 1 × 1 and row to row 2 × 2 feet) following the method used for cotton sowing.

16.5.2.9 Moringa World Trade and Prospects in Pakistan

Moringa has multimillion USD world trade for its different products. Today, the moringa market is estimated at more than 4 billion USD, which is expected to cross the 7 billion USDby 2020, while growing at 9% per annum. India has the major market share, which is more than 90% of the world trade (Figure 16.1). The commercialization of moringa products in Pakistan is still informal and

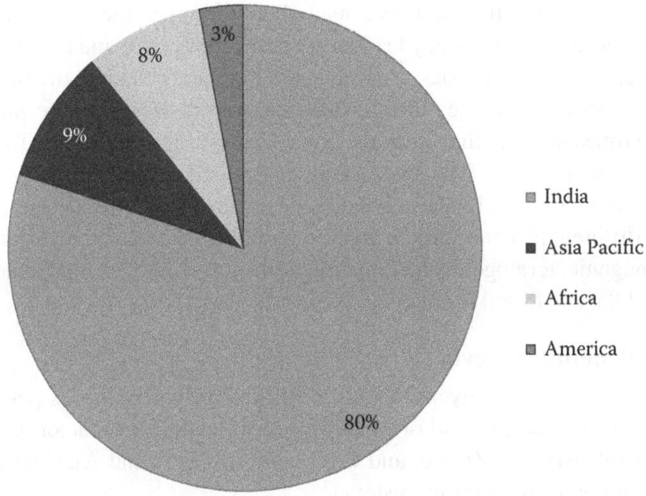

FIGURE 16.1 Share of different countries in the world moringa trade during 2015. (From Anonymous, 2015. http://www.prunderground.com/the-2015-global-moringa-meet-updates-moringa-export-market-trend-upward-integration-opportunities/0064082/ [accessed June 25, 2016].)

neglected. However, recently there has been an increasing awareness and many moringa products have been launched as herbal supplements. Through the efforts of Alternate Crops Lab, University of Agriculture, Faisalabad, moringa is becoming popular as a crop for fodder and feed and natural source of biostimulant. There are huge opportunities in Pakistan to popularize moringa for cattle fattening, fodder supplement for enhancing milk production, poultry feed, and so on. Since the plant is native to Pakistan, it has a wide variety germ plasm, which needs to be screened. Local moringa landraces have much more yield potential than the post popular Moringa varieties PKM1 and PKM2. Lab studies proved that local moringa strains have better biomass, seed, oil, and other product potential. To effectively exploit the existing market potential, their commercialization should eventually become more structured and formalized.

16.5.3 Stevia

For a healthier lifestyle, many alternatives are being researched to replace regular table sugar. Stevia leaf is a healthy 0 calorie alternative sweetener. Stevia contains no added chemicals, making it a natural sugar substitute. Stevia is much sweeter compared to sugar and its extract can taste 200–300 times sweeter than table sugar. In addition to its sweetness, its leaves also contain antioxidants, minerals like potassium, zinc, magnesium, and vitamin like B3. Recently stevia has replaced a fair quantity of U.S. tobacco.

16.5.3.1 Botanical Description

Stevia (*Stevia rebaudiana* (Bertoni) Berton) is a perennial crop which grows up to the height of 65–80 cm, with sessile and oppositely arranged leaves. Different species of stevia contain several sweetening compounds, with *S. rebaudiana* being the sweetest of all. Stevia is a semihumid subtropical plant and can be grown easily, like any other vegetable, even in the kitchen garden. The required soil pH range varies from 6.5 to 7.5 and can be grown in well-drained sandy loams and red soils. Saline soil is unfavorable for its cultivation (Goyal et al., 2010).

16.5.3.2 Significance and Uses

Stevia is not a new discovery. This crop was cultivated commercially in Paraguay for the first time in 1964 and after that it has been introduced in several countries. Now *S. rebaudiana* is successfully grown under a wide variety of climatic locations and conditions in the world. Stevia is a gift from nature and is also called sweet leaf, honey leaf, sweet herb, candy leaf, and honey yerba. These sweet herb leaves have been used for centuries as sweeteners to counterbalance the bitter taste of various plants. There sweet tasting glycosides like rebaudiosides and stevioside are present in its leaves, which are 100–300 times sweeter than sucrose. Even shade-dried leaves are (10–15 times) sweeter than sucrose and can be used raw. The sweetening compound of this plant is zero with no caloric value of glycemic index. Stevioside and rebaudioside are its stable products at wide ranges of pH and temperature in different pharmaceutical and food products. Extracted diterpene glycoside is safe and is free of carcinogenic, teratogenic, and mutagenic effects. They are nonfermentative and do not change the taste and flavor of food products in which they are used (Singh et al., 2014).

16.5.3.3 Health Benefits of Stevia

Stevia based products and steviol glycosides have minimal or zero calories per serving. For those who are looking to manage diabetes and/or want to control weight, these factors could have a positive effect. Various minerals (such as Zn, Fe, and Ca), vitamins (B, C and A), electrolytes (K and Na), other elements, and proteins are present in stevia.

Stevia contains several antioxidants and sterols compounds such as, tannins, triterpenes, and flavonoids. Several polyphenols, flavonoid, phytochemicals, and antioxidants are present in stevia. For example, chlorogenic acid, quercetin, kaempferol, isoquercitrin, caffeic acid, and isosteviol are widely reported. The risk of pancreatic cancer can be reduced up to 23% using kaempferol. In the extract of stevia, the presence of glycosides has been found which dilates blood vessels, and increase urine output and sodium excretion. Stevia help in lowering blood pressure at higher doses than as a sweetener (Tandel, 2011). The cardiotonic action of the plant normalizes and regulates heartbeat and blood pressure (Gupta et al., 2013).

The extracts of stevia have a strong bactericidal action against a wide range of pathogenic bacteria, including *coli* strains and certain *Escherichia*. Furthermore, stevia may have antioxidant, antiproliferative, and antimutagenic properties (Gupta et al., 2013). As steviol glycoside has no calories, it can reduce dental problems. As it doesn't affect blood sugar levels it is more suitable for diabetics (Duran et al., 2001).

16.5.3.4 Production Technology

Stevia seeds are very small and have short life with very low viability. It is a highly heterozygous species and as a result shows variation in plant growth when raised from seeds. The true to type plants are not produced and a constant reselection is needed in mother-seed plots to get true to type plants (Rank and Midmore, 2006). Stevia is mainly propagated vegetatively by stems on raised beds. The plant population is maintained around 28 to 30 thousand per acre with a 20–25 cm plant distance. Rooting is usually enhanced by using rooting hormones like synthetic auxins. Stevia grow well in drained fertile sandy loam or loam soils with high organic matter and with acidic to neutral pH (6–7). It grows well under continuously wet conditions under partial shade. To obtain plantlets from stem cuttings, appropriate soil moisture with a temperature range of 24–35°C is required during first 2 weeks. Two to three shoots are produce in stevia plants depending on the number of buds available on stem segments. To produce a good number of leaves for harvest these shoots produce multiple shoots. Water stress and temperature extremes are not suitable for vegetative growth. Stevia responds well to fertilizers the and recommended dose is 70:35:45 kg ha^{-1} of N, P, and K, respectively.

After 4 months of planting, the leaves of stevia plants are ready for first harvesting and after every 3–4 months' time interval, subsequent harvesting can be performed. Depending upon the land type, climatic conditions, and variety one can get three to four commercial harvests in a year. The glycoside contents are highest in the young actively growing shoot sections and leaves, being less in the overmatured leaves. The harvesting of leaves should be done before the onset of flowering or immediately after flower bud formation as total glycoside content of the leaves start to decrease with the onset of flowering (Singh et al., 2014).

16.5.3.5 Market Potential

Steviol glycosides from stevia have been accepted as natural sweeteners by all leading health organizations including the European Food Safety Authority (EFSA), the European Union (EU), United States Food and Drug Administration (USFDA), European Commission's Scientific Committee on Foods (SCF), FAO/WHO Expert Committee on Food Additives (JECFA) and Food Safety and Standards Authority of India (FSSAI) (Singh et al., 2014). Up to 20% of world sugar usage can be replaced with stevia according to WHO (Singh et al., 2014).

Commercially, stevia is cultivated in Japan, Canada, Brazil, the UK, Australia, China, Belgium, Spain, Taiwan, Israel, South Korea, and Thailand. The major world producers and exporters of diterpene glycosides are Japan and China. Stevioside is used in many food items including teas, soft drinks, and cereals in Japan. With a future potential for expansion stevia is currently a niche market as in recent years' stevia has generated considerable media attention and interest. It may prove to be profitable for small-scale growers who are willing to cultivate stevia and develop a market through local farmer's markets, or for wholesale to smaller distributors and other direct markets. Pakistan, which has the largest diabetic population in the world, is among the leading consumers of cane sugar and stevia is ideally poised to make significant contributions in satisfying the local demand of natural low-calory sweeteners. Malaysia, Japan, and Korea sweetener market contain 40% of *S. rebaudiana* extract. Stevia is a recent introduction in Pakistan and has yet to establish its market. Since there is growing international market for Stevia, it can also be a potential future cash crop for Pakistan.

16.5.4 Teff

Teff [*Eragrostis tef(Zucc.) Trotter*] is an annual grass indigenous to Ethiopia and Eritrea. It is an ancient, self-pollinated warm season annual grain species. It has been recently included in the list of the super foods of the twenty-first century. Like quinoa, teff's international popularity is rapidly growing (Collyns, 2013). It is cultivated for its grain, which is used to make porridge or flour. Because it lacks gluten in its grain, teff can be used to produce gluten-free products, especially for those people who have allergies to gluten.

16.5.4.1 Nutritional Value

In addition to being gluten-free, teff is rich in minerals (P, Cu, Al), thiamine, and is a very good source of protein, amino acids, carbohydrates, and dietary fiber. Ca, Fe, and Zn contents of teff are better than wheat, maize or rice. It is recommended as staple food for diabetic patients as it helps in blood sugar control (Zung, 2004). Teff grain is used to make the Ethiopian staple bread *INJERA*. Teff grain is very tiny and comes in a variety of colors, from light tan to dark brown, pale white to ivory white, dark brown to reddish-brown purple (Gamboa and van Ekris, 2008). In the highlands of Ethiopia white teff is preferred for growing as a staple grain. It requires the most rigorous growing conditions, has the mildest flavor, and is the most expensive form of teff. Brown teff, the least expensive form but has the highest iron content. Mixed teff has even higher iron content (Stallknecht, 1997).

Depending on variety, teff is ready for harvest two to five months after sowing. Its grain is the smallest in the world, and for one kernel of wheat the teff 150 grains are needed (Assefa et al., 2001; Gamboa and van Ekris, 2008). Due to very small grain size of teff, the postharvest loss is thereby resulting in relatively low yield (around 1.2 tha^{-1}). Teff is susceptible to lodging, and this could account for up to 30% of the potential loss of teff yields (Fufa et al., 2011).

16.5.4.2 Taxonomy

Teff belong to family Poaceae, sub-family *Eragrosteae*, and member of the tribe *Eragrosteae*. Over a wide geographic range, both annuals and perennials species of *Eragrostis*with 300 reported species. The classification of *Eragrostis* species based on the characteristics of spikelets, culms, pedicels, panicles, lateral veins, flower scale colors and flowering scales. Recently, on the basis of numerical taxonomy techniques, biochemistry and cytology, including seed protein electrophoretic patterns and leaf flavonoids, the taxonomy of teff has been clarified (Bekele and Lester, 1981).

Teff is chasmogamous annual cereal, C_4, and a self-pollinated crop. It has elbowing types, with fibrous root system, and with mostly erect stems. The plant possesses many tillers (lateral offshoots originating from the base of the stem) and a large crown. The plant grows in tufts and has a fine stem. The panicle contains spikelets from 190 to 1410 and can be either loose or compact in form. The grain of teff ranges in color from reddish purple or white to brown and is very small (1.0–1.7 mm in length). There are 2–12 dark brown or white flowers on the spikelets.

16.5.4.3 Production Technology

Well-drained and clay soils are favorable for the cultivation Teff. Teff also thrives in both drought stressed and water logged soil conditions. It has a short growing season with temperature range 10–27°C and with rainfall 450–550 mm (Roseberg et al., 2005). Teff flowers best during 12 hours of daylight as it is sensitive to long days (Roseberg et al., 2005). Teff can grow at altitudes where many other crops cannot grow. Moreover, it can be grown at sea level to as high as 3000 m altitude, with maximum production occurring at about 1800–2100 m (Stallknecht, 1997).

The ideal period for teff growing is when rainfall is reliable and well distributed. The rainfall between 300–500 mm per growing season is adequate for majority of the cultivars. Seed rate is recommended at 5–8 kg/ha. Recommended fertilizers rate is 60:25 kgN and P_2O_5 ha^{-1}. One to two hand weedings are needed to make crop fields weed free. Before planting of 1–2 weeks, presowing herbicides should be applied, while at early tillering the postemergence herbicides should be applied. The crop maturation period is 60–70 days. Compared to other cereals, teff is relatively free of plant diseases (Bekele, 1985).

16.5.4.4 Industrial Demand

Currently Ethiopia and Eritrea are the major producers of teff. It is relatively recent introduction in Europe and other parts of the world as gluten free super food. However, Ethiopian teff yields have not met the increased demands, thus leading to ever increasing prices. According to the Ethiopian Grain Trade Enterprise (EGTE), from 2007 to 2008, the price of teff skyrocketed above the 1000 USD/metric ton mark, which is four times the 2000–2008 average of 250 USD/metric ton (USDA FSA, 2013). In 2010, the price of teff fell, but was remained above 700 USD/metric ton (USDA FSA, 2013). Many Ethiopian families who switched to other cereals as substitutes now feel hardships (Demeke et al., 2011). Teff retail prices in Addis Ababa had reached over 800 USD/ metric ton by the end of 2012 (USDA FSA, 2013). Still, Ethiopian's preferred staple cereal is teff, as evidenced by demand reflected in the persistently high prices in recent years.

Due to increasing demands for teffin in the United States, the teff production has exploded widely and teff is currently grown in at least 25 states across the nation (Davidson and Laca, 2009). Teff has recently been introduced in Pakistan by Alternate Crops lab, University of Agriculture, Faisalabad, which showed its potential as kharif crop. Currently, there is no alternative of rice and cotton. Teff can be an interesting candidate being alow input, hardy, and high export value crop in Pakistan.

16.5.5 Amaranth

Amaranth (*Amaramthus* spp.) is a herbaceous annual with upright growth. The center of origin of amaranth is the Americas and Europe. Chinese have been growing amaranth for more than 400 years as vegetable plant, although it is grown for different purposes such as vegetable, ornamental, or seed in different parts of the world since long (O'Brien and Price, 1983, revised Yarger, 2008).

Since past times, it is known for its use as a leafy vegetable and seed in human nutrition. Its seed are used in making flour or are popped like popcorn. Now again, the amaranth seed is being recognized as a high quality and nutritious food supplement along with cereals (Pedersen et al., 1987). It is also being considered as a new millennium crop with a great pharmaceutical and nutraceutical value. In the Indo-Pak subcontinent, it is commonly known as Chulai and its leaves are used as an additive when cooking the traditional dish "Saag" (Rastogi and Shukla, 2013).

16.5.5.1 Systematics and Morphology

Taxonomically, the genus *Amaranthus* belongs to the family Amaranthaceae. The genus has more or less 70 species with a cosmopolitan distribution and with a great deal of morphological diversity (Costea and DeMason, 2001). Most of the species are wild with a weedy behavior, although *Amaranthus cruentus*, *A. hypochondriacus* and *A. caudatus* are taken as essential grain species. In Pakistan, Qureshi et al. (2009) reported on *A. ovalifolius* from Tehsil Chakwal with a note on its medicinal use. However, Erum et al. (2012) collected 13 genotypes of amaranth belonging to two species *A. hypochondriacus* and *A. tricolor* from Mensehra (KPK), Gilgit (Gilgit-Baltastan), and Mirpur (AJK) areas of Pakistan.

Amaranth is a hardy annual plant species. Morphologically, the amaranth plant is broad leaved with the leaf color ranging from green to pink and variegated in different species. Although variable from species to species, the plant height of amaranth ranges from 30 to 130 cm. The species collected from Pakistan showed 5–13 branches per plant. The seed yield per plant turns out to be highly variable (9–129 g/plant) in material collected from Pakistan (Erum et al., 2012).

16.5.5.2 Nutritional, Nutraceutical, and Pharmaceutical Value

Because amaranth does not belong to the family Poaceae, and its seed is not regarded as a grain, it is considered as a pseudocereal. Both vegetative parts and seed of amaranth species are consumed due to good nutritional profiles. The nutritional profile from the germplasm collected from different areas of Pakistan showed that seed contained 87–192 mg/mL carbohydrates, 15%–30% fats and 35–102 μg/mLproteins (Erum et al., 2012). However, the detailed analysis showed that the seed of amaranth contains a wealth of health and nutritionally important constituents. Carbohydrates are 65% with major proportion as starch (57%) and dietary fiber (7%). The fats are 7% with a major proportion of polyunsaturated (3%) fatty acids. Total proteins comprise 13% showing the presence of all essential amino acids with glutamic acid, glycine, aspartic acid, serine, and arginine as the major ones. The caloric content of amaranth seed is 371 kcal.

The seed contains almost all important nutraceuticals such as vitamins, proteins, and fats. Among vitamins, vitamin B6, B5, and B9 are the predominant ones. Amaranth seed is a rich source minerals such as phosphorus, iron, magnesium, and manganese. Since amaranth seed is gluten free, it can be used as a wheat-substitute for those who suffer allergy from gluten (Cooper, 2012). Concerning its pharmaceutical importance, the amaranth seed is used in the treatment of cardiovascular diseases, hypertension, lowering cholesterol, and reducing blood pressure. The plant is also used in gynecological troubles, treating anemic patients, and controlling diarrhea (Rastogi and Shukla, 2013).

16.5.5.3 Production Technology

As regards production technology, the seed rate for amaranth is about 2 kg ha^{-1} for direct sowing. The seed requires a 20–25°C temperature for germination, which lasts 8–10 days, while for growth over 20°C is optimal with an average germination rate of 80%. The seeds remain viable for about one year. For better germination, the seed should be sown at about 1–1.5 cm depth using sowing drill. At the time of sowing, there should be enough moisture in soil, alternatively light irrigation will be required after sowing. Seminal roots may go deep up to 1 m with a spread of around 30 cm, while taproot may attain a depth of 1.5–2.0 m. However, this depends upon the species used for cultivation (Anonymous, 2018). Recommended plant-to-plant distance is 30–45 cm while row-to-row distance is about 30 cm with a plant density of about 200,000 ha (Apaza-Gutierrez et al., 2002; Erum et al., 2012). Since the amaranth is a cross-pollinated crop, an isolation distance of about 400 m is recommended for maintaining genetic purity of lines/genotypes (Brenner et al., 2013).

Although the amaranth can grow well in marginal soil with a pH of 6–7, it responds to fertilizers. For obtaining optimal seed yield of about 2 t ha^{-1}, 80 kg ha^{-1} of N, 80 kg ha^{-1} of P, and 40 kg ha^{-1} of K is recommended (Apaza-Gutierrez et al., 2002). The amaranth requires frequent irrigation at 4–6 days intervals. For weed control, two hand-hoeings are sufficient while no specific weedicides are prescribed. Likewise, insecticides may be decided according to incidence of insect attack depending upon the insect fauna of the area where the crop is sown (Brenner et al. 2013). In Bolivia, amaranth is prone to the attack by chewing and sucking insects, armyworm, and cowpea aphid (O'Brien and Price, 1983, revised Yarger, 2008); the same may therefore apply to other parts of the world too.

16.5.5.4 Harvesting, Storage, and Marketing

The individual leaves, preferably side leaves, can be picked as and when needed, but should be done when the plants are 25–30 tall. For marketing as a leafy vegetable, the amaranth plants can be pulled and sent to market. For harvesting seed of amaranth, remove a few seeds and chew to check whether they have passed dough stage. Then cut the heads with mature seeds and dry under shade. When the grains are about to shatter from the heads, put a few heads in the cloth bags and beat on the floor for few times or strike the bag gently with a stick so that seeds are not damaged. Transfer the threshed seeds to a bucket. To remove most of the chaff from the threshed seed, winnowing will be required, which may be done with a fan. To remove the chaff completely, the seed can be put in the water in a bucket where the seeds will settle and the chaff will float. Remove the seed and dry to 9%–12% moisture for safe storage (O'Brien and Price, 1983; Yarger, 2008).

Since amaranth has is yet being cultivated in Pakistan, it is only imported and available in selected supermarkets. However, looking at the production cost to seed sale benefits in the United States, the cost of production is on an average USD 100/acre while gross return is USD 320/acre. So, there is average net return of USD 220/acre, which shows a cost/benefit ratio of 1:2.2 (Cooper, 2012).

In view of its increasing demand, amaranth is increasingly becoming an export commodity from developing countries to the EU; Germany and Netherlands being the biggest importers. Over the past 3 years, South Africa has become the biggest exporter of amaranth to the EU with a total export volume of 709 tones. The import price of amaranth is 4–6 Euro/kg, which is reasonably better than for many other commodities. For packing purposes, strong paper bags or polypropylene bags with a capacity of 25 kg are often used, but this depends upon the importers preference (Anonymous, 2015).

16.5.6 Chia

16.5.6.1 Systematics and Morphology

The plant chia belongs to the genus *Salvia*, specific epithet is *hispanica* and family name is Lamiaceae. DNA sequence data has shown that the chia is paraphyletic in origin (Walker et al., 2004). The largest centers of origin of chia are tropical America and southwest Asia (Hedge, 1992). As mentioned in the flora of Pakistan, chia is not endemic to Pakistan, although seven other species of the genus are found (Anonymous, 1996).

Chia is a shrubby plant and grows up to height of 1–1.5 m. The stem is squared and hairy. Its leaves are simple, petiolate (petiole attached to the stalk), oval, serrated, and grow on opposite sides. These leaves are dark green, ranging between 10 and 15 cm long and 3–5 cm wide. The inflorescence is a pedicelled flower (stalked), found in groups of six or more flowers gathered on the rachis or main axis of the inflorescence. The flower is hermaphroditic and flowering occurs in July and August in the northern hemisphere. Flowers colors are purple or white and are produced in numerous clusters in a spike at the end of each stem. Fruit is an indehiscent achene. Seeds are oval shaped and very small, approximately 1.5 mm wide × 2 mm long; seed color depends upon the variety, and may be plain white, brown, black, or mottled (Anonymous, 2018).

16.5.6.2 Nutritional, Nutraceutical, and Pharmaceutical Value

The word chia means 'oily.' The plant has gained great importance due to its specific economical and medicinal properties (Hussain et al., 2011). Chia has a caloric value of 436 kcal/100 g seed. Total carbohydrates are 7.72%, and dietary fiber 34%. Total fats are ~31%, with very low saturated fatty acids (3.3%) while polyunsaturated fatty acids comprise 24% of total fatty acids (omega-3 and omega-6 being 18% and 6%); cholesterol is absent. Seed proteins are ~17% while niacin (59%), thiamine (54%), and riboflavin (14%) are the major vitamins. The major seed minerals are Mn, P, Mg, Ca, Fe, and Zn (USDA, 2016).

Among the pharmaceuticals, the chia seed contains an array of phenolics, terpenoids, and antibiotics, which make chia a medicinally important species (Baricevic and Bartol, 2000; Jamboonsri et al., 2012). The seeds of chia have a neutral taste and can be used in preparation of fruit juice, yogurt, pastries, and ice cream (Anonymous, 2018). Chia seeds contains mucilage representing soluble fiber and a source of hydrocolloids with different functional properties for food industry products; for example, as a thickener and stabilizer, as an emulsifier and for high water retention capacity in the formation of foam (Hernández, 2012).

Due to these phytochemical constituents, the chia seed is used to treat a large number of ailments including allergies, angina, enhance athletic performance, used in cancer treatment, heart attack, coronary heart disease, hormonal/endocrine disorders, hypertension, stroke and vasodilatation, and hyperlipidemia. The seed also shows antioxidant, anticoagulant, and antiviral effects (Ulbricht et al., 2009; Hamidpour et al., 2014).

16.5.6.3 Production Technology

The chia plants can be grown in normal fields or pots. Its production is more feasible in the tropics (Kaiser and Ernst, 2016). A mature chia plant can attain the size of a large bush or small tree. Since chia grows faster than other herbs. Also, chia is not a ground-hugger like mint and will grow much taller than even the biggest parsley, sage, or rosemary plants; therefore, sufficient space has to be provided for Chia allowing the aerial parts to expand before it flowers (Yeboah et al., 2014).

While conducting field trials at different location in Argentina, Coates and Ayerza (1996) used 6 kg ha^{-1} seed rate using a seed plotter with a row spacing of 70–80 cm while there were 47–92 plants m^{-2} at different locations. Sowing was done in January, 1995. Normal irrigation and pest control practices as for other crops were adapted. The average crop duration at various locations lasted for 160 days. The average seed yield at six locations was 665 kg/hectare. Adopting these field sowing practices in Ghana, Yeboah et al. (2014) reported the Chia seeds were spread in the drills and covered lightly to enhance germination. In another set of experiment, the seedlings were transplanted at three planting densities (10,000, 20,000, and 40,000 plants ha^{-1}). Comparison of the two sowing methods showed that direct sown plants always gave greater seed yield than transplanted ones (2605 kg ha^{-1}) However, planting density of 40,000 plants/hectare gave the highest average seed yield (2329 kg ha^{-1}).

Alternate Crop Lab have recently conducted some initial trials. The optimum sowing time was in November and crops matured in May–June. However, the crop sown in December did not produced seeds and is still green where regular irrigation was applied. The average yields harvested are around 800 kg ha^{-1}. Hand sowing on ridges was better than broadcast and seedling transplanting. Weed management was a very critical issue. More research is needed to develop local production technology.

16.5.6.4 Harvesting, Storage, and Marketing

Harvesting of chia seed may be started as soon as most of the petals have fallen off the flower and seeds reach the nail denting stage. Coates and Ayerza (1996) used commercial combines for harvesting chia seeds, with modifications made to improve performance. Chia seed can last up to 5 years if stored in a cool and dry place (Anonymous, 2015).

Chia has a number of medicinal, confectionery, feed, food, and other uses. Because of these benefits, the chia market is gradually expanding. The use of chia seed flour up to 5% in making bread products was approved by the European Commission. Chia seeds or its oil are used for different applications such as bars, breakfast cereals, fruit juices, cookie snacks, cakes, and yoghurts around various countries of the world including Australia, New Zealand, the US, Canada, Chile, and Mexico (Borneo et al., 2010). Regarding the chia trade, in 2008, Argentina contributed approximately 4% of the world's production (Lema, 2010). Although chia seed has been commercialized for a long time in Argentina, there is yet a great need to promote the chia trade in the international market to meet world requirements (Coates and Ayerza, 1996).

According to the U.K. Department of Agricultural Economics, although the initial expenditure of Chia production may be similar to row crop, the net returns are likely to be greater. For instance, per acre cost of soybean production is expected to be about US$ 470, while this cost is expected to be much less for Chia production (Kaiser and Ernst, 2016). In Pakistan, a tiny bag of quinoa could cost Rs. 2000, while 100 g of Chia seeds may cost up to Rs. 400. Although quite costly, the cost is not a deterrent for those seeking healthier lifestyles (Anonymous, 2015).

16.5.7 Soybean

The soybean (*Glycine max* L.) belongs to family Fabacae and is a native legume to East Asia. Among the legumes, the soybean is valued for its high protein (38%–45%) and (20%) oil contents. Its oil consists of 85% unsaturated fatty acid containing two essential fatty acids, linolenic acid, and linoleic acid; whereas its protein contains one of the limiting amino acids, lysine. Like other legumes, it has ability of biological nitrogen fixation and thus helps in improving the soil's fertility status.

Soybean may be cultivated during the spring and/or autumn in regions with mean temperatures of 20–30°C. However, it does not grow well at temperatures below 20°C and above 40°C. Although soybean can be raised in a range of soils, it grows best in moist alluvial soils with good organic content. Soybean is one of the few genetically modified food crops with introduced glyphosate-tolerance. This enabled its cultivation in conservation tillage systems.

In the Indian subcontinent, initial work on soybean was initiated during the 1930s in the rainfed areas of prepartition Punjab. Later, plantation were started in Tandojam, Sindh in 1960. Nonetheless, soybean cultivation remained very limited and showed even a declining trend because of inadequate market demand and competition with the major crops. Production could be sustained only in the areas with a well-developed processing industry.

Soybean may be planted as spring (Zaid Rabi) and/or autumn (Kharif) crop in areas that remain fallow during part of the year. This includes (i) riverine lands which are flooded during summer from June to September but are dry during the winter from November to May; (ii) cotton fallow areas where no crop is grown between two crops of cotton from December to May; (iii) rain fed areas which are kept fallow before wheat planting, which are available from June to October; (iv) areas under September-planted and spring-planted sugarcane available for intercropping of soybean; and (v) dobari lands in Sindh. In this regard some proposed crop rotations are listed below:

- Wheat-sorghum/millet-fallow-soybean-wheat
- Cotton-soybean-cotton
- Wheat-soybean-wheat
- Intercropping soybean with maize, sorghum, cotton, or sugarcane.

16.6 CHALLENGES, OPPORTUNITIES, AND PROSPECTS OF CROP DIVERSIFICATION

For crop diversification, a multicomponent approach, such as modern technologies, improved research, institutional infrastructure, and economic support is necessary. There is need to overcome constraints that reduce the productivity of crop diversity.

16.6.1 Emerging Technology and Crop Diversification

Crop diversification depends significantly upon the development of technology along with climate, land and socioeconomic factors. The technology includes seed, fertilizer, draught power, which can alter the benefit of one crop in relation to another. For example, in India, irrigation is an important technology that can influence cropping patterns. Multiple cropping systems and higher productivity may be achieved through irrigation introduction. More diversification in cropping patterns occurred in areas where irrigation was introduced, especially if there was a better process of water delivery (Behera et al., 2007).

Due to high population pressure, with less productivity from agriculture and small land size for promotion of cultivation, the available option is use of modern technologies for expansion of crop diversification. These will consist of agriculture mechanization, balanced crop nutrition, planting materials, irrigation, and utilization of new technologies like organic farming and protected cultivation, and the use of improved seeds. The utilization and development of technologies should be encouraged through capacity building, proper policies, and appropriate mechanism implementation (Behera et al., 2007). Out of traditional commodities, the diversification process is triggered by factors like quick technology change in agricultural production in an extensive variety of commodities including pulses, cereals, and high value crops.

Genetic engineering and biotechnology in crops improve the quality and yield of many important crop plants by focusing on quality traits and primary productivity. Due to these emerging technologies, the diversification in favor of crops, such as oilseeds, pulses, vegetables, and fruits will be the future focus. GIS, usage of information technology, and decision support systems lead to crop diversification.

16.6.2 Research and Developmental Support for Crop Diversification

For agricultural diversification, research efforts are required to produce innovative solutions to new difficulties related with unknown and alternative production enterprises. Investment in extension and research system may be crop specific (like increase in yield, improved quality characteristics, short duration cultivars, and tolerance to pest stress) or general system level research like tillage systems and land management that permit change in cropping patterns and water management at farm level to facilitate growing different crop varieties within a season (Barghouti et al., 2004).

16.6.3 Institutional and Infrastructure Developments towards Crop Diversification

A multicomponent approach, including various particular investment regions, is required for diversification initiatives. Science and technology, irrigation and drainage, policy, institutional environmental and rural infrastructures are a few examples. Development of infrastructural facilities like irrigation facilities, use of electricity for agriculture, motorable roads and markets infrastructures are the important components that regulate the variation in cropping patterns and consequently the profitability and extent of crop diversification (Kumar and Gupta, 2015).

The implementation of crop technologies also depends upon infrastructure and institutional factors along with limited resource-based factors. It was reported that it is not the farm size but the infrastructure characteristics, such as motorable roads and irrigation, access to market, profitability, and extent of diversification through high-value crops such as off-season vegetables on which success depends (Chand, 1995). Estache (2003) reported that infrastructures facilitated access of underdeveloped areas and poorer individuals to central economic activities, therefore permitting them to be more productive. Ashok and Balasubramanian (2006) also presented the significance of infrastructures in terms of diversification. They concluded that irrigation, markets, and access to motorable roads determine the profitability and success of diversification through higher value crops. Infrastructure systems and input markets are still insufficient in many areas of developing world. In various cases, capital markets, input unavailability, support services, communication and transporting systems, and irrigation and drainage systems can limit crop diversification.

Roads and markets are the main elements which affect the status of crop diversification. A chief factor that determines the pattern—if not the pace—of diversification is the market,. For example, in India and many other underdeveloped countries, a higher elasticity supply for many crops in relation to price is suggested (Vyas, 1996). In India, the crop diversification in Kharif and Rabi seasons proved advantageous over traditional cropping system as determined in terms of higher income, more efficient use of field resources and labor, better wages for field workers and reduced pest attack (Behera et al., 2007). The studies report that domestic prices fluctuate less compared to international prices (Nayyar and Sen, 1994). Therefore, if the small and marginal farmers are present in agriculture sector, it is necessary to insulate the domestic producers from the effects of international price fluctuations. However, price is one part of the market impact on the cropping patterns. Institutional arrangements and market infrastructure also have great importance. The producers do not get appropriate signals due to lack of communication facilities, poor transport systems, and lack of proper information. This is also correct for the institutional arrangement. For agricultural producers, cropping decisions are determined through delivery systems for credits and inputs (Behera et al., 2007).

There is a need of adopting important measures to improve institutional arrangements that can facilitate both production and market reform for the mechanisms of diversification of high value produce. An immediate measure can be appropriate infrastructures, developing roads, proper markets, and promoting the private sector involvement for processing and value addition. For crop diversification, a precondition is to reform domestic markets in favor of high-paying crops and products. The main principle of all programs for fruitful diversification can be determined by market demand (Behera et al., 2007).

In developing countries, in crop diversification pursuits, quick economic development is associated with increased demand for high-value commodities, availability of new technologies, increasing the role of private sector, supportive role of government, consistent supply chain management, better food safety and quality and trade liberalization are important elements of growth. At the same time, risk of fluctuating prices, declining market size, land rights and suitability, difficulties in the supply of skilled labor force, and degrading irrigation infrastructures are main debacles in accelerating the crop diversification (Benziger, 1996; Braun, 1995; Dorjee et al., 2003; Joshi et al., 2007; Shahbaz et al., 2017).

16.7 CONSTRAINTS IN CROP DIVERSIFICATION

Crop diversification implies adding or fitting new crops species or new varieties to the existing crop species. In the agricultural context, crop diversification implies reallocation of the farmers' resources in terms of land, capitals, and farm mechanics for growing new crops. This is very challenging because of the fact that the farmers have become accustomed to the crops species they have been growing for decades or centuries. Several studies are available in the literature that focus on the issues in existing crop diversification with respect to the economic condition of the farmers (Bowman and Zilberman, 2013) or land held and response towards crop diversification (Dube et al., 2016). Any new crop will find its place in the existing pattern only when it has special attraction for the farmers. So, the challenges in crop diversification are multipronged are discussed in the following subsections.

16.7.1 Receptivity towards Diversification

Introduction of a new crop in the existing agro-ecosystems is only possible if there is enough attraction to it from the farmers. This is because the new crops introducted in a specific area might not have been assessed for the agro-based practices in that area. The farmers will be receptive to adopt new materials only when the level of risk is minimal, because the farmers are usually risk-averse

(Bowman and Zilberman, 2013). So, sharing proper knowledge on the production technology, fitness in the existing agroecosystem, environmental factors, consumer and market demand, and cost/benefit ratio is pivotal to make the farmers receptive to certain new crop being introduced in any area. The receptiveness towards crop diversification among small landholders may be reduced due to a greater risk involved (Bacon et al., 2012; Dube et al., 2016).

16.7.2 Climate Change Issues

The main purpose of crop diversification remains to cultivate a greater number of crop species on a given piece of land to generate more income. However, such an activity is always prone to prevailing climatic conditions. Climate change is always a critical constraint in crop diversification. Major climatic factors affecting crop diversification include water excess or scarcity, heat and frost, ionic strength in the soil solution, and pest flora and fauna of the area (Anonymous, 2018). These factors are likely to influence the growth and phenology of the new crop materials in any area (Farooq et al., 2011). This would eventually change the quality of agricultural produce, and thus having great consequences for food security (Schmidhuber and Tubiello, 2007; Kang et al., 2009).

Unpredicted changes in climatic conditions are among the major bottlenecks in the diversification due to noncompatibility of a crop in an area. The monoculture of crops may confront greater risks of unforeseen climatic changes leading eventually to severe declines in productivity brought on by the emergence of pests and the sudden onset of frost or drought. Another important factor linked to climate change is the reduced availability of rechargeable water for growing crop. This problem is more serious in developing countries (Akanda, 2010).

16.7.3 Farmland Holding

Farmland holding is an important constraint that limits crop diversification. Based on land holding, farms are classified as large, medium, and small (Sichoongwe et al., 2014). Small farm holders are more on a stake since they have small and fragmented land holdings and avoid all costs, including resources needed to adopt the cultivation of new crops; if the crop fails these farmers will be at complete loss. In addition, the cropping pattern on small farms is determined by household-food needs, and food crops, which occupy 4/5 of the cropped area. On the contrary, farmers with large landholdings are at an advantage since they can spare more land and resources for crop diversification (Jha, 2001). Thus, small landholding is another constraint in the diversification.

16.7.4 Marketability of Produce and Consumer Demand

Financial requirements of diversification revolve around the costs involved in researching the species to be planted and training in the management of diversified systems. Prior knowledge of market niche must be gained before a decision is taken to cultivate new crop species. Preliminary feasibility studies and market research need to be considered for financial requirements. Infrastructure (such as transport and storage) and marketing costs should also be worked out empirically (Anonymous, 2018). Increase in average incomes is always relevant to increase demand of novel and valuable goods; the income elasticity of these products is also typically high. This means that even a small change in income is likely to result in a proportional change in demand for any one product (Barghouti et al., 2004). Since most of the newly introduced crop species are either exotic or are not grown widely, an improper availability of market and consumer demand might also be a substantial constraint.

16.7.5 Socioeconomic Implications

Another important constraint is the socioeconomic status of the farmers involved in crop diversification. Socioeconomic issues are hinged on profitability. Main factors in this regard are nonavailability of land and inputs, fluctuating markets trends, and unstable prices for new crops (Adriano, 1989). Although all these factors are interactive, the small landholders are more vulnerable and have to be more careful in selecting any new crop, especially after attaining sufficient knowledge of the production technology, market trends, and marketability of the product.

16.8 CONCLUSION

The existing agriculture system around the globe seems to be unable to fulfill the future food requirements and food security, including the nutritional quality of food. At present, the monoculture cropping system leads to severe problems such as yield instability, nonjudicious and wasteful use of agricultural inputs and resources (land and water), poor household income of farmer communities, and resultantly poor economic growth. To strengthen the agro-ecosystem, crop diversification offers practicable solutions to achieve the food demand and food security around the globe, most specifically in the developing economies. Crop diversification means the efficient utilization of agricultural resources and giving better options to farmers to grow different types of crops which fit well in the existing cropping systems. Inclusion of commercial crops in agro-ecosystems will stabilize the income of farmers and improve the utilization of resources. Crop diversification, brought about in a managed fashion (growing staple and commercial crops), will ensure improved farmer incomes, better nutrition, food security, and more importantly, the ecological security of nature.

REFERENCES

Abdulkarim, S.M., K. Long, O.M. Lai, S.K.S. Muhammad and H.M. Ghazali. 2005. Some physico-chemical properties of MorFRinga oleifera seed oil extracted using solvent and aqueous enzymatic methods. *Food Chem.* 93: 253–263.

Abro, A.A. 2012. Determinants of crop diversification towards high value crops in Pakistan. *Int. J. Buss, Mgt. Eco. Res.*, 3: 536–545.

Acharya, S., S.P. Basavaraja, H. Kunnal, L.B. Mahajanashetti and A.R.S. Bhat. 2011. Crop diversification in Karnataka: An economic analysis. *Agric. Econ. Res. Rev.*, 24: 351–358.

Adriano, M.S. 1989. Implications for policy of the studies on profitability of irrigated non-rice crop production: A synthesis. *Proc. National workshop on Crop Diversification in Irrigated Agriculture in the Philippines.* IIMI, Sri Lanka, 134–142.

Afzal, M.I. and M.A. Iqbal. 2015. Plant nutrients supplementation with foliar application of allelopathic water extracts improves wheat (*Triticum aestivum* L.) yield. *Adv. Agri. Biol.*, 4: 64–70.

Ahamed, N.T., R.S. Singhal, P.R. Kulkarni and P. Mohinder. 1998. A lesser-known grain, Chenopodium quinoa: Review of the chemical composition of its edible parts. *Food Nutr. Bull.*, 19: 61–70.

Akanda, A.I. 2010. Rethinking crop diversification under changing climate, hydrology and food habit in Bangladesh. *J. Agri. Env. Int. Dev.*, 104: 3–23.

Ali, M. and U. Farooq. 2003. Diversified consumption to boost rural labor productivity: Evidence from Pakistan. Asian Vegetable Research and Development Center, Discussion Paper.

Anonymous, 1996. Flora of Pakistan. Available at: www.eFlora.org

Anonymous, 2015. http://www.prunderground.com/the-2015-global-moringa-meet-updates-moringa-export-market-trend-upward-integration-opportunities/0064082/ (accessed June 25, 2016)

Anonymous, 2018. Amaranth Growing and Harvest Information. Available at: http://veggieharvest.com/vegetables/amaranth.html (accessed February 20, 2018)

Anwar, F., M. Ashraf and M.I. Bhanger. 2005. Interprovenance variation in the composition of *Moringa oleifera* oilseeds from Pakistan. *J. Am. Oil Chem. Soc.*, 82: 45–51.

Apaza-Gutierrez, V., A. Romero-Saravia, F.R. Guillen-Portal and D.D. Baltensperger. 2002. Response of grain amaranth production to density and fertilisation in Tarija, Bolivia. In: Janick, J., Whipkey, A. (eds.) *Trends in New Crops and New Uses.* ASHS Press, Alexandria, 107–109.

Ashok, K.R. and R. Balasubramanian. 2006. Role of infrastructure in productivity and diversification of agriculture. A Research Report, SANEI, Islamabad, Pakistan.

Assefa K., H. Tefera, A. Merker, T. Kefyalew F. Hundera. 2001. Quantitative trait diversity in tef [Eragrostis tef (Zucc.) Trotter] germplasm from Central and Northern Ethiopia. *Genet. Res. Crop Evol.* 48: 53–61.

Ayerza R. 2011. Seed yield components, oil content, and fatty acid composition of two cultivars of moringa (*Moringa oleifera* Lam.) growing in the Arid Chaco of Argentina. *Ind Crops Prod.*, 33: 389–394. doi: 10.1016/j.indcrop.2010.11.003.

Bacon, C.M., C. Getz, S. Kraus, M. Montenegro and K. Holland. 2012. The social dimensions of sustainability and change in diversified farming systems. *Ecol. Soc.*, 17(4): 41. http://dx.doi.org/10.5751/ES-05226-170441

Barghouti, S., S. Kane, K. Sorby and M. Ali. 2004. Agricultural diversification for the poor: Guidelines for the practitioners. Agriculture and Rural Development, Discussion Paper 1, World Bank, Washington D.C.

Baricevic, D. and T. Bartol. 2000. The biological/pharmacological activity of *Salvia* genus V., Pharmacology. In: S.E. Kintzois (Ed). *Sage: The Genus Salvia.* Harwood Academic Publishers, Abingdon Marston, New York, USA. pp. 143–184.

Basra, S.M.A., S. Iqbal and I. Afzal. 2014. Evaluating the response of nitrogen application on growth, development and yield of quinoa genotypes. *Int. J. Agric. Biol.*, 16: 886–892.

Basra, S.M.A. and C.J. Lovatt. 2016. Exogenous applications of moringa leaf extract and cytokinins improve plant growth, yield, and fruit quality of cherry tomato. *HortTechnology*, 26: 327–337.

Basra, S.M.A., W. Nouman, H.U. Rehman, M. Usman and Z.H. Nazli. 2015. Biomass production and nutritional composition of *Moringa oleifera* under different cutting frequencies and planting spacings. *Int. J. Agric. Biol.*, 17: 1055–1060.

Bazile, D., D. Bertero and C. Nieto. 2015. *State of the Art Report on Quinoa around the World in 2013.* Food and Agriculture Organization of the United Nations (FAO) & CIRAD (Centre de coopération internationale en recherche agronomi quepourle développement), Rome.

Behera, U.K., A.R. Sharma and I.C. Mahapatra. 2007. Crop diversification for efficient resource management in India: Problems, Prospects and Policy. *J. Sustain. Agri.*, 30: 97–127.

Bekele, E. 1985. A review of research on diseases of barley, teff, and wheat in Ethiopia. In: T. Abate (ed.). *A Review of Crop Protection Research in Ethiopia. Proc. First Ethiopian Crop Prod. Symp.* Dept. Crop Protection, Inst. Agr. Res., Addis Ababa, Ethiopia, pp. 79–108.

Bekele, E. and R.N. Lester. 1981. Biochemical assessment of the relationship of *Eragrostis teff* (Zucc) Trotter with some wild *Eragrostis* species (Gramineae). *Ann. Bot.*, 48: 717–725.

Benziger, V. 1996. Small fields, big money: Two successful programs in helping small farmers making transition to high value added crops. *World Dev.*, 24: 1681–1693.

Berger, M.R., M. Habs, S.A.A. John and D. Schmahi. 1984. Toxicological assessment of seeds from *Moringa oleifera* and *Moringa stenopetala*, two highly efficient primary coagulants for domestic water treatment of tropical waters. *East African Med. J.*, 61: 712–717.

Biswas, A.K., T.S. Hoqueand and M.A. Abedin. 2016. Effects of moringa leaf extract on growth and yield of maize. *Prog. Agric.*, 27: 136–143.

Borneo, R., A. Aguirre and A.E. León. 2010. Chia (*Salvia hispanica* L) gel can be used as egg or oil replacer in cake formulations. *J. Am. Dietetic Assoc.*, 110: 946–949.

Bowman, M.S. and D. Zilberman. 2013. Economic factors affecting diversified farming systems. *Ecol. Soc.*, 18(1): 33.

Braun, J.V. 1995. Agricultural commercialization: Impact on income diversification processes and policies. *Food Policy*, 20: 187–202.

Brenner D.M., W.G. Johnson, C.L. Sprague, P.J. Tranel and B.G. Young. 2013. Crop–weed hybrids are more frequent for the grain amaranth 'Plainsman' than for 'D136-1'. *Genet. Resour. Crop Evol.*, 60: 2201–2205.

Chand, R. 1995. Agricultural diversification and small farm development in western Himalayan region. *National Workshop on Small Farm Diversification: Problems and Prospects*, NCAP, New Delhi, India.

Chapin, F.S., B.H. Walker, R.J. Hobbs, D.U. Hooper, J.H. Lawton, O.E. Sala and D. Tilman. 1997. Biotic control over the functioning of the Ecosystem. *Science*, 277: 500–504.

Coates, W. and R. Ayerza. 1996. Production potential of Chia in northwestern Argentina. *Ind. Crops Prod.*, 5: 229–233.

Collyns, D. 2013. Quinoa brings riches to the Andes. Retrieved from http://www.theguardian.com/world/2013/jan/14/quinoa-andes-bolivia-peru-crop

Cooper, C.C. 2012. Gluten free and healthy – dietitians can help reverse nutrition deficiencies common in celiac disease patients. *Today's Dietitian*, 14: 24.

Costea, M. and D.A. DeMason. 2001. Stem morphology and anatomy in *Amaranthus* L. (Amaranthaceae): Taxonomic significance. *J. Torrey Bot. Soc.*, 128: 254–281.

Cowger, C. and R. Weisz. 2008. Winter wheat blends (mixtures) produce a yield advantage in North Carolina. *Agron. J.*, 100: 169–177.

Culver, M., T. Fanuel and A.Z. Chiteka. 2012. Effect of moringa extract on growth and yield of tomato. *Greener J. Agric. Sci.*, 2: 207–211.

Dahot, M.U. 1988. Vitamin contents of the flowers of Moringa oleifera. *Pak. J. Biochem.*, 21(1–2): 21–24.

Davidson, J. and M. Laca. 2009. *Grain Production of 15 teff Varieties Grown in Churchill County, Nevada during 2009.* University of Nevada Cooperative Extension. Retrieved from https://www.unce.unr.edu/publications/files/ag/2010/fs1036.pdf

De Bruin, A. 1963. Investigation of the food value of quinoa and canihua seed. *J. Food Sci.*, 29: 872–876.

Delaveau, P. and P. Boiteau. 1980. Huiles a interêt pharmacologique, cosmetologique et dietetique. 4. Huiles de Moringa oleifera Lamk. et de M. drouhardii Jumelle. *Plantes Medicinales et Phytotherapie*, 14: 29–33.

Demeke A.B., K. Alwin and Z. Manfred. 2011. Using panel data to estimate the effect of rainfall shocks on smallholders' food security and vulnerability in rural Ethiopia, *Climatic Change*, 108: 185–206.

Dillard, C.J. and J.B. German. 2000. Review Phytochemicals: Nutraceuticals and human health. *J. Sci. Food Agric.*, 80: 1744–6.

Dini, I., G.D. Tenore and A. Dini. 2005. Nutritional and anti-nutritional composition of Kancolla seeds: An interesting and underexploited andine food plant. *Food Chem.*, 92: 125–132.

Dorjee, K., S. Broca and P. Pingali. 2003. Diversification in South Asian Agriculture: Trendsand Constraints, ESA Working Paper No. 03-15, Agricultural and Development Economics Division, Food and Agriculture Organization of the United Nations, Rome, Italy.

Dube, L., R. Numbwa and E. Guveya. 2016. Determinants of Crop Diversification amongst Agricultural Co-operators in Dundwa Agricultural Camp, Choma District, Zambia. *Asian J. Agric. Rural Dev.*, 6: 1–13.

Duran, S.A., M.P.N. Rodriguez, K.A. Cordon and J.C. Record. 2001. Stevia (*Stevia rebaudiana*), non-caloric natural sweetener. *Rev. Chil. Nutr.*, 39: 203–206.

Erum, S., M. Naeemullah, S. Masood, A. Qayyum and M.A. Rabbani. 2012. Genetic divergence in amaranthus collected from Pakistan. *JAPS*, 22: 653–658.

Estache, A. 2003. *On Latin America's Infrastructure Privatization and Its Distributional Effects.* The World Bank, Mimeo, Washington, DC.

Fahey, J.W. 2005. Moringa oleifera: A review of the Medical evidence for its nutritional, Therapeutic and prophylactic properties. Part 1. *Trees for Life J.*, 2005: 1–5.

Farooq, M., H. Bramley, J.A. Palta and K.H.M. Siddique. 2011. Heat stress in wheat during reproductive and grain-filling phases. *Crit. Rev. Plant Sci.*, 30: 1–17.

Foidl, N., H.P.S. Makkar and K. Becker. 2001. The potential of Moringa oleifera for agricultural and industrial uses. In: Lowell J. F. (ed.) *The Miracle Tree: The Multiple Uses of Moringa.* CTA, Wageningen, The Netherlands, pp. 45–76.

Freiberger, C.E., D.J. Vanderjagt, A. Pastuszyn, R.S. Glew, G. Mounkaila, M. Millson and R.H. Glew. 1998. Nutrient content of the edible leaves of seven wild plants from Niger. *Plant Foods for Hum. Nutr.*, 53: 57–69.

Fufa, B., B. Behute, R. Simons and T. Berhe. 2011. Strengthening the Tef value chain in Ethiopia. Available at: http://www.fao.org/sustainable-food-value-chains/library/details/en/c/243930/ (accessed February 20, 2018).

Fuglie, L. 2000. New uses of Moringa studied in Nicaragua. ECHO's Tropical Agriculture Site. Available at: http://www. echotech.org/network/modules.php?name=News&file=article&sid=194

Fuglie, L.J. 1999. *The Miracle Tree: Moringa oleifera: Natural Nutrition for the Tropics.* Church World Service, Dakar, Senegal.

Fuglie, L.J. 2001. *The Miracle Tree: Moringa Oleifera: Natural Nutrition for the Tropics.* Church World Service, Dakar, 1999. p. 68. Revised in 2001 and published as The Miracle Tree: The Multiple Attributes of Moringa, p. 172.

Fuglie, L.J. 2005. *The Moringa Tree: A Local Solution to Malnutrition.* Church World Service in Senegal.

Gamboa, P.A. and L. van Ekris. 2008. Survey on the nutritional and health aspects of teff (Eragrostis Teff). Javeriana.edu.co. Retrieved from http://educon.javeriana.edu.co/lagrotech/images/patricia_arguedas.pdf

Garcia, M., Raes, D. and Jacobsen, S.E. 2003b. Evapotranspiration analysis and irrigation requirements of quinoa (*Chenopodium quinoa*) in the Bolivian highlands. *Agric. Water Manage.*, 60: 119–134.

Garcia, M., D. Raes, S.-E. Jacobsen and T. Michel. 2007. Agroclimatic contraints for rainfed agriculture in the Bolivian Altiplano. *J. Arid Environ.*, 71: 109–121.

Gassenschmidt, U., K.D. Jany, B. Tauscher and H. Niebergall. 1995. Isolation and charaterization of a flocculating protein from *Moringa oleifera* Lam. *Biochimica et Biophysica Acta*, 1243: 477–481.

Gilani, A.H., K. Aftab, F. Shaheen et al. 1992. Antispasmodic activity of active principle from Moringa oleifera. In: Capasso F., Mascolo N. (eds). *Natural Drugs and the Digestive Tract.* EMSI, Rome, pp. 60–63.

Goyal, S.K., Samsher and R.K. Goyal. 2010. Stevia (*Stevia rebaudiana*) a bio-sweetener: A review. *Int. J. Food Sci. Nutr.*, 61: 1–10.

Gupta, E., S. Purwar, S. Sundaram and G.K. Rai. 2013. Nutritional and therapeutic values of Stevia rebaudiana: A review. *J. Med. Plants Res.*, 7: 3343–3353.

Hamidpour, M., R. Hamidpour, S. Hamidpour and M. Shahlari. 2014. Chemistry, pharmacology, and medicinal property of sage (salvia) to prevent and cure illnesses such as obesity, diabetes, depression, dementia, lupus, autism, heart disease, and cancer. *J. Tradit. Complement Med.*, 4: 82–88.

Haque, T. 1996. Diversification of Small Farms in India: Problems and Prospects. In: Haque T. (ed.) *Small Farm Diversification: Problems and Prospects.* National Centre for Agricultural Economics and Policy Research, New Delhi, India.

Hariadi, Y., Marandon, K., Tian, Y., Jacobsen, S.-E. and Shabala, S. 2011. Ionic and osmotic relations in quinoa (*Chenopodium quinoa* Willd.) plant grown at various salinity levels. *J. Exp. Bot.*, 62: 185–193.

Hayami, Y. and K. Otsuka. 1995. Beyond the Green Revolution: Agricultural Development Sterategy into the new country. In Jock R.A. (ed.) *Agricultural Technology: Policy Issues for the International Community.* CAB International, World Bank, Wallingford, Oxon, UK.

Hedge, I.C. 1992. A global survey of the biogeography of the Labiatae. In: Harley R.M. and Reynolds T. (eds.) *Advences in Labiatae Science.* Royal Botanic Gardens, Kew, UK, pp. 7–17.

Hernández, L.M. 2012. Mucilage from Chia Seeds (*Salvia hispanica*): Microestructure, Physico-chemical Characterization and Applications in Food Industry. Doctoral Thesis submitted to Pontificia Universidad Catolica De Chile.

Hooper, D. and P.M. Vitousek. 1997. The effect of plant composition and diversity on ecosystem processes. *Science*, 277: 1302–1305.

Hussain, A., F. Anwar, T. Iqbal and I. Bhatti. 2011. Antioxidant attributes of four Lamiaceae essential oils. *Pak. J. Bot.*, 43: 1315–1321.

Iqbal, M.A. 2011. Response of canola (*Brassica napus* L.) to foliar application of moringa (*Moringa oliefera* L.) and brassica (*Brassica napus* L.) water extracts. MSc thesis, Department of Agronomy, University of Agriculture, Faisalabad, Pakistan.

Jacobsen, S.-E. 2011. The situation for quinoa and its production in southern Bolivia: From economic success to environmental disaster. *J. Agron. Crop Sci.*, 197: 390–399.

Jacobsen, S.-E., C. Monteros, J.L. Christiansen, L.A. Bravo, L.J. Corcuera and A. Mujica. 2005. Plant responses of quinoa (*Chenopodium quinoa* Willd.) to frost at various phenological stages. *Eur. J. Agron.*, 22: 131–139.

Jacobsen, S.-E., C. Monteros, L.J. Corcuera, L.A. Bravo, J.L. Christiansen and A. Mujica. 2007. Frost resistance mechanisms in quinoa (*Chenopodium quinoa* Willd.). *Eur. J. Agron.*, 26(4): 471–475.

Jacobsen, S.-E., Mujica, A. and Jensen, C.R. 2003. Resistance of quinoa (*Chenopodium quinoa* Willd.) to adverse abiotic factors. *J. Expt. Bot.*, 54(Suppl. 1): i21.

Jahangeer, A. 2011. Response of maize (*Zea mays* L.) to foliar application of three plant water extracts. MSc thesis, Department of Agronomy, University of Agriculture, Faisalabad, Pakistan.

Jahn, S.A.A. 1986. Proper use of African natural coagulants for rural water supplies-Research in the Sudan and a guide to new projects. GTZ Manual No. 191.

Jamboonsri, W., T.D. Phillips, R.L. Geneve, J.P. Cahill and D.F. Hildebrand. 2012. Extending the range of an ancient crop, *Salvia hispanica* L. A new ω3 source. *Genet. Resour. Crop Evol.*, 59: 171–178.

Jensen, C.R., S.E. Jocobson, M.N. Andersen, N. Nunez, S.D. Andersen, L. Rasmussen and V.O. Mugensen. 2000. Leaf gas exchange and Water relation Characteristics of Field quinoa (*Chenopodium quinoa* Willd.) during soil drying. *Eur. J. Agron.*, 13: 11–25.

Jha, D. 2001. Agricultural Research and Small Farms. *Indian J. Agric. Econ.*, 56(1): 1–23.

Jian Y.Q. and M. Kuhn. 1999. Characterization of *Amaranthus cruentus* and *Chenopodium quinoa* starch. *Starch–Stärke*, 51: 116–120.

Joshi, P.K., A. Gulati and R. Cumming Jr. 2007. *Agricultural Diversification and Smallholders in South Asia.* Academic Foundation, New Delhi, India.

Kahan, D. 2013. *"Managing Risk in Farming" Farm Management Extension Guide.* Food and Agriculture Organization of United Nations, Rome, Italy.

Kaiser, C. and M. Ernst. 2016. *Chia.* Cooperative Extension Service, University of Kentucky College of Agriculture, Food and Environment. Available at: https://www.uky.edu/Ag/CCD/introsheets/chia.pdf (accessed 7 April, 2016).

Kamran, M., Z.A. Cheema, M. Farooq and A.U. Hassan. 2016. Influence of foliage applied allelopathic water extracts on the grain yield, quality and economic returns of hybrid maize. *Int. J. Agric. Biol.*, 18: 577–583.

Kang, Y., S. Khan and X. Ma. 2009. Climate change impacts on crop yield, crop water productivity and food security – a review. *Prog. Nat. Sci.*, 19: 1665–1674.

Kiær, L.P., I.M. Skovgaard and H. Østergård. 2009. Grain yield increase in cereal variety mixtures: A meta-analysis of field trials. *Field Crops Res.*, 114: 361–373.

Kumar, S. and S. Gupta. 2015. Crop diversification in India: Emerging trends. *Int. J.Curr. Res.*, 7: 17188–17195.

Lema, D. 2010. Growth and productivity in Argentine agriculture. *Conference on Causes and Consequences of Global Agricultural Productivity Growth*, Washington DC, USA.

Leonard, M. and P. Rweyemamu. 2006. *Report on Workshop "Moringa et Autres Végétaux à Fort Potentiel Nutritionnel: Stratégies, normes et marchés pour un meilleur impact sur la nutrition en Afrique".* 16–18 Novembre 2006, Accra, Ghana.

Li, C., X. He, S. Zhu, H. Zhou, Y. Wang, Y. Li, J. Yang et al. 2009. Crop diversity for yield increase. *PLoS ONE*, 4: e8049.

Lorenz, K. and L. Coulter. 1991. Quinoa flour in baked products. *Plant Foods Hum. Nutr.*, 41: 213–223.

Mahapatra, I.C. and U.K. Behera. 2004. Methodologies of farming systems research. In: Panda, D., Sasmal, S., Nayak, S.K., Singh, D.P., and Saha, S. (eds.) *Recent Advances in Rice-Based Farming Systems.* Central Rice Research Institute, Cuttack, Orissa, pp. 79–113.

Makkar, H.P.S., G. Francis and K. Becker. 2007. Bioactivity of phytochemicals in some lesser-known plants and their effects and potential applications in livestock and aquaculture production systems. *Animal*, 1: 1371–1391.

Makonnen, E. A. Hunde and G. Damecha. 1997. Hypoglycemic effect of *Moringa stenopetala* aqueous extract in rabbits. *Phytoth. Res.*, 11: 147–148.

Mandloi, M., S. Chaudhari and G.K. Folkard. 2004. Evaluation of natural coagulants for direct filtration. *Environ. Technol.*, 25: 481–489.

Manzoor, M., F. Anwar, T. Iqbal and M.I. Bhnager. 2007. Physico-chemical characterization of Moringa concanensis seeds and seed oil. *J. Am. Oil Chem. Soc.*, 84: 413–419.

Mehta, K., R. Balaraman, A.H. Amin, P.A. Bafna and O.D. Gulati. 2003. Effect of fruits of *Moringa oleifera* on the lipid profile of normal and hypercholesterolaemic rabbits. *J. Ethnopharmacol.*, 86: 191–195.

Middleton, Jr. E., C. Kandaswami and T.C. Theoharides. 2000. The effects of plant flavonoids on mammalian cells: Implications for inflammation, heart disease, and cancer. *Pharmacol. Rev.*, 52: 673–751.

Miean, K.H. and S. Mohamed. 2001. Flavonoid (myricetin, quercetin, kaempferol, luteolin and Apigenin content of edible tropical plants. *J Agri Food Chem.* 49: 3106–3112.

Morton, J.F. 1991. The horseradish tree, *Moringa pterigosperma* (Moringaceae). A boon to arid lands. *Econ. Bot.*, 45: 318–333.

Mughal, M.H., G. Ali, P.S. Srivastava and M. Iqbal. 1999. Improvement of drumstick (*Moringa pterygosperma* Gaertn.) – a unique source of food and medicine through tissue culture. *Hamdard Med.*, 42: 37–42.

Mujica, A. 1994. Andean grains and legumes. In: Hernando Bermujo, J.E., Leon, J. (eds) *Neglected Crops: 1492 from a Different Perspective*, FAO, Rome, pp. 131–148.

Muluvi, G.M., Sprent, J.I., Soranzo, N., Provan, J., Odee, D., Folkard, G., McNicol, W.J. and Powell, W. 1999. Amplified fragment length polymorphism (AFLP) analysis of genetic variation in *Moringa oleifera* Lam. *Mol. Ecol.*, 8: 463–470.

Nadkarni, A.K. 1976. *Indian Materia Medica.* Popular Prakashan, Bombay, pp. 810–816.

Naeem, S., L.J. Thomson, S.P. Lawler, J.H. Lawton and R.M. Woodfin. 1994. Declining biodiversity can affect the functioning of ecosystems. *Nature*, 368: 734–737.

Naseer, M.Z. 2016. Future of Agriculture Lies in Diversification. agrihunt.com. (accessed June 27, 2016)

Nayyar, D. and A. Sen. 1994. International trade and agricultural sector in India. In: Bhalla, G.S. (ed.) *Economic Liberalization and Indian Agriculture.* Institute for studies in Industrial Development, New Delhi, India.

Nikkon, F, Z.A. Saud, M.H. Rehman and M.E. Haque. 2003. In vitro antimicrobial activity of the compound isolated from chloroform extract of *Moringa oleifera* Lam. *Pak. J. Biol. Sci.*, 22: 1888–1890.

Nouman, W., M.T. Siddiqui, S.M.A. Basra, H. Farooq, M. Zubair and T. Gull. 2013. Biomass production and nutritional quality of *Moringa oleifera* as field crop. *Turk. J. Agric. For.*, 37: 410–419.

Nyein, M.M. and T. Aye. 1997. The use of *Moringa oleifera* (dan-da-lun) seed for the sedimentation and decontamination of household water. Part II: Community-based study. *Myanmar Health Sci. Res. J.*, 9(3): 163–166.

O'Brien, G.K. and Price, M.L. 1983. Amaranth: Grain & vegetable type. Echo Technical Note. Durrance Rd, North Ft. Myers, FL 33917, USA.

Olsen, A. 1987. Low technology water purification by bentonite clay and *Moringa oleifera* seeds flocculation as performed in sudanese village: Effects of Schistosoma Mansoni cericariae. *Water Res.*, 21: 81–92.

Pal, S.K., P.K. Mukherjee and B.P. Saha. 1995. Studies on the antiulcer activity of *M. oleifera* leaf extract on gastric ulcer models in rats. *Phytother. Res.*, 9: 463–465.

Palada, M.C. and L.C. Chang. 2003. Suggested Cultural Practices for Moringa. AVRDC pub # 03-545. Available at AVRDC website at www.avrdc.org

Palada, M.C., L.C. Chang, R.Y. Yang and L.M. Engle. 2007. Introduction and varietal screening of drumstick tree (Moringa spp.) for horticultural traits and adaptation in Taiwan. *Acta. Hort.*, 752: 249–253.

Pasko, P., H. Barton, P. Zagrodzki, S. Gorinstein and M. Folta. 2009. Anthocyanins, total polyphenols and antioxidant activity in amaranth and quinoa seeds and sprouts during their growth. *Food Chem.*, 115: 994–998.

Pedersen, B., L. Hallgren, I. Hansen and B.O. Eggum. 1987. The nutritive value of amaranth grain (*Amaranthus caudatus*) 2. As a supplement to cereals. *Plant Foods Hum. Nutr.*, 36: 325–334.

Qaiser, M. 1973. Moringaceae. In: Nasir, E. and Ali, S.I. (eds.) *Flora of West Pakistan. 38, 1973.* Dept. of Bot. Univ. Karachi, Karachi, Pakistan. pp. 1–4.

Qureshi, R, A. Waheed, M. Arshad and T. Umbreen. 2009. Medico-ethnobotanical inventory of tehsil Chakwal, Pakistan. *Pak. J. Bot.*, 41: 529–538.

Rady, M.M., C. Bhavya Varma and S.M. Howladar. 2013. Common bean (*Phaseolus vulgaris* L.) seedlings overcome NaCl stressas a result of presoaking in *Moringa oleifera* leaf extract. *Sci. Hort.*, 162: 63–70.

Rady, M.M., G.F. Mohamed, A.M. Abdalla and H.M.A. Yasmin. 2015. Integrated application of salicylic acid and *Moringa oleifera* leaf extract alleviates the salt-induced adverse effects in common bean plants. *Int. J. Agric. Technol.*, 11(7): 1595–1614.

Ramachandran, C., K.V. Peter and P.K. Gopalakrishnan. 1980. Drumstick (*Moringa oleifera*): A multipurpose Indian vegetable. *Econ. Bot.*, 34(3): 276–283.

Rank, A.H. and J.D. Midmore. 2006. An intense natural sweetener-laying the ground work for a new rural industry; May 2006 RIRDC Publication No 06/020 RIRDC Project No UCQ-17A.

Rastogi, A. and S. Shukla. 2013. Amaranth: A new millennium crop of nutraceutical values. *Crit. Rev. Food Sci. Nutrit.*, 53: 109–125. 10.1080/10408398.2010.517876

Reich, P.B., J. Knops, D. Tilman, J. Craine, D. Ellsworth, M. Tjoelker, T. Lee et al. 2001. Plant diversity enhances ecosystem responses to elevated CO_2 and nitrogen deposition. *Nature*, 410: 809–810.

Repo-Carrasco, R., Espinoza, C. and Jacobsen, S.E. 2003. Nutritional value and use of the Andean crops Quinoa (*Chenopodium quinoa*) and Kaniwa (*Chenopodium pallidicaule*). *Food Rev. Int.*, 19: 179–189.

Reyes, S.N. 2006. Moringa oleifera and Cratylia argentea: Potential fodder species for ruminants in Nicaragua. *Doctoral thesis*. ISSN 1652-6880, ISBN 91-576-7050-1. Available at: http://diss-epsilon.slu.se/archive/00001027

Roseberg, R.J., N. Steve, S. Jim, C. Brian, R. Kent and S. Clint. 2005. Yield and quality of teff forage as a function of varying rates of applied irrigation and nitrogen. Retrieved from http://oregonstate.edu/dept/kbrec/sites/default/files/documents/hort/teff.pdf

Ruales J. and B.M. Nair. 1994. Properties of starch and dietary fiber in raw and processed quinoa (*Chenopodium quinoa* Willd.) seeds. *Plant Foods Hum. Nutr.*, 45: 223–246.

Ruckmani, K., S. Kavimani, R. Anandan and B. Jaykar. 1998. Effect of *Moringa oleifera* Lam on paracetamol-induced hepatoxicity. *Indian J. Pharm. Sci.*, 60: 33–35.

Ruffino, A.M.C., M. Rosa, M. Hilal, J.A. Gonzalez and F.E. Prado. 2010. The role of cotyledon metabolism in the establishment of quinoa (*Chenopodium quinoa*) seedlings growing under salinity. *Plant Soil.*, 326: 213–224.

Schmidhuber, J. and F.N. Tubiello. 2007. Global food security under climate change. *PNAS*, 104: 19703–08.

Shahbaz, P., I. Boz and S. Haq. 2017. Determinants of crop diversification in mixed cropping zone of Punjab Pakistan. *Direct Res. J. Agric. Food Sci.*, 5: 360–366.

Sichoongwe, K., L. Mapemba, D. Ng'ong'ola and G. Tembo. 2014. The determinants and extent of crop diversification among smallholder farmers. IFPRI Working Paper 05, International Food Policy Research Institute.

Siddhuraju, P and K. Becker. 2003. Antioxidant properties of various solvent extracts of total phenolic constituents from three different agro climatic origins of drumstick tree (*Moringa oleifera* Lam.) leaves. *J. Agric. Food Chem.*, 51: 2144–2155.

Singh, B., J. Singh and A. Kaur. 2014. Agro-production, processing and utilization of *Steviarebaudiana* as natural sweetener. *J. Agri. Engg. Food Tech.*, 1: 28–31.

Singh, S., S.P. Mishra, P. Singh and R.S. Prasad. 2013. *Moringa oleifera* leaf extract as biostimulant for increasing pea yield. *Indian For.*, 139: 562–563.

Smale, M. and A. King. 2005. *Genetic Resource Policies. What is Diversity Worth to Farmers? Briefs 13–18.* International Food Policy Research Institute and the International Plant Genetic Resources Institute. Source- Science Tech, Entrepreneur, VOL.12/N0.10, October 2004.

Somali, M.A., M.A. Bajnedi and S.S. Al-Faimani. 1984. Chemical composition and characteristics of *Moringa peregrina* seeds and seed oil. *J. Am. Oil. Chem. Soc.*, 61: 85–86.

Stallknecht, G.F. 1997. New crop fact sheet: Teff. Hort. purdue. Retrieved from http://www.hort.purdue.edu/newcrop/cropfactsheets/teff.html

Sutar, N.G., C.G. Bonde, V.V. Patil, S.B. Narkhede, A.P. Patil and R.T. Kakade. 2008. Analgesic activity of seeds of *Moringa oleifera* Lam. *Int J Green Pharm.*, 2: 108–110.

Sutherland, J.P., G.K. Folkard and W.D. Grant. 1989. Seeds of Moringa species as naturally occurring flocculants for water treatment. *Sci. Tech. Dev.*, 7: 191–197.

Szolnokim, T.W. 1985. *Food and Fruit Trees of the Gambia*. Bundesforschungsanst. für Forst- und Holzwirtschaft, Hamburg.

Tandel, K.R. 2011. Sugar substitutes: Health controversy over perceived benefits. *J. Pharmacol. Pharm.*, 2: 236–243. doi: 10.4103/0976-500X.85936

Tilman, D., C.L. Lehma and K.T. Thomson. 1997. Plant diversity and ecosystem productivity: Theoretical considerations. *Proc. Nat. Acad. Sci. USA*, 94: 1857–1861.

Ulbricht, C., W. Chao, K. Nummy, E. Rusie, S. Tanguay-Colucci, C.M. Iannuzzi, J.B. Plammoottil, M. Varghese and W. Weissner. 2009. Chia (*Salvia hispanica*): A systematic review by the natural standard research collaboration. *Rev. Recent Clin. Trials*, 4: 168–174.

UN News Centre. 2013. http://www.un.org/apps/news/story.asp?NewsID=44184#.V8FEHvl97IU

USDA. 2016. *Risk Management/Risk in Agriculture*. United States Department of Agriculture, Economic Research Service, Washington, DC, USA.

USDA FSA. 2013. Ethiopia grain and feed annual report. USDA. Retrieved from http://gain.fas.usda.gov/Recent%20GAIN%20Publications/Grain%20and%20Feed%20Annual_Addis%20Ababa_Ethiopia_5-24-2013.pdf

Vyas, V.S. 1996. Diversification of agriculture: Concept, rationale and approaches. *Indian J. Agric. Econ.*, 51: 636–643.

Walker, J.B., K.J. Sytsma, J. Treutlein and M. Wink. 2004. Salvia (Lamiaceae) is not monophyletic: Implications for the systematics, radiation, and ecological specializations of Salvia and tribe Mentheae. *Am. J. Bot.*, 91: 1115–1125.

Wright, K.H., O.A. Pike, D.J. Fairbanks and S.C. Huber. 2002. Composition of *Atriplexhortensis*, sweet and bitter *Chenopodium quinoa* seeds. *Food Chem. Toxicol.*, 67: 1383–1385.

Yachi, S. and M. Loreau. 1999. Biodiversity and ecosystem productivity in a fluctuating environment: The insurance hypothesis. *Proc. Nat.al Acad. Sci. USA*, 96: 1463–1468.

Yarger, L. 2008. Amaranth: Grain and Vegetable Types. Echo Technical Note, Florida. pp. 1–14.

Yasmeen, A, S.M.A. Basra, R. Ahmad and A. Wahid. 2012. Performance of late sown wheat in response to foliar application of *Moringa oleifera* Lam. leaf extract. *Chil. J. Agric. Res.*, 72: 92–97.

Yasmeen, A., S.M.A. Basra, M. Farooq, H. Rehman and N. Hussain. 2013. Exogenous application of moringa leaf extract modulates the antioxidant enzyme system to improve wheat performance under saline conditions. *Plant Growth Reg.*, 69: 225–233.

Yeboah, S., E. Owusu Danquah, J.N.L. Lamptey, M.B. Mochiah, S. Lamptey, P. Oteng-Darko, I. Adama, Z. Appiah-Kubi and K. Agyeman. 2014. Influence of planting methods and density on performance of Chia (*Salvia hispanica*) and its suitability as an oilseed plant. *Agric. Sci.*, 2: 14–26.

Zung, A. 2004. Type 1 diabetes in Jewish Ethiopian immigrants in Israel: HLA class II immunogenetics and contribution of new environment. *Hum Immunol.*, 12: 1463–8. Retrieved from http://www.ncbi.nlm.nih.gov/pubmed/15603874

17 High Value Horticultural Crops

Iftikhar Ahmad, Saeed Ahmad, Khurram Ziaf, M. Muzammil Jahangir, and Raheel Anwar

CONTENTS

17.1 HIGH VALUE HORTICULTURE—CURRENT STATUS

Pakistan has a rich topography and climate along with high variations in soil, on which a large range of horticultural crops, such as fruits, vegetables, ornamentals, medicinal/ aromatic plants, and spices are grown. In Pakistan, horticultural production is mainly high-value fruits, vegetables, and ornamentals. The tropical-to-temperate climate of the country allows growing 40 different kinds of vegetables, 21 types of fruits, and many ornamental and medicinal plants. Pakistan is a fascinating country in its horticultural wealth, where most of the fruits consumable by humans are grown. Fruits are low in calories, fat, and are a source of simple sugars, fiber, and vitamins, which are essential for good human health. Strawberry is a nonclimacteric small fruit with high visual appeal and desirable flavor (Cherian et al. 2014). Its global production is 4.5 million tonnes (FAOSTAT 2013) and world trade volume is 1.4 million tonnes (FAOSTAT, 2011). It is a good source of vitamins, K, Mg, Ca, Na (Khan et al. 2010; Wasim et al. 2012), and phytonutrients like anthocyanin, flavonoids, phenolics, ellagitannins, and catechins (Mahmood et al. 2012a; Ornelas-Paz et al. 2013; Seeram 2008). These phytonutrients exhibit anti-inflammatory, antioxidant and anticarcinogenic properties that lead to the reduction in oral maladies (Seeram 2008). The strawberry is not only relished as fresh fruit but also used in various processed foods including jams, jellies, custards, ice creams, and pies.

The nutritional value of strawberry fruit is well recognized and has led to its commercial production in many countries including Pakistan (Mahmood et al. 2012a). In Pakistan, this high-value crop is mainly cultivated in Swat and its peripheral areas. Due to low chilling requirements, it can be grown in tropical and subtropical regions. At 800 m above sea level, strawberry is cultivated as a perennial but below this altitude it is grown successfully as an annual crop. That is why, for the last 2 decades, its cultivation across the river belt areas of Punjab and Sindh is gaining popularity. At present Islamabad, Lahore, Charsada, and Mardan are the major strawberry producing areas.

It is also grown in other parts of Pakistan like Mansehra, Haripur, Abbottabad, Peshawar, Khairpur, Sakhar, and some parts of central and Southern Pakistan like Gujrat, Sialkot, Jehlum, Chakwal, Multan, and Muzaffargarh (Khan 2013). At present, strawberry production in Pakistan is 609 tonnes from 179 hectares of rich land resources (MNFSR 2016). Moreover, other minor fruits such as ber and fig are also neglected by the researchers, however, comparison of their nutritional profile—especially of ber—with major fruits shows their enhanced mineral and vitamin C contents.

In Punjab, total marginal land is 6.67 million hectares, which is approximately 3% of the total cultivated land (Khan 1998). Most of the agronomic and horticultural crops cannot be grown successfully in these marginal areas and are considered as wastelands. Therefore, it is very important to encourage growers to establish ber orchards on such lands. It is the need of the time to enhance the establishment of ber orchards on a scientific basis on marginal lands, where other crops cannot be grown, to improve the socioeconomic status of the farmers and enable them to uplift the nutritional status of the nation.

An efficient, productive, and profitable horticulture sector can play a vital role for accelerating overall economic growth, alleviate poverty and transform the country towards agro-based industrialization. Overall, the agriculture sector accounts for 20.9% of GDP and 43.5% of employment, including strong backward and forward linkages. During recent years, a considerable increase in fruit and vegetable production has been noticed, which also has increased exports and the availability of the produce for local consumption. However, there are still several high potential horticultural enterprises to be explored to further improve exports.

Among these potential areas, production of minor fruit crops such as fig, ber, olive, strawberry, and so forth, even on marginal lands offer higher returns compared to several major fruit crops. In vegetable production, indigenous seed production in combination with screening of elite germplasm for biotic and abiotic stress tolerance, processing of vegetables, and mushroom production offer opportunities for improvement and increasing growers' profitability. For ornamentals, cut flower production and optimizing their postharvest handling technology for distant markets, crop diversification, and extraction/preparation of different value added by-products are potential enterprises to alleviate poverty of the farming community. Moreover, Pakistan is rich in native or naturalized flora that have a high value for use in various pharmaceuticals, as well as in daily culinary uses. These all need to be explored in order to utilize the full potential to uplift our agriculture under the prevailing circumstances.

This chapter will provide information about the potential of high-value horticultural crops, which can be explored to increase horticultural production in the country for local as well as export markets. Moreover, it will describe the current status of high-value horticultural crop production, its scope, and the way forward for future expansion.

17.2 MINOR FRUIT CROPS

Exploring the potential of underutilized fruits could be a major gateway to uplift the economy of Pakistan. Minor fruits are usually fast growing plants, some of which start to produce fruits after 2 to 4 years of planting. These include strawberry, ber, fig, pomegranate, loquat, litchi, olive, jaman, papaya, falsa, mulberry, etc. The annual production of minor fruits in Pakistan is very low due to the small area of established orchards (Tables 17.1 through 17.4). There is dire need to increase this area under cultivation of these minor fruit crops by convincing the farmers and providing them true-to-type plants of these fruits. This, in turn, will lead to boost the export potential as well as the availability of value-added products on long-term basis, which will ensure welfare and increase economic development in country. Minor fruits are comprised of many antioxidants such as polyphenolic flavonoids, vitamin-C, and anthocyanins. Ber fruits have more calcium, phosphorus, iron, and ascorbic acids as compared to apple (Pasternak et al. 2016). These compounds serve as shields in the human body against oxidative stress, diseases, and cancers. Minor fruits are grown on small areas and are neglected due to low production volumes, difficult postharvest management, and unawareness of advanced marketing techniques. Lack of applied research by researchers further aggravates the situation. However, the

TABLE 17.1
Area of Minor Fruits in Pakistan (Hectares)

	2004–2005	2005–2006	2006–2007	2007–2008	2008–2009	2009–2010	2010–2011	2011–2012	2012–2013	2013–2014	2014–2015
Pomegranate	13,262	13,283	13,555	13,510	13,494	12,986	12,952	12,802	1120	11,155	9434
Jaman	1337	1282	1338	1343	1258	1208	1196	1209	1241	1122	1114
Litchi	455	449	445	450	441	431	410	286	295	278	274
Phalsa	1363	1394	1250	1119	1266	1250	1226	1163	1157	1131	1193
Walnut	1354	1455	1497	1292	1297	1254	1225	1225	1241	1686	1686
Ber	3039	3152	3905	4470	5200	5356	5965	5893	5291	5266	5129
Loquat	1407	1429	1472	1501	1678	1604	1502	1534	1567	1477	1415
Mulberry	844	822	818	624	724	625	550	545	510	543	459
Fig	198	195	164	162	162	145	139	129	125	122	58
Strawberry	0	0	0	0	78	79	82	84	92	171	179

TABLE 17.2
Production of Minor Fruits in Pakistan (Tonnes)

	2004–2005	2005–2006	2006–2007	2007–2008	2008–2009	2009–2010	2010–2011	2011–2012	2012–2013	2013–2014	2014–2015
Pomegranate	49,904	50,109	48,074	56,629	61,090	55,276	49,997	48,589	46,081	45,318	42,648
Jaman	7607	7578	7712	7991	7930	7254	7392	7536	7398	6407	6364
Litchi	2842	2917	2895	2971	2903	2854	2567	1736	1811	1666	1644
Phalsa	13,319	14,867	15,195	12,722	12,748	11,854	10,306	10,640	9926	10,094	14,831
Walnut	5375	5491	4666	4167	4574	4457	4373	3991	3902	3851	4063
Ber	17,288	17,874	23,225	25,291	28,079	28,066	29,618	28,377	25,634	25,309	24,635
Loquat	10,042	10,171	10,688	10,479	13,159	10,141	8577	8731	9304	9002	8823
Mulberry	4249	4221	3717	2950	3293	3085	2578	2615	2325	2530	2100
Fig	729	730	720	773	741	658	621	525	494	500	459
Strawberry	0	0	0	0	274	250	284	292	312	591	609

TABLE 17.3
Area of Minor Fruits in Punjab (Hectares)

	2004–2005	2005–2006	2006–2007	2007–2008	2008–2009	2009–2010	2010–2011	2011–2012	2012–2013	2013–2014	2014–2015
Pomegranate	2301	2298	2278	2220	2208	2045	1901	1747	1614	1535	1444
Jaman	1141	1151	1148	1146	1127	1108	1088	1087	1058	938	933
Litchi	445	449	445	450	441	421	410	286	295	278	274
Phalsa	605	638	630	600	600	598	536	577	561	541	568
Ber	1377	1403	1406	1388	1390	1385	1380	1365	1289	1268	1223
Loquat	736	753	764	779	763	766	776	814	815	778	767
Mulberry	394	359	347	339	338	321	253	252	221	212	208

TABLE 17.4
Production of Minor Fruits in Punjab (Tonnes)

	2004–2005	2005–2006	2006–2007	2007–2008	2008–2009	2009–2010	2010–2011	2011–2012	2012–2013	2013–2014	2014–2015
Pomegranate	16,723	16,671	16,453	15,733	15,667	14,582	13,304	11,859	11,038	10,325	9762
Jaman	7045	7196	7187	7443	7546	7265	7061	7140	6800	5809	5780
Litchi	2842	2917	2895	2971	2903	2754	2567	1736	1811	1666	1644
Phalsa	2657	2786	2754	2548	2553	2416	2341	2381	2306	2256	2387
Ber	10,123	10,500	10,380	10,059	10,266	10,211	10,126	9965	9566	9248	8792
Loquat	4505	4585	4687	4266	3990	4032	4114	4414	4451	4181	4131
Mulberry	1714	1551	1499	1497	1507	1456	1214	1184	1068	1033	987

nutritional status of some minor fruits is better than for apple and citrus. Moreover, some minor fruits (ber, fig, falsa) can be successfully grown on marginal lands. The production of these fruits can prove a supplementary food security due to sustained production. Value addition in minor fruits (like ber, falsa, papaya, jaman, fig, pomegranate, mulberry, loquat, litchi, olive, strawberry, etc.) is expected to impact positively on income generation of farmers and entrepreneurs but also on nutritional food security of the country. There are more than 90 processing plants for citrus, 28 for mango, and two for date processing in Pakistan yet none are available for minor fruit crops. Incentives to encourage processing of minor fruits and their exports are also lacking.

Unfortunately, there are very few planned orchards for only some of these minor fruits such as ber, falsa, and papaya in the country, which also have been established during recent years. In such a situation, there is a need to establish certified nurseries for propagating true-to-type saplings of these fruit crops and to motivate farmers for plantation of these neglected fruits, especially on marginal lands. Moreover, the production technology of these crops has changed over time. Falsa strains having higher pulp ratio with small seeds need to be developed. Plants are being grown adopting the technology of High Density Plantation (HDP) and plant heights are maintained by pruning according to plant spacing. Awareness about the production technology of these plants should be increased among farmers. The total losses from harvest to the point of consumption are as high as 30%–40%, which account for millions of rupees. About 10%–15% of fresh fruits are shriveled, which lower market value and consumer acceptability. Enhancement in postharvest processing and preservation technologies are critical strategies to add value to the minor fruits. The best way to reduce fruit wastage and to utilize delicate commodities during seasonal gluts is through processing. Once postharvest processing and preservation technologies are developed for these minor fruits, then areas under cultivation and yields can be increased, which will boost their export potential and availability of value-added products as well. This will play a major role to improve the socioeconomic status of farmers and of the country. Pakistan produces a huge amount of guava but export is negligible; the same situation applies to the Indian plum (Ber), jambol (Jaman), and strawberry. The UAE and China are big markets for the export of both fresh and processed ber products. Ber fruit products such as jams, pickles, and chutneys can increase the income of farmers.

Currently, production and postharvest management of strawberries in Pakistan severely lacks scientific recommendations by research and extension institutes. Strawberry runners (seedlings) are generally purchased at Rs. 2–4 per runner from nurseries in Swat. Strawberry planting time in Punjab province ranges from mid-October to mid-November and depends highly upon availability of resources and the convenience of farmers. Likewise, its harvesting time also fluctuates from February to late April. Average yield of strawberry in Punjab province ranges from 2–4 tonnes per acre (Akhter 2013). It is generally harvested during early morning or late evening hours at over 3/4th red stage to fully ripe stage, graded into different sizes, and immediately packed into baskets for marketing. Strawberry is mostly packed in 10–11 kg mulberry or plastic baskets lined with newspaper and cushioned with grass and weeds. During the evening, packed strawberries are transported in open-top trucks (for distant markets) or rickshaws (for nearby markets) to markets where strawberry baskets are auctioned in wholesale markets at an average price of Rs. 90–120 per kg. Due to improper harvesting methods and rough packing, handling and transport, postharvest losses in strawberry are estimated to be more than 40% from farm to wholesale market. Since, strawberry is a very delicate fruit, even slight mishandling causes pathogens, especially fungal infections. Frequent use of insecticides and fungicides during production, and even just before harvesting, results in higher amounts of health-hazardous residues in fruit. Moreover, poor postharvest practices and absence of cool chain systems further lead to the very short shelf life (2–3 days) of strawberries.

Crop planting time strongly influences planting success, plant growth, stress resistance, yield potential, and fruit quality. Planting time of strawberry depends on its day length requirement, which plays an important role in determining flower production and fruit set on strawberry plants. So, it is imperative to standardize planting time of strawberry cultivars grown in Punjab province. This may also provide a better chance to explore the possibility of extending production and harvesting time of strawberry without compromising fruit yield and quality.

The strawberry production business is growing rapidly among farming communities due to high profitability, described in Table 17.5, which clearly indicates its relative advantage over other horticultural crops. This profit margin can further be enhanced by increasing production and reducing postharvest losses. Production may still be enhanced by growing strawberry in Punjab plains closer to river belts and applying proper amounts of N, P, and K fertilizers, as well as foliar application of micronutrients—after consultation with the Agriculture Extension Department. Major reasons of high postharvest losses (>40%) in strawberry include improper harvesting, packing, and handling during supply chain. There is need to organize regular training for all stakeholders of the strawberry supply chain about proper harvest and handling of this delicate crop. It is important to harvest strawberry at the mature red stage and pack them in small size packings ($\leq$ 1 kg). Due to lack

TABLE 17.5
Feasibility Estimate for Strawberry Production (Per Acre)

Description	Cost (Rs.)
Soil preparation	10,000
Plastic mulch (Rs. 230/kg)	8000
Walking tunnel (wire, polythene, etc.)	45,000
Nursery plants (50,000 runners per acre)	100,000
Transplantation (manual)	6000
Weed management (manual)	40,000
Irrigation (every 3 days in October–February, every alternate day in March–May from canal water or tube well)	40,000
Insect, pest, and disease management	30,000
Mineral nutrient management (urea, DAP, SOP, NP fertilizers, etc.)	45,000
Harvesting (manually @ Rs. 250/day/person from February to mid-May)	50,000
Packing (Rs. 35/ mulberry or plastic basket)	45,000
Others unforeseen expenses	50,000
Subtotal: Expenses	**469,000**
Yield (kg/acre)	17,500
Whole-sale price (Rs./kg)	80
Subtotal: Sale	**1,400,000**
Net profit (total sale minus total cost)	**931,000**

Note: Commercial production of strawberries is mostly practiced without tunnels. Expenses, yields and net profits greatly vary with household income, farm size, farming practice, geological and climatic conditions, and disease incidence. Fruit yield per acre varies between 10–20 tonnes/acre.

of awareness, farmers harvest all sorts of strawberry fruit (half red or immature, full red, mature and deep red, or overmature) and pack them in 10–12 kg mulberry or plastic baskets lined with newspaper and cushioned with weeds. This sort of harvesting and packing causes serious postharvest losses. Since, our current supply chain lacks cooling systems for this perishable commodity, farmers have to sell the harvested produce at comparatively low prices to the traders and middle men in wholesale markets. With the establishment of cool chain systems (from farm to consumer), strawberry fruit quality could be maintained, and is postharvest losses could be further reduced, which would result in income generation for all stakeholders of the strawberry industry.

Product quality, fruit nutritional status, and consumer acceptability of the strawberry fruit are greatly influenced by maturity stage and ripeness (Pelayo-Zaldivar et al. 2005; Kafkas et al. 2007). Sugars (mainly glucose and fructose), organic acids (mainly citric acid and ascorbic acid), phenolics, and flavonoids accumulate in strawberry fruit with ripening (Mahmood et al. 2012a,b), thus ripening regulates fruit quality. Therefore optimum maturity stage at harvest helps retain nutritional and functional properties of the strawberry fruit for longer periods during storage. Since, most growers in Punjab province of Pakistan harvest strawberry without considering maturity, it would be intriguing to explore impact of maturity stage on fruit quality and shelf life of strawberry fruit. Fruits picked either too early or too late in their season are more susceptible to postharvest physiological disorders (Kader 1999). Susceptibility of strawberry fruit to fungal decay is also accelerated under the highly humid subtropical climate of Punjab, Pakistan. Market life of strawberry may also be significantly extended by improving packaging and developing cool system in its local supply chain.

Even though, strawberry is successfully grown at different locations of Pakistan, comprehensive research on various aspects of strawberry production and postharvest management is still in its infancy stage. There has been some research studies conducted on growth media (Ayesha et al. 2011; Khalid et al. 2013; Tariq et al. 2013), planting density (Tariq et al. 2013), tunnel production (Qureshi et al. 2012), potassium application (Ahmad et al. 2014), mulching (Kaska et al. 1988), phenolic profiling at different maturity stages (Mahmood et al. 2012a), sugars (Mahmood et al. 2012b), growth regulators (Qureshi et al. 2013) but there is still huge potential of research on many aspects of strawberry production and postharvest management (Figure 17.1).

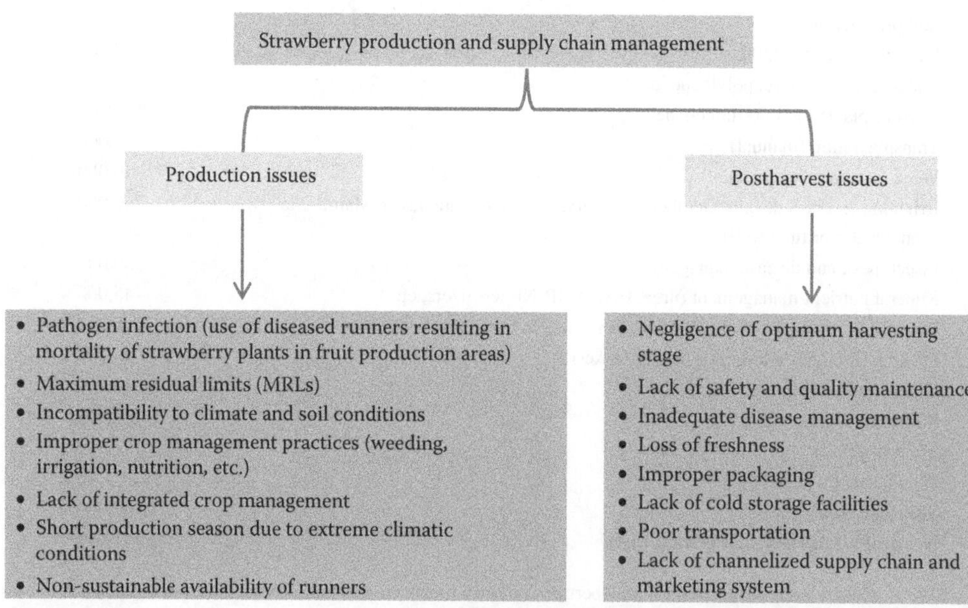

FIGURE 17.1 Major preharvest and postharvest issues related to strawberry production and supply chain management in Pakistan. (Adapted from Bhat and Stamminger, 2015.)

17.3 VALUE ADDED OLERICULTURE

Olericulture is the branch of horticulture that deals with cultivation of vegetable crops. Vegetables are an essential component of human diet because they provide fiber, mineral elements, carbohydrates, proteins, and vitamins. These are also rich source of antioxidants and thus beneficial for human health. Vegetables are perishable in nature because they continue to respire and therefore, metabolic activities continue until they perish. Perishability of vegetable crop restricts their transport to distant places, especially overseas, because the quality deteriorates quickly with the passage of time. Moreover, glut in the market further aggravates their deterioration because produce take more time to be sold. Due to gluts, farmers receive very low prices for their produce. Further, high prices during off-season due to shortage of vegetables is a problem for consumers. Value addition is therefore mandatory to overcome these problems in vegetable crops. Allah has blessed Pakistan with diverse climatic conditions, which permit the cultivation of a variety of vegetable crops at different times of the year. Annual vegetable production in Pakistan, excluding potato, during 2014–2015 was 3116.81 thousand tonnes on an area of 385.58 thousand hectares (MNFSR 2016). Contribution of fruit and vegetables to total value addition in the agriculture sector is about 11%, with export values of about 400 million USD (MNFSR 2013). This share can be further increased manyfolds through interventions which are discussed in the next subsections.

17.3.1 Ecological Zonning

As mentioned earlier, diversity of climatic conditions permits the cultivation of vegetables at different times of years in different areas of the country, for example, cabbage, cauliflower, tomato, and other vegetables requiring mild growing temperature can be successfully grown in the Soan valley of district Khushab, Murree, Quetta, Abbotabad, and northern hilly areas during summer season. In addition to this, use of early, mid, and late season varieties in these areas can further extend the availability of vegetables in the market during the year. So, zoning of vegetable producing areas can be replanned and implemented by the government of Pakistan to meet the demand of off-season vegetables in the plain areas of Punjab and also in Sindh. Moreover, ecological zoning can be helpful in vegetable seed production because specific environmental conditions are required for seed production of elite varieties of different vegetable crops. For example, cabbage seed cannot be produced in subtropical as well as tropical climatic zones, but various pockets in KPK, Gilgit Baltistan, and Balochistan can be used for this purpose.

17.3.2 Processing and Preservation of Vegetables

Overproduction is a big problem facing the farmers of Punjab due to early supply of vegetables from Sindh, resulting in about 25%–30% of the produce wasting each year (Din et al. 2011). Sometimes, farmers even do not get a price that can compensate for packaging (sacks/bags) and transportation and therefore, they dispose of their produce in the field. Many vegetable crops have great potential of value addition, which can help the farmers to save their produce and earn a good return. For example, raw tomato can be processed into paste and ketchup on small scale by local farming groups. This processing is easy to handle for layman and cheap as well. Similarly, some vegetables can be dried and converted into different forms. Nowadays, bitter gourd, that has several health benefits, is dried and converted into powder form and pickled, and used by the patients of diabetes and high blood pressure. Dried bitter gourd slices can also be used for cooking during off-season. Freezing of peas is a very old practice in many countries including Pakistan. Similarly, other vegetables can be processed in different ways and can get good prices in the market during off-season. Conventional processing methods for some vegetables are given in Table 17.6. Government and NGOs may facilitate the training of farming communities for valued processing methods of vegetables.

TABLE 17.6
Processing Methods for Value Addition of Various Vegetable Crops

Crop	Processing Method
Onion	Slicing and frying
Tomato	Paste and ketchup
Carrot	Slicing and freezing
Cucumber	Slicing and freezing; fermentation in brine
Peas	Blanching, followed by freezing or canning in brine solution
Cabbage	Shredding followed by dry salting
Okra	Drying followed by grinding to a powdered form
Coriander	Blanching followed by dehydration to make powder which can be used for chutney production
Mentha (Mint)	Blanching followed by dehydration to make powder which can be used for chutney production
Spinach	Salting and packaging
Turnip	Slicing and suspending in thread followed by sun drying
Radish	Drying to make powder

Besides local processing units at grower level, national, and multinational companies can also be involved in processing of vegetable crops at large scale and the processed products can be marketed at national level or exported abroad. In recent years, some companies have been processing vegetables but mostly limited to tomato ketchup. Mitchell's Pakistan Limited has started processing of Brassica (sarsoon) saag and ready-to-use saag is available in the market. Similar processing can be adopted for other vegetables, which can decrease losses and increase farmers income.

17.3.3 Nursery Business in Olericulture

Vegetable growers in Pakistan are now adopting modern vegetable production techniques. Production of transplants using plug culture and common nursery raising techniques are becoming popular day by day. Moreover, the trend is changing and instead of raising vegetable nurseries, farmers have now started purchasing ready to transplant nursery from nurserymen specialized in vegetable nursery production. This practice has solved many problems such as adulteration of seed, low germination, and nonuniform germination of the seed. Farmers can purchase good quality uniform sized seedlings from such nurserymen. However, this nursery is usually raised in field beds instead of plug trays. There are several problems associated with nursery raising directly in the field. First of all, crop stand is not uniform, and therefore there are great variations in size of the seedlings. Even if the stand is uniform, weeding becomes difficult because some vegetables, such as onions, are highly sensitive. If their leaves are slightly damaged during weeding, they cannot recover. These problems can be overcome by raising nursery in lines. Another problem associated with raising nursery in soil beds is the high mortality rate as compared with nursery raised by using plug culture. Therefore, plug culture can provide uniform sized seedlings, ease in weeding, and other cultural practices during nursery raising, as well as transplanting with minimum mortality rates. Moreover, nursery raising was limited to those vegetables that are considered easy to transplant such as tomato, chilies, cauliflower, cabbage, onion, etc. Cucurbits like cucumber, bitter gourd, bottle gourd have not been yet grown commercially using transplants, for which plug technology is the only way to raise nursery. Plug culture can be popularized among the nurserymen and they must be trained for using plug trays of various sizes. This technology is particularly useful for hybrid varieties for which each seed cost 5 to 10 rupees (for tomato and cucumber) (Figure 17.2). Although the vegetable nursery business is at its infancy stage, it has great potential and therefore needs promotion by the government and private sector.

FIGURE 17.2 Vegetable nursery raising in plug trays for best quality uniform sized seedling production.

17.3.4 Seed Production

Seed is one the inputs in production of any crop that determines its success or failure. High quality seed is the prerequisite for high yield of vegetable crops. For last few decades, there has been a trend to grow hybrid cultivars with high yield potential. Use of hybrid seed escalates the production cost to a great extent as compared to seed of open pollinated varieties (OPs). According to an estimate, total vegetable seed requirement in Pakistan during 2009–10 was 5500 metric tonnes, of which only 36% were locally produced seeds (seed produced at national level both by private companies and farmers). Meanwhile, the remaining 64% of the required vegetable seed was imported from different countries, including China, the United States, India, the Netherlands, Korea, etc. (Table 17.7). The cost of this imported seed was very high (Rs. 1085 million) (Hussain 2011). During 2013–2014, import of vegetable seed was further increased to 5177 metric tons in comparison with 482 metric tons of locally produced seed, which is only 8% of the total seed available in the market (Rana 2014). This volume of imported vegetable seed is increasing every year. Beside high cost, imported seed is a big source of diseases, especially viruses. Moreover, imported seed is also not properly evaluated for acclimatization, which results in crop failure or low production (Personal communication). So, local production of seed can decrease the cost of production and also can prevent inflow of various diseases as well as reduce the burden on economy. Development of a strong local seed industry will also provide job opportunities at various scales. Fortunately, a few determinate and indeterminate

TABLE 17.7
Seed Distribution in Pakistan during 2014–2015 (Metric Tonnes)

Crop	Public Sector	Private Sector	Imported Hybrid Seed/OPV[a]	Total
Vegetables	12.54	52.46	4709.20	4774.20
Potato	34.50	29.00	4217.15	4280.65
Onion	0.25	1.10	34.52	35.87

Source: FSC&RD, Ministry of National Food Security and Research.

[a] Open Pollinated Varieties.

hybrids of tomato, developed by Ayub Agricultural Research Institute and the Nuclear Institute of Agriculture and Biology, are in the pipeline for commercial availability, a step towards indigenous vegetable seed production.

Due to the availability of seed production pockets in various areas of Pakistan, vegetable seed production can be started in conjunction with the breeding staff of government organizations, such as NARC, NIAB, and AARI, which have developed hybrids and high yielding OPVs of some vegetables, especially tomato, bitter gourd, okra, onion, etc. So, the seed of these crops can be multiplied by the progressive growers or national private companies for commercialization of high yielding varieties. At present, the Punjab Seed Corporation is producing and marketing seeds of some vegetable crops but its production share is marginal (Rana 2014). Availability of seed of local high yielding varieties will decrease the import of vegetable seeds and thus, will also reduce the cost of production. At the same time, multinational companies can be invited to establish their research and seed multiplication units in Pakistan, which will also decrease the cost of seed.

17.4 MUSHROOM CULTURE

There are several mushroom species being grown wild in different regions of Pakistan during different seasons. Some of these mushrooms can be collected, dried, and exported to different countries to earn foreign exchange. Potential mushrooms for cultivation in different areas of Pakistan include button mushrooms (Figure 17.3), oyster mushrooms (Figure 17.4), king oyster mushrooms, and milky mushrooms.

17.4.1 Potential Agricultural Waste Materials for Mushroom Cultivation

Pakistan is mainly an agricultural country. Different agricultural waste materials are available for mushroom cultivation in Pakistan, such as wheat straw, rice straw, sorghum straw, soybean straw, pea straw, carrot straw, radish straw, turnip straw, fenugreek straw, corncobs, sunflower straw, corn leaves, cotton ginning waste, coconut pith, banana leaves, barley straw, coconut coir, banana pseudostems, bean straw, mustard straw, citrus fruit peels, corn fiber, legume straws, sawdust, cereal straw, sugarcane bagasse, chopped tree bark, vegetable biomass, etc. These agricultural waste materials are easily available in different areas of Pakistan. All of these agricultural waste materials are relatively cheap and sometimes available free of cost. All of these agricultural waste materials can be used for the successful cultivation of mushrooms.

Cotton gin waste is normally composed of leaves, sticks, soil particles, burs, other plant materials, cotton lint, mote, etc. A slight difference in the proportions of the components exists between different mechanical harvest methods. In general, as a whole, cotton gin waste is regarded as a lignocellulosic substrate, that is, a material generally comprised of cellulose, lignin, and hemicellulose, etc. One of the biggest problems faced by the cotton ginning industry in Pakistan and rest of the world is cotton gin waste management. It has been estimated that ginning one bale (227 kg) of spindle harvested

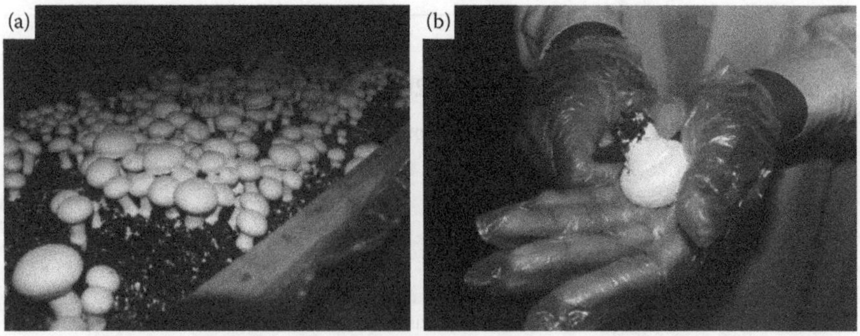

FIGURE 17.3 Button mushroom production: (a) ready to harvest and (b) stem trimming.

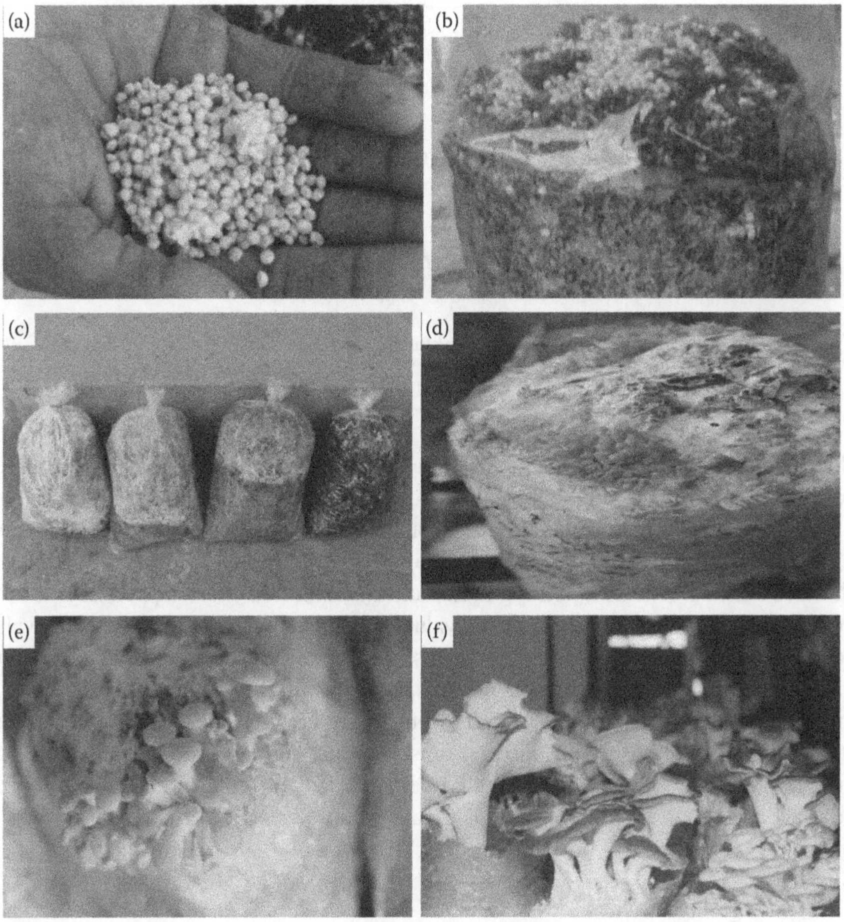

FIGURE 17.4 **(See color insert.)** Oyster mushroom production: (a) spawn (seed), (b) spawned bag, (c) stages of spawn running in tied bags, (d) pinhead initiation, (e) fruit body formation, and (f) cropping.

seed cotton lint approximately produces 37 to 147 kg of waste. On average, about 12 million bales are ginned annually in Pakistan, the amount of cotton gin waste produced in Pakistan could be approximately close to 3.5 billion pounds per year. Disposal of such huge amounts of cotton gin waste is an alarming issue as may cause serious threats to the environment. At the moment, the global cotton industry is trying to reduce their cotton gin waste by alternative options for handling this cotton gin waste as a by-product, because it has multiple uses.

Use of cotton gin waste as a growing media for oyster mushroom cultivation could be a viable option in this regard. Besides reducing environmental pollution, it can also provide food to the masses and can help in ensuring food security in food deficient countries like Pakistan. Additionally, growing of oyster mushrooms on cotton gin waste can be helpful in creating additional jobs for the village-based communities and can improve their livelihood. Different steps of button and oyster mushroom production technology have been demonstrated in Figure 17.4.

17.4.2 Mushroom Value Addition: Current Scenario

The Pakistan mushroom industry is still predominantly production and trade of fresh produce. Almost the entire domestic trade is in the fresh form, while all imports are in the preserved form (canned) (Table 17.8). Current era is characterized by greater awareness of quality and the demand for the

TABLE 17.8
Import and Export Volume and Value of Mushrooms from Pakistan for 2013

Element	Item	Quantity (Tonnes)	Value (1000 US$)
Export	Canned mushrooms	1	2
	Mushrooms and truffles	520	802
Import	Canned mushrooms	569	283
	Mushrooms and truffles	7	20

Source: FAO, 2013. http://www.fao.org/statistics/en/ FAOSTAT Date: Wed Apr 27 09:45:29, CEST 2016.

readymade or ready-to-make food products. Value can be added to the mushrooms at various levels and to varied extents, right from grading to the readymade snacks or the main-course items. Improved and attractive packaging, which may be called the secondary value-addition, is important but is a totally neglected area in the mushroom supply chain. Packaging of mushrooms is still done in unprinted plain polypouches in Pakistan. Whereas, packing in attractive and labeled overwrapped trays is in vogue in the developed countries. Small growers may add value by grading and packaging. Potential value-added products of mushrooms include mushroom biscuits, mushroom bread, mushroom cakes, mushroom pickles, mushroom chutneys, mushroom nuggets, mushroom soup powder, mushroom ketchup, mushroom chips, ready-to-eat mushroom curry, mushroom cream, mushroom porridge, mushroom jams, fast food items like mushroom burgers, mushroom cutlets, and mushroom pizza. Moreover, mushroom value addition in canned, dehydrated, freeze dried, or frozen form can also be utilized for its year-round availability in the country as well as for export purposes to international markets.

17.5 VALUE ADDED FLORICULTURE

Floriculture is presently considered as the most lucrative agro-enterprise for making higher profits. However, being the utmost perishable horticultural commodity, there remains several bottlenecks in its proper marketing and getting maximum return by the producers. Therefore, value addition is an important arena for proper utilization of ornamental crops either fresh or in processed form. Value addition is a process of increasing the economic value and appeal of the flower commodities through processing or improved presentation, which in turn increases profitability of the growers and stakeholders. There are many value-added floral products available in the market, which have enhanced the profitability in competitive markets. These include branded packaged floral products (Figure 17.5a), floral ornaments, dehydrated flowers, and buds popular for making potpourri (Figure 17.5b), improved soilless substrates for high quality ornamentals (Figure 17.5c), various by-products such as essential oils, flavors, pharmaceutical compounds, pigments and natural dyes, rose water (Figure 17.5d), rose jam, perfumes (Figure 17.5e), and insecticidal compounds, using most advanced supercritical fluid extraction technology (Figure 17.5f).

In Pakistan, commercial floriculture is still in infancy stage due to several reasons, which include unavailability of improved plant material, poor infrastructure for quality production, lack of suitable substrates for quality ornamentals production for export, lack of cool-chain facility for export of cut flowers and potted plants, unavailability of trained manpower about innovative production and processing technologies, poor knowledge about international standards and value addition, etc. However, during recent years, floriculture has emerged as a major diversification option in the agribusiness and a rapid increase in floriculture production has been followed (Ahmad et al. 2010). This situation holds good future prospects, being one of the most profitable components of diversified horticultural industry. There is an increasing trend for areas under cut flower production, particularly

FIGURE 17.5 Value addition of ornamentals: (a) modern packaging of cut flowers, (b) potpourri, (c) soilless substrates for quality plant production, (d) rose water, (e) oil based perfumes, and (f) supercritical fluid extraction plant.

bulbous flower crops such as gladiolus, lilies, tuberoses, tulips, etc. Moreover, exotic species including delphinium, lisianthus, marigold, snapdragon, stock, sunflower, and zinnia have been introduced by the University of Agriculture, Faisalabad. Specialty cut flowers in the local markets and their commercial production would diversify the currently available flower products in the markets. Demand for quality cut flowers has increased several folds during last few years on account of change in lifestyle and increase in flower use in various social and religious festivities. Therefore, it is need of the time to increase the value and appeal of floriculture products to ensure high premiums to growers and to diversify floral products available to satisfy consumers' aesthetic needs.

17.5.1 Value Addition of Cut Flowers, Foliage, and Potted Ornamentals

Cut flowers, foliage, and potted ornamental plants are major floriculture components whose share is more than 3/4th of the global floriculture trade. In developed countries, there are only 5%–10% losses during postharvest handling of chain of cut flowers. While in Pakistan, these losses are about

20% during winter and 30%–40% during summer because of unavailability of cool-chain handling facility for the perishable flowers. For cut flowers, value can be added by selection of improved cultivars, which have high demand in global markets, have no prickles, can withstand shipping/transportation, and have reasonably longer vase life of more than a week. These may be produced in greenhouses under controlled atmospheric conditions for year-round production of good quality flower stems, while outdoors for seasonal production, if resources are limited. Production of quality stems/plants also depends on substrate in which plants are raised. Therefore, use of advanced soilless substrates not only ensures uniform nursery production but also the best quality flower and final plant production (Ahmad et al. 2012). Unfortunately, most of commercial production of ornamentals in the country is being done using soil, which result in poor quality production due to several limiting factors including salinity and soil borne diseases. Moreover, plants grown in soil cannot be exported due to quarantine issues. To overcome this issue, efforts are being made to develop low cost sustainable soilless substrates using local components for nursery production. Most important way to add value is adoption of proper postharvest handling technology, which is completely unavailable to the farming community in Pakistan. Value/price of the flower stems may be enhanced by proper handling at low temperature, flower dyeing with different colors (rainbow roses are getting popularity all around the world), improved branding/labeling, and specialized innovative packaging to ensure good quality after transportation.

For nursery production of ornamentals, plug culture is the best option, which can ensure the maximum transplanting success with almost no mortality and good quality uniform crop stand. It also allows more protection at nursery stage and flexibility in time to transplant. Similarly, for potted plants, use of plastic or biodegradable pots, locally available organic substrates such as rice hulls, coco coir, biochar or spent mushroom compost, proper labeling along with care tips would add value and increase profitability. Similarly, different bioactive compounds and essential oils having high demand in local and international markets can be extracted from various ornamental crops such as roses (*Rosa centifolia* and *R. damascena*), jasmine. (*Jasminum* spp), tuberose (*Polianthes tuberosa*), Murva (*Murraya exotica*), etc. Different components of value addition of ornamentals have been discussed with possible outcomes in Tables 17.9 and 17.10.

17.6 VALUE ADDED MEDICINAL PLANTS

Crude drugs play a vital role in the pharmaceutical industry. A drug has both active and inactive constituents; however, research has proved that only active constituents can be used as medicine. Crude drugs, besides their use in natural form are also used for extraction of active constituents and other products. For example, fennel is a carminative, while opium is hypnotic. Natural products can be used to impart color, taste, and odor to various dishes. For example, pectin is isolated from lemon peels, while cinnamon is used as flavoring agent. Honey is used as sweetening agent. Some medicinal plants have bioactive compounds, which have great healing power against heart diseases and cancers. For example, Digitalis plants and their drugs (Digoxin and Digitoxin) are used in curing heart diseases. Catharanthus plants and their drugs (Vincristin and Vinblastin) are used to cure cancers. The majority of medicinal plants and their derivatives are considered as a source of affordable healthcare.

In Pakistan, there is no coordinated approach for the collection, cultivation, and sale of many high value medicinal plants commonly used in various folk recipes and medicines such as ispaghol (*Plantago ovata*), saunf (*Foeniculum officinalis*), linseed (*Linum usitatissimum*), Kawar Gandal (*Aloe vera*), and mullathi (*Glycyrrhiza Glabra*) (most consumed/marketed medicinal plants) (Figure 17.6). Major constraints include lack of a database describing distribution of medicinal plants, nonscientific methodology, approach to medicinal plants collection and improper method of selection, harvesting, drying, curing, preservation, packing, garbling, and storage in which losses can be as high as 70%. Moreover, difficulty in predicting the marketability of plant extracts and market preferences for natural extracts are also major bottlenecks in the development of the natural medicine industry.

TABLE 17.9
Possible Value Addition Components of Floriculture Crops

Possible Value Addition Component	How?	Possible Outcome(s)
Improved handling and marketing of cut flowers	Optimal postharvest handling and promotion Establishing cool chain facility from farms to markets and airports Branding/labeling Innovative packaging Flower dyeing with natural dyes	Flowers with good quality and long lasting vase life Longer shipment durability Diversification at retail
Uniform nursery production with good quality saplings	Plug culture Soilless substrates	Good quality nursery with minimum transplant mortality Improved transportation/shipping Seed saving (lower cost of production)
Value addition of potted plants	Innovative pots (plastic or biodegradable) Organic substrates (coco coir, rice hulls, spent mushroom compost) Labeling	Quality enhancement Acceptable in global markets Easier shipping Promotion and branding
Flower forcing	Off-season flower production Production in winter when high demand in international markets	Higher returns from exports
Cut foliage longevity	Waxing of cut foliage and potted plants	Improved glossy appearance of the produce with minimum water loss
Diversification of cut flower industry	Introduction of new potential species and cultivars which have high market demand Optimizing production and postharvest handling protocols for new crops	Provision of more choices to not only producers but also to the consumers to fulfill their aesthetic needs

TABLE 17.10
Value Addition of Flower Crops Through Dehydration and Processing

Possible Value Addition Component	How?	Possible Outcome(s)
Dehydration of flower petals and buds	Controlled solar dehydration	Potpourri Saving extra produce
Floral ornaments	Garlands Floral bangles and crowns Buttonholes Flower bouquets	Value addition of loose flowers More income per unit area/production
By-products development	Essential oils Phytochemicals and plant pigments extraction Rose jam Rose water Rose eye drops Pharmaceutical compounds	Processed high value products/pigments such as xanthophylls, luteins, anthocyanins, etc. Essential oil of marigold, tuberose, basil, rose, jasmine, etc. Products for textile dying and pharmaceutical industries Value added rose products such as rose jam, rose water and eye drops
Flower dyeing	Flower tinting Sulphuring (color removal) Dyeing	Supply of specific color flowers for specific occasions (Valentine's day—red) to increase profitability

FIGURE 17.6 **(See color insert.)** High Value medicinal plants of Pakistan: (a) Kawar Gandal (*Aloe vera*), (b) ispaghol (*Plantago ovata*), (c) linseed (*Linum usitatissimum*), and (d) mullathi (*Glycyrrhiza Glabra*).

According to World Health Organization (WHO) estimations, traditional medicines, mostly plant drugs, can fulfill medicinal needs of 80% of the world population. Moreover, natural products being nonnarcotic have fewer side effects as compared to synthetic drugs. Basically, drugs are used mainly for preventive (prophylactic) or curative (therapeutic) purposes. Natural sources give a number of useful drugs, which are difficult to produce commercially by synthetic or chemical and microbiological means as active constituents. Therefore, the only way to produce these drugs is through cultivation of respective medicinal plants.

Cost of treatment using medicinal plants is quite low compared to the use of synthetic drugs. Scientists use natural compounds as models or prototypes for synthetic drugs because they have pharmacological activities similar to the original compounds. Some natural sources are converted into useful product by means of chemical or microbiological treatment. After treatment, these are converted into potent drugs because they have compounds with little or no activity themselves. Therefore, these drugs are very useful in the field of medicine.

17.6.1 Commercial Production of Medicinal Plants

There is a long history of commercial production of medicinal plants during eras of ancient civilization. Plant growth and development, and often the nature and quantity of secondary metabolites, are affected by temperature, rainfall, aspect length of day, and altitude. Such effects have been studied by growing particular plants in different climatic areas and observing variations. Certain drugs are obtained from commercially cultivated plants only. These include cardamoms, Indian hemp, ginger, peppermint, and spearmint being used for oil production. Others include Ceylon cinnamon, linseed, fennel, cinchona, and opium. While for some drugs, both wild and commercially cultivated plants are used. Consumption of medicinal plants is widespread and increasing due to the increasing demand of herbal plants. Commercial production of medicinal plants is the main source of raw material. The high selling medicines that have large market shares such as *Ginkgo biloba* (maidenhair tree) or *Hypericum performatum* (common St. John's wort) are mostly under production by large European and American herbal commercial companies. But medicinal traits of these plants can be improved by plant breeding methods and molecular markers and can also assist in this regard. Use of tissue culture and genetic transformation techniques can also help to alter the pathway for the production of target metabolites.

Pharmaceutical companies are dependent on the natural resources (mainly plants) for drug development. This importance of medicinal plants has compelled modern science to produce plant species on a commercial scale. During the last few decades, production of high-value metabolites and phytochemicals by tissue culture, and in vitro systems for growing tissues, organs, or protoplasts has increased. Moreover, the power to confine collection to species, varieties, or hybrids that have desired phytochemical characteristics such as aconite, cinnamon, fennel, cinchona, labiates drugs, and valerian has improved. But there is still much to accomplish to explore this naturally available high-value resource in Pakistan.

REFERENCES

Ahmad H, Sajid M, Ullah R, Hayat S and Shahab M. 2014. Dose optimization of potassium (K) for yield and quality increment of strawberry (*Fragaria* × *ananassa* Duch) Chandler. *Amer. J. Exp. Agric.* 4(12): 1526–1535.

Ahmad I, Ahmad T, Gulfam A and Saleem M. 2012. Growth and flowering of Gerbera as influenced by various horticultural substrates. *Pak. J. Bot.* 44(SI1): 291–299.

Ahmad I, Dole JM, Khan MA, Qasim M, Ahmad T and Khan AS. 2010. Present status and future prospects of cut rose production in Punjab (Pakistan). *HortTechnology* 20(6): 1010–1015.

Akhter M. 2013. Dynamics of strawberry marketing system in Punjab. M.Sc. (Hons.) Diss., University of Agriculture, Faisalabad, Pakistan.

Ayesha R, Fatima N, Ruqayya M, Qureshi KM, Hafiz IA, Khan KS and Kamal A. 2011. Influence of different growth media on the fruit quality and reproductive growth parameters of strawberry (*Fragaria ananassa*). *J. Medicinal Pl. Res.* 5(26): 6224–6232.

Cherian S, Figueroa CR and Nair H. 2014. Movers and shakers' in the regulation of fruit ripening: a cross-dissection of climacteric versus non-climacteric fruit. *J. Exp. Bot.* 65(17): 4705–4722.

Din A, Parveen S, Ali MA and Salam A. 2011. Safety issues in fresh fruits and vegetables—A review. *Pak. J. Food Sci.* 21(1–4): 1–6.

FAOSTAT. 2011. Food and Agriculture Organization. (Available online with updates) http://faostat.fao.org/site/339/default.aspx. Assessed on March, 27, 2015.

FAOSTAT. 2013. Food and Agriculture Organization. (Available online with updates) http://faostat.fao.org/site/339/default.aspx. Assessed on March, 27, 2015.

Hussain A. 2011. Status of Seed Industry in Pakistan. Available at: http://siteresources.worldbank.org/PAKISTANEXTN/.../statusofseedindustr.

Kader AA. 1999. Fruit maturity, ripening, and quality relationships. *Acta Hort.* 485: 203–208.

Kafkas E, Kosar M, Paydas S, Kafkas S and Baser KHC. 2007. Quality characteristics of strawberry genotypes at different maturation stages. *Food Chem.* 100: 1229–1236.

Kaska N, Shah AH, Khan DA and Khokar KM. 1988. Strawberry production under low polythene tunnels with different mulch system in Islamabad. *Pak. J. Agric. Res.* 9(4): 543–548.

Khalid S, Qureshi KM, Hafiz IA, Khan KS and Qureshi SU. 2013. Effect of organic amendments on vegetative growth, fruit and yield quality of strawberry. *Pak. J. Agric. Res.* 26(2): 104–111.

Khan MA. 1998. Chanda beel: wildlife and biodiversity in peril. In: Gain P (ed) *Earth Touch*, Vol. 4. pp. 17–20. The Society for Environment and Human Development, Dhaka, Bangladesh. Freshwater prawn farming in Bangladesh. Aquacul Res 39: 806–819.

Khan MH. 2013. Published in Dawn Newspaper. http://www.dawn.com/news/1102694.

Khan MN, Sarwar A, Bhutto S and Wahab MF. 2010. Physicochemical Characterization of the Strawberry Samples on Regional Basis Using Multivariate Analysis. *Int. J. Food Properties* 13: 789–799.

Mahmood T, Anwar F, Abbas M and Saari N. 2012a. Effect of maturity on phenolics (phenolic acids and flavonoids) profile of strawberry cultivars and mulberry species from Pakistan. *Intl. J. Mol. Sci.* 13: 4591–4607.

Mahmood T, Anwar F, Abbas M, Boyce MC and Saari N. 2012b. Compositional variation in sugars and organic acids at different maturity stages in selected small fruits from Pakistan. *Intl. J. Mol. Sci.* 13: 1380–1392.

MNFSR. 2013. *Fruit, Vegetable and Condiments Statistics of Pakistan 2011–12*. Ministry of National Food Security and Research, Economic Wing, Govt. of Pakistan, Islamabad, Pakistan.

MNFSR. 2016. *Fruit, Vegetable and Condiments Statistics of Pakistan 2014–15*. Ministry of National Food Security and Research, Economic Wing, Govt. of Pakistan, Islamabad, Pakistan.

Ornelas-Paz J, De J, Yahia EM and Gardea A. 2013. Changes in external and internal colour during postharvest ripening of "Manila' and "Ataulfo" mango fruit and relationship with carotenoid content determined by liquid chromatography–APcI$^+$-time-of-flight mass spectrometry. *Postharvest Biol. Technol.* 50: 145–152.

Pasternak D, Nikiema A, Ibrahim A, Senbeto D and Djibrilla I. 2016. How demosticated *Zizphus mauitiana* (Lam) spread in the Sahel region of Africa and Ethiopia. *Chronica Hortic.* 56(1): 21–25.

Pelayo-Zaldivar C, Ebeler S and Kader A. 2005. Cultivar and harvest date effects on flavor and other quality attributes of California strawberries. *J. Food Qual.* 28: 78–97.

Qureshi KM, Hassan F, Hassan Q, Qureshi US, Chughtai S and Saleem A. 2012. Impact of cultivation systems on growth and yield of strawberry (*Fragaria ananassa*) cv. "Chandler." *Pak. Agric. Res.* 25(2): 129–135.

Qureshi KM, Chughtai S, Qureshi US and Abbasi NA. 2013. Impact of exogenous application of salt and growth regulators on growth and yield of strawberry. *Pak. J. Bot.* 45(4): 1179–1185.

Rana MA. 2014. The Seed Industry in Pakistan: Regulation, Politics and Entrepreneurship. Working Paper No. 019, Pakistan Strategy Support Program.

Seeram NP. 2008. Berry fruits for cancer prevention: Current status and future prospects. *J. Agric. Food Chem* 56: 630–635.

Tariq R, Qureshi KM, Hassan I, Rasheed M and Qureshi US. 2013. Effect of planting density and growing media on growth and yield of strawberry. *Pak. J. Agric. Res* 26(2): 113–123.

Wasim M, Khalid N, Asif A, Arif M and Zaidi JH. 2012. Elemental characterization of strawberry grown in Islamabad by k_0-instrumental neutron activation analysis and atomic absorption spectrophotometry and its dietary assessment. *J. Radioanal Nucl. Chem.* 292: 1153–1159.

18 Forestry and Range Management in Pakistan
Present Potential and Way Forward

Muhammad T. Siddiqui, Muhammad F. Nawaz, Rashid A. Khan, and Zahoor H. Khan

CONTENTS

18.1 PRESENT FORESTRY SCENARIO

The estimated forest area of Pakistan is 4.2 million ha, which is only 4.8% of the total land area (87.98 million ha). All four provinces have different ecological conditions and forests also vary with the changes in the ecology. Variability of forest area is evident as given in Table 18.1. Pakistan is comparatively a forest deficient country with only 0.05 ha of forest per capita as compared to a world overage of 1.0 ha (FAO, 2010). There are many issues hindering the proper forest management but deforestation, injudicious forest management practices, lack of incentives for afforestation, lack of wood-based industries, faulty timber markets, weak institutional infrastructure, and so on are some of the major problems regarding forestry in Pakistan. It has been estimated that more than one third of the original cover has disappeared, having grave consequences for human livelihoods, ecosystems, and environmental concerns (FAO, 2005).

In Pakistan, the estimated deforestation rate is approximately 39,000 ha/year (FAO, 2009). If this injudicious practice of forest logging goes unabated, it is feared that within the next 30–40 years, Pakistan will lose most of its forest area. Being a forest deficient country, protection and conservation of its forest resources is urgently required. Moreover, deforestation and degradation can have local impacts such as barren watersheds, more floods and landslides, or broader impacts, like global warming, desertification, land use changes and global climate shift (TEEB, 2010). Such menaces can result in substantial loss of financial resources needed for forest protection and conservation (UNEP, 2011).

Although there is an increasing trend regarding forest cover expansion in developed countries over the last decade; however, vast area of these forests is largely over exploited resulting in degradation and deterioration. The situation has been aggravated by environmental pollution, impacts of climate change, wildfires, introduction of exotic invasive species, and fragmentation. (Naeem et al., 1994; Roberts et al., 2009). Consequently, forest habitats continues to degrade and their biodiversity is suffering, which can affect forest ecosystem functioning and forest productivity

TABLE 18.1
Forest Area by Province (ha)

	AJK	Baluchistan	Northern Area	KPK	Punjab	Sindh	Total
Forest area	275	592	666	1684	608	399	4224
Total area	1330	34,719	7040	10,179	20,626	14,091	87,980
% age tree cover	20.7	1.7	9.5	16.6	2.9	2.8	4.8

Source: Forestry Sector Master Plan 1992–2018. FSMP. [online] available: http://www.environment.gov.pk/pub-pd... t3-Chp%203.pdf.

TABLE 18.2
Forest Area by Legal Classification and Province (ha)

Forests Category	AJ&K	Baluchistan	Northern Area	KPK	Punjab	Sindh	Total
Reserved	567	707	–	98	337	292	2001
Protected	–	378	67	629	2747	726	4547
Un-classed	–	–	–	7	115	25	147
Resumed lands	–	–	–	33	9	5	47
			Private Forest				
Guzara forests	–	–	–	550	–	–	550
Communal forests	–	–	2982	–	–	–	2982
Section 38 areas	–	1	–	26	9	–	36
Chos Act areas	–	–	–	–	1	–	1
Private plantations	–	–	–	159		–	159
Miscellaneous	–	–		53	42	–	95
Total	567	1086	3049	1555	3260	1048	10565

Source: FAO 2010 Global forest resources assessment. *Main report forestry paper 163*. Food and Agriculture Organization of the United Nations, Rome.

(Butchart et al., 2010; Norris, 2012). So, conservation as well as restoration of forests and their ecosystems is of prime importance to ensure the survival of life on Earth in a sustainable manner (Lambin and Meyforoidt, 2011).

Table 18.1 shows that 40% of the forest area is composed of gymnosperm species and scrub forest in the northern part of the country. The remainder of the area includes riverain forests along the banks of major rivers, irrigated plantations, mangrove forests, and trees planted on agriculture fields. It is well understood that forest resources of the country cannot cope with the increasing demand of wood and wood products of a population growing at the rate of 2.6%. Furthermore, under the prevailing arid and semiarid climate of the country, it is not possible to expand public forest areas at a rate that could meet the forest product demands of the society.

In Pakistan, according to FAO (2010), forests have been classified based on vegetation composition, legal status, function, and ownership. Detail of the areas of state and private forest subcategories in the provinces is given in Table 18.2.

18.1.1 State Owned Forest

Major portions of forest land (85%) are state owned. These are legally further classified into the following four classes discussed in the next subsections.

18.1.1.1 Reserved Forest

Pakistan inherited and accepted the Forest Act of 1927 under which reserved forests are in the control of the Forest Department. Unless permitted through special permission, local dwellers generally do not have rights and privileges on these forests.

18.1.1.2 Protected Forests

According to the Forest Act 1927, in contrary to "Reserved Forests" local dwellers have limited rights and concessions for livestock grazing, cutting of grass, collection of dead and drywood, and so on from "Protected Forests."

18.1.1.3 Unclassed Forests

State-owned forestlands controlled by provincial forest departments, which are neither put under the category of reserved nor protected forests are called unclassed forests.

18.1.1.4 Resumed Lands

There are some lands which have been surrendered by native landowners under the land reforms act of 1959 following the ceiling on the limit of land ownership. Concerned landlords preferred to retain cultivated lands and surrender the wooded lands that were over the fixed ceiling. The Forest Department manages these lands under the category of "resumed lands."

18.2 PRIVATELY OWNED FORESTS

It's a wider classification term entailing the forests in private ownership. The five categories are as follows:

18.2.1 Guzara Forests

Under this category, a reasonable afforested area close to public dwellings was defined to fulfill the domestic needs of native people. Such forests were defined when state forests were reserved at the first land ownership settlement in Hazara, NWFP (now KPK), 1872. Ownership of such wooded land is given to local dwellers either individually, or communally as village Shamilat. It is the responsibility of these local dwellers to maintain these wooded patches.

18.2.2 Communal Forests

They come under the category of Guzara forests. These forests provide a broad spectrum of social services and are essentially owned by the inhabitants of an entire village. These forests are mostly found in the Pothwar region of Punjab province.

18.2.3 Chose Act Area

The sensitive and easily erodable lands, liable to natural calamities, are put under Chose act areas. These are privately owned lands, which can cause damage to public buildings and installations because of erosion or earthquakes and taken over by the state under the Chose Act, 1900. After stabilization, original owners may be returned the ownership of these areas.

18.2.4 Section 38 Areas

A mutual agreement between private owners and the Forest Department for afforestation and management can be promulgated for a period ranging from 10 to 20 years under section 38 of the Pakistan Forest Act, 1927.

18.2.5 Farm Forest Areas

Growing trees on private farmlands either in linear fashion or compact blocks on farmlands owned individually or by a family. Such areas are found throughout the country both in rain fed and irrigated farming areas of Pakistan.

18.3 STRATEGIES FOR COPING WITH DEFORESTATION

Unprecedented deforestation has resulted in degradation of huge areas of wooded lands. To check the pace of deforestation, certain practical steps must be taken to restore forest ecosystems and make the environment healthy and stable. Following are major actions to be considered by all concerned:

18.3.1 Afforestation

According to FAO (2009), wood demand around the world is projected to increase twofolds by the year 2030 and nearly six times by 2060 because of expected increasing consumption for fuel wood (Harrington, 1999). In addition, because of regional economic shifts, population rise, energy shortage, polices, and regulations on environment, the consumption and utilization of forest products will also continue to rise.

Afforestation/plantation forestry can help to fulfill the growing need for forest products, thus reducing population pressure on natural forest resources. These plantations provide multiple benefits, such as enhanced secondary growth and high-density biodiversity and offer a scope for mosaic and landscape forest restoration (Raunikar et al., 2010).

Although, in Pakistan, the deforestation rate is high for natural forests but in developing countries, there has been an increasing trend in plantation forests. For example, in 1990 the planted area was 234,000 ha, and an increasing trend was observed in the following years like in 2000 it increased to 296,000 ha, in 2005, 318,000 ha, and in 2010 it reset to 340,000 ha. The annual change rate was in the range of 2.38%, 1.44%, and 1.35% from 1990 to 2000, 2000 to 2005, and 2005 to 2010, respectively (FAO, 2011). Furthermore, it was estimated that out of total wood volume 240,002,000 m^3 planted on 85,221 $\times$ 10^3 ha in various ecological zones about 135,572,000 m^3 was for timber wood purpose (Table 18.3).

A Household Energy Strategy Study (HESS) demand survey, conducted in four provinces, revealed that about 1.251 billion trees were planted on agricultural lands during 1990–1991 and about 0.108 billion trees were harvested. Among planted trees, about 0.132 billion were fruit trees. Results of this survey about felling and planting different trees in various provinces of Pakistan are described in Table 18.4.

TABLE 18.3
Estimates of Standing Wood Volume (000 m^3) in Different Agro-Ecological Zones of Pakistan

Zone	Area ($\times$ 10^3 ha)	Wood Fuel	Timber	Total Wood Volume
Desert	25,234	1857	286	2143
Semiarid	25,383	7857	6000	13,857
Natural forest	12,997	61,143	43,143	104,286
Barani	2788	4429	429	4858
Irrigated lowlands (Sindh/Balochistan)	4461	4429	4429	7715
Irrigated lowlands (Punjab/KPK)	3186	14,286	14,286	28,572
Irrigated highlands (Sindh /Baluchistan)	1987	4714	4714	8857
Irrigated highlands (Punjab/KPK)	9185	36,857	36,857	69,714
Total	85,221	135,572	135,572	240,002

Source: Archer G, 1993 Pakistan Household Energy Strategy Study (HESS). Planning and Development Division. Government of Pakistan.

TABLE 18.4
Tree Planting and Felling Rates in Various Provinces of Pakistan

	Punjab	Sindh	KPK	Balochistan	Total
Trees planted (billion)	0.609	0.0651	0.4416	0.1354	1.2511
% of plantings	49.7	5.2	35.3	10.8	100
Fruit trees (billions)	0.0569	0.0261	0.0371	0.0125	0.1325
Nonfruit trees (billions)	0.5521	0.0390	0.4045	0.1229	1.1186
% nonfruit trees	90.6	59.9	91.6	90.8	89.4
Trees planted/acre	3	1	22.2	9.9	4.4
Total farmers (million)	3.396	1.117	0.800	0.245	5.558
% of farmers planting trees	42.9	26.6	48.4	51.8	40.8
Trees felled (billions)	0.0542	0.0204	0.0163	0.0175	0.1083
Planting/felling ratio	11.2	3.2	27.1	7.7	11.6

Source: Archer G, 1993 Pakistan Household Energy Strategy Study (HESS). Planning and Development Division. Government of Pakistan.

18.3.2 Training of Foresters and Forest Management

Forest department is using traditional methods to manage forest resources. In fact, before the division of subcontinent, there was 25% of the total area under forests and British Government made the policy to protect these forests. So, foresters were given complete training, such as in military service, and were even equipped with weapons for use against timber mafias. After the division of Pakistan and India, most of the forests went to India and a meager forest area was given to the newly born state of Pakistan. However, despite facing a scarcity of wood and wood products, a viable forest policy addressing future needs was not framed. Protection forests were cleared to meet the demands of wood without considering future stability of the ecosystem. The forest law of 1927 was valid until the 1990s, where only Rs. 200 fine and 6-month prison terms were proposed for forest related offences. Forests followed the same old policies which resulted in insignificant improvement. Even today, all the foresters in the PFI (Pakistan Forest Institute, Peshawar) are trained in almost a similar way. Up-to-date teaching and research-oriented curricula should be introduced to train the officers to address the changing national and international environmental scenarios. The forest sector needs to have professional competent personnel and merely persons with pseudo military training. It is strongly recommended that staff like Forest Guards and Forest Block Officers be given good training because they spend the most of their time in the field; however, all officers should be competent professionals and trained researchers. Other options are to recruit the trained foresters (officers) through a competitive selection process from all the relevant academic institutions of the country. These recruits would then receive intensive professional training of 6 month in a well reputed institution.

18.3.3 Promotion of Participatory Approach

Forest Departments rarely involves the local people to plant trees in their forest area. In many foreign countries, on special events, local people are invited to plant the trees in their name and take care of them throughout their lives. Globally, the concepts of social forestry, agroforestry, community forestry, and urban forestry are more attractive than state forestry. It is recommended that participation of local people from all walks of life should be ensured in tree planting and post planting operations. This will develop a sense of ownership and participation in the people. The tasks of protecting and caring for trees will also be shared through participatory approaches.

18.3.4 Strong Collaboration among Forest Department, Universities/Colleges, and Research Institutes

In Pakistan, poor collaboration exists between forest departments, agricultural and forestry universities, and research institutions. A national taskforce should be organized on a collaborative basis taking all stakeholders on board. It is strongly recommended that a task force consisting of forest professionals, forest department officers, academicians, public representatives, social activists, and farmers be constituted under the governance of the Chief Minister of the province to boost the plantation of trees on a large scale.

18.3.5 Provision of Funds and Facilities

Adequate facilities and funds are essential for afforestation, reforestation, and protection of forests. An adequate budget with strong auditing and evaluation is mandatory for this purpose.

18.3.6 Planned Forestry

In Pakistan, tree planting has been a regular feature over the last 60 years. During every planting season, the forest departments distribute millions of tree seedlings to the people. Quite a number of the plants, however, fail to survive or grow because common people lack adequate knowledge of proper site preparation, species selection, planting techniques, and follow-up care.

It is a common notion that a tree will grow anywhere or everywhere. When selecting a tree for planting, one should look around and see which trees are thriving locally. The choice of suitable species is critical to success. The following points should be considered while planning plantation:

- Suitability of climate, minimum annual rainfall, and its seasonal distribution
- Suitability of soil, such as soil texture, structure, fertility, pH, porosity, and drainage
- Wind velocity and frequency, particularly in shallow soils and desert
- Value as fuel, fodder, and forage
- Rate of growth and requirement of subsequent aftercare
- Susceptibility to insects, diseases, parasites, etc.
- Market requirements

It is a false notion that the job of the tree planter is finished as soon as the tree is put into the soil. Baby plants need care and to be looked after until its establishment. Therefore, tending and aftercare is the most important part of the whole operation. Watering at proper intervals, weeding, hoeing, staking, and so forth are also important operations.

18.3.7 Covering Forest Area by Regeneration/Afforestation

After logging of mature trees, forests can be regenerated either through natural or artificial regeneration. Artificial regeneration is carried out by seed sowing, or by planting directly in fields the nursery raised seedlings. The estimated rate of annual regenerated area is about 28,500 ha. Afforestation of newly areas is carried out at the rate of 2300 ha/year. If the actual scenario persists, it would require more than 20 years of afforestation on 1 million ha area. Afforestation programs should likely be accelerated by eliminating physical, financial, and legal constraints. In economic planning, a low priority has been given to the management of natural resources. Its importance should be included in economic planning to reorient it because forestry is about long-term planning and requires a number of years before harvesting trees. Use of appropriate and suitable inputs can increase the growth of trees and wood production per unit area in both natural forests and irrigated plantations.

18.3.8 Caring of Existing Stock

Growing forests are vulnerable to multiple risks that may range from natural threats such as insects, disease, and wild animal infestations to anthropogenic threats such as grazing, lopping, and illegal cutting. These factors can cause a great damage to existing stocks of forests and their growth. It is reported that damage caused by biotic factors, especially anthropogenic and livestock, is greater than for all other factors. It is a known that due to poor socioeconomic conditions, people living in adjoining areas to forests are forced to use forest resources to fulfill their daily needs. They can exploit the forests for timber, fuel, fodder, and/or cultivation. Provision of alternatives to forest resources can only help the local dwellers to survive without damaging forests. If caring for existing stocks is neglected then the success of new afforestation projects will not be beneficial on a sustainable basis.

18.3.9 Sustainable Forest Management

For the continuation of services provided by forest ecosystems, proper scientific management, regeneration of existing forests, and afforestation of new areas is indispensable. Pakistan has been blessed with ecologically diverse forests ranging from alpine pastures in the north to mangroves in the south. So, different management schemes should be used accordingly. Various management techniques such as cleaning, pruning, weeding, and thinning are essential to improve the timber quality, enhance the growth of trees, maintain a balance in species composition, and reduce the intraspecies competition. These techniques should be employed based on scientific principles.

18.3.10 Management Plans

Only about 45 forests have management plans while the rest are managed under ad hoc conditions. Preparation of management plans for all existing forests is very important and crucial. Moreover, forests are now regarded more as biodiversity reserves, abodes of wildlife, sources of oxygen, carbon sinks, noise absorbers, and so on and not just as a source of wood. The management plans should be written by a group of multidisciplinary scientists rather than an individual. A planning commission form number 1 (PC1) should also be prepared to provide financial support to the working plan document.

18.3.11 Need of Data Bank

In contrast to developed countries, Pakistan has very limited data pertaining to forests, forestry, and related disciplines. So, in the absence of current and latest data, the provincial and federal government have to rely on old figures. A central data bank in all provinces is therefore mandatory for future planning and development in forestry.

18.3.12 Demarcation of Forest Areas

To curb encroachments on forestlands it is essential that all the forest areas are accurately defined and boundary pillars be constructed to demarcate forest boundaries. To control forest damage, forest areas without defined boundaries should be given top priority, as boundaries are essential for better administrative management of forests.

18.3.13 Livestock Routes

During the winter season, a comsiderable number of livestock (in millions) moves from high hills and alpine pastures to plains with or without the local dwellers. During the summer, the livestock moves in opposite direction. These livestock herds consist mostly of sheep and goats. During their passage within forests, in either direction, this livestock cause great damage to saplings and seedlings through trampling, grazing, and browsing. Demarcation of livestock routes and limiting livestock movements to these routes is crucial for forest growth and regeneration.

18.3.14 Afforestation of Saline and Waterlogged Soils

According to reports, it is estimated that in the Indus basin about a 1.5 m ha area is waterlogged and about a 5.5 m ha area is saline. These areas are mostly the result of anthropogenic agricultural activities and are without vegetation. Scarce vegetation is disappearing at an alarming rate due to its removal by man for fuel or fodder purposes. These areas require high mechanical and financial inputs to ameliorate them. However, these areas can be ameliorated on a sustainable basis through biological approaches such as afforestation. Planting of suitable trees on these degraded areas will provide multiple benefits to uplift the socioeconomic conditions of local dwellers as well as ameliorate soil fertility and productivity. These areas, if managed on long-term basis, cannot only serve as timber mines but can also create job opportunities for local people.

18.3.15 Ban on Forest Land Leasing

Cultivation leases given in the irrigated plantations and the riverain forests have also resulted in the decrease of the forestry covered area and reduction of tree growth. First, water supply, which is meant to grow trees, is diverted to crop cultivation. Moreover, people who take the forest lands on lease are very influential people and do not comply with the terms and conditions of the agreement. So, they do not plant trees or regenerate the forests on these lands according to given schedule. Several times, it has been notices that unlawful possession continues for years and ultimately the lessees become owners of the land.

18.3.16 Use of Substitute Materials

There are many substitute wood and nonwood materials available in the market. Many wooden products are now made from wood substitutes derived from wood reduced to homogenous particles by chipping or chemicals. Metallic frames of doors and windows, steel furniture, electric transmission poles, cement concrete railway sleepers, iron beds, and so on are now quite common in Pakistan. This is welcomed, although not as environmentally friendly in term of the use of nonrenewable and nondegradable resources, because it has reduced population pressure on natural forests of Pakistan. So, it is the need of time to promote the use of wood substitutes. Similarly, wood cement composites like cement-bonded particleboard and wood-cement blocks are very useful in fabricating low-cost housing and in the building industry for paneling. Though low quality wood and manufacturing wastes are used to make these products they are resistant to many hazards: termites, fire, weather elements, and fungi.

In Pakistan, people living in hilly areas are using huge quantities of good quality timber wood for their house construction. These areas are heavily infested with winter snow and rains that cause damage to these houses and they need repeated repairs. It would be useful to provide CGL sheets on subsidized rate. Provision of these mentioned materials in Murree Hills and Azad Kashmir resulted in remarkable successes towards decreasing the dependence of local population on forests. Extension of these projects is needed in other hilly areas.

Sustained supplies of kerosene oil and gas cylinders in the hilly areas, or even supply of firewood from plains, should be very helpful to save the valuable coniferous and broad-leafed tree wood often used for heating and cooking. Improvement in the design of firewood stoves could also save a lot of fuel.

18.4 WOOD PRODUCTION IN CROPLANDS (AGROFORESTRY)

Agroforestry is an important part of the routine life of the population of Pakistan, especially, the rural population. The term agroforestry is derived from two words, agro and forestry. Agro means "agricultural crops" and forestry means "tree plantations." So it can be defined as the plantation of trees along the farmlands: Trees are grown with the farm crops to enhance benefits from used land. Many tree crop combinations are used in different sites and in different environments to get maximum benefits from the land.

There are different systems of agroforestry like agri-silviculture, boundary plantation, block plantation, energy plantation, alley cropping, agri-horticulture, agri-silvi-horticulture, agri-silvi-pasture, silvi-olericulture, horti-pasture, horti-olericulture, silvi-pasture, forage forestry, shelter-belts, wind-breaks, live fence, horti-apiculture, aqua-forestry, riparian buffer strips, and homesteads. Agroforestry also is used for nontimber forest products. While they clearly offer economic and ecological advantages, these systems also involve complex interactions, which complicate their management. When designing an agroforestry enterprise, one should research the marketing possibilities and include the agroforestry system in the complete business plan of the farm.

Agroforestry provides services and numerous products to the society and millions of people in the world rely on forests for their livelihood, especially in developing countries. Agroforestry provides timber and fuel wood and fodder for livestock. People collect fruits and medicinal plants from agroforestry. Honey, gums, resins, and number of other products are provided to forests dwellers and people living adjacent to farms. Agroforestry also generates a many job opportunities for native people. Timber harvesting, conversions, and transportation to sale points create jobs at all levels. Similarly, forest food, trade of medicinal plant, and many other nonwood products have a high value around the globe.

At present, 90% firewood and 65% furniture wood comes from agroforestry (finding of the Forest Department) without any government incentive and support. This figure indicates the existing potential of the farmlands, which could be further explored by incentivizing the agroforesters.

18.5 SUGGESTIONS FOR THE IMPROVEMENT OF WOOD PRODUCTION IN CROPLANDS

18.5.1 Compatible Tree Species for Croplands

In the present situation, farmers/agroforesters (private sector) and the state foresters (government sector) grow almost same tree species in irrigated plantations and farmlands, and sell their produce in the same timber markets on a competitive basis. Of course, there is a considerable difference between public and private entrepreneurs in terms of inputs and labor. Ultimately, when wood from state forests flood the market with a prefixed price per cubic foot (which is usually lower than the open market price), private foresters cannot get reasonable prices for their produce. At the government level, there is ambiguity and a lenient attitude regarding wood price fixing. Middle men are active and loot the growers. Since dealing with government enterprises, the state foresters are safe but the agroforesters do not cope with the vicious circle of the middle men and "Arthies" and get marginal profit for their produce. This unfavorable situation is very unfortunate and continuously discourages agroforesters to grow trees in croplands.

The second important factor is that both forester groups (agroforesters and state foresters) are planting fast growing trees with less gestation period in different situations, but use the same species for wood production. In fact, the state foresters can wait longer for the final felling of their forest crop while agroforesters need early incomes and want short rotations of tree crops. High quality timber trees with long rotation should be planted in the state forests for long-term environmental stability and site conservation and should not be allowed to grow fast growing trees and leave it to poor farmers to grow such trees with reasonable returns.

The third factor is that agroforesters want to plant deciduous and less shade casting trees to save their crops from shade effects, therefore block forestry should be promoted in state forests because their prime produce is wood.

The fourth factor is that state foresters get four times more irrigation water than a common farmer, so it should be mandatory for the state foresters to grow high-quality wood like Shisham, Kikar,

Frash, etc. Presently, the situation is reversed, in spite of possessing highly trained manpower, ideal situations, quality land, many times more irrigation water, and lavish planting facilities, the Forest Department is mainly planting eucalyptus, poplar, and simal in the state-owned forests. Currently, most of the forestlands under their control are either unplanted or supporting substandard—as well as unwanted—tree species like mesquite, etc.

As a principle for agroforestry practices, crop-compatible trees should be grown in croplands but due to lack of knowledge and vision, our farmers plant all kinds of tree species without knowing the market demand and have to face colossal losses. In some areas, farmers don't even grow trees in croplands especially in cotton and rice growing zones to avoid shade effects. In fact, there are tree species available which can safely/successfully be planted in such shade nontolerant cropping zones with little after planting cares.

In order to avoid competition between state foresters and agroforesters for wood production, the following tree species are recommended on farmlands and the Forest Department should not be allowed to grow these trees in the irrigated forest plantations.

1. Poplar: *Populus deltoides*
2. Simal: *Bombax ceiba*
3. Eucalyptus: *Eucalyptus camaldulensis* (for problem lands)

The tree species listed above should be declared "farmer trees" by law to boost wood production in the farmlands and the Forest Department should concentrate on high quality woods like Shisham, Kikar, Frash, etc. The crux of the matter is that fast growing species require minimum tending operations, inputs, and manual labor for their establishment and development; therefore, the Forest Department willfully grows such trees in irrigated forest plantations to show their efficiency and pseudo-performance.

18.5.2 Managerial Changes in the Forest Department

In the present situation, all highly diversified forests/rangelands, that is, high hill forests, alpine ranges, coastal forests, plateau scrub forests, irrigated forest plantations, riverain plantations, avenue/linear plantations, desert/semi desert rangelands, and even social forestry practices in croplands are under the administrative control of the Forest Department. During the last sixty years, experience shows that instead of any improvement, these forest covers are declining and degrading fast throughout Pakistan. Despite lavish budgets, trained manpower, comparable infrastructure, research grants, and mobility facilities the Forest Department could not prove its worth and the negative trends in the forestry sector are still progressing. A few years ago, the Chief Minister of Punjab appointed forestry experts to review the progress of three important irrigated forest plantations in Punjab. All three inquiry reports pointed out serious discrepancies and other managerial blunders. It was further stated that big chunks of land were unplanted, mesquite cover was rampant and thousands of domestic animals belonging to the forestry personnel and adjoining communities were found illegally grazing in these plantations. Misuse of irrigation water, illegal cutting of timber, abuse of funds, and forest land-grabbing cases were common. Resultantly, many senior officers were dismissed, and financial penalties were imposed. But can these actions bring back the vanished vegetative cover? Under the circumstances, the only reliable option is to change the administrative and managerial setup of the Forest Department. The viable recommendations in this regard are as follows:

According to spatial diversity, there are two major types of forests/range covers, that is:

1. Hill forests/ranges (high hill forests, alpine ranges, and plateau forest scrubs)
2. Forests/ranges in the plains (irrigated forest plantations, desert/semidesert ranges, riverine forests, avenue plantations, and agroforestry zones)

It is recommended that the management of high hill forests, alpine ranges, and plateau scrubs be kept under the control of the Forest Department. Since these productive/protective forest lands are very important as watershed areas and play a key role in the conservation and management of dams, rivers, and irrigation systems, which needs continuous watch, ward, and technical supervision. Managerial staff of the Forest Department is always proud of the quality education and training in the fields of forestry, range, and watershed sciences they receive from the Pakistan Forest Institute (PFI) Peshawar. The task of managing these zones should be given to Forest Department so that it can make the best use of the existing staff and facilities.

The management of forests/rangelands in the plains (irrigated forest plantations, riverain forests, avenue plantations, and desert/semi desert rangelands) should be transferred to the Agriculture Extension Department. In spite of huge infrastructure and funding, the duties of this department are not well defined. This additional duty of managing forests and rangelands in plains will enable them to show their worth practically on the ground using their intellectual skills. The network of Agriculture Extension Department possesses much better access to the farmer through field assistants and managerial staff who are highly qualified (MSc Hons or PhD) with special training in agro-forestry/range sciences. Therefore, it is recommended that forestry/range management in the plains including the irrigated forest plantations, forestation in agricultural lands, and management of desert/semidesert rangelands should be assigned to them. These recommendations are debatable and could be refined in order to achieve the targets of greening the country given by the Prime Minister of Pakistan and Chief Minister of Punjab.

18.5.3 Participatory Management of Avenue Plantations along Roads, Canals, Drains, and Railway Tracks

In the present situation, all flanking areas along roads, canals, drains, and railways are hardly planted anywhere in the country. During the previous decades, command and control of these lands shifted between departments, therefore, these avenues remained neglected and unplanted. Since plantation on such sites directly affect the crops of the adjoining farmers, therefore discourage the planting efforts to keep the places accessible for grazing of livestock and other activities. Unless they have domestic needs, the avenue belts cannot be successfully planted. It is very important to manage the linear plantation of thick tree cover with adjoining farmers. According to this recommendation, the farmers along these linear structures should be allowed to plant trees on the flanks of roads, canals, drains, and railway tracks adjoining their farmlands. As an incentive, these farmers should be relaxed from paying water charges and land revenues of the adjoining fields (acres). Local staff of the Agriculture Extension Department should supervise these tree belts, post planting care, and felling of trees. In case of successful plantations, each adjoining farmer should be given half of the harvested wood as an incentive. This participatory approach to afforestation and free-of-cost management of the avenue plantations should be considered to frame rules in collaboration with the Revenue Department. If efficiently developed, this system can be very successful for managing these "timber mines" for wood production on a self-help basis without any additional government funding. The arrangement of competitions and distribution of prizes among farmers participating in avenue plantation exercises can enhance their interest and income.

18.5.4 Compulsory Planting of Trees in Croplands by Law

Shortage of tree cover is an evident and a persistent issue in the country. Previous experience shows that despite sincere efforts, tree-planting campaigns by the Forest Department could not bring a visible change in the landscape of the country. At farmlands, due to labor division and an inherited know-how about growing woody plants, each farmer can easily plant and care for trees on his farm land. Farmlands have the potential to contribute to wood production without a considerable decrease in farm produce. Farmers should be convinced and bound (to some extent) to grow at least

20–25 trees per acre throughout the country. And as an incentive, government may curtail water rates or land revenue from those framers who will complete the given assignment within specified time. The Agriculture Extension Department or Revenue Department can be assigned to check the recommended number of trees per acre and help implement the decided incentives for successful farmers. This is the most viable option to positively sensitize farming communities for planting trees in croplands. In the current scenario of climate change, the establishment of forest cover is highly important to meet the wood needs and mitigate environmental complications and pollution hazards in the country.

18.6 RANGELANDS IN PAKISTAN

Rangelands are considered a significant and potential natural resource of Pakistan. It constitutes the largest land use (58%–62%) of the country, which extends throughout Pakistan. It supports the livestock industry by providing range forage to 5% buffaloes and cows and 60% sheep and goat, and 40% camels, horses, donkeys, and mules. In addition to forage, it provides habitat for wildlife, a regular supply of clean fresh water free of sediments to the dams responsible for power generation and crop production, supports recreational activities like hunting, fishing, camping, hiking, etc. It also enriches biodiversity due to a wide range of flora and fauna, it moderates the climate and cleans the environment while providing many other commodities of cultural, ecological, and economic value to the society, particularly to pastoral communities living in and around these rangelands. But unfortunately, the present status of these rangelands is very poor due to overexploitation by human and livestock populations, which are increasing tremendously. Therefore, the major factors for alarm are the depletion of range vegetation, the deterioration of soils, the reduced efficiency and productivity of range livestock and wildlife. The productivity of rangelands has declined 10%–50%. This adverse situation strongly suggests probing into the matter very carefully and analyzing the situation with national interest and patriotism along comprehensive scientific lines. There is need to formulate policies and develop strategies, which would be compatible under local ecological conditions of each rangeland of the country.

18.6.1 Classification of Rangelands

Rangelands can be divided into two broad categories depending on climatic zones.

18.6.1.1 Arid and Semiarid Rangelands

These are the rangelands located in areas where the annual rainfall is less than 30 cm per annum. These areas are characterized by low and erratic precipitation, high temperatures, deep and brackish underground water, unstable sand dunes, scanty vegetation, and hot dry winds. Overgrazing and scarcity of water turns these landscapes into desert. Here, the pastoral population is ill-clad, lives in medieval habitations, and their children are often anemic; their livestock is often in poor health as well. Thal, D.G.Khan, and Cholistan in Punjab; Thar, Registan in Sindh; and Central Kalat, Swabi, Makran, Chagi Kharan in Blochistan are some examples of arid and semiarid rangelands. These are poor rangelands and their present production is far below than their potential due the reasons pointed out above.

18.6.1.2 Alpine and Moist Temperate Rangelands

Moist temperate rangelands are most productive such as Swat, Hazara, Malakand, Upper and Lower Dir, of the Khyber Pukhtunkha province. In addition, the southwestern tracts of KPK, including D.I.Khan, Kohat, Bannu and tribal territories are not productive rangelands whereas Gilgit, Sakrdu are present in sub Alpine zone. These are lush green meadows supporting a large number of livestock. The most productive rangeland is Pothowar, which is in Punjab province, and consists of the entire districts of Attock, Rawalpindi, Chakwal, Jhelum, and hilly areas of Gujrat (Pubbi hills).

Overgrazing by nomads results in the suppression and uprooting of palatable grasses and shrubs. Range site becomes compacted by heavy grazing and browsing pressure. The nomads come down from the hills and mercilessly devastate the vegetation.

Water erosion on steep slopes is alarming and creates havoc, especially on slopes which have been overgrazed and trampled heavily.

18.6.1.2.1 Improvement Operations

- Improvement of these rangelands is possible by controlled grazing, especially in the rainy season, and planned grazing with proper stocking rates is the key to sustainable range management.
- Slopes should be covered with vegetation throughout the year. The government should provide alternate feeding material to livestock. Substitutes for fuelwood should be provided like kerosene oil, LPG, and so on to save vegetation in fragile areas.
- Provision of community services to the local people will convince them to take care of range lands.
- Local people should be made aware of the importance of the ranges that are sources of livestock rearing. Livestock has grown into an industry producing meat, milk, wool, hides, skin, and their byproducts.

18.7 RECOMMENDATION/STRATEGIES FOR DEVELOPMENT OF RANGELANDS IN PAKISTAN

18.7.1 Establishment of an Independent Authority

For the sake of conservation and development of rangelands there is a dire need of establishment of an independent authority/organization vested with authority, responsibility, and accountability with the mandate for exploiting and increasing the productive potential of rangelands in the country. Its field offices should be in every rangeland of each province. Adequate funds must also be provided for effective and visible improvement of rangeland areas.

18.7.2 Provision of Trained Staff

Lack of professionally trained staff having higher education degrees in range science is also one of the reasons for not achieving targets of rangeland management and development. Since the creation of Pakistan until now, other than range experts, professionals like foresters, animal husbandry experts, botanists, and so on are given assignments for short periods to conduct surveys and give suggestions to provincial and federal governments for the development of rangelands. A majority of these experts don't differentiate between deserts and rangelands. Therefore range science should be strengthened by including it in the curriculum of the agricultural universities and colleges so that this important natural resource can be properly exploited and managed.

18.7.3 Provision of Infrastructures in Range Stations

The task or targets of range development and improvement infrastructure for residence of staff and for range development management and improvement purposes like offices, stores, sheds, machinery and water tanks etc is an unavoidable need. Schools, dispensaries, markets, transport, and community colleges should be established, so that, range staff can work efficiently and lead life comfortable life.

18.7.4 Establishment of National Range Institutes and Schools

It is a bitter fact that, as compared to other disciplines, there is an acute shortage of rangelands experts and trained/skilled technicians in the country. Along the lines of forest schools, there must be range schools for imparting professional education and practical training. It must be kept in mind that rangelands comprise of a vast area of the country and thousands of technicians and skilled labor are required. It is only possible if rangeland schools and institutes are opened in sufficient numbers.

18.7.5 Special Incentives for Range Staff

It is a well known fact that the environment of rangelands is very harsh and discouraging for people who work there. Therefore their salary packages should be very attractive. All other facilities like residence and transport should be provided. Rangelands are mostly located in remote areas, therefore hardship allowances must be granted.

18.7.6 Constituting Pastoral Societies for Awareness

Research on socioeconomic conditions of pastoral communities/graziers has revealed the fact that almost all graziers are illiterate and unaware of the horrible consequences of overgrazing and unplanned grazing. Local people should be involved in decision-making by organizing societies to solve their problems and convince them to cooperate with government officials for the improvment of range resources. No success can be achieved without the help and collaboration of the local people.

18.7.7 Alleviation of Poverty of Pastoral Community

As already mentioned, the socioeconomic conditions are low, moreover middlemen also exploit the situation and purchase live animals and animal products at very low prices. The government should establish sale points and purchase live animals and animal products like milk at encouraging rates. On the other hand, graziers should be involved in various development projects. They should be paid their due share from the income of the projects if possible. When graziers are prosperous and rich, they will avoid overgrazing, which is the root cause of the deterioration of rangelands.

18.7.8 Provision of Basic Needs of Animals

Herdsmen need fodders, straws, concentrates for complementary and supplementary feed, and water for animals during famine periods. All these basic commodities or basic needs must be provided to avoid any considerable loss to graziers. In fact, graziers are the key figure for the successful completion of any project of range development, improvement, and management. So, they must be satisfied in all respects.

18.7.9 Range Soil and Water Conservation

Soil and water are the two major factors of paramount importance and play key role in all range management operations required for increasing range productivity. Unfortunately, there is acute shortage of water in range areas, especially southern rangelands of the country where annual rainfall never or seldom exceeds 300 mm. Soil conservation and its stability is an important prerequisite for promoting water conservation and range vegetation. For example, it is crucial for harrowing, cultivation, deep ploughing, strengthening field boundaries, field leveling, pitting, trenching, ridging, water spreading, making large water storage, and aquifers.

18.7.10 Water Harvesting

Rainwater harvesting is very necessary for irrigation maintenance, water conservation structures such as dykes, check dams, and stable soil. To increase soil fertility, uprooting of undesirable range plants, stubbles and ploughing for burning will enhance the water retention ability of the soil. In addition, mulching, green manuring, loosening, trapping, and mixing sand in heavy clayey soils are also very important in this regard.

18.7.11 Model Ranching

Seeing is believing proverb should be applied to convince the graziers. Develop a model ranching for demonstration. This will convince the graziers to achieve the maximum range production and its conservation for the welfare of rangeland and their consumers both herbivores and omnivores.

18.7.12 Safari Parks

Establishment of safari parks is gaining high popularity all over the world. This should also be practiced in Pakistan as one of the major use of rangelands for the public. Various kinds of wildlife should be kept for the amusement of visitors to wild lands. For example, Cholistan has immense potential to be developed into a good safari park.

18.7.13 Introduction of Best Exotic Forage Varieties (Grasses and Legumes) and Animal Breeds

For increasing high potential of range vegetation and livestock imports and establishment of high yielding, palatable, and nutritious exotic varieties of range grasses and legumes through artificial reseeding is must. Availability of palatable vegetation will ensure the high productive potential of range livestock.

REFERENCES

Archer G. 1993 Pakistan Household Energy Strategy Study (HESS). Planning and Development Division. Government of Pakistan.

Butchart SHM, Walpole M, Collen B. 2010 Global biodiversity: indicators of recent declines. *Science* 328: 116–1168.

FAO. 2005 *Microfinance and Forest Based Small Scale Enterprises.* Food and Agriculture Organization of the United Nations, Rome.

FAO. 2009 *State of the World's Forests.* Food and Agriculture Organization of the United Nations, Rome.

FAO. 2010 Global forest resources assessment. *Main Report Forestry Paper 163.* Food and Agriculture Organization of the United Nations, Rome.

FAO. 2011 *State of the World's Forests 2011.* Food and Agriculture Organization of the United Nations, Rome.

Forestry Sector Master Plan. 1992–2018. FSMP. [online] available: http://www.environment.gov.pk/pub-pd...t3-Chp%203.pdf.

Harrington CA. 1999 Forests planted for ecosystem restoration or conservation. *New For* 17: 175–190.

Lambin EF, Meyfroidt P. 2011 Global land use change, economic globalization and the looming land scarcity. *Proc Natl Acad Sci* 108: 3465–3472.

Naeem S, Thompson LJ, Lawler SP. 1994 Declining biodiversity can alter the performance of ecosystems. *Nature* 368: 734–737.

Norris K. 2012 Biodiversity in the context of ecosystem services: the applied need for systems approaches. *Phil Trans R Soc B* 367: 163–169.

Raunikar R, Buongiorno J, Turner JA, Shushuai Z. 2010. Global outlook for wood and forests with the bioenergy demand implied by scenarios of the intergovernmental panel on climate change. *J For Policy Econ* 12: 48–56.

Roberts G, Parrotta J, Wreford A. 2009. Current adaptation measures and policies. In: Seppala R, Buck A, Katila P (eds) *Adaptation of Forests and People to Climate Change- a Global Assessment Report.* IUFRO world series, volume 22.

TEEB. 2010 *The Economics of Ecosystems and Biodiversity: mainstreaming the Economics of Nature: a Synthesis of the Approach, Conclusions and Recommendations of TEEB*, Progress Press, Malta.

UNEP. 2011 Towards a green economy: pathways to sustainable development and poverty eradication. A synthesis for policy makers. http://www.unep.org/greeneconomy/portals

19 Biotechnology and GM Crops

Muhammad Sarwar Khan and Faiz Ahmad Joyia

CONTENTS

19.1 INTRODUCTION

Exploding populations, predominantly of nations with developing economies, will require an estimated 70% increase of food production by 2050. Biotechnology has been employed for 3 decades to create genetically modified (GM) crops, especially with two traits: Insect resistance and herbicide tolerance. Consequently, crop yield has increased benefiting farmers of developed as well as developing countries (Delaney, 2015). Biotechnology is an applied science meant for altering the genetic makeup not only of plants but animals, and microorganisms as well for value addition. Generally defined, agricultural biotechnology comprises tissue culture techniques, mutation breeding, recombinant DNA technology, the use of molecular markers for breeding, and genetic modification using transgenes to develop GM crops. The pace of OMICS has escalated with robust methods for genome sequencing and gene discovery (Anthony and Ferroni, 2012). Scientists have developed genetically engineered organisms, including bacteria and fungi, to yield expensive molecules, plant resistance against biotic and abiotic stresses, animals with enhanced quality and quantity of milk, meat, and eggs. The possible benefits of this technology are being harnessed in the food, feed, and industry. However, certain groups—both in the scientific community as well as common public, have raised their concerns about its potential implications on human and animal health, global biodiversity, and ecosystems. In Pakistan adoption of the biotech crops is overwhelming and cotton grown on major areas of cotton growing belt is genetically modified. A large number of biotech teaching and research institutes have been established, which are on their way to producing their own products in the form of GM crops (Malik, 2014).

19.2 ROLE OF BIOTECHNOLOGY IN AGRICULTURE

Genetic engineering of living organisms has been there for more than 3 decades. While genetically modified plants have now become widely adopted and commercially accessible for more than 20 years. Commercialization of GM crops for the last 2 decades has revealed they have provided significant economic, environmental as well as social advantages to all its stakeholders especially farming communities in both the developed as well as developing countries. Such a prompt adoption of the technology by farmers reflects their belief in the possible benefits of technology. In the year 2016, a total of 18 million farmers planted 185.1 million hectares (457.4 million acres) of biotech crops in 26 countries with an increase of 3% or 5.4 million hectares (13.1 million acres) from 2015. Out of these 18 million, around 17 million were small scale, resource-poor farmers. Impact analysis for the 20 years of commercial deployment of GM crops has shown that their cultivation has been beneficial for farmer and consumers, producing broader welfare gains and positive effects for the environment and human health. GM crops have increased the agricultural productivity and reduced the need of insecticides and fungicides (ISAAA, 2016). It is possible to transfer genes conferring resistance against biotic as well as abiotic stress conditions. Transgenic plants now can endure harsh environmental conditions like drought, salinity, heat, frost, etc. GM plants have become efficient users of available nutrients and water. Moreover, numerous crops have been engineered for nutrient biofortification in order to fight malnutrition. Production of numerous biopharmaceuticals is another value addition of GM plants and the use of plants parts as edible therapeutics exhibit promising results (Khan et al., 2016) not only for humans (Kwon et al., 2013) but for livestock as well (Jacob et al., 2013).

19.2.1 International Scenario

Biotech crops are being cultivated in 26 countries of all six continents. The United States, Brazil, Argentina, India, Canada, China, Paraguay, Pakistan, South Africa, Uruguay, Bolivia, the Philippines, Australia, Burkina Faso, Myanmar, Mexico, Spain, Colombia, and Sudan are the important countries where about 18 million farmers are growing biotech-crops, out of which 90% farmers are small and resource-poor farmers. Each of these countries is growing biotech crops on areas of more than 50,000 hectares. In 2016, developing countries surpassed industrial countries in cultivating biotech crops with a share of 54% and a cumulative area of 99.6 million hectares out of the total area of 185.1 million hectares under biotech crops (Figure 19.1). A significant rise in the area under cultivation of GM soybean in Brazil and Argentina and GM cotton in Pakistan, Myanmar, and Sudan has contributed in this increased share of developing countries. In 2016, Pakistan was in the 8th position in growing biotech crops globally with an area of 2.9 million hectares (ISAAA, 2016).

Biotechnology is the use of living organisms, their processes, and products for the benefits of mankind. Keeping in view this generalized definition, biotechnology has been in place for over 30,000 years when humans started taming animals, domesticating plants, and selecting the best ones

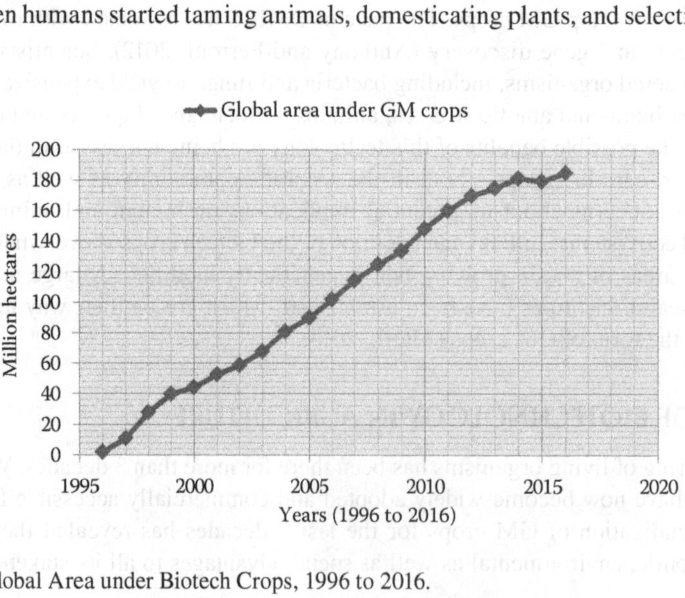

FIGURE 19.1 Global Area under Biotech Crops, 1996 to 2016.

for their benefit. Later, the concept of breeding started and organisms with desirable characters were bred to obtain new generations having good characters of both parental lines.

Figure 19.2 illustrates the timeline of development of biotech crops. Biotechnology started around 32,000 years ago, when cave dwellers started taming dogs. Hence, the dog is supposed to be the pioneer animal selected artificially by our ancestors. After domestication, there was artificial selection to increase docility, making the dogs the first ever animals tamed by human beings (Zimmer, 2013).

In case of plants, the record of the earliest artificial selection dates back to 7800 BC in the southwestern Asia, where early varieties of wheat were domesticated (Balter, 2013). Similarly, corn emerged as a crop by long-term breeding and artificial selection of a wild grass teosinte, which originally had small ears with a small number of kernels. A parallel course of domestication gave birth to numerous crops like sugarcane, rice, sorghum, millets, beans, bananas, apples, and many others.

Recently, the procedure of selective breeding progressed into the idea of genetic engineering. Innovators, enthused by the world's utmost critical issues of food, health, and human wellbeing, paved the way for GMOs, which led not only to incredible benefits, but also raised certain terribly imperative concerns.

Engineering of first successful genetically modified (GM) bacteria proved to be a breakthrough in 1973 (Cohen et al., 1973). One year later, Jaenisch and Mintz utilized a similar procedure in animals by introducing foreign DNA into mouse embryos (Jaenisch and Mintz, 1974).

The new technology opened up countless avenues, however, a multitude of concerns were raised about its potential ramifications on health, biodiversity, and environment. Hence, the research was halted; experts of science, law, and government sat together, debated the safety of genetic engineering at the Asilomar Conference in 1974 and concluded that the Genetic Engineering (GE) projects should be permitted to continue with explicit regulations in place (Berg et al., 1975). Hence, development of the first biosafety guidelines was initiated. In 1980, the Supreme Court in the United States permitted the patenting of GE bacteria giving legal proprietorship rights over GMOs. That legal decision incentivized the development of GMOs for large multinational companies (MNCs) who earned huge profits out of such GMOs. In 1982, the US-FDA approved the first human medicine "insulin" produced by genetically engineered bacteria (Altman, 1982).

In 1992, after 5 years of rigorous health and environment related testing, Calgene's FlavrSavr tomato was approved by the USDA for commercialization. In 1996, the first Bt crop was approved by the U.S. EPA after thorough testing. The same year first glyphosate-resistant (herbicide-resistant) soybean, a product of Monsanto, was approved for commercial release. This herbicide-resistance technology has now been extended to many other crops, including corn, canola, and sugar beets. Development of golden rice in 2000 was another milestone to fight against the deficiency of vitamin A in humans (Ye et al., 2000).

The same year, the U.S. FDA granted approval to the first biological product, ATryn, for the treatment of human blood clotting disorders. There is a long list of biotech products that are at various stages of approval, trials, and production. More recently, Bt brinjal has been deployed for commercial cultivation in developing countries like Bangladesh (Choudhary et al., 2014) and India (Shelton, 2010).

19.2.2 National Scenario

19.2.2.1 Development of GM Crops in Pakistan

Rapid progress in the field of biotechnology has produced remarkable impacts on human society in the fields of agriculture, industry, medicine, and environment. Regarding agriculture, scientists have developed protocols to develop novel crop varieties requiring lower inputs by resisting biotic and abiotic stresses and providing higher yields with healthier nutritional qualities, longer shelf life, and reduced postharvest losses.

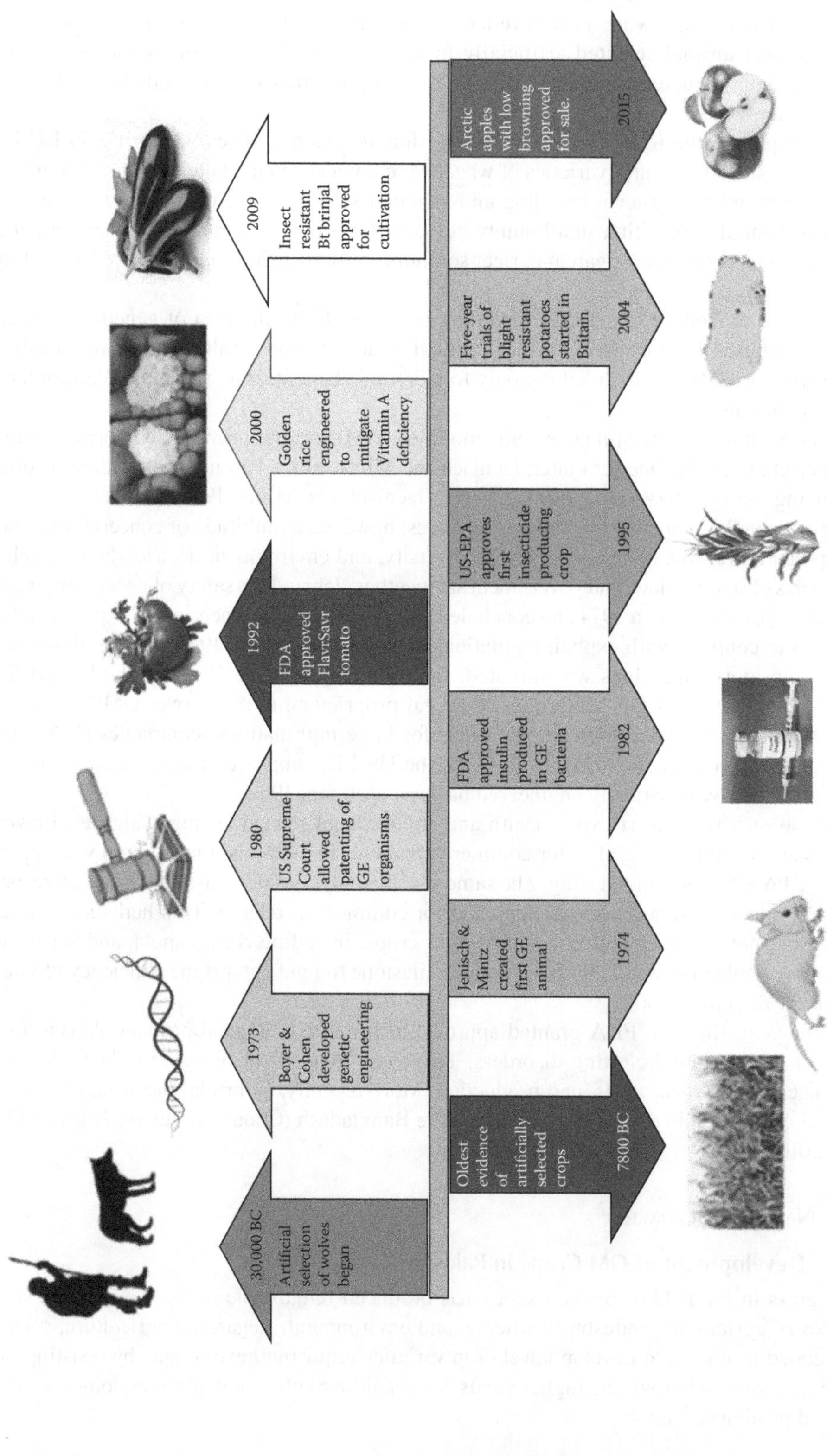

FIGURE 19.2 Timeline of Biotech Crops Development.

Taking in account the steady developments in the field of biotechnology, it was incorporated as one of the six priority areas of science and technology by the National Commission of Science and Technology in Pakistan. As a result, the Mid-Term Development Framework (2005–2010) of the Planning Commission of Pakistan included an Action Plan to improve the research and development of biotechnology in Pakistan. Therefore, development of biotech departments in numerous universities and research institutes has been encouraged. More than 2.0 Billion PKR has been expended since last few years through numerous national funding agencies including the Higher Education Commission of Pakistan (HEC), Ministry of Science and Technology (MoST), and Ministry of Agriculture (MoA). Those funds have been spent in improving the research facilities and training the manpower to undertake biotech-based research and development (RD) activities, particularly in agriculture, industry, health, and environment.

Currently, more than 600 scientists, with more than 400 indigenous as well as foreign PhDs in biotechnology and allied subjects have been trained and employed in universities and research institutes. At present, almost all major public sector universities in all provinces have developed their biotech departments and institutes, producing more and more skilled manpower in the field of biotechnology. Likewise, numerous private universities have been encouraged to progress in the field of biotechnology. Resultantly, numerous private universities are also progressing in this field.

Some institutions, in Pakistan, have now entered the phase where they have some deliverable products in the form of goods and services. Although, most of the accomplishments are in the field of agriculture, yet other fields are also progressing. Initially, tissue culture was extensively established and commercialized for various crops. Salient products are virus-free potato and banana. Similarly, the technology has been used to grow exotic orchids and cut flowers, which are being sold are high prices in the country. Tissue culture also holds a great potential in micropropagation of important forest trees and medicinal plants.

Besides tissue culture, serious efforts are being made for genetic engineering of important crops to improve numerous traits. Initially, most of the research activities using modern biotechnological approached were focused on cotton and sugarcane, which are amongst the top five significant crops in Pakistan. Initially, some insect resistant (Bt) cotton varieties were brought into Pakistan through smuggling and spread among farmers. By virtue of their insect resistance ability, those varieties were adopted by farmers very quickly and spread over the major cotton growing belt of Pakistan (Sharma, 2009). The Pakistan Biosafety Rules were approved and came into force in 2005. Since then, all the major biotech institutes have devised their Institutional Biosafety Committees (IBCs) and more than twenty submissions, from public as well as private sector institutes have been made for evaluation and approval by the Technical Advisory Committee (TAC) and the National Biosafety Committee (NBC). Originally, two crops—cotton and sugarcane, were focused to employ biotechnology in Pakistan. Soon after the approval of biosafety rules, the TAC approved and recommended three proposals to NBC for approval. Out of three, only the proposal on sugarcane received approval with minor observations (Khan et al., 2011). Now, the development of GM cotton is at an advanced stage in Pakistan. Currently, many national seed companies are also dynamically engaged in GM cotton seed multiplication. The Pakistan Atomic Energy Commission (PAEC) provided 40,000 Kg of seeds of Bt cotton varieties in 2005. Those varieties performed well as far as insect resistance was concerned and subsequently got overwhelming acceptance in the farming community (Gabol et al., 2012). Furthermore, numerous renowned multinational companies, namely Monsanto, Syngenta, Bayer, and Pioneer, are also planning to market their seeds with endorsement from the NBC. After devolution of the subject environment under the 18th amendment of NBC itself under legal complications; however, it is believed that the enactment of the Pakistan Biosafety Act 2015, such legal disputes may be resolved.

Later on, work on developing GM rice, wheat, and groundnut was started, which are at advanced stage of development. Brassica, chickpea, chilies, cucurbits, potato, tobacco, and tomato have recently been taken up by various RD institutes and universities. Keeping in view the imperative problems of agriculture in Pakistan, key emphasis is on engineering tolerance to salinity and drought in addition to fungus and insect resistance. More recently, efforts are being made to enhance the nutritional quality of major staple crops, including wheat through biofortification. Although, GM plants of many important crops have been obtained, field trails and subsequent commercial release has been hampered due to delays in approval of Biosafety Guidelines.

19.2.2.2 Awareness and Acceptance of GM Crops

The aim of plant genetic manipulation is yield enhancement as well as quality improvement. Yield is subjective to various factors like insect-pests, soil, and climatic conditions. Substantial improvements have been accomplished by improving irrigation, pesticide application,s and the addition of fertilizer. However, most of these interventions increase the input costs, especially for small-scale farmers in developing countries (Sanchez, 2002) like Pakistan. Hence, the development of transgenics has provided us with innovative breakthroughs to cope with abiotic as well as biotic stresses. A growing body of knowledge, based upon evidences gathered from on farm surveys, indicates that the acceptance of GM crops in developing countries is overwhelming (Paarlberg, 2000; Carpenter, 2010). The major reason for such an ambitious acceptance is the considerable reductions in pest infestations with minimum input cots on insecticides. Consequently, crop yields have been improved and farmers are getting more profits out of the same pieces of land.

Nevertheless, even with several administrative as well as research efforts during the last decade, Pakistan could not make commercial approval of any genetically modified cotton variety until 2010. On the other hand, the delay in approval for commercialization has resulted in the unregulated adoption of Bt cotton varieties in Pakistan. According to the Pakistan Agricultural Research Council (PARC), in 2007 almost 60% of cotton growing area in Pakistan was under Bt cotton (PARC, 2008). However, in 2011 the proportion increased to 85%. Under these circumstances, numerous Bt cotton varieties were developed locally through backcrossing of transgenic cotton expressing *CryIAc* and was disseminated without any formal monitoring framework. It raised numerous apprehensions at national as well as international levels. In 2010, Pakistan became the twelfth country worldwide to formally approve commercial cultivation of Bt cotton after the United States, China, India, Australia, South Africa, Brazil, Argentina, Columbia, Mexico, Costa Rica, and Burkina Faso. In 2010, some indigenously developed Bt cotton varieties based on Monsanto event MON531 and one hybrid expressing the fusion gene *cryIAc* and *cryIAb* were approved for commercial cultivation.

During the last few decades, numerous inventions and innovations in biological sciences gave rise to biotechnology, which has become one of the fastest growing areas of science. Biotechnology has made a profound impact on human life. Therefore, this century may appropriately be designated as "Century of Biology," anticipating that these developments will be more beneficial than electricity and computers for humans. Agricultural biotechnology remained an authoritative, but controversial technology. Although farmers continue to adopt transgenic crops, it is a fact that the eventual success or failure of agricultural biotechnology will unavoidably be influenced by consumer opinion.

There have been a large number of publically and privately-funded studies in developed countries for examining consumer awareness about biotech-based food products. Conversely, there has been no study about the consumer awareness and consumer perception about GM crops/foods in Pakistan. As a result, the policy makers are left making decisions about the likely future of GM agricultural products and biotech-based industry in Pakistan without consistent and competent data. Hence, there is an urgent need of such studies to be initiated in Pakistan.

19.2.2.3 Role of Public and Private Stakeholders

Framers, public as well as private RD organizations, the Ministry of Climate Change, and Federal and Provincial Environmental Protection Agencies are the major stakeholders to biotech crops in Pakistan. GMOs being the patentable entities, in many parts of the world, have spurred a tremendous amount of private sector biotechnology research as 75% of all patents globally in agricultural biotechnology are held by the private sector, mostly by a few large MNCs.

In Pakistan, more than 60 biotech institutes (Table 19.1) with about 600 scientists in biotechnology and related fields are working. Although, few of them have reached a stage where they have some deliverable products. However, most of them have numerous products in the pipeline, which will be in the commercialization stage in next few years, mostly in the area of agricultural biotechnology. It is expected that there would be strong competition between products of national biotechnology institutes and those of multinational companies. It is on one hand, necessary for local biotech institutes to be prepared for this tough competition while, on the other hand, it would be necessary for policy makers to devise policies safeguarding the local biotech-based products and ensuring healthy competition.

With the increasing body of awareness about the importance of biotechnology and the availability of funding resources for biotechnology, numerous private teaching institutes in Pakistan have started dedicated biotech programs and developed their biotech centers. Forman Christian College (FCC), Lahore University of Management Sciences (LUMS), University of Lahore (UOL), and University of South Asia (USA) in Lahore while Faisalabad Institute of Research Science and Technology (FIRST) in Faisalabad are key examples in this regard. It is expected that more and more private universities will get involved in teaching and research of biotechnology in the near future.

The situation of the biotech-based industry in Pakistan is very disappointing because this industry in Pakistan is almost zero. With the establishment of biotechnology teaching and research institutes, sufficient biotech products as well as skilled manpower will be available in near future. Hence, establishment of biotech-based industry in Pakistan will be direly needed in the near future. In the present scenario, biotechnology has become a main provider to the development of Small and Medium Enterprise (SME) by establishing start up companies. Such activities should be facilitated by vigorous contribution of the government agencies through financial support. So as to achieve this, it is recommended that some venture capital fund should be established with a fair contribution from the government, represented by the Higher Education Commission (HEC) and the Ministry of Science and Technology (MoST), and the private sector, represented by the Federal Chamber of Commerce and Industry. Such an activity would help in commercialization of biotechnology.

19.2.2.4 Enforcement of Biosafety Regulations

Acceptance and adoption of transgenic crops by farmers is noteworthy. However, most of the genetically modified food and feed crops have not been approved for commercial cultivation so far in Africa, Asia, and the Middle East. Among developing countries, there are few exceptions GM production like South Africa and the Philippines (transgenic maize), and Argentina (transgenic maize and soybean) after appropriate approval by regulatory bodies. The foremost reason of slow approval procedure is that regulators in developing countries frequently follow highly conservative cautionary practices while deciding about the approval of a biotech crop. Likewise, uncertain apprehensions regarding biosafety of transgenic crops for humans and environment, together with potential constraints arising from international trade policies of certain dominant countries/regions of the world have also been influential in this respect.

Currently, in Pakistan, Bt-cotton is the only biotech crop being grown with approval of commercial release. However, there are numerous biotech crops developed by Pakistani biotech institutes of the public sector which are waiting for approval of commercial release by regulatory authority. Similarly, different MNCs have also applied for approval of many products.

Pakistan is a party to the Cartagena Protocol on Biosafety (CPB). The National Biosafety Centre was established as a project with an objective to cater the obligations of Cartagena Protocol on Biosafety and implementation of Pakistan Biosafety Rules and National Biosafety Guidelines as they provide the necessary management and regulatory framework for the GMOs. Pakistan Biosafety Rules were approved and came into force in 2005. These regulations established a system to evaluate the health and environmental safety of GMOs prior to release for commercial use.

The NBC issued its first approvals for Bt cotton varieties in 2010 but did so for the majority of them for a limited duration of 3 years. Meanwhile, the Punjab Seed Council (PSC) began issuing its

TABLE 19.1
List of Universities/Centers/Institutes Involved in Biotech Teaching and/or Research

S. No.	Institute/Department	Organization/University	City
	Public Sector Universities		
1	Centre of Agricultural Biochemistry and Biotechnology (CABB)	University of Agriculture	Faisalabad
2	Department of Bioinformatics and Biotechnology	Government College University	Faisalabad
3	National Centre for Excellence in Molecular Biology (NCEMB)	University of the Punjab	Lahore
4	Institute of Biochemistry and Biotechnology (IBB)	University of the Punjab	Lahore
5	School of Biological Sciences (SBS)	University of the Punjab	Lahore
6	Department of Microbiology and Molecular Genetics	University of the Punjab	Lahore
7	Department of Biotechnology and Microbiology	Lahore College for Women University	Lahore
8	Department of Biotechnology	Kinnaird College for Women	Lahore
9	Industrial Biotechnology Institute	Government College University (GCU)	Lahore
10	Institute of Biochemistry and Biotechnology	UVAS	Lahore
11	Department of Human Genetics and Molecular Biology	Uiversity of Health Sciences Lahore	Lahore
12	Department of Biological Sciences	Quaid-i-Azam University	Islamabad
13	Institute of Bioinformatics	Quaid-i-Azam University	Islamabad
14	Department of Biotechnology	Quaid-i-Azam University	Islamabad
15	Atta-ur-Rehman School of Biosciences (ASAB)	National University of Sciences and Technology (NUST)	Islamabad
16	Institute of Molecular Biology and Biotechnology	Baha-ud-Din Zakariya University	Multan
17	Department of Biochemistry and Biotechnology	Islamia University	Bahawalpur
18	Department of Biochemistry	University of Gujrat	Gujrat
19	Department of Biotechnology	University of Sargodha	Sargodha
20	Department of Biotechnology	University of Karachi	Karachi
21	Center for Molecular Genetics	University of Karachi	Karachi
22	Department of Microbiology	University of Karachi	Karachi
23	Dr. Punjwani Center for Molecular Medicine and Drug Research (PCMDR)	University of Karachi	Karachi
24	Hussain Ebrahim Jamal (HEJ) Research Institute of Chemistry	University of Karachi	Karachi
25	Dr. A.Q. Khan Institute of Biotechnology and Genetic Engineering (KIBGE)	University of Karachi	Karachi
26	National Center for Proteomics	University of Karachi	Karachi
27	Institute of Biotechnology and Genetic Engineering (IBGE)	University of Sindh	Jamshoro
28	Institute of Biotechnology and Genetic Engineering (IBGE)	The University of Agriculture Peshawar	Peshawar
29	Center for Animal Biotechnology	Veterinary Research Institute, KPK	Peshawar
30	Center of Biotechnology and Microbiology	University of Peshawar	Peshawar

(Continued)

TABLE 19.1 (*Continued*)
List of Universities/Centers/Institutes Involved in Biotech Teaching and/or Research

S. No.	Institute/Department	Organization/University	City
31	Institute of Biochemistry	University of Balochistan	Quetta
32	University Institute of Biochemistry and Biotechnology	PMAS-Arid Agriculture University	Rawalpindi
33	Department of Biotechnology and Genetic Engineering	Kohat University of Science and Technology, Kohat	Kohat
34	Department of Genetics	Hazara University	Mansehra
35	Department of Biotechnology	University of Science and Technology Bannu	Bannu
36	Department of Biotechnology	Shaheed Benazir Bhutto University, Sheringal, Dir	Upper Dir
37	Department of Biotechnology	University of Malakand, Chakdara, Dir	Malakand
38	Department of Biotechnology/ Microbiology	Sarhard University of Science and Information Technology	Peshawar
39	Faculty of Natural and Applied Sciences	Preston University	
40	Department of Bio-Sciences	COMSATS	Islamabad, Sahiwal, Abbottabad
41	Department of Biotechnology	SardarBahadur Khan Women University	Quetta
42	Department of Biotechnology	Mir Pur University of Science & Technology	Azad Kashmir
43	The Department of Biotechnology	The University of Azad Jammu & Kashmir	Azad Kashmir
	Private Sector Universities		
1	Department of Biology, Syed Babar Ali School of Science and Engineering	Lahore University of Management Sciences (LUMS)	Lahore
2	Department of Biological Sciences	Forman Christian (FC) College	Lahore
3	Institute of Molecular Biology and Biotechnology	The University of Lahore	Lahore
4	Department of Biotechnology	University of South Asia (USA) Lahore	Lahore
5	Akhuwat-FIRST	Akhuwat Faisalabad Institute of Research Science and Technology (FIRST)	Faisalabad
6	Faculty of Life Sciences	University of Central Punjab	Lahore
	Research Institutes/Centers		
1	National Institute for Biotechnology and Genetic Engineering (NIBGE)	Pakistan Atomic Energy Commission (PAEC)	Faisalabad
2	Agricultural Biotechnology Institute	Ayub Agricultural Research Institute (AARI)	Faisalabad
3	Biotechnology and food Research Center	Pakistan Council for Scientific and Industrial Research (PCSIR) Laboratories	Lahore
4	Cytogenetics Section	Central Cotton Research Institute (CCRI)	Multan
5	Agricultural Biotechnology Institute (ABI)	National Agricultural Research Center (NARC)	Islamabad
6	National Institute of Genomics and Advanced Biotechnology (NIGAB)	National Agricultural Research Center (NARC)	Islamabad
7	Biomedical and Genetic Engineering Division	Dr. A.Q. Khan Research Laboratories	Islamabad

own approvals for cultivation only in the province of Punjab. The PSC issued and renewed approvals in 2010, 2011, and 2013, but it was not until 2014 that the NBC convened a session again to approve a new set of GM varieties. It is not clear whether the PSC approvals were in line with the NBC review process or not but the regulatory uncertainty and confusion started after promulgation of the 18th amendment in the constitution. National Biosafety Committee (NBC) has the responsibility to approve or reject cases related to GM crops in Pakistan.

In order to implement the Pakistan Biosafety Rules 2005 in the country, the National Biosafety Centre should be promoted to nondevelopment status. It should be empowered to enforce biosafety rules in Pakistan. The NBC's limited capacity to conduct biosafety evaluations delayed the federal government's ability to respond towards approval of new biotech crops in order to promote effective and safe deployment of this technology. Furthermore, a National Biosafety Laboratory should be established and permitted to work independently. Along with these immediate developments, a country-wide mass awareness campaign about GMOs, as well as capacity building of Pakistan the National Biosafety Centre at least up to provincial level, should be initiated.

Regulations regarding crop seed approval, certification, and sale need to be revisited. Existing regulation "the Pakistan Seed Act 1976," do not practically ensure the farmer's right to guaranteed seed quality, its purity, and gene expression levels upon visual inspection prior to purchase. Hence, farmers have to rely upon the information provided by the retailers. Moreover, the structure of Pakistan's seed system is poorly designed to manage the private sector's increasing share in the seed market. Briefly, there is a need to expedite the approval process and strengthen its implementation to encourage foreign investment.

19.2.2.5 Need to Strengthen the Technology

Biotechnology has emerged as the fastest growing field in life sciences with inestimable benefits for mankind. During the last decade, biotechnology has attracted substantial financial support from the Government of Pakistan. Government has taken many initiatives to reorient the field of biotechnology in order to shift the economy of Pakistan towards a knowledge-based economy.

Hence, biotechnology was incorporated in the six priority areas of Science and Technology Policy of the country. To formulate a national policy and to excel in this emerging field, a National Commission on Biotechnology was established in 2001 in order to advise the Ministry of Science and Technology (MoST) for monitoring new developments at the national level. Consequently, a National Policy and Action Plan were formulated, which later became an important part of the Mid Term Development Framework (2005–2010) of the Planning Commission of Pakistan.

Accordingly, an amount of more than 2.0 Billion PKR was expended to develop state-of-the-art infrastructure and capacity building for undertaking RD in the field of biotechnology. Currently, the needed expertise in biotechnology and all associated fields are available in the country. Inspite of all these efforts, current investment in biotechnology remains subcritical. This has been further compounded by the current economic crisis, resulting in the suspension of a number of biotechnology related developmental projects. Hence, with an already developed infrastructure and the available expertise, any further investment into this sector will be most fruitful than ever before. Nevertheless, it is essential to keep on appraising the status of biotechnology RD in the country, its potential for commercialization, and comparing it with the most recent international developments in biotechnology, and its allied fields like bioinformatics, nanobiotechnology, systems biology, synthetic biology, and biomemitics.

19.3 ROLE OF BIOTECHNOLOGY IN RURAL DEVELOPMENT

Biotech crops hold greater potential for smallholder farmers in developing countries by improving crop productivity. Around 18 million smallholder farmers are already benefiting from biotech crops in the developing countries (Anthony and Ferroni, 2012; ISAAA, 2016) including Pakistan.

Cultivation of GM crops by such a huge proportion of resource-poor farmers in rural areas attest to the multiple advantages they have been deriving for the last 20 years. Salient benefits include: increased productivity contributing towards global food, feed, and fiber security; self-sufficiency on nations' arable land; conservation of biodiversity; prohibiting deforestation; mitigating climate change associated complications, and improving economic, health, and social benefits. Only 27% of the total geographical area of Pakistan is arable, which is diminishing rapidly due to an exploding population, expanding urbanization, rural to urban migration, increasing soil erosion, and dwindling water resources. In such a perplexing scenario, development of GM crops has persuasively demonstrated extraordinary benefits in terms of increased per hectare yield and lesser use of costly chemical pesticides. It has resulted overwhelming adoption of biotech cotton by Pakistani farmers for last 14 years (2002–2016).

Cotton contributes ~8% in the value addition of agriculture sector generating nearly 3.2% of GDP. About two-thirds of Pakistan's export earnings are derived from this crop where a large textile sector including ginning, weaving, and clothes manufacturing greatly relies upon domestic cotton production. Furthermore, a huge farming population depends on cotton as a major source of their livelihood. Commercial cultivation of Bt cotton in Pakistan was permitted officially in 2009. However, it was being cultivated widely long before that permission. Ali and Abdulai (2010) reported about 20% yield enhancement due to Bt-cotton adoption. They estimated a positive and statistically significant effect of Bt cotton adoption on household income and welfare while reducing the chances of adopters being poor to be 11%–14% lesser than that of nonadopters. According to another study in 2011, Bt cotton, in Pakistan, was grown on about 2.6 million hectares, which was almost 81% of the total cotton growing area of Pakistan. Evaluation impact studies revealed an average yield gain of more than 12.6%, ≈$20/ha, an average insecticide cost saving and seed premium of about $14/ha–$15/ha. On the basis of this study, the calculated increase of an average farm in 2011 was $57.4/ha and a total of farm income gain was estimated upto $149 million in 2011 (ISAAA, 2011). Since 2009, there has been a cumulative increase of $334.2 million in farm income attributed to Bt cotton adoption.

Ali and Abdulai (2010) revealed that the productivity gains from Bt cotton were more for small farmers than that of the medium or large farmers. Though, apparently the income gains of large farmer appears to be greater, yet adoption tends to help small farmers out of poverty but shows no statistically substantial influence on poverty level of medium and large farmers.

The direct effects of new agricultural technologieson poverty reduction are the productivity benefits enjoyed by the farmers who actually adopt the technology. These benefits usually manifest themselves in the form of higher farm incomes.

19.4 FUTURE PROSPECTS

Agriculture is crucial for attaining food, feed, and fiber security (Borlauge, 2007). After its creation, Pakistan faced a major food security threat in 1950s. Later, the "Green Revolution" resulted in self-sufficiency of major staple foods. However, this was only possible for cereal crops like wheat and rice. Pakistan still has a huge import of numerous food items which include edible oil, pulses, tea, and other commodities, posing a continuous threat to the nations' food security. The population has also increased, thus threatening even the sufficiency in cereal crops.

With the current population at over 200 million, Pakistan is the sixth largest country and its population is expected to reach 250 million by the year 2022. About half of the Pakistani population has been estimated to be food insecure due to deficiency of calories, proteins, vitamins and minerals. This food insecurity is associated with health insecurity and malnutrition. As there is limited scope to bring new land under cultivation, the major increase in production must be attained through application of modern technologies which increase production per unit area, decrease cost of production by enhanced efficiency of input utilization, and increase the nutritive value of food being produced in the country.

Modern plant breeding exploits the knowledge of molecular biology coupled with the tools of biotechnology to improve crop production. The knowledge on genetic composition of crop plants is used to characterize the relevant gene sequences, identify the desirable alleles, develop markers to detect those alleles in a segregating population, and use the transformation techniques to insert the genes in a more targeted and efficient way. The crop production enhancement expected through cutting edge biotechnological interventions is often termed as "Gene Revolution."

Keeping in view the overall challenges and opportunities related to agricultural biotechnology, the following recommendations are suggested to the policy makers for future biotechnology policy.

- It is recommended that genes and gene-carrying expression cassettes instead of transgenic seeds should be procured from sources including Monsanto, Bayer Crop Sciences, and others to strengthen our indigenous RD activities to develop climate resilient high-yielding cotton and other crops. We are at par with the international community as far as technology and expertise are concerned. Where we are deficient, is the availability of genes or gene expression cassettes since genes and their regulatory sequences are patented by either universities abroad or multinational companies. So, it is strongly recommended to procure genes instead of final product like Bollgard-II or any other crop seed. The procurement of the later would be much costlier since developing transgenic plants in developed countries involves costly inputs compared to their development in the developing countries.
- The policy should be devised to encourage technology adoption by small-scale farmers, which will help improve their farm production and revenues while improving their living standards. Favorable policies include enhancing their access to information in order to reduce their hesitations towards adoption of new technologies and formal credit in order to overcome liquidity constraints.
- The policy should be devised to improve human capital of small-scale farmers in the form of schooling, health facilities, better infrastructure, and advanced extension services.
- Empowerment of agricultural biotechnology in all agricultural universities and colleges through ample funding, human resource development, and provision of infrastructure.
- New posts for young biotechnologists should be created in research and extension system, to enable strengthening of the research and extension system with the modern tools of biotechnology.
- Both the agricultural universities and agricultural research institutes should be provided with basic infrastructure for agricultural biotechnology research.
- Research in agricultural biotechnology should be encouraged to solve the major agricultural problems, mainly in improving biotic and abiotic stress resistance, improved input responsiveness, and better nutrition qualities.
- Greater emphasis should be laid on crop genomics, physiology and crop–pest molecular interactions. Human resource development in these areas would be essential for development of molecular information-based breeding program and selection of desirable plants.
- The public institutions should actively contribute to the development of transgenic crops and biotechnological crops along with the national industry to avoid the public perception against the multinational companies as the only owners of biotechnological products.
- All agricultural biotechnology projects must strongly consider the sustainability and environmental safety aspects of their research and products. The protection of environment and ecosystem should be a top priority. The biosafety aspects of all agriculture related biotechnology products should be thoroughly tested before their release. A strong and neutral biosafety body should be devised for this purpose, which should be compatible with international regulations.
- There is an urgent need to train manpower in interdisciplinary fields of science including bioinformatics, agri-nanobiotechnology, systems biology, synthetic biology, and use of ICT in agriculture.

- Nonconventional uses of transgenic crops, including production of plant-based biopharmaceuticals should be also funded on a priority basis.
- Serious efforts are needed to produce biopharmaceuticals for humans as well as livestock at an industrial scale.
- Development of a biotech-based industry should be encouraged and strengthened so that job opportunities can be created for young graduates of agri-biotechnology. Moreover, the potential of industry associated with biotechnology related commodities, products, and instruments should be exploited.
- Special efforts should be made to encourage public–private partnerships in this domain.
- Efforts should be made to increase public awareness about the products and benefits of the agricultural biotechnology.

REFERENCES

Ali, A. and A. Abdulai. 2010. The adoption of genetically modified cotton and poverty reduction in Pakistan. *J. Agri. Eco.* 61: 175–192.

Altman, L. K. 1982. A newinsulin given approval for use in the US. *The New York Times* October 1982. http://www.nytimes.com/1982/10/30/us/a-new-insulin-given-approval-for-use-in-us.html. Accessed July 11, 2016.

Anthony, V. M. and M. Ferroni. 2012. Agricultural biotechnology and smallholder farmers in developing countries. *Curr. Opin. Biotechnol.* 23: 278–285.

Balter, M. 2013. Farming was so nice,it was invented at least twice. *Sciencemag.* http://www.sciencemag.org/news/2013/07/farming-was-so-nice-it-was-invented-least-twice. Accessed July 11, 2016.

Berg, P., D. Baltimore, S. Brenner, R. O. Roblin and M. F. Singer. 1975. Summary statement of the Asilomar conference on recombinant DNA molecules. *Proc. Natl. Acad. Sci. USA* 72(6): 1981–1984.

Borlaug, N. 2007. Feeding a hungry world. *Science* 318(5849): 359.

Carpenter, J. E. 2010. Peer-reviewed surveys indicate positive impact of commercialized GM crops. *Nature Biotechnology* 28(4): 319–321.

Choudhary, B., K. M. Nasiruddin and K. Gaur. 2014. The Status of Commercialized Bt Brinjal in Bangladesh. ISAAA Brief No. 47. ISAAA: Ithaca, NY.

Cohen, S. N., A. C. Chnag, H. W. Boyer and R. B. Helling. 1973. Construction of biologically functional bacterial plasmids *In vitro. Proc. Natl. Acad. Sci. USA* 70(11): 3240–3244.

Delaney, B. 2015. Safety assessment of foods from genetically modified crops in countries with developing economies. *Food Chem. Toxicol.* 86: 132–143.

Gabol, W. A., A. Ahmed, H. Bux, K. Ahmed, K. Mahar and D. Laghari. 2012. Genticaly modified organisms (GMOs) in Pakistan. *Afr. J. Biotechnol.* 11: 2807–2813.

ISAAA Brief 43. 2011. Executive Summary Global Status of Commercialized Biotech/GM Crops: ISAAA Brief No. 43. ISAAA: Ithaca, NY.

ISAAA. 2016. Global Status of Commercialized Biotech/GM Crops: 2016. ISAAA Brief No. 52. ISAAA: Ithaca, NY.

Jacob, S. S., S. Cherian, T. G. Sumithra, O. K. Raina and M. Sankar. 2013. Edible vaccines against veterinary parasitic diseases: Current status and future prospects. *Vaccine* 31: 1879–1885.

Jaenisch, R. and B. Mintz. 1974. Simian Virus 40 DNA sequences in DNA of healthy adult mice derived from pre-implantation blastocysts injected with Viral DNA. *Proc. Natl. Acad. Sci. USA* 71(4): 1250–1254.

Khan, M. S., S. Ali and J. Iqbal. 2011. Developmental and photosynthetic regulation of *Bacillus thuringiensis* δ-endotoxin reveals that engineered sugarcane conferring resistance to "dead heart" contains no toxins in cane juice. *Mol. Biol. Rep.* 38: 2359–2369.

Khan, M. S., G. Mustafa, S. Nazir and F. A. Joyia. 2016. Plant molecular biotechnology: Applications of transgenics. In Khan, M.S., Khan, I.A and Barh, D. (Eds.) *Applied Molecular Biotechnology: The Next Generation of Genetic Engineering.* CRC Press Taylor & Francis, USA, pp. 61–89.

Kwon, K., D. Verma, N. D. Singh, R. Herzog and H. Daniell. 2013. Oral delivery of human biopharmaceuticals, autoantigens and vaccine antigens bioencapsulated in plant cells. *Adv. Drug Delivery Rev.* 65: 782–799.

Malik, K. A. 2014. *Biotechnology in Pakistan Status and Prospects.* Pakistan Academy of Sciences, Islamabad, Pakistan.

Paarlberg, R. 2000. Genetically modified crops in developing countries: Promise or Peril. *Environment: Science and Policy for Sustainable Development*, 42(1): 19–27.

PARC. 2008. Status of Cotton Harboring Bt Gene in Pakistan. Institute of Agri-Biotechnology & Genetic Resources, National Agricultural Research Centre, Pakistan Agricultural Research Council, Islamabad.
Sanchez, P. A. 2002. Soil fertility and hunger in Africa. *Science* 295: 2019–2020.
Sharma, D. 2009. The illegal way to promote BT cotton in Pakistan, June 20, 2009. Available at: http://devinder-sharma.blogspot.com/2009/06/bt-cotton-seeds-smuggled-from-india.html. Last Accessed February 15, 2018.
Shelton, A. M. 2010. The long road to commercialization of Btbrinjal (eggplant) in India. *Crop Prot.* 29: 412–414.
Ye, X., S. Al-Babili, A. Kloti, J. Zhang, P. Lucca, P. Beyer and I. Potrykus. 2000. Engineering the provitaminA (β-Carotene) biosynthetic pathway into (Carotenoid-Free) rice endosperm. *Science* 287(5451): 303–305.
Zimmer, C. 2013. From fearsome predator to man's best friend. *New York Times* May 2013.

20 Plant Diseases of Major Crops and the Way Forward for Their Management

Muhammad Amjad Ali, Amjad Abbas, Muhammad Atiq, Nasir Ahmad Rajput, Khalid Naveed, and Nazir Javed

CONTENTS

20.1 INTRODUCTION

Agriculture is the backbone of Pakistan's economy. It accounts for over 20% in the national GDP. Crop plants are a very important part of the agricultural system of the country. Crop plants faced different abiotic and biotic stresses in different agro-ecological zones of the country. Plant pathogens are major threat to agriculture and result in reduction of yield of important food and fiber crops. The less developed countries are facing new challenges with the emergence of new diseases because of gap in research and less empowered laws, that is, quarantine and seed laws. Plant diseases caused by certain plant pathogens are very important dynamics in the economy and social state of affairs of a country. Plant diseases have resulted in several famines in the world, which in turn have resulted in casualties in the millions and significant socioeconomic impacts in the world. The plant diseases of economically important crops are a serious menace to the economy of Pakistan. This fact can be highlighted as annual losses caused by only wheat (*Triticum aestivum*) diseases in Pakistan are estimated at around 3.6 billion rupees (http://www.nifa.org.pk/CBDpathologyGroup.htm). Similarly, the single disease of cotton (*Gossypium hirsutum*), cotton leaf curl disease, has caused a loss of US$ 20 billion equivalent to 500 million bales in a period of 5 years from 1992 to 1997. Moreover, the diseases of cotton, rice, citrus, and mango not only result in serious yield losses but also have a significant impact on the economy of Pakistan in terms of foreign exchange earning losses. As a consequence, the agricultural economy is disturbed. This is not only an economic issue but also threatens the country's population due to food insecurity, malnutrition, poverty, and imbalanced nutrient uptake.

In this chapter, we provide an overview of the important diseases related to economically important crop plants from Pakistan's perspective. The current status of diseases like wheat rusts, cotton leaf curl disease (CLCuD), rice blast, red rot of sugarcane, corn stalk rot, mango anthracnose, citrus blemishes and greening diseases, wilt and blight diseases of chickpea, plant parasitic nematodes of vegetables and tree declines of shishum, mango, guava, and citrus are discussed in this chapter. Moreover, a brief account of the practices to control these diseases and the way forward to manage these diseases is given.

20.2 COTTON LEAF CURL DISEASE

Cotton is the backbone of the textile industry of the country. It accounts for 8.6% of the value added in agriculture and about 1.7% to GDP (Anonymous, 2010). Cotton not only fulfills the lint needs of the national textile industry but also contributes towards the edible oil production (Ali and Khan, 2007; Ali et al., 2008; Ali and Awan, 2009). However, cotton is attacked by different diseases and insect pests. Cotton leaf curl disease is the most important yield constraint in cotton production in the Indian-subcontinent after the attack of sucking and chewing insects. It is caused by a monopartite begomovirus with associated beta- and alpha-satellite molecules. The disease caused huge losses over a period of 5 years from 1992–1997 (Figure 20.1) and it remains a persistent challenge to the agricultural scientists. In Pakistan, the disease was first found in 1967 in the district of Multan but was ignored as it was not a problem at that time. However, the disease appeared in epidemic form in 1992 resulting in a loss of 5 million cotton bales to Pakistan cotton industry. Cotton is grown on more than 15% of cultivated land area in Pakistan and provides 60% of foreign exchange earnings. The textile industry provides employment to many people. According to a recent survey conducted in 2015, CLCuD is causing a loss of 3 million cotton bales to the cotton industry each year.

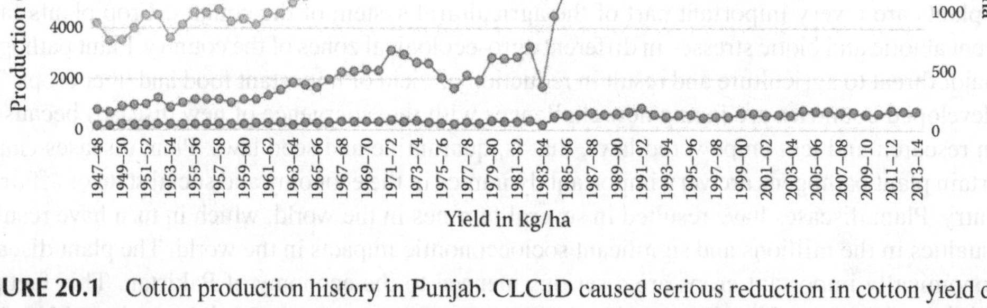

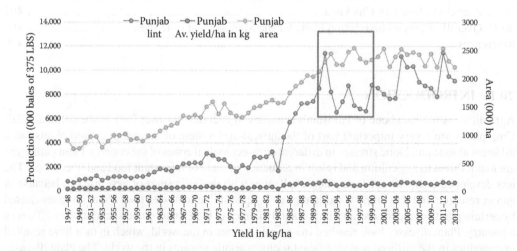

FIGURE 20.1 Cotton production history in Punjab. CLCuD caused serious reduction in cotton yield during 1992–1997 in the province. (From Anonymous. 2014–2015. *Agricultural Statistics of Pakistan, Ministry of Food, Agriculture and Cooperative*, Islamabad.)

Symptoms of CLCuD on infected cotton plants include upward and downward curling of leaves, thickening of veins, stunted of plants, and formation of leaf enations on the underside of leaves. Severely stunted plants produce fewer bolls and the quality of lint is affected. Affected plants produce poor quality lint and the yield is reduced drastically (Briddon et al., 1990). Genome size of CLCuV is 2.8 kb with associated small size (1.4 kb) betasatllite that has a role in disease development. CLCuD is currently a problem in India, Pakistan, China, and certain African countries where a major proportion of cotton is grown (Briddon and Markham, 2000). So far, no variety with complete resistance towards all CLCuV species is available. It has become a difficult challenge to manage due to the diversity of begomovirus species associated with the disease (Amrao et al., 2010). Resistant varieties developed through conventional breeding have failed due to the emergence of resistance breaking strains (Mansoor et al., 2003). There is threat of disease spreading to other parts of the world, which are currently free from disease including North and South Americas. The worldwide trade of plant materials can transport CLCuD to these regions. China was free from CLCuD until 2008, when the causal agent of the disease was found in China. Assumptions are that this spread of CLCuD to China occurred along with the trade of ornamentals from Pakistan to China (Cai et al., 2010). However, in Africa and India different begomovirus species are associated with CLCuD. In Africa, a different species known as *Cotton leaf curl Gezira virus* is associated with infected cotton plants (Idris and Brown, 2000). In Pakistan, the major species associated with the disease is *Cotton leaf curl Multan virus* (CLCuMV). CLCuD is transmitted by the whitefly *Bemisia tabaci* in persistent-circulative manner and a single infected whitefly can inoculate several cotton plants.

No big difference was seen in the incidence of CLCuD between 2014 and 2015, where important yield declines were experienced in the Punjab province of Pakistan (Figure 20.2). This demonstrated that the cotton decline in 2015 was not due to the incidence of CLCuD in cotton. The main cause of cotton decline was heavy rainfalls coupled with lack of agronomic interventions and re-emergence of pink boll worm.

In order to control the disease, management of whitefly vector is very important. Different insecticides are recommended to control the whitefly vector which includes Imidacloprid, Methamedophos and Score to manage the whitefly. Efforts should be made to introgress several genes in the same variety, which will offer durable resistance to the disease. Quarantine measures should be implemented to prevent the spread of disease to new regions, which are currently free from

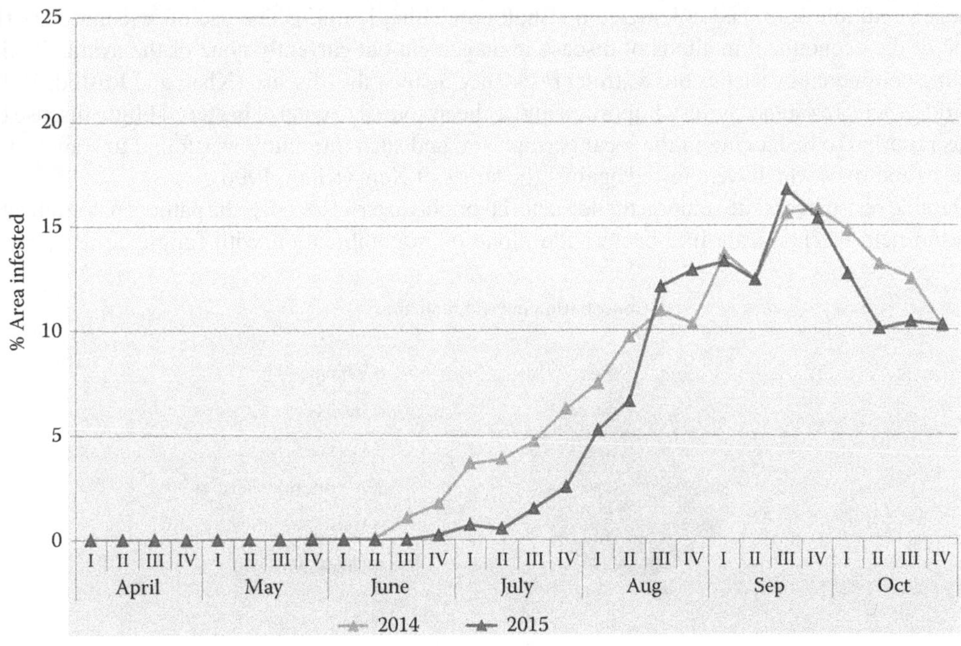

FIGURE 20.2 Week wise incidence of CLCuD (% area infested) in cotton during 2014–2015.

disease. Insecticides should be applied to avoid resistance build up in the insects. Pathogen-derived resistance offers an alternative long-term control in which different genes of virus genome can be targeted to stop virus replication. RNAi technology can be used to target different genes of CLCuV so that virus transcription can be interfered resulting in no disease.

20.3 BACTERIAL BLIGHT OF COTTON

Bacterial blight caused by *Xanthomonas campestris* pv. *malvacearum* (Xcm) is one of the important disease causing heavy losses to the cotton crops. It is a silent disease, which does not completely eradicate the plant but reduces its overall growth and quality of the products. It was proposed as the major disease by Hussain and Tahir (1993) because of its high frequency in the core cotton areas and secondary cotton growing regions of the country. The target of this disease on the plant is diversity in the foliar regions, as well as internal physiological damage to the crop plants (Casson et al., 1977). In Pakistan, the disease was reported the first time in the Multan region in the year 1965 (Ali, 1968) when 50% disease incidence was recorded (Hussain and Ali, 1975). *Xanthomonas* is a genus of diverse plant pathogens and is considered among the top ten damaging bacteria to the plants (Mansfield et al., 2012). More than 30 pathogenic races of Xcm are reported in the world (Huang et al., 2008), while in Pakistan only four races of the pathogen are documented out of which race 18 is more prevalent and virulent in its action (Hussain et al., 1985). During 2015, Punjab province of Pakistan experienced a huge cotton production decline and it was 33% less as compared to the previous year. The reduction in the yield was caused by the development of insect pest, emergence of diseases, and the selection of unauthorized seed. Meanwhile, the environmental conditions were more fluctuating than during the previous year. Global warming resulted in unpredicted rains during the month of April and high temperatures with moist conditions invited the reemergence of old insect and microbial pests. The timing of phenological events like bud formation, flowering, square formation and boll formation was also disturbed due to these uneven environmental fluctuations. Area under crop was not changed largely but the per unit area yield was severely diminished last year. A survey among the farmers concluded that 5% of the yield reduction in 2015 was caused by bacterial blight (Figure 20.3).

The chlorophyll contents are damaged due to chlorosis caused by bacterial blight and continuous spread of the pathogen and 50% reduction in the yield was reported under favorable conditions of pathogen multiplication and colonization (Bhutta and Bhatti, 1983). The use of resistant varieties is one of the economical methods of disease management but currently none of the available high yielding commercial varieties has a strong resistance against this disease (Khan and Rashid, 1997). Natural genetic resistance is very important and a cheap source to control bacterial blight disease but it was reported to be lacking in the local germplasm, and such immunity was found to some extent in the exotic lines which were tested against the stress of Xcm (Khan, 1996).

Durable resistance is the solution to the bacterial blight disease. Usually, the pathogen is eradicated with the help of chemicals like bactericide alone or in combination with fungicide as spray or

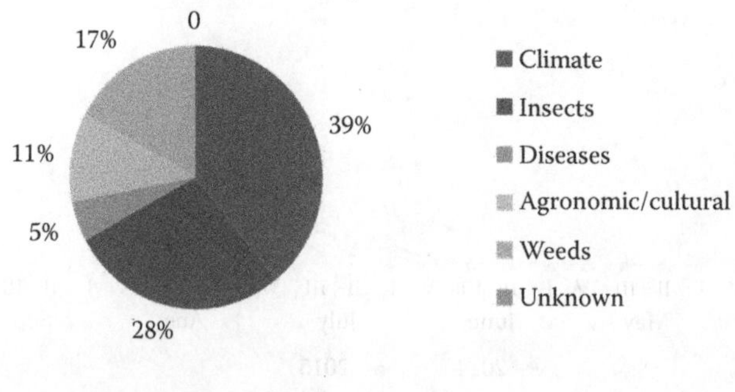

FIGURE 20.3 Percentage contribution of different factors involved in cotton decline 2015. (Taken from Cotton Decline 2015 Report, UAF.)

seed treatment which limits its spread to some extent but does not eradicate it completely (Khan and Ilyas, 1999; Hussain and Tahir, 1993). Khan and Ilyas (1999) reported that Agromycine-100, streptomycine sulphate, and carboxin were more effective to control bacterial blight disease as compared to Pencozeb and Nemispore, while Tricyclazole were found in efficient to combat Xcm under lab trials. The varietal response to blight disease was different in the changing environmental conditions in a controlled experiment reported by Khan and Ilyas (1999). We can conclude from the reports that, since the bacterium was found to survive in the soil for 6 months and 3 months in trash (Verma et al., 1977). The infection at post anthesis stage clearly show that the pathogen has the ability to survive in adverse temperatures of core areas of cotton in Pakistan, for example, during the month of June and July which have average high temperatures above 45°C.

Sajid et al. (2013) reported the efficacy of plant extracts in a comparison study with three chemicals and found different concentrations effective against the growth of Xcm under *in vitro* trials. Three chemicals (Plant Protector, Agrimycine, and Copper oxy Chloride) and plant extracts (from *Citrullus colocynthis, Nicotiana tobaccum,* and *Curcuma lunga*) were tested. *Nicotiana tobaccum* extract at concentrations of 10% was reported as the most effective dose to combat the pathogen with a relatively higher inhibition zone.

In Pakistan, the most important thing regarding plant pathology research is to strengthen the present effort to combat the bacterial blight of cotton by fresh data of comprehensive surveys in the so-called core cotton growing areas. Similar efforts are needed in the secondary crop areas of the country. More than 30 pathogenic races of Xcm are reported in the world while in Pakistan a study based on morphological and biochemical testing proved the existence of only four pathogenic races. There is a need for comprehensive surveys and reporting about the existence of multiple races of the pathogen based on modern molecular tools. This would determine the existence and abundance of a particular pathogenic race specific to an area, which will be helpful to engineer resistant cultivar.

The virulence of the pathogen, due to its multiple races, is an important topic to investigate and devising cultivars from the existing local germplasm with some exotic blood for the development of better disease resistant genotypes. Genes of virulence from the local identified strains should be identified along with the corresponding resistance genes in the germplasm of cotton. This would help to generate lines with single resistance (*R*) genes for combating the pathogen. Such genes in cotton crop could be incorporated into the high yielding cultivars with broad spectrum of disease resistance. Cultivars specific to a particular district having special conditions and varying adaptation as compared to other areas should be devised to tackle the disease.

Beside existing chemicals, new chemistries are needed to trial for better management of the disease. Biocontrol agents, especially plant growth promoting rhizobacteria (PGPR) and endophytic bacteria, which might be helpful for the biological management of the disease, should be tested to manage the disease along with existing popular approaches like chemicals. There is a need to convert this research into products for community service to uplift the status of the farming community involved in the growing of cotton.

20.4 RUST DISEASES OF WHEAT

Bread wheat is the main source of food in the world as it feeds around 35% of the world's population (Dreisigacker, 2004), and over 75% of the world's population consumes wheat as part of their daily diet. Pakistan was the 8th leading producer of wheat after the EU, China, India, the United States, Russia, and Canada in 2013 (Economic Research Service, USDA; Figure 20.4). Wheat is grown on a large area in the country and it covers most of the cultivated areas, followed by cotton and rice. The area under wheat cultivation was 9.199 million ha, production was 25.98 million tonnes, and grain yield was 2824 kg ha^{-1} (Pakistan Bureau of Statistics, 2014). However, there is a very large gap between New Zealand (which is giving the highest production in terms of per unit area) and Pakistan (32nd position in terms of production per unit area) (USDA, 2014). This yield gap is due to several biotic and abiotic factors, out of which wheat rusts are the most important in terms of yield losses.

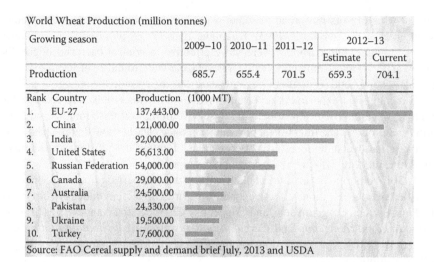

World Wheat Production (million tonnes)

Growing season	2009–10	2010–11	2011–12	2012–13	
				Estimate	Current
Production	685.7	655.4	701.5	659.3	704.1

Rank	Country	Production (1000 MT)
1.	EU-27	137,443.00
2.	China	121,000.00
3.	India	92,000.00
4.	United States	56,613.00
5.	Russian Federation	54,000.00
6.	Canada	29,000.00
7.	Australia	24,500.00
8.	Pakistan	24,330.00
9.	Ukraine	19,500.00
10.	Turkey	17,600.00

Source: FAO Cereal supply and demand brief July, 2013 and USDA

FIGURE 20.4 Status of Pakistan in wheat production in the world. (Adapted from Wheat in Pakistan, A Status Paper, NCW, PARC.)

Wheat crops are attacked by several diseases like rusts, smuts, bunts, mildews, septoria leaf blotch, and root cyst nematodes. However, the rusts are the most devastating diseases in wheat. They are a serious menace to wheat productivity worldwide and in Pakistan. There are three types of rusts in wheat, that is, leaf rust, stipe rust, and stem rust. Leaf and stripe rusts are the most prevalent rust diseases in Pakistan. Stripe rust, which is caused by *Puccinia striiformis* f. sp. *tritici* fungus, has been associated with wheat in cool, temperate regions of the world, including Asia, Europe, the Middle East and Africa (Line, 2002). It produces symptoms on the host plant which include the appearance of yellow uredia (spore mass) on the surface of leaf in the form of stripes or lines sometimes present on the leaf sheath, stems, and heads. Black spores (telia) are produced in stripes as the crop mature, these black spore are enclosed by the leaf outer surface known as epidermis (Smiley and Cynthia, 2003). In Pakistan around 70% of the total area under wheat cultivation is pruned by yellow rust disease. The rusts have appeared in epidemic forms in 1947–1948, 1953–1954, 1958–1959, 1977–1978, and 1992–1993 causing severe yield losses (Hassan et al., 1979; Ahmad et al., 1991; Khan and Mumtaz, 2004; Afzal et al., 2007; Afzal et al., 2008). Early incidence of stripe rust may result into 100% yield losses in wheat depending on the degree of susceptibility of the host plants. This rust disease results in 10%–70% yield losses (Chen, 2005). Stripe rust has destroyed cereal production throughout the world with fast systematic infection of the plants resulting in defoliation and shriveled kernels.

Yield losses due to yellow rust can be prevented or reduced by growing resistant varieties and with timely fungicide application. Development of resistant cultivars is a very lengthy process, while use of fungicides has health and environmental concerns. Wheat rust has been mostly controlled by chemotherapy using various fungicides. Huge costs are involved in the use of fungicides, which is an increasing burden on the finances of medium growers in the developing countries (Chen, 2005). It is estimated that the use of fungicides to control stripe rust in Australia costs AU$40 million in one year (Wellings and Kandel, 2004). The application of fungicides has not only health and environmental concerns but also results in contamination of agroproducts (Tao et al., 2014). Moreover, the continuous use of fungicides to control stripe rust has resulted in the development of fungicide-resistance in different strains of stripe rust (Chen, 2005).

Leaf rust in wheat is caused by the fungus *P. triticina*. It is the most important rust disease after yellow rust and causes serious yield losses (Park et al., 2007). This disease has the ability to cause yield losses up to 50% due to its high frequency and widespread prevalence (Huerta-Espino et al., 2011). The leaf rust fungus is adaptive to a variety of climatic conditions and mainly affects grain attributes like number of grains per spike, thousand grain weight, caused by interruption in photosynthesis in the upper plant leaves (Din et al., 2017). Pakistan has experienced leaf rust epidemics in 1948, 1954, and 1978. However, the epidemic of this disease in 1978 resulted in a 10% yield loss producing a national loss of US $86 million (Hussain et al., 1980).

The stem rust disease of wheat is caused by *P. graminis* f. sp. *tritici*. After the so-called "green revolution" during the 1960s and the 1970s, most of the cultivars in the world were found resistant against stem rust. This resistance was largely conditioned by the presence of *Sr31* gene, which was introgressed into the bread wheat from rye (*Secale cereale*) (McIntosh et al., 1995). Around 80% of the wheat lines from developing countries are protected by the *Sr31* gene against stem rust (Iqbal et al., 2010). The breakdown of resistance due to *Sr31* has led to the evolution of race Ug99 in Uganda during the year of 1999 (Pretorius, 2000). Local races of stem rust have important impacts on wheat yields and the economy of Pakistan. However, experiments were conducted in Kenya and all the major wheat varieties from Pakistan were found to be susceptible to Ug99 (Anonymous, 2005).

This is alarming for the world and Pakistan that very few wheat lines have shown partial resistance to the Ug-99 race. Still, there is no report of the Ug99 race in Pakistan, however, continuous travel of people and trade from Ug99 infested areas/countries to Pakistan may result in the spread of the disease in the country. In the absence of any resistant source against this race, this strain of rust will result in a serious menace to wheat production in the country. Hodson et al. (2011) reported that the Inqlab 91 cultivar is highly susceptible to Ug99. The problem in Pakistan is most of the areas under wheat cultivation are either covered by Inqlab 91 or its derivatives like Sehar-2006. Moreover, according to Khoury et al. (2008), Pakistan is at potential risk of invasion of Ug99 (Figure 20.5).

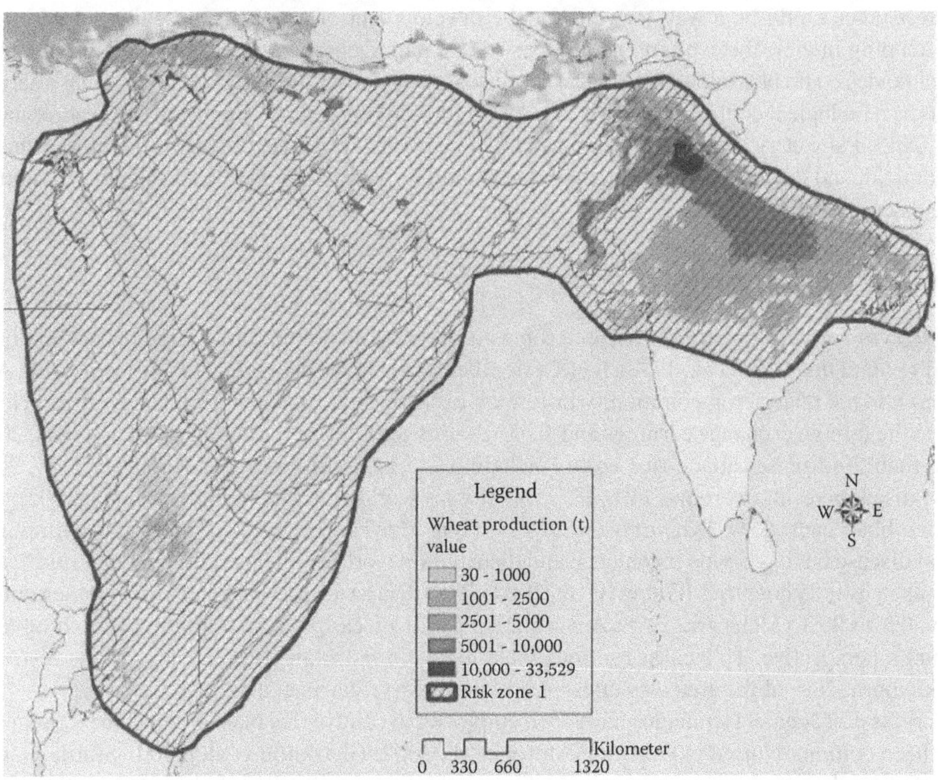

FIGURE 20.5 **(See color insert.)** Area of the world at risk from Ug99 infestation. (Reproduced from Khoury W, Cressman K, and Yahyaoui A. 2008. Ug99 status, management and prevension. International Plant Protection Convention, Special session, Rome, April 9, 2008 (https://www.ippc.int/static/media/files/publications/en/1217860549444_Ug99_CPM_3_side_event_9Apr08.pdf).)

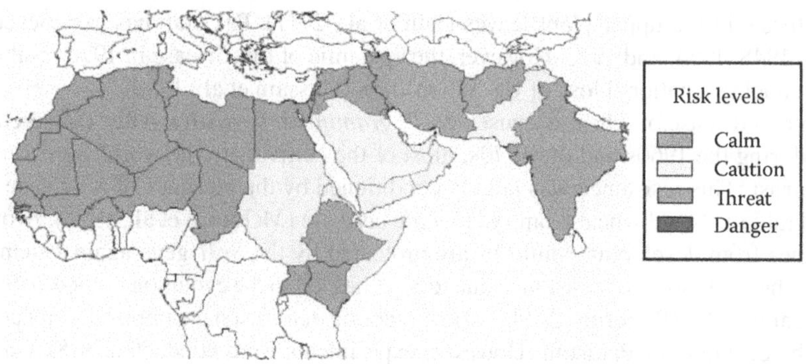

FIGURE 20.6 Risk levels of Ug99 spread in Pakistan and its neighboring countries. (Reproduced from Khoury W et al. 2008. *Ug99 status, management and prevension.* International Plant Protection Convention, Special session, Rome, April 9, 2008 (https://www.ippc.int/static/media/files/publications/en/1217860549444_Ug99_CPM_3_side_event_9Apr08.pdf).)

Nonetheless, to date, the risk level of a Ug99 invasion in Pakistan is mild compared to other related countries in the region like Iran (Khoury et al., 2008; Figure 20.6) A combination of stem rust resistant genes like *Sr13*, *25*, *36*, *37* and *44* should be very carefully used against local stem rust race. Similarly, the CIMMYT lines from Ug-99 screening nurseries could be a handful to develop resistance against this dangerous disease (Iqbal et al., 2010). The strategies like fast breeding for rust resistance could be a way forward for the development of rust resistant cultivars with more pace keeping in view the evolving pathotypes of rust pathogens. Moreover, the potential of various biofungicides could be exploited to control rust diseases in wheat. Similarly, use of different microbial agents as a biological control strategy is also a way forward for controlling rust diseases in wheat (Li et al., 2013; Pang et al., 2016). However, development of rust resistant cultivars is the most effective, economical, and ecologically safe method that should be adopted by Pakistani scientists to manage these diseases.

20.5 RICE BLAST DISEASE

Rice (*Oryza sativa*) is a very important crop as it provides staple food to a large proportion of the world's population (Luo et al., 1998). It has a significant role in the economy of Pakistan as well. After cotton, it is the major crop commodity in foreign exchange earnings for the country. It contributes 15% to the foreign exchange earnings and 1.3% towards the GDP of the country (Anonymous, 2009). More than 50 diseases attack rice crops, including 21 fungal diseases (Jabeen et al., 2011, 2012). These diseases result in around 14%–18% yield losses worldwide (Mew and Gonzales, 2002).

Rice blast caused by *Magnaporthe oryzae* (syn: *Pyricularia oryzae*) is the most important fungal disease having a wide prevalence and distribution worldwide. It is estimated that this disease prevails in over 85 countries (Gilbert et al., 2004). This disease causes enormous yield losses ranging from 50% to 90% (Mehrotra, 1998; Agrios, 2005). This disease is very important according to the country's perspective. This causes serious yield losses in Pakistan (Hafiz, 1986). In the district of Sheikhupura, 20% of the area was infested with rice blast (Shazia et al., 2003).

Various management strategies have been employed to control rice blast. For instance, fungicides have been commonly used in Pakistan (Anwar et al., 2002; Ghazanfar et al., 2009). Similarly, plant extracts have been applied to manage the disease (Iftikhar et al., 2010; Babar et al., 2011). Garlic juice and neem extracts have shown good promise to minimize the disease incidence (Slusarenko et al., 2008; Amadioha, 2000). Moreover, antagonistic microorganisms also have been utilized, that is, *Trichoderma* spp. (Ouazzani et al., 1998). Recently, Hajano et al. (2012) used all these strategies in Pakistan and reported that they could be used for controlling the disease.

Although the disease could be controlled by the use of cultural practices, chemotherapy, and biotechnological methods (Ribot et al., 2008), the use of host plant resistance is the most economical and environmentally safe strategy (Khan et al., 2001; Haq et al., 2002). Host plant resistance is still vulnerable to breach due to the outbreak of new virulent races of the pathogen in the present scenario of changing climate. This demands the development of rice cultivars with broad resistance, that is, multigenic resistance against emerging races of blast disease in addition to climate resilience.

20.6 CORN STALK ROT

The disease is present wherever corn is grown and is caused by many soil borne fungi. Different species of *Fusarium* and other fungi involving *Gibberella* species are involved in the development of the disease (Gilbertson et al., 1985). Symptoms of stalk rot are similar in cases of *Fusarium* and *Gibberella*; however, the infections can be differentiated by looking at the inner stalk color of infected plants. *Fusarium verticilloides* is responsible for corn stalk rot in hot and dry areas (Doohan et al., 2003). Symptoms of corn stalk rot on infected corn plants include rotting of inner stem, crown, and internodes. On the inner portions of the stem, pink to tan colored discoloration develops. Stalk rot results in 5% yield losses in corn fields each year, reaching up to 10%–20% in severe infections.

Infected plants suffer malnutrition during the seed filling stage as there is insufficient movement of water in vascular bundles and poor translocation to upper parts of the plants. Kernels on infected plants are poorly filled and yield is reduced drastically. Diseased plants may die prematurely resulting in yield reduction. Infection with corn rot fungi occurs right after flower formation and during the process of pollination. Warm and dry weather is favorable for disease development in addition to insect injury, which provides an entry passage for the fungus into the plant.

The development of disease-resistant hybrids is needed for the management of corn stalk rot. Management of insects is also important in order to prevent injury on healthy plants. Maintaining proper plant population in fields is also an important criteria for disease management. Healthy crops will have resistance to the invading pathogens and management of disease will become easier. Stress and insect injury are two common reasons for disease occurrence on corn plants, therefore plants should be given a healthy environment and proper nutrition.

20.7 RED ROT OF SUGARCANE

Sugarcane, because of its fast growth and high sugar contents, is cultivated on a large area in Pakistan, mostly in the subtropical regions of the country. The crop is grown on an area of 1.13 million ha in Pakistan with a production of 63718.52 thousand tons of cane biomass annually. Still the average yield per unit area in Pakistan is around 56.50 tonnes/ha (Govt. of Pakistan, 2013), which is far less compared to many countries across the globe. Red rot of sugarcane is a major disease of sugarcane and results in low production, causing heavy losses each year in Pakistan. This disease is caused by the fungus *Colletotrichum falcatum* (Sharma and Tamta, 2015).

Red rot can reduce cane weight from 29% to 83%; similarly it may decrease the juice content of the cane from 24% to 90% in addition to a reduction in sugar recovery from 31% to 75% at various infection levels (Munir et al., 1986). However, during the years 2003–2006, the greatest loss to sugarcane crop was brought by red rot disease epidemic. During this epidemic condition of the disease, the loss in cane weight was about 29.07% and resulted in a 30.8% loss in sugar recovery (Alvi et al., 2008). The problem of red rot is faced by many countries like the United States, India, Australia, Thailand, and Pakistan and reduces the recovery of sugar from 5% to 10% on an average (Viswanathan and Samiyappan, 2002). There is a strong need to monitor the core areas of Pakistan's cane production to evaluate the prevalence and disease incidence. The disease devastates the quality and quantity of sugarcane. The pathogen hydrolysis the stored sugar in the standing crop by capitalizing its sugar degrading enzymes, and glucose and fructose molecules are formed which alter the final sucrose quality (Sehtiya et al., 1993). It is aptly called the "Cancer" of sugarcane (Khan et al., 2011).

In Pakistan, the occurrence of red rot of sugarcane has resulted in extensive damage in recent past and has attained the status of the most devastating and sustained problem in sugarcane (Chaudhry et al., 1999). A variety of management strategies have been adopted in the past, for instance, use of fungicides (McGrath, 2004; Subhani et al., 2008). Although the role of fungicides is significant in modernization and changing the status of agricultural crop is commendable (Mehta, 1971), they have several health and environmental concerns which cannot be overlooked. After independence, Pakistan imported several sugarcane cultivars from India, such as L-54 and L-118. These lines were very susceptible to red rot of sugarcane under Pakistani environmental conditions, which resulted in epidemics of red rot during 2003–2006 and caused serious losses in terms of cane weight and sugar recovery (Alvi et al., 2008). This demands the development of resistant cultivars against this dangerous disease, which is the most economical and environmentally safe method.

With the increasing trend of organic farming, beneficial microbes are being identified and applied to tackle different biotic and abiotic stresses including plant diseases. For instance, plant growth promoting rhizobacteria (PGPR) are gaining attention because of their attributes as biofertilizers, bioprotectants, and soil remediation agents (Vessey, 2003; Tank, 2009). This strategy could be adopted to control red rot disease of sugarcane as well (Hassan et al., 2014). Various QTLs conferring resistance against red rot could be found in diverse sugarcane germplasm to develop high yields and disease-resistant cultivars. As crossbreeding is difficult in Pakistan due to inability of the plants to bear true flowers and seeds, somaclonal variations could be utilized for developing disease free clone/variants of the crop.

20.8 MANGO ANTHRACNOSE

It is a fungal disease caused by *Colletotrichum gloeosporioides*, which causes severe yield losses in mango (*Mangifera indica*) production each year. Another fungus, *C. autumn* is involved, however, and the former one is mainly associated with the disease in humid environments. *Colletotrichum autumn* is responsible for minor infections in sporadic cases. The disease is particularly a problem in wet weather during blossoming of trees. Disease symptoms appear on leaves, branches, and fruits as sunken dark brown to black spots, which increase in size at later stages of disease expanding to the entire leaves and fruits. The spots on ripe fruits penetrate deep into tissues making them unfit for human consumption. Spots are normally present near the margins, which extend towards the center and cover the entire leaves. Fungus produces spores during humid conditions which are carried to healthy trees by wind and rain splashes (Arauz, 2000). Anthracnose causing fungi infect both young tissues and mature fruits on trees. Anthracnose is considered a postharvest disease and causes up to 15% losses in mango crop each year in humid seasons. Disease spots on leaves greatly reduce the photosynthetic production, thus hindering chlorophyll synthesis and ultimately reducing yields (Wang, 2009). Anthracnose fungi cause latent infections in mango trees in which the fungus penetrates the green fruits where the symptoms do not develop until the fruits are ripe. The incidence of mango anthracnose can reach 100% in severe cases in humid and wet conditions.

Dipping mango fruits in hot water (52°C) for 5 minutes, in which fungicides have been added, is a recommended strategy. As the rainy season is conducive for disease development, the crop should be managed to avoid the rainy season. Potential biocontrol agents are available for management of mango anthracnose which provide efficient disease management (Patino-Vera et al., 2005). Routine monitoring of the orchards is recommended in order to manage the disease, also fungicide sprays with mancozeb are recommended at 14 days interval. Emulsion coatings are used for disease management (Diaz-Sobac et al., 2000).

20.9 PLANT PARASITIC NEMATODE OF VEGETABLES

Plant-parasitic nematodes are obligate biotrophic parasites causing serious yield reductions in a broad range of crop plants such as tomatoes, wheat, soybean (*Glycine max*), potatoes and sugar beet, etc.

(Ali et al., 2015). The annual yield losses caused by plant-parasitic nematodes have been estimated at over $157 billion worldwide (Abad et al., 2008). Plant parasitic nematodes can cause up to 20% yield losses in individual crops, which can be devastating for low-income farmers in developing countries (Atkinson et al., 1995). Out of several economically important nematode species, root-knot and cyst nematodes, within the family Heteroderidae, are especially dangerous. Root-knot nematodes induce galls or knots comprising of several giant cells on plant roots and fetch nutrients from the vasculature (Jones and Payne, 1978). Among the root-knot nematodes, *Meloidogyne incognita* is the most dangerous one, which parasitizes almost all cultivated plants including vegetables (Abad et al., 2008).

These nematodes are present everywhere on earth ranging from the polar regions to the deep oceans. Several nematode species parasitize vegetable crops. A comprehensive survey of Punjab province was carried out by Anwar and McKenry (2012) to find out the frequency of plant parasitic nematodes on about 20 of the most prevalent vegetable crops. The finding is that root knot nematode *Meloidogyne incognita* (90%) was the most prevalent nematode species in all these vegetables followed by *Pratylenchus penetrans* (30.2%) and *Tylenchorhynchus clarus* (29%) (Anwar and McKenry, 2012). Similarly, they investigated the crop losses incurred by these nematodes in these vegetables and found that on an average these parasitic worms cause 23% yield losses in vegetable crops in Punjab, Pakistan. However, working on four commercially grown vegetables crops (eggplant, okra, tomato, and cucumber), Shakeel et al. (2012) reported 32.5% yield losses caused by mainly 5 species of nematodes.

In case of potato, golden cyst nematodes (*Globodera rostochiensis*) usually feed on roots and generally produce aboveground symptoms like leaf yellowing, necrosis and patchy stunted growth leading to huge crop losses. Moreover, the potato tubers are also seriously affected due to these nematodes, which results in poor quality of the tubers. This nematode not only invades potato but also parasitize tomato and other solanaceous crops, which might result in huge development of inoculum. However, they are often neglected due to lack of obvious aboveground symptoms and lack of knowledge about plant-parasitic nematodes in farming community. *Globodera rostochiensis* has already resulted in huge losses to the country's economy as the shipments were sent back by Russia due to the presence of eggs and larvae of potato cyst nematodes in the tubers. This is alarming situation as Pakistan exports 40%–50% of its potato solely to Russia (http://www.pakistantoday.com.pk/2013/10/25/business/pakistan-approaches-russia/). The occurrence of potato cyst nematode in Pakistan was first reported in 1986 in northern areas of Pakistan by Khan and colleagues but no further work was carried out to survey core potato growing areas of the country and the management of this nematode to avoid its spread.

Several management strategies have been adopted to cope with this nematode species in various crop plants. These includes use of various chemical nematicides, resistant cultivars (Sorribas et al., 2005), cultural practices, use of biochemicals in the form of plant extracts, or use of biological control agents like antagonistic fungi and bacteria (Dababat and Sikora, 2007; Adam et al., 2014). However, more often the plant parasitic nematodes are controlled by the application of chemical nematicides. These chemicals are very costly and impose an extra financial burden on low-income farmers (Sorribas et al., 2005). On the other hand, the use of nematicides available has serious environmental and health concerns. Per hectare cost of a nematicide is as much as US$ 500 per year and can only be acceptable if the crops have high value in the market. Moreover, transgenic strategies like modification of gene expression of different defense related genes in the feeding sites like giant cells and synytia induced by parasitic nematodes could be established and employed (Ali et al., 2013a,b; 2014a,b). So, there is a need to explore alternative strategies for management of plant parasitic nematodes, which are environment friendly as well.

20.10 TREE DECLINES IN PAKISTAN

For the past few decades, certain forest and horticulture tree species in Pakistan are facing remarkable rates of decline or death with displayed foliar symptomology, reduced radial growth, and increased mortality. The most severely affected species are trees/shrubs, which include shisham (*Dalbergia sissoo*), mango (*Mangifera indica*), citrus (*Citrus spp.*), and guava (*Psidium guajava*). Although trees of all ages are affected by decline, the problem has been most severe among the older age-classes. The causes of tree decline are not completely known, but there are many natural (biotic and abiotic) and management related factors that are thought to stress or injure the trees.

20.10.1 Shisham Decline

Soil-borne injury on shisham tree, associated with fungal pathogens, has been reported since early 1956 in Pakistan (Khan, 1989). Typical symptoms are that the branches start withering and disappearing from tip to downward, because of virulent pathogens which destroy the root system completely and result in the death of the tree. The affected shisham leaves loose their green color and turn brown at the later stages. Under severe conditions, the twigs start dying one after another, causing the decline of the entire tree (Bakhshi, 1954; Bajwa et al., 2003; Rajput et al., 2008). 30% of shisham trees were affected by decline disease during a survey of different areas of Punjab (Bajwa et al., 2003). However, Arshad et al. (2008) observed that 70% of the shisham tree were affected in the Punjab. Moreover, Pathan and coworkers reported that 40% of the trees have been affected by dieback disease in Sindh (Pathan et al., 2007).

Many strategies have been applied for management of shisham decline in Pakistan and worldwide. For instance, Bajwa and Javaid (2007) used Benomyl and Ridomil to control mycelial growth of *F. solani*, one of the causal pathogen of shisham decline. Moreover, they recommended Benomyl to control *D. sissoo* wilt (Bajwa et al., 2003). Furthermore, high concentrations of Ridomil Gold and Dithane M-45 significantly reduced the mycelial growth of *F. solani* and enhanced the inoculated seedling growth of *D. sissoo* (Rajput et al., 2012). Similarly, Purohit et al. (1998) and Mehrotra et al. (1998) reported various fungicides were most effective against seed mycoflora of *D. sissoo*. Neem is an evergreen tree with many uses, the most important use for as antagonistic to fight against crop pests and diseases without any harmful effects on environment. Neem products showed significant reduction of *F. solani* in the growth of shisham seedlings (Rajput et al., 2011). In addition, *D. sissoo* were initially developed as disease free plant through micropropagation (Zaidi et al., 2016). However, the suitable control for shisham decline can be adopted through genetic engineering to develop resistant shisham trees.

For better understanding of shisham decline, we need to understand different functional branches of the plant immune systems targeted by biotic and abiotic factors through adaptation of genetic engineering techniques. By using these innovative strategies, we can develop disease-resistant woody tree. The present approach could be employed to obtain genetically improved and resistant shisham trees against decline disease. It is important to develop a screening program suitable for wood production to ensure a continued supply of this valuable. Resistance genes analogues could be identified and detailed information about vegetative clones and seed producing orchards of shisham could be made available for the future using these technologies.

20.10.2 Citrus Decline

Citrus fruits belong to the Rutacease family and are grown in both tropical and subtropical regions of the world. The worldwide annual production of citrus fruits was estimated 115.53 million tons in 2010–2011 (FAO STAT, 2012). In total, 142 countries produce citrus fruits, including China, Brazil, United States, Mexico, India, Nigeria, Pakistan, etc. Pakistan is in 10th position for citrus production in the world. The area under citrus cultivation in Pakistan is 204.07 thousand hectares and 2.334 million tonnes production per year (FAO STAT, 2012). Citrus production is facing serious problems due to the attack of a number of diseases like *Phytophthora* gummosis, slow decline, vascular wilt, root and stem rots, feeder root-rot, dry root rot, dieback, twig blight, and citrus canker (Armstrong and Armstrong, 1975; Farih et al., 1981; Bender et al., 1982; Valle, 1987; Nemec et al.,

1989; El-Borai et al., 2002). Citrus nematode (*Tylenchulus semipenetrans*) is widely distributed in all the citrus groves of Punjab province. Although different species of nematode have been found in association with citrus roots worldwide (Verdejo and McKenry, 2004; Duncan, 2005), *Tylenchulus semipenetrans* has been mostly associated with citrus slow decline.

Association of a great variety of plant pathogens and nematodes are commonly perceived to be indirect, the result of alteration in the host, such as localized or systemic responses to wounding (Westerlund et al., 1974: Moorman et al., 1980; Carter, 1981; Hillocks, 1986; El-Borai et al., 2002). Since the last few years, most of the citrus growing areas in Punjab province were found to suffer from the infestation of citrus decline disease. Since then, the citrus decline has become one of the most severe problems in citrus growing region of Pakistan, especially of the Punjab province. A survey of different areas demonstrated serious loss of citrus because of these diseases (Akhtar and Ahmed, 1999). Hundred percent incidence of *Fusarium* association with citrus decline was found in Jhang and Faisalabad followed by 25% in Sargodha. However, incidence of *Phytophthora and Diplodia spp.* association with citrus decline was 5% and 8.33% in Sargodha respectively. (Ali et al., 2014a,b).

Citrus decline is caused by several soil-borne plant pathogens at seedling stage in nurseries and mature trees in fields. In this regard, different management strategies are needed. In the early 1990s, the use of metalaxyl in citrus nurseries was used extensively, resulting in the development of resistant isolates of *Phytophthora nicotianae* in Florida (Timmer et al., 1997). The growers ranging from 74.16 to 90.8% had awareness about fungicides against diseases; however, 37 to 45% adopted the use of all these fungicides (Ashraf et al., 2014). Another study indicated that *T. semipenetrans* results in reduction of *P. nicotianae* infection in citrus seedlings and mitigate the disease caused by the fungus. (El-Borai et al., 2002). A confirmation of a synergistic effect between these two pathogens (*F. semitectum* and *T. semipenetrans*) proved that a synergistic relation should be a part of an integrated control strategy to management of citrus decline (Safdar et al., 2013). Moreover, the nematicidal plant extracts could be used to enhance the vegetative growth of citrus seedlings and reduced the nematode population (Ayazpour et al., 2010).

Citrus decline is one of the major problems throughout the citrus growing areas of the world. Citrus production can be increased through nurseries run on a scientific and professional basis. Certified citrus nurseries are needed to solve the problems of citrus decline. Pathogen vector relationships should be further investigated. The citrus market is also facing a setback because of shortages of cold storage and market manipulation by middlemen. Suitable fungicide or bioagents may be applied at the nursery level. Moreover, genomic-based diversity has not been explored for citrus nematode species from Pakistan. So it is very important to study its genetic diversity in the citrus fields of Punjab, which is the main citrus producing province. Different antagonistic endophytic microbes could be employed for controlling the nematodes below the threshold level. Moreover, different advancements used for the improvement of citrus plants could be adopted to manage various citrus diseases, including citrus decline (Ali and Nawaz, 2017).

20.10.3 Mango Sudden Decline

Mango sudden decline (MD) is a complex disorder of mango (*Mangifera indica* L.) that is caused by several fungi, insects, and different soil factors. The disease is widespread in all mango producing areas worldwide and results in huge yield losses (Iqbal and Saeed, 2012). The disease is also known as mango sudden death syndrome (MSD) as the affected trees are killed very rapidly due to the disease. MSD has become a persistent problem in mango production areas worldwide. In Pakistan, mango is an important summer fruit crop that is grown on large acreages in Sindh, Punjab, and Khyber Pakhtunkhwa. India is the largest mango producing country followed by China, Thailand, and Mexico. In Pakistan, more than 300 mango varieties are grown with an annual production of 1.40 million tonnes. Around 7%–10% of total mangoes produced in Pakistan are exported to other countries, which earn up to US$ 20 million. Pakistan occupies the 5th position among mango producing countries and provides 4% of the world's mango share.

Several fungal pathogens are involved in mango decline, including *Fusarium solani* and *Diplodia theobromae*, which causes the blockage of phloem vessels. Blockage of phloem vessels results in a reduced food supply to the roots and other lower parts of the plant, eventually resulting in plant death (Masood et al., 2011). Bark beetle is involved in the transmission of fungi from infected plants to healthy plants during their feeding process. In Oman, *Ceratocystis fimbriata* is found associated with disease in association with *D. theobromae* and *C. amanuensis*. MSD has been found in more than 60% of mango trees in Oman (Al-Adawi et al., 2003, 2006).

Disease symptoms include sudden wilting of diseased trees, which starts from one side of tree and covers the entire tree at later stages of disease. Infected trees secrete amber-colored gum from their trunks and branches and vascular tissues become discolored. The oozy material secreted from the stem of infected trees attracts the insects, which feed on the gum and carry germs along with their legs. Several research institutes and universities have been trying to find a cure for the disease. Studies show that application of a balanced amount of fertilizer is helpful in controlling the disease. Application of potash fertilizer can help to control the disease.

Spray of Talstar should be done to control bark beetles at the rate of 200 cc/100 litres of water. Several systemic fungicides are also helpful for disease management in addition to contact fungicides (Khanzada et al., 2005).

Regular monitoring of mango orchards should be done for timely control of this disease. Early detection can be very helpful in managing the disease and to prevent the further spread of associated fungi to nearby growing healthy plants. Before establishing a new orchard, nursery sources should be checked to be free from disease. Care should be taken during farm operations not to injure the trees so that the pathogen is not spread to healthy trees. Tree trunks should be pasted with bordeax mixture and other copper based fungicides. For eco-safe management of this mango disease, breeding efforts should focus on introducing resistant mango varieties, which offer durable resistance to the disease. Gene pyramiding can be applied to combine multiple genes with minor effects for resistance against the disease.

20.10.4 Guava Decline

Guava has the second highest content of vitamin C. It is also very rich in pectin soluble fiber, which cleans and rinses our intestine walls. A guava weighing 100 g would contain 260 mg of vitamin C, which is four times greater than in an orange. Guava is also rich in carotenoids and potassium. Its traditional use against diarrhea, gastroenteritis, and other digestive complaints has been validated in numerous clinical studies. In Pakistan, guava is one of the most gregarious fruit trees, its annual production is 512.3 thousand tonnes and occupies the 3rd position among major fruits grown in the country, after citrus, and mango (Anonymous, 2012). Guava decline is caused by many fungal pathogens such as *Botryodiplodia theobromae*, *F. oxysporum* f.sp. *psidii*, *P. parasitica* and *F. solani* f.sp. *psidii* (Bokhari, 2009; Safdar et al., 2015). Root knot nematodes are also considered an important part of the disease complex coupled with different fungal pathogens (Gomes et al., 2011). It is estimated that guava decline diseases is responsible for high yield losses in Pakistan. Guava yield reduction were recorded between 2003–2004 (8920 kg per hectare) and 2008–2009 (8223 kg per hectare) due to tree decline (Anonymous, 2010). Moreover, Safdar and coworkers reported a 36% disease incidence of guava trees affected by decline during a survey of different areas of Punjab (Safdar et al., 2015).

Guava decline and wilt diseases have been spread epidemically during the last few years and have caused tremendous losses. Timely applications for the control of guava decline have been recommended by various workers using chemical and biocontrol agents. For instance, Dwivedi (1992) controlled the Guava wilt disease through *Trichoderma* spp. and *Streptomyces chibaensis*. Additionally, seed oil of *Foeniculum vulgare* was reported to control wilt diseases (Dwivedi, 1993). In Pakistan, a combination of Topsin M sprays and the antagonists *Trichoderma harzianum* and *Arachniotus* sp. added in soil amended with wheat straw controlled the decline of guava (Ansar et al., 1994). Currently, carbendazin application has resulted in reduction of fungal infection in guava trees and suppressing of the decline disease with significant enhancement in vegetative growth of plants (Safdar et al., 2015).

Farmers of guava cultivation face many problems for various reasons, including poor communication system and lack of governmental support. Marketing of the guava fruit is also facing a setback for shortage of cold storages and manipulation by middlemen. The growers have also been being deprived of fair price of the fruit for the last several years. There are two major factors, intercropping in both summer and winter seasons and physical beating of flowers, are responsible for decline in a long way losses to guava yields. Also a number of inappropriate fertilizers are responsible for decline in crop. In this regards, productivity of guava yield can be inverted through selection of proper fertilizers, sufficient irrigations and by better management practices.

20.11 CITRUS GREENING DISEASE

Huanglongbing/citrus greening caused by *Candidatus liberibacter* asiaticus (CLas) is a widespread and destructive disease of tropical and subtropical areas. It is potential threat to commercial or global citrus industries (Islam et al., 2012). It is estimated that more than 60 million trees are destroyed due to this disease worldwide. Three million trees of citrus discarded due to this disease have been reported in Indonesia. Approximately 3 million HLB-affected sweet orange trees were removed in Brazil. This disease has caused 30%–100% losses worldwide especially in Africa (Iftikhar et al., 2014). Disease incidence reported in various citrus cultivars is given in Figure 20.7 (Ahmad et al., 2011). The disease progressed in the orchard relatively fast, reaching more than 95% incidence in 3 to 13 years after onset of the first symptoms (Gottwald, 2010).

Chemotherapy, thermotherapy, cultural, and biological control methods are used to manage citrus greening. Strict application of quarantine measures help to minimize the spread of this disease (Stover and McCollum, 2011; Iftikhar et al., 2014). Seven weeks application of $ZnSO_4 + MnSO_4$ (3:1) is effective for induction of resistance and growth in the infected trees (Ahmad et al., 2011). Tetracycline hydrochloride and penicillin carbendazin are effective in controlling the disease. Dimethoate EC (44%), melathion EC (50%), and carbofuran FP (40.64%) show economically good control of *Citrus psylla*. Fungal mycelia *Paecilomyces fumosoroseus* and *Hirsutella citriformis* are used to control or kill the vector (Iftikhar et al., 2014). Methyl salicylate-based traps to control vector has been formulated for commercial exploitation (Stelinski et al., 2013).

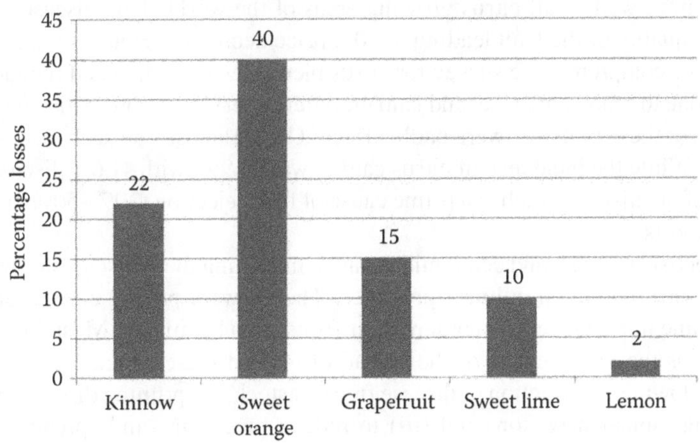

FIGURE 20.7 Percentage losses caused by citrus greening disease in various citrus cultivars. (Data taken from Ahmad K, et al. 2011. *Afr. J. Microbiol. Res.* 5: 4967–4979.)

It is necessary to determine pathogen seasonal transmission for developing proper management strategies. Location based confirmation of dimethoate and monocrotophos is necessary for their effectiveness. Similarly, evaluation of different traps to control *C. psyilla* is mandatory in Pakistan to develop sound strategy against citrus greening disease. Replacement of outdated insecticides with new formulation for controlling vector of citrus greening disease could be handy to control disease and stop development of fungicide resistance in the pathogens. Investigation of biochemical markers that may be helpful for detecting resistant source and for management of disease infected plants. Test plots of citrus plants should be established to evaluate new methods of scouting and therapy. Exploitation of DNA sequence of different citrus cultivars may be powerful tool in future to develop strong management strategies against citrus greening and vector. It is the need of the hour to establish in vitro evaluation techniques to study bacterium genome, which will provide new approach for its successful management. Support of small-scale projects to develop different strategies to avoid potential threats of citrus greening in future could be very important.

20.12 CITRUS BLEMISHES

More than 96% citrus produces in Punjab while Sargodha district contributes 43% in citrus production. Kinnow (Mandarin) is the most popular variety grown in Sargodha. Pakistan is ranked the 10th in citrus production worldwide. Its total production in Pakistan is 2.334 million tons and export of citrus is 592 thousand tons. Area under cultivation of citrus in Pakistan is 195 million hectares (FAO STAT, 2012) and is the 6th largest producer of Kinnow (Syed, 2010).

Many factors contributing towards its low quality and quantity of citrus fruits but the most important limiting factors are diseases and insects (Cernadas and Benedetti, 2009). Sudden climatic changes favor the emergence of different diseases, like citrus canker, scab, and mellanose, and so forth, which badly affect the quality and quantity of fruit. These diseases produce blemishes on fruit and become a leading threat to the citrus industry as it reduced the fruit quality, which in turn results in decrease in foreign exchange of the country. Among these diseases, citrus canker is the most destructive one (Ware, 2015) and playsa foremost role in decreasing international exports of citrus of Pakistan. During 2015, the European Union rejected our citrus consignments due to citrus canker disease and Pakistan had to bear huge economic losses (Pervaiz, 2015). Citrus canker causes 5%–30% losses worldwide (Spreen et al., 2003), while in Pakistan it ranges from 30%–40% (GOP, 2013).

Incidence of citrus melanose, caused by *Diaporthe citrii*, increased from 16% to 22.6% in different citrus producing areas of the world from 2003 to 2010, and it now becomes a potential threat in Pakistan, as well as all citrus growing areas of the world. This disease results in serious reductions in the quality of the fruit leading to 80% price decreases (Hyun et al., 2013). Malik et al. (2014) conducted a comprehensive survey for citrus blemishes in Punjab and recoded that 47.5% of blemishes were due to scab, melanose, and canker. After observing 33 orchards, they concluded that 77% of the fruits of these orchards were badly affected by melanose and scab diseases with a 15.6% disease severity. While the incidence of citrus canker was 10.9%, with 41.6% disease severity, it is a potential cause of blemishes, which is a prime cause of fruit rejection (20%–50%) and is responsible for decreased exports.

The most effective, reliable, and economical way of managing these diseases is the use of resistant cultivars along with integrated cultural practices. The growers paying careful attention to their orchards is of prime importance to safeguard their fruits from blemishes. Most farmers are absentee landlords, which is the main reason of the spread of these diseases. Incidence of diseases can be minimized by following: (i) selection of disease free rootstock, (ii) pruning disease infected branches and twigs during monsoon season, and (iii) to reduce inoculums and spread of these diseases with applying copper-based fungicides periodically along with insecticides (Das, 2003). Among chemicals, Score was effective against *E. fawacettii* (Malik et al., 2014), while copper oxichloride and copper hydroxide work against canker and scab (Das, 2003). Amistar was also effective against melanose (Nelson, 2008). In citrus fruit assay, Topsin-M, Nativo, and Score exhibited prominent results against melanose/scab (Rehman et al., 2013).

Citrus cultivars with enhanced levels of antifungal/antibacterial activity screened and should be improved through different genetic/molecular approaches. Development of biochemical markers for reliable identification of scab, melanose, and canker are needed. These markers will help breeders to develop resistant varieties and chemist to develop bactericides and fungicides which will be more effective against these diseases.

Identification of genes responsible for production of antimicrobial compounds will be helpful to understand biochemical defense and management of these diseases. There is a need to investigate depository of secondary compounds of fungicides from higher plants, which may serve as template for new formulations of fungicides/bactericides. Because new strains of bacterium and fungi associated with canker, scab, and melanose are emerging continuously due to mutations, it is necessary to investigate and identify these strains for proper management.

It is necessary to investigate the most economical, quick, and reliable methods to create systemic acquired resistance in citrus plants against citrus diseases. The government should announce benefits for farmers who produce blemishe-free citrus fruits. The government should develop a mobile expert team for each citrus growing area to facilitate awareness of farmers about blemishes, management practices, packing materials, etc. Strict application of quarantine measures is necessary to avoid the import of any disease affected citrus germplasm in Pakistan. The government should support farmers for the establishment of concrete walls/fences to decrease wind speeds to avoid citrus blemishes, as wind facilitates the dispersal of the pathogens of canker, scab, and melanose in orchards. Development of citrus nursery in canker-free areas like Layya and Bhakkar is also helpful to manage this disease.

20.13 CHICKPEA WILT

Chickpea (*Cicer arietinum* L.) belonging to legume family Fabaceae is an autogamous diploid crop with chromosome number $2n = 2x = 16$. Pakistan is the world's third largest producer of chickpea with 8.7% of world total chickpea production (FAO STAT, 2012). Chickpea is an important pulse crop, and it is highly affected with *Fusarium* wilt disease every year, caused by *F. oxysporum* f. sp. *ciceris* (*Foc*). The crop is grown in different arid and semiarid areas of Punjab and Sindh provinces. *Foc* attacks plants during their seedling and flowering stages. Losses up to 61% and 43% are reported whether the attack occurs at the seedling or postanthesis stage, respectively. On average, 10%–50% yield losses are observed every year due to *Foc* pathogen attacks on chickpea (Khan et al., 2002; Subhani et al., 2011). Haware (1990) reported that annual yield losses of 10%–15% are a regular.

Foc is assumed to contain various strains, as globally it has eight pathogenic strains, that is, 0, 1A, 1B/C, 2, 3, 4, 5 and 6, but very little is known about those present in Pakistan, or about their geographical distribution. At least 8 races of this fungus have been reported (Haware and Nene, 1982). Races 0, 1A, 1B/C, 2, 3, 4, 5, and 6 have been described to date. Races 1A, 2, 3, 4, 5, and 6 are wilting pathotypes. While races 0 and 1A/B are yellowing pathotypes (Ahmad et al., 2014).

However, no information is available on the geographical distribution of these races in Pakistan, though variation in isolates of the fungus, collected from different sites, have been noticed.

It is widely distributed among the chickpea growing areas in Pakistan (Khan et al., 2002). Among fungal diseases in Pakistan, *Fusarium* wilt is the most serious disease and has significant economic importance. For the last 4 to 5 years, due to the prevalence of drought conditions in the country, this disease has emerged as a disastrous problem for chickpea. It has high saprophytic ability in the soil. The pathogen can survive for a long period during which it undergoes different biological competition and environmental stresses, which may lead to the existence of physiologic race (Bendre and Barhate, 1998). Soil temperatures between 22 and 26°C are considered best for *Foc* infection (Ayyub, 2001).

Wilting in chickpea can be observed 20–25 days after sowing. The disease appears at the seedling (September–October) and reproductive stages (February–March) under field conditions. The infected plants do not show external rotting and look healthy. When splitting their roots vertically from the collar region downward, brown discolorations of the vascular tissues appear. The main symptoms of the disease are drying and yellowing of leaves from the base on upward, improper branching, drooping of petioles and rachis, browning of vascular bundles, withering and wilting of plants (Bendre and Barhate, 1998).

The most important and reliable way to overcome the losses due to wilt is the use of resistant cultivars. It was reported that resistance in chickpea cultivars was due to a single recessive allele. The allele was found at the same locus among all the six tested resistant lines in subsequent generations after crossing with single a susceptible parent (Mahmood et al., 2011). The pathogen is susceptible to fungicides related to different chemical classes. Subhani et al. (2011) tested a batch of six fungicides against the pathogen using poison food technique under in vitro trials and found that all of the fungicides were able to reduce the growth of mycelia by more than 70% at a concentration of 5 ppm. Similarly, plant extracts also have efficacy to control the growth of the *Foc* pathogen. Mukhtar (2007) tested the extracts of four plant species (*Azadaracta indica*, *Ocimum sanctum*, *Parthenium hysterophorus* and *Datura metel*) at various concentrations to measure their antifungal strength. It was concluded that among these extracts *D. metel* was active at a 40% concentration and inhibited the growth of fungi up to 80% under in vitro trials, which was almost similar to the fungicide Benomyl used as a control in the experiment.

Plants have a complex genetic makeup in which some genes of resistance are activated upon priming with some biochemicals. A similar approach was adopted by Sarwar et al. (2005) who used salicylic acid to induce resistance in chickpea plant against necrotrophic fungi. External application of salicylic acid at a concentration of 1.0–1.5 mM as a seed treatment significantly reduced *Foc* damage in chickpea plant which was susceptible nature genetically. Mahmood et al. (2015) checked the comparative efficacy of biocontrol agents and some selected chemicals and concluded after dual cultures of antagonists with *Foc* that *Pseudomonas flourescens* suppressed the growth of *Foc* up to 70%, followed by *Trichoderma harzianum* with 63.95% suppression. Among the chemicals, carbendazim was more effective to reduce the growth of *Foc*. Similar results were also observed in greenhouse experiments.

Foc is main pathogen of chickpea and has a great impact on the productivity of crops in Pakistan. Eight pathogenic races with varying degrees of virulence are reported in the world causing wilt diseases; however, Pakistani isolates remain poorly understood, lacking virulence profiling, geographical distribution, and relative frequency of abundance. Previously, preliminary reports had been documented by several researches that suggested the prevalence of this pathogen in the country, but molecular-level research was needed. Normally, chickpea competes during the sowing season with major crops like wheat but—due to less output in the form of products—most farmers have shifted to growing of wheat. This occasional sowing of chickpea gives a crop rotation, but for this fungus the long persistency of its spores is reported and disease epidemic can be seen even after a patch of six years. The fungus produces conidiospores, which can persist for 5 to 6 years in the absence of chickpea crop. So practically, crop rotation is of little use in the control of this devastating pathogen. The available cultivars vary in susceptibility and we do not have varieties offering complete resistance. The possible reason for this has been the over focus of investment of energy and resources on the major staple food crops like wheat, maize and rice and so on during the past few decades. Naturally, in self-pollinated crops, the genetic material remains conserved as there are little chances of genetic reshufflings, as compared to cross-pollinated crops. So efforts are required to develop resistant cultivars along modern and innovative lines leading to high yields. For this purpose marker assisted breeding and selection based on molecular markers is essential and should be adopted by the research institutes.

In the present era, several disease management strategies have been employed for chickpea, such as chemical and biological controls. After screening of available germplasms, resistant varieties are being produced from time to time but the necessary variability is lacking, therefore, there is a need to introduce exotic genotypes. Other possible efforts could be genetic transformation of disease resistant genes from distant relatives for the improvement of yield potential of the crop. This could be achieved after optimizing transformation procedure in this crop in the country.

20.14 ASCOCHYTA BLIGHT OF CHICKPEA

Chickpea is one of the most important food legumes, pulse crops, which accounts for 80% of the total area under pulses. The prodcution is the highest among all pulses. Due to its high protein content, it is often regarded as the alternate to poultry meats. The protein of chickpea is of superior quality and varies from 19% to 21% depending upon the type of cultivar, while it also contains a larger portion of carbohydrates (up to 60%) (Roy et al., 2001).

During 2014–2015, the crop was grown on an area of 960 thousand hectares with total production of 484 thousand tons. While in the last year, the production was 399 thousand tons from an area of 950 thousand hectares (Anonymous, 2014–2015). The better production in 2014–2015 is considered due to favorable environmental conditions. The other main factor is deployment over more area as compared to last year. The crop suffered from many diseases including wilt and blight caused by *F. oxysporum* f.sp. *ciceris* and *Ascochyta rabiei*. The later is the most damaging disease. In the areas with low temperatures, chickpea crop is more vulnerable to blight disease (Bokhari et al., 2011). Blight disease attacks the foliar parts of the plant and could be managed by the application of chemicals (Malik et al., 1991). The chickpea crop in Pakistan is mainly grown in the Thal region, which depends on the rainfall for moisture needs of the plant with low input seed conditions, so application of the fungicide as seed treatment is not practiced (Chaudhry et al., 2006). The main emphasis is on the need of genetic resistance of the crop against blight disease, which is reported based on the artificial screening experiments are spread of natural inoculum of the fungus during the growing season of the crop (Ilyas et al., 2007). Ali et al. (2013a,b,c) investigated the natural resistance of chickpea cultivars against blight disease by conducting trials on many local germplasm containing ten genotypes. They looked for the sources of resistance against this fungal pathogen and could not find a single complete resistant line. The study was based on correlation and combining ability experiments and revealed partial resistance in some of the lines tested.

The plant's natural resistance, in the form of resistance conferring genes, is considered the safest, and most reliable source of resistance, which reduces further costs of production compared to other management methods. Ahmad et al. (2013) reported the existence of genetic resistance in chickpea germplasm in a disease screening experiment in field conditions and found that out of 48 lines tested, only 10 lines showed resistant response against *A. rabiei*. They further recommended including these lines in advanced chickpea breeding programs.

Singh and Reddy (1991) proposed that resistance against blight disease in chickpea was controlled by a single dominant or recessive gene. Ali et al. (2011) gave different opinion, based on a molecular marker studies, involving role of three dominant and one recessive gene with independent segregation. In other independent studies (Collard et al., 2003; Hina et al., 2008), influence of additive gene action, interallelic gene interactions, and role of QTL was established, which indicates that inheritance of resistance against Ascochyta blight in chickpea is not simply due to independent segregation of dominant or recessive genes. Therefore, a comprehensive study to find out the number of genes of chickpea involved in resistance to Ascochyta blight and genetic interaction of these genes to engineer resistant chickpea cultivars in future. The disease could be managed by alternative strategies other than chemicals. Use of alternative crops or not sowing chickpea in fields that have a history of blight disease could be helpful to minimize the risks of heavy invasion of the pathogen. Another way involves continuous monitoring of sick plants and residues having early symptoms of the disease. Burying them will also be beneficial to combat the blight invasions and would be good prevention measure at the initial stage (Navas-Cortes et al., 1995; Kaiser, 1997; Gan et al., 2006).

Chickpea blight is an important disease causing collapse of crop yields during many growing seasons in Pakistan. There are many research questions which need to be addressed. Geographical distribution of the *Ascochyta* spp. and its virulent types demands a comprehensive survey for 3–5 growing seasons amongst the core areas of chickpea cultivation. For research purposes, disease screening nursery should be establish in the key chickpea cultivation areas for real time screening of available germplasms. There is a need for international collaboration in the field of pulse research, to build capacity of researchers and to search for new ways and techniques to handle these challenging issues.

The availability of certified healthy seed is a top priority to enhance per acre yield of pulses. Cultivars that respond well to low inputs are needed as most of the crop are cultivated in arid zones and depend on rainfall as irrigation. Seed treatment with fungicides is mostly neglected by the farmers, which should be improved by providing pre-basic to the farmers. Genetic transformation and integration of disease resistant genes in the chickpea genome is big task and there are very few reports about the transformation of the chickpea plant by using tissue culture technique but the reproducibility of these experiments is questionable worldwide. So there it is a challenge and the responsibility for our plant biotechnologists to clear the way for efficient plant transformation in the chickpea plant so that in future high yielding and adapted local cultivars can be spared from this fungal disease.

20.15 OVERALL STRATEGIES AND ACTION PLANS FOR THE MANAGEMENT OF DISEASES

Strategies	Action Plans	Responsible Institutes
	Short-Term Strategies	
Disease monitoring and surveillance	• Spatiotemporal estimation of disease incidence • Determination of disease severity • Estimation of yield losses in major diseases	Pest warning department, Agri-Extension department, R&D institutions
Seed health improvement	• Proper seed certification and registration • Seed treatment for seed and soil borne diseases • Development of infrastructure for seed health testing • Seed health testing using classical and modern tools • Avoiding introduction of new pathogens by properly enforcing quarantine laws	Seed certification department, Pest warning department., Quarantine department
Judicious and recommended use of chemicals	• Identification of proper chemical pesticides • Optimization of dose and method of application • Recommendation of proper dose	R&D institutions, Pest warning department, and Agri-Extension department
Outreach programs regarding plant diseases	• Farmer's problem-oriented surveillance at plant health level • Establishment of the research activities applied in the field by keeping in mind the need of farming community • Proper feedback to the farmers through outreach programs	R&D institutions, Pest warning department, and Agri-Extension department
	Mid-term strategies	
Disease forecasting	• Correlation of disease incidence with environmental factors • Development of models for disease forecasting based of previous information • Dissemination of knowledge among the farming community using extension and information and communication technologies (ICTs).	R&D institutions, Pest warning department, Agri-Extension department, Meteorological department. and SUPARCO

Continued

Strategies	Action Plans	Responsible Institutes
Generation of provincial databases for plant pathogens and tested germplasm by using diagnostic labs	• Division of plant pathogens into different categories • Maintenance of the cultures of plant pathogens along with their molecular data (i.e., sequence of 18S or 16S genes) • Maintenance of full catalogue of the pathogens • Characterization of available germplasm against various pathogens in different crops • Pathogen zoning according to crop distribution zones	R&D institutions, Pest warning department, Agri-Extension department
Biological control strategies	• Identification of indigenous biological control resources • Evaluation of biocontrol agents against different plant diseases • Recommendation of these biological control agents for targeted controlling different diseases	Public and private R&D institutions
Development of disease resistant GMOs	• Identification of resistance genes against various diseases • Procurements of genes property rights from developers • Genetic engineering for the development of GMO plants resistant to various diseases • Evaluation of the GMO events in greenhouse and field condition	R&D institutions, provincial and federal Governments.
Integrated Disease Management	• Identification of potential approaches for different diseases • Optimization of these approaches to devise IDM package • Recommendation to the farming community	Public and private R&D institutes, Agri-Extension department
	Long-term strategies	
Resistance breeding programs	• Collection and introduction of plant germplasm • Screening of plant germplasm against different diseases • Crossing plans for development of disease resistant varieties coupled with high yield	Public and private R&D institutes of plant breeding and pathology
Development of collaborations, infrastructure, financial and technical human resources	• Promotion of new areas of research through competitive research grants • For commercialization of research, development of linkages with industry • Collaboration with national and international scientists • Spacious laboratories and classrooms will be needed to accommodate maximum number of scientists, students, and trainees	R&D institutions, provincial and federal Govts.

REFERENCES

Abad P, Gouzy J, Aury JM, Castagnone-Sereno P, Danchin EGJ, Deleury E, Perfus-Barbeoch L et al. 2008. Genome sequence of the metazoan plant-parasitic nematode *Meloidogyne incognita*. *Nat. Biotechnol.* 26: 909–915.

Adam M, Heuer H, and Hallmann J. 2014. Bacterial antagonists of fungal pathogens also control root-knot nematodes by induced systemic resistance of tomato plants. *PloS One* 9: e90402.

Afzal SN, Haque MI, Ahmedani MS, Bashir S, and Rehman A. 2007. Assessment of yield losses caused by *Puccinia striiformis* triggering stripe rust in the most common wheat varieties. *Pak. J. Bot.* 39: 2127–2134.

Afzal SN, Haque MI, Ahmedani MS, Rauf A, Munir M, Firdous SS, Rehman A, and Ahmad I. 2008. Impact of stripe rust on kernel weight of wheat varieties sown in rainfed conditions of Pakitan. *Pak. J. Bot.* 40: 923–929.

Agrios GN. 2005. *Plant Pathology* (5th edition). Elsevier-Academic Press, San Diego, CA. pp. 922.

Ahmad K, Sijam K, Hashim H, Rosli Z, and Abdu A. 2011. Field assessment of calcium, copper and zinc ions on plant recovery and disease severity following infection of huanglongbing (HLB) disease. *Afr. J. Microbiol. Res.* 5: 4967–4979.

Ahmad S, Khan MA, Sahi ST, and Ahmad R. 2013. Evaluation of chickpea germplasm against *Aschochyta rabiei* (Pass). *Lab. J. Anim. Plant Sci.* 23: 440–443.

Ahmad S, Rodriguez A, Sabir F, Khan R and Panah M. 1991. *Economic losses of wheat crops infested with yellow rust in highland Balochistan.* MART/AZR Project Research Report # 67. ICARDA Quetta. pp. 15.

Ahmad Z, Mumtaz AS, Ghafoor A, Ali A, and Nisar M. 2014. Marker Assisted Selection (MAS) for chickpea *Fusarium oxysporum* wilt resistant genotypes using PCR based molecular markers. *Mol Biol Rep.* 41: 6755–6762.

Akhtar MA, and Ahmed I. 1999. Incidence of citrus greening disease in Pakistan. *Pak. J. Phytopathol.* 11: 1–5.

Al-Adawi AO, Deadman ML, Al-Rawahi AK, Al-Maqbali YM, Al-Jahwari AA, Al-Saadi BA, Al-Amri IS, and Wingfield MJ. 2006. Aetiology and causal agents of mango sudden death decline syndrome in the sultanate of Oman. *Eur. J. Plant Pathol.* 116: 245–254.

Al-Adawi AO, Deadman ML, Al Rawahi AK, Khan AJ and Al-Maqbali YM. 2003. Diplodia theobromae associated with sudden decline of mango in the sultanate of Oman. *Plant Pathol.* 52: 419.

Ali, M. 1968. Pests and diseases of cotton in Multan. *Ann. Prog. Rep. Cotton Res. Sta., Multan*: pp. 14–15.

Ali MA, Abbas A, Azeem F, Javed N, and Bohlmann H. 2015. Plant-nematode interactions: From genomics to metabolomics. *Int. J. Agric. Biol.* 17: 1071–1082.

Ali MA, Abbas A, Kreil DP, and Bohlmann H. 2013a. Overexpression of the transcription factor RAP2.6 leads to enhanced callose deposition in syncytia and enhanced resistance against the beet cyst nematode *Heterodera schachtii* in Arabidopsis roots. *BMC Plant Biol.* 13: 47.

Ali MA, and Awan SI. 2009. Inheritance pattern of seed and lint traits in *Gossypium hirsutum* L. *Int. J. Agric. Biol.* 11: 44–48.

Ali MA, and Khan IA. 2007. Assessment of genetic variation and inheritance mode of some metric traits in cotton (*Gossypium hirsutum* L.) *J. Agric. Soc. Sci.* 4: 112–116.

Ali MA, Khan IA, and Awan SI, Ali S and Niaz S. 2008. Genetics of fibre quality traits in cotton (*Gossypium hirsutum* L.). *Aust. J. Crop Sci.* 2(1): 10–17.

Ali MA, and Nawaz MA. 2017. Advances in lime breeding and genetics. In: Khan M.M., R. Al-Yahyai and F. Al-Said (ed) *The Lime: Botany, Production and Uses*, CAB International, Oxfordshire, UK. pp. 37–53.

Ali MA, Plattner S, Radakovic Z, Wieczorek K, Elashry A, Grundler FMW, and Bohlmann H. 2013b. An Arabidopsis ATPase gene involved in nematode-induced syncytium development and abiotic stress responses. *Plant J.* 74: 852–866.

Ali MA, Wieczorek K, Kreil DP, and Bohlmann H. 2014a. The beet cyst nematode *Heterodera schachtii* modulates the expression of WRKY transcription factors in syncytia to favour its development in Arabidopsis roots. *PloS One* 9: e102360.

Ali Q, Ahsan M, Tahir MHN, Farooq J, Waseem M, Anwar M, and Ahmad W. 2011. Molecular markers and QTLs for *Ascochyta rabiei* resistance in chickpea. *Int. J. Agro Vet. Med. Sci.* 5: 249–270.

Ali Q, Iqbal M, Ahmad A, Tahir MHN, Ahsan M, Javed N, and Farooq J. 2013c. Screening of chickpea (*Cicer arietinim* L.) germplasm agaisnt aschochyta blight (*Ascochyta rabiei* (Pass.) Lan.) correlation and combining ability analysis for various quantitave traists. *J. Plant Breed Crop Sci.* 5(6): 103–110.

Ali SR, Snyder J, Shehzad M, and Khalid AR. 2014b. Associations among fungi, bacteria, and phytoplasma in trees suffering citrus decline in Punjab. *J. Agric. Technol.* 10: 1343–1352.

Alvi AK, Iqbal J, Shah AH, and Pan YB. 2008. DNA based genetic variation for red rot resistance in sugarcane. *Pak. J. Bot.* 40: 1419–1425.

Amadioha, AC. 2000. Controlling rice blast *in vitro* and *in vivo* with extracts of *Azadirachta indica. Crop Prot.* 19: 287–290.

Amrao L, Amin I, Shahid S, Briddon RW, and Mansoor S. 2010. Cotton leaf curl disease in resistant cotton is associated with a single begomovirus that lacks an intact transcriptional activator protein. *Virus Res.* 152:153–63.

Anonymous. 2005. Expert Panel on the Stem Rust Outbreak in Eastern Africa. Sounding the Alarm on Global Stem Rust: An assessment of race Ug-99 in Kenya and Ethiopia and the potential for impact in neighboring regions and beyond. Mexico City: International Maize and Wheat Improvement Center (CIMMYT). (September 8). Available at http://www.cimmyt.org/english/wps/news/2005/aug/pdf/Expert_Panel_Report.pdf

Anonymous. 2009. *Economic Survey of Pakistan, Government of Pakistan, Finance Division, Economic.* Adviser's Wing, Islamabad, Pakistan.

Anonymous. 2010. *Pakistan Statistical Yearbook 2010.* Federal Bureau of Statistics, Ministry of Economic Affairs and Statistics, Government of Pakistan, Islamabad.

Anonymous. 2012. *Pakistan Statistical Yearbook 2012.* Federal Bureau of Statistics, Ministry of Economic Affairs and Statistics, Government of Pakistan, Islamabad.

Anonymous. 2014–2015. *Agricultural Statistics of Pakistan, Ministry of Food, Agriculture and Cooperative,* Islamabad, Pakistan.

Ansar M, Saleem A, and Iqbal A. 1994. Cause and control of guava decline in Punjab (Pakistan). *Pak. J. Phytopathol.* 6: 41–44.

Anwar A, Bhat GN, and Singhara GN. 2002. Management of sheath blight and blast in rice through seed treatment. *Ann. Plant Protect. Sci.* 10: 285–287.

Anwar SA, and Mckenry MV. 2012. Incidence and population density of plant-parasitic nematodes infecting vegetable crops and associated yield losses. *Pak. J. Zool.* 44: 327–333.

Arauz LF. 2000. Mango anthracnose: Economic impact and current options for integrated management. *Plant Dis.* 84: 600–611.

Armstrong GM, and Armstrong JK. 1975. Reflections on the wilt *Fusaria. Annu. Rev. Phytopathol.* 13: 95–103.

Arshad HMI, Khan JA, and Jamil FF. 2008. Screening of rice germplam against blast and brown spot disease. *Pak. J. Phytopath.* 20(1): 52–7.

Ashraf S, Khan GA, Ali S, Iftikhar M, and Mehmood N. 2014. Managing insect pests & diseases of citrus: On farm analysis from Pakistan. *Pak. J. Phytopathol.* 26: 301–307.

Atkinson HJ, Urwin PE, Hansen E, and Mcpherson MJ. 1995. Designs for engineered resistance to root-parasitic nematodes. *Trends Biotechnol.* 13: 369–374.

Ayazpour K, Hasanzadeh H, and Arabzadegan MS. 2010. Evaluation of the control of citrus nematode (*Tylenchulus semipenetrans*) by leaf extracts of many plants and their effects on plant growth. *Afr. J. Agric. Res.* 5: 1876–1880.

Ayyub MA. 2001. Evaluation of Chickpea Germplasm, Fungitoxicants, Organic and Inorganic Materials for the Management of ilt (*F. oxysporum* f. sp. *ciceriss*). *PhD thesis*, University of Agriculture, Faisalabad.

Babar LK, Iftikhar T, Khan HN, and Hameed MA. 2011. Agronomic trials on sugarcane crop under Faisalabad conditions, Pakistan. *Pak. J. Bot.* 43: 929–935.

Bajwa R, Arshad J, and Saleh A. 2003. Extend of shisham (*Dalbergia sissoo* Roxb.) decline in Sialkot, Gujaranwala, Lahore and Sargodha districts. *Mycopathology* 1: 1–5.

Bajwa R, and Javiad A. 2007. Integrated disease management to control shisham (*Dalbergia sissoo* Roxb.) decline in pakistan. *Pak. J. Bot.* 39: 2651–2656.

Bakhshi BK. 1954. Wilt of Shisham (*Dalbergia sissoo* Roxb.) due to *Fusarium solani* Sensu. Snyder and Hansen. *Nature* 174: 278–291.

Bender GS, Menge JA, Ohr HD, and Burns RM. 1982. Dry root rot of citrus: Its meaning for the grower. *Calif. Citroger.* 67: 249–254.

Bendre NJ, and Barhate BG. 1998. A souvenir on Disease Management in Chickpea. M.P.K.V., Rahuri during December 10, 1998.

Bhutta AR, and Bhatti MAR. 1983. Incidence of bacterial blight of cotton and reaction of different cultivars to *Xanthomonas campestris* pv. *malvacearum*. *Pak. Cottons.* 27: 75–78.

Bokhari AA. 2009. Studies on Guava Decline and Disease Management. *PhD thesis*, University of Agriculture, Faisalabad, Pakistan. pp. 44.

Bokhari AA, Ashraf M, Rehman A, Ahmad A, and Iqbal M. 2011. Screening of chickpea germplasm agaisnt Aschochyta blight. *Pak. J. Phytopathol.* 23: 5–8.

Briddon RW, and Markham PG. 2000. Cotton leaf curl virus disease. *Virus Res.* 71: 151–159.

Briddon RW, Mansoor S, Bedford ID, Pinner MS, Saunders K, Stanely J, Zafar Y, Malik KA and Markham PG. 2001. Identification of DNA components required for induction of cotton leaf curl disease. *Virology* 285: 234–243.

Briddon RW, Pinner MS, Stanley J, and Markham PG. 1990. Geminivirus coat protein replacement alters insect specificity. *Virology* 177: 85–94.

Cai JH, Xie K, Lin L, Qin BX, Chen BS, Meng JR, and Liu YL. 2010. Cotton leaf curl multan virus newly reported to be associated with cotton leaf curl disease in China. *New Dis. Rep.* 20: 29.

Carter WW. 1981. The effect of *Meloidogyne incognita* and tissue wounding on severity of seedling disease of cotton caused by *Rhizoctonia solani*. *J. Nematol.* 13: 374–376.

Casson ET, Richardson PE, Brinkerhoff LA, Gholson RK. 1977. Histopathology of immune and susceptible cotton cultivars inoculated with *Xanthomonas campestris* pv. *malvacearum*. *Phtopathology.* 67: 195–196.

Cernadas RA, and Benedetti CE. 2009. Role of auxin and gibberellin in citrus canker development and in the transcriptional control of cell-wall remodeling genes modulated by *Xanthomonas axonopodis* pv. *citri*. *Plant Sci.* 177: 190–195.

Chaudhry MA, Ilyas MB, and Malik KB. 1999. Identification of physiological strains of red rot of sugarcane in Pakistan. *Proc. 2nd Nat. Conf. P.P. UAF.*, Faisalabad, Pakistan, pp. 253–257.

Chaudhry MA, Muhammad F, and Afzal M. 2006. Screening of chickpea germplasm against *Fusrium* wilt. *J. Agric. Res.* 44: 307–312.

Chen XM. 2005. Epidemiology and control of stripe rust [*Puccinia striiformis* f. sp. *tritici*] on wheat. *Can. J. Plant Pathol.* 27: 314–337.

Collard BCY, Pang ECK, Ades PK, and Taylor PWJ. 2003. Preliminary investigations of QTL associated with seedlings resistance to Ascochyta blight from *Cicer echinospermum*, a wild relative of chick pea. *Theor. Appl. Genet.* 107: 719–729.

Dababat AA, and Sikora RA. 2007. Induced resistance by the mutualistic endophyte, *Fusarium oxysporum* strain 162 toward *Meloidogyne incognita* on tomato. *Biocontrol Sci. Technol.* 17: 969–975.

Das AK. 2003. Citrus canker: A review. *J. Appl. Hort.* 5: 52–60.

Diaz-Sobac R, Perez-Florez L, and Vernon-Carter EJ. 2000. Emulsion coatings control fruit fly and anthracnose in mango (*Mangifera indica* cv. Manila). *J. Hort. Sci. Biotechnol.* 75: 126–128.

Din GM, Ali MA, Naveed M, Abbas A, Anwar J, and Tanveer MH. 2017. Effect of leaf rust disease on various morpho-physiological and yield attributes in bread wheat. *Pak. J. Phytopathol.* 29: 117–128.

Doohan FM, Brennan J, and Cooke BM. 2003. Influence of climatic factors on *Fusarium* species pathogenic to cereals. *Eur. J. Plant Pathol.* 109: 755–768.

Dreisigacker S. 2004. Genetic Diversity in Elite lines and Land races of CIMMYT Spring Bread Wheat and Hybrid Performance of Crosses among Elite Germplasm. PhD thesis, Faculty of Agriculture, University of Hohenheim, Gemany.

Duncan LW. 2005. Nematode parasites of citrus. In: Luc M., R.A. Sikora and J. Bridge (eds.) Plant parasitic nematodes in subtropical and tropical agriculture. 2nd ed. CABI Publishing, Wallingford, UK, pp. 593–607.

Dwivedi SK. 1992. Effect of culture filtrate of soil microbes of pathogens inciting wilt disease of guava (*Psidium guajava* L.) under *in vitro* conditions. *Natl. Acad. Sci. Lett.* 15: 33–35.

Dwivedi SK. 1993. Fungitoxicity of *Foeniculum vulgare* seed oil used against a guava wilt pathogen. *Natl. Acad. Sci. Lett.* 16:207–208.

El-Borai FE, Duncan LW, and Graham JH. 2002. Infection of citrus roots by *Tylenchulus semipenetrans* reduces root infection by *Phytophthora nicotianae*. *J. Nematol.* 34: 384–389.

FAO STAT. 2012. Citrus fruit fresh and processed. Food and agriculture organization of the United Nations. Annual statistics. CCP:CI/ST/2012.

Farih A, Menge JA, Tsao PH, and Ohr HD. 1981. Metalaxyl and efosite aluminum for control of Phytophthora Gummosis and root rot on citrus. *Plant Dis.* 8: 654–657.

Gan YT, Siddique KHM, Mac Leod WJ, and Jayakumar P. 2006. Management options for minimizing the damage by Aschochyta blight (*Aschochyta rabiei*) in chickpea (*Cicer arietinum* L.). *Field Crops Res.* 97: 121–134.

Ghazanfar MU, Wakil W, Sahi ST, and Saleem-il-Yasin. 2009. Influence of various fungicides on the management of rice blast disease. *Mycopath.* 7: 29–34.

Gilbert MJ, Soanes DM, and Talbot NJ. 2004. Functional Genomic Analysis of the Rice Blast Fungus Magnaporthe grisea. *Appl. Mycol. and Biotechnol.* 4: 331–352.

Gilbertson RL, Brown WM, and Ruppel EG. 1985. Prevalence and virulence of Fusarium species associated with stalk rot of corn in Colorado. *Plant Dis.* 69: 1065–1068.

Gomes VM, Souza RM, Mussi-Dias V, Silveira SF, and Dolinski C. 2011. Guava decline: A complex disease involving *Meloidogyne mayaguensis* and *Fusarium solani*. *J. Phytopathol.* 159: 45–50.

GOP. 2013. *Economic Survey of Pakistan 2012–2013.* Ministry of Finance, Government of Pakistan, Islamabad, Pakistan.

Gottwald TR. 2010. Current epidemiological understanding of citrus huanglongbing. *Ann. Rev. Phytopathol.* 48: 119–139.

Hafiz A. 1986. *Plant Diseases*. Directorate of Publication, Pakistan Agricultural Research Council, Islamabad, Pakistan. pp. 552.

Hajano J, Lodhi AM, Pathan MA, Khanzada MA, and Shah GS. 2012. *In vitro* evaluation of fungicides, plant extracts and bio-control agents against rice blast pathogen *Magnaporthe oryzae* Couch. *Pak. J. Bot.* 44: 1775–1778.

Haq IM, Fadnan M, Jamil FF, and Rehman A. 2002. Screening of rice germplasm against *Pyricularia oryzae* and evaluation of various fungitoxicants for control of disease. *Pak. J. Pythopath.* 14: 32–35.

Hassan MH, Afghan S, Hassan Z, and Hafeez FY. 2014. Biopesticide activity of sugarcane associated rhizobacteria: Ochrobactrum intermedium strain NH-5 and *Stenotrophomonas maltophilia* strain NH-300 against red rot under field conditions. *Phytopathol. Mediterr.* 53: 229–239.

Hassan SF, Hussain M, and Rizvi SA. 1979. Wheat diseases situation in Pakistan. Paper presented at National Seminar of Wheat Research and Production, August 6–9, Islamabad.

Haware MP, and Nene YL. 1982. Races of *Fusarium oxysporum* f. sp. *ciceri*. *Plant Dis.* 66:809–810.

Haware MP. 1990. Fusarium wilt and other important diseases of chickpea in the Mediterranean area. *Options Mediterr. Ser. Semin.* 9: 163–166.

Hillocks RJ. 1986. Localized and systemic effects of root-knot nematode on incidence and severity of *Fusarium* wilt in cotton. *Nematologica* 32: 202–208.

Hina A, Iqbal N, Haq MA, Shah TM, Atta BM, and Hameed A. 2008. Detection of QTLs for blight resistance in chickpea genotypes with DNA based markers. *Pak. J. Bot.* 40: 1721–1728.

Hodson DP, Nazari K, Park RF, Hansen J, Lassen P, Arista J, Fetch T et al. 2011. Putting Ug99 on the map: An update on current and future monitoring. In: *Proceedings of BGRI Technical Workshop*, June 13–16, 2011, St. Paul, Minnesota, USA.

Huang X, Zhai J, Luo Y, and Rudolph K. 2008. Identification of a highly virulent strain of *Xanthomonas axonopodis* pv. *malvacearum*. *Eur. J. Plant Pathol.* 122: 461–469.

Huerta-Espino J, Singh R, German S, McCallum B, Park R, Chen W, Bhardwaj S, and Goyeau H. 2011. Global status of wheat leaf rust caused by *Puccinia triticina*. *Euphytica* 179: 143–160.

Hussain T, and Ali M. 1975. A review of cotton diseases of Pakrstan. *Pak. Cott.* 19: 71–86.

Hussain M, Hassan SF, and Kirmani MAS. 1980. Virulence in Puccinia recondite Rob.ex. Desm. f. sp. tritici in Pakistan during 1978 and 1979. *Proceedings of the 5th European and Mediterranean Cereal Rust Conference*, Bari, Italy. 179–184.

Hussain T, Mehmood T, Ali L, Bhatti NN, and Ali V. 1985. Resistance of some cotton lines to bacterial blight in Pakistan. *Trop. Pest Manag.* 31: 73–77.

Hussain T, and Tahir M. 1993. Chemical control of bacterial blight of cotton. *Pak. J. Phytopathol.* 5: 119–121.

Hyun J-W, Yi P-H, Hwang R-Y, and Moon K-H. 2013. *Aspect of Incidence of the Major Citrus Diseases Recently*. pp. 102–107. Korean Society of Plant Pathology. http://koreascience.or.kr/article/ArticleFullRecord.jsp?cn=SMBRCU_2013_v19n2_102

Idris AM, and Brown JK. 2000. Identification of a new, monopartite begomovirus associated with leaf curl disease of cotton in Gezira, Sudan. *Plant Dis.* 84: 809.

Iftikhar T, Babar LK, Zahoor S, and Khan NG. 2010. Best irrigation management practices in cotton. *Pak. J. Bot.* 42: 3023–3028.

Iftikhar Y, Saeed R, Umbreen S, and Muhammad Awais Z. 2014. Huanglongbing: Pathogen detection system for integrated disease management—A review. *J. Saudi Soc. Agric. Sci.* 15: 1–11.

Ilyas MB, Chaudhry MA, Javed N, Ghazanfar MU, and Khan MA. 2007. Sources of resistance in chickpea germplasm against *Ascochyta* blight. *Pak. J. Bot.* 39: 1843–1847.

Iqbal MJ, Ahmad I, Khanzada KA, Ahmad N, Rattu A, Fayyaz M, Ahmad Y, Hakro AA, and Kazi AM. 2010. Local stem rust virulence in Pakistan and future breeding strategy. *Pak. J. Bot.* 43: 1999–2009.

Iqbal N, and Saeed S. 2012. Isolation of mango quick decline fungus from mango bark beetle *Hypocryphalous magniferae* (Coleoptera: Scolytidae). *J. Anim. Plant Sci.* 22: 644–648.

Islam MDS, Glynn JM, Bai Y, Duan YP, Coletta-Filho HD, Kuruba G, Edwin L, Civerolo EL, and Lin H. 2012. Multilocus microsatellite analysis of "Candidatus Liberibacter asiaticus" associated with citrus Huanglongbing worldwide. *BMC Microbiol.* 12: 39.

Jabeen R, Iftikhar T, Ashraf M, and Ahmad I. 2011. Virulence/aggressiveness testing of *Xanthomonas oryzae* pv. *oryza*isolates causes blb disease in rice cultivars of Pakistan. *Pak. J. Bot.* 43: 1725–1728.

Jabeen R, Iftikhar T, and Batool H. 2012. Isolation, characterization, preservation and pathogenicity test of *Xanthomonas oryzae* pv. *Oryzae* causing BLB disease in rice. *Pak. J. Bot.* 44: 261–265.

Jones MGK, and Payne HL. 1978. Early stage of nematode-induced giant-cell formation in roots of Impatiens balsamina. *J. Nematol.* 10: 70–84.

Kaiser WJ. 1997. Inter and International spread of Aschochyta pathogen of chickpea, faba bean and lentil. *Can. J. Plant Pathol.* 19: 215–224.

Khan AH. 1989. *Pathology of Trees*, Vol. II. University of Agriculture, Faisalabad.

Khan MA. 1996. Relationship of *Xanthomonas campestris* pv. *malvacearum* population to development of symptoms of bacterial blight of cotton. *Pak. J. Phytopathol.* 8: 152–155.

Khan MA, and Rashid A. 1997. Identification of resistant sources from cotton germplasm against bacterial blight and leaf curl virus disease. *Pak. J. Agri. Sci.* 34: 26–31.

Khan MA, and Ilyas MB. 1999. Cotton germplasm response of slow blighting against *Xanthomonas campestris* pv. *malavacearum* and slow curling against CLCuV infection. Proc 2nd Nat. conf. Plant Pathology. Sep. 27-29, University of Agriculture, Faisalabad, Pakistan, pp. 138–139.

Khan H, Awais M, Raza W, and Zia A. 2011. Identification of sugarcane lines with resistance to red rot. *Pak. J. Phytopathol.* 23: 98–102.

Khan IA, Alam SS, Haq A, and Jabbar A. 2002. Selection for resistant to wilt in relation with phenols in chickpea. *Int. Pieonpea Newslett.* 9: 19–20.

Khan JA, Jamil FF, Cheema AA, and Gill MA. 2001. Screening of rice germplasm against blast disease caused by *Pyricularia oryza*. In: *Proc. National Conf. of Plant Pathology, Held at NARC*, October 1–3, Islamabad, pp. 86–89.

Khan MA, and Mumtaz H. 2004. Combining yellow rust resistance with high yield in grain wheat. In: Abstracts. Second Regional Yellow Rust Conference for Central & West Asia and North Africa, March 22–26, 2004, Islamabad, Pakistan, pp. 28.

Khanzada MA, Lodhi AM, and Shahzad S. 2005. Chemical control of *Lasiodiplodia theobromae*, the causal agent of mango decline in Sindh. *Pak. J. Bot.* 37: 1023–1030.

Khoury W, Cressman K, and Yahyaoui A. 2008. *Ug99 status, management and prevension*. International Plant Protection Convention, Special session, Rome, April 9, 2008 (https://www.ippc.int/static/media/files/publications/en/1217860549444_Ug99_CPM_3_side_event_9Apr08.pdf).

Li H, Zhao J, Feng H, Huang L, and Kang Z. 2013. Biological control of wheat stripe rust by an endophytic *Bacillus subtilis* strain E1R-j in greenhouse and field trials. *Crop Prot.* 43: 201–206.

Line RF. 2002. Stripe rust of wheat and barley in North America: A retrospective historical review. *Annu. Rev. Phytopathol.* 40: 75–118.

Luo Y, Tang PS, Febellar NG, TeBeest DO. 1998. Risk analysis of yield losses caused by rice leaf blast associated with temperature changes above and below for five Asian countries. *Agric. Ecosys. Environ.* 68: 197–205.

Mahmood K, Saleem M, and Ahsan M. 2011. Inheritance of resistance to Fusarium wilt in chickpea. *Pak. J. Agri. Sci.* 48: 55–58.

Mahmood Y, Khan MA, Javed N, and Arif J. 2015. Comparative efficacy of fungicides and biological control agents for the management of chickpea wilt caused by *Fusarium oxysporum* f. *sp. ciceris. J. Anim. Plant Sci.* 25: 1063–1071.

Mansoor S, Amin I, Iram S, Hussain M, Zafar Y, Malik KA, and Briddon RW. 2003. Breakdown of resistance in cotton to cotton leaf curl disease in Pakistan. *Plant Pathol.* 52: 784.

Malik A, Khan IA, Rehman A, Yasin M, Bashir H, Ahsan M, Aleem S, Saleem B, and Khan AA. 2014. *Kinnow quality issues and strategies for improvement.* Survey Report and Citrus Blemishes Resource Guide.

Malik MR, Iqbal SM, and Malik BA. 1991. Economic loses of *Ascochyta* blight in chickpea. *Sarhad J. Agric.* 8: 765–768.

Mansfield J, Genin S, Magori S, Citovski S, Anum MS, Ronald S, Dow M et al. 2012. Top 10 plant pathogenic bacteria in molecular pathology. *Mol. Plant Pathol.* 13: 614–629.

Masood A, Saeed S, Da Silveira SF, Akem CG, Hussain N, and Farooq M. 2011. Quick decline of mango in Pakistan, survey and pathogenicity of fungi isolated from mango tree and bark beetle. *Pak. J. Bot.* 43: 1793–1798.

McGrath MT. 2004. What are Fungicides? The Plant Health Instructor. http://www.doi.org/10.1094/PHI-I-2004-0825-01

McIntosh, RA, Welling CR, and Park RF. 1995. *Wheat rusts: An Atlas of Resistance Genes.* CSIRO, Melbourne, Australia; Kluwer Acad. Publ., Dordrecht, the Netherlands.

Mehrotra MD, Punam S, and Singh P. 1998. Study on seed borne fungi of some forest trees and their management. *Indian J. Forest.* 21: 345–354.

Mehrotra RS. 1998. *Plant Pathology*. Tata MC Graw-Hill Co. Ltd., New Delhi.

Mehta PR. 1971. Role of fungicides in modernizing agriculture in India. *Indian Phytopath.* 24: 235–246.

Mew TW, and Gonzales P. 2002. *A Handbook of Rice Seedborne Fungi*. International Rice Research Institute, Los Banós, Philippines. pp. 83.

Moorman GW, Huang JS, and Powell NT. 1980. Localized influence of *Meloidogyne incognita* on *Fusarium* wilt resistance of fluecured tobacco. *Phytopathology* 70: 969–970.

Mukhtar I. 2007. Comparison of phytochemical and chemical control of *Fusarium oxysporum* f. sp. *ciceri*. *Mycopathology* 5: 107–110.

Munir A, Roshan A, and Fasihi SD. 1986. Effect of different infection levels of red rot of sugarcane on cane weight and juice quality. *J. Agric. Res.* 24: 129–131.

Navas-Cortes JA, Trapero-casas A, and Jimens-Diaz RM. 1995. Survival of *Didymella rabiei* in chickpea stra debris in Spain. *Plant Pathol.* 44: 332–339.

Nelson S. 2008. Citrus Melanose. Department of Plant and Environmental Protection Sciences. Plant Disease November 2008 P-59.

Nemec S, Zablotowice RM, and Chandler JL. 1989. Distribution of *Fusarium* spp. and selected microflora in citrus soil and rhizospheres associated with healthy and blight diseased citrus in Florida. *Phytophylactica* 21: 141–146.

Ouazzani TA, Mouria A, Douira A, Benkirane R, Mlaiki A, and El-Yachioui M. 1998. *In vitro* effect of pH and temperature on the ability of *Trichoderma* spp., to reduce the growth of *Pyricularia oryzae*. *J. CAB Direct* 96: 19–24.

Pang F, Wang T, Zhao C, Tao A, Yu Z, Huang S, and Yu G. 2016. Novel bacterial endophytes isolated from winter wheat plants as biocontrol agent against stripe rust of wheat. *Bio. Control* 61: 207–219.

Park RF, Wellings CR, and Bariana HS. 2007. Preface to global landscapes in cereal rust control. *Aust. J. Agric. Res.* 58: 469.

Pathan MA, Rajput NA, Jiskani MM, and Wagan KH. 2007. Studies on intensity of shisham dieback in Sindh and impact of seed-borne fungi on seed germination. *Pak. J. Agric. Agril. Engg. Vet. Sci.* 23: 12–17.

Patino-Vera M, Jimenez B, Balderas K, Ortiz M, Allende R, Carrillo A, and Galindo E. 2005. Pilot-scale production and liquid formulation of *Rhodotorula minuta*, a potential biocontrol agent of mango anthracnose. *J. Appl. Microbiol.* 99: 540–550.

Pervaiz S. 2015. Fruit, vegetable fail to enter Eurpeon Union. A Report. http://thedailystar.net/business.

Pretorius ZA, Singh RP, Wagoire WW, and Payne TS. 2000. Detection of virulence to wheat stem rust resistance gene Sr31 in Puccinia graminis f. sp. tritici in Uganda. *Plant Dis.* 84: 203.

Purohit M, Jamaluddin, and Mishra GP. 1998. Studies on germination and seedborne fungi of some forest tree species and their control. *Indian Forest.* 124: 315–320.

Rajput, NA, Pathan MA, Jiskani MM, Rajput AQ, and Arain RR. 2008. Pathogenicity and host range of *Fusarium solani* Mart. Sacc., causing dieback of Sisham (*Dalbergia sissoo* Roxb.). *Pak. J. Bot.* 40: 2631–2639.

Rajput NA, Pathan MA, Lodhi AM, Dou D, and Rajput S. 2011. Effect of neem (*Azadirachta indica*) products on seedling growth of shisham dieback. *Afr. J. Microbiol. Res.* 5: 4937–4945.

Rajput NA, Pathan MA, Lodhi AM, Dou D, Tingli L, Arain MS, and Rajer FU. 2012. *In vitro* evaluation of various fungicides against *Fusarium solani* isolated from *Dalbergia sissoo* dieback. *Afr. J. Microbiol. Res.* 6: 5691–5699.

Rehman A, Malik AU, Yasin M, Ahsan M, Bashir H, Alam MW, Saleem B, Riaz K. 2013. Strategies for improving the Kinnow scab/malanose management plan: A major export quality issue in Pakistani citrus. https://www.apsnet.org/meetings/Documents/2015_meeting_abstracts/aps2015abP302.htm

Ribot C, Hirsch J, Balzergue S, Tharreau D, Notteghem JH, Lebrun MH, and Morel JB. 2008. Susceptibility of rice to the blast fungus, *Magnaporthe grisea*. *J. Plant Physiol.* 165: 114–24.

Roy A, Das K, Kumar J, Rao BV. 2001. A new chickpea variety for Hills zone of Assam, India. *Int. Chickpea Pigeonpea News Lett.* 8: 6–7.

Safdar A, Javed N, Khan SA, Safdar H, Haq IU, Abbas H, and Ullah Z. 2013. Synergistic effect of a fungus, *Fusarium semitectum*, and a mematode, *Tylenchulus semipenetrans*, on citrus decline. *Pak. J. Zool.* 45: 643–651.

Safdar A, Khan SA, and Safdar MA. 2015. Pathogenic association and management of *Botryodiplodia theobromae* in guava orchards at Sheikhupura district, Pakistan. *Int. J. Agric. Biol.* 17: 297–304.

Sajid M, Rahid A, Ehetisham-ul-haq M, Javed MT, Jamil H, Mudassar M et al. 2013. *In vitro* evaluation of chemicals and plant extracts against colony growth of *Xanthomonas axonopodis* pv. *malvacearum* causing bacterial blight of cotton. *Eur. J. Exp. Biol.* 3: 617–621.

Sarwar N, Hayat Zahid Ch M, Haq I, and Jamil FF. 2005. Induction of systemic resistance in chickpea against Fusarium wilt by seed treatment with salicylic acid and bion. *Pak. J. Bot.* 37: 989–995.

Sehtiya HL, Phawan AK, Virk KS, and Dendsay J. 1993. Carbohydrate metabolism in relation to *Colletotrichum falcatum* in resistant and susceptible sugarcane cultivars. *Indian Phytopathol.* 46: 83–85.

Shakeel Q, Javed N, Iftikhar Y, Haq IU, Khan SA, and Ullah Z. 2012. Association of plant parasitic nematode with vegetables crops. *Pak. J. Phytopathol.* 24: 143–148.

Sharma R, and Tamta S. 2015. A review on red rot: The "Cancer" of sugarcane. *J. Plant Pathol. Microbiol.* S1: 003.

Shazia I, Ahmad I, and Ashraf M. 2003. A study on fungi and soil borne diseases associated with rice-wheat cropping system of Punjab province of Pakistan. *Pak. J. Biological Sci.* 6: 1–6.

Singh KB, and Reddy MV. 1991. Advances in disease resistance breeding in chickpea. *Adv. Agron.* 45: 191–222.

Slusarenko AJ, Patel A, and Portz D. 2008. Control of plant diseases by natural products: Allicin from garlic as a case study. *Eur. J. Plant Pathol.* 121: 313–322.

Smiley R, and Cynthia MO. 2003. Information on wheat-stripe rust (yellow rust), an online guide to plant disease control, Oregon State University Extension. http://plant-disease.ippc.orst.edu

Sorribas F, Ornat C, Verdejo-Lucas S, Galeano M, and Valero J. 2005. Effectiveness and profitability of the *Mi*-resistant tomatoes to control root-knot nematodes. *Eur. J. Plant Pathol.* 111: 29–38.

Spreen TH, Zansler ML, and Muraro RP. 2003. The costs and value loss associated with Florida citrus groves exposed to citrus canker. *Proc. Flor. State Agric. Soc.* 116: 289–294.

Stelinski LL, Ali JG, Alborn HT, Mann R, and Pelz-Stelinski K. 2013. U.S. Patent Application 13/774,112.

Stover ED, and McCollum G. 2011. Incidence and severity of huanglongbing and Candidatus *Liberibacter asiaticus* titer among fieldinfected citrus cultivars. *HortScience* 46: 1344–1348.

Subhani MN, Chaudhry MA, Khaliq A, and Muhammad F. 2008. Efficacy of various fungicides against sugarcane red rot (*Colletotrichum falcatum*). *Int. J. Agri. Biol.* 10: 725–727.

Subhani NS, Sahi ST, Husain S, Ali A, Iqbal J, and Hameed K. 2011. Evaluation of various fungicides for the control of gram wilt caused by *Fusarium oxysporum* f. sp. *ciceris*. *Afr. J. Agri. Res.* 6: 4555–4559.

Syed R. 2010. Seedless kinnow export: Pakistan to meet European countries demand. *Daily Times*, July 15, 2010.

Tank MS. 2009. Enhancement of plant growth and decontamination of nickel-spiked soil using PGPR. *J. Basic Microbiol.* 49: 195–204.

Tao A, Pang F, Huang S, Yu G, Li B, and Wang T. 2014. Characterisation of endophytic *Bacillus thuringiensis* strains isolated from wheat plants as biocontrol agents against wheat flag smut. *Biocontrol Sci. Technol.* 24: 901–924.

Timmer LW, Graham JH, and Zitko SE. 1997. Metalaxyl-resistant isolates of *Phytophthora nicotianae*: Occurrence, sensitivity, and competitive parasitic ability on citrus. *Plant Dis.* 82: 2254–2261.

USDA. 2014. Economic Research Service, USDA, USA.

Valle N del. 1987. Citrus blight in Cuba, a review. In: *Proc. 1st Conf. Int. Symp. of Citrus Canker, Declinio/Blight and Similar Diseases*. Sao Paulo, Brazil. 1977. Zinc and water-soluble phenolic levels in the wood for the diagnosis of citrus blight. pp. 274–284.

Verdejo LS, and Mckenry MV. 2004. Management of citrus nematode, *T. semipenetrans*. *J. Nematol.* 36: 424–432.

Verma JP, Nayak ML, and Singh RP. 1977. Survival of *Xanthonionas malvacearum* under North Indian conditions. Ind. *Phytopath.* 30: 361–365.

Vessey JK. 2003. Plant growth promoting rhizobacteria as biofertilizers. *Plant Soil* 255: 571–586.

Viswanathan R, and Samiyappan R. 2002. Induced systemic resistance by fluorescent Pseudomonads against red rot disease of sugarcane caused by *Colletotrichum falcatum*. *Crop Prot.* 21: 1–10.

Wang J. 2009. The infection processes of Colletotrichum turnatum on Lentil [graduate thesis]. Saskatoon: Crop Development Centre, Department of Plant Sciences, University of Saskatchewan.

Ware M. 2015. Oranges: Health Benefits, Nutritional Information. http://www.medicalnewstoday.com/articles/272782.php

Wellings CR, and Kandel KR. 2004. Pathogen dynamics associated with historic stripe (yellow) rust epidemics in Australia in 2002 and 2003. In: *Proceedings of the 11th International Cereal Rusts and Powdery Mildews Conference*. August 2004, John Innes Centre, Norwich, UK. European and Mediterranean Cereal Foundation, Wageningen, Netherlands, Cereal Rusts and Powdery Mildews Bulletin, Abstr, A2.74. http://www.crpmb.org/icrpmc11/abstracts.htm

Westerlund FU Jr., Campbell RN, and Simble KA. 1974. Fungal root rot and wilt of chick-pea in California. *Phytopathology* 64: 432–436.

Zaidi SA, Ijaz S, Khan AI, and Rana IA. 2016. Development of source independent micropropagation system in *Dalbergia sissoo* Roxb, as a basis for germplasm conservation and disease free plants production. *Mol. Plant Breed.* 15: 1–12.

21 Trends in Sustainable Management of Emerging Insect Pests

*Muhammad Jalal Arif, Waqas Wakil,
Muhammad Dildar Gogi, Rashad Rasool Khan,
Muhammad Arshad, Muhammad Sufyan,
Ahmad Nawaz, Abid Ali, and Shahid Majeed*

CONTENTS

21.1 INTRODUCTION

Every organism in nature is striving for its survival and interacts with other species and all components of the ecosystem. These interactions include those between pests and humans, other plants, and animals. These interactions can cause many problems such as competition for space, food, and water; injury to plants, property, or animals; spread of endemic or epidemic diseases or nuisances. In any interactive agro-ecosystem, there are many forces and factors that regulate the pest population and its control (Figure 21.1). A comprehensive knowledge of these factors helps in deciding the effective strategies to cope with emerging or prevailing pest conditions (Buurma 2008; Heong et al. 2008; Pedigo and Rice 2009; Schowalter 2011). Effective pest management largely depends on a knowledge of the technology being used and biological knowledge of the pest against which the technology is being practiced (Figure 21.1). A knowledge of the technology helps to select and impliment appropriate pest management tools (e.g., insecticides, equipments). Knowledge of the biological aspects of a pest's biology highlights the appropriate place (where), timing (when), and procedure/method (how) for efficient application of any technology and allows for the economically effective management of any pest (Buurma 2008; Heong et al. 2008; Pedigo and Rice 2009). This knowledge is imperative for strengthening the impact of pest management strategies, reducing the risk of crop failure by pests, minimizing threats to environmental stability and human health, reducing operational costs and improving crop profitability by minimizing the intensity of inputs (Knipling 1979; Inayatullah 1995; Norris et al. 2002; Trivedi 2002; Sorby et al. 2005; Dhaliwal et al. 2006; Pedigo and Rice 2009; Alam 2010; Jha 2010). Most researchers suggest three basic characteristics or essentials of sustainable pest management (Figure 21.2). A detailed knowledge and feasibility of these characteristics/elements lays the foundation of a successful, efficient, economical, and ecofriendly pest management program for any pest, crop, locality, and time-period (Geier et al. 1973; Knipling 1979; Dhaliwal et al. 2006; Buurma 2008; Heong et al. 2008; Pedigo and Rice 2009; Alam 2010; Schowalter 2011). Effective and sustainable insect pest management also depends on economic decision levels, which are indispensable for determining the course of action, ensuring sensible pesticide applications, reducing significant economic damage, safeguarding the profits of the producer, and conserving the environmental quality in any pest situation (Knipling 1979; Inayatullah 1995; Norris et al. 2002; Sorby et al. 2005; Dhaliwal et al. 2006; Pedigo and Rice 2009; Alam 2010; Jha 2010).

The changing scenarios of cropping schemes/patterns, climate, ecosystems, habitat, intensified cultivation systems, constant niches for pest multiplication, inputs (e.g., imbalance fertilizers), pesticide usage, and conventional low inputs varieties to input intensive high yielding varieties/hybrids have resulted in a shift in the status of a pest and loss of intensity in time and space. Expansion in the host horizon of pests, development of resistance to pesticides, resurgence of primary and secondary pests, and replacement of major pests with the minor pests have been reported in different countries in agro-ecosystems, forest-ecosystems, orchard ecosystems, and other natural and man-made situations (Prakash et al. 2014; Rai et al. 2014; Boulanger et al. 2016). Examples from India include occurrence of chili gall midge (*Asphondylia capparis*) in Tamil Nadu and Andhra Pradesh; *Phenacoccus solenopsis* in brinjal, tomato, okra, and cucurbits; *Henosepilachna vigitioctopunctata* and *Epilachna dodecastigma* on cowpea and bitter gourd; *Sphenaeches caffer* (plume moth) in bottle gourd (Rai et al. 2014). Drastic geographical variations in the incidence and species composition of different borers, hoppers, bugs, leaf-folders, mites, and so forth on various crops have been experienced across Asia (Sain and Prakash 2008; Prakash 2013; Prakash et al. 2014; Rai et al. 2014). The composition and distribution of species, niche size, range shifts, and habitat adaptations have also been predicted to be hampered among tephritid fruit fly species due to severe variability in future climate scenarios (McKenney et al. 2007; FAO 2008; Masembe et al. 2016). In Pakistan, some insect pests have resurged in severity while some minor pests have become major pests with some considered as invasive species. For example, cotton stainer, *Dysdercus koenigii* F. (Hemiptera: Pyrrhocoridae), dusky cotton bug, *Oxycarenus* spp. (Hemiptera: Lygaeidae), and stink bugs (Hemiptera: Pentatomidae) have emerged as a great threat to cotton in Pakistan due to changes in cotton cultivation systems (Shah 2014). Maize stem borer (*Chilo partellus* (Swinhoe); Pyralidae: Lepidoptera), wheat aphid, fruit flies (*Bactrocera* spp.; Tephritidae: Diptera), armyworm (*Spodoptera* spp.: Noctuidae; Lepidoptera), Citrus psylla (*Diaphorina citri* Kuwayama (Phyllidae: Homoptera), pink bollworm (*Pectinophora gossypiella* (Saunders); Gelechiidae: Lepidoptera), and mango mealybug (*Drosicha mangifera* (Green): Margarodidae; Homoptera) are resurgent while red palm weevil has invaded. These pests have emerged as future threats to agriculture productivity.

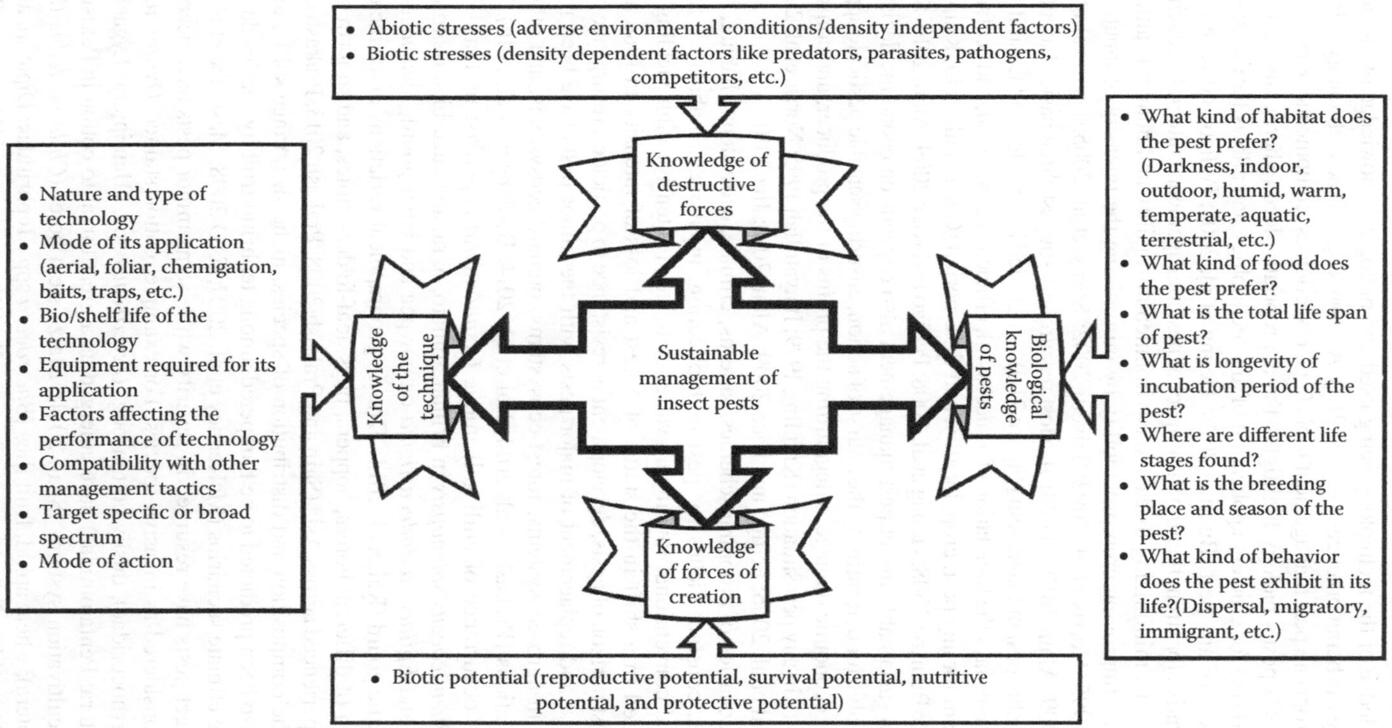

FIGURE 21.1 Schematic diagram demonstrating some prerequisite knowledge and information that lay the foundation of successful and sustainable pest management program of insect's pests.

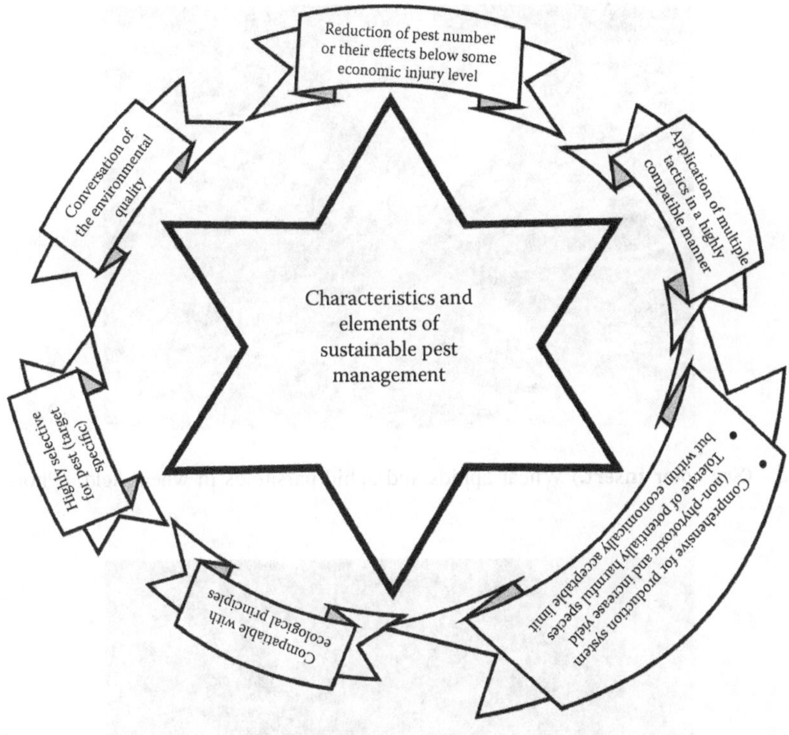

FIGURE 21.2 Diagrammatic illustration of the characteristics and elements of a sustainable pest management program in any ecosystem.

Among emerging cotton insect pests, cotton pink bollworm caused 4.5 million bales losses in 2015–2016. Overall, whitefly caused 30%–40% yield losses in cotton, sugarcane, and vegetables. Cropping patterns of wheat and oil seed crops are heavily infested by aphid species, and wheat aphid species are potential threats for stagnant wheat yields in Pakistan. Wheat aphids were reported to cause 35%–45% and 20%–80% yield losses directly and indirectly in wheat crop. Twenty to 75% and 80% yield losses were recorded as a result of fruit fly damage in melon and guava, respectively. Six percent of the total volume of Pakistani export mangoes (including 0.9%–2.4% during the previous 4 years) were rejected due to heavy fruit fly infestations. Other field crops like maize and sorghum were infested by maize stem borers which caused 50%–70% yield losses. Fruits, citrus, and dates are important crops. Citrus pyslla acts as a transmission vector for citrus greening disease and there is need to measure the exact yield losses in Pakistan citrus. To better understand the importance of citrus psylla, 10% losses were estimated (8257 lost jobs among the citrus work force) and caused a reduction of $ 2.7 billion (U.S.) revenue and $ 1.8 billion (U.S.) economic activities in the citrus industry. Heavy infestation of date palm weevil in datcs resulted in 10%–20% yield losses (last time estimated in 1992) (Figures 21.3 through 21.7).

This chapter focusses on the pest status, biology, seasonal incidence, distribution, host range, damage pattern, losses, flight behavior, and management of *C. partellus*, wheat aphid, fruit flies (*Bactrocera* spp.), *Spodoptera* spp., *D. citri*, *P. gossypiella*, *D. mangifera*, and red palm weevil, which have been reported as emerging future threats to agriculture productivity. In addition, future research areas of these emerging pests are also discussed. This chapter also accentuates the SWOT analysis (Figure 21.8) of the insect pest management situation in Pakistan and highlights the strengths, weeknesses, opportunities, and threats (SWOT) associated with insect pest management scenarios in Punjab, Pakisan. Various issues, strategies and short, medium, and long-term action-plans/ways-forward (Figure 21.9) associated with effective and sustainable insect pest management are also summarized.

FIGURE 21.3 (**See color insert.**) Wheat aphids and aphid parsitoids in wheat field. (Photo credit: M.J. Arif, UAF.)

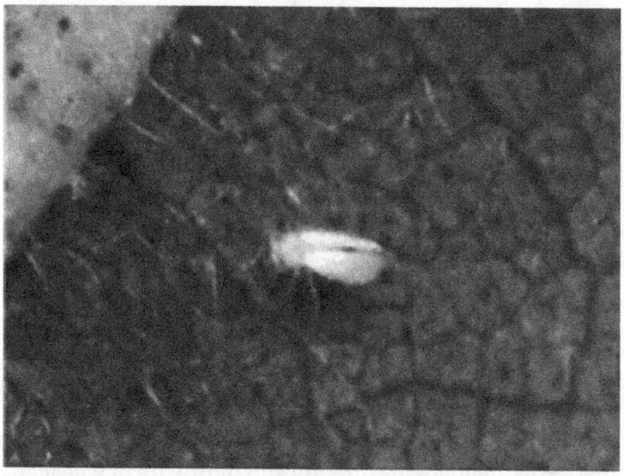

FIGURE 21.4 Adult of *Bemisia tabaci.* (Photo credit: M.J. Arif, UAF.)

FIGURE 21.5 (**See color insert.**) Leaf enation due to CLCuV transmitted by cotton whitefly. (Photo credit: M.J. Arif, UAF.)

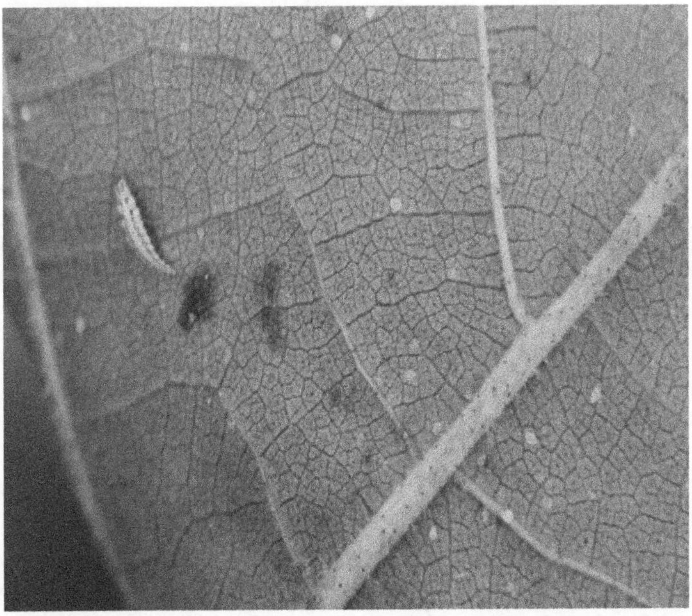

FIGURE 21.6 **(See color insert.)** Larva of *Chrysoperla carnea* feeding on whitefly nymphs. (Photo credit: M.J. Arif, UAF.)

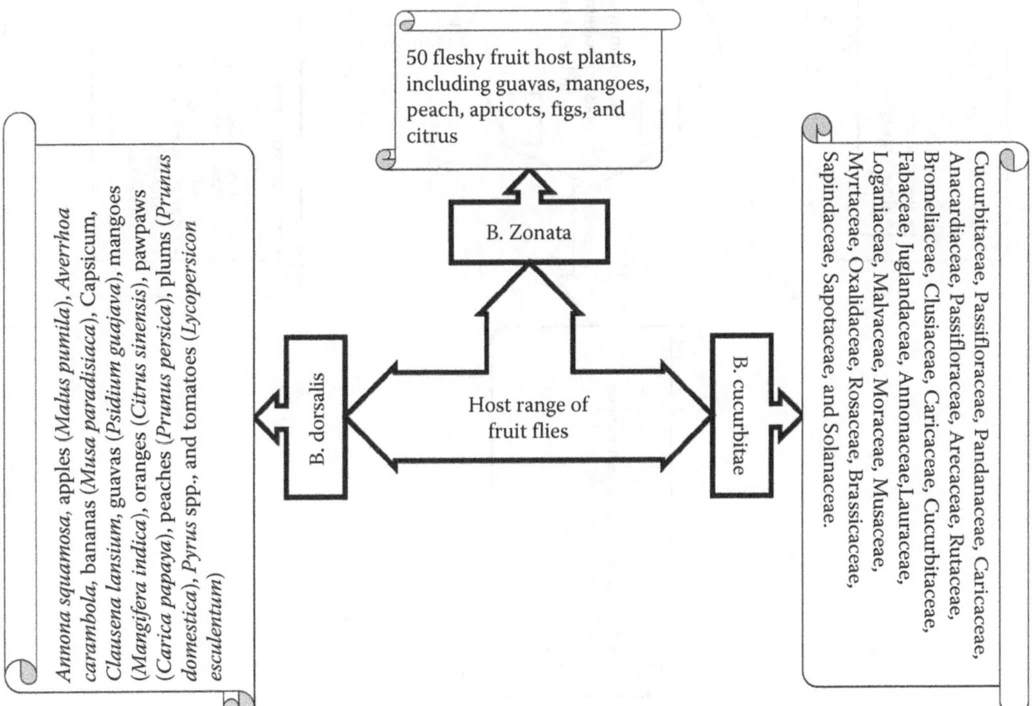

FIGURE 21.7 Host range of B. cucurbitae, B. zonata, and B. dorsalis.

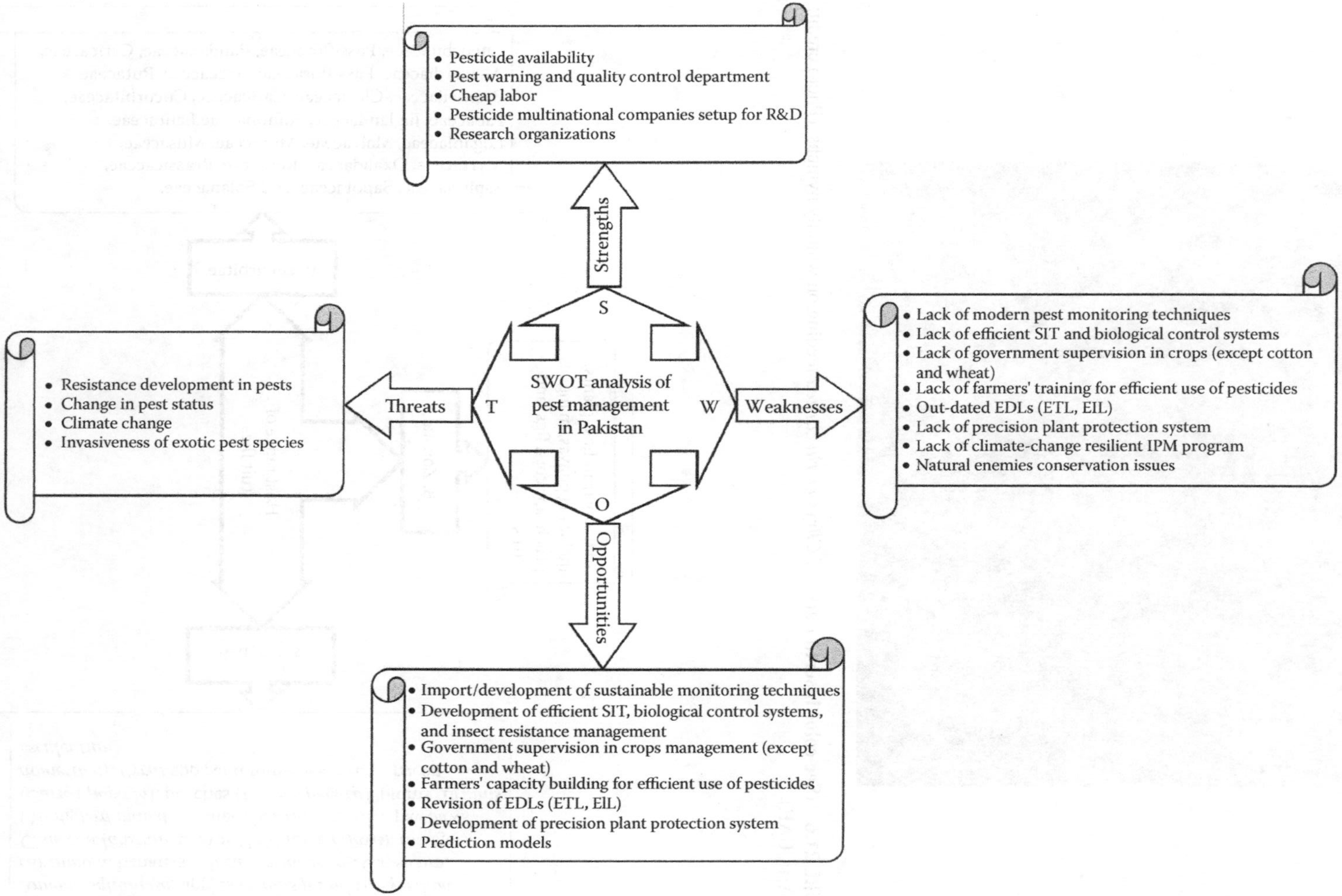

FIGURE 21.8 Diagrammatic illustration of SWOT analysis of pest management in Pakistan.

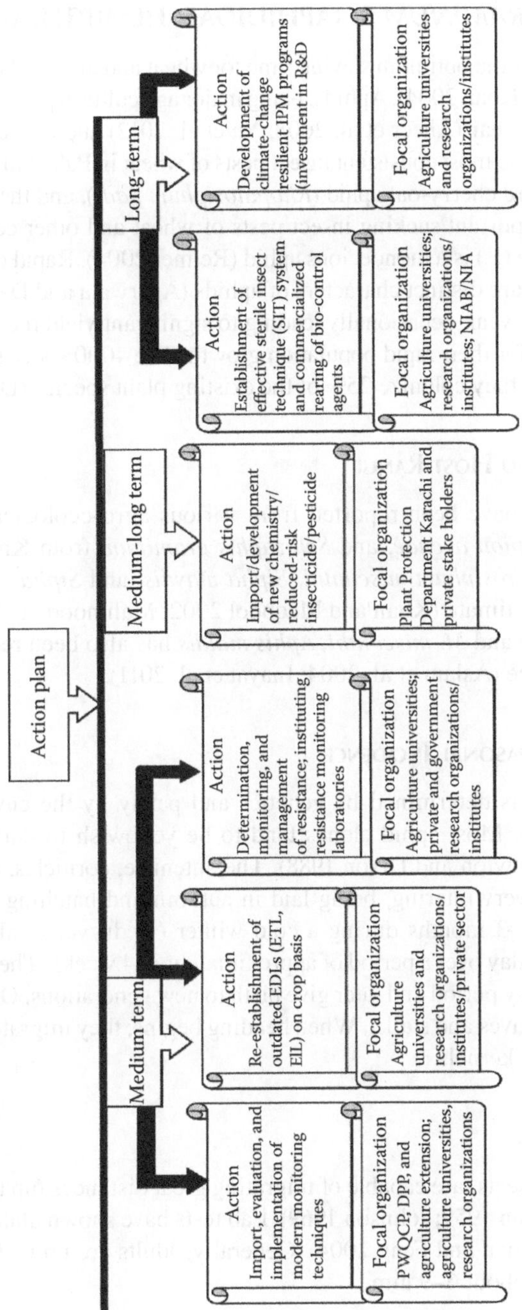

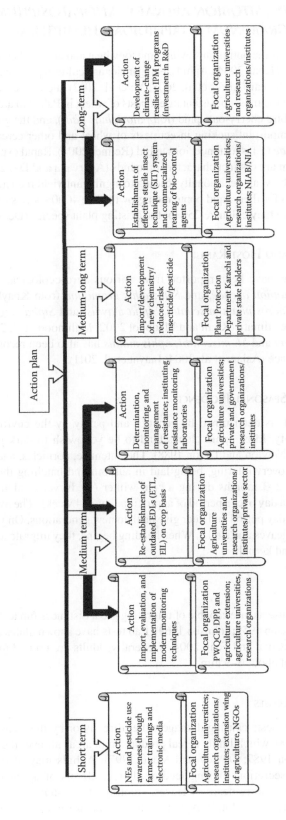

FIGURE 21.9 Diagrammatic illustration of action-plan/way forward for sustainable pest management in Pakistan.

21.2 WHEAT APHIDS, *SITOBION AVENAE* F., *RHOPALOSIPHUM PADI* L., *SCHIZAPHIS GRAMINUM* R. (APHIDIDAE: HEMIPTERA)

Generally, mid-November is the optimum sowing time for wheat and any further delay results in yield reductions of 50 kg/ha/day (Khan 2004). Aphids, being major agricultural pests, cause significant yield losses to wheat plants each year (Aheer et al. 2008; Yu et al. 2012) and are responsible for stagnant yields in Pakistan. Among the most consistent insect pests of wheat in Pakistan are: The English grain aphid (*Sitobion avanae*), bird cherry-oat aphid (*Rhopalosiphum padi*), and the greenbug (*Schizaphis graminum*). Aphids are important sucking insect pests of wheat and other cereals, vegetables, and fruit crops. *Sitobion avenae* (F.) is a monoecious aphid (Reimer 2004). Rapid exponential population growth and polymorphism are distinct characters of aphids (Agarwala and Das 2012). Cereal aphids infest wheat throughout the year, occasionally leading to significant yield reduction when biotic and abiotic factors are optimal for their rapid population growth. The 4000 species of aphids live mostly in temperate regions where they colonize 25% of the existing plant species (Dedryver et al. 2010).

21.2.1 Distribution and Host Range

Several species of aphids have been reported from various agro-ecological zones of Pakistan; *Rhopalosiphum padi*, *Sitobion avenae,* and *Schizaphis graminum* from Khayber Pakhtun Khaw (KPK) province with, *Macrosiphum miscanthi*, *Sipha maydis,* and *Sipha elegans* from northern hilly areas with temperate climate (Khan and Maqbool 2002; Mahmood et al. 2002; Khan 2005b). In addition to *S. graminum* and *M. miscanthi*, *Aphis maidis* has also been recorded from the plain areas of the Punjab province (Aslam et al. 2004; Inayat et al. 2011).

21.2.2 Biology and Seasonal Incidence

The body color of aphids is determined by genetics and partly by the environment, especially depending on food quality. Most aphid clones tend to be yellowish to dark green or brown, or sometimes orange-pink (Newton and Dixon 1988). The antennae, cornicles, and some parts of the legs are black. Eggs are overwintering, being laid in autumn and hatching the following spring, generally diapausing for 2–3 months during a cold winter (Dedryver et al. 1998). Each female produces several nymphs a day over a period of approximately 3–4 weeks. The nymphs pass through four instars over an 8–12-day period and later give birth to new generations. On young cereal plants, these aphids colonize the leaves and stalks. When heading begins, they migrate towards the ears and settle among the bracts and kernels.

21.2.3 Behavior

Aphids, like many other insects, are capable of migrating great distances (up to 1300 km) by means of wind (Elton 1925; Dickson 1959; Johnson 1969). Lab tests have shown alates are able to fly up to 9 km in a 5-hour period (Chen and Feng 2004). Generally, adults are up to 3 mm in length. Alate aphids have a wingspan of about 6–9 mm.

21.2.4 Damage and Losses

Cereal aphids are serious pests causing damage either directly or by the transmission of viruses (Fiebig and Poehling 1998), which limits cereal production in areas all over the world (Vickerman and Wratten 1979; Dixon 1987; D'Arcy and Mayo 1997). Aphids may also cause damage by injecting toxic salivary secretions during feeding. Direct effects of aphid feeding on cereals include yellowing and premature death of leaves, stunting of the stems, and reduction in grain size (van Emden and Harrington 2007). Aphid feeding on plant sap causes significant reduction in grain protein (Ba-Angood and Stewart 1980). Aphids suck the cell sap from leaves and shoots which results in curling, chlorosis, and distortion of leaves and thus stunted growth (Kindler et al. 1995; Akhtar and Khaliq 2003). Indirectly, the secretion of honeydew on leaves interferes with photosynthetic and respirational functions and consequently boosts leaf senescence (Bardner and Fletcher 1974). Annually, aphids cause significant yield losses of crop plants (Yu et al. 2012). Extent of direct losses by aphids ranges between 35% and 40% (Kieckhefer and Gellner 1992), while indirect losses vary between 20% and 80% (Marzocchi and Nicoli 1991; Rossing et al. 1994). Several studies have documented grain yield decreases of up to 16.4% by feeding of 7.2 aphids per tiller (Aheer et al. 1993), at seedling stage (2–3 leaves); 50% reduction in yield was reported with mean densities of 25–30 aphids per stem (Kieckhefer and Kantack 1988). Yield losses up to 50% and 36%, respectively, were reported in autumn and spring infestation (Kuroli and Nemeth 1987) and aphid infestation at a 1% level decreased the yield by 0.50% (Karren 1993). Losses caused by *R. padi* and *S. graminum* were 19% and 31% at booting stage and 15% and 20% at anthesis stage, respectively (Voss et al. 1997).

21.2.5 Nonchemical Control

The use of intercropping (e.g., pea, mung bean, oil seed rape, wildflower strips) as a strategy for increasing the biodiversity of natural enemies in fields has the potential to reduce pest damage. Judicious use of fertilizers is an important factor for aphid control. In order to reduce populations of aphids and increase in populations of its parasitoids, it is wise to use mixed wheat cultivars to control wheat pests by intercropping of a field cultivar with wheat cultivars with different levels of resistance to aphids. Conservation of chrysopids, syrphids, coccinellids, and aphid parasitoids (See Figure 21.1) should be encouraged at heavy aphid infestation stages (heading, flowering, and booting stage). Iqbal (2008) observed higher wheat yields and consumption of aphids by chrysopids as compared to coccinellids. Growing of canola for intercropping or as refuge crop in wheat, and conservation of natural enemies at wheat heading and flowering stages is encouraged for the control of aphids.

21.2.6 Chemical Control

Seeds of the neem tree, *Azadirachta indica*, A. Juss. (L.) have extractable compounds like azadirachtin (Butterworth and Morgan 1968), a tetranotriterpenoid limonoid known to have antifeedant and growth disruptive effects on more than 540 insect species (Mordue and Blackwell 1993). Many studies reported its effectiveness for managing different insect pests (Schmutterer 1990; Lowery et al. 1993; Walter 1999; Liang et al. 2003; Nathan et al. 2005). Iqbal et al. (2011) observed varying effects of different botanicals against wheat aphid populations where the maximum level of aphid mortality (65.69%) was observed by orange peel extract followed by garlic (57.91%) and tobacco (57.90%). Aziz et al. (2013), on the basis of cost benefit ratio analysis, recommended the use of neem (*Azadirachta indica*) seed kernel extract against *S. avenae* on wheat as compared to imidacloprid. Sprays of the entomopathogenic fungi *Verticillium lecanii* could be used to optimize an integrated pest management strategy against cereal aphids (Aqueel and Leather 2013). Except in Pakistan, pesticides are being used against aphids on wheat, and these pesticides are known to have negative effects on key natural enemies like coccinellids, chrysopids, and aphid parasitoids (Desneux et al. 2007). Aphids can be managed successfully with either neonicotinoid insecticides like imidacloprid applied as a seed treatment (Ahmed et al. 2001; Royer et al. 2005; Zhang et al. 2015) or foliar applications (Shahzad et al. 2013). Nevertheless, frequent and nonjudicious use of insecticides and their absorption through grains endangers the health of consumers and farm workers and poses threats to environmental stability (Lalah and Wandiga 1996). The use of pesticides is delayed or not preferred in Pakistan due to its side effects on human health. Recently, lectin genes have been engineered with substantial success into wheat plants for aphid control viz. *Galanthus nivalis* agglutinin against *S. graminum*, *S. avenae* and *Rhopalosiphum padi* (Stöger et al. 1999;

Liang et al. 2004; Xu et al. 2004) and *Pinellia ternata* agglutinin and cry1Ac against *S. graminum* (Yu and Wei 2008).

21.2.7 Future Research

- Aphids versus transgenic crops engineered for resistance via a nontoxic mode of action could be an efficient alternative strategy.
- There is a need to focus on the following aspects in the future:
  - Predator—prey relationships
  - Push and pull strategy
  - Refuge crops
  - Fertility of soil (fertilizer)
  - Economics of pheromones
  - Insecticide resistance

21.3 WHITEFLY *BEMISIA TABACI* (GENNADIUS) (ALEYRODIDAE: HOMOPTERA)

The cotton, potato, or silver whitefly, *Bemisia tabaci* (Gennadius) is a significant insect pest of vegetables and agronomic crops, especially in the subtropical and tropical regions and some ornamentals worldwide (Fan et al. 1998; Wan et al. 2009). It is in the family Aleyrodidae of the large suborder, Sternorrhyncha (Brown et al. 1995). A study by De Barro et al. (2011) reported that silver leaf whitefly is a complex with 24 structurally identical species. Species occur around the world in tropical, subtropical, and to the lesser extent in temperate habitats. Adults and immatures do not survive cold temperatures (Greenberg et al. 2000). Directional adult flight is finite but winds may carry flying adults over prolonged distances. Whitefly is moved worldwide with plant trade. It was reported during 1996–1999; attacking almost 160 plant species including fruits, vegetables, cash crops, ornamentals, forest trees, and weeds in Pakistan (Attique et al. 2003). In Pakistan, its invasion is documented from approximately more than 100 plants belonging to 24 families. There are 12 generations in a year and hence it remains active in all seasons. A single female carrying virus can infect many host plants. It damages the plant in various ways, that is, by continuously sucking the cell sap, which results in 50% boll reduction, by secreting honeydews on which sooty mold develops (Ahmad et al. 2002) and also acts as a vector of leaf curl virus (CLCUV) (Nelson et al. 1998). The average yield loss was reported to be 38.7% during 1993 in Pakistan caused by CLCUV (Khan and Khan 1995) and still this disease persists in agricultural fields, which threatens the cotton-based economy.

21.3.1 Distribution and Host Range

Whiteflies are cosmopolitan in distrubution and are reported from all continents except Antarctica. They are present in fields in most of southern Europe from Portugal to Turkey, and also in Slovakia and Ukraine. They are restricted to greenhouses in Western, Central, and Northern Europe. They are not known from Ireland, Finland, and the United Kingdom. They are polyphagous pests with a very broad range of host plants including many herbaceous and some woody plants like cucurbits, cole crops, tomatoes, cotton, peppers, crape myrtle, hibiscus, lantana, and roses. It has been reported during 1996–1999 that whitefly has 160 host species, which belong to 113 genera of 42 families including fruit, crops, forest trees, ornamentals, and weeds from Pakistan (Attique et al. 2003). In Pakistan, it was recorded in 104 plants belonging to 24 families. Cotton, squash, cabbage, cucumber, melons, tomatoes, okra, beans, eggplants, and many ornamental plants are the crops that supported numbers of silver leaf whitefly (Brown and Bird 1992). Important crops harboring the vector during the "Rabi" season include sunflowers, tomatoes, cucurbits, rape-seed, and pepper. CLCUV can be transferred by grafting and through whiteflies to soybean, cowpea, cotton, and okra (Hameed et al. 1994). The presence of B-type of *B. tabaci* was reported by Hameed and Khalid (1996) in Pakistan.

21.3.2 Biology and Seasonal Incidence

It remains active throughout the year and has more than 12 generations. Whitefly populations are prevalent throughout the year in orchards, on vegetables, and on ornamentals, which provide best support to this insect vector for survival. The insect is polyphagous, abundant throughout most of the year, transmits CLCUV very efficiently, produces 11–15 generations yearly and it is also a phloem feeder (Ali et al. 1995). It possesses high reproduction potential; a single female can lay more than 300 eggs in her lifetime. Female whiteflies can reproduce without mating (parthenogenesis) and produce only male progeny. Initially eggs are white in color and change to brown at the time of hatching, within 5–7 days. After the egg stage, whiteflies develop through four instars. The first instar, generally called a crawler, is 0.3 mm in size and develops to 0.6 mm till the fourth instar. In the first instar, the body is greenish and flat (Brown et al. 1995). During the immature stage, the whiteflies have a milky white appearance and become immobile from their feeding site. These immature whiteflies use their mouthparts to injure the plant and consume the plant's juices at the feeding site. At the pupal stage, their eyes turn a deep-red color, the body turns yellow, and overall they become thick. After the completion of development, adult whiteflies gain a size approximately four times that of the egg, possessing light yellow bodies and white wings (Brown et al. 1995).

21.3.3 Behavior

In the absence of an olfactory response (Mound 1962) whiteflies rely mostly on vision for navigation and orientation. The literature on the visual sensitivity of silver whiteflies reports that these insects are intensely attracted to the UV wavelength (Mound 1962). Therefore, intrusion with UV-vision may result in interruption of orientation and dispersal processes of whiteflies. The free-flight behavior of *B. tabaci* was investigated and a flight tendency was comparable from 06.00 to 19.00 hours (Blackmer and Byrne 1993). The propensity to take off proportion exhibiting phototactic orientation and flight duration varies with the age of the whitefly. Host quality also affects the timing of flight behavior (Blackmer and Byrne 1993).

21.3.4 Damage and Losses

Heavy infestations cause significant losses in yield, ranging from 20% to 100%, and depend upon the crop and season. The direct feeding by mature and immature whiteflies may decrease the strength and growth of the host plant, initiate chlorosis, irregular ripening, and induce physiological disorders. The secretion of honeydew by nymphs, serves as a substrate for the development of black sooty mould on leaves and the bolls of cotton. Photosynthetic activity decreases as a result of the mould and hence yield is reduced and the crop remains unmarketable. Immature B biotype whiteflies also induce phytotoxicity reactions in pest-ridden plants. They also act as a vector of approximately 38 plant diseases (Hussain et al. 1991). Whitefly-transmitted gemini viruses evoke 40 diseases of vegetables and fiber crops around the world (Fauquet and Fargette 1990). A list of 61 host plants of CLCUV was reported by Mirz et al. (1994). Pumpkin has been reported as the host on which whitefly progeny survived better out of six virus-infected plant species (Costa et al. 1991).

21.3.5 Nonchemical Control

Through cultural control methods, different planting areas can restrict *B. tabaci* infested plants. Growing various host plants distant from each other will reduce infestation rates and damage. Hence, the best control is to increase the distance, by as much as possible, and increase the time interval between host crops. Better sanitation in winter and spring plants is also necessary to maintain and control whitefly populations. Weeds and host crop by-products must be removed as soon as possible to prevent infestation. Mulch covers of silver/aluminum can prevent adult silver leaf whitefly. Therefore, placement of reflective polyethylene mulch on the seedbeds at sowing time can significantly reduce the colonization of whiteflies. The use of trap crops has also proven to be an effective control method. Due to the whiteflies attractiveness to squashes, it can act as a trap crop for the silver leaf whitefly (Schuster 2004). Tomato yellow leaf curl virus infestations can be checked and restricted when squash serves as a trap crop (Schuster 2004). To control whitefly infestations, biological methods are also proposed, and can be used in collaboration with chemical insecticides. The washing of crop plants, especially the undersides of the leaves, help to reduce many plant pests and makes their control more effective. The color yellow attracts whiteflies, so yellow sticky paper can serve as traps to check infestations. Dead leaves can be removed, burnt, or carefully thrown in closed bins to prevent future infestations and disease spreading. Many entomophagous agents proved to be operative in managing whitefly invasions, including big-eyed bugs, green lacewings, phytoseiid mites, minute pirate bugs, ladybirds, and damsel bugs. An efficient controlling of whiteflies might also be achieved by applying microbial-based biopesticides such as *Beauveria bassiana* (effective for nymphs and adults) and *Isaria fumosorosea*. The larvae of green lacewings have ravening cravings. They attack whiteflies, and other pests, like thrips, aphids, mealybugs, leafhopper nymphs, spider mites, scales, and moth eggs. Ladybird beetles eat mostly insect eggs, but also feed on scale insects and aphids.

21.3.6 Chemical Control

Chemical insecticides are generally applied in order to manage whitefly and aphid colonies in crops (Nzanza et al. 2011). The waxy shelters produced by whiteflies resist effectiveness of chemicals. Contact with the pest's immovable nymphal and pupal phases and also the movable and outstanding reproductive capability of aphids contribute to the decreased efficacy of chemical control (van Lenteren 1990; Lowery et al. 1993; James 2003; Yang et al. 2010). This insect pest has developed resistance to a number of insecticides including growth regulators (IGRs), pyrethroids, carbamates, and organophosphates due to frequent use of pesticides (Cahill et al. 1994; Horowitz et al. 1994; Jazzar and Hammad 2003). Thus, significant strides are made for the development of more eco-friendly technologies for managing insect pests (Dimetry et al. 1996; Jazzar and Hammad 2003). Plants with insecticidal properties like neem (*Azadirachta indica*) and wild garlic (*Tulbaghia violacea*) are effective against unique target pest species and might be used in integrated management programs (Markouk et al. 2000; Tare et al. 2004; Ateyyat et al. 2009). Extracts of the neem plant serve as antifeedants and repellents, inhibit oviposition and moulting, and reduce growth and development causing high mortality in whiteflies and aphids (Coudriet et al. 1985; Prabhaker et al. 1989; Liu and Stansly 1995; Mitchell et al. 2004; Kumar et al. 2005; Kumar and Poehling 2006). Though, all portions of the neem plant have traits for pest management, most definitive results are gained with neem seed and oil, due to high concentrations of azadirachtin (Rovest and Deseo 1991; Dimetry and Schmidt 1992; Dimetry et al. 1996). Insecticides used for whitefly management usually possess neonicotinoid substances as active compounds: Thiamethoxam, clothianidin, imidacloprid, acetamaprid and dinotefuran. The use of insecticides in different groups in rotation can be more effective in checking resistance. Insecticides like chlorfenapyr, pyriproxyfen, clothianidin, and dinotefuran are in the same insecticide family.

21.3.7 Future Research

Under a collective pest management program, different methods are used to manage *B. tabaci*. From trial experimentation, researchers have noticed that sugar apple oil (Lin et al. 2009), and other natural oils, have the same power as other insecticides with the advantage of being environmentally friendly

(Schuster 2009). Through cultural and mechanical control methods the insects are managed by trap crops and using man-made traps. The most essential and environment friendly method of controlling the silver leaf whitefly is the use of biological control agents including predators, parasites, and pathogens. *Eretmocerus*, a wasp species is an effective natural enemy of whiteflies (Hoddle et al. 1999). Most of the predators of *B. tabaci* eat the insides of the pest, while pathogens transmit deadly viruses (Hoddle et al. 1999).

Strategies include:

- Never allow CLCUD susceptible varieties/hybrids to be permitted for cultivation.
- Create facilities for early sowing before the end of April by providing irrigation and enforcing a ban on sowing after 7th May.
- Insect pest management must be based on sticky traps, reflective sheets, suction traps, and soap emulsions of neem oil, castor oil, fish oil rosin soap, and insect growth regulators.
- Appropriate spray methods must be used to ensure that the spray fluid covers the undersurface of leaves.
- Fields and vicinity must be kept weed free

21.4 PINK BOLLWORM, *PECTINOPHORA GOSSYPIELLA* (SAUNDERS) (LEPIDOPTERA: GELECHIIDAE)

The pink bollworm *Pectinophora gossypiella* Saunders is one of the key lepidopteron pests of cotton and is distributed in scattered locations throughout southern Europe, Africa, the Middle East, Asia, and Australia. In the New World it occurs from the southern United States to Argentina, including the Caribbean (Hill 1975). The other known species is *P. malvella* Zeller, recorded in Africa and southern Europe. Adults of *P. gossypiella* are small, grayish brown to dark-brown, measuring about 12–20 mm across the wings. The hind wings are broader than the fore wings and silvery grey with darker, shimmery hind margins. Wings are covered with poorly defined black spots, which are held folded over the back when not in use. The head is reddish brown pale in color with iridescent scales. The basal segment of antennae bears pectin of five or six long, stiff, hair-like scales that are brown in color. Young larvae of pink bollworm are white caterpillars with dark brown heads. At maturity, they become about 0.5 inches long with wide transverse pink bands on the dorsum. The larvae feed on the blossoms, lint, and seeds of cotton and cause serious losses to the crop.

21.4.1 Distribution and Host Range

Pectinophora gossypiella was first described from larvae recovered from infested cotton bolls in India in 1843 (Noble 1969). It has since become one of the most destructive pests in cotton production areas worldwide. The insect has been taxonomically labeled under numerous other generic names and the complete synonymy was described by Common in 1958. The origin of pink bollworm (PBW) remains cryptic but the diversity of pest species found in Pakistan (Cheema et al. 1980) seems to support an Indo-Pakistan origin (Ingram 1994). However, the infestation and spread of pink bollworm in different countries was documented during various years. For example, in Australia the pest was first recorded in 1911 (Wilson 1972), while in Sri Lanka, Burma (Pearson 1958), Malaysia (Lefroy 1906a,b), and China it was before 1918 (Hunter 1918). In Africa, it was first reported in Tanzania (Vosseler 1904) and Egypt about 1906–1907 (Willcocks 1916), while in Sudan about 1914–1915 (Ripper and George 1965). However, the insect did not reach Malawi until 1939 (Smee 1940), though it appeared in Zimbabwe in the late 1950s (Whellan 1960). In the Western Hemisphere it has been distributed through cotton seeds shipped from Egypt to Brazil, Mexico, the West Indies, and the Philippine Islands between 1911 and 1913. The insect was transported to the United States through cotton seeds dispatched from Mexico to Texas oil mills in 1916 (Spears 1968). *Pectinophora gossypiella* is an oligophagous pest with about 7 families, 24 genera, and 70 species having been recorded as its alternate hosts (Noble 1969). Pink bollworm sometimes also prefers to feed on okra over cotton especially at the end of the season when the cotton boll's surface becomes hard (Khidr et al. 1990). The contribution of weeds and other vegetation towards pink bollworm fluctuation is still unclear; however, it is not considered a major factor influencing cotton infestations (Noble 1969). The fact is that the pink bollworm is essentially limited to cotton fields and this can assist in management approaches.

21.4.2 Biology and Seasonal Incidence

Pink bollworm mating occurs in the summer and usually begins the second night of the adult life (Henneberry and Leal 1979). The female lays small eggs that are difficult to detect without a lens. The adult's life span is about 2–3 weeks and a single female lays nearly 100–200 eggs. The eggs are deposited singly or in small clusters on any part of the green cotton bolls or its calyx or even in the flower. The eggs hatch in 3–5 days depending on environmental conditions. Newly hatched larvae are very small, glassy white with a light brown head and thoracic shield. The larvae pass through four instars in 12–18 days; the pupal stage take 6–8 days (Butler and Henneberry 1976b) and takes place in the ground about 50 mm below the soil surface (Saunders 1843). Adult emergence starts in late March and continues until late July and early August (Wene et al. 1961; Watson and Larsen 1968; Watson et al. 1970; Rice and Reynolds 1971; Slosser and Watson 1972a; Fye 1979a). Under optimal conditions, the life cycle is completed in 25–31 days (Green and Lyon 1989). Adults are nocturnal in nature and females begin egg laying a day or two after emergence. The eggs laid by these moth populations produce first generation larvae that enter cotton flower buds (squares) and become mature about the time the cotton flowers open. Under optimum (cool, dry) conditions, *P. gossypiella* larvae undergo diapause in small cocoons in partially opened bolls in cotton lint, stored seeds, or in the soil. In favourable conditions, which may vary from area to area, diapausing larvae emerge from March to April but may remain quiescent for up to 2.5 years (Metcalf and Metcalf 1993).

21.4.3 Behavior

One of the most important means of detecting and monitoring of insect pest population fluctuation levels is the use of pheromone traps and other semiochemicals (Asaro et al. 2004; El-Sayed et al. 2006). For dispersal and mating, insects usually fly randomly over large areas in upward and downward directions (Reynolds et al. 2007; Byers 2012a). Male moth response towards pheromone baited traps is always complex and often involves behaviors that effect the males finding the pheromone source (Carde 1979). The change in behavior starts with the male fanning his wings, when the moth first identifies lures from a distance, this is often called activation. It then progresses from flights with upward and downward movements after a series of interaction. The pink bollworm moth can travel long distances in search of susceptible crops. In most cases, moth flights take place at night when temperatures exceed 50° F and wind speed is below 10 mph. Moth flight behaviors stem from a complex assimilation of visual responses from drifts with a chemically altered program of contrary direction (Baker et al. 1985). The flight behavior of the pest can be manipulated for pest management strategies (e.g., mating).

21.4.4 Damage and Losses

Among the cotton bollworm complex, the pink bollworm assumed major pest status in the recent past (Ghosh 2001). Newly hatched larvae tunnel into the buds and flowers in the early spring, while later in the season they attack the bolls as they become available. A single larva can destroy several buds and flowers eating the seeds and spoiling the lint badly. In most cases, affected bolls fail to open or may open in a very awkward fashion indicating the attack of the pest. The pest hibernates in between the seeds and causes double seed formation, the best indication of the damaged seeds. First and second instar larvae are sometime difficult to see against the white lint of the bolls. There are several factors, such as weather conditions, that decide the extent of infestations in a particular year during the cotton production season. Warmer than normal winters and wetter than normal summers often result in increased problems from the pink bollworm. The pink bollworm emerged as a major pest in cotton growing areas of Pakistan in 2015–2016 and caused the country about a 4.5 million bale losses during the season according to Government officials (Amin 2016). Globally, pink bollworm has become economically the most damaging pest of cotton and is known to cause about 2.8%–61.9% loss in seed cotton yield, 2.1%–47.10% and 10.70%–59.20% loss in oil content and in normal opening of bolls, respectively (Patil 2003). Amongst the bollworm complex, the pink bollworm *Pectinophora gossypiella* (Saund.) and the spotted bollworm *Earias insulana* (Boisd.) are the most damaging pests in Pakistan and cause significant yield reductions (Abro et al. 2004; Hamed and Nadeem 2010). As transgenic cotton offered a high level of resistance against the bollworm complex (Kranthi 2002; Kranthi and Kranthi 2004), however, *Bt* cotton is more vulnerable to climatic variations and attack by sucking insect pests, which are outside the resistance-claim purview.

21.4.5 Nonchemical Control

The pink bollworm's migratory nature and reproductive ability present a challenge for cotton growers in their efforts to manage this pest. Additional to cultural techniques, synthetic sex pheromones have been employed extensively in the monitoring and control of *P. gossypiella*. Trapping with the pheromone gossyplure has been widely used and resulted in a significant reduction of pest populations in some studies (Gao et al. 1992). The potential of pheromones for pink bollworm behavioral control was assessed in various studies (Shorey 1976; Gaston et al. 1977). Therefore, careful monitoring of pheromone traps can provide useful information for estimating the level of the adult populations and subsequent produced economic infestations of larvae in the bolls. The impact of natural enemies to manage pink bollworm populations in cotton is still not well understood. It has been difficult to quantify and predict the effect of natural enemies because of complex biological and ecological relations occurring within and between natural enemies and pest-insect hosts. However, there are some predators and parasitoids in the orders Coleoptera, Hymenoptera, Neuroptera, Hemiptera, and Dermaptera that help to manage *P. gossypiella* (Orphanides et al. 1971; Legner and Medved 1979; Naranjo et al. 1992a,b). In addition to other control methods, sterile insect technique has also been successful in keeping pink bollworm infestations at low levels (Frisbie et al. 1989). Sterile insect technique involves the rearing and sterilization of thousands of moths, which are released into cotton fields, where they then contest with the native existing male population for mating. Climatic cycles may trigger insect population cycles through changes in the insect or in the host. The development of different predicting phonological, population dynamic, deterministic, and stochastic models help to anticipate their responses and management practices.

21.4.6 Chemical Control

The control of pink bollworm through synthetic chemicals is always a difficult task as the pest feeds inside the cotton bolls. However, in some studies, frequent application of selective organophosphate insecticides after one week intervals has delayed pink bollworm infestations and increased yields (Tollefson 1987). However, early-season control with insecticides must be carefully evaluated against the possibility of outbreaks of secondary pests and preserving the natural enemy populations.

21.4.7 Future Research

- Development of prediction models
- Behavioral control with pheromones (e.g., monitoring, mating disruption, mass trapping)

- Transgenic cotton
- Augmentative biological control
- Sterile insect technique

21.5 ARMYWORM, *SPODOPTERA* SPP. (NOCTUIDAE: LEPIDOPTERA)

The armyworm, also known as tobacco caterpillar, *Spodoptera* spp. (Lepidoptera: Noctuidae) are polyphagous herbivores that are cosmopolitan in distribution. The genus *Spodoptera* (Lepidoptera: Noctuidae) contains around thirty species (Pogue 2002; Ellis 2004), which are considered among the most economically significant pests of major crops. In Pakistan, armyworm attack has been recorded on cotton, tomato, cabbage, soybean, tobacco, cauliflower, radish, carrot, onion, etc. (Maree et al. 1999). Out of thirty species, *Spodoptera littoralis*, *Spodoptera litura*, *Spodoptera exigua*, *Spodoptera fugepera* are common in South Asia. *Spodoptera. litura* and *S. exigua* are common species in Pakistan. These insects play a major role in damaging agriculture crops with economic losses from 10% to 30% (Ferry et al. 2004). Recently, it has been reported that *Spodopetra* spp. infested more than 290 species of plants belonging to 99 families (Wu et al. 2004). Under favorable conditions, due to increased populations of *Spodopetra* and their gregarious behavior, they are called "armyworm." Due to resistance development, *Spodopetra* has become a major pest in South Asia (Armes et al. 1997; Kranthi et al. 2001, 2002). In 2003, an outbreak of armyworm occurred in Pakistan in the cotton belt and ended with a major loss of the cotton crop.

21.5.1 Distribution and Host Range

Armyworms are generalist pests of various economic crops and vegetables. *Spodopetra* spp. are distributed throughout the tropical to temperate regions such as the Middle East, East Asia, Australia, and the Pacific islands (Zhang and Zhao 1996). Due to good flying abilities, long distance migration is possible where climatic conditions are favorable. Several species, such as *S. exigua*, *S. frugiperda* and *S. litura*, can migrate 100–1000 km in several weeks, taking nocturnal flight in the downwind direction (Gatehouse 1997; Saito 2001; Westbrook 2008). In order to improve reproductive fitness (i.e., longevity and fecundity), *Spodopetra* have the capacity to locate suitable food from alternate host plants (Song et al. 2007; Wenninger and Landolt 2011). Armyworms have an ability to perceive the chemical signals emitted by host plants though olfactory oragns like antenna (Hansson et al. 1999; Stensmyr et al. 2001). Various behavioral and electrophysiological studies indicate that armyworms also have olfactory receptors to detect favorable ubiquitous volatiles from suitable host plants (Saveer et al. 2012). This accurate perception and detection allows switching among the 290 host plants, which include major crops, vegetables, and ornamental plants.

21.5.2 Biology and Seasonal Incidence

The reproduction and development of *Spodoptera* vary on the quality of the host-plant, which may reduce growth of larvae, pupae, and cause lower size and weight of adults. Knowledge of the biology of *Spodoptera* is the key step towards the development of control strategies. Two to five days postemergence, *Spodoptera* females lay 50–300 eggs in masses preferably on the lower surfaces of leaves. Within 6 to 8 days, a single female can lay around 1500 to 2500 eggs. Eggs are spherical in shape and color changes from pearly green to black with time. Larvae emerge from the eggs within 3–4 days (Chari and Patel 1983). A newly hatched larva is tiny, blackish green with a distinct black band on the first abdominal segment. Fully-grown larvae are firm and smooth with dispersed short setae. In later instars, larvae switch their color from dark grey to dark brown marked with yellow dorsal and lateral stripes. After 15–30 days, larvae shift to redish brown pupae around 18–22 mm in length. Within 2 weeks, they switch to a whitish yellow, 14–18 mm adult. A life cycle is complete in an average of 30–60 days.

21.5.3 Behavior

Like other moths, *Spodoptera,* behavior is mainly based on the chemical communication and olfactory function (Kanzaki and Shibuya 1992; Hansson 1995). Male *Spodoptera* use pheromones to track females over long distances (Baker 1989), while female *Spodoptera* use odor bouquet to identify preferable oviposition sites (Renwick 1989). Both sexes, by using olfactory cues, find food sources in the form of nectar-rich flowers (Gabel et al. 1992). The flight capability of *S. litura*, contributes considerably to its pest status across the globe. The flight behavior is mainly dependent on wind speed and temperature (Westbrook 2008). Long-distance migration has been most prominent in insects and birds (Alerstam 2006; Holland et al. 2006; Newton 2008). Many noctuid moths, including *Spodoptera* (Chapman et al. 2008a,b, 2010) perform seasonal migrations by switching between summer and winter ranges separated by thousands of kilometres, by undertaking nocturnal flights at altitudes of hundreds or thousands of metres above the ground. For example, the movement of the African armyworm moth, *Spodoptera exempta* and *Spodoptera frugiperda* may be a strategy used to locate hosts in areas of recent rain (Rose et al. 1985; Johnson 1987).

21.5.4 Damage and Losses

Spodoptera larvae cause significant damage by feeding on leaves and bolls of cotton plants leading to complete stripping of the plants (EPPO/CABI 1997). Normally, larvae are leaf feeders and may act as a cutworm with crop seedlings resulted into the stunted development and fruit may be small or late to develop (USDA 2005). *Spodoptera* feeds on the lower side of leaves ended with feeding scars and skeletonization of leaves. Early larval stages feed gregariously, however, later stages are solitary in feeding. Because of this pest's feeding activities, holes and bare sections are later found on leaves, young stalks, bolls, and buds. These sporadic pests of many crop plants causes economic losses around 26%–100% (Dhir et al. 1992).

21.5.5 Nonchemical Control

Management of *Spodotera* includes the hand picking of egg masses and caterpillars as well as ploughing fields after harvest. Avoid weeds in cotton fields (i.e., Itsit and Jantar). These weeds attract armyworm and ultimately cause infestations in the major crops. In addition, biological control has been considering an alternative to traditional means of insect-pest control, especially in the context of environmental protection and food safety (De Clercq et al. 2003; Zanuncio et al. 2006). Biological control could include biocontrol agents, entomopathogenic fungi (EPFs), entomopathogenic nematodes (EPNs), and viruses.

21.5.6 Chemical Control

Resent data shows that around thirty-four insecticides are used to control *Spodoptera* spp. (Saleem et al. 2008). Approximately 6–10 applications of insecticides are made in cotton, and 2–4 applications in other crops such as okra and cauliflower. Diversified groups of pesticides have been used like organophosphates and pyrethroids, as well as the newer chemistries, spinosad, abamectin, emamectin benzoate, indoxacarb, lufenuron, and methoxyfenozide in cotton in Pakistan. The inappropriate use of these pesticides without pest scouting is hazardous to health (Ahmad and Arif 2010). The sporadic outbreaks of *Spodoptera* in Pakistan are due to the prevalence of resistance to conventional and new chemistry insecticides (Ahmad and Arif 2010).

21.5.7 Future Research

- Revisit host range and distribution patterns
- Impact of climate change must be studied

- Investigate host plant associations
- Investigate the migration of *Spodoptera* among Indo-Pak continents

21.6 MAIZE STEM BORER, *CHILO PARTELLUS* (SWINHOE) (PYRALIDAE: LEPIDOPTERA)

Different kinds of biotic and abiotic factors are responsible for hampering production and productivity of maize in major maize producing countries. Among the major biotic factors, *C. partellus* is reported as a key insect pest of maize in eastern and southern Africa and South Asia (Kfir et al. 2002; Ahmed et al. 2007) where it attacks all growth stages of maize and causes severe yield losses (Tamiru et al. 2011). About 47 species of the genus *Chilo* have been reported by various authors from different regions (De-Prins and De-Prins 2016). It is native to Asia and now has been distributed in many countries where sorghum maize and/or its host are cultivated. It has a wide distribution as a devastating pest in different types of agro-ecological zones due to its effectual and proficient colonizing potential (Tamiru et al. 2011; Asmare et al. 2011; Asmare 2012). A fractional displacement of native stem borers by *C. partellus* has been documented in many maize-growing countries. For example, *C. partellus* has displaced *C. orichalcociliellus* and *B. fusca* from various areas of Kenya and proportionally the incidence of *C. partellus* increased from 3% to 91% in a period of 6–7 years, 1986–1992 (Wondimu 2013). Morphologically, spotted stem borer (*C. partellus*) has much resemblance with many other species of *Chilo*. However, diagnostic characters of the male and female genitalia can be used for their true identification (Chernoh 2014). Although maize borer severely damages maize and grain sorghum, it has the potential to devastate pearl millet, finger millet, sugarcane, rice, wheat, foxtail, and various grass species, including Sudan grass and Napier grass (Kfir et al. 2002; Matama-Kauma et al. 2008). In different countries of Africa, three species of stem/stalk borers severely infest maize and demonstrate similar behavioral and development patterns. These species include African maize stalk borer (*Busseola fusca* Fuller) (attacks maize and sorghum); the spotted stem borer (*Chilo partellus* Swinhoe) (attacks maize, sorghum, millet, rice and sugarcane) and the African pink stem borer (*Sesamia calamistis* Hampson) (attacks maize, sugarcane, millet, and rice) (Chernoh 2014).

21.6.1 Distribution and Host Range

Chilo partellus has a worldwide distribution. It has established as major pest of maize in Yeman, Sri Lanka, India, Afghanistan, Pakistan, Japan, Thailand, Nyasaland, Nepal, Taiwan, Laos, Iraq, Bangladesh, Indonesia, and Vietnam (CABI 2007; EPPO 2015). It is also widely distributed as a notorious maize pest in southern sub-Saharan and most eastern countries of Africa including East Africa, South Africa, Mozambique, Botswana, Sudan, Comoros, Cameroon, Lesotho, Zambia, Swaziland, Eritrea, Ethiopia, Somalia, Uganda, Kenya, Tanzania, and Malawi (CABI 2007; EPPO 2015). Distribution of *C. partellus* in Western and European countries has rarely been reported (Arabjafari and Jalali 2007). However, a few reports highlight and support its occurrence in Australia (Ampofo and Saxena 1989) as well as in some provinces of the east mediterranean region of Turkey in 2014–2015 (Bayram and Tonğa 2015). In 2010, *C. partellus*, was reported as an invasive pest in Israel on sorghum and corn and had widely spread all over northern Israel by 2011 (Ben-Yakir et al. 2013). Agro-ecological zones having dry coastal areas with dry mid-altitude and moist-transitional as well as moist mid-altitude have been reported as ideal zones for its distribution (Muhammad and Underwood 2004). *C. partellus* has been reported as a oligophagous pest, which mainly infests maize and sorghum; however, its feeding has also been documented on various other host plant species (Arabjafari and Jalali 2007; EPPO 2015). The plant species reported as hosts in various countries include *Eleusine coracana* (finger millet), *Megathyrsus maximus* (Guinea grass), *Pennisetum glaucum* (pearl millet), *Rottboellia compressa*, *Setaria italica* (foxtail millet),

Sorghum bicolor subsp. *Verticilliflorum*, Poaceae (grasses), *Vossia cuspidate*, *Hyparrhenia rufa* (Jaragua grass), *Oryza sativa* (rice), *Pennisetum purpureum* (napier/elephant grass), *Saccharum officinarum* (sugarcane), *Sorghum bicolor* (sorghum), *Sorghum halepense* (Johnson grass), and *Zea mays* (maize) (Arabjafari and Jalali 2007; CABI 2007; Matama-Kauma et al. 2008; EPPO 2015).

21.6.2 Biology and Seasonal Incidence

Chilo partellus remains active from March–April to October–November. However, it spends the rest of the months of the year in hibernation in maize stubble, stalks or unshelled cobs in the form of full-grown larva. Hibernating full-grown larva come out from hibernation and pupate in March. From these pupae, adult moths emerge at the end of March or in early April. The moths become active at night, mate just after emergence and female moths deposit flat, oval shaped yellowish eggs in the form of overlapping clusters on the upper or underside of the leaves mainly near the mid-ribs or in the leaf-sheath (Muhammad and Underwood 2004; Tamiru et al. 2012; Lella and Srivastav 2013; Wondimu 2013). The whorl stage of maize is the preferable stage for female moths for oviposition (Tamiru et al. 2012). The fecundity of each female moth during a life span of 2–12 days is about 300 eggs (10–80 batches of overlapping eggs), which hatch out in 4–10 days depending upon the temperature (4–5 days in summer months) (Muhammad and Underwood 2004; Tamiru et al. 2012; Lella and Srivastav 2013; Wondimu 2013). Approximately 80%–85% of the deposited eggs hatch successfully into larvae (Tamiru et al. 2012; Wondimu 2013). The first two instar preferably feed in the whorl while mid-late instars leave the whorl, bore inside the stem, feed there and grow for 2–3 weeks till full grown. The mid-late instar larvae may also come out of the whorl, crawl down the maize stem outwardly/externally and then bore into the maize stem just above the internode or cobs (Tamiru et al. 2012; Lella and Srivastav 2013; Wondimu 2013). The full-grown larvae pupate inside the stem for 6–14 days but before pupation, larvae make holes for adult emergence (Muhammad and Underwood 2004; Tamiru et al. 2012; Lella and Srivastav 2013; Wondimu 2013). At the cob formation and tasselling stages, the older larvae inside the stem may enter the ears through the shank or reach tassels and damage the ears or tassels (Lella and Srivastav 2013; Wondimu 2013). Depending upon temperature and other factors, one generation may take 25–60 days to complete the life cycle (Muhammad and Underwood 2004; Tamiru et al. 2012; Wondimu 2013). During a year, maize stem borer may complete five or more (5–8) successive and overlapping generations based on climatic conditions, as well as availability of maize and other host crops/plants (Tamiru et al. 2012; Lella and Srivastav 2013; Wondimu 2013). However, the larva of the last generation undergoes winter hibernation from mid-October/November to mid-February/March. The spring maize crop is damaged by the first two or three generations while summer maize is attacked by the fourth to eighth generation (Kfir et al. 2002; Muhammad and Underwood 2004; Tamiru et al. 2012; Lella and Srivastav 2013; Wondimu 2013). The attack of the larvae of this pest result in destruction of growing point, stem breakage, and lodging, disruption of nutrient translocation, stunting and direct damage to tasselling as well as ears (Polaszek 1998; Kfir et al. 2002).

21.6.3 Behavior

A comprehensive knowledge of the behavior and flight of *C. partellus* is of paramount importance for its effective and successful control with semiochemicals. The timing of flight, migration and dispersal of *C. partellus,* like others insects, depend upon its mating behavior (Pats and Wiktelius 1989, 1992). The female moth except exhibits almost no flight activity on eclosion night. The mated female moths show peak flight activity at their peak oviposition activity period (before midnight). However, the peak flight activity period of male moths coincides with the eclosion of female moths and occurs after midnight (Pats and Wiktelius 1992). Comparatively, mated female moths exhibited longer flight periods than unmated moths and the flight for egg-deposition lasts for three nights. Irrespective of the sex and the mating condition, the moths can fly with a speed of 0.9 m/s (Pats and Wiktelius 1989, 1992).

21.6.4 Damage and Losses

Chilo partellus attacks all growth stages including vegetative, grain-filling, earing, silking, and tasselling stages of the maize crop. At the early vegetative growth stage, its damage appears in form of a series of "pin holes" in lines on younger leaves and/or "window panes" (patches of transparent leaf epidermis) in older leaves. The penetration of larvae through a central growing point into the stem results in the development of "dead heart" (dead and died central growing shoot). Its attacks at the tasselling stage results in the damage and spoilage of the androecium parts of the tasseling and ultimately impairs pollination processes as well as grain formation. If its attack appears on earing and silking stage, the larvae cut the silks of cubs (which are actually the styles of the gynoecium) and results in the impairment of pollination as well as grain formation. In the case of severe outbreaks of its population at cob-formation and grain-filling stage, the infesting larvae damage the grains and ultimately reduce the grains yield (Mwimali 2014). The attack of the larvae result in destruction of the growing point, stem breakage and lodging, disruption of nutrient translocation, stunting and direct damage to tasselling as well as ears (Polaszek 1998; Kfir et al. 2002). The losses and damage potential of *C. partellus* varies with altitude, moisture gradient, plant species, season, geographical areas, stage of plant growth, and plant protection practices. Dejen et al. (2014) reported that *C. partellus* demonstrated 1%–100% damage in South Wollo zone at elevation range of 1492–2084 m; 2%–26% in North Wollo zone at elevation range of 1850–2044 m; and 84%–99% in Oromia administrative zone at elevation of 1400–1669 m on cultivated maize and sorghum. A substantial loss of 20%–25% by *C. partellus* has been reported in Kenya in dry transitional and lowland tropical zones; whereas, about 10% in moist mid-altitude and dry mid altitude zones and less than 1% damage in highland tropics and moist transitional zones. Gupta et al. (2010) documented *C. partellus* as one of the most destructive insect pests of maize and sorghum at altitudes below 1500 m above sea level. Yield losses of 24%–75%, 10%–50%, 80%–88% and 13%–16% by this pest have been documented by Kumar (2002) in India, Farid et al. (2007) and Mashwani et al. (2011) in Peshawar valley of Pakistan, Van den Berg (2009a,b) in Africa and Patra et al. (2013) in Meghalaya-India, respectively. *Chilo partellus* has also been declared responsible for varying degrees of grain yield losses ranging from 24% to 36% in different agro-ecological zones of India (Bhanukiran and Panwar 2000; Patra et al. 2013) and 25% to 40% at different pest densities and phenological stage of the crop in Africa (Khan et al. 1997). Farid et al. (2007) reported varying degrees of damage at various plant growth stages of maize in Peshawar Valley of Pakistan. They reported 8%–20%, 10%–22%, 10%–21%, 9%–22% and 8%–22% damages on vegetative, grain-filling, earing, silking and tasselling stages of the maize crop, respectively. Depending upon the type and nature of the crop, 60% and 80% losses are in grain-maize and fodder-maize, respectively while 88% losses in sorghum have been documented due to *C. partellus* (Jalali and Singh 2003). As a polyphagous and key-pest, *C. partellus* has the potential to induce an estimated grain yield loss of 80% and maximum stalk damage in sorghum when it is 20 days old whereas, no significant loss is experienced when the sorghum crop is attacked soon after emergence (i.e., 6 days) (Van den berg 2009a,b). Mashwani et al. (2015) reported 15%–27% stubble infestation and 3%–10% stalks infestation in maize crop cultivated in Mardan, Pakistan. Overall, *C. partellus* is responsible for 50%–70% yield losses in maize and sorghum in various geographical regions of the world where these crops are cultivated either for grain or for fodder purposes (Mutyambai et al. 2014; Smith 2015).

21.6.5 Nonchemical Control

Destruction of stems and stubble of the previous crop, volunteer crop plants and/or alternative host removal as well as deep ploughing of crop residues are good crop hygiene practices which not only reduce carryover of stem borers from one growing season to the next but also limit the severe attacks on young crops early in the next growing season. Slashing maize and sorghum stubble and stems proves very effective in destroying 70% of the population of the *C. partellus* carry-over in stubble and stems. Subsequent disc ploughing results in the destruction of an additional 24% of the carry-over population. Tillage practices directly (by mechanical injury to larvae/pupae and/or burying them deep in soil) or indirectly (by exposing larvae/pupae to natural enemies and/or adverse weather) suppress substantial carryover populations of stem borer (more than 80%) (Kfir et al. 2002; Wondimu 2013; Khanzada et al. 2015).

- Partial burning of crop debris disrupts the life cycle and destroys hibernating larvae inside stalks and stubbles. This practice results in 95% destruction of hibernating larvae (Wondimu 2013; Chernoh 2014; Khan et al. 2015a,b; Khanzada et al. 2015).
- Manipulation of sowing dates is a good damage avoiding practice, which helps to evade peak adult activity periods (Kfir et al. 2002; Wondimu 2013; Chernoh 2014; Khanzada et al. 2015).
- Intercropping maize with beans, cowpea, or molasses grass (*Melinis minutiflora*) is an effective technique for reducing spotted stem borer attacks. Some volatile allelochemicals are emitted from such crops, which not only repel stem borers but also attract their parasitoids (Wondimu 2013; Khanzada et al. 2015).
- Planting some highly preferred hosts like Napier grass (*Pennisetum purpeum*) and Sudan grass, *Sorghum* (*vulgare sudanense*) as trap crops is also suitable for the management of maize stem borer (Khan et al. 1997; Wondimu 2013).
- In Africa, planting trap crops like Napier grasses around fields (to trap moths), intercropping *Desmodium* in maize field (to repel stem borers), removal of volunteer maize and grassy weeds (which act as host stem borers), rotation of maize crop with noncereal crops like legumes, deep ploughing, and burning of stubble after harvesting to reduce carryover of diapausing larvae and pupae are the most effective preventive nonchemical measures (CABI 2007; Chernoh 2014; Khanzada et al. 2015).
- Depending upon the geographical area, different types of natural enemies, especially entomophagous parasitoids prevail in the maize agro-ecosystem and suppress the population of *C. partellus* to varying degrees by natural parasitism. Natural enemies, which control *C. partellus* successfully and efficaciously in Tanzania, Kenya, Uganda, South Africa, and Pakistan include *Bacillus thuringiensis kurstaki* (a larval pathogen) and *Cotesia flavipes* (a larval parasitoid) (CABI 2007; Asmare 2012). According to Divya et al. (2009), *Cotesia flavipes* is responsible for 29% parasitism and *Tetrastichus* sp. is accountable for 2% pupal parasitism during kharif season in November, whereas *Sturmiopsis inferens* is responsible for 28% parasitism during rabi-summer crop in February. Periodical release of *Cotesia flavipes* should be made for the augmentation of parasitoid populations and suppression of maize borer.
- Use of tolerant maize varieties like *Bt*-maize proves very effective against borers. *Bt* maize provided effective and economical protection against the stem borer even under its severe outbreak by producing insecticidal protein (Fischer et al. 2015).
- Monitoring in young crop on the basis of pin holes (on funnel leaves and stem), window panes on open leaves, dead hearts, and stem borer larvae in stem is recommended weekly to assess the population buildup and decide upon control measures. Direct control of *C. partellus* should be practiced if 8%–12% plants have these damage symptoms (Chernoh 2014; Khanzada et al. 2015).
- *Chilo partellus* moths exhibit peak flight from mid March to mid May. During these months, as well as summer months (both seasons), numerous light traps should be installed in the fields for attracting and trapping flying moths. The traps should be emptied weekly (Farid et al. 2007).
- Staple 80–100 *Trichogramma* parasitoid egg cards per acre on leaves opposite to the sun (Farid et al. 2007).

- Spraying beneficial fungi based products (*Aspergillus flavus, Metarhizium anisopliae, Beauveria bassiana*) in the whorl of the plant proves effective for the control of maize borer (Chernoh 2014; Khanzada et al. 2015).

21.6.6 Chemical Control

Seed dressing before sowing with imidaclorid 70 WS and thiamethoxam 70 WS @ 5 gm kg^{-1} seed can reduce the infestation of *C. partellus* upto 97% and 88%, respectively. Similarly, granular insecticides like Monomehypo®-5G at 7 kg acre^{-1}, carbofuran 3G at 8 kg acre^{-1} and cartap 3G at 9 kg acre^{-1}, and foliar insecticides like imidacloprid 25 WP at 100 gm acre^{-1}, trichlorofon 80 SP at 1 gm L^{-1}, Chlorpyriphos 40EC at 5 mL L^{-1}, fipronil 35 SC at 5 mL L^{-1} and Deltaphos® 36EC (Deltamethrin + Triazophos, at 0.60 L acre^{-1}) can reduce infestations by 80%–90% (Mashwani et al. 2011). The recommended and effective method for application of granular and foliar insecticides is the whorl application. However, chemigation can also be practiced for these insecticides. According to Saleem et al. (2014), carbofuran 3G at 8 kg acre^{-1}, Fipronil 4G at 12 kg/ha, Abamectin at 500 mL/100 L, and profenofos 40% EC at 500 mL acre^{-1} should be used for direct control of maize borer when it infestation is at ETL (8%–12% infested plants). Khan et al. (2015a,b) recommended integrated application of cultural (removal of crop residues and irrigation with 20 days interval), mechanical (destruction of crop residues/stubbles before planting and removal of infested plants in early stage of the crop), biological (release of the *Trichogramma* @ 20,000 parasitoids/acre) and chemical control (funnel application of Confidor 70 WP @ 250 gm/acre at sowing time and Furadon 3G @ 2 g/plant at 6–8 leaf stage) for attaining effective control of maize borer and cost-effective higher yield of maize crop. Application of a mixture of neem powder with dry clay or sawdust in a ratio of 1:1 in the funnel/whorl of the plant proves very effective against maize stem borer. By the whorl application method, 1500–2000 plants can be treated with 1 kg of neem powder (Wondimu 2013; Chernoh 2014).

21.6.7 Future Research

- Identification of species of borers attacking maize crops is lacking and research should be carried out on this aspect at the national level.
- In Pakistan, exploration, multiplication, and manipulation of parasitoids of *C. partellus* is direly needed.
- Fundamental, molecular and applied research on parasitoid venom injected by the parasitoid into the body of *C. partellus* larva for paralyzing and killing should be carried out. Such research will help in sequencing and identifying venom (toxic peptides) transcripting genes and then their insertion into the maize plant by biotechnological approaches for developing GM-maize.
- Research on the characterization and preparation of commercial formulations of active peptides present in parasitoid venom is required.
- Research on isolation, identification, multiplication, and entomopathogenic potential of entomopathogenic fungi against *C. partellus* should also be carried out.
- There is also the need to explore semiochemicals for *C. partellus* for its ecofriendly management.

21.7 FRUIT FLIES, *BACTROCERA DORSALIS, B. ZONATA,* AND *B. CUCURBITAE* (TEPHRITIDAE: DIPTERA)

Tephritidae, with 4000 species (Fletcher 1987; White and Elson-Harris 1994; Drew and Romig 2000), is one of the largest and most diversified families of Acalyptrate: Diptera, and is divided into three subfamilies: Dacinae, Trypetinae, and Ceratitinae, of which the Dacinae has, traditionally been subdivided into two main genera, *Bactrocera* and *Callantra,* and consists of the most predominant, economically important, and biologically interesting dacine fruit flies (Christenson and Foote 1960). These belong to a little explored fruit fly guild with concealed feeding habits inside the fruit (Novotny et al. 2005) and have been widely explored in the tropical and subtropical regions of the world as agricultural pests (White and Elson-Harris 1994; Drew and Romig 1997; Clarke et al. 2001; Leblanc et al. 2001). The genus *Bactrocera* contains, approximately 11 subgenera and 48 species, distributed, primarily, in Southeast Asia, the South Pacific, and Australia (Syed 1969; White and Elson-Harris 1992). These species are considered economically important phytophagous dipteran pests of fruits throughout the world (Christenson and Foote 1960; Robinson and Hooper 1989), preferentially those of the families Cucurbitaceae (White and Elson-Harris 1992; Saelee et al. 2006), Anacardiaceae (*Dracontomelon* and *Mangifera* spp.), Moraceae (*Artocarpus* sp.), Oxalidaceae (*Averrhoa* sp.), Sapotaceae (*Manilkara* sp.) (White and Elson-Harris 1992), Solonaceae (Tan and Lee 1982; White and Elson-Harris 1992), and Asclepiadaceae as larval hosts (Drew et al. 1982; Munro 1984) in South and Southeast Asia. Out of these 48 species, *B. cucurbitae*, *B. dorsalis*, *B. oleae*, *B. tryoni*, *B. ciliates,* and *B. zonata* are regularly causing great reductions in both quality and quantity of yield and inflict economic losses to fruits and vegetables (Nagappan et al. 1971; Bateman et al. 1976; Cavalloro 1983; DFID crop protection programme 2004–2005), which are key sources of nutrition and of marketable produce for millions of poor people in South Asia (DFID Crop Protection Program 2004–2005).

21.7.1 Distribution and Host Range

The melon fruit-fly, *Bactrocera cucurbitae* (Coquillett) (Insecta: Diptera: Tephritidae: Dacinae), is indigeous to India (Eta 1985), and was detected in 1943 in the Northern Mariana Islands and was eliminated from these islands by releasing sterile male flies in 1963 (Steiner et al. 1965; Mitchell 1980), but it became established again in 1981 (Wong et al. 1989). In 1982, it was observed in Nauru and eradicated in 1999, but it became re-established in 2001 (Hollingsworth and Allwood 2002). It had also been reported in Hawaii, but it has not been recorded in the continental United States (Weems and Heppner 2001). It is now widespread throughout the world (Eta 1985). *Bactrocera cucurbitae* is native to the Asian region, but has potentially been established in other temperate and tropical regions. It is geographically distributed in Asia (Iwahashi 1977; Fischer-Colbrie and Busch-Petersen 1989; Anonymous 1993), Africa, North America (Carey and Dowell 1989), Oceania (Australia) (Cunningham 1989b), Papua New Guinea, the Solomon Islands (Eta 1986) and has been recorded almost exclusively on commercial cucurbits and solanaceous vegetables (Tan and Lee 1982; White and Elson-Harris 1992), especially, both tropical (e.g., *Momordica charantia*) and temperate species of Cucurbitaceae (Weems 1964; CABI and EPPO 2003). However, courgettes (*Cucurbita pepo*), cucumbers (*Cucumis sativus*), melons (*Cucumis melo*) (CABI and EPPO 2003), and tomato (*Lycopersicon esculentum* Miller) (Ranganath and Veenakumari 1996) are the main potential and confirmed hosts. This species is distributed widely, up to the elevations of 2000 m, on 4 plant families (Cucurbitaceae, Passifloraceae, Pandanaceae, Caricaceae), and 12 species (*Carica papaya*, *Coccinia grandis*, *Cucurbita maxima*, *Cucumis satilvus*, *Diplocyclos palmatus*, *Lagenaria siceraria*, *Lufa acutangula*, *Momordica chtrrantia*, *Pandanus odoratissimus*, *Passifora edulis*, *Strychnos nux-vomica*, *Trichosanthes cucumerina*, etc.) (Tsuruta et al. 1997). But, according to Carroll et al. (2004), more than 20 plants families serve as host plants for melon fruit flies (Figure 21.7). Batrocera. *dorsalis* is a complex of 75 species (Clarke et al. 2005) and occurs in the Northern Hemisphere and Southeast Asia. It has been reported in 40 countries of Africa, some countries of North America and Oceania but has not been recorded in the E.U. *Batrocera dorsalis* occurs on a wide range of fruit crops (Figure 21.7) (Clausen et al. 1965; Koyama 1989; Drew and Hancock 1994; De Meyer et al. 2012; Mguni 2013; CABI 2015; EPPO 2015; De Villiers et al. 2016). *Batrocera zonata* originated in South and Southeast Asia. It has a wide distribution in many states of India and Pakistan as well as in many countries of Africa. *Batrocera zonata* originated in South and Southeast Asia. It has a wide distribution in many states of India and Pakistan as well as in many countries of Africa. It has also been documented from the Near East, Israel, Egypt, and Libya and some islands in the Indian Ocean like Mauritius and Reunion. The distribution of *B. zonata* has also been reported from several countries in the Arabian Peninsula (Oman, Saudi Arabia, United Arab Emirates, and Yemen), the Gezira region in Sudan, Sub-Saharan region (Meyer et al. 2007), North America (California) (Carey and Dowell 1989), Moluccas (Amboina) (White and Elson-Harris 1994), Sumatra (CABI/EPPO 2001), United Arab Emirates, and Switzerland but to lesser extent from southern Europe. It attacks more than 50 fleshy fruit host plants, including guavas, mangoes, peach, apricots, figs, and citrus (CABI 2015; EPPO 2015).

21.7.2 Biology and Seasonal Incidence

The seasonal occurrences of *B. zonata, B. dorsalis, and B. cucurbitae* vary with type of host fruit and difference in region. In East Asia, population trends of these tephritids exhibits little variation in different regions. The lowest incidence of these tephritids has been reported in November–February followed by a gradual increase in populations until it reaches its peak in March–May. Their populations then show a declining trend in June–July followed by an increasing trend again in August. Another peak population trend occurs in September. This variation in its incidence is attributed to difference in temperature and relative humidity of these months. A positive association between population fluctuation of these tephritids and low temperature as well as humidity has been reported by different researchers (Qureshi et al. 1975; Clarke et al. 2001; Mohamed 2002; Dhillon et al. 2005; Mahmood and Mishkatullah 2007; Dale and Patel 2010; Alim et al. 2012; Draz et al. 2016). These species complete seven to eight generations in a year with varying generation duration. A generation period of 28–37, 33–35, 21–31, 30–50, 60–70, and 90–130 days have been reported by various researchers for spring, spring–summer overlapping, summer, autumn, autumn–winter overlapping, and winter–spring overlapping generations, respectively (Audemard and Millaire 1975; Richmond et al. 1983; Clarke et al. 2001; Dhillon et al. 2005; Shehata et al. 2008; Khalil et al. 2010; Alim et al. 2012; Delrio and Cocco 2012; Vayssières et al. 2015; Draz et al. 2016). Development threshold and thermal constant of these tephritids required to develop from egg to adult are 11.84°C and 487.92 thermal units, respectively (Draz et al. 2016).

21.7.3 Behavior

Tephritid fruit flies belonging to the genus *Bactrocera* are very strong fliers and possess long distance migratory and dispersal capacities. Long-distance flight enables these flies to migrate and disperse inter- and intraregionally in Asia, Africa, and other regions of the globe (Yan 1984; Fletcher 1989; Zhu and Qiu 1989; Tan and Serit 1994; Liang et al. 2001; Chen and Ye 2007; Chen et al. 2007; 2015; Froerer et al. 2010; Wan et al. 2011). For example, *B. dorsalis* exhibits extensive spreading and migratory potential after fruit harvests in the radius 20–97 km, even over water, depending on the direction of the wind. It can fly 20 km upwind and 97 km downwind (Yan 1984; Zhu and Qiu 1989; Tan and Serit 1994; Chen et al. 2007). Female flies are better flyers than males and their flying capacity (speed and distance) decreases with age (Sharp et al. 1975; Chen et al. 2007, 2015; Hao et al. 2013).

21.7.4 Damage and Losses

In Pakistan, the most dominant and prominent fruit fly species include *Bactrocera zonata, B. cucurbitae, B. dorsalis, Myiopardalis pardalina, Carpomiya inompleta, C. vesuviana, Dacus ferrugincus,* and *D. diversus*, which cause qualitative, quantitative, and economical losses in fruits and vegetables (Srinivasan 1959; Mote 1975; Abdullah and Latif 2001; Alyokhin et al. 2001; Abdullah et al. 2002; Stonehouse et al. 2002; Armstrong 2003; Follet and Armstrong 2004; Panhwar 2005; Ekesi and Billah 2007). Among these species, *Bactrocera dorsalis, B. zonata,* and *B. cucurbitae* are the most destructive fruit fly pests of fruits and vegetables, and A1 quarantine pests (OEPP/EPPO 1983;

Ravikumar and Viraktamath 2007). These fruit fly species cause both qualitative and quantitative losses in fruits and vegetables (Srinivasan 1959; Lall and Singh 1969; Mote 1975; Rabindranath and Pillai 1986). Their high reproductive potential, scarcity of natural enemies, and wide host range make them the most significant limiting factors in obtaining a quality production of fruits (Bachrouch et al. 2008; Gogi et al. 2010). Female fruit flies damage the fruit by ovipositing their eggs in fruit skins (Hollingsworth and Allwood 2002). Larvae feed on pulp inside the fruits, which makes them nonconsumable (Ye and Liu 2005). The maggots damage fruits (Dhillon et al. 2005) or flowers (Narayanan 1953; Weems and Heppner 2001), which develop improperly and fall down or rot on the plant (Dhillon et al. 2005). Pseudo-punctures and brown resinous deposits, produced by female flies on fruits, reduce its market value (Fletcher. 1987; Miyatake et al. 1993). *Batrocera cucurbitae* is a serious issue for growers of horticultural crops in the India and Pakistan, where it causes enormous damage (Agarwal et al. 1987; Singh and Singh 1998). According to Agarwal et al. (1987), about 50% of the cucurbitaceous crops are partly or entirely damaged/infested by *B. cucurbitae* in India. In Pakistan, it was reported that *D. cucurbitae* normally causes 20%–75% damage to melon production (Siddiqui and Ashraf 2002). *Batrocera zonata* is the most dominant (94.7% prevelance) species in guava, mango, peach, apricot, fig, and citrus (Ekesi and Billah 2007; Ravikumar and Viraktamath 2007). It causes 80% losses in guava (Kafi 1986) and 3%–100% in different fruits (Karar et al. 2016). In addition to direct damages and losses, infestation of fruits with fruit flies is a key trade obstacle causing severe monetary losses to farmers (Kafi 1986; CABI and EPPO 2003). *Batrocera dorsalis* and *B. zonata* are the most destructive fruit fly pests of fruits like mango, guava, and citrus (Ravikumar and Viraktamath 2007) and the second most in cucurbit crops. The consignments of fruits infested with fruit fly are often destroyed when they enter the international market. In 2013, almost all mango shipments/consignments from Pakistan were disposed of by the U.K. quarantine department because of the detection of fruit flies inside the fruits (Zaheer 2013). A warning to Pakistan has been issued by the EU, and mango consignments from Pakistan have been rejected and destroyed due to fruit fly contamination (Khan 2014; Mirza 2014). The U.K.'s Food and Environment Research Agency (FERA) reported that 6% of the total volume of Pakistani mangoes was denied entry whereas, 0.9% to 2.4% were rejected in the previous 4 years due to the presence of fruit fly maggots (FreshFruitPortal.com 2013). According to the Euro-Fruit Tribune (2014), approximately 2.8% of the 1652 lots of Pakistani mangoes were found to contain fruit fly and rejected from the U.K. market. Similar threats are also significant in terms of the export of citrus and vegetables. There is a need to address this issue of fruit flies contamination; otherwise, fruit exports, especially mango and citrus will be totally banned in European countries and other international markets in the coming years. This will lead to the loss of huge amounts of foreign exchange. In developing countries like Pakistan, the management of fruit flies totally depends upon cover sprays of insecticides, which not only cause ecological backlashes in fruit flies against insecticides and lethality to nontarget beneficial arthropods (Williams et al. 2003; Yee 2007) but also increase the cost of production and leave toxic residues in fruits and vegetables (EPA 2005; Klungness et al. 2005; Gogi et al. 2010). Fruits with residue levels higher than the defined MRLs are rejected for export to the international market under quarantine and SPS barriers (Gogi et al. 2007). In Pakistan, farmers blindly used insecticides on their fruit crops to make their fruit crops insect-pests free. This injudicious application of insecticides leaves toxic residues in fruits below or above MRL limits. A variable intensity of fruit samples having pesticide residues above MRL has been reported from different provinces of Pakistan. Tahir et al. (2009) reported 13.8% and 41% of the total fruit samples from KPK having residues above and below MRL limits, respectively. According to Khan (2005a), 17% and 70% of total fruit samples from Punjab were found to have residues above and below MRL limits, respectively. Similarly, Parveen et al. (2011) documented about 22% and 62.5% of total samples from Sindh, which had pesticide residues above and below MRL, respectively. Pakistan's export consignments of fruits having residues above MRL have been rejected by the quarantine departments of various countries. From 2004 to 2011, more than 140 consignments of fruits from Pakistan were confiscated by the quarantine departments of countries of the E.U. due to the detection of pesticide residues above MRL.

21.7.5 Nonchemical Control

Fruit infestation can and checked and minimized by holding paper bags over fruits before the onset of fruit flies. The timing of fruit bagging depends on the type of fruits (Fang and Chang 1987; Omar and Hashim 2004; Panhwar 2005). Bagging cucumber fruits, 3 days after anthesis and retained for 5 days proved significantly effective in the reduction of fruit infestation by fruit flies (Akhtaruzzaman et al. 1999). Fruits can be saved from fruit fly attack if fruits are wrapped over by shrink-wrap film (Jang 1990). Fang (1989a) and Jaiswal et al. (1997) also reported that bagging bitter gourd fruits, as a noninsecticidal method for the control of melon fruit flies, successfully increased the yield and net income by 45% on the bitter gourd and 58% on the angled luffa. Similar reports have been documented by Panhwar (2005), who concluded that bagged fruits helped in controlling the infestation of fruits before ripening and reducing the population build-up in fruit fly hotspot areas. However, this technique reduces the quality of fruit if bags of poor quality are used. Phytosanitary measures like the inspection of fruits from fruit flies hotspot countries for symptoms of infestation and destruction of suspected consignments may be effective for fruit-fly management. At the fruit ripening transit phase, forced hot-air treatment (Armstrong et al. 1995) or hot-vapor treatment of fruits at 44.5°C for 8.75 h (USDA 1994) or 46°C for 30 min (Iwata et al. 1990) is suffient for killing larvae developing inside the fruits. In the packinghouse, holes in the packing boxes are covered with a mesh to prevent fruit fly infestation. Fruits are also inspected twice for symptoms of fruit fly infestation; once, during harvest at the farm, and again during the packing and grading processes (Omar and Hashim 2004). Many advanced technologies have been developed and tried all over the world to control fruit flies (Tan 2000a,b). However, control of fruit flies from postoviposition to adult eclosion has been completely ignored in many of these fruit-fly management programs (Klungness et al. 2005). Disposal of infested and damaged fruits from orchards/fields is highly encouraged by various quarantine agencies for complete sanitation of fruit-flies breeding sources (Allwood 2000; Dowell et al. 2000; Seewooruthun et al. 2000; Klungness et al. 2005). Collection and burying of all un-harvested fruits and vegetables, deep into the soil, as a field sanitation measure, can be effective to break the reproduction cycle and reduce the population increase of fruit flies (Klungness et al. 2005). According to Panhwar (2005), maggots of fruit flies, feeding inside the fruits, can be killed and development of overlapping generations of fruit flies can be reduced/suppressed effectively, by collecting and burying all infested fruits, 4–5 feet deep in the soil. Similarly, another safe sanitary method to kill larvae, inside the fruits, is to irradiate the fruits (Panhwar 2005), though some people resist this technique because of the threat of radiation contamination of the fruit and vegetables. Tilling of crops, after the final harvest, along with the removal of infested fruits from the cropping system proves effective in checking the build-up of fruit-fly populations and, ultimately, in reducing fruit-fly infestations (Pandey 2004). However, Klungness et al. (2005) reported that the tilling of fruits into the ground only partially reduced the eclosion of melon fruit flies. Simple smashing of fruits, also, does not sufficiently reduce the rate of eclosion of melon fruit flies. Three methods: placing culled fruit in the augmentoria, burying the fruit 0.46 m underground, or placing fruit on the screen under and over 0.7 m, beyond the fruit pile, can be practiced as new sanitation approaches. Burying fruit 0.15 and 0.30 m deep only partially, whereas, burying fruit 0.46 m deep, fully prevents the adult fly eclosion. Burying of the edges of screen between the infested fruit and the ground prevents the escape of 90.2% of enclosing melon flies. Installation of augmentoria, a new sanitation approach, entraps all adult flies emerging from infested fruits placed inside the structure (Klungness et al. 2005). Fruit flies should be continually monitored, using pheromone/bait traps (Bateman 1982), which have been used for luring and controlling fruit flies (Bhutani 1975; Perdomo et al. 1976; Gupta and Verma 1979; Lee 1988; Omar and Hashim 2004). With the increasing apprehensions over severe environmental deterioration and human health concerns by pesticides sprays for fruit-fly control, a Bait-Application Technique (BAT) has been established, which attracts fruit flies of both genders. This technique has been proved more efficient than the male-only attracting method (Sabine 1992). Protein-baits have been developed from yeast autolysate, in Queensland (Smith and Nannan 1988; Sabine 1992) and from yeast protein, in Malaysia by the

Malaysian Agricultural Research and Development Institute (MARDI) (Vijaysegaran 1989; Loke et al. 1992). These baits have been tested and proved to be the most successful in controlling fruit flies in star-fruit (Vijaysegaran 1989; Loke et al. 1992), soursops, chili (Sabine 1992), angled luffa (*L. acutangula*), and bitter gourd (*M. charantia*) (Chinajariyawong et al. 2003). Good control of fruit flies was achieved in an integrated pest management (IPM) programme, in Brisbane when protein hydrolysate based baits were sprayed (Sabine 1992). Rice and his coworkers (2003) used a McPhail bait (food lure) trap from 2001 to 2002 for detecting fruit flies that are not attracted to specific lures (Rice et al. 2003). Trap of yellow color, baited with an ammonium bicarbonate, food-attractant lure (Suterra LLC [Consep], Bend, Ore.) plus an olive fly spiroketal pheromone lure (Vioryl S.A., Athens, Greece), attracts more fruit flies compared to a trap baited only with the ammonium bicarbonate food-lure, which also attracts reasonable numbers of fruit flies (Longo and Benfatto 1982; Rice et al. 2003). Fruit flies can be controlled effectively, through using food-baits mixed with the insecticides, which even more powerfully attract and kill fruit flies (DFID crop protection program 2004–2005). Protection of fruits from fruit flies is possible by using a bait-spray, prepared by mixing insecticides like malathion with protein hydrolysate (Roessler 1989; Wharton 1989; Sabine 1992; Omar and Hashim 2004). Bait-sprays are more effective, economical, and ecofriendly than cover sprays because the former can be deployed as a spot-treatment to attract and kill fruit flies with minimal impact on natural enemies (Wharton 1989). Different types of baits have been reported by various researchers for the management of fruit flies. For example, application of Pinnacle-bait (autolysed yeats bait,) or Thai-bait (brewery-yeast-bait), in angled-luffa and bitter gourd plots considerably reduced the percentage of damaged fruits, as compared to untreated plots, and yielded 81.57% higher production in the angled-luffa plot, and 59.98%–67.22% higher production in the bitter gourd plots when compared with the control (Chinajariyawong et al. 2003). Satisfactory control of *B. tau* is achieved on angled-luffa when a mixture of molasses and fenvalerate (molasses bait) is applied on the crop as a bait-spray (Saikia and Dutta 1997). The application of molasses mixture with malathion (Limithion 50 EC) and water (1:0.1:100 v/v/v) gave a good control of melon fly in cucumber, causing an effective and significant reduction in the fruit infestation (Akhtaruzzaman et al. 2000). Spray application of the fenitrothion-bait, prepared by coadministration of fenitrothion (0.025%) and protein hydrolysate (0.25%), on a bitter gourd crop reduced fruit fly damage approximately 5 fold, causing 8.7% damage as compared to 43.3% damage in the untreated control (Gupta and Verma 1982). Research organizations are now focusing on the development of female attractive lures for area-wide control of fruit flies (Cheng et al. 2003). Similar efforts were made by Nasiruddin et al. (unpublished data), who tested pheromone-traps and mashed sweet-gourd (MSG) bait-traps in Bangladesh and reported MSG-bait attracted both sexes of fruit flies. They further reported that both bait-traps, resulted in about 60%–70% and 70%–85% less fruit damage and 50%–80% and one- to threefold higher yields in cucumber and pumpkin, respectively when compared with the untreated ones. GF-120 Naturalyte Fruit Fly Bait is the only sprayable bait which is marketed and used at 10–20 fl oz. per acre per spray after dilution in ratio of 1:1.5–1:4 with water. Dilution GF-120 is sprayed, in large droplets, once a week for effective control of fruit flies in various fruit and vegetables (Thomas and Mangan 2005). Yee (2007), in Washington State revealed that fresh aplication of spinetoram and GF-120 were significantly effective and toxic for *Rhagoletis pomonella*. However, during a postapplication period of 7–14 days, their efficacy was reduced due to their breakdown or absorption into plant parts under bright, warm, and dry weather. It has been reported that it needs to be applied every 7 days for an effective control of *R. pomonella*, in central Washington and New York (Reissig 2003; Yee 2007), while for only 3 days in California against walnut husk fly, *Rhagoletis completa* Cresson, (Van Steenwyk et al. 2003). Likewise, it also proved effective in reducing infestation by the apple maggot and blueberry maggot in Michigan (Pelz et al. 2005). According to a project report compiled by Smith during 2002–2005 (Smith 2005), application of GF-120 bait demonstrated 100% control of cherry fruit fly from 2003 to 2005, and no larval infestation was observed in 12,000 fruits collected during 2005 from bait-treated locations. Melon fruit flies are attracted to some plants like sorghum and maize, where they feed on the inflorescence of the crop. This behavior of the melon fruit fly has been used as a management strategy (Dhillon et al. 2005). Experiments conducted along similar lines, in Hawai and India, revealed that GF-120 was found to be highly effective against protein-hungry melon fruit flies when sprayed on border crops like sorghum and maize, surrounding the main cucurbit crop. However, this technique was found ineffective in inhibiting protein-satiated female flies from settling on the major cucurbit crop (Prokopy et al. 2003; Dhillon et al. 2005). Flies are mostly attracted to yellow and green colors, and this is used to develop colored mechanical traps. In these traps, flies are trapped by glue- or tape-panels. At present, different types of mechanical devices and/or baits like yellow-sticky-panel-traps, McPhail type bottle-traps, and lure and kill traps are the most attractive for fruit flies and remain effective for about 2 weeks (Mangan and Moreno 2002).

Fruit flies can be monitored (Chen et al. 1995) and economically controlled by using pheromones like methyl euginol, cue-lure (Hooper 1978; Pawar et al. 1991; Permalloo et al. 1998; Seewooruthun et al. 1998; Vargas et al. 2000; Omar and Hashim 2004; Shelly et al. 2004), or parapheromones (DFID crop protection program 2004–2005). Parapheromones (chemical sexual lures), soaked into wooden blocks with a small quantity of insecticides, which attract and kills male fruit flies (DFID crop protection programme 2004–2005) can be used effectively to control fruit flies. Pawar et al. (1991) reported cue-lure traps are comparatively more efficient than food-bait-traps for melon fruit flies (*B. cucurbitae*). Different types of lure-traps are used for trapping, monitoring, and attract-an-kill of the melon fruit flies (Pawar et al. 1991; Permalloo et al. 1998; Seewooruthun et al. 1998). For area-wide control of *B. cucurbitae*, male cuelure B1® can be used effectively, because it proves 4–9 times higher effective in attracting *Ceratitis capitata* males than trimedlure® (Mau et al. 2003b). Methyl eugenol, raspberry ketone (RK), and cue-lure have also been used for attracting males of *B. cucurbitae* (Beroza et al. 1960; Drew and Hooper 1983; Tan and Lee 1982; Ramsamy et al. 1987; Zaman 1995). Raspberry ketone and cue-lure act as sex-pheromones attracting male fruit flies (Kuba and Sokei 1988; Tan 1993; Tan and Nishida 1996; Tan 2000a,b). In addition, zingerone/butanone has also been reported attractive to males of *Bactrocera* species (Tan and Nishida 2000). Melon fruit flies have also been suppressed by planting *O. sanctum* plants as border crops and spraying with a protein-based bait (Roomi et al. 1993). Control programs against *Bactrocera* species use ME- and CL-baited traps in order to detect and suppress pest populations (Gilbert and Bingham 2002). These baits consist of a toxicant mixed in a male attracting pheromone. Steiner et al. (1965) successfully eradicated *B. dorsalis* from Rota Island by installing ME and toxicant soaked fiberboard blocks, soaked with a toxicant and ME. Toxicant baited CueLure has also proven effective in suppressing *B. cucurbitae* in Miyako Island (Kuba et al. 1996). A reasonable number ranging from 2.36 to 4.57 flies/trap/day, were captured when the trichlorfon-bait traps were installed in a field of bitter-gourd (Chowdhury et al. 1993). Coadministration of lures has been reported to reduce the number of traps/acre, as well as the associated labor and the amount of pesticide used (Shelly et al. 2004); however, the results are asymmetric, inconsistent, and contradictory (Hooper 1978; Vargas et al. 2000; Shelly et al. 2004). The data, available in the literature, show that cuelure reduces the trap-capture of ME-responding species (Hooper 1978; Vargas et al. 2000). Whereas, the data documented on *B. cucurbitae*, by Hooper (1978) from Taiwan, show nearly twice the number of males captured in traps, baited with mixture of methyl euginol and cuelure, when compared with the traps baited with CL alone. Vargas et al. (2000) also conducted similar experiments and finally documented that a mixture of lures having 25% cuelure attracted as many male *B. cucurbitae* as trapped by cuelure alone. Similarly, different commercially available pheromone-baits like Flycide® (with 85% cue-lure content), Eugelure® (20%), Eugelure® (8%), Cue-lure® (85% + naled), Cue-lure® (85% + diazinon), and Cue-lure® (95% + naled) have been reported to control fruit flies effectively (Iwaizumi et al. 1991). Methyl euginol and Cue-Lure is effective when used with Male Inhalation Technique (MAT) (Afzal and Javed 2001; Dhillon et al. 2005; Stonehouse et al. 2007).

Many cultural practices have been used to reduce the damage of fruit flies (Wong et al. 1984; Sheo et al. 1990; Joshi et al. 1995; Borah 1996; Omar and Hashim 2004; Teixeira and Isaacs 2007; Gogi 2009). Infestation of fruit flies in vegetables can be minimized considerably by adapting different cultural practices, from sowing to harvesting (Khan and Manzoor 1992). Borah (1996) reported that sowing seasons and varieties significantly reduced fruit infestation by *B. cucurbitae*. Similarly, Joshi et al.

(1995) reported a significant variation in fruit infestation of bitter gourd among different sowing times. Agarwal et al. (1987) suggested cultivation practices for the destruction of melon fruit-fly pupae, which are present in the soil. Panhwar (2005), in her article on the control of fruit flies, reported that ploughing the field, 2–3 times, effectively killed the soil aboding larvae and/or pupae, by exposure to the sun and/or other natural enemies and helped to suppress population of fruit flies in the field. Melon fruit flies can be attracted and suppressed by planting maize, as a border-crop, along with the application of protein-bait (Dhillon et al. 2005). Fruit flies spreading from infested to uninfested areas can be interrupted through tight quarantine import barriers (Bayart et al. 1997; Dhillon et al. 2005) as well as through a cold treatment, at 1.1±0.6°C, for 12 days (Armstrong et al. 1995) and heat treatment, at 40°C, for 24 h (Jang 1996) of fruits, infested with eggs and larvae of tephritid fruit flies. Bait stations of different shapes, and colors are good attractants for male and female fruit flies (Alyokhin et al. 2000; Montell 2009). Oriental fruit flies show more attraction to green than yellow (Hardie 1986). Males and females of *Bactrocera tryoni* are more attracted to a white-sphere rather than than orange-, red-, green-, yellow-, or black spheres (Drew et al. 2003). Many types of baits with specific lures are used against fruit flies in Hawaii (Vargas et al. 2008, 2009). Spherical shape objects/traps of yellow, white, or orange color can be used for effective and efficient capturing of various fruit flies (Rousse et al. 2005; Pinero et al. 2006).

21.7.6 Chemical Control

Chemical control of fruit flies is relatively ineffective (Dhillon et al. 2005). In the past, fruit flies were controled with contact insecticides (Bhutani 1975; Perdomo et al. 1976; Gupta and Verma 1979; Lee 1988; Saikia and Dutta 1997; Omar and Hashim 2004). However, these insecticides impose severe health hazard impacts, particularly when such treated fruits and vegetables are consumed unwashed (Wong et al. 1984; Sheo et al. 1990). Conventional insecticides, especially organophosphates (Neilson and Maxwell 1964; Neilson and Sanford 1974; Mohammad and AliNiazee 1989; Reddy 1997; Yee 2007) and pyrethroids (Borah 1997) have been used to control various fruit-fly species, through direct cover sprays or through insecticide mixed-bait systems. Malathion is usually used to develop a bait spray with protein hydrolysat for fruit-fly management (Roessler 1989; USDA 1994). Malathion (0.5%) is reported more effective than carbaryl (0.2%) and quinalphos (0.2%) against melon fruit flies on some cucurbit crops (Bhatnagar and Yadava 1992). A similar report has been published by Srinivasan (1991), who found a significant reduction in melon fruit-fly infestation/damage, when either 0.05% fenthion or 0.1% carbaryl was applied at 50% appearance of the male flowers and repeated, at 3 days, after fertilization. Carbofuran granules, when applied @ 1.5 kg a.i./ha, at the time of sowing, vining, and flowering, reduced 83.35% of fruit infestation in melon by the fruit fly in bitter gourd (Thomas and Jacob 1990). Application of dicrotophos at 600 g a.i./ha and trichlorfon at 1920 g a.i./ha effectively controlled *B. cucurbitae* in a muskmelon crop (Chughtai and Baloch 1988). Talpur et al. (1994) conducted experiments to compare the effectiveness of two insecticides, formathion and trichlorfon. They reported that formathion caused a greater reduction in fruit fly infestations than trichlorfon. Diflubenzuron has also been reported to control the melon fly effectively, causing significant reduction in fruit infestation (Mishra and Singh 1999). Similarly, treatment of pumpkin, with carbofuran @ 1.5 kg a.i./ha, at 15 days after germination, accounted for low fruit infestation and high yields (Borah 1998). However, some of the OPs and OCs insecticides, like malathion, dichlorvos, phosphamidon, and endosulfan have been reported to be moderately effective against the melon fly (Agarwal et al. 1987). According to Chou et al. (2002), weekly application of insecticides, like dimethoate and dibrom, did not give an adequate control of melon fly in Hawaiian crops, rather, allowed 15%–30% or higher fruit infestation. Field and laboratory tests of new insecticides, like, imidacloprid, indoxacarb, pyriproxyfen, spinosad (85% spinosyn A and 15% spinosyn D), thiacloprid, and thiamethoxam against the apple maggot, *Rhagoletis pomonella* (Walsh) (Diptera: Tephritidae) revealed that imidacloprid reduced the oviposition the most, whereas, imidacloprid and spinosad were the most toxic (Reissig 2003). Teixeira and Isaacs (2007) concluded that chemical control of blueberry maggot, under organic standards, in Michigan, included an application of kaolin clay-mineral- particle film, neem, pyrethrum, and spinosad products. An area-wide management of fruit flies includes RSIT (Klassen 2005), which is not economical and practically convenient in developing countries like Pakistan. It is being replaced with new lethal systems and other methods like chemosterilization (Navarro-Llopis et al. 2004) and biopesticides (Dolinski and Lacey 2007). Chemosterilants with IGR mode of action are effective alone and in combination because they impose effects on the reproductive system, natality, growth as well as on metamorphosis and successfully induce sterility in fruit flies with successful results under both in vivo and in vitro conditions (Riddiford and Truman 1978; Alam et al. 2001; Magoc et al. 2005). Chemosterilants like lufenuron (Alemany et al. 2008; Bachrouch et al. 2008), apholate (Wendell and Ruth 1964), hexamethylphosphoramide, and hexamethylmelamine (Chang et al. 1964) tepa, hempa tretamine and ethanesulfonates, (Chance et al. 1969) and biopesticides like fungi (*B. bassiana* & *M. anisopliae*) (Mangan and Moreno 2007) and Bt isolates (Lacey et al. 2001) have been studied as spray against a wide range of fruit flies but for their safe, practical, and effective application. Bait stations have also been proposed (Mangan and Moreno 2007; Robert et al. 2009). Spray applications (Lacey and Goettel 1995) and autodissemination (dissemination of pathogens among target pest populations by using attractive materials/devices) (Maniania 2002; Dimbi 2003; Dimbi et al. 2003; Scholte et al. 2004; Maniania et al. 2006; Vega et al. 2007) are the main methods used for the introduction of microbial agents in fruits and vegetable agro-ecosystems against fruit flies (Talaei-Hassanloui et al. 2007). Microbial infection in fruit flies alters mating and attraction behaviors (Schaechter 2000) and sterility (Dimbi et al. 2009). Kovanci and Kovanci (2006) reported that application of *B. bassiana* as biopesticide alone and with IPM models, suppressed cherry fruit flies in the field without imposing ecotoxicological effects. Various botanical insecticides induce various behavioral and physiological disorders/abnormalities (Schmutterer and Singh 1995; Singh and Singh 1998) in fruit flies. For example, crude extracts as well as enriched extracts of neem-seed kernel deter/repell many dipterans (Chiu 1985; Rice, et al. 1985; Zebitz 1987; Anonymous 1992), especially tephritid flies (Singh and Srivastava 1983; Sombatsiri and Tigvattanont 1984; Areekul et al. 1988; Naumann and Isman 1995; Chen et al. 1996) from oviposition. Similar experiments, carried out by Singh and Singh (1998) revealed that EtOH NSK, neem oil and EtOH Oil, and Acet. DNSKP deterred *Bactrocera cucurbitae* (Coq.) at all concentrations. Ranganath et al. (1997) documented both of the neem-based treatments (Neem oil at 1.2% and neem cake at 4.0% concentration) to be as effective as dichlorvos against the melon fruit fly, *Bactrocera cucurbitae*. Rajapakse and Ratnasekera (2007) concluded that under three intensive systems, that is, integrated, chemical and organic agriculture, excellent control of the fruit fly *Bactocera cucurbitae* can be achieved by integration of neem-based products with predatory ants in an organic system. However, some reports have shown that pure azadirachtin does not deter fruit flies from oviposition (Saxena and Rembold 1984; Singh and Singh 1998). Neem-seed kernels have been reported to have some other volatile and nonvolitile compounds, which actually deter *Bactrocera cucurbitae* and *B. dorsalis* from oviposition (Saxena and Basit 1982; Balandrin et al. 1988; Naumann and Isman 1995; Singh and Singh 1998). Like neem-based botanicals, some other botanicals have also been reported to affect the behavior and physiology of fruit flies. Of these, an extract of *Acorus calamus* has been observed to reduce the adult longevity from 119.2 days to 26.6 days, when 1 mL of 0.15% concentrated extract of *A. calamus* is mixed with 1 g sugar and fed continuously to the adult melon fruit flies (Nair and Thomas 1999).

21.7.7 Future Research

- There is a need to work on pheromone nanotechnology for economic development and application of pheromone-based technologies.
- No fruit-fly special insecticide, except trichlorfon, is available and resistance in fruit flies has been reported against it. There is a dire need to evaluate some soft and new chemistry insecticides and their ecofriendly application techniques.

- Pheromones are species specific. But only methyl euginol is available in Pakistan and is being recommended for all fruit-fly species in Pakistan. Work on other pheromones for other fruit flies of economic importance is needed.
- Lures and toxicants for detection and control programs are not available for all concerned species. Existing and future lures and toxicants should be evaluated by application techniques that comply with current environmental mandates.
- Severe lacunae in sanitation measurement at farmer's fields which embolden fruit flies breeding and many overlapping generations during fruiting season. There is a need to organize campaigns for the awareness of the farmers in this regard.

21.8 CITRUS PSYLLA, *DIAPHORINA CITRI* (KIWAYAMA) (PSYLLIDAE: HEMIPTERA)

The Asian citrus psyllid (*Diaphorina citri* K.) is a sap-sucking, hemipteran insect in the family Psyllidae. It is called Asian Citrus Psyllid because of its origin. It is an important pest of citrus in several countries because it acts as a vector to transmit phloem-limited bacteria (*Candidatus Liberibacter* spp.) to citrus. This bacterium is responsible for citrus greening disease (huanglongbing) of citrus. Citrus greening disease is considered by some to be the world's most serious disease of citrus (McClean and Schwartz 1970; Bove 2006). The psyllid has slowly spread throughout southern Asia and many other countries of the world. The name *citrus psylla* sometimes applied to both *Diaphorina citri* K. and *Trioza erytreae* D. (a psyllid pest of citrus in Africa). Therefore, to avoid confusion, *D. citri* should be referred to as the Asian citrus psyllid and *Trioza erytreae* D. as the African citrus psyllid. There are six *Diaphrina* species reported on citrus but these are nonvector species of relatively little importance. *Diaphorina citri* and *Trioza erytreae* D. are the only known vectors of citrus greening disease (Halbert and Manjunath 2004a,b).

21.8.1 Distribution and Host Range

The Asian citrus psyllid ranges primarily in tropical and subtropical Asia. Geographical origin of this pest is thought by some to be southern Asia, probably India. The psyllid is reported from the following Asian countries: Thailand, Nepal, Hong Kong, Ryukyu Islands, China, India, Myanmar, Taiwan, Philippine Islands, Malaysia, Indonesia, Réunion, Mauritius, Sri Lanka, Pakistan, Afghanistan, and Saudi Arabia (Mead 1977; Halbert and Manjunath 2004a,b). During the 1940s, the psyllid was known to be in South America in Brazil (Mead 1977) in the states of São Paulo and Parana. *Diaphorina citri* invaded the West Indies during the 1990s, and is reported from Abaco Island, Grand Bahama Island, and the Cayman Islands (Halbert and Núñez 2004). In 1998, it was found in Florida (Tsai and Liu 2000). During the last decade, the psyllid was found in Puerto Rico, Cuba, The Dominican Republic, Mexico, Venezuela, and Argentina (Halbert and Núñez 2004; Pluke et al. 2008). In June 2010, many areas or even entire states of the United States were quarantined due to the presence of the Asian citrus psyllid (USDA 2010c). The host range of Asian citrus psyllid includes 25 genera in the family Rutaceae, although not all of these are good hosts (Halbert and Manjunath 2004a,b). In 2004, a list of 59 plant species were reported as hosts by Halbert and Manjunath. The most common or preferred hosts includes citrus (Citrus spp), orange jasmine (*Murraya paniculata* [L.] Jack) orange boxwood (*M. koenigii* [L.].Sprengel), Chinese boxthorn (*Severinia buxifolia* [P.]), and other related species of the family Rutaceae (Mead 1997; Halbert and Manjunath 2004a,b). Due to the restricted host range, monitoring efforts for this pest should be focused on citrus and closely related plants species. The Asian citrus psyllid acts as a vector for citrus greening disease, which also has been found in some noncitrus species. But, this disease is a direct economic problem only in citrus. Some preferred host plants are ornamentals (orange jasmine and Chinese boxthorn) and regulatory efforts to limit the spread of the psyllid and disease includes ornamental plants and negatively affects the ornamental plant industry.

21.8.2 Biology and Seasonal Incidence

A living *Diaphorina citri* adult appears dusty due to the presence of whitish waxy secretions on the body. Adult citrus psyllids are small in size (2.7–3.3 mm long) with a mottled brown body and light brown head. They are active, jumping/flying insects, which can readily fly short distances upon disturbance. The forewing of the Asian citrus psyllid is broadest in the apical half. They are mottled in color and have a brown band extending around the periphery of the outer half of the wing. The antennae have black tips with two small, light brown spots on the middle segments. Nymphs feed exclusively on new growth and are green or yellowish-orange in color. The 1st nymphal instar is light pink with red compound eyes and 0.3 mm long and 0.17 mm wide. During the 2nd instar, the rudimentary wing pads are visible on the dorsum and size increases to 0.45 mm long and 0.25 mm wide. Antennal segmentation starts at the 3rd instar, which also has well-developed wing pads and averages 0.74 mm long and 0.43 mm wide (Tsai and Liu 2000). Mesothoracic wing pads extended towards compound eyes during 4th instar period and the nymph averages 1.01 mm in length and 0.7 mm in width (Tsai and Liu 2000). The number of setae is one and two on each antenna during the 3rd and 4th instars, respectively (Husain and Nath 1927). The 5th instars averaged 1.6 mm long and 1.02 mm wide and have three setae on each antenna while the metathoracic wing pads reach the fourth abdominal segment. The color varies among mature nymphs from bluish green to pale orange (Tsai and Liu 2000). The average size of an egg of *D. citri* was reported to be 0.31 mm long and 0.14 mm wide (Tsai and Liu 2000). Eggs are elongated, almond-shaped, thicker at the base, and tapering toward the distal end. Newly laid eggs are pale-yellow, turn yellow, and finally bright orange before hatching. There are two distinct red eyespots at egg maturity. The adult female of *D. citri* places eggs on plant tissue with the long axis vertical to the surface. Adults of *D. citri* are found resting or feeding on leaves. They feed with their heads down almost touching the leaf surface. Their bodies are held at a 45° angle from the leaf surface due to the shape of their heads. The psyllid is a sap-sucking insect and inserts its mouthparts into plant tissue. Adults can feed on young stems and on leaves during all stages of development while the nymphs always feed on the new growth. The lifecycle includes an egg stage, five nymphal instars, and adults. Adult males and females find each other for mating in-part using substrate-borne vibrational sounds (Wenninger et al. 2008a). But there is some evidence of the release of sex pheromones for mating by the female as well (Wenninger et al. 2008b). Adult psyllids mate multiple times with different partners. Activities like mating, oviposition and other movements are restricted to daylight hours (Wenninger and Hall 2007). Females continuously lay eggs throughout their lives if young leaves are present. They lay 500 to 800 or more eggs over a period of 2 months. The egg is anchored to plant tissue and large numbers of eggs may be found on a single flush shoot. During feeding, the nymphs continuously secret honeydew and a thread-like waxy substance from circumanal glands (Tsai and Liu 2000). A black sooty mold often develops on the honeydew deposited on the lower leaves. Nymphs pass through five instars. The total life cycle requires from 15 to 47 days, depending upon the season. Adults may live for two to several months. There is no diapause throughout the year. Populations are low in winter (the dry season) with nine to ten generations a year. However, 16 generations have been observed in field cages.

21.8.3 Behavior

Adults of *Diaphrina citri* are active, jumping/flying insects. The psyllid flies pretty well in the absence of wind. The longest flight duration observed was 47 minutes for females and 49 minutes for males. Similarly, the maximum distance covered was 978 m for females and 1241 m for males. The psyllids behave differently at different temperatures. For example, at 25°C, adult life span is between 40 and 44 days. The maximum adult longevity ranges from 51 days at 30°C to 117 days at 15°C. Adults feed and mate only during daylight hours (Hall 2008). The ideal temperature for psyllids is between 20°C and 30°C and at this temperature a single female psyllid can lay between 300 and 750 eggs but if the temperature increases to above 32°C egg laying capacity decreases to fewer than 70 eggs per female (Rogers and Stansly 2012).

21.8.4 Damage and Losses

Asian citrus psyllid has piercing-sucking mouthparts used for feeding on plants. It directly damages citrus and related ornamental plants (Halbert and Manjunath 2004a,b). Damage results from the withdrawal of large quantities of sap from the foliage. During feeding, they inject salivary toxin that stops terminal elongation and causes malformation of leaves and shoots. An immature (nymph) can cause permanent malformation of a citrus leaf in less than 24 h feeding on that leaf. They severely damage citrus flush resulting in the abscission of leaves and shoots or malformation of leaves. The most serious damage of Asian citrus psyllid is the transmission of the phloem-limited bacteria (*Candidatus Liberibacter* spp.) that causes citrus greening disease (Huanglongbing) in citrus (McClean and Schwartz 1970; Bove 2006).

21.8.5 Management

A citrus plantation with guava at a 1:1 tree ratio is very effective to control *Diaphrina citri* because guava reduces infestations of the psyllid on citrus. Toxins associated with guava might exist that negatively affect the biology of the psyllid, which results in reduction of psyllid reproduction in citrus (Hall et al. 2008). Several chemicals effectively control the adults and nymphs of *Diaphrina citri*. Insecticides such as chlorpyrifos, imidacloprid, aldicarb, carbaryl, fenpropathrin, and dimethoate are effective against this pest (Browning et al. 2005). The important pesticides with recommended doses against citrus psylla are Polytrin-C 440 EC 125 mL/hac, Bifenthrin 10 EC 50 mL/hac, Actara 25 WP 25 g/hac, Imidacloprid 70 WP 25 g/hac. Natural enemies for biological of *D. citri* such as syrphids, chrysopids, and different species of parasitic wasps, such as *Tamarixia radiate* W. are effective against *D. citri* (Hall et al. 2008). The parasite *Diaphorencyrtus aligarhensis* S. is also an important biocontrol agent of *D. citri*.

21.8.6 Future Research

Scientific research on the disease complex is greatly needed for developing Asian citrus psyllid management strategies. However, disease-resistant citrus trees have been developed but such varieties are not yet available (Hall et al. 2013). Chemical control can cause health hazards to nontarget organisms. The skillful application of biocontrol could be very effective at managing citrus psyllid. Nursery stock must be propagated in an approved greenhouse structure. Infested trees at retail nurseries are subject to quarantine action. An integrated management approach by using different control measures could help to manage the Asian citrus psyllid and to improve the citrus industry.

21.9 MANGO MEALYBUG, *DROSICHA MANGIFERAE* (MARGARIDIDAE: HOMOPTERA)

Mango mealybug, is a pest of mango crops. Pakistan is a significant exporter of mangoes (Balal et al. 2011) and this fruit has commercial importance and is grown in Punjab and Sindh provinces on large acreages. Mango has faced insect pest problems for many years (Mohyuddin and Mahmood 1993; Masood et al. 2009). Among these insect pests, mango mealybug is a serious pest. This pest was reported in the subcontinent in 1907 and caused serious losses to mango. Mango mealybug hinders the mango sector from realizing its full potential by posing serious threats to the exploitation of foreign markets and, as such is helping to jeopardize the lucrative trade in fresh fruits from the region. When this pest is introduced into the orchard, it is very difficult to control its populations (Green 1908); however a number of control practices are used to manage this pest. Approximately 20 species of mango mealybugs are recorded in Southeast Asia of which *Drosicha mangiferae* (G.), *Drosicha stebbingi* (G.), and *Rastrococcus iceryoides* (G.) are the most destructive pest of mango. *Drosicha mangiferae* (G.) is the serious pest of mango orchards in Pakistan. This is a polyphagous, dimorphic, and destructive pest (Karar et al. 2006) and severe infestations have a significant economic impact on mango producers and traders. Mango mealybug infestation losses varied from 53% to 100% reduction of total production depending on the variety, the site of the orchard, and the period of harvest (Nebie et al. 2016).

21.9.1 Distribution and Host Range

Worldwide, mango mealybug constitutes a major threat to mango production, causing heavy preharvest and postharvest losses and curtailing expansion of both domestic and international trade of fruits (Osman et al. 1989). This pest is considered an important phytosanitary threat and infestation of fruits results in the rejection of exports or sold locally with significant economic implications as this could seriously jeopardize the viability and future access to these markets (Pieterse et al. 2010). In the tropics, the problem is aggravated by the prevailing humid warm climate, which is conducive for overlapping fruiting patterns, resulting in overlapping generations of mealybug populations and the potential for year-round infestation, though at varying degree of severity. *Drosicha mangiferae* is not only a pest of mango but also some 70 other plants (Tandon and Lal 1978). Although this pest is considered a main pest of mango, in heavy infestation areas, it also attacks plum, peach, papaya, many native plants and weeds, and every species of citrus (Khan 2001). Many other species of related mango pests also attack in different regions of the world. *Rastrococcus iceryoides* (Green), also an invasive species, is reported in Tanzania, , and Malawai. Mango mealybug is also found in Indonesia, Malaysia, India, Sri Lanka, and Pakistan (Muniappan et al. 2012). *Rastrococcuc spinosus* is also found in southern areas of Pakistan (Mehmood et al. 1980) but is not considered as economically important compared to the *D. mangiferae* (Karar et al. 2006).

21.9.2 Biology and Seasonal Incidence

Mealybug appears in March–April and is univoltine. Males are winged insects, which do not cause damage to the plants, short lived, and only mate with the female. Female mealybug starts to lay eggs in April–May. Purple colored eggs are laid in a sac surrounded by a waxy thread like mass. Eggs hatch in December–January and nymphs start to climb on the tree to reach the fruiting bodies of the mango plant. Nymphs go through three instars. First instars complete development in 45–71 days, 2nd instars in 18–38 days, and 3rd instars in 15–26 days. The whole life cycle completes in 78–135 days. Environmental factors—especially temperature and relative humidity, directly and indirectly affect the tritrophic association of host plants, mealybug, and their natural enemies and sometimes results in mealybug outbreaks (Kontodimas et al. 2004; Walton and Pringle 2005; Chong and Oetting 2006; Nakahira and Arakawa 2006; Gutierrez et al. 2008). The host plant characteristics may also hinder or favor reproduction, development, and survival of mango mealybugs (Leru and Tertuliano 1993; Yang and Sadof 1995; Nassar 2007).

21.9.3 Behavior

Sexual dimorphism exists in mealybug populations. Adult female mango mealybugs are large and wingless while males are small and winged. Females are recognized by their flat shape with white flocculent wax covering. Nymphs are pink and brown in color and flat in shape. Both adult and immature are highly pestiferous and have a strong preference for mango. Many studies have treated the dispersal activity in the newly emerged crawlers or 1st instar nymphs and adults. However, adult females and other nymphal stages move only short distances but also can be moved a few meters to several kilometers by wind, air currents, water, human, domestic and wild animals, and ants (Kosztarab and Kozár 1988; Ranjan 2006).

21.9.4 Damage and Losses

Mango mealybug is a serious pest of mango in Asia. Immatures and females suck the cell sap from different parts of the plants like tender leaves, inflorescences, shoots, and fruit peduncles (a stalk like part through which any part of the plant attach). Its excessive feeding results in drying of inflorescence and fruit drop (Tandon and Lal 1978). This pest also secretes honeydew on the plant (Pruthi and Batra 1960). The growth of sooty mold occurs due to this secretion (Smith et al. 1997), which creates disturbances in the photosynthetic activity of leaves and makes the fruits unmarketable. Mealybug infestations disturb the normal morphological and physiological process of the affected plants, which results in the fall of leaves and spikes, delay of flowering, and slows the emission of new branches. Some studies also reported significant reduction in weight and size of fresh mango fruits (Ivbijaro et al. 1991; Tobih et al. 2002). This pest causes more than 50% yield loss to all genotypes of mango. Due to persistency of this pest, many farmers in the Pakistan have uprooted mango orchards (Karar et al. 2006).

21.9.5 Nonchemical Control

There are a number of nonchemical control methods for the management of mango mealybug. Sticky traps are used as a mechanical control method. These traps are installed on the base of the plants. Mealybugs stick to the trap while climbing on the mango tree. Female mealybugs lay eggs into the soil. Its eggs start to hatch in the end of February. Nymphs crawl and climb on the tree and start to suck sap. At this point, a polythene sheet can be installed at the tree stem in such a way that its upper margin establishes like a funnel around the tree stem and the lower margin scrumps up and is fixed to the stem. Mud is used to cover the lower margin to prevent the escape of nymphs from cracks and crevices. This trap is attached at a height of 3–4 feet from the ground. This technique is more effective than sticky traps because nymphs make a bridge while sticking on the trap and provide a way for their companions to climb on the tree and approach fruits. Some times mealybugs escape through cracks under the bands. This can be minimized by spreading parathion dust below the band on the tree. As mealybug lay eggs into the soil, one should plough the soil with a turning plough to expose eggs to the sun and natural enemies like predators including birds. Ploughing of soil during May–June results in the destruction of egg masses due to summer heat. Application of irrigation by a flooding method in October kills the eggs in the soil. Biological control with natural enemies is cost-effective and a most recommended method in many countries. *Monochillus sexmaculatus*, *Rodolia fumida*, and *Suminius renardi* are important biological control agents of mealybug nymphs. *Beauveria bassiana*, an entomopathogenic fungus, effectively controls the nymph of this pest. Foliar application of *Beauveria bassiana* at 5 g/mL per liter of water effectively controls the population of mealybugs. *Gyranusoidea tebygi* and *Angyrus mangicola* were introduced into Africa from India to control mango mealybug (Muniappan et al. 2012). However, several factors affect natural enemies, which include the cryptic behavior of mealybugs, tending of mealybugs by ants, interference, and intraguild predation (Chong and Oetting 2007). Host plant resistance is the innate property of a plant to withstand damage and tolerate it. It inhibits the growth of the insect due to morphological characters, good defense systems (antixenosis), and tolerance. This technique has been most useful in IPM programs during the last 2 decades. This technique works with the combination of various tactics, which makes the environment unfavorable for insect growth (Dhaliwal and Singh 2004). Less susceptible mango varieties use multiple IPM tactics in a compatible manner. This has been proving most useful in eliminating pest populations to the extent possible.

21.9.6 Chemical Control

There are many insecticides available for mango mealybug control. However, the extensive use of these chemicals is objectionable because of their hazardous impact on human health and the environment, and the risk of insecticide resistance development in mealybug is a concern (Flaherty et al. 1982). In addition to health and environmental hazards caused by pesticides, chemical control does not generally provide adequate control for mealybugs in the long term, owing to their typical waxy body cover, their cryptic behavior, and clumped spatial distribution pattern (Franco et al. 2009). It is not viable to have a spraying program that can manage the mealybugs on infested plants (Sagarra and Peterkin 1999). Moreover, repeated insecticide use, especially broad-spectrum chemicals, has been reported to adversely impact the mealybug's natural enemies (Walton and Pringle 1999). However, some insecticides, that is, Chlorpyrifos 40% w/v at 50 mL 10 L of water, Spintoram 25% w/w at 7.5 mL/10 L water, and Sulfoxaflor 50% w/w at 7.5 mL/10 L water can be used as chemical control substances.

21.10 RED PALM WEEVIL, *RHYNCHOPHORUS FERRUGINEUS* (OLIVIER) (COLEOPTERA: CURCULIONIDAE)

The Red Palm Weevil (RPW), *Rhynchophorus ferrugineus* (Olivier) (Coleoptera: Curculionidae), is the most destructive and voracious feeder of date palm worldwide (Wakil et al. 2015). The RPW is an important invasive pest, which has invaded and is fully established in more than 50% of the date palm growing areas of the world. It has a high fecundity (Faleiro 2006), is capable of living and interbreeding in the same tree for several years (Rajamanickam et al. 1995; Avand-Faghih 1996), can fly long distances (Wattanapongsiri 1966), and is tolerant to a wide range of climatic conditions due to its hidden habit in palm trees.

21.10.1 Distribution and Host Range

The pest is native to South Asia, Southeast Asia, and Melanesia, where it had been a key pest of coconut palms (Lefroy 1906a,b; Brand 1917; Viado and Bigornia 1949; Nirula 1956). It is distributed within 35° North and 15° South latitudes. In the southern hemisphere it is reported from Papua New Guinea, Indonesia, and Tanzania. In the northern hemisphere the pest is present in the Philippines, Thailand, Burma, India, Pakistan, and the Arabian countries of Iraq, Saudia Arabia, UAE, and others (Ramachandran 1998). After the 1980s, the weevil was recorded from North Africa (1992) into southern Europe (1994), and was reported for the first time in the south of Spain (Cox 1993; Barranco et al. 1995). Eventually it reached North America in 2009 (Roda et al. 2011). The pest is polyphagous in nature and found in different climates and farming systems. It has been recorded attacking greater than 40 palm species in 23 genera and has caused extensive mortality of the Canary Island date palms (*P. canariensis* Chabaud) in the Mediterranean and date palms (*P. dactylifera* L.) in the Middle East and North Africa (Murphy and Briscoe 1999).

21.10.2 Biology and Seasonal Incidence

Adult RPW are large insects, 2–3 cm long and 1 cm wide (Giblin-Davis et al. 2013) with the male a little smaller than the male. They exhibit variability in coloration that is commonly composed of an orange thorax with varying amounts of discrete black mottling, and black and orange striations on the elytra (Rugman-Jones et al. 2013). Seasonal flight activity patterns are different in different countries when captured with pheromone traps. A high peak activity was reported during April–May, followed by a smaller peak during October–November in Al-Hassa and Al-Qateef regions of Saudi Arabia (Vidyasagar et al. 2000). The peak active period during April–June was recorded in Israel in pheromone traps and decreased with the onset of cold winter (Soroker et al. 2005). A similar trend of activity was reported in Egypt by El-Garhy (1996). However, also in Egypt, El-Sebay (2003) reported two active periods, the first during April and the second during November, but no relationship was observed between seasonal population fluctuations and weather factors. Two flight peaks were recorded from Amria district (Egypt) during two years of pheromone trapping of RPW; the 1st peak was recorded in the 1st half of September (mean number 0.13 weevil/trap), the 2nd peak occurred in the 2nd half of March (Hashem 2016). Al-Saoud and Ajlan (2013) reported highest catches in March and April, while this number gradually decreased in September when pheromones were installed in date palm plantations in Al Rahba (U.A.E.). Previous studies have found similar patterns in weevil populations (Abraham et al. 1999a,b; Vidyasagar et al. 2000; Al-Saoud et al. 2010). In Goa, Faleiro (2005) reported peak activity in October and November and minimal activity in June and July. Faleiro (2005) further revealed a correlation between prevailing temperature and rainfall. The weevil infestation in Japan was recorded between the temperature ranges of 30°C and 40°C, even in winter season. El-Lakwah et al. (2011) conducted field experiments in Wardan and Abu-Ghalep villages (Egypt). They found the lowest adult population was recorded during December and January. The population showed four peaks each year. The following four peaks of emergence were recorded during 2009: 2nd week of April, 1st week of June, 1st week of August, and 2nd week of November. In 2010, the following four peaks were also recorded: 4th week of March, 3rd week of June, 3rd week of July, and 2nd week of November. High seasonal activity of RPW in the region as noticed from pheromone trap captures is between April–May and again from October–November. However, from the weevil management point of view the first peak (April–May) is important as most of the eggs laid by female weevils during this period hatch into damage-inflicting grubs, while most of the eggs laid during the second seasonal peak are caught in the winter and fail to hatch (Alhudaib et al. 2008). In Bangalore, South India, weevil activity was observed year-round but the maximum trap catches were recorded during June–August and October–December (Chakravarthy et al. 2014).

21.10.3 Life History

The female commences egg laying 1–7 days after mating in scooped small holes on tender parts of young palms. In mature trees, females lay eggs in leaf axils, wounds, injury caused by rhinoceros beetles or disease, cuts on trees, or most preferably in exposed plant tissues. A single female can lay 58–531 eggs during her life span, which incubate for 1–6 days before hatching (Abraham et al. 2002). A whitish-yellow apodous (legless) larva hatches from the egg, which live for 25–105 days. Larval development requires 3–9 larval instars (Jaya et al. 2000). These variations in larval instars may be due to rearing conditions and food host. The last instar larvae pupate in a cocoon made up from chewed fiber; pupation lasts for 11–45 days. After emergence, the adult weevil still remains in the pupal cocoon for an extra 4–12 (8 days on an average) days before exiting (Menon and Pandalai 1960a,b). It is believed that during this period, the weevil completes its sexual maturity (Hutson 1922). A week after emergence, females start egg laying which lasts for 8–10 weeks and stops before 10 days of dying. Adult weevils live for about 2–3 months, feeding on palms, mating multiple times, and laying eggs. It has been theoretically assumed that in the absence of biotic and abiotic factors a single pair of RPW can give rise to more than 53 million progeny in four generations (Menon and Pandalai 1960a,b). However, reports based on successive research deny this highly unusual multiplication (Rahalkar et al. 1972; Avand-Faghih 1996; Esteban-Duran et al. 1998; Cabello 2006). The high reproductive rates may be due to continuous and high ovipositional rates year around with a few months of relatively higher fecundity. This may be the case in countries with a hot humid climate where temperature is suitable for RPW development throughout the year as mentioned by Salama et al. (2002). If a fraction of the brood survives, they may cause substantial threats to the host plants.

21.10.4 Behavior

During flight initiation, weevils first wipe their rostrum with the prothoracic legs, pump their pygidium, lift their elytra, and extend their wings. Just before flight, weevils raise the mesothoracic legs above the pronotum. Weevils preparing for a subsequent flight frequently skip the first two preflight events (Weissling et al. 1994). Adult weevils are strong fliers and capable of flying long distances to find hosts in widely separated areas (Abraham et al. 2002). A weevil can sustain constant flight for 900 m (~900 yards) and can cover up to 7 km (~4.3 miles) in 3–5 days (Abbas et al. 2006). Research revealed that most adults make short flights (<100 m); 46% of adults are able to perform medium- or long-distance flights (from 100–5000 m, and more than 5000 m, respectively), which was not influenced by size and sex of individuals. The beetle remains active during early hours of the day and the last few hours before sunset; it has also been recorded that RPW can fly up to 900 m during 2 days (Nirula 1956). Diurnal activity of the weevil shows activity from midnight-0600 h in India and between 18:00–08:00 h in Sri Lanka (Faleiro and Satarkar 2003). Activity period differs in Goa (India) due to weather; the monsoon in Goa is restricted between June and September while Sri Lanka receives rain almost throughout the year. This suggests that weather parameters play an important role in determining RPW activity (Faleiro 2006).

21.10.5 Damage and Losses

In palms, mortality is caused by internal larval feeding as larvae spend their entire life concealed within the palm making early detection of infestations difficult (Giblin-Davis et al. 2013). Internal larval feeding of 2–3 generations for 1–2 years can cause trunk collapse or mortality of the apical growing areas (Giblin-Davis et al. 2013). Under sever infestations, date palm yields may be reduced up to 90%. The larvae (sometimes over 80/tree) can tunnel in the trunk, weaken the tree, and bringing about its decline and even breakage under pressure (such as strong winds). Infestations may remain undetected until damage is extensive and even upright infested trees have to be destroyed lest they fall and damage other trees. The estimated economic losses in the KSA reached 1%–5% date palm plantations (El-Sabea et al. 2009) and the estimated cost of damage ranged from 5.18 to 25.92 million USD, respectively. They further observed that in a single year up to three complete generations can take place inside one single palm and indirect losses increase this figure several fold. The curative treatments of young palms with infestation levels of 1%–5% may range from $20.73 to $103.66 million US, respectively (El-Sabea et al. 2009). Due to the high infestation, losses have been reported from 10 tons to only 0.7 ton per ha in the gulf region (Gush 1997). Recent reports regarding the occurrence of RPW in California put the annual spending (70 million USD) on ornamental palms at risk (Hoddle and Hoddle 2015). In Pakistan, 10%–20% loss has been reported in different date palm varieties due to this voracious pest (Baloach et al. 1992). Resultantly, these annoying pests cause direct losses worth thousands of millions of dollars annually in money involved in management efforts to reduce populations below economic threshhold levels (Simberloff et al. 2013).

21.10.6 Nonchemical Control

Among nonchemical approaches, early detection of RPW is a key component to avoid extensive damage to date palms, and pest dispersal to neighboring areas (Faleiro et al. 1998; Hallett et al. 1999; Faleiro 2006). Moreover, agronomic practices such as phytosanitary measures, field sanitation, host removal, eradication, and host disposal can also greatly reduce population build up. The deployment of microwave radiation to the host plant can effectively control all the developmental stages of RPW by increasing the tree temperature without harming the host plant (Massa et al. 2011). Uses of semiochemicals play a key role in managing RPW populations and are a convenient tool for IPM programs. The general methods for monitoring that are usually deployed are mass trapping, attract-and-kill, attract-and-infest, and push-pull (Cook et al. 2007; Witzgall et al. 2010). Host plant resistance (HPR) also influences the degree of damage done by insects (Painter 1951). Sterile insect technique (SIT), within an area-wide integrated pest management (AW-IPM) approach has proven to be a powerful control tactic for the creation of pest-free areas or areas of low pest prevalence (Simmons et al. 2009). A vast array of RPW natural enemies including vertebrates, insects, mites, entomopathogenic bacteria, entomopathogenic fungi (EPNs), entomopathogenic nematodes (EPNs), yeasts, and viruses are reported to infest RPW across many countries of the world. Entomopathogens EPNs and EPFs are commonly used against RPW under laboratory, field and semifield conditions successfully in many parts of the world (Faleiro 2006).

21.10.7 Chemical Control

The most commonly used control treatments for this pest are chemical insecticides such as diazinon, imidacloprid, phosmet, and phosphine (Llácer et al. 2010). Pesticides are applied with numerous application methods, including wound dressing, frond axil filing, fumigation, injection, and spraying and are being tried for the control of RPW infestations (Al-Rajhy et al. 2005). During the early stage of RPW infestation, damage symptoms are difficult to detect, in this regard intentions are generally imposed on preventive control measures. This is challenging because of the concealed nature of RPW but chemical insecticides can play an important role (Llácer et al. 2010). For preventive treatments, the following measures must be kept under consideration: injection of palm trunk with chemical insecticides during onset of infestation (Wickremasuriya 1958; Murphy and Briscoe 1999); soaking of palm trunk with insecticides (Abraham et al. 1998); treatments with repellents and leaf axil filling insecticide dusts such as (BHC) and sand mixture (Anonymous 1956); drenching of infested tree crown with insecticides (Kurian and Mathen 1971); and insecticidal treatments of pruned off shoots immediately after removal (Faleiro 2006). Besides chemical insecticides, fumigants such as aluminium phosphide are very effective against cryptic stages of this pest. Good results were obtained under laboratory and field experiments (Mesallam 2010). Llácer and Jacas (2010) suggested that under containment, even low dosages of aluminium phosphide are able to kill all developmental stages of RPW. They found complete mortality of RPW when phosphine fumigant was applied on infested canary palm crowns (Llácer and Jacas 2010).

21.10.8 Future Research

- Early detection of infestation of RPW
- Current tree departure and postentry quarantine protocols
- Use of semiochemicals and trapping systems
- Use of biological control agents
- Use of chemical control
- Identification of resistance and induce plant defenses

21.11 CONCLUSION

Agricultural and horticultural products face a number of constraints due to insect pests and the new emergence of any indigenous or invasive species always increases the threat. Pest problems can be correlated with environmental hazards, economic losses, and food security, which lead towards their potential control. Among different control measures, chemical control of insect pests is quick and rapid, which is why it is adopted most by the farmers. However, heavy use of chemical treatments causes environmental damage, harms nontarget organisms, disturbs beneficial fauna, and also leads to the development of insecticide resistance. Thus, a comprehensive knowledge of host-plant association, growth of pests, and feeding is essential for designing efficient and sustainable management strategies. Pest control tactics are changing to humble approaches with fewer hazards; an example would be biological control. Integration of all the possible, approachable, and safe meaures with an understanding of pest life cycles and behavior could prove efficient to manage pests. These programs should be designed to produce a pest-free environment with less environmental contamination in present crops, and also to reduce the potential for damage in subsequent crops. All future directions should be evaluated to cope with these originating and emerging pests.

For most insect pests, monitoring techniques are a prerequisite for effective management as insect populations can fluctuate dramatically over time, especially in changing climatic conditions. Climatic cycles may trigger insect population cycles through changes in the insect or in the host. Lacking predictive modeling and monitoring techniques is a major limitation for the effective development of pest management strategies. Globally adopted pest monitoring techniques are not being imported and marketed by pesticide industries in Pakistan. Traditional insecticides are available, however, the people involved in their application are not properly trained and or not well informed about their appropriate application methodology and strategy. Currently, involvement of agricultural organizations in combating the cotton crises under the effectual and competent supervision of government has resulted in a better crop situation this year (2016–2017). But unfortunately, this kind of supervision is lacking in other economic crops. Pest economic decision levels are too outdated and need to be updated according to the changing scenario of pest status, pest-ecology, and climate. The Sterile Insect Technique (SIT) has also been successful at keeping pink bollworm infestations at low levels in developed countries. Briefly, SIT involves the rearing and sterilization of insects, which are released in the fields where they then contest with the native existing wild male population for mating. Functional, purposeful, and serviceable biological control systems for these insect pests is also currently lacking. A practical attention is needed for biodiversity conservation which has been disturbed due to the insecticide treadmill.

The government should initiate legislation to enhance the vigilant monitoring and implementation of rules and regulations regarding crop pests, pesticides, and their application methods. The capacity building of farmers should be carried out on a regular basis through provincial extension departments for efficient use and management of resources during crop production. The agriculture universities should also take part in the training of farmers by interaction with their postgraduate students by giving them different tasks during their study. In addition, they can also start campaigns by focusing on different crop growing areas on a seasonal basis. All the active departments should focus the 4Rs' principles (right time of sowing, right dose of pesticides, right time of crop management application, and right method of application) for efficient use of crop management resources. New chemistry/reduced-risk insecticides adopted globally should be imported, tested, and recommended for field applications. The government should require pesticide companies to import and market globally successful pest monitoring. Similarly, the determination/monitoring of insecticide resistance development in insects and its mitigation should be assessed by using alternate chemical compounds, their combinations, synergists, additives or alternate methods of pest control. Resistance development in transgenic crops should also be estimated on a regular basis especially in cotton.

Economic decision levels (EDLs) are now outdated due to changes in various inputs including climate. Therefore, research organizations and agriculture universities should be given the task to reestablish economic injury levels (EILs) and economic threshold levels (ETLs) of all the major and emerging pests. The conservation of natural enemies in the environment for the biological control of insect pests is another issue, which can be solved through training of farmers by provincial extension departments. Conservation can be obtained by construction of artificial structures, provision of supplementary food in the field, stipulation of alternative hosts in the absence of host plants, improvement of pest-natural enemy synchronization by developing natural enemies, commercial mass-rearing laboratories, and dissemination (releases) in fields during pest infestation, and modification of adverse agricultural practices, which cause damage to natural enemies.

The import and dissemination of the latest transgenic crops and insect pest-control tactics (e.g., SIT [Sterile Insect Technique] establishment for fruit flies and pink bollworm pests) will be effective for sustainable management of all pests. The ever-changing climate (considered to be a major threat for the emergence and invasion of insect pests) should be considered during management decisions. In conclusion, using the combination of all the above-mentioned practices, stakeholders and organizations should develop flexible and climate-change resilient sustainable integrated pest management programs.

ACKNOWLEDGMENTS

The Red Palm Weevil and fruit fly sections of this chapter were supported through grants on RPW by the Higher Education Commission, Islamabad (Pakistan).

REFERENCES

Abbas MST, Hanounik SB, Shahdad AS, Al-Bagham SA. 2006. Aggregation pheromone traps, a major component of IPM strategy for the red palm weevil, *Rhynchophorus ferrugineus* in date palms (Coleoptera: Curculionidae). *J. Pest Sci.* 79: 69–73.

Abdullah K, Akram M, Alizai AA. 2002. Nontraditional control of fruit flies in guava orchards in D. I. Khan. *Pak. J. Agric. Res.* 17: 195–196.

Abdullah K, Latif A. 2001. Studies on bait and dust formulation of insecticides against fruit fly (Diptera; Tephritidae) on melon (*Cucumis melon*) under seemi arid conditions of D. I. Khan. *Pak. J. Biol. Sci.* 4: 334–335.

Abraham VA, Al-Shuaibi, MA, Faleiro JR, Abuzuhairah RA, Vidyasagar PSPV. 1998. An integrated management approach for red palm weevil, *Rhynchophorus ferrugineus* Oliv., a key pest of date palm in the Middle East. *Sultan Qabus Univ. J. Sci. Res. Agric. Sci.* 3: 77–84.

Abraham VA, Faleiro JR, Nair CPR, Nair S S. 2002. Present management technologies for red palm weevil *Rhynchophorus ferrugineus* Olivier (Coleoptera: Curculionidae) in palms and future thrusts. *Pest Manag. Hort. Ecosyst.* 8: 69–82.

Abraham VA, Nair SS, Nair CPR. 1999a. A comparative study on the efficacy of pheromone lures in trapping red palm weevil, *Rhynchophorus ferrugineus* Oliv. (Coleoptera: Curculionidae) in coconut gardens. *Indian Coconut J.* 30: 1–2.

Abraham WR, Stro$mpl C, Meyer H et al. 1999b. Phylogeny and polyphasic taxonomy of *Caulobacter* species. Proposal of Maricaulis gen. nov. with Maricaulis maris (Poindexter) comb. nov. as the type species, and emended description of the genera Brevundimonas and Caulobacter. *Int J Syst Bacteriol* 49: 1053–1073.

Abro GH, Syed TS, Tunio GM, Khuhro MA. 2004. Performance of transgenic *Bt* cotton against insect pest infestation. *Biotechnol.* 3: 75–81.

Afzal M, Javed H. 2001. Evaluation of soaked wooden killer blocks for male annihilation (MA) on fruit fly Bactrocera spp. (Diptera: Tephritidae). *Online J. Biol. Sci.* 1(7): 577–579.

Agarwal ML, Sharma DD, Rahman O. 1987. Melon fly and its control. *Indian Hortic.* 32(3): 10–11.

Agarwala BK, Das J. 2012. Weed host specificity of the aphid, *Aphis spiraecola*: Developmental and reproductive performance of aphids in relation to plant growth and leaf chemicals of the Siam weed, *Chromolaena odorata*. *J. Insect Sci.* 12(24): 1–13.

Aheer GM, Ali A, Munir M. 2008. Abiotic factors effect on population fluctuation of alate aphids in wheat. *J. Agric. Res.* 46: 367–371.

Aheer GM, Rashid A, Afzal M, Ali A. 1993. Varietal resistance/susceptibility of wheat to aphids, *Sitobion avenae* F. and *Rhopalosiphum rufiabdominalis* Sasaki. *J. Agric. Res.* 31: 307–311.

Ahmad M, Arif MI. 2010. Resistance of beet armyworm, *Spodoptera exigua* (Lepidoptera: Noctuidae) to endosulfan, organophosphorus and pyrethroid insecticides in Pakistan. *Crop Prot.* 29: 1428–1433.

Ahmad M, Arif MI, Ahmad Z, Denholm I. 2002. Cotton whitefly (*Bemisia tabaci*) resistance to organophosphate and pyrethroid insecticides in Pakistan. *Pest Manag. Sci.* 58: 203–208.

Ahmed N, Kanan H, Inanaga S, Ma Y, Sugimoto Y. 2001. Impact of pesticide seed treatments on aphid control and yield of wheat in the Sudan. *Crop Prot.* 20: 929–934.

Ahmed S, Rehman OU, Ahmad F, Riaz MA, Hussain A. 2007. Varietal resistance in maize against chemical control of stem-borer, shoot-fly and termites in Sahiwal, Punjab, Pakistan. *Pak. J. Agric. Sci.* 44(3): 493–500.

Akhtar IH, Khaliq A. 2003. Impact of plant phenology and coccinellid predators on the population dynamic of rose aphid *Macrosiphum rosaeiformis* Das (Aphididae: Homoptera) on rose. *Asian J. Pl. Sci.* 2: 119–122.

Akhtaruzzaman M, Alam MZ, Ali-Sardar MM. 2000. Efficiency of different bait sprays for suppressing fruit fly on cucumber. *Bull. Inst. Trop. Agr. Kyushu Univ.* 23: 15–26.

Akhtaruzzaman M, Alam MZ, Sardar MA. 1999. Suppressing fruit fly infestation by bagging cucumber at different days after anthesis. *Bangladesh J. Entomol.* 9(1–2): 103–112.
Alam MA. 2010. *Encyclopedia of Applied Entomology.* Anmol Publications Pvt. Ltd., New Delhi, India.
Alam, MJ, Funaki Y, Motoyama N. 2001. Distribution and incorporation of orally ingested cyromazine into house fly eggs. *Pestic. Biochem. Physiol.* 70: 108–117.
Alemany A, Gonzalez A, Juan A, Tur C. 2008. Evaluation of a chemosterilization strategy against *Ceratitis capitata* (Diptera: Tephritidae) in Mallorca island (Spain). *J. Appl. Entomol.* 132: 746–752.
Alerstam T. 2006. Conflicting evidence about long-distance animal navigation. *Science* 313: 791–794.
Alhudaib K, Arocha Y, Wilson M, Jones P. 2008. First report of a 16SrI "Candidatus Phytoplasma asteris" group phytoplasma associated with a date palm disease in Saudi Arabia. *Plant Pathology* 57: 366.
Ali M, Ahmad Z, Tanveer M, Mahmood T. 1995. *Identification and Characterization of Virus in: "Cotton Leaf Curl Virus in Punjab during 1991–92".* CLCuV Proj. Ministry Food, Agric. & Livestock. Govt. Pak. Asian Dev. Bank. pp. 7–11.
Alim MA, Hossain MA, Khan M, Khan SA, Islam MS, Khalequzzaman M. 2012. Seasonal variations of melon fly, *Bactrocera cucurbitae* (Coquillett) (Diptera: Tephritidae) in different agricultural habitats of Bangladesh. *J. Agric. Biol. Sci.* 7(11): 905–911.
Allwood A. 2000. Regional approaches to the management of fruit flies in the Pacific. In: Tan K-H (ed.), *Area Wide Control of Fruit* Flies and Other Insect Pests. Sinaran Bros., Sdn. Bhd., Penang, Malaysia, pp. 439–448.
Al-Rajhy DH, Hussein HI, Al-Shawaf AMA. 2005. Insecticidal activity of carbaryl and its mixture with piperonylbutoxide against red palm weevil *Rhynchophorus ferrugineus* (Olivier) (Curculionidae: Coleoptera) and their effects on Acetylcholinesterase activity. *Pak. J. Biol. Sci.* 8: 679–682.
Al-Saoud AH, Ajlan A. 2013. Effect of date fruits quantity on the numbers of red weevil, Rhynchophorus ferrugineus (Coleoptera: Curculionidae) captured in aggregation pheromone traps. *Agric. Biol. J. North Am.* 4(4): 496–503.
Al-Saoud AH, Al-Deeb MA, Murchie AK. 2010. Effect of color on the trapping effectiveness of red palm weevil pheromone traps. *J. Entomol.* 7(1): 54–59.
Alyokhin AV, Messing RH, Duan JJ. 2000. Visual and olfactory stimuli and fruit maturity affect trap captures of oriental fruit flies (Diptera: Tephritidae). *Microscopy Res. Techniq.* 55: 57–67.
Alyokhin AV, Mille C, Messing RH. 2001. Selection of pupation habitats by oriental larvae in the laboratory. *J. Insect Behave.* 14: 57–67.
Amin T. 2016. Pink bollworm causes huge cotton loss. Business Recorder published on March 5, 2016. https://fp.brecorder.com/2016/03/2016030522716/.
Ampofo JKO, Saxena KN. 1989. Screening methodologies for maize resistance to *Chilo partellus* (Lepidoptera: Pyralidae). *CIMMYT. Toward Insect Resistant Maize for the Third World: Proc. of the International Symposium on Methodologies for Developing Host Plant Resistance to Maize Insects.* CIMMYT, Mexico, D.F.
Anonymous. 1956. Red palm weevil: The hidden enemy that work from within. *Coconut Bull.* 10: 77–81.
Anonymous. 1992. *Azadiractin.* Tech. Bull. AgriDyne Technologies Inc., Salt Lake City, UT, USA.
Anonymous. 1993. Eradication of the melon fly from Japan. *Quart. Newsletter APPPC* 36: 4–5.
Aqueel MA, Leather SR. 2013. Virulence of *Verticillium lecanii* (Z.) against cereal aphids; does timing of infection affect the performance of parasitoids and predators? *Pest Manag. Sci.* 69(4): 493–498.
Arabjafari KH, Jalali SK. 2007. Identification and analysis of host plant resistance in leading maize genotypes against spotted stem borer, Chilo partellus (Swinhoe) (Lepidoptera: Pyralidae). *Pak. J. Biol. Sci.*, 10: 1885–1895.
Areekul S, Sinchaisri P, Tiogvatananon S. 1988. Effect of Thai plant extract on the oriental fruit fly III. Attractancy Test. *Kasetssart J. Nat. Sci.* 22(2): 160–164.
Armes NJ, Wightman JAF, Jadhav DR, Ranga Rao GV. 1997. Status of insecticide resistance in Spodoptera litura in Andhra Pradesh, India. *Pestic. Sci.* 50: 240–248.
Armstrong JW. 2003. Quarantine security of bananas at harvest maturity against Mediterranean and Oriental fruit flies (Diptera: Tephritidae) in Hawaii. *J. Econ. Entomol.* 94: 302–314.
Armstrong JW, Hu BKS, Brown SA. 1995. Single-temperature forced hot-air quarantine treatment to control fruit flies (Diptera: Tephritidae) in papaya. *J. Econ. Entomol.* 88: 678–682.
Asaro C, Cameron RS, Nowak JT, Grosman DM, Seckinger JO, Berisford CW. 2004. Efficacy of wing versus delta traps for predicting infestation levels of four generations of the Nantucket pine tip moth (Lepidoptera: Tortricidae) in the Southern United States. *Envir. Entomol.* 33: 397–404.
Aslam M, Razaq M, Ahmad F, Faheem M, Akhter W. 2004. Population of aphid (*Schizaphis graminum* R.) on different varieties/lines of wheat (*Triticum aestivum* L.). *Int. J. Agric. Biol.* 6 (6): 974–977.

Asmarc D. 2012. Distribution, species composition, phenology and management of stemborers on sorghum (Sorghum bicolor L.) in north eastern Ethiopia. PhD thesis, Harmaya University. pp. 24–34.

Asmare D, Getu E, Azerfgne F, Ayalew A. 2011. Efficacy of some botanicals on stem borers, *Busseola fusca* (Fuller) and *Chilo partellus* (Swinhoe) on sorghum in Ethiopia under field conditions. *Biopestic. Int.* 7(1): 24–34.

Ateyyat MA, Al-Mazra'awi M, Abu-Rjai T, Shatnawi MA. 2009. Aqueous extracts of some medicinal plants are as toxic as Imidacloprid to the sweet potato whitefly, *Bemisia tabaci*. *J. Insect Sci.* 9: 15–20.

Attique MR, Rafiq M, Ghaffar A, Ahmad Z, Mohyuddin AI. 2003. Hosts of *Bemisia tabaci* (Genn.) (Homoptera: Aleyrodidae) in cotton areas of Punjab, Pakistan. *Crop Prot.* 22(5): 715–720.

Aubert B, Quilici S. 1988. Monitoring adult psyllas on yellow traps in Reunion Island. In: Garnsey SM, Timmer LW, Dodds JA (eds.), *Proceedings of the 10th Conference of International Organization of Citrus Virologists.* International Organization of Citrus Virologists, Riverside, CA, pp. 249–254.

Audemard H, Millaire G. 1975. Le piegeage du carpocapce sexual de syntheses: Primers results utilisables pour *L. estimation* des populations conduite de la lutte. *Ann. Zool. Ecol. Ani.* 7(1): 61–80.

Avand-Faghih A. 1996. The biology of red palm weevil, Rhynchophorus ferrugineus Oliv\. (Coleoptera: Curculionidae) in Saravan region (Sistan and Balouchistan Province, Iran). *Appl. Entomol. Phytopathol.* 63: 16–18.

Aziz M, Ahmad M, Nasir M, Naeem M. 2013. Efficacy of different neem (*Azadirachta indica*) products in comparison with imidacloprid against english Grain aphid (*Sitobion avenae*) on Wheat. *Int. J. Agric. Biol.* 15: 279–84.

Ba-Angood SA, Stewart RK. 1980. Sequential sampling for cereal aphids on barley in southwestern Quebec. *J. Econ. Entomol.* 73(5): 679–681.

Bachrouch O, Mediouni-Ben JJ, Alimi E, Skillman S, Kabadou T, Kerber E. 2008. Efficacy of the lufenuron bait station technique to control Mediterranean fruit fly (medfly) *Ceratitis capitata* in citrus orchards in Northern Tunisia. *Tunisian J. Pl. Prot.* 3: 5–45.

Baker TC. 1989. Sex pheromone communication in the Lepidoptera: New research progress. *Experientia* 45: 248–262.

Baker TC, Willis MA, Phelan PL. 1985. Optomotor anemotaxis polarizes self-steered zigzagging in flying moths. *Physiol. Entomol.* 9: 365.

Balal RM, Khan MM, Shahid MA, Waqas M. 2011. *Mango Cultivation in Pakistan.* Institute of Horticultural Sciences, University of Agriculture-38040, Faisalabad, Pakistan. Available Online at http://agrihunt.com/horti-industry/293.html

Balandrin MF, Lee SM, Klocke JA. 1988. Biologically active volatile organosulphur compounds from seeds of the neem tree, *Azadirachta indica* (Meliaceae). *J. Agric. Food Chem.* 36: 1048–1054.

Baloach HB, Rustamani MA, Khuro RD, Talpur MA, Hussain T. 1992. Incidence and abundance of date palm weevil in different cultivars of date palm. *Proc. of 12th Cong. Zool. Pak* 12: 445–447.

Bardner R, Fletcher K. 1974. Insect infestations and their effects on the growth and yield of field crops: A review. *Bull. Entomol. Res.* 64: 141–160.

Barranco P, De La Pena J, Cabello T. 1995. Un Nuevo curculiónido tropical para la fauna Europa, *Rhynchophorus ferrugineus* (Olivier 1790), (Curculionidae: Coleoptera). *Boletin de la Asociación Espanola de Entomolgia* 20: 257–258.

Bateman MA. 1982. Chemical methods for suppression or eradication of fruit fly populations. In: Drew RAI, Hooper GHS, Bateman MA (eds.), *Economic Fruit Flies of the South Pacific Region*, 2nd ed. Queensland Department of Primary Industries, Brisbane, Australia, pp. 115–128.

Bateman MA, Boller EF, Bush GL, Chambers DL, Economopoulos AP, Fletcher BS. 1976. Fruit flies. In: Delucchi VL (ed.), *Studies in Biological Control*, vol. 1. Cambridge Univ. Press, pp. 11–49. 304pp.

Bayart JD, Phalip M, Lemonnier R, Gueudre F. 1997. Fruit flies. Results of four years of import control on fruits in France. *Phytoma* 49: 20–25.

Bayram A, Tonğa A. 2015. First report of *Chilo partellus* in Turkey, a new invasive maize pest for Europe. *J. Appl. Entomol.* 140(3): 236–240.

Ben-Yakir D, Chen M, Sinev B, Seplyarsky V. 2013. *Chilopartellus* (Swinhoe) (Lepidoptera: Pyralidae) a new invasive species in Israel. *J. Appl. Entomol.* 137: 398–400.

Beroza M, Alexander BH, Steiner LF, Mitchell WC, Miyashita DH. 1960. New synthetic lures for the male melon fly. *Science* 131: 1044–1045.

Bhanukiran Y, Panwar VPS. 2000. In vitro efficacy of neem products on the larval of maize stalk borer. *Ann. Pl. Protec. Sci.* 8: 240–242.

Bhatnagar KN, Yadava SRS. 1992. An insecticidal trial for reducing the damage of some cucurbitaceous fruits due to *Dacus cucubitae* Coq. *Indian J. Entomol.* 54: 66–69.

Bhutani DK. 1975. Insect pests of fruit crop and their control—Mango. *Pesticides*, 7(3): 36–42.

Bindra OS, Sohi BS, Batra RC. 1974. Note on the comparative efficacy of some contact and systemic insecticides for the control of citrus psylla in Punjab. *Ind. J. Agri. Sci.* 43: 1087–1088.

Blackmer JL, Byrne DN. 1993. Flight behaviour of *Bemisia tabaci* in a vertical flight chamber: Effect of time of day, sex, age and host quality. *Phys. Ent.* 18(3): 223–232.

Borah RK. 1996. Influence of sowing seasons and varieties on the infestation of fruit fly *Bactrocera cucurbitae* (*Dacus cucurbitae*) in cucumber in the hill zone of Assam. *Indian J. Entomol.* 58(4): 382–383.

Borah RK. 1997. Effect of insect incidence in cucumber (*Cucumis sativus*) in hill-zone of Assam. *Indian J. Agric. Sci.* 67(8): 332–333.

Borah RK. 1998. Evaluation of an insecticide schedule for the control of red pumpkin beetle and melon fruit-fly on red pumpkin in the hill-zone of Assam. *Indian J. Entomol.* 60: 417–419.

Boulanger Y, Gray DR, Cooke BJ, DeGrandpre L. 2016. Model-specification uncertainty in future forest pest outbreak. *Global Change Biol.* 22(4): 1595–1607.

Bove JM. 2006. Huanglongbing: A destructive, newly-emerging, century-old disease of citrus. *J. Plant Pathol.* 88: 7–37.

Brand E. 1917. Coconut red weevil. Some facts and fallacies. *Trop. Agric. Mag. Ceylon Agric. Soc.* 49: 22–24.

Brown JK. 1994. Current status of *Bemisia tabaci* as a pest and virus vectorin agro-ecosystems worldwide. *FAO Pl. Prot. Bullet.* 42(1–2): 3–32.

Brown JK, Bird J. 1992. Whitely-transmitted geminiviruses and associated disorders in the Americans and Caribbean basin. *Pl. Dis.* 76: 220–225.

Brown JK, Frohlich DR, Rosell RC. 1995. The Sweet potato or Silver leaf whiteflies: Biotypes of *Bemisia tabaci* or a species complex? *Annu. Rev. Entomol.* 40(1): 511–534.

Browning HW, Childers CC, Stansly PA, Peña J. 2005. Soft-bodied insects attacking costa lima AM da 1942. *Homopteros. Insetos do Brazil* 3: 1–327. Esc. Na. Agron. Min. Agr. Foliage and Fruit. In 2005 Florida Pest Management Guide. Available at http://edis.ifas.ufl.edu/CG004

Burd JD, Elliott NC, Reed DK. 1996. Effects of the Aphicides Gaucho'and CGA-215944 on feeding behavior and tritrophic interactions of Russian wheat aphids. *Southwest Entomol.* 21: 145–152.

Butler GD, Henneberry TJ. 1976b. *Temperature-Dependent Development Rate Tables for Insects Associated with Cotton in the Southwest.* US Department of Agriculture, Washington DC, ARS-W-38, 36pp.

Butterworth JH, Morgan E. 1968. Isolation of a substance that suppresses feeding in locusts. *Chem. Commun.* 1: 23–24.

Buurma J. 2008. Stakeholder involvement in crop protection policy planning in the Netherlands. ENDURE – RA3.5/SA4.5 Working Paper. LEI Wageningen UR, The Hague, The Netherlands.

Byers JA. 2012a. Estimating insect flight densities from attractive trap catches and flight height distributions. *J. Chem. Ecol.* 38: 592–601.

Cabello TP. 2006. Biology and population dynamics of red palm weevil in Spain. *Proceedings of the 1st International Workshop on Red Palm Weevil*, November 28–29, 2005, IVIA, Valencia, Spain (in press).

CABI. 2007. *Selected Texts for Chilo partellus.* CAB International, 2007. Crop Protection Compendium, 2007 Edition. CAB International, Wallingford, UK.

CABI. 2015. *Crop Protection Compendium.* CAB International, Wallingford, UK. Available at http://www.cabi.org/cpc (accessed June 2016)

CABI/EPPO. 2001. *Bactrocera zonata.* Distribution Maps of Plant Pests, Map No. 125. Wallingford, UK: CAB International.

CABI and EPPO. 2003. EPPO Data Sheets on Quarantine Pests: Bactrocera cucurbitae. EPPO A1 list: No. 232. EPPO quarantine pest. pp. 1–6. Available at http://www.eppo.org/QUARANTINE/insects/Bactrocera_cucurbitae/DACUCU_ds.pdf (accessed November 17, 2008)

Cahill M, Byrne FJ, Denholm I, Devonshire A, Gorman K. 1994. Insecticide resistance in *Bemisia tabaci. Pestic. Sci.* 42: 137–138.

Carde RT. 1979. Behavioral responses of moths to female-produced pheromones and the utilization of attractant-baited traps for population monitoring. In: Rabb RL Kennedy GG (eds.), *Movements of Highly Mobile Insects: Concepts and Methodology in Research.* North Carolina State University Press, Raleigh, pp. 286–315.

Carey JR, Dowell RV. 1989. Exotic fruit pests and California agriculture. *California Agric.* 43: 38–40.

Carroll LE, Norrbom AL, Dallwitz MJ, Thompson FC. 2004. Pest fruit flies of the world—larvae. Version: 13th April 2005. http://delta-intkey.com/

Cavalloro R. 1983. Fruit flies of economic importance. In: Cavalloro R (ed.), *CEC/IOBC Symposia*, 1982, Athens, Greece, Rotterdam. Balkema, Germany, 642pp.

Chakravarthy VSK, Reddy TP, Reddy VD, Rao KV. 2014. Current status of genetic engineering in cotton (*Gossypium hirsutum* L): An assessment. *Critical Rev. Biotechnol.* 34: 144–160.

Chance LE, Degrugillier M, Leverich AP. 1969. Comparative effects of chemosterilants on spermatogenic stages in the house fly I. Induction of dominant lethal mutations in mature sperm and gonial cell death. *Mutation Res* 7(1): 63–74.

Chang SC, Terry PH, Borkovec AB. 1964. Insect chemosterilants with low toxicit for mammals. *Science*. 144: 57–58.

Chapman JW, Nesbit RL, Burgin LE, Reynolds DR, Smith AD, Middleton DR, Hill JK. 2010. Flight orientation behaviors promote optimal migration trajectories in high-flying insects. *Science* 327: 682–685.

Chapman JW, Reynolds DR, Hill JK, Sivell D, Smith AD, Woiwod IP. 2008a. A seasonal switch in compass orientation in a high-flying migrant moth. *Curr. Biol.* 18: 908–909.

Chapman JW, Reynolds DR, Mouritsen H, Hill JK, Riley JR, Sivell D, Smith AD and Woiwod IP. 2008b. Wind selection and drift compensation optimise migratory pathways in a high-flying moth. *Curr. Biol.* 18: 514–518.

Chari MS, Patel SN. 1983. Cotton leaf worm *Spodoptera litura* Fabricius its biology and integrated control measures. *Cotton Dev.* 13(1): 7–8.

Cheema MA, Muzaffar N, Ghani M. 1980. Biology, host range and incidence of parasites of *Pectinophora gossypiella* (Saunders) in Pakistan. *Pakistan Cottons* 24: 37–73. DC.

Chen C, Feng MG. 2004. *Sitobion avenae* alatae infected by Pandora neoaphidis: Their flight ability, post-flight colonization, and mycosis transmission to progeny colonies. *J. Invert. Pathol.* 86: 117–123.

Chen CC, Dong YJ, Cheng LL, Hou RF. 1996. Deterrence effect of neem seed kernel extracts on oviposition of the oriental fruit fly (Diptera: Tephritidae) on guava. *J. Econ. Entomol.* 89: 462–466.

Chen HD, Zhou CQ, Yang HPJ, Liang GQ. 1995. On the seasonal population dynamics of melon and Oriental fruit flies and pumpkin fly in Guangzhou area. *Acta. Phytophyacica. Sinica* 22(4): 348–354.

Chen M, Chen P, Ye H, Li JP, Ji QE. 2014. Morphological structures and developmental characteristics of the ovaries of *Bactrocera dorsalis. J. Environ. Entomol.* 36: 1–5.

Chen M, Chen P, Ye P, Yuan R, Wang X, Xu J. 2015. Flight capacity of *Bactrocera dorsalis* (Diptera: Tephritidae) adult females based on flight mill studies and flight muscle ultrastructure. *J Inset Sci* 15(1): 132–141.

Chen P and Ye H. 2007. Population dynamics of *Bactrocera dorsalis* (Diptera: Tephritidae) and analysis of factors influencing populations in Baoshanba, Yunnan, China. *Entomol. Sci.* 10: 141–147.

Chen P, Ye H, Mu QA. 2007. Migration and dispersal of the oriental fruit fly, *Bactrocera dorsalis* in regions of Nujiang River based on fluorescence mark. *Acta. Ecologica Sinica.* 27: 2468–2476.

Cheng EY, Kao CH, Chiang MY, Hwang YB. 2003. Modernization of oriental fruit fly control in Taiwan: The planning and execution of an area-wide control project. *Proceedings of the Workshop on Plant Protection Management for Sustainable Development: Technology and New Dimension.* September 4, 2003, Taichung, Taiwan. pp. 49–66.

Chernoh E. 2014. *Maize Stalk Borers.* Plantwise Knowledge Bank. Available at www.plantwise.org/knowledgebank and BIONET-EAFRINET http://keys.lucidcentral.org/keys/v3/eafrinet/index.htm (accessed June 20, 2016)

Chinajariyawong A, Kritsaneepaiboon S, Drew RAI. 2003. Efficacy of protein bait sprays in controlling fruit flies (Diptera: Tephritidae) infesting angled luffa and Bitter gourd in Thailand. *Raff. Bull. Zool.* 51(1): 7–15.

Chiu SF. 1985. Recent research findings on Meliaceae and other promising botanical insecticides in China. *Z. Pflkrankh. Pflanzenschutz* 92: 320–329.

Chong JH, Oetting RD. 2006. Influence of temperature and mating status on the development and fecundity of the mealybug parasitoid, *Anagyrus* sp nov nr *sinope* Noyes and Menezes (Hymenoptera: Encyrtidae). *Environ. Entomol.* 35: 1188–1197.

Chong JH, Oetting RD. 2007. Intraguild predation and interference by the mealybug predator *Cryptolalemus montrouzieri* on the parasitoid *Leptomastix dactylopii. Biocont. Sci. Technol.* 17: 933–944.

Chou MY, Mau RFL, Pandey RR. 2002. Encouraging results from the fruit fly suppression program in Oahu. In: Sugano J, Hiraki C (eds.), *HAW-FLYPM Newsletter, Coop. Ext. Serv., Col. of Trop. Agric. and Human Resources.* Univ. of Hawaii.

Chowdhury MK, Malapert JC, Hosanna MN. 1993. Efficiency of poison bait trap in controlling fruit fly, *Dacus cucurbitae* in Bitter gourd. *Bangladesh J. Entomol.* 3: 91–92.

Christenson LD, Foote RH. 1960. Biology of fruit flies. *Annu. Rev. Entomol.* 5: 171–192.

Chrysanthus TM, Samira M, Sunday E, Prem G. 2010. Distribution, host plant and abundance of the invasive mango mealybug *Rastrococcus iceryoides* and its associated natural enemies in Africa. *ESA 58th Annual Meeting*, December 12–15, San Diego, California, USA.

Chughtai CG, Baloch VK. 1988. Insecticidal control of melon fruit-fly. *Pak. J. Entomol. Res.* 9: 192–194.

Clarke AR, Allwood A, Chinajariyawong A, Drew RAI, Hengsawad C, Jirasurat M, Kong-Krong C, Kritsaneepaiboon S, Vijaysegaran S. 2001. Seasonal abundance and host use patterns of seven Bactrocera Macquart species (Diptera: Tephritidae) in Thailand and Peninsular Malaysia. *Raff. Bull. Zool.* 49(2): 207–220.

Clarke AR, Armstrong KF, Carmichael AE, Milne JR, Raghu S, Roderick GK, Yeates DK. 2005. Invasive phytophagous pests arising through a recent tropical evolutionary radiation: The *Bactrocera dorsalis* complex of fruit flies. *Ann. Rev. Entomol.* 50: 293–319.

Clausen CP, Clancy DW, Chock QC. 1965. *Biological Control of the Oriental Fruit Fly (Dacus dorsalis Hendel) and Other Fruit Flies in Hawaii.* United States Department of Agriculture, Technical Bulletin No. 1322, 102 pp.

Cook SM, Khan ZR, Pickett JA. 2007. The use of push-pull strategies in integrated pest management. *Ann Rev Entomol* 52: 375–400.

Coombe PE. 1981. Wavelength behavior of the whitefly *Trialeurodes vaporariorum* (Homoptera: Aleyrodidae). *J. Comp. Physiol.* 144: 83–90.

Coombe PE. 1982. Visual behavior of the greenhouse whitefly, *Trialeurodes vaporaiorum. Physiol. Entomol.* 7: 243–251.

Costa HS, Brown JK, Byrne DN. 1991. Life history traits of the whitefly, *Bemisia tabaci* (Homoptera; Aleyrodidae) on six virus infected or healthy plant species. *Envir. Entomol.* 20(4): 1102–1107.

Coudriet DL, Prabhaker N, Meyerdirk DE. 1985. Sweet potato Whitefly (Homoptera: Aleyrodidae): Effects of neem-seed extract on oviposition and immature stages. *Environ. Entomol.* 14: 776–779.

Cox ML. 1993. Red palm weevil, *Rhynchophorus ferrugineus* in Egypt. *FAO Plant Protection Bulletin* 41: 30–31.

Crey JR, Dowell RV. 1989. Exotic fruit pests and California agriculture. *California Agri.* 43(3): 38–40.

Cunningham RT. 1989b. Control; insecticides; male annihilation. In: Robinson AS, Hooper G (eds.), *World Crop Pests 3(B). Fruit Flies; Their Biology, Natural Enemies and Control.* Elsevier, Amsterdam, Netherlands, pp. 345–351.

D'Arcy CJ, Mayo M. 1997. Proposals for changes in luteovirus taxonomy and nomenclature. *Archives Virol.* 142: 1285–1287.

Dale NS, Patel RK. 2010. Population dynamics of fruit flies (*Bactrocera spp.*) on guava and its correlation with weather parameters. *Current Biotica.* 4(2): 245–248.

De Barro PJ, Liu S-S, Boykin LM, Dinsdale AB. 2011. *Bemisia tabaci*: A statement of species status. *Annual Review of Entomology* 56(1): 1–19.

De Clercq P, Peeters I, Vergauwe G, Thas O. 2003. Interaction between Podisus maculiventris and Harmonia axyridis, two predators used in augmentative biological control in greenhouse crops. *BioControl* 48: 39–55.

De Meyer M, Mohamed S, White IM. 2012. Invasive fruit fly pests in Africa. A diagnostic tool and information reference for the four Asian species of fruit fly (Diptera, Tephritidae) that have become accidentally established as pests in Africa, including the Indian Ocean Islands. Available at http://www.africamuseum.be/fruitfly/AfroAsia.htm (accessed June 2016).

De Villiers M, Hattingh V, Kriticos DJ, Brunel S, Vayssières JF, Sinzogan A, Billah MK, Mohamed SA, Mwatawala M, Abdelgader H, Salah FEE, De Meyer M. 2016. The potential distribution of *Bactrocera dorsalis*: Considering phenology and irrigation patterns. *Bull. Entomol. Res.* 106: 19–33.

Dedryver CA, Le Gallic JF, Gauthier JP, Simon JC. 1998. Life-cycle in the cereal aphid Sitobion avenae F.: Polymorphism and comparison of life history traits associated with sexuality. *Ecol. Entomol.* 23: 123–132.

Dedryver CHA, Le ralec A, Fabre F. 2010. The conflicting relationships between aphids and men: A review of aphid damage and control strategies. *Comptes Rendus Biologies* 333(6–7): 539–553.

Dejen A, Getu E, Azerefegne F, Ayalew A. 2014. Distribution and impact of *Busseola fusca* (Fuller) (Lepidoptera: Noctuidae) and *Chilo partellus* (Swinhoe) (Lepidoptera: Crambidae) in Northeastern Ethiopia. *J. Entomol. Nematol.* 6(1): 1–13.

Delrio G, Cocco A. 2012. The peach fruit fly, *Bactrocera zonata*: A major threat for Mediterranean fruit crops? *Acta. Hort.* 940: 557–566.

Dembilio Ó, Jaques JA. 2015. Biology and management of red palm weevil. In: Wakil W, Faleiro JR, Miller TA (eds.), *Sustainable Pest Management in Date Palm: Current Status and Emerging Challenges, Sustainability in Plant and Crop Protection.* Springer International Publishing, Switzerland, pp. 13–36.

De-Prins J, De-Prins W. 2016. Afromoths, online database of Afrotropical moth species (Lepidoptera). Available at www.afromoths.net (accessed June 22, 2016)

Desneux N, Decourtye A, Delpuech JM. 2007. The sublethal effects of pesticides on beneficial arthropods. *Ann. Rev. Entomol.* 52: 81–106.

DFID, crop protection programme. 2004–2005. Lord of the fruit flies. 1–26.

Dhaliwal GS, Singh R. 2004. *Host Plant Resistance to Insects: Concepts and Applications.* Panima Publication, New Delhi, 578p.

Dhaliwal GS, Singh R and Chhillar BS. 2006. *Essentials of Agricultural Entomology.* Kalyani Publishers, Ludhiana, New Delhi, India.

Dhillon MK, Naresh JS, Singh R, Sharma NK. 2005. Influence of physico-chemical traits of bitter gourd, *Momordica charantia* L. on larval density and resistance to melon fruit-fly, *Bactrocera cucurbitae* (Coquillett). *J. Appl. Entomol.* 129(7): 393–399.

Dhir BC, Mohapatra HK, Senapati B. 1992. Assessment of crop loss in ground nut due to tobacco caterpillar, *Spodoptera litura* (F.). *Indian J. Plant Prot.* 20: 215–217.

Dickson RC. 1959. Aphid dispersal over southern California deserts. *Ann. Entomol. Soci. Amer.* 52: 368–72.

Dimbi S. 2003. Evaluation of the potential of hyphomycetes fungi for the management of the African tephritid fruit flies *Ceratitis capitata* (Weidemann), *Ceratitis cosyra* (Walker) and *Ceratitis fasciventris* (Bezzi) in Kenya. *PhD thesis*, Kenya University, Nairobi.

Dimbi S, Maniania NK, Ekesi S. 2009. Effect of Metarhizium anisopliae inoculation on the mating behavior of three species of African Tephritid fruit flies, *Ceratitis capitata*, *Ceratitis cosyra* and *Ceratitis fasciventris. Biol. Contr.* 50: 111–116.

Dimbi S, Maniania NK, Lux SA, Ekesi S, Mueke JK. 2003. Pathogenicity of *Metarhizium anisopliae* (Metsch.) Sorokin and *Beauveria bassiana* (Balsamo) Vuillemin, to three adult fruit fly species: *Ceratitis capitata* (Weidemann), C. rosa var. fasciventris Karsch and *C. cosyra* (Walker) (Diptera: Tephritidae). *Mycopathol.* 156: 375–382.

Dimetry NZ, Gomaa AA, Salem AA, Abd-El-Moniem ASH. 1996. Bioactivity of some formulations of neem seed extracts against the whitefly. *Anz. Schädlingsk. Pflanzenschutz, Umweltschutz* 69: 140–141.

Dimetry NZ, Schmidt GH. 1992. Efficacy of Neem Azal-S and Margosan-O against the bean aphid, *Aphis fabae* Scop. *Anz. Schädlingsk. Pflanzenschutz, Umweltschutz* 65: 75–79.

Divya K, Marulasiddesha KN, Krupanidhi K, Sankar M. 2009. Population dynamics of spotted stem borer, *Chilo partellus* (Swinhoe) and its interaction with natural enemies in sorghum. *Indian J. Sci. Technol.* 3(1): 70–74.

Dixon AFG. 1987. Cereal aphids as an applied problem. *Agric. Zool. Rev.* 2: 1–57.

Dolinski C, Lacey LA. 2007. Microbial control of arthropod pests of tropical tree fruits. *Neotrop. Entomol.* 36(2): 161–79.

Dowell RV, Siddiqui IA, Meyer F, Spoungy EL. 2000. Mediterranean fruit fly preventative release programme in southern California. In: Tan K-H (ed.), *Area Wide Control of Fruit Flies and Other Insect Pests.* Sinaran Bros., Sdn. Bhd., Penang, Malaysia, pp. 369–375.

Draz KA, Tabikha RM, El-Aw MA, El-Gendy IR, Darwish HF. 2016. Population activity of peach fruit fly *Bactrocera zonata* (Saunders) (Diptera: Tephiritidae) at fruits orchards in Kafer El-Shikh Governorate, Egypt. *Arthropods*, 5(1): 28–43.

Drew RAI, Hancock DL. 1994. *The Bactrocera dorsalis Complex of Fruit Flies in Asia.* Bulletin of Entomological Research: Supplement Series. Supplement No. 2. CAB International, Wallingford, UK.

Drew RAI, Hooper GHS. 1983. Population studies of fruit flies (Diptera: Tephritidae) in South-East Queensland. *Oecologia* 56: 153–159.

Drew RAI, Hooper GSH, Bateman MA. 1982. *Economic Fruit Flies of South Pacific Region*, 2nd ed. Qld. Dept. Primary Ind., Brisbane, Australia, 139pp.

Drew RAI, Prokopy RJ, Romig MC. 2003. Attraction of fruit flies of the genus Bactrocera to colored mimics of host fruit. *Entomol. Exper. Appl.* 107: 39–45.

Drew RAI, Romig MC. 1997. Overview—Tephritidae in the Pacific and Southeast Asia. In: Allwood AJ, Drew RAI (eds.), *Management of Fruit Flies in the Pacific, a Regional Symposium, ACIAR Proceedings No. 76*, October 28–31, 1996, Nadi, Fiji, pp. 46–53. Australian Centre for International Agricultural Research, Canberra.

Drew RAI, Romig MC. 2000. Tephritid taxonomy into the 21st Century research opportunities and application. In: Tan RH (ed.), *Area-Wide Control of Fruit Flies and Other Insect Pests. Proc. Int. Conf. of Insect Pests, May 28–June, 1998 and 5th. Int. Symp. on Fruit Flies*, June 1–5, 1998. Penerbit Universiti, Malaysia, pp. 677–692.

Ekesi S, Billah MK. 2007. *A Field Guide to the Management of Economic Important Tephritid Fruit Flies in Africa.* ICIPE, Nairobi, Kenya.

El-Garhy ME. 1996. Field evaluation of the aggregation pheromone of the red palm weevil, *Rhynchophorus ferrugineus* in Egypt. *Brighton Crop Protection Conference: Pests and Diseases*, vol. 3, pp. 1059–1064.

El-Lakwah FAM, El-Banna AA, Rasha A, El-Hosary, El-Shafei WKM. 2011. Population dynamics of the red palm weevil (*Rhynchophorus Ferrugineus* (oliv.) on date palm plantations in 6th october governorate. *Egypt. J. Agric. Res.* 89(3): 1105–1116.

Ellis SE. 2004. *New Pest Response Guidelines: Spodoptera*. USDA/APHIS/PPQ/PDMP. Available at http://www.aphis.usda.gov/ppq/manuals/

El-Sabea MR, Faleiro JR, Abo-El-Saad MM. 2009. The threat of red palm weevil *Rhynchophorus ferrugineus* (Oliv) to date plantations of the Gulf region of the Middle-East an aconomic prospective. *Out Looks Pest Manag.* 20(3): 131–134.

El-Sayed AM, Suckling DM, Wearing CH, Byers JA. 2006. Potential of mass trapping for long-term pest management and eradication of invasive species. *J. Econ. Entomol.* 99: 1550–1564.

El-Sebay Y. 2003. Ecological studies on the red palm weevils Rhynchophorus ferrugineus Oliv. (Coleoptera: Curculionidae) in Egypt. *Egyptian J. Agric. Res.* 81: 523–529.

Elton CS. 1925. *The Dispersal of Insects of Spitzbergen*. Transactions American Entomological Society, London, pp. 289–299.

EPA (Environmental Protection Agency). 2005. Malathion; Revised risk assessments, notice of availability, and solicitation of risk reduction options. *Federal Register* 70: 55839–55842.

EPPO. 2015. PQR – EPPO database on quarantine pests. Available at http://www.eppo.int (accessed June 2016).

EPPO/CABI. 1997. *Spodoptera littoralis* and *Spodoptera litura*. In: Smith IM, McNamara DG, Scott PR, Holderness M (eds.), *Quarantine Pests for Europe*, 2nd ed. CAB International, Wallingford, UK, pp. 518–525.

Esper J, Buntgen U, Frank DC, Nievergelt D, Liebhold A. 2007. 1200 years of regular outbreaks in alpine insects. *Proc. R. Soc. B.* 274: 671–679.

Esteban-Duran J, Yela JL, Beitia-Crespo F, Jimenez-Alvarez A. 1998. Biologia del curculionido ferruginoso de las palmeras *Rhynchophorus ferrugineus* (Olivier) en laboratorio y campo: Ciclo en cautividad, peculiaridades biologicas en su zona de introduccion en Espana y metodos biologicos de deteccion y posible control (Coleoptera: Curculionidae: Rhynchophorinae). *Boletín de Sanidad Vegetal – Plagas* 24: 737–748.

Eta CR. 1985. *Eradication of the melon fly from Shortland Islands (special report)*. Solomon Islands Agricultural Quarantine Service, Annual Report. Ministry of Agriculture and Lands, Honiara.

Eta CR. 1986. *Review—eradication of the melon fly from Shortland Islands, Western Province, Solomon Islands*. Annual Report of Solomon Islands Agriculture Quarantine Service, 1985, pp. 14–23. Solomon Islands Agricultural Quarantine Service, Honiara, Solomon Islands.

Euro-Fruit Tribune. 2014. Fruit flies plagues Pakistani mangoes. Downloaded on July 29th, 2014. Available at http://www.fruitnet.com/eurofruit/article/159004/pakistani-mango-exports-refused-uk-access

Faleiro JR. 2005. Pheromone technology for the management of red palm weevil *Rhynchophorus ferrugineus* (Olivier) (Coleoptera: Rhynchophoridae) –A key pest of coconut, Technical Bulletin No.4, ICAR Research Complex for Goa, 40pp.

Faleiro JR. 2006. A review of the issues and management of the red palm weevil *Rhynchophorus ferrugineus* (Coleoptera: Rhynchophoridae) in coconut and date palm during the last one hundred years. *Int. J. Trop. Insect Sci.* 26: 135–154.

Faleiro JR, Abraham VA, Al-Shuaibi MA. 1998. Role of pheromone trapping in the management of red palm weevil. *Indian Coconut J.* 29: 1–3.

Faleiro JR, Satarkar VR. 2003. Diurnal activity of red palm weevil, *Rhynchophorus ferrugineus* Olivier in coconut plantations of Goa. *Insect Environ.* 9: 63–64.

Fan Y, Fan P, Frederick. 1998. Dispersal of the broad mite, Polyphagotarsonemus latus (Acari: Tarsonemidae) on *Bemisia tabaci* (Homoptera: Aleyrodidae). *Exp. Appl. Acarol.* 22(7): 411–415.

Fang MN. 1989a. Studies on using different bagging materials for controlling melon fly on Bitter gourd and sponge gourd. *Bull. Taichung Dist. Agric. Improv. Stn.* 25: 3–12.

Fang MN, Chang CP. 1987. Population changes, damage of melon fly in the Bitter gourd garden and control with paper bag covering method. *Plant Prot. Bull. Taiwan* 29: 45–51.

FAO. 2008. *Climate-Related Transboundary Pests and Diseases*. Technical background document from the expert consultation held on 25 to 27 February 2008. FAO, Rome.

Farid A, Khan, MIN, Khan A, Khattak SUK, Alamzeb, Sattar A. 2007. Studies on maize stem borer, *Chilo partellus* in Peshawar Valley. *Pakistan J. Zool.* 39(2): 127–131.

Fauquet CM, Fargette D. 1990. African cassava mosaic virus: Etiology, epidemiology and control. *Pl. Dis.* 74: 404–411.

Ferry N, Edwards MG, Gatehouse AMR. 2004. Plant-insect interaction: Molecular approaches to insect resistance. In: Sasaki T, Christou P (eds.), *Biotechnology*, vol. 15, pp. 155–161.

Ficbig M, Poehling HM. 1998. Host-plant selection and population dynamics of the grain aphid *Sitobion avenae* (F.) on wheat infected with Barley Yellow Dwarf Virus. *Bull. IOBC/WPRS.* 21: 51–62.

Fischer K, Van den Berg J, Mutengwa C. 2015. Is Bt maize effective in improving South African smallholder agriculture? *South African J. Sci.* 111(1/2): 1–2.

Fischer-Colbrie P and Busch-Petersen E. 1989. Pest status; temperate Europe and west Asia. In: Robinson AS, Hooper G (eds.), *World Crop Pests 3(A). Fruit Flies; Their Biology, Natural Enemies and Control.* Elsevier, Amsterdam, Netherlands, pp. 91–99.

Flaherty DL, Peacock WL, Bettiga L, Leavitt GM. 1982. Chemicals losing effect against grape mealybug. *California Agriculture* 36: 15–16.

Fletcher BS. 1987. The biology of Dacine fruit flies. *Annu. Rev. Entomol*, 32: 115–144.

Fletcher BS. 1989. Life history strategies of tephritid fruit flies. In: Robinson AS, Hooper G (eds.), *World Crop Pests vol. 3B: Fruitflies Their Biology, Natural Enemies and Control.* Elsevier Science Publishing Company Inc., New York, pp. 195–208.

Follet JA, Armstrong JW. 2004. Revised irradiation doeses to control melon fly, Mediterranean fruit fly and oriental fruit fly (Diptera: Tephritidae) and a genric dose for Tephritid Fruit Flies. *J. Econ. Entomol.* 97: 1254–1262.

Franco JC, Zada A and Mendel Z. 2009. Novel Approaches for the Management of Mealybug Pests. In: Ishaaya I, Horowitz AR (eds.), *Biorational Control of Arthropod Pests: Application and Resistance Management.* Springer International Publishing AG, Netherland. pp. 233–278.

FreshFruitPortal.com. 2013. Fruit fly hitting Pakistani mango export potential. Downloaded on 25 July, 2013. Available at http://www.freshfruitportal.com/2013/07/25/fruit-fly-hitting-pakistani-mango-export-potential/?country=pakistan

Frisbie RE, El-Zik KM, Wilson LT. 1989. *The Future of Cotton IPM. Integrated Pest Management Systems and Cotton Production.* John Wiley and Sons, Inc., New York, USA, pp. 413–428.

Froerer KM, Peck SL, McQuate GT, Vargas RI, Jang EB, McInnis DO. 2010. Long-distance movement of *Bactrocera dorsalis* (Diptera: Tephritidae) in Puna, Hawaii: How far can they go? *Am. Entomol.* 56: 88–94.

Fye RE. 1979a. *Insect Diapause: Field and Insectary Studies of Six Lepidopterous Species.* U.S. Department of Agriculture, Science and Education Administration ARR-W-7, Washington, 51pp.

Gabel B, Thiéry D, Suchy V, Marion-Poll F, Hardsky P, Farkas P. 1992. Floral volatiles of *Tanasetum vulgare* L. attractive to *Lobesia botrana* (Den et Schiff.) females. *J. Chem. Ecol.* 18: 693–701.

Gao ZR, Zhao HY, Jiang YF. 1992. A study on the occurrence, damage and control of the pink bollworm in Henan Province [in Chinese; summary in English]. *Plant Protect.* 18(4): 29–30.

Gaston LK, Kaae RS, Shorey HH, Sellers D. 1977. Controlling the pink bollworm by disrupting sex pheromone communication between adult moths. *Sci.* 196: 904–905.

Gatehouse AG. 1997. Behavior and ecological genetics of wind-borne migration by insects. *Ann. Rev. Entomol.* 42: 475–502.

Geier PW, Clark LR, Andersen DJ, Nix HA. 1973. *Insects: Studies in Population Management.* Ecological Society of Australia, Canberra.

Ghosh SK. 2001. G.M. crops: Rationally irresistible. *Current Sci.* 6: 655–660.

Giblin-Davis RM, Faleiro JR, Jacas JA, Peña JE, Vidyasagar PSPV. 2013. Biology and management of the red palm weevil, *Rhynchophorus ferrugineus.* In: Peña JE (ed.), *Potential Invasive Pests of Agricultural Crop Species.* CAB International, CABI Wallingford, Oxfordshire, pp. 1–34.

Gilbert AJ, Bingham RR. 2002. *Insect Trapping Guide*, 9th ed. State of California, Department of Food and Agriculture, Sacramento.

Gogi MD. 2009. Mechanisms of resistance and management for melon fruit fly, *Bactrocera cucurbitae* (Coquillett) (Diptera: Tephritidae) in bitter gourd, *Momordica charantia. PhD thesis,* Department of Entomology, University of Agriculture, Faisalabad, pp. 1–6.

Gogi MD, Ashfaq M, Arif MJ, Khan MA, Ahmad F. 2007. Co-administration of insecticides and butanone acetate for its efficacy against melon fruit flies, *Bactrocera cucurbitae* (Insects: Diptera: Tephritidae). *Pak. Entomol.* 29(2): 111–116.

Gogi MD, Ashfaq M, Arif MJ, Sarfraz RM, Nawab NN. 2010. Investigating phenotypic structures and allelochemical compounds of the fruits of *Momordica charantia* L. genotypes as sources of resistance against *Bactrocera cucurbitae* (Coquillett) (Diptera: Tephritidae). *Crop Protect.* 29(8): 884–890.

Goldsmith TH. 1994. Ultraviolet receptors and color vision: Evolutionary implications and dissonance of paradigms. *Vision Res.* 34: 1479–1487.

Gottwald TR, Aubert B, Zhao XY. 1989. Preliminary analysis of citrus greening (huanglongbin) epidemics in the People's Republic of China and French Reunion Island. *Phytopath.* 70: 687–693.

Government of Pakistan. 2015. *Crops Area and Production (by Districts)*. Statistics Division Federal Bureau of Statistics (Economic Wing), Islamabad, Pakistan.

Green EE. 1908. Remarks of Indian scale insects (Coccidae) Part-III with a catalogue of all species hitherto recorded from the Indian Continent. *Memo Dept. Agric. India Entomol. Ser.* 2: 15–46.

Green MB, Lyon DJ de B. 1989. *Pest Management in Cotton*. Ellis Horwood Limited, Chichester, West Sussex, UK, 259pp.

Greenberg SM, Legaspi BC, Jones WA, Enkegaard A. 2000. Temperature-dependent life history of *Eretmocerus eremicus* (Hymenoptera: Aphelinidae) on two whitefly hosts (Hymenoptera: Aleyrodidae). *Envir. Entomol.* 29: 851–860.

Gunn A, Gatehouse AG. 1993. The migration syndrome in the African armyworm moth, *Spodoptera exempta*: Allocation of resources to flight and reproduction. *Physiol. Entomol.* 14: 419–427.

Gupta JN, Verma AN. 1979. Relative efficacy of insecticides as contact poisons to the adults of melon fruit-fly, *Dacus cucurbitae* (Coq.). *Indian J. Entomol.* 41: 117–120.

Gupta JN, Verma AN. 1982. Effectiveness of fenitrothion bait sprays against melon fruit-fly, *Dacus cucurbitae* Coquillett in Bitter gourd. *Indian J. Agric. Res.* 16: 41–46.

Gupta S, Handore K, Pandey IP. 2010. Effect of insecticides against *Chilo partellus* (Swinhoe) damaging *Zea mays* (maize). *Int. J. Paras. Res.* 2(2): 4–7.

Gush H. 1997. Date with disaster. *The Gulf Today*. September 29. p. 16.

Gutierrez AP, Daane KM, Ponti L, Walton VM, Ellis CK. 2008. Prospective evaluation of the biological control of vine mealybug: Refuge effects and climate. *J. App. Ecol.* 45: 524–536.

Halbert SE, Manjunath KL. 2004a. Asian citrus psyllids (Sternorrhyncha: Psyllidae) and greening disease of citrus: A literature review and assessment of risk in Florida. *Fla. Entomol.* 87(3): 330–353.

Halbert SE, Manjunath KL. 2004b. Asian citrus psyllids (Sternorrhyncha: Psyllidae) and greening disease of citrus: A literature review and assessment of risk in Florida. *Fla. Entomol.* 87(3): 330–353.

Halbert SE, Núñez CA. 2004. Distribution of the Asian citrus psyllid, *Diaphorina citri* Kuwayama (Rhynchota: Psyllidae) in the Caribbean basin. *Fla. Entomol.* 87(3): 401–402.

Hall DG. 2008. Biology, history and world status of *Diaphorina citri*. *Proceedings of the International Workshop on Huanglongbing and Asian Citrus Psyllid*, held on May 7–9, 2008, at Hermosillo, Sonora, Mexico. pp. 1–11.

Hall DG, Gottwald TR, Stover E, Beattie GAC. 2013. Evaluation of management programs for protecting young citrus plantings from huanglongbing. *Hort. Sci.* 48: 330–337.

Hall DG, Hentz MG, Adair RC. 2008. Population ecology and phenology of *Diaphorina citri* (Hemiptera: Psyllidae) in two Florida citrus groves. *Environ. Entomol.* 37(4): 914–24.

Hall DG, Richardson ML, Ammar ED, Halbert SE. 2013. Asian citrus psyllid, *Diaphorina citri*, vector of citrus huanglongbing disease. *Entomol. Exp. Appl.* 146: 207–223.

Hallett RH, Oehlschlager AC, Borden JH. 1999. Pheromone trapping protocols for the Asian palm weevil, *Rhynchophorus ferrugineus* (Coleoptera: Curculionidae). *Int. J. Pest Manag.* 45: 231–237.

Hamed M, Nadeem S. 2010. Prediction of pink bollworm (*Pectinophora gossypiella* (Saunders) population cycles in cotton by accumulating thermal units in the agro-climate of Faisalabad. *Pakistan J. Zool.* 42: 431–435.

Hameed S, Khalid S. 1996. Occurrence of B-biotype of Bemisia tabaci in Pakistan. *Proceedings of Brighton Crop Protection Conference, Pests and Diseases*, held on November 18–21, 1996 in British Crop Protection Council, London, UK. pp. 81–85.

Hameed S, Khalid S, Haq E, Hashmi AA. 1994. Cotton leaf curl disease in Pakistan caused by a whitefly transmitted geminivirus. *Pl. Dis.* 78(5): 529.

Hansson BS. 1995. Olfaction in Lepidoptera. *Experientia* 51: 1003–1027.

Hansson BS, Larsson MC, Leal WS. 1999. Green leaf volatile-detecting olfactory receptor neurones display very high sensitivity and specificity in a scarab beetle. *Physiol. Entomol.* 24(2): 121–126.

Hao YN, Jin M, Wu YQ, Gong ZJ. 2013. Flight performance of the orange wheat blossom midge (Diptera: Cecidomyiidae). *J. Econ. Entomol.* 106: 2043–2047.

Hardie RC. 1986. The photoreceptor array of the dipteran retina. *Trends Neurosci.* 9: 419–423.

Hashem MH. 2016. Genetic variations among the red palm weevil *Rhynchophorus ferrugineus* populations collected from Egypt. *Egypt. J. Genet. Cytol.*, 45: 33–45.

Henneberry TJ, Leal MP. 1979. Pink bollworm: Effect of temperature, photoperiod and light intensity, moth age and mating frequency on oviposition and egg viability. *J. Econ. Entomol.* 72: 489–492.

Hensel LL, Grbić V, Baumgarten DA, Bleecker AB. 1993. Developmental and age-related processes that influence the longevity and senescence of photosynthetic tissues in Arabidopsis. *Plant Cell* 5: 553–564.

Heong KL, Escalada MM, Huan NH, Ky-Ba VH, Quynhm PV, Thiet LV, Chien. 2008. Entertainment-education and rice pest management: A radio soap opera in Vietnam. *Crop Prot.* 27: 1392–1397.

Hill D. 1975. *Agricultural Insect Pests of the Tropics and Their Control.* Cambridge University Press, Cambridge, UK. p. 516.

Hobbs P, Morris ML. 1996. *Meeting South Asia's future food requirements from rice-wheat cropping systems: Priority issues facing researchers in the post-green revolution era.* NRG Paper No. 96-01. CIMMYT, Mexico DF, 46pp.

Hoddle MS, Hoddle CD. 2015. Evaluation of three trapping strategies for the palm weevil, *Rhynchophorus vulneratus* (Coleoptera: Curculionidae) in Sumatra, Indonesia. *Pak Entomol.* 27: 73–77.

Hoddle MS, Sanderson JP, Van Driesche RG. 1999. Biological control of *Bemisia argentifolii* (Hemiptera: Aleyrodidae) on poinsettia with inunudative releases of *Eretmocerus eremicus* (Hymenoptera: Aphelinidae): Does varying the weekly release rate affect control? *Bull. Entomol. Res.* 89: 41–51.

Holland RA, Wikelski M & Wilcove DS. 2006. How and why do insects migrate? *Science* 313: 794–796.

Hollingsworth R and Allwood AJ. 2002. *Melon Fly.* SPC Pest Advisory Leaflets, Australian Center for International Agricultural Research, Australia, pp. 1–2.

Hooper GHS. 1978. Effect of combining methyl euginol and cuelure on the capture of male tephritid fruit flies. *J. Aust. Entomol. Soc.* 17: 189–190.

Horowitz AR, Forer G, Ishaaya I. 1994. Managing resistance of *Bemisia tabaci* in Israel with emphasis on cotton. *Pestic. Sci.* 42: 113–122.

Hunter WD. 1918. The pink bollworm problem in the United States. *Bull. State Plant Board Florida* 2: 139–158.

Hussain A, Saleem A, Khan WS, Tariq AH. 1991. Vector, whitefly (*B. tabaci*). In: *Cotton Leaf Curl Viruses, the Problem, Disease Situation, Research Update and Control.* Public. Directorate of Agric. Inf., Lahore, 7p.

Husain MA, Nath D. 1927. *The citrus psylla (*Diaphorina citri, *Kuw.) [Psyllidae: Homoptera].* Memoirs of the Department of Agriculture in India. Entomolo. Series. vol. 10, no. 2, 27pp.

Hutson JC. 1922. The Red Weevil of Palm Weevil (*Rhynchophorus Ferrugineus*). *Trop Agr* 9(4): 249–254.

Inayat TP, Rana SA, Rana N, Ruby T, Sadiqui MJI, Abbas MN. 2011. Predation rate in selected coccinellid (coleoptera) predators on some major aphidid and cicadellid (Hemipteran) pests. *Int. J. Agric. Biol.* 13: 427–430.

Inayatullah C. 1995. *Training Manual: Integrated Insect Pest Management.* Entomological Research Laboratories & NARC Training Institute, Islamabad.

Ingram WR. 1994. *Pectinophora* (Lepidoptera: Gelechiidae). In: Matthews GA, Tunstall JP (eds.), *Insect Pests of Cotton.* CAB International, Wallingford, UK, pp. 107–149.

Iqbal J. 2008. IPM of aphids on wheat, *Tricticum asetivum* (L.) in Punjab, Pakistan. PhD thesis, Department of Agri. Entomology, University of Agriculture, Faisalabad, Pakistan.

Iqbal MF, Kahloon MH, Nawaz MR, Javaid MI. 2011. Effectiveness of some botanical extracts on wheat aphids. *J. Anim. Plant Sci.* 21(1): 114–115.

Ivbijaro MF, Udensi N, Ukwela UM, Anno-Nyako FV. 1991. Geographical distribution and host range in Nigeria of the mango mealy bug, *Rastrococcus invadens* Williams, a serious exotic pest of horticulture and other crops. *Insect Sci. App.* 13: 411–416.

Iwahashi O. 1977. Eradication of the melon fly, *Dacus cucurbitae*, from Kume Island Okinawa, with the sterile insect release method. *Res. Pop. Eco.* 19: 87–98.

Iwaizumi R, Sawaki M, Kobayashi K, Maeda C, Toyokawa Z, Ito M, Kawakami T, Matsui M. 1991. A comparative experiment on the attractiveness of the several kinds of the cue-lure toxicants to the melon fly, *Dacus cucurbitae* (Coquillett). *Res. Bull. Pl. Protect. Serv. Japan* 27: 75–78.

Iwata M, Sunagawa K, Kume K, Ishikawa A. 1990. Efficacy of vapour heat treatment on netted melon infested with melon fly, *Dacus cucurbitae. Res. Bull. Pl. Protect Serv. Japan* 26: 45–49.

Jaiswal JP, Gurung TB, Pandey RR. 1997. Findings of melon fruit-fly control survey and its integrated management, 1996/97, Kashi, Nepal. Lumle Agriculture Research Centre Working Paper, 97/53, pp. 1–12.

Jalali SK, Singh SP. 2003. Determination of release rate of natural enemies for evolving bio-intensive management of *Chilo partellus* (Swnhoe) (Lepidoptera: Pyralidar). *Shashpa* 10(2): 151–154.

James RR. 2003. Combining azadirachtin and *Paecilomyces fumosoroseus* (Deuteromycotina: Hyphomycetes) to control *Bemisia argentifoli* (Homoptera: Aleyrodidae). *J. Econ. Entomol.* 96: 25–30.

Jang EB. 1990. Fruit fly disinfestation of tropical fruits using semipermeable shrink-wrap film. *Acta. Hortic.* 269: 453–458.

Jang EB. 1996. Systems approach to quarantine security: Postharvested application of sequential mortality in the Hawaiian grown "Sharwil" avocado system. *J. Econ. Entomol.* 89: 950–956.

Jaya S, Suresh T, Sobhitha-Rani RS, Sreekumar S. 2000. Evidence of seven larval instars in the red palm weevil, *Rhynchophorus ferrugineus* Olivier reared on sugarcane. *J. Entomol. Res.* 24: 27–31.

Jazzar C, Hammad EAF. 2003. The efficacy of enhanced aqueous extracts of *Melia azedarach* leaves and fruits integrated with the *Camptotylus reuteri* releases against the sweet potato whitefly nymphs. *Bull. Insectol.* 56: 269–275.

Jha LK. 2010. *Applied Agricultural Entomology.* New Central Book Agency (P) Ltd., Kolkata Pune, Delhi, India.

Johnson CG. 1969. *Migration and Dispersal of Insects by Flight.* Methuen, London, 763pp.

Johnson SJ. 1987. Migration and life history strategy of the fall armyworm, *Spodoptera frugiperda* in the Western Hemisphere. *Ins. Sci. Appl.* 8: 543–549.

Joshi VR, Pawar DB, Lawande KE. 1995. Effects of different training systems and planting seasons on incidence of fruit flies in Bitter gourd. *J. Maharashtra Agric. Uni.* 20(2): 190–291.

Kafi A. 1986. Progress and problems in controlling fruit fly infestation. RAPA, angkok, 16–19 December, 1986. In: Ysegaran SV, Ibrahim AG (eds.). *Fruit Flies in the Tropics, Proc. Ist. Int. Symp.*, 14–16 March, Malysia, 21pp.

Kanzaki R, Shibuya T. 1992. Olfactory processing pathways of the insect brain. *Zool. Sci.* 9: 241–264.

Karar H, Arif MJ, Saeed S, Sayyed HA. 2006. A threat to Mango. *DAWN Sci-tech. World*, December 23, 2006.

Karar H, Saeed S, Naeem-Ullah U, Shakeel-ur-Rehman, Abbas MA, Ayyaz A, Sadiq H, Qayyum M, Ahmad M. 2016. Production of quality and cosmetic valued mangoes and management of fruit fly (Tephritidae: Diptera). *Pak. Entomol.* 38: 95–98.

Karren JB. 1993. The Russian wheat aphid in Utah. In: University US (ed.), *An update.* Utah State University Extension, USA.

Kfir R, Overholt WA, Khan ZR, Polaszek A. 2002. Biology and management of economically important lepidopteran cereal stem borers in Africa. *Ann. Rev. Entomol.* 47: 701–731.

Khalil AA, Abolmaaty SM, Hassanein MK, El-Mtewally MM, Moustafa SA. 2010. Degree-days units and expected generation numbers of peach fruit fly *Bactrocera zonata* (Saunders) (Diptera: Tephritidae) under climate change in Egypt. *Egypt. Acad. J. Biol. Sci.* 3(1): 11–19.

Khan AS. 2014. Mango consignment found infested with fruit fly. *DAWN*, downloaded on June 22nd, 2014. Available at http://www.dawn.com/news/1114336 (accessed August 6, 2014)

Khan BA. 2005a. Studies on the residues of commonly used insecticide on fruit and vegetables grown in NWFP-Pakistan, *PhD thesis*, NWFP Agriculture University, Peshawar.

Khan IA, Khan MN, Akbar R et al. 2015a. Assessment of different control methods for the control of maize stem borer, *Chilo partellus* (Swinhoe) in maize crop at Nowshera-Pakistan. *J. Entomol. Zool. Studies* 3(4): 327–330.

Khan IU, Nawaz M, Said F, Sohail K, Subhanullah. 2015b. Integrated pest management of maize stem borer, *Chilo partellus* (Swinhoe) in maize crop and its impact on yield. *J. Entomol. Zool. Studies* 3(5): 470–472.

Khan LI, Manzoor U. 1992. Control of melon fly, *Dacus cucurbitae* (Diptera: Trypetidae) on melon in Pakistan. *Trop. Pest Manag.* 38(3): 261–264.

Khan MA. 2001. Control of insect pest of mango. *Proceedings of International Mango Workshop*, 27 February 1 March. Direct Agric-Multan Region, Punjab, 224p.

Khan MA. 2004. *Wheat Crop Management for Yield Maximization.* Annual Research program, Arid Zone Research Institute, Bhakkar, Pakistan.

Khan SA. 2005b. Studies on aphid distribution pattern and their natural enemies in wheat and maize crop. *PhD thesis*, Department of Plant Protection, Agricultural University, Peshawar, Pakistan, pp. 61–62.

Khan SM, Maqbool R. 2002. Varietal performance of wheat (*Triticum aestivum*) against wheat aphid (*Macrosiphum miscanthi*) and its chemical control with different doses of insecticides. *Asian J. Plant Sci.* 1(2): 205–207.

Khan WS, Khan AG. 1995. Strategies for increasing cotton production. *National seminar held at Agric. House*, Lahore. April 26–27.

Khan ZR, Overholt WA, Hassana A. 1997. Utilization of agricultural biodiversity for management of cereal stem borers and striga weed in maize-based cropping systems in Africa -a case study. Available at http://www.cbd.int/doc/casestudies/agr/cs-agr-cereal-stemborers.pdf (accessed May 23, 2016)

Khanzada SR, Nyamwasa I, Khanzada MS. 2015. *Pest Management Decision Guide for Maize Stem Borer, Chilo partellus on Maize.* Plantwise-CABI, Available at www.plantwise.org (accessed June 21, 2016)

Khidr AA, Kostandy SN, Abbas MG, El-Kordy MW, El-Gougary QA. 1990. Host plants, other than cotton, for the pink boll worm *Pectinophora gossypiella* and the spiny boll worm *Earias insulana*. *Agric. Res. Review (Cairo)* 68: 135–139.

Kieckhefer R, Gellner J. 1992. Yield losses in winter wheat caused by low-density cereal aphid populations. *Agron. J.* 84: 180–183.

Kieckhefer R, Kantack B. 1988. Yield losses in winter grains caused by cereal aphids (Homoptera: Aphididae) in South Dakota. *J. Econ. Entomol.* 81: 317–321.

Kindler SD, Springer TL, Jensen KBL. 1995. Detection and characterization of the mechanisms of resistance to Russian wheat aphid (Homoptera: Aphididae) in tall wheat grass. *J. Econ. Entomol.* 88: 1503–1509.

Klassen W. 2005. Area-wide integrated pest management and the sterile insect technique. In: Dyck VA, Hendrichs J, Robinson AS (eds.), *Sterile Insect Technique. Principles and Practice in Area-Wide Integrated Pest Management.* Springer, Dordrecht, The Netherlands, pp. 39–68.

Klungness LM, Jang EB, Mau RFL, Vargas RI, Sugano JS, Fujitani E. 2005. New sanitation techniques for controlling tephritid fruit flies (Diptera: Tephritidae) in Hawaii. *J. Appl. Sci. Environ. Mgt.* 9(2): 5–14.

Knipling. 1979. *The Basic Principles of Insect Population Suppression and Management.* US Government Printing Office, Department of Agriculture, Washington, DC, USA, Agriculture Handbook No. 512.

Kontodimas DC, Eliopoulos PA, Stathas GJ, Economou LP. 2004. Comparative temperature-dependent development of *Nephus includens* (Kirsch) and *Nephus bisignatus* (Boheman) (Coleoptera: Coccinellidae) preying on *Planococcus citri* (Risso) (Homoptera: Pseudococcidae): Evaluation of a linear and various nonlinear models using specific criteria. *Environ. Entomol.* 33: 1–11.

Kosztarab M, Kozár F. 1988. *Scale Insects of Central Europe.* Dr. W. Junk Publishers, Dordrecht.

Kovanci OB, Kovanci B. 2006. Reduced-risk management of Rhagoletis cerasi flies (host race *Prunus*) in combination with a preliminary phenological model. *J. Insect Sci.* 6: 1–10.

Koyama J. 1989. Pest status; south-east Asia and Japan. In: Robinson AS, Hooper G (eds.), *World Crop Pests 3(a) Fruit Flies; Their Biology, Natural Enemies and Control.* Elsevier, Amsterdam, Netherlands, pp. 63–66.

Kranthi KR. 2002. Modalities of Bt cotton cultivation in India, its pros and cons including resistance management and potential ecological impact. *Nation. Sem. on Bt Cotton Scenario with Special Reference to India*, May 23, 2002. UAS, Dharwad, Karnataka, pp. 26–50.

Kranthi KR, Kranthi NR. 2004. Modelling adaptability of the cotton bollworm, *Helicoverpa armigera* (Hubner) to Bt cotton in India. *Current Sci.* 87: 1096–1107.

Kranthi KR, Jadhav DR, Kranthi S, Wanjari RR, Ali SS, Russell DA. 2002. Insecticide resistance in five major insect pests of cotton in India. *Crop Prot.* 21: 449–460.

Kranthi KR, Jadhav DR, Wanjari RR, Ali SS, Russell D. 2001. Carbamate and organophosphate resistance in cotton pests in India, 1995 to 1999. *Bull. Ent. Res.* 91: 37–46.

Kuba H, Kohama T, Kakinohana H, Yamagishi M, Kinjo K, Sokei Y, Nakasone T, Nakamoto Y. 1996. The successful eradication programs of the melon fly in Okinawa. In: McPheron BA, Steck GJ (eds.), *Fruit Fly Pests: A World Assessment of Their Biology and Management.* St. Lucie Press, Delray Beach, FL, pp. 543–550.

Kuba H, Sokei Y. 1988. The production of pheromone clouds by spraying in the melon fly, Dacus cucurbitae Coquillett (Diptera: Tephritidae). *J. Ethol.* 6: 105–110.

Kumar H. 2002. Resistance in maize to larger grain borer, *Prosphanus truncates* (Horn) (Coleoptera: Bostrichidae). *J. Stored Prod. Res.* 38: 267–280.

Kumar P, Poehling HM. 2006. Persistence of soil and foliar azadirachtin treatments to control sweet potato whitefly *Bemisia tabaci* Gennadius (Homoptera: Aleyrodidae) on tomatoes under controlled (laboratory) and field (netted greenhouse) conditions in the humid tropics. *J. Pestic. Sci.* 79: 189–199.

Kumar P, Poehling HM, Borgemeister C. 2005. Effects of different application methods of Neem against Sweet potato Whitefly *Bemisia tabaci* Gennadius (Homoptera: Aleyrodidae) on Tomato plants. *J. Appl. Entomol.* 129: 489–497.

Kurian C, Mathen K. 1971. Red palm weevil-hidden enemy of coconut palm. *Indian Farming* 21: 29–31.

Kuroli G, Nemeth I. 1987. Aphid species occurring on winter wheat, their damage and results of control experiments. *Növényvédelem* 23: 385–394.

Lacey LA, Frutos R, Kaya HK, Vail P. 2001. Insect pathogens as biological control agents: Do they have a future? *Biol. Control* 21: 230–248.

Lacey LA, Goettel MS. 1995. Current developments in microbial control of insect pests and prospects of the early 21st century. *Entomophaga* 40: 3–27.

Lalah JO, Wandiga SO. 1996. The persistence and fate of malathion residues in stored beans (*Phaseolus vulgaris)* and maize (*Zea mays*). *Pestic. Sci.* 46: 215–220.

Lall BS, Singh BN. 1969. Studies on the biology and control of melon fly, *Dacus cucurbitae* (Coq.) (Diptera: Tephritidae). *Labdev J. Sci. Tech.* 7: 148–153.

Leblanc L, Balagawi S, Mararuai A, Putulan D, Clarke AR. 2001. *Fruit Flies in Papua New Guinea*. Pest advisory leaflet No. 37, Secretariat of the Pacific Community- Plant Protection Service, Fiji, 12pp.

Lee, WY. 1988. *The Control Programme of the Oriental Fruit Fly in Taiwan*. Spec. Publ. No. 2, Entomol. Soc. Rep. China, pp. 51–60.

Lefroy HM. 1906a. *Indian Insect Life*. Superintendent of Government Printing, Calcutta.

Lefroy HM. 1906b. *Indian Insect Pests*. Office of the Superintendent of Government Printing, India, p. 346.

Legner EF, Medved RA. 1979. Influence of parasitic Hymenoptera on the regulation of pink boll-worm, *Pectinophora gossypiella*, on cotton in the lower Colorado Desert. *Envir. Entomol.* 8: 922–930.

Lella R, Srivastav CP. 2013. *Screening of maize genotypes against stem borer Chilo partellus* L. in kharif season. *Int. J. Appl. Biol. Pharmaceut. Technol.* 4(4): 394–403.

Leru B, Tertuliano M. 1993. Tolerance of different host-plants to the cassava mealybug *Phenacoccus manihoti* Matile-Ferrero (Homoptera, Pseudococcidae). *Intern. J. Pest Manag.* 39: 379–384.

Liang F, Wu JJ, Liang GQ. 2001. The first report of the test on the flight capability of oriental fruit fly. *Acta Agric. Univ. Jiangxi.* 23: 259–260.

Liang G-M, Chen W, Liu TX. 2003. Effects of three neem-based insecticides on diamondback moth (Lepidoptera: Plutellidae). *Crop Prot.* 22: 333–340.

Liang H, Zhu YF, Zhu Z, Sun DF, Jia X. 2004. Obtainment of transgenic wheat with insecticidal lectin from snowdrop (*Galanthus nivalis* agglutinin: GNA) gene and analysis of resistance to aphids. *Acta Genet. Sin.* 31: 189–194.

Lin CY, Wu DC, Yu JZ, Chen BH, Wang CL, Ko WH. 2009. Control of silverleaf whitefly, cotton aphid and kanzawa spider mite with oil and extracts from seeds of sugar apple. *Neotrop. Entomol.* 38: 531–536.

Liu TX, Stansly PA. 1995. Deposition and bioassay of insecticides applied by leaf dip and spray tower against *Bemisia agentifolii* nymphs (Homoptera: Aleyrodidae). *Pestic. Sci.* 44: 317–322.

Liu YC. 1993. Pre-harvest control of oriental fruit fly and melon fly. *Plant Quarantine in Asia and the Pacific-Report of APO Study Meeting*, March 17–26, 1992, Taipei, Taiwan Republic of China, pp. 73–76.

Llácer E, Jacas JA. 2010. Efficacy of phosphine as a fumigant against *Rhynchophorus ferrugineus* (Coleoptera: Curculionidae) in palms. *Spanish J. Agri.* Res. 8: 775–779.

Llácer E, Dembilio Ó, Jacas JA. 2010. Evaluation of the Efficacy of an Insecticidal Paint Based on Chlorpyrifos and Pyriproxyfen in a Microencapsulated Formulation against *Rhynchophorus ferrugineus* (Coleoptera: Curculionidae). *Journal of Economic Entomology* 103: 402–408.

Loke WH, Tan KH, Vijaysegaran S. 1992. Semiochemicals and related compounds in insect pest management-Malaysian experiences. In: Kadir AASA, Barlow HS (eds.), *Pest Management and the Environment in 2000.* CAB International, Wallingford, UK. Arizona Press, Tucson, AZ, USA, pp. 111–126.

Longo S and Benfatto D. 1982. Observation on olive fly (Dacus oleae (Gmel.))-population dynamics in Sicily. In: Balkema AA (ed.), *Proc CEC/IOBC Int. Symp., Fruit Flies of Econ Importance*, Athens, Greece. Rotterdam, pp. 612–615.

Lowery DT, Isman MB, Brard NL. 1993. Laboratory and field evaluation of neem for the control of aphids (Homoptera: Aphididae). *J. Econ. Entomol.* 86: 864–870.

Lubbock J. 1882. *Ants, Bees and Wasps: A Record of Observations on the Habits of the Social Hymenoptera (new edition, based on the 17th, 1929).* Kegan Paul, Trench, Trubner, London.

Magoc L, Yen JL, Hill-Williams A, McKenzie JA, Batterham P, Daborn PJ. 2005. Cross-resistance to dicyclanil in cyromazine-resistant mutants of Drosophila melanogaster and Lucilia cuprina. *Pestic. Biochem. Physiol.* 81: 129–135.

Mahmood K, Mishkatullah. 2007. Population dynamics of three species of genus *Bactrocera* (Diptera: Tephritidae: Dacinae) in BARI, Chakwal (Punjab). *Pak. J. Zool* 39(2): 123–126.

Mahmood R, Poswal MA, Shehzad A. 2002. Distribution, host range and seasonal abundance of *Sipha* sp. (Homoptera: Aphididae) and their natural enemies in Pakistan. *Pak. J. Biol. Sci.* 5(1): 47–50.

Mangan RL, Moreno DS. 2002. *Development of a Bait System to Control Tephritid Fruit Flies.* USDA-Agricultural Research Service Kika de la Garza. Subtropical Agricultural Research Center Crop Quality and Fruit Insects Research Unit, Texas.

Mangan RL, Moreno DS. 2007. Development of bait stations for fruit fly population suppression. *J. Econ. Entomol.* 100(2): 440–450.

Maniania NK. 2002. A low-cost contamination device for infecting adult tsetse flies, Glossina spp., with the entomopathogenic fungus Metarhizium anisopliae in the field. *Biocontr. Sci. Technol.* 12: 59–66.

Maniania NK, Ekesi S, Odulaja A, Okech MA, Nadel DJ. 2006. Prospects of a fungus-contaminated device for the control of tsetse fly *Glossina fuscipes* Fuscipes. *Biocontr. Sci. Technol.* 16: 129–139.

Maree JM, Kallar SA, Khuhro RD. 1999. Relative abundance of *Spodoptera litura* F. and Agrotis ypsilon Rott. on cabbage. *Pakistan J. Zool.* 31: 31–34.

Markouk M, Bekkouche K, Larhsini M, Bousaid M, Lazrek HB, Jana M. 2000. Evaluation of some Moroccan medicinal plant extracts for larvicidal activity. *J. Ethanopharmacol.* 73: 293–297.

Martinez SS, de Carvalho AOR, Vieira LG, Nunes LM, Bianchini A. 2000. Identification, geographical distribution and host plants of *Bemisia tabaci* (Genn.) Biotypes (Homoptera: Aleyrodidae) in the state of Paranov Brazil. *Anais-da-Sociedade-Entomologica-do-Brasil.* 29(3): 597–603.

Marzocchi L, Nicoli G. 1991. The principal pests of wheat. *Informatore Fitopatologico*; 41: 29–33.

Masembe C, Isabirye BE, Rwomushana I, Nankinga CK, Akol AM. 2016. Projections of climate-induced future range shifts among fruit fly (Diptera: Tephritidae) species in Uganda. *Plant Protect. Sci.* 52: 26–34.

Mashwani MA, Ullah F, Sattar S, Ahmad S, Khan MA. 2011. Efficacy of different insecticides against maize stem borer, *Chilo partellus* Swinhoe (Lepidoptera; Pyralidae) at Peshawar and Swat valleys of Khyber Pakhtunkhwa, Pakistan. *Sarhad J. Agric.* 27(3): 459–465.

Mashwani MA, Ullah F, Ahmad S, Sohail K, Usman M, Shah SF. 2015. Infestation of maize stem borer, *Chilo partellus* (Swinhoe) in maize stubbles and stalks. *Acad. J. Agric. Res.* 3: 94–98.

Masood A, Saeed S, Sajjad A, Ali M. 2009. Life cycle and biology of mango bark beetle, *Hypocryphalus mangiferae* (Stebbing), a possible vector of mango sudden death disease in Pakistan. *Pak. J. Zool.* 41: 281–288.

Massa R, Caprio E, De Santis M, Griffo R, Migliore MD, Panariello G, Pinchera D, Spigno P. 2011. Microwave treatment for pest control: The case of *Rhynchophorus ferrugineus* in *Phoenix* canariensis. *Bulletin OEPP.* 41(2): 128–135.

Matama-Kauma T, Schulthess F, Pierre LRB, Mueke J, Ogwang JA, Omwega C. 2008. Abundance and diversity of lepidopteran stem borers and their parasitoids on selected wild grasses in Uganda. *Crop Prot.* 27: 505–513.

Mau RFL, Sugano JS, Jang EB. 2003b. Farmer education and organization in the Hawaii area wide fruit fly Pest Management program. In: Inamine K (ed.), *Recent Trends on Sterile Insect Technique and Area-Wide Integrated Pest Management: Economic Feasibility, Control Projects, Farmer Organization and Dorsalis Complex Control Study.* Research Institute of Subtropics, Okinawa, Japan, pp. 47–57.

McClean APD, Schwartz RE. 1970. Greening of blotchy-mottle disease in citrus. *Phytophylactica* 2: 177–194.

McKenney DW, Pedlar JH, Lawrence K, Campbell K, Hutchinson MF. 2007. Potential impacts of climate change on the distribution of North American trees. *Bio Sci.* 57: 939–948.

Mead FW. 1977. *The Asiatic Citrus Psyllid, Diaphorina citri Kuwayama (Homoptera: Psyllidae).* Entomology Circular 180. Florida Department of Agriculture and Consumer Services, Division of Plant Industry. 4pp.

Mead FW. 1997. Asiatic citrus psyllid, Diaphorina citri Kuwayama. EENY-33. UF/IFAS Featured Creatures. EENY-33. http://creatures.ifas.ufl.edu/citrus/acpsyllid.htm

Mehmood R, Mohyuddin I, Kazimi S. 1980. *Restrococcus spinosus* (Homoptera; Pseudococcidae) and its natural enemies in Pakistan. *Proceedings of 1st Pakistan Congress Zoology.* Tandojam, pp. 291–294.

Menon KPV, Pandalai KM. 1960a. *A Monograph of the Coconut Palm.* Indian Central Committee, Ernakulam, India, 384pp.

Menon KPV, Pandalai KM. 1960b. Pests. In: *The Coconut Palm. A Monograph.* Indian Central Committee, Inrankulam, South India, pp. 261–265.

Mesallam T. 2010. Effect of different date palm varieties on some biogical aspects of the red palm weevil, *Rhynchophorus ferrugineus* (Oliv.) (Coleoptera, Curculionidae). *PhD thesis*, Fac. Agric., Zagazig Univ., 350pp.

Metcalf RL, Metcalf RA. 1993. *Destructive and Useful Insects.* McGraw Hill Inc., New York.

Meyer MD, Mohamed S, White IM. 2007. *Invasive Fruit Fly Pests in Africa.* Royal Museum for Central Africa, Tervuren, Belgium. Available at http://www.africamuseum.be/fruitfly/AfroAsia.htm (accessed May 2016).

Mguni C. 2013. *Notification of Bactrocera invadens Detection in Bindura District (Mashonaland Central Province) of Zimbabwe.* Ministry of Agriculture, Mechanisation and Irrigation Development. Available at https://www.ippc.int/sites/default/files/documents/20130423/1360934298_ippc_secretariat_bi_declaration_2013042321%3A20.pdf (accessed May 2016)

Mirz JH, Ahmad W, Ayub MA, Khan O, Ahmad S. 1994. *Studies on the Identification, Transmission and Host Range of CLCuV Disease in the Punjab with Special Reference to Its Control.* Final Res. Deptt., UAF.

Mirza I. 2014. Compaign against fruit fly to avert mangoes' export ban to EU. *Business Recorder*, downloaded on May 6th, 2014. Available at http://www.brecorder.com/agriculture-a-allied/183:pakistan/1180060:campaign-against-fruit-fly-to-avert-mangoes-export-ban-to-eu?date=2014-05-06/ (accessed August 6, 2014).

Mishra PN, Singh MP. 1999. Studies on the ovicidal action of diflubenzuron on the eggs of *Dacus (Bactrocera) cucurbitae* Coq. damaging cucumber. *Ann. Pl. Prot. Sci.* 7: 94–6.

Mitchell PL, Gupta R, Singh AK, Kumar P. 2004. Behavioural and developmental effects of neem extracts on *Clavigralla scutellaris* (Hemiptera: Heteroptera: Coreidae) and its egg parasitoid, *Gryon fulviventre* (Hymenoptera: Scelionidae). *J. Econ. Entomol.* 97: 916–923.
Mitchell WC. 1980. Verification of the absence of oriental fruit and melon fruit-fly following an eradication program in the Mariana Islands. *Proc. Hawaiian Entomol. Soc.* 23: 239–243.
Miyatake T, Irabu T, Higa R. 1993. Oviposition punctures in cucurbit fruits and their economic damage caused by the sterile female melon fly, *Bactrocera cucurbitae* Coquillett. *Proceedings of the Association of Plant Protection*, vol. 39, Kyushu, pp. 102–105.
Modarresi M, Mohammadi V, Zali A, Mardi M. 2010. Response of wheat yield and yield related traits to high temperature. *Cereal Res. Commun.* 38: 23–31.
Mohamed AM. 2002. Seasonal abundance of peach fruit fly, *Bacterocera zonata* (Saunders) with relation to prevailing weather factor in Upper Egypt. *Assuit J. Agric. Sci.* 33(2): 195–207.
Mohammad AB, Aliniazee MT. 1989. Malathion bait sprays for control of apple maggot Diptera: Tephritidae). *J. Econ. Entomol.* 82: 1716–1721.
Mohyuddin AI, Mahmood R. 1993. Integrated control of mango pests in Pakistan. *Acta. Hort.* 341: 467–487.
Montell C. 2009. A taste of the drosophila gustatory receptors. *Curr. Opin. Neurob.* 19(4): 345–353.
Mordue A, Blackwell A. 1993. Azadirachtin: An update. *J. Insect. Physiol.* 39: 903–924.
Mote UN. 1975. Control of fruit fly (*Dacus cucurbitae*) on bitter gourd and cucumber. *Pesticides* 9: 36–37.
Mound LA. 1962. Studies on the olfaction and colour sensitivity of *Bemisia tabaci* (GENN.) (Homoptera, Aleurodidae). *Entomol. Exp. Appl.* 5: 99–104.
Muhammad L, Underwood E. 2004. The maize agricultural context in Kenya. In: Hilbeck A, Andow DA (eds.), *Environmental Risk Assessment of Genetically Modified Organisms: A Case Study of Bt Maize in Kenya*, vol. 1. CABI Publishing, Wallingford, Oxford shire, UK, pp. 1–2.
Muniappan R, Shepard BM, Carner GR. 2012. *Arthropod Pest of Horticultural Crops in Tropical Asia.* Gutenberg Press Limited, Tarxien, Malta, 99 pp.
Munro KH. 1984. A taxonomic treatise on the Decidae (Tephritoidea: Diptera) of Africa. Entomological Memories of the Department of Agriculture. *Repub. South Africa* 61: 1–313.
Murphy ST, Briscoe BR. 1999. The red palm weevil as an alien invasive: Biology andprospects for biological control as a component of IPM. *Biocontrol* 20: 35–45.
Mutyambai DM, Midega CA, Bruce TJ, van den Berg J, Pickett JA, Khan ZR. 2014. Behaviour and biology of *Chilo partellus* on maize landraces. *Entomol. Exper. Appl.* 153: 170–181.
Mwimali MG. 2014. Genetic analysis and response to selection for resistance to two stem borers, *Busseola fusca* and *Chilo partellus*, in tropical maize germplasm, *PhD dissertation*, African Centre for Crop Improvement, School of Agriculture, Earth and Environmental Sciences, College of Agriculture, Engineering and Sciences, University of KwaZulu-Natal Republic of South Africa.
Nagappan K, Kamalnathan S, Santharaman T, Ayyasamy MK. 1971. Insecticidal trials for the control of the melon fruit-fly, Dacus cucurbitae Coq. infesting snake gourd, *Trichosanthes anguina. Madras Agric. J.* 58: 688–690.
Nair S, Thomas J. 1999. Effect of *Acorus calamus* L. extracts on the longevity of *Bactrocera cucurbitae* Coq. *Insect Environ.* 5: 27.
Nakahira K, Arakawa R. 2006. Development and reproduction of an exotic pest mealybug, *Phenacoccus solani* (Homoptera: Pseudococcidae) at three constant temperatures. *Appl. Entomol. Zool.* 41: 573–575.
Naranjo SE, Gordh G, Moratorio M. 1992a. Inundative release of Trichogrammatoidea for biological control of pink bollworm. In: Cotton A (ed.), *College of Agriculture Report.* University of Arizona, Tucson, pp. 110–116.
Naranjo SE, Gordh G, Moratorio M. 1992b. Biology and behavior of Trichogrammatoidea bactrae, an imported parasitoid of pink bollworm. In: Herber DJ, Richter DA (eds.), *Proceedings of the Beltwide Cotton Conference.* Memphis: National Cotton Council, USA, pp. 920–922.
Narayanan ES. 1953. Seasonal pests of crops. *Indian Farming* 3(4): 8–11 and 29–31.
Nassar NMA. 2007. Cassava genetic resources and their utilization for breeding of the crop. *Genet. Mol. Res.* 6: 1151–1168.
Nathan SS, Kalaivani K, Murugan K, Chung PG. 2005. Efficacy of neem limonoids on *Cnaphalocrocis medinalis* (Guenée) (Lepidoptera: Pyralidae) the rice leaffolder. *Crop Prot.* 24: 760–763.
Naumann K, Isman MB. 1995. Evaluation of neem (*Azadirachta indica* A. Juss.) seed extracts and oils as oviposition deterrents to noctuid moths. *Entomol. Exp. Appl.* 76: 115–120.
Nava DE, Torres MLG, Rodrigues MDL, Bento JMS, Parra JRP. 2007. Biology of *Diaphorina citri* (Hem., Psyllidae) on different hosts and at different temperatures. *J. Appl. Entomol.* 131: 709–715.
Navarro-Llopis V, Sanchis-Cabanes J, Ayala I, V. Casana-Giner, Primo-Yúfera E. 2004. Efficacy of lufenuron as chemosterilant against Ceratitis capitata in field trials. *Pest Manag. Sci.* 60(9): 914–20.

Nebie K, Nacro S, Otoidobiga LC, Dakouo D, Somda I. 2016. Population dynamics of the mango mealybug *Rastrococcus invadens* Williams (Homoptera: Pseudococcidea) in Western Burkina Faso. *Amer. J. Exper. Agric.* 11(6): 1–11.
Neilson WTA, Maxwell CW. 1964. Field tests with a malathion bait spray for control of the apple maggot, *Rhagoletis pomonella*. *J. Econ. Entomol.* 57: 192–194.
Neilson WTA, Sanford KH. 1974. Apple maggot control with baited and unbaited sprays of azinphos-methyl. *J. Econ. Entomol.* 67: 556–557.
Nelson MR, Nadeem A, Ahmad W, Orum TV. 1998. Global assessment of cotton viral diseases. In *Proceedings of Beltwide Cotton Conference*, San Diego, CA. January 5–9, 1998. National Cotton Council, Memphis, TN, pp. 161–162.
Newton I. 2008. *The Migration Ecology of Birds*. Academic Press, Elsevier, London, UK.
Newton C, Dixon AFG. 1988. A preliminary study of variation and inheritance of life history traits and the occurrence of hybrid vigour in *Sitobion avenae* F. Hemiptera: Aphididae. *Bull. Entomol. Res.* 78: 75–83.
Nirula KK. 1956. Investigations on the pests of coconut palm. Part IV. *Rhynchophorus ferrugineus*. *Indian Coconut J.* 9: 229–247.
Noble LW. 1969. *Fifty Years of Research on the Pink Bollworm in the Unites States*. U.S. Dept. Agric. Agric. Res. Serv. Handb, p. 357.
Norris RF, Caswell-Chen EP, Kogan M. 2002. *Concepts in Integrated Pest Management*. Prentice-Hall of India Private Limited, New Delhi, India.
Novotny V, Clarke AR, Drew RAI, Balagawi S, Clifford B. 2005. Host specialization and species richness of fruit flies (Diptera: Tephritidae) in a New Guinea rain forest. *J. Trop. Ecol.* 21: 67–77.
Nzanza B, Marais D, Soundy P. 2011. Response of tomato (*Solanum lycopersicum* L.) to nursery inoculation with *Trichoderma harzianum* and *Arbuscular mycorrhizal* fungi under field conditions. *Acta Agr. Scand.* B-SP: 1–8.
OEPP/EPPO. 1983. Data sheets on quarantine organisms No. 41, Trypetidae (non-European). *Bulletin OEPP/EPPO, Bulletin* 13(1): 1–3.
Omar HM, Hashim N. 2004. *Technical Document for Market Access to Star Fruit (Carambola) (Averrhoa carambola L.; Oxalidaceae)*. Crop Protection and Plant Quarantine Services Division, Department of Agriculture, Kuala Lumpur, Malaysia, p. 8.
Orphanides GM, Bartlett BR, Dawson LH. 1971. *Bracon kirkpatricki* Wlkn: Laboratory life tables and releases against pink bollworm in southern California cotton fields. *Bollettino del Laboratorio di Entomolo. Agric. Filippo Silvestri, Portici* 28: 135–144.
Osman M, Bin S, Chettanachitara C. 1989. Postharvest insects and other pests of rambutan. In: Lam PF, Kosiyanchinda S (eds.), *Rambutan: Fruit Development, Postharvest Physiology and Marketing in ASEAN*. ASEAN Food Handling Bureau, Kuala Lumpur, Malaysia, pp. 57–60.
Painter RH. 1951. *Insect Resistance in Crop Plants*. MacMillan, New York, 520 pp.
Pair SD, Raulston JR, Sparks AN, Westbrook JK, Dounce GK. 1986. Fall armyworm distribution and population dynamics in the southeastern states. *Florida Entomol.* 69: 468–487.
Pandey RR. 2004. Plan your sudex borders for melon fly control. In: Hiraki C (ed.), *HAW-FLYPM Newsletter*. Coop. Ext. Serv., Col. of Trop. Agric. and Human Resources, Univ. of Hawaii, February 2004. Available at http://www.fruitfly.hawaii.edu/
Panhwar F. 2005. *Mediterranean Fruit Fly (Ceratitis capitata) Attack on Fruits and Its Control in Sindh, Pakistan*. Digitalverlag GmbH, Germany. Available at http://www.chemlin.de/publications/documents/Mediterranean_fruit_fly.pdf (accessed September 24, 2008)
Papp M, Mesterházy Á. 1993. Resistance to bird cherry-oat aphid (*Rhopalosiphum padi* L.) in winter wheat varieties. *Euphytica* 67: 49–57.
Parveen Z, Masud SZ, Khuro MI, Kausar N. 2011. Organophosphate pesticide residues in fruits. *Pak. J. Sci. Res.* 31: 53–56.
Patil SB. 2003. Studies on management of cotton pink bollworm *Pectionophora gossypiella* (Saunders) (Lepidoptera: Gelechiidae). *PhD thesis*, University of Agricultural Sciences, Dharwad, India.
Patra S, Rahman Z, Bhumita P, Saikia K, Thakur NSA. 2013. Study on pest complex and crop damage in maize in medium altitude hill of Meghalaya. *Bioscane* 8(3): 825–828.
Pats P, Wiktelius S. 1989. Tethred flight of *Chilo partellus* (Swinhoe) (Lepidoptera: Pyralidae). *Bull. Entomol. Res.* 79(1): 109–114.
Pats P, Wiktelius S. 1992. Diel flight periodicity of *Chilo partellus*. *Entomol. Exp. Appl.* 65: 165–170.
Pawar DB, Mote UN, Lawande KE. 1991. Monitoring of fruit fly population in Bitter gourd crop with the help of lure trap. *J. Res. Maharashtra Agric. Uni.* 16: 281.
Pearson EO. 1958. *The Insect Pests of Cotton in Tropical Africa*. Empire Cotton Growing Corporation and Commonwealth Institute of Entomology, London.

Pedigo LP, Rice ME. 2009. *Entomology and Pest Management*, 6th ed. PHI Learning Private Limited, New Delhi, India.

Pelz KS, Isaacs R, Wise JC, Gut LJ. 2005. Protection of fruit against infestation by apple maggot and blueberry maggot (Diptera: Tephritidae) using compounds containing spinosad. *J. Econ. Entomol.* 98: 432–437.

Perdomo AJ, Nation JL, Baranowski RM. 1976. Attraction of female and male fruit lies to food-baited and male-baited traps under field conditions. *Environ. Entomol.* 15: 1208–1210.

Permalloo S, Seewooruthun SI, Joomaye A, Soonnoo AR, Gungah B, Unmole L, Boodram R. 1998. An area wide control of fruit flies in Mauritius. In: Lalouette JA, Bachraz DY, Sukurdeep N, Seebaluck BD (eds.), *Proceedings of the Second Annual Meeting of Agricultural Scientists*, August 12–13, 1997. Food and Research Council, Reduit, Mauritius, pp. 203–210.

Pieterse W, Muller DL, Jansen-van-Vuuren B. 2010. A molecular identification approach for five species of Mealybug (Hemiptera: Pseudococcidae) on citrus fruit exported from South Africa. *African Entomol.* 18(1): 23–28.

Pinero JC, Jacome I, Vargas R, Prokopy RJ. 2006. Response of female melon fly, *Bactrocera cucurbitae*, to host-associated visual and olfactory stimuli. *Entomol. Exp. Appl.* 121(3): 261–269.

Plant Health Australia. 2009. *A Threat Specific Contingency Plan for Spotted Stem Borer (Chilo partellus).* Kalang Consultancy Services Pty Ltd, Australia, pp. 1–26.

Pluke RWH, Qureshi JA, Stansly PA. 2008. Citrus flushing patterns, *Diaphorina citri* (Homoptera: Psyllidae) populations and parasitism by *Tamarixia radiata* (Hymenoptera: Eulophidae) in Puerto Rico. *Fla. Entomol.* 91: 36–41.

Pogue MG. 2002. A world revision of the genus *Spodoptera* Guenée (Lepidoptera: Noctuidae). *Mem Am Entomol Soc.* 43: 1–202.

Polaszek A. 1998. *African Cereal Stem Borers: Economic Importance, Taxonomy, Natural Enemies and Control.* CABI, Wallingford, UK, 530pp.

Prabhaker N, Toscano NC, Coudriet DL. 1989. Susceptibility of the immature and adult stage of the sweet potato white fly (Homoptera: Aleyrodidae) to selected insecticides. *J. Econ. Entomol.* 82: 983–988.

Prakash A. 2013. Changing pest scenario in rice. Key-note address on rice, pest and climate change. In: *National Symposium on. Man, Animal and Environmental Interaction in the Perspective of Modern Research*, March 8–9, 2013, North Bengal University, Darjeeling, West Bengal, India, pp. 3–5.

Prakash A, David BV, Bambawale OM. 2014. *Plant Protection in India: Challenges and Research Priorities.* AZRA Publications, Cuttak-India, 170pp.

Prokopy RJ, Miller NW, Pinero JC, Barry JD, Tran LC, Oride L, Vargas RI. 2003. Effectiveness of GF-120 fruit fly bait spray applied to border area plants for control of melon flies (Diptera: Tephritidae). *J. Econ. Entomol.* 96(5): 1485–1493.

Pruthi HS, Batra HN. 1938. A preliminary annotated list of fruit pests of the North-West Frontier Province. *Misc. Bull. Imp. Council Agric. Res. India.* 19: 10–12.

Pruthi HS, Batra HN. 1960. Some important fruit pests of North West India. *ICAR Bull.* 80: 1–113.

Qureshi ZA, Ashraf M, Bughio AR, Siddiqui QH. 1975. *Population Fluctuation and Dispersal Studies of Fruit Fly Dacus zonata.* International Atomic Energy Agency (the IAEA and the FAO of the United Nations), Vienna, Austria, pp. 201–206.

Rabindranath K, Pillai KS. 1986. Control of fruit fly of Bitter gourd using synthetic pyrethroids. *Entomon.* 11: 269–272.

Rahalkar GW, Harwalkar MR, Rananvare HD. 1972. Development of red palm weevil, *Rhynchophorus ferrugineus* Oliv. on sugarcane. *Ind. J. Ent.* 34: 213–215.

Rai AB, Halder J, Kodandaram MH. 2014. Emerging insect pest problems in vegetable crops and their management in India: An appraisal. *Pest Manag. Hortic. Ecosyst.* 20(2): 113–122.

Rajamanickam K, Kennedy JS, Christopher A. 1995. Certain components of integrated management for red palm weevil, *Rhynchophorus ferrugineus* F. (Curculionidae: Coleoptera) on coconut. *Mededelingen Faculteit Landbouwkundige en Toegepaste Biologische Wetenschappen Universiteit Gent* 60: 803–805.

Rajapakse R, Ratnasekera D. 2007. The management of the major insect pests *Bactocera cucurbitae* (Diptera: Tephritidae) and Aulacaphora spp. (Coleoptera: Scarabaeidae) in cucurbits in Southern Sri Lanka under three intensive systems: Integrated, chemical and organic agriculture. III International Symposium on Cucurbits held on September 11, 2005 in Townsville (Australia). *ISHS Acta Hort.* 731: 303–309.

Ramachandran CP. 1998. Biotypic variability among four populations of red palm weevil, *Rhynchophorus ferrugineus* Oliv. from different parts of India. *Coconut Res. Develop. (CORD)* 14: 26–41.

Ramsamy MP, Rawanansham T, Joomaye A. 1987. Studies on the control of *Dacus cucurbitae* Coquillett and *Dacus demmerezi* Bezzi (Diptera: Tephritidae) by male annihilation. *Revue. Agricole. Sucriere ltle Mauriee* 66: 1–3.

Ranganath HR, Suryanarayana MA, Veenakumari K. 1997. Management of melon fly *Bactrocera (Zeugodacus) cucurbitae* in cucurbits in South Andaman. *Insect Environ.* 3: 32–33.

Ranganath HR, Veenakumari K. 1996. Tomato (*Lycopersicon esculentum* Miller): A confirmed host of the melon fly, *Bactrocera cucurbitae* Coquillett. *Insect Environ.* 2: 3.

Ranjan R. 2006. Economic impacts of pink hibiscus mealybug in Florida and the United States. *Stoch. Environ. Res. Risk Assess.* 20: 353–362.

Ravikumar P, Viraktamath S. 2007. Attraction of female fruit flies to different protein food baits in guava and mango orchards. *Karnataka J. Agric. Sci.* 20(4): 745–748.

Reddy AV. 1997. Evaluation of certain new insecticides against cucurbit fruit-fly (*Dacus cucurbitae* Coq.) on bitter gourd. *Ann. Agric. Res.* 18(2): 252–254.

Reimer L. 2004. Clonal diversity and population genetic structure of the grain aphid *Sitobion avenae* (F.) in central Europe. Ph.D. Dissertation, University of Gottingen, Germany, pp. 9–11.

Reissig WH. 2003. Field and laboratory tests of new insecticides against the apple maggot, *Rhagoletis pomonella* (Walsh) (Diptera: Tephritidae). *J. Econ. Entomol.* 96: 1463–1472.

Renwick JAA. 1989. Chemical ecology of oviposition in phytophagous insects. *Experientia* 45: 223–228.

Reynolds AM, Reynolds DR, Smith AD, Svensson GP, Lofstedt C. 2007. Appetitive flight patterns of male *Agrotis segetum* moths over landscape scales. *J. Theor. Biol.* 245: 141–149.

Rice MJ, Saxton S, Esmail AM. 1985. Antifeedant phytochemical blocks oviposition by sheep blow fly. *J. Aust. Entomol. Soc.* 24: 16.

Rice RE, Phillips PA, Stewart-Leslie J, Sibbett GS. 2003. Olive fruit fly populations measured in Central and Southern California. *California Agric.* 57(4): 122.

Rice RE, Reynolds HT. 1971. Distribution of pink bollworm larvae in crop residues and soil in Southern California. *J. Eco. Entomol.* 64: 1451–1454.

Richmond JA, Thomas HA, Hattachargya HB. 1983. Predicting spring flight of Nantucket pine tip moth (Lepidoptera: Olethreutidae) by heat unit accumulation. *J. Econ. Entomol.* 76: 269–271.

Riddiford LM, Truman JW. 1978. Biochemistry of insect hormones and insect growth regulators. In: Rockstein M (ed.), *Biochemistry of Insects*. Acad. Press, New York, pp. 307–357.

Ripper WE, George L. 1965. *Cotton Pests of the Sudan: Their Status and Control.* Black well Scientific Publications, Oxford.

Robert RH, Lavallee SG, Schnell E, Midgarden DG, Epsky ND. 2009. Laboratory and field cage studies on female-targeted attract-and-kill bait stations for *Anastrepha suspensa* (Diptera: Tephritidae). *Pest Manag. Sci.* 65(6): 672–677.

Robinson A, Hooper G. 1989. *World Crop Pests: Fruit Flies; Their biology, Natural Enemies and Control.* vol. 3. Elsevier, Amsterdam.

Roda A, Kairo M, Damian T, Franken F, Heidweiller K, Johanns C, Mankin R. 2011. Red palm weevil (*Rhynchophorus ferrugineus*), an invasive pest recently found in the Caribbean that threatens the region. *Bulletin OEPP.* 41(2): 116–121.

Roessler Y. 1989. Control; insecticides; insecticidal bait and cover sprays. In: Robinson AS, Hooper G (eds.), *World Crop Pests 3(B). Fruit Flies; Their Biology, Natural Enemies and Control.* Elsevier, Amsterdam, Netherlands, pp. 329–336.

Rogers ME, Stansly PA. 2012. *Biology and Management of the Asian Citrus Psyllid, Diaphorina citri Kuwayama.* Citeseer, Florida Citrus.

Roomi MW, Abbas T, Shah SAH, Robina S, Qureshi AA, Hussain SS, Nasir KA. 1993. Control of fruit flies (Dacus spp.) by attractants of plant origin. *Anzeiger fur Schadlingskunde, Aflanzenschutz Umwdtschutz* 66: 155–157.

Rose DJW, Page WW, Dewhurst CF, Riley JR, Reynolds DE, Pedgley DE, Tucker MR. 1985. Downwind migration of the African armyworm moth, Spodoptera exempta, studied by mark-and-capture and by radar. *Ecol. Entomol.* 10: 299–313.

Rossing W, Daamen R, Jansen M. 1994. Uncertainty analysis applied to supervised control of aphids and brown rust in winter wheat. Part 2. Relative importance of different components of uncertainty. *Agr. Syst.* 44: 449–460.

Rousse P, Duyck PF, Quilici S, Ryckewaert P. 2005. Adjustment of Field Cage Methodology for Testing Food Attractants for Fruit Flies (Diptera: Tephritidae). *Ann. Entomol. Soc. Am.* 98(3): 402–408.

Rovest L, Deseo KV. 1991. Effectiveness of neem seed kernel extract against *Leucoptera rnalifoliella* Costa, (Lep., Lyonetiidae). *J. Appl. Entomol.* 111: 231–236.

Royer T, Giles K, Nyamanzi T, Hunger R, Krenzer E, Elliott N, Kindler S, Payton M. 2005. Economic evaluation of the effects of planting date and application rate of imidacloprid for management of cereal aphids and barley yellow dwarf in winter wheat. *J. Econ. Entomol.* 98: 95–102.

Rugman-Jones PF, Hoddle CD, Hoddle MS, Stouthamer R. 2013. The lesser of two weevils: Molecular-genetics of pest palm weevil populations confirm *Rhynchophorus vulneraturs* (Panzer 1798). as a valid species distinct from *R. ferrugineus* (Olivier 1790), and reveal the global extent of both. *PLoS ONE* 8: 1–15.
Sabine BNE. 1992. *Pre-harvest Control Methods.* International Training Course Fruit Flies, MARDI, Kuala Lumpur. 4th–15th May 1992. 20pp.
Saelee A, Tigvattananont S, Baimai V. 2006. Allozyme electrophoretic evidence for a complex of species within the Bactrocera tau group (Diptera: Tephritidae) in Thailand. *Songklanakarin J. Sci. Technol.* 28(2): 249–259.
Sagarra LA, Peterkin DD. 1999. Invasion of the Carribean by the hibiscus mealybug, Maconellicoccus hirsutus Green [Homoptera: Pseudococcidae]. *Phytoprotection* 80(2): 103–113.
Saikia DK, Dutta SK. 1997. Efficacy of some insecticides and plant products against fruit fly, *Dacus cucurbitae* Coq. on ridge-gourd, *Luffa acutangula* L. *J. Agric. Sci. Soc. North East India* 10(1): 132–135.
Sain M, Prakash A. 2008. *Major Insect Pests of Rice and Their Changing Scenario, in Rice Pest Management.* AZRA Publications, Cuttak-India, pp. 7–17.
Saito O. 2001. Flight activity of three *Spodoptera spp., Spodoptera litura, S. exigua and S. depravata*, measured by a flight actograph. *Physiol. Entomol.* 2: 112–119.
Salama HS, Hamdy MK, El-Din MM. 2002. The thermal constant for timing the emergence of red palm weevil *Rhynchophorus ferrugineus* (Oliv.) (Curculionidae: Coleoptera). *Anzeiger für Schädlingskunde.* 75: 26–29.
Saleem MA, Ahmad M, Ahmad M, Aslam M. 2008. Resistance to selected organochlorine, carbamate and pyrethroid in Spodoptera litura (Lepidoptera: Noctuidae) from Pakistan. *J. Econ. Entomol.* 101: 1667–1675.
Saleem Z, Iqbal J, Khattak SJ, Khan M, Muhammad N, Iqbal Z, Khan FU, Fayyaz H. 2014. Effect of different insecticides against maize stem borer infestation at Barani Agricultural Research Station, Kohat, KPK, Pakistan during Kharif 2012. *Int. J Life Sci. Res.* 2(1): 23–26.
Saunders WW. 1843. Description of a species of moth destructive to the cotton crops in India. *Trans. R. Entomol. Soc. Lond.* 3: 284–285.
Saveer AM, Kromann SH, Birgersson GR, Bengtsson M, Lindblom T, Balkenius A, Hansson BS, Witzgall P, Becher PG, Ignell R. 2012. Floral to green: Mating switches moth olfactory coding and preference. *Proc. R. Soc. B Biol. Sci.* 2791737: 2314–2322.
Saxena KN, Basit A. 1982. Inhibition of oviposition by volatiles of certain plants and chemicals in the leaf hopper, *Amrasca devetans* (Distant). *J. Chem. Ecol.* 8: 329–338.
Saxena KN, Rembold H. 1984. Orientation and ovipositional responses of Heliothis armigera to certain neem constituents. *Proc. 2nd Int. Neem Conf.*, Rauischholzhausen, Germany, 1983, pp. 199–210.
Schaechter E. 2000. Weired and wonderful fungi. *Microbiolgy Today* 27: 116–117.
Schmutterer H. 1990. Properties and potential of natural pesticides from the neem tree, *Azadirachta indica. Annu. Rev. Entomol.* 35: 271–297.
Schmutterer H, Singh RP. 1995. List of insect pests susceptible to neem products. In: Schmutterer H (ed.), *The Neem Tree Azadirachta Indica A. Juss. and Other Meliaceous Plants.* CH Publications, Weinheim, Germany. pp. 326–365.
Scholte EJ, Knols BGJ, Takken W. 2004. Autodissemination of the entomopathogenic fungus *Metarhizium anisopliae* amongst adults of the malaria vector *Anopheles gambiae. Malaria J.* 3: 45–50.
Schowalter TD. 2011. *Insect Ecology: An Ecosystem Approach.* Academic Press, London NW1 7BY, UK.
Schuster DJ. 2004. Squash as a trap crop to protect tomato from whitefly-vectored tomato yellow leaf curl. *Int. J. Pest Manag.* 50: 281–284.
Seewooruthun SI, Permalloo S, Gungah B, Soonnoo AR, Alleck M. 2000. Eradication of an exotic Fruit Fly from Mauritius. In: Tan K-H (ed.), *Area Wide Control of Fruit Flies and Other Insect Pests*, Sinaran Bros., Sdn. Bhd., Penang, Malaysia, pp. 389–393.
Seewooruthun SI, Sookar P, Permalloo S, Joomaye A, Alleck M, Gungah B, Soonnoo AR. 1998. An attempt to the eradication of the oriental fruit fly, *Bactrocera dorsalis* (Hendel) from Mauritius. In: Lalouette JA, Bachraz DY, Sukurdeep N, Seebaluck BD (eds.), *Proceedings of the Second Annual Meeting of Agricultural Scientists*, August 12–13, 1997. Food and Research Council, Reduit, Mauritius, pp. 181–187.
Shah SIA. 2014. Cotton stainer (*Dysdercus Koenigii*): An emerging serious threat for cotton in Pakistan. *Pak. J. Zool.* 46(2): 329–335.
Shahzad MW, Razaq M, Hussain A, Yaseen M, Afzal M, Mehmood MK. 2013. Yield and yield components of wheat (*Triticum aestivum* L.) affected by aphid feeding and sowing time at Multan, Pakistan. *Pak. J. Bot.* 45: 2005–2011.
Sharma-Natu P, Sumesh K, Lohot VD, Ghildiyal M. 2006. High temperature effect on grain growth in wheat cultivars: An evaluation of responses. *Indian J. Plant Physi.* 11: 239–245.

Sharp JL, Chambers DL, Haramoto FH. 1975. Flight mill and stroboscopic studies of oriental fruit flies1 and melon flies, including observations of Mediterranean fruit flies. *Proc. Hawaiian Entomol. Soc.* XXII: 137–144.

Shehata NF, Younes MWF, Mahmoud YA. 2008. Biological studies on the peach fruit fly, *Bactrocera zonata* (Saunders) in Egypt. *J. Appl. Sci. Res.* 4(9): 1103–1106.

Shelly TE, Pahio E, Edu J. 2004. Synergistic and inhibitory interactions between methyl eugenol and cue lure influence trap catch of male fruit flies, *Bactrocera dorsalis* (Hendel) and *B. cucurbitae* (Diptera: Tephritidae). *Florida Entomol.* 87(4): 480–486.

Sheo ST, Vargas RI, Gilmore JE, Kurashima RS, Fujimoto MS. 1990. Sperm transfer in normal and gamma-irradiated, laboratory-reared Mediterranean fruit flies (Diptera: Tephritidae). *J. Econ. Entomol.* 83: 1949–1953.

Shorey HH. 1976. Application of pheromones for manipulating insect pests of agricultural crops. In: Yuskima T (ed.) *Proceedings of the Symposium on Insect Pheromones and Their Applications.* National Institute of Agriculture, Nagoaka and Tokyo, pp. 45–72.

Siddiqui QH, Ashraf M. 2002. Significance of moisture percentage and depth levels of pupation substrate in the quality production of *Bactrocera zonata. Pak. J. Bio. Sci.* 5(12): 1311–1312.

Simberloff D, Martin JL, Genovesi P. 2013. Impacts of biological invasions: What's what and the way forward. *Trends Ecol. Evol.* 28: 58–66.

Simmons DR, Robertson AE, McKay LS, Toal E, McAleer P, Pollick FE. 2009. Vision in autism spectrum disorders. *Vision Research* 49(22): 2705–2739.

Singh RP, Srivastava BG. 1983. Alcohol extract of neem (*Azadirachta indica* A. Juss.) seed oil as oviposition deterrent for *Dacus cucurbitae* (Coq.). *Indian J. Entomol.* 45: 497–498.

Singh S, Singh RP. 1998. Neem (*Azadirachta indica*) seed kernel extracts and azadirachtin as oviposition deterrents against the melon fruit-fly (*Bactrocera cucurbitae*) and the oriental fruit fly (*Bactrocera dorsalis*). *Phytoparasitica* 26(3): 1–7.

Skelley LH, Hoy MA. 2004. A synchronous rearing method for the Asian citrus psyllid and its parasitoids in quarantine. *Biol. Control.* 29: 14–23.

Slosser JE, Watson TF. 1972a. Influence of irrigation on overwinter survival of the pink bollworm. *Environ. Entomol.* 1: 572–576.

Smee C. 1940. *Report of the Entomologist, 1939.* Department of Agriculture, Nyasaland.

Smith D, Beattie GAC, Broadley R. 1997. *Citrus Pests and Their Natural Enemies: Integrated Pest Management in Australia.* Information series Q197030. Queensland Department of Primary Industries, Brisbane.

Smith D, Nannan L. 1988. Yeast auto lysate bait sprays for control of Queensland fruit fly on passion fruit in Queeensland. *Queensland J. Agric. Ani. Sci.* 45(2): 169–177.

Smith RHS. 2015. *Chilo partellus* (Swnhoe): Plant pest of Middle East. Available at http://www.agri.huji.ac.il/mepests/pest/Chilo_partellus/ (accessed June 20, 2016)

Smith TJ. 2005. *Organic Cherry Fruit Fly Control with Spinosad (Enrust, GF-120 bait), Compared to a Conventional Provado Standard and an Untreated Check.* Washington State University Extension, North Central Washington. Available at http://www.ncw.wsu.edu/treefruit/documents/2005CFFResults TJSmithOrganic.pdf (accessed September 23, 2008)

Sombatsiri K, Tigvattanont S. 1984. Effects of neem extracts on some insect pests of conomic importance in Thailand. *Proc. 2nd Int. Neem Conf.*, Rauischholzhausen, Germany, 1983, pp. 95–100.

Song ZM, Li Z, Dian-Moli Xie BY, Xia JY. 2007. Adult feeding increases fecundity in female *Helicoverpa armigera* (Lepidoptera: Noctuidae). *European J. Entomol.* 104: 721–724.

Sorby K, Fleischer G, Pehu E. 2005. *Integrated Pest Management in Development: Review of Trends and Implementation Strategies.* Agriculture and Rural Development Working Paper 5, World Bank, Washington, DC. Downloaded from website: http://documents.worldbank.org/curated/en/2003/04/2455449/integrated-pest-anagement-development-eviewtrendsimplementation-strategies (accessed December 25, 2016)

Soroker V, Blumberg D, Haberman A, Hamburger-Rishad M, Reneh S, Talebaev S, Anshelevich L, Harari AR. 2005. Current status of red palm weevil infestation in date palm plantations in Israel. *Phytoparasitica* 33(1): 97–106.

Spears JF. 1968. The westward movement of the pink bollworm. *Bull. Entomol. Soc. Am.* 14: 118–119.

Srinivasan K. 1991. Pest management in cucurbits: An overview of work done under AICVIP. In: *Group Discussion of Entomologists Working in the Coordinated Projects of Horticultural Crops*, 28–29 January 1991. Central Institute of Horticulture for Northern Plains, Lucknow, Uttar Pradesh, India, pp. 44–52.

Srinivasan PM. 1959. Guard your Bitter gourd against the fruit fly. *Indian Farming* 9: 8.

Steiner LF, Harris EJ, Mitchell WC, Fujimoto MS, Christenson LD. 1965. Melon fly eradication by over flooding with sterile flies. *J. Econ. Entomol.* 58: 519–522.
Stensmyr M, Larsson M, Bice S, Hansson B. 2001. Detection of fruit- and flower-emitted volatiles by olfactory receptor neurons in the polyphagous fruit chafer Pachnoda marginata (Coleoptera: Cetoniinae). *J. Comp. Physiol. A Neuroethol. Sens. Neural Behav. Physiol.* 187(7): 509–519.
Stöger E, Williams S, Christou P, Down RE, Gatehouse JA. 1999. Expression of the insecticidal lectin from snowdrop (*Galanthus nivalis* agglutinin; GNA) in transgenic wheat plants: Effects on predation by the grain aphid *Sitobion avenae*. *Mol. Breed.* 5: 65–73.
Stonehouse JM, Mahmood R, Poswal A, Mumford JD, Baloch KN, Chaudhary Z, Makhdum M, Mustafa AHG, Huggett D. 2002. Farm field assessments of fruit flies (Diptera: Tephritidae) in Pakistan: Distribution, damage and control. *Crop Prot.* 21: 661–669.
Stonehouse JM, Mumforda JD, Vergheseb A et al. 2007. Village-level area-wide fruit fly suppression in India: Bait application and male annihilation at village level and farm level. *Crop Protect.* 26: 788–793.
Subhan F, Khan M. 2004. Effect of different planting date, seeding rate and weed control method on grain yield and yield components in wheat. *Sarhad J. Agric.* 20: 51–55.
Syed RA. 1969. *Studies on the Ecology of Some Important Species of Fruit Flies and Their Natural Enemies in West Pakistan.* CIBC, Commonwealth Agriculture Bureau, Farnham Royal, Slough, UK, 12pp.
Tahir D, Lee EK, Oh SK, Tham TT, Kang HJ, Jin H, Heo S, Park JC, Chung JG, Lee JC. 2009. Determination of the organophosphorus pesticide in vegetables by high-performance liquid chromatography. *Am-Eurasian J. Agric. Environ. Sci.* 6(5): 513–519.
Talaei-Hassanloui R, Kharazi-Pakdel A, Goettel MS, Little S, Mozaffari J. 2007. Germination polarity of *Beauveria bassiana* conidia and its possible correlation with virulence. *J. Invertebr. Pathol.* 94: 102–107.
Talpur MA, Rustamani MA, Hussain T, Khan MM, Katpar PB. 1994. Relative toxicity of different concentrations of Dipterex and Anthio against melon fly, *Dacus cucurbitae* Coq. on bitter gourd. *Pak. J. Zool.* 26: 11–12.
Tamiru A, Bruce TJA, Woodcock CM, Caulifield CJ, Midega CAO, Ogol CKPO, Mayon P, Birkett MA, Pickett JA, Khan ZR. 2011. Maize land races recruit egg and larval parasitoids in response to egg deposition by a herbivore. *Ecol. Lett.* 14:1075–1083.
Tamiru A, Getu E, Jembere B, Bruce TJA. 2012. Effects of temperature and relative humidity on the development and fecundity of *Chilo partellus* (Swinhoe) (Lepidoptera: Crambidae). *Bull. Entomol. Res.* 02: 9–15.
Tan KH. 1993. Ecohormones for the management of fruit fly pests: Understanding plant-fruit fly-predator interrelationships. *Proceedings of the International Symposium of Insect Pests, Nuclear and Related Molecular and Genetic Techniques.* IAEA, Austria, pp. 495–503.
Tan KH. 2000a. *Area Wide Control of Fruit Flies and Other Insect Pests.* Sinaran Bros., Sdn. Bhd. Penang, Malaysia, p. 782.
Tan KH. 2000b. Sex pheromone components in defense of melon fly, Bactrocera cucurbitae against Asian house gecko, *Hemidactylus frenatus*. *J. Chem. Ecol.* 26: 697–704.
Tan KH, Lee SL. 1982. Species diversity and abundance of Dacus (Diptera: Tephritidae) in five ecosystems of Penang, West Malaysia. *Bull. Entomol. Res.* 72: 709–716.
Tan KH, Nishida R. 1996. Sex pheromone and mating competition after methyl eugenol consumption in the *Bactrocera dorsalis* complex. In: McPheron BA, Steck GJ (eds.), *Fruit Fly Pests: A World Assessment of Their Biology and Management.* St. Lucie Press, Delray Beach, Florida, pp. 147–153, 320.
Tan KH, Nishida R. 2000. Mutual reproductive benefits between a wild orchid, Bulbophyllum patens, and Bactrocera fruit flies via a floral synomone. *J. Chem. Ecol.* 26: 533–546.
Tan KH, Serit M. 1994. Adult population dynamics of *Bactrocera dorsalis* (Diptera: Tephritidae) in relation to host phenology and weather in two villages of Penang Island, Malaysia. *Environ. Entomol.* 23: 267–275.
Tandon PL, Lal B. 1978. The Mango coccid, *Restrococcus iceryoides* Green (Homoptera; Coccidae) and its natural enemies. *Curr. Sci.* 13: 46–48.
Tare V, Deshpande S, Sharma RN. 2004. Susceptibility of two different strains of Aedes aegypti (Diptera: Culicidae) to plant oils. *J. Econ. Entomol.* 97: 1734–1736.
Teixeira L, Isaacs R. 2007. Options for organic management of blueberry maggot. *The New Agricultur Network on-line Newsletter* 4(5): June 27, 2007. Available at http://www.new-ag.msu.edu/issues07/6-27.htm (accessed September 23, 2008)
Thomas C, Jacob S. 1990. Bioefficacy and residue dynamics of carbofuran against the melon fruit-fly, *Dacus cucurbitae* Coq. infesting bitter gourd, *Momordica charantia* L. in Kerala. *J. Entomol. Res.* 14: 30–34.
Thomas DB, Mangan RL. 2005. Non target impact of spinosad GF-120 Bait sprays for control of the Mexican fruit fly (Diptera: Tephritidae) in Texas citrus. *J. Econ. Entomol.* 98: 1950–1956.

Tobih FO, Omoloye AA, Ivbijaro MF, Enobakhare DA. 2002. Effects of field infestation by *Rastrococcus invadens* Williams (Hemiptera: Pseudococcidae) on the morphology and nutritional status of mango fruits, *Mangifera indica* L. *Crop Prot.* 21: 757–761.

Tollefson S. 1987. High yielding, short-season cotton production in Arizona. In: Herber DJ, Richter DA (eds.), *Proceedings of the Beltwide Cotton Production Research Conference.* National Cotton Council of America, Memphis, pp. 73–75.

Trivedi PC. 2002. *Plant Pest Management.* Aavishkar Publishers Distributors, Diamond Publishing Press, Jaipur, India.

Tsai JH, Liu YH. 2000. Biology of *Diaphorina citri* (Homoptera: Psyllidae) on four host plants. *J. Econ. Entomol.* 93(6): 1721–1725.

Tsai JH, Wang JJ, Liu YH. 2000. Sampling of *Diaphorina citri* (Homoptera: Psyllidae) on orange Jessamine in southern Florida. *Fla. Entomol.* 83(4): 446–459.

Tsuruta K, White IM, Bandara HMJ, Rajapakse H, Sundaraperuma SAH, Kahawatta SBMUC, Rajapakse GBJP. 1997. A preliminary note on the host-plants of fruit flies of the *Tribe dacini* (Diptera, Tephritidae) in Sri Lanka. *Esakia* 37: 149–160.

Ugarte C, Calderini DF, Slafer GA. 2007. Grain weight and grain number responsiveness to pre-anthesis temperature in wheat, barley and triticale. *Field Crops Res.* 100: 240–248.

USDA. 1994. *Treatment Manual.* USDA/APHIS, Frederick, USA.

USDA. 2005. *New Pest Response Guidelines,* Spodoptera. United States Department of Agriculture, Animal and Plant Health Inspection Service, p. 82.

USDA. 2010c. *National Quarantine: Citrus Greening and Asian Citrus Psyllid.* Plant Health. (4 December 2014).

Vaishampayan SM, Kogan M, Waldbauer GP, Woolley JT. 1975a. Spectral speciÞc responses in the visual behaviour of the greenhouse whitefly, *Trialeurodes vaporaiorum* (Homoptera: Aleyrodidae). *Entomol. Exp. Appl.* 18: 344–356.

Vaishampayan SM, Waldbauer GP, Kogan M. 1975b. Visual and olfactory responses in orientation to plants by the greenhouse whitefly *Trialeurodes vaporaiorum* (Homoptera: Aleyrodidae). *Entomol. Exp. Appl.* 18: 412–422.

Van den Berg J. 2009a. Can Vetiver grass be used to manage insect pests on crops? *Third International Conference on Vetiver,* October 6–9, 2003, Guangzhou- China, China Agriculture Press, 572pp.

Van den Berg J. 2009b. Case Study: Vetiver grass as component of integrated pest management systems. Available at www.vetiver.org/ETH_WORKSHOP_09/ETH_A3a.pdf; https://www.researchgate.net/publication/239588464 (accessed June 19, 2016).

Van Emden HF, Harrington R. 2007. *Aphids as Crop Pests.* CABI, Wallinford, UK, p. 768.

Van Lenteren JC. 1990. Biological control in a tritrophic system approach. In: Peters DC, Webster JA, Chlouber CS (eds.), *Aphid-Plant Interactions: Populations to Molecules.* USDA-ARS & Oklahoma State University, Stillwater, OK,pp. 3–28.

Van Steenwyk RA, Zollbrod SK, Nomoto RM. 2003. Walnut husk fly control with reduced risk insecticides. In: Beers B (ed.), *Proceedings of the 77th Annual Western Orchard Pest and Disease Management Conference,* Portland, Oregon. Washington State University, Pullman, 5pp.

Vargas RI, Pinero JC, Mau RFL, Stark JD, Hertlein M. 2009. Attraction and mortality of oriental fruit flies (Diptera: Tephritidae) to SPLAT-MAT- methyl eugenol with spinosad. *Entomol. Exp. Appli.* 131: 286–293.

Vargas RI, Stark JD, Hertlein M, Mafra-Neto A, Coler R, Pinero JC. 2008. Evaluation of SPLAT with spinosad and methyl eugenol or cue-lure for "attract-and-kill" of oriental and melon fruit flies (Diptera: Tephritidae) in Hawaii. *J. Econ. Entomol.* 101: 750–768.

Vargas RI, Stark JD, Kido MH, Ketter HM, Whitehand LC. 2000. Methyl eugenol and cue-lure traps for suppression of male oriental fruit flies and melon flies (Diptera: Tephritidae) in Hawaii: Effects of lure mixtures and weathering. *J. Econ. Entomol.* 93(1): 81–87.

Vayssières JF, De Meyer M, Ouagoussounon I, Sinzogan A, Adandonon A, Korie S, Wargui R, Anato F, Houngbo H, Didier C, Bon HD, Goergen G. 2015. Seasonal abundance of mango fruit flies (Diptera: Tephritidae) and ecological implications for their management in mango and cashew orchards in Benin (Centre & North). *J. Econ. Entomol.* 108(5): 2213–30.

Vega FE, Dowd PF, Lacey LA, Pell JK, Jackson DM, Klein MG. 2007. Dissemination of beneficial microbial agents by insects. In: Lacey LA, Kaya HK (eds.), *Field Manusl of Techniques in Invertebrate Pathology,* 2nd ed. Springer, Dordrecht, pp. 127–146.

Viado GBS, Bigornia AE. 1949. A biological study of the Asiatic palm weevil, *Rhynchophorus ferrugineus* Oliv. (Curculionidae: Coleoptera). *Philipp. Agric.* 33: 1–27.

Vickerman GP, Wratten SD. 1979. The biology and pest status of cereal aphids (Hemiptera: Aphididae) in Europe: A review. *Bull. Entomol. Res.* 69: 1–32.

Vidyasagar PS, Al-Saihati PV, Al-Mohanna AA, Subbei AI, Abdul Mohsin AM. 2000. Management of red palm weevil *Rhynchophorus ferrugineus* Olivier, A serious pest of date palm in Al- Qatif, Kingdom of Saudi Arabia. *J. Plant. Crops* 28: 35–43.

Vijaysegaran S. 1989. An improved technique for fruit fly control in carambola cultivation using spot sprays of protein baits. *Seminar Belimbing Dayamaju dan Prospeks*, July 8–9, 1989. Kuala Lumpur, Malaysia, 12pp.

Voss TS, Kieckhefer RW, Fuller BW, McLeod MJ, Beck DA. 1997. Yield losses in maturing spring wheat caused by cereal aphids (Homoptera: Aphididae) under laboratory conditions. *J. Econ. Entomol.* 90: 1346–1350.

Vosseler J. 1904. *Einige feinde der baumwolkulturen in Deutsh-Ostafrika. [Some Enemies of the Cultuvation of Cotton in german East Africa].* Mitteilungen Aus Den Biologisch, Landwirtschaftlichen Institut Amani, No. 18.

Wakil W, Faleiro JR, Miller TA. 2015. *Sustainable Pest Management in Date Palm: Current Status and Emerging Challenges, Sustainability in Plant and Crop Protection.* Springer International Publishing, Switzerland, 429p.

Walter JF. 1999. Commercial experience with neem products. In: Hall FR, Menn JJ (eds.), *Method in Biotechnology. Biopesticides*, vol. 5. Humana Press, Totowa, New Jersey, pp. 155–170.

Walton VM, Pringle KL. 1999. Effects of pesticides used on table grapes on the mealybug parasitoid *Coccidoxenoides peregrinus* (Timberlake) (Hymenoptera: Encyrtidae). *South African J. Enol. Viticult.* 20: 31–34.

Walton VM, Pringle KL. 2005. Developmental biology of vine mealybug, Planococcus ficus (Signoret) (Homoptera: Pseudococcidae), and its parasitoid *Coccidoxenoides perminutus* (Timberlake) (Hymenoptera: Encyrtidae). *African Entomol.* 13: 143–147.

Wan FH, Zhang GF, Li SS, Luo C, Chu D. 2009. Invasive mechanism and management strategy of *Bemisia tabaci* (Gennadius) biotype B: Progress Report of 973 Program on Invasive Alien Species in China. *Sci. China Life Sci.* 52: 88–95.

Wan XW, Nardi F, Zhang B, Liu YH. 2011. The oriental fruit fly, *Bactrocera dorsalis*, in China: Origin and gradual inland range expansion associated with population growth. *PLoS One* 6: 225–238.

Watson TF, Barnes KK, Fullerton DG. 1970. Value of stalk shredders in pink bollworm control. *J. Econ. Entomol.* 63: 1326–1328.

Watson TF, Larsen WE. 1968. Effects of winter cultural practices on the pink bollworm in Arizona. *J. Eco. Entomol.* 61: 1041–1044.

Wattanapongsiri A. 1966. *A Revision of the Genera Rhynchophorus and Dynamis (Coleoptera: Curculionidae).* Departament of Agriculture Science Bulletin, Bangkok, Thailand, 328p.

Weems HV. 1964. *Melon Fly (Dacus cucurbitae Coquillett) (Diptera: Tephritidae).* Entomology Circular, Division of Plant Industry, Florida Department of Agriculture and Consumer Services, No. 29, 2p.

Weems HV, Heppner JB. 2001. *Melon Fly, Bactrocera cucurbitae Coquillett (Insecta: Diptera: Tephritidae).* Florida Department of Agriculture and Consumer Services, Division of Plant Industry and Fasulo TR, University of Florida. University of Florida Publication EENY- 199, FL.

Weissling TJ, Giblin-Davis RM, Gries G, Gries R, Perez AL, Pierce HD, Oehlschlager AC. 1994. Aggregation pheromone of palmetto weevil, *Rhynchophorus cruentatus* (F.) (Coleoptera: Curculionidae). *J. Chem. Ecol.* 20: 505–515.

Wendell WK, Ruth RP. 1964. Effect of the chemosterilisant Apholate on the synthesis of cellular components in developing house fly eggs. *Bioch. J.* 92: 353–357.

Wene GP, Sheets LW, Woodruff HE. 1961. Emergence of overwintered pink bollworm in Arizona. *J. Eco. Entomol.* 54: 192.

Wenninger EJ, Hall DG. 2007. Daily timing of and age at mating in the Asian citrus psyllid, *Diaphorina citri* (Hemiptera: Psyllidae). *Fla. Entomol.* 90: 715–722.

Wenninger EJ, Hall DG. 2008. Importance of multiple mating to female reproductive output in *Diaphorina citri. Phys. Entomol.* 33: 316–321.

Wenninger EJ, Hall DG, Mankin RW. 2008a. Vibrational communication between the sexes in *Diaphorina citri* (Hemiptera: Psyllidae). *Ann. Entomol. Soc. America.* 102(3): 547–555.

Wenninger EJ, Landolt PJ. 2011. Apple and sugar feeding in adult codling moths, Cydia pomonella: Effects on longevity, fecundity, and egg fertility. *J. Insect Sci.* 11: 1–11.

Wenninger EJ, Stelinski LL, Hall DG. 2008b. Behavioral evidence for a female-produced sex attractant in in *Diaphorina citri* (Hemiptera: Psyllidae). *Entomol. Exp. App.* 128: 450–459.

Westbrook JK. 2008. Noctuid migration in Texas within the nocturnal aeroecological boundary layer. *Integr. Comp. Biol.* 48: 99–106.

Wharton RH. 1989. Control; classical biological control of fruit-infesting Tephritidae. In: Robinson AS, Hooper G (eds.), *World Crop Pests 3(B). Fruit Flies; Their Biology, Natural Enemies and Control.* Elsevier, Amsterdam, Netherlands, pp. 303–313.

Whellan JA. 1960. Pink bollworm (*Platyedra gossypiella*) in the Federation of Rhodesia and Nyasaland. *FAO Plant Prot. Bull.* 8: 113.

White IM, Elson-Harris MM. 1992. *Fruit Flies of Economic Significance. Their Identification and Bionomics.* International Institute of Entomology, Center for Agriculture and Biosciences International, Wallingford, Oxon, London, UK, 601pp.

White IM, Elson-Harris MM. 1994. *Fruit Flies of Economic Significance: Their Identification and Bionomics.* C.A.B International, Wallingford, 601pp.

Wickremasuriya CA. 1958. An important injection technique for coconut palm with special reference to the control of *Rhynchophorus ferrugineus* F. *Ceylon Coconut Quart.* 9: 40–54.

Willcocks FC. 1916. *The Insect and Related Pests of Egypt. Vol. 1. The Insect and Related Pests Injurious to the Cotton Plant. Part L. The Pink Bollworm.* Sultanic Agricultural Society, Cairo.

Williams T, Valle J, Vinuela E. 2003. Is the naturally derived insecticide Spinosad® compatible with insect natural enemies? *Bio. Sci. Technol.* 13: 459–475.

Wilson AGL. 1972. Distribution of pink bollworm, *Pectinophora gossypiella* (Saund.), Australia and its status as a pest in the Ord irrigation area. *J. Aus. Ins. of Agric. Sci.* 38: 95–99.

Witzgall P, Kirsch P, Cork A. 2010. Sex Pheromones and their impact on pest management. *J. Chem. Ecol.* 36: 80–100.

Wondimu M. 2013. Management of *Chilo partellus* (Swinhoe) (Lepidoptera: Crambidae) through horizontal placement of stalks and application of *Jatropha curcas* on maize (*Zea mays* L.) in central rift valley of Ethiopia. *MSc thesis*, Department of Plant Science, School of Graduate Studies, Haramaya University, Ethiopia.

Wong TTY, Cunningham RT, McInnis DO, Gilmore JE. 1989. Spatial distribution and abundance of *Dacus cucurbitae* (Diptera: Tephritidae) in Rota, Commonwealth of the Northern Mariana Islands. *Environ. Entomol.* 18: 1079–1082.

Wong TTY, Kobayashi RM, Whitehand LC, Henry G, Zadig DA, Denny CL. 1984. Mediterranean fruit fly (Diptera: Tephritidae): Mating choices of irradiated laboratory reared and untreated wild flies of California in cages. *J. Econ. Entomol.* 77: 58–62.

Wu CJ, Fan SY, Jiang YH, Yao HH, Zhang AB. 2004. Inducing gathering effect of taro on *Spodoptera litura* Fabricius. *Chinese J. Ecol.* 23: 172–174.

Xu QF, Tian F, Chen X, Hou WS, Li LC, Du LP, Xu HJ, Xin ZY. 2004. Inheritance of sgna gene and insect-resistant activity in transgenic wheat. *Acta Agron. Sin.* 30: 475–480.

Yan QT. 1984. Study on *Dacus dorsalis* Hendel (Diptera: Trypetidae) on Okinawa. *Chin. J. Entomol.* 4: 107–120.

Yang J, Sadof CS. 1995. Variegation in *Coleus blumei* and the life history of citrus mealybug (Homoptera: Pseudococcidae). *Environ. Entomol.* 24: 1650–1655.

Yang NW, Li AL, Wan FH, Liu WX, Johnson D. 2010. Effects of plant essential oils on immature and adult sweet potato whitefly, *Bemisia tabaci* biotype B. *Crop Prot.* 29: 1200–1207.

Ye H, Liu JH. 2005. Population dynamics of the oriental fruit fly, Bactrocera dorsalis (Diptera: Tephritidae) in the Kunming area, southwestern China. *Insect Sci.* 12: 387–392.

Yee WL. 2007. Attraction, feeding and control of *Rhagoletis pomonella* (Diptera: Tephritidae) with GF-120 and added ammonia in Washington State. *Florida Entomol. 90* (4): 665–673.

Yee WL, Lacey LA. 2005. Mortality of different life stages of *Rhagoletis indifferens* (Diptera: Tephritidae) exposed to the entomopathogenic fungus *Metarhizium anisopliae. J. Entomol. Sci.* 40: 167–177.

Yu XD, Pickett J, Ma YZ, Bruce T, Napier J, Jones HD, Xia LQ. 2012. Metabolic engineering of plant-derived (E)-β-farnesene synthase genes for a novel type of aphid-resistant genetically modified crop plants. *J. Integr. Plant Biol.* 54(5): 282–299.

Yu Y, Wei ZM. 2008. Increased oriental armyworm and aphid resistance in transgenic wheat stably expressing *Bacillus thuringiensis* (Bt) endotoxin and *Pinellia ternate* agglutinin (PTA). *Plant Cell Tiss. Org. Cult.* 94: 33–44.

Zaheer F. 2013. *UK's Quarantine Department Destroying All Shipments of Pakistani Mangoes.* The Express Tribune, July 3rd, 2013 and downloaded from website: http://tribune.com.pk/story/571524/uks-quarantine-department-destroying-all-shipments-of-pakistani-mangoes/ (accessed August 6, 2014)

Zaman M. 1995. Assessment of the male population of the fruit flies through kairomone baited traps and the association of the abundance levels with the environmental factors. *Sarhad J. Agric.* 11: 657–670.

Zanuncio JC, Lemos WP, Lacerda MC, Zanuncio TV, Serrão JE, Bauce E. 2006. Age-dependent fecundity and fertility life tables of the predator Brontocoris tabidus (Heteroptera: Pentatomidae) under field conditions. *J. Econ. Entomol.* 99: 401–407.

Zebitz CPW. 1987. Potential of neem seed kernel extracts in mosquito control. *Proc. 3rd Int. Neem Conf.*, Nairobi, Kenya, 1986, pp. 555–573.

Zhang P, Zhang X, Zhao Y, Wei Y, Mu W, Liu F. 2015. Effects of imidacloprid and clothianidin seed treatments on wheat aphids and their natural enemies on winter wheat. *Pest Manag. Sci.* 72: 1141–1149.

Zhang SM, Zhao YX. 1996. *The Geographical Distribution of Agricultural and Forest Insects in China*. China Agriculture Press, Beijing, China.

Zhu YI, Qiu HT. 1989. The reestablishment of *Dacus dorsalis* Hendel (Diptera: Tephritidae) after flee eradication on Lanbay Island. *J. Econ. Entomol.* 9: 217–230.

FIGURE 3.1 Indus River along with its tributaries. (From FAO-Aquastat 2012. Irrigation in Southern and Eastern Asia in figures: AQUASTAT Survey—2011. Rome.) [3]

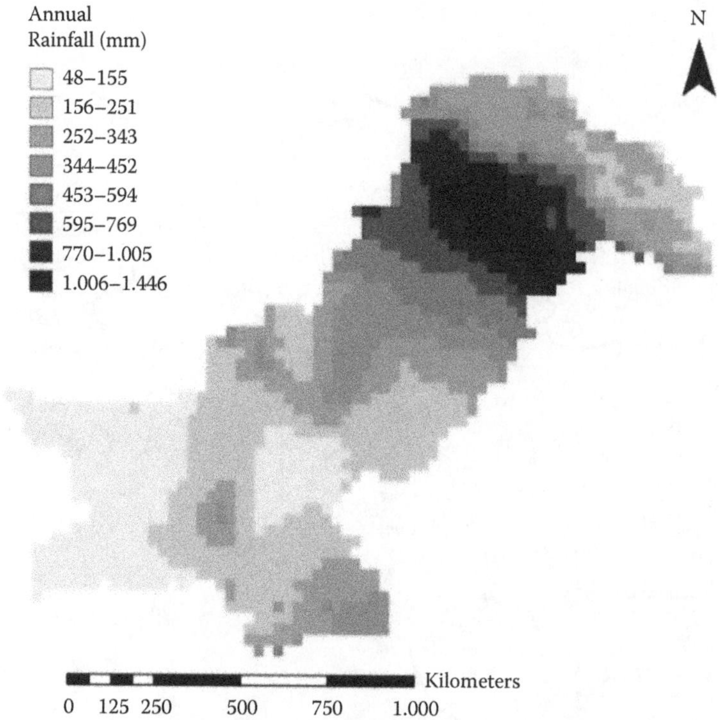

FIGURE 3.5 Regional variations in average annual rainfall based on TRMM rainfall data. (From Shahid MA, Boccardo P, Garcia WC, Albanese A and Cristofori E 2013. Evaluation of TRMM satellite data for mapping monthly precipitation in Pakistan by comparison with locally available data. In: *III CUCS Congress—Imagining Cultures of Cooperation: Universities Working to Face the New Development* Challenges, Turin, Italy.) [15]

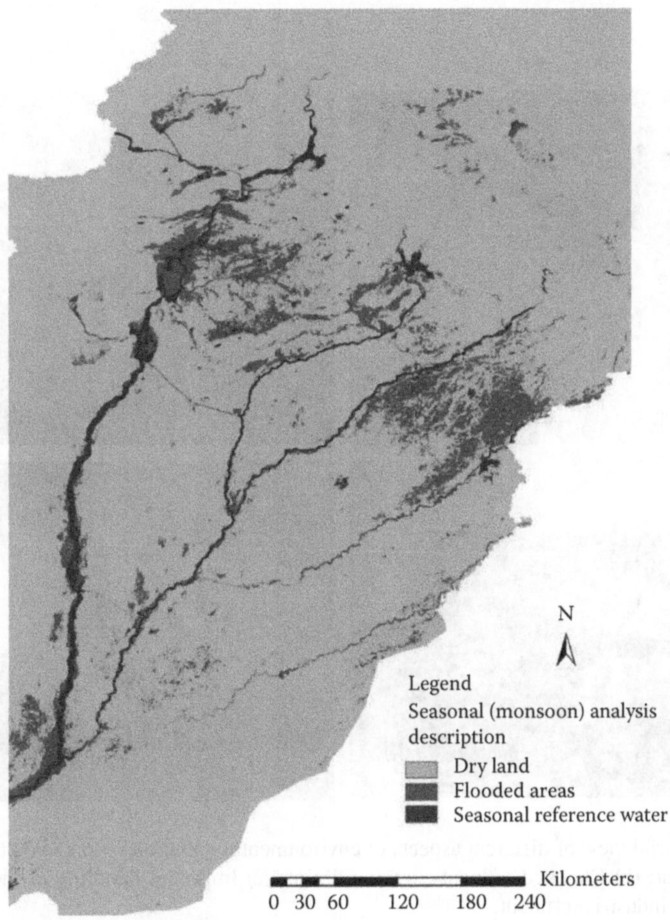

FIGURE 3.7 Flood inundation map for Upper Indus Basin, Pakistan. (From Shahid MA 2015. Geoinformatic and Hydrologic Analysis using Open Source Data for Floods Management in Pakistan. *PhD thesis*, DOI: 10.6092/polito/porto/2604981.) [18]

FIGURE 5.1 Pictorial view of different aspects of environmental pollution in Pakistan: (a) Treated effluent from treatment plant; (b) Chocked effluent-carrying drains; (c) Improper handling of industrial waste; (d) Untreated & mixed industrial effluent.

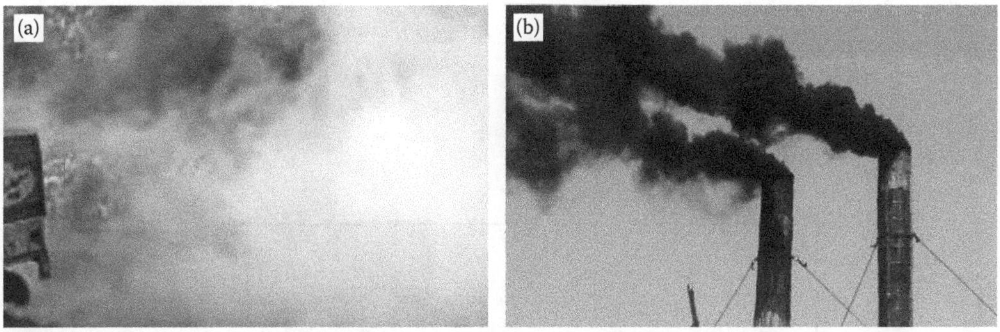

FIGURE 5.3 Industrial and vehicular air pollution sources: (a) Vehicular pollution on roads; (b) Industrial air pollution.

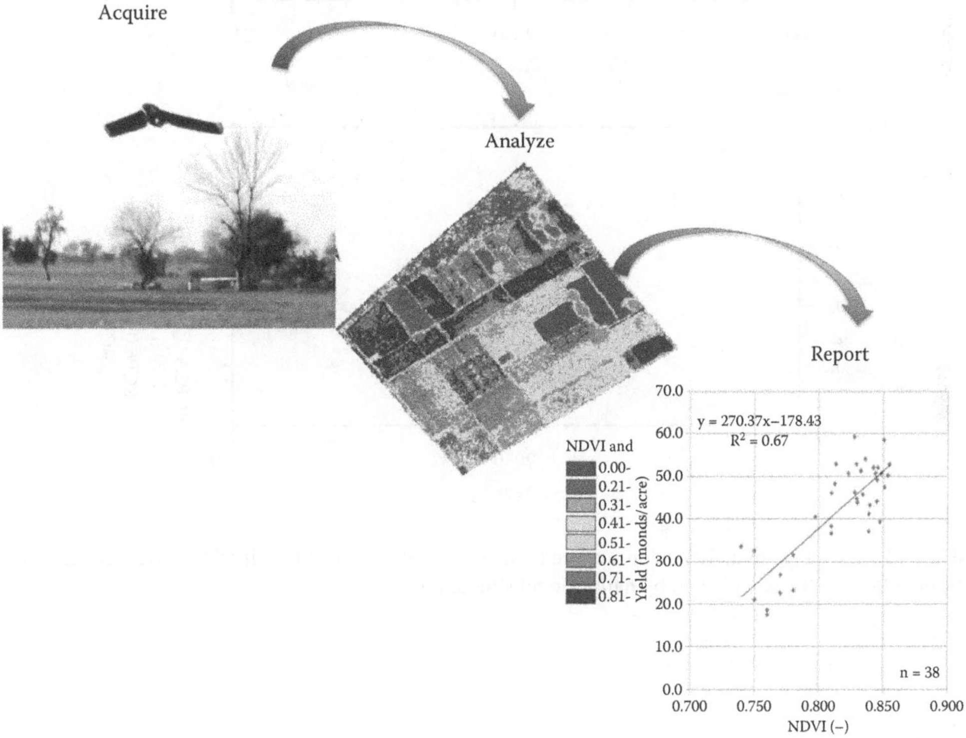

FIGURE 8.2 Information flow from data acquisition to crop health monitoring and yield estimation using UAV system.

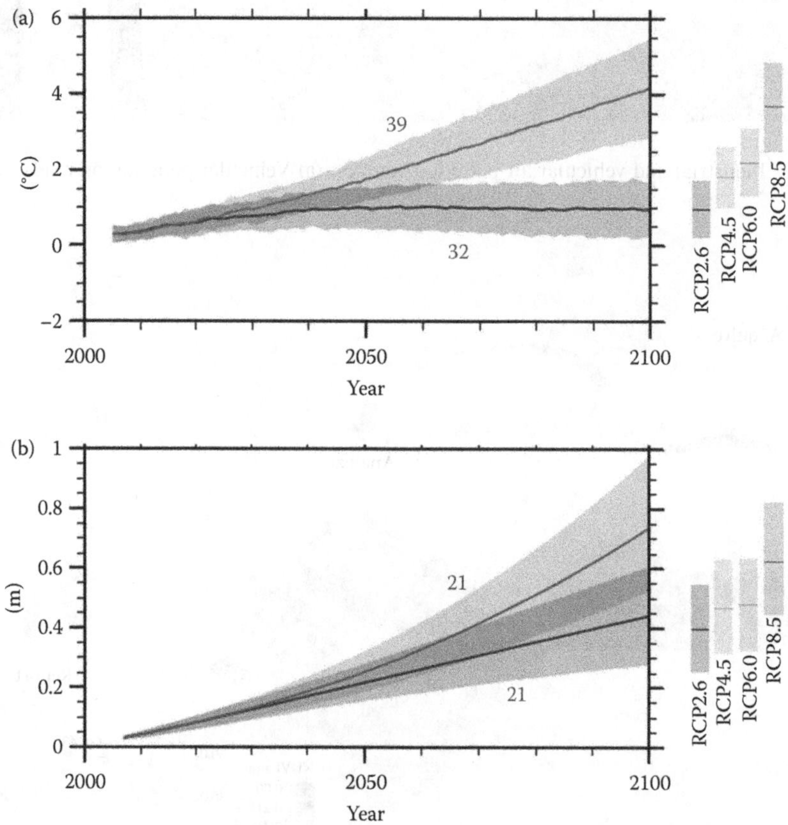

FIGURE 13.1 Change in global mean surface temperature (a) and variations in global average sea level rise (b) from 2006 to 2100 as projected by multi-model simulations.

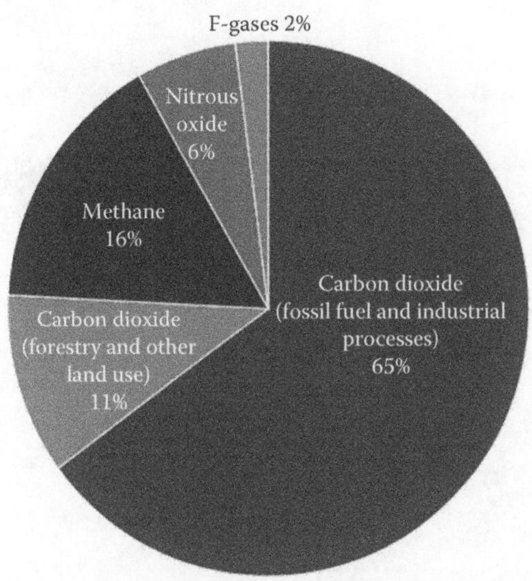

FIGURE 13.2 Global GHGs emission during year 2010. (From IPCC, 2014)

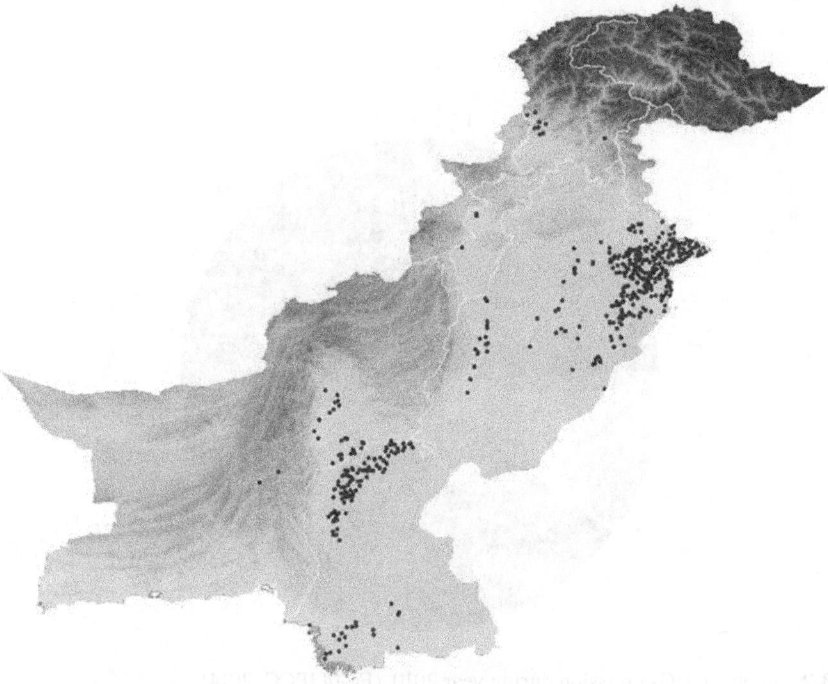

FIGURE 15.8 Rice growing areas of Pakistan. Each green dot denotes 5000 ha of rice. (From http://ricepedia.org/index.php/pakistan.)

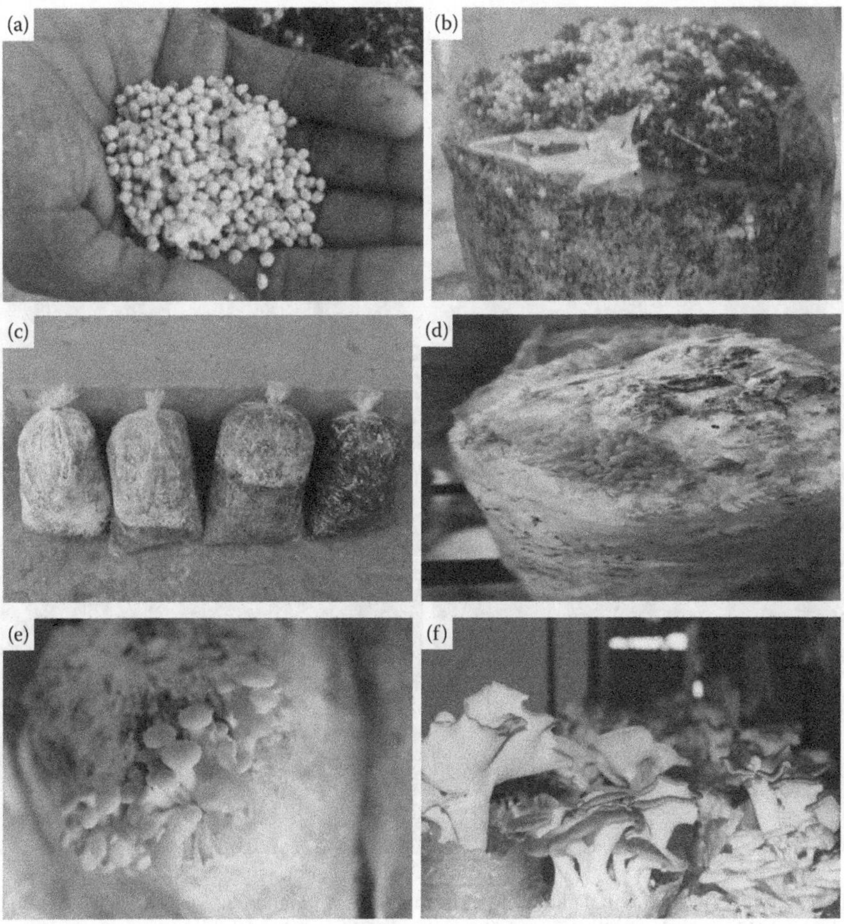

FIGURE 17.4 Oyster mushroom production: (a) spawn (seed), (b) spawned bag, (c) stages of spawn running in tied bags, (d) pinhead initiation, (e) fruit body formation, and (f) cropping.

FIGURE 17.6 High Value medicinal plants of Pakistan: (a) Kawar Gandal (*Aloe vera*), (b) ispaghol (*Plantago ovata*), (c) linseed (*Linum usitatissimum*), and (d) mullathi (*Glycyrrhiza Glabra*).

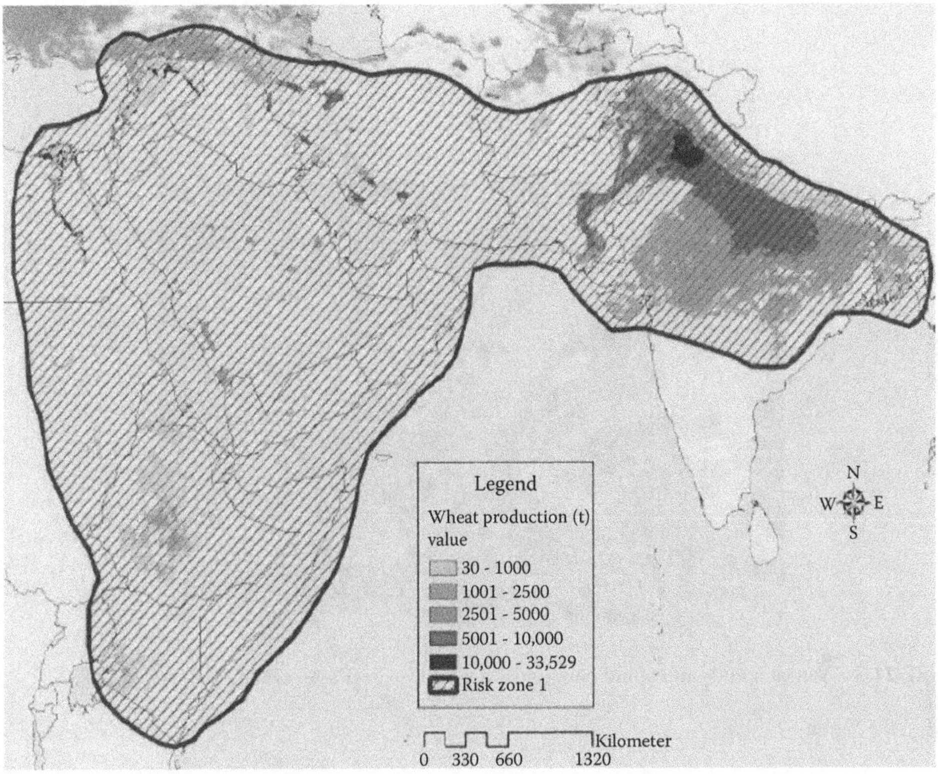

FIGURE 20.5 Area of the world at risk from Ug99 infestation. (Reproduced from Khoury W, Cressman K, and Yahyaoui A. 2008. Ug99 status, management and prevension. International Plant Protection Convention, Special session, Rome, April 9, 2008 (https://www.ippc.int/static/media/files/publications/en/1217860549444_Ug99_CPM_3_side_event_9Apr08.pdf).)

FIGURE 21.3 Wheat aphids and aphid parsitoids in wheat field. (Photo credit: M.J. Arif, UAF.)

FIGURE 21.5 Leaf enation due to CLCuV transmitted by cotton whitefly. (Photo credit: M.J. Arif, UAF.)

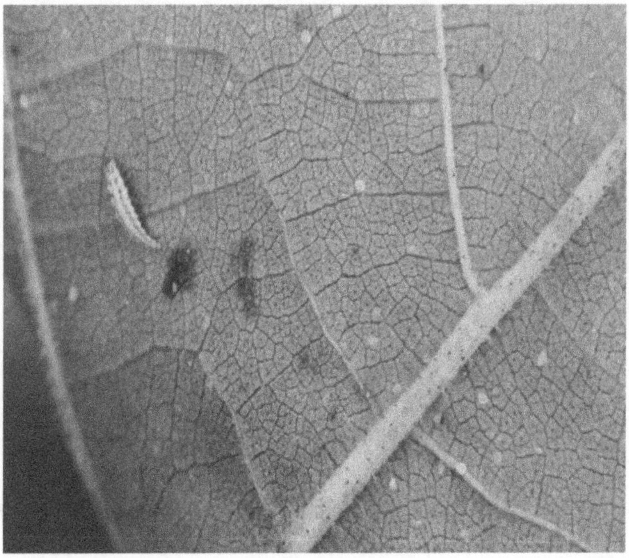

FIGURE 21.6 Larva of *Chrysoperla carnea* feeding on whitefly nymphs. (Photo credit: M.J. Arif, UAF.)

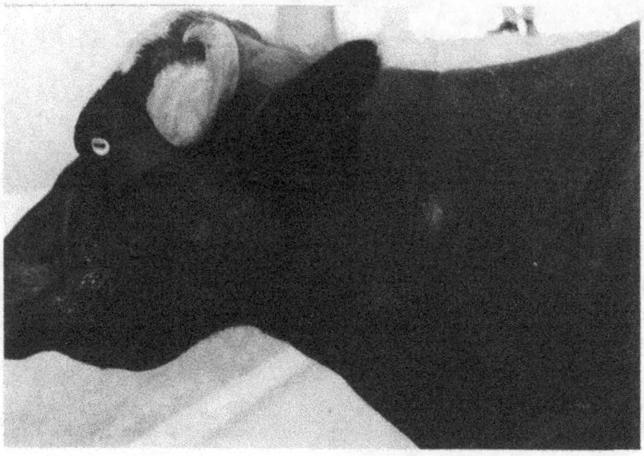

FIGURE 26.3 Tuberculosis positive buffalo indicated by swelling on the neck in tuberculin test.

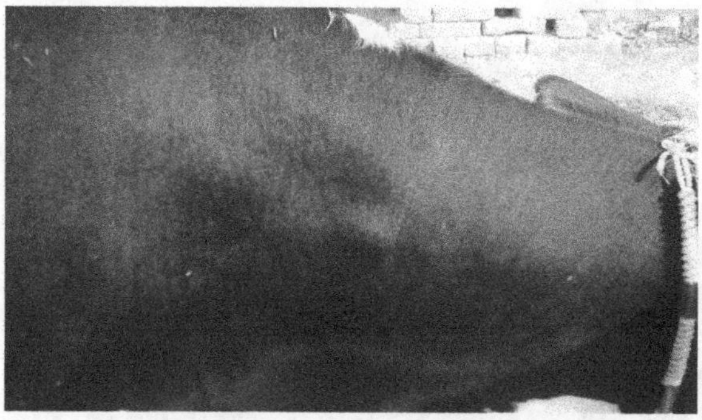

FIGURE 26.4 Tuberculosis positive cattle indicated by swelling on the neck in tuberculin test.

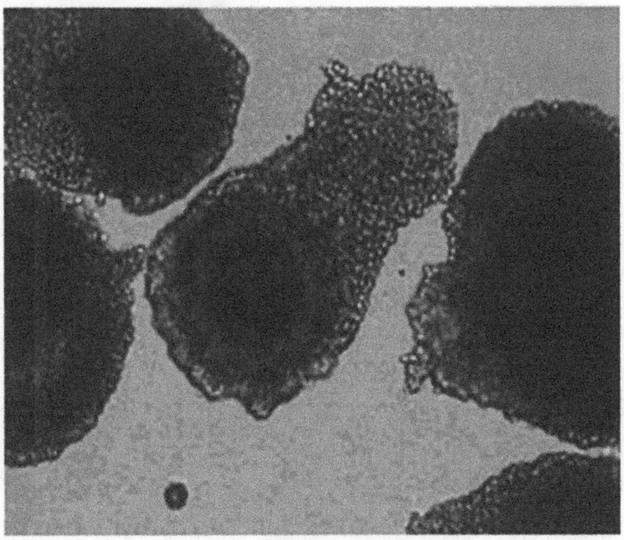

FIGURE 26.7 Harvested oocytes under microscope.

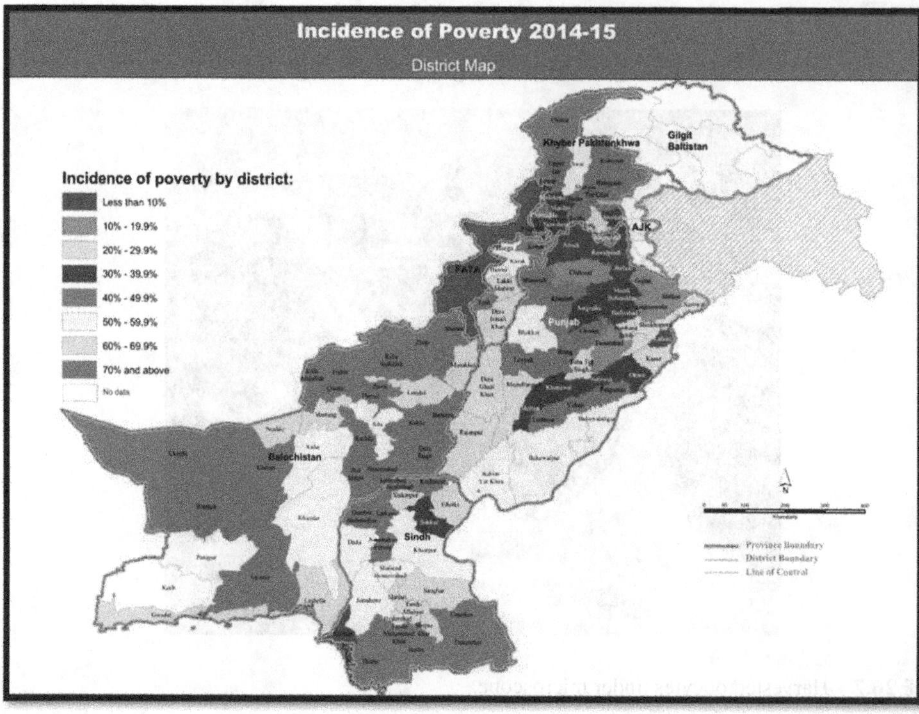

FIGURE 30.3 District wise population in Pakistan.

22 Status of Pesticides and Their Alternatives for Crop Sustainability in Pakistan

Amjad Abbas, Muhammad Amjad Ali, Arbab Ahmad, Safdar Ali, Amer Habib, Luqman Amrao, and Abdul Rehman

CONTENTS

22.1 INTRODUCTION

Pesticides are the chemicals that have the potential to check the population of living agents which could harm the normal stay of crop plants in the field. These agents are regarded as agriculture pests and include insects, microbial pathogens, weeds, or rodents (Bearden, 2010). The chemicals that are used to control the pest population to a certain level are called as pesticides. These chemicals repel, retard, or even kill these pests. But their toxicity is not totally under control, there are meager chances of harmful effects of these chemicals on the health of humans. Introduction of these pesticides, which include organophosphates (1960), carbamates (1970), and pyrethroids (1980); herbicides and fungicides (1970s–1980s) (Hayes and Laws, 2013) have increased the production of almost all crops by reducing pre and postharvest losses.

Rapid population growth rates in Pakistan signify greater demands of agricultural products every year. This could be achieved by increasing the area under cultivation and or increasing per unit area production from our cultivated lands. There is always a limit to obtaining new areas of cultivation, so there is need to take measures for enhanced production from available areas (Stoytcheva, 2011). Additionally, the high growth rate of our population is threatening the agricultural sector in two ways; by converting the fertile lands into urban and semiurban colonies and imposing the pressure of enhanced and sustainable growth on agriculture. The advent of the "green revolution" of 1960s in Pakistan exploited the genetic potential of crop plants in the right direction and achieved success with other nations, but increases seemed to stagnate afterwards. Crop yield is a complex phenomenon and depends on many natural factors, some of which are hardly manageable, while others in the form of proper and timely inputs along with proper pest management could be effective for high production and profitability. Insects, weeds, and microbial pathogens are the main obstacles to achieve higher production of any crop. These are called plants pests and compete with crop plants for inputs and deteriorate the yield in the field or in storage. To overcome the problem of pests, chemicals of both organic and inorganic nature are used. The definition of pesticides proposed by FAO is "any substances used as solely or in mixture to prevent, destroy, repel or extenuate any sort of pest both from animal and plant side which causes harm or interfere in the field, storage, processing and transport of agriculture products."

The pesticide sector was privatized in 1980s and this business grew well due to the extraordinary liberties regarding policies of import and registration of pesticide companies in the country. Introduction of synthetic pesticides were effective and widely adapted to control insect and weed pests. The demands of these chemicals gradually increased every year. Alternative methods of pest management were tried from research labs to the fields but demands for efficacy and urgency of chemical pesticides was much higher than proposed by these approaches. Additionally, the ease in application, readiness in use, and easy availability made chemical pesticides a top priority for growers in planning their cultivations with objectives of profitability (Haque, 1991; Khooharo et al., 2008).

Existing checks in pesticide policy are based on the agriculture pesticide ordinance of 1971 and its later amendments, which do not make any significant differences in the regulation of this important commodity. The problems of growers, suppliers of pesticides, and other stakeholders are not solved completely by these amendments of the existing laws regarding legality and implementation to achieve regularity in the existing system. The pesticides rules and regulation address the usual complaints from interested parties and penalties are fully mentioned (Ullah et al., 2015). As an example, pesticide traders put the responsibility of fitness of these chemicals on the manufacturers when they face legal actions according to existing laws. On the other hand, when we analyze the net return from this business, then these traders enjoy maximum profits from the sales. Another important point is imposition of small financial penalties and fines when someone violates the pesticide act, which need to be revised according to current market situation. Such amendments in the existing laws would discourage such violations in pesticide manufacturing and sale, and will save the farming communities from being exploited. A first information report (FIR) nomination is also suggested by traders against companies involved in this business, this would be helpful to improve this important sector. Furthermore, pesticide inspectors avoid awkward legal processes because these cases are often only lightly sanctioned, and fines light as compared to the payback benefits gained by the traders (Khooharo, 2008).

22.2 HISTORY OF MODERN SYNTHETIC PESTICIDES

The era of synthetic pesticide was started between 1940 and 1960. An important pesticide, DTT, was discovered during that time and was banned later. Similarly, a second important group of pesticides organophosphate insecticides came from Germany, and these were considered very effective against several pests. Meanwhile, phenoxyalkanoic acid group was used for the first time in Great Britain as commercial herbicide. In the mid 1940s carbamates, considered as the first generation herbicides were also discovered in the United Kingdom. These were found very handy in the control of weeds pests of commercial crops. Meanwhile, the organochlorines (which are mostly used as insecticide) were introduced and an important example of this group "chlordane" was formulated for the first time in Switzerland. Between 1950 to1955 herbicides with urea origin were discovered in America, such as fungicides marketed under the tradename Captan and Glyodin; malathion was also introduced at the time.

1950–1960 is considered the early period of pesticide invasion. In the 1950s many new pesticides and herbicides were introduced. Particularly, triazine herbicides which are actually ammonium-based herbicides. This was the early development area of herbicide introduction. Research in subsequent years resulted in the many new chemical formulations for insect and weed control. On an average, 13 new pesticides per year were discovered and tried up to 1980. This discovery was achieved on the basis of AI types and their miscibility in water or other solutions. In later years, the number of new AI s introduced in the pesticide market varied. In the years 1997 and 1998, there were 19 and 13 pesticides with novel chemistry discovered and applied, respectively (Gianessi and Marcelli, 2000).

22.3 CLASSIFICATION OF PESTICIDES

Pesticides are diverse in their chemistry, mode of action, origin, and effectiveness. There are different classification systems of pesticides. In the table, the classification is very broad and relies on target organism. Most of the pesticides are of organic nature but inorganic, synthetic, natural plant products, and biological organism are also used. The information regarding pesticides classification in Pakistan is given in Table 22.1 while pesticides with their particular chemical group is enlisted in Table 22.2.

Pesticides are important as they are used in the field and under postharvest storage conditions. Pesticides were subsidized by the government before 1980, through the Ministry of Agriculture, which was later privatized and the subsidized discount rate was waived off (Farid-u-ddin, 1985). After 1980, the consumption of the pesticides was increased by 70 times. But it was interesting that 80% of all of the pesticides were used on cotton crop as insecticides (Tariq, 2002).

The main issue of chemical consumption is an increase in the cultivation of cotton crop in an environment of subtropical climate with diversity of local and acclimatized insect pest of distant localities. There are hundreds of different insect pest which use cotton crops as hosts. Cotton leaf curl virus, which is considered as threat to world cotton, is dispersed from plant to plant through white flies, and more than four species of white flies are involved in this transmission (Sattar et al. 2013).

TABLE 22.1
Classification of Pesticides

Pesticide	Target Organism	Type	References
Insecticides	Very diverse in nature and effective against insect pests of the crop plants	Inorganic, organic, synthetic, biological, and natural plant products. Organic and synthetic are more common	Saeed et al. (2016), Mamoon-ur-rashid et al. (2012), Tayyib et al. (2005)
Herbicides	They target weeds in the field of crop plants, selective in action against weed pest. Little harm to main crop	Could be inorganic, organic, or syntheticchemicals	Hussain et al. (2013), Ali et al. (2004), Naseer-ud-din et al. (2011)
Fungicides	Fungal pathogens are very common. These chemicals arrest various physiological process. Both foliar and soil applications are practiced	Could be inorganic, organic, biological, or natural plant products	Khan et al. (2004), Mahmood et al. (2008), Sultana and Ghaffar (2010)
Nematicides	Root Knot nematodes are common and could be managed, soil sterilization is	Could be inorganic, organic, biological, or natural plant products	Ahmad et al. (2004), Irshad et al. (2012), Hussain et al. (2011)
Rodenticides	Used to control the population of rats	Inorganic and organic rodenticides are more common	Durr-i-Shahwar et al. (1999), Khan et al. (2012), Khan (2007)

TABLE 22.2
Common Pesticide and Their Groups

Common Name	Chemical Class
Acephate	Organophosphorus
Diazinon	
Chlorpyrifos	
(DDVP) Dichlorvos	
Parathion	
Malathion	
Propetamphos	
Aldrin	Organochlorine
Dieldrin	
Heptachlor	
Benzene hexachloride	
Carbon tetrachloride, perchloromethane	
Chlordane	
Chlorinated camphene	
Chlorinated naphthalene	
Edolan-U	
Endosulfan	
Endrin	
Ethylan	
(DDT) Dichlorodiphenyltrichlorethane	
Pentachlorophenol	
(LPCP) Pentachlorophenyl laureate	
Methoxychlor	
Mitin FF	
Ammonium fluosilicate	Inorganic
Arsenic trioxide, arsenous acid	
Silica gel	
Sodium arsenite, arsenous acid	
Sodium fluoride	
Sodium fluoroacetate	
Sodium fluorosilicate	
Thallium sulfate	
Zinc hexafluorosilicate	
Zinc phosphide	
Zinc hexafluorosilicate	
Zinc phosphide	
Phosphine	
Phosphorus, white	
Copper acetoarsenite	
Cryolite	
Boric acid	
Calcium arsenate	
Lead arsenate	
Mercuric chloride	
Carbaryl	Carbamate
Carbofuran	
Bendiocarb	
Dithiocarbamte	

(*Continued*)

TABLE 22.2 (*Continued*)
Common Pesticide and Their Groups

Common Name	Chemical Class
Propoxur	
Carbon disulfide	Misc.fumigant
Chloropicrin	
Ethylene dibromide	
Ethlene dichloride	
Ethyl formate, formic acid ethyl ester	
Ethylene oxide	
Formaldehyde, formalin	
Hydrogen cyanide, hydrocyanic acid	
Allethrin	Synthetic pyrethroid
Cyfluthrin	
Phenothrin	
Permethrin	
Piperonylbutoxide	
Resmethrin	
Tetramethrin	
Camphor	Botanical
Derris, Cube	
Strychnine	
Ryania	
Sabadilla	
Pyrethrins	
Red Squill	
Nicotine	
kerosene	Petroleum distillate
Methyl bromide	Halogenated hydrocarbon
Paradichlorobenzene	
Ortho dichlorobenzene	
Perchloroethylene	
Trichloroethylene	
Naphthalene, white tar	Simple hydrocarbon
Thymol	Phenol
Ortho phenyl phenol	
Carbolic acid, hydroxybenzene	
Carbon dioxide, dry ice	Inert gas
Nitrogen	
Propargite	Organosulfite
Sulfurylflouride	Inorganic fumigant
(TEPP) Tetraethyl pyrophosphate	Misc.fumigant
Sulfur dioxide, sulfurous anhydride	
Warfarin	Coumarin

These insect pests have become resistant, thus the amount of chemicals needed had to be increased in quantity and frequency. The problem of white fly, which was proved as vector of cotton leaf curl virus in Pakistan in early 90s, established a trend of higher consumption of insecticides against white fly and was further extended to American bollworm, pink boll worm, and spotted boll worm. According to recent reports, 74% of all pesticides used are insecticides and 80% of this amount is used to manage or control the population of three or four main insects in cotton crops. The second class of pesticides, which are very common in Pakistan, are herbicides, accounting for 14% of the total pesticides consumed (Khan, 1998). These chemicals are dangerous but are becoming popular. The crops are grown without a break period and there is no sanitation of the field for the seeds of unwanted plants like weeds. Additionally, poor farming practices include the use of unauthorized, adulterated, and self-maintained seed of two of the very important cereal crops, that is, wheat and rice. These weed plants compete with main crops and require a high usage of herbicides. Fungal pathogens of different species are also well adapted in our region. The fungal diseases of wheat, rice, potato, tomato, vegetables, and fruits, both in the field and storage conditions, are being managed by the use of various chemicals, both of organic and inorganic nature. These are called fungicides and they account about 9% of total pesticides consumed in Pakistan. The nature of herbicides is different as they are used to suppress the growth or vital pathways of the weed plant development. There is an immense need for care in the handling and dose determination of these chemicals because overuse or careless application could damage the crop plants. The weed issue is of an economical nature in wheat and rice fields. Since these are main cereals and are considered basic food crops, a lot of money is being spent on herbicides (Khooharo et al., 2008). It is interesting to note that the dominant type of pesticide has been synthetic pyrethroids, which accounts for approximately 70% of all pesticides.

22.4 THE PESTICIDE INDUSTRY IN PAKISTAN

In Pakistan, the history of pesticides dates back to the early 1950s. At that time, the main problem was swarms of locust attacks affecting a large portion of cereal and other crops, and the surrounding vegetation. To handle this problem, pesticides were imported for the first time in the country by the government and only 4% of the farmers were convinced to use these chemicals. This was the first adoption of chemical management of the pests and the start of the pesticide business in the country, which became an industry in the 1980s (Nasira, 1996). Abundant use of pesticides was started in 1992 to control the population of white flies.

Major use of pesticide is in the form of insecticides, which have resulted in a twofold increase of our cotton production over the last three decades; however, use of pesticides in 2003 was 4–5 times higher than before, resulting in a very high cost of production for this crop (Figure 22.1) (Khooharo et al. 2008).

After 1980, the pesticide sector was privatized and some incentives were offered for investors to start their business as local manufacturing under the Agricultural Pesticide Ordinance of 1971 and

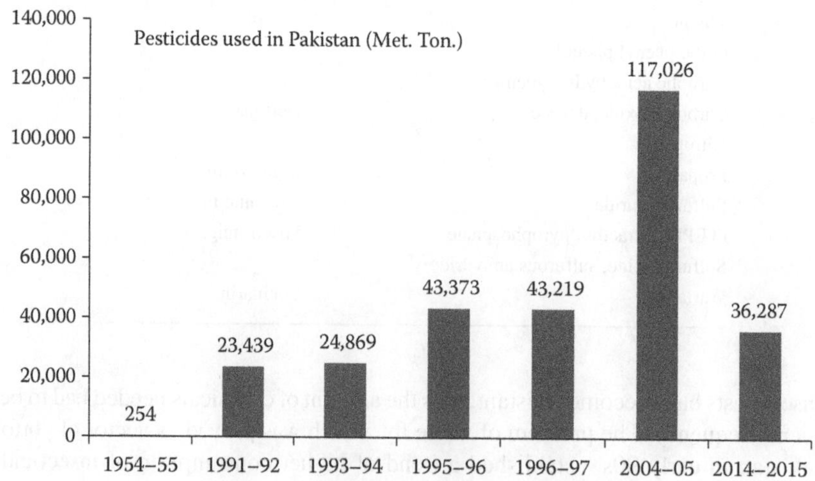

FIGURE 22.1 History of pesticide use in Pakistan.

Agriculture Pesticide Rules of 1973. For the local manufactures, there were very clear instructions regarding the nature of AI s, pesticide application procedures, environmental risk assessments, prior testing, appropriate labeling, permission and license for launching new formulations, renewal of licenses after three years, residual effects, safe storage, and transportation procedures. This was further patterned by appointing a special inspector to monitor the whole procedure chain of market supply, distribution, and application in the field. Pesticide laws were very carefully designed but the enforcement regarding production, distribution, and application were not adequate, which resulted in adulteration of pesticide contents, sales of expired products, and wrong labeling. Another important concern, which was not addressed appropriately, was no requirement for training regarding safety of farmers and their families (Yasmin, 2003). At the moment, there are pesticide suppliers, wholesalers, and retailers working across the country, but they don't give the necessary safety training to farmers; unfortunately no such arrangements are being offered by the government either (Ahad et al., 2001). At the time of nationalization of institutes in Pakistan, two private pesticide manufacturing factories were working as DDT-Nowsera and BHC & DTT-Kala Shah Kaku. After nationalization, these factories were closed in the 1970s (Mushtaque, 1999).

To meet the local needs, the government facilitated the growers to directly import pesticides. There was no regulation for testing these pesticides in Pakistan prior to their distribution during 1990s when the demands for these chemicals were very high. Such non-stringent import rules resulted in huge amount of imports of active ingredients of various pesticides in bulk into Pakistan through shipments. After arrival of imports, the pesticides were labelled and packed in Pakistan. The amount of active ingredients in some pesticides was found to be less and also their quality questionable. It is often seen that these pesticides are substandard, which forces farmers to increase the number of sprays to get control of pest. The products offered by multinational companies have higher prices but have better efficacy.

After 18th amendment in the constitution in national assembly, agriculture is a federal as well as a provincial subject. To assess the AIs, substandard products, and adulteration in pesticide contents, the Punjab Government took an initiative to establish a Provincial Pesticide Reference Laboratory, Kala Shah Kaku, Sheikhupura under the umbrella of Ayyub Agricltural Research Institute in Faisalabad. This lab provides a platform for establishing quality criteria of high standard pesticides, their testing for AIs, dose determination, and other regulatory provisions for the effective chemical control of plant pests (Tariq, 2002).

22.5 PUBLIC AND PRIVATE SECTOR RESEARCH ON PESTICIDES

In Pakistan, pesticide research regarding application and residual effects is carried out by both the government and private registered companies. In the private sector, research activities cost an estimated 26.5 million Rupees in the year 1987, while the public sector spent 41.2 million. Universities, mainly from the public sector and the plant protection department, conduct experimental field and greenhouse trials. In the private sector, indigenous and international or multinational pesticides companies conduct research on pesticides which also favor the development of their business. The data shows that domestic companies invested less on pesticide research as compared to multinational companies. The research objectives, however, mainly support development of product and marketing of these products after approval from the plant protection department. This information is further used by different research institutions from the public sector in the formulation of policies, necessary law amendments, and regulatory setups. (Ahmad, 1987). There are different multinational pesticide companies working in the country like Burma shell, Ciba-Geigy, Dow agro chemicals, Hoechest, FMC, Sandoz, ICI, Four brother, FMC, etc. Pesticide research at an advanced level involves sophisticated techniques of analytical chemistry and histopathology. To establish such costly facilities with trained manpower demands investment and involvement of the public sector. Since, multinational and local companies work with an objective of profit and firms cannot sacrifice a loss of credit by shifting their resources towards the research side. Additionally, there is no incentive from the government side for these companies to conduct research and evaluate chemicals (Khooharo, 2008). The main research focus of pesticides companies has been the adaptive research trials, effectiveness of their products in field conditions, testing and methods of spraying, and analytical or statistical procedures to evaluate the effectiveness of these chemicals. The economics or profit ratio-based research is also important in the pesticides business for enhanced acceptability of these chemicals for high profits. Little attention has been given on the residual effects of pesticides, harms, and hazards to human, birds, and animal populations as well as detecting micro trace levels of these toxins in raw and processed food. There is also a gap in the areas of product improvement based on the defects identified in existing chemicals by experts or farmers (Ali and Iqbal, 2005). Research studies on various aspects of pesticides such as use, residual effects, and efficacy were expected from the companies working in the private sector, including many local, national, and international sectors. Some international and very few local companies conducted trials on pest management, product demonstration, and product improvement (Khooharo, 2008).

The overall research budget in Pakistan is very small. Moreover the share of resources of agriculture research is just 0.02% of GDP in the year 1990. Even in 2006, the research budget of agriculture is same, which shows the nonserious attitude toward such an important issue. This amount is small compared to other countries. As an example, India spends 1.8% of its GDP.

22.6 CERTIFICATION, LICENSING, MONITORING, AND EVALUATION OF PESTICIDES

22.6.1 Pesticide Registration

The Federal Ministry of Food, Agriculture and Livestock is responsible for policy development and the marketing of pesticide applications. There are two regulations on pesticide provided by the Ministry Ordinance, that is, 1971 Agricultural Pesticides (APO, 1971) and agricultural pesticides, APO 1973 standards 1971 and 1973 Pesticide Rules were proposed and implemented under the FAO guidelines. These regulations state information policy regarding the import, manufacture, certification and sale of pesticides in Pakistan.

APO 1971 has been revised to incorporate the import rules of pesticides and penalties on defaulters (Jabbar and Mallick, 1994). According to revised 1971 APO, the process of local registration of pesticides is exempt: (1) if the pesticide in question is imported under the same generic name and the country of origin, and (2) if a pesticide registration and it is completed in the member countries of organization of Economic Cooperation and Development (OECD) and Chinese, with the brand name. The OECD was established in 1961 for economic policy cooperation in industrialized countries. There are 29 member countries, including Australia, Canada, France, Germany, Japan, South Korea, United Kingdom and United States, among others. As permitted by this provision, most pesticides brand and generic base materials are imported routinely in Pakistan from China, Germany, the United States, and other member countries of the OECD.

Before this scheme, the registration process in Pakistan was complex and cumbersome. For example, to register a product, an applicant usually needs to wait for a long period of 3 years. There was a monopoly of few companies and the rates of pesticides were high. The policy was revised, the pesticides were imported by dropping the verification process because they were tested in the exporting countries after meeting all conditions. The non-stringent import policies have resulted in the import of new chemical pesticides in the domestic market, which created a price competition.

22.6.2 Monitoring and Evaluation of Pesticides

For quality control of pesticides, pesticide inspectors have the competencies needed to take samples of pesticides. Assistant Plant Protection Officers (APPO), Deputy District Officers (DDO) Agriculture and District Officer (DO) Agriculture are in charge of such controls. Currently, the number of these inspectors is not enough. There are too many pesticides to be tested and in most cases, there is a lack of training of these officials. In addition, judicial proceedings are not effective enough to punish the defaulters. As a result, adulteration of pesticides is a very common phenomenon. Some of the pesticides contain only half of the AIs mentioned on their labels. In the worst case, some products have been sold containing less than 3% of the ingredients shown on the label.

A few banned pesticides such as DDT, aldrin, dieldrin may still have originated in Pakistan's black market. There have been a few occasions when environmentally unsafe pesticides were banned and its registration was cancelled; however, some of these pesticides are still available in very few local market due to lack of compliance with the legislation. There is a small gap in the implementation of the APO, 1971, and adequate coordination between the regulatory authorities is also needed.

The lack of a well-defined appropriate legislation and regulatory infrastructure has caused serious problems causing the misuse of pesticides. Often farmers complain that there is no attention to their complaints about adulteration. Similarly, pesticide dealers do not want the blame being put on them. According to the pesticide dealers, adulterated pesticides are provided by importers and distributors, and authorities unjustifiably put pressure on them although they are not responsible for the problem. There is also an assumption that multinational companies are influential in the departmental check process because the department of agriculture lacks advanced scientific facilitates to point out all violations.

22.6.3 Pakistan and the Rotterdam Convention at the United Nations in New York

Pakistan became the member of the Rotterdam Convention at the United Nations in New York on the 9th of September, 1999. The main objective of the Rotterdam Convention was to share information on standards and international trade regulations on chemicals and pesticides. Prior Informed Consent (PIC), which was developed by members, was to receive advance information on chemicals and hazardous and restricted pesticides. PIC countries would be facilitated in their decision-making for trade in pesticides. PIC is supposed to serve as an important tool for the import and export of pesticides to ensure quality among member countries.

22.7 SHORTCOMINGS OF PESTICIDE USAGE

The so-called ideal pesticide is meant for killing or repelling target pests without any detrimental effects to nontarget living communities in the ecosystem. Because these ideals are not met by most pesticides, pesticide use is controversial. It seems impossible today to get the higher crop yields without the application of pesticides (Aktar et al., 2009). The dependency of the farmers on these chemicals and their injudicious use has resulted in hazardous effects on the environment and human health, where farmers, their children, and fieldworkers are the most endangered individuals (Hough, 2014).

Pesticides may cause chronic or acute poisoning. The former type of poisoning badly affects the nervous system in addition to the skin by damaging the immune system and inducing allergic reactions, and induces allergic symptoms. Acute poisoning could be due to mild skin irritation, which sometimes may lead to a complex systemic disease that can be lethal. Pest control strategies, based on chemicals, have disrupted the agricultural ecosystem by killing of nontarget and ecofriendly organisms such as predators, parasites, and birds (Eddleston et al., 2002).

Each year, about 2.23 million women are exposed to pesticide poisoning. It is estimated that the woman worked 5 hours less per year due to the incidence of pesticide poisoning resulting in an approximate monetary loss of Rs. 60 million (Khan et al., 2011). While adding the costs associated with the treatment of health, the total loss is estimated at Rs. 765 million. Moreover, it is estimated that around 10,000 farmers are poisoned each year by the injudicious use of insecticides in the cotton growing areas of Pakistan (Tariq et al., 2007).

Similarly, higher level of pesticide residues in blood and milk makes farmers and their families more vulnerable to pesticide poisoning. The urban population is also victim of pesticide poisoning, because of consumption of vegetables and fruits with residue levels above the standards of the World Health Organization (Azmi et al., 2006). Additionally, pest problems have become more severe due to improper and indiscriminate use of pesticides as the pests have developed resistance against widely used chemicals.

Depletion of pollinators and soil fauna was attributed to the overuse of pesticides; however, they were of the view that they are not conducting studies to quantify the losses (Khan et al., 2002). In some case studies, the estimate of externalities of biodiversity, with special reference to the economic evaluation was unsatisfactory as rigorous econometric modeling and calculations were needed.

These facts, on one hand emphasize the use of proper dosages of pesticides, and on the other hand these underline the use of alternative strategies. Pesticides dosage is very important for control of pest population below the economic threshold level at which they are unable to harm the crop plants. If the pesticide dosage is applied at lower concentrations than the recommended ones, then the economic losses could not be controlled efficiently. Additionally, extra use of pesticides has injurious effects on ecosystems, the host crop itself, and human life. So, utmost care is essential while adjusting AIs for field applications. Pesticides are available in the market as wetable or soluble powder and liquid concentrations. Liquid concentrations are widely used as they are easy to handle. Water is mainly used as diluent in the application of pesticides. Kerosene is also used as diluent in limited cases. Granules and powder pesticides are used as such. The amount of actual toxicant or AIs is expressed in pounds per gallon, % by weight in case of liquid or powder pesticide respectively.

Since, AIs are actual toxicant and are a small portion of pesticides. To determine the total amount of pesticide per acre and obtain the desired concentration of AI per acre can be calculated according to Smith (2005);

$$\text{WP, G, or D per acre in (pounds)} = \frac{\text{lbs of active ingredient required}}{\text{\% a.i in WP, G, D}} \times 100$$

In the equation (WP) is wettable powder, (D) is dust or (G) is granules, lbs is pounds, and a.i. stands for AI . Percentage of AI is labeled on all pesticides products. Other units of weight like grams can be used instead of lbs.

If the 50WP is mentioned as label of any wetable powder pesticide product then it means it contains 50% AI and it can be simply calculated for any type of weight for example in 15 lbs of pesticide then to determine active AI in any amount of pesticide the following ratio equation can also be used

$$\frac{\text{50 lbs of a.i}}{\text{100 lbs of WP}} = \frac{\text{x lbs of a.i}}{\text{15 lbs of WP}}$$

or

$$\text{100 lbs of WP} \times \text{x lbs of a.i.} = \text{15 lbs of WP} \times \text{50 lbs of a.i}$$

or

$$\text{X lbs of a.i.} = \frac{\text{15 lbs of WP} \times \text{50 lbs of a.i}}{\text{100 lbs of WP}}$$

$$\text{X} = 7.50\ \text{lbs}$$

To determine total amount (pints) of liquid pesticide per acre required to obtain desired concentration of AI per acre can be calculated as

$$\text{Concentration of liquid pesticide per acre} = \frac{\text{desired amount (lbs) of a.i} \times 8}{\text{amont of a.i per gallon of liq.con.}}$$

In the above equation the number "8" can be replaced by 1, 4, or 128 to obtain answer in gallons, quarts or fluid ounces respectively. One gallon contains 4.55 liters.

22.8 ALTERNATIVE WAYS OF PEST CONTROL

Intensive farming system largely depend on chemical control through the use of chemical pesticides, herbicides, fungicides, and nematicides. These chemicals have been divided mainly in three categories based on their mode of actions; that is, fumigants, organophosphates, and carbamates. The use of this strategy is very common all over the world. These chemicals are very costly and impose an extra financial burden on low-income farmers (Sorribas et al., 2005). Due to high cost and health hazards, chemical pesticides are losing their value with time, thus paving the way towards the use of cultural practices, resistant crop varieties, biological control, and transgenic strategies for biotic stress management.

22.8.1 Cultural Methods

Management of insect pests and disease is necessary to sustain economically viable crop productivity and to guarantee food security for low-income farmers. For this purpose, various cultural strategies could be adopted to minimize the effect of these biotic factors to a certain level. For insect and disease management, cultural practices like crop rotation, inter/trap cropping, flooding, solarization, managed use of fertilization, alteration of sowing date and sanitation and phyto-sanitation measures have largely been used.

Trap cropping has widely been used for management of insect pests in various crops (Reviewed by Hokkanen 1991). For instance, strips of alfalfa were interplanted with cotton as a trap crop for lygus bugs (Family: Miridae). The alfalfa, which attracts lygus bugs more strongly than cotton, is usually treated with an insecticide to kill the bugs before they move into adjacent fields of cotton. Increasing the soil temperature by solarization by summer ploughing has been proved to be valuable technique for controlling nematodes (Katan, 1981). Similarly, alteration in sowing date (early sowing) has been helpful for managing aphid populations on wheat (Wains et al., 2010). However, this strategy is valid only for areas with long periods of uninterrupted sunshine. Similarly, flooding is also helpful for nematode control because it fills the soil pore spaces with water and oxygen supply to nematodes becomes limited (MacGuidwin, 1993). Moreover, one of the oldest and most key methods for managing nematodes is crop rotation. It has been proved to be an effective strategy to control nematodes, especially for those species that have a narrow host range, that is, cyst nematodes (Westphal, 2011). However, it has limitations against the species like *M. incognita* which have a broad range of host plants, parasitizing around 3000 plant species (Abad et al., 2003).

22.8.2 Manipulation of Host Plant Resistance

The use of resistance crop varieties is another way to control pests and diseases. In the past, the utilization of natural host resistance from various crop species is a preferred approach for pest management because it is environmentally safe and a cost-saving option compared to chemical control (Williamson and Kumar, 2006). For instance, this strategy has largely been employed for enhancement of rust resistance in wheat. Different variants of resistance genes are present in diverse germplasm accessions against leaf rust (*Lr* genes), yellow rust (*Yr* genes) and stem rust (*Sr* genes) which have been utilized to develop rust resistant wheat cultivars against all types of rust diseases (Reviewed by Khan et al., 2013). These genes work on the principle of gene for gene interaction with virulence (*Avr* gene) from the pathogenic race of rust and its corresponding *R* gene in wheat. The main limitation with this approach in the emergence of different rust races, which overcome the effect of R gene in the wheat cultivars which were resistant before. For instance, the emergence of virulent race against an establish yellow rust gene (*Yr27*) was observed in Pakistan and India which eliminated the mega cultivars like PBW343 and Inqilab 91 harboring *Yr27* gene based resistance (Khan et al., 2013).

Moreover, Fuller et al. (2008) have reviewed the use of host resistance against various plant parasitic nematodes. They have further argued that the plant species have significant variation for nematode resistance based on accumulation of genes responsible for resistance or susceptibility (Fuller et al., 2008). Recently, aphid resistance was incorporated in bread wheat and resulted in substantial improvements in grain yield per plant (Wains et al., 2014).

22.8.3 Biological Control

Use of natural enemies of insect pests and diseases as predators and parasites has been practiced widely to manage pest populations below the economic threshold level. This is an alternative to the use of chemical pesticides, which is not only natural way to control insect pests and diseases but also lack health and environmental hazards. Several predators and parasites have been used to control various insect pests in different crop plants, that is, *Crysoperlacarnea* and entomopathogenic nematodes are important examples of biological insect management. *Crysoperlacarnea* has widely been used to control dangerous insect pest in different crop plants (reviewed by Tauber et al., 2000). Tauber et al. (2000) have discussed in detail the systematics and dynamics of Crysoperla commercialized field applications for insect control in the end of twentieth century. Similarly, *Trichogrammachilonis*is largely being used in Pakistan to control various sugarcane borers under field conditions recently (Nadeem and Hamed, 2011; Saljoqi and Walayati, 2013).

Antagonistic relationship between plant pathogens and other microorganisms is being widely exploited these days. Most of the soils have diverse microbial community to operate antagonistic to the pathogen, thus leading to specific inhibition of a particular disease. These antagonists are used as biological control agents (BCAs) for controlling different diseases caused by bacteria, fungi, and nematodes. For instance, the fungal parasites are the most common antagonists of plant parasitic nematode (Kerry, 1980; Kerry, 1988; Stirling, 1991). The main limitation in using these biological agents is their application on large scale needs a lot of resources in space and time.

22.8.4 Plant Extracts and Biopesticides

Several allelochemicals or botanical extracts have shown their promise to control various pests. These are attractive alternatives to chemicals pesticides of synthetic nature for pest management as they have very little or even no detrimental effect on the environment or human health (Koul and Walia, 2009). Different botanical extracts have successfully been used against harmful insects in various crop plants. For instance, residual and contact toxicity of over 30 plant extracts was examined on the larval instars of the Colorado potato beetle, which concluded that various botanical extracts were lethal to the beetle larvae (Gokce et al., 2006). Likewise, many botanical extracts and essential oils have been used as biofungicides or nematicides to manage fungal and nematode diseases of crops. Gray mold disease of tomato (caused by *Botrytis cinerea*) was successfully control by the application of extracts from the giant knotweed, *Reynoutriasachalinensis*in the greenhouse. Moreover, the allelochemicals have been widely utilized for their weedicidal activity as well. For example, allelopathic extracts from four different plant species, that is, *Sorghum bicolor*, *Morus alba*, *Echinochloacrusgalli*, and *Withaniasomnifera* were used to assess their weedicidal potential against wild oat (*Avenafatua* L.) and canary grass (*Phalaris minor* Ritz.) in wheat (Jabran et al., 2010). The results showed that the extract from *Morusalba*was was most effective in controlling these weeds followed by *Withaniasomnifera, Echinochloacrusgalli,* and *Sorghum bicolor.*

However, limitations of these botanicals lie in their regulatory barriers and availability of competing products of microbial origin and detrimental effects on nontarget organisms (Koul and Walia, 2009). So, it is important to organize natural sources available for the plant extracts, develop quality control and regulatory mechanisms followed by the adoption of standardization strategies to apply them in the field.

22.8.5 Transgenic Approaches

Genetically modified organisms (GMOs) were mainly developed to minimize the use of synthetic chemicals. For instance, *Bt* gene coding for the *Bt* toxin were taken from Gram-positive, soil-dwelling bacterium *Bacillus thuringiensis* and was incorporated in cotton to avoid chewing by caterpillars of different bollworms (Perlak et al., 1990; Benedict et al., 1996; Jenkins et al., 1997). Bt cotton resulted in management of bollworms below the threshold level in addition to several economic, social, and environmental benefits (Edge et al., 2001). They further concluded that *Bt* cotton provides an effective method for the control of lepidopteran insects in cotton, which is environment friendly and safer to humans as compared to conventional broad-spectrum chemical insecticides, making *Bt* cotton the best emerging tool for integrated pest management.

Transgenic plants have largely been used in the past to enhance nematode resistance in plants. The heterologous expression of R genes in different plants has been carried out. For instance, Mi (resistance against *M. incognita*) gene from tomato, $Hs1^{pro\text{-}1}$ from sugar beet against *H. schachtii*, *Gpa*-2 from potato against *G. pallida,* and *Hero* from tomato against *G.rostochiensis* (Reviewed by Fuller et al., 2008). Another main strategy was the targeted silencing of known nematode effector proteins in plants through RNAi technology (Fuller et al., 2008). Unlike these strategies, some recent research has suggested that nematode resistance could be enhanced in plants by modifying the expression of particular genes in syncytia (Klink and Matthews, 2009; Ali et al., 2013a, b, 2014, 2015). Similarly, for controlling weeds, glyphosate resistance genes were introduced from Agrobacterium spp. (strain CP4), which codes for an enzyme called 5-enolpyruvylshikimate-3-phosphate (EPSP) synthase resulting in glyphosate tolerant lines in soybean (Padgette et al., 1995).

22.9 CONCLUSION

Pakistan is an agricultural country and there is a need to improve its systems of agriculture production through the distribution of necessary inputs at subsidized rates, selection of markets for good returns to the farmers, development of transportation facilities, and crop insurance in case of epidemics or natural disasters.

The problem of diseases is getting more attention in an era of free movement of people between continents and high risk of environmental fluctuations. As in case of wheat, which is chief food grain crop in the country, there are reports about the potential risks of Ug99 rust from the neighboring nations due to free movement between the borders. The quarantine department of any country should safeguard against the possible risks imposed by unchecked imports of infected seeds or food grains. This department has a responsibility to detect and stop such risks at their source. This is compounded by the increasing trend in the use of pesticides for the control of insects, weeds, and microbial pathogens.

Environmental changes, which are due to many reasons such as deforestation, urbanization, industrialization, and fossil fuel dependency, have imposed an increase of 1–2°C on average temperatures in last decade. Such temperature changes have invited nonseasonal rains, and a good example of this is the failure of cotton crops in the country during 2015–2016. This resulted in the emergence of new microbial and insect pathogens along with the reemergence of older more vigorous ones. So, there is need to return to basics and redesign our breeding programs and carry out more selections from the existing germplasm to adapt to such environmental extremes and to minimize the use of pesticides. All the while, natural bioproducts should be explored as well.

Breeding against diseases requires rigorous selections under different inoculums strengths of the pathogens. For that we need the most virulent races or strains. This data on the diversity of pathogens demands surveys of the distant areas of the country for the collection of reliable information and the geographical distribution of microbial pathogens. Such efforts would result in real-time screening of our germplasm. For more reproducibility, local research stations could be engaged in disease-screening program for area-specific availability of genotypes. The research on the various aspects of plant disease resistance should be streamlined nationwide and the researchers should know the research objectives of other teams to avoid redundancy, which is actually prevailing at this time, in order to save valuable resources and time.

REFERENCES

Abad P, Favery B, Rosso MN, Castagnone-Sereno P. 2003. Root-knot nematode parasitism and host response: Molecular basis of a sophisticated interaction. *Mol. Plant Pathol.* 4(4):217–224. doi: 10.1046/j.1364-3703.2003.00170.x

Ahad K, Hayat Y, Ahmad I. 2001. Capillary chromatographic determination of pesticide residues in Multan Division. *The Nucleus* 38(2):145–149.

Ahmad M. 1987. *Agricultural research in pakistan's private sector. Management of Agricultural Research and Technology Project.* The Pakistan Agricultural Research Council, Islamabad and the USAID Mission to Pakistan, Islamabad.

Ahmad MS, Mukhtar T, Ahmad R. 2004. Some studies on the control of Citrus nematode (*Tylenchulussemipenetrans*) by leaf extracts of three plants and their effects on plant growth variables. *Asian J. Plant Sci.* 3:544–548.

Aktar MW, Sengupta D, Chowdhury A. 2009. Impact of pesticides use in agriculture: Their benefits and hazards. *Interdiscip Toxicol.* 2: 1–12.

Ali M, Sabir S, Ud-din QM, Ali MA. 2004. Efficacy and economics of different herbicides against narrow leaved weeds in wheat. *Int. J. Agric. Biol.* 6:647–651.

Ali MA, Abbas A, Azeem F, Javed N, Bohlmann H. 2015. Plant-nematode Interactions: From Genomics to Metabolomics. *Int. J. Agric. Biol.* 17(06):1071–1082.

Ali MA, Abbas A, Kreil DP, Bohlmann H. 2013a. Overexpression of the transcription factor RAP2.6 leads to enhanced callose deposition in syncytia and enhanced resistance against the beet cyst nematode Heterodera schachtii in Arabidopsis roots. *BMC Plant Biol.* 13:47.

Ali MA, Plattner S, Radakovic Z, Wieczorek K, Elashry A, Grundler FM, Ammelburg M, Siddique S, Bohlmann H. 2013b. An Arabidopsis ATPase gene involved in nematode-induced syncytium development and abiotic stress responses. *Plant J.* 74:852–866.

Ali MA, Wieczorek K, Kreil DP, Bohlmann H. 2014. The beet cyst nematode Heterodera schachtii modulates the expression of WRKY transcription factors in syncytia to favour its development in Arabidopsis roots. *Plos One* 9(7):e102360.

Ali S, Iqbal M. 2005. Total factor productivity growth and agricultural research and extension: An analysis of pakistan's agriculture, 1960–1996 [with comments]. *Pak. Dev. Rev.* 44(4):729–746.

Azmi MA, Naqvi S, Azmi MA, Aslam M. 2006. Effect of pesticide residues on health and different enzyme levels in the blood of farm workers from gadap (rural area) Karachi-Pakistan. *Chemosphere* 64(10):1739–1744.

Bearden DM. 2010. *Environmental Laws: Summaries of Major Statutes Administered by the Environmental Protection Agency.* DIANE Publishing, Washington, DC, USA.

Benedict, JH, Sachs ES, Altman WD, Deaton WR, Kohel RJ, Ring DR, Berberich SA. 1996. Field performance of cottons expressing transgenic CryIA insecticidal proteins for resistance to *Heliothisvirescens* and *Helicoverpazea* (Lepidoptera: Noctuidae). *J. Econ. Entomol.* 89(1):230–238.

Durr-i-Shahwar, Beg MA, Mushtaq-ul-hassan M, Khan AA. 1999. Inhibiting rodent depredations. I. Distribution and abundance of rats in a wheat rice based crop land. *Pak. J. Agric. Sci.* 36:3–4.

Eddleston M, Karalliedde L, Buckley N, Fernando R, Hutchinson G, Isbister G, Konradsen F et al. 2002. Pesticide poisoning in the developing world—A minimum pesticides list. *The Lancet* 360(9340):1163–1167.

Edge JM, Benedict JH, Carroll JP, Reding HK. 2001. Bollgard cotton: An assessment of global economic, environmental, and social benefits. *J. Cotton Sci.* 5:121–136.

Farid-u-ddin A. 1985. Review of agro-pesticide consumption and its impact on crop protection in Pakistan. *Pak. Agric.* 7: 28.

Fuller VL, Lilley CJ, Urwin PE. 2008. Nematode resistance. *New Phytol.* 180(1):27–44.

Gianessi LP, Marcelli MB. 2000. *Pesticide Use in Us Crop Production: 1997.* National Center for Food and Agricultural Policy, Washington, DC.

Gokce A, Whalon ME, Çam H, Yanar Y, Demirtaş İ, Gören N. 2006. Plant extract contact toxicities to various developmental stages of Colorado potato beetles (Coleoptera: Chrysomelidae). *Ann. Appl. Biol.* 149(2):197–202.

Haque H. 1991. Imported generic pesticides need to be checked before marketing. PAPA Bulletin (Pakistan).

Hayes WJ, Laws ER. 2013. *Classes of Pesticides.* Elsevier Science, San Diego, USA.

Hokkanen HMT. 1991. Trap cropping in pest management. *Annu. Rev. Entomol.* 36(1):119–138. doi: 10.1146/annurev.en.36.010191.001003

Hough P. 2014. *The Global Politics of Pesticides: Forging Concensus from Conflicting Interests.* Taylor & Francis Publishers, Ann Arbor, Michigan, USA.

Hussain MA, Mukhtar T, Kayani MZ. 2011. Efficacy evaluation of *Azadirachtaindica*, *Calotropisprocera*, *Daturastramonium* and *Tageteserecta* against root-knot nematodes *Meloidogyne incognita. Pak. J. Bot.* 43:197–204.

Hussain Z, Marwat KB, Munsif F, Samad A and Ali K. 2013. Evaluation of various herbicides and their combinations for weed control in wheat crop. *Pak. J. Bot.* 45(1):55–59.

Irshad U, Mukhtar T, Ashfaq M, Kayani MZ, Kayani SB, Hanif M, Aslam S. 2012. Pathogenicity of citrus nematode (*Tylenchulussemipenetrans*) on *Citrus jambhiri. J. Anim. Plant Sci.* 22(4):1014–1018.

Jabbar A, Mallick S. 1994. *Pesticides and Environment Situation in Pakistan.* Sustainable Development Policy Institute, Islamabad.

Jabran K, Farooq M, Hussain M, Hafeez-ur-Rehman, Ali MA. 2010. Wild oat (*Avena fatua* L.) and canary grass (*Phalaris minor* Ritz.) management through allelopathy. *J. Plant Protec. Res.* 50(1):41–44.

Jenkins JN, McCarty Jr. JC, Buehler RE, Kiser J, Williams C, Wofford T. 1997. Resistance of cotton with bendotoxin genes from *Bacillus thuringiensis* var. kurstaki on selected lepidopteran insects. *Agron. J.* 89:768–780.

Katan J. 1981. Solar heating (Solarization) of soil for control of soilborne pests. *Annu. Rev. Phytopathol.* 19(1):211–236.

Kerry B. 1980. Biocontrol: Fungal parasites of female cyst nematodes. *J. Nematol.* 12(4):253–259.

Kerry B. 1988. Fungal parasites of cyst nematodes. *Agric. Ecosyst. Environ.* 24(1–3):293–305. doi: 10.1016/0167-8809(88)90073-4

Khan, MS. 1998. Pakistan crop protection market. *PAPA bulletin.* 9: 7–9.

Khan AA. 2007. Comparitive evaluation of twoanticoagulant and two acute rodenticide in sugarcane fields. *Sarhad J. Agric.* 23(3):713–718.

Khan AA, Munir S, Hussain I. 2012. Evaluation of in-burrow baiting technique for control of rodents in groundnut crop. *Pak. J. Zool.* 44(4):1035–1039.

Khan MA, Iqbal M, Ahmad I, Soomro MH. 2002. Economic evaluation of pesticide use externalities in the cotton zones of Punjab, Pakistan. *Pak. Dev. Rev.* 41(4 Part II):683–698.

Khan MH, Bukhari A, Dar ZA, Rizvi SM. 2013. Status and strategies in breeding for rust resistance in wheat. *Agric. Sci.* 4(6):292–301.

Khan MR, Khan SM, Mohiddin FA. 2004. Biological control of Fusarium wilt of chickpea through seed treatment with the commercial formulation of *Trichodermaharzianum* and/ or *Pseudomonas fluorescens. Phytopathology.* 43:20–25.

Khan MS, Shah MM, Mahmood Q, Hassan A, Akbar K. 2011. Assessment of pesticide residues on selected vegetables of Pakistan. *Pak. J. Chem. Soc.* 33(6):816–821.

Khooharo AA. 2008. A study of public and private sector pesticide extension and marketing services for cotton crop. PhD thesis, Sindh Agriculture University, Tando Jam, Pakistan.

Khooharo AA, Memon RA, Mallah MU. 2008. An empirical analysis of pesticides marketing in Pakistan. *Pak. Eco. Social Rev.* 46(1):57–74.

Klink VP, Matthews BF. 2009. Emerging approaches to broaden resistance of soybean to soybean cyst nematode as supported by gene expression studies. *Plant Physiol.* 151(3):1017–1022.

Koul O, Walia S. 2009. Comparing impacts of plant extracts and pure allelochemicals and implications for pest control. *CAB Rev.: Perspect. Agric., Vet. Sci., Nutr. Nat. Resour.* 4(49):1–30.

MacGuidwin AE. 1993. *Management of Nematodes. Potato Health Management.* APS Press, St. Paul, MN.

Mahmood Y, Khan MA, Iqbal M. 2008. Evaluation of various fungicides against powdery mildew disease on peas. *Pak. J. Phytopathol.* 20(2):270–271.

Mamoon-ur-rashid M, Khattak MK, Abdullah K. 2012. Evaluation of botanical and synthetic insecticides for the management of cotton pest insects. *Pak. J. Zool.* 44(5):1317–1324.

Mushtaque A. 1999. Is there potential for producing major pesticides in Pakistan? First Pakistan Agric. Business Conference, April 26th – 28th 1999. Internet WWW page, at URL: http://www.pakagribiz.com/FMC.ppt (February 15, 2003).

Nadeem S, Hamed M. 2011. Biological control of sugarcane borers with inundative release of *Trichogramma chilonis* (Ishii) (Hymenoptera: Trichogrammatidae) in farmer fields. *Pak. J. Agric. Sci.* 48(1):71–74.

Naseer-ud-din GM, Shahzad MA, Nasrullah HM. 2011. Efficacy of various pre and post-emergence herbicides to control weeds in wheat. *Pak. J. Agric. Sci. Vol.* 48(3):185–190.

Nasira, N. 1996. Invisible Farmer Khoj, Lahore.

Padgette SR, Kolacz KH, Delannay X, Re DB, LaVallee BJ, Tinius CN, Rhodes WK, Otero YI et al. 1995. Development, identification, and characterization of a glyphosate-tolerant soybean line. *Crop Sci.* 35(5):1451–1461.

Perlak FJ, Deaton RW, Armstrong TA, Fuchs RL, Sims SR, Greenplate JT, Fischhoff DA. 1990. Insect resistant cotton plants. *Biotechnology* 8:939–943.

Saeed R, Razzaq M, Raffiq M, Naveed M. 2016. Evaluating insectide spray regimes to manage cotton leafhopper, *Amrascadevstans* (Distant): Their impact on natural enemies, yield and fiber characteristics of Transgenic *Bt* Cotton. *Pak. J. Zool.* 48(3):703–711.

Saljoqi A, Walayati WK. 2013. Management of Sugarcane Stem Borer *Chilo infuscatellus* (Snellen) (Lepidoptera: Pyralidae) Through *Trichogramma chilonis* (Ishii) (Hymenoptera: Trichogrammatidae) and Selective Use of Insecticides. *Pak. J. Zool.* 45(6):1481–1487.

Sattar MN, Kvarnheden A, Saeed M, Briddon RW. 2013. Cotton leaf curl disease—An emerging threat to cotton production worldwide. *J. Gen. Vir.* 94:695–710.

Smith P. 2005. Pesticides rate and dosage calculations. Extension Entomologist. 833–840. http://www.caes.uga.edu/c.../Com_Pesticide_Safety.pdf

Sorribas F, Ornat C, Verdejo-Lucas S, Galeano M, Valero J. 2005. Effectiveness and profitability of the Mi-resistant tomatoes to control root-knot nematodes. *Eur. J. Plant Pathol.* 111(1):29–38.

Stirling GR. 1991. *Biological Control of Plant-Parasitic Nematodes.* CAB International, Wallingford, UK.

Stoytcheva M. 2011. Pesticides—Formulations, effects, fate. In: *Pesticides in Agricultural Products: Analysis, Reduction, Prevention*, Shokrzadeh M, SaeediSaravi SS (Eds), Intech Open Access Publisher, Rijeka, Croatia, 225–242, ISBN: 978-953-307-532-7.

Sultana N, Ghaffar A. 2010. Effect of fungicides, microbial antagonists and oilcakes in the control of Fusariumsolani, the cause of seed rot, seedling and root infection of bottle gourd, bitter gourd and cucumber. *Pak. J. Bot.* 42(4):2921–2934.

Tariq, MA 2002. Need to tap agriculture sector. Daily Dawn, Economic and Business Review.

Tariq MI, Afzal S, Hussain I, Sultana N. 2007. Pesticides exposure in Pakistan: A review. *Environ. Int.* 33(8):1107–1122.

Tauber MJ, Tauber CA, Daane KM, Hagen KS. 2000. Commercialization of predators: Recent lessons from green lacewings (Neuroptera: Chrysopidae: Chrosoperla). *Am. Entomol.* 46(1):26–38.

Tayyib M, Sohail A, Shazia, Murtaza A, Jamil FF. 2005. Efficiency of some new chemistry insecticides controlling the sucking insect pests and mites on cotton. *Pak. Entomol.* 27(1):63–66.

Ullah R, Khan MZ, Ullah K, Butt KM. 2015. Model farm services center approach: An implication to boost farmer's yield. *Agric. Sci.* 6(9):953.

Wains MS, Ali MA, Hussain M, Anwar J, Zulkiffal M, Sabir W. 2010. Aphid dynamics in relation to meteorological factors and various management practices in bread wheat. *J. Plant Protec. Res.* 50(3):385–392.

Wains MS, Jamil MW, Ali MA, Hussain M, Anwar J. 2014. Germplasm screening and incorporation of aphid resistance in bread wheat (*Triticum aestivum* L.). *J. Anim. Plant Sci.* 24(3):919–925.

Westphal A. 2011. Sustainable approaches to the management of plant-parasitic nematodes and disease complexes. *J. Nematol.* 43(2):122–125.

Williamson VM, Kumar A. 2006. Nematode resistance in plants: The battle underground. *Trends Genet.* 22(7):396–403.

Yasmin T. 2003. Pesticide poisons: Women at risk. Farming Outlook. January – March 2003. 39, Street, I-8/2, Islamabad, Pakistan.

Section III

Animal Production and Health

23 Livestock Production
Status and Policy Options

M. Sajjad Khan, S. A. Bhatti, S. H. Raza, M. S. Rehman, and F. Hassan

CONTENTS

23.1 LIVESTOCK RESOURCES AND THEIR UTILIZATION

The livestock in Pakistan comprises buffaloes, cattle, sheep, goat, camels, and equines. A small population of yaks is found in the northern areas. The wild species of sheep and goats are found in mountainous regions. These species from diverse agroecological zones are used depending on breeds and their utilization. The trend in the livestock population is positive (Figure 23.1).

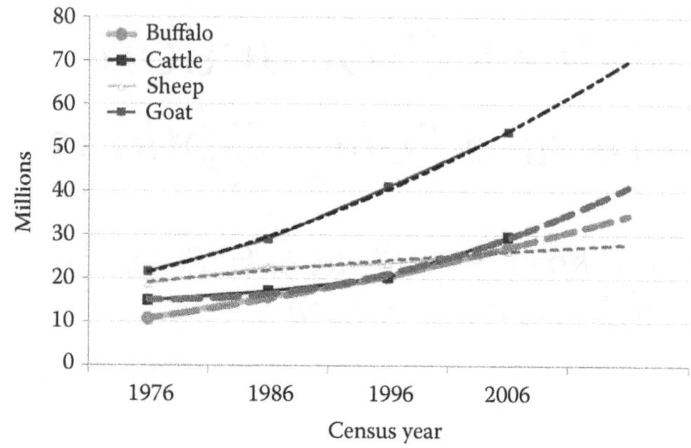

FIGURE 23.1 Trend in livestock population.

23.1.1 Buffaloes

There are 37 million buffaloes in Pakistan (GOP, 2016); Nili–Ravi and Kundhi are the main breeds. Nili and Ravi are still recognized as different breeds although their merger as Nili–Ravi was officially accepted in 1963. Azi-Kheli, a less known breed of buffalo is found in Swat valley, Khyber Pakhtunkhwa. About 50% of the buffalo population is nondescript as these animals do not fulfill the exact phenotypic characteristics of a particular breed.

Buffaloes are reared for milk as they contribute 65% of the milk produced in the country. Buffalo milk is preferred over cow milk because of its high fat and nutrient value. Buffalo meat contributes 32% of the red meat produced in the country. The distribution of buffaloes across provinces is not uniform and resembles statistics of the human population. The general production system is low-input subsistence small farming. Cattle in and around big cities are mainly inhabited by buffaloes. Landhi Cattle Colony, Karachi, inhabits a quarter million buffaloes.

Buffaloes, in their second to fifth lactation, are purchased from central and southern Punjab and taken to various settlements around big cities from Karachi up to Azad Jammu and Kashmir and Gilgit-Baltistan. Most of these buffaloes are slaughtered after completing lactation.

Performance recording and genetic selection programs are absent except for Nili–Ravi buffaloes on a limited scale. Accuracy of this recording system is low and limited number of bulls (and daughters per bull) combined with delayed assessments has yielded limited genetic gains. These programs should be executed with complete technical and budgetary reinforcements. Government farms have few thousand breeding buffaloes for research and production of bulls for artificial insemination (AI). Currently, AI coverage in buffaloes in Punjab is less than 10% and at a lesser rate in Sindh and other provinces. Government farms do not have pure Nili or Ravi breeds and, therefore, semen of these breeds is not available.

23.1.2 Cattle

The cattle population in the country is 43 million. Sahiwal and Red Sindh are internationally known tropical dairy cattle breeds. Cholistani and Tharparkar are important breeds surviving as dairy breeds under resource-limited environments. Achai, Lohani, and Gibrali with their small size are best suited under subsistence production setups. Nondescript cattle are grouped as Desi, Bhagnari, Dajal, Dhanni, Kankraj, and Rojhan are draft breeds. Hissar and Hariana are in low number and are used in Ox races. Cattle have been reared for bullock production for agricultural operations; this role is diminishing gradually. Cattle, including crossbreds, share one-fourth of the milk and about 28% of the red meat produced in the country. The total milk supplied by cows is 19.4 million tons.

Of the total cattle in Pakistan, 42% are in Punjab. The group size in dairy cattle is small as 76% of cattle are reared in herd size of less than 10, raised for subsistence in rural irrigated areas. Dairy cattle crossbreeds of Friesian and Jersey are raised with buffaloes to fill the milk supply gap during summer. Purebred Holsteins are being imported in thousands for high-input systems. In Punjab, importers are private entrepreneurs and military farms. Their influx has been encouraged by softer loans. Crossbreds can produce milk twice that of Sahiwal at a reduced age at first calving and (sometimes) better calving intervals as well. Yet, productivity is highly variable to feed/fodder accessibility and health cover. Profitability of pure exotic dairy breeds has not been evaluated and documented; however, the closure of purebred farms raised question on their success story in the country. Crossbreeding of nondescript cattle with Friesian in irrigated and with Jersey in rain-fed areas has been allowed. Semen for the crossbreeding program has been imported as well as produced locally.

Cattle crossbreeding for beef production has been trialled. For this purpose, the semen of Charolais and Simmental has been imported. A new beef breed, Narimaster, was developed by crossing Australian Droughtmaster (AD) and Bhagnari. The population of this breed, however, could not surpass few hundred animals as growth rate in the breed is not quite different from local breeds. Punjab has permitted the import of semen of Brahman, Charolais, Simmental, and Angus breeds. Low meat cost in the local market discourages crossbreeding for beef production. Veal production for export purposes has not been tried; however, it could be an option under the enhanced communication system of CPEC.

The performance recording and genetic selection is underway for Sahiwal breed under Research Center for Conservation of Sahiwal Cattle, Jhang. However, the base population has not reached the required 30,000 to test 150 bulls per year. Currently, 20–30 bulls are tested every year, limiting the selection intensity to achieve any tangible genetic gain in productivity. In general, AI coverage is available for 10%–15% of cattle population and is directed to crossbreeding rather than pure breeding. The use of embryo transfer technology is still at the experimental level. The Breeding Services Authority in Punjab has recently started regulating semen production of livestock in Punjab. Conservation programs are missing for breeds other than Sahiwal and require immediate attention.

23.1.2.1 Upgradation Strategy of Nondescripts through Crossbreeding

Crossbreeding has been looked as a quick fix to improve the performance of locally available low yielding animals in developing countries. Crossing of *Bos taurus* and *Bos indicus* breeds has been attempted in many countries. Generally, improved animals have been obtained in the F1 generation. Back crossing with either of the purebred local or purebred exotic breed, however, resulted in consequent loss of productivity due to adaptability problems and low resistance to diseases and parasites. First generations have been better due to maximum hybrid vigor, but subsequent generations usually performed poorer due to recombination and inbreeding losses.

In Pakistan, the craze for crossbreeding started with the inception of AI in the early 1970s with a vision of upgradation of local cows to improve milk yield and other economic characteristics. However, performance in the first crosses were better as Holstein and Jersey crosses with Sahiwal produced more than 70% of their local dams, whereas Holstein × nondescript produced 42% more milk than their dams. Research based on data from 48 herds (from six countries) reports that crossbreds produced 2.2 times more milk in the same herd as did the native cattle. The results for later generations were poor, but all these ventures were not as fruitful and sustainable over a long term as envisioned as in the case of Pakistan. The reasons for poor results were lack of implementation of proper mating plan after first generation, availability of semen of different crossbred grades, sustainable funding problems for long-term programs, environment adaptability and noncompliance, and/or unavailability of proper national breeding policy. Based on national and international experience, it is emphasized that we have to establish a stable crossbreeding system to maintain exotic level of inheritance between 50% and 75%.

Two crossbreeding strategies namely grading up of nondescript cattle and continuous F1 production can, therefore, be opted. Nondescript cattle would be inseminated with Holstein Friesian and Jersey semen for production of 50% bulls. For production of 75% bulls, F1 females would be back crossed with exotic bull semen. The selected 50% and 75% bulls would be used on main cattle population. Exotic inheritance level would fluctuate between 50% and 75%. The use of Holstein semen is recommended in irrigated areas while Jersey in hilly areas as per Punjab livestock breeding policy. Whenever, crossbreeding is used to upgrade any nondescript population, it should not be haphazard. Performance recording is the basic need for breed improvement efforts and must be ensured through the institutions mandated for this.

The second strategy which could be opted for enhancement of productivity in cattle is continuous production of F1. Bulls of exotic breeds would be used on females of the local population to produce F1. For this purpose, large local populations have to be maintained for crossbreeding. All the F1 females would become part of commercial herds/corporate farms. Nucleus herd would be maintained in the public sector while registration of progressive crossbred herds in the field would be carried out. Selected F1 males will be kept at calf rearing centers for growth evaluation at selected nucleus station. After final selection, male claves would be shifted to semen production unit for production of semen of selected bulls. Effective utilization of these two strategies could result in enhancement of productivity of local cattle in Pakistan to meet increasing demands of milk in the country.

23.1.2.2 Development of New Dairy and Beef Breeds

The development of synthetic dairy cattle breed is an option in the present scenario of dairy production setup in Pakistan to meet the increasing demands of milk and milk products. There are many success stories of synthetic dairy breed development around the globe, but in the Indian subcontinent, Sunandini is a good example. Sunandini cattle was developed in Kerala, India, using Brown Swiss, Holstein, and Jersey crosses with local nondescript cattle ultimately leading to a synthetic cattle breed. The Sunandini breed, which actually evolved from a group of crossbred animals through selection for economic traits, have greatly influenced the dairy economy of the state of Kerala. Another example of developing dairy synthetic breed by utilization of crossbred population of Military dairy farms of India is "Frieswal." It was developed using Friesian and indigenous Sahiwal cattle breeds, possessing an inheritance of about 62% Friesian and 38% Sahiwal, capable of producing 4000 kg milk with 4% butter fat in 300 days lactation period under a national project that was launched in the year 1985 at Military Farm, Meerut. Considering the experience of development of Frieswal and Sunandini, it is quite possible to develop a synthetic dairy breed in Pakistan.

A reasonable population of crossbred dairy cattle is present with only military dairy farm facilities. These farms have suitable housing, feeding, and management facilities apart from health coverage. Milk is also processed to make powder and cheese. These farms, therefore, with all the facilities and base population of crossbreds can be used to develop a synthetic dairy breed. Only implementation of a sustainable breeding program along with strict selection is required for this purpose. Raising of bull calves and their periodic evaluation to produce progeny tested bulls of different grades would be required to meet the demands for breeding of crossbred females. After successive generations of performance recording, selection, and evaluation, uniform breed characteristics could be opted for new dairy synthetic breed. A dairy breed thus developed could be a suitable animal for dairy production, especially for smallholder dairy production and corporate setups.

There is no beef breed in the country. Cattle and buffaloes are mainly bred for milk and draught purpose, and culled animals and surplus calves are the main source of beef, with cattle and buffaloes having equal share. Buffalo calf fattening have been conducted at various institutional facilities with encouraging results. Cattle crossbreeding to produce a beef breed was initiated at Sibi in 1960s. It involved AD and indigenous breeds such as Bhagnari, Tharparkar, and Sahiwal. Crossbreds with 62.5% AD were named as Narimaster. Although, Narimaster was evolved as a synthetic beef breed, their performance was not very impressive and, therefore, did not get acceptance at the farmer level. Crossbreeding of indigenous cattle with Charolais in 1970s and with Simmental, Hereford, and Angus in the 1990s has been conducted at experiment stations with limited success. Besides these experimental projects on beef crossbreeding, situation regarding beef production has not been very promising as no concept of beef grades is prevalent along with poor and ill-practiced market channel. In this scenario, fattening of surplus cow and buffalo calves is a viable option which fits into routine dairy production setup and can better utilize existing resources. However, for experimental purposes, crossbreeding could be opted to evaluate different grades of crossbreds for their growth potential and feed efficiency. This may be helpful in future to target international beef market, especially Middle East beef halal market where beef quality and grade matters with better economic returns.

23.1.3 Sheep and Goats

The population estimates for sheep breeds in the country are 30 million. There are at least 25 breeds and distribution across provinces is very heterogeneous with Balochistan having 40% of the population. Sheep breeds can be divided into thin-tailed or fat-tailed. Sheep raised in the cooler regions of the country keep on moving from cooler to hotter regions (and back) in search for free grazing lands. All sheep breeds produce coarse fleece and are used for producing mutton. They are more commonly raised in mixed flocks with goats. Wool from breeds, for example, Kaghani and Kari is finer than the other breeds but not sufficiently fine to meet the domestic needs and fine wool is, therefore, imported. Sheep population has not indicated growth rate like that of goats and the reasons for slow growth are deteriorating ranges, diminishing demand of wool due to diversified synthetic fiber and availability of (preferred) goat meat.

Development of new synthetic breeds (to produce better finer wool (and ore meat) was attempted in 1960s and 1970s, and resulted in Baghdale, Ramghani, Pak-Awassi, and Pak-Karakul breeds. Some public institutions still keep some typical animals of these breeds. Crossbreeding with American Rambouillet was attempted in KP to produce animals of medium fine wool and to some extent continues till date with a nucleus flock of Rambouillet at Jaba. Experiment stations with governments still distribute rams to interested ranchers but to an exceptionally limited extent. Experimental AI has been attempted in few breeds, for example, with liquid semen in Lohi and use of frozen semen is quite recent (Asim, 2015). There is no breed improvement program for any sheep breed except upgradation of Kaghani with Rambouillet in KP. Breed association/societies do not exist.

There are around 70 million goats in Pakistan. There are at least 25 breeds in the four provinces and including GB and AJK will add another dozen to this number. Population estimates are available for one-third of the breeds and in the absence of any pure breeding and breed improvement program reduction in purebreds is easy to believe. With the exception of Angora (imported in the 1960s) for producing mohair (now limited to a research station), there are no major import of goats. Recently, however, South African Boer has been imported for high-input mutton production and semen of Saanan has also been used by PARC experimentally.

Major breeding objective for all breeds is goat meat (mutton). The average milk production is 2–3 L per day. Breeds such as Beetal can easily produce more than 6 L a day intended from Saanan (even if purebreds succeed) and their fattened males can easily surpass 200 kg as seen in goat shows. Teddy and Barbari are quite prolific. Documentation of Beetal for its different strains (Faisalabadi, Makhi-Cheeni, Nuqri, Gujrati, Nagri, and RY Khan) has reported wide within breed diversity (Khan and Okeyo, 2016).

The goat population is more evenly distributed among the four provinces than other species. Punjab, Sindh, Balochistan, and KP have 37%, 24%, 21%, and 18% population, respectively. The annual rate of increase in goat population is most astounding among all the ruminant species (Figure 23.1). These population trends and the actual number of animals have recently been criticized. The actual numbers may be lower than these estimates. Mainly fresh meat is sold. Thousands of goats are slaughtered on religious events, particularly on Eid. Grazing alone and grazing with provision of some fodder is the most widely recognized arrangement of feeding. Concentrate feeding is restricted to animals that are displayed at shows. Some areas specifically produce sacrificial goats under high-input system. There is no specific conservation program for any of the goat breeds. Punjab government has recently distributed Nuqri bucks in the hometract of the breed. Few programs use goats to reduce poverty but impacts are yet to be seen. Beauty competitions shows now involve goats. Milk competitions are also held and generally winners are Beetals. Breeding is for the most part natural yet endeavors to freeze goat semen have been fruitful at the University of Agriculture Faisalabad. KP has launched AI in goats using Beetal (from Punjab) and results are encouraging. Government livestock farms are source of superior males yet their contribution in the improvement of breeds is negligible. Government farms raise about one-third of indigenous breeds but wide-scale performance recording and genetic improvement are still awaited. Lessons learnt from past and production potential parameters indicate that both in sheep and goats, we should focus on within breed selection and crossbreeding has no place.

23.1.4 Camels

Camel population has been reported as static for the last decade (around 1 million), but general belief among technocrats is that it may be on a decline. Almost all the population belong to one humped (dromedary type) camel. The presence of two humped camel (bactrian type) has also been reported in GB. Uncertainty exists with respect to number of breeds (or strains isolated by locations) as earlier reports just had four dromedary breeds (Bikanari/Mahra/Merecha, Sindhi, Bagri/Booja, and Mountain) while Isani and Baloch (2000) reported that there were 20 breeds of dromedary camel in Pakistan.

Camel is part of socioeconomic culture of arid, mountainous, and coastal regions of the country. Balochistan has 44% of the camel population. Camel meat and milk are not favored food items for people residing in irrigated areas where buffalo/cattle milk is in abundance but keepers raising camels savor these items sufficiently. Gamekeepers consume fresh, raw, or soured camel milk. Camels are also sacrificed at Eid ul Azha all over Pakistan. The other major utility of camel is as a pack animal. Camel hides are also used for making decorations such as table lamps. Hair is utilized for making mats and clothes. Camel racing and dancing is a piece of rural fun. The potential to export camels to gulf countries both for meat and milk is great.

23.1.5 Equines

The horse population has been diminishing over the years and is estimated at half a million at present. Balochi, Heerzai, Waziri, Makra, Pak—thoroughbred breeds have been reported as indigenous breeds. Different families also raise few strains (Kakka, Kajlan, Morna, Shien, etc.) and they try keeping the relationship of animals close to a specific male or a female (linebreeding). Facilities with armed force raise thoroughbred as well as draft breeds such as Suffolk, Percheron, and so on, for producing mules. Horses are also imported for racing. Close to the Afghan outskirt in KP and Balochistan, Afghan stallions are also found in areas close to Afghanistan both in Balochistan and KP.

Raising and riding horses is a much older tradition than the historical backdrop of Pakistan. Horses have always been an image of power, trustworthiness, and honor. With the declined usage of horses for "Yakka" (Tonga) or two-wheeled carriage, horse-drawn carts can still be seen in at least small towns. Horses are still utilized for individual or family rides yet tent pegging and horse dancing are essential events at animal shows/weddings and specifically arranged horse shows and tournaments. Horse meat is not consumed in Pakistan. Crossbreeding with exotic thoroughbreds is a general practice for producing polo and racing horses. Army's Mona Depot (at Sargodha) is the major facility for horse breeding and research.

Pakistan has some 5 million donkeys with a positive population trend in spite of their decreasing utility as pack animals in major cities. Mules with an estimated population of more than 2 million are mainly used as pack animals. Donkeys are also used for furrowing (alongside a bullock) in sub-hilly/barani regions. As livestock and agriculture farming go hand in hand, pulling (with female donkey), of fodder for the bullocks/cows/buffaloes, other agricultural produce or family articles is not uncommon in villages. Under climate change scenario, donkeys (along with camels) are likely to become more important as they can consume high-fiber feed and can tolerate harsh summers without major care. International efforts have started to see if more milk and meat can be harvested from this socially ignored species in future.

23.1.6 Yak

Yaks are found in GB. Official statistics (GOP, 1996) reported a population of 17,000 with Skardu district being the most populated one. A few reports show that a small feral population is also available in GB. Yak is a multi-purpose animal that provides milk, meat, leather, hair, and manure to people living in harsh cold areas. Yaks are also used as pack animals and adds to the esthetic value of GB. Crossbreeding with cattle (such as Holsteins and Jerseys) has threatened this species but in the absence of any legislation to stop unabated crossbreeding seems impossible. It requires immediate attention of policy makers and other stakeholders.

23.1.7 Application of Genomics and Other Biotechnologies in Livestock Improvement

Genomic selection is a new era in genetic improvement, it allows breeders to estimate the genetic merit of an animal at birth and then improve the estimated breeding value of an animal once phenotypic information is available.

In Pakistan, where it is difficult to collect large quantities of high-quality phenotype and pedigree data on farm animals over long periods of time, genomic selection has the potential that dramatically enhance genetic improvement of animals. Low-cost tests for high-throughput genotypes are being developed. It will allow an animal's entire genome to be inferred from a relatively small number of genetic markers. Breeding value can, therefore, be estimated at a very early age. Detection of thousands of single-nucleotide polymorphisms (SNPs) and continuous reduction in SNP genotyping cost can potentially benefit our breeders and the dairy industry.

Genomic selection reduces the need for recording all breeding animals and helps in traits that are otherwise difficult to measure. For example, traits that are often difficult or impossible to measure in elite animals, such as disease resistance (requires disease challenge), female fertility (sex limited), and longevity (needs to wait for long). So practically to apply genomic selection, instead of recording traits on the entire population, both a large reference population with many recorded traits and information of the genome are required. The priority area would be creation of reference population where animals are recorded with authenticity.

The use of genomics can speed up genetic improvement and increase levels of production quickly but it has scientific, economic, and political challenges. Considerable increases in public and private research funding will be required to develop and utilize novel tools and collection of required data on appropriate animals. Advances in livestock genomics have major implications for increasing food output as well as improving human health.

23.1.8 Way Forward

- Documentation of indigenous livestock breeds is a pre-requisite to assess their production potential under optimum conditions of management. Less optimum performance and resource-poor conditions and comparison with exotics can mislead policies.
- Prioritization of potential breeds with respect to their hometracts and type is important in national/provincial breeding policies.
- Within country registration mechanism is required for available and new breeds to encourage new breed development and help in material transfer protocol agreements with external buyers.
- Performance recording is the key for genetic selection of sires/bulls and should be expanded to widen selection base and allow reasonable selection intensity for tangible genetic

gain. Progeny testing programs should be protected through legislation so that political and administrative changes do not drag them back wasting all the previous efforts and investment.

- Distribution/culling of pedigreed heifers (and other performance recorded animals) to support poverty alleviation programs should be banned to allow performance recording programs to yield progress.
- Species such as camel and donkeys and adapted cattle breeds such as Sahiwal, Red Sindhi, and Cholistani will become more important under climate change scenario and specialized institutions should be developed to further target productivity enhancement that can lead to conservation of indigenous breeds through improved utilization.
- Small ruminants have been left alone to develop. Each province should prioritize breeds for development and technologies such as AI can help multiply better/improved genetics through access of better germplasm to marginalized farmers. Failed crossbreeding attempts in the past should provide enough lessons to focus on within breed selection.
- Identification of high-prolific breeds will become important with respect to climate change scenario under low/traditional-input systems in tropics. Modern cutting-edge technologies like high-throughput genomic selection may be targeted in exploration of such characters.
- Cap on milk and meat prices is likely to kill indigenous dairy breeds when inputs have no limits to hike. Processed milk and milk product businesses should have limits to use imported powder. Duties on powder milk and imported whey must be and rationalized to stop indigenous dairying to further deteriorate.
- Development of disease-free farms/zones for meat export especially around northern route of China–Pakistan Economic Corridor will help improve sustainable utilization of indigenous livestock resources and in export earnings in future.
- Veal production for export market to neighboring countries in the north and west remains to be tested and might be a good opportunity under the improved communication network of CP.

23.2 LIVESTOCK FEEDING AND NUTRITION

The importance of proper nutrition in animal feeding cannot be overemphasized. Properly fed animals have higher productivity and are more economical to maintain than underfed animals (Figure 23.2). Underfeeding reduces the growth rate of animals and thus increases the age at puberty and age at

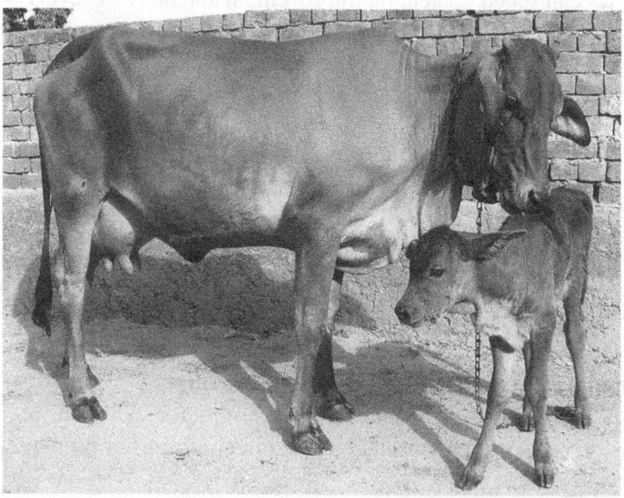

FIGURE 23.2 A well-fed heifer that calved at the age of 29 months.

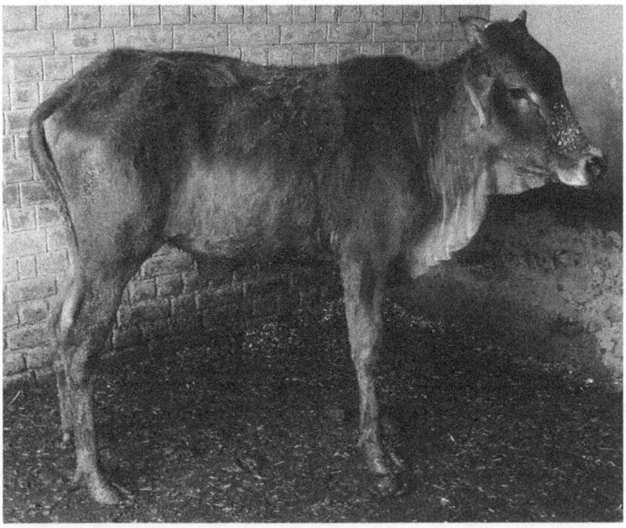

FIGURE 23.3 Underfed cow heifer; age 4 year, weight 158 kg.

calving (Figure 23.3). Underfed animals are a liability for the owner and are burden on land resources. The underfed heifer, at the age of 4 years, still needs to grow and calve, whereas a well-fed heifer has already calves at the age of 2.5 years. Providing the required quantity and quality of feeds can enhance livestock production up to 50% from exiting genetic potential of animals.

23.2.1 Current Nutritional Status of Livestock in Pakistan

In Pakistan, nutritional requirements of animals are mainly met through fodder crops, shrubs, grasses, and agro-industrial wastes. Livestock are getting 50.7%, 37.8%, 6.1%, 2.4%, and 3% of their total nutrient requirements from crop residues, fodders, cereal by-products, oilcakes, and other wastes, respectively. It has been reported that livestock are getting only 75% of the required amount of total digestible nutrients (TDN) and there was 60% shortage of digestible crude protein. The existing available feed resources can only fulfill the maintenance requirements of animals. With the rising livestock population and squeezing land for fodder production, this gap is likely to increase in the years to come.

23.2.2 Fodder Supply Situation

As mentioned earlier, livestock are getting major share of their nutrients from crop residues and fodders. Fodder supply round the year is not uniform (Figure 23.4). There are two "official" fodder scarcity periods during the year; one in summer (May–June) and the other during winter (December–January). However, actual availability of fodder to the livestock during other months of the year is not abundant either. The time and duration of scarcity period can shift to either side depending upon rains, onset of winter and water availability for the fodder crops.

Reasons for short supply of fodder to the livestock are described below:

- Fodder is not a priority crop for the farmers. Fodder is sown on the land that is available in between the harvesting time of one cash crop and sowing time of the other. Therefore, fodder calendar recommended by the agronomists for round the year supply of fodder is merely a theoretical solution.
- Cut and carry system is a hurdle in continuous fodder supply. In a commercial livestock farming operation, daily cut and carry system is not feasible for feeding livestock. Unexpected rains can either delay sowing of fodder at proper time or destroy the germination

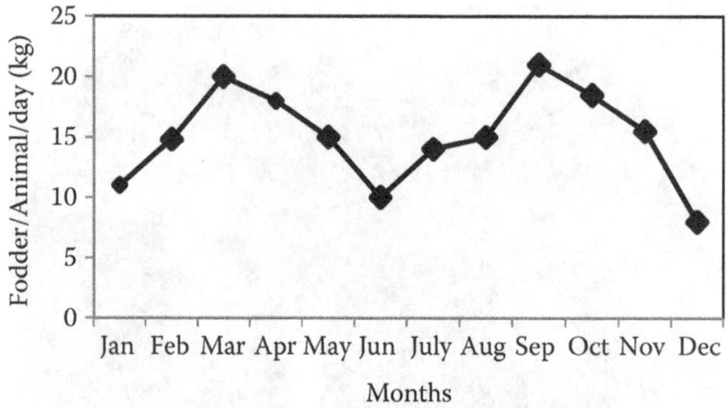

FIGURE 23.4 Fodder availability in Pakistan with two scarcity periods.

if received just after sowing. Delayed monsoon can limit water supply to the fodder crop and decrease per acre yield and lower its feeding value (Figures 23.5 and 23.6). Frost in winter can destroy the whole Berseem crop. The closure of canals in canal-irrigated areas can affect the per acre yield of fodder due to short supply of water at proper time. Shortage of electricity supply to the tube wells can limit water supply to the fodder crops badly affecting expected per acre yield. Even if fodder is sown according to the requirements of the animals at the farm, or even 10% more than their requirements (following fodder calendar) and quality seed is used, fertilizer is applied according to recommendations and water availability was not limiting, rains were beneficial to the fodder crops and fodder was available in the field in plenty, even then, it does not reach the animals at all times. This is mainly because of labor issues, nonavailability of transportation, and in order machinery required for harvesting and chopping. Extreme weather (severe hot and cold, wind storms) can hamper in fodder harvesting, if done manually, and adds to another excuse for not transporting fodder from the field to the farm.

23.2.3 Way Forward

- Collective efforts of all the stakeholders are required to address the problem of feed shortage in the country. The role of each stakeholder for enhancement of feed resources needs to be understood and specified. Government has a major role to solve this issue. The provision of seed of high-yielding fodder varieties through local research or import of seed

FIGURE 23.5 Sorghum crop in the month of August due to shortage of water as a result of no rains.

FIGURE 23.6 Making hay can give more cuts of alfalfa than with cut and carry system.

of high-yielding fodder varieties at farmer's doorstep and ensuring quality of fodder seed available in the market can be better handled when government is on the steering wheel with all the other stakeholders in the van. This is the most important single issue, if handled, can improve fodder supply and thus improve nutritional status of the animals. The maintenance and development of range lands can also be better handled by the government organizations for increased biomass from the same land. Thus, government needs to focus on how to use these utilized land resources to provide more nutrients to the livestock. The availability of machinery for hay and silage should be developed at local level. However, in the absence of such machinery, government should encourage import of such machinery by private sector, so that land-use efficiency could be increased to grow more fodder on the same piece of land. Provision of water to water-scarce areas should also be arranged by the government. Judicious use of feed resources through balanced ration and the use of biotechnological innovations for maximizing the nutrient availability from the same feed resources lies on the shoulders of the farmers, for which they need to be given education. Creating awareness among farmers is the job of public-run institutions whether it be government departments or educational institutions (universities). Besides, farmer organizations can also play roles in creating awareness on modern feeding practices.

- The preservation of fodder in the form of hay and (or) silage offers a solution for irregular supply of fodder for livestock. Making silage makes the land free for another crop and increases the nutrient availability from the same piece of land (Figures 23.7 and 23.8). Hay making from alfalfa crop can give one to two more cuts of quality fodder compared with cut and carry system (Figure 23.6). With small land holdings, buying machinery for hay and silage making is an issue, however, can be solved with rented machinery for the purpose, as is done for threshing of wheat and paddy.
- Using healthy seeds of high-yielding fodder varieties can increase per acre fodder yield. However, nonavailability of quality seed for fodder crops at farmers' doorstep is a big hurdle.
- Better agronomic practices can increase the per acre fodder yield; these include proper land preparation, sowing at proper time, providing the required water, fertilizers, and pesticides.
- Harvesting fodder crop at proper time can provide more nutrients from the same dry matter: With increasing age of plants, dry matter yield per acre increases but at the same time quality of fodder goes down. Fodder crops should be harvested at a stage when they have maximum nutrients rather than weight. For harvesting maximum nutrients, grasses should be harvested when they are 50% heading stage and legumes should be harvested before budding stage.

FIGURE 23.7 Fodder cutting and chopping for silage making.

- Chemical/physical/biological treatment of low-quality roughages can enhance nutrient availability from the same dry matter. Straws are major feed resources for livestock in the country; they have low nutritional values because of higher fiber and low protein contents. Nutritional values of straws can be increased by chemical treatments. However, this solution is only of an academic importance in the absence of farmer-friendly technology for chemical treatment of straws.
- Use of ionophores, prebiotics, buffer salts can improve the efficiency of feed utilization by livestock. These products are available in the market with many brand names.
- Range lands are potential feed reservoirs for livestock in Pakistan. Proper management of range land can provide fodder to the small ruminants. Range management, conservation, and improvement policies are required to utilize range land for nutrient supply to the livestock. (Range management is not discussed here, as it is more related to forestry and discussed there.)

FIGURE 23.8 Silage being prepared in a bunker.

- Providing a nutritionally balanced ration can enhance the output of animals. Haphazard feeding to livestock can result in wastage of nutrients, which if fed in proper proportion can be better utilized. For example, feeding berseem and (or) maize fodder separately to livestock will result in energy- and protein-deficiency, respectively. However, if fed in combination, with a proper proportion can provide a balanced ration for livestock. Similar is the situation with other agro-industrial feed ingredients for livestock. Thus, with a judicious feeding management, a lot of feed resources, which otherwise are being utilized inefficiently, can be used very effectively with more output from them.
- Cholistan area and other such lands that are otherwise fertile but are unproductive because of no water can provide fodder to livestock if water availability is ensured. This is easier said than done; however, for long-term solution of the problem, attention must be paid to make this land productive.

23.3 LIVESTOCK MARKETING

Efficient marketing system is vital for the viability of any business venture. Efficient marketing system boosts production and brings more investments and technologies for the growth of respective sector. This chapter explains present status and way forward for livestock marketing system.

23.3.1 Live Animal Marketing

Marketing of live animals (dairy, meat, wool, draught, or other purposes) is very important in livestock production. The traditional livestock markets in the country are neither buyer friendly nor seller friendly. Traditional marketing system exploits the one who has less tacit knowledge of the market by the ones who are more familiar with the system. Buying and selling is not based on any transparent criteria but depends upon mutual agreement between the two parties with little surety of the worth of the animal being sold or purchased. It is not possible for a buyer to be certain about the worth of an animal he is paying the price for. Market administrations are less interested in the welfare of the animals brought for sale, or protecting the interests of both buyers and sellers, but are more interested in money making. Livestock markets are generally held at district headquarters or other important hubs for livestock on specified dates on monthly, quarterly, or weekly basis. Livestock beoparis (the person who buys and sells livestock) buy animals from livestock farmers at their doorstep and bring them in the livestock market to sell others. These beoparis earn profit from marketing of livestock, which otherwise could have gone to the pockets of livestock farmers. Thus, in the present livestock marketing system there is exploitation at every level.

23.3.2 Milk Marketing

Milk marketing system is more complex than livestock marketing. There are more than one marketing system in milk marketing. These are: urban production (diminished), peri-urban, rural production, and subsistence). Almost 80% milk is produced in rural area. Of the total milk produced, 3%–4% enters through formal channels and rest 97% by informal players using a multi-tiered layer of agents. Out of the total milk available for human consumption, only 30%–40% enters the urban market, the rest is used locally (only 3%–4% processed). The country is deficit in milk supply, therefore, dry milk and whey powder worth US$ 130 million are imported each year. Brief description of milk marketing channels is described as follows.

23.3.2.1 Informal Milk Collection

Under this system, a closely interlinked network of milk collection agents called katcha dodhi (100 L/day), pacca dodhi (400–800 L/day), contractors (1500–3000 L/day) are involved.

With the involvement of middlemen the price increases, quality decreases, and consumers get inferior products at a high price (80%–100% higher than farm gate price). More than 70% milk comes from smallholders; these smallholders are unable to negotiate with companies or collectors on milk prices. They have to accept what is offered. In most cases, milk price is less than production cost.

23.3.2.2 Formal Milk Collection System

This milk collection system is run by different processing companies and is described as follows:

- Progressive farmers or direct from farmers: Under this system, milk is collected directly from farmer on fat% basis. Milk-processing companies supply milk canes, chemical, and other necessary equipments; milk is collected by company vehicle and payment is made on weekly basis
- Village milk collection: This system is similar to the one described above, except that there is a commission agent involved for collection of milk from the smallholder milk producers in a village
- Mini contractor: Under this system, milk is collected from farmers and dodhis on fat% basis and supplied to the milk-processing companies and payment is made on weekly basis.
- Hilux contractor: Under this system, company provides no facilities; milk is received at collection centers and payment is done on a weekly basis.

The main issues identified in milk marketing are the absence of integrated cold chain, adulteration, scattered farming making milk collection and transportation difficult, no farm gate price fixing (local authorities fix the milk prices but ignore the increasing prices of feed inputs), perishable nature, and small quantity of milk (unable to negotiate).

23.3.2.3 Cooperative Milk Marketing

The largest milk marketing system is cooperative marketing. In major dairy countries, it is the most effective milk marketing system. The best example is India where Gujarat Cooperative Milk Marketing Federation has more than 80,000 villages and 12 million farmers as members and is handling more than 10 million liters milk daily. Such a system had been tried in Pakistan, however, it could not sustain due to political interference. For a country like Pakistan, cooperative system of milk marketing is the best choice as it gives negotiation power to small farmers.

23.3.3 Meat Marketing

The marketing of animals for meat production is shown in Figure 23.9. The flow diagram explains the entry channels and their share in animal trade. Middlemen and beoparies play a major role in meat animal trade. There are many layers of traders between producers and consumers. The beoparies/butchers (or their agents) purchase animals from villagers, butcheries, from different areas, and households, and transport them to nearby markets. Where large beoparies or contractors/traders purchase and send to main markets for further trade. The commission agents/traders sell these animals to local butchers and wholesalers who partially export these animals to KPK or Afghanistan. Wholesale butchers purchase animals from main markets, slaughter them and sell to small butchers for retail consumers.

The meat industry in Pakistan has enormous potential. Pakistan produces 3.7 million tons of meat annually; out of which 53%, 18%, and 29% comes from beef, mutton, and poultry, respectively. The demand of halal meat is increasing day by day even in non-Muslim countries. We ought to

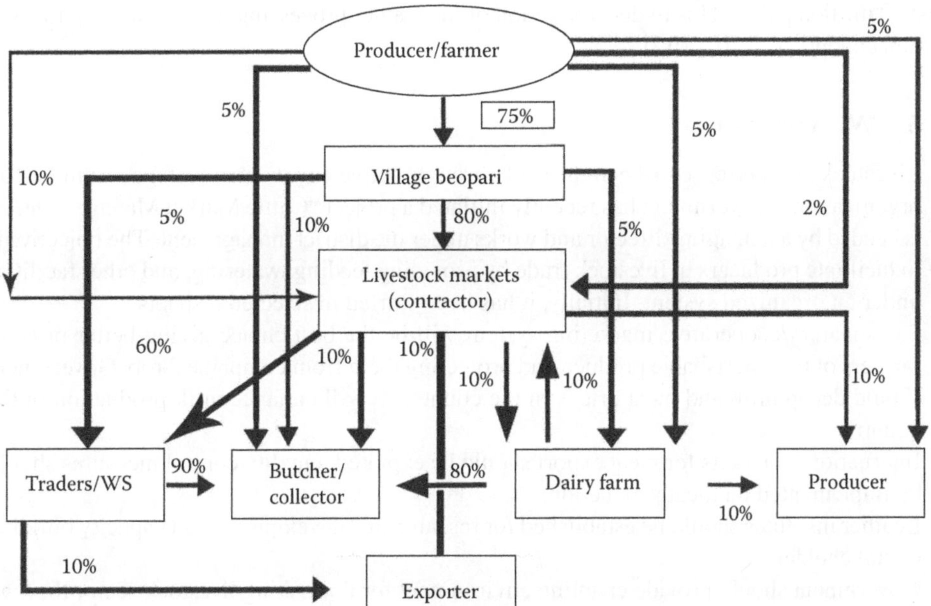

FIGURE 23.9 Marketing of live animals.

find our share in it. Pakistan can access to 470 million consumers in Middle East, Central Asia, and Europe.

23.3.4 Leather Marketing

Leather industry (tanning, leather footwear, leather garments, gloves, shoe uppers, and leather goods) in Pakistan is the second most dynamic sector after textile industry; it contributes 5% in GDP, 7% in export and provides employment to 0.2 million people (Muhammad, 2008). Pakistan produces 15.34 and 51.87 million hides and skins, respectively. Leather industry is mainly situated at Karachi, Sialkot, Kasur, and Lahore. Leather industry is export oriented (90%) in the form of leather products/finished goods (US$ 1.0 billion). Leather export has a growth rate of 11% per annum.

There are 2500 tanneries/footwear making units. Growth in the leather industry is hampered due to inability to cope with changing trends, rigorous customers' requirements, and leather fashion industry. Poor flaying techniques, pre- and post-animal and hides/skins management deteriorate leather quality. The main constraints of leather industry are animal smuggling, unavailability of raw skins, pre-slaughter management (warble fly, mange, etc., MDF, 2015), insufficient expertise in marketing, designing and product development, old machinery, lack of research and development, energy and water crises (270–300 kW for 100 ft^2 needed), expensive dyes and other chemicals (Zaman, 2006), the absence of infrastructure for hides collection, high cost on water purchase and government rules about water treatment and environment pollution. These constraints increase the cost of production. Unskilled labor worsens the situation by causing damage to leather and producing poor goods. Pakistan has capacity to produce 90 million ft^2 of leather but is working on 67% of its capacity. Presently, 461 leather garments and apparels making units are producing 28% less than the capacity, 524 footwear units are producing 100% less than the capacity (100 million vs. 200 million). Gloves manufacturing units are also producing 100% less than the capacity (5 million against 10 million pairs). This under-utilization of the leather processing units is making the leather sector uneconomical (Saif, 2012).

23.3.5 Way Forward

- Livestock marketing can be improved through active involvement of government-run organizations. Government has recently initiated a project "Cattle Market Management." It is headed by a managing director and works under the district management. The objective is to facilitate producers in livestock trade by providing feeding, watering, and other facilities under an organized system. Initially, it has been started in selected districts.
- Participatory/cooperative marketing system will be the best choice giving better price to farmers of their perishable produce and protecting them from companies' rob. Government should decap milk and meat prices in the country; it will enhance milk production in the country.
- International markets for meat export should be explored. Quality control measures should be implemented on meat production.
- Leather institutes should be established for research and development and capacity building of stakeholders.
- Government should provide enabling environment for flourishing livestock sector through policy-making rather than itself entering into business (e.g., selling stomach powder). Formulation of policies and their implementation, creating enabling environment for people to do business, monitoring and evaluation, licensing, research on vaccines, and other livestock issues and (education) capacity building are the domains of governments while doing businesses should be left to private sector.

23.4 ANIMAL PRODUCTION EDUCATION AND CAPACITY BUILDING

Education is a process of bringing desirable changes into the behavior of human beings. These changes must be desirable to the society at large. The education is effective when it results in changes in knowledge, skill, attitude, and action. Education can be formal or informal. Informal education is generally out of the classroom. Extension education is a kind of informal education, which aims at teaching the people living especially in rural areas how to raise their standard of living by their own efforts using their own resources of manpower and materials with the minimum assistance from government. The broader function of extension work is to help people to solve their own problems through the application of scientific knowledge is now generally accepted.

Animal production scientists act as innovators and leaders in respective fields of livestock production. Currently, such livestock experts are being produced in the institutes of higher learning (colleges and universities). They have also gotten education from universities of advanced countries including Europe, the United States, Canada, Australia, China, and others. In a lower tier, livestock assistants are needed to act as implementers of the animal production knowledge. Such manpower is being produced in training institutes specially established for this purpose and in some of the universities in the country. Livestock farmers and other stakeholders should have updated knowledge in modern husbandry practices. Creation of awareness among livestock farmers on modern husbandry practices is being done through extension services provided by provincial livestock department and other institutions.

23.4.1 Institutes and Organizations Involved in Animal Production Education

More than 10 universities are offering a 5-year degree, DVM, with 80:20 emphasis on health and production. Out of these, the University of Agriculture, Faisalabad and Sargodha University offer

BSc agriculture with animal science as a major subject. BSc Hons (animal husbandry) degree with focus on animal production alone was abandoned in 2003.

There are other institutes/centers that are involved in the provision of short-term, nondegree training on livestock production and health. Under this diploma course, theoretical and practical training is provided in various disciplines. Diploma certificates are also awarded.

Pakistan has a large network of NGOs (both registered as well as unregistered). Many NGOs address poverty reduction issues and livestock is focused as one of the tools to alleviate poverty in rural areas.

Although major emphasis of field activities of the provincial livestock departments is on health coverage, however, the literature on animal production in the form of brochures, leaflets, and booklets is published by these departments and is available to the interested livestock farmers free of cost. Mobile vans have also been introduced in limited areas.

Livestock and Dairy Development Department Punjab has established a helpline for the farmers. Farmers can call at this toll-free number to get the required information.

Information technology has revolutionized the system of information delivery. With the progress in information technology, government, semigovernment, private organizations have put animal production information on their websites. Information on these websites is available to everyone who has access to Internet. Local and internal organizations are providing information on most of the animal production aspects. It is all on an individual's palm in a cellular phone with Internet facility. Localized information is available on these websites. There are other websites too and are growing day by day. Other than these websites, farmers share information on Facebook and Twitter accounts. However, the use of information technology is limited to literate farmers only. Extension messages can be broadcasted through SMS on mobile phones. This service is on a limited scale, however, is likely to grow and may take a leading role in spreading extension messages.

23.4.2 Current Field Situation of Livestock Production Extension

In the field, the extension service is practically an animal health delivery system. One of the major issues in the animal production education in the country is wrong direction of the extension services provided by the provincial livestock departments. Provincial livestock departments are mostly dominated by veterinarians. Main focus of the extension staff is on health issue of the livestock and not animal production. There has always been realization that animal production education is of utmost importance for the development of livestock, however, all on ground efforts are made to promote veterinary education. Since the field staff is dominated by the veterinarians, they are the "livestock expert" educating the livestock farmers on animal production. The para-veterinary staff is generally interested in money making by providing treatment outside the hospital boundaries rather than educating farmers on animal production education. This practice has resulted in general ignorance of livestock farmers about modern husbandry practices. Some projects were initiated in the past by provincial livestock departments to focus on animal production education but actually ended up with focus on health cover.

23.4.3 Issues in Animal Production Education and Extension

The biggest issue in the animal production education and extension is nonavailability of animal production experts. Before 2003, there were two independent degree programs; one of them was offered in animal production and the other on veterinary care. However, the "intellectuals" of the livestock sector thought it was wastage of resources. Thus, they were successful in combining the two degrees into one 5-year composite degree of DVM with 80:20 courses of animal production and health. Thus, the BSc (Hons) degree in animal husbandry, with focus on animal production at the University of Agriculture, Faisalabad was stopped. The DVM graduates of the new 5-year degree programs are neither veterinarian nor animal production experts. When they go into the field, they are/will not be able to deliver animal production education to the livestock farmers because of their superficial knowledge in animal production.

With veterinarian-dominated thinking, there is no hope ever that this country will ever be able to produce graduates with sole focus on animal production. In the absence of right manpower in the field of animal production, all the other steps will be no more than a mere lip service. This is major cause of lack of ownership in breed development programs where veterinary experts feel they have been misplaced as they cannot earn extra income by extending their expertise to needy farmers. Those who are forced to join such projects try finding clients all day for whom they can extend veterinary service. Hunting such clients does not need documentation. Moreover, public relations (PR) earned can help them continue veterinary service as a main job at the expense of public money. Data quality and authenticity is, therefore, at stake. Millions of rupees spent may go waste. Erroneously ranked bulls further disseminated through AI pull back any genetic progress otherwise possible. Provincial livestock departments mainly focus on health coverage, therefore, there is no real system of information delivery to the livestock farmers with focus on livestock production.

One major problem of animal production education in Pakistan is the absence of breeders' organizations in the country. Some NGOs have organized village communities but the main binding force is the availability of micro-credits and some infrastructure development. Therefore, there is no emphasis on capacity building of village organizations in animal production education. The few organizations, visible otherwise are weak and lack capacity to bring any tangible change. Another few are single person shows. The other side of the picture is that government wants them to remain weak so that they have little say in its affairs.

23.4.4 Way Forward

Animal production education and extension services can be improved through the following measures:

- Producing the right manpower: Need for producing right manpower for animal production education must be realized. Correcting the direction of provincial livestock departments to develop a real infrastructure, with major focus on animal production education is the first step in this direction. Without changing the veterinarian-dominated mindset, it is not possible to produce the right manpower that can deliver information on animal production.
- Development of an independent infrastructure for animal production extension: An independent infrastructure is required with the sole responsibility of creating awareness among farmers on animal production aspects. This infrastructure must be provincially controlled and have offices throughout the province. Within this infrastructure, all modern means of communication should be employed to reach the farmers. The role of information technology in this era may be more than any other means of communication.
- Specialized institutions for farmers training: There should be specialized training institutes for capacity building of livestock farmers. This role can be assigned to the existing para-veterinary and extension training centers.
- Role of universities in extension service and capacity building: Universities are already publishing extension material on livestock production. Besides, universities should be involved in capacity building of livestock farmers through short training courses of one to two weeks duration depending on the farmer's training needs. They may act as catalysts but should not try to replace the government setups.
- Livestock shows, farmer days, seminars, workshops: Livestock shows, farmer days, seminars, workshops are effective means of education and capacity building. Livestock shows play a major role in breed improvement and knowledge sharing among farmers. Farmer days, seminars, workshops organized by provincial livestock departments, farmer organizations, and universities are very effective means of extension services and capacity

of farmers. They should regularly be organized. In milk and beauty competitions, however, discouraging "animal rentals" is necessary, that is, owners should be real. Furthermore, show records are least desirable in breed developments as they are "environmentally induced." Now with enough know-how spread by these shows, animals with real genetic potential for lactation performance, for example, should be decorated.

- IT-based learning: Information technology is the cheapest and fastest way of information delivery in the present age. As described earlier, information technology as a single source or combined with other means will take a leading role in extension services in the years to come. We need to concentrate on how to use this technology effectively and efficiently for animal production education and for capacity building of all stakeholders.

23.5 ANIMAL WELFARE ISSUES

The situation on animal welfare is very poor in the country. Horses, donkeys, and mules are the most mistreated animals. Animal welfare is rejected by the majority considering it a nonissue. Different metropolitan bodies kill stray animals by poisoning or shooting them, which is costly as well as inhumane. Such pitiless practices are prohibited under the act (the Prevention of Cruelty to Animals Act, 1890). Controlling stray dog population and vaccinating them against rabies to eradicate the danger of virus transmission is an achievable target and it's simply an issue of government priorities and esteeming life. The administration can build facilities for this purpose or use the current ones. The private backing ought to likewise be looked for. Such an activity would likewise give a chance to veterinary students to learn and harness their abilities.

The Prevention of Cruelty to Animals Act 1890 made it a criminal offense to be brutal to animals and kill them with cruelty. Punishments range from Rs 50 to Rs 500 and detainment sometimes are a part of the act. The 1890 demonstration was altered thrice by the British, however, progressive governments in Pakistan did not change the law. It applies to all the four provinces and to Gilgit-Baltistan since 1959 and to Islamabad since 1981. The punishments, however, remain the same and the law has, thus, lost its impediment impact. Provinces should, therefore, establish new laws and entrust its enforcement to either the wildlife or livestock department. The 1890 act particularly manages domestic animals and it is imperative that governments update the law as per the present necessities so as to cover all parts of animal abuse.

Another important issue is slaughtering of meat animals. Millions of animals are slaughtered for food annually in the country. The slaughter process begins most often with food animals crowded in inadequate trucks with little protection from the elements and transported long distances without water over harsh roads. Since the distances involved often are quite substantial and the management of the animals during this process is poor, transportation has deleterious effects that result in significant food losses. Moreover, slaughter facilities generally lack sanitation facilities or veterinary care in the country. Some traditional methods of handling, processing, and marketing of meat undermine quality, and poor sanitation leads to considerable loss of product as well as to the risk of food-borne diseases.

A high percentage of animal slaughter takes place in rural areas under very primitive conditions that do not meet even minimal technical and hygienic requirements. Because of the level of bacterial contamination, meat produced under such conditions can deteriorate easily and lead to food poisoning. Many of the large-scale slaughterhouses in the country are in poor condition. These are usually located in or around large cities. Even the abattoirs that have been designed specifically to supply meat to the expanding centers of urban population, all too often are unsatisfactory from a hygienic viewpoint. Once the meat leaves the abattoir, its hygienic quality also is influenced by careless and poor handling. Carcasses, quarters, unwashed offal, and other items are placed together on the floor of the market or on dirty concrete or wooden tables in meat shops, increasing the microbiological contamination of the meat.

REFERENCES

Asim M. 2015. Semen quality and pregnancy rate using artificial insemination in Lohi and Kajli sheep breeds. *MSc (Hons) Thesis*, University of Agriculture Faisalabad.

GOP (Government of Pakistan). 1996. *Livestock Census 1996*. Agricultural Census Organization, Statistics Division, Government of Pakistan, Lahore.

GOP (Government of Pakistan). 2016. Pakistan Economic Survey 2015–16. (www.finance.gov.pk)

Khan M.S. and A.O. Okeyo. 2016. Judging and Selection in Beetal Goats. GEF-UNEP-ILRI FAnGR Asia Project, University of Agri., Faisalabad, Pakistan.

Market Development Facility (MDF). 2015. Inclusive sector growth strategy. Leather.

Muhammad K. 2008. Leather sector crises in Pakistan. *J. Agric. Res.*, 44(3):229–236.

Saif O.B. 2012. *Leather Sector Analysis*. Pakistan Inst. of Trade and Development. Ministry of Commerce. Pakistan.

Zaman Q. 2006. Diagnostic study of Korengi tanneries zone (Korengi Leather cluster). UNIDO (United Nation Industrial Dev. Org.).

24 Poultry Production
Status, Issues, and Future Prospect

Pervez Akhtar, Umar Farooq, and M. Sajjad Khan

CONTENTS

1. *Short-term policies*: Diseases control, improving marketing channels, formation of rules and regulations, establishment of poultry export zones, formation of poultry export and import policy, development of markets for other than chicken breeds, consumer education/awareness.
2. *Medium-term policies*: Increase poultry processing, explore new feed ingredients. Production of bio-tech grains, Implementation of rules and laws.
3. *Long-term policies*: Development of commercial and rural chicken breeds, waste management, value addition, consumer education, research and development, and disaster management.

24.1 PREAMBLE

Poultry production is one of the most vibrant sectors of agriculture in Pakistan and it is playing a key role in providing quality food to the nation at reasonable prices. In Pakistan, poultry production has emerged through two management systems, that is, (i) commercial poultry and (ii) rural poultry. Before the 1960s, however, rural poultry was the only source of meat and eggs in the country. But because of the fact that rural chicken were only capable of producing around 0.7 kg of meat in 4–5 months and on an average 30 eggs a year (Sahota and Bhatti, 2003a,b), it was not possible to fulfill the nutritional needs of the ever growing population employing local breeds for production. So exotic chickens specifically produced, bred, and selected either for meat or eggs were imported in the country and laid the basis for the commercial poultry production. The commercial poultry production in Pakistan although led by the private sector was established through public-private partnership. The first commercial chick hatchery was established by the Pakistan International Airline (PIA) in collaboration with Shaver Poultry Breeder Farms from Canada, and the first poultry feed mill was established by the Lever Brothers (Pvt.) Ltd. Nevertheless, after introduction of exotic breeds, the industry grew rapidly over the years; in the early seventies the annual growth rate was 20% to 30% but the pace slowed down to 10%–15% per annum in the eighties, and then sustained its current annual growth rate, that is, 8%–10% (GOP 2014, Economic survey 2015–2016). The history of development of the poultry industry has been extensively discussed previously (Hussain et al. 2015). Nonetheless, a brief overview is given in Table 24.1 for a better understanding of the driving forces (i.e., government policies, developments, and constraints) of the poultry sector growth in the past. Table 24.1 refers to the depression in growth during certain periods due to overlooking of certain problems related to intensive production, both on part of industry and government. Initial developments had been concentrated in specific areas without taking consideration of environmental changes, availability of raw/quality feed ingredients (i.e., most of the feed mills underproduce due to limited supplies of ingredients), marketing issues (i.e., commercial chicken meat and eggs were perceived artificial by consumer compared with Desi/indigenous products), and disease risks causing overall production decreases in terms of number of birds, meat, and total number of eggs.

Although limited supply of feed ingredients and marketing issues had slowed down the pace of growth, the emergence of diseases remained a continuous feature and limiting factor in growth of the industry such that outbreak of hydro pericardium syndrome in broilers and broiler breeder flocks in 1990's, infectious bursal disease in broilers, parent stocks and layer flocks in 1991, and avian influenza (bird flu) later in the periods of 1995, 1999, and 2006–2007 etc. are some of the examples in this regard. The avian influenza did not caused much lose in terms of mortality but the damage done to sales and consumption of poultry products was huge as demand for poultry products fell down to one of the lowest levels in the history of the industry. Due to avian influenza in 2004, a 40% drop in production was observed, and the 2007 attack forced shutting down 50% of the breeder farms in the country, which resulted in a significant reduction in day old chick supply. Pakistan was hit twice by the highly pathogenic avian influenza strains H5N1 and H7N3, and low pathogenic avian influenza strains H7N3 and 22 (FAO, 2011, Hussain et al. 2015). In addition to disease outbreaks, other factors which contributed to slowing down the pace of industry such as natural disasters including the earthquakes and flooding which did unprecedented damaged to industry.

The government played pivotal role in the early development of industry. Briefly, many favorable governmental intervention such as suitable liberal financing and credit, income tax exemption, duty free import of grandparent and parent flocks, machinery related to poultry farms and hatcheries, and import of raw feed ingredients such as soybean meal, sunflower, and maize helped the industry survive and grow rapidly over the years. In addition, establishment of poultry research institutes in 1978 in Karachi and Rawalpindi, with assistance from UNDP/FAO funds, provided research services to poultry farmers and the creation of the Federal Poultry Board in 1979—to establish an effective link between industry and government—was also useful. It is worth mentioning despite favorable government interventions, unfavorable policies caused tremendous damages too; for example in 1996, imposition of a sales tax on poultry feed led to closing of around 40% sheds as the cost of production became too high. Other example in this regard are the marriage act of 1997, which banned the serving of food at marriages, and that led to drop in production by almost 40%.

Despite the fact that many efforts and huge investments from both government and private sectors have been made to help the industry grow well over the years, no concrete efforts are carried out at present to counter those issues which have been major causes of poor growth in the past. So, similar problems persist today, that is, disease outbreaks and their prevalence, shortage of raw feed ingredients (i.e., grains, etc.) for feed milling, the challenges of shortages of land, water resources, energy/fuel, high cost of production, unstable market conditions, global warming, and disaster management. These problems will grow in the future and if the poultry sector wants to flourish it must develop and work through a comprehensive plan keeping in view these future needs, demands, supplies, and constraints.

24.2 DYNAMICS OF POULTRY PRODUCTION

The current contribution of the poultry sector in the total meat production and overall GDP is 30% and 1.4%, respectively, while its share in agriculture and livestock value added products is 6.9 and

TABLE 24.1
Development of Poultry Industry: Policy Role and Constrains

Phase	Description	Developments	Government Role	Constraints
Phase-I (1965–1970)	Introductory phase	Establishment of chick hatcheries, feed mills, easy availability of drugs and vaccines	• Series of major policy decisions • Permission to import parent stock • Poultry farming income was made income tax free • Lease of state owned land for poultry farming • Meatless days for red meat • Not serving red meat in parties with more than 150 guests • Establishment of directorate of poultry production for extension services	• Shortage of grains for feed milling • Lower market price for poultry products due to consumer biasness for indigenous Poultry products
Phase-II (1971–1975)	Boom phase	Establishment of animal research institute, increase in production followed by export of poultry products	• Exemption of sales tax • A ban on poultry export	• Discontinuation of poultry exports • Continued consumer resistance
Phase-III (1975–1980)	Institutional development	Establishment of institutions in Government sector, and Pakistan Poultry Association in private sector	• Federal Poultry Board • Poultry research institute • Imposition of income tax	• Limited supply of feed ingredients • Marketing issues • Disease issues
Phase-IV (1980–1985)	Depression phase	Decrease in overall growth, and relocation of the industry		• Climatic stress • Disease issues • Marketing issues
Period from 1991 to 2000	Disease out breaks and re-emergence of industry	High prevalence of disease such as hydro pericardium syndrome, Gumboro	• Conditioned withdrawal of Income tax • Imposition of sales tax and withdrawal • Tax levy on import of parent stock, raw material, and machinery • Imposition of 10% electricity tariff on farm and hatcheries • Prohibition of wasteful expenditure on marriages ordinance	• High cost of production • Halt modernization
Period from 2001 to 2016	Modernization and investment	Adoption of poultry controlled housing technology	• ZTBL sada bahar loaning scheme for poultry • Zero custom duty on poultry vaccines, feed additives, and coccidiostates • Announcement of poultry export zones in Karachi and Faisalabad • Establishing of Avian Influenza laboratories • National program for the control and preservation of Avian Influenza • Poultry development policy • Zero Percent custom duty on poultry farm, hatcheries, and feed mill equipment • Sales tax exemption to uncooked poultry meat • Credit assistance for poultry under rural development scheme • Duty free import of poultry processing machinery • Lift of ban of export to Saudi Arabia	• Natural disasters e.g. flood and earth quake • Disease out breaks • Band on export • Limited supply of feed ingredients

11.7%, respectively (Economic Survey 2015–2016). The current poultry status is given in Table 24.2 while provincial statistics are given in Figure 24.1. At present over 190 billion rupees worth of agricultural produce (including 70% of total grains produced in the country) and agro-byproducts are being used in poultry feeds (PPA, 2016). Poultry is the largest consumer of oil seed meals, wheat bran, rice polishing, corn gluten, broken rice, guar meal, animal byproducts, and fishmeal in the country. The consumption of over 70% of grains though create significant competition between humans, poultry farming and other livestock farming, yet use of agro-byproducts of these grains in poultry feed makes a significant contribution to the reduction in prices of other human foods such as rice, wheat, and sugar, etc. However, at present the proportion of agro-byproducts in relation to whole grains in poultry feed formulation is pretty low and needs to be enhanced in order to decrease poultry feed costs. Therefore, future efforts should focus on increased used of agro-byproducts in the feed through advancement in knowledge of food science, nutrition, and biotechnology. This will not only decrease the competition for grains between human, other livestock and poultry but will also help making grains available at economical prices for human consumption.

The poultry is the second largest industry in Pakistan, though contrary reports are available on the overall investment in this sector as the official and nonofficial statistics vary significantly, that is, Rs. 200 billion versus Rs.750 billion per annum, respectively (Economic survey, 2015–2016; PPA, 2016). The variation in figures, however, could be due to the methods of estimation adopted in the valuation process, for example, economic surveys rely on statistical calculations from the 2006–2007 census (Economic Survey of Pakistan, 2006–2007) rather than real time data while nonofficial data may be exaggerated as per the industry's perspective. However, it is obvious that over the past 10 years lots of new investments have taken place in this sector hence, the worth and potential of the industry is much

TABLE 24.2
Current Status of Poultry Production in Pakistan (Economic Survey vs. Pakistan Poultry Association)

Type	Units	Economic Survey 2015–2016	Pakistan Poultry Association Statistics
Domestic Poultry	Million nos.	**84.6**	
Cocks	-do-	11.2	
Hens	-do-	40.9	
Chicken	-do-	32.4	
Eggs	-do-	4,090	
Meat	000 tonnes	115.2	
Duck, Drakes & Duckling	Million nos.	**0.46**	
Eggs	-do-	20.3	
Meat	-do-	0.6	
Commercial Poultry	Million nos.	**56.9**	
Layers	-do-	45.6	60
Broilers	-do-	874	1404
Breeding stock	-do-	11.2	2.225
Day old chicks	-do-	912.9	1560
Eggs	-do-	12,077	18,000
Meat	000Tonnes	1,054	2,250
Total Poultry			
Day old chicks	Million nos.	945	
Poultry birds	-do-	1,016	
Eggs	-do-	16,188	
Poultry meat	000Tonnes	1,170	

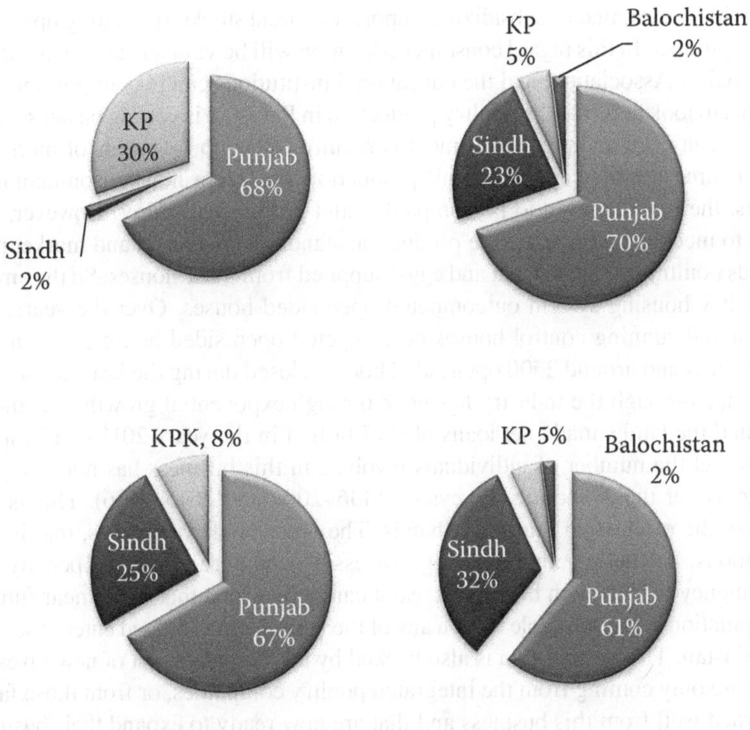

FIGURE 24.1 Province wise distribution of Broiler breeder, layer breeder, commercial broiler, and layers.

more what might be anticipated for policy interventions and decision-making relying on the figure provided by economic surveys. Such discrepancies in statistics may result in over citation of certain issues and undermine certain other problems that might be associated with the future survival and growth of the industry. Therefore, exact statistics need to be established. For this purpose, industry representative bodies and the government may work in collaboration with educational institutions to work out long term and comprehensive data reporting and processing strategies in this regard.

Commercial poultry in Pakistan encompasses chicken broilers for meat and chicken layers for egg production. The term poultry includes many species, yet except chicken, none of these species have been commercialized on the scale of chickens in Pakistan. Other poultry species like turkeys, ducks, and quails, which are extensively grown on a commercialized scale across the globe have limited commercial production in Pakistan. The reason behind such discrepancy is unclear. Nevertheless, the possible causes might be the personal liking/disliking and the difficult and risky scenario of creating markets for new poultry species in Pakistan. This might be true as it has been the case with commercial chicken production in early periods of development in the country where its consumption was hampered by personal myths and lack of information about processes of artificial incubation, hatching of chicken, feeding, and feed manufacturing. Certain myths about broiler meat and commercial chicken eggs exist even today such as broiler meat is produced by supplementation of hormones and steroids, broiler can hardly stand on its feet, walk and run so it produces nutritionally low quality meat and eating of poultry meat and eggs in hot weather conditions is not good for health etc. Despite the difficult scenario of entry of other poultry species to commercial production in Pakistan, species like turkeys, ducks, and quails have great potential to bridge the gap of protein deficiency in the country. These species have some advantages over chicken such as they have a better ability to survive against some of the most prevalent diseases. Nevertheless, turkeys and ducks have their presence in rural production where they mostly feed on kitchen wastes, while quails are produced commercially on a limited scale. There is also great potential for export of these species: therefore, there is a need to get these species in the commercial production system. The public sector can play leading role in this regard, that is, subsidizing imports of parent stocks and setting up of support prices for the finished product. In this regard consumer education will be vital and as well local governments, the Pakistan Poultry Association, and the educational institutions can play important roles.

Another area to look at is that the poultry production in Pakistan is getting business of the rich with passing years because lot of initial investment is required to coop the needs of modern production system. About three decades ago, when all production was from nonenvironmentally controlled poultry houses, the markets used to be compatible and profits were fairer. However, with the shift of production to mechanized houses, the production standards increased and market preference got skewed towards poultry products (meat and eggs) supplied from these houses. So the environmentally controlled poultry housing system outcompeted open sided houses. Over the years, the high cost of construction and running control houses outcompeted open sided house communities from the production business and around 2500 open sided houses closed during the last decade. It is also clear from the fact that although the industry has gone through exponential growth over the last decades (Table 24.3), and the banks made net loans of Rs.7 billion in the years 2011–2012 especially in the farming sector, yet the number of individuals involved in this business has not changed much (1.5 million people as per the economic surveys of 2006–2007 and 2015–2016). This is a situation of consolidation of the production into fewer hands. The other poultry ventures, that is, feed milling, breeder operations, hatcheries, and meat/egg processing units are already expensive business that require lot of money to start with businesses. So it can be foreseen that in the near future more than 95% of the population will not be able to own any of the "production" related enterprises of the poultry business in Pakistan. This assumption is also backed by the fact that most of new investments in the poultry sector are only coming from the integrated poultry companies, or from those farmers having previously earned well from this business and that are now ready to expand their business. Some of the investments are from abroad in the form of technical support to poultry feed millers, hatcheries, and disease control to some extent. So future poultry production is going to be in the hands of the elite or upper middle class of the country.

Current analysis of industry reveals three type of investors, (*i*) the main poultry stakeholders/investors which are the main source of investment in the industry and (*ii*) fresh investors which only do secure-investments and leave when circumstances are not favorable—but their number are small; and (*iii*) the investors who are involved in allied fields, that is, sale, purchase, and distribution of feed ingredients, medicine, vaccines, and feed additives suppliers, etc. This scenario demands careful policy interventions for future production where the areas which increase cost of production should be subsidized as most of the consumers will be the low income communities. Taxation may be exercised on income generated from the poultry enterprises. Future policy making should

TABLE 24.3
Growth of Poultry Sector (2006–2016)

Category	Unit	2006–2007	2015–2016	Percent Increase[a]	
Day old chicks	Million no	401	912	127	127
Layers (farming)	Million no	23.8	45	89	91
Broiler (farming)	Million no	315.8	874	177	176
Breeding stocks	Million no	7.1	11.2	58	57
Poultry meat production	000 tonnes	480	1054	120	119
Egg production	Million no	5222	12077	131	131
Human population	Million no	161.5	189.8	18	17.5

[a] Calculated from difference of Economic Survey 2015–2016 and 2006–2007.

focus on the nature of the business and the communities involved in this business in order to make comprehensive strategies for sustainable growth of the industry.

In order to achieve a suitable policy framework, however, there is also need to understand the commercial chicken production process and related issues. For example commercial broilers are produced through so-called four way crosses that essentially means that the broilers are derived from four different grandparent lines including pure lines or great grandparents, grandparents, parent flocks and at the end commercial broilers. Each step in production, from the bottom to the top, is expensive, sophisticated, and highly technical. Pakistan import grandparent stocks and rest of the crosses are carried out in the country. Any problematic flock in the chain, or any extra expense on top hierarchy (such as taxes, duties or any other costs), increases cost of production to the bottom level. This is clear from the fact that one grandparent meat type female is the source for production of around 3600 broilers or 3.78 metric tonnes of meat while one egg type female on the top level is a source of production of around 12,000 eggs. Therefore, the topmost birds in the pedigree are much costlier, hence running these enterprises are highly expensive. The other major areas of poultry investment are the poultry feed manufacturing and chick hatcheries. The policy decisions required to take into account that the initial costs on construction and purchase of equipment are not the only costs involved, and the running expenses are much more sensitive in determining the cost of the finished products (i.e., chick and feed). The running costs involved repair maintenance, purchase of raw material, electricity, duty, and taxes, etc. The chick hatchery business heavily depends on availability of electricity and shortfalls may increase cost exponentially.

In addition, feed is one of the most expensive commodities for poultry production. The demand for grains will grow in future with increases in human and livestock/poultry population. Therefore, in addition to taxation policies, there is also need to develop mechanism for responsible production where growers will share responsibility of replenishing resources such as land, production of grains, water, and other liabilities. Similarly, poultry production and its supplies will not be risk free unless local meat and egg producing breeds are developed. Simply relying on import of exotic chicken resources for meat and egg production will not be enough. Strong policy work will be required for development and propagation of local breeds for commercial production.

24.3 MARKETING ISSUES

The focus of this section is on marketing issues related to poultry production. The reason to focus on marketing issues is that the current marketing system, due to involvement of intermediaries, is one of the major factors responsible for low profit margins. Currently, only 5%–6% broilers are sent to proper slaughter houses for processing to sell as fresh/frozen meat or value added ready to cook meat products. More than 95% of broilers are transported live to wholesale markets through middlemen (Arthi) and wholesalers, and then they are brought into the cities and sold to the retailers. The marketing channels and cost and profit margins from farmer to retailer are given in Figure 24.2 (Cholan, 2007). It indicates that from commercial farmers to the consumer, the chicken go through commission agents and retailers who fetch maximum profits on minimum investment. The high profit margins on the retailer's end deprive both farmers and consumers from getting fair prices for their products. Another ambiguity in this type of marketing system is that the chicken are transported to long distances and the vehicles which are used for transportation are not equipped with environmental control system, so the transportation losses are high (0.25%–1% chicken die, some get injured, traumatized, or muscle fatigued, etc.) (Mukhtar et al. 2012).

For development of market price control mechanisms for poultry products one needs to keep in mind that poultry are perishable products. They cannot be held back or stored longer once they reach the marketand. Due to poorly established marketing channels the prices are highly exploitable. Generally, meat demand decreases in the months of Zul Hajj and Safer of the Islamic calendar, while in months of Muharam and Ramadan there may be fair prices, and then in months of Shaban, Shawwal, and Rabi-ul-awwal the demand will be high because of the wedding season. Moreover,

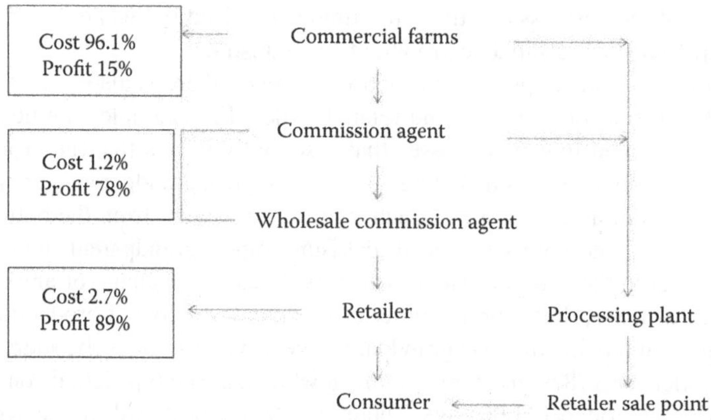

FIGURE 24.2 Marketing channels cost and profit margins from farm to retailer.

the climatic changes, summer and winter, per capita income, and price of supplementary foods such as beef, mutton, and fish also affect supply and demand of poultry meat and eggs. The poultry egg prices go down in summer due to consumer perception that eggs are not good to eat during summer and that the schools holidays are observed during summer. Some other factors in this regard are natural disasters that can kill birds or damage infrastructure, and disease outbreaks such as avian influenza in 2004 and 2007 etc.

Other issues are that there are few functionaries that control the price and create conditions where farmers get forced to sell their products at maneuvered prices. Furthermore, unsuitable conditions of low market prices, any disease condition, and lower than normal weight of birds at market age may lead to increased commission charges by the middlemen. Unluckily, there is no agency to check such unfair commission rates. Besides missing direct linkages between producers and consumers and lack of investment to develop infrastructure are some of the other reasons for lower profits at producer's end.

In the recent years, the government has decided to intervene in price control by fixing poultry meat and egg prices; However, this has so far not contributed to price stability positively as demand and supply fluctuate significantly. This is clear from the sale prices over the last 10 years for meat and eggs (Figures 24.3 and 24.4). As per these figures, the increase in prices from 2006–07 to 2010–2011 were high because of the attacks of avian influenza in 2004 and 2007 which reduced the supply and demand initially; in the later period demand went high but the supplies took time to fill the gaps. From 2010–2011 to 2015–2016, nevertheless, the percent increase in egg and meat prices were very

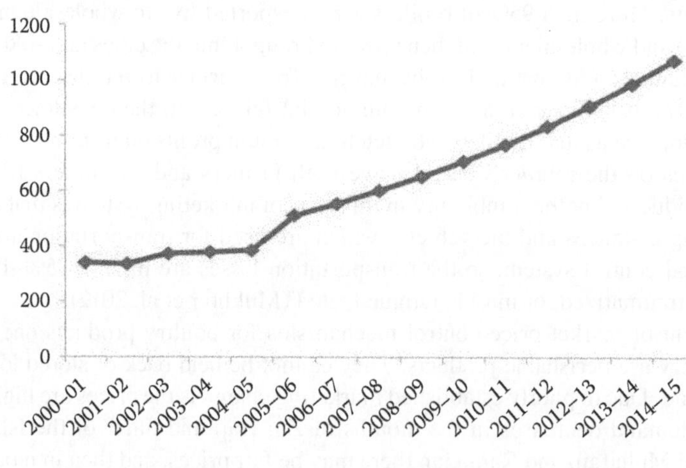

FIGURE 24.3 Yearly increase in meat production (000 tonnes).

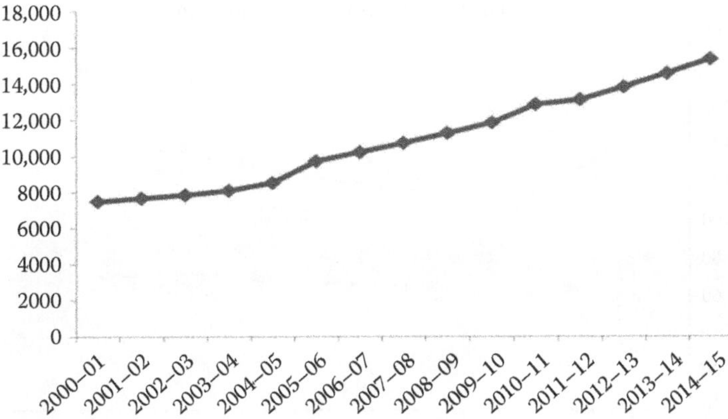

FIGURE 24.4 Yearly increase in egg production (million Nos.).

less than the previous 5 years (Figures 24.5 through 24.8). This might be attributed to the fact that the industry consolidated during this period and overall farm size and production increased due to advancement in technologies, for example, environmental controlled poultry housing. During the same period, per capita income increased wich not only increased demand for poultry, meat, and egg in big cities but also in small towns and villages playing an important role in balancing supply and demand of poultry products. Moreover, use of poultry meat in fast-food and in the local restaurant chains increased tremendously during this period and triggered this growth. Nevertheless, the early increase in production led to overproduction during the middle period due to which farm rates for meat and eggs dropped and resulted in the closing down of around 40% of the farms in Islamabad, Pothohar, and Rawalpindi regions. This situation can be analyzed from the fact that in April and May 2016 chicken meat sale prices hit 300/kg mark and this was due to the shortage which were created by low prices in the previous months. The farm rate of 1-kilogram live chicken for selected months, that is, July-2015 to May-2016 is given in Figure 24.9 (PPA, 2016). The lower than meat production cost for most of the months forced the shutting down of many farms resulting in shortages observed in the later months (PPA, 2016). So The government interference for price control of poultry finished products will not be fruitful unless a different system of marketing (i.e., through slaughter house and cold chain storage)

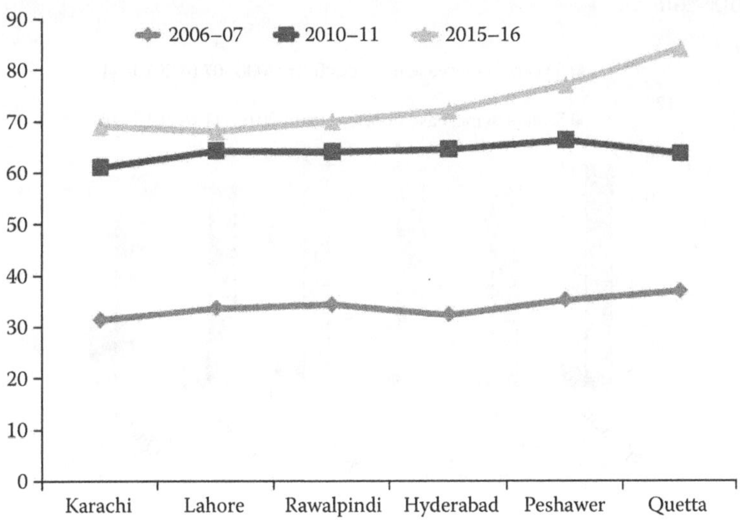

FIGURE 24.5 Yearly increase in wholesale per dozen egg prices in big cities of Pakistan.

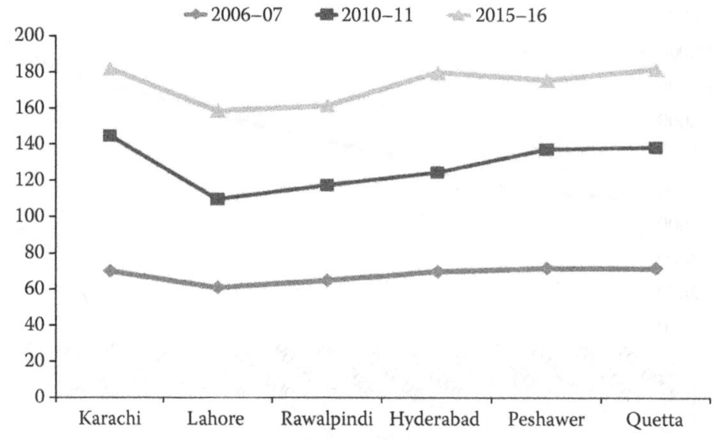

FIGURE 24.6 Yearly increase in wholesale meat prices in big cities of Pakistan.

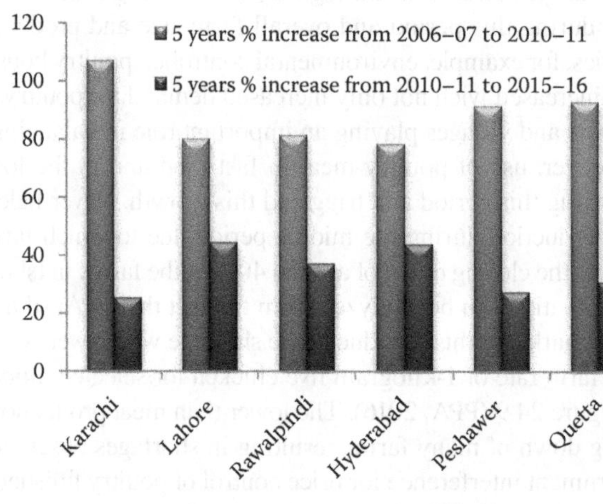

FIGURE 24.7 Comparison of increase in egg prices in big cities between 2006–2007 to 2010–2011 vs 2010–2011 to 2015–2016.

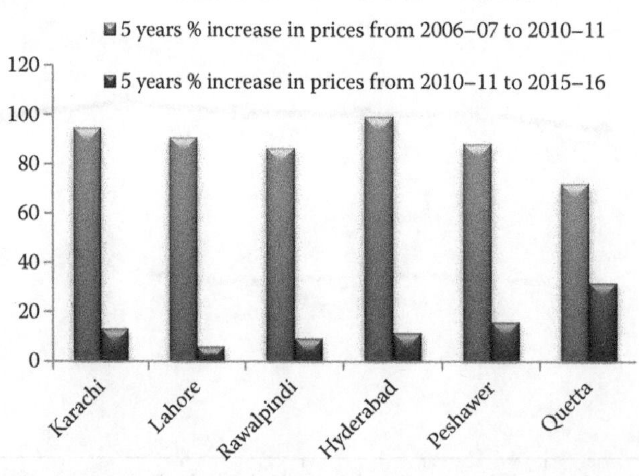

FIGURE 24.8 Comparison of increase in meat prices in big cities between 2006–2007 to 2010–2011 vs 2010–2011 to 2015–2016.

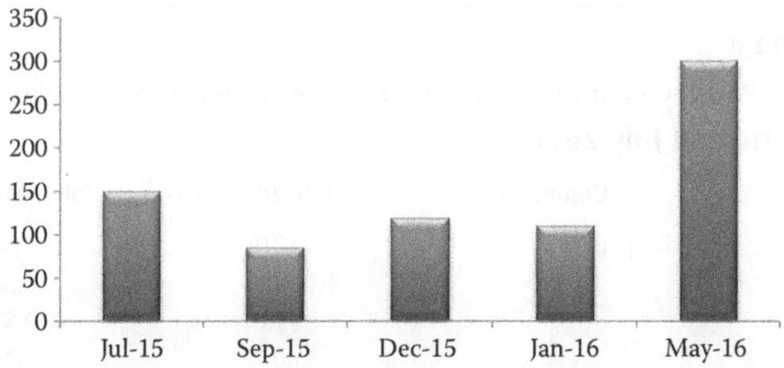

FIGURE 24.9 Fluctuation in per kg live weight broiler market prices (selected months) from July-2015 to May-2016.

is introduced, and contributing factors at each level of production that is hatchery, feed mills, poultry farms, and the factors that contribute to demand and supply are addressed.

24.4 POULTRY MEAT EXPORT

Pakistan produces 1170 thousand tonnes of poultry meat annually. The sources of poultry meat are rural chicken, commercial broiler, spent layers, parent stocks, ducks, ducklings, and drakes. Ducks, drakes, and ducklings are, however, contributing 1% to the total meat production while 99% of the poultry meat comes from chicken (Economic survey 2015–2016). The annual growth of poultry meat is 8.2%. Yearly increases in chicken, mutton, and beef production from 2012–2013 to 2014–2015 is given in Figure 24.10. The figure showed that, though there has been a steady increase in production of meat from all three sources, the rate of increase was higher for chicken meat production for the year 2014–2015 compared with beef and mutton. The poultry meat processing was first introduced in Pakistan in 1980 where a poultry meat processing plant of 250 birds per hour was established in Poultry Research Institute in Rawalpindi with the help of United Nation Development Program. Soon after, two major companies started their meat processing plants with a capacity of 6000 bird per hour, soon followed by 5 to 6 other small companies. However, this effort could not grow further as there was limited consumer acceptance for frozen/processed poultry meat in Pakistan. At present, around 176 abattoirs are operating in the country but are mostly involved in processing meat of large animals.

The current worth of the global halal meat product market is 2.1 trillion dollars, 16% of the total world trade. Australia, Brazil, Canada, Indonesia, India, Malaysia, Philippines, Thailand, New Zealand and the United States are the world's leading suppliers for Halal meat products. Among these, Brazil and France are the biggest Halal chicken exporters in the Middle East. Pakistan's share is only 2.9% in the global meat export and that is mostly from beef and mutton. Since 2014 exports of beef and mutton has shown positive growth while on other hand the poultry export

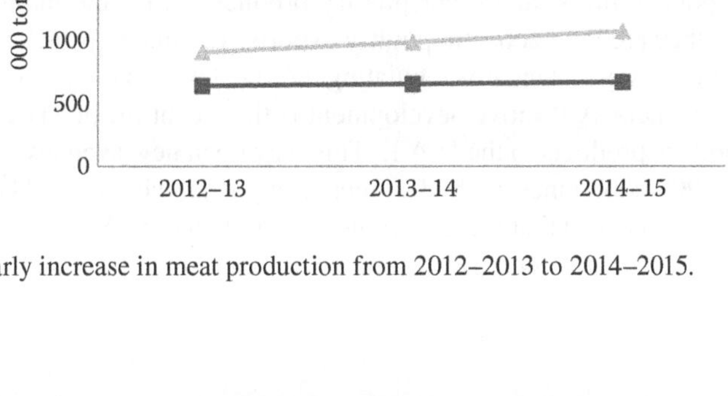

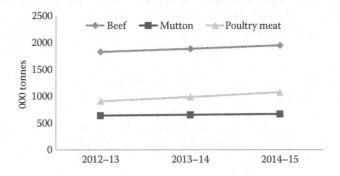

FIGURE 24.10 Yearly increase in meat production from 2012–2013 to 2014–2015.

TABLE 24.4
Pakistan Poultry Products Exports (Million Rupees) for the Years July 2013–2014 and July 2014–2015

Country	Commodity	July 2013–2014	July 2014–2015
Afghanistan	Live poultry	219	214
Malaysia		–	0.0025
Oman		1.07	2.8
Tajikistan		0.7	1.2
Afghanistan	Chicken weight up to 185 g	0.38	–
Bangladesh		5.0	–
Oman		–	0.77
Tajikistan		–	1.23
Afghanistan	Fertilized chicken egg	50.9	0.48
Jordan		9.07	74.3
Oman		32.5	75.81
Saudi Arabia		–	211.5
Tajikistan		–	2.667
Yemen		–	1.788
Afghanistan	Fresh eggs	184.9	2.372
Oman			1.938
Total		504.6	591.7

has shown a negative trend decreasing from Rs. 2.08 billion in 2010–2011 to Rs. 365 million in 2011–2012. Country-wise exports of chicken products from Pakistan for the years 2013–2014 and 2014–2015 are given in Table 24.4. Pakistan major exports have been to Afghanistan and to the Middle-East; however, there exist many more opportunities as the global halal market is growing at rate of 6.9% annually. The global Muslim population is around 1.83 billion but get more than 80% of their meat supplies from nonMuslim countries. Another future market for export of chicken meat is China as its meat imports are increasing steadily (imports have grown from 1.7 billion in 2009 to 6.1 billion dollars by 2014). Pakistan should explore this market though the China-Pakistan Economic Corridor.

Pakistan produces meat that is trusted for being Halal in Muslim countries, hence there exists great opportunities for export of poultry products but there is need to take concrete steps to avail these opportunities. Some sound steps have already been taken by the government with the establishment of the Pakistan Halal Development Agency and by providing some exemptions of import taxes on slaughter house related equipment and on income. However, the private sector should take the lead in investing in poultry slaughterhouses and exports. The poultry sector in the country is well organized and chicken production standard are not less than in any other advanced country. Some of the companies engaged in poultry exports are now meeting all international standards. A few others are primarily involved in red meat exports and have the capacity to diversify their business into poultry exports. In 2010, about 20 trading companies in the country were involved in exports of grains, red meat, poultry birds, and other poultry products. Now, the number has risen to 70 (industry sources) but they are least active in poultry exports. The main obstacles are: the absence of a centralized certification system, standards regulating the Halal meat industry, and awareness on part of the farmers and consumers. A Positive development is the recent lift of THE long-drawn ban on export of Pakistani poultry products to the U.A.E. This may open new avenues as Saudi Arabia only imports little less than 900,000 tonnes of chicken meat every year, which could be a good market for future exports. Thus, there is need that the government should take steps to increase poultry exports and encourage the private sector to step forward in meeting international standards for exports.

24.5 RURAL POULTRY PRODUCTION

Almost every rural household raise domestic poultry. Rural poultry consist of 84.5 million birds and its contribution to total egg and meat production in Pakistan is 25% and 9.8%, respectively. The comparison of rural and commercial chicken egg and meat production is given in Figures 24.11 and 24.12, respectively.

Indigenous chicken is predominantly Desi (local), mainly raised as a village scavenging bird. Among desi, the most important birds include Aseel that are raised on a high input system as a pet/game bird and its major utility is cock fighting and meat. There exist few dozen strains of Aseel depending upon their color and fighting patterns. Aseel's potential for egg production is 20–40 eggs per year (Usman et al. 2014). In order to improve rural chicken production, new breeds, that is, Fayoumi and Rhode Island Red were introduced in Pakistan and were crossbred (Ashraf et al. 2003). The purebred were also raised in semicommercial setups or under rural subsistence setups for production of meat and eggs. Breeder farms and hatcheries were built around big cities with the focus to produce and distribute these chickens.

Under the breed improvement program, the first successful development was the production and distribution of Lyallpur Silver Black breed in mid 1960s by the University of Agriculture Faisalabad; however, due to lack of continuation in the research and development, this breed became eventually extinct. Recently, a new attempt has been successful in producing a egg laying backyard chicken at the University of Agriculture Faisalabad. "UniGold" is reported to have the capacity to produce more than 200 eggs per year under local conditions. The production data and status of some other of such breeds—developed under different projects, are given in Table 24.5. The data showed isolated and short-term efforts of breed improvement through cross-breeding, but the actual efforts on commercialized or distinct breed development were missing for serious consideration in future programs.

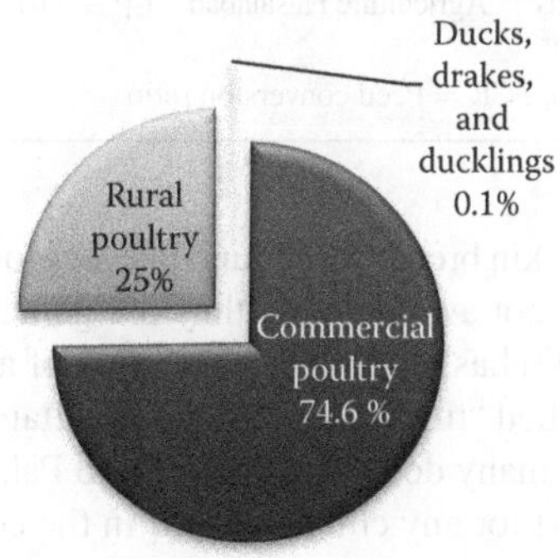

FIGURE 24.11 Share of rural and commercial poultry in egg production.

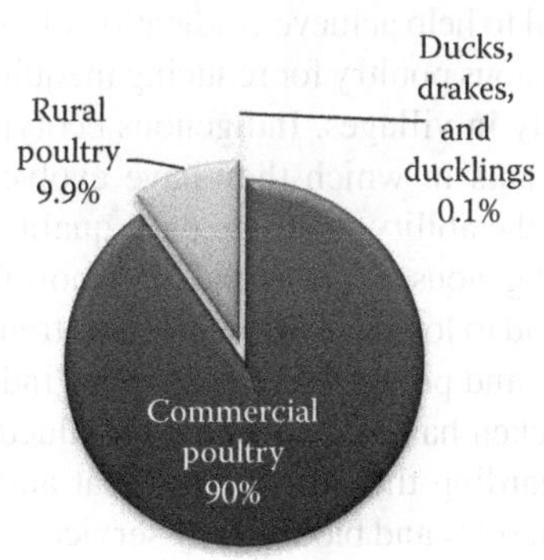

FIGURE 24.12 Share of rural and commercial poultry in meat production.

TABLE 24.5
Attempts for Local Breed Development, their Production Performance, and Status

Breed	Cross	Executing Agency	Production	Status
Lyallpur silver black	White Cornish White leghorn New Hampshire	UAF 1965–1966	BW = 1200–1350 EP% = 49 FCR = 4.96	Very few (Ashraf et al. 2003)
Necked neck cross bred	Necked neck White leghorn	Poultry and wildlife program, National Agriculture Centre NARC Islamabad	BW = 1461 EP% = 66	Reasonable number of chicken http://www.parc.gov.pk/index.php/en/contact-us
Necked neck × white leghorn	Necked neck × White leghorn	Pakistan Agriculture Research council	BW = 1210 EP (40 w) = 185 Viability = 92%	80,000 chicken were distributed in different district (http://www.parc.gov.pk)
Desi × White leghorn	Desi × White leghorn	Pakistan Agriculture Research council	BW = 1192 EP (40 w) = 190 Viability = 93%	
Rhode Island red × Fayoumi	RIR male × Fayoumi female and Fayoumi male × RIR female	Poultry research institute Rawalpindi	BW = 1213–1260 FCR = 4.6–5.60	Reasonable number of chicken (Khawaja et al. 2012)
(Rho de Island red × Fayoumi) × white leghorn	(Fayoumi female × RIR male) × White leghorn	Poultry research institute Rawalpindi	BW = 1253 FCR = 4.46	Reasonable number of chicken (Khawaja et al. 2013)
UniGold	Synthetic having Aseel and other local breeds	University of Agriculture Faisalabad	BW = 1200 g EP = 210	M.S Khan (personal communication)

BW = Body weight, EP = Egg production, FCR = Feed conversion ratio.

Domestic ducks (Mallard and Pekin breeds) are found in every town of the country. The breed-wise population estimates of ducks are not available but they are generally raised for subsistence. Food and Agriculture Organization (FAO) has reported a population of about 4 million ducks in Pakistan. Similarly, guinea fowls (locally called "titri") are common in villages and towns. Also, nineteen wild ducks species, many pigeons, and many doves are common to Pakistan (Roberts, 1991).

There are no conservation effort for any chicken breed in the country. It is generally agreed that the population of the naked-neck breed has decreased over the years mainly because eggs produced by commercial layers are more easily available. For species other than chicken, authentic population estimates are not available.

Rural poultry has great potential to help achieve eradication of extreme poverty and hunger. There is a great potential in the use indigenous poultry for reducing inequities in access to resources, income generation, and nutrition especially in villages. Indigenous ecotypes have unique combinations of adaptive traits to local environments in which they have evolved. As such, as they are disease resilience, heat tolerant, and have the ability to utilize poor quality scavenge feeds.

The poor productivity of the indigenous chickens and high mortality rates due to disease outbreaks and predation have increasingly lead to losses of diversity. This trend has also worsened over time by the negative perception of farmers and policy makers' towards indigenous chicken. In addition, the importance of the indigenous chicken has often been undervalued, with the focus mainly directed towards the apparent benefits regarding the quantity of meat and number of egg produced while limited attention given to the social roles and biodiversity services which are key factors in the in-situ conservation of indigenous livestock species as pointed out by institutions like FAO. An institution committed to the development and improvement of chicken lines, for commercial and rural poultry, needs to coordinate research and development on chickens with a repository of various breeds and ecotypes. Development of (and propagation of) new strains of indigenous chicken, such as UniGold, will be an important objective of the said program.

The universities and other public sector institutions have provided much needed support in the form of research, development, and trained human resources. They have provided much needed work to improve indigenous breeds and development of high producing crossbreds, yet due to lack of funding and noncoordinated efforts, there has been little success in breed development.

24.6 CONSUMER DYNAMICS, FUTURE TRENDS, AND PREDICTIONS

Pakistan is the sixth most populated country in the world with an estimated population of 195 million. The population growth is 1.89% (Economic survey, 2015–2016). The current per capita consumption of meat and eggs are 7 kg and 60–65 eggs per annum, respectively, which have increased steadily over the years. However, the developed world is consuming 41 kg of meat and over 250–300 eggs per capita per year. According to WHO the average daily requirement of protein is 27 g per person, whereas, in Pakistan the available protein per person is 17 g, out of which 5 g is from poultry; the gap here is around 788,000 tonnes of meat (Memon, 2012; Hussain et al. 2015). Pakistan falls among those nine countries where half of the world population growth is expected to come from during 2015 to 2050. The world food production by 2050, thus it needs to be raised by 70% of its current production (Johnson and Jorgensen, 2006).

The demand for meat and eggs in Asia and Pacific countries will increase by 56% from 2011 to 2021. By 2021, the consumers will depend on poultry meat to fulfill 68% of their protein needs (FAO, 2011). A considerable increase in poultry production will be required which will demand more resources. In the past, supply of grains has constantly been on the lower side of the demand. So to satisfy increasing and changing demands for animal foods, sustaining natural resources will be one of the major challenges faced by highly populated countries. The following future challenges may be anticipated.

- Future production systems will be more intensive with warmer, more humid, and more disease-prone environments.
- Poultry production will change from market-oriented to an increasingly integrated enterprise.
- Competition for common property resources such as land, grains, and water will increase.
- Large-scale industrial production will be associated with environmental and public health risks.
- Chances of emergence of new diseases will increase.

These challenges will raise crucial global and national public-policy issues which must be addressed. Therefore, there is need to realize and understand the changes likely to unfold and implement development plans accordingly.

24.7 WAY FORWARD

To meet the demand of egg and poultry meat, substantial improvement in productivity is required. It necessitates long-term strategies with all stakeholders on board. The commercial poultry sector is totally dependent on imported genetic resources in the form of parent and grandparent flocks. When investment in the sector is in billions, there must be a strategy to reduce dependency on imports of parent stock. Many developing countries have successfully produced their own breeds. For rural poultry, cutting edge technologies should be adopted to assist conventional practices. This will require genetic improvement of indigenous birds. A separate research and development institution should be established tasked to develop sustainable egg laying, meat producing, and dual-purpose breeds for backyard poultry. Emerging issues such as climate change, transboundary diseases, and environmental degradation, etc., can be better addressed keeping in view international obligations. The presently squeezed poultry science discipline must flourish to address the country's needs.

The following strategies and approaches are suggested for (i) commercial, and (ii) rural poultry production in the country:

1. Commercial setup
   a. Regulations for meat processing facilities: Special subsidies to promote and facilitate slaughter house operations.
   b. Being an agricultural activity, poultry should have similar taxation levies as other agriculture commodities.
   c. Regulate poultry marketing channels in order to minimize profits margins of middlemen. Long-term policies are needed to shift the culture from live market to slaughterhouse.
   d. Research for exploring new sources of raw ingredients for feed or up-gradation of low quality agro-byproducts is required. Feed millers may be bound to fulfill 70% of their grain's requirement by self-cropping through rehabilitation of barren/un-used land and also invest in research and development of grain production and agro-byproduct usage in poultry.
   e. Pakistan Poultry Association has to take responsibility of developing new commercial egg and meat type breeds for the country.
   f. Special incentive should be given for entry of other poultry species such as ducks, turkies, and Japanese quails in the commercial production chain.
   g. Establishment of hi-tech labs, specifically for local vaccine production, for various diseases of poultry. Production of biotech grains for disease prevention should be encouraged, Disaster management cells need to be established to help in disease outbreak management and other natural calamities.
   h. Establishment of poultry export zones can be helpful. Consideration should be given to export of processed meat, eggs, and other poultry products
   i. The government should increase funding for higher education in the field of poultry science and related fields, and universities should play their role in research and development.
   j. Implementation of rules and laws governing production of healthy and hygienic poultry feed and meat. Pakistan Poultry Association should promote consumer education on poultry meat and egg consumption.

2. Rural poultry
   a. Identification and characterization of indigenous chicken germplasm, both phenotypically and genetically, which is first step for conservation and effective utilization of indigenous genetic resources.
   b. Establishment/strengthening of poultry research institutes for development, propagation, and further improvement of specialized egg and dual type breeds for different agro-ecological zones.
   c. Development of standards/procedures for raising chicks, pullets, and adult birds under semi commercial and scavenging conditions. Push for capacity building of rural poultry farmers to improve production by learning poultry husbandry practices.
   d. Assessment of the potential of indigenous poultry genetic resources for both intensive and backyard farming.
   e. Systemic up-gradation in indigenous breeds through selective breeding for genetic improvement and conservation.
   f. Use of genomics and other biotechnologies for improvement of local chicken genetic resources. Development of technologies for production of antibiotic-free and organic poultry products, and nutritionally enriched meat and eggs.
   g. Alternative use of poultry litter and waste as fertilizer, for production of biogas and electricity, and for other purposes.
   h. Research and development initiatives for optimization of vaccination of rural chicken against various diseases to decrease mortality.

REFERENCES

Ashraf M, Mahmood S, Khan MS and Ahmad F. 2003. Productive behavior of Lyallpur Silver Black and Rhode Island Red breeds of Poultry. *Int. J. Agri. Biol.* 5(3):384–387.

Cholan TZ. 2007. Marketing margins of broiler in Azad Jammu Kashmir: Challenges and opportunities. *Sarhad. J. Agric.* 23(1): 157–168.

Economic Survey of Pakistan. 2006–2007. *Government of Pakistan, Finance division*. Economic Advisory Wing Islamabad, Pakistan.

Economic Survey of Pakistan. 2015–2016. *Government of Pakistan, Finance division* Economic Advisory Wing Islamabad, Pakistan.

FAO. FAO representation in Pakistan. 2011. Pakistan and FAO: Achievements and success stories.

GOP (Government of Pakistan). 2014. *Economic Survey of Pakistan*. Ministry of finance, Government of Pakistan.

Hussain J, Rabbani I, Aslam S and Ahmad HA. 2015. An overview of Poultry industry in Pakistan. *W. Poult. Sci.* 71(4) 689–700.

Johnson I, Jorgensen S. 2006. The Road to 2050 sustainable development for 21st century (2006). The International Bank for reconstruction and development/The World Bank, Washington, DC (Report Number 36021).

Khawaja T, Khan SH, Mukhtar N, and Parveen A. 2012. Comparative study of growth performance, meat qualityand haematological parameters of Fayoumi, Rhode Island Red and their reciprocal crossbred chickens. *Ital. J. Anim. Sci.* 11:211–216.

Khawaja T, Khan SH, Mukhtar N, Parveen A and Ahmed T. 2013. Comparative study of growth performance, meat quality and haematological parameters of three-way crossbred chickens with reciprocal F1 crossbred chickens in a subtropical environment. *J. Appl. Anim. Res.* 41(3) 300–308.

Memon NA. 2012. Poultry: Country's second-largest industry. *Exclusive on Poultry*, Nov–Dec 2012.

Mukhtar N, Khan SH and Khan RNA. 2012. Structural Profile and emerging constraints of developing poultry meat industry in Pakistan. *W. Poult. Sci. J.* 4(68):749–757.

PPA (Pakistan Poultry Association). 2016. Present Status of Poultry Sector. notes/pakistan-poultry-association/pakistan-poultry-industry-poultry-industry-has-very-strong-roots-in-pakistan-as-/609998695692006

Roberts, TJ. 1991. *The Birds of Pakistan. Vol 1. Non-Passeriformes*. Oxf. Uni. Press, Oxford.

Sahota AW and Bhatti BM. 2003a. Productive performance of Desi field chickens as affected under deep litter system. *Pak. J. Vet. Res.* 1(1):35–38.

Sahota AW and Bhatti BM. 2003b. Growth performance of different varieties of Desi generation-1 chickens maintained under deep litter system. *Pak. J. Vet. Res.* 1(1):46–49.

Usman M, Zahoor I, Basheer A, Akram M and Muhammad A. 2014. Aseel Chicken- A preferable choice for cost-effective and sustainable production of meat type poultry in tropics. *Sci. Int.* 26(3):1301–1306.

25 Inland Fisheries and Aquaculture in Pakistan

Muhammad Javed and Khalid Abbas

CONTENTS

25.1 BRIEF HISTORY OF INLAND FISHERIES IN PAKISTAN

Geographically, Pakistan consists of three major regions; the mountainous North, where three of the world's greatest mountain ranges (Himalayas, Karakoram, and Hindukush) meet; the vast but thinly populated plateau of Balochistan in the southwest; and the plains of the Indus River and its tributaries in Punjab and Sindh provinces. Pakistan is located in south Asia, between 23°42′ and 36°55′ N and 60°45′ and 75°20′ E, and is bordered by China on the north-east, India on the east, Iran to the south-west, Afghanistan on the north-west, and the Arabian Sea in the south. Pakistan has variety of ecosystems, from the northern mountains, to the relatively hotter plains of the Indus Valley, and a southern coast region with temperate conditions. Generally, the rainfall is limited, ranging from 130 mm/yr in the northern areas of the lower Indus plains to 890 mm/yr in the Himalayan region. The Monsoon rains are often in late summer. The average national rainfall is almost 760 mm/yr. The Indus River, flowing from mountain ranges of the Himalaya and Karakoram for 2500 km to the Arabian Sea, determines the agriculture in Pakistan. The Indus River system comprises of the largest irrigation system in the world and serves to irrigate the 1.4 million ha of the cultivated land.

The geographic position of Pakistan makes it rich with diverse climatic conditions and both marine and freshwater resources. The freshwater fisheries resources include rivers, lakes, streams, canals, and huge reservoirs created by the construction of dams and barrages. The origin of all rivers flowing into Pakistan is India who initiated constructing dams on these rivers, thereby restricting the entry of water into Pakistan. As the economy of Pakistan is mainly based on agriculture, a need for the development of a comprehensive irrigation system was a priority. In early seventies, after the Indus Basin Treaty, construction of barrages and dams were completed across the main rivers, primarily for irrigation and hydropower generation. It brings great benefits to the country but simultaneously it castrated the ecosystem and natural habitats for the aquatic life thriving therein. The migration of anadromous fish for breeding was restricted by the construction of the Sukkur and Kotri barrages because fish ladders were not designed to accommodate the behavior and habitat of the fish and were ineffective. Construction of huge water reservoirs resulted in the disappearance of sufficient natural spawning grounds for fish. In water bodies, aquatic vegetation flourished while huge areas of land became water-logged or saline. As a result, fish breeding activity/breeding behavior changed. The natural fish seed stocks have been depleted over the years due to aquatic pollution. Thus evolved the need to develop and adopt artificial breeding technology for rehabilitation of fish stocks.

To control the problem of excessive aquatic vegetation, an exotic fish species, the grass carp (*Ctenopharyngodon idella*), was imported from China and introduced into the pond culture system in Pakistan. It mainly feeds on grass and other aquatic weeds/vegetation. Other fish species, namely silver carp (*Hypophthalmichthys molitrix*), bighead carp (*Aristichthys nobilis*), Gulfam (*Cyprinus carpio*), and tilapia species have also been imported to maximize fish production in the country.

The practice of artificial fish breeding began with the adoption of techniques like hypophysation and extraction of hormones from pituitary glands, their preservation, and use for fish breeding. Research studies on fish breeding behavior started along with its physiological functions and maturity. Conditions required for breeding, and water quality monitoring were also studied during the mid 1960s (DOF 2005). The department made a breakthrough in sixties and succeeded in induced spawning of fish, thus happas were erected in flowing waters to breed different fish species in 1967 (DOF 2005). However, the response to hypophysation remained poor. In 1973, the department of Punjab Fisheries started the establishment of Nursery Units along with fish hatcheries. At the start, there were no major progress and only a single species, *Cirrhinus mrigala* was propagated through artificial spawning in 1974 (DOF 2005). Then the Chinese system of induced spawning was adopted by the department of Punjab Fisheries and success was acheived in breeding *L. rohita*. Investigations on fish seed and fry rearing, fish diseases, prophylactic measures, and water quality management continued. During the year 1982, the department of Zoology and Fisheries, University of Agriculture developed "Fisheries Research Farms" and provided training on fish biology and ecology, fish nutrition, water and soil chemistry, fisheries management and aquaculture (DOF 2005).

In areas away from the sea, inland fisheries used to serve as a major source of animal protein for the ever-increasing human population. In the current scenario, human attitudes towards inland water resources have changed in face of high demand of fish and the growing knowledge of the environment's role in human well-being. This adaptation change is integral to the idea for sustainability of ecosystems and using the potential of available freshwater resources to meet demands. The old exploitative models are being replaced with evolving priorities for forestry, agriculture, and fisheries for long-term food security through sustainable ecosystems. Generally, the increased pressure on bioresources has led humans to make policy for conservation and sustainability. Inland fisheries and aquaculture are immensely related to the geographical landscape and socio-economic status of population in a given area. Currently, these sectors are being reframed with regard to the political, economic, and demographic evolution of human society.

25.2 CURRENT STATUS OF FISHERIES

Amidst the national economy, the fisheries sector plays a substantial role. Although, the main component is the marine fishing sector, with a 57% contribution to the overall national fish production, the rest of the production is based on inland fisheries and aquaculture systems (GOP 2015). The inland fishery, based on rivers, reservoirs and irrigation canals, is generally of subsistence nature, involves some 180,000 people, and about 20,000 smaller fishing crafts (FAO 2009). During recent years, the government has focused on the aquaculture subsector and developed numerous fish hatcheries and training centers to facilitate fish farming communities. Pond fish culture is being practiced in the provinces of Punjab, Sind, and Khyber Pakhtunkhawah (KPK) on a limited scale, involving such fish species as common carp, Chinese carps, trout, and some other carps alongside the indigenous major carps such as *Labeo rohita, Cirrhinus mrigala,* and *Catla catla*. The culture of marine shrimp species (*Penaeus merguiensis* and *P. indicus)* was started on pilot project scale, but could not be brought to commercial scale for want of seed of shrimp, technical expertise, infrastructure, and required facilities (MinFAL 2006).

Due to the huge potential for development of the sector, the government plans to develop the sector in future. The government has designed a policy document for development of aquaculture and fisheries in Pakistan (MinFAL 2005). This document highlights the role of aquaculture and fisheries sectors in achieving the Pakistan's goals of development for economical fish yields in a broader perspective. The three policy goals set by the government are (*i*) increased growth of economy, (*ii*) poverty alleviation and (*iii*) food security. The production statistics of inland fisheries and aquaculture in Pakistan, over the last decade, are summarized in Table 25.1 and Figure 25.1):

25.3 INVENTORY OF INLAND AQUATIC RESOURCES IN PAKISTAN

Nature has endowed Pakistan with huge water resources, given in Table 25.2. Freshwater, brackish water, and marine water resources could be utilized for development of economically viable and ecosystem friendly aquaculture systems (Figure 25.2).

25.4 ICHTHYOGRAPHIC REGIONS OF PAKISTAN

On the basis of distribution patterns of different fish species, Pakistan has been categorized into following five ichthyographic regions:

25.4.1 Northern Mountainous Region

This region comprises the northern mountainous areas of Pakistan and Kashmir above 1500 maltitude, including the northern areas (Gilgit, Skardu and Diamer), upper parts of Chitral, Kaghan valleys, and Swat. The fish fauna is mostly high Asian (Central Asian) and mainly comprises snow trouts (Schizothoracinae), the catfish genus of *Glyptosternum,* and loaches (*Noemacheilus*) (Akhtar et al. 2014). Some south Asian forms of the genera; *Tor, Labeo, Garra, Puntius, Botia, Glyptothorax* and *Ompok* have also been referred (Jan et al. 2016).

25.4.2 Yagistan Region

This region was previously named as the northwestern mountainous region. After the old tribal name of this area, it was renamed as the Yaghistan region. Its altitude is between 1000 and 1500 m and

TABLE 25.1
Production Statistics of Pond Fisheries and Inland Catches in Pakistan

Production (Tonnes)

Year	2001	2002	2003	2004	2005	2006	2007	2008	2009	2010	2011	2012	2013	2014	2015
Pond fisheries	57,632	66,898	72,978	76,583	80,547	121,740	130,000	131,005	133,900	135,202	137,500	139,850	142,712	143,900	146,000
Inland catches	122,468	114,030	92,794	93,687	94,644	140,000	100,000	96,995	94,600	93,067	91,009	89,004	86,021	85,000	82,021

Source: FAO 2009. *Fishery and Aquaculture Country Profile: Islamic Republic of Pakistan*. Rome, Italy; GOP 2015. *Economic Survey of Pakistan—2014–2015*. Government of Pakistan, Islamabad.

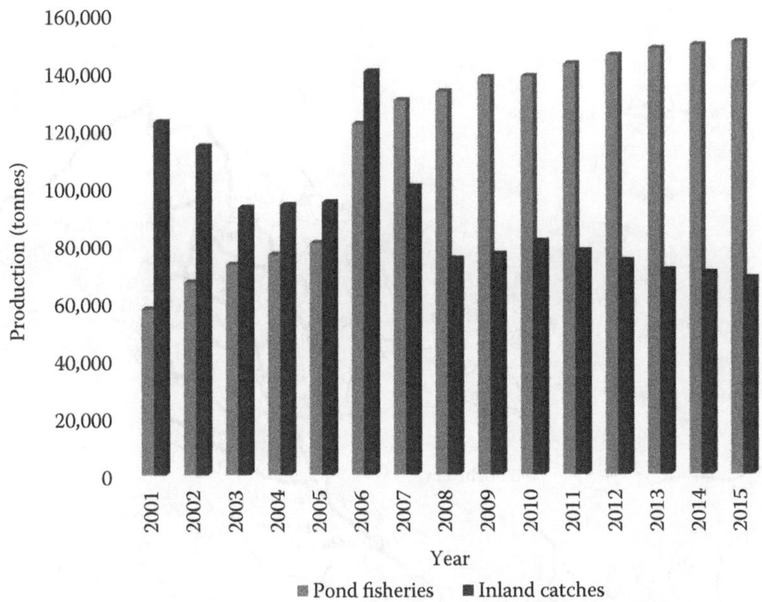

FIGURE 25.1 Graph showing the relative fish production from pond fisheries and inland catches.

TABLE 25.2
Inventory of Pakistan's Aquatic Resources by Province (Area in ha)

Resource	Sind	Punjab	Blochistan	KPK	Total
Rivers and streams	160,000	2,940,000	–	2408	3,102,408
Canals, drains, and abandoned canals	321,340	23,700	–	1763	346,803
Lakes	110,000	6700	4047	6362	127,109
Dams and reservoirs	97,000	65,800	6070	26,800	195,670
Waterlogged areas	3,000,000	30,000	–	1,600	3,031,600
Indus delta	700,000*	–	–	–	700,000
Flood irrigated area	1,000,000	–	–	–	1,000,000
Fish farming systems	49,170	10,400	100	560	60,230
Marine Waters	352 km × 12 N.Miles	–	1,129 × 12 N.Miles	–	1481 × 12 N.Miles

Source: MinFAL 2005. *National Policy and Strategy for Fisheries and Aquaculture Development in Pakistan.* Ministry of Food, Agriculture and Livestock, Government of Pakistan, Islamabad.

surrounded by the Koh Safaid range in the north, the Suleman range in the east, the Marri-Bugti hills in the south, and the central Brohui range in the southwest. In the northwest, it extends up to Afghanistan with areas drained by the rivers Kurram, Tochi, Gomal and their tributaries. The fish fauna is a mixture of south Asian, high Asian and west Asian forms, but south Asian forms predominates.

25.4.3 Aba-Sinh Region

This region comprises the southern areas of the Malakand division, the vale of Peshawar, submountainous Hazara, adjoining areas of the Kashmir and Punjab, north of the Kala Chita and Koh Safaid ranges. It extends up to southeastern parts of Afghanistan drained by the river Kabul

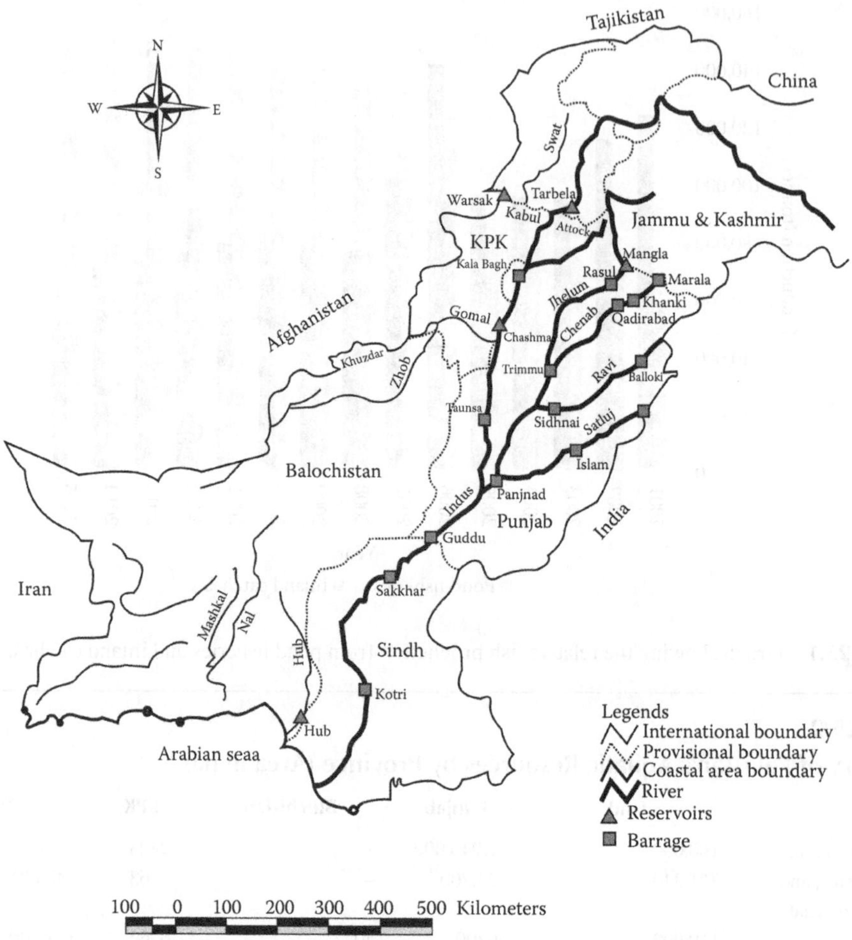

FIGURE 25.2 Network of rivers in Pakistan.

and the tributaries. This area is dominated by the south Asian fish fauna, although some high Asian fishes are also found.

25.4.4 Mehran Region

Previously, this region was known as the Indus plain region. It includes the Indus plain, adjoining hills, that is, the Salt range and Kohat hillls, Sindh-Baluchistan hills in the southwest, and the Pothohar plateau in the north. The south Asian fish fauna dominates this region except for some west Asian forms. Rarely, the genus *Schizothorax* is also observed.

25.4.5 Gedrosian Region

This region comprises Baluchistan plateau, Hala and west of central Brohui ranges. Its northern part is drained by the rivers of Mashkel and Lora, ending into salt lakes of Hamun-e-Mashkel, respectively. Its southern part is drained by the rivers of Dasht, Hingol and their tributaries which fall into Arabian Sea. The fish fauna of this region is a mixture of south Asian and west Asian (Lora drain) forms.

Presently, the fish fauna in many of these regions is threatened due to anthropogenic activities which have polluted the water due to indiscriminate discharge of untreated industrial effluents and city sewage. Due to drought, this situation is further worsened and has caused lowering of subsoil and groundwater levels. In aquatic environments, it also has contributed to higher concentration of pollutants. The construction of barrages and dams across major rivers, aimed for irrigation and hydropower generation, have also destroyed natural fish habitats and breeding grounds. These structures have also hampered the upstream migration of fish.

25.5 MANAGEMENT STRUCTURE OF FISHERIES IN PAKISTAN

The inland fisheries are a provincial subject in Pakistan. The inland fishing grounds, which have been so far exploited, fall within the administrative control of provincial fisheries departments. The fisheries administrative structure of Provincial Governments is as follows:

- Directorate of Fisheries Khyber Pakhtunkhawah (Head office in Peshawar) headed by a Director
- Directorate of Fisheries, Punjab (Head office in Lahore) headed by a Director General
- Directorate of Fisheries, Sindh (Head office in Karachi) headed by a Director
- Directorate of Fisheries, Balochistan (Head office in Pasni) headed by a Director.

The Water and Power Development Authority (WAPDA) is another federal agency with fisheries related functions and has its own Directorate of Fisheries to develop and manage fisheries in the six major reservoirs with a total surface area of about 100,000 ha under its administrative control (FAO 2009).

The following are the common management measures:

- Prohibition of harmful or destructive fishing gears for fishing
- Regulation of mesh-size limits for the fish nets
- Closed areas and closed seasons.

For the shrimp fishery, the recruitment period is the closed season. Shrimp trawling is prohibited in nursery areas. The fish catch of major carps is also banned during June-July, that is the breeding season of carps.

25.5.1 Inland Subsector

In Pakistan, freshwater catch fishery is dominated by the Indus river system. The fish fauna in the northern part of the Indus River system is cold-water type while the warm water fish species dominate the middle and southern parts of the system. So different rivers of the same zone may have the same fish fauna and great differences exist between various zones of one river. Hydrological alterations, resulting from the development of the Indus basin, has greatly modified riverine hydrology. As a result, riverine fish populations have been adversely influenced by environmental and anthropogenic factors mostly.

Of the total inland fish production, 80% comes from reservoirs and rivers (FAO 2009). The riverine fishery is managed and operated by the fisheries departments of provinces. The regulatory laws, that is, restricting catch by fish-size and enforcing closed seasons, are promulgated by the provinces. But due to lack of proper fisheries management plans and coordination, the development of inland fisheries is limited.

In Pakistan, 14 million ha are irrigated (FAO 2009). There are three major reservoirs, 19 barrages, 12 interlink canals, and 43 independent irrigation command areas. The total length of the main canals is 58,500 km. About 79% of the total wheat crop comes from irrigated fields. None of the major dams include fish passes, but some barrages have fish passes, which are largely nonfunctional. Most pumping stations have no fish protection devices and where these are present they do not function well . In the year 2001, 122,468 tonnes of fish were captured from rivers, reservoirs, natural lakes, and irrigation canals; 57,632 tonnes were produced in aquaculture, and about 75% of these were produced in fish ponds fed by irrigation canals and groundwater (FAO 2009). There is a potential for enhancing fish production in irrigation and multipurpose reservoirs, and also inlargely unexploited brackish waters. Abandoned irrigation canals are used for fish production and managed as fishponds. Flood control compartments, ranging from 10 to 5000 ha, are naturally stocked with fish during floods but drained a few months later, which does not allow the fish to grow to market size. With proper management, these structures could be better utilized by stocking with appropriate fish species.

25.5.1.1 River Fisheries

In rivers, both carp and catfish species breed in the natural waters. But these fish species are under tremendous pressure from water pollution and struggle for survival. Keeping themselves away from enemies, the fish have to find food by effort. During Monsoon floods in June and July, carp species such as Rohu, Thaila, and Mirgal ascend water currents and streamlets to lay the eggs in shallow waters and then return to rivers. In the shallow pools, young ones emerge within 24 hours and grow to fry and fingerling stages. After a month or so, with the next flood, they find their way to rivers by ascending the streams. In their third year, they become adult and repeat the reproductive process (Rafique and Khan 2012). River fisheries have the following characteristics:

- Greater variation of depth, width, and speed of water
- There is a great range of water temperatures
- To thrive against enemies, life is hard
- Searching for food is difficult
- Breeding grounds have to be located by the fish
- Natural barriers have to be surmounted
- Fish face a variety of water pollutants

25.5.1.2 Canal Fisheries

In canals, the life of fish is difficult compared to life in the rivers. Actually, any fish that enters the canal is fated to perish. This is due to closure of canals, twice a year, for desilting. Almost all the water is drained from canals; therefore, most of the fish die. However, a small number of fish is caught and sold in the market while most of the fish iare eaten away by their enemies. A small number of fish which survive remain somewhere in the deeper waters and cannot go back into the river due to the absence of fish ladders in canals.

25.5.1.3 Lake Fisheries

Lake fisheries are of recent origin in Pakistan. In Sindh Province alone there are more than one hundred natural lakes of various sizes expanding over a total area of about 100,000 ha. Among them is Kinjhar Lake (12,000 ha), Halijee Lake (1800 ha), and Manchar (16,000 ha), which are significant with regard to fish production. There are about 2000 fishing families' living on Lake Manchar alone. Moreover, Bakar Lake, which is a cluster of small lakes, covers an area of about 40,000 ha. In Punjab, natural lakes extend over about 7000 ha. Some lakes such as Uchhali (943 ha), Nammal (480 ha), Kalar-Kahar (100 ha), Jahlar (100 ha), Khabaki (283 ha), and Kharral (235 ha) are brackish (FAO 2009). Mangala Lake (District Jhelum) is reputed for common carp (*Cyprinus carpio*) and Mahaseer (*Tor putitora*). Both these lakes are full of waterweeds. Apart from the principle 26 carp fish species, the lakes have *Channa* fish (Murrel) as well as the big *Walago* in large numbers. Additionaly, small dams of various areas varying from 100 to 1000 acre are also present. They are regularly stocked with carp fish fry and a fairly good yield is obtained. The characteristics of lake fishery include (*i*) greater depth and expanse of water; (*ii*) at the farther end there is main river current, entering into the lake but the main reservoir is quite; and (*iii*) the depth, temperature, and food provide different habitats for the fish, however, it takes quite a long time for a lake to provide necessary conducive environment for fish life, usually a decade.

TABLE 25.3
Fish Species Commonly Cultured in Pakistan

Sr. No.	Common Name	Scientific Name
1.	Calbans	*Labeo calbasa*
2.	Grass Carp	*Ctenopharyngodon idella*
3.	Gulfam	*Cyprinus carpio*
4.	Mahaseer	*Tor putitora*
5.	Mori or Mirgal	*Cirrhinus mriagala*
6.	Mulee	*Wallago attu*
7.	Paree	*Notopterus notopterus*
8.	Rohu	*Labeo rohita*
9.	Silver Carp	*Hypophthalmichthys molitrix*
10.	Thaila	*Catla catla*
11.	Tilapia	*Orecochromis sp.*

Source: Mirza MR and IA Sandhu 2007. *Fishes of the Punjab, Pakistan*. Polymer Publications, Pakistan.

The entire process of ecosystem stability depends upon the flow of river water into the lake/dam that may become slow and ultimately loses its entity. With the passage of time, the silting process increases and enhances the transparency of water. Afterwards, phytoplankton populations will start developing from a sparse community to dense growth. The zooplankton will follow suit. Phosphates in the water are depleted as they are used up by the planktonic biomass. Forming their own stratification, dissolved gases will be formed. Carbon dioxide is used up and oxygen is produced at the surface. Due to density of phytoplankton, all sunlight is absorbed at the upper layers of water. Deeper waters will have less light and therefore less photosynthetic activity. Oxygen may be depleted resulting in fish mortality.

25.5.1.4 Small Dams

In the past four decades, six large water reservoirs have been established by the development of dams and barrages. These reservoirs provide about 250,000 ha area for fish production (FAO 2009). Several smaller reservoirs also add to the fish production system. Although the major reservoirs and barrages mainly account for fish production, a gradual decline in fish catch has been recorded over the years. Currently, the fishery operation in reservoirs is managed by the WAPDA through fish exploitation limits in different seasons resulting in minimum size of fish landings. Over the past 38 years, the Small Dams Organization of the Irrigation and Power Department of the Punjab Government has created a number of small dams in the province to provide irrigation to agricultural lands. At present there are 32 dams in operation. Currently, the fishery management in the small dams/reservoirs in Punjab province is controlled by the Department of Fisheries and has been decentralized to District Officers. The major management functions are carried out by the Department of Fisheries in small dams/reservoirs by stocking the fish seed and issuing permits for their harvesting. The following indigenous and exotic species dominate the fish stocks in small dams (Table 25.3):

25.6 AQUACULTURE SUBSECTOR

Despite huge potential for development, the aquaculture industry in Pakistan is still in itsjuvenile stage. During the last decade, it has gained great momentum. Since 2000, aquaculture production has rapidly increased from about 35,000 tonnes in 2001 to reach over 146,000 tonnes during 2015 (GOP 2015).

25.6.1 Cold Water Fisheries

Cold water in the mountainous areas of KPK, AJK, Balochistan, and Northern Areas provides a unique opportunity for aquaculture. At present, there are two cold-water fish species, that is, rainbow trout (*Oncorhynchus mykiss*) and brown trout (*Salmo trutta*), which are cultured and propagated successfully. On a commercial scale, intensive raceways culture of trout is being practiced in Dir, Swat, Chitral, AJK, Hazara in KPK ,and Northern Areas. The province of KPK has comparatively fewer trout farms than in Swat, Dir, Chitral, Malakand, Mansehra, FATA, and Northern Areas.

25.6.2 Warm Water Fisheries

Despite diversification in its environment and being blessed with large fresh, brackish, and marine waters, Pakistan only has carps cultured in inland waters. Carp farming is done in earthen ponds, mostly using semiintensive culture systems with very little inputs. Although Pakistan is rich in fish fauna, the commercial fish production is based on only seven warm water species. In the Indus delta, trials of shrimp culture could not turn successful due to unavailability of shrimp seed and feed from the local market. Freshwater carp culture is the main aquaculture in the provinces of Punjab, KPK, and Sindh.

As the aquaculture is in its infancy, the sector is suffering from poor management. Two projects funded by the Asian Development Bank (ADB) have helped in bringing reforms to strengthen the institutional structure and extension services through the establishment of model fish farms, fish hatcheries and juvenile production, and human resource development (MinFAL 2006). Because of attention from the government, aquaculture has received substantial investments, and facilities are available now which can provide the basis for a significant expansion of the industry in future.

Except for trout culture in the KPK province and the northern areas, almost all aquaculture is practiced in ponds stocked with various carp species. Despite huge potential, coastal aquaculture operations in Pakistan are almost nonexistant. As per recent estimates, about a 60,500 ha area is covered by fish ponds (Punjab, 10,500 ha; Sindh, 49,170 ha; KPK, 560 ha), and others (Balochistan, Azad Jammu Kashmir, and Northern Area, 240 ha). So far, about 13,000 fish farms, varying considerably in size, have been established throughout Pakistan (MinFAL 2005). The fish farm average size ranges from 5 to 10 ha. The correct information about the number of people engaged in fish farming and allied activities is not available as fish culture, in most of the areas in Pakistan, is a side activity to agriculture. The current surveys report that about 50,000 persons are employed in the fisheries sector either directly or indirectly (GOP 2015).

In Sindh Province, districts Badin, Thatta, and Dadu are the hotspots of fish farming as these are located close to the Indus river. Thatta and Badin areas are suitable for fish farming due to waterlogged floodplains. In the Punjab Province, most of the fish farms are developed in irrigated areas that have abundant rainfall and alluvial soil. For these reasons, about 3/4th of all the fish farms in the Punjab are located in Districts Gujranwala, Sheikhpura, and Attock.

In KPK Province, carp farming is practiced in Dera Ismail Khan, Kohat, Mardan, Abbotabad, and Swabi districts through semiintensive culture system. In Pakistan, a typical carp polyculture system has an appropriate ratio of warm water fish species as follows: Rohu (30%–35%), Mrigal (15%–20%), Catla (10%–20%), Silver Carp (15%–20%), and Grass Carp (15%–20%). However, intensive fish farming has not been started yet in the country due to unavailability of least cost balanced feed and limited expertise for the management of intensive culture system.

25.7 ROLE OF FISH HATCHERIES IN ENHANCEMENT OF FISHERIES

Before 1970, the fish seed was mainly collected from the natural water sources. During 1974 and onwards, success in the experimental trails of induced breeding of carp species made a breakthrough

TABLE 25.4
Distribution of Freshwater Fish Hatcheries in Pakistan

	Provinces				
Type of Hatchery	**Punjab**	**Sindh**	**KPK**	**Baluchistan**	**AJK and Others**
Government hatcheries	14	5	6	1	2
Private hatcheries	76	31	10	–	–
Trout hatcheries	1	–	8	1	7

Source: Ayub M 2007. Freshwater fish seed resources in Pakistan. In: Bondad-Reantaso MG (ed) *Assessment of Freshwater Fish Seed Resources for Sustainable Aquaculture*. FAO Fisheries Technical Paper. No. 501. FAO, Rome.

to promote pond fish culture. As such, development of both warm water and cold-water aquacultures were brought about through establishment of fish hatcheries and nurseries in different provinces, as shown in Table 25.4.

The public and private fish hatcheries are major sources of fish seed in the country. These hatcheries could be further categorized based on the fish species being used for spawning purposes. It could be carp-, trout- and mahseer-specific hatchery, etc. At present, there are 28 public carp hatcheries located in different districts being run by provincial fisheries departments whereas 117 hatcheries are owned by the private sector. There are 17 trout-specific hatcheries to enhance fish culture in cold water resources of the country. For propagation of Mahseer fish, there is one hatchery operational in the semi-cold water area of Malakand, in KPK province, while another Mahseer-specific hatchery is being established under a developmental project in Attock District, Punjab (Ayub 2007).

In carp hatchery, management of fish seed comprises two distinct stages as: (*i*) nursery ponds-based rearing of postlarvae to the fry phase; and (*ii*) fry to fingerling phase reared in earthen ponds. Under both conditions, the most important functions are maintenance and preparation of nurseries, stocking and rearing of ponds, which principally guarantees the health, survival, and growth of fish seed stocks. Growth performance, survival rate, and general health indices of the fish seed stock are properly monitored to assess quality of the seed. Segregation of seed of different fish species is advised as mixing of seed of related species creates management problems for farmers who stock poly culture units as per specified stocking ratios of different species of fish. For the sale of fish seed, no special markets exist in the country. Fish farmers purchase fish seed directly from hatcheries for transport to their farms. When fish seed are transported over a long distance by road, seeds are packed in polyethylene bags with enough oxygen.

25.8 CONSTRAINTS IN THE DEVELOPMENT OF FISHERIES

During the last several decades, aquaculture in Pakistan has developed well, particularly in Punjab Province, as compared to catch fishery. The important impediments still faced by aquaculture sector are: (*i*) inadequate supply of quality fresh water in some areas, particularly in lower Sindh, coupled with declining levels of groundwater which renders it unaffordable for fish farmers; (*ii*) inadequate supply of quality seed of culturable fish species; and (*iii*) competition due to carps imported from Myanmar. In inland water resources, unchecked fish exploitation has led to depletion of stocks. Use of dynamite, toxic chemicals, and gases as means of harvesting fish in the past have caused the killing of nontarget species and fish of undesirable size. Such activities have catastrophic effects on natural fish populations. Despite stocking natural water bodies, catches of some commercially valuable fish species such as Kalbans, Rita, Mhasheer, and numerous other cat fish species have declined remarkably. The stocks of freshwater fish resources have not been assessed over the last few years. Changes in freshwater species composition, determination of losses, knowing the maximum sustainable yield, and implementation of adapted management could not be realized due to lack of information on the subject. The menace of pollution in inland water bodies throughout the country has been escalating, especially close to urban populations and industries which have a negative influence on fish catches and the livelihood and health of fishing communities.

25.8.1 Research and Development

In the country, a number of universities, institutions, and research organizations are engaged in the subjects related to fisheries and aquaculture teaching and research but most of the research does not pertain to the pragmatic problems of the aquaculture and fisheries industries (FAO 2009). The development of a fisheries sector in Pakistan demands substantial research support. The information related to aquaculture and fisheries and related data in Pakistan is inadequate, inaccurate ,and unauthentic to large extent. This is reflected by inefficient data collection with regard to fisheries along the coastal line in Balochistan. It is more or less the same for other provinces when comes to reliable data collection systems. In most of cases, data on fisheries of Sindh are simple estimates and subjected to personal judgments. Unreliable data collection systems for aquaculture and fisheries impedes management of these sectors. Organizational structure of fisheries institutions, especially in all the provinces, needs to be strengthened and widened for specialized working; the research and development wings of the fisheries departments are not on par with the needs of the sector, and require up-gradation and streamlining.

25.8.2 Post-Harvest Issues

Improper fish handling and lack of appropriate preservation facilities on fishing vessels and at landing centers result in poor quality fish raw material for further processing and marketing. It is estimated that almost 70% of the captured seafood is deteriorated before reaching the end user or/and processing centers. Instead of producing fishmeal, the fishermen leave the by-catch from trawl fisheries to perish. Despite launching a program for the up-gradation of fish holding and handling setup at the fishing points under the sponsorship of SMEDA and the Fishermen's Cooperative Society, the lack of commitment and technical guidance led to a failure. The same unhygienic conditions prevail in the preprocessing industry where peeling of shrimps is performed in an unclean manner (SMEDA 2014). The freshwater fishery products are facing similar issues of postharvest losses. Such drawbacks limit the accessibility of fish products to local consumers and degrade the image and position of Pakistan among international markets also.

25.8.3 Fish Marketing

The fish marketing chain is same as that of other agricultural products. In the fish market, wholesalers purchase products from farmers and then sell them to retailers and consumers through commission-based agents. Farmed fish is most often sold either at the fish farm, through agents, or ice-packed fish is transported to markets for sale through auction systems. Buyers can be public, wholesalers, retailers, agents of processing plants, and/or exporters. Fish markets are quite prevalent in Sindh but are limited to selected cities in Punjab. All markets are controlled by the local district administration. The required facilities, that is, proper cold storage, hygiene, and adequate communication links are lacking in most fish markets. Mostly, aquaculture products are consumed at the local level. The Rohu fish has a good acceptability in local markets; the marketable size is usually from 2 kg up to a maximum of 3 kg. The price of a fish tends to decline if more than 3 kg. The price of fish also depends upon species, freshness, and the supply–demand status in the market.

25.8.4 Provision of Quality Fish Seed

Awareness of genetic issues pertaining to artificial propagation of fish in Pakistan is low. The induced breeding of major carps at large scale in private and public fish hatcheries have influenced the genetic integrity of indigenous, as well as exotic, fish species. Generally, technical knowledge about fish genetics and acquaintance with the possible ecological consequences of inbreeding is very poor. This situation has led to negative selection, inbreeding depression, interspecific hybridization, and genetic introgression in hatcheries as well as natural stocks of commercially important species (Alam and Islam 2005; Simonsen et al. 2005).

The fish seed of major carps and some exotic species are being produced at large scale through traditional spawning in hatcheries to meet increasing demands for aquaculture products. The hatchery operations are driven primarily by profit, and as such, hatchery managers focus on quantity of fish seed instead of quality. They usually keep a small size of broodstock of relatively small-sized fish to lower production costs. This has resulted in low survival rates, growth performance, resilience against diseases, and environmental stress etc. The poor performance of the hatchery-produced fish seed is caused primarily by genetic deterioration through negative selection, inbreeding, hybridization at various scales, poor broodstock keeping in hatcheries, or a combination of all these factors (Sultana et al. 2015).

25.8.5 Environmental Degradation

Currently, Pakistan is confronted with environmental issues including climate change, thermal and chemical water pollution and waste disposal, water logging and salinity, irrigated agriculture, biodiversity loss and many associated phenomena (GOP 2011). Pakistan is one of the most vulnerable countries to the impacts of climate change. Environmental degradation is a global issue in the current age. Environmental hazards are potentially influencing the economic, national, and social arenas. Fast economic growth and human intervention in the natural system are the key factors behind this degradation. Remarkable increases in the frequency and extent of extreme weather conditions, along with flood and droughts, are causing threats to fish in Pakistan. Major issues include the projected recession of the Karakoram-HinduKush-Himalayan glaciers due to global warming, deadly water influx into the Indus river system, elevated siltation of major dams and reservoirs due frequent floods, and higher temperatures causing stressful conditions for fish life, particularly in arid and semiarid areas. All these factors are resulting in decreased agricultural productivity, decline in forest cover, restricted natural migration of adversely influenced fish species, and increased influx of saline water to the Indus delta thus damaging the coast line mangroves, which serve as the breeding grounds for marine fish.

The freshwater fish have been threatened by a wide range of pollutants/factors, but genetic diversity of many species have declined mainly due to anthropogenic interruptions which include unplanned introduction of exotic fish species, impoundment of rivers, habitat degradation, water quality deterioration, and overexploitation. Anthropogenic intervention is the most important factor contributing to loss of genetic resources of fish populations in the wild (Abbas et al. 2010). The hydrological alterations and increased human interventions during the past many years have resulted in loss/reduction of spawning, breeding, and nursing grounds in natural water bodies. The construction of dams has interrupted the natural migration of many fish species and thus interferes with their life cycles (Agosthinho et al. 2008). Pollution reduces rates of survival of fish spawn. Among other human interventions, siltation due to deforestation and the merciless catching of brood fish during the breeding season contribute to declines of natural fish populations in the wild (Collares-Pereira and Cowx 2004). Moreover, traditional breeding operations in hatcheries and natural fish population size reductions have detrimental genetic consequences and cause the loss of growth potential, disease resistance, and stress tolerancc. Since 1998, there has been a consistent decline in fish production of 2% per year from natural sources (MinFAL 2005). Alarming declines in productivity in so many major fish stocks have been observed due to overfishing. Wild fish stocks have been heavily overfished, resulting in a noticeable leveling of fish landings during the last decade (FAO 2014).

25.9 OPPORTUNITIES AND FUTURE ROAD MAP

The economy of Pakistan has gained a boost during the fiscal year 2004–2005 with an annual GDP growth of 8.4%, the fastest rate of the last two decades (FAO 2009). The exceptional growth rate owed to factors such as macroeconomic policies, renewed confidence from the private sector, a growing domestic demand, fiscal discipline, and competitive exchange rates. According to the estimates of FAO, aquaculture is contributing increasingly to world fish production. It was pointed out that an additional 37 million tonnes/year of fish will be required by the year 2030 to sustain the present pace of fish consumption. An anticipated increase of two billion people annually means that aquaculture will be expected to produce nearly 85 million tonnes of annual fish to maintain current per capita consumption (FAO 2014).

Due to government attention and two ADB-funded program, aquaculture has expanded at a remarkable pace over the years. The Punjab province has become able for mass production of eggs and juvenile fish, and is providing the seed restocking programs and the aquaculture production both Punjab and other parts of the country. The total production of fish seed by Punjab FOD was 91 million during 2014–2015 (GOP 2015). Coastal aquaculture has not become a success story despite public and private sector efforts. In almost all maritime countries, marine aquaculture has developed tremendously and provides raw materials as exportable commodities. Pakistan has not been able to compete with other nations having major aquaculture sectors providing alternative and dependable sources of raw materials for exports. The unavailability of cost-effective commercial fish feed has also rendered he taquaculture industry less viable. However, some experimental feeds have been prepared and utilized very effectively by the University of Agriculture, Faisalabad at the experimental level (Hussain et al. 2011; Nazish and Mateen 2011). Small fish caught, as a by-catch from shrimp trawlers as well as fish waste, are used as raw materials for preparation of fishmeal on industrial scale. About 189,134 tonnes of small pelagic fish was obtained as by-catch which yielded 42,230 tonnes of fish meal according to the reports of FAO (2009). Some progressive fish farmers are using fishmeal and/or trash fish in aquaculture operations, but it is not a common practice.

Aquaculture in Pakistan began as a sideline and small-scale business for crop growers. However, with the establishment of public and private fish hatcheries, the smaller fish farms are turning into larger units facilitated by business investment into this sector. This is, however, area specific and confined to regions close to large cities such as Lahore and Multan in Punjab, and Thatha and Badin in Sindh where people own large areas of land. It is a common trend in Pakistan to manage large water bodies and establish large farms. Among the fishing communities, the participation of women in fisheries sector is routine. Women usually do not work in the business of independent companies due to social taboos. However, their participation in aquaculture activities, when it is a family enterprise, helps for fish feeding, pond management and surveillance when the farm is close to the residence. All the goals of aquaculture and fisheries cannot be achieved without capacity-building. The following are recommendations for this aim:

25.9.1 Integrated Fish Farming

The concept of fish culture integrated with crops and livestock production systems in Pakistan is not common. Very few technical efforts and little financial investment have been made to materialize this concept. The obvious reasons include the diverse ecological zones, a typical public religiosity, lack of education on part of farmers as well as consumers, and poor acquaintance with the latest technologies and advances in production systems. Investment plays a major role in the development of any sector. In livestock and fisheries industries, people objected that livestock fecal matter was recycled for fish pond productivity, and this discouraged eating habits at first. Small farmers can enhance their income significantly, with little extra inputs, by practicing integrated fish farming. Particularly among the rural population, women could also participate in poultry, livestock, and fish production activities. The increasing pressure to maximize quality food production with minimal input cost and energy conservation led to the idea of fish culture integrated with agriculture and animal husbandry. Despite a long history of practice in China and other Asian countries, the concept of integrated fish farming system was finally introduced to Pakistan. The integration of fish culture with cash crops and/or livestock carries great potential for enhancing production of animal proteins, creating opportunities of employment for the rural people, and to uplift their socio-economic status (Akhtar 1995). New horizons of increasing production per unit area have been opened through integrated fish farming for its ability to replace major inputs of high cost with animal manures and other agricultural by-products. It is a farming system of multiple commodities, with efficient waste recycling as the key feature, and fish culture as the major activity (Figure 25.3).

25.9.2 Cage Culture in Lakes and Reservoirs

Culturing fish in cages is the most viable option to a wide range of open water systems such as reservoirs. It can effectively exploit the natural productivity of such water bodies, and thereby share the burden of overexploitation on other water resources. The technology involved is very simple and based on locally available resources to construct and operate the cage, which renders it socially, economically, and environmentally acceptable. Due to the low-trophic level feeding habit of carp, carp fingerlings could be reared with minimum influence on the environment. Intensive culture of carp and catfish species makes wise use of resources, both natural and cost-effective artificial diet for fish feeding. Cage culture also eliminates predation-based losses and helps in prophylactic steps to address disease outbreaks, which ensures very high survival rates of fish. Manpower is effectively used, as the routine monitoring, management, and harvesting are very simple, time-efficient, easier, and perfect. High yields could be achieved with minimal inputs due to intensive culture of fish in

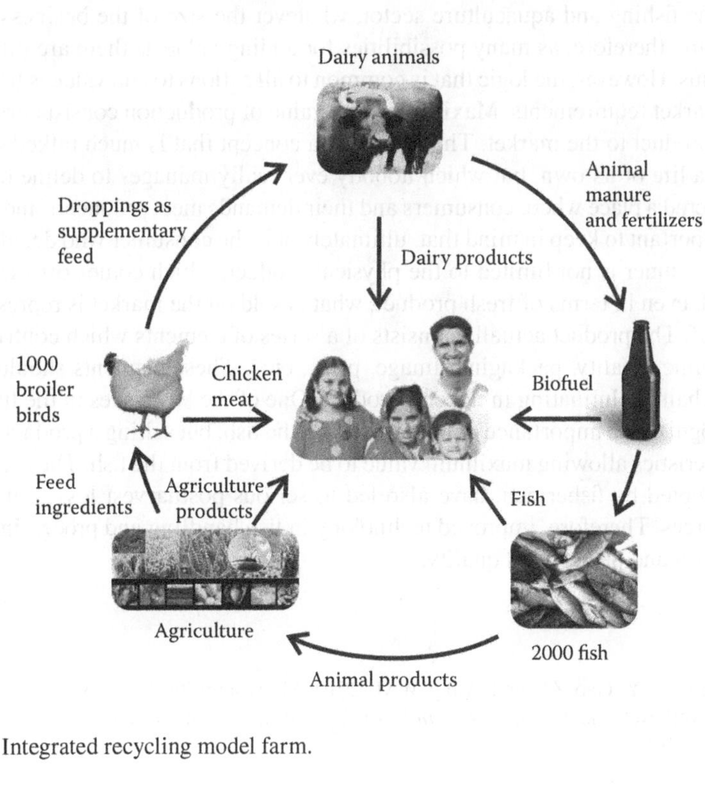

FIGURE 25.3 Integrated recycling model farm.

cages. Water reservoirs are mostly designed for multiple purposes, including supplying water for drinking, as such cage culture is suitable, causing minimal pollution and sustaining the ecological status of water bodies. Rearing of fry-fingerlings in cage culture requires less cost compared to pens and nurseries. An array of eight cages can provide fish seed sufficient to stock a water body covering 200 ha. In one year, three crops of fingerlings can easily be obtained.

25.9.3 Establishment of Fish Feed Industry in the Country

Good nutrition has a pivotal role for the production of healthy and high quality fish. Aquaculture is rapidly growing quality food production industry, and about 50% of all fish consumed in the world comes from aquaculture. In fish farming, cost of fish feed is critical because it represents 40%–50% of total production costs. Furthermore, high amounts of feed ingredients from marine sources have raised both environmental and economic concerns. Thus, it is imperative for the aquaculture industry to focus on the development of least-cost and ecofriendly local ingredients for the preparation of fish feed. During recent years, huge advancements have been made in the area of fish nutrition and new balanced commercial diets have been developed for optimal fish growth and health. The development of new species-specific diet formulae helps the aquaculture industry to cope with the growing demand of safe, affordable, and quality fish. However, the formulation of least-cost and nutritionally balanced fish feed needs the availability of raw ingredients to cater for energy, feeding, species-specific metabolic requirements and curtail the cost of feed. In Pakistan, farmers are facing a serious problem regarding the availability of nutritionally balanced formulated/processed feed as this industry has not been established yet. Therefore, a fish feed industry should be launched through public-private partnership, thus providing an impetus to the aquaculture business in Pakistan.

25.9.4 Postharvest Processing and Value Addition

It is important to realize that the issue of maximizing the value derived from production applies to all levels of the fishing and aquaculture sector, whatever the size of the business or the species targeted. There are, therefore, as many possibilities for adding value as there are different types of production systems. However, the logic that is common to all actions to add value is to ensure that the products meet market requirements. Maximizing the value of production consists first and foremost of tailoring the product to the market. The market is a concept that is much talked about, and that appears to have a life of its own, but which nobody ever really manages to define or comprehend. Market is considered a place where consumers and their demands meet producers and their products. However, it is important to keep in mind that, ultimately, it is the consumer who decides what to buy. The choice of consumer is not limited to the physical product, which comes off the boat or out of the pond. Indeed, even in terms of fresh product, what is sold on the market is represented by more than the fish itself. The product actually consists of a series of elements which contribute to its sale (availability, volume, quality, packaging, image, price, etc.). These elements should be developed along the value chain, culminating in the end product. One of the key issues in the fishing sector is, therefore, to recognize the importance of not just selling the fish, but selling a product that comprises different characteristics allowing maximum value to be derived from the fish. The traditional fishing technologies, adopted by fishermen, have also led to serious postharvest losses; it often involves wastage of resources. Therefore, improved technology in fish handling and processing will increase the shelf life of fish and preserve its quality.

REFERENCES

Abbas K, Zhou XY, Li Y, Gao ZX and Wang WM. 2010. Microsatellite diversity and population genetic structure of yellowcheek, *Elopichthys bambusa* (Cyprinidae) in the Yangtze River. *Biochem. System. Ecol.* 38: 806–812.

Agosthinho AA, Pelicice FM and Gomes LC. 2008. Dams and the fish fauna of the Neotropical region: Impacts and management related to diversity and fisheries. *Braz. J. Biol.* 68: 1119–1132.

Akhtar N. 1995. *Sustainable Fisheries: The Pakistan National Conservation Strategy Sector Paper.* Environment & Urban Affairs Division, Government of Pakistan, Islamabad.

Akhtar N, Khan S and Saeed K. 2014. Exploring the fish fauna of River Swat, Khyber Pakhtunkhwa, Pakistan. *World J. Fish Mar. Sci.* 6: 190–194.

Alam MS and Islam MS. 2005. Population genetic structure of *Catla catla* (Hamilton) revealed by microsatellite DNA markers. *Aquaculture* 246: 151–160.

Ayub M. 2007. Freshwater fish seed resources in Pakistan. In: Bondad-Reantaso MG (ed) *Assessment of Freshwater Fish Seed Resources for Sustainable Aquaculture.* FAO Fisheries Technical Paper. No. 501. FAO, Rome, pp. 381–394.

Collares-Pereira MJ and Cowx IG. 2004. The role of catchment scale environmental management in freshwater fish conservation. *Fish. Mang. Ecol.* 11: 30–312.

DOF. 2005. *Fish and Fisheries in the Punjab–A Manual.* Department of Fisheries, Punjab, Lahore.

FAO. 2009. *Fishery and Aquaculture Country Profile: Islamic Republic of Pakistan.* FAO, Rome, Italy, pp. 1–18.

FAO. 2014. *The State of World Fisheries and Aquaculture: Opportunities and Challenges.* United Nations Organizations, Rome, Italy.

GOP. 2011. *National Climate Change Policy.* Ministry of Environment, Government of Pakistan, Islamabad.

GOP. 2015. *Economic Survey of Pakistan—2014–2015.* Government of Pakistan, Islamabad.

Hussain M, Hussain SM, Afzal M, Raza SA, Hussain N and Mubarik MS. 2011. Comparative study on the effect of replacement of maize gluten with rice bran (3:1 and 1:3) feed supplement on fish growth in composite culture after yearling stage. *Pak. J. Agri. Sci.* 48: 321–326.

Jan A, Rab A, Ullah R, Shah H, Haroon, Ahmad I, Younas M and Ullah I. 2016. Current scenario and threats to ichthyo-diversity in the foothills of Hindu Kush: Addition to the checklist of coldwater fishes of Pakistan. *Pak. J. Zool.* 48: 285–288.

MinFAL. 2005. *National Policy and Strategy for Fisheries and Aquaculture Development in Pakistan.* Ministry of Food, Agriculture and Livestock, Government of Pakistan, Islamabad.

MinFAL. 2006. *Aquaculture and Shrimp Farming in Pakistan.* Ministry of Food, Agriculture and Livestock, Government of Pakistan, Islamabad.

Mirza MR and IA Sandhu. 2007. *Fishes of the Punjab, Pakistan.* Polymer Publications, Pakistan.

Nazish N and Mateen A. 2011. Winter growth of carps under different semi-intensive culture conditions. *Pak. Vet. J.* 31: 134–136.

Rafique M and Khan NUH. 2012. Distribution and status of significant freshwater fishes of Pakistan. *Rec. Zool. Surv. Pak.* 21: 90–95.

Simonsen V, Hansen MM, Mensberg M, Sardar MRI and Alam MS. 2005. Widespread hybridization among species of Indian major carps in hatcheries, but not in the wild. *J. Fish. Biol.* 67: 794–808.

SMEDA. 2014. *Inland Fish Farming: Pre-Feasibility Study.* Ministry of Industries & Production, Government of Pakistan, Lahore.

Sultana F, Abbas K, Xiaoyun Z, Abdullah S, Qadeer I and Hussnain RU. 2015. Microsatellite markers reveal genetic degradation in hatchery stocks of *Labeo rohita. Pak. J. Agri. Sci.* 52: 775–781.

26 Livestock and Poultry Health
Issues and Way Forward

M. Tariq Javed, Ghulam Muhammad, Nazir Ahmad, Laeeq Akbar Lodhi, Zafar Iqbal, Sajjad-ur-Rahman, Ahmad Din Anjum, Faqir Muhammad, Zafar Iqbal Qureshi, Muhammad Sohail Sajid, Muhammad Kashif Saleemi, Farah Deeba, Bilal Aslam, Aisha Khatoon, Muhammad Imran, Muhammad Imran Arshad, and Rizwan Aslam

CONTENTS

26.1 INTRODUCTION

The livestock sector is an integral part of agriculture. Its role is pivotal towards the socioeconomic development of the rural community. The 2016–2017 Economic Survey of Pakistan indicates that nearly 8 million families are involved in livestock raising, and are deriving more than 35% of their income from livestock production related activities (Pakistan Economic Survey 2016–2017). This sector has the potential to play an important role in poverty alleviation and foreign exchange earnings for the country. It is further documented that livestock contributes approximately 58.33% of the agriculture and 11.3% in the GDP (Pakistan Economic Survey 2016–2017). The total population of cattle, buffaloes, sheep, goat, camel, and horses are 44.4, 37.7, 30.1, 72.2, 1.1 and 0.4 million heads, respectively. The gross milk production is 56,080 thousand tones, which witnessed an increase of 1752 thousand tones from last year (Pakistan Economic Survey 2016–2017). The meat production is 4061 thousand tones, which witnessed an increase of 188 thousand tones over the last year.

Poultry sector is one of the vibrant segments of the livestock sector in Pakistan. This sector provides employment to over 1.5 million people. The investment in poultry industry is more than Rs. 700 billion (Pakistan Economic Survey 2016–2017). Poultry today has been a balancing force to keep check on the prices of lamb and beef, and is serving as the backbone of the agriculture sector. Poultry meat contributes 31% of the total meat production in the country. The oultry sector has witnessed a growth of 5–10%, which reflects its inherent potential. The poultry has contributed 1.4% of GDP, while its contribution in agriculture and livestock value added stood at 7.1% and 12.2%, respectively. The poultry value added has shown an increase of 7.7% (Pakistan Economic Survey 2016–2017).

Veterinarians have played a key role in the development of the livestock and poultry sector although diseases still pose a serious threat. The public and private sectors have engaged veterinarians to deal with diseases of various kinds, but this fight against diseases is an ongoing phenomenon. Hereafter, the importance of health-related issues affecting both livestock and poultry with suggestions to move forward have been ignored out so that this sector can play a much-improved role in the build-up of the economy and health of the nation.

26.2 SIGNIFICANCE OF ANIMAL DISEASES IN THE ECONOMIES OF DEVELOPED AND DEVELOPING COUNTRIES

Productivity of dairy animal's hinges on a variety of such factors such as their genetic potential, nutrition, management, reproductive efficiency, and the health status. Diseases adversely affect productivity and therefore sap farmer's profit in several ways, including but not limited to the following:

1. Mortality
2. Retarded growth
3. Suboptimal production
4. Lowered reproductive efficiency
5. Reduced market value of the animal products

According to the FAO, partial costing estimates of the impact of animal disease hover around 17% of turnover within the livestock sector of the developed world and 35–50% in the developing world (Bishop et al. 2009). There is limited local data on pecuniary losses associated with animal diseases in Pakistan. Reproductive disorders of dairy animals reportedly caused an annual loss of 11.5 million tons of milk in 2005, which translated into a loss of 120 billion rupees in buffaloes alone in Punjab (Younas 2006).

Animal diseases in many instances are also communicable to human (zoonosis) and the prevalence of infectious diseases in human is inversely related to IQ_s (intelligence quotients) of different nations (Daniele and Ostuni 2013).

26.3 MISSION OF THE VETERINARY SERVICE

The traditional mission of veterinary services have been to protect animal health only. Over the years, however, this thinking has radically changed and now control of risks along with the safety of the food chain of animal origin and animal welfare are also considered as essential components of veterinary services (OIE Terrestrial Code Commission 2008). The following are the current globally accepted objectives of official veterinary services in any country (Afzal 2009): (1) An effective control of animal diseases together with the capability to rapidly detect and diagnose these diseases; (2) minimizing and controlling the risks all along the food chain (veterinary public health) including zoonosis and food borne diseases; and (3) protection and promotion of animal welfare.

26.4 ISSUES RELATED TO ANIMAL HEALTH AND THEIR PROPOSED SOLUTIONS

26.4.1 Unreliable Animal Population and Other Relevant Data Required for Apportioning of Financial Resources and Planning

Reliable data related to the animal population and other aspects forms the basis for apportioning of financial resources for improving the health of animals in the country. However, there are some contradictions in these data in Pakistan. For example, according to data from Pakistan Economic Surveys from 1976 to 1997, the fodder crops cover the 16%–19% of the country's total cropped area. With minor fluctuations, this area decreased from 2.6 million ha in 1976/1977, to 2.45 million ha in 2005/2006, or about 1.6% per annum. Paradoxically, however, animal population in the country has witnessed an increasing trend (Pakistan Economic Survey 2015–2016). Nobody in his/her right mind would believe that animal population would register an increase with a concomitant decrease in area under fodder. Unreliable data create difficulties in the allocation of financial resources for animal health and other domains of animal agriculture. It is suggested that a fresh census of livestock to be conducted to correctly appraise the livestock population in the country.

26.4.2 Misplaced Focus of National Veterinary Service

The current veterinary service in Pakistan is treatment centric. The practice of curative medicine (treatment), sometimes called "Fire brigade" practice of veterinary medicine, is far costlier than the practice of preventive (proactive) medicine. For example, the cost-benefit ratio of prevention of mastitis is generally around 1:9 whereas, treatment of mastitis is generally disappointing. Taking stock of the higher cost effectiveness of preventive veterinary medicine than that of the curative medicine, the government of Punjab has recently started according preference to preventive veterinary medicine practices (e.g., vaccination, deworming, farmers training and education using print and electronic media) over the treatment of sick animals. The government should divert its animal health, budgetary resources to preventive health programs and should pick up the entire tab of preventive vaccination.

26.4.3 Weak Mechanism of Technology Transfer to the Farmers and Continuing Education

Several research institutes and universities in public sector conduct research on animal health related issues. However, the findings of their investigations are not relayed to the institutions engaged in rendering animal health related services. According to FAO, one of the major constraints to realizing the potential for increased animal productivity in Pakistan is the woeful lack of a mechanism to transfer existing technology from government research institutes to farmers (Anonymous 1987). It is suggested that all research institutes and universities should establish special cells within their administrative ambits for communicating the findings of their research investigations to provincial Livestock and Dairy Development Departments, dairy farmers, NGOs, and other agencies involved in the provision of animal health related services. Another problem relates to lack of a setup for updating the skills of animal health personnel. It is suggested that Livestock and Dairy Development Departments of all the provinces should establish a system of planned training of all in-service veterinarians and paraveterinary staff. It is further suggested that all in-service veterinarians and auxiliary personnel must undergo 3–6 months training before moving on to a higher rung of their professional career. The faculty of the veterinary schools can be utilized to this end.

26.4.4 Difficulties in Adoption of Technologies Due to Small Herd Size

One of the hallmarks of the dairying in Pakistan is extremely small herd size; 84% of farmers own only 1–4 animals/farmer (FAO 2011). Mass adoption of animal health related technologies is ostensibly difficult under this scenario. India has addressed this problem by creating dairy cooperatives. One of these dairy cooperatives (Amul) spurred the "White Revolution" and made India the world's largest producer of milk and milk products (https://en.wikipedia.org/wiki/Amul). Amul model can be replicated *mutatis mutandis* in Pakistan for mass adoption of animal health related technologies by smallholder dairy farmers.

26.4.5 Disease Diagnosis: Quality Issues and Reporting

Proper disease diagnosis is of utmost importance in veterinary health. Currently, in Punjab several disease diagnosis labs are working at district level. In these laboratories, most of the simpler diagnostic approaches are insufficient, including microbial isolation and identification, postmortem examinations, etc. The diagnosis relies on many things, the most significant of which is the use of diagnostic tests. Few diagnostic agents (produced by VRI) are also used, including Mallein, Tuberculin (Avian and mammalian PPD), the *Brucella abortus* agglutination antigen (Milk ring test, Rose Bengal plate test antigen, SAT antigen), and Salmonella pullorum stained antigen.

There is a lack of technical and technological facilities at district diagnostic laboratories. These district disease diagnostic laboratories may be converted into sample collection centers, to provide services to dairy and livestock farmers at local levels. There is a lack of reference laboratories in the country to perform correlation with the previous disease outbreaks in the area.

Remedial measures taken could be to provided rapid diagnostic kits to the district diagnostic laboratories. Molecular biology and serology tools should be provided to the district disease diagnostic laboratories. Clinical microbiology and diagnostic skills of the field veterinarians, especially disease investigation officers, should be built up by quarterly training programs. Such programs can be initiated in collaboration with the Institute of Microbiology/Department of Pathology at UAF and the Livestock and Dairy Development Department, Punjab. Several mobile training workshops (disease outbreak and investigation training) can also be initiated at remote and intensive livestock farming areas. In addition to traditional diagnostic laboratories, referenced diagnostic laboratories should also be maintained that can provide international recognition of the disease data. To facilitate reference diagnostic laboratories a culture bank facility should be initiated at the Institute of Microbiology (IOM) at UAF moreover, provincial disease diagnostic laboratories must be launched in the IOM, to provide quality milk and meat services from the joint forum of L&DD Department-UAF.

26.4.6 Parasitic Threats/Issues

Parasitic problems increase with intensification of production and the lack of attention to strict sanitation. In general, parasites can be divided into internal (worms, protozoa, infection with larval forms of arthropods, etc.) and external parasites (ticks, mites, infestation with larvae of arthropods, etc.). The major challenges of parasites to animal health and production in Pakistan include gastrointestinal worms/fascioliasis (grazing livestock), coccidiosis (poultry and small ruminants), ticks and blood parasites (large animals), and warble fly infestations (rangeland livestock).

Production losses due to parasitism may vary from one parasite to the other. Gastrointestinal nematodes, for example, may interfere with food absorption and thus adversely affect animal productivity (Soulsby 1982). In tick infestation, losses may be due to loss of blood (Hunter and Hooker 1907), tick paralysis (Varma 1993), transmission of different diseases (Ram et al. 2004; Teglas et al. 2005; Jonsson 2006; Bazarusanga et al. 2007), treatment costs (Figure 26.1) (Sutherst et al. 1982; Horn 1987; Cobon and Willadsen 1990), lowered milk yield (Jonsson et al. 2001), etc. Protozoa may also tax animal production due to poor growth, and reduction in body weight, milk, egg, and wool (Bhatia et al. 2004).

The prevalence of parasites has been reported from 10% to 90% in different agro-ecological zones of Pakistan in various species of animals (Siddiqi and Jan 1986; Khan et al. 1988, 1989, 1993, 2006; Pal and Qayyum 1992, 1993; Iqbal et al. 1993, 2005a, b; Maqsood et al. 1996; Sajid et al. 1999, 2008, 2009; Abbas et al. 2006; Raza et al. 2007; Asif et al. 2008).

Though, sporadically estimated on a small scale, the reported parasitic problems inflict heavy economic losses in the form of lowered production and mortality of the animals. Economic losses due to parasites have been estimated based on the losses of meat, milk, wool, quality of hides/skins, mortality of animals, and cost of treatments/management. In Pakistan, Iqbal et al. (1993) have estimated a loss of Rs. 31.43 million due to hemonchosis in Faisalabad, Pakistan. Likewise, Razzaq et al. (2012) estimated losses of Rs. 894.3 million annually, due to mortality of sheep and goats owing to worm infestations.

The economic losses have been estimated in billions in other countries due to similar parasitic diseases. For example, US$ 64.12 million in Kenya due to fascioliasis and hemonchosis (Mukhebi et al. 1985; Kithuka et al. 2002) and about US $ 800 million per year globally due to avian

FIGURE 26.1 Treatment of parasite infected animal through dipping solution.

coccidiosis (Williams 1998). Likewise, US$ 7 losses per animal due to tick infestation at the global level (McCosker 1979) has been reported. In Tanzania, economic losses due to tick-borne diseases resulting from production/mortality losses, treatment, and control costs associated with tick-borne diseases have been estimated at US$ 364 million (Kivaria 2006).

26.4.6.1 Strategies for Sustainable Parasite Control

The current parasite treatment and control in Pakistan largely relies on the use of synthetic chemicals (for all types of parasites); however, a small proportion of the farmers use medicinal plants/herbs (particularly for parasites of the gut) as a routine practice as "*masaullas*"/physic balls for multiple benefits. Vitamin-mineral supplementation/feed additives are also used by a large segment of the farmers as an adjunct/supportive treatment to reduce the production losses associated with parasites. There is no systematic/formal parasite control program in place in Pakistan. Animal health workers, including qualified veterinarians and para-vets lack proper knowledge/training on administration of antiphrastic as to their dosage, indication, etc. Consequently, antiphrastic are used indiscriminately and often are underdosed leading to the development of resistance.

In contrast to practices in Pakistan, animal health workers and farmers are far more aware of the options for parasite control in different situations. They have knowledge and practice for management of parasites, especially for worm infections. For example, the use of copper oxide particles for the control of gastrointestinal parasites, biological control using fungi and arthropods, grazing management, target-selective treatment, food supplementation, feed additives, etc. They go for more specific treatments based on an effective monitoring of parasitic burden through FAMACHA Card (Figure 26.2). Another practice is breeding of parasite-resistant breeds for the control of parasites of poultry and livestock.

26.4.6.2 Way Forward

The remedial measures, which may be considered important, are the followings: (*i*) adoption of integrated parasite control (IPC) practices. IPC is the application of two or more than two methods/options together in the control of parasites. These options include use of chemicals (all types of parasites), botanicals (gastrointestinal worms), vaccination (mainly for protozoal diseases), development of parasite resistant breeds of animals (worms and external parasites), nutritional supplementation to enhance resilience in the animals against parasites (all types of parasites), grazing management (gastrointestinal parasites and those having intermediate hosts), strategic and tactical use of different options, and biological/manual/mechanical control (larvae of worms and/or external

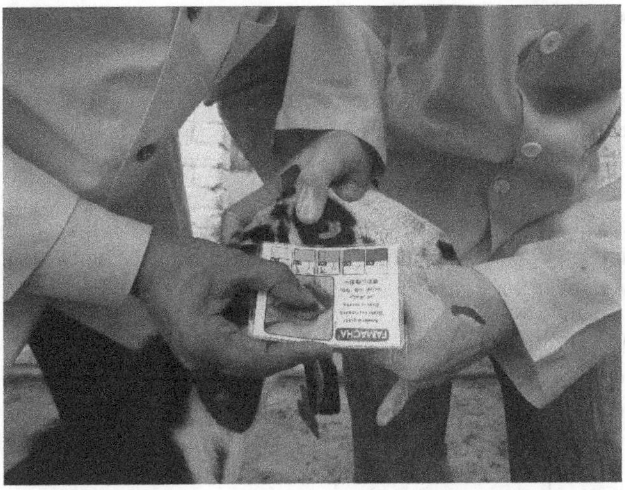

FIGURE 26.2 Targeted selective treatment through FAMACHA card.

parasites like ticks); (*ii*) development of modules and training/continuing education of the animal health workers on IPC options and practices; (*iii*) sensitization and awareness programs for the farming community and general public using electronic/print media, seminars, symposia, and so on; and (*iv*) investments to be made in research and development for start of the art diagnostic facilities, drug discovery, country medicine, parasite surveillance programs, capacity building, academia-industry linkages, outreach, etc.

26.4.7 Climate Change: Control of Vector Borne Diseases

Most of the arthropod borne infections travel worldwide and have a direct relation with the incidence of Dengue fever, Zika virus, etc. It is imperative to reduce the occurrence of such contagious infections through the rigorous control of the vector, keeping a surveillance on the flow of newly emerging infections through the regular monitoring of migratory birds to reduce the disease risk in the country.

GIS-based disease forecasting may be promoted to curtail the spread of diseases through strict quarantine measure around the endemic area. In the high-risk areas, veterinarian team may be sent in advance to prepare farmers about the possible adoption of management and health practices to safeguard precious animal resources.

In the face of natural and man-made disasters, the risk of disease spread may also deviate and it requires an immediate response from the veterinary health services. Therefore, an expert opinion may be provided to the farmers through information and communication technologies (ICTs) and media system so that further animal health losses may be avoided through timely intervention at the farm and proper vaccine coverage may be implemented in the domestic and wild animals of the affected area.

26.4.7.1 Issues and Remedies

The issues include: (1) climate change temperature, humidity and global warming; (2) increase in vector density and associated disease morbidity, for example, Dengue fever, West Nile virus infection, Bovine ephemeral fever, *Francisella talurensus* (talurenia); and (3) tick borne diseases, for example, Spirochetosis in poultry.

The remedial measures possibly adopted include: (1) preparation of disease preparedness and response strategies; (2) early warning systems; (3) vaccination in flooded areas; (4) establishment of buffered zones; and (5) restriction of animal movement policies for epidemic areas.

26.4.8 One Health Approach to Control Zoonotic Diseases

Traditionally, every type of disease has been thought to embrace three basic determinants, that is, host, agent, and environment. However, owing to recent global expansion of the host spectrum, the disease scenario is changing. The animal-human-environment interaction is playing a role development of new disease and development of zoonotic pathogens. This process of zoonosis has led to several emerging and reemerging diseases in livestock, poultry, and humans as well. This one health approach integrates veterinarians, medical physicians, researchers, environmental, and social scientists. Ultimately, these professionals are uniting in a common platform to better understand the high risk of infection prevailing in the community. Moreover, such efforts can lead to appropriate biosurveillance of various diseases.

26.4.8.1 Issues and Remedies

The issues include: (1) recurrent pandemic outbreaks of reemerging diseases such as Bird-flu, Ebola virus, Zika virus can lead to major economic losses in livestock and poultry and to the public health; (2) quarantine issues are also a source of zoonosis and reemergence of diseases; (3) improper or unrestricted import and export of the livestock/poultry and their products are also a major source of infection; (4) horizontal and vertical transmission of antimicrobial resistance leads to the emergence of multidrug resistant bacterial infections and ultimately contribute to health and economic losses all over the country; and (5) the reemergence of diseases can lead to public health risk of disease spread and occupational risk.

The remedial measures could be: (1) national disease surveillance system should be initiated which may provide epidemiological investigation of all the common and rare infections prevalent throughout the country; (2) in addition, antimicrobial drug residue monitoring system should be developed which can control the emergence of multidrug resistant microorganisms; (3) training programs and awareness seminars should be launched for professionals to properly regulate working conditions. Adaptability to this new concept can be increased by proper advocacy and awareness programs; and (4) risk assessment data of all types of healthcare units should also be maintained to prevent outbreaks and the chance of zoonosis.

26.4.9 The Disease Surveillance and Quarantine System

The country has very weak disease surveillance and quarantine system. This system should be the backbone of the livestock disease eradication and control program. This system not only provides information on current disease status of all the diseases of animals in different areas of the country, but also provide important information on the emergence of new diseases and the changing face of existing diseases.

The quarantine system provides the disease-protected areas where animals are kept before moving or when enter a new area. This unit plays an important role in the development of livestock of any country.

The disease surveillance system should be a central provincial unit, but working on multiple ends. There is need to have a data collection and reporting system. To collect and use veterinary health surveillance data, the central unit staff works closely with divisional and local veterinary health departments; experts in the field and institutes; stakeholders should have its own team of collection of data. This unit should develop and implement consistent standards, tools, training, and technology to help ensure that disease-reporting systems is working well at tertiary level. The surveillance system should also have a close look at the animals being slaughtered, any death occurring in animals or birds anywhere in the homes, farms, fields, and even in the jungles. There is need to have a network of laboratories and those should also be linked to the animal disease surveillance unit.

26.4.9.1 Lack of Information on a Full Spectrum of Prevalent Diseases and Their Epidemiology

Notwithstanding that epidemiology of animal diseases has been identified as one of the three national health research imperatives for animals in Pakistan (Afzal and Usmani 2005), the full spectrum of diseases is yet unknown. A national project entitled "Epidemiology of Major Livestock Diseases in Pakistan" executed in the 1990s focused largely on a handful of clinically important diseases. Glaring omissions in this project included such economically important diseases as infectious bovine rhinotracheitis, parainfluenza 3, rota/corona virus infections, bovine respiratory syncytial virus, bovine viral diarrhea, Q fever, and ephemeral fever. Although, these diseases are currently important in dairy animals in Pakistan, information on their prevalence and their epidemiology is extremely sketchy. The ongoing trend of establishing large dairy farms with the import of exotic breeds of dairy cows underpins the need to know the complete spectrum of prevalent and emerging diseases of livestock. The government should, therefore, launch a new epidemiological project with a view to determine the complete spectrum of dairy animal diseases with greater emphasis on those currently important diseases which were overlooked in the 1990s.

26.4.9.2 Lack of a Proper Setup to Intercept the Spilling Over of Contagious Animal Diseases from Neighboring Countries

Pakistan currently lacks a setup to intercept the spilling over of contagious animal diseases from neighboring countries (Saudi Arabia, Iran, Afghanistan, and India). Rift Valley fever is a highly fatal viral infection of sheep and other ruminants that has been reported in Saudi Arabia and Yemen (Gould et al. 2006). It can spill over to livestock in Pakistan with potentially tragic consequences. The government should setup an animal surveillance unit to keep a close watch on infectious diseases of animals in the contiguous counties and prepare a contingency plan to intercept their spilling over to livestock populations in Pakistan.

26.4.9.3 Transmission of Animal Diseases through Unrestricted Movement of Animals

Unrestricted movement of animals from one area to another is the main way to spread infectious diseases, particularly FMD and PPR. Outbreaks of FMD occur particularly at the time of Eid-ul-Azha, when a large number of animals move from one geographical locale to another (Anjum et al. 2006). There is a need to register every animal in the province kept at home or on a farm (Javed 2007). Registering means that every animal in the country should be numbered and identified (data fed on computers). Necessary tests of diseases should be carried out and the data stored. The provincial livestock department may be asked to do this job. A nominal fee for registration may be charged, but all laboratory tests for specified diseases should be free.

There is need to ban movement of all kinds of animals across villages, cities, etc. The movement of animals from one place to another may be allowed under supervision and certification. When animals are supposed to be moved by request of the owners, a fresh testing against a set of diseases should be done at the owner's expense. This certificate should be carried along with the animal and compulsory data should be entered into the computer at the new place. Every district or administrative area should have the right to reject any animal with a specific disease entering their jurisdiction.

26.4.9.4 The Current Disease Control/Eradication Program

Pakistan has successfully eradicated rinderpest disease, which is a milestone. Currently, Pakistan is trying to control the PPR and FMD diseases. In disease eradication, the animals are not vaccinated for the disease to be eradicated but when animals with that disease are identified, they are killed and buried deep or incinerated. In disease control programs, different measures are adopted, which include management, biosecurity, effective treatment, and vaccination. Through these procedures the impact of disease in animals is kept to a minimum and thus diseases are kept under check.

There is need to expand this system of eradication and control to target more diseases in the program. Bovine tuberculosis and glanders are diseases for which developed countries started eradication programs in late 1800s. Many countries have eradicated this diseases from their animals and are now declared bovine tuberculosis and glanders free. Currently, the world is addressing the paratuberculosis and brucellosis, the programs in different countries for eradication are in full swing. Furthermore, the other steps taken by developed countries were to supply only pasteurized milk to their people to prevent infection from TB infected animal through the milk to humans. However, the pasteurization process also helped to store the milk for longer durations, which was an added advantage. Unfortunately, nothing has yet been done in Pakistan for zoologically important diseases (tuberculosis, paratuberculosis, brucellosis, and glanders). The data suggest that these diseases exist in the country and their magnitude is increasing. The reports before 2001 indicated less than 10% prevalence of bovine tuberculosis (Amin et al. 1992; Rehman 2001). However, later an increase in animal and herd level prevalence of tuberculosis has been reported, which at some farms was beyond 10% (at animal level), with 100% herd prevalence (Javed et al. 2010, 2011). Similar is the situation for brucellosis and other diseases (Gul and Khan 2007). The best way to move forward in Pakistan is now, when the disease incidence is low, then less number of animals are to be sacrificed. Furthermore, dairy or beef farming is just in its infancy stage. The farmers should be educated right now rather than later when their farm are plagued with the disease-causing organisms.

26.4.9.4.1 The diseases of significant importance for control or eradication

There are a number of diseases that are of significant importance in the animal sector and need to be eradicated or controlled. Tuberculosis (Figures 26.3 and 26.4), brucellosis, and paratuberculosis are diseases of significant importance to Pakistan caused by bacteria, and diseases like rabies, Q fever, and Cowpox are of significant importance to Pakistan, which are caused by viruses. These diseases in animals are also responsible for production losses, reproduction losses, and making animals economically unsuitable. Diseases that cause heavy economic loses include FMD, PPR, HS, and black quarter along with some others whose status is yet not very clear, for example, Q fever, West Nile Virus disease, etc. There are still some other diseases of high significance those are metabolic, deficiency diseases, and some other caused by multiple disease agents like mastitis.

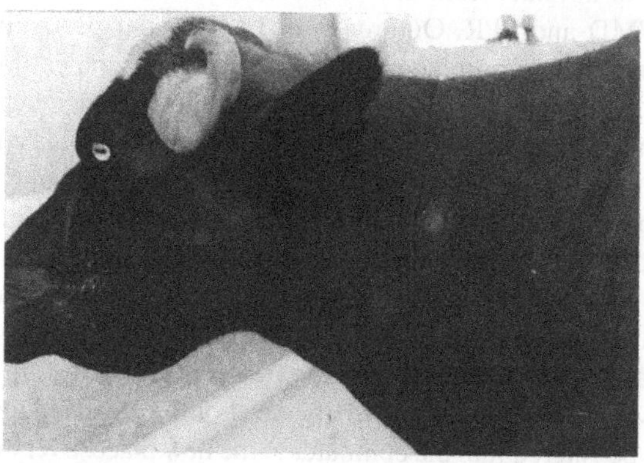

FIGURE 26.3 **(See color insert.)** Tuberculosis positive buffalo indicated by swelling on the neck in tuberculin test.

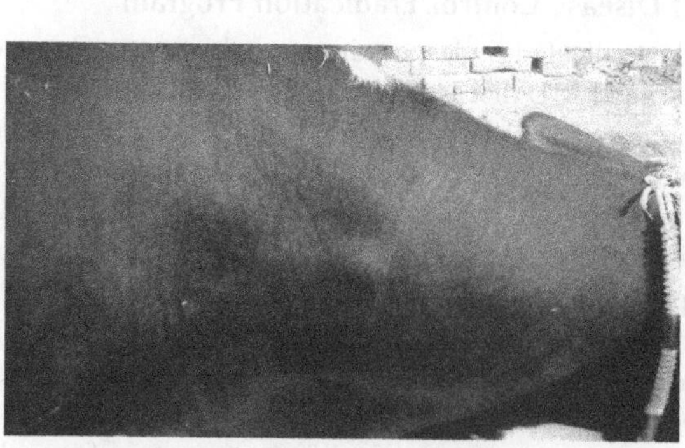

FIGURE 26.4 **(See color insert.)** Tuberculosis positive cattle indicated by swelling on the neck in tuberculin test.

26.4.9.4.2 *Disease Eradication and Control Program*

To eradicate tuberculosis and other diseases, stamping out methods are being used all over the world (Javed et al. 2014). Stamping-out methods involve the eradication of bovine tuberculosis by killing all the infected animals and require compensation to farmers, otherwise they will be reluctant to participate. The components of disease eradication program are: (1) setting up a task force to put up suggestions for necessary early regulations, framework, goals, and objectives to be achieved; (2) making a setup/body at the provincial level with its wings in each division and going down to the sistrict or Tehsil level, and setting up the necessary infrastructure, including reference lab, provincial, divisional, and district labs, internet resource setups, printing units, etc; (3) setting up control program by the provincial body with further necessary legislation, TORs, and SOPs at different levels/bodies, etc; (4) setting up the infrastructure of stakeholders with necessary legislation, TORs, and SOPs, etc; (5) creating an awareness among stakeholders and general public from time to time through electronic and print media; and (6) revisiting the control program after 5 years and making changes if necessary.

The other way of controlling a disease is by vaccination. Everyone should comply with vaccination programs. The program must be maintained for a long period of time, that is, 5 to 10 years. The surveillance of the diseases in animals is important so as to know where a specific disease is present at high magnitude, so that effective measures can be taken to minimize the spread of disease to humans. This surveillance will have to be a continuous process covering all animals. Slaughterhouse surveillance is also important.

Testing of animals for disease require comprehensive and highly equipped laboratory setups for disease eradication and control programs. The laboratory setups are very important and should also act as source of information to all concerned about diseases in the country. There is need to have a reference laboratory for every important disease of animals. Every district laboratory should be responsible for doing initial testing of samples received from villages. The sample collection can be done by sample collection units, like livestock dispensaries/hospitals and artificial insemination centers. Every district laboratory should send portions of sample to every reference laboratory for testing of samples with state-of-art testing procedures and confirmation of results. The reference labs should be responsible for storing disease data with full details and should be the source of information to the government for those specific diseases.

26.4.10 Vaccines and Drugs

26.4.10.1 Insufficient Quantum of Vaccines Produced and Their Questionable Efficacy

According to the Livestock Vision 2025 document prepared by the University of Veterinary and Animal Sciences, Lahore in 2015, only about 15–25% of the animal population in Pakistan is vaccinated. Only 10–15 million doses of foot-and-mouth disease (FMD), *peste des petits ruminants* (PPR), and hemorrhagic septicemia (HS) vaccines are being produced as against 100 million doses required. This insufficient vaccination coverage leads to colossal economic losses in terms of unacceptably high morbidity and mortality as well as reduced productivity. Current dairy animal vaccination programs are generally directed only against HS, FMD, black quarter, and anthrax. Large dairy farms, managing imported exotic dairy cows, also vaccinate against infectious bovine rhinotracheitis, parainfluenza 3, bovine respiratory syncytial virus, bovine viral diarrhea, ephemeral fever, and leptospirosis by using imported or smuggled vaccines. Owing to their exorbitantly high prices, the indigenous production of these vaccines is obviously in order to make them affordable by resource-poor smallholder dairy farmers.

Incidence of vaccination failures with vaccines produced in the public sector is unacceptably high. Some locally manufactured vaccines of questionable efficacy undermine the confidence of farmers in the effectiveness of vaccination as a prophylactic tool (Muhammad 2012). It is proposed that an independent Vaccine Testing Authority be established and the quantum of vaccines produced locally be scaled up to commensurate with the size of animal population in the country.

26.4.10.1.1 Indigenous Vaccine Production Policy

In Punjab, Veterinary Research Institute (VRI), Lahore is the sole indigenous vaccine-producing institute. Currently, VRI is producing two major types of vaccines: Bacterial and viral vaccines both for livestock and poultry. The bacterial vaccines of VRI are: hemorrhagic septicemia vaccine, black quarter vaccine, Listeriosis vaccine, Contagious Caprine Pleuropneumonia vaccine, Enterotoxaemia-cum-lamb dysentery vaccine, Anthrax spore vaccine, and Spirochetosis vaccine. While the viral vaccines of livestock and poultry include rinderpest vaccines, foot and mouth disease vaccines, antirabies vaccines, sheep pox vaccines, goat pox cell culture vaccines, Newcastle disease vaccines, Fowl Pox vaccines, and Hydropericardium Syndrome vaccines.

In addition to vaccines, some of the antisera are also produced. These are obtained from hyperimmuned buffalo calves. Foot and mouth disease antisera are the most commonly used products for transient immunity.

26.4.10.1.2 Issues and Remedies

Vaccine shortage is one of the main hurdles to the disease control programs started in Punjab. Strain specific vaccines are not currently produced in VRI. Vaccine failure is due to the availability of low quality vaccines and poor storage conditions. There is no proper facility of vaccine storage and distribution, especially in remote areas.

The remedial measures could be that (1) the exploitation of local vaccine production should be initiated by using the local strains of the pathogens; (2) vaccine efficacy testing centers should be established with the task to test production batches and to regulate the quality of the vaccine after distribution to remote vaccine distribution units; (3) vaccine surveillance and monitoring centers should be activated with a task of ensuring comprehensive vaccination programs; (4) one shot vaccine is the newest concept to decrease the economic burden of the small farmers.

26.4.10.2 The Role of Drugs in the Treatment and Control of Animal Diseases

Several animal diseases are treated and controlled with the help of drugs. This is accomplished through the general principle that drugs modify the already existing physiological processes of animal bodies. Interference in these physiological processes with the drugs needs careful and wise use, especially in food producing animals.

In human medicine, while launching a drug in the market, its clinical efficacy and therapeutic effects are balanced with respect to the risks of drug resistance and host toxicity. Unlike human medicine, veterinary medicine should be given additional consideration, especially when drugs are used in food producing animals because of drug residue issues in edible animal tissues such as milk, meat, and eggs. The ultimate consumers of these animal foods are human beings. Hence, rational use of drugs is of utmost importance in veterinary medicine (Sanders 2007).

26.4.10.2.1 Injudicious Use of Drugs

Many veterinary drugs are being used injudiciously in Pakistan. The use of dewormers is a particular case in point. Research conducted on gastrointestinal parasites of sheep and goats in most of the countries of the world, including Pakistan (Saeed et al. 2010) has shown that many parasites have developed resistance to almost all classes of currently available dewormers (anthelmintics). This is partly because of injudicious use of dewormers. About 20% of animals (mainly young animals and those which are underfed) in a flock or herd harbor 70–80% of the parasites. Modern recommendation is that this group (i.e., animals carrying a high number of parasites) should be identified, treated with dewormers and kept away from the rest of the flock/herd, which is not treated. In other words, deworming of all animals in a flock/herd is unnecessary. Contravening this recommendation, the Livestock and Dairy Development Departments in different provinces regularly conduct mass anthelmintic treatment campaigns whereby animals are treated with dewormers regardless of their internal worm status.

In order to slow down the speed of development of anthelmintic resistance in *Hemonchus contortus*, South African scientists have developed a method for Targeted Selective Treatment whereby only anemic sheep and goats in a flock are selected for treatment against this parasite. This method is called FAMACHA© (diminutive word for FAffa Malan CHArt) anemia guide. In this method, the color of the mucous membrane of the lower eyelid of sheep or goat is matched to the FAMACHA© eyelid color score (1–5) system printed on a card. The use of FAMACHA© card helps to identify sheep or goats that are suffering from severe anemia (score 3–5) and need immediate deworming and possibly supportive treatment. The FAO has been supporting the validation and worldwide application of FAMACHA as a part of the Targeted Selective Treatment Strategy in sheep and goat for reducing the development of anthelmintic resistance. However, by its nature, the application of FAMACHA is limited to a few blood sucking parasites (particularly *Hemonchus contortus* and liver fluke) which produce anemia. Several other important parasites are not targeted in this system. South African scientists (Bath et al. 2010) have introduced a more comprehensive system (called Five Point Check© abbreviated as 5•√©). Adoption of the FAMACHA© and Five Point Check© system-based deworming for sheep and goat will not only check the development speed of anthelmintic resistance but would also avoid frittering away financial resources of government and farmers. It is strongly recommended that L&DD Departments should shun the practice of blanket anthelmintic treatment by adopting the FAMACHA© and Five Point Check© systems.

Tragic consequences have been reported due to self-treatment of animals using over-the-counter drugs like ivermectin (Muhammad et al. 2004). Self-treatment of animals by the farmers, due to over-the-counter availability of drugs, is one of the important reasons for overdosing/under dosing leading to toxicities, lack of response to the drugs used, as well as development of resistance to antibiotics and anthelmintics. The use of oxytocin for milk let down, particularly in buffalo in Pakistan is rampant (Khan et al. 1987; Wet et al. 2010). Administration of oxytocin induces several undesirable changes in milk composition (Hameed et al. 2010, 2016). The sale of oxytocin in India has been banned under section 12 of the "Prevention of Cruelty to Animals Act," "The Food and Consumable Substances Adulteration Act," and "The Drug Control Laws." Oxytocin in India is a scheduled substance, meaning thereby that this drug can only be supplied with the prescription of a registered medical or veterinary practitioner. It is suggested that in Pakistan too, oxytocin should be granted the status of a scheduled drug and made available only through registered veterinary practitioners.

In Pakistan, the most common irrational use of drugs includes: (1) use of too many drugs per patient (polypharmacy); (2) inappropriate and unnecessary use of antibiotics in food producing animals; (3) overuse of injections instead of oral medications; (4) prescribing without clinical guidelines; (5) quackery or self-medication by animal owners; and (6) failure to observe proper drug withdrawal times in animals.

The followings are well-established ways to estimate the type and degree of irrational use of drugs: (1) cumulative drug used data to identify costly drugs with lower efficacy; (2) Defined Daily Dose (DDD) methodology to compare drug use among institutions and regions; (3) focused drug evaluations to know the problems related to the use of specific drugs in specific diseases; and (4) the qualitative procedures established in social science, such as focused group discussions, in-depth interviews and structured observation/questionnaires).

Unwanted effects of irrational use of drugs include idiosyncratic or allergic reactions, anaphylactic shock, multiple organ dysfunction, mutagenesis, carcinogenesis, and teratogenesis. Thus, in veterinary clinical practice, veterinary physicians must be very careful in considering the frequency of application, dose of the drugs, and side effects, especially in the case of banned drugs (Ćupić and Dobrić 2003; Giguere et al. 2006).

26.4.10.2.2 Core Interventions to Promote Rational Use of Drugs

The core interventions include, (1) establishment of a multidisciplinary national body to formulate and implement policies on drug use; (2) strict observation of clinical guidelines; (3) identification and use of a national essential drugs list; (4) constituting drug use monitoring committees in the districts and hospitals; (5) inclusion of problem-based pharmacotherapy trainings in undergraduate curricula; (6) continuing in-service veterinary medical education as a licensure requirement; (7) supervision, audit, and feedback; (8) public education about veterinary drugs, especially following recommended drug withdrawal times; (9) avoidance of perverse financial incentives; and (10) government priority to ensure availability of essential veterinary drugs and staff.

26.4.10.2.3 Responsibilities of Drug Regulatory Authorities

The responsibilities of drug regulatory authorities include: (1) establishment of clinical guidelines for veterinary health institutions; (2) selecting cost-effective and safe drugs (hospital/ district veterinary drug formulary); (3) implementing and evaluating strategies to improve drug use in consultation with antibiotic and infection control committees; (4) providing in-service staff education/trainings; (5) controlling promotional activities of pharmaceutical companies; (6) monitoring and prevention of adverse drug reactions; and (7) providing advice on drug quality, cost and availability.

26.4.11 Fading Focus on Complementary and Alternative Veterinary Medicine

Hippocrates, Father of modern medicine coined a phrase "*Vis medicatrix*" means nature is the physician of diseases, presently recognized as Complementary and Alternative Medicine (CAM) in human medicine and Complementary and Alternative Veterinary Medicine (CAVM) in veterinary medicine.

26.4.11.1 Complementary and Alternative Veterinary Medicine: Global Status vis-à-vis Pakistan

In livestock farming, antibiotics, and growth promoters are handed out like sweets. At lower rates of ingestion, the effects are usually subclinical. Excessive use of antibiotics and growth promoters in animals reduces their fertility and immune status. Moreover, if drug residual levels increase in the food chain, these drugs pose a threat to human medicine. We are polluting the environment so veterinarians should make-up their mind to save future generations (Knapton 2016). The Indus valley civilization is one of the foremost contributors in the history of development of Veterinary Science. In Pakistan, 66% population lives in rural areas. Around 70%–80% population uses complementary alternative medicine (CAM) as a first line of treatment. Among various therapeutic modules of CAM, the herbal and homeopathic systems are the most prevalent (Figure 26.5; Shaikh and Hatcher 2005). So, it can be generalized that livestock farmers prefer complementary alternative veterinary medicine (CAVM) over the allopathic system for their animals. Studies in Europe and Pakistan concluded that allopathic medicine and CAVM systems provide similar levels of protection against infectious diseases in livestock (Spranger 2000; Kenyanjui and Sheikh-Ali 2009). Affordability is one of the most important virtues of the CAVM system. Drawbacks to modern veterinary practice include questionable quality of allopathic drugs, the development of chemoresistance in livestock and user-friendly effects such as high antibiotic and hormone residues in the milk and other animal products (Monteiro et al. 1998; Deeba et al. 2009). Excessive use of oxytocin and bovine somatotropin hormone are banned in Pakistan due to their adverse effects in dairy animals and humans (Ilyas 2015). Organic foods (meat, milk) come from animals given no antibiotics or hormones. Pakistan cannot afford to neglect organic farming. It is not a new concept for Pakistan, because our predecessors used CAVM for their animals and were organic farmers. After CAVM adoption, it doesn't mean that we are following western countries, but revitalizing our traditional medicine. By fulfilling the increasing demand of organic products for health-conscious consumers Pakistan can earn heavy foreign exchange through export of organic foods. In the long run Pakistan will be one of the suppliers of organic products.

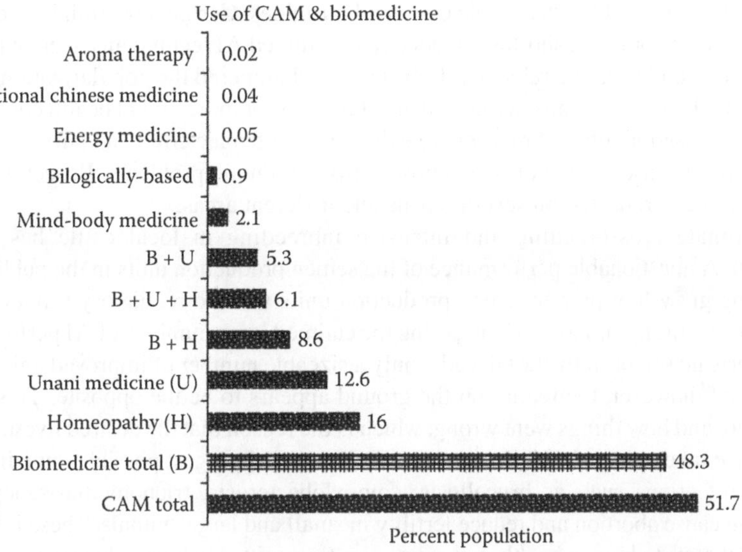

FIGURE 26.5 Use of complementary and alternative medicine and biomedicine. (From Hussain SS et al. 2009. *J. Alter. Complement. Med.* 15: 545–550.)

26.4.11.2 Suggestions

Complementary and Alternative Veterinary Medicine basic courses should be incorporated in DVM curriculum in veterinary institutes to prepare a new generation of veterinarians well equipped with traditional health care system. Collection and compilation of traditional veterinary knowledge to protect Intellectual Property Rights Pakistani farmers draw on over a millennia of knowledge and experience of CAVM. Priority research areas in CAVM (Phytotherapy and homeopathy, etc.) should be funded without discrimination. Scientific validation of already available products (efficacy/safety/quality) is required to win the confidence of consumers and provide accessibility to adulteration-free affordable complementary and alternative therapeutic modules. Resuscitation of CAVM will spur organic livestock farming in Pakistan.

26.4.12 Reproductive Herd Health Management

The production of livestock is directly related to reproduction. The data for last 3 years has recorded an increase in the number of cattle, buffalo, sheep, and goat population and so is the corresponding increase in the milk and meat production (Pakistan Economic Survey 2016–2017). The real question is the per head increase in the productivity of animals. The late puberty, long calving intervals, low conception rate, and high incidence of reproductive disorders in female animals have hampered the rate of reproduction. Since reproduction has a bearing on production, this has resulted in low per head productivity in the presence of inadequate nutrition, poor breeding policy, poor management and husbandry, high losses due to diseases, and marketing problems.

Artificial insemination (AI) has proven a useful technology to improve the milk herd in several countries of the world. Denmark started AI in 1936 through a "Cooperative Cattle Association" and the other countries followed this practice. In a short span of 20 years, Denmark, Japan, and Israel were using AI to breed over 90% of their cattle population while the United States were using it for over half of its population by that time. Artificial Insemination was started in Pakistan in 1954 as "Experimental Trials in AI" and later, with development of the independent Directorate of AI, it spread rapidly in the whole of the Punjab. Unfortunately, even 55 years after its introduction, AI is used for less than 30% of the breedable cows and buffaloes. Certain constraints, including shortage of genetically superior bulls, shortage of adequately trained AI technicians, lack of infrastructure to educate farmers, and lack of a reliable AI service have hampered the popularization of AI in Punjab (Lodhi 1997). The veterinarians in the field also failed to combine sexual health control and AI. This resulted in increased number of reproductive disorders in female cows and buffaloes and the spread of reproductive diseases, some of which pose serious zoonotic problems. Brucellosis, for example, is one that has been reported on several farms and different areas.

Indiscriminate crossbreeding and intensive inbreeding in local cattle has caused genetic deterioration. A questionable performance of the semen production units in the public sector and the mushrooming growth of private semen production units with no regulatory framework have added fuel to the deteriorating situation. Considering the claimed large numbers of AI performed by both the public and private sector in the last decade, only a sizeable number of improved animals should have been produced; however, the reality on the ground appears to be the opposite. This is an unending story as where and how things went wrong, which is the reason that the desired livestock development could not have taken place. It is better to identify issues and suggest possible remedial measures.

Specific infections such as brucellosis, compylobacterosis, trichomoniasis, leptospirosis, and listeriosis can cause abortion and reduce fertility in small and large animals. These infections can also pose serious threats to human health as they are also transmitted to human beings through contact with infected animals or consumption of their products. However, there are no official reports regarding the occurrence of these infections among domestic animals in Pakistan, except brucellosis for which a few reports by different workers are available. Even for brucellosis, the incidence varies widely from region to region and from species to species. For example, its prevalence has been reported to vary from 0.5% in cattle on government livestock farms (Nasir et al. 1999) to 6.9% and 6.6% in cattle and buffaloes of the Rawalpindi/Islamabad region (Ali et al. 2013) and 9.33% in goats from Azad Kashmir (Din et al. 2013).

Nonspecific uterine infections are the major cause of endometritis, which is one of the most common reproductive disorders of cows and buffaloes (Akhtar et al. 2009) and is the main cause of repeat breeding (animal is not conceived and coming to heat again and again) and reduced fertility in these dairy animals. These infections gain entrance into the uterus at the time of service or parturition. Noninfectious problems like anestrous and silent heat are also among the causes of infertility in the buffaloes, while cystic ovarian degeneration affects fertility in cattle.

26.4.12.1 Assisted Reproductive Technologies

Besides AI, other modern assisted reproductive technologies are being applied for reproductive management in animals. These include: (1) multiple ovulation and embryo transfer (MOET); (2) in vitro embryo production (IVEP); (3) sexing of sperm/embryos; (4) intracytoplasmic sperm injection (ICSI); and (5) animal cloning.

26.4.12.1.1 Multiple Ovulations and Embryo Transfer

Multiple ovulations and embryo transfer (MOET) is a reproductive biotechnology by which the widespread use of superior females is made possible through induction of multiple ovulations. These days MOET has been well established in advanced countries, and more than 80% of embryos produced for commercial purposes are obtained through this technique. In this technique, donor females are subjected to multiple ovulation treatment using pituitary (FSH) or placental (eCG) gonadotropins during the luteal phase. The donor is inseminated with two doses of semen from proven sires at the interval of 12 hours. Embryos are recovered through flushing of the uterus around 7 days after insemination and are transferred to the recipient after necessary treatment, so that donor and recipient are at the same stage of the cycle (Dawuda et al. 2002). The main use of MOET is in a breeding program, where this technique is applied to increase the number of young-ones obtained from animals of high genetic potential. In MOET program, a central breeding herd is established, from which necessary data is obtained for the genetic selection in order to produce a herd of high genetic potential to improve the overall performance of the herd (Bogh and Greve 2009).

26.4.12.1.2 In Vitro *Embryo Production*

In vitro embryo production (IVEP) consists of the oocyte collection, in vitro maturation of collected oocytes (IVM), in vitro fertilization of oocytes with capacitated spermatozoa (IVF), and in vitro culture of the presumptive zygote (IVC; Bogh and Greve 2009). The IVEP has emerged as a useful technique to overcome the limitations of MOET. Oocytes are generally collected from the ovaries of animals after slaughter. But ovum pick-up through ultrasound guided follicular aspiration has emerged as a repeatable procedure to get oocytes from live animals in large numbers. This technique does not affect the normal productive and reproductive cycles of the donors. Oocytes can be collected from any female starting from 6 to 3 months of pregnancy, and even 2–3 weeks after parturition. Ovum pick-up can be done twice a week, or sporadically for many weeks or even months, without any after effects (Figures 26.6 and 26.7).

26.4.12.1.3 *Sexing of Sperm/Embryos*

For ages, farmers have had a desire to choose the sex of their livestock offsprings. These days, the new technologies make it possible to control the sex of the offspring through using sexed semen or embryo. Sexing of sperm or embryo has a number of positive points (Kakar et al. 2012): (1) farmers can believe in the importance of AI in the genetic improvement of their livestock; (2) farmers can produce the calf of their desired sex, male or female; and (3) there is no need of application of any additional reproductive management technology with AI, such as estrus synchronization.

26.4.12.1.4 *Intracytoplasmic Sperm Injection*

In Intracytoplasmic Sperm Injection (ICSI), a single sperm is directly injected into the egg in order to fertilize it. The procedure for ICSI is similar to IVF, but instead of performing fertilization in a dish, the expert selects sperm from the sample and a single sperm is injected directly into each egg. The fertilized egg (embryo) is then transferred to the recipient female. This technique is used as a treatment of infertility of male origin.

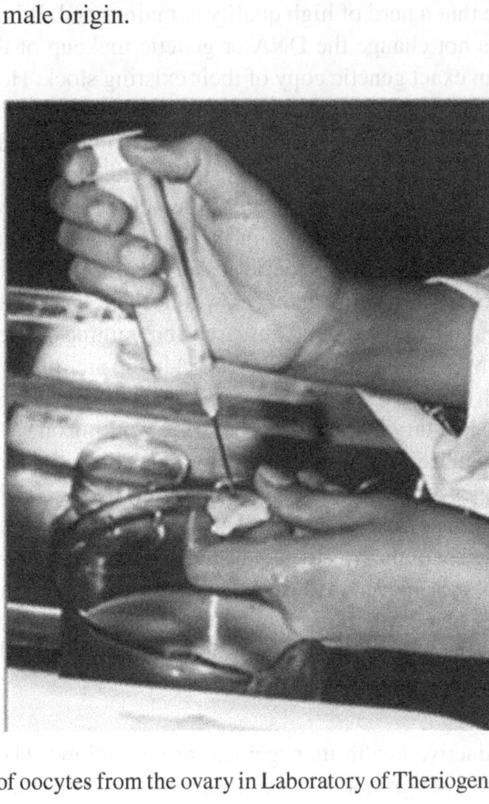

FIGURE 26.6 Harvesting of oocytes from the ovary in Laboratory of Theriogenology Department, University of Agriculture Faisalabad.

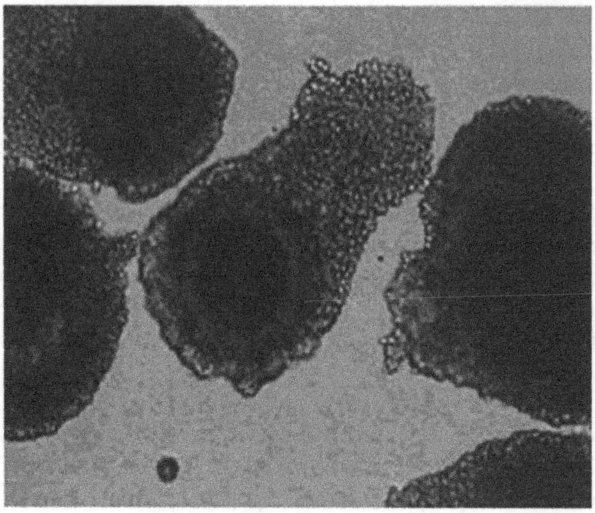

FIGURE 26.7 (**See color insert.**) Harvested oocytes under microscope.

26.4.12.1.5 Animal Cloning

Animal cloning is the recent technology in which an exact copy of a superior animal is produced from a single cell. With this technique, the nucleus of the ovum containing DNA is removed and a mature cell having DNA is injected into the empty egg, which is then subjected to electric stimulus. The developing embryo is transferred to another female which will give birth to a young resembling genetically to the donor of the cell. This is the reliable technique to reproduce animals of high genetic make and to ensure that a herd of high quality is maintained. It is also useful for saving the endangered species. It does not change the DNA or genetic makeup of the animal. But it enables the breeders to reproduce an exact genetic copy of their existing stock. Healthy animals of superior germ plasm are produced; thus, it minimizes expenses on antibiotics, hormones, or other chemicals.

These assisted reproductive biotechnologies are widely used in many countries of the world including the United States, the United Kingdom, Canada, Australia, etc. However, the livestock industry in Pakistan is still lagging behind the world. Relatively little attention has been paid to the application of various modern reproductive biotechnologies in the improvement of our dairy animals through modifying their genetic make up. Research on various aspects of IVEP has been conducted in bits and pieces by individual workers at various institutes. However, there is no well-organized institute with modern infrastructure and equipment for quality research and training on reproductive biotechnologies in the country. The work on MOET and ovum pick-up is yet in the preliminary stages. The first calf was reportedly produced through embryo transfer a few years ago at the Buffalo Research Institute, Pattoki. At present, researchers at the Military Dairy Farm, Okara are working on various aspects of MOET and in vitro embryo production. Future development in the livestock industry depends upon the application of these emerging technologies in the production system of our livestock. Use of technologies such as MOET, in vitro produced embryos, sexed semen, or embryo and animal cloning can open new avenues for the research and development of the livestock industry in the country. However, the objectives of all these biotechnologies cannot be achieved in the true sense unless artificial insemination is popularized in the country. Moreover, the objective of improving livestock production cannot be achieved without farmers' education and awareness regarding the usefulness of these technologies.

26.4.12.2 Issues Identified

The issues affecting reproductive health management are as follows: (1) Lack of human resource specific to support various programs. (2) Shortage of superior bulls/superior germ plasm.

(3) Deficiency of green fodder and grasses and other nutritional elements. (4) Late maturity, low conception rate, long calving intervals and reproductive disorders in female cows and buffaloes. (5) Absence of proper surveillance of reproductive diseases. (6) No or little check on the emerging private semen production units. (7) Questionable performance of public sector semen production units. (8) Absence of refresher courses for Veterinarians and AI technicians. (9) Failure to show the benefits of AI. (10) Lack of properly trained professional AI staff. (11) Low conception rates with frozen semen, especially in buffaloes. (12) Lack of sexual health management with AI. (13) Lack of knowledge among livestock farmers about significance and applications of modern reproductive biotechnologies. (14) Lack of well-organized research and training institutes on modern assisted reproductive biotechnologies.

26.4.12.3 Possible Remedies/Suggestions

(1) Ensure availability of superior bulls to be used for AI and for natural breeding until the AI is made popular and acceptable. (2) Human resource should be developed for specific programs. (3) Increase the availability of green fodder and grasses through increasing areas under fodder crops. Agro-forestry be encouraged. Barren lands may be developed and made useful for animal fodders. (4) A technical and performance audit of all the Semen Production Units (SPU) whether in the public or private sector should be carried out. The SPUs and their roles may be redefined. (5) The veterinarians in the field may be made responsible to maintain optimum fertility so that the livestock enterprise functions as efficiently and profitably as possible. (6) Regular refresher courses for veterinarians and AI technicians be held every 3 years. No AI technician should be allowed to work in the field, unless trained and registered by L&DD department. (7) The breeds should be developed through breed specific and region-specific programs. (8) Formation of Breed Association for improvement of indigenous breeds. The particular task of this association would be production of quality males. (9) Round the clock breeding services should be provided to farmers at their door steps. There should be a shift from a stationary clinic to a mobile facility. (10) Adequate attention should be paid for the reproduction of equines and camels, which, despite huge potential have been ignored and their population has been stagnant for the last several years. (11) Emphasis on small ruminants should be on increasing body weight, improving reproductive efficiency, reducing disease and mortality, better nutrition, and breeding. (12) Frequent and regular serological examination of the herd for presence of specific infections. (13) Awareness among farmers about the importance of hygienic measures at the time of parturition. (14) Farmers education on benefits of the application of modern reproductive biotechnologies in livestock. (15) Establishment of a well-organized Centre of Excellence for Reproductive Biotechnologies with modern infrastructure and equipment for quality research and training in the country.

26.4.13 Indifference of the Pakistani Society to Animal Welfare

Pakistani society by in large is indifferent to animal welfare, although Islam enjoins us to be kind to the mute creatures of the universe. Print and electronic media may be utilized for changing the mind-set of the society in this regard.

26.4.14 The Undesirable Situation of Rampant Quackery

One of the hallmarks of the livestock scenario in Pakistan is the unhindered practice of medicine by nonprofessionals. Despite a sizeable number of veterinary graduates being unemployed, there is no room for the practice of veterinary medicine by unqualified veterinarians. Pakistan Veterinary Medical Council and L&DD Departments should take all measures to curb the practice of veterinary medicine by quacks and charlatans.

26.5 POULTRY HEALTH ISSUES AND SOLUTIONS

Poultry farming is an integral part of the Pakistan agricultural system and is one of the most technologically advanced sectors of agriculture.

26.5.1 Significance of the Poultry Industry

Poultry meat has a low cost of production compared to sheep and goat meat. Although it is a cheaper animal protein in the Pakistanis' diet, yet its consumption is only 7 kg/capita/annum (as compared to 40 kg in the developed world) and only 65–70 eggs/capita/annum (as compared to 300 eggs in the developed world). As per standards of the World Health Organization, the daily requirement of animal protein in the human diet is 69.5 g/capita/day, out of which 27.5 g must be animal source protein. In Pakistan, the availability of animal protein is 17.4 g/capita/day. Three per cent of the babies born in Pakistan are mentally retarded because their mothers don't consume the minimum recommended amount of protein while carrying the baby.

The current population of Pakistan is 192,458,864, based on the latest United Nations estimates of 2016. It is estimated to grow to 309,639,865 in 2050 (www.Worldometers.info). In the face of a geometric progression of human population, a viable and progressive poultry industry would perhaps be a dire need in the future to bridge the yawning supply-demand protein gap.

26.5.2 Systems for Poultry Production

Two basic systems exist for poultry production: (1) The small-scale production sector includes rural poultry having small flocks of mixed species and undescript breeds. This is a low input system using poor housing, low or nonexistent feed supplementation and health inputs. There is high mortality up to 60–80%. There is low output (eggs and meat), almost insufficient for household consumption or local trade. Veterinary services do not exist in this sector. (2) The industrial sector includes commercial poultry. These are large units with hybrid breeds for specialized production (layers or broilers) and uses high inputs (infrastructure and feed) (Figures 26.8 and 26.9). Veterinary services are provided by public veterinarians, private experts in poultry diseases, as well as nonveterinarians. The sector aims national and international markets. Biosecurity management, diagnostic facilities, vaccination and other prophylactic measures, and therapy are highly variable among units.

FIGURE 26.8 Control shed for broilers in Faisalabad.

FIGURE 26.9 Fully automated control house for layers in Lahore.

It is speculated that the poultry industry will face several problems and challenges in the future, in addition to existing ones, as mentioned hereafter.

26.5.3 Issues and Challenges of Poultry Industry in Pakistan

Rising feed costs are mainly related to limited availability of raw ingredients, mainly agro-wastes. The current supply of feed ingredients is insufficient to meet the requirements of Poultry Feed Industry, which will lead to the suboptimal performance of the sector; and one can foresee that the requirement of feed ingredients will increase exponentially. If an adequate supply chain is not assured for the future, increased costs of production are anticipated in the face of the "low buying power" of average Pakistanis. Changes in global energy structure, that is, maize for biofuel and soy for biodiesel, are making international maize and soy no longer attractive. How do we cope?

Currently, in general, advice to poultry farmers is provided by private veterinarians or paraveterinarians or poultry medicine representatives to promote the sale of feed or medicine. This increases the cost of production with little or no gain to farmers and promotes the injudicious use of medicine in poultry. Noteworthy, the meager disease diagnosis facilities provided by the government are also withdrawn in the province of Punjab, in the near past, by the closure of the Government Poultry Diagnostic Laboratories.

A "notifiable disease" refers to a disease that poses a risk for international spread, or zoonotic potential, or significant spread within a naive population, or an emerging disease. There are 14 notifiable poultry diseases in the OIE-Code, but in the European Union there are 3 including HPAI, LPAI, and Newcastle disease. No lessons have been learned from the havoc of Gumboro disease in 1980s, Avian influenza in 1990s, and yet more to come. No legislation is active in Pakistan on reporting any devastating disease in poultry.

The marketing channels of broilers and eggs are predominantly in the unorganized sector. The market instability is mainly due to overproduction, little processing, and storage facilities (power deficit), as well as poor coordination among stakeholders. Development in the use of spent hens is also required.

The development of antibiotic resistant bacteria, common in both animals and humans, is a continuous public health hazard. Supplementation of poultry feed with antibiotic growth promoters (AGPs) improves performance, but can also increase the prevalence of drug-resistant bacteria.

Varying production costs in various regions of the world will lead to an increase in the global movement of poultry and poultry products. The Pakistan poultry industry is not oriented for this.

In the future, global cooperations and trade will force governments to harmonize the existing legislations related to trade and animal disease control, etc. Poultry diseases are, and will remain, a major challenge to the industry. Once an outbreak of a given disease occurs, it may have a significant negative impact on trade within the country and exports.

Currently, the animal welfare in developed countries is given high importance, while this is not the case in rapidly developing poultry sector in Pakistan. The collaboration between researchers and the poultry industry is missing at the moment. Lastly, poor record keeping and disclosure may keep the industry planning process down and regulatory bodies ineffective.

26.5.3.1 Suggestions

The feed costs can be stabilized by increasing budgetary allocations to agriculture. The government should make an "Agricultural policy for the poultry industry" in consultation with the stakeholders. There is a need to secure a reasonable subsidy for maize and soy production targets for the poultry industry—feed-millers should be partners with growers. The grants should be provided to universities and research institutes for converting local agriculture wastes to support the poultry feed industry, that is, use of nontraditional feed material to reduce feed cost and using nanotechnology (nanotech feed mix) to reduce feed cost.

The role of "Disease Diagnosis Laboratories" should be redefined to provide the farmer a faster, more sensitive and more accurate diagnosis of infectious diseases; early interventions can then become a reality. Training and certification programs for poultry care and formal or on-the-job training opportunities for professionals should be encouraged.

Legislation should be developed to enforce reporting devastating diseases to the diagnostic laboratories in the area. Criteria for reporting may include rapid spread, or high mortality, or important economic impact, or zoonotic potential. The lab staff should confirm the disease and implement appropriate control measures. There is also a need to develop good management practices (GMPs) and focus should also be given to interfarm distances, these will help in the control of diseases.

The poultry industry should develop a National Strategic Poultry Plan. The Pakistan Poultry Association should be involved in self-pegulatory issues, registration, licensing, and imposing levies to each stakeholder to generate funds for marketing, research, solving problems, and dealing with emergencies. Strategic planning should evaluate where we stand. and where we are going. Information systems and databases should be developed for planning purposes. The public sector should develop a policy for streamlining the marketing of poultry. A balance should be maintained between large-scale production and rural poultry towards lower cost of production. Incentives should be given for value added products and storage facilities (cheap power) to stabilize markets of meat and eggs.

Regulation and standardization of poultry production for export is required. There is a need to adopt a standard operating procedure (SOP), quality standardization, and trade regulation.

To address animal welfare, "Codes of Practice" should be developed for each type of poultry. The public sector may be involved in "On-farm Inspections." Producers should document their management program accordingly. Researchers and academia must communicate current knowledge and management practices to farmers to address animal welfare concerns and problems.

It should be mandatory for the private sector to develop a liaison with veterinary universities/ scientific research institutions for quality control. Short trainings on various aspects of poultry farming will encourage farmers to share their experiences with academia and researchers and seek guidance. Finally, encourage the culture of record keeping and record disclosure among farmers.

26.5.4 Poultry Farming Practices: Food Safety Issues

It has been observed with great concern that poultry farming practices are not in line with human health concerns (Qamar et al. 2014). There is a need to look forward and address this concern as there is no data available on the human health concerns associated with untoward practices adopted by poultry farmers. These practices result in meat and eggs of low quality, especially with reference to drug residues, mycotoxins, and other biochemical substances. There is considerable talk of antibiotic resistance faced by human health authorities for which one-way poultry farming practices are to be blamed. The medicines at poultry farms are given even on the day of sale of broiler flocks for human consumption. Similarly, the layer farmers are using antibiotics and other drugs inadvertently to keep their flocks free of diseases, or as a treatment regimen. This causes contamination of eggs with drugs in the form of residues, which are ultimately consumed by humans. In this way both broiler and layer farmers are responsible for drugs and other kind of residues like mycotoxins, insecticides, and pesticides making their way to the human table. There is a need to introduce pharmacist's statutory rules for the sales, distribution, manufacture, and dispensing of veterinary pharmaceuticals to reduce the development of antibiotic resistance and residues in meat. No self-medication, only the medication on the veterinarian's prescriptions.

Another important threat coming from poultry is the zoonosis Salmonella, *E. coli,* Influenza etc. These organisms are reaching consumers through poultry products. The household women and their loved ones are vulnerable when handling contaminated eggs or when using half cooking practices. The broiler meat sale points in live bird markets are also considerable sources of these organisms, soiling the meat during processing of broilers.

Apart from poultry farmers, the feed sellers are also contributing to the production of low quality meat and eggs. One of the important reasons is the presence of mycotoxins in meat and eggs, which are come partly from the feed mills, because the grains are contaminated with mycotoxins during storage. Also, perhaps some feed millers purchase low-quality grains already contaminated with mycotoxins and fungal metabolites. Although, part of the blame should be put on the poultry farmers as feed storage facilities at poultry farms are not ideal and contaminate feed with mycotoxins, etc.

There is a need for legislation and implementation protocols relating to poultry farming practices. The legislation should address the sale of broiler meat after withholding of the flock where medicines, including antibiotics, are used for at least one week. The legislation should also include the penalties in the form of a ban of the poultry farming practices for a certain period along with fine and/or imprisonment of a certain duration. The legislation is also required on sale of unwholesome meat in the live bird market. This checking mechanism is necessary to ensure the sale of wholesome meats. The layer farmers should be educated and advised not to sell their eggs in the market for human consumption at the time when their flocks are under treatment with various medicines like antibiotics and other drugs. If found guilty, such farmers may not be allowed to be involved in poultry farming for a certain period and may be fined as well.

Legislation is required to put a ban on sale practices being adopted by meat and egg sellers. There is also a need to start treatment of meat and eggs with detergents and other antiseptics/disinfectants like vinegar, etc. The fresh meat and egg sale points must be routinely checked for hygienic standards and practices. The people not following recommendations should be forbidden to running such businesses for certain durations. They may be allowed to run a business again, but repeated breach may result in cancellation of their license. Besides, they must be fined for the first, second and so on breaches and or imprisonment for a certain period accordingly.

26.6 STRATEGIES TO COMBAT MYCOTOXINS IN POULTRY AND LIVESTOCK

Mycotoxicosis is a disease condition produced by the mycotoxin contamination of feed stuffs and finished poultry and livestock feeds. This is an important noninfectious disease condition in animals, birds, and even in humans. The contamination of agricultural grains (Figure 26.10) with fungi result in mycotoxin production in feedstuffs under favorable environmental conditions (Figure 26.11). This contamination occurs both at preharvest and postharvest level. Mycotoxins are secondary metabolites produced by fungi. These are known immunosuppressive substances that expose birds

FIGURE 26.10 Fungal contamination of corn.

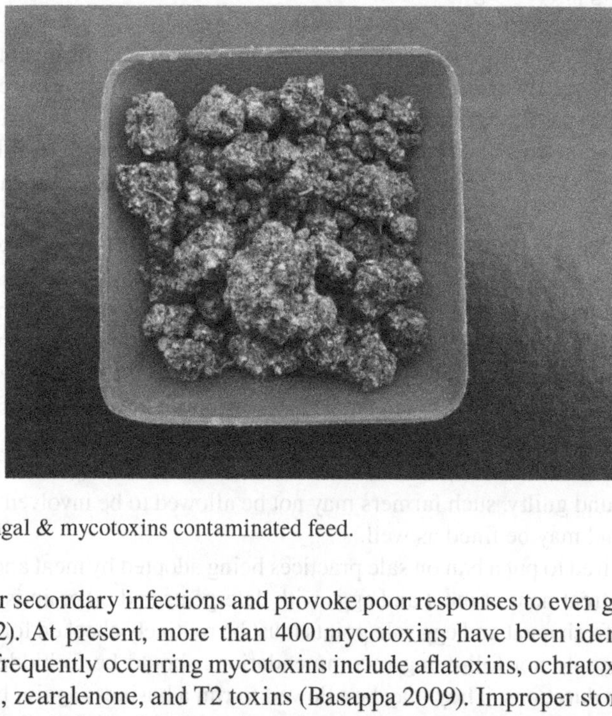

FIGURE 26.11 Fungal & mycotoxins contaminated feed.

and animals to other secondary infections and provoke poor responses to even good quality vaccines (Hassan et al. 2012). At present, more than 400 mycotoxins have been identified, however, the most common and frequently occurring mycotoxins include aflatoxins, ochratoxins, deoxynivalenol, fumonisins, patulin, zearalenone, and T2 toxins (Basappa 2009). Improper storage of poultry feeds may also lead to accumulations of mycotoxins to levels injurious to animals and birds (Figures 26.12 through 26.14). Aflatoxins (AF) are produced by *Aspergillus flavus* and *Aspergillus parasiticus*. The most important subgroups of aflatoxins include aflatoxin B1, B2, G1, and G2. Among these, AFB1 is the most potent hepatotoxic, hepatocarcinogen, and immunosuppressive, classified as a group 1 carcinogen for humans by the International Agency for Research on Cancer (Anonymous 1993). The metabolites of AFB1 in milk samples are known as aflatoxin M1 (AFM1). Ochratoxins (OT) are mainly produced from *Aspergillus ochraceous* (Saleemi et al. 2010). Ochratoxins are divided into three subgroups known as ochratoxin A, ochratoxin B, and ochratoxin C. The OTA toxicity results in degenerative changes in vital organs like kidneys and liver (Figure 26.12). Both ochratoxins and aflatoxins may also accumulate in eggs and meats of the poultry, entering the food chain at that point (Hussain et al. 2010).

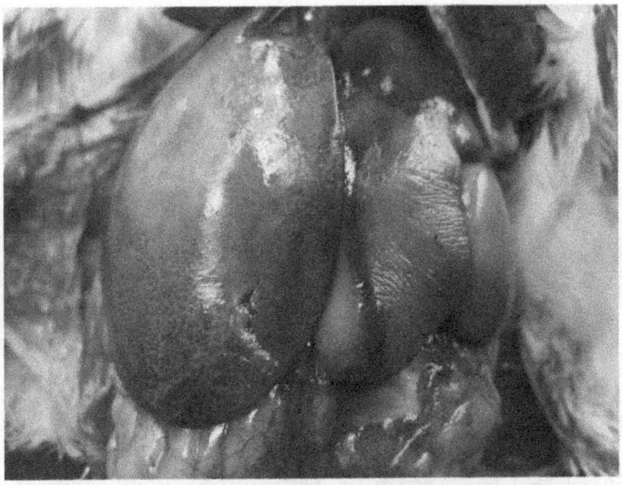

FIGURE 26.12 Congested and swollen liver in aflatoxins affected bird.

26.6.1 Mycotoxins Problem in Pakistan

The most important problem faced by the world poultry industry, after high feed prices, are mycotoxins. At present, the Pakistan poultry industry (Poultry Medicine and Feed Mills directly) is importing mycotoxins binders from different countries including the United States, The United Kingdom, Latin America, Europe, and Turkey etc. A huge investment is leaving Pakistan for the import of these toxin binders. About 3.0–4.0 billion rupees are spent annually on the import of feed additives and toxin binders to reduce the losses in feed in poultry feed manufacturing. In Pakistan, scattered reports indicate the AF/AFB1 contamination of different foods, including cereals, chicken meat, etc. (Saleemi et al. 2012). Aflatoxin M1 (a metabolite of AFB1) has been reported in dairy milk (Hussain et al. 2008). Ochratoxins have been detected in broiler meat (Hussain et al. 2010). Therefore, it is a serious health issue faced by the poultry and livestock industry in the country.

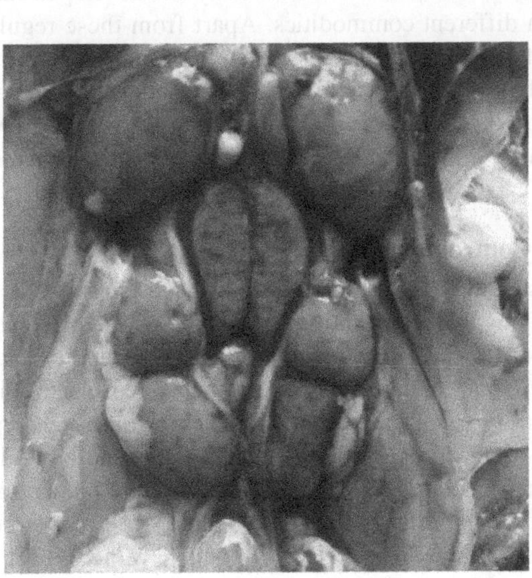

FIGURE 26.13 Swollen kidneys in ochratoxins affected bird.

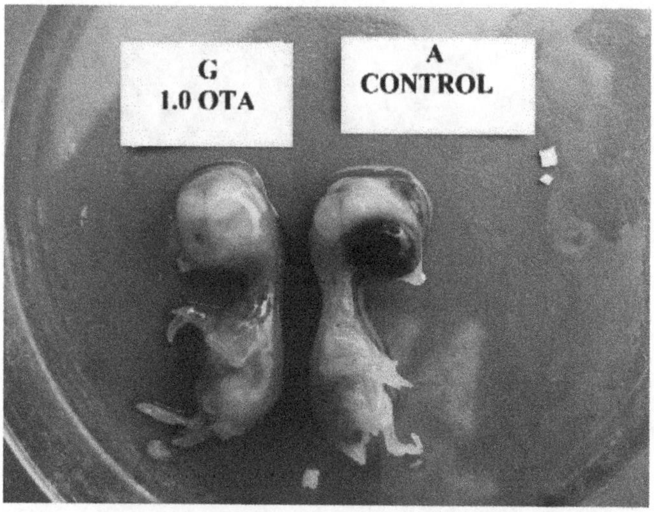

FIGURE 26.14 Embryotoxicity of ochratoxin A: one eye missing (in left) chick embryo.

26.6.2 Solutions and Way Forward

Strategies to control mycotoxins are occasionally being divided into three main categories, that is, physical, chemical, and biological. Discussing about the physical strategies, physical separation of contaminated seeds from sound kernels (FAO 1991), simple shifting to separate contaminated material using UV light (Basappa and Shanta 1996). As far as chemical methods are concerned, the use of ozone, diatomaceous earth (DE), organic acids, ammonia, and antioxidant agents are important methods to minimize the toxicity of different mycotoxins. Biological methods for the control of mycotoxins involve the use of different microorganisms (probiotics), which are used for the treatment of different conditions, including fungal alterations.

For the development and implementation of standards for different food and feed ingredients, the Pakistan Standard and Quality Control Authority (PSQCA), along with its Food and Agriculture Division is present nationally, but up until now this body had developed standards for aflatoxin only among all mycotoxins in different commodities. Apart from these regulations, certain countries have banned the import of different commodities from Pakistan. The fact of the matter in this regard is that PSQCA still needs to develop strict regulatory standards for all hazardous mycotoxins to improve and facilitate the export of different commodities. Apart from developing standards, these regulatory measures should be implemented in all steps of production to minimize mycotoxin levels in end products.

26.6.2.1 Pre-Harvest Control Measures

The following measures should be adopted at preharvest levels to avoid the entry of mycotoxins into the food chain. (1) Identifying the potent antifungal peptides from different sources and transferring them to susceptible crops can result in transgenic (fungal resistant) crops. (2) Use of antibodies in plants is another strategy to control mycotoxins at pre-harvest levels. (3) Plants should be prevented from exposure to different types of infections. After harvesting fungal spores remain in the soil, so, before the cultivation of new crops, old seed heads, stalks and other debris should be removed from the soil. (4) Poor nutrients or water deficiency may lead to plant stress, which could result in increased susceptibility to fungal infections. To avoid plant stress proper nutrients and water supply should be ensured and planting should be done keeping in view the soil pH and proper plant spacing. (5) Inoculation of atoxigenic fungi of the same genus in standing crops should be introduced for competitive inhibition to eliminate the chances of fungal contamination.

26.6.2.2 Postharvest Control Measure

Production of mycotoxins during storage and transport (postharvest) depends mainly upon the amount of inoculum present, temperature, humidity, moisture content, and insect activity. Fungal infections usually occur before harvest, but can also occur after harvesting from dormant fungal spores in silos. Rodents or insects might also transport these. During storage, water content must be below 14% or else high moisture content might promote fungi growth. Stored feed should be given good aeration at high environmental temperatures. Feedstuffs should be stored at a relative humidity less than 80% and at temperatures below 20°C. Same control measures should also be adopted during transportation of feed or feed ingredients. One important method in this regard is the Hazard Analysis Critical Control Point (HACCP) system, which should be adopted throughout the food chain, that is, from preharvest until the offering of feed to animals/poultry.

26.6.2.3 Control of Mycotoxins in Finished Poultry and Livestock Feeds

Mycotoxins usually grow in the form of pockets and removing these pockets from the feed can reduce the overall load of mycotoxins from the lot. Before offering feed to animals/poultry one should know the levels of different mycotoxins in the feed, and maximum tolerable levels for each mycotoxin should be strictly kept in mind. Use of appropriate mycotoxin binders and antioxidants should be ensured.

Use of mycotoxin binders and antioxidants in feed can ultimately decrease the carryover of mycotoxins into biological systems. Different microorganisms have the tendency to adsorb or detoxify the mycotoxins within the body, hence the use of different probiotics can also minimize the toxicity in animals/poultry.

26.7 RESEARCH IMPERATIVES

The financial resources of the government are scarce and there is no room for funding poor research projects. There is a need to have about 80% of solution-oriented research aimed at local problems, about 10% may be focused on the development of diagnostic kits/biologicals/reagents, etc., and about 10% should be basic research of applied nature focusing on local problems. Basic research cannot be neglected altogether, which is being carried out in most research prioritization programs. Whether it is Punjab Agriculture Research Board (PARB), Agriculture Linkage Program (ALP) or other funding agencies, may ensure that research on native animals. for example, buffalo may be funded, which is of basic type to generate useful data about the animal, for example, how the immune system works against diseases. The basic research also ultimately helps to frame ways and means to control certain diseases. Such immune mechanisms are linked to host genetics.

There is considerable evidence now that a majority of the diseases which animals face have a genetic link. These diseases could be due to microorganisms or they may be metabolic in nature. In diseases caused by pathogens, there is a clear evidence of an association of host factors, including innate and humoral immune responses. A lot of work has been done now in humans and animals, especially with reference to tuberculosis (Qi et al. 2015). Many markers have been identified which has been found associated with susceptibility or resistance of humans and animals to tuberculosis. Likewise, work on other diseases with reference to genetic susceptibility and resistance indicate that the genes are important and play a significant role in the occurrence of diseases. Scientists in China have produced a transgenic cattle herd genetically resistant to tuberculosis, which has been achieved by using mouse *SP110* gene in bovine macrophages (Tuggle and Waters 2015). Similarly, work is underway to develop a parasite (Trypanosome) resistant cattle in Africa, which is a big problem in African cattle. Such work has not been done in Pakistan and is missing both in livestock and poultry.

We need to carry out research on our local animals and need to develop local breeds which are genetically resistant to tuberculosis and other diseases. We need to establish a separate unit to carry out such studies. Therefore, we need to improve the genetics of our animals for host resistance to diseases, but also for improved production of our dairy and meat animals. Similarly, work should be done to target poultry and develop breeds by taking genes even from crows, eagles, owls, etc. to make our local breeds resistant to New Castle disease, which is deadly and devastating as well as against other similar diseases. There is a need to establish a center of excellence in the province or in the country and use a multipronged approach to look for breeds of animals resistant to certain diseases for transgenesis.. Likewise, such a center will also help to produce animals with required levels of certain types of proteins and other useful nutrients in their milk, meat, or eggs. Such milk, meat and eggs will not only help to protect local human population from certain diseases, but may also be used as a treatment source; for example, we can increase the level of vitamin A or carotenoids in buffalo milk, which can be used for infant feeding. Similarly, there are many other minerals and vitamins, and so on, which can be added into milk or meat or even in eggs though gene manipulation. Therefore, funds should be allocated to establish a center of excellence in animal genetics to meet the future demands of the people of Pakistan.

26.8 UNCHARTERED MISSION OF VETERINARY EDUCATION

There are 12 veterinary schools in Pakistan, which are churning out around 600 veterinarians per year, which is at least five times what can be employed in the public and private sectors. As a remedial measure, all veterinary schools in Pakistan should strive to include courses aimed at training the budding veterinarians in entrepreneurship skills so that at least half of the graduating veterinarians can start their own businesses, which would then have the capacity to absorb other unemployed veterinarians.

26.9 A BULLETED BRIEF

- A fresh census of livestock in the country should be conducted to correctly appraise the livestock population in the country.
- The Government should divert its animal health budgetary resources to preventive health programs and should pick up the entire tab of preventive medication.
- All research institutes and universities should establish special cells within their administrative ambits for communicating the findings of their research investigations to provincial Livestock and Dairy Development Departments, livestock farmers, NGOs, and other agencies involved in the provision of animal health related services.
- Amul model can be replicated *mutatis mutandis* in Pakistan for mass adoption of animal health related technologies by smallholder dairy farmers.
- Semen production units, semen quality, and AI technicians need special attention. A technical and performance audit of all Semen Production Units (SPUs), whether in public or private sectors, should be carried out.
- Regular refresher courses for veterinarians and paraveterinary staff should be held every 3 years.
- Rapid diagnostic kits should be provided at the district diagnostic laboratories.
- A well-organized Center of Excellence for Reproductive Biotechnologies with modern infrastructure and equipment for quality research and training should be established in the country.
- Adoption of integrated parasite control (IPC) practices by the farmers should be encouraged.
- Sensitization and awareness programs should be arranged for farming communities and general public through electronic/print media, seminars, symposia, etc. on livestock and poultry problems along with their solutions.
- Investments should be made in research and development of state of the art diagnostic facilities, drug discovery, country medicine, parasite surveillance programs, capacity building, academia-industry linkages, outreach, etc.
- Establishment of comprehensive disease surveillance and quarantine network, netted with disease reference laboratories at provincial level against important poultry and livestock diseases is proposed. A comprehensive and regular epidemiological survey, including all

important diseases should be carried out under this umbrella to ascertain the low and high disease burden zones and clear zones against a set of diseases.

- Disease control and eradication program against important diseases should be started simultaneously.
- Animal numbering systems and animal movement under restriction needs special attention.
- The concept of one health needs special attention, which will help to focus on alternative treatment and control strategies. Legislation on farming practices, meat sale, egg sale, and on live bird markets should be made with reference to food safety.
- Development of indigenous vaccines should be given priority. The Veterinary Research Institute and such other departments should be given special tasks and their facilities should be improved accordingly. One-shot vaccine is the newest concept to be initiated to decrease the economic burden on small farmers.
- The vaccines and drugs should be monitored routinely for quality, and a special cell may be created to take up this task.
- Establishment of a multidisciplinary national body to formulate and implement policies on drug use, clinical guidelines and preparation of the national essential drugs list. Government should give priority to ensure availability of essential veterinary drugs and staff.
- Inclusion of problem-based pharmacotherapy trainings in undergraduate curricula.
- Preparation of disease preparedness and response strategies and early warning system.
- Basic courses on Complementary and Alternative Veterinary Medicine should be incorporated in DVM curriculum in veterinary institutes to prepare a new generation of veterinarians that are well equipped with traditional health care system.
- Collection and compilation of traditional veterinary practices to protect Intellectual Property Rights. Resuscitation of complementary alternative veterinary medicine will spur organic livestock farming in Pakistan.
- Mycotoxin control in the crop field, postharvest, during feed storage, and in live animals should be ensured and local toxin binders need to be developed and tested.
- The government should make "Agriculture policy on Poultry Industry" in consultation with the stakeholders. The poultry industry should develop a National Strategic Poultry Plan.
- Regulation and standardization of poultry production for export is required.
- Legislation should be developed for reporting devastating diseases in poultry. Criteria for reporting may include rapid spread, high mortality, important economic impact, or zoonotic potential.
- The role of "Disease Diagnosis Laboratories" should be redefined to provide farmers a faster, more sensitive, and more accurate diagnosis of infectious diseases so that early interventions become a reality.
- Researchers and academia must communicate current knowledge and management practices to address animal welfare concerns and problems.
- There is need to encourage the culture of record keeping and record disclosure.
- A state of the art research institute on host genetics and diseases with focus on the development of genetically resistant breeds to different stresses, including disease-stress and value-added animal product development through host genetic manipulation should be established.
- Animal health and production are two sides of the same coin. To maximize the animal's productivity, a holistic view should be taken and animal health integrated with optimum nutrition, genetic improvement, and optimum management.

REFERENCES

Abbas RZ, Iqbal Z, Akhtar MS, Khan MN, Jabbar A and Sindhu ZD. 2006. Anticoccidial screening of Azadirachta indica (neem) in broilers. *Pharmacologyonline* 3: 365–371.

Afzal M. 2009. Improving veterinary service in Pakistan. *Pak. Vet. J.* 29: 206–210.

Afzal M and Usmani RH. 2005. Research and development needs of dairy sector in Pakistan. *Sci. Tech. Develop.* 24: 34–37.

Akhtar MS, Farooq AA and Inayat S. 2009. Treatment of first degree endometritis by cloprostenol and estradiol in Cholistani cows. *J. Anim. Plant Sci.* 19: 20–21.

Ali S, Ali Q, Abatih EN, Ullah N, Muhammad A, Khan I and Akhter S. 2013. Sero-prevalence of *Brucella abortus* among dairy cattle and buffaloes in Pothohar plateau, Pakistan. *Pakistan J. Zool.* 45: 1041–1046.

Amin S, Khan MA, Hashmi HA, Khan MS, Ahmad I and Bhatti MA. 1992. Detection of buffalo Tuberculosis by using short thermal test and isolation of causal organisms from lymph nodes. *Buffalo J.* 8: 83–87.

Anjum R, Hussain M, Zahoor AB, Irshad H and Farooq U. 2006. Epidemiological analyses of foot and mouth disease in Pakistan. *Intl. J. Agri. Biol.* 8: 648–651.

Anonymous. 1987. *Report of the FAO/Asian Development Bank Cooperation Programme.* Pakistan Livestock Sector Study, Phase I Report. As Pak 39, FAO, Rome, Italy. 1: 55–87.

Anonymous. 1993. *Monographs on the Evaluation of Carcinogenic Risks to Humans: Some Naturally Occurring Substances, Food Items and Constituents, Heterocyclic Aromatic Amines and Mycotoxins.* International Agency for Research on Cancer, Geneva, 56: 489–521.

Asif M, Azeem S, Asif S and Nazir S. 2008. Prevalence of gastrointestinal parasites of sheep and goats in and around Rawalpindi and Islamabad. *Pak. J. Vet. Anim. Sci.* 1: 14–17.

Basappa SC. 2009. *Aflatoxins Formation, Analysis and Control.* Alpha Science International Ltd., Oxford, UK.

Basappa SC and Shanta T. 1996. Methods for detoxification of aflatoxins in foods and feeds. A critical appraisal. *J. Food Technol.* 33: 95–107.

Bath GF, van Wyk JA and Malan FS. 2010. Targeted selective treatment of sheep using the Five Point Check ©. *Egyptian J. Sheep Goat Sci.* 5: 369–374.

Bazarusanga T, Vercruysse J, Marcotty T and Geysen D. 2007. Epidemiological studies on theileriosis and the dynamics of *Theileria parva* infections in Rwanda. *Vet. Parasitol.* 143: 214–221.

Bhatia BB, Pathak KML and Banerjee DP. 2004. *A Text Book of Veterinary Parasitology.* Kalyani Publishers, Ludhiana, New Delhi, India.

Bishop S, de-Jong M and Gray D. 2009. *Opportunities for Incorporating Genetic Elements Into The Management Of Farm Animal Diseases: Policy Issues.* Commission On Genetic Resources for Food and Agriculture: A report. ftp://ftp.fao.org/docrep/fao/meeting/015/aj629e.pdf

Bogh IB and Greve T. 2009. Assisted reproduction. In: Noakes DE, Parkinson TJ and England GCW (eds) *Veterinary Reproduction and Obstetrics.* 9th Ed., Saunders Elsevier, London, UK, pp. 855–894.

Cobon GS and Willadsen P. 1990. Vaccines to prevent cattle tick infestation. In: GC Woodrow and MM Levine (eds), *New Generation Vaccines.* Marcel Dekkar, New York, pp. 109–142.

Ćupić V and Dobrić S. 2003. Sadašnje stanje i perspektive u razvoju antimikrobnih lekova. *Veterinarski žurnal Republike Srpske.* 2: 36–42.

Daniele V and N Ostuni. 2013. The burden of disease and the IQ of nations. *Learn. Individual Diff.* 28: 109–118.

Dawuda PM, Scaramuzzi RJ, Lees HJ, Hall CJ, Peters AR, Drew SB and Wathes DC. 2002. Effect of timing of urea feeding on the yield and quality of embryos in lactating dairy cows. *Theriogenology* 58: 1443–1455.

Deeba F, Muhammad G, Iqbal Z and Hussain I. 2009. Appraisal of ethno-veterinary practices used for different ailments in dairy animals in peri-urban areas of Faisalabad. *Intl. J. Agric. Biol.* 11: 535–554.

Din AMU, Khan SA, Ahmad I, Rind R, Hussain T, Shahid M and Ahmad S. 2013. A study on the sero-prevalence of brucellosis in human and goat population of district Bhimber, Azad Jammu and Kashmir. *J. Anim. Plant Sci.* 23(Suppl 1): 113–118.

FAO. 1991. *Food Nutrition and Agriculture—Food for the Future.* Edited by JL Albert, R Tucker, N Roland, H Gigli and M Criscuolo, No 1, pp. 1–55.

Giguere S, Prescott JF, Baggot JD, Walker RD and Dowling MP. 2006. *Antimicrobial Therapy in Veterinary Medicine.* 4th Ed. Iowa State University Press, Ames, Iowa, USA.

Gould EA, Higgs S, Buckley A and Gritsun TS. 2006. Potential arbovirus emergence and implications for the United Kingdom. *Emerg. Infect. Dis.* 12: 549–555.

Gul ST and Khan A. 2007. Epidemiology and epizootology of brucellosis: A review. *Pak. Vet. J.* 27: 145–151.

Hameed A, Anjum FM, Zahoor T and Jamil A. 2010. Consequence of oxytocin injections on minerals concentration in Sahiwal cow milk. *Pak. J. Agri. Sci.* 47(2): 147–152.

Hameed A, Anjum FM, Zahoor T, Zia-ur-Rahman, Akhtar S and Hussain M. 2016. Effect of oxytocin on milk proteins and fatty acid profile in Sahiwal cows during lactation periods. *Turkish J. Vet. Anim. Sci.* 40: 163–169.

Hassan ZU, Khan MZ, Saleemi MK, Khan A, Javed I and Noreen M. 2012. Immunological responses of male White Leghorn chicks kept on ochratoxin A (OTA)-contaminated feed. *J. Immunotoxicol.* 9: 56–63.
Horn S. 1987. Ectoparasites of animals and their impact on the economy of South America. *Proc., 23rd World Veterinary Congress*, Montreal, Canada.
Hunter WD and Hooker WA. 1907. Information concerning the North American fever tick. *USDA Bureau Entomol. Bull.* 72: 1–87.
Hussain A, Khan MZ, Khan A, Saleemi MK, Hameed MR, Ul-Hassan Z, Javed I, Hussain T and Ahmed I. 2010. Ochratoxin A (OTA) residues in tissues of commercial broilers in Pakistan. *Proceedings of The World Mycotoxin Forum -6th Conference*, 8–10 November 2010, Noordwijkerhout, the Netherland. p. 111.
Hussain I, Anwar J, Munawar MA and Asi MR. 2008. Variation of levels of aflatoxin M1 in raw milk from different localities in the central areas of Punjab, Pakistan. *Food Cont.* 19: 1126–1129.
Hussain SS, Malik F, James H and Hamid A. 2009. Trends in the use of complementary and alternative medicine in Pakistan. *J. Alter. Complement. Med.* 15: 545–550.
Ilyas F. 2015. Banned hormones still in use in dairy business despite health hazards. http://www.dawn.com/news/1209783, Accessed 28-05-16
Iqbal Z, Akhtar M, Khan MN and Riaz M. 1993. Prevalence and economic significance of Haemonchosis in sheep and goats slaughtered at Faisalabad abattoir. *Pak. J. Agri. Sci.* 30: 51.
Iqbal Z, Jabbar A, Akhtar MS, Muhammad G and Lateef M. 2005a. Possible role of ethno- veterinary medicine in poverty reduction in Pakistan: Use of botanical anthelmintics as an example. *J. Agric. Soc. Sci.* 1: 187–195.
Iqbal Z, Lateef M, Jabbar A, Muhammad G and Khan MN. 2005b. Anthelmintic activity of *Calotropris procera* (Ait.) Ait. F. flowers in sheep. *J. Ethnopharmacol.* 102: 256–261.
Javed MT. 2007. Future of Livestock. *The Nation "Money Plus" Magazine*, January 15, 2007, pp. 18–19.
Javed MT, Aziz-ur-Rehman and Qamar M. 2014. Updates on bovine tuberculosis in cattle and buffaloes, speculated future of disease and control program for Pakistan. Part 2. *The News and Views*, 15–21 November, (10), 10.
Javed MT, Irfan M, Ali I, Farooqi FA, Wasiq M and Cagiola M. 2011. Risk factors identified associated with tuberculosis in cattle at 11 livestock experiment stations of Punjab Pakistan. *Acta Trop.* 117: 109–113.
Javed MT, Shahid AL, Farooqi FA, Akhtar M, Cardenas GA, Wasiq M and Cagiola M. 2010. Association of some of the possible risk factors with tuberculosis in water buffalo around two cities of Punjab Pakistan. *Acta Trop.* 115: 242–247.
Jonsson NN. 2006. The productivity effects of cattle tick (*Boophilus microplus*) infestation on cattle, with particular reference to *Bos indicus* cattle and their crosses. *Vet. Parasitol.* 137: 1–10.
Jonsson NN, Davis R and Witt MD. 2001. An estimate of the economic effects of cattle tick (*Boophilus microplus*) infestation on Queensland dairy farms. *Austr. Vet. J.* 79: 826–831.
Kakar MA, Kakar E, Shahwani MN, Jan M, Raza AM, Hassan J, Saeed M, Babar S and Baloch SK. 2012. Reproductive biotechnologies in dairy industry in Pakistan. *J. Anim. Plant Sci.* 22(2 Suppl.): 84–86.
Kenyanjui MB and Sheikh-Ali M. 2009. Observations on cattle dairy breeds in Pakistan: Need to curb unseen economic losses through control of mastitis and endemic diseases. *JAEID* 103: 155–172.
Khan BB, Abdullah M, Ahmad N, Akram M and Ahmad Z. 1987. Use of oxytocin for milk ejection in buffaloes and cows in and around Faisalabad. *Pak. J. Agri. Sci.* 24: 36–44.
Khan MN, Hayat CS, Chaudhry AH, Iqbal Z and Hayat B. 1989. Prevalence of gastrointestinal helminths in sheep and goats at Faisalabad abattoir. *Pak. Vet. J.* 9: 159.
Khan MN, Hayat CS, Iqbal Z, Hayat B and Naseem A. 1993. Prevalence of ticks on livestock in Faisalabad (Pakistan). *Pak. Vet. J.* 13: 182–184.
Khan MN, Iqbal Z, Sajid MS, Anwar M, Needham GR and Hassan M. 2006. Bovine hypodermosis: Prevalence and economic significance in southern Punjab, Pakistan. *Vet. Parasitol.* 141: 386–390.
Khan MQ, Hayat CS, Ilyas M, Hussain M and Iqbal Z. 1988. Effect of haemonchosis on body weight gain and blood values in sheep. *Pak. Vet. J.* 8: 62.
Kithuka JM, Maingi N, Njeruh FM and Ombui JN. 2002. The prevalence and economic importance of bovine fasciolosis in Kenya—An analysis of abattoir data. *Ond. J. Vet. Res.* 69: 255–262.
Kivaria FM. 2006. Estimated direct economic costs associated with tick-borne diseases on cattle in Tanzania. *Trop. Anim. Hlth. Prod.* 38: 291–299.
Knapton S. 2016. Antibiotics for animals to be restricted under government plans to beat drug resistance. http://www.telegraph.co.uk/science/2016/05/13/antibiotics-for-animals-to-be-restricted-under-government-plans/ Accessed 26-05-2016

Lodhi LA. 1997. Artificial Insemination services with reference to livestock improvement. Paper presented at *workshop on Breeding Status and Future Policies/requirements for increased Productivity*. April 29–30, 1997, Lahore, Pakistan.

Maqsood M, Iqbal Z and Chaudhry AH. 1996. Prevalence and intensity of hemonchosis with reference to breed, sex and age of sheep and goats. *Pak. Vet. J.* 16: 41.

McCosker PJ. 1979. Global aspects of the management and control of ticks of veterinary importance. In: *Recent Advances in Acarology II*. Academic Press, New York, NY, USA, pp. 45–53.

Monteiro AM, Wanyangu S, Kariuki DP, Bain R, Jackson F and McKellar QA. 1998. Pharmaceutical quality of anthelmintics sold in Kenya. *Vet. Rec.* 142: 396–398.

Muhammad G. 2012. Vaccination based prophylactic and metaphylactic disease control plans. In: Ullah Z, Avais M, Riaz A, and Sattar A (eds) *Proc. Seminar on Road Map for Development of Dairy Sector of Pakistan*, March 19, 2012. University of Veterinary and Animal Sciences, Lahore, Pakistan, pp. 19–20.

Muhammad G, Jabbar A, Khan MZ and Saqib M. 2004. Use of neostigmine in massive ivermectin toxicity in cats. *Vet. Human Toxicol. USA* 46: 28–29.

Mukhebi AW, Gitunu M, Kavoi J and Iroha J. 1985. Agropastoral systems of Southern Machakos District of Kenya. Technical Paper No. 7, Socioeconomics Division, Kiboko National Range Research Station, Ministry of Agriculture and Livestock Development, Kenya.

Nasir AA, Shah MA and Rashid M. 1999. Current status of brucellosis in cattle at various government livestock farms in Punjab. *Intl. J. Agri. Biol.* 1(4): 337–338.

OIE Terrestrial Code Commission. 2008. *Terrestrial Animal Health Code 2008*. Section 3. Quality of Veterinary Services, Office International des Epizooties, Paris, France.

Pakistan Economic Survey. 2016–2017. Ministry of Finance, Government of Pakistan, Islamabad. http://www.finance.gov.pk/survey_1415.html

Pal RA and Qayyum M. 1992. Breed, age and sex-wise distribution of gastro-intestinal helminths of sheep and goats in and around Rawalpindi region. *Pak. Vet. J.* 12: 60.

Pal RA and Qayyum M. 1993. Prevalence of gastrointestinal nematodes of sheep and goats in upper Punjab, Pakistan. *Pak. Vet. J.* 13: 138–141.

Qamar M, Javed MT, JA Khan and B Aslam. 2014. Poultry farming practices and human health concerns. *Veterinary News Views* (8): 10.

Qi H, Sun L, Wu X, Jin Y, Xiao J, Wang S, Shen C, et al. 2015. Toll-like receptor 1(TLR1) Gene SNP rs5743618 is associated with increased risk for tuberculosis in Han Chinese children. *Tuberculosis* 95: 197–203.

Ram H, Yadav CL, Banerjee PS and Kumar V. 2004. Tick associated mortality in crossbred cattle calves. *Indian Vet. J.* 81: 1203–1205.

Raza MA, Iqbal Z, Jabbar A and Yaseen M. 2007. Point prevalence of gastrointestinal helminthiasis in ruminants in Southern Punjab, Pakistan. *J. Helminthol.* 81: 323–328.

Razzaq A, Islam M, Ahmad S, Shideed K, Shomo F and Athar M. 2012. Prevalence of internal parasites in sheep/goats and effective economic de-worming plan at upland Balochistan, Pakistan. *Afri. J. Biotech.* 11: 12600–12605.

Rehman M. 2001. Prevalence of tuberculosis in cattle around Lahore using intradermal test. *M.Sc. (Hon.) Thesis*. Department of Clinical Medicine and Surgery, College of Veterinary Sciences Lahore, UAF.

Saeed M, Iqbal Z, Jabbar A, Masood S, Babar W, Saddiqi HA, Yaseen M, Sarwar M and Arshad M. 2010. Multiple anthelmintic resistance and possible contributory factors in Beetal goats in an irrigated area (Pakistan). *Res. Vet. Sci.* 88: 267–272.

Sajid MS, Anwar AH, Iqbal Z, Khan MN and Qudoos A. 1999. Some epidemiological aspects of gastro-intestinal nematodes of sheep. *Intl. J. Agric. Biol.* 1: 306–308.

Sajid MS, Iqbal Z, Khan MN and Muhammad G. 2008. Point prevalence of hard ticks infesting domestic ruminants of lower Punjab, Pakistan. *Intl. J. Agric. Biol.* 10: 349–351.

Sajid MS, Iqbal Z, Khan MN and Muhammad G. 2009. In vitro and in vivo efficacies of Ivermectin and Cypermethrin against the cattle tick Hyalomma anatolicum anatolicum (Acari: Ixodidae). *Parasitol. Res.* 105: 1133–1138.

Saleemi MK, Khan MZ, Khan A and Javed I. 2010. Mycoflora of poultry feeds and mycotoxins producing potential of *Aspergillus* species. *Pak. J. Bot.* 42: 427–434.

Saleemi MK, Khan MZ, Khan A, Javed I, Hassan ZU, Hameed MR and Mehmood MA. 2012. Occurrence of toxigenic fungi in corn and corn-gluten meal from Pakistan. *Phytopathologia Medeter.* 51: 219–224.

Sanders P. 2007. Veterinary drug residue control in the European Union. *Technologija Mesa.* 1: 59–68.

Shaikh BT and Hatcher J. 2005. Complementary and alternative medicine in Pakistan: Prospects and limitations. *eCAM* 2(2): 139–142. doi: 10.1093/ecam/neh088.

Siddiqi MN and Jan AH. 1986. Ixodid ticks Ixodidae of N.W.F.P. Pakistan. *Pak. Vet. J.* 6: 124–126.

Soulsby EJL. 1982. *Helminths, Arthropods and Protozoa of Domesticated Animals*. English Language Book Society, Bailliere Tindall, London, UK.
Spranger J. 2000. Testing the effectiveness of antibiotic and homeopathic medication in the frame of herd reorganization of subclinical mastitis in milk cows. *Br. Homeopath. J.* 89(1): S62.
Sutherst RW, Jones RJ and Schnitzerling HJ. 1982. Tropical legumes of the genus *Stylosanthes* immobilize and kill cattle ticks. *Nature* 295: 320–321.
Teglas M, Matern E, Lein S, Foley P, Mahan SM and Foley J. 2005. Ticks and tick-borne disease in Guatemalan cattle and horses. *Vet. Parasitol.* 131: 119–127.
Tuggle CK and Waters WR. 2015. Tuberculosis-resistant transgenic cattle. *Proc. Natl. Acad. Sci. USA* 112(13): 3854–3855.
Varma MRG. 1993. Ticks and mites (acari). In: Land RP and Crosskey RW (eds) *Medical Insects and Arachnids*. Chapman and Hall, Ltd., London, UK, pp. 11–35.
Wet JD, Iqbal MU and Umer MS. 2010. *Housing, Shed Cooling and Milking Machines*. Nestle Sarsabz Farm and Training Centre, Renala, District Okara, Pakistan. p. 33.
Williams RB. 1998. Epidemiological aspects of the use of live anticoccidial vaccines for chickens. *Int. J. Parasitol.* 28: 1089–1098.
Younas Z. 2006. *Poverty Focused Investment Strategies for the Punjab*. Punjab Resource Management Program (PRMP), Planning & Development Department, Government of the Punjab Lahore, Pakistan. http://prmp.punjab.gov.pk/sites/prmp.pitb.gov.pk/files/PFIS_MAIN_STRATEGY.pdf

27 Malnutrition in Children and One Health

Mian Kamran Sharif, Masood Sadiq Butt, Muhammad Kashif Saleemi, and Muhammad Imran Arshad

CONTENTS

27.1 BACKGROUND

27.1.1 Global Situation

Malnutrition is a universal issue, nevertheless developing countries are significantly hit by this menace. Individuals with increased nutrient requirements such as children, pregnant and lactating women, and old age population are more vulnerable to undernutrition. Globally, it is widespread in the form of low weight, stunting, and wasting whereas children in developing developed countries are becoming overweight and obese. Currently, 870 million people (12.5% of world population) are malnourished and 852 million of them are from emerging countries. Globally, approximately 178 million kids under 5 are affected by stunting and 55 million are being wasted (Akhtar et al., 2013). Undernutrition diminishes a nation's economic progression by at least 8% because of direct productivity losses, losses via poorer cognition, and reduced schooling. The current annual rate of reduction in childhood malnutrition is 2.1% whereas the World Health Assembly has called for a 40% reduction in the number of stunted children under-5 by 2025, which can only be achieved with a 3.9% annual reduction. The highest number of malnourished children reside in South Asia followed by eastern and western Africa. It is expected that there will be a substantial decrease in the prevalence of stunting because of strategies and policies being adopted by South Asian countries. Likewise, global prevalence of wasting was 8% (52 million children under 5) in 2011 (Morris et al., 2008). The local government and international organizations have been focusing on deficiencies of vitamin A, zinc, and iron for many decades now (Akhtar et al., 2011). Among micronutrient deficiencies, iron is at the top affecting about 20% global individuals (Huma et al., 2007). Iron deficiency is extensively prevalent in cereal, especially among wheat consuming peoples due to presence of phytates which limit iron bioavailability. In these countries, various strategies including diet diversification, supplementation, and food fortification are being implemented at limited levels to tackle this problem (Ahmed et al., 2014).

27.1.2 Pakistani Context

Pakistan has been facing malnutrition for decades; it is now where it is severest in the region. According to the recent Global Hunger Index (2016), Pakistan is ranked 11th among 118 countries with respect to its undernourished population (22%) and prevalences of higher rates of stunting (45%), wasting (10.5%), and child mortality (8.1%) (IFPRI, 2016). According to a World Bank report, malnutrition is about 3% GDP of developing nations. Neighboring countries of Pakistan are in a better position on the basis of adolescent and adult nutritional status, uptake of micronutrients, and prevalence of stunting. Malnutrition is causing 33% annual deaths among children. Likewise, these children have nine times greater mortality rates than those of healthy children thriving on balanced diets. According to the latest National Nutrition Survey of Pakistan, core maternal and childhood nutrition indicators showed little change with respect to macro- and micronutrient deficiencies. However, only the iodine status has improved because of universal salt iodization, however, the vitamin A status has worsened. Among children under 5, 43.7% were stunted and 15.1% were wasted, whereas 31.5% were underweight reflecting poor status of health and nutrition in the country (GOP, 2011). Overall, urban areas have lower prevalences of stunting (36.9%) than rural communities (46.3%). Similarly, the share of wasted children is high (16.1%) in rural areas compared to municipalities (12.7%). Besides stunting and wasting, about one third of the children (31.5%) are underweight with higher proportions in villages (33.3%). The proportion of wasted children was lower in urban areas than rural counterpart. Vitamin and mineral deficiencies were also prevalent in children (vitamin A: 54.0%; vitamin D: 40%) (Fe: 43.8%, Anaemia: 61.9% and zinc: 39.2%). There were large variations regarding the prevalence of anaemia in different regions of the country and urban rural communities. The lowest anaemia rates was found in northern areas (41%), and the highest in Sindh province (72.5%). Likewise, frequency of severe anaemia was relatively lower in urban areas (3.6%) than in rural localities (5.5%). Comparatively, low ferritin levels (26.8%) were in non-pregnant women in Punjab with slight high levels (26.8%) in urban population than that of rural areas (26.6%). Considerable differences were found at the provincial level, ranging from 28.9%–43.4% in KPK to Balochistan, respectively. Malnutrition in Pakistan is mainly due to lack of clean water, hygiene and sanitation systems, inflation, poverty, illiteracy, improper feeding and weaning practices, poverty, and lack of interest by the officials. However, it is worth mentioning that wealthier families are also challenged by the malnutrition either due to excessive calories in the diet or improper eating habits.

27.2 NUTRITIONAL REQUIREMENTS IN CHILDREN

Fetal development, infancy, and adolescence are life stages described by fast growth and maturity of body organs and systems. The disparities in the quantity and quality of nutrients obtained by pregnant mothers or infants have shown powerful and long-lasting impacts on developing tissues and organs, as evidenced by numerous studies. The classic example is of the Dutch famine, which severely hit western areas during winter season in 1944–1945. The food supplies were blocked by German forces for approximately 6 months. Consequently, the nutritional status of pregnant women was compromised during the first and in some cases third trimester having pronounced impacts on birth weights of newborn babies. Birth weights among babies affected by famine in the third trimester were around 250 g less than those born before or conceived after the food shortage (Roseboom et al., 2001, 2011). The real nutritional atmosphere faced by the fetus is far more complicated depending upon the mother's health, daily intake, physical activity, fetal growth rate, and placental function. According to recent studies, nutritional insults occuring during pregnancy affect adult offsprings (Samuelsson et al., 2008). Animal studies have proven that malnutrition runs in the family.

Growth and development in infancy is at its peak. However, the accumulative impacts which occur next are noteworthy. Subsequently, the pace towards adulthood speed up extremely when child enters in teen hood. The fast growth during the first year directly reflects the nutrient intake, which is assessed by health professionals measuring weight and height at specific intervals. A healthy full-term infant doubles its weight after 5–6 months, and triples it over the next 6 months, reaching to 9–11 kg. The increase in length is, however, comparatively slower than weight, adding about 10 inches during the first year. In subsequent years children gain 4–5 kg weight and grows 4–5 inches in length. Because of rapid growth and development, the energy requirements of the infant based on body weight are about twice than that of adult. Adults require only 40 Kcal per kg of body weight whereas infant need 100 Kcal per kg of body weight. After 6 months, the infant's energy requirements drop due to slow growth rate. The energy requirements of infant are derived from the average requirements of health and full-term infant nourishing on exclusive breast feeding. Subsequently solid foods are introduced alongwith breast milk to cater the growing requirement. The major source of energy for all body cells and especially brain is from carbohydrates. Relative to the size of the body, the size of an infant's brain is greater than that of an adult's hence it uses comparatively 60% more glucose. Fat is mainly contributed by the breast milk and standard formula. Protein being building block of body tissues and component of the most body fluids is considered essential for growth and development. An infant's needs for most of the vitamins and minerals are more than double than those of an adult. These recommendations are established from the average amount of nutrients consumed by flourishing infants breastfed by health and well-nourished mothers. Infants have comparatively more percentage of water in body than adults. To replace fluid losses in a healthy infant, breast milk or infant formula usually delivers enough water except in case of hot environment infants need supplemental water. After 12 months of age, cow or buffalo milk become a primary source of most of the nutrients Other traditional foods should be provided in variety and amounts appropriate for total energy needs. Besides provision of nutritious foods parents are also responsible for safe and loving environment in which the children may grow and develop (Rolfes et al., 2015).

27.2.1 Childhood

Each year from age 1 to adolescence, a child normally gains height by 2–3 inches and weight by 2–3 kg. The optimum growth can be assessed through growth charts. The positive or negative changes in weight gain proportion to height may reflect overeating, inactivity and inadequate food intake leading to malnutrition. A healthy child can stand alone and start toddling after one-year of age, walk, and run by 2, and can jump and climb with confidence by 3 years. Bones and muscles increase in mass and density to make these accomplishments possible. For this purpose, they need optimum energy and strong bones and muscles. Children's energy intake varies extensively from meal to meal; however, total daily intake remains constant. Individual children's energy needs vary widely, depending on physical activity and growth phase. By age 10, an active child needs

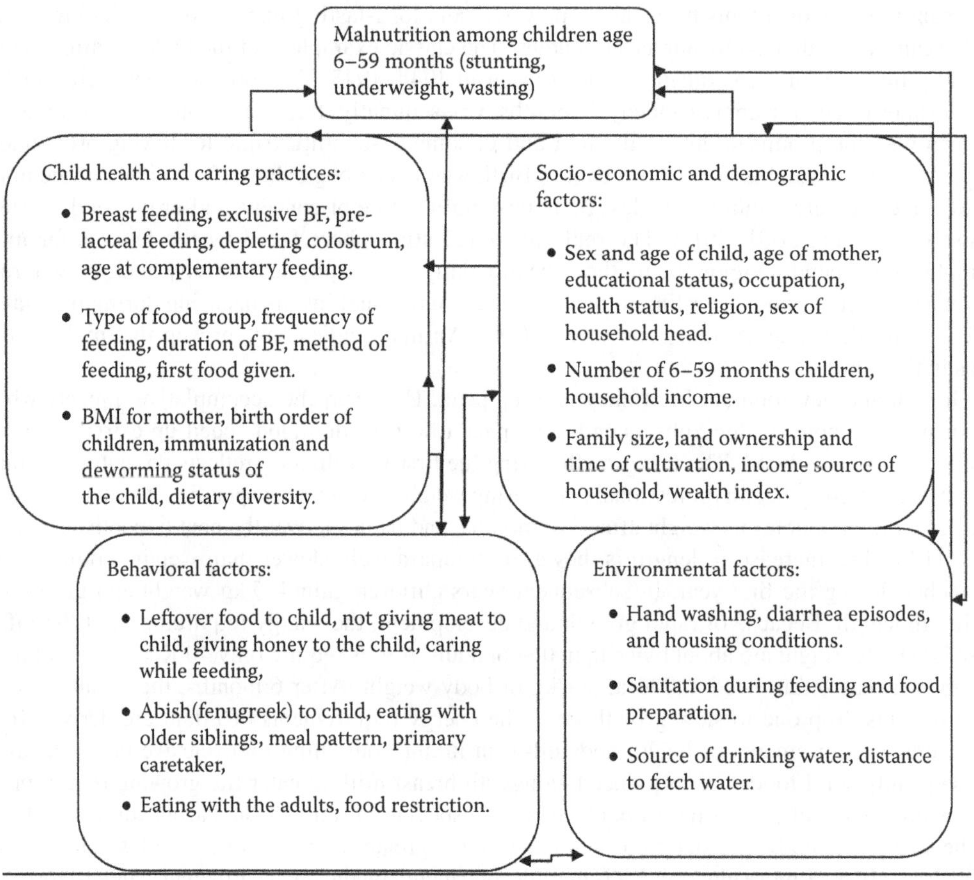

FIGURE 27.1 Contributory factors resulting in malnutrition among the children under 5.

about 2000 Kcal a day. Children in developed countries are becoming overweight and obese due to greater availability of junk food and low physical activity, whereas in developing countries they are underweight, stunted, wasted, and deficient in vitamins and minerals due to socioeconomic and demographic factors, behavioral and environmental factors, and poor child health and care practices (Figure 27.1).

Carbohydrate recommendations for children are based on glucose use by the brain. These are same for children after 1 year and adults. Dietary fiber should be adequate to prevent constipation and regulating bowl movement. Up till now no RDA for total fat has been established, however DRI Committee recommends fat intake sufficient to provide 30%–40% energy in kids from 1 to 3 years of age and 25% to 35% in children 4 to 18 years of age. Like energy needs, total protein requirements urge slightly with age, however when compared with child's body weight there is slight decline in protein requirement. Protein recommendations should be based on protein quality and maintaining positive nitrogen balance. The micronutrient needs of children increase with age, which can be fulfilled providing balanced diets. Children's foods must supply 7 to 10 mg of iron per day to prevent iron deficiency anaemia. This can be achieved through the provision of snacks d meals having iron rich foods including whole-grain products, meats, poultry, eggs, fish, pulses and fortified products. The daily requirements of Vitamin D (10 micrograms) can be met by drinking vitamin–D fortified milk and eating fortified foods. Generally, supplements are not recommended for children. Dietary supplements on the market today include many herbal products that have not been tested for safety and effectiveness in children (Rolfes et al., 2015).

27.3 MALNUTRITION IN CHILDREN

27.3.1 Types of Malnutrition

Malnutrition is the chief health issue worldwide and is widespread in developing countries. Three South Asian countries, that is, Pakistan, India, and Bangladesh are the home of more than half of the world's malnourished children. In Pakistan, it has been resilient since many years. It extends throughout the lifecycle affecting health and economics of the individuals negatively. There are two main types of malnutrition, that is, protein energy malnutrition (PEM) and micronutrient deficiencies. The brief description of each is given below:

27.3.1.1 Protein-Energy Malnutrition (PEM)

Macronutrient deficiency refers to a lack of carbohydrates, protein, and fat, which are consumed in ample quantities to fuel the body, normal growth and development, and maintenance of health. PEM is a major macronutrient deficiency prevalent in children consuming inadequate quantities of protein and energy to fulfill body requirements. The occurrence of PEM is much higher in South Asia compared with subSaharan Africa where more than 90 million children under 5 are underweight. There are significant discrepancies among countries, Sri Lanka having the lowest rate and Bangladesh having the highest. Underweight, stunting, and wasting are the result of PEM. There are several reasons for PEM in children and may include inadequate food supply, poor quality food, malabsorption issues, poor personal hygiene and sanitation, chronic ailment, and gastrointestinal illness. In Pakistan, 43.7% children are affected by stunting followed by low weight (31.5%) and wasting (31.5%). Stunted children fail to reach their genetic potential. The widespread prevalence of infection coupled with low intake or absence of nutrients and stress greatly disturb length and weigh t in children. Globally, 178 million children are facing stunting and about one third of them are from developing nations. Overall, the occurrence is the highest in Asia affecting one in every two children. The major symptoms of stunting among the children are reduced height, low weight, chubby appearance, and delayed bone growth. Stunting is more prevalent in children with low protein intake in proportion to total calorie, frequent infections early in life and hormone changes prompted by stress. Wasting typically occurs due to shortfall in total calories due to inadequate intake of nutrients or poor food quality. The affected children fail to gain weight or lose weight swiftly due to inadequate dietary intake or frequent infection. Initially body fat stores are depleted due to severe calorie deficiency followed by death in extreme cases. Children with wasting may exhibit symptoms like low muscle and fat mass, fatigue, drastic weight loss, thinness, and poor wound healing. Shortage of food and poor quality, weakened immune system, frequent infections, and starvation are the major causes of wasting which can be reversed by improving the supply of energy, proteins, and nutrients required for the synthesis of fat, tissues, and muscles. However, quick refeeding of wasted children may result in life-threatening refeeding syndrome.

27.3.1.2 Hidden Hunger

Micronutrients, that is, vitamins and minerals are required for the proper functioning of enzymes, hormones, and numerous other substances. Vitamins are organic substances that must be supplied in smaller amounts to maintain health. Similarly, minerals are not only required for body building but also for maintaining electrolyte balance in body, blood pressure, normal functioning of nerve impulse and many as part of enzymes and hormones. The non-optimum quantities of vitamins and minerals in body disturb the normal functioning of mental and body (Bowman and Russell, 2006). The deficiencies of vitamin A, D, B_{12}, folic acid, iron, zinc, and iodine are now well established across the globe. Iron is an essential components of hemoglobin, responsible for carrying oxygen in the body. It is critical for the normal growth of the central nervous system. Iron deficiency is the most common nutritional disorder in the developing countries as well as in affluent societies affecting almost all segments of the society (Yip, 2000). According to WHO, about two billion humans are affected by anemia and a majority of them are living in developing countries. In some regions, the prevalence of anemia is much higher. It affects more than 47% of all preschoolers globally resulting in impaired cognitive and physical development, compromised immune responses, poor academic performances, and behavioral problems (Grantham-McGregor and Ani, 2001; Stoltzfus et al., 2009).

Zinc is an essential mineral and an important component of more than 200 enzymes involved in protein and cell division. It is vital for growth and development, the gastrointestinal tract, and the immune system. It helps in appetite, neural function, skin integrity, wound healing, and testicular maturation. Zinc deficiency increases the chances of death due to diarrhea, pneumonia, and malaria by 13%–21% (Caulfield and Black, 2009). According to WHO, 20% of the global population is zinc deficient whereas about two billion have insufficient iodine for nourishment. Iodine deficiency is a major public health problem throughout the world. Vitamin A and its allied compounds are essential for strong immunity, better vision, and normal functioning of the body. Vitamin A deficiency (VAD) contributes towards several health disparities like xerophthalmia, night blindness, and some others while its excess causes hypervitaminosis (Diaz et al., 2003; Underwood, 2004). According to the WHO, globally 254 million children under 5 are vitamin A-deficient, and 15%–60% among them are from Africa and Southeast Asia. Severe deficiencies of this vitamin have resulted in night blindness among huge number of preschool children (Long et al., 2007). Moreover, vitamin A deficiency increases the risk of mortality by 20%–24% due to measles, diarrhea and malaria (Rice et al., 2009). Likewise, folate deficiencies result in more than 200,000 severe birth defects among expectant mothers each year (Long et al., 2007).

27.3.2 Causes, Signs, and Symptoms

Food inequity, poverty, sociocultural beliefs, nutrient deficient soils, illiteracy, body losses, frequent infections, incorrect feeding practices, and lack of access to age appropriate foods are the major contributory factors resulting in malnutrition. During infancy, appropriate complementary foods, vaccination, supplementation and deworming is required for rapid growth and development. Subsequently, intense physical, psychosocial and cognitive development in adolescence require accelerated demands for nutrients as this is the stage where they gain up to 50% of their adult weight, 50% of their adult skeletal mass and >20% of adult height during this period. Likewise, the nutritional choices of the adults are greatly affected by sedentary life styles, imbalanced diets, urbanization, industrialization and sociocultural norms. Undernutrition is not merely related to food, there are numerous causes for it. The UNICEF conceptual framework (Figure 27.2) categorizes 3 levels of causes of undernutrition: immediate, underlying and basic, whereby factors at one level influence other levels. Immediate causes operating at the individual level are the result of disease or lack of dietary intake. In general, poor nutrition can result in reduced immunity to infections. This can raise the probability of an individual getting an infection or increase its duration or severity. Infection can result in loss of appetite, increased nutrient requirements and decreased absorption of nutrients consumed. This trigger further weight loss and reduced resistance to further infection. This vicious cycle needs to be broken by treatment of infection and improved dietary intake. Whether an individual gets enough food to eat or whether she/he is at risk of infection is mainly the result of factors operating at the household and community level, that is, household food insecurity, inadequate care (breastfeeding, appropriate complementary feeding, hygiene & health seeking behavior) and unhealthy household environment and lack of health services. The third category of the underlying causes of undernutrition refers to those related to poor public health. This includes factors relating to the health environment, exposure to disease, and access to basic health services. The health environment is affected by access to clean and safe water, sanitation, quality of shelter, presence of malarial breeding sites, and consequent level of cold, stress, and overcrowding. For example, a child who is sick is dependent on his/her mother to take him/her to a health facility yet her time is dependent on the work she must do in the fields for the upcoming harvest. The third level of factors identified by the conceptual frameworks are considered basic causes. These refer to what resources are available and how they are used. Overcoming entrenched poverty and underdevelopment requires

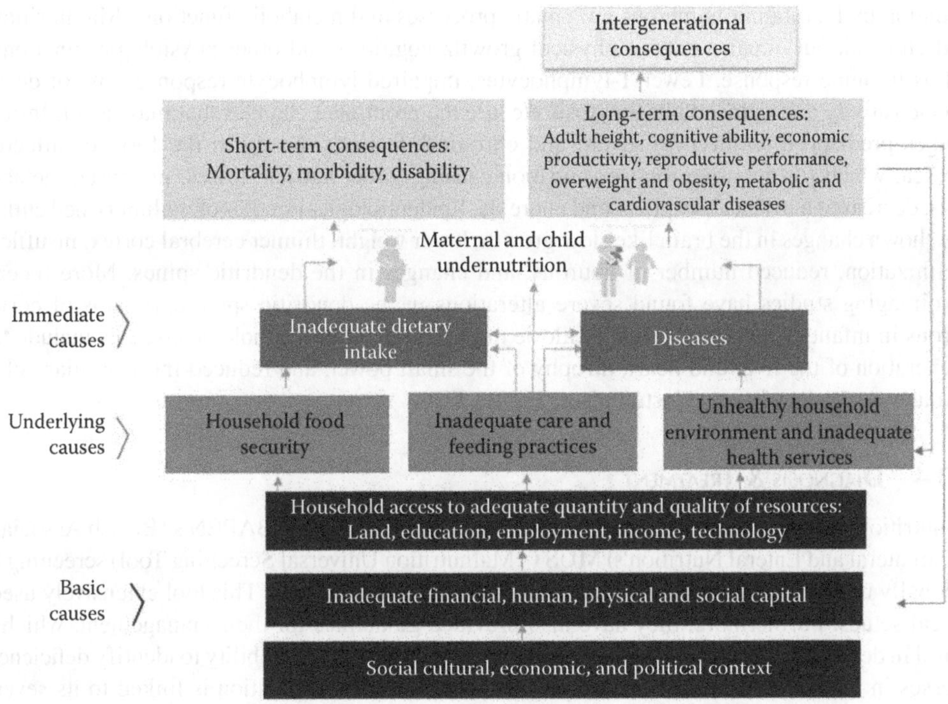

FIGURE 27.2 The UNICEF conceptual framework.

resources and inputs. If the basic causes of undernutrition are to be addressed, greater and better-targeted resources and better collaborations are needed. There is a clear difference between signs and symptoms; sign is something detected by doctor or other people (e.g., rash) whereas symptom is something felt and reported by the patient (e.g., pain). The signs and symptoms of malnutrition include:

- Respiratory failure or difficulties in breathing
- Hypothermia
- Complications associated with surgery
- Weakened immune system
- Depression
- Feeling cold
- Poor healing of wounds
- Delayed recovery from illnesses
- Frequent infections
- Reduced muscle and tissues mass
- Irritability
- Tiredness, fatigue, or apathy
- Poor fertility and low sex drive

27.3.3 Pathophysiology

Malnutrition virtually results distress in every organ system. In human body, the amino acids for the synthesis of proteins and other allied compounds are provided by the dietary proteins. Similar, energy is required for various physiological and metabolic functions in the body. Furthermore, micronutrients in the form of vitamin and minerals have dual role, that is, bodybuilding and regulation by facilitating numerous enzymatic processes and metabolic functions. Malnutrition in children results in impairments in physical growth, cognitive and other physiologic functions as well as immune response. Fewer T-lymphocytes, impaired lymphocyte response, loss of delayed hypersensitivity and reduced phagocytosis etc. are the prominent changes that may occur. Immune changes predispose children to severe and chronic infections mostly in the form of infectious diarrhea, which further compromises nutrition causing direct nutrient losses, increased metabolic needs, decreased nutrient absorption, and anorexia. Epidemiological studies of malnourished children have shown changes in the brain like slow growth, lower weight, thinner cerebral cortex, insufficient myelinization, reduced number of neurons, and changes in the dendritic spines. More recently, neuroimaging studies have found severe alterations in the dendritic spine apparatus of cortical neurons in infants with severe protein-calorie malnutrition. Other pathologic changes include fatty degeneration of the liver and heart, atrophy of the small bowel, and reduced intravascular volume leading to secondary hyperaldosteronism (Figure 27.3).

27.3.4 Diagnosis & Treatment

Malnutrition complications can be stopped only after quick diagnosis. BAPEN's (British Association for Parenteral and Enteral Nutrition's) MUST (Malnutrition Universal Screening Tool) screening tool is globally used for identification of undernourished and obese adults. This tool extensively used in clinical setups. Furthermore, they have also provided guidelines for their management, which are helpful in developing care plans. The only limitation of MUST is the inability to identify deficiencies/excesses in vitamin and/or mineral intake. The treatment of malnutrition is linked to its severity. National Institute for Health and Clinical Excellence, UK (NICE) have framed guiding principles

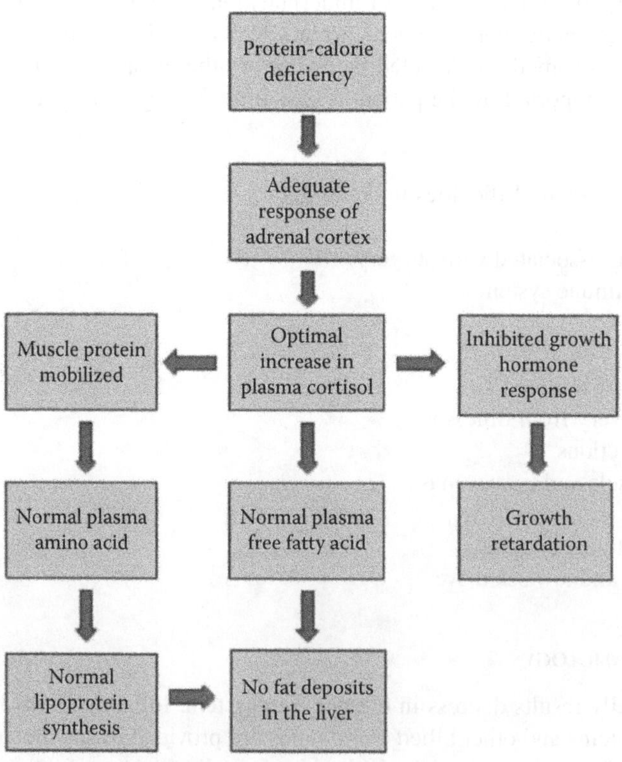

FIGURE 27.3 Pathophysiology of malnutrition.

for malnutrition treatment. Artificial nutritional support is available for those patients who are unable to meet sufficient nutrition through oral route.

27.3.5 Consequences of Malnutrition in Early Life

Inadequate intake of protein, calories, and other nutrients in malnourished children results in weight loss, illness, and developmental delays. The first few years of life are crucial for optimal growth and development in children, and nutrient insult at this stage results in immune and growth related complications making them more susceptible to communicable ailments. Weakened immune function in children is mainly due to deficiencies of micronutrients particularly iron, zinc and vitamin A. Initially child is contracted with infection followed by infection of gastrointestinal system accelerating nutrient deficiencies due to poor absorption. Subsequently, nutrient deficiency coupled with infection further results in growth retardation. Furthermore, deficiency in one nutrient may lead to a deficiency in another one. Growth and cognition are compromised by the short term implications of malnutrition giving the way to long-term problems. Resultantly bone grow this limited resulting in stunting and negative impact on brain development leads to learning disabilities, decreased IQ scores, attention deficit disorder, memory deficiency, reduced social skills, impaired school performance, reduced language development, and problem-solving abilities.

27.4 ONE HEALTH AND NUTRITION

27.4.1 Global Scenario

The "One health" concept has become a movement now. The idea of one health is to improve the human health at societal level by preventing the human health challenges and minimizing the risk factors related to humans, animals, and environments. This concept promotes multisectoral and collaborative approaches. It needs involvement of society as a whole. One health recognizes that humans, animals, and their environments are interconnected and these are all important pillars of this triangle. The common theme of one health is collaboration across sectors and to assess the direct and indirect effects on human health. It necessitates that all people working in their specific areas should make proper use of resources for the betterment of society. For strengthening the one health concept in a proper way there is need to balance efforts of different groups and networks including animal health experts, human physicians, and environmental and social scientists. The one health concept elucidates that human health is dependent upon animals and ecosystem health. Because humans have very close contact with animals and environment. We can quote many examples of zoonotic diseases in our daily life, for example, rabies, avian influenza, bovine spongiform encephalopathy, salmonella, etc. Therefore, combined efforts are required from people working in public health sectors, veterinary health, and environmental health. This is the basic theme of the one health concept. This concept was started 4 years ago and has received much encouragement, becoming since a fast growing movement. This one health movement is officially owned and recognized by international agencies including EU, USDA, CDC, WHO, FAO, OIE, etc. The current one health movement attracted the international community unexpectedly after outbreaks of highly pathogenic avian influenza in 2005. It changed the international scenario and people are thinking of cross-sectoral collaborations for serious health risks to humans. Different groups and institutes are continuously arranging conferences, meetings, and symposia to increase awareness among different sectors. The major conferences have been held in Winnipeg (Manitoba, Canada, March 2009), Hanoi (Vietnam, April 2010), and Stone Mountain (US, May 2010), as well as the first international One Health scientific congress, which took place in Melbourne, Australia, in February of 2011. In Georgia, at the 1-Health expert meeting, scientists were convinced that one health should not be "possessed" or "mastered" by any one organization or institution; that one health should remain flexible and comprehensive; and that one health can be promoted by various institutions, but it should not be institutionalized. It was also decided that One Health Global Network should be initiated to improve coordination and collaboration. Similarly, the U.S. National Academy of Sciences (NAS), in collaboration with the Pakistan Academy of Sciences, started a pilot project of one health involving Pakistan scientist from different sectors to promote the concept of one health (Arif et al., 2014).

27.4.2 Role in Child Health and Nutrition

There is strong need of a one health concept implementation in Pakistan to mitigate child malnutrition. The strong multisectoral efforts are necessary to improve child nutrition. All the partners including veterinarians, human health physicians, environmental scientists, and social scientists should play their key role to tackle this important national issue.

27.5 STRATEGIES FOR IMPROVED NUTRITION

Malnutrition is not just about food, or just about health care, but is an issue that requires a multisector response. Ensuring good nutrition for children is not the task of just one group or government ministry. Instead, it involves the activities of a wide range of groups and government bodies working together to achieve results. For example, children can receive the nutritious foods, but poor environmental sanitation can lead to diarrhea and other childhood infections, which limit their ability to absorb nutrients from the food, resulting in a vicious cycle of malnutrition and disease. The following strategies can be adopted to mitigate malnutrition especially among the children.

1. School nutrition programs
2. School health programs
3. Food multimix/composite flours
4. Targeted fortified products
5. Diet diversification
6. One health.

27.5.1 School Nutrition Programs

Meeting the nutrition and education needs of children is critical to supporting their healthy growth and development. Historically, in 1790 Germany stated the first-ever school feeding program, soon followed by France and England (1800s). In the United States, Ellen H. Richards introduced the first school-feeding program in the city of Boston in 1894. The purposes of these programs were to feed hungry and underfed children, create employment opportunities for unemployed individuals, and create a market for surplus agricultural commodities. Subsequently, in 1930s the federal government started financial support for school lunch program. National School Lunch Act (NSLA) played an important role to carry school lunch programs on a permanent basis. NSLA emphasize that participating schools should provide nutritious lunches by meeting the requirements set by USDA consistently with the Dietary Guidelines for Americans. After the approval of the Child Nutrition Act in 1966, school breakfast programs were launched to fill the gap for children attending school after travelling long distances by bus. Evidence-based studies have proven the effectiveness of these programs, as reflected in better classroom performances. In the Unites States, the National School Lunch Program is providing meals to more than 30 million children to meet at least one third of the recommended daily intakes for energy, protein, iron, calcium, vitamin A and C, and focusing on variety of food choices. Likewise, School Breakfast Programs are available in more than 80% of the nation's schools that offer school lunch, and close to 9 million children participate in it (Table 27.1).

TABLE 27.1
Foods and Beverages That Meet Recommended School Food Standards

Preferred foods for all students	
Foods	Whole grains, fruits, vegetables and related combination products and non-fat and low-fat milk products that are limited to 200 Kcal or less per serving are considered alongwith: • Not more than 35% of total energy from fat; <10% from saturated fats; trans fat-free (≤0.5 g per serving) • 35% or less energy from sugars • 200 mg or less intake of sodium per portion as packaged
Beverages	• Normal water • Low-fat and nonfat milk (soy beverages and flavored milk having <22 g sugars per 8 oz serving • 100% fruit juice (4 oz for elementary/middle schools & 8 oz for high schools) • Caffeine-free
Snacks for high school students after school	
Foods	Snack foods with no more 200 Kcal per portion as packaged and: • No more than 35% energy from fat; <10% from saturated fats and free from *trans* fats • 35% or less energy from sugars • 200 mg or less intake of sodium per portion as packaged
Beverages	Nonnutritive-sweetened, noncaffeinated, nonfortified beverages with <5 Kcal per portion as packaged

Source: Stallings A and Yaktine AL 2007. *Nutrition Standards for Foods in Schools: Leading the Way Toward Healthier Youth*. National Academies Press, Washington DC, USA.

27.5.2 School Health Programs

School nutrition programs have traditionally been viewed as a method to reduce the vulnerability of poor children by encouraging regular attendance and maximum enrollment. School atmosphere and the food served at school can influence children's diet for better or for worse. This is one of the oldest concepts first adopted by developed countries with the notion that improved health conditions and created conducive environments to motivate students for learning. Afterwards, numerous developing countries like India, Sri Lanka, Bangladesh, Iran, the Philippines, Thailand, Malaysia, and most recently Pakistan have launched health-related initiatives in the schools. The most common interventions in schools include supervised hand washing with soap, tooth brushing, biannual deworming, giving a ration on each school day, health screening and remedial measures, health and nutrition education, providing a safe and supportive environment in school, capacity building for health screening, nutritional interventions like mid-day meals and provision of iron-folate tablets. In India, during lunch breaks, kids are served with *khichdi*, a dish made of rice mixed with lentils and vegetables. In certain parts of India, children are sent to school with empty bellies. The mid-day meal is their first meal of the day and their only regular source of vegetables and lentils and—in some states with better lunch menus—eggs (Figure 27.4). The costs of all these activities are comparatively low but benefits are high. Because of these interventions, there has been a reduction in infectious diseases (30%–50%), dental cavities (40%–50%), and prevalence of helminths (80%). Likewise, there was increase in school attendance (20%–25%) and a 20% reduction in the number of students below-normal height and weight.

FIGURE 27.4 School children in Haryana, India, eat rice and kadhi, a curry made with onions, garlic, yogurt, and fritters made with chick pea flour.

In Pakistan, the only negligible resources are spent on health and education. In such situation, there is dire need of such interventions at country level to mitigate malnutrition. Historically, several elements of school health programs have been delivered in a fragmented manner in 1970s and 1980s, mainly focusing on health screening by the doctors, which were abandoned due to the lack of interest by health professionals. Recently, the Punjab Health Sector Reforms Programme (PHSRP), in collaboration with the National Commission for Human Development, launched School Health Programs in Punjab province to improve and increase school enrolment, attendance, health, nutrition, and learning performance for more than 34 million children enrolled in 228,304 educational institutions. Initially, School Health Programs with financial assistance of the Bill and Melinda Gates Foundation benefitted 1.86 million students enrolled in 23,266 primary schools of 17 districts. Considering the importance of the program towards improved health and education for the children, Pakistan needs a similar nationwide comprehensive school health program also addressing school environment, education, health services, and nutrition of the children.

School nutrition programs are not a new in Pakistan. Tawana Pakistan was a multifaceted pilot nutrition and social development project (Sept. 2002 to June 2005) investing Rs. 3.6 billion to combat malnutrition and increase school enrolment among primary school girls. The cost of noon meal was USD 0.12/child (meal from 3 food groups providing 600 Kcal, deworming medicine, and multi-micronutrient supplements). Consequently, wasting, low weight, and stunting were decreased by 45%, 22%, and 6%, respectively. However, overall the program failed to achieve its targets due to mismanagement

27.5.2.1 Tackling Malnutrition Through Snacks Feeding Program—Summary Of Feasibility Study of Corn Based Extruded Snacks for School Nutrition Programs

Recently, the National Institute of Food Science & Technology, University of Agriculture, Faisalabad has developed rice and corn based extruded crisps by adding soy & chick pea flours to augment protein contents. Additionally, these products have been fortified with micronutrients. These crispy ingredients can be used by adding some seasoning and can also be converted into sweet nuribars.

The basic composition (Table 27.2) and feasibility study of corn-based extruded snacks for school nutrition programs is given in below:

• GDP of Pakistan	US $269.97 billion (Pak Rs. 28,098.48 billion)
• GDP loss due to malnutrition	3%–4% (US$ 8–10; 842–1123 billion Rs.)
• Expenditures on health	0.45% of GDP (Pak. Rs. 133.9 billion)
Expenditures on nutr. interventions are very negligible	
• NSP (Sindh)	4.1 billion Rs.
• BNPMC (Balochistan)	1.5 billion Rs.
• BISP (Rs. 18,800/- per annum)	115 billion Rs. (Waseela-e-Sehat: Life & Health Insurance)

- Processing and packaging cost = Rs. 1.5
- Total price of 100 g snacks = Rs. 11.13
- 1 day serving cost = 11.13/2 = Rs. 5.57
- Malnourished children in Pakistan = 66.1%
- Total numbers of malnourished children = 40.66 million (4,06,60,000)
- Cost of feeding them for one day = 40,660,000 × 5.57 = Rs. 226,476,200 (226.4762 million rupees)
- Cost for 1 year = 226.4762 × 365 = 82,663.813 million rupees (82.663,813 billion)
- Total cost of the project = 82.663,813 billion
- If program efficiency is 20% than covered losses = 0.20 × 842.9544–0.20 × 1123.9392 = Rs. 168.59–224.7878 billion
- Actual benefit = benefitted amount from according to the program efficiency-cost of project 168.59–82.66 = 85.93 & 224.7878–82.66 = 142.13

Actual benefit = 85.93–142.13 billion/year; this benefit is 0.3%–0.5% of Pakistan's GDP can be saved by this program with only 20% program efficiency.

27.5.3 Food Multimix/Composite Flours

Food-based approaches are considered sustainable and cost-effective ways to improve food and nutrition security and decrease the incidences of micronutrient deficiencies. A preferrable approach is food-to-food fortification or the food multimix (FMM) concept in which a composite recipe or diet is formulated carefully choosing the constituents considering their individual strengths to complement each other and provide an enriched composite product. For example, vitamin C from the citrus family promotes absorption of non-haem (form of iron other than derived from hemoglobin and myoglobin) iron in plant-based foods (cereals). A food multimix approach is considered equally useful to meet the energy and micronutrient requirements of vulnerable segments of the population through an empirical process. The recipes thus formulated are subsequently used for the manufacturing of numerous food products (porridge, soups, cakes, bread, muffins) commonly used in the region.

TABLE 27.2
Basic Composition and Cost Analysis

Ingredients	Amount in 100 g	Unit Price (Rs.)	Cost of 100 g
Soybean	15	130/kg	1.95
Chickpea	15	80/kg	1.2
Corn	70	45/kg	3.15
Nutrient Premix	DV (635 mg)	5200/kg	3.30
Table salt	1	25/kg	0.025

Another advantage of FMM approach is the development of food products considering specific physiological and clinical needs of the people. The micronutrient status of children can be improved using this novel food-to-food fortification technology. FMM approach has long-lasting impacts in society due to sustainability, cultural, and sensoric acceptability and cost effectiveness. Food recipes are normally developed to meet 40% of daily energy requirements and >50% mineral and vitamin supplies depending on age. However, in occasional instances, the need for limiting nutrients can be met through the addition of food fortificants in the form of premixes. The knowledge gained and experiences learned from FMM approaches will ensure and improve the nutritional value of foods and the health of the people (Zotor et al., 2015).

27.5.4 Targeted Fortified Products

Food fortification is a cost-effective, flexible, and generally acceptable approach to improve the nutrients intake in the vulnerable segments. Processed foods are often fortified with certain micronutrients accessible to large number of peoples thus playing pivotal role in prevention of deficiencies and disorders (Fiedler et al., 2008). Fortification of foods with micronutrients is an effective method of increasing micronutrient intakes in populations. The advantages of fortification over the distribution of vitamin and mineral supplements, or commodity foods, include little change in dietary habits, delivery of nutrients to large population segments, better calibration of nutrients to avoid excessive intakes and possibility of toxicity, multiple nutrient fortifications, low cost of adding nutrients to foods, and profitable economic activity for smallholder farmers (Allen et al., 2006). Moreover, food fortification of staple foods and conversion into shelf-stable and easy-to-eat food can be an effective strategy for combating micronutrient malnutrition as well as to provide emergency and humanitarian assistance in response to natural or manmade disasters. In Pakistan, only a few fortified food products like fortified wheat flour, ghee & vegetable oil, common salt and milk are available in the market. In AJK, all eleven flour mills have started iron fortification in wheat flour with the support of the Micronutrient Initiative and the World Food Program. Likewise, Global Alliance for Improved Nutrition is helping to do the same in Punjab. Universal Salt Iodization (USI) Programs are being implemented through public-private partnership models and are supporting the government in 110 districts to benefit almost 174 million people countrywide. There is dire need to enhance the availability of fortified food products, especially for school children in Pakistan, to alleviate micronutrient deficiencies in the country.

27.5.5 Diet Diversification

Agronomic practices, climatic changes, cultural, ecological, and socioeconomic conditions have changed current dietary patterns, ultimately dictating food choices of the masses. The nutritional needs of the consumers can effectively be met if adequate quantities and array of foods are available. For the purpose, various strategies have been suggested by the international agencies to ensure household and community requirements for diet diversification. For this purpose, quantity and variety of food commodities can be enhanced by escalating activities for food diversification. The FAO has suggested the following strategies to ensure balanced diets capable of meeting nutritional requirements (FAO, 1997, 2004):

- Cultivation of new crops with more nutrients
- Rearing of small ruminants/livestock
- Home/kitchen gardening by exploiting traditional crops
- Improved food preservation methods to reduce waste and curtail postharvest
- More consumption of fishery and forestry products
- Nutrition education of the masses for consumption of healthy and nutritious food throughout the year

- Accelerate activities to generate extra income
- Establishment of small and medium scale agro-processing industries

Focusing on dietary diversification for optimal health is need of the time to prevent objectionable disparities which may limit the consumption of some nutrients. The inclusion of nutrient dense foods in daily diet may be helpful in correcting nutrient deficiencies in the target groups. Foods like fruits & vegetables, pulses, green leafy vegetables etc. are considered health. When these are use in various combinations these complement each other. For example, fruits rich in vitamin C besides proving the activity of vitamin C also accelerate the absorption of ionic iron. Likewise, in societies where cereal based diets are used as staple foods, deficiencies of iron, zinc and vitamin A are widespread which can be mitigated through diet diversification. The addition of plant foods rich in carotenoids in staple diets can significantly improve the recommended levels of vitamin A. For example, daily use of cooked carrots (50 g) can provide approximately 500 mg retinol equivalents. Likewise, use of palm oil can be helpful in the provision of carotenoids in many countries. Therefore, daily use these foods may provide 100% or more of the daily requirements for retinol equivalents. Likewise, the consumption of 20–25 g chicken liver or fish may provide more than the recommended vitamin A nutrient density for virtually all age and sex groups.

27.6 RECOMMENDATIONS

- Malnutrition and hunger cannot be eliminated alone by the governments; there is shared responsibility of all stakeholders including food producers, processors, consumers, policy makers, enforcement bodies, and the civil society for coordinated actions to tackle the menace.
- Household food security must be addressed especially considering the vulnerable segments of the society especially children, pregnant and lactatiing women, and the aged.
- Climate change is hampering food production and quality of commodities; planning and adaptations must be done accordingly to maintain the food production levels; R&D facilities should be strengthened to mitigate the negatives impacts of this change.
- Capacity building and skill development of rural communities should be carried out to produce more domestic food, reduce postharvest losses, and create value addition.
- Prices of food commodities should be stabilized to ensure availability of foods at an affordable cost.
- Nutrition education and awareness campaigns should be launched involving print and electronic media for raising awareness and policy advocacy.
- Short and long-term strategies such as enhancing the productivity of major crops, food diversification, targeted fortification and supplementation, and school health and nutrition programs should be promoted to mitigate the problems associated with malnutrition.

REFERENCES

Ahmed A, Ahmad A, Khalid N, David A, Sandhu MA, Randhawa MA and Suleria HAR. 2014. A questionmark on iron deficiency in 185 million people of Pakistan: Its outcomes & prevention. *Cri Rev Food Sci Nutr.* 54(12): 1617–1635.

Akhtar S, Anjum FM and Anjum MA. 2011. Micronutrient fortification of wheat flour: Recent development and strategies. *Food Res Int.* 44: 652–659.

Akhtar S, Ismail T, Atukorala S and Arlappa N. 2013. Micronutrient deficiencies in South Asia: Current status and strategies. *Trends Food Sci Technol.* 31: 55–62.

Allen L, de-Benoist B, Dary O and Hurrell R. 2006. *Guidelines on Food Fortification with Micronutrients.* World Health Organization/Food and Agricultural Organization, Geneva, Switzerland.

Arif GM, Arooq S, Nazir S and Sathi M. 2014. Child malnutrition and poverty: The case of Pakistan. *PakDev Rev.* 53(1): 29–48.

Bowman BA and Russell RM. 2006. *Present Knowledge in Nutrition*. International Life Sciences Institute, Washington DC, USA.

Caulfield LE and Black RE. 2009. Zinc deficiency. In: Ezzati M, Lopez AD, Rodgers A and Murray CJL (ed) *Comparative Quantification of Health Risks: Global and Regional Burden of Disease Attributable to Selected Major Risk Factors*. World Health Organization, Geneva.

Diaz JR, Cagigas AD and Rodriguez R. 2003. Micronutrient deficiencies in developing and affluent countries. *Eur J Clin Nutr*. 57(1): 70–72.

FAO. 1997. *Agriculture Food and Nutrition for Africa—A Resource Book for Teachers of Agriculture*. Food and Agriculture Organization of the United Nations, Rome, Italy.

FAO/WHO. 2004. Vitamin and mineral requirements in human nutrition. Report of a Joint FAO/WHO Expert Consultation on Human Vitamin and Mineral Requirements. World Health Organization, Geneva, Switzerland.

Fiedler JL, Sanghvi TG and Saunders MK. 2008. A review of the micronutrient intervention cost literature: program design and policy lessons. *Int J Health Plann Manag*. 23: 373–397.

GOP. 2011. *National Nutrition Survey of Pakistan 2011*. Nutrition Wing, Cabinet Division, Government of Pakistan, Islamabad-Pakistan.

GOP. 2016. *Economic Survey 2015–16*. Economic Advisor Wing, Finance Division, Government of Pakistan, Islamabad-Pakistan.

Grantham-McGregor S and Ani C. 2001. A review of studies on the effect of iron deficiency on cognitive development in children. *J Nutr*. 131: 649S–668S.

Huma N, Rehman SU, Anjum FM, Murtaza MA and Sheikh MA. 2007. Food fortification strategy—preventing iron deficiency anemia: A review. *Cri Rev Food Sci Nutr*. 47(3): 259–265.

IFPRI. 2016. *2016 Global Hunger Index: Getting to Zero Hunger*. International Food Policy Research Institute, Bonn Washington, DC, USA.

Long KZ, Rosado JL and Fawzi W. 2007. The comparative impact of iron, the B-complex vitamins, vitamins C and E, and selenium on diarrheal pathogen outcomes relative to the impact produced by vitamin A and zinc. *Nutr Rev*. 65(5): 218–232.

Morris SS, Cogill B and Uauy R. 2008. Effective international action against undernutrition: Why has it proven so difficult and what can be done to accelerate progress? *Lancet*. 371: 608–621.

Rice AL, West KP and Black RE. 2009. Vitamin A deficiency. In: Ezzati M, Lopez AD, Rodgers A and Murray CJL (ed) *Comparative Quantification of Health Risks: Global and Regional Burden of Disease Attributable to Selected Major Risk Factors*. World Health Organization, Geneva.

Rolfes SR, Pinna K and Whitney E. 2015. *Understanding Normal and Clinical Nutrition*. Thomson and Wadsworth, USA.

Roseboom TJ, Painter RC, van Abeelen AF, Veenendaal MV and de Rooij SR. 2011. Hungry in the womb: What are the consequences? Lessons from the Dutch famine. *Maturitas*. 70: 141–145.

Roseboom TJ, van der Meulen JH, Ravelli AC, Osmond C, Barker DJ and Bleker OP. 2001. Effects of prenatal exposure to the Dutch famine on adult disease in later life: An overview. *Mol Cell Endocrinol*. 185: 93–98.

Samuelsson AM, Matthews PA, Argenton M, Christie MR, McConnell JM, Jansen EH, Piersma AH et al. 2008. Diet-induced obesity in female mice leads to offspring hyperphagia, adiposity, hypertension, and insulin resistance: A novel murine model of developmental programming. *Hypertension*. 51: 383–392.

Stallings A and Yaktine AL. 2007. *Nutrition Standards for Foods in Schools: Leading the Way Toward Healthier Youth*. National Academies Press, Washington DC, USA.

Stoltzfus RJ, Mullany L and Black RE. 2009. Iron deficiency and anaemia. In: Ezzati M, Lopez AD, Rodgers A and Murray CJL (ed) *Comparative Quantification of Health Risks: Global and Regional Burden of Disease Attributable to Selected Major Risk Factors*. World Health Organization, Geneva.

Underwood BA. 2004. Vitamin A deficiency disorders: International efforts to control a preventable pox. *J Nutr*. 134(1): 231–236.

Yip R. 2000. Significance of an abnormally low or high hemoglobin concentration during pregnancy: Special consideration of iron nutrition. *Am J Clin Nutr*. 72(1): 272S–279S.

Zotor FB, Ellahi B and Amuna P. 2015. Applying the food multimix concept for sustainable and nutritious diets. *Pro Nut Soc*. 74(4): 505–516.

Section IV

Agricultural Incentives for Farmers

28 Gender Dimensions of Agriculture

Status, Trends, and Gap

Farkhanda Anjum, Muhammad Iqbal Zafar, Kanwal Asghar, and Ayesha Riaz

CONTENTS

28.1 INTRODUCTION

Agriculture as a significant contributor to economic growth fights against hunger, undernourishment and helps to improve livelihood. About 60% of the Pakistani population lives in rural areas. The majority of this rural population derives an income from agriculture as farmers, wage laborers, and service providers either directly or indirectly. The agriculture sector employs approximately 44% of the total labor force with a contribution of 21% to GDP (GOP, 2015b). Agriculture is the central economic activity that engages the most people in rural areas in Pakistan.

Women of rural areas working in the agricultural sector play a vital role in the well being of the household, particularly in spite of being overburdened, susceptible, and under-rewarded. Within conservative household units, wives are presented mostly as family workers whose financial interests are in harmony with the interests of their husband, and whose labor is subsumed to his. The following recurring issues are significant for understanding the family dynamics in different locations: (i) Married women face vulnerable situations of losing family assets at the death of their husbands, or in the case of separation and divorce; (ii) husbands will reduce their contributions as production and income of wife increases, (iii) husbands will take over enterprises of their commercially successful wives, (iv) cultural and family values affect women's work in public fields.

Women of Pakistan are busy in agricultural through their participation as farm operators and livestock managers. Women spend about 12–14 hours daily in agricultural activities. The work of women is generally ignored, unpaid, and not counted economically. Women also lack access to microcredits, agricultural knowledge, agricultural skills, and extension services. There is a requirement to improve female's education and offer on-farm educational opportunities and training.

In developing countries, women are the backbone of the country's economy. It is a valid economic argument that there should be investments on women in the agricultural sector as farmers, in farm-processing, and marketing. However, female farmers do not have equal access to assets and resources; this considerably limits their capacity to enhance productivity. They are mostly disadvantaged when they try to secure land tenure privileges, own livestock, access economic services, receive technical services and other resources that could increase their output. Women perform a fundamental role in food security, livestock development, natural assets and environmental administration, water conservation, marketing, and nonfarm income generation. It is concluded on the basis of inclusive literature that females are instrumental in Agricultural development in developing nations.

BOX 28.1 KEY CONCEPTS OF GENDER

Gender has become a central point for agricultural improvement and development, hence it is important to understand the terminology.

Gender roles *are behaviors, domestic responsibilities and everyday tasks of men and women considered appropriate by society.*
Gender relations define as *the manner in which the society delineates rights, duties and identities men and women in relation to one another.*
Gender discrimination *is every omission or limitation rooted on gender roles and relations that stop an individual from full human rights.*
Gender equality *refers to both women and men having equal rights, prospects, and entitlements in their public and political lives.*
Gender equity *refers to equality and neutrality in dealing with men and women in terms of rights, profit, responsibility and opportunities.*
Gender mainstreaming *is a worldwide recognized approach for accomplishing gender equality.*

In the development of agriculture, the work of women appears to be mostly limited to the labor of a wife on the field of a husband, as unpaid labor. A study on Bangladeshi female residents in London and Dhaka indicate that women cannot get any sustainable benefit by participating in innovative opportunities to keep themselves as capital is scarce (Kabeer, 2000).

Focus on gender may increase agricultural output and livestock system, and improvement in food security and sustenance. If women's access to productive assets can be increased on the same level with men, the yields would increase by 20%–30%. In response to this, the agricultural production in developing countries would increase by 2.5%–4%, reducing the total number of the starving populace by 12%–17%, which is approximately 100–150 million people (FAO, 2010). When women have equal access to resources, they may produce yield equal to that of men. But as there are gender disparities in acquiring resources like seed, technology and other inputs, uplifting women to the level the men's access to the resources is a significant chance to improve the general productivity. Women are more conscious than men in utilization of their income to develop the wellbeing of their family and community. Particularly in the case of children, for women it would have greater significance to invest in their education, nutrition, and health (Njuki and Sanginga, 2010).

BOX 28.2 AGRICULTURAL ISSUES AND ANALYSIS

- The women of South Asia are comparatively more involved in agricultural paid employment as compared to other regions, mainly expected the result of women's lesser property rights and other resources than in nearly all other regions, and tied with growing landlessness.
- According to "The International Labor Organization, 2007" about 59% of the total female labor force in South Asia is workings as contributing household workers.
- The rural women worker in the agriculture sector often tolerate harsh working conditions; they are busy in different activities nearly for 12 hours a day without weekly rest with less facilities and delayed wages.
- The agricultural workforce is full-time employed less than 6 months a year and is frequently forced to migrate to other areas for blue-collar jobs during the off-season.

28.1.1 Global Position of Pakistan about Gender Issues

In Pakistan, women have less advantage from economic developments compared to men. Low social indicators together with inadequate income-generating prospects have increased considerably the poverty level of women more than that of men. The sociocultural restriction on mobility of women is an important factor preventing their access to development prospects. Socially driven restrictions are transformed with the passage of time and there are extensive differences in the way of manifestation of social constraints between different regions of Pakistan. For instance, the gender parity index of both primary and secondary education has improved gradually over recent years (from 0.85 for primary education in 2005 to 0.87 in 2015) as attitudes toward female education have also changed. But cumulative national figures cover severe regional differences. However, additional education cannot necessarily translate into women employment due to other constraints, like workplace discrimination and physical insecurity. The considerably higher percentage of jobless women (25–34 years) having the same credentials as men are largely illiterate. With slow increases, the rate (26%) of women's labor force involvement in Pakistan is the lowest in the region. The maternal mortality rate (MMR) is also evidence of poor women's health. Ratios have become poorer over the last 5 years regardless of considerable investments in basic health and emergency obstetric and neonatal care services and MMR has contributed to the inequitable sex ratio of 106 men to every 100 women in 2015 (World Economic Forum, 2015) Figure 28.1.

Pakistan also has low ranking in other supplementary events of gender equality and women empowerment. In 2011, Pakistan ranked 133 among 135 countries in the Global Gender Gap Index, whereas in 2015 according to the Global Gender Report by the World Economic Forum, Pakistan ranked 144 among 145 countries (The Global Gender Gap Report, 2015). There are sociocultural, geographical, and political determinants of low women empowerment in Pakistan. Evidence confirms that if female farmers in the developing world possessed the same resources as men do like land, better seed varieties, modern technologies, and improved farming practices, production could swell by 30% per family and the countries could witness an increase of 2.5%–4% in agricultural productivity (Bill and Melinda Gates Foundation, 2015).

Country Score Card

	Rank	Score	Sample average	Female	Male	Female-to-male ratio
ECONOMIC PARTICIPATION AND OPPORTUNITY	143	0.330	0.592			
Labour force participation	140	0.30	0.67	26	86	0.30
Wage equality for similar work (survey)	88	0.61	0.60	—	—	0.61
Estimated earned income (PPP US$)	140	0.19	0.54	1,503	8,000	0.19
Legislators, senior officials, and managers	124	0.03	0.27	3	97	0.03
Professional and technical workers	122	0.28	0.64	22	78	0.28
EDUCATIONAL ATTAINMENT	135	0.813	0.946			
Literacy rate	136	0.66	0.89	46	70	0.66
Enrolment in primary education	134	0.87	0.93	67	77	0.87
Enrolment in secondary education	124	0.74	0.64	32	43	0.74
Enrolment in tertiary education	99	0.98	0.92	10	10	0.98
HEALTH AND SURVIVAL	125	0.967	0.957			
Sex ratio at birth (female/male)	1	0.94	0.92	—	—	0.95
Healthy life expectancy	131	1.02	1.04	57	56	1.02
POLITICAL EMPOWERMENT	87	0.127	0.230			
Women in parliament	72	0.26	0.27	21	79	0.26
Women in ministerial positions	141	0.00	0.24	0	100	0.00
Years with female head of state (last 50)	26	0.10	0.20	5	45	0.10

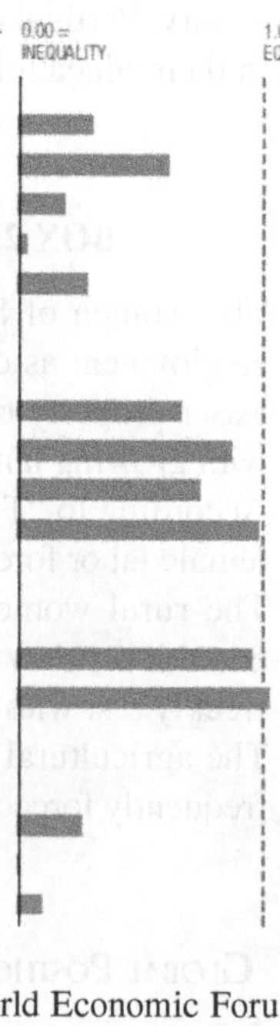

FIGURE 28.1 Global Gender Gap: Country Score Card (Pakistan). (From World Economic Forum. 2015. The Global Gender Gap Report-2015. World Economic Forum. Available at: http://reports.weforum.org/global-gender-gap-report-2015/)

28.2 GENDER AND FOOD SECURITY

Food Security at the individual, household, national, regional, and global levels is achieved when all people, at all times, have physical, social, and economic access to sufficient, safe, and nutritious food to meet their dietary needs and food preferences for a healthy and active life. According to FAO estimation, about 795 million people are undernourished globally, and every year more than 3 million children die due to malnutrition before than their fifth birthday. About 780 million people are from developing regions and their overall share of undernourishment is 12.9% of the total population. It means that just over one in every nine people in the world is unable to consume enough food for an active and healthy life. Deficiency of micronutrients affects around two billion million people, results in poor growth, blindness, imminent severity of infections, and sometimes leads to casualty (FAO, 2015b) (See Figure 28.2).

Gender-based disparities along the chain of food production, "from farm to platter," slow down the attainment of proper food security. Maximizing the effects of agricultural growth on food security requires enhancing women's roles as rural farming producers along with the key caretakers of their family unit. Key objective of food security is the consistence agricultural development and economic and social progress.

Food safety and agriculture are significantly connected to each other. Agriculture, whether familial or international, is the only way of food supply equally in favor of direct consumption as well as raw stuff for finished food stuff. Agricultural production establishes food accessibility. The constancy of access to foodstuff by production and purchase is administrated through domestic policies, comprising policies on social well-being and agricultural investment preferences that lessen threats (e.g.; droughts) in the agro production cycle. Food security would turn into a reality simply when the agricultural zone becomes vibrant.

There are vital gender dimensions in food security. The gender disparity in agriculture—established by females' inadequate access to dynamic resources, inputs, technical services, and education—contributes significantly to the situation of food and farming production today. According to FAO (2011) estimates, by addressing the limitations faced by females in agriculture, one could provide for as many as 100–150 million malnourished inhabitants in the world. Hence, by adopting a gendered approach to advancement one could facilitate and categorize gender specific limitations to address key development challenges.

28.2.1 The Role of Women in Food Security

Women are vital for the transformation of vibrant agriculture yield into nutritional well being for their families. Mostly farmers as a source of income not only grow food crops but also cultivate

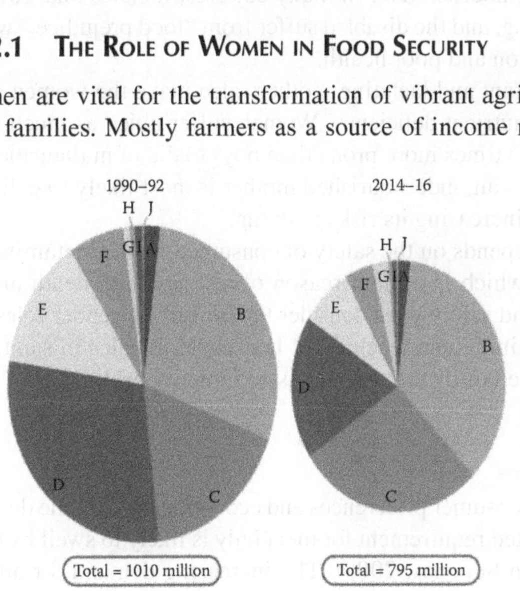

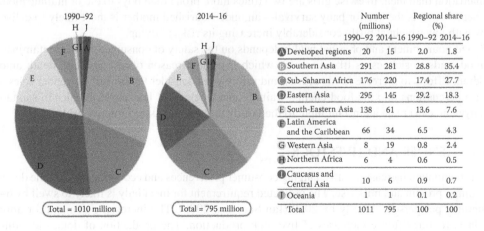

	Number (millions) 1990–92	2014–16	Regional share (%) 1990–92	2014–16
A Developed regions	20	15	2.0	1.8
B Southern Asia	291	281	28.8	35.4
C Sub-Saharan Africa	176	220	17.4	27.7
D Eastern Asia	295	145	29.2	18.3
E South-Eastern Asia	138	61	13.6	7.6
F Latin America and the Caribbean	66	34	6.5	4.3
G Western Asia	8	19	0.8	2.4
H Northern Africa	6	4	0.6	0.5
I Caucasus and Central Asia	10	6	0.9	0.7
J Oceania	1	1	0.1	0.2
Total	1011	795	100	100

FIGURE 28.2 Numbers and shares of undernourished people by region, 1990–1992, 2014–2016. (From Food and Agriculture Organization. 2015b. *State of Food Insecurity*. FAO.)

commercial crops along with their family. Whenever females have income, they spend most of it on food and for their children's requirements. Generally, household females are responsible the preparation of food. Household females provide food security for their family units. A direct relation exists between agriculture and food security, especially for small farmers who depend on agricultural farming for their subsistence. Similar to other developing nations, in Pakistan the majority of rural females are engaged in farming. They are involved in different agricultural farm activities like land preparation, threshing, collecting manure, seed preparation, weeding, and harvesting. Rural females are also responsible for the clean-up, drying, and storage of agro - grains. Looking after farm animals is another duty that is generally done by rural females. They collect, slice and serve silage, clean up the stable and process the farm animal products. Additionally, they carry out a range of household chores like cooking, house cleaning, washing, gathering fire stalks, fetching water, and looking after children and aged family members. Household females always try to make sure that all family members get an appropriate share of foodstuff. So far, their participation in prolific activities relevant to livestock and agricultural farming is underestimated and considered as regular housework. In Pakistan rural areas experience food insecurity and this food insecurity is significantly higher among females. Yet the household females' role in family units' food and nutritional security is under-reported, regardless of this reality that their contribution is significant to Pakistan's rural economy.

28.2.2 Food Security Issues and Analysis

In developing nations, rural females and males play a variety of tasks for the assurance of food and nutritional security for their family units and neighborhoods. Usually, rural males are responsible for cultivating primarily field crops while females grow and prepare mainly the foodstuff, which are consumed in their family, and take care and manage small livestock, which are a source of protein for their family unit.

Thus, women play a key role in nutritional food security, dietetic diversity, and the health of their children. Females are more liable to utilize their earnings on foodstuff and their children's demands—this is empirically shown as a child's survival probability increases up to 20% when the household budget is run by the mother. Having a sufficient availability of foodstuff does not necessarily translate into sufficient levels of nourishment. In many cultures, females and girls eat after the family males . The females, the ailing, and the disabled suffer from "food prejudice," which is the major cause of unrelieved undernutrition and poor health.

The physiological requirements of expectant and lactating mothers also make the women more vulnerable to undernourishment and micronutrient deficiency. Women suffer twice as much from malnutrition than men, likewise girls are two times more prone than boys to die of malnutrition. A mother's health is the key for baby survival—an undernourished mother is most likely to deliver a baby with low weight at birth, considerably increasing its risks of dying.

Good nourishment and physical health depends on the safety of consumed food. Contamination of food leads to sicknesses like diarrhea, which is a main reason of sickness and death among children. Efforts to increase food security and safety must consider the prevailing gender roles and their repercussions into the food chain—training and teaching of household females in sanitation and hygiene can create an instant input to the family unit as well as community health.

28.3 GENDER AND LIVESTOCK

As the population increases, with changing consumer preferences and economic growth, the demand for animal protein and milk rise. The projected requirement for meat only is likely to swell by 6–23 kilograms per person globally by 2050 (Ian Scoones, 2008). The increasing demand for animal products requires future increases of livestock production. The production of domestic animals provides not only a source of earning and employment creation but also provide the better food and nutritional security with different production methods along diverse value chains (like meat, live animals, eggs and dairy). In a number of countries, livestock production now contributes about 80% of agricultural gross domestic product (World Bank, 2007).

Livestock production is the second key segment of the agriculture sector in Pakistan. With the contribution of approximately 50% of agricultural GDP and in general 11% to national GDP, it is an important economic activity. Livestock rearing is mainly a subsistence activity to fulfill the need of basic food of the family to enhance farm incomes. Almost each household in rural areas of Punjab, Pakistan possess some farm animals. The involvement of Government for growing livestock production is not sufficient but the private sector in general should encourage investing in production, marketing and processing and livestock products (GOP, 2008). The major proportion of the female population of rural Pakistan is engaged in livestock production. To take care of the animals is common activity carried out by rural females. These activities involve fodder collection, cleaning sheds and processing of animal products. Unfortunately the female participation in livestock management is underestimated and considered as housework. The women's role in the domestic routine matters is traditional. The women in rural areas have extremely hectic work schedules which includes livestock and poultry care without help from males.

The livestock, as a subsector of agriculture in Pakistan, is also important to rural livelihood particularly for small farmers and rural people without lands. Livestock characterizes the prime agricultural opportunity that is based on farm gate value and livestock total production contribute about 55.4% to agricultural GDP and 11.9% to total GDP 16. Almost 7 million rural households depend on the livestock subsector for their survival and the greater part of them are women. Women who carry out the most of the care and management of livestock obtain very slight returns because of their restricted mobility and having practically no access to markets. Livestock production has a tremendous potential for economic growth and poverty reduction in Pakistan as a short-term activity (FAO, 2015a,b).

28.3.1 Livestock Sector Issues

There are a number of gender issues which are essential to the discussions of agricultural sources of revenue. The following are the key issues faced by women currently involved in the livestock sector, as shown in Figure 28.3.

28.3.1.1 Access to Livestock and Other Assets

Managing assets like land, livestock, agricultural equipment, and water has a direct effect on whether men and women can shape life-enhancing income generating strategies. However, there is a general custom of the husband's family getting animals and other property from the widow and children after the death of the husband. This has a direct impact on females and their children in the form of losses as potential income, food, and safety assurance. The loss of assets and resources has an effect on women's capability to manage and take economic advantage from livestock. Poultry, on the other hand, is an almost widespread exception throughout the world because women are inclined to have extra control over poultry.

28.3.1.2 Roles and Responsibilities in Livestock Management

Universally, women and men provide labor for various livestock-related roles. Yet, the role of gender is undefined and it fluctuates for different socioeconomic, environmental, and health-related grounds. So far at times of labor shortage, females of family could and do perform "men's" responsibilities, such as herding and watering animals. However, men rarely performed "women's" farm duties, except in the cases where they have the opportunity to get control over property and resources (Hill, 2003).

28.3.1.3 Decision Making

Women might control and be involved in household decision-making processes; particularly in the case of livestock and livestock products. Only in the case of poultry may they sell or replace their poultry without their husband's consent. Women perform a significant role in livestock rearing,

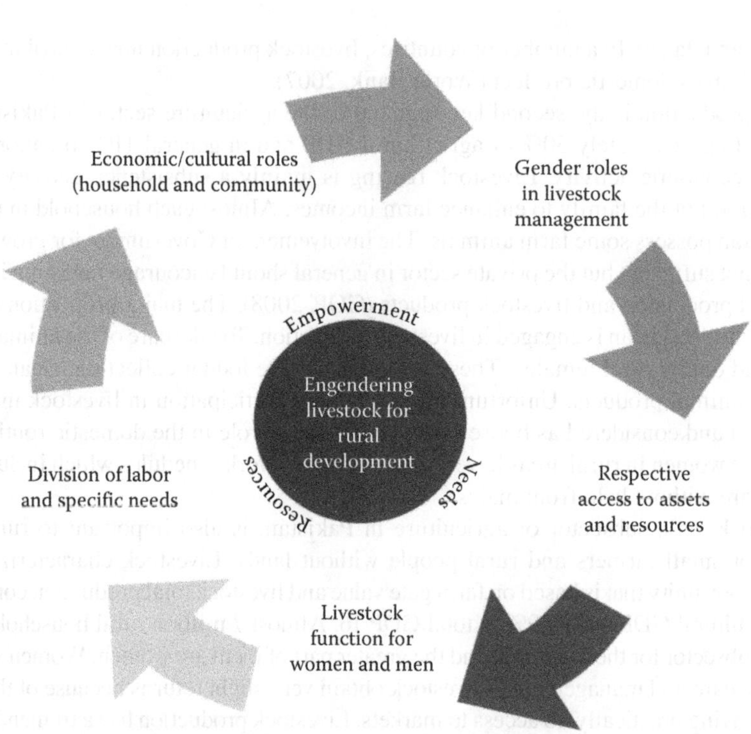

FIGURE 28.3 Key issues related to gender and livestock. (From International Fund for Agricultural Development (IFAD). 2010. IFAD's Performance with regard to Gender Equality and Women's Empowerment December 2010 Report No. 2324.)

management, and marketing, playing a role of care providers and birth supporters. Women are also engaged in the production of milk, while most of them do not manage its sale and products. Recognizing and supporting the roles female's as owners, processors, and consumers of livestock products while escalating the power of decision-making and their competence are important aspects in encouraging women's empowerment and therefore providing an approach to enable women of rural areas to alleviate poverty.

28.3.1.4 Economic and Cultural Roles

Women's roles in livestock production varies from one region to another, and the division of livestock ownership among men and women strongly depends on social, economic, and cultural factors. Usually, it is based on the kind of animals they hold. In several societies, for instance, cattle and bigger animals are possessed by men, whereas smaller animals—like goats, sheep, and poultry kept in the courtyard of the house are mostly managed by women. Although raising small domestic animals for household income, the ownership and management of these animals often passes over to men.

28.3.1.5 Lack of Access to Livestock Services and Technical Knowledge

Due to restructuring of the livestock sector, the services structure has also changed. With the increasingly privatized system of services, women are facing uneven challenges compared to men in receiving information and services regarding livestock management. Women's have less access to the market, livestock services, new technologies, updated information, and microcredit, which decreases their capability to improve output and profit. There are also other important productivity limitations such as deficient technical knowledge on feeding, animal health, reproduction, and lack of training institutions for small farmers, particularly for women in Pakistan. Besides, women are also exploited by the middlemen due to their lack of understanding of market regulations. Some causes perceived for this situation are the lower level of female literacy in Pakistan and the lack of formal education on livestock services to improve management of animals.

28.3.2 Pillars of Gender Mainstreaming and Women Empowerment in Livestock

The following are the key pillars of mainstreaming gender in livestock enterprises (see Figure 28.4).

28.3.2.1 Improve Household Wellbeing

Concentrating on gender in livestock rearing means to recognize different livelihood requirements, priorities, benefits, and restrictions of men and women. It means maximizing the available social capital through engaging all household members as agents of poverty reduction. Men and women are extremely more likely to take part in efforts to improve their livestock production if they perceive that profits (such as better productivity, food safety, income generation, and fewer diseases) compensate the expenses (like, time, effort, social responsibility). The management, processing, and marketing of domestic animal products creates more income compared to other activities women are likely to be engaged in, and reaps benefits for the entire family.

28.3.2.2 Improve Social Protection

Mainstreaming gender into livestock programs and development is an essential social protection measure. Doing so builds assets at the individual, household, and community levels through reducing vulnerability and increasing the opportunities of men, women, boys, and girls. Women throughout the world utilize income produced from poultry and livestock production on social commodities such as school fees of their children, health expenses,and other resources for their family. Particularly, livestock initiatives are appropriate to protect those who are in vulnerable position and are forced to take risks to secure food, income, shelter, clothing, and other necessities (Figure 28.4).

28.3.2.3 Asset Ownership

Livestock is such an asset that females can easily possess. It is usually easier for the rural women in developing nations to acquire assets through inheritance compared to purchasing farmland or new

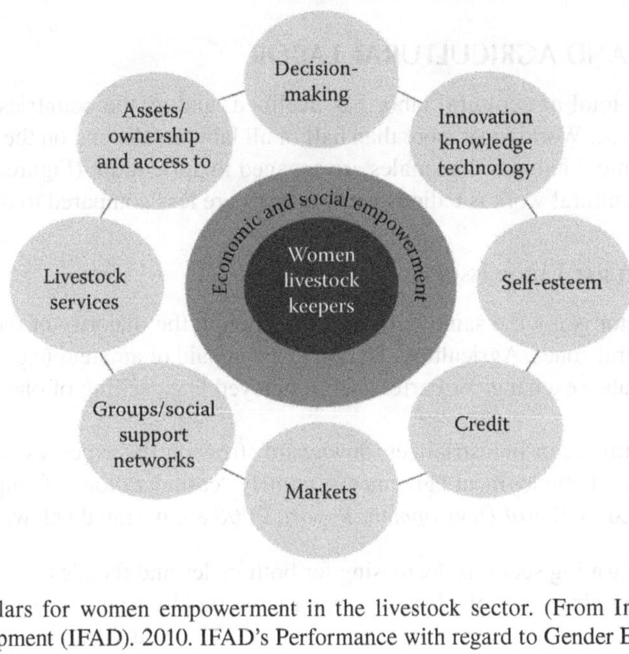

FIGURE 28.4 Pillars for women empowerment in the livestock sector. (From International Fund for Agricultural Development (IFAD). 2010. IFAD's Performance with regard to Gender Equality and Women's Empowerment December 2010 Report No. 2324.)

financial assets. The ownership of assets increases the bargaining power of women, their decision-making power in households and domestic expenditures on the education and health of their children. Interventions regarding the improved access to livestock, which protect women from dispossession, stealing, or unfortunate death afterwards facilitate women in reducing poverty.

28.3.2.4 Livestock Value Chains

Women's involvement at every level of the value chain is affected by different factors, including access to assets, their expertise, capability to manage, and limited mobility. Women have less secure rights of land, labor, and resources and also lack services from formal economic institutions. As animal production becomes more commercialized, women might not be capable to compete and get equal benefits.

28.3.2.5 Income Management

Women's capacity to handle their earnings is very important to the survival of their household. Studies revealed that females spend about 90% of their income on their household, whereas men spend only 30%–40%, yet the whole income is not adequate to meet family needs. The income managed by women may raise their bargain power, diminish violence on women and enhance the dietary condition of their family.

28.3.2.6 Decision-Making and Empowerment

The ownership of livestock increases women's decision-making and financial power in both the family and society. Livestock is a source of income and it creates a chance to get credit (the sale of small ruminants can provide an emergency source of cash for medical treatment or school fees, while daily milk presents the usual flow of hard cash income frequently used to obtain food and other household items).

28.3.2.7 Self-Esteem

Possessing, controlling, and gaining advantage from animal production enhances the self-esteem of women and strengthens their responsibility as producers and income creators within the family and community.

28.4 GENDER AND AGRICULTURAL LABOR

In many countries, total agricultural labor has declined, and as the countries industrialize, this tendency will continue. Worldwide, more than half of all laborers depend on the agricultural sector. In various regions, more females than males are engaged in agriculture (Figure 28.5).

Most of the agricultural work is tedious while returns are less compared to the other sectors.

28.4.1 Agricultural Labor Issues

The agriculture sector is not the same in rural areas, though the majority of the agricultural tasks are carried out in rural zones. Agricultural labor may be unpaid or amateur (e.g.; on-farm kin labor), paid-in-kind (e.g., labor exchange or barter), self-employed (e.g., selling of one's own products), or gross/wage labor.

As the agricultural sector industrializes, downward stress on the expenses of employment leads towards more informal employment agreements, namely "casualization" of employment. Potential dimensions identified by *World Development Report, 2008* are narrated below:

- Labor in the farming sector is decreasing for both males and females.
- Males are migrating out of the farming sector more rapidly from some areas.
- Young rural females who migrate from their areas in search labor are predominantly susceptible to abusive agreements and working situations, and poor environments.

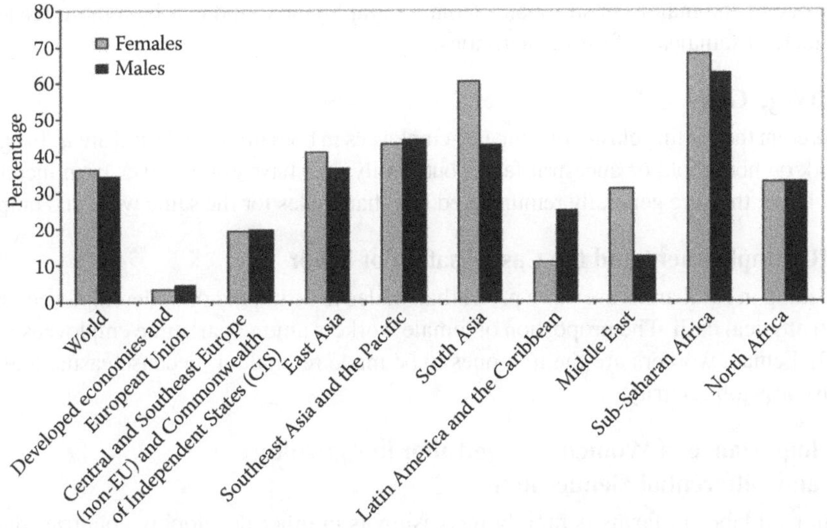

FIGURE 28.5 Percentages of women and men in agriculture by region, 2007. (From ILO 2006.)

28.4.1.1 Trends in Agricultural Labors by Sex

In spite of the significance of farming activities, which are not presented in national statistics, because of unpaid/ amateur and informal features, the contribution of women to the farming labor force has considerably increased, while male's share has noticeably decreased (see in Figure 28.6).

Women engaged in agricultural activities (comprising two thirds of all working women in developing countries) work onerously as mature/unpaid family workforce in livestock and crop production, together with postharvest activities.

28.4.1.2 Women's Time Allocation

Within the "reproductive economy" females are the key workers: Raising children, preparing food, maintaining households, and looking after the sick and old parents as well as indigent relatives. In rural areas, women are more constrained because these activities are more tedious due to the scarcity of basic facilities, for example, electrical energy and water. Lack of sufficient child and parental care

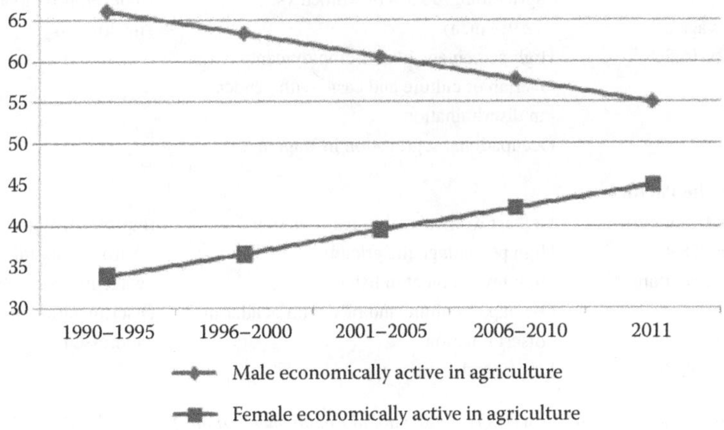

FIGURE 28.6 Economically active male and female labor force.

symbolizes one of the major constraints to female's employment and can be a main reason for the greater obstacle of females in farming activities.

28.4.1.3 Wage Gaps

Females represent the leading cluster of "unpaid" employees in both urban and rural areas. In agriculture, females work on household or ancestral farms but hardly ever have control over farm income. When females are hired, they are generally remunerated less than males for the same work and outputs.

28.4.1.4 Unemployment and the Casualisation of Labor

Females residing in rural areas are more prone than males to be unpaid or underemployed and having no access to physical cash. The proportion of female workers among part-time employees is growing (ILO, 2003). Female workers are the first ones to be made redundant, because casual and seasonal workers have less job security.

28.4.1.5 Importance of Women's Wage Labor in Agriculture and Differential Gender Role

In Pakistan, paid labor in farms is rapidly increasing as in other developing countries due to high population growth rates; however, low earnings in farm activities are taking out males from the agricultural sector and pushing more females into it; basically since females have less flexibility compared to males. In agriculture, females are mostly concentrated in high labor crops like legumes and vegetables where most of the activities are done manually, apart from plowing and occasionally planting, while males are involved in mechanized agricultural activities (plowing, planting, combine harvesting, etc.) which involve limited workers for specific times, permitting flexibility to find labor other than in agriculture (Table 28.1).

TABLE 28.1
Regional Characteristics and Key Issues of Women's Agricultural Labor

Regions	Characteristics of Women's Agricultural Labor Force	Key Issues for Women's Agricultural Labor
South Asia		
Employment to population ratios: Women: 31.4%; Men: 78.1%	High percentage of informal agricultural labor	Unequal access for women in formal sector employment
Working women in agriculture (2007): 60.5%	Higher percentage of women in agriculture (60.5% of women vs. 42.9% men)	Few legal protections
Working women in wage and salaried jobs (2007): 15.5%	High percentage of self-employment	Undeveloped labor market institutions
	Overlap of culture and caste with gender in discrimination	
	Occupational segregation in wage market	
Southeast Asia and the Pacific		
Employment to population ratios: Women: 62.5%; men: 78.4%	Highest women's labor participation	Improvement in work conditions in Agro processing and agricultural wage markets needed
Working women in agriculture (2007): 43.4%	High percentage in agriculture	Discrimination in all forms to be addressed
Working women in wage and salaried jobs (2007): 39.2%	High involvement in fisheries	
	Overlap of culture and race with gender in discrimination	
	Large gender wage gap	

Source: International Labour Organization (ILO). 2008. *Global Employment Trends for Women, March 2008*. ILO, Geneva; World Bank. 2007. *World Development Report 2008: Agriculture for Development*. World Bank, Washington, DC.

28.4.1.6 Violence, Health, and Safety

The increasing prevalence of females in casual low-paid labor with inadequate security shows the way to other buses. Workplace violence and sexual harassment are more common under these situations. Male administrators have power over decisions regarding work performance, and consequently the payment for the "task."

Females encounter health risks such as backaches and pelvic harm during weeding and rice cultivation. Agricultural work can be onerous for both sexes, but to the degree that females are involved in specific agricultural activities, they face more exposure to various health risks.

Health hazards in the rapidly developing horticulture industry encompass the exposure to noxious products because of improper training and lack of protective clothing, deplorable hygienic conditions, and long working hours. Every year no less than 170,000 agricultural employees lose their lives due to workplace incidents, and some 40 thousand of these are from exposure of toxic pesticides (ILO, 2003). As rural females predominate in most of these agricultural activities, they are more exposed.

28.4.1.7 Women Empowerment in the Rural Labor Force

As in Pakistan, most rural females are family farm workforce in developing countries, however, they have poor control and access over most of the economic resources and other opportunities. Therefore, their inputs towards the agricultural growth are limited and only partially contribute to socioeconomic development. For poverty reduction and economic growth, it is important to take effective steps towards women's empowerment and reducing the gender gaps, which would benefit each nation.

Gender disparities in all labor markets are commonly available. Reducing labor disparities establishes good development insight. Latin American economies showed both declines in poverty and boosts in economic growth by increasing females' workforce participation (Table 28.2). If the wages of males and females became equal, a 6% increase of growth is possible (Tzannatos, 1999) (Figure 28.7). Empirical facts demonstrate that females spend more than males for the betterment of children; hence, better job opportunities and wages for females would not only boost the existing economic development, but also have intergenerational effects (see relationships in following Figure 28.7).

28.5 RURAL FINANCE

Rural finance is about giving financial services for all people living in rural areas. The vital role of services regarding finance in the development of rural livelihood is widely accepted around the world. A great emphasis was placed on establishing an "inclusive financial sector" in recent years. This means helping the whole range of financial institutions for providing funds and capital for pro-poor development. Rural finance then denotes the variety and array of financial services available and approachable in rural areas, not only linked to agriculture, but also financial issues related to the development of nonagricultural areas (World Bank, 2009).

BOX 28.3 KEY CONCEPTS OF RURAL FINANCE

There is significant confusion regarding rural finance in Pakistan. Mostly it was considered as synonymous with agricultural finance or microfinance. The State Bank of Pakistan in 2008 defined in its report that rural finance encompasses the following:

- *Rural credit*: Rural credit refers not only to agriculture credit but also includes the obligation of credit related to all rural business and agricultural related activities.
- *Saving mobilization*: It is as important for rural development and accelerated agriculture as the provision of credit, but most often a neglected aspect of rural finance.

- *Payment system*: It includes the remittances, transfer, and payment of funds.
- *Insurance services*: Availability for insurance for various risk management purposes is a vital part of the finance sector and availability and dissemination of insurance in the rural sector is a fundamental part of the rural finance market.

Rural finance covers a wide range of products. These includes loans of different amounts, with changing conditions and time frames, in order to pursue an array of activities related to livelihood which includes both agricultural and non-agricultural activities. Different types of saving services in order to encounter different needs help to manage not only day-to-day domestic cash flows but also to build assets for long term development. Policies related to insurance are effective to decrease risk and liability. Procedures related to leasing are helpful for building assets. Services related to remittance

TABLE 28.2
Gender Gaps in Rural Wages

Region	Sector	Type of Employment	Wage Unit	Women's Wage as % of Men's
Sub-Saharan Africa				
Ghana	Agriculture	Self-employment	Hourly	65
Senegal	Agriculture (NTAEs)	Wage	Hourly	Similar to men's
Kenya	Agriculture	Wage	Hourly	93
Tanzania	Agriculture	Wage	Monthly	69
South Africa	Agriculture	Wage	Hourly	84
South Asia				
Afghanistan	Agriculture	Wage	Daily	50
	Nonagriculture	Wage	Daily	20–30
Bangladesh	Agriculture (fry catchers and sorters)	Wage		64
India	Agriculture	Casual wage	Daly	69
	Agriculture	Regular wage	Daly	79
	Nonagriculture	Casual wage	Daly	65
	Nonagriculture	Regular wage	Daly	57
Pakistan	Agriculture (sugar)	Wage	Daly	50
Southeast Asia				
China	Nonagriculture	Wage	Monthly	68
Vietnam	Agriculture	Wage	Hourly	73
Latin America				
Costa Rica	Agriculture	Self-employment	Hourly	53
El Salvador	Agriculture	Wage	Hourly	82
	Agriculture	Self-employment	Hourly	63
	Nonagriculture	Self-employment		70
Mexico	Agriculture (NTAEs)	Wage	Daily	88
	Avocado		Daily	99
	Mango		Daily	78
	Cucumber		Daily	97
	Flowers		Daily	80

Source: Food and Agriculture Organization. 2015. *State of Food Insecurity*. FAO.

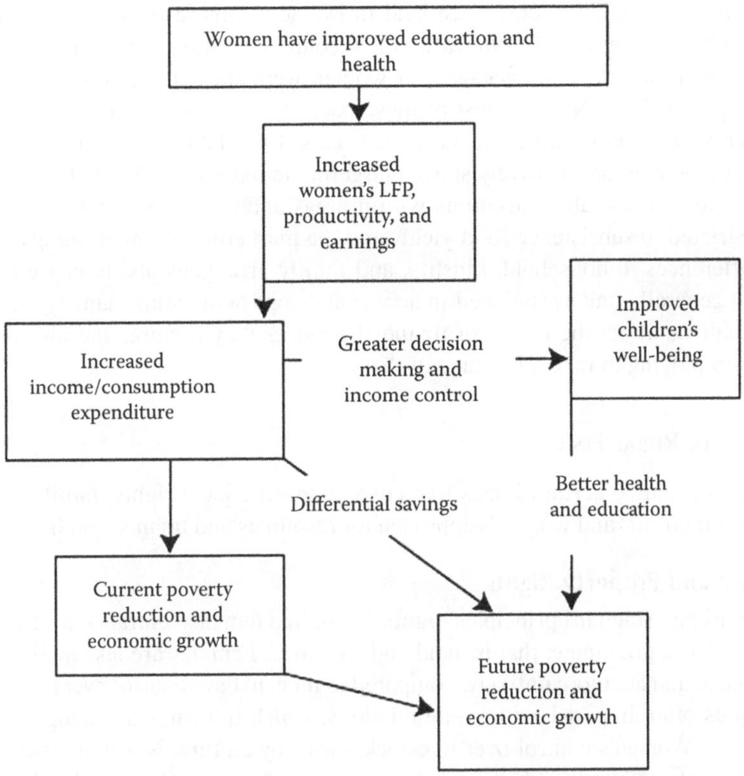

FIGURE 28.7 Relationship between Women Labor Force Participation (LFP), Poverty, and Economic Growth. (Based on Morrison et al 2007. Gender-Equality, Poverty and Economic Growth. World Bank Policy Research Working Paper No. 4349, Gender and Development Group, Poverty Reduction and Economic Management Network, World Bank, Washington, DC, September.)

transfer enable migrants to send more money to their home for investment. This money in turn is utilized by family members not only in raising the standards of livelihood but also in asset building and consumption, thus decreasing the vulnerability and dependence of the family. Bill payment services and pension schemes are also an effort to diminish long-term helplessness (World Bank, 2009).

Women are an essential force not only for the development of rural areas but also to the enhancement of national economies. Almost 43% of the work force related to agriculture worldwide comprises females. This range accedes to almost 70% in many developing countries. Women work hard and longer hour's side by side men and also act as caregivers who look after and help their family. Moreover, many rural females work as entrepreneurs and investors, who devote and dedicate their earnings for the benefit and welfare of their families as well as to society. But in spite of that much efforts and little progress, most rural women are struggling. They usually have to face more hindrances compared to men in attaining access to public services, decent employment opportunities, social protection, and access to market and other institutions.

If women are given the same opportunity and access to resources as men, they can increase farm yields. When women are empowered in reference to economics and social status they surely prove themselves as agents of change for both economic and social progress. Moreover, they play their effective role in sustainable development (FAO, 2012).

Gender issues and problems related to financial services are always a matter of concern. From the first decade of 1970s most of the women's movements in various countries started showing their interest to what extent the women were capable of attaining access to and get advantage from poverty focused credit. The concern to women's access to credit was greatly emphasized in 1975 at the time of first International Women's Conference held in Mexico. This resulted in the establishment of Women's World Banking system. In 1985, after the second International Women's Conference, many income generating programs and packages for women were started not only by the public sector but were also sponsored by NGOs. Most of these packages were related to savings and credit. In the 1990s, microfinance programs of the Grameen Bank, Foundation for International Community Assistance, and ACCION progressively started targeting females (IFAD, 2009).

While there are considerable variations within areas and crop types and economic activities, women are restricted to subsistence food yields and the marketing of low-profit articles. There are noteworthy differences in household, kinship, and family arrangements. Because of this, women are also almost generally underprivileged in access and control of assets, namely land and income. Such discriminations affect the nature of financial services they require, and also the means they are capable of employing to take advantage of them.

28.5.1 Issues in Rural Finance

It is necessary to examine certain issues like context-specific legal rights, family responsibilities, social and cultural norms, and women's approach for resources and finance and how they acquire it.

28.5.1.1 Land and Property Rights

Lawful directions and standard principles regularly confine females' entrance to resources that can be acknowledged as a guarantee; that is, land and livestock. Females are less inclined to have land titled to their name and are more unlikely, compared to men, to have control over an area. Prejudiced inheritance rights often impart land to male relations, which in turn, leave daughters and widows at a disadvantage. Women's control over livestock varies by culture. Normally, men are in charge for the purchase, sale and trade of big animals, as cattle and horses. On the other hand, women are inclined to have claim of small animals like poultry, goats and sheep. In areas where male members are considered as the central breadwinner, women's capacity to offer family possessions act as a guarantee. Moreover, their incentives to devote in productive activities are affected by family dynamics which are generally to prioritize the male's investments (World Bank, 2007).

28.5.1.2 Cultural and Social Taboos and Family Responsibilities

Sociocultural norms can have deep effects on the nature of economic deeds in which women can partake in resources and technologies accessible to them, people, agencies and organization with which they can interrelate, the areas and places they can visit, available time and the control they can exercise over their own money. In certain areas where sociocultural norms and taboos act as hindrances to women's mobility, their interactions with males and their ability to receive formal education and schooling, women's approach to information, agencies and markets is constricted and reduced. Moreover, when women are not allowed from imingling and interacting with males other than in the family context, it limits their participation in agricultural and financial training. In that regard, when they have the information they are less equipped to process it. Cultural customs and family setup can also restrict women's ability to have control over the savings they own and the semiliquid resources they possess.

28.5.1.3 Psychological and Behavioral Differences

Due to psychological characteristics or influence of sociocultural norms, males and females tend to demonstrate variances in their behavior pattern. Women are reluctant to take risks compared to men. Women are likely to waive activities that grant higher returns if such activities demand high risks.

28.5.1.4 Organizational Discrimination

The access of women to financial resources in rural areas is hampered by unfair lending practices adopted by financial institutions which consider women unattractive customers. Rural women are also denied financial assistance because of lack of knowledge of the concerned institutions (Fletschner, 2009). The range and level to which financial institutions offer financial help to women and the conditions applied for such assistance vary significantly. Rural women are less privileged because financial institutions rarely fund the activities usually controlled by women. Institutions also do not entertain female guarantors for rural women, which is also a disadvantage.

28.5.1.5 Economic Empowerment: An Overview

The increase in microfinance system since the 1990s has remarkably enhanced women's approach to facilities related to loans and savings. The improved access to microfinance facilities has been perceived as contributing to poverty reduction, financial sustainability, increased well-being, and socio-political empowerment for women, thus highlighting and meeting goals of empowerment and gender equality. Increased access to microfinance facilities can help to attain economic empowerment of women. Moreover, their part in household financial issues can improve. This financial stability can enable females to initiate their own economic activities, participate more in current activities, obtain assets and raise their standard in household economical activities (IFAD, 2009).

Various efforts have been made by the Government of Pakistan and non-Government Organizations to promote women empowerment in rural areas and one of these efforts is the microfinance intervention. The various studies conducted in this field have confirmed that targeting women is an active way to fight poverty because women are likely to employ their earnings and incomes in food, education and health care of their children. These act as investments in the future and participate towards long-term economic developments not only for their families but for the communities and eventually for their countries (Meenu et al., 2011).

Women empowerment, along with gender mainstreaming, have to be central to rural finance interventions if rural microfinance targets are to contribute towards the development and economic growth of the poor in society. Women make up around half the population in rural areas. Any developmental scheme and strategy which ignores women is a limited and partial strategy.

Rural finance interventions related to women's empowerment can help in poverty reduction. Specific consideration to women in deprived households is needed in order to achieve the Millennium Development Goals regarding poverty reduction. Women are not only the overwhelming majority of the underprivileged people; they are more likely to spend additional wages in the health and nutritive status of the family and also for their children's education. Thus, targeting women has a greater positive effect on children and domestic poverty reduction, which can be calculated in terms of nutritive status, and well being. Financial growth is another aspect. Gender equality is a vital element of economic growth, which enables females to become real economic performers.

In Pakistan, between the 1970s and 1990s, rural economics were mostly governed by the Agricultural Development Bank of Pakistan, later known as Zaraee Taraqqiati Bank Limited. The ADBP delivered finances for investment in agriculture, including working money, equipment, tube wells, and agriculture industry. In spite of the Bank's system of administered credit, based on mobile credit officers who were both male and female, it was supposed to deliver widespread outreach. In practice, few farmers, the landless and females were aided by the ADBP (IFAD, 2007).

At present, many public and private organizations are working for women's uplift through microfinance. Among them are the First Women Bank, Khushali Bank, and various leading NGOs like AKRSP, PRSP Akhuwat and Aurat Foundation, etc. Under the public sector, many programs were launched for the economic enablement of women like Jafakash, Aurat established in 2005 under the National Fund for Advancement of Rural Women, to benefit 23,000 rural women around Pakistan (ADB, 2008). Benazir Income Support Program is another example.

In spite of this broadly accepted notion, rural financial packages were largely designed and executed with the male as the proposed consumer and failed to identify that females are energetic and productive economic managers with their own monetary needs and limits (Fletschner, 2009).

On average, males have 3.5 times higher chances of obtaining finance and credit in comparison to females in rural areas of Pakistan (Zaidi et al., 2007). Another reason is women's dearth of mobility. Their inability to go to the bank or to the meetings of NGOs, illiteracy, lack of understanding of official procedures and forms, or interaction with administrative staff outside their specific community are other drawbacks.

Several recent innovations employed in many countries can expand women's access to monetary and financial resources. In an effort to tackle some of the constraints, many novel and innovative services and delivery modules have been introduced in many countries. These measures include the following:

- Technological innovations that develop access to current financial services. Technological innovations include prepaid cards for distributing loan payments. Mobile phone schemes for loan payments and transferring of cash, making it easier for females to have access to money without travelling far or less interaction with males (Duncombe and Boateng, 2009). Technologies such as biometric smart cards facilitate and empower women to attain control over that can access their savings accounts in the banks (Quisumbing and Pandofelli, 2010).
- Changes in product design according to women's preferences and restraints such as housing loans.
- Establishment of innovative products like micro insurance, for example, health insurance. Most micro insurance plans cover pregnancy and birth-related expenses.
- Self-help groups are recognized as an effective path for linking women with financial institutes. Self-help groups of females work at the village level. All the savings are collected from every member and deposited in local rural banks, or loaned to the other members of the group. After the self-help group has proved that it has the capability to collect loans, the rural banks offer extra money that group fellows use usually for agriculture purposes (World Bank, 2009).
- Client-focused loans are another innovation. Majority of the loans of Grameen Bank in Bangladesh had a 1-year period with a weekly repayment schedule. The Grameen II, loans are now accessible with changing terms and schedules. Although these loans are granted for business use, the members utilize them for what they desire which in turns helps to stabilize weak and fragile livelihoods (IFAD, 2009).

28.6 LAND OWNERSHIP

The major factor contributing to poverty is nonownership of land. Approximately less than 50% of rural families own their land. About 2.5% of families own almost 40% of land. This misrepresentation contributes to the increased poverty rates in rural areas of Pakistan. The link between lack of land and vulnerability to poverty is a fact. Thus the almost total lack of access for females leaves them at risk. In the rural area of Punjab, a meager 4% of females possess some land. In urban areas, a very low percentage of women own land, and this is mainly due to inheritance instead of buying by the females themselves (ADB, 2008).

Females' right of inheritance of land is defined under Islamic law, which is also recognized as such under the civil laws in Pakistan. The Constitution of Pakistan guarantees basic rights for every citizen without any discrimination on the basis of gender. In the Pakistani Constitution of 1973, it is stated that all citizens are equal according to law (Article 25(1), and are enabled to equal protection of law. Article 25(2) states that there shall be no discrimination on the basis of sex alone (Kokab, 2012). Muslim Family Laws Ordinance 1961 and the 1962 Act of Muslim Personal Law Shariat Application are also established laws for inheritance of land, property, and agriculture and are at par with the orders of the Holy Quran. There are differences between Sunni and Shia laws. But both acknowledge women's right in inheritance.

Individual Women	Inheritance Right
As Mother	1/6th of X
As wife	• 1/8th of X if she has any children • 1/4th of X if she does not have any children

As Daughter	• ½ of brother's share, if she has brothers
	• Women having sisters
	  • Sunni/Hanafi—2/3rd of X is divided equally among themselves
	  • Shia….total inheritance is equally divided among themselves
	• Women having no siblings
	  • Sunni/Hanafi—1/2 of X
	  • Shia- All property (X)

• X in above table means "all wealth of the deceased person after settlement of mortgages and loans".

Females have equal rights to access, own, and control land since housing and property are recognized under international law. But in Pakistan, the presences of discriminatory laws, unclear policies, patriarchal customs and social norms are still hampering women from enjoying their rights. There are numerous factors responsible for this unjust scenario in the country.

28.6.1 Expected Dependency on Parents after Marriage

Mostly, females do not claim their right of inheritance due to the fears and uncertainties prevailing in their marital life. They feel that at some point during their lives, they might have to go to their parent's house. Parents, on the other hand also do not want to give inheritance rights to their daughters, because they feel that their burden will be born by their sons in the long run and don't want to anger them.

28.6.2 Women Low Social Status

Women's low social status is another factor acting as a barrier to their right to property. Lesser socialization, unrecognized productive labor, and above all, lack of education in Pakistani communities often make them hesitant to demand their legitimate rights (World Bank, 2009).

28.6.3 Lack of Awareness about Their Land Rights

Women's rights to land and their role in the family finance and economy are interrelated. Rights to land enhance women's negotiating ability within the household; it not only result in enlarged provision of household resources to their children and themselves but also amplifies household prosperity. Women with land and property rights and ownership are less exposed to domestic violence (Panda and Agarwal, 2005).

Acquired rights and land ownership also empower women to contribute more meritoriously in their communities as well as in civil society. Enabling women's involvement in extrahousehold institutes lessens men's supremacy at the community level decision-making. It also helps to formulate women's administrative abilities, social capital, and networks. Women with acquired land rights are considered more vigorous and energetic fellows of their communities (World Bank, 2009). The majority of the people, especially females in Pakistan, are illiterate and they do not have knowledge and awareness of their rights. Many women lack information, confidence, and resources to obtain what they are legally entitled to (Awaz Foundation Pakistan, 2010).

28.6.4 Social Customs and Norms

The prevailing trends in Pakistani society compel women not to demand their rights; it is an understood custom among families not to respect the rights of their women (Awaz Foundation

Pakistan, 2010). Cultural biases and prohibitions regarding women's land ownership are most of the times more powerful than laws, and do not permit females to own land. Women's' land rights are typically inseparable from their status in their spouse's family.

In such societies, where social practices are dominant, land rights are usually defined by sociocultural institutions (World Bank, 2009).

28.6.5 Lack of Knowledge of Legal Issues

Another constraint is women's low level of literacy related legal issues, This includes understanding legal language and legal drafts (World Bank, 2009). Generally, in legal terms women often inherit and own land, but in reality they do not have control over how the land operations are managed and have no access to the income derived from its use. Overall, women are not aware of the system of civil law which guards their rights to land. Land registration systems, processes regarding taking possession of land, and settlements of land disputes through legal formalities are some of the issues. To deal with such issues, women have to rely on male members.

28.7 REVIEW OF POLICIES AND STRATEGIES REGARDING GENDER EMPOWERMENT

Gender empowerment is a key to the development of the country. For women's empowerment Pakistan has taken very important steps as follows:

- National Commission on the Status of Women (NSCW) in 2000
- The current reports of the National Commission on the Status of Women on the Hudood Ordinances (2003)
- Women in Public Sector Employment (2004),
- Qisas & Diyat (2005)
- Home-Based Women Workers (2005)
- National Policy for Development and Empowerment of Women (NPDEW), articulated formally in 2002
- Women development departments established to plan tactical interventions to fill the identified gaps
- Women's political participation in Assemblies through National Programme for Women's Political Participation has also been achieved, it provides an incorporated approach to encourage women's equal involvement in decision-making at all levels of governance and it also creates public knowledge and support for women's improved participation in policies
- GRAP - Gender Reform Action Plan - focuses on budgetary procedure to narrow the gap in civil expenditure, reorganization of national mechanism dealing with gender issues.
- IT policies 2000 highlight the approach to assist and encourage the training and employing of women in this sector to facilitate increases in employment
- Parliament approved a bill on Honor Killing "Criminal Law Act 2004"

Considering the issues of gender mainstreaming into policy, development projects and programs would be the key plan strategies to encourage gender equality in Pakistan. On one hand, it would develop women's status and on the other contribute to social and economic development of the country. The key mechanism for gender integration include gender concerns in all developmental plans and projects; stability in the policy frameworks for female; gender central points in every ministry; to establish the linkage development; provision of practical support through the Ministry of Women Development to focal points of gender in further ministries.

28.8 MOVING AHEAD TOWARDS ATTAINING SUSTAINABLE TRANSFORMATIVE CHANGE FOR WOMEN

Agriculture is the fundamental source of national economics and rural females play vital roles in the production of agricultural goods. Gender relationships are the key to understand the way family farm work is structured within the family unit and beyond, the means by which assets and resources like finance, land, equipment, labor, are managed, and the approach through which decisions are implemented. The possible sustainable improvements and poverty cutbacks through social and economic development would not be accessible unless a proper effort is made by devoted government and development organizations regarding gender mainstreaming and women's empowerment.

Primarily because of confusing definitions, females' rural economic activities and responsibilities are more or less considered informal, hence not entirely captured and documented by the government. This point usually not adequately presented during conventions because it is not discussed. These deeds are not even taken into account by official statistics, particularly those deeds that are temporary, seasonal, or performed at household level. This direct failure to take into account female economic activities in many research studies skews research on employement and trends in the national economy.

REFERENCES

Asian Development Bank. 2008. Releasing Women's Potential Contribution to Inclusive Economic Growth: Country Gender Assessment for Pakistan. Available at http://www.adb.org/

Awaz Foundation Pakistan. 2010. Denial of women's right of inheritance. Enhancing their Vulnerability to Domestic &Societal Violence. Available at: http://awazcds.org.pk/Downloads/rstudies/Inheritance%20 Rights.pdf

Bill and Melinda Gates Foundation. 2015. Agricultural Development: Creating Gender-Responsive Agricultural Development Programs. http://www.gatesfoundation.org/What-We-Do/Global-Development/ Agri.-Development/

Duncombe, R. and Boateng, R. 2009. A review of evidence on mobile use by micro and small enterprises in developing countries. *Journal of International Development*, 22(5): 641–658.

Fletschner, D. 2009. Rural women's access to credit: Market imperfections and intra household dynamics. *World Development*, 37(3): 618–631.

Food and Agriculture Organization (FAO). 2011. *The State of Food and Agriculture 2010–11. Women in Agriculture closing the Gap for Development.* FAO of United Nations, Rome.

Food and Agriculture Organization. 2012. *UN Launches New Program to Empower Rural Women and Girls.* Rome: FAO. Available at: http://www.fao.org/news/story/en/item/158377/icode/

Food and Agriculture Organization (FAO). 2015a. *The State of Food and Agriculture 2015. Social Protection and Agricultur.* Rome: FAO of United Nations, Rome. Available at: www.fao.org/publications/SOFA/2015/en.

Food and Agriculture Organization. 2015b. *State of Food Insecurity.* FAO of United Nations, Rome.

Govt. of Pakistan. 2008. *Pakistan Economic Survey.* 2007–08 Finance Division, Ministry of Finance, Islamabad, Pakistan.

Govt. of Pakistan. 2015a. *Pakistan Economic Survey 2014–15.* Ministry of Finance, Islamabad, Pakistan. Available At: www.Fianace.gov.pk.

Govt. of Pakistan. 2015b. *Pakistan Economic Survey.* 2015–16 Finance Division, Ministry of Finance, Islamabad, Pakistan.

Hill, C. 2003. *Livestock and Gender: The Tanzanian Experience in Different Livestock Production Systems.* A Glance at LinKS: LinKS Project Case Study No. 3. Rome: FAO.

Ian Scoones. 2008. *The Growing Demand for Livestock.* ID21 Insights 72, February, Institute of Development Studies, Brighton, www.id21.org.

IFAD International Fund for Agricultural Development. 2009. Gender and rural microfinance: Reaching and empowering women. Available at: https://www.ifad.org.

IFAD International Fund for Agricultural Development. 2007. Working Paper on Rural Finance* https://www. ifad.org/documents/10180/7bd8eb94-d9b0-4f97-a3e9-5eb33144cbde

International Fund for Agricultural Development (IFAD). 2010. IFAD's Performance with regard to Gender Equality and Women's Empowerment December 2010 Report No. 2324.

International Labour Organization (ILO). 2003. Decent Work in Agriculture. Background Paper for International Worker's Symposium on Decent Work in Agriculture, Geneva, September 15–18.

International Labour Organization (ILO). 2006. *Global Employment Trends Model.* ILO, Geneva.

International Labour Organization (ILO). 2007. The Informal Economy: Enabling Transition to Formalization. Background document for the Tripartite Interregional Symposium on the Informal Economy: Enabling Transition to Formalization, ILO, Geneva, November 27–29. Available at: www.ilo.org/public/english/employment/policy/events/informal/download/back-en.pdf.

International Labour Organization (ILO). 2008. *Global Employment Trends for Women, March 2008.* ILO, Geneva.

Kabeer, N. 2000. *The Power to Choose: Bangladeshi Women and Labor Market Decisions in London and Dhaka.* VERSO, London/ New York.

Kokab, J. 2012. Women Rights and Women Protection Bill in Pakistan. Available at: http://www.slideshare.net/katikokabl/women-right-and-women-protection-bill-in-pakistan-15112739

Meenu, S., Arora and Reeder. 2011. Women empowerment through microfinance interventions in the commercial banks. *International Journal of Economics and Research*, 2(2): 35–45.

Morrison, A., Raju, D. and Sinha, N. 2007. Gender-Equality, Poverty and Economic Growth. World Bank Policy Research Working Paper No. 4349, Gender and Development Group, Poverty Reduction and Economic Management Network, World Bank, Washington, DC, September.

Njuki, J. and P. Sanginga. 2010. *Gender and Livestock: Issues, Challenges and Opportunities.* International Livestock Research Institute (ILRI). International Development Research Center, Canada.

Panda, P., and B. Agarwal. 2005. Marital violence, human development and women's property status in India. *World Development*, 33(5): 823–850.

Quisumbing, A. S. and L. Pandofelli. 2010. Promising approaches to address the needs of poor female farmers; resources, constraints and interventions. *World Development*, 38(4): 581–592.

Tzannatos, Z. 1999. Women and labor market changes in the global economy: Growth helps, inequalities hurt and public policy matters. *World Development*, 27(3): 551–69.

World Bank. 2007. *World Development Report 2008: Agriculture for Development.* World Bank, Washington, DC.

World Bank, Food and Agriculture Organization, and International Fund for Agricultural Development. 2009. *Gender in Agriculture Sourcebook.* The International Bank for Reconstruction and Development / The World Bank 1818 H Street, NW. Washington, DC. Available at: www.worldbank.org

World Economic Forum. 2015. The Global Gender Gap Report-2015. World Economic Forum. Available at: http://reports.weforum.org/global-gender-gap-report-2015/

Zaidi, S. A., H. Jamal, S. Javeed and S. Zaka. 2007. *Social Impact Assessment of Microfinance Program.* European Union/Pakistan Financial Services Sector Reform Programs, Islamabad.

29 Population Planning and Labor in Pakistan

A. A. Maan, Izhar A. Khan, and N. Farah

CONTENTS

29.1 THE BACKGROUND AND SIGNIFICANCE OF POPULATION PLANNING IN PAKISTAN

The population of Pakistan was about 32.5 million at the time of independence in 1947. By 1988, the population of Pakistan had become 102 million, exclusive of three million Afghan refugees. In 1998, the total population was 130.58 million (Population and Housing Census of Pakistan, 1998). Today, Pakistan is among the most populous countries in the world. Furthermore, Pakistan has the highest growth rate within the most populous countries (Hakim, 2000). The high population growth rate has dampening implication for socioeconomic development of the country. These implications affect all aspects of life including housing, health, employment, education, and economics (Afzal, 2006). Pakistan's educational record is particularly poor. Pakistan is among the countries with the lowest literacy rates. The total number of children enrolled in primary schools is just half of the total number of school-aged children. This percentage is even lower among female children compared to male children. Furthermore, the dropout and repeater rates in schools in Pakistan are among the highest in the world. Half of total primary school students leave before completing 5 years. On average, students complete only about 2 years of schooling (PDHS, 2014; Population and Housing Census of Pakistan, 2017). The low literacy rates and poor school attendance demonstrate high fertility and low use of contraception. Previous data showed that almost all female helpers at Family Welfare Centers were illiterate. Many of the Family Welfare Assistants had not completed the recruitment criteria of 10 years schooling.

The reasons for illiteracy in Pakistan lie in the socioeconomic milieu of the society. The first important reason may be the traditional setup of the society. In many traditional cultures, parents rarely prefer to send children to school. Unemployment and underemployment of educated youngsters discourage parents to provide education to their children. Probably, poverty is the most important impediment to schooling and attaining education in Pakistan. Even if these difficulties are removed, there is an insufficient number of schools for the ever-growing number of school-age children. This situation will become more difficult in the future, as more and more children will be entering the school-going age group. The overall economy of a country is also important to determine the socioeconomic status of its people. Income is one of the important prerequisites to attain the basic necessities of life. Pakistan is struggling to transform its economy from agriculture to industry. About 65% of the population is living in rural areas. The agriculture sector engages more than a half of the labor force. Furthermore, agriculture contributes 24% of the Gross Domestic Product (GDP) and most of the country's export (Hakim, 2003; Government of Pakistan, 2014). The annual per capita income is very low. The average monthly income of a rural household is much less than in an urban household. The annual GDP growth rate is also discouraging (Anjum, 2001; PDHS, 2014). The situation demonstrates the poor economic conditions in Pakistan compared to other countries.

The influence of the sociodemographic structure of population to determine the level and make the economic activities sustainable among people in any country is crucial. One cannot ignore the influence of population growth on the economy of Pakistan. It is not just about population size but also about its structure. The population structure in Pakistan may be more important when its economic development is considered (Anjum, 2001; Hakim, 2003). More than half of the Pakistani population is comprised of children and elderly people who are economically dependent. There is a feeling that the socioeconomic conditions of Pakistan cannot be improved without population control (Rehman, 2006). The most widely used approach to control population growth is through the establishment of national population planning programs. Population planning programs exist in many countries including Pakistan. Pakistan is among the pioneer countries to initiate population-planning services.

29.2 POPULATION PLANS IN PAKISTAN: A CRITICAL ANALYSIS

The Family Planning Association of Pakistan (FPAP), with the help of International Planned Parenthood Federation (IPPF), launched a family planning program in the country during the early 1950s (See Appendix 29.A). This program received limited financial support from the government as part of the government's First Five Year Plan 1955–1960, (Government of Pakistan, 1955). The aim was to slow down the population growth rate and popularize the small family norm through using contraceptives. An independent family planning program was introduced in the public sector in the Third Five Year Plan (1965–1970). Since then, many attempts and changes have been made yet little has been achieved. Achieving an increase in the acceptance and continued use of contraception to slow down the high population fertility rates have been the top priority objectives of the Pakistani programs since the first 5 year plan (Rukanuddin, 2001; Hakim, 2003).

29.2.1 First Five Year Plan (1955–1960)

Soon after the emergence of Pakistan, the population problem was recognized as needing consideration and action by those in the government and outside government as well. However, the government had to solve many other social and political problems. Up to the First Five Year Development Plan (1955–1960), some agencies and individuals became successful in convincing the government to provide support for the development of population welfare activities by private organizations. While the Government of Pakistan was unable to introduce the family welfare program, it did provide half a million rupees (Government of Pakistan, 1955) to assist the family planning activities introduced by FPAP.

Strengths

- FPAP carried on the family welfare activities with the support of IPPF.
- They organized a number of voluntary associations in different cities.
- They worked to reduce the importance of the large family norm.
- They provided family planning services in some cities through clinics.
- It encouraged to take the first few steps to tackle the population issue.

Weaknesses

- This small-scale approach could not solve the problem of overpopulation.

29.2.2 Second Five Year Plan (1960–1965)

The government of Pakistan launched its own family planning project as part of the Second Five Year Plan 1960–1965 (Government of Pakistan, 1960).

Strengths

- A large amount of thirty million rupees was proposed.
- The family planning activities were made the responsibility of the Health Department.
- The distribution of contraceptives was made through public sector hospitals and dispensaries across the country.
- A National Research Institute of Family Planning was established.
- Research was started on different features of the family planning program.
- Training institutes were set up to develop and improve research skills.
- The trainers of research produced and publicized FP related material.
- They showed audio-visual films in the villages and cities.

- They shared training and research facilities for publicity material.
- A Pak–Swedish Project was initiated to promote communication skills.
- National family Planning Boards were established for success of the program.

Weaknesses

- The experience of utilizing the Health Department was not successful.
- Only clinic-based services were provided.
- No efforts were made for community participation.
- No efforts were made to motivate potential service users.
- No follow-up of those who attended clinics.

At the end of this five year plan, the government decided to start a full-fledged, independent, and single-dimensional program to make it a success. During the period of this plan, FPAP became more popular and effective and attained the favor of many influential officials of the government and people in society.

29.2.3 Third Five Year Plan (1965–1970)

Strengths

- A new structure of family planning program was designed and implemented.
- A Family Planning Commissioner was appointed to control the activity.
- A Central Family Planning Council was established to implement and administer the program in coordination with the Commissioner.
- A Family Planning Board was established in each province on the similar lines of the Central Council.
- As a district level, District Family Planning Boards were designed (Government of Pakistan, 1965; Hakim, 2003).
- In this plan, money was allocated for publicity, communication, and training.
- Training and refresher courses were arranged for TBAs, family planning, and medical personnel.
- Other strengths of the program included.
- An independent program.
- A clinical approach.
- Political support and awareness among couples.

Weaknesses

- The weaknesses of the program included.
- The lack of training for TBAs.
- A lack of appropriate supervision.
- A dependency on health personnel.
- No evaluation of service supply.
- No evaluation of the achievement of stated goals was made.
- No impact upon high population growth rate, the program continued to 1968 when it was discontinued following the dismissal of the then government.

29.2.4 Fourth Five Year Plan (1970–1975)

This program, in terms of policy, can be divided into three periods (Government of Pakistan, 1970).

29.2.4.1 The Early Period (1970–1973)

- The program faced financial and operational problems during this period.

29.2.4.2 The Late Period (1973–1975)

Strengths

- The presence of a multipurpose program.
- Change of name to population planning.
- The presence of contraceptive choice (condom and pill supplied at home and IUD at clinic).
- The development of more training institutes and a policy and research centre.
- The greater emphasis on the relationship between population size and national resources.
- The strategy for supervision introduced into service provision.
- Direct contact with husbands.
- The initiation of a multiphase media campaign.
- The use of community contraceptive distributors (shopkeepers and villagers).
- The training and involvement of employees of different governmental departments in the program.

Weaknesses

- Poor year-to-year planning and implementation.
- A poor implementation phase.
- The presence of inexperienced workers.
- Poor selection and appointment procedures.
- Failure of the CMs at national level.

29.2.4.3 The Extra Period (1975–1977)

- The policy described in the "late period" continued for another 2 years.

29.2.4.4 Mid to End of 1977

Strengths

- Program was federalized.

Weaknesses

- Program activities and structure were severely reduced.
- The television and radio campaign were ceased.
- The motivation program was stopped.

29.2.5 Fifth Five Year Plan (1978–1983)

- The population plan for this time can be divided into two periods.

29.2.5.1 The Early Period (1978–1980)

Weaknesses

- After the change of government in 1977, the program collapsed.
- Some program activities were revived in 1978 but without publicity.

- These limited activities continued up to 1980 without any apparent achievement.
- During 1980, the population planning was transferred from the Ministry of Health to the Ministry of Planning and Development.
- The program (and the employees associated with it) was taken over by the government.

Strengths

- There were signs that the population planning program would be revived on the appointment of the Advisor on Population Welfare during 1980.

29.2.5.2 The End Period (1980–1983)

Weaknesses

- A new plan for 1980–1983 was designed.
- The results were discouraging for the administrators and employees of the population-planning program.
- The working staff was reduced to about half their previous number.
- The functioning and coverage by the program was reduced in scale.

However, the government subsequently did not feel satisfied with all the changes made in the program. The reason may be that the decisions made were not made on an agreed basis but in order to please some segments of the society. A second reason was the pressure by the Aid Agencies to give priority to a population-planning program (Government of Pakistan, 1978).

Strengths

- The presence of a multipurpose and multiphase program.
- A strong IEC program (in place of CMS).
- A decentralized, project-oriented and broad-base program.
- A new Division of Population Welfare within the Ministry of Planning and Development.
- The creation of "Family Welfare Centers."
- Awareness campaign about family planning methods.
- Most of the staff eventually made their adjustment.
- Staff numbers again began to grow.

29.2.6 Sixth Five Year Plan (1983–1988)

Strengths

- A mass-media campaign was again planned and launched during 1986.
- Special innovative elements were introduced in the fields of communication and supply in the period 1986–1988.
- Services were supplied through dais, hakeems, barbers, NGOs and by the social marketing of condoms.

Weaknesses

- The results were not encouraging. The reasons for these poor results may be found in the need of continuous supplies and regular supporting messages in the mass media.
- The long disruption made it difficult to restart the program.
- The communication processes were difficult to reestablish.

- The time-to-time disruption of the program appears as a major obstacle in convincing and familiarizing people that the small family norm and adoption of contraception are acceptable goals.
- Frustration among the staff transferred from central to provincial institutions.
- A lack of field supervision.
- A lack of coordination between the relevant departments.
- Poor follow-up.
- Restricted and unclear public advertisements.
- The absence of strong and open political support (Government of Pakistan, 1983).

29.2.7 Seventh Five Year Plan (1988–1993)

Strengths

- To lower the population growth rate through widening the idea of birth control.
- Strengthening the service delivery and motivational mechanisms.
- To make the attitudes of the masses more favorable.
- To increase the use of contraception from 12.9% (estimate during 1987–1988) to 23.5% by the 1st year of the plan.
- To provide maternal and child health care services.
- To lower the Crude Birth Rate (CBR) from 42.3 to 39.0 per thousand.
- More consideration was given to clinical contraception such as the IUD, injectable, and surgery.
- Field-supervision and the IC component of the program were strengthened.
- Involvement of NGOs, Hakims, and Social marketing was also not neglected.
- Once again, the involvement of other government departments was encouraged.
- The importance of the TBAs was also again recognized.
- The involvement of registered medical practitioners was also encouraged.

Weaknesses

- Though the multisectoral approach remained, its influence was not stressed as much as it had been during the Sixth Plan (Government of Pakistan, 1987).
- However, despite these efforts the performance of the program remained poor.

29.2.8 Eighth Five Year Plan (1993–1998)

Strengths

- Village Based Male and Female Workers provided family planning services.
- Mobile Service Units provided services in rural areas in liaison with the Village Based Workers.
- The number of Family Welfare Centers was increased.
- More centers for RH Services were established.
- Training centers were equipped with more facilities.
- More public and private hospitals started to provide RH Services.
- Social Marketing of Contraceptives continued with the Govt. support.
- NGOs were encouraged as previous to make the plan a success.
- Training programs were enhanced and strengthened.
- IEC campaign was made broader and more effective.
- Community participation was encouraged and ensured.

- Supervision and management were made more effective and successful (Government of Pakistan, 1993).

29.2.9 Ninth Five Year Plan (1998–2003)

Strengths

- Most of the steps of the previous plan continued.
- First Population Policy of Pakistan was developed.
- Master Degree in Population Sciences was initiated in Pakistani universities.

Weaknesses

- This plan, as previously, could not make any major dents in the situation (Government of Pakistan, 1998).

29.2.10 Tenth Five Year Plan (2003–2008)

- Most of the steps of the previous two plans continued.
- The important initiatives included first youth policy in the country (Government of Pakistan, 2003).

29.2.11 Eleventh Five Year Plan (2008–2013)

Strengths

- To develop devolution plan and establish a local govt. system.
- To meet the unmet need of couples who want to limit/postpone childbearing.
- To increase accessibility of services in village and urban slums.
- To improve outreach and facilitate referrals in the health sector.
- To double primary enrolment rates for the coming 15 years.
- To provide social protection to 15 million never school-going children by 2015.
- To enhance also the secondary school enrolments dramatically.
- To meet the skilled and nonagricultural job needs.
- To enhance the youth labor force contribution to the economy.
- To facilitate to transfer from agricultural to nonagricultural employment.
- To increase female labor force opportunities in paid jobs.
- To subsidize food and education for poorer households.
- To create saving and investment strategies for emerging labor force and changing household work system (Government of Pakistan, 2008).

29.2.12 Twelfth Five Year Plan (2013–2018)

- General elections were held.
- Decentralization of important ministries such as ministry of population welfare and ministry of youth development took place.
- Population Policy (Government of Punjab) included strategies:

Strengths

- To converge service availability at community level.
- To fulfill couples unmet-need for contraception.
- To stress upon quality of services.

- To ensure contraceptive commodity security.
- To improve advocacy.
- To ensure work on demand generation and social mobilization.
- To stress upon human resource development (HRD).
- To make sure the support of development partners and private entities.
- To carry on monitoring and evaluation.
- To ensure research and metrics.
- To focus on men's mobilization for responsible parenthood.
- To prioritize issues of adolescents and youth.
- To work for departmental restructuring,
- To coordinate to consolidate the services.
- To ensure proper utilization of funds.
- To improve governance and development issues.
- To promote small family as a right.
- To improve service delivery.
- To focus behavior change by effective communication through modern techniques (Government of Punjab, 2013).

29.3 LABOR/LABOR FORCE

Choosing a specific occupation, job or employment can be seen as an important determinant of socioeconomic status of an individual or family in society. It is generally the trend that men are engaged in jobs having better salaries and status as compared to women. This is an important social issue that needs investigation and data in order to understand its reasons and minimize this gender-gap. The available empirical evidence shows that there are several characteristics of individuals that stand behind their selection for a particular occupation. The major characteristics include ban and discrimination for women to join certain labor force groups, but also available opportunities, expected cost to join a particular job of choice, interest in a particular occupation, personality traits, capacities, cognitive abilities, norms and values of society, education levels, skills, and training. Moreover, women's priority is to go for nonpecuniary characteristics of jobs. This orients women's choices towards low-paying occupations. Therefore, their representation is low in high-paying professions. These social, cultural, and personal characteristics vary by gender and have serious implications in the decision-making process for joining a particular labor force-group for earning a livelihood. Furthermore, the entry into so-called female-jobs is easy for women. Therefore, they concentrate in these occupations. Resultantly, this oversupply of women for these jobs turns them into the low-paid and low-earning individuals of society (Tabassum, 2017).

29.4 WOMEN LABOR FORCE IN PAKISTAN

The literature on Pakistani situation also shows that workingwomen are concentrated in low paying positions while men in high paying jobs. Education, experience, and other personal characteristics have been proven important correlates of a women's career. However, discrimination against women has been reported as an important barrier for their entry into high-paid occupations. Due to the growing interest and need of women in the labor market, there is a need to bring out programs to enhance opportunities for women employment by increasing their chances of recruitment and employment in the existing labor-force structure and system (Tabassum, 2017).

29.4.1 Rural Women and Agriculture-Labor

The economic participation of women can be accessed through various indicators including activity rates, employment status, and representation of women across occupations and the vectors

(See Appendix 29.B). Unfortunately, labor force data considerably underestimate the extent of female participation. There are significant differences in the activity rates of rural and urban women. In most studies, occupation is used as an explanatory variable in wage equations, which does not reveal the choice of an individual, rather that explains the variation in wages. Based on the definition of work including the expanded categories, it is found that women's daily work burden is greater than for male family members. Women face a number of constraints in the labor market which create wide gender differentials at the workplace. Low levels of education among women, lack of employment opportunities in the formal sector, and sociocultural taboos against work outside the home, particularly in rural areas, limit women's participation in labor markets. In rural areas, women are mostly engaged in the agricultural sector and contribute largely to the rural economy. Many of the various agricultural activities carried out by women are usually not counted as work. The invisibility of their contribution to agricultural work limits the benefits that women can derive from their economic activities. Most of them work under poor work conditions and receive low remuneration. Above all, women lack control over financial and other productive resources derived from their economic activity. Lack of economic autonomy and the limited mobility of women act as social barriers, which prevent them from responding to the dynamics of labor market. Women living in rural areas, who are traditionally conservative, get married at an early age due to sociocultural pressures. The early age at marriage exposes women to child bearing and family formation, which limit options for their economic participation. Delay in marriage increases the chances of women's participation in education and work, which eventually affects their status in society. It is argued that the prime responsibility of women is to look after domestic affairs and their economic participation leaves little time for household responsibilities. Pakistan is basically a male-dominated society where most of the family decisions are taken by male members of the household. The head of the family is empowered to decide about the education and occupational life of women. However, it is found that women who are older, better educated, head of the household, or coming from smaller, better-off urban families are more empowered to take decisions on their own about their employment. Furthermore, in Pakistan, it is perceived that Islam strictly forbids the employment of women, which is debatable. The role of media, also, is not encouraging to promote the status of women and it projects women in a way that does not help their cause. Some orthodox employers never let women rise to better positions because of the understanding that women are not as competent and efficient as their male counterparts, especially for technical and managerial positions. However, research studies have shown that women are as competent, hard working, and efficient as male members of the society (Tabassum, 2017).

29.4.2 Rural Women and Livestock Labor

In developing countries, there is a huge demand for livestock-production which is expected to double in future. Livestock, in many developing countries including Pakistan, is used for many benefits such as income-enhancement of rural people, stability of social activity, agricultural and environmental sustainability, improvement in crop-farming, and crop-productivity.

29.4.2.1 The Situation in Asia and Africa

In the developing world, rural women were engaged in exclusive duties, that is, making animals sheds, making and collection of dung cakes and manure collection. In some cases, men work out of the house (fodder-cutting, etc.) and women are involved in almost all livestock activities, that is, care, production, and marketing. Furthermore, the level of women's involvement in all these activities varies. It has also been reported that women participate in about 90% of activities. Mostly, they are engaged in cleaning animal sheds, milking animals, making dung cakes, fodder cutting, and chopping. In Africa and Asia, 70% of the women are engaged in livestock activities, but fewer were involved in making animal-houses, animal marketing, production, and slaughtering. In livestock related tasks, women with low or no education are involved. Therefore, they have fewer skills to adopt new technologies, little or no training, and have attended fewer workshops. Resultantly, this reduces efficiency and productivity (Akmal and Sajida, 2004; Khushk and Hisbani, 2004; PARC, 2004; Rasheed, 2004; Farinde and Ajavi, 2005; Javed et al., 2006; Hashmi, 2009; Nirmala et al., 2012; Taj et al., 2012; Munawar et al., 2013; Government of Pakistan, 2014).

Observation of the incidence and treatment of epidemic diseases among animals have had positive impact upon animals' health. Women recognized the importance of immunization against livestock diseases for livelihood-sustainability. Problems about animals' reproductive-system and production were controlled and productivity improved, especially in milk animals. Training for animals' health-management and programs for confirmed accessibility of veterinary facilities were launched. In this way building and improving women's achievements and aspirations were made possible. Small steps of an organization improved women's social-status, self-confidence and self-motivation (Miller, 2001; Devendra and Chantalakhana, 2002; Nielsen and Heffernan, 2002; Arshad et al., 2010).

29.4.2.2 Situation in Pakistan

Literature shows that women's are mostly engaged in soft work rather than hard work compared to men (See Appendix 29.B also). A gender-based division of labor has been observed, both in agriculture and livestock sectors. Mostly, women participate in work inside their houses, usually milking and feeding animals. On the other side, men usually work outside the homes, that is, selling and marketing milk and milk products. A study about the production of milk and different levels of dairy herds showed that the average production of milk was through indigenous methods of livestock production. Milk was the major source of national economy. Milk was produced in different ways, that is, in the household and in the courtyard of farm. A significant role and involvement of women in livestock-related activities has been reported. However, this role was always undervalued due to gender-discrimination. The research results showed different difficulties in data collection about women's importance and roles, which have significant contribution in national economy. Furthermore, women's role in livestock care and management in regional statistics are rarely observed. In Pakistan, high constraints were faced by the programs working for the improvement of livestock production. These constraints included policy and market problems, nutritional issues, and inadequate support services. The women of rural Pakistan played significant roles in managing their livestock and poultry farming. They were also involved in doing all kinds of domestic work. However, they were always neglected in important decision-making of family as well as livestock issues. Managing their livestock and poultry farming activities were always considered the women's responsibility and a part of domestic work. Therefore, the major duties of their domestic work are ignored. They are not able to focus upon their domestic duties including about childcare and household maintenance (Niamir-Fuller, 2008; Arshad et al., 2013; Maan and Khan, 2015).

It has been found in different studies that women faced more problems in livestock-care and management as compared to men. They have had fewer opportunities for participating in programs for training and education in livestock care and management issues. Livestock was found as second source of income for families, especially when involved in poultry and buffalo farming. Livestock is also an insurance during crop-failure and drought. During these times, a large number of women earned enough income for their families from livestock farming. Improved livestock-stock and production have had positive effects on domestic income. Mostly, rural women have better knowledge about livestock care and management activities because they routinely care for their animals. Women engaged in livestock-farming earned more than their men. Rural women are empowered by having money spend it for household expenses like childcare and food (Amuguni, 2001; Flolmaan et al., 2005; Niamir-Fuller, 2008; Nadeem et al., 2012; Arshad et al., 2013; Maan and Khan, 2015).

29.4.3 Education, Women, and Labor Force

Women having education are important for a country. These women can play a significant role in the socioeconomic development of a country like Pakistan. They can contribute for their development, and that of the nation, through participating in the labor force. Education is considered the key factor for acceptable and respectable jobs and careers. Investment in the education of women and providing them job facilities is the most effective means of raising the general level of development and promoting sustainable development. According to several studies by International Agencies, including the World Bank, UNESCO and the United Nations Development Program, positive correlations have been found between improved educational opportunities for women in developing countries and their labor force participation. One study has concluded that three variables (education, fertility and urbanization) affected the female labor force participation, but the effect of education was more significant. Furthermore, this accounted for more than two-third of the variation in the level of female participation in the labor force. Another study analyzed the determinants of female labor force participation. Empirical results of the study indicated that higher levels of education significantly influenced the female labor force participation. Furthermore, household per capita income was correlated with wage employment, but self-employed women appear to be unaffected. One study on the topic has pointed out the problems educated women have to face in performing economic activities; it also examined the categories of jobs offered for women. It was concluded that in Pakistan women were not encouraged to work with men. It was suggested that women should be provided with job facilities in-line with their skills and qualification (Faizunnisa and Ikram, 2003; Khalid, 2006; Tabassum, 2017).

Some other researchers studied the factors determining work participation and labor supply decision in the urban areas of Pakistan. Empirical results indicated that labor force participation (LFP) rate rose with increases of expected earnings, wages, and the level of education. The economic factors affecting female labor force participation, household income, and female education have been discussed in the literature. Household income is a primary determinant of women's entry into the labor market. It is inversely related to women's supply in the labor market. A number of variables for household income such as husband's occupational status, husband's wage rate, and adult male earners have been used in the studies. The empirical results supported the hypothesis that husband's occupational status, husband's wage rate and number of adult male earners was inversely related to the percentage of women entering in the labor market. While the effects of educational levels on women's labor force participation rate were different, illiterate women's participation was 44%, having primary and middle educational-level women's participation rate was above 30%, having matric level education women's participation rate was 60% and at F.A. and B.A. level participation rate was 50%. It has been concluded that the number of women at home has a positive effect on female labor force participation. The determinants of female labor supply have also been discussed in the published literature. The data were collected in the form of 166 earning and 219 nonearning women living in the poor settlement of Rawalpindi. The major factors affecting female labor supply included education, household income, widowed household, household size and composition. The effect of household income and presence of children had negative impact on female labor supply. The changes over time in the level and pattern of women's employment in Pakistan have also been studied. These changes have been analyzed in the context of supply and demand factors influencing women's participation in the labor market. The increased participation in the labor market of women at the top and bottom end of the socio-economic scale was investigated. Women from the more privileged classes in Pakistan have been able to acquire university education and take up professional jobs, but women from the poorest strata are pushed into labor market due to desired economic necessity. They have to be engaged in low paid and unskilled jobs in the informal sector due to cultural restrictions, household responsibilities and low level of education and skills (Aliya and Shahnaz, 2001; Tabassum, 2017).

Another socioeconomic factors that influence women's labor market participation decisions have also been studied. It has been concluded that women's wage rate and education were positively correlated with women labor force participation (WLFP) rate, where as marital status, the number of children, and age were negatively correlated with WLFP rate. Important factors influencing women's decision to enter the labor force have stressed that the level of education had a great influence on women's decision-making ability, especially if they have received some form of postgraduate education. Other factors, which influence the decision, are income level of husband, family system, and urban settings. Literature has also connected two important aspects of women's decisions regarding their participation in economic activities and how these decisions are made. The study concluded that women labor force participation could be enhanced with the improvement in their educational level, lower fertility rates, and better care of children. It has been revealed that most of the women labor force (generally in low skills, low paid economic activities) belongs to families located in rural areas, or low-income families. The study explained that age, better education, responsibilities, and family environment were important in impacting the female labor force participation decisions.

29.4.3.1 Education, Rural Women, and Labor

It has been found that most of the rural women started working at an age of less than 15 years and they were forced to work to meet their large families' basic needs. Most of them were illiterate, unskilled, and had poor health as well as low pay. They worked as much as women in the formal sector but were paid much less than those working in the formal sector. The study suggested that rural women's advancement should be facilitated with education, training, hospital, reasonable wage, raw material, transportation, and marketing facilities, etc. to play a useful role in the prosperity and development of the country. Another study determined the factors affecting the optimum time allocation between market and housework of females in rural Pakistan. The study examined, whether women's decisions not to work outside the home were influenced by factors such as social norms, economic constraints, lack of relevant education and training, nonavailability of job opportunities, and low wages. The study showed that women who participated in market activity also spent longer hours on home production as compared to their nonparticipating counterparts. It was also found that education does not have a significant effect upon the market time allocation. This could be due to the characteristics of the sample, which had very few educated women. The study concluded that the labor force participation decision of rural women was mostly influenced by cultural taboos (constraints) and job opportunities (Aliya and Shahnaz, 2001; Tabassum, 2017).

At the end, it can be said that the female labor force in developing countries like Pakistan is economically exploited and underestimated. Social, cultural, and economical barriers hinder their full and proper participation. These studies also concluded that there is a positive correlation between education and female labor force participation. Another inference drawn is that increase in women's education generally leads to increase in their labor force participation and, resultantly, in their earnings. Higher education costs more and needs to be more utilized.

29.5 CHILD LABOR

The incidence of child labor is very high in Pakistan. Although, child labor is often harmful for the children, there are situations where its alternatives may offer even deeper poverty both for the children and their families. It is commonly believed that child labor often comes at the expense of child schooling. Child enrolment may be enhanced by child subsidies, but its impact in curbing child labor remains ambiguous. Since the main cause of child labor is poverty of the households, it is sometimes argued that the incidence of child labor can be lowered by higher adult wages. The government may intervene in the market to curb child labor, but its success largely depends on how this intervention takes place. For instance, it can be argued that compulsive measures meant for elimination of child labor can sometimes be justified when incentives and regulations are combined. The government of Pakistan has banned employment of children below the age of 14 years since 1991 and has made it an offence punishable by imprisonment and fine. It is important to realize, however, that such state interventions run the risk of reducing the already limited choices available to the child. If this legislation were implemented effectively, it would mean that the child could neither work nor go to school. The point here is that to tackle this complex problem, different policy instruments are required for the state, not only to address the obvious market failure, but also to tackle the efficiency and distribution considerations of such services. Increasingly, it is being realized that the issue of child labor can be understood better by gaining insights on the supply side factors that influence household decision-making.

29.5.1 Household Characteristic and Child Labor

Burki and Shahnaz (2001) have investigated the supply side determinants of child schooling and child work. About 60% of rural and 76% of urban children attended school on full-time basis. Moreover, children who combined work with school were only 1% or less in either rural or urban areas. However, full-time child workers both in rural and urban areas were 12% each. The proportion of nowhere children located in rural areas was more than urban areas. Some studies related education and employment of parents with child labor decision. It has been argued that investment on children in larger households is reduced. Each child has a different probability of being engaged in work depending upon the age of siblings in the same household. It doubles if compared with such children located in urban areas. The age of the head of household indicates the stage in lifecycle, which is expected to influence the household decisions on child labor. Only 4%–8% of the households in rural or urban areas were headed by females, which are expected to have its ramifications on household decisions. The proportion of literate household heads was much higher in urban than in rural areas. Further, this proportion was highest for children going to school and second highest for children combining school with work in both rural and urban areas. By contrast, the proportion of literate heads was found to be lowest for the category of children involved in child labor. Like literate mothers, literacy of household head also had positive effect on school and work choices of households for females. The full time child worker had relatively higher family size than school going children in both rural and urban areas (PAP, 2006, 2007, 2010). However, in urban areas the largest family size was found to be for child workers and nowhere children. The highest incidence of child labor was found in the province of Punjab, irrespective of rural or urban areas. Similarly, Sind and NWFP provinces had second and third highest incidence of child labor while the lowest incidence of child labor was found in Baluchistan (PAP, 2006, 2007, 2010). This distribution of child workers seems to be quite consistent with the distribution of business and agricultural activities in the respective provinces. The probability of a child's attending school and combining school with work for wages increased at a decreasing rate while the age of the child was not a statistically significant determinant of the probability of child work. Enrolment in rural schools peaked at 10.63 years while school and work choice peaked at 12.58 years. This evidence is consistent with peak enrolments of 11 years (Asif et al., 2001).

Life cycle of the head of household does matter at least for rural household choices for work. It was found (PAP, 2006, 2007, 2010) that the probability of child labor and child combining school with work falls as the head of household gets older in rural areas of Pakistan. The gender of the household head is statistically significant in schooling and school with work choices, which suggest that female headed household have very high probability of sending their children to school and relatively much lower positive probability to combine school with work. Like the effect of literate mothers on child schooling, the probability of a child going to school increases by 22.4% if head of the household is literate. The presence of older siblings increases the chances of attending school and decreases the chances of doing full time work. The negative effect of presence of younger siblings on schooling and positive effect on work for wages is very likely to be explained by greater demand for children for home production, such as caring the young children, or the need for supplementing household income by working full time. The residence of household also matters in child schooling and work decisions; the strongest effect was found in Punjab. The high incidence of child workers in Punjab can be attributed to a greater proportion of labor force involved in agricultural production with labor-intensive farming practices, which are high in demand for female and child labor (Basu, 2000; Dessy, 2000).

29.5.1.1 Gender of the Child and Labor

In most South Asian countries, including Pakistan, boys are considered an asset because they are supposed to live with their parents and to support them financially in their older age. However, girls are considered a liability because they are supposed to live with their in-laws after their marriage in a joint family system or with their husband in a nuclear family. Hence, it is more likely that education of daughter may not involve any financial return to poor parents. Therefore, male children get preference in schooling. May be for the same reasons, it is preferred that females do not work at places other than child's own home due to social, cultural, and ethical issues. It has been observed that the gender of the child renders powerful influence on schooling and work decisions of the household. For instance, female children in rural areas were 20% less likely to choose full-time schooling than their male counterparts (PAP 2006, 2007, 2010), which is representative of the national composition of female literacy and enrolment patterns in Pakistan. The evidence that females from rural areas are 36% more likely to help in home production also supports this argument. It has also been observed that females are about 13% less likely to accept child labor. Only a negligible proportion is less likely to combine school with work. These results are in line with the customs and traditions in rural areas of Pakistan where most female children assist only in home production for the reasons described earlier. As expected, very few females were attending school or are economically as active than their male counterparts. Female child laborers as a proportion of males were found to be more than double in rural than in urban areas. Similarly, the proportion of females who were neither going to school nor were economically active was much higher in rural than in urban areas. These enrolment and work patterns observed were representative of the national literacy and labor force participation rates (Chaudhry and Jabeen, 2007; Khan et al., 2010).

29.5.1.2 Mother's Profile and Child Labor

The parents' characteristics play an important role in household decisions about their children. More importantly, mother's education has a positive influence on the probability of the child going to school and negative implication on the probability of going for work for wages. More specifically, our evidence (PAP, 2006, 2007, 2010) showed that school enrolment of children was positively associated with the education level of their mothers. Quite consistent with these findings, mothers with primary, secondary or higher education were less likely to allow their children to become child laborer. It implies that mothers who have been to school try that their offspring also get this opportunity. Children of working mothers were 12% less likely to attend school than children of nonworking mothers. Mothers' employment has a strong effect on household decisions on girls schooling or work choices. More specifically, we found (PAP, 2006, 2007, 2010) that girls with employed mothers were about 12% more likely to become child workers and about 4% less likely to go to schools as compared with girls of nonworking mothers. It has also been found that more than 96% of mothers of those children who were economically active or were away from schools were illiterate. Moreover, it also appeared that educated mothers were more likely to send their children to school in rural as well as urban areas. In other words, investment on mothers' education can significantly affect child schooling and child work patterns. Another thesis is that child labor does not necessarily affect child schooling. Rather, the important to consider is the difference in schooling only, part-time schooling with work, full time work and no schooling and no work choice. Further important to be considered is characteristics of the child, the mother and the household. Studies suggest a positive influence of mother's education on child schooling and a negative impact on child work while mother's employment was found to be negatively associated with child schooling and positively with child work. The presence of siblings/other children in the age group (0–4) years within the household ass found to have a negative implication for child schooling (Ravallion and Wondon, 2000; Ray, 2000; Ranjan, 2001; PAP, 2006, 2007, 2008, 2010; Batool and Khan, 2010).

29.5.2 Community Issues and Adolescent/Child Labor

The coexistence of declining school enrolments and rising labor force of adolescents and children, especially in the rural areas of Pakistan, is not a cheerful situation. It has been argued that schooling and child labor are two sides of the one problem, which interprets the reasons for not going to school. Going to school or for labor results from the decision-making process to send the child to school or to work. Furthermore, factors responsible to increase children's working hours generally tend to decrease their hours of study. However, hours of study seem to be more affected by community variables, such as electricity supply and distance to water as compared to hours of work. The association between child education and child work provides mixed evidence of substitution and complementarily. Again, the educational attainment of working children is significantly lower than that of nonworking children. The issue of community level influences on schooling and work activity of Pakistani children has been presented in the literature. Child characteristics such age and gender, household characteristics such as household structure and socioeconomic resources, school characteristics such as same-sex public/private school within 1 km, co-education school and no school, labor market characteristics such as returns to education and infrastructure characteristics such household access to electricity, natural gas connection, household level water supply, and paved/unpaved roads were used as community characteristics in a study. The effect of all these community factors on the likelihood that youth engage in work and/or schooling was examined. There is literature available on the health, education, work, marriage, and child bearing of adolescents in Pakistan. It has been argued that the demand for children and adolescents' work in the household may inhibit their ability to attend school, complete school, and perform well in school. A clear negative relationship between schooling and work participation exists among children and adolescents in Pakistan (Durrani, 2000; Lovcnbalk et al., 2003; PAP, 2006, 2007, 2010; Haque and Sultana, 2010). However, it would be premature to argue that all working children are not attending school because they are working. It is quite possible that children are working because of less availability of schools, because of dropping-out from school due to dissatisfaction with quality of education at school, or work offers them more valuable skills and experience than schooling. A strong influence of parental education, household income and status on schooling investment has been found. It has also been revealed that more than a half (56.9%) of the adolescent males and females aged (10–14) were still enrolled at up to the primary level. This phenomenon is more obvious in the rural areas where 63.1% of all male and female adolescents currently enrolled at up to the primary level belonged to the (10–14) age group. A distribution was made by taking male and female adolescents who completed either up to primary level or above primary level of schooling in the past. It was found that more than three-fourth (79.9%) of the adolescents have had completed only up to primary level of schooling. However, 43.8% of the adolescents who worked, were those who were previously enrolled in school. It was further found that majority (58.3%) of the adolescents were working in an agricultural occupation. This occupational pattern is reenforced when we have a look upon the occupational distribution of the 48.2% working adolescents who were never enrolled in school. Again, a higher proportion (75%) of these adolescents were engaged in an agricultural occupation (PAP, 2006, 2007, 2010; Sabir and Aftab, 2007; Government of Pakistan, 2009).

29.5.2.1 Poverty and Adolescent/Child Labor

The underlying reason for relationship between child/adolescent labor and schooling can also be seen in terms of the existence of poverty in society (See Appendix 29.B also). The correlation of child labor and low school enrolments can be seen as a result of the poverty of income and opportunity. The empirical evidence on the issue shows a negative association between labor and schooling of children through the intervening of poverty (Shahnaz and Bukhari, 2001; Chaudhry et al., 2006). The data support that the supply of child labor cannot be analyzed apart from schooling and population characteristics. A reasonable solution to the problem of child labor can be found in the reduction of poverty, improvement in the socioeconomic status of women, accessibility, and quality of schooling.

The contribution of poverty for establishing a link between work and education has been stated as poverty is the greatest single force pushing the flow of children into the workplace. Acute need makes it nearly impossible for households to invest in their children's' education and the price of education can be very high as the so-called "free" public education is in fact very expensive for a poor family. Macroeconomic crises slow down the economic growth and a rise in adult unemployment, leading to transitory or persistent chronic poverty (Majeed, 2007). The social impact of economic crises, thus, deteriorates the human capital of poor people. There, it creates an inverse relationship between child/ adolescent schooling and labor. This extends support to the hypotheses of falling school enrolments to save household schooling expenditures and rising child labor to supplement household earnings. Therefore, it is reasonable to expect that parents would not send their children to work if their own wages were higher or employment prospects were better. A study by Arif (2001) links the rise in poverty in Pakistan during the 1990s to a fall in the ability of households to enroll their children in school. He found a large gap in primary school enrolment, health status/access to health care facilities and housing conditions between the poor and nonpoor as well as between the rural and urban households. He proposed that particular attention should be directed towards reducing the rural–urban and gender disparities in primary school enrolment. The role of poverty is also crucial to lower school attendance and increases drop out. This is due to the perception that schooling is of lesser advantage due to its cost and the need for children to earn. It has been also found, that for Pakistan, there exists a positive association between child labor hours and poverty, and a negative association between child schooling and poverty. Again, various shapes of poverty, such as low household income, low occupational status, and low educational status of the household head play an important role in constraining the demand for children's schooling (PAP, 2007, 2010).

In short, the evidence presented above reflects a great concern about the prevailing situation regarding the link between child/adolescent labor, child/ adolescent schooling. Slow economic growth, widespread adult unemployment, insecure and low wage jobs, and macroeconomic policy reforms all work jointly to lower the economic value of children schooling and raise the economic value of child labor. As the economic function of adults weakens due to unstable macroeconomic conditions, the economic function of children increases. This macro-micro connection brings out the crucial role played by the dual poverty of income and opportunity in determining the choice between child schooling and child labor. A two-fold effect of the poverty of income and opportunity is hidden in the inability of parents to finance investments in child education and in the dire need for any income that child labor can bring to supplement the poor household incomes. The parental choice leading to lower school enrolment is made in the short-run at a point in time. However, it has many long-run implications for future economic growth and poverty. Poverty may be an important cause of this labor-schooling relationship in the short-run. But, it certainly emerges as an even more serious consequence of this labor-schooling association in the long run. There is no dearth of evidence on the rote of developed human resources in promoting and sustaining economic growth. Investments in human capital (schooling and health) of children and adolescents are the only means of ensuring productive and healthy adult working individuals. We cannot be expected to build up our human capital stock by compromising on the promise of economic rewards emerging from child/adolescent schooling in the long run for the benefits of child/adolescent labor in the short-run.

29.6 CONCLUSION AND POLICY DIMENSIONS

29.6.1 Labor Policy Issues

The policy message from the above is that the state has a definite role to play in reversing the existing link between work and education of children. No single policy measure can work to resolve this tedious issue. It is a coordinated effort of several dimensions that will help improve the socio-economic indicators of development in Pakistan. The most important direction for policy is that people have to be protected from the damaging social implications of macroeconomic crises in the country. Furthermore, it is difficult to reverse negative implications upon the human capital of already underprivileged people of a country having a slow economic growth. Poor people are caught in a vicious cycle of poverty that leads to further poverty. It is well documented that educated children have higher lifetime earnings than illiterate ones. Furthermore, their health and safety improve by not entering the labor force market at an early age.

29.6.1.1 Policy Message for Rural Labor in Pakistan

From a policy perspective, if increasing enrolment rate in the country is a fundamental objective, then rural children, especially girls, who are neither attending school nor are going to work, should be the first targets of policy makers. Such children are more than double in numbers in rural compared to urban areas. This stresses, from a policy point of view, that efforts to increase enrolment in rural schools must gain further insights on the issue. Social and cultural factors may be responsible for the conservative attitude of parents toward female education. The fact that most females help in housekeeping also explains the observed pattern. Part of the reason for nonenrolment in schools may be poor quality of schools, distance or travel time to the schools, and other indicators of school price. Mothers' employment has a negative effect on child schooling but a positive effect on female child labor. Moreover, mothers' employment has much stronger effects on female child work in rural areas. Therefore, there is a need to understand the close link between adult-female labor-markets and female-child labor markets, especially in agricultural labor-markets. It is difficult to gauge that reducing the demand for female child labor, through the implementation of the laws prohibiting child labor in the agricultural markets, would enhance or not enrolment in schools. It is because there is an equal demand for female children in home production. It is suggested to make awareness strategies in the community effective and sustainable by involving university teachers and students in the country. The quality of schools should also be improved through occasional revisions of curriculum and hiring of qualified, hardworking, and dedicated teachers. It can play an effective and useful role in decreasing dropout rates, especially of female students. Households with educated parents (especially mothers) are better places to appreciate the need and to understand the benefits of educating their children. Hence, they are more likely to enroll their children in school irrespective of their rural or urban background. Voluntary groups of parents, especially mothers, can be used for a community campaign for advocating the benefits of sending children to school rather for labor.

29.6.1.2 Policy Message for Provinces of Pakistan

On the basis of residence of household, there are different policy measures at province level. The probability of becoming a child laborer is found to be remarkably higher in Punjab than all other provinces. This probability is second highest for Sindh and third highest for NWFP. It can be attributed to a very high concentration of small scale, informal and cottage industries as well as labor-intensive farming practices, which are high in demand for child labor. This confirms the existence of inequalities in incidence of child labor on a regional basis which deserves appropriate policy consideration. It should also be pointed out that the effect of residence of household on the probability of going to school or combining school with work is found to be nonexistent or relatively much weaker in both rural and urban areas. These results confirm the existence of equality in availability and quality of postprimary institutions meant for children in age cohorts of 10–14 years. However, the only instance where this effect has been found to be strong is in urban NWFP, which shows that children from urban NWFP are about 10% less likely to attend school on a full-time basis. This unequal access to postprimary institutions in NWFP should be a cause for concern for policy makers.

29.6.2 Population Planning and Policy

The population planning and policy in Pakistan is one of a shifting policy from one strategy to another; from the direct family planning program to multidimensional population welfare approach; from the only nonclinical contraceptives to a concentration on clinical methods; from the use of Traditional Birth Attendants (TBAs) to the employment of male and female motivators under Continuous Motivation System (CMS); from CMS workers back to TBAs; from centralized to decentralized program; from single sector to multispectral approaches and from purely administrative-to-top-down-to-community-participation approaches. Add the ups and downs of political support, and the difficulties of maintaining consistency in the population planning and policy become apparent.

The Government (2013–2018) has once again announced decentralization of the Population Welfare activities since 2013. This has again disturbed the whole structure and system of population planning and policy, including youth policy, as this department/ministry has also been devolved and handed over to the provinces. The draft policy (Government of Punjab, 2016) has been announced while other provinces have yet to design and announce it. The following suggestions are made by the authors of this chapter to be considered for incorporation in this population policy to be implemented in Punjab.

Empowering and enhancing the status of departments through granting authority for resolving different family issues and conflicts relevant to population welfare and development programs. For instance, by putting issues like nikah, divorce, separation, and other family disputes and activities within this department; ensuring political commitment; involving religious and community leaders; introducing the incentives for couples; incentives for FP workers; improving FWC's infrastructure & facilities; initiating pre-marital counseling; ensuring quality of care; ensuring HRD; having impact assessment and continuity of result-giving programs; ensuring strict implementation of five-year plans; initiating responsible parents, especially mothers associations and implementing adolescents & youth policy.

APPENDIX 29.A: POPULATION PLANNING AND LABOR FORCE IN PAKISTAN (SWOT ANALYSIS)

Strengths

- Young Population
- Energetic
- Active and Productive
- Learning Potential
- Working Potential
- Increasing enrolment in Higher Education

Weaknesses

- Shifting from one to another
- Shaky Political support
- Unrealistic targets and ineffective plans
- Poor implementation
- Political recruitment and transfers
- Misuse of resources
- Poor national economy
- Poor drinking water
- Poor health facilities
- High infant, child, and maternal mortality
- Poor nutrition
- Education—Illiteracy
- Out-of-school children—Female children
- Underemployment and unemployment
- Crimes

- Unplanned—Unbridled city growth
- Safety–Security issues
- Low life expectancy
- Poverty
- Social injustice
- Beggary
- Human smuggling
- Child and women labor

Opportunities

- Entrepreneurship
- Skill development
- Police/Military/Security Agencies
- Recruitment and training to fight terrorism
- Women Participation

Threats

- High population growth rate
- Extremism—Terrorism -Tribalism
- Food safety
- Sectarian—Ethnic conflicts
- Diseases—Incidence/Prevalence and Treatment
- Violation of human rights
- Squeezing land holdings
- Water and sanitation
- Irrigation Water distribution
- Doubling population (Population Bomb)
- Per capita income
- Corruption
- Under development
- Hollow—Shell institutions
- Frustration—Tension—Social conflicts
- Dishonesty—Corruption
- Dissatisfaction—Hopelessness—Suicides

APPENDIX 29.B: POPULATION PLANNING AND LABOR IN PAKISTAN

History

- Population planning shifting quickly from one strategy to another
- Trends of labor force, particularly women and child labor, show that rural, gender, poverty, community and household issues are major determinates

Quick Facts

- The 6th most populous country with highest growth rate
- Health and education indicators remain low in South-Asia
- Socioeconomic indicators for women are the lowest
- The 146th position on UN 2013 Human Development Index
- A quarter of the population below the poverty line

- Poverty and predominantly (80%) is rural
- Rural-agriculture economy makes one fifth of the national economy
- Rural agriculture labor force is highest in number in Punjab
- Poor agricultural and rural transformation and rural poverty are interconnected
- Many of the poorest depend on nonfarm income sources
- Small and landless farmers-especially women make the most-affected segment

UN-World Bank Statistics

- GNI per capita, Atlas method (current US$-2014) 1,410.0
- Population, total (2014) 185,132,926.0
- Rural population (2014) 114,221,461.0
- Number of rural poor (million, approximate-2014) 30,839,794.5 (IFAD, 2015–16)

APPENDIX 29.C: STRATEGIC ACTION PLAN

A. **Agriculture and rural transformation (Long term plan—10 years)**
(i) Focusing on agriculture transformation by focusing on small farmers by providing loans, quality water, quality inputs, and genuine price for ensuring higher production and resultantly higher income to reduce rural poverty. (ii) Promoting rural transformation through poverty reduction by focusing on nonfarm initiatives with the microfinance support to enable the rural poor to invest more in their nonfarm enterprises for higher production and higher profit.

B. **Population planning (Medium term plan—5 years)**
(i) Reinterpretation of religious perspectives on population planning. (ii) Enhancing the status of population department by giving authority for nikkah, and solving divorce, separation, and other family disputes. (iii) Premarital counseling initiative by the population department. (iv) Incentives in kind for couples planning family and for employees with additional increments and cheap health services on priority.

C. **Child & Women Labor Force (Medium term plan—5 years)**
(i) Making primary enrolment compulsory by compensating children and controlling corruption. (ii) Skill development for primary and middle pass and technical training for matriculate pass rural youth
(Short term plan—2 Years)
(i) Law enforcement for girl-children working as domestic workers. (ii) Elimination of discrimination for women agricultural labor force regarding wage, working hours, and differential attitudes at work place. (iii) Making agricultural marketing accessible and safe for women labor force. (iv) Skill development for women labor force and awareness about their rights.

REFERENCES

Afzal, M. 2006. Population and economic development in Pakistan: An econometric investigation. Paper Published in *Population at the Crossroads of Development Seventh Annual Population Research Conference Proceedings*, November 28–30, Peshawar, Pakistan.

Akmal, N. and S. Sajida. 2004. *Women and Livestock Management in Sindh*. Pakistan Agricultural Research Council, Islamabad.

Aliya, K. and L. Shahnaz. 2001. Trade-off patterns between work and education of adolescents: Evidence from PIUS 1998–1999. Paper published in *Pakistan Population Stabilization Prospects 2nd Annual Population Research Conference Proceedings*, October 31 – November 2, 2001, Islamabad.

Amuguni, M. 2001. A Gender Study Focusing on the Turkana and Pokot of" North West Kenya. Prepared for the Community-based Animal Health and Participatory Epidemiology Unit (CAPE) of IHC Program for the Pan-African Control of Epizootics (PACE) of the Organization of African Unity/Inter-African Bureau for Animal Resources (OAU/IBAR).

Anjum, M. I. 2001. Population-development nexus: Pakistan's progress, problems and policy options. Paper Published in *Pakistan's Population Issues in the 21st Century Proceedings*, October 24–26, 2000, Pakistan.

Arif, G. M. 2001. Recent Rise in Poverty and its implications for the poor households in Pakistan. Paper presented at the *16th Annual General Meeting of the Pakistan Society of Development Economists*, January 22–24, 2001, Islamabad.

Arshad. S., S. Muhammad, and I. Ashraf. 2013. Women's participation in livestock farming activities. *The Journal of Animal & Plant Sciences* 23(1): 304–308. ISSN: 1018-7081.

Arshad, S., S. Muhammad, M. A. Randhawa, I. Ashraf, and K. M. Choudhry. 2010. Rural women's involvement in decision-making regarding livestock management. *Pakistan Journal of Agricultural Sciences* 47(2): 1–4.

Asif, A. Z., T. Syed, and S. Irfan. 2001. Population growth, resources scarcity and environment security: The case of Pakistan. Paper published in *Pakistan Population Stabilization Prospects 2nd Annual Population Research Conference Proceedings*, October 31 – November 2, 2001, Islamabad.

Basu, K. 2000. The intriguing relation between adult minimum wage and child labor. *Economic Journal* 110: 50–60.

Batool, T. and A. H. Khan. 2010. Determinants of female labor force participation: A case study of District Khushab. Paper Published in *Population Dynamics and Security: Public Policy Challenges, Tenth Annual Population Research Conference*, March 9–11, 2010, Islamabad, Pakistan.

Burki, A. and L. Shahnaz. 2001. The implications of household factors for children's time use in Pakistan. Paper presented at the *IPAD International Conference on Child Labor in South Asia*, October 15–17, New Delhi.

Chaudhry, I. S. and T. Jabeen. 2007. The determinants of labor force participation of higher educated women in Pakistan: A logit model analysis. Paper Published in *Population and Regional Development Nexus, Eighth Annual Population Research Proceedings*, December 17–18, 2007, Islamabad, Pakistan.

Chaudhry, I. S., S. Malik, and A. Hassan. 2006. The impact of socio-economic and demographic variables on poverty: A village study. Paper Published in *Population at the Crossroads of Development Seventh Annual Population Research Conference Proceedings*, November 28–30, Peshawar, Pakistan.

Dessy, S. E. 2000. A defense of compulsive measures against child labor. *Journal of Development Economics* 62: 261–273.

Devendra, C. and C. Chantalakhana. 2002. Animals, poor people and food security: Opportunities for improved livelihoods through efficient natural resource management. *Outlook on Agriculture* 31(3): 161–175.

Durrani, V. L. 2000. Adolescent Girls and Boys in Pakistan: Opportunities and Constraints in the Transition to Adulthood. Islamabad: Population Council (Research Report No 12).

Faizunnisa, A. and A. Ikram. 2003. What is female labor force participation rate in Pakistan? Paper Published in *Population Research & Policy Development in Pakistan 4lil Conference Proceedings*, December 9–11, 2003, Faisalabad.

Farinde, A. J. and A. O. Ajavi. 2005. Training needs of women farmers in livestock production: Implications for rural development in Oyo state of Nigeria. *Journal of Social Science* 10(3): 159–164.

Flolmaan, F., L. Rivas, N. Urbina, B. Rivera, L. A. Giraldo, S. Guzman, M. Martinez, A. Medina, and G. Ramirez. 2005. The role of livestock in poverty alleviation: An analysis of Colombia. *Livestock Research for Rural Development*, 17(1).

Government of Pakistan. 1955. Planning Commission/Planning and Development Division. First Five-Year Plan (1955-60). Islamabad.

Government of Pakistan. 1960. Planning Commission/Planning and Development Division. Second Five-Year Plan (1960-65). Islamabad.

Government of Pakistan. 1965. Planning Commission/Planning and Development Division. Third Five-Year Plan (1965-70). Islamabad.

Government of Pakistan. 1970. Planning Commission/Planning and Development Division. Fourth Five-Year Plan (1970-75). Islamabad.

Government of Pakistan. 1978. Population Division. Fifth Five Year Plan (1978–1983). Ministry of Planning and Development, Islamabad.

Government of Pakistan. 1983. Planning Commission. The Sixth Five Year Plan (1983–1988). Islamabad.

Government of Pakistan. 1988. Planning Commission. The Seventh Five Year Plan (1988–1993). Islamabad.

Government of Pakistan. 1993. Planning Commission. The Eight Five Year Plan (1993–1998). Islamabad.

Government of Pakistan. 1998. Planning Commission. The Ninth Five Year Plan (1998-2003). Islamabad.

Government of Pakistan. Population and Housing Census of Pakistan. 1998. Population Census Organization, Statistics Division, Islamabad.

Government of Pakistan. 2003. Planning Commission. The Tenth Five Year Plan (2003–2008). Islamabad.

Government of Pakistan. 2008. Planning Commission. The Eleventh Five Year Plan (2008–2013). Islamabad.

Government of Pakistan. 2009. *Labor Force Survey of Pakistan*. Federal Bureau of Statistics, Islamabad.

Government of Pakistan. 2014. *Economic Survey of Pakistan*. Finance Division, Economic Advisors Wing Islamabad.

Government of Punjab. 2013. Ministry of Population Welfare. The Twelfth Five Year Plan (2013–2018). Lahore.

Government of Punjab. 2016. *Draft of Population Policy*. Ministry of Population & Welfare 2016. Lahore.

Hakim, A. 2000. Population change and development prospects: Demographic issues in Pakistan. Paper Published in *Pakistan's Population Issues in the 21M Century Proceedings*, October 24–26, 2000, Pakistan.

Hakim, A. 2003. Pakistan's population 1998–2028: Where are we pleading? Paper Published in *Population Research & Policy Development in Pakistan 4th Conference Proceedings*, December 9–11, 2003, Faisalabad.

Haque, M. and M. Sultana. 2010. Lapping potential in youth: Policy challenges and prospects. Paper Published in *Population Dynamics and Security: Public Policy Challenges. Tenth Annual Population Research Conference*, March 9–11, 2010, Islamabad, Pakistan.

Hashmi, M. S. 2009. Women in agricultural development. *The Dawn*, May 19, 2009. Online Available: http://dawn.com

Javed, A., S. Sadaf, and M. Luqman. 2006. Rural women's participation in crop and livestock production activities in Faisalabad – Pakistan. *Journal of Agriculture and Social Sciences* 2(3): 150–154.

Khalid, U. 2006. Educational performance and labor force participation of teenagers in Pakistan. Paper Published in *Population at the Crossroads of Development Seventh Annual Population Research Conference Proceedings*, November 28–30, Peshawar, Pakistan.

Khan, T. M., A. A. Maan, M. I. Zafar, I. Ashraf, and M. A. Khan. 2010. Awareness and practices of women's rights: An initial step towards women's empowerment. Paper Published in *Tenth Annual Population Research Conference*, March 9–11, 2010, Islamabad, Pakistan.

Khushk, M. A. and S. Hisbani. 2004. *Rural Women at Work*. The Daily Dawn, Islamabad, Pakistan.

Lovenbalk, J., D. Hjarne, A. A. Taoutaou, O. Mertz, M. Dirir, P. M. Dyg, K. M. Lassen, and M. Sehested. 2003. Opportunities and Constraints for Agricultural Intensification in Communities Adjacent to the Crocker Range National Park Sabah, Malaysia, P: 6. ASEAN Review of Biodiversity and Environmental Conservation (ARBEC).

Maan, A. A. and I. A. Khan. 2015. *Socio-Cultural Factors Affecting Women's Participation in Livestock Management in Punjab Pakistan*. HEC, Islamabad.

Majeed, M. T. 2007. Trade, poverty and employment: Empirical evidence from Pakistan. Paper Published in *Population and Regional Development Nexus. Eighth Annual Population Research Proceedings*, December 17–18, 2007, Islamabad, Pakistan.

Miller, A. B. 2001. Empowering women to achieve food security: Rights to Livestock. IFPRI Policy brief 4 of 12.

Munawar, M. U. Safdar, M. Luqman, T. M. Butt, M. Z. Y. Hassan, and M. F. Khalid. 2013. Factors inhibiting the participation of rural women in livestock production activities..!. *Agricultural Research*, 51(2): available at: www.jar.com.pk

Nadeem, N., M. I. Javed, I. Hassan, W. Khurshid, and A. Ali. 2012. Total factor productivity growth and performance of livestock sector in Punjab, Pakistan. *Journal of Agricultural Research* 50(2): 279–287.

Niamir-Fuller, M. 2008. *Women Livestock Manager in the Third World: Focus on Technical Issues Related to Gender Roles in Livestock Production*. IFAD, Rome.

Nielsen, L. and C. Heffernan. 2002. *Motivation and Livestock-based Livelihoods: An Assessment of the Determinants of Motivation among Restocked Households in Kenya*. FAO, Rome. www.fao.org/docs/eims/upload/agrotecIV1928/r7402_bsas_111.pdf

Nirmala, G., D. B. V. Ramana, and B. Venkateswarlu. 2012. Women and scientific livestock management: Improving capabilities through participatory action research in semi-arid areas of south India.

PARC. 2004. Pakistan Agriculture Research Council, Annual Report; Finance Division.

PDHS. 2014. *Pakistan Developing Health Survey*. National Institute of Population Studies, Islamabad.

Population and Housing Census of Pakistan. 2017. *Population Census Organization*. Statistics Division, Govt. of Pakistan, Islamabad.

Population Association of Pakistan. 2006. Population at the crossroads of development. *Seventh Annual Population Research Conference Proceedings*, November 28–30, 2006, Peshawar, Pakistan.

Population Association of Pakistan. 2007. Population and regional development nexus. *Eight Annual Population Research Proceedings*, December 17–18, 2007, Islamabad, Pakistan.

Population Association of Pakistan. 2008. Population dynamics and security: Public policy challenges. *Ninth Annual Population Research Conference*, December 2–4, 2008, Lahore, Pakistan.

Population Association of Pakistan. 2010. Population, peace and development. *Tenth Annual Population Research Conference*, March 9–11, 2010, Islamabad, Pakistan.

Ranjan, R. 2001. Child labor, child schooling and their interaction with adult labor: Empirical evidence for Peru and Pakistan. *World Bank Economic Review* 14(2): 347–367.

Rasheed, T. 2004. Women's Participation in Livestock Care and Management Activities in Rural Sialkot. *MSc thesis*, Department of Rural Sociology, University of Agriculture Faisalabad.

Ravallion, M. and Q. Wondon. 2000. Does child labor displace schooling? Evidence on behavioral responses to an enrolment subsidy. *The Economic Journal* 110: 158–174.

Ray, R. 2000. Child labor, child schooling and their interaction with adult labor empirical evidence for Peru and Pakistan. *World Bank Economic Review* 14(2): 347–367.

Rehman, N. 2006. Impact of population policy on population indicators: 1990–2006. Paper Published in *Population at the Crossroads of Development Seventh Annual Population Research Conference Proceedings*, November 28–30, Peshawar, Pakistan.

Rukanuddin, R. 2001. Uptake of family planning in Pakistan: Trends and emerging issues. Paper Published in *Pakistan's Population Issues in the 21st Century Conference Proceedings*, October 24–26, 2001, Pakistan.

Sabir, M. and Z. Aftab. 2007. Capitalizing the demographic dividend. Paper Published in *Population and Regional Development Nexus, Eighth Annual Population Research Proceedings*, December 17–18, 2007, Islamabad, Pakistan.

Shahnaz, L. and A. A. Bukhari. 2001. An empirical analysis of child labor and child schooling in rural areas of Pakistan. Paper published in *Pakistan Population Stabilization Prospects 2nd Annual Population Research Conference Proceedings*, October 31–November 2, 2001, Islamabad.

Tabassum, N. 2017. *Women in Pakistan Status in Socio-Cultural and Politico-Legal Domains*. HEC, Islamabad.

Taj, S., A. Bashir, R. Shahid, and H. Shah. 2012. Livestock development through micro credit: A hope for poor resource women in rural areas of Faisalabad, Punjab. *Journal of Agricultural Research* 50(1): 135–143.

30 Rural Poverty

Saria Akhtar, Muhammad Iqbal Zafar,
Shabbir Ahmad, and Naima Nawaz

CONTENTS

30.1 INTRODUCTION

This chapter on rural poverty tries to explain what is poverty? How it affects the different dimensions of the lives of the rural people? What happens to the nations when poverty keeps on thriving for a long time? What are crucial factors which are the real cause of rural poverty? What are the major contributions of the Pakistani government to eradicate poverty from its masses? What can people do for themselves to get out of the vicious mire of poverty?

The eradication of poverty has long been recognized as one of the challenges for human society as more human beings have been suffering from this chronic deprivation today than ever before in history. During the last century, a locus of poverty and hunger concentrated in third world countries, particularly in rural areas. According to the conflict theorists' poverty thrives in a nation when exploitation becomes the order of the day. In rural areas of Pakistan such as Punjab and Sindh, feudal lords exploit the labor of peasants and tenants causing the poor to remain poor. In urban areas, according to the conflict theory, the same is done by industrialists to promote poverty among people. It is high time now that we define poverty. For this purpose, the suitable definitions of poverty are given below:

- A state or condition in which you find yourself unable to afford the basic necessities of life is called poverty.
- When the basic needs such as food, clothing, and shelter are not property met despite your best efforts due to unemployment, natural disasters, and other reasons is called poverty.

30.2 TYPES OF POVERTY AND DIFFERENT CONCEPT OF RURAL POVERTY

30.2.1 Absolute Poverty

Absolute poverty is the life-threatening sort of poverty where you are unable to provide for yourself basic provisions of life such as wholesome food, clean and pure water, and shelter and housing facilities. The absolute kind of poverty is found in the underdeveloped countries of the world such as Ghana, Somalia, and most of the African countries. Absolute poverty is highly prevalent in the developing and underdeveloped nations of the world. There is lawlessness and civil war-like conditions in these countries. Absolute poverty invites disease, unemployment, hunger, lack of education, and absence of infrastructure and facilities for telecommunication and transportation.

FIGURE 30.1 It is a graphic picture of absolute poverty. The shacks, unpaved streets, no drainage facility, children without clothes, and poor people can be seen.

Figure 30.1 shows a grim picture of a locality somewhere in Pakistan. It is a graphic picture of absolute poverty. It shows shacks, unpaved streets, no drainage facility, children without clothes, and jobless people.

30.2.2 Relative Poverty

Relative poverty is actually an indirect type of poverty in which you are poor but in relation to the other people living around you. It can be explained with the help of the following: a family can be considered poor if it cannot afford vacations, or cannot send its children to the university. It is self-assumed poverty and the people adopt unfair means in order to come out of this poverty. It is not poverty but a sense of inferiority complex in the people that why do they have not have the modern means enjoyed by other people? It is a mad race in which people run like horses in order to get what other people have. It is the most dangerous type of poverty, which destroys the nations due to mass corruption in the departments of governance.

30.2.3 Situational Poverty (Transitory)

Suppose you are a seasonal worker who is doing well while your employment lasts and when the season is over you find yourself unemployment. This is the case for cotton pickers, wheat harvesters, laborers working in sugar mills, private teachers (without job in the months of June, July, and August), and worker in the kilns. The situational poverty is also called transitory poverty. In this type of poverty, the people or families are poor because of some misfortunes like earthquakes, floods, or a serious illness.

30.2.4 Generational or Chronic Poverty

Generational poverty is a capricious gift which is bestowed by one generation to the other without any hope to come out of the vicious circle of poverty. It is the result of ignorance, lethargy, fatalism, and traditional means of cultivation, and poor heath of the people. It is also called chronic because you do not have any hope of coming out this trap in spite of your best efforts.

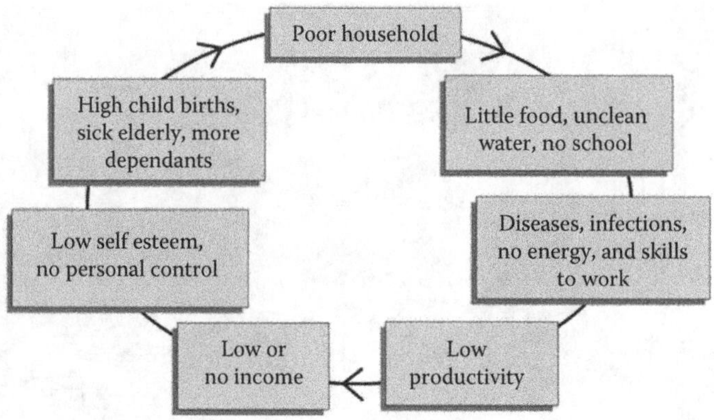

FIGURE 30.2 The vicious cycle of poverty.

30.3 THE VICIOUS CYCLE OF POVERTY

This is a marvel utilized frequently by financial researchers. It just means destitution sires neediness. It is an idea that outlines how neediness causes destitution and traps individuals in neediness unless an outer intercession is connected to break the cycle.

Let's take a glance at this situation with a family in absolute poverty (Figure 30.2). An exceptionally poor family with children has almost nothing to eat and has no access to social welfare offices where they can seek help, if any. Subsequently, the children are malnourished, and they have no access to social welfare opportunities. They are in this way not able to go to class (regardless of the possibility that there is a school in the following town). They grow up with no education or any skill development and cannot contribute to income-earning activities. Their guardian dies from preventable infections as a consequence of no access to health facilities, and they are entangled in the trap of never-ending poverty divulging into fatalistic schizophrenia. As the kids grow up, they find spouses who are just on the same level of neediness as them, and together have children who are just as poor. They hand over this condition to their kids, who will likewise experience childhood under the same conditions.

It takes intervention from governments, social welfare associations, and relatives to help to get the young people to do some sort of income-earning activities to acquire some wages. Without that, this cycle will proceed for ages and is hard to escape.

This idea can likewise be connected to nations and bigger economies, in spite of the fact that the elements might be marginally distinctive.

30.4 TECHNIQUES UTILIZED AT DIMINISHING POVERTY

Neediness can't be totally killed, as it is, to a great extent, created by human components. Over the previous years there has been a considerable measure of Poverty Alleviation Programs intended to soften the cycle of destitution of numerous families and groups on the planet. The outcome is astounding, however, there is still a great deal to be accomplished.

Poverty reduction includes the vital utilization of infrastructures, for example, instruction, financial advancement, social welfare, and wage redistribution to enhance the employment of the poor masses by governments and globally recognized associations. They likewise go for relinquishing social and legitimate boundaries to promote developmental attitudes among poor people.

30.4.1 Training

Quality training engages individuals to exploit opportunities around them. It helps the youth to get quality learning, essential information, and particular aptitudes to understand their hidden potential.

Preparing and developing instructional materials, building schools and training institutions, and providing training are essential components of poverty eradication programs. The technical and vocational institutions in Pakistan are working to provide training to the youth so that they can indulge themselves in the entrepreneurship activities.

30.4.2 Wellbeing, Sustenance, and Water

Numerous projects designed by the government provide children facilities at school and give them comfort. This urges parents to send their children to schools and keep them there. If the children have food to eat, clean water to drink, money to pay for their education, obtain opportunities for recreation, and part-time job opportunities—they perform well in their respective domains.

30.4.3 Procurement of Aptitudes and Training

The young and ready to-work in the groups are given aptitudes to homestead work or other financial action, which helps them win cash to bring home the bacon and deal with their families.

30.4.4 Salary Redistribution

It is critical that the legislature augments its advancement projects, for example, streets, spans, and other financial offices to rustic regions, to make it simple for merchandise and administrations and homestead produce to move to and from the cultivating groups.

With a touch of exertion in the territories said above, it won't take long to see genuine changes in the living states of the group.

30.4.5 Resource Distribution and Poverty Alleviation

Assets are financial or profitable component required to finish a movement, or as intends to attempt an undertaking and accomplish craved result. Three most fundamental assets are area, work, capital; different assets incorporate vitality, enterprise, data, aptitude, administration, and time.

Pakistan is a nation of differences, with expanded help having superb high mountain ranges snow-secured tops, endless ice sheets, and the between mountain valleys in the north. Flooded fields in the Indus bowl stand out from stark betrays and tough rough levels in southwest Baluchistan. The nation is dry and semi-parched with generous variety in temperature relying on the geology and portrayed by mainland kind of atmosphere. Throughout the years since autonomy the regular assets of the nation (land and water) have been bridled which thus made it conceivable to bolster the developing populace which more than quadrupled amid the previous sixty years.

30.5 DISTRIBUTION OF POPULATION BY GENDER AND AREA 2014–2015

Over population and excessive abuse of the resources by the public has not been completely figured out by the analysts and the policymakers (Table 30.1). Given the way that sound living communities

TABLE 30.1
Population Distribution in Pakistan

	Total			Rural			Urban		
Population	Total	Male	Female	Total	Male	Female	Total	Male	Female
Total	100.00	51.00	49.00	65.21	33.13	32.08	34.79	17.87	16.93

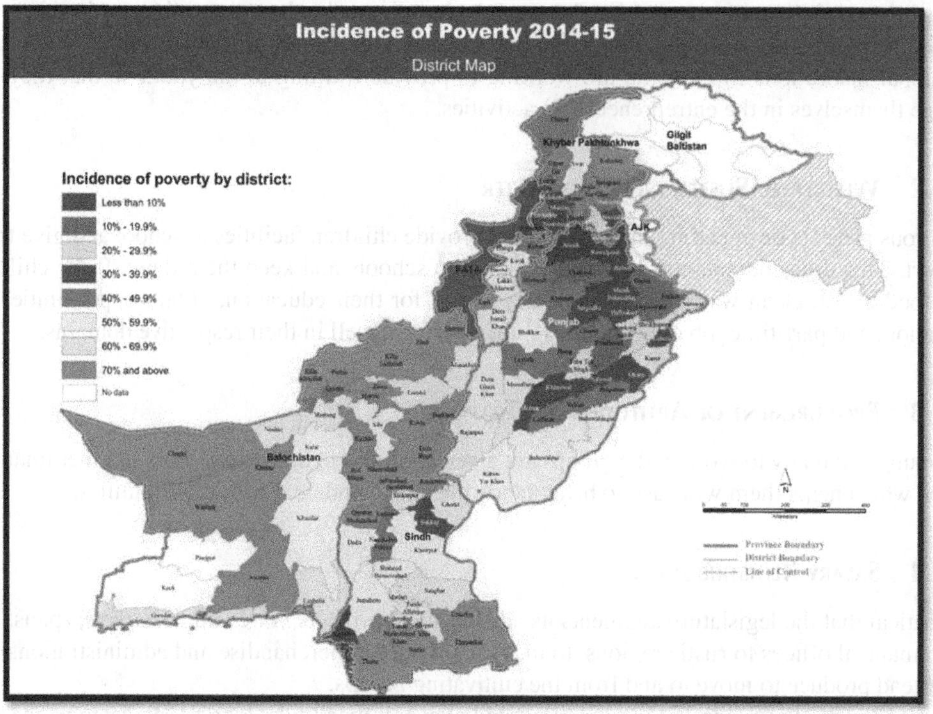

FIGURE 30.3 **(See color insert.)** District wise population in Pakistan.

deliver the prerequisites forever highlight the pivotal linkages between the general public and eco-frameworks. The unpredictable connections between administration of distinctive sources and survival system of poor are not completely inspected and explored in Pakistan. This is regardless of the way that poor depend more on common resources than the rich. Lamentably few, assuming any, exploration tried in Pakistan has led to unwind the nexus between poverty and typical resources.

Here is district wise poverty map of Pakistan (Figure 30.3).

Poverty reduction, and resources administration have not been analyzed in a related system of the environment. Improvement and public development should be emphasized in a proper fashion. It should be highlighted so that the general public may gain confidence from the ongoing policies of the government.

Twenty percent of the world's population consumes more than 80% of the world's resources. Commentators of globalization help us to remember this bad form of poverty which has prevailed in our country in the shape of poverty and prolonged hunger in the poor people without discrimination. The principle purpose behind that 20% consuming 80% of the resources is that they create 80% of resources. The 80% expend just 20% since they deliver just 20% of resources. The issue is that numerous individuals are poor, not that specific individuals are rich.

There is no unambiguous answer when requesting the components that cause poverty. Poverty stays a standout amongst the most complex issues on the planet with not a single clear determination to be seen. The disparity crevice between the rich and poor keeps on developing, while very nearly three billion individuals live on under $2.50 a day. Yet, why does poverty exist? That is by all accounts the million-dollar address that has not been satisfactorily replied over the previous decades. In spite of the fact that there is no straightforward response to this inquiry, there are numerous variables that add to this difficulty. Unequal conveyance of resources and at last riches is one of the main considerations.

It is a known fact across the world that the rich are getting richer and the poor are getting poorer. There are rising variations among the distribution of resources in creating nations that sustain the cycle of poverty. In many areas, the rich get a bigger pay, while the poor get a shockingly small part. On the off chance that riches are not redistributed, or wages are not made to be more equivalent, poverty will dependably remain an enormous issue. Poverty, brought about by the unequal dispersion of resources wins inside a general public or between social orders.

Albeit frequently ignored, natural debasement can hugely affect poverty rates and the prosperity of individuals. If resources are drained because of environmental change, regular debacles and deforestation, then natives will probably be living in poverty. Ecological issues have prompted deficiencies of water, sustenance and materials for lodging, and also other key assets. Without these things, the poor would not just stay poor, additionally build their odds of unexpected passing. Obviously, poverty is an issue that requirements to hold our aggregate consideration and goad our aggregate activity.

Unequal distribution of resources on the planet is due to the presence of unending desire in some parts of the world. Particularly among the younger population, desire can be fatal or have grave outcomes. Lately, the value surge is another reason for unequal dissemination of resources on the earth. It is unlikely that individuals in rich nations will go hungry during a food emergency. Here, individuals spend a small measure of their pay on sustenance. While in developing and underdeveloped countries, people spend up to 80% of their pay on sustenance. At times, with soaring prices, the under privileged would not have the capacity to purchase.

Results on estimation of poverty show that poverty levels are higher in provincial territories than in urban areas. Given that the provincial populace represents 66% of the aggregate, a dominant part of the poor live in rural Pakistan.

Four regions, for the most part are characterized on the premise of ethnicity, which additionally matches with regular resource gifts, frequently order the nation. In this manner, Punjab and Sindh regions have rich agrarian resource base and additionally both are more fashioned to provide agrarian products to the inhabitants of the country in abundance as compared with the rest of the provinces and the federally administrated areas.

30.5.1 Land

Land changes, particularly resources creation for landless workers, are resources for poverty alleviation. There was a solid relationship amongst poverty and area possessions, in that destitution increases as ownership diminishes. Some assessments recommend that around half of every single rural family unit claim no area while the big landholders own more than 33% of all developed area. This may clarify why around 40% of Pakistanis live in poverty.

The bolster costs profited extensive landowners while sharecroppers had neither the motivating force nor the ability to build their efficiency. They pushed on the requirement for rural arrangements to concentrate on small farmers. 57 percent of poor farmers who owed cash to landowners developed area for nothing while 14 percent did likewise for just Rs. 28 every day. Pakistani state had 2.6 million sections of land of cultivable area which if partitioned into five-section of land parcels could make resources for 58 percent for each of the nation's 897,000 sharecroppers.

The remaining 42% could be offered credits to buy their own asset that when assessed would be worth around Rs. 4 billion. Then again, arrive changes that are vital for social and business sectors.

This indicates that landlessness among the rural population is the prime reason for poverty in Pakistan. A high convergence of landownership and out of line tenure contracts are real deterrents to rural development and reducing poverty. Along these lines both rural development and easing out poverty can be accomplished, if land disparity is diminished and the occupants are secured by implemented tenure contracts. Land redistribution would bring about expanded productivity, more interest to work, and lessened poverty. Landless poor are subjected to non-horticultural sources of salary. In rural economy, subsistence is primarily occasional and is based on low wages, leaving

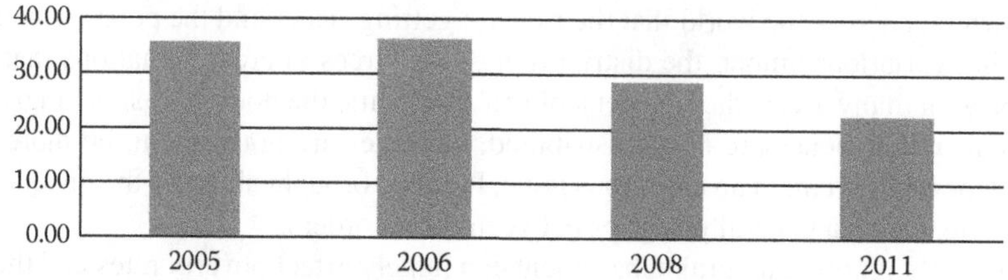

FIGURE 30.4 Number of individuals (in millions) living below $1.25/day in Pakistan.

many landless families in poverty. In this setting, occupation programs for provincial open works can be critical in decreasing rural poverty. It is, along these lines, it was proposed to start country wide open works projects and scale up the current projects (Figure 30.4). By and large, the number of individuals living beneath $1.25 is decreasing over the time. In any case, the purpose for this decrease is not due to less number of individuals in poverty but rather due to increment in the ostensible wages at work.

Better training and social services will be valuable to decrease poverty. Increment in training lessens the rate of poverty, poverty crevice and poverty seriousness. Access to enhance water sources and sanitation helps people and family units to break the endless loop of poverty. Since training and well-being are the principle segments of human capital, any change in instruction and well-being area would prompt create human capital. Arrangement of human capital in the economy may have its immediate effect on the profitability of the people. Increment in the normal went to years of tutoring expands the extent of the work power. In addition, the effectiveness per unit of proficient laborers increments. This animates the development of the economy. More accessible instruction offices empower the work power to embrace new cutting-edge methods of creation. Expanded profitability of work power empowers the economy to create more proficient and beneficial items. In this way build training expands the financial development of the economy which at last prompts lessen poverty. Women are the most powerless amongst poor people. Increment in human capital through better well-being training administrations for women can be useful for upgrading their profitability. Female procuring increments with the expansion in instruction and well-being administrations for them.

There is a critical need in Pakistan economy that measures ought to be taken to expand the mobilization rate keeping in mind the end goal to ease poverty. There is a lack of schools and well-being focuses in remote provincial territory. There is a critical need of expanding the training advancement use as a rate of GDP according to sanctions by the United Nations. The formation of non-ranch openings for work is attractive by extending the formal and specialized training in the new rural zones. Mobilization of the provincial youth is guaranteed by ensuring that schools record undertakings of individuals to serve the country. Procurement of better medical facilities and good administration would lessen poverty.

Although diminishing rural poverty has been the key motivation for financial changes in Pakistan, rural poverty is still on the rise. The reasons for rural poverty are intricate and multidimensional. The rural poor entirely differ both in the issues they confront and the conceivable answers for these issues. The local poor are not homogeneous as far as their characteristics or financial situations or their sources of salary are concerned.

30.6 MAIN FACTORS CONTRIBUTING TOWARDS POVERTY

Under unfavorable circumstances, rural profitability development could deliver unreasonable results for poor people. It is proposed that different components may compel the potential for efficiency increments to decline poverty (Figure 30.5).

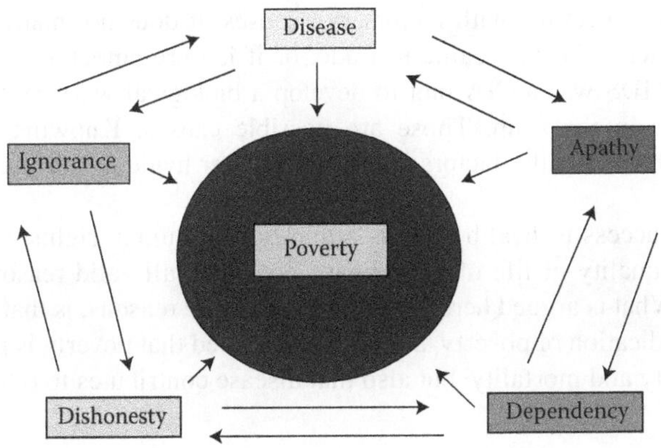

FIGURE 30.5 Factors of poverty.

The explanation of these factors is given underneath.

30.6.1 Ignorance

Ignorance means having a lack of information, or lack of knowledge. It is different from stupidity which is lack of intelligence, and different from foolishness which is lack of wisdom. The three are often mixed up and assumed to be the same by some people.

"Knowledge is power," goes the old saying. Unfortunately, some people, knowing this, try to keep knowledge to themselves (as a strategy of obtaining an unfair advantage), and hinder others from obtaining it. Do not expect that if you train someone in a skill, or provide some information, that the information or skill will naturally trickle or leak into the rest of a community.

It is important to determine what the information is that is missing. Many planners and good-minded persons who want to help a community become stronger think that the solution is education. But education means many things. Some information is not important to the situation. It will not help a farmer to know that Romeo and Juliet both died in Shakespeare's play, but it would be more useful to know which kind of seed would survive in the local soil, and which would not.

The training in this series of community empowerment documents includes (among other things) the transfer of information. Unlike a general education, which has its own history of causes for the selection of what is included, the information included here is aimed at strengthening capacity, not for general enlightenment.

30.6.2 Disease

When a community has a high disease rate, absenteeism is high, productivity is low, and less wealth is created. Apart from the misery, discomfort, and death that result from disease, it is also a major factor in poverty in a community. Being well (well-being) not only helps the individuals who are healthy, it contributes to the eradication of poverty in the community.

Here, as elsewhere, prevention is better than cure. It is one of the basic tenets of PHC (primary health care). The economy is much healthier if the population is always healthy; more so than if people get sick and must be treated. Health contributes to the eradication of poverty more in terms of access to safe and clean drinking water, separation of sanitation from the water supply, knowledge of hygiene and disease prevention—much more than clinics, doctors and drugs, which are costly curative solutions rather than prevention against disease.

Remember, we are concerned with factors, not causes. It does not matter if tuberculosis was introduced by foreigners who first came to trade, or if it were autochthonic. It does not matter if HIV that carries AIDS was a CIA plot to develop a biological warfare weapon, or if it came from green monkeys in the soup. Those are possible causes. Knowing the causes will not remove the disease. Knowing the factors can lead to better hygiene and preventive behavior, for eradication.

Many people see access to health care as a question of human rights, the reduction of pain and misery, and the quality of life of the people. These are all valid reasons to contribute to a healthy population. What is argued here, additionally to those reasons, is that a healthy population contributes to the eradication of poverty, and it is also argued that poverty is not only measured by high rates of morbidity and mortality, but also that disease contributes to other forms and aspects of poverty.

30.6.3 Apathy

Apathy is when people do not care, or when they feel so powerless that they do not try to change things to improve conditions.

Sometimes, some people feel so unable to achieve something, they are jealous of their family relatives or fellow members of their community who attempt to do so. They seek then to bring these people down to their level of poverty. Apathy breeds apathy.

Sometimes apathy is justified by religious precepts, "Accept what exists because God has decided your fate." That fatalism may be misused as an excuse. It is OK to believe God decides our fate, if we accept that God may decide that we should be motivated to improve ourselves. "Pray to God, but also row to shore," a Russian proverb, demonstrates that we are in God's hands, but we also have a responsibility to help ourselves.

We were created with many abilities: to choose, to cooperate, to organize in improving the quality of our lives; we should not let God or Allah be used as an excuse to do nothing. That is as bad as a curse upon God. We must praise God and use our God-given talents.

In the fight against poverty, the mobilizer uses encouragement and praise, so that people (i) will want to and (ii) learn how to—take charge of their own lives.

30.6.4 Dishonesty

When resources that are intended to be used for community services or facilities, are diverted into the private pockets of someone in a position of power, there is more than morality at stake here. In this training series, we are not making a value judgment that it is good or bad. We are pointing out, however, that it is a major cause of poverty. Dishonesty among persons of trust and power. The amount stolen from the public, that is received and enjoyed by the individual, is far less than the decrease in wealth which results for the public.

The amount of money that is extorted or embezzled is not the amount of lowering of wealth to the community. Economists tell of the "multiplier effect." Where new wealth is invested, the positive effect on the economy is more than the amount created. When investment money is taken out of circulation, the amount of wealth by which the community is deprived is greater than the amount gained by the embezzler. When a Government official takes a 100-dollar bribe, social investment is decreased by as much as a 400 dollar decrease in the wealth of the society.

It is ironic that we get very upset when a petty thief steals ten dollars' worth of something in the market, yet an official may steal a thousand dollars from the public purse, which does four thousand dollars' worth of damage to the society as a whole, yet we do not punish the second thief. We respect the second thief for her or his apparent wealth, and praise that person for helping all her or his relatives and neighbors. In contrast, we need the police to protect the first thief from being beaten by people on the street.

The second thief is a major cause of poverty, while the first thief may very well be a victim of poverty that is caused by the second. Our attitude, as described in the above paragraph, is more than ironic; it is a factor that perpetuates poverty. If we reward the one who causes the major damage, and punish only the ones who are really victims, then our misplaced attitudes also contribute to poverty. When embezzled money is then taken out of the country and put in a foreign (e.g., Swiss) bank, then it does not contribute anything to the national economy; it only helps the country of the offshore or foreign bank.

30.6.5 Dependency

Dependency results from being on the receiving end of charity. In the short run, as after a disaster, that charity may be essential for survival. In the long run, that charity can contribute to the possible demise of the recipient, and certainly to ongoing poverty.

It is an attitude, a belief, that one is so poor, so helpless that one cannot help one's self, that a group cannot help itself, and that it must depend on assistance from outside. The attitude, and shared belief is the biggest self-justifying factor in perpetuating the condition where the self or group must depend on outside help. There are several other documents on this web site, which refer to dependency. When showing how to use the telling of stories to communicate essential principles of development, the story of Mohammed and the Rope is used as a key illustration of the principle that assistance should not be the kind of charity that weakens by encouraging dependency, it should empower. The community empowerment methodology is an alternative to giving charity (which weakens), but aids, capital and training aimed at low income communities identifying their own resources and taking control of their own development—becoming empowered. All too often, when a project is aimed at promoting self-reliance, the recipients, until their awareness is raised, expect, assume and hope that the project is coming just to provide resources for installing a facility or service in the community. Among the five major factors of poverty, the dependency syndrome is the one closest to the concerns of the community mobilizer.

30.6.6 Population Growth

Population development can likewise influence the net effect of expanded vocation opportunities and profitability picks up on destitution decrease. In their multisectoral development model, Irz and Roe (2000) exhibit that a base rate of profitability development is important to counter populace development and stay away from the "Malthusian trap," whereby populace development outpaces per capita financial development.

30.6.7 Technology

When chances to build, profitability depend on enhanced advances or mechanical development, a few elements may confine the advantages to poor people. The poor face numerous requirements that breaking point innovation appropriation in this manner, while there are numerous potential advantages of enhanced innovation for poor people, they don't generally appear (Irz et al. 2001; Thirtle et al. 2001). Mechanical change can likewise affect the dissemination of salary, particularly where reception is uneven (crosswise over family units or districts). Innovation alone will be deficient to lessen destitution without framework and training (Thirtle et al. 2001). Innovation will probably produce positive advantages for the poor where starting resource and salary disparity is lower and related foundation and social administrations are created all around (Thirtle et al. 2001).

30.6.8 Asset and Income Distribution

Poverty lessening relies on upon the generation and utilization of multipliers coming about because of expanded horticultural efficiency. In any case, where wages, resources, and area appropriation are unequally distributed, most of the advantages will be in favor of the higher middle class and the additional resources created will move towards foreign manufactured goods for the best interest of the community, as opposed to privately delivered, work intensified products and administrations (Thirtle et al. 2001). Where imbalance between the top and base salary quintiles is more prominent, the pay impact of agricultural development is more grounded for the most astounding quintile than the least. Where starting salary disparity is small, agrarian development adds to a change in pay dispersion whereby the versatility of poverty to agricultural development decreases progressively with each higher wage quintile (Mellor 1999). Also, lacking access to arrive compels the potential for poverty lessening through smallholder driven farming advancement.

30.6.9 Market Access

While base is regularly expected to empower market access and subsequently profitability, small-scale level confirmation gives clashing discoveries to the bearing of causality. Utilizing joined World Bank Living Standards Measurement Survey information from Tanzania, Guatemala, and Vietnam, Rios et al. (2008) demonstrate that family units with higher profitability tend to take part more in business sectors. There is no proof of an opposite causal linkage whereby market access would prompt higher efficiency. They reason that interests in business sector access base give just insignificant, assuming any, upgrades in agricultural efficiency. Improvements in farm structure and capital, then again, can possibly build profitability and business sector investment (Rios et al. 2008). Then again, Rao et al. (2004) contend that since a considerable number of the determinants of agricultural profitability are generally altered, for example, the amount of area and work, separation to center markets, and atmosphere, proceeded with advancement in efficiency originates from expansions area and work and through diminishing transport costs by means of upgrades in base, for example, streets and ports (Rao et al. 2004).

Right now, 1.22 billion individuals are living in acute poverty. Since the World Bank's foundation in 1944, their main goal has been to make "a world free of neediness" (World Bank 2015). Be that as it may, this suggests a conversation starter, in what manner can nations ease poverty? A general comprehension of the components that have gone into reducing poverty in the past is a great way to begin.

30.7 SOCIAL INJUSTICE AND CONTROL SYSTEM

Social injustice' is an idea identifying with the apparent ignominy or foul play of a community in its divisions of rewards and punishments. The idea is from those of equity in law, which could possibly be viewed as good as anything else.

The United Nations exposure on human rights says, "Each man is a joint inheritor of all the normal assets and of the forces," developments and potential outcomes gathered by our predecessors. He is entitled, inside the measure of these resources and without refinement of race, color, caste and creed, convictions or conclusions, to the food, shelter and medicinal consideration expected to understand his full conceivable outcomes of physical and mental advancement from birth to death. Despite the different and unequal characteristics of people, all men might be considered totally equal according to the law, similarly imperative in social life and similarly qualified for the admiration of their kindred men.

Inhabitants of Pakistan are unconscious of the social equity idea and its consequences for society. The brutal reality is that the poor are getting poorer and the wealthier are getting richer. The elite and exclusive class of Pakistan has never advanced the idea of social democracy that can provide incorporate benefit to the public. Whereas all the efforts tried to motivate the elite and distinguished class to rise depriving the general public without solid class framework, approaches and projects to diminish unemployment, welfare society, appreciation of work regardless of their level of instruction, admiration of individuals independent of their family foundation, society of trustworthiness, an unequivocally dynamic duty framework, liberal government procurement of non-money advantages, for example, training, well-being and lodging, legit and autonomous judiciary, devoted society where premiums of collectivity are preeminent, guideline introduction, constitution is viewed as a hallowed report, religious resilience, genuine common and military organization, which act inside their spaces, military finds no room in legislative issues, elevated expectation, public segment instruction, provincial joining, acknowledgment of Pushto, Baluchi, Panjabi and Sindhi as official dialects alongside Urdu.

The patterns toward inequality in the public arena are not inescapable and can be altered by sound and glaring strategies considering a hearty variant of the idea of "equivalent worth." Measures which enhance the financial status of minimal well off, expansion job, decrease disparity, and "acculturate" the work environment are prone to create critical changes for every one of us. On the other hand, lack of involvement, immaturity and apathy to the destiny of our close citizens are sure to prompt superfluous and noteworthy social separation. It's every one of an issue of what we are set up to do.

30.7.1 Egality

Egalitarian groups will be gatherings of individuals who have lived respectively, with populism as one of their center qualities. An expansive meaning of libertarianism is "equivalent access to assets and to basic leadership power."

In Pakistan, individuals living in provincial regions are not treated the same as individuals living in urban areas. Country regions do not have the offices of well-being, instruction, advancement, and security. Indeed, even where training is given in the rural zones the guidelines are not same concerning urban zones. There exists distinctive medium of guidelines out in the open and restricted area. Territorial dissimilarity is available in the structure as standard of instruction shifts in various areas of Pakistan. In FATA, the education rate is lamentable constituting 29.5% in males and 3% in females.

Five categories such as: elitism is productive, avoidance is important, preference is characteristic, eagerness is great, and despondency is unavoidable.

The picture underneath is from an UNDP report. It demonstrates the unequal conveyance of world wage and represents why 3 billion individuals on this planet have no response to the exceptionally essential necessities of life (Figure 30.6).

The procurement of social equity and the poverty issues couldn't be tended to without preparation of resources. Although, administration is attempting to grow the duty base to prepare more resources for social area improvement and to take measures for fracture of riches.

Just 21% of the aggregate assessment potential is being completed while the duty hole stays at 79%. Our policymakers need to move their center and take dynamic choices if they are keen on making the public prosperous. A reasonable dissemination of assets would prompt social advancement that at the

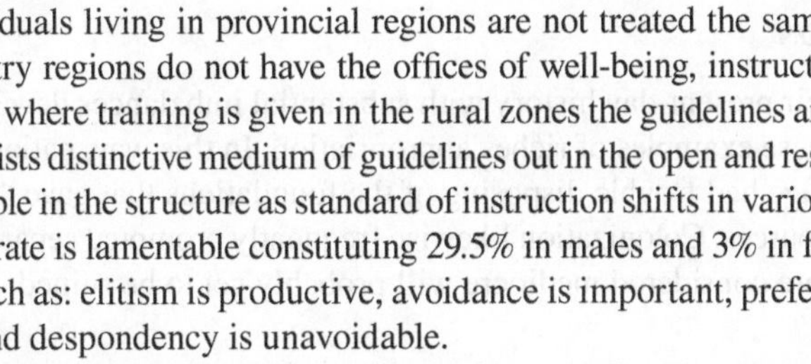

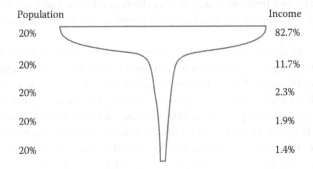

FIGURE 30.6 Unequal conveyance of world wages.

appointed time wipe off the hazard of poverty from the nation. It is the center obligation of the state to deal with the resource preparation and support the spending. There exists gender discrimination to all the poverty-related issues. Women get less sustenance, well-being and training; consequently, they endure more in the wake of crises and fiascos. Most of the women live at the lower ebb; we must embrace discourage gender discrimination from all the spheres of life such as social, political, economic, and religious to ensure the safe passage to women so that they may be able to discharge their duties well in an integrated fashion.

There is a need to encourage the adolescent to spread the message of social equity. We must diminish our spending to guarantee asset designation for the penniless and poor. Free enterprise has expanded consumerism and unemployment at the same time. Sixty-four percent of our populace is gaining under USD 2 a day while around 80% don't get the nourishment of 3000 calories. With an expanding group of rich, destitution has increased manifolds in nation.

30.7.2 Unequal Distribution of Wealth

Everywhere throughout the world, the rich keep on getting wealthier and the poor keep on getting poorer. There are rising differences among assets in nations, which sustains the cycle of neediness. In many spots, the rich get the biggest offer of salary, while the poor get a shockingly smaller treatment. If riches are not redistributed or livelihoods are not made more equivalent, neediness will remain a preponderant issue for the foreseeable future.

30.7.3 Colonization

Nations that started their present-day history with substantial imbalances develop into social orders that keep on holding these examples of riches appropriation. In this way, nations that accomplished colonization and bondage had trouble disposing of the foundations that gave "higher class natives" with more cash and resources. Colonization likewise frequently prompted separation, in this manner, the individuals who were considered mediocre will probably get to be ruined.

30.7.4 Discrimination

Separation, bigotry and bias all fit together to constitute a prime reason for neediness. In Pakistan, discrimination has turned into a typical abhorrence that fundamentally originates from the class distinctions in the public. In this nation, where individuals are living on two extremes, it has turned out to be exceptionally hard to dispose of separation. In any case, it is troublesome, however, not difficult to dispose of this social abhorrence, which is making a wide divide amongst individuals and creates a sentiment loathing in their souls.

Separation is being drilled in the public in different structures. Those individuals who are rich or have built up a higher position in the public eye don't care about the discouraged class. They detest these needy individuals and frequently act towards them in total barbaric manner. This conduct of the rich produces awful feelings in the hearts of poor people as the crevice between these two classes becomes wider.

Another critical variable is the training procedure of Pakistan, which has a massive gap between the substitutes of private foundations and open establishments. This gap causes a feeling of inadequacy among the individuals who can't stand to take confirmation in an English medium school or who don't have a solid family foundation. All things considered, this immense difference can be decreased to a greatest degree by making changes in the training framework by presenting a solitary course for private and open organizations. The law for everyone is another achievement, which will help in containing class segregation. On the off chance that the administration begins dealing with these issues, the class segregation will begin diminished and it will bring forth a happier society.

Massive social welfare irregularities exist inside and between the territories along the lines of class, provincial urban separation, sexual orientation, position and religion. With expanding neediness and high unemployment, an individuals' buying power concerning medicinal services is quickly being reduced.

30.7.5 Poor Governance

Administration does not sufficiently mirror the legislature, but rather likewise the common social orders, systems or markets that activity control over the administration of a nation's social and financial resources for improvement. If debasement and political precariousness are widespread in these organizations, then the state will come up short in satisfying its obligations regarding the residents and stay frail. Thusly, high rates of poverty are generally found inside nations with degenerate pioneers, frail state foundations, and no guideline of law.

30.7.6 Environmental Degradation

Albeit frequently disregarded, ecological degradation can significantly affect destitution rates and the prosperity of individuals. If assets are exhausted because of environmental change, debacles and deforestation, then residents will probably be living in poverty. Natural issues have prompted deficiencies of water, nourishment, lodging, and additionally other vital assets. Without these, the poor won't just stay poor, but increase their odds of sudden death. Clearly, poverty is an issue that requires holding our collective consideration and goading our aggregate activity.

The government is completely dedicated to take after a maintained poverty-reducing technique and allot at least 4.5% of GDP to social and poverty-related issues. The government organized 17 professional impoverished areas through the medium-term expenditure framework (MTEF) in the PRSP-II. The amount of GDP consumed on the poor segments in 2010–2011 remained at 13.24%. In 2011–2012, these were 11.55% of GDP and in 2012–2013, 13.10% of GDP. Between 2013 and 2014, added up to Rs. 1,934.95 billion, which was 14.16% of GDP to reduce poverty from the poor people of Pakistan.

Benazir income support program (BISP) is going on extinguishing gripping rural poverty through the provision of financial aid to the masses. The present government has supported the BISP and has additionally expanded the stipend to Rs. 1200/month and after that to Rs. 1500/month furthermore expanded BISP financial budget to Rs. 97 billion in 2014–2015 from Rs. 75 billion in 2013–2014. All out use of the BISP amid the current monetary year is anticipated to cross Rs. 90 billion. The quantity of BISP recipients has increased from 4.6 million in 2013–2014 to 5.0 million before the end of this budgetary year. BISP is required to select 500,000 youngsters in school during the current monetary year, under its Waseela-e-Taleem activity. The administration has expanded the month-to-month stipend under the Waseela-e-Taleem activity from Rs. 200 to Rs. 250 every month for each youngster.

The Pakistan Poverty Alleviation Fund (PPAF) additionally contributes help in the form of micro-credits, water, dry season moderation, training, social welfare, and crisis reaction intercessions. From July 2014 to March 2015, the PPAF has figured out how to dispense Rs. 9.8 billion to its different on-going ventures. Under the 18th established Amendment, the collection of Zakat has been declined to the Provinces/Federal Areas. An aggregate sum of Rs. 4778.18 million has been distributed in the mass amongst the regions and other regulatory territories for the year 2014–2015. Pakistan Bait-Ul-Mal (PBM) is likewise attempting to reduce of neediness by giving help to the down and out, dowagers, vagrants, invalids, decrepits, and other poor persons through various activities. From July 2014–March 2015, Pakistan Bait-Ul-Mal (PBM) dispenses Rs. 2.28 billion to its central activities.

The legislature ought to tax all these rich and spend these assets on poor people. This will lead the country to a welfare state. The social orders, which do not take measures for the welfare and improvement of poor people, ends up contributing lawlessness, mass-scale riots, and social injustice in the long run.

The visually impaired after of the free enterprise has expanded the social disasters and wrongdoings in the society. We require strong social equity components to guarantee the procurement of rights to all. The generous spending is around 5% of the aggregate spending plan. We can illuminate our need-related issues if the administration figures out how to really spend no less than 15% for the financial backing on the welfare of poor people. One percent individuals have involved 90% of the resources which are not utilized for poor people and masses but rather to extend the abundance of the rich ones. We must change our public on the premise of the social equity.

30.8 MILLENNIUM DEVELOPMENT GOALS TOWARDS POVERTY REDUCTION

Featuring the exertion of destitution easing is the United Nation's Millennium Development Goals (MDGs). The primary objective of the MDGs is to "kill great neediness and appetite." The MDGs were established in 2000 at the United Nations Millennium Summit and the foundation set out to have the objective accomplished by 2015. The G8, or the "Gathering of 8," demonstrates the gathering of eight nations driving the world, six of which are viewed as lasting individuals. The account pastors of each G8 nation have vowed to offer cash to the World Bank, International Monetary Fund (IMF), and the African Development Bank (AfDB) to reduce destitution. (Joined Nations 2014).

The primary objective of the MDGs is to "Destroy compelling neediness and craving." This objective is subdivided into three targets, including: (1) Halve, somewhere around 1990 and 2015, the extent of individuals living on under $1.25 every day; (2) Achieve nice vocation for ladies, men, and youngsters; (3) Halve, somewhere around 1990 and 2015, the extent of individuals who experience the ill effects of craving. Starting 2013, the focus of this objective has been come to, yet the total of the principal objective still has far to go. (Joined Nations 2014).

30.8.1 Delicate Public–Private Partnership

Open private associations (PPPs) are courses of action amongst government and private part substances with the end goal of giving open framework, group offices and well-being administrations. Such associations are described by the sharing of venture, danger, obligation and prize between the accomplices. The purposes behind building up such organizations change yet by and large include the financing, outline, development, operation and upkeep of open base and administrations.

The fundamental rationale for building up organizations is that both the general population and the private segment have remarkable qualities that furnish them with focal points parts of administration or venture conveyance. The best organization plans draw on the qualities of both general society and private segment to set up reciprocal connections. The parts and obligations of the accomplices may change from task to extend. For instance, in some anticipates, the restricted area accomplice will have noteworthy inclusion in all parts of administration conveyance, in others, just a minor part.

While the parts and obligations of the private and open division accomplices may vary on individual overhauling activities, the general part and obligations of government don't change. Open private organization is one of various methods for conveying open foundation including well-being administrations. It is not a substitute for solid and compelling administration and basic leadership by government. In all cases, government stays capable and responsible for conveying well-being administrations and ventures in a way that secures and assists the general population interest.

In Pakistan, different horticultural augmentation and provincial improvement projects were started to quicken farming segment. Agribusiness expansion administrations are considered as the prime vehicle to spread data which support innovation exchange, human asset and social capital advancement, savvy and ecologically practical approaches. It additionally reinforces the limit working of ranchers and enhances their job. Be that as it may, people in general horticultural expansion administrations part running lacking assets, bureaucratic methodologies, inadequately inspired staff and overlooking the ladies and poor ranchers in basic leadership process. The propensities show the movement towards privatization.

The private augmentations are developing while concentrating on wealthier ranchers. Their basically reason for existing is to have most extreme benefit. Thusly, the study is intended to satisfy the correspondence and information hole, with a specific end goal to safeguard institutional consistency towards expansion administrations in view of the past studies made by various schools of thought. In this respect writing surveyed precisely over on the idea of open and private augmentation framework at worldwide and in addition at national point of view, with a specific end goal to discover the downsides of existing expansion framework. Through this article, we contribute while talking about on institutional consistency and cognizance to guarantee present day horticultural augmentation administrations for enhancement of agribusiness area. Moreover, we prescribe arrangement measures for the feasible augmentation administrations in the creating nations like Pakistan.

Administration of Pakistan dispatched different augmentations cum group improvement and rural projects in a steady progression. They recorded V-AID-P, BDS, RWP, IRDP, PWP, Traditional Agriculture Extension System, and TV as principle rural expansion programs. (Abbas et al. 2009). Each system had its own qualities, shortcomings and most was bound to customary straight methodologies. The major reason behind annuls of various augmentation projects were: absence of powerful linkage between associated offices, absence of prepared specialized staff, absence of operational assets, absence of participatory methodologies, deficient responsibility and assessment strategies, political flimsiness, debasement, non-helpful bureaucratic conduct, absence of comprehension of agriculturist's needs, and top-down arranging.

In Pakistan, different augmentation cum-group improvement programs at national level were propelled occasionally considering the routine direct approach keeping in mind the end goal to upgrade ranch creation and elevate the expectation for everyday comforts of provincial groups (Abbas et al. 2009). All these projects however had restricted achievement and were shut down in a steady progression because of deficient bolster administrations, top-down progressively style, absence of contact between framework performers, incapable and temperamental information supply framework, absence of portability and brought together administration. Moreover, we talk about significant expansion programs/approaches started by the legislature as takes after.

30.9 MAJOR EXTENSION APPROACHES TOWARDS POVERTY REDUCTION

Including the effort of poverty facilitating is the United Nation's Millennium Development Goals (MDGs). The central goal of the MDGs is to "crush convincing destitution and voracity." The MDGs were built up in 2000 at the United Nations Millennium Summit and the establishment set out to have the target fulfilled by 2015. The G8, or the "Social event of 8," shows the get-together of eight countries driving the world, six of which are unchanging people. The record ministers of each G8 country have vowed to offer money to the World Bank, International Monetary Fund (IMF), and the African Development Bank (AFDB) to lessen desperation. (Joined Nations 2014).

The central target of the MDGs is to "Devastate extraordinary destitution and desiring." This goal is subdivided into three specific targets, including: (1) Halve, some place around 1990 and 2015, the degree of people living on under $1.25 consistently; (2) Achieve reasonable occupation for women, men, and adolescents; (3) Halve, some place around 1990 and 2015, the degree of people who encounter the evil impacts of longing. Beginning 2013, the key center of this target has been come to, yet the aggregate of the primary target still has far to go. (Joined Nations 2014).

30.9.1 Sensitive Public–Private Partnership

Open private affiliations (PPPs) are arrangements amongst government and private division components with the deciding objective of giving open structure, bunch workplaces and prosperity organizations. Such affiliations are depicted by the sharing of endeavor, risk, commitment and prize between the assistants. The clarifications behind setting up such affiliations change yet all things considered incorporate the financing, plan, advancement, operation and upkeep of open structure and organizations.

The concealed reason for setting up associations is that both the all-inclusive community and the private zone have exceptional characteristics that give them central focuses specifically parts of organization or undertaking movement. The best affiliation blueprints draw on the characteristics of both general society and private portion to develop corresponding associations. The parts and commitments of the assistants may vary from errand to amplify. Case in point, in some expects, the private range assistant will have essential relationship in all parts of organization movement, in others, only a minor part.

While the parts and commitments of the private and open division associates may shift on individual modifying exercises, the general part and commitments of government don't change. Open private association is one of different techniques for passing on open structure including prosperity organizations. It is not a substitute for strong and effective organization and fundamental initiative by government. In all cases, government stays careful and in charge of passing on prosperity organizations and assignments in a way that secures and encourages individuals as a rule interest.

In Pakistan, distinctive cultivating increase and common progression ventures were begun to revive agribusiness part. Agriculture development organizations are considered as the prime vehicle to scatter information which invigorate development trade, human resource and social capital progression, adroit and biologically sensible methodologies. It moreover strengthens the cutoff working of agriculturists and improves their occupation. Nevertheless, individuals when all is said in done green increase organizations fragment running lacking resources, bureaucratic approaches, inadequately convinced staff and ignoring the women and poor agriculturists in fundamental administration process. The inclinations demonstrate the development towards privatization.

The private extensions are creating while focusing on wealthier farmers. Their generally aim is to have most amazing advantage. Along these lines, the present study is planned to fulfill the correspondence and data gap, with a deciding objective to ensure institutional consistency towards increase organizations considering the past studies made by different schools of thought. In this regard composing surveyed intentionally over on open and private enlargement structure at worldwide and furthermore at national perspective, remembering the deciding objective to discover the weaknesses of existing extension system. Through this article, we contribute while analyzing on institutional consistency and reasonability to ensure present day cultivating expansion organizations for improvement of agribusiness range. Furthermore, we propose approach measures for the sensible increase organizations in the making countries like Pakistan. Governing body of Pakistan dispatched distinctive developments cum bunch change and provincial tasks reliably. They recorded Traditional Agriculture Extension System, and TV as essential agrarian growth programs. (Abbas et al. 2009). Every venture had its own qualities, deficiencies and most was bound to standard straight techniques. The fundamental purpose for invalidates of different extension tasks were: nonappearance of convincing linkage between collaborated divisions, nonattendance of arranged specific staff, nonappearance of operational resources, nonattendance of participatory systems, inadequate obligation and evaluation methodologies, political frailty, degradation, non-pleasant bureaucratic behavior, nonappearance of understanding of agriculturist's needs, and top-down orchestrating.

In Pakistan, diverse expansion cum-bunch change programs at national level were pushed occasionally in light of the routine direct approach remembering the final objective to update farm creation and lift the desire for regular solaces of nation gatherings (Abbas et al. 2009). All these ventures however had obliged accomplishment and were closed reliably because of lacking support organizations, top-down logically style, nonappearance of contact between system on-screen characters, insufficient and tricky data supply structure, nonattendance of compactness and united organization. In addition, we look at genuine expansion programs/approaches began by the organization as takes after furthermore confront activity blockage, natural corruption air and water contamination.

30.9.2 Poverty Reduction Programs in Pakistan

Presently we will examine neediness easing developers in Pakistan and its qualities and shortcoming. From the most recent 19 years, the administration of Pakistan has contributed distinctive sort arrangements to reduce Poverty among the masses. Because of These arrangements diminishment in neediness has been seen at various levels from 56% in 1992 to 38% in 2003 (IMF Country Report No. 05/307 August 2005). For destitution diminishment, a progression of project has been eaten during recent decades in Pakistan. This was chosen by government which kept in perspective the rustic advancement and concentrated on accomplishing the national flourishing. In the mid-fifties, the country ate its first significant group improvement program known as Village Aid.

30.9.2.1 Village Aid (1952–1959)

Village Aid (Agricultural and industrial development program) was launched to contribute social administration to the general population on the premise of their felt needs. It was a far reaching rustic improvement exertion in view of inspiration, self-improvement and independence. The methodology was multi-dimensional going for advancement of material and HR through the participation of the general population and dynamic backing of the state. While the system encouraged the execution of advancement ventures as for enhancement of crop cultivation, general well-being of the people, training etc.

30.9.2.2 Rural Works Program (1961–1969)

Overall results of the village Aid Program were not satisfactory because it led to the wastage of money and extreme efforts which were invested to bear away the palm in the scenario of this program. As we know that every failure leads to the way of success. So, keeping in mind, a new program, with better understanding bearing the lesson learnt from the Village-Aid program was launched with the twin objectives of increasing the productivity, development of rural areas and providing better jobs to the unemployed rural manpower by engaging them in labor intensive schemes such as construction of farm to market roads, bridges and culverts, water channels etc.

30.9.2.3 Basic Democracy System 1959–1968 (BDS)

The reign of Field Marshal General Ayyub Khan in 1959 he experimented a political approach which was called Basic Democracies to secure victory in the upcoming Presidential Elections.

In 1962, a new structure was once promulgated as a product of that oblique optionally available approach. Ayyub khan did not sense that a sophisticated parliamentary democracy was as soon as well matched for Pakistan. Alternatively, the Basic democracies, because the person administrative fashions have been called, were purported to provoke and train a in large part illiterate population within the working of presidency via giving them limited illustration and associating them with resolution making at a Degree Commensurate with Their Capability. Basic democracies have been involved with out a greater than nearby government and rural progress. They have been imagined furnishing a -manner channel of communique between the Ayyub khan regime and the usual ladies and men and allow social exchange to move slowly. The Basic democracies consisted within the 5 tiers. The lowest but essential tier was composed of union councils, one every for companies of villages having an approximate general population of 10,000. Each union council comprised ten at once elected contributors and 5 appointed people, all known as ordinary democrats. Union councils were dependable for regional agricultural and community development and for rural regulation and order maintenance; they have been empowered to impose local taxes for community projects. these powers, although, had been greater than balanced at the local stage via using the reality that the controlling authority for the union councils was once the deputy commissioner, whose immoderate reputation and normally paternalistic attitudes maximum commonly elicited obedient cooperation instead than demands.

The subsequent tier consisted of the tehsil (sub district) councils, which accomplished coordination offerings. Above them, the district (zilla) councils, chaired by means of the use of the deputy commissioners, had been composed of nominated authentic and nonofficial contributors, along with the Chair-persons of union councils. The district councils were assigned both obligatory and non-obligatory functions concerning education, sanitation, neighborhood lifestyle, and social welfare. Above them, the divisional advisory councils coordinated the events with representatives of presidency departments. The quality viable tier consisted of 1 development advisory council for every province, chaired with the aid of the governor and appointed with the resource of the president. The city areas had a comparable arrangement, below which the smaller union councils were grouped together into municipal committees to take part in similar duties. In 1960, the elected participants of the union councils voted to verify Ayyub khan's presidency, and underneath the 1962 constitution they common an electoral training to select the president, the countrywide meeting, and the provincial assemblies.

The technique of popular democracies did now not have time to take root or to fulfill Ayyub khan's intentions earlier than he and the technique fell in 1969. Whether or now not a brand-new class of political leaders organized with a few administrative information could have emerged to switch those educated in British constitutional regulation was once in no way found out. And the method did not provide for the mobilization of the rural populace round institutions of national integration. The authority of the civil provider was augmented within the fundamental democracies, and the power of the landlords and the colossal industrialists in the west wing went unchallenged.

30.9.2.4 Benazir Income Support Program 2009

Benazir Income Support Program (started by the Peoples Party in the reign of Asif Ali Zardari) was launched by the government of Pakistan (BISP) in 2009. The major objective of this program is to lend a helping hand to the poor, down trodden and vulnerable people. A poverty census using Poverty Score Cards (PSC) was conducted to identify poor house-holds. Valuable information on the various characteristics of the households as well as its assets was collected. Through the Poverty Score Card Survey, eligible households by the employment of Proxy Means Test (PMT) were identified that determines welfare status of the household on a scale between 0 and 100. It was a major innovation to employ Poverty Score Card Survey and Proxy Means Test (PMT).In October, 2010 this survey was started and has been completed across Pakistan except in two major Agencies of FATA. It was a great innovation employed by the data collectors that they used PMT in the identification of eligible household below a poverty cut-off score. In 2005–2006, using PSLM, the PSC based on PMT was developed. It was latter updated using PSLM 2007–2008. This data not only provided poverty profile of each household but provided useful data on 12 key indicators also. The collected data included the following information about the households such as education, agricultural landholding size, and livestock ownership, type of housing and toilet facilities, child status, and household assets.

This poverty survey has the subsequent critical functions:

First 7.7 million households have been recognized who were living under cut-off rating of 16.17.

Second, it created a huge and reliable countrywide registry of the socio-economic fame of around 27 million households across Pakistan. Third, it uses GPS to map the information of the complete country for choice making (for instance, to reply to herbal screw ups and different emergencies). Furthermore, it validates the concentrated-on process thru 1/3 celebration assessment.

30.9.2.4.1 Objectives

The simple objective of BISP is to combat poverty with the aid of using multiple social safety gadgets designed to deliver a sustainable nice exchange in the lives of persistently excluded and deprived households.

30.9.2.5 Five Point of Junejo Regime (1985–1988)

In December 31, 1985, Junejo announced his five-point program for the development of rural areas. For the implement of these five points for the prosperity of the country many schemes and programs were planned and proposed. One of the Nai Roshini Scheme was successfully launched. A five-point program for socio-economic acceleration was introduced and funds were mobilized from domestic and international sources for financing the program. The main theme of the five points was to assist for the prosperity of less developed areas and improvement of kachi abadi, Link roads etc.

30.9.2.6 Waseela-E-Haq (2009)

Benair Income Support Program (BISP) had launched Waseela-e-Haq scheme in September 2009 to provide and enhance small business and entrepreneurship among poorest of the poor to come out of poverty trap. It was meant to promote self-employment among women beneficiaries or their nominated able bodied members of their families as a mean to improve their livelihood.

Waseela-e-Haq was a targeted scheme of providing interest free loan amounting up to Rs. 300,000 in two or more installments to the randomly selected beneficiary families already receiving the cash transfers under BISP. The loan for Waseela-e-Haq was meant for establishing businesses. For Waseela-e-Haq initiative, BISP beneficiaries had been sub-divided into sub-groups of 3,000 each (Commune System). Every month one beneficiary out of each commune was selected randomly through computerized balloting. An amount of Rs. 2205 Million has been disbursed among 13,455 beneficiaries.

30.9.2.6.1 Targets

The fundamental objective of Waseela-e-Haq to offer and beautify small commercial enterprise and entrepreneurship among underprivileged people to come out of poverty circle. Waseela-e-Haq is purely supposed to promote self-employment among girl's beneficiaries or their nominees to improve their livelihood.

30.9.2.7 Khushali Bank for Microfinance

Microfinance is an institute which was established for the extreme poor peoples. Microfinance supply loans consumer, credit, saving, pension, insurance and money transfer services financial services to the poor. The main objectives of the Khushhali Bank were to eradicate extreme poverty and absolute hunger. And its objective was to achieve universal primary education throughout the entire rural areas of Pakistan.

30.9.2.8 Punjab Rural Support Program (1997-Onwards)

The main aim of this program is to develop the rural areas of Pakistan. Furthermore, PRSP also launched other various useful and multi-dimensional Support and subsidies programs for the rural uplift,socio-economic welfare of the people including other schemes such as micro credits given to people welfare. PRSP is currently working in 26 nominated district of the Punjab province with the core program and through other intervention in partnership with the government and other Donors. PRSP's mission shall be best achieved through organized the poor household. Creating solidarity among groups. PRSP believes supporting the people and harnessing their potential.

30.9.2.9 Baluchistan Rural Support Program BRSP 1991

Baluchistan rural support program is the major program which was launched to create awareness among the people of Baluchistan province like the other rural support programs. The major objective of this multi- dimensional program is to alleviate poverty particularly from the rural areas and improve the quality of life of the rural poor people by yoking the potential of people to manage their own development through their own institutions. Fundamental target of this project enhances the living condition and personal satisfaction. Under PRSP numerous system began Social Mobilization is a hypothesis of social union that is being promoted by BRSP to reduce the sufferings of minimized groups of Baluchistan. "Social" has been gotten from society, which implies aggregate or comprehensive and "activation" remains for collecting or sorting out assets to bolster goal. Along these lines, the hypothesis of "social assembly" signifies "aggregate endeavors of sorted out assets to accomplish particular destinations." Besides, BRSP additionally accepts on cooperation and limit working of burdened groups to guarantee their possession in their general improvement motivation. Social Organization for just about two decades, improvement scholars and professionals have attracted regard for the significance of social foundations or "social capital" as a method for upgrading human abilities and flexibility. As indicated by 21 years of experience of BRSP, social activation is discovered a standout amongst the most appropriate methodologies for country improvement; along these lines, it is being considered as bedrock of whole advancement motivation of BRSP. The key improvement of social preparation is to arrange groups in various levels of gatherings. Since early January 2010, BRSP has received three level structure of institutional improvement, for example, group association (CO) at group level, Village Organization (VO) at town level and Local Support Organization (LSO) at union gathering level. This structure is being viewed as the most proficient and viable intends to get to and extend effort to minimized groups. When people group are sorted out in given gatherings then delegates from every gathering are being prepared so they may arrange, compose, actualize and control their own particular improvement with compelling authority aptitudes. Accordingly, these composed and capacitated groups are being considered as a social capital as well as give stages to all other improvement activities at the group level for instance well-being, training, water, sanitation, cover, sustenance, agribusiness, social insurance, sexual orientation mainstreaming, vocation and other base advancements.

30.9.2.10 Aga Khan Rural Support Program (1982-Onwards)

The Aga Khan Rural Support Program popularly known as (AKRSP 2012) is a non- profit and private organization established by the Aga Khan Foundation in 1982 to accelerate the individual satisfaction of the poor villagers cum peasants of Gilgit-Baltistan and Chitral regions. It is the major objective of AKRSP to improvethe approaches to raise the living standard of the general population and their innate capacities and faculties. It depends on the conviction that vulnerable groups can possibly arrange and deal with their own improvement, once they are sorted out and gave access to fundamental aptitudes and capital. The association's proclivity for a participatory methodology discovered much backing in Shoaib Sultan, the establishing General Manager of AKRSP.

It is based on the above mentioned broader areas of interventions AKRSP enhanced the social development process of its targeted areas, its major outcomes in terms of institutional and community development include:

- Establishment of the First Microfinance Bank in private sector.
- 9 Rural Support Programs in Pakistan.
- 8 RSPs under AKDN across Asia, Africa and the Middle East.
- Core values such as "community participation" and "community development" internalized by various government and non-government program.

30.10 WORLD REPORT ON POVERTY

The World Bank team's mission is engraved on a stone at our Washington D C Headquarters: "Our Dream is a globe free from Poverty." This mission underpins all our analytical, operational, and convoking work in more than 145 client nations, and is bolstered by means of our objectives of ending severe poverty within an iteration and selling shared prosperity in a sustainable method throughout the globe. The UNO is also working to mitigate poverty from Pakistan by providing financial assistance through WHO, UNHR, UNDP etc.

30.10.1 An Overview of Poverty Forecast of South Asia

Region	1990	1999	2011	2012	2015
South Asia	50.6	41.8	22.2	18.8	13.5

An overview of poverty forecast of south Asia indicates that a gradual decline has been observed from 1990 to 2015 as indicated in the table mentioned above. There has been checked advancement on diminishing destitution during the past years. The enclosure accomplished the essential Millennium progress reason objective—to lessen the 1990 neediness cost in half of through 2015—five years forward of time table, in 2010. In October 2015, the World bank has anticipated for the essential time that the nature of men and women living in amazing destitution used to be expected to have fallen under ten percent. Despite this development, the quantity of individuals living in extreme poverty all around stays inadmissibly high.

Pakistan has foremost key gifts and advance abilities. The nation is set at the junction of South Asia, essential Asia, China and the Middle East and is as an outcome at the support of a provincial business sector with a considerable people, enormous and various resources, and undiscovered skills for exchange.

Pakistan's creating working age populace offers the nation with a capabilities demographic profit however in addition with the basic assignment to outfit adequate offerings and employments.

Pakistan faces tremendous budgetary, administration and insurance requesting circumstances to gather solid change results. The determination of contention inside the fringe zones and security challenges all through the nation affect all elements of life in Pakistan and obstructs progress.

Different administration and change environment signs embrace that profound upgrades in administration are fancied to unleash Pakistans change capacities.

Money related improvement is indicating manifestations of supported recovery helped by means of falling ware and fuel costs, expanded life accessibility and extended assurance circumstances. preparatory actualities for the essential portion of monetary year 2016 the big-time development quickening on the lower back of better entertainment in enormous scale assembling and generation, the last being driven particularly through utilizing start of China Pak Economic Corridor (CPEC) framework and power assignments.

The advanced business execution is anticipated to adjust, to some degree, the atmosphere mishaps found inside the agribusiness district. besides, offerings quarter is foreseen to likewise develop drove by means of the money related zone, considerable auto retail salary, enhanced port side interest, and better telecom wage; even though wholesale and retail change is however to progress.

Pakistan positions 147 out of 188 nations in the 2015 human advancement record (HDI) with greatest markers diminish than most global areas in south Asia. Inspire admission to preparing stays low and of whole expense for essential preparing is a considerable number of the most minimal inside the worldwide. In Fiscal year 2014, open spending on training got to be 2.1% of GDP which thinks about the wonderful, negative instructing and becoming more acquainted with results and inadequate base.

Despite the way that a few territories (Punjab) have made progress in bringing down the gender discrimination in almost all walks of life particularly in the education sector to bring women forward in the job sector. It has also been estimated that spending on well-being got to be 0.8% of GDP in fiscal year 2014, making Pakistan one of the least spenders worldwide. Well-being results have enhanced yet at a drowsy pace in the meantime as wholesome outcomes have no more progressed during the last quite a while, and include even disintegrated for some signs. The 2011 national nourishment overview expected that the rates of child hindering have no more altered because 1965 with 45% of adolescents being hindered, 16% of Pakistani children under 5 experience the ill effects of intense unhealthiness.

Economic indicators in the first half of FY17 suggest that pressures are mounting for both fiscal consolidation and external balances. The current account deficit will more than double in FY19 from 1.1% of GDP in FY16. Reserves are forecast to be around $18 b by FY19, still well above three months of imports. The fiscal deficit will widen from 4.5% of GDP in FY16 to 5.1% in FY18, and will decline slightly to 4.9% in FY19. Pakistan has also embarked on an ambitious structural reforms program. Implementation record has been mixed. There were early successes in taxation, the financial sector, the business environment (at both the national and provincial levels), and the electricity sector. However, significant reforms undertaken in the electricity sector have stalled since the Government stopped privatization a year ago.

The quantity of people around the destitution line stays extreme, which prompts limitless defenselessness to neediness. This customary decrease in destitution has come upon disregarding spans of moderate blast. Blast rate bends outline the prepared ghastly nature of Pakistan's blast with utilization of the poorest growing speedier than propose utilization. In any case, the drivers of this procedure are not surely knowing. Enormous settlements, concentrated amongst very terrible family units, are an urgent source, yet developing "shrouded" urbanization and an expanding casual locale will be similarly fundamental.

30.11 UN REPORTS ON POVERTY

The 2013 report positioned Pakistan 146 out of 187 worldwide areas on a human advancement record, equivalent to Bangladesh and only a stage in front of Angola and Myanmar. "Pakistan has most likely the least interests regarding tutoring and health–it burns through 0.8% of its GDP on prosperity and 1.8% on training," the United Nations said in an announcement, including that 49% of the people lives in destitution. Typically, legislators who are presently on the campaign trail utilized this chance to recommend the report's systems and promise brings up in prosperity and tutoring spending plans after the record used to be propelled.

The Senator Razina Alam of the Pakistan Muslim League-N (PML-N) promised to change over the training procedure, maintaining astounding instruction must be given to adolescents all through the nation. "We will extend assets for tutoring and no less than four percent of GDP can be assigned by means of the 12 months 2018," she specified. "In the well-being division, we can make a triple grow inside the funds by means of 2018," she conveyed.

Shafqat Mehmood, data secretary for the Pakistan Tehreek-e-Insaf, expressed his festival would triple spending on instruction and raise spending on prosperity five-fold. It is to be expressed that "The issue of administration is a genuine one in Pakistan as there's an absence of focus toward reacting to people groups' issues," specified Mehmood.

Then again, an assigned political hooligan said, "If voted into force, we will grow each the prosperity and tutoring spending plans to be 5% of GDP for each part," expressed Muttahida Qaumi development (MQM) pioneer Farooq Sattar. Sattar expressed his get-together would impose primitive boss, control defilement and enhance open part divisions to raise cash for social advancement.

Moreover, this report is humiliating and educational in the meantime, he included.

The PPP didn't go to the dispatch of the record, with coordinators saying its agent drop their participation on the last moment.

In accordance with its statement, the PPP expanded the tutoring spending plan with the guide of 196 for every penny to $78 million for 2012–2013 "In our ensuing term, we will prompt a grow in state spending on prosperity to 5 rate of united govt. spending by method for the highest point of our next day and age," the statement specified.

In the meantime, tending to the occasion, Lars-Gunnar Wigemark, Ambassador of the European Union to Pakistan expressed, "We should be focused on human advance and make this improvement supportable for future eras."

UNDP nation Director Marc-André Franche specified, "Expedient human improvement progress in South Asian countries is controlling a verifiable movement in worldwide flow. Pakistan can as a result pick up learning of from unmistakable nations on upgrade the capacities and productiveness of its people. It ought to embrace more intercessions to downsize destitution, expand base and bolster administration."

30.12 STRATEGY FOR THE FUTURE

The procedure for the future ought to along these lines concentrate on (a) Creating and reinforcing synergetic linkages amongst urban and rustic regions and (b) Managing Urbanization.

30.12.1 Rural Urban Integration

More than half of the rural families in Pakistan are landless and consequently looking for better employments, higher wages and getting a capacity to dispatch a segment of this salary to their families back home. Rural –urban relocation and urban-abroad movement would thusly remain a fundamental piece of their adapting procedure. It would be sentimental vision to envision that in perspective of such capable push and force components relocation to urban territories will retreat. Despite what might be expected, the electronic media now accessible generally in even the remotest towns of the urban residents further fortifies this pattern. It is prudent to consider connecting and incorporating the country urban economies as opposed to wishing the driving forces for relocation will leave with rustic growing as it were. Rural profitability increases through effective water use, land leveling, high yielding seeds, build utilization of motorization and broadening into worth included items would assist lessen the interest in work. Provincial urban relocation is a key element for the basic monetary change.

Agrarian division is itself experiencing a noteworthy intra sectoral change. Major and minor products shape a littler extent of farming segment esteem added today contrasted with ten years back and the offer of domesticated animals, dairy, fisheries and agriculture has been step by step rising. The developing dairy industry is an illustrative case of the synergetic linkages amongst urban communities and towns. The enormous majority of the milk is created in the provincial regions however is devoured in the urban communities. Milk gathering, transportation, preparing and sanitization, bundling and retail circulation frame a whole production chain which is shared among the farming, business and administrations parts profiting both the rustic and urban portions of the populace. Meat and poultry utilization is likewise on a common ascent and the area of abattoirs and poultry ranches close to the urban communities would encourage add to this between linkages. More than a substantial portion of the worth expansion in these subsectors happens after these items leave the ranch.

Interests in agribusiness Research and Development, soil and water administration, veterinary and creature cultivation administrations, stockpiling and cool chains, refrigerated transportation and so forth are essential to understand the supply capability of horticulture. Interests in streets, logistics, handling, bundling and sorted out retailing will reinforce the consistent stream of speculations will likewise expand the salaries of the agriculturists which, thusly, will push the interest in customer products delivered in the urban communities. Cell telephones and satellite, digital TV stations are starting to blue the refinements between the urban and rustic ways of life.

Land Markets don't work well since area titles are not clear, documentation and records stay inadequate and are liable to control, and the control of the unimportant civil servants is overpowering. Eighty percent of the case in the courts of Pakistan relates to land question. If area markets function admirably, land will be portable amongst clients and dispensed profitably.

30.12.2 Overseeing Urbanization

The endless loop of improbable arranging, unsuitable execution, powerless requirement of ordinances and principles, poor budgetary circumstance of the City Governments, divided limits and covering locales has, in Pakistan, tilted the equalization of externalities agglomeration and scale economies towards the negative Zone. Step by step instructions to break this endless loop is the significant test for overseeing urbanization. The beginning stage ought to be the devolution of authoritative and monetary independence for overseeing the urban communities—both huge cities and in addition Intermediate urban areas.

The eighteenth amendment has reverted powers from the Federal to the Provincial budgetary assets. Be that as it may, this devolution will stay inadequate and impenetrable to the necessities of the basic nationals until such time that an arrangement of solid District Governments is set up. The limits of the District Governments must be manufactured with the goal that they can satisfy their obligations. Institutional changes will likewise reach out to Land Markets, Labor Markets and conveyance of fundamental administrations.

The administration of the vast urban areas ought to be only in the space of the Metropolitan Governments headed by an elected Mayor. Expert arranging, Land allotment and use, open transport, lodging, framework advancement, parks and comforts, water, sewerage and strong waste transfer ought to fall solely under the control of these MGs. For middle of the road urban areas, City District Governments with the imperative authoritative forces and monetary assets would be the proper vehicle. Without these representing components, Katchi-Abadis and ghettos, deficient city administrations, significant foundation crevice, poor and costly transport would keep on persisting. The covering and parallel wards of different bodies inside the space of the city or metropolitan governments have made discontinuity. They will all must be brought under the Master Planning order of the Metropolitan government. They will have the capacity to bear on the metropolitan capacities inside their individual ward yet submitting to the principles, standards, and benchmarks, determined in the Master Plan.

The second real test is the assembly and supportability of urban open funds. Urbanization in Pakistan has ended up synonymous with in formalization of the economy. The huge majority of the drifters from the provincial territories are invested in the unplanned economy. This represents a significant issue for urban open accounts. As the expanded convergence of populace alongside normal development forces a weight for growing water, sanitation, sewerage, open transport and other urban framework benefits the extent of those paying assessments and client charges for these administrations continues declining. In other comparable circumstances, they vote by feet by relocating to rural areas where the taxation rate is lower. In Pakistan, the inclination is to dodge the expenses by paying off the authorities or utilizing their impact and associations. This further decreases the development of general expense accumulation while the uses continue rising influencing the monetary circumstance of the urban areas. City Bonds, an oft utilized instrument for financing uneven long growth foundation ventures, can't be summoned due to the poor money related soundness of the city Governments. Hence the future efficient arranged development of urban areas is compelled by absence of financing suitability.

The third issue relates to the bless and recognition of the Master arrangement. End-all strategy readiness and execution for all the major urban settlements in the nation is the key element of urban administration. Producing new spaces or reviving existing spaces for retaining future inflows into urban areas requires interest in base and effective conveyance of essential open administrations and would remain a tolerating imperative.

The issue emerges when the arrangements are decreased to bits of paper and the execution is more in infringement instead of in adherence to the arrangement. Authorization of by-laws, rules, zoning laws, occasional happens because the uncontrolled debasement, acts of neglect, nepotism and partiality winning in the Building Control Authorities. Unless straightforwardness and great administration are presented and honed no measure of admonishment would achieve any adjustments in the present arrangement of Urban sprawl, squatter and ghetto settlements, deficiencies of supplies of water and sewerage, weak old urban transport framework.

30.12.3 Territorial Growth Poles

The third segment of this system is another conceivable activity that needs some thought. This is to utilize the fringe zones between the neighboring nations as development shafts. Gravity models recommend that the lower is the separation between two urban areas the bigger is the stream of merchandise and administrations between them. Fringe exchange both sides of Kashmir has as of now been permitted. The time has come to expand the thought of Regional Growth Poles to another fringe locale.

Local Integration can in this manner be utilized for building up the retrogressive zones of the nation and overseeing organized urban development that will come about because of financial resurgence and boost gave by these territorial mix Economic ties will likewise at last straightforwardness political strains assuming any and will be harbinger for serene and congruous relations between the neighboring nations.

The relocation is a dynamic procedure and it impacts the development of economy. The procedure of relocation may happen because of number of reasons and elements. The relocation might be inside and outside. The present study examines and clarifies the elements in charge of relocation in Pakistan. This study is restricted to clarify the interior relocation in Pakistan and reasons that female instruction is a noteworthy determinant of the inside movement. The study utilizes the information from Labor Force Survey 2010–2011. Information is factually dissected on the premise of appropriation of vagrants by their qualities and on the premise of relocation reasons. It catches the effect of age, instruction and different elements on the relocation procedure. The determinants of movement are evaluated with the assistance of logistic relapse model. Here ward variable of relocation model is dichotomous. Along these lines, logistic relapse model is figured to know the effect of female instruction and different components on inside movement in Pakistan. The positive and massive impact of training on movement for both guys and females is found and it demonstrates that relocation is a human capital speculation.

30.13 CONCLUSION

Poverty is condition when your basic needs are not to be satisfied despite your best efforts due to inflation, unemployment, bad governance, political instability, frequent disasters such as earthquake, flood, bad crops, wars, civil wars and many more factors, load shedding, foreign debts, mass corruption and injustice. There are many types of poverty such as absolute poverty, relative poverty, generational or transitory poverty and situational poverty. Pakistan is basically an agricultural country and her major part of population is living in the rural areas. There is lack of world class infrastructure, the fastest means of telecommunication, health facilities, educational facilities, and other basic facilities of life. That is why the people of rural areas are facing poverty. They are unable to live a life which can be called a life. They are entangled in the vicious circle of poverty. They are unable to break this circle and continue to live in the abject poverty for generation to generation. There are many factors which contribute to their poverty such as superstition, astrology, ignorance, fatalism, Peer-ism, lack of industries in the village, gambling, prostitution, betting, lethargy, and traditional style of cultivation. They are always looking to the govt. to have the reasonable solution of their problems but the govt. is busy in her unproductive ventures not heeding to the basic problems of the rural masses. In the rural areas of Pakistan poverty is on the go but reasonable efforts are still lacking. There is much empirical evidence for poverty reduction through increases in agricultural productivity. Much of the literature suggests that this effect occurs through the impact on real household incomes, however there are multiple, complex pathways linking agricultural productivity to real income changes that respond to various market forces. There is convincing evidence for indirect poverty reduction through employment generation, rural non-farm multiplier effects, and food prices effects, however contextual factors determine whether market forces resolve most favorably for the poor. Furthermore, the resulting equilibrium in agricultural and labor markets may affect poor net food buying households differently than poor net food producers. The available evidence supports the theories that when farm incomes and the real wage rate increase and the rural non-farm economy grow, real household incomes increase and the percentage of the population living below international poverty lines decreases. Nutritional status or other aspects of well-being, such as health measures and education, may also improve. However, initial asset endowments, and land assets in particular, are significant determinants of households' ability to access and effectively use productivity enhancing knowledge and technologies. Poor households face barriers to technology adoption and market access. In sum, the importance of productivity to agricultural sector growth and to poverty reduction is complex and depends on a variety of contextual factors including the initial distribution of poverty, asset endowments, strength of market linkages and the extent and nature of the poor's participation in the agricultural sector.

Public and private partnership joint venture processes are emerging to solving the issues of poor farming community by coordinating both public and private extension activities which will influence farmers. Hence, in the structural paradigm changing from top-down system to participatory extension in agriculture extension is emerging. The extension methods and competency level of extension field staff were the important components in convincing the farming community for adoption of sophisticated agriculture technologies and recommendations.

The communication interventions are the significant channels and bridge between technology disseminators and technology adopters. There is also need to encourage holistic approaches. Promote group-based extension programs and strategies it is therefore suggested that field-level training should be arranged for farmers to enhance the capacity building of the respondents. Farmers are first and last in the ladder of agricultural development, therefore, it is suggested that rather than selective approach toward better farmers the private extension services should utilize a participatory approach the farmers having small and medium size land holdings.

REFERENCES

Aga Khan Rural Support Program (AKRSP). 2012. Young Community Leadership Development Program (YCLDP).

Irz, X., Lin, L., Thirtle, C., and Wiggins, S. 2001. Agricultural productivity growth and poverty alleviation. *Development policy review*, 19(4): 449–466.

IMF. 2005. Country Report No. 05/307 August 2005.

World Bank. 2015. Poverty Forecasts. http://www.worldbank.org/en/publication/global-monitoring-report/poverty-forecasts-2015.

31 Rural–Urban Migration

N. Farah, Izhar A. Khan, and A. A. Maan

CONTENTS

The people's movement within and across regions is greater than ever before. Traditional spheres of human activity are rapidly transformed by globalization and migration. Migration both international and internal has increasingly diversified livelihood patterns in rural areas and most rural families are no longer attached to farming activities only (IFAD 2008). Throughout the world migration is a continuing demographic process within one's region, country, or beyond under the form of both voluntary and involuntary movement (IOM 2016). Currently, internal migration has a leading share (80%) in global migration, and the flow of this movement is mostly from rural areas to urban centers (Qin et al. 2014). Such inclinations have produced a rise in the population of urban centers across the globe, which will host 69% of the population by 2030, from 10% in 1900% and 50% in 2009 (Grimm et al. 2008).

Previously, the human migration literature has extensively focused on socioeconomic determinants, while the current literature explores various causation factors behind migration decisions. These causation factors include access to finance, economic opportunities, social capital, governance, and environmental conditions and research has found that migration decisions are mostly economic and not purely social (Mazumdar 1987; Irfan 1986; Kolev 2013). These push and pull factors arise as a result of a country's sociopolitical, ecological, and fiscal conditions, and push people to migrate from unjust areas to stable places (Black and Sward, 2008). Besides these, various personal and household factors and the pressures of population growth often contribute to an individual's decisions to move. Migration is usually a household decision and is an extremely complicated procedure that must considers the potential benefits, family, social, and economic costs and changes in the division of labor; migration policies and rules; availability of networks; and many other aspects. A heterogeneous mix of factors may affect mobility and decisions to migrate such as wage differentials between areas of origin and destination; the local employment situation; distance to areas of dynamic economic activity; total labor supply within the household; the potential migrant's level of education, labor market experience, access to assets, age, sex, etc. Moreover, these factors vary widely among countries and regions (Waddington and Wheeler 2003).

Migration within one's own country/region exceeds migration beyond boundaries as a UNDP 2009 report indicates that globally, 740 million people migrate within the country while 214 million people move across the boundaries. In developing countries specifically, the bulk of people are moving internally and 800 million people have migrated from rural to urban areas during the last 50 years, and this trend is gradually swelling (FAO 2008). This huge influx of migrants is considerably remodeling the existing socioeconomic structures of rural societies, and the efforts to tackle with poverty need to consider these changing social and economic realities and incorporate them into innovative strategies for stimulating rural development (IFAD 2008).

31.1 MIGRATION: THE PAKISTAN'S SCENARIO

Pakistan has long been a nation defined by its countryside. This is where the majority of the population is based; where the largest industries are ensconced and where some of the chief political power centers are anchored. Today, however, this tradition is endangered and the country is urbanizing at 3% annual rate, which is the fastest pace in South Asia (Planning Commission 2013), and 20% of this urban growth is due to rural–urban migration (Hasan and Raza 2009).

In Pakistan, the concept of internal migration is not a new phenomenon. Today, rural–urban migration for new livelihoods, better social services, escape from conflicts, insecurity, and natural disasters push more people towards cities. The rapid growth of the country's population also highlights the rising trend of urbanization. But during different time periods, the volume and nature of internal migration has changed and so has its effects on migrant households (Arif 2005) and on economy (Naseem 1981).

Table 31.1 shows the current trends of urban and rural population changes in Pakistan. It is projected that until 2050, the urban population will be more than 50% of the total population of Pakistan.

In Pakistan, internal migration has become a major policy concern and has been identified both as a determinant of economic growth and modernization on one hand, and responsible for urban deprivation and a destroyer of traditional rural life, on the other hand. This tension is not unusual for a country undergoing rapid socioeconomic transition—from a low income agrarian to middle income industrial. To a great extent migration, industrialization, and urbanization are a single symbiotic process, and the underpinning forces are hard to resist.

TABLE 31.1
Proportion of Urban and Rural Population and Annual Rate of Change

Urban (Thousands)			Rural (Thousands)			Urban Population (%)			Average Annual Rate of Change %
2014	2030	2050	2014	2030	2050	2014	2030	2050	2010–2050
709,12	107,880	155,747	114,221	123,864	115,335	38.3	46.6	57.5	1.1

Source: UN 2014. *Urban and Rural Areas*. Department of economic and social affairs. Population Division. United Nations. New York. https://www.unpopulation.org

31.1.1 Migration Trends & Patterns

During the 1960s the "Green Revolution" technologies in rural areas of Pakistan transformed subsistence agriculture into a capital-intensive system creating a demand for money, which, obviously, village economies failed to generate for poor people in villages. Simultaneously, demand for unskilled labor, as a result of industrialization in urban areas, sped up the process of rural–urban migration. Population growth pressure on land coupled with high costs and non-profitability of agricultural production forced small landowners into daily wage labor or seasonal work to secure their livelihoods; cities offered them improved economic opportunities (Hasan 2010).

Recently, the net migration rate of Pakistan is –1.4 and it is ranked 160th in the world (CIA 2015) with 5,935,176 people's outward migration from Pakistan and 3.05% of all Pakistanis lived outside their country of origin in 2015 (IOM 2016) (Figure 31.1).

Recent data shows that 8% of the Pakistani population moved from their places of origin and mostly (64%) move within the country and are settled in cities from rural areas. Most of the internal movement is towards the large cities and 25% of them settled in the four largest cities, that is, Karachi, Lahore, Faisalabad, and Rawalpindi (MHHDC 2014).

31.1.1.1 Migration Flows

Rate of internal migration in Pakistan has increased with the passage of time and the flow is mostly from rural to urban areas than from urban to urban (Hamid 2010). The share of rural-to-urban migration is 26.2% of the total flow of internal migrants. Among the more obvious reasons to move are marriages, movement with parents, with spouse, or with son/daughter. Job search, transfer, and business are the other major reasons for migration in Pakistan. Education is the least likely reason to move, according to the Labour Force Survey 2013–2014 (GoP 2014).

It is estimated by Irfan (1986) that in Pakistan, 42% of all movement are within districts, 39% beyond districts, and 19% across provincial boundaries. A more recent study by Arif (2005) estimates rural-to-urban migration constitutes 40% of total internal migration. Labour Force Survey 2014–2015 showed a descending sequence in rural–urban migration as Baluchistan (43.9%), Punjab (25.1%), Khyber Pakhtunkhwa (KPK) (21.1%) and Sindh (13.9%). It is noticeable that most of the migration is towards districts with high population densities. Facts document that during the last 10 years, out of 63% of the people who migrated towards cities, 56% of them moved to the provincial or federal capital (Mahmud et al., 2010). The reason behind this is the high development in these cities and the presence of family networks. In light of the recent 7th NFC awards, provincial capitals will see heavy concentration of migrants as these districts will now receive a disproportionately higher share of development funds. (Mahmud et al., 2010). The recent trends in migration are reflected in the Labour Force Survey (2012–2013), which shows decreasing migration rates in provinces, that is, (68.7%) in Punjab, (20.7%) in Sindh, (10.0%) in KPK, and (0.6%) in Baluchistan. Except in Sindh province, the rate of migration increased for all provinces (GoP 2014).

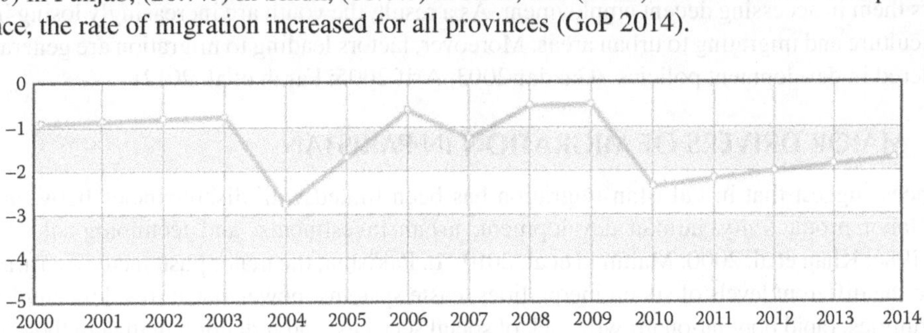

FIGURE 31.1 Trends of migration in Pakistan. (CIA 2015. CIA World Factbook 2015. Retrieved from https://www.cia.gov/library/publications/download/download-2015/index.html).

31.1.1.2 Gender Perspective of Migration

In the movement of population within and across boundaries, a recent change is the feminization of migration. Among the complex reasons of women's migration are both social and economic factors. The literature from Southeast Asia as well as from South Asia exposes the changing patterns of the internal migration that are more urban oriented and female oriented and now like single male migration; single woman also move out for employment and education. In east and Southeast Asia, mostly young and unmarried women are rural–urban migrants and they concentrate in the "mega cities" (Guest, 2003).

In Pakistan, mostly males migrated from rural to urban areas, that is, 30.3% while females were less mobile, that is, 23.5% and the same trend is pronounced in all provinces. Marriage is found to be the major reason for migration of women in Pakistan, and they mostly migrate to join their spouses or move with their parents. Though, current qualitative research on migration has shown the rising tendencies of independent movement of women in some districts of central Punjab in search of education as well as employment opportunities (GoP 2014). The male out-migration affects women too, which adds to their socioeconomic vulnerability in rural Pakistan. Evidence showed that in rural areas of Pakistan, as a result of out-migration of males in poor farmer households, the females left behind faced the additional burdens of family matters traditionally addressed by males.

31.1.1.3 Youth and Internal Migration

Migration is a solution against restricted employment opportunities, mainly for the youth in rural areas. Rural migration out of agriculture is generally related to ambitions and perceptions of the youth. Currently, a general tendency prevails among most young people to be negatively inclined towards farm life, caused by the nature of the labor performed, limited earnings, lack of mobility, and low social status (Leavy and Smith 2010). It has been observed in a World Bank report (2006), that the 12–24-year age group constitutes almost 33% of the total migrants from developing countries, and they are 40% more likely than aged people to migrate from villages to cities, or to move across urban regions. While explaining the higher propensity of rural youth for internal migration (FAO 2015) argued that the lack of decent employment opportunities and poor working conditions among youth are the unquestionably major reasons for youth out migration.

Pakistan has a young age population structure with a 60% of young people of working age group, i.e., 15–29 years (GoP 2015). This vibrant group is the chief source to promote the economic growth of the country through its productive capacity. However, 52.4% of youth are jobless and their unemployment rate is 22%, which is much greater than the adult unemployment rate in Pakistan. Besides rising trends of unemployment for urban youth, they also have to work under poor conditions and work as unpaid family worker. Young rural people worked at their own farms, daily wages or as seasonal workers, and have low wages and insufficient social protection and no job security. All this hinders them in accessing decent employment. As a result, the youth are increasingly losing interest in agriculture and migrating to urban areas. Moreover, factors leading to migration are generally not considered in development policies. (Gazdar 2003; Arif 2005; Farah et al. 2012).

31.2 MAJOR DRIVERS OF MIGRATION IN PAKISTAN

Evidences suggest that in Pakistan migration has been linked with discrepancies between rural-urban labor productivity, human development, urban investments, and technological advances (Irfan 1986; Khan et al. 2000; Mahmud et al. 2010). In Pakistan, the major push factors in rural areas include the different levels of social inequalities, caste systems, power structures, less employment opportunities, rapid population growth, lack of social services, and reduction in the landholding size and per capita share in agricultural labor. The urban pull factors, on the other hand, are expanded economic opportunities, higher wages, improved social amenities like drinking water, sanitation, transportations, advanced communications, educational opportunities, freedom of marriage, and improved facilities of health and recreation for people (Farooq et al. 2005).

31.2.1 The "Push:" Mainly Declining Opportunities in Agriculture

The rural people in Pakistan are mainly pushed out due to pressures of rapid population growth, low landholdings, poverty, food insecurity, and lack of social and cultural opportunities. The main stay of rural economy is agricultural production and is vulnerable to more risks and adversely influenced by water shortages, high input prices, energy crises, floods and marketing problems along with many other factors. In these conditions, poor rural households depend on migrant family members who send money for necessities, and to start a business or continue expansive farming. Besides this, numerous forms of ecological degradation, like desertification, persistent droughts, and salinization are some other factors which might depreciate agricultural livelihoods and push people to the cities. Climatic variations can alter livelihoods and possibly increase urbanization (Mueller et al. 2014). Also, not owning property and lack of access to capital define rural poverty (Naveed and Ali 2012) and promote migration. Pakistan has a history of agricultural migration from arid areas to irrigated regions, especially in the province of Sindh and southern Punjab.

High cost of agricultural inputs and falling prices of agricultural commodities has been a recent trigger for rural–urban migration in Pakistan. Small farmer also suffer water shortages and cannot afford the high expenses of tube well irrigation, hence production declines and farmers are pushed to leave farming and move towards cities.

Water-logging in some parts of Pakistan is also an important trigger for the migration of affected people. The uncontrolled and excessive utilization of irrigation water has resulted in waterlogging and salinization in some parts of Pakistan. This leads to declines in crop yield pushing poor rural families to leave.

Lack of decent employment is another factor pushing rural people towards cities. Situations of surplus labor arising from scarcity of cultivated land, inequitable land distribution, low agricultural productivity, high population density, and the concentration of the rural economy frequently lead to an increase in outmigration. Usually, people's mobility is due to access better economic opportunities elsewhere. This link can be confirmed from patterns of migration that show movement of people from labor abundant rural areas of North West Frontier Province and Punjab to urban centers of Punjab and Sindh. Mostly out-migration flows are from regions of the country with low and uncertain incomes. The mechanical poverty-migration link can be best understood by some of the facts on which labor markets operates in the country. First, sex discrimination exists in labor markets and there are fewer opportunities for female migrants. A general lack of employment opportunities for females often restrict them to low status jobs like domestic workers. Second, the formal public sector has less demand for workers as compared to private and the informal sectors, and here social networks play a crucial role in getting jobs for migrants. Last, in cities migrants are often offered cheap labor such as construction work on daily wages or the socially disregarded occupations (Gazdar 2003).

Land ownership as a determinant of migration is an important push factor. Increasing landlessness due to certain factors in agrarian communities compels the families to adopt alternate livelihood strategies; one of which is migration. Limited gainful employment in the agriculture sector is another major impediment in rural areas. There are four ways in which land can impact internal migration: (*i*) as wealth; (*ii*) as an investment opportunity; (*iii*) as employment; and (*iv*) through ownership inequality. (Leah 2005) found a negative effect of landholding size on out-migration for smaller landholders, while a positive effect was found on out-migration for larger landholders. Due to ambiguous land titles, incomplete and manipulated documentation and land records, land markets do not function well and the control of petty bureaucrats is overwhelming. In the courts of Pakistan, 80% of the cases concern land disputes.

The seasonal nature of agricultural employment pushes many people towards cities in search of employment. Agricultural farming offers only seasonal employment and does provide sufficient income to sustain family households for an entire year, the only solution is to move to secure a livelihood.

31.2.1.1 Climate Change Induced Migration: A New Dimension

During the last 20 years, new intuitions have been added concerning climate change impacts on population movement. The assessment report of Intergovernmental Panel on Climate Change (IPCC) reflects that during the twenty-first century, climate change will enhance the extent of human displacement and migration. It is projected that 25 million to 1 billion people will migrate by 2050 due to climate change effects like shifts in rainfall, heat stress, and water scarcity (Myers 2005; Lovell and London 2007; Parry 2007). Mostly, the residents from rural communities of arid and semiarid regions will move. (Qin et al. 2014). It is not surprising that migration is considered by the latest scientific research as the chief transformational adaptation to climate change (Hoermann et al. 2010; Massey et al. 2010; McLeman and Hunter 2010; Qin et al. 2014).

Currently, Pakistan is in the top countries influenced by climate change and in rural areas, changes in average temperature and precipitation have affected agricultural productivity. Particularly in the winter season, crops bear significant heat stress that affects agricultural productivity (Qin et al. 2014). Wheat is the most consumed Pakistani staple food and its production may drop by 5%–25% due to climate change (Sultana et al. 2009). Similarly, Majid and Zahir (2014) found that the drought in the provinces of Sindh and Punjab significantly impacts the yield of different crops like wheat, cotton, sugarcane, and rice. They identified that this decline in crop yields adversely affects poor farmers, and pushes them to move somewhere else for seeking alternative livelihoods. Another similar study was conducted by Mueller et al. (2014), which found, to a lesser extent, that heat stress was a determinant of internal migration in Pakistan. They concluded the benefits of migration may be greater than the moving costs for poor rural people, specifically those who move as a result of extreme heat waves/heat stress.

31.2.1.2 Migration Due to Conflict

According to Mehboob-ul-Haq Human Development Centre's report (2014), increasing urbanization in South Asia is also characterized by forced migration including natural disasters and conflicts. Internal conflicts in Afghanistan have resulted in increased urbanization in Pakistan as a large number of Afghan refugees enter Pakistan. The conflicts in Swat, Wazirstan, and other tribal areas have also pushed many Internally Displaced Persons (IDP's) to the cities. In Karachi, the law and order situation with increasing target killings are pushed many people, particularly businessmen, to move to Punjab province or leave Pakistan. Natural disasters like earthquake (2005, 2010, 2012) and a series of floods (2010, 2012, 2013, 2014 & 2015) every year have added to the number of internal migrants.

31.2.2 The "Pull:" Often Urban-Based Opportunities

The main pull factor viewed by development economists during the 1950s were the growing demands for skilled and unskilled labor by "modern industrial complexes" and the rural–urban wage gap. The people's aspirations to acquire new skills is the other important pull factor of migration. The surplus and unemployed labor of areas surrounding cities are attracted towards urban areas. Rapid economic growth is thus linked with urbanization and the speed of urbanization is, in turn, augmented by migration.

Through urbanization, moving goods, peoples and ideas become cost effective which expedites agglomeration economies with enhanced production of manufacturing and facilitating growth of services sector. As a result, various employment opportunities are generated and attract people from nearby rural communities to improve their average incomes. A portion of their income is sent to their families living in villages as remittances and help to upgrade their living standards. This reduces the economic and social disparities among leading and lagging regions. The excess and underemployed labor from the farms migrate towards cities which help in raising agricultural productivity and produces comparatively better incomes for the left behind farm workers. As a result, many nonfarm jobs are also generated from such enriched farm incomes. Absorption of rural migrants in industrial and service sector also generates higher incomes in the towns and cities. The aggregate demand effect on the economic production is therefore increased (Arif 2005).

31.3 IMPACT OF MIGRATION IN PAKISTAN

31.3.1 Migration and Rural Development

Despite migration, farming is still the mainstay in rural areas. Among rural households, family members with higher opportunity cost tend to migrate and the old, uneducated households stay behind to manage farm operations resulting in labor shortages in rural areas (Ohajianya 2005; Paris et al. 2009). The productivity of the old-aged labor also affects agriculture productivity. However, income received from migrants can have positive impact on crop production and the net impact on crop productivity is negative (Rozelle et al. 1999). Migration has a positive impact on nonagricultural income and a negative impact on agriculture income. The nonmigrant households earn higher agricultural incomes compared to migrant households (Zahonogo 2011), as migration positively affects nonagriculture income and negatively affects agriculture income.

31.3.1.1 Remittances

For many poor rural families, remittances constitute an effective coping strategy to tackle with adversities like low farm productivity and the related risks of farm activities. Further, remittances serve to minimize the adverse impacts on food security through insurance to recover or counter emergency conditions (Lucas 2005, 2006). Migration in Pakistan is largely to do with economic opportunities and benefits to individuals, families, communities and the national economy. However, evidence are not available for the contribution of internal migration for rural development, this is understandable, given that savings remitted by Pakistanis working abroad constitute the largest single source of foreign exchange earnings for the country. The importance of these remittances has varied, however. In the early 1980s, for example, the flow of remittances was equivalent to around 10% of the Gross National Product (GNP). Currently, remittances are thought to be around 4% of GNP, or US$ 2.4 billion. In 2013, Pakistan ranked seventh among the top ten countries receiving migrant remittances in the developing world (US$ 14.6 billion) (GoP 2014).

31.3.1.2 Agricultural Development

Mahmud et al. (2010) elaborated on the facts that in Pakistan, people mostly invested their remittances in the agriculture sector because a majority of the migrants belonged to rural areas. Location is not only the single factor for investment in agriculture but this is the mind-set, lack of education, lack of exposure or lack of awareness etc. The migrants who were not investing were those people who had a few month or years duration of migration. It is evident that agriculture remained the preferred sector of investment by the migrants. Therefore, it can be argued that migration has a direct impact on agricultural development. People mostly invest in sowing crops. The livestock and farming sectors play a vital role in poverty reduction in Pakistan. So people prefer to invest in this sector to enhance their income. The migrant's money was used for the purchase of livestock and machinery and this is contributed in the overall agricultural development of Pakistan.

Imran et al. (2016) found a negative impact of rural–urban migration on crop productivity of cotton in Muzaffargarh district. Cotton production requires intensive use of workforce from sowing to harvesting and lack of labor has affected cotton productivity.

31.3.1.3 Poverty Reduction

Mansuri (2006) and, Arif and Chaudhry (2015) found significantly positive impact of international migration on school attainment and reduction in child labor in rural Pakistan. Besides this, international migration exerts a positive effect on accumulation of human capital particularly, for girl's access to education, hence, helps in lowering gender inequalities to a substantial degree.

In Pakistan, remittances sent by internal migrants have been mainly used for better living standards and improvement in the migrant's families. The social sector does not get any investment through the remittances but these can help the communities in the time of crises and natural disasters like earthquakes, floods, and droughts. In addition, the important effects of remittances result in limiting the importance of agricultural incomes, which can lead local rural communities to lease their land. This leads to the emergence of a new underclass of landless laborers who now work as tenants on migrants' farms (Hasan 2010).

31.3.2 Migration & Urbanization

In Pakistan, urban slums and periurban areas of large cities are the preferred destinations for the rural-urban migrant families, which results in unregulated urban growth and deviations from the city's development plans (Kugelman 2014). It is estimated that the population residing in the slum areas of only the four largest cities, that is, Karachi, Lahore, Faisalabad, and Gujranwala constitutes about 50% of the urban slum population in Pakistan (Ghani 2012). This unplanned urban expansion usually results in the conversion of prime agricultural lands into urban infrastructures, mostly housing schemes and slums. (Zaman 2012) showed that in Lahore district 94% of the total area was under agricultural use in 1972, but in 2010 it was reduced to 29.5%. This loss of agricultural land will directly affect the livelihood of those associated with agriculture, especially the poorest of the poor (Figure 31.2).

31.3.2.1 Migrants and Urban Poverty

Migrants are often blamed for increasing urban poverty, but it is not true that all migrants are poor. The general pursuit of migrants is to seek formal employment and education and they are often the better-off rural residents. However, in many cases migrants make up a disproportionate share of the urban poor and face some difficulties like accessing proper housing and basic social amenities. These rural migrants usually work in low-paid, long hours and unsafe jobs, like the majority of the urban poor, and are exposed to various environmental threats due to the unavailability of basic infrastructures in their low-income and informal settlements that are situated in hazardous areas where land is available at cheap rates. Further, most of these poor migrants must send money back to home to support their families and repay debts, this makes it difficult to invest for better housing and education in the city (Tacoli 2015).

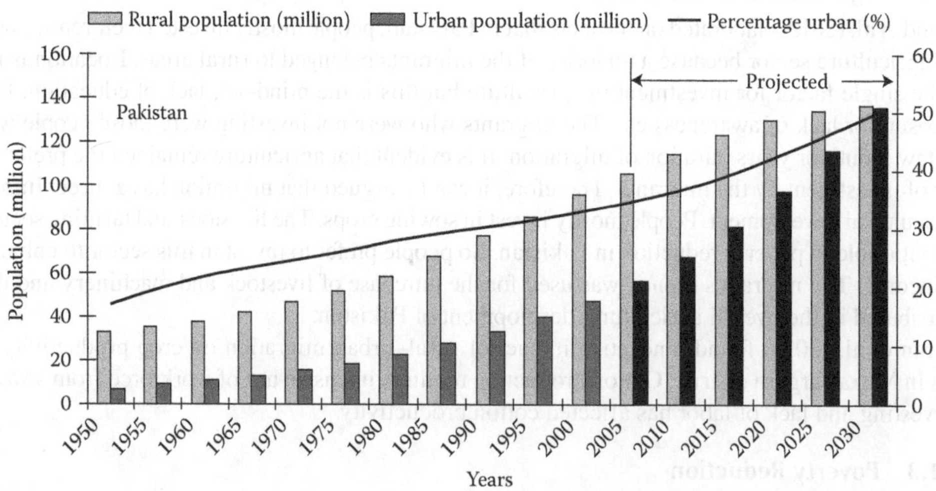

FIGURE 31.2 Urban growth projection of Pakistan. (Asian Development Bank 2006. *Urbanization and Sustainability in Asia: Case Studies on Best Practice Approaches to Sustainable Urban and Regional Development 2006.* Asian Development Bank, Manila, Philippines).

31.3.2.2 Rural–Urban Integration

Unplanned and unregulated urban expansion results in changes of agricultural areas to residential and industrial zones which bring changes to the livelihoods of people living along the rural–urban fringes. Most often, the rich are able to adopt urban attitudes and shed off rural attitudes, the poor are slower in doing so and mostly small farmers lose their land (Tacoli 2003). However, urbanization also creates opportunities by providing means of access to basic amenities and diversifying periurban economies. The accumulative effects of succession and dominance factors have made land progressively scarce for periurban farmers. But a location near a major, growing metropolitan area is often attractive to farmers. Easy access to food markets is of critical economic value to any farming operation. Moreover, diversified nonfarm job opportunities are also produced as result of these linkages (Mahmoud et al. 2003).

Thus urbanization creates employment opportunities for periurban dwellers and provides markets, trading opportunities, and access to various services and infrastructure. The areas are consequently transformed into complex monetized urban economies and are integrated into the urban system. As a result, simple and rural agrarian economies take on urban characteristics owing to the trickledown effects of urban development. As rural areas become urbanized, multiple livelihood sources evolve. In other words, land use changes from agricultural to nonagricultural result in diversified livelihood sources (Aberra and King 2005).

In the major urban centers of Pakistan, a similar trend has been found, where cities not only provide many employment opportunities for rural residents, but also provide markets for agricultural products. It is more evident in Lahore City and its peripheries; the road infrastructure has made urban centers easily accessible and has helped rural people to find better job opportunities and to sale their food and agricultural commodities in urban markets. Another important urban corridor is near the twin cities of Islamabad and Rawalpindi. Here it has linked the surrounding rural communities and populations to the adjoining cities of Jhelum, Chakwal, and Attock and has provided employment opportunities in the services sector. A third urban corridor links Sialkot, Gujrat, and Gujranwala where light manufacturing industries have provided employment opportunities for rural people (ADB 2005).

31.4 REVIEW OF EXISTING MIGRATION & URBANIZATION POLICIES OF PAKISTAN

The issue of migration has received little attention in Pakistan's rural and urban development plans. The recent trends of migration show that most of the rural–urban migration is from deprived areas towards those area which are highly concentrated and developed. The huge gap between rural and urban development is pushing more people towards cities in search of better opportunities. The overview of different national plans reveals urban biased policies and the lack of attention towards rural development.

Migration Policy: There is no internal migration policy of Pakistan. A National Emigration Policy exists in the form of a draft, but the focus of this policy is only international migration and it exclusively discusses the issues of immigrants and remittances and has no focus on internal and specifically rural–urban migration.

Five Years Plans: Throughout the history of Pakistan, the Five-Year Plans fail to present any workable strategy to curb the rural–urban migration and to manage rapid urban growth. The recently announced 11th Five Years Plan of Pakistan has no significant planning to manage the rising issues of rural urban migration, urbanization, and related problems.

Vision 2025: There is no concrete planning for managing large influx of rural–urban migration in Vision 2025. The idea of smart cities was presented but rural and urban development for the small city centers is ignored. Rural youth and rural women's problem was not addressed in order to curb the huge out-migration of villages.

Official Migration Data: Most of the data about internal migration is based on the 1998 census and the annual Labour Force Survey's reports, which do not present the real picture. After a gap of 17 years, the government conducted the last Population Census in 1998, but it failed to provide any data about the place of birth, therefore it is difficult to measure the exact direction of migration flows. The current residence of rural and urban population is mentioned in census questionnaires but it does not ask about the migrant's place of origin. Due to lack of such data, the rate and flow of migration could only be measured with substantial errors (Reza 2003).

31.5 MOVING FORWARD: STRATEGIC DIRECTIONS

To deal with migration three different approaches are used worldwide, which can be classified as negative and manipulative preventive policies. These policies were adopted by different countries at different times, as China followed a negative approach such as imposed relocation from urban to rural areas, bulldozing of slums, banning new migrants into cities and trying to have control over rural–urban migration (Parnwell 1993). The manipulative approach, on the other hand, considers migration as inevitable for development; as there is a capability of value addition from the redirection of movement streams. Venezuela adopted a manipulative approach during the 1970s through policies enabling urban, industrial, and administrative decentralization. Preventive policies encourage a reduction in "push" factors at home through an expansion of the rural labor market and availability of land for farming thus making urban areas less attractive for migrants. Malaysia attempted to follow a preventive approach by adopting its rural urbanization scheme. (Afsar 2003; Farooq et al. 2005).

Stopping rural people from migrating to cities only make poverty less visible; however, as an adaptation measure a combination of preventive and manipulative measures can work to keep control over movements of people from villages to cities. The following are some of the initial recommendations based on preventive and manipulative approaches, for future national planning of rural–urban development in Pakistan: (See Table 31.2)

- *Development of Small/Intermediate City Centers:* The large flux of urbanization cannot be controlled but it can be managed. Therefore, one way is developing small intermediate cities or towns, which can act as linkage between large urban areas and rural areas by providing access of people to urban markets (Jamal and Ashraf 2004; Hussain 2014). These centers can also be helpful in lowering the load of big cities through facilitating additional 'Pull factors," such as better education and health care services. These intermediate cities may also help lessen the rural–urban income gaps through markets which can provide employment to seasonal laborers and rural landless people. Migration to small city centers creates enhanced income and employment, such as urban and periurban agriculture and other nonfarm occupations; however, this movement to small city centers is highly affected by macroeconomic policies. These nonfarm activities and the development of small urban centers may influenced by trade policies. Particularly, export-oriented industries offer various employment opportunities for rural people to reduce the flow of migration (Tacoli 2013).
- *Establishment of Agro-Based Industries in Rural Areas:* In order to provide job opportunities nearby and minimize migration towards cities, agro-based industries should be established in rural areas. Agricultural processing units will not only open new avenues for earning but will also ensure food security and help prevent perishable commodity losses through value addition. There should be direct interventions at the household level to build human capital through vocational training and skill development. People must be given adequate training to equip them with alternative strategies to ensure sustainable livelihood development. The most important is to develop women's skills for the sustainability of their livelihoods.
- *Promoting Rural–Urban Linkages:* Rural–urban linkages play a key role for income and employment. Yet, the importance of such linkages is not recognized (for many reasons) in national development plans and trade policies. There is a need to investigate institutional obstacles, infrastructural problems, and barriers in trade which block the integration of villages

TABLE 31.2
Pathway to Manage the Rural–Urban Migration in Punjab

Timeline	Target	Strategy
Long-Term	Development of small/intermediate cities & towns	Along the motorway intermediate cities can be developed, in different agricultural processing zones.
		Improvement of existing towns should be focused. An example of BastiMalook 35 km from Multan City can be taken as pilot case.
Medium-Term	Provision of decent employment in rural areas	Establishment of agri. processing units in rural areas, cottage industries, and rural women entrepreneurship in agriculture and livestock.
	Promoting rural–urban linkages	Improvement in rural roads, processing and packaging, storage, and systematized marketing especially in dairy sector.
	Development of migration plans from a youth lens	Provision of skill training programs long with credit facilities to youth to establish value-addition enterprises especially in the areas like livestock, horticulture, and fisheries as well as nonfarm sector.
Short-Term	Monitoring of internal migration	National registration system to count the population mobility data within the country and it should be reinforced by legislative amendments for making it compulsory for migrants to register themselves.

and cities. This impedes the process of rural development and economic growth. It is required to focus countryside and city centers into regional planning for creating enabling environment for expansion of business networks and exchange of knowledge between cities and villages. In Pakistan, migration will remain an important coping strategy for poor landless rural households and it would be a fantasy to think that—in the presence of such influential push and pull factors—rural-urban migration will recede. On the contrary, we should integrate and link urban and rural economies rather than try to reverse the urge of migration through rural development only. The growing dairy industry of Pakistan is an example of synergetic linkages between cities and rural areas, as most of the milk is produced in the villages and periurban areas but is sold in the cities. The focus on development of the agricultural and dairy value chain would benefit both rural and urban populations. Improvement of rural roads, processing and packaging, storage and systematized marketing will increase the flow of investments and enhance the incomes of the farmers; in turn, farmers will demand for consumer goods produced in the cities.

- *Development of Migration Plans from a Youth Lens:* Decent rural employment may be a crucial as a part of augmentative opportunities for rural youth to enable them to stay in their home communities, and also offer those that have migrated with the choice of returning. The approach entails coordinated efforts by national and provisional governments, development partners, and also the private sector to create capacities for rural youth and supply them with the required resources, skills, and technologies. In Pakistan, most of the agriculture-related industries and rural nonfarm economies are based on agriculture, hence, rural youth involvement in services sector and entrepreneurship for value added growth of agriculture sector and improvement in income cannot be neglected. Therefore, the government ought to support the youth in acquiring new skills to set up agro-based enterprises, as there are many business opportunities in the areas of livestock, horticulture, and fisheries as well as in the nonfarm sector. The government should provide the youth will credit facilities along with technical support, which would economically benefit rural communities.
- *Monitoring of Internal Migration:* The patterns of internal migration, along with its drivers and outcomes, are difficult to fully comprehend due to lack of valid country-level

information; therefore, much of the existing data comes from case studies. This deficiency obstructs effective monitoring of patterns and flows of internal migration and hinders informed decision-making. It is time to create a National Registration System to produce population mobility data and reinforce it by legislative amendments making compulsory for migrants to register.

- *Managing Climate Change:* A recent study by the Sustainable Development Policy Institute reflects that in the prevailing sociopolitical and financial situation, climate change is likely to aggravate development challenges (SDPI 2016). A future climate change scenario indicates that the rising trend of heat waves in most of the wheat-producing areas will adversely affect the production of wheat, which will increase the vulnerability of the rural landless people. The higher migratory flows towards the cities will add to the already prevailing problems of urban population explosion. There is a need to manage the consequences of climate change through proper policies promoting adaption measures other than migration.
- *Improved Service Delivery:* Under the 18th amendment of the constitution of Pakistan, the revival of local governments plays important role for curbing rural-to-urban migration in two ways. On one hand, more funds are allocated to megacities and provincial capitals in the absence of local governments (Mahmud et al., 2010), and a limited funds trickle down to district, tehsil, union, and village levels. On the other hand, the resultant balanced power distribution will permit local governments to make decisions about fund allocations to directly solve the local social issues. This can be acheived by improving service delivery for rural and local populations. Now, the local governments should take steps for better and improved service delivery to both rural and urban areas. The most important step is the introduction of new forms of village organization and group village people into larger units like *cooperatives*, which would help in the easy delivery of infrastructure, agricultural inputs, and other services. Simultaneously, improved service provision is also required in the urban slums, which absorb the major influx of informal migrants from rural areas.
- *Future Research:* For a better understanding of the link between rural–urban migration with rural/urban growth and transformation, more focused and in-depth research on different aspects of internal migration is required. Analysis of the impacts of different policies on outcomes of internal migration and evaluation of how policies have facilitated internal migration is needed. Also, how does the composition of the migrant population (age and gender) influence the impact of migration on households, recipients, and sending communities?

31.6 CONCLUSION

Overall, internal migration is perceived as a desirable process that spreads the advantages of economic agglomerations to marginalized areas. Moreover, in most cases, internal mobility does not seem to cause overpopulation or unemployment where people settle. Thus, permitting free mobility within countries is not just ethically justified, it also results in a stronger economic base for all citizens, regardless of where they live. However, in some circumstances, internal migration can reduce the well-being of migrants and their families. More research is needed to better understand the relationship between internal migration and well-being of the people, as well as the resulting policy implications.

REFERENCES

Aberra E and King R. 2005. Additional Knowledge of Livelihoods in the Kumasi Peri-Urban Interface (KPUI), *Ashanti Region*, Development Planning Unit, and University College London, Ghana.

ADB. 2005. *Islamic Republic of Pakistan: Preparing the mega-city Sustainable Development Project.* Technical Assistance Report. Manila: ADB. http://www.adb.org/sites/default/files/projdocs/2005/38408-PAK-TAR.pdf

Afsar R. 2003. Dynamics of Poverty, Development and Population Mobility: The Bangladesh case. *Ad Hoc Expert Group Meeting on Migration and Development, Organized by the Economic and Social Commission For Asia And The Pacific*, Bangkok 27–29 August.

Arif GM. 2005. Internal migration and household well-being: Myth or reality. in Hisaya Oda (ed.) *Internal Labour Migration in Pakistan*. Institute of Developing Economies, Japan External Trade Organisation, Chiba, Japan.

Arif R and Chaudhry A. 2015. The effects of external migration on enrolments, accumulated schooling and dropouts in Punjab. *Applied Economics*, 47(16):1607–1632.

Asian Development Bank. 2006. *Urbanization and Sustainability in Asia: Case Studies on Best Practice Approaches to Sustainable Urban and Regional Development 2006*. Asian Development Bank, Manila, Philippines.

Black R and Sward J. 2008. Measuring the Migration-Development Nexus: An Overview of Available Data. Development Research Centre on Migration, Globalization and Poverty.

CIA 2015. CIA World Factbook 2015. Retrieved from https://www.cia.gov/library/publications/download/download-2015/index.html

FAO. 2008. *Rural Population Change in Developing Countries: Lessons for Policymaking*. Rome.

FAO. 2015. Decent Rural Employment and Youth employment. http://www.fao.org/rural-employment/en/

Farah N, Zafar MI, Naima N. 2012. Socio-economic and cultural factors affecting migrationbehavior in District Faisalabad. *Pak. J. Life Soc.Sci.* 10(1):28–32.

Farooq M, Mateen A and Cheema MA. 2005. Determinants of Migration in Punjab: A case study of Faisalabad metropolitan. *Journal of Agriculture and Social Sciences*. 1(3):280–282.

Gazdar H. 2003. A review of Migration issues in Pakistan. Paper Presented in *Regional Conference on Migration, Development and Pro-Poor Policy Choices in Asia*. 22–24 June, Dhaka, Bangladesh.

Ghani E. 2012. Urbanization in Pakistan: Challenges and Options. *Pakistan Institute of Development Economics (PIDE) Working Paper.*

GoP. 2014. *Labour Force Survey 2012–13*. Statistics Division. Pakistan Bureau of Statistics. Govt of Pakistan. Islamabad.

GoP. 2015. *Pakistan Economic Survey 2014–15*. Finance Division of Government of Pakistan, Islamabad.

Grimm NB, Faeth SH, Golubiewski NE, Redman CL, Wu J, Bai X and Briggs JM. 2008. Global change and the ecology of cities, *Science*, 3195864: 756–760.

Guest P. 2003. Bridging the gap: Internal migration in Asia, Population Council Thailand, *Paper Prepared for Conference on African Migration in Comparative Perspective*, Johannesburg, South Africa, 4–7 June.

Hamid S. 2010. Rural to urban migration in Pakistan: The gender perspective. *PIDE Working Paper No. 201056*.

Hasan A. 2010. Migration, small towns and social transformations in Pakistan. *Environment and Urbanization*, 22(1): 33–50.

Hasan A and Raza M. 2009. *Migration and Small Towns in Pakistan. International Institute of Environment and Development (IIED) Working Paper 15*. IIED, London. http:// pubs.iied.org/pdfs/10570IIED.pdf

Hoermann, B, Banerjee, S and Kollmair, M. 2010. *Labour Migration for Development in the Western Hindu Kush-Himalayas: Understanding a Livelihood Strategy in the Context of Socioeconomic and Environmental Change*. International Centre for Integrated Mountain Development, Kathmandu. Available at: http://lib.icimod.org/record/8050/files/attachment_695.pdf

Hussain I. 2014. Urbanization in Pakistan. *Keynote Address Delivered at South Asia Cities Conference and Pakistan Urban Forum Held at Karachi on January 9, 2014*.

IFAD. 2008. *International Migration, Remittances and Rural Development*. International Fund for Agricultural Development. Rome.

Imran M, Bakhsh K, Hassan S. 2016. Rural to urban migration and crop productivity: Evidence from Pakistani Punjab. *Mediterranean Agricultural Sciences* 29(1): 17–19.

IOM. 2016. *Glossary on Migration, International Migration Law Series No. 25*, New York.

Irfan M. 1986. Migration and development in Pakistan: Some selected issues, *The Pakistan Development Review*, XXV(4).

Jamal Z and Ashraf M. 2004. Development of Intermediate-size towns: An alternative form of urbanization, *Quarterly Science Vision*, 9(1–2).

Khan AH, Shehnaz L and Ahmed AM. 2000. Determinants of internal migration in Pakistan: Evidence from the Labour Force Survey, 1996–97, *The Pakistan Development Review*, 39(4): 695–712.

Kolev A. 2013. 'Labour migration and development: A critical review of a controversial debate,' in Cazes, S. and Verick, S. (eds) *Perspective on Labour Economics for Development*, International Labour Office, Geneva.

Kugelman M. 2014. *'Pakistan's Runaway Urbanization: What Can Be Done?'* Wilson Center, Washington DC. ISBN: 978-1-938027-39-0.

Leah KV. 2005. Land ownership as a determinant of international and internal migration in Mexico and internal migration in Thailand. *The International Migration Review*, 39(1): 141–172.

Leavy J and Smith S. 2010. Future farmers: Youth aspirations, expectations and life choices, Discussion Paper 013, *Future Agricultures.*

Lovell J and London M. 2007. Climate change to make one billion refugees-agency, *Reuters*, 13 May 2007 http://www.reuters.com/article/latestCrisis/idUSL10710325

Lucas Robert EB. 2005. *International migration regimes and economic development.* Executive summary of the report prepared for the Expert Group on Development Issues, Swedish Ministry of Foreign Affairs. Department of Economic and Social Affairs, United Nations, New York, http://www.un.org/esa/population/meetings/thirdcoord2004/P22_AnnexIII.pdf.

Lucas Robert EB. 2006. Migration and rural development. Background Paper presented at the *Conference Beyond Agriculture: The Promise of a Rural Economy for Growth and Poverty Reduction.* Rome: FAO.

Mahmoud, B, Samana C, Bitrina D, Gouro D, Fred L, David O and Enoch O. 2003. Changing rural–urban linkages in Mali, Nigeria and Tanzania. *Environment and Urbanization*, 15(1): 13–24.

Mahmud M, Musaddiq T and Said F. 2010. Determinants of internal migration in Pakistan—Lessons from existing patterns, *Pakistan Development Reivew* 49(4): 593–607.

Majid H and Zahir H. 2014. Climate Change and the impact on farmer productivity: The role of socioeconomic vulnerability in rural Pakistan. *Climate Change in Pakistan: Working Paper Series #1.*

Mansuri G. 2006. Migration, School Attainment and Child Labor: Evidence from Rural Pakistan. Policy Research Working Paper; No. 3945. World Bank, Washington, DC.

Massey DS, Axinn WG and Ghimire DJ. 2010. 'Environmental change and out migration: Evidence from Nepal,' *Population and Environment*, 32(2–3): 109–136.

Mazumdar D. 1987. 'Rural-urban migration in developing countries,' in E.S Mills (ed.) *Handbook of Regional and Urban Economics*, Elsevier, Amsterdam, 2, pp. 1097–1128.

McLeman RA and Hunter LM. 2010. Migration in the context of vulnerability and adaptation to climate change: Insights from analogues. *Wiley Interdisciplinary Reviews: Climate Change*, 1(3): 450–461. doi: 10.1002/wcc.51

MHHDC. 2014. *Human Development in South Asia 2014: Urbanization: Challenges and Opportunities.* MahbubulHaq Human Development Centre, Lahore p. 21.

Mueller V, Gray C and Kosec K. 2014. Heat stress increases long-term human migration in rural Pakistan, *Nature Climate Change*, 4(3): 182–185.

Myers N. 2005. Environmental refugees: An emergent security issue, *Paper for the 13th Economic Forum, Organisation for Security and Cooperation in Europe*, Prague, May 2005, pp. 23–27.

Naseem SM. 1981. *Underdevelopment, Poverty and Inequality in Pakistan.* Vanguard Publication. Ltd.

Naveed A and Ali K. 2012. *Clustered Deprivations: District Profile of Poverty in Pakistan.* Sustainable Development Policy Institute, Islamabad.

Ohajianya DO. 2005. Rural-urban migration and effects on agricultural labor supply in Imo state, *Nigeria. Int. J. Agri. Rural Dev.*, 6: 111–118.

Paris TR, Luis J, Villanueva D, Rola-Rubzen MF, Chi TTN, Wongsanum C. 2009. Labour out migration on rice farming households and gender roles: Synthesis of findings in Thailand, the Philippines and Vietnam. Paper Presented at the *FAO-IFAD-ILO Workshop on Gaps, Trends and Current Research in Gender Dimensions of Agricultural and Rural Employment: Differentiated Pathways out of Poverty*, Rome, Italy.

Parnwell M. 1993. *Population Movements and the Third World.* Routledge Publishers. London.

Parry ML. 2007. 'Climate change 2007: Climate change impacts, adaptation and vulnerability,' *Contribution of Working Group II to the Intergovernmental Panel on Climate Change Fourth Assessment Report, 4*, Cambridge University Press, UK. Available at: https://www.ipcc.ch/pdf/assessmen t-report/ar4/wg2/ar4-wg2- chapter7.pdf

Planning Commission of Pakistan. 2013. *Pakistan 2025, One Nation-One Vision.* Govt. of Pakistan. Islamabad.

Qin D, Plattner GK, Tignor M, Allen SK, Boschung J, Nauels A and Midgley PM. 2014. Climate Change 2013: The physical science basis, in T. Stocker (ed.), *Contribution of Working Group I to the Fifth Assessment Report of the Intergovernmental Panel on Climate Change, Cambridge University Press Cambridge*, UK, and New York, p. 1535.

Reza A. 2003. Understanding urbanisation, in S. Akbar Zaidi (ed.) *Continuity and Change: Socio-Political and Institutional Dynamics in Pakistan*, City Press, Karachi.

Rozelle S, Taylor JE and Debrauw A. 1999. Migration, remittances, and agricultural productivity in China. *The American Economic Review*, 2: 287–291.

SDPI. 2016. Climate induced rural-to-urban migration in Pakistan. *Sustainable Development Policy Institute*. Islamabad. Pakistan. Working paper.

Sultana H, Ali N, Iqbal MM and Khan AM. 2009. Vulnerability and adaptability of wheat production in different climatic zones of Pakistan under climate change scenarios, *Climatic Change*, 94(1–2): 123–142.

Tacoli C. 2003. The links between urban and rural development. *Environment and Urbanization*, 15(1): 3–12.

Tacoli C. 2013. The Potential of Rural–urban Linkages for Sustainable Development and Trade. *Swiss national centre for competence in research. Working Paper No 2013/37*

Tacoli C. 2015. Stopping rural people going to cities only makes poverty less visible, and stripping migrants of rights makes it worse. *Blogs* 15 March, 2015. International Institute of Environment and Development.

UN. 2014. *Urban and Rural Areas*. Department of economic and social affairs. Population Division. United Nations. New York. https://www.unpopulation.org

UNDP. 2009. "Overcoming barriers: Human mobility and development," *Human Development Report 2009*, UNDP, New York.

Waddington H and Wheeler RS. 2003. *How Does Poverty Affect Migration Choice?: A Review of Literature*. Institute of Development Studies, University of Sussex, Brighton, UK.

Zahonogo P. 2011. Migration and agricultural production in Burkina Faso. *African J. of Agri. Res.*, 7: 1844–1852.

Zaman KU. 2012. Urbanization of arable land in Lahore City in Pakistan: A case-study, *European Journal of Sustainable Development*, 1(1): 69–83.

32 Rural Development

Tanvir Ali, Babar Shahbaz, Muhammad Iftikhtar, Ijaz Ashraf, Shoukat Ali, Ghazanfar Ali Khan, Aqeela Saghir, and Muhammad Saleem Mohsin

CONTENTS

32.1 BACKGROUND INFORMATION

Rural development is a process that helps rural people to modify positively their lifestyles towards a better living. Despite the being the most urbanized country in South Asia,* Pakistan is still predominantly a rural country where about two-thirds of the total population lives in rural areas. The recent trends indicate more than 3% annual increase in urbanization. The rapid urbanization is as a result of the structural transformation of the economy and rural–urban migrations. Until the 1970s, rural development and agricultural development were considered two sides of the same coin. However, in subsequent years the concept of rural development changed from just agricultural growth and increasing farm production to improvement of living standards and providing water, food security, health, and education services for the rural communities (Israr et al., 2009). In Pakistan, the rural sector's performance and well-being is greatly associated with the development of the agriculture sector, which also influences highly the overall national economic growth. Over the last decade, somewhat stagnant and declining trends in agricultural growth (Figure 32.1) have challenged rural development and agricultural development programs and policies in Pakistan.

In the past, many rural development interventions were not backed by a strong institutional base and were devoid of participatory approach in the decision-making and implementation process. However, despite these policy weaknesses, the agriculture sector continued to survive; however, over the past few years agricultural growth has started declining and we are now witnessing the agrarian economy in a critical position, particularly since the last few years. Associated with multiple problems including land and water resource degradation and weak rural development policies, the decline in agriculture growth has raised concerns about sustainability of this sector and eventually the rural economy of Pakistan. Rural dwellers still have less access to facilities and resources required for rapid development. Rural development remains a matter of great concern, as most of the people in Pakistan live in rural areas or have a rural background.

32.2 RURAL DEVELOPMENT PROGRAMS BEFORE INDEPENDENCE

In preBritish villages, the occupation and trade had been established in such a way that each village had become a self-sufficient economic unit. It was not a planned effort but poor and unsafe means of communication pushed the rural communities to be self-dependent. The economic life of the village was centered on agriculture and all supportive occupations were part and parcel of each village. For example, each village had a carpenter who made wooden agricultural implements, a blacksmith who furnished iron implements, a weaver who made cloth for farmers.

The concept of agricultural value addition was materialized by oilmen, weavers, and leather-workers. Oilmen extracted oil from sarsoon, linseed, til, etc. Villagers used sarsoon oil especially as cooking oil in addition to desi ghee (a value added product of milk). Every home had a spinning

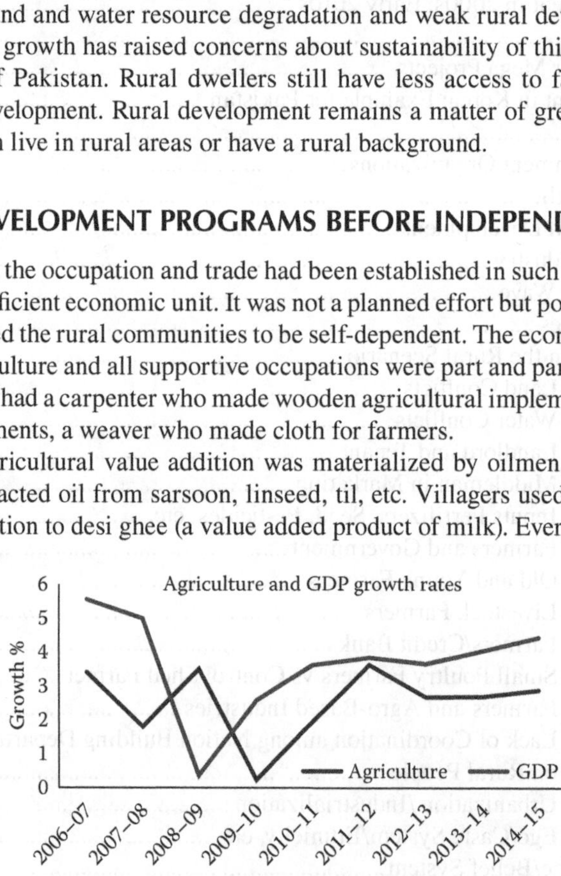

FIGURE 32.1 Trends in agriculture growth and GDP growth.

* It is estimated that approximately 58 million people reside in cities in Pakistan.

wheel to spin cotton-yarn. Spinning was an inherent duty of women who gave cotton-yarn to weavers for weaving clothes. Leather workers processed the hides of animals to make leather. It was used to make a variety of shoes for villagers. The other essential artisans of the village were potters, barbers, messengers, night watchmen, *Brahman* priests (for Hindus), *molvis* (for Muslims), etc. The services of such village servants were returned after every 6 months at the time of harvest in the form of specific crop shares.

The most important economic component of a village was a *bania* who always established a retail shop for the purchase of agricultural commodities and sale of daily life items. He also served as a lender to extend interest-based income for the villagers to fulfill their ceremonial and other needs.

The social life of the villagers largely centered around the well, which was excavated in the center of the each village by the efforts of the villagers. This water well was a symbol of development because the well served as an exclusive source of water on which the life of crops, animals, and humans depended.

In the preBritish era, because a canal system was not established, most of the villages were settled around banks of rivers or in the territories where subsoil water was fit for drinking and irrigation.

Villagers grew the crops according to their needs because there was no concept of money. Community exchanged goods and services (barter system) without use of money.

But, in the British-era, introduction of currency and canal systems promoted the commercialization of the rural settings. The villages started to grow crops for the market to earn money.

The idea of a canal irrigation system was reinforced by periodic famines and the interest of the British to promote British commerce in India.

The British government started to establish canals in 1830; however, the canal system in West Punjab was established in 1886 when canal colonies of the west Punjab were developed. These colonies include Lyallpur, Sargodha, Montgomery, and Multan. Later, the colonization was extended in the princely state Bahawalpur through the canal system. Most of the colonists came from Ludhiana, Amritsar, and Jullundur.

The colonists grew crops of market value, which were exported to England and other countries. But, in this cycle, the agricultural and rural development remained stagnant and villagers still suffered from periodic famine. About 60 million people were affected in the famine of 1876–1878, and more than 5 million people died due to starvation. This huge loss pushed the government to constitute a commission to inquire and report on the matter.

In Punjab, the government became interested in agricultural and rural development when the Famine Commission published its report in 1881. Consequently, Punjab Agricultural college and research institute was established in Lyallpur in 1906 with the objective to promote agricultural education, research, and extension for agriculture and rural development. The college released high yielding wheat varieties especially C-518 and C591. These varieties changed wheat husbandry in Punjab, NWFP (now Khyber Pakhtoonkhwa), and Sindh. Similarly varieties of other crops were also released, which had great impacts on the rural economy.

In 1920, the Indian National Congress launched rural development programs with the agenda to increase agricultural production, promote use of local made goods, sensitize rural people for unity and improvement of social fabrics.

It was a developmental as well as a political move of the congress against British rule. Under these circumstances, the government of India planned to develop agriculture and rural communities and consequently Royal commission on agriculture was constituted in 1926. The objective of the commission was to analyze the existing conditions of agriculture and rural areas and to give recommendation for development. The recommendations of the commission strengthened the developmental efforts in the country.

Meanwhile, F.L. Brayne who was appointed as Deputy Commissioner of Gurgaon (now a district in eastern Punjab) launched a rural reconstruction scheme in his district. Brayne observed that there was a prehistoric system of agriculture in the villages and people were ignorant and degraded. The customs of the rural dwellers were opposed to moral, social, or material progress.

Brayne stated the objectives of the scheme to "jerk the villagers out of his old grooves, convince them that improvement is possible by combating climate, diseases and pests"

Seven objectives of the program were to:

1. Improve farming
2. Clean the village
3. Make houses lighted and airy
4. Take precautions against epidemic
5. Humanize women (sending girls to school and forbidding child marriage)
6. Stop waste (marriage and postfuneral expenditures)
7. Beatify the home

To achieve the stated objectives, Brayne heavily depended on the intensive use of propaganda. He believed that "there is no habit or custom that cannot be undermined by propaganda and no new method that cannot be popularized with propaganda."

The contact methods used by Brayne were:

1. Singing parties
2. Magic shows
3. Printed materials
4. Competitions
5. Demonstration
6. Exhibitions
7. Shows, and
8. Dramas

All such methods were designed to undermine the unwanted habits. Moreover, Brayne established a school of rural economy in 1925.

The objective of the school was "to convince the villagers that village uplift was a complete remedy for all the ills of village life."

The contents covered in the school included: scouting, practical agriculture, health and sanitation, first aid, child care, and village work. Brayne was considered a pioneer of the Rural Reconstruction program in India. This program set an example for later rural development efforts by the governments in India and Pakistan. The Brayne approach got recognition but its sustainability was poor after Brayne left the district; people had adopted many innovations just out of fear of the deputy commissioner.

32.3 RURAL DEVELOPMENT PROGRAMS AFTER INDEPENDENCE

Pakistan became independant in 1947. Its focus on rural development was the need of the time as a vast majority of people lived in villages at that time. Most of the villages were devoid of facilities, which were available to cities.

32.3.1 Rural Development during the 1950s

The economy of Pakistan was agrarian (with almost zero industrial base) when it got independence in 1947 and a majority of the farmers had small landholdings while a minority of big and powerful landlords were present in rural areas. The annual growth rate during this period (1950–1960) was low because agriculture was the main driver of the economy of the country and the yield/production of major field crops was very low (Chaudhry et al., 1996). This low yield is attributed to poor performance of the agriculture sector throughout the 1950s (Mahmood et al., 2008). In view of the backwardness of agriculture during this era, the government of Pakistan started to develop agriculture using modern crop production and protection technologies through rural extension work. The priority of the then government was to develop the agriculture sector (Mahmood et al., 2008) because the country was dependent on imported commodities to meet its food requirements (Hussain, 2004). In order to develop agriculture along modern crop management lines, the government of Pakistan started different programs during this era of stagnant agricultural growth. The first program was initiated in the early 1950s. It was called Village AID (Village Agricultural and Industrial Development Programme) (Lodhi et al., 2006). Community development, that is, provision of basic facilities to the rural community (education, health, recreational facilities, and rural infrastructure) and the promotion of agro-based rural industry (cottage industry) were the main themes of rural development strategies during this period (Sobhan, 1968). Under this program, the development unit called the "Development Area (DA)" consisted of about 150 villages. An official called "Development Officer" was held responsible for the overall development in the area. He was supported by VAID workers (having one year training in development process). In each village, there was a village council (10–15 members) responsible to undertake small development projects. The VAID worker acted as the advisor. The District Advisory Committee functioned under the chairmanship of the District Commissioner (DC). Its function was to supervise and coordinate the work of the constituent village councils.

The policy of providing employment opportunities to rural people for raising their income through different projects was also adopted by state authorities during the 1950s (Cheema, 1980). The investments of the state in rural infrastructure and the introduction of modern capital in rural economies of many developing countries including Pakistan in 1950s paved the way for a dual economy. The notion of modern rural capital was used to develop agriculture on up to date modern scientific lines and adopt improved agricultural technologies in order to increase productivity. To involve the rural youth in the rural development process an important step was taken by the state authorities by launching the rural youth movement in the 1950s under the banner of V-AID system. It also gave birth to *Chandtara* (mon-star) youth clubs (Mallah, 1997). Different community-based village organizations were instituted to ensure participation of local rural people (Abbas et al., 2009). In the late 1950s a new system of governance was initiated under the name Basic Democracies System (BDS) to solve rural community problems related to agriculture, education, infrastructure, and sanitation. This program was basically planned to bring both the elements of community and political development closer to each other at the local grass root level (Luqman, 2004). The focus of the first Five-Year Development plan (1955–1960) was attaining high national level income through growth in agriculture and its subsectors (Haider, 2011). With the start of this Five-Year Development plan, a new era of economic development was started based on a scientific approach (Ghafoor, 1987).

32.3.2 Rural Development during the 1960s

Due to the structural transformation in the rural economy in the 1960s and fundamental changes in policies and objectives of the state regarding the rural development paradigm, diversification was started in the rural economy of Pakistan. During this period, emphasis was given to nonfarm income activities along with farm income, which are essential for rural development (Gill et al., 1999). The agriculture sector was also given special importance in rural development strategies in the 1960s (Khan et al., 2011a). Increased availability of irrigation water and provision of subsidies on agricultural inputs played an important role in bringing tangible breakthrough in agricultural production (Islam, 1996). Agricultural technology transfer which was the prime duty of agricultural extension also played a significant role in adoption of new and improved agricultural technologies and boosting the agricultural production in 1960s (World Bank, 2007). The positive and significant impact was found in agriculture sector during this period due to the green revolution (Davidson and Ahmad, 2003). The idea of the green revolution was to increase the income of rural people through increases in farm production and increased provision of employment opportunities (Ahmad et al., 2004). High yielding varieties were introduced in Pakistan during the era of green revolution. This revolution resulted in increase in yield of two major food crops (wheat and rice) of Pakistan. An intensive program was also initiated in early 1960s for uplifting the standards of living of rural people with the name of Rural Works Programme (RWP) (Gill et al., 1999). Development of rural infrastructure and provision of employment opportunities to rural community at their door steps was the major feature of rural development themes of 1960s (Abbas et al., 2009). Mechanized farming was also introduced during this period under the strategy of rural development through agricultural development (Ahmad et al., 2004). In late 1960s Farm Guide Movement was also introduced in the country for encouraging involvement of rural youth in agriculture (Naeem, 2005).

32.3.3 Rural Development during 1970s

In 1970s especially during Bhutto regime (1972–1977) there was reversal of old rural development strategy. In this reversal of development strategy in which growth was mainly led by private sector shifted to nationalization and creating state monopoly in export of agricultural produces especially cotton and rice (Islam, 1996). In this regime preference was given to large scale nationalization of different sectors including agriculture and increased role of the state in welfare of the people (Tahir and Ali, 1999). These rural development strategies were mainly based on state led agricultural policies as major portion of economic growth mainly depended upon agriculture (Hamid, 2008). Under the preview of new development strategy Integrated Rural Development Programme (IRDP) was initiated and the development of agriculture sector was the main theme of this program (Luqman, 2004). Development practitioners emphasized on trickle-down theory of rural development covering multiplicity of rural development activities during this period (Sobhan, 1968). In this rural development strategy special importance was given to small and marginalized farmers through disbursement of agricultural credit (loans) by the state in order to promote high yielding varieties (Abbas et al., 2009). The credit served as an instrumental in spreading new agricultural technology through agricultural extension wing. But the agricultural growth slowed down during this period and the government changed their priority in development strategy which focused on industrial growth (Mahmood et al., 2008). The basic force behind the focus on rural industrialization was the idea which confirmed the contribution of industrialization to rural development by giving direct support to agricultural advancement (United Nations, 1978). In order to fulfill the basic needs of rural people, that is, schools, health centers, roads, clean drinking water, women training centers, etc., government launched a program with the name of People's Works Programme (PWP) through participation of local community in early 1970s. This notion of satisfaction of basic human needs up to minimum standards of living in rural areas was the poverty focused approach towards rural development (United Nations, 1978). The rural development strategies of this period played an important role in declining rural poverty (Tahir and Ali, 1999). The supervised agricultural credit scheme for small farmers was started during this decade.

32.3.4 Rural Development during 1980s

During the late 1980s (1988), a new experiment was tried to involve private sector in agricultural delivery system in order to raise the income level of rural people and to attain food security at national level. During this period special attention was given to agriculture and rural development including provision of welfare services to the rural community under the Structural Adjustment Programme (SAP) reforms in different sectors including agriculture (Anwar, 1996; Islam, 1996). The program was launched in order to minimize the economic imbalances in the country. Under this program there was shift from industrialization to farming of cash crops and increase in export of agricultural commodities (Bhutta, 2001). But due to the SAP unemployment and inflation had increased and slow economic growth during this period lead to increase in rural poverty (Khan et al., 2011). Some other research studies also proved that due to SAP poverty and income inequality increased in the country (Kemal, 1994; Jaffery et al., 1995). Due to the failure of state-led rural development policies in 1980s and the space given by the public sector in the development of rural people and to alleviate rural poverty, was filled by mushrooming of different civil society organizations commonly known as non-governmental organizations (NGOs) (Bennett, 1998; Kalim and Salahuddin, 2011; Banks and Hulme, 2012). The main focus of these organizations was to develop rural poor and marginalized people socially and economically on sustained basis as poverty is more severe in rural areas than urban areas (Rehman and Ismail, 2012). In this era due to the high poverty rate among rural women, attention was given to their development and also to involve them in the development process (Afzal, 2009). The performance of agriculture sector remained stagnant during this period, which was mainly responsible for high poverty rate in rural areas (Asghar et al., 2012).

32.3.5 Development during 1990s

During 1990s there was also increase in rural poverty as during this period government reduced subsidies on agricultural inputs like fertilizers, pesticides, seeds, electricity, etc. (Bhutto and Bazmi, 2007). Political instability and poor economic growth during this period was also responsible for rise of rural poverty (Asghar et al., 2012). Numerous research studies proved that there was an increasing trend of rural poverty in this era (Intizar, 2004; Orden et al. 2006; Omer and Sarah, 2008). In 1990s different rural support programs (RSPs) were also originated in the form of government assisted NGOs having mandate to promote rural development in Pakistan using participatory approach and also empowerment of rural poor (Riaz et al., 2012a,b). They served as an intermediate between state departments and NGOs (Kalim and Salahuddin, 2011). During this era microcredit in rural areas was given to rural community for agricultural purpose (Waheed, 2009). The microcredit schemes launched in rural areas of the country under the umbrella of different NGOs, RSPs, private and public sector banks had some impact on poverty reduction (Latif et al., 2011). The idea of bringing improvement in the well-being of rural people and in broader sense "human development" covered the rural development initiatives of 1990s through the "well-being approach" of rural development instead of "technocratic approaches". Under the theme of well-being rural development approach different programs like Tameer-e-Watan Programme, Social Action Programme (SAP), and Khushal Pakistan Programme were started to bring improvement in the rural infrastructures and provision of basic human needs to rural communities (Azizi, 1999; Khan and Khan, 2001; Government of Pakistan, 2005). The theme of sustainable rural development gained importance during this period due to the heavy criticism of the green revolution and previous rural development policies in which emphasis was only given to production without keeping in mind ecological imbalances (Hussain, 1982; Ali and Byerlee, 2002; Ahmad et al., 2004). In the 1990s social sector development policy became the central part of rural development strategy in which greater importance was given to human development (European Commission Pakistan, 2007). In the early 1990s gender and development became an integral part of rural development process in all the developing countries including Pakistan by adopting gender mainstreaming approach (IFAD, 2011; Grigorian, 2007).

32.3.6 Development in 2000s: Early 2010s

In 2000, the Millennium Development Goals were adopted by the government under the umbrella of overall development strategy for sustainable human development and rural poverty reduction (Arif and Farooq, 2012a,b).

One of the most significant initiatives during the 2000s was "Devolution of Power Plan" introduced by the military government of General Parvez Musharraf in 2001. A new system of local governance was announced that has five fundamentals, or 5Ds;

- Decentralization of governance
- Deconcentration of management functions

- Devolution of political power
- Diffusion of the power-authority nexus, and
- Distribution of resources to the lower level (i.e., district level)

The crux of this system was that the local governments are answerable to people and rural development should be initiated at the lowest level of administrative tier (i.e., union council) (Figure 32.2).

In this system special seats were allocated for women, peasants, laborers, and minorities. There was provision of citizen community boards (CCBs) through which the local people could form groups and submit their development plans for direct funding by the government. This system remained intact until the regime of General Musharraf, but afterwards the political government gradually abolished it because of heavy criticism by the elected parliamentarians. In 2010, the 18th amendment in the constitution of Pakistan added another dimension to the development paradigm when many functions of the federal government were devolved to the provincial governments. An important rural development program launched in 2003 was the Tamir-e-Watan program. This program covers interventions relating to telecommunication, education, furl, electricity, roads and bridges, drinking water, sanitation and health, etc. Each member of the national assembly and Senators, were allocated Rs. 10 million yearly and they identified and recommended development projects for execution.

During the last 2–3 years, there has been increasing emphasis on the development of rural areas. Some of the most recent initiatives of the current government include the "Khadim-e-Punjab Rural Road Program" in Punjab which is a mega project aimed at the rehabilitation and repair of rural roads to increase market accessibility. The Prime Minister of Pakistan has announced the mega

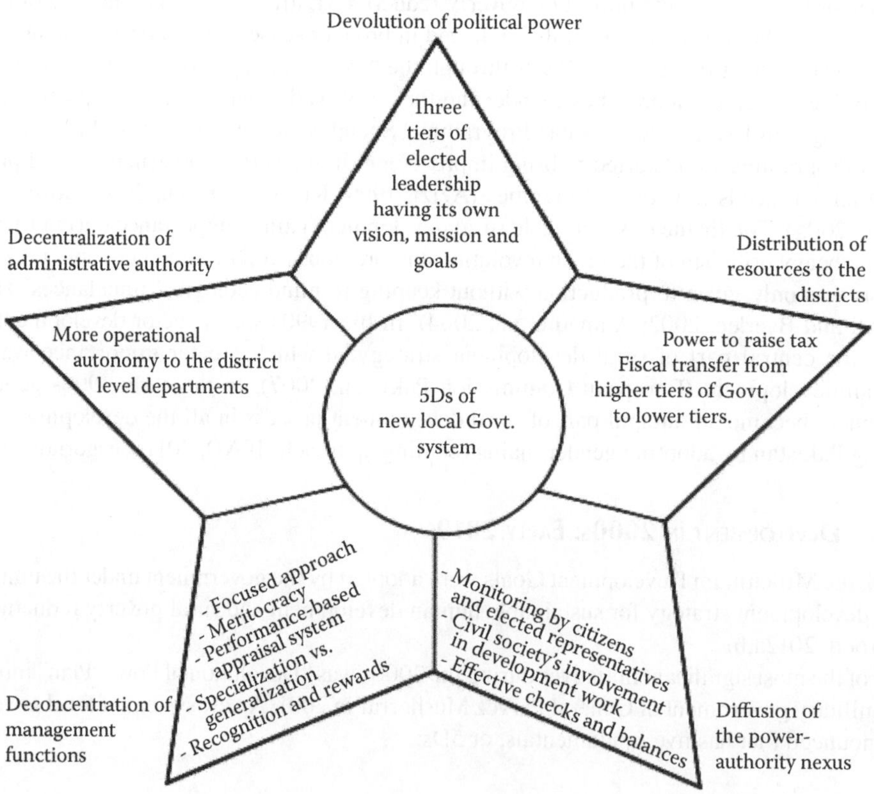

FIGURE 32.2 The local government plan of General Musharraf of Pakistan (NRB, 2005).

Kissan Package in 2016 with emphasis on providing relief to the farming community of Pakistan with agricultural credits, cash payments (zari relief) to cotton and rice growers, and increased access to agricultural inputs (particularly fertilizer).

32.4 CURRENT STATUS

Given the fact that nearly 70% of the population of the country resides in rural areas, the government is devoting lots of resources to empower local communities through various development initiatives as a part of overall national integrated rural development programs including;

1. Primary and secondary education
2. Technical and vocational education
3. Professional and skill development in agricultural farm and nonfarm activities
4. Community health and hygiene
5. Establishment of women skill development centers (WSDC)

Besides those mentioned above, other programs exist for the overall betterment of rural areas, the government announced various agricultural and rural development initiatives from time to time. The present government has recently announced a kissan package for the overall improvement of the agriculture sector and rural life. The following are some of the initiatives of the present government

1. Prime minister's kissan package
2. Prime minister's youth program including interest free loan (IFL) scheme
3. Tunnel farming development program
4. District road resource management project
5. Construction of farm to market roads
6. World Bank assisted training of local government elected leadership and state functionaries project
7. Pilot project for model village development

The present government has recently announced an unprecedented Kissan Package with 341 billion rupees allocated for the development of the agriculture sector and overall rural development. Out the total 341 billion rupees, 25 billion rupees were allocated to provide fertilizers at subsidized rates. Furthermore, it was decided to give cash support of 34 billion rupees to almost 1.6 million farmers in the country to compensate crop damages due to floods. The package also included crop insurance and support for the installation of solar tube wells, etc. Similarly, import duties on agricultural machinery were significantly reduced. This package is appreciated by many farmers and agricultural scientists; however, there are some critics of this package. According to them input subsidies and cash support have created suspicions that who would really benefit from such policies? Under the recent Kissan Package, it was decided that all farmers of the country would benefit from this package. Many issues have been reported concerning the distribution of this relief package. A farmer reported that his cotton and rice crops were destroyed consecutively for two seasons and he remained totally deprived of any relief announced by the government. He said that people/ farmers having good ties with government officials and local politicians are getting the support. Such viewpoints are very rare but they do exist according to some experts, cash support is not a sustainable option but it is mere a part of to avoid crop losses in future. According to such critics the short-term solutions in terms of input subsidies would not be a substitute for developing overall agriculture sector.

In recent programs, focus is given on short-term incentives; for example, cash support, input subsidies, etc., while there is limited emphasis on given for overall resource management. Being an agricultural country, agricultural policy should be the most important policy subject in Pakistan, but it has seldom been taken seriously for the past few decades. During the green revolution, Pakistan was amongst the early adopters of the new agricultural technology. The adoption of improved agricultural technology was largely aided by government support. As the result of government support, agricultural growth jumped to 3.8% form just 1.8% and to 6% during the 3rd Five-Year development plan. The highest growth rate of 11% was recorded during the years 1967–1968. However, this momentum could not be sustained and the agricultural growth plummeted 1.9% in the next decade after the green revolution era. But, the post-green revolution policies and their impacts on agricultural growth and rural development have rarely been critically investigated. The lack of a holistic rural development approach and political instability have resulted into the present agricultural crisis. Poor water resource development has also badly affected agricultural development and eventually rural development in the country. The Asian Development Bank has warned in its Asian Development Outlook 2013 report that Pakistan is close to being classified as a "water scarce" country. The government is interested in dealing with current water issues, however, long-term water resource planning is not addressed in agricultural and rural development programs.

32.5 PARADIGM SHIFT FOR MEGA PROJECTS

During past, many mega projects planned and implemented either related to agriculture or overall rural development in the country. But, we see a paradigm shift as there are more mega projects related to the development of cities rather than villages. During the 1950s, a mega project was initiated under the first technical assistance agreement signed between Pakistan and the United States in 1952 with the purpose to increase agricultural production. It was named Village AID program which started in 1952 and ended in 1961. In the 1960s most of the government funds were invested in the development of villages through Basic Democracies. It was a system of developing rural areas through local government institutions. The local council called union council (the basic tear of BDS) was held responsible for preparation of development schemes. It was supervised by the DC of the district. During the 1970s (1971–1978) the mega project "Integrated Rural Development Programme (IRDP)" totally focused on rural development. It was based on joint efforts by farmers, governmental departments, and local organizations to make Markaz (community) a unit of integrated development.

After 1980 there was a paradigm shift from state-led to community led rural development programs. A booklet entitled, "Guiding principles for people's participation projects: Design, operation, monitoring, and ongoing evaluation was published by FAO in 1983. It served as a guideline for participatory rural development projects. During 1989–1990 about twenty-two billion rupees (33% of PSDP) were allocated for rural development—rural health, electrification, farm to market roads, safe drinking water, and sanitation facilities. It included mega projects like USAID resources management projects and ADP farm to market roads projects. In the 1990s, different projects were started under Tameer-e-Wattan and Social Acton Programs. Later on, there was more focus on urban development rather than rural development. The mega projects of urban roads, motorways, higher education (universities in cities) remained the focus. Recently, the government has focused on providing help to farmers through Prime Ministers Kissan Package. But still, overall rural development needs more attention by the government. While critically reviewing the Five-Year Plans and developments in the country, we find out that actions did follow plans.

The 10th Five-Year Plan emphasized food security, rural industrialization, social mobilization, women empowerment, and construction of farm to market roads for overall rural development. The plans look good on papers but the good actions need to be taken as per plans. The latest Five-Year plan (11th) has emphasized upon the development of human and social capital. It focuses on education, health, women empowerment, poverty reduction, partnership, reforms in energy and communication sector. But, the main root causes of most of the evils such as lawlessness, corruption, favoritism, and nepotism, still need serious and honest efforts to reach the planned destination. As long as the rules remain for the poor only and the rich find ways to get escaped, no vision has the potential to be turned to reality. Many visions emerged and failed. We remain blaming others for the failure but, in reality we are not touching the real cause, that is, the law and order situation in the country.

32.6 RURAL DEVELOPMENT IN KOREA: EXAMPLE FOR PAKISTAN

Rural development in South Korea is often taken as example for developing countries. Korea was a very poor and undeveloped nation when it became independent in 1945; immediately after independence there was the Korean War (1950–1953). During the late 1950s, South Korea started its rural development programs with little success, but in the early 1970s a mega rural development initiative (Saemaul Undong) initiated by the government completely changed that situation and in short period of time Korea surpassed not only Pakistan but also almost all of the third world countries in terms of rural development. Now Korea is one of the highly developed nations of the world. In contrast Pakistan, which had also made huge investment in rural development, is still struggling to achieve rural development.

Korea is located in Southeast Asia, has a population of about 50 million with less than 100 thousand square kilometers of land area. Its literacy rate is almost 100% and GDP per capita is among the highest in the world.

At the time of independence in 1945, the Korean economy was largely based on agriculture. The agriculture was controlled by feudal landowners, and rural communities were extremely poor and hungry. The main staple food was rice, which was in very short supply during the 1950s (KREI, 2010); the main focus of the then Korean government was to ensure the food supply. The agriculture sector was characterized with low productivity, outdated technology, lack of inputs, and bad infrastructure (ibid). However, after the war, the country got large funding (mainly form the Unites Sates). During this period the government invested heavily on ensuring the supply of inputs (fertilizers and seeds) and capacity building in rural areas (Reed, 2010). But the land was in the hands of few landlords and majority of the farmers were tenants therefore the need for land reform was recognized by Korean government during 1950s. Consequently, a process of comprehensive land reforms was started. There were two major parts in the land reform process: (1) prohibition of tenant farming and, (2) an upper ceiling limit of three hectares (Park, 1998). Thus, in a short period of time landlord-tenant relationships were broken and a new class of farmers emerged during 1951 to 1958 (KREI, 2010). Therefore, the effect of the land reform was substantial and a majority of the farmers became owners, and rural society became more homogeneous. This homogeneity played a crucial role in the participatory rural development initiatives of the 1960s and 1970s.

The establishment of farmer cooperatives was a key policy of the Korean government along-with land reforms during the 1950s; as well, Agricultural Cooperative Bank and village cooperatives were developed during this decade. However, rural development initiatives of South Korea created incredible momentum during early 1960s when General Park Chung-Hae became president and straightaway the expansion of the industrial and engineering base of the country became the focus of the government. The first 5-year economic development plan was introduced in 1962 and consequently the rate of growth in Korea became more than 7% (Hong, 2013). While the period during 1960 to 1970 was a period of speedy industrialization in Korea, there was comparatively slow development in the agricultural sector. The Rural Development Administration was established in 1962 to foster agricultural research and extension, investments in inputs supply, irrigating facilities, and marketing were also made during this period (Reed, 2010).

The second 5-year plan was started in 1967 and achieving self-sufficiency in rice was outlined as one of its most important goals. During the 1970s, a massive rural development program "Saemaul Undong" (New Village Movement) was introduced as an integrated rural development project intended for social mobilization in rural areas and development/transformation of the rural infrastructures. Later, on Saemaul Undong (a Korean rural development movement) encompassed the whole paradigm of rural development policies (Reed, 2010). The Saemaul Undong became the institutional and ideological framework for 3rd 5-year economic and development plan started in 1972 and eventually all rural development interventions were brought into the framework of Saemaul Undong. There was an enormous increase in the investment on agriculture and several new programs were announced in the third 5-year plan like rural electrification, irrigation, communication, marketing, etc. Many authors agreed that the impacts Saemaul Undong on rural society in South Korea were tremendous and the country entered into the golden age during late 1970s as far a rural development is concerned.

Some of the key success factors of the Saemaul movement were the self-help spirit, the homogenized rural society after land reforms, rural cooperatives, industrial development, active social mobilization, and full support by the government. The new village movement (Saemaul Undoing) completely transformed rural areas of Korea into modern and production oriented centers. Rural infrastructure was also greatly improved, traditional thatched huts were converted into modern houses, roads and bridges were built, the self-help spirit was promoted, and household income was considerably increased due to higher agricultural productivity. Many researchers have agreed that during Saemaul Undong Korean villages became almost unrecognizable (Douglas, 2013) and by the end of 1970s not a single village was categorized as underdeveloped.

There are three main component of Saemaul Undong, that is, diligence, self-help, and cooperation and these three components were promoted through the country and made a marked impact on Korean society as a whole. One of the salient features of Saemaul Undong was the emergence of a new generation of egalitarian village leaders who replaced the traditional landlords and local chiefs. The Saemaul Undong also brought rural women to the mainstream and it was compulsory to have one woman leader in each village (Douglas, 2013). The basic factor of success of Saemaul Undong (or New Village Movement) was the solid contribution of government and abolition of the feudal system. President Park himself took a deep interest in all of the activities of Movement and his passion was another factor that tremendously encouraged rural developmental efforts. President Park visited personally most of the villages where integrated rural development projects were under way and participated in many activities himself. Rewards and incentives was another strategy to motivate the self-help efforts of villagers.

The Korean version of the green revolution took place during the 1970s when the country became self-sufficient in rice in 1975 and this self-sufficiency gave spirited support to the Saemaul Undong. Many researchers called self-sufficiency in rice a big achievement because rice is the main food staple. The agricultural extension services played the most important role in the green revolution and the Rural Development Administration used contemporary approaches to educate rural people about production technologiesof high yielding rice varieties. The dynamic role of farmers' cooperatives was also an important factor in rural development in Korea.

President Park Chung-Hee was assassinated in 1979, followed by political unrest in country that put the economy of the country into a crisis but by then Korea had achieved marvelous developments in the rural sector. However, the contribution of the agriculture in the national GDP began to drop in the 1980s from 26% in 1970 to about 15% in 1980 and less than 8% in 1990. Similarly, the rural-to-urban migration rate significantly increased. The income of rural households was also reduced and consequently the government's agricultural policy was shifted from increasing rice productivity to emphasis on high-value crops and nonfarm income opportunities. Accordingly, a holistic plan was developed in the mid 1980s to improve the rural economy. The main features of the program were establishment of rural industrial units to increase nonfarm income. Integrated Rural Development Approach was the key focus of 5th 5-year plans in 1982–1986. Major structural changes were also introduced in cooperatives, which were democratized. Emphasis was made on tunnel farming for the production of high-value cash crops and this initiative brought a second green revolution in the country.

A notable feature of Korean rural development policy after 2000 is the sensitization on trade liberalization and global climate change issues. For instance, a green growth strategy was highlighted in agricultural policies in 2007 and the first 5-year plan for Environmental Friendly Agriculture was announced in 2001, and then a second 5-year plan for environmentally friendly farming was announced in 2006. There was a significant reduction of agrochemicals, such as pesticides and chemical fertilizers by 2010. Village tourism has been promoted during the recent years as an alternate income source for rural communities. Rural tourism is very attractive for urban citizens who have little interaction with rural areas. These programs are claimed to have an inordinate potential to improve rural livelihoods and the preservation of the cultural heritage of the country (for details see, Shahbaz et al., 2014).

32.7 OPPORTUNITIES

32.7.1 Nongovernment Organizations

There are a number of nongovernment organizations (NGOs) currently working for the overall development of the rural poor. Due to bad intentions and nonethical acts of some NGOs the word "NGO" has become a bad name, but still many NGOs have good record of decent work in rural development. We personally know some NGOs that have recently been distributing foods items among the rural poor during this Ramadan. Some other NGOs are involved in free education of rural children. Such NGOs are a great opportunity for the government to get their participation in government initiatives for rural development. In order to involve NGOs in development, a fair monitoring and evaluation system is a basic requirement.

Various stakeholders and sectors are working at micro as well as macro levels to make people prosperous and developed. NGOs is the top leading name nowadays. NGO is defined in two ways; one is in the broader sense while the other is in a narrow sense.

The failure of the Pakistani government in delivering basic services to its poor has given birth to various NGOs which are serving these disadvantaged segments of the population (ADB 1999). The roots of various NGOs in Pakistan can be traced back to 1947, when the country got its freedom from the United Kingdom. Initially, these NGOs had put all their efforts for the rehabilitation of migrants from India after partition. A rapid increase in the number of these NGOs was noted in the 1970s and 1980s in Pakistan; the major factor behind this mushrooming growth was a continuous inflow of donations and generous funding from international donors as a result of its alignment with the United States and its allies. During this period, these NGOs have played a key role in creating awareness among the general public regarding many social and economic fields such as basic human rights, overpopulation, women empowerment and the provision of many basic necessities of life in the remote areas of the country.

The services offered by NGOs during the 1990s were expanded to include basic health services, education, awareness campaigns, income generating skills, water and sanitation facilities through the provision of credit facilities, microfinance programs, micro infrastructure projects (MIPs), and targeted poverty alleviation programs (ADB, 1999). These NGOs thus started to place more emphasis on different community-based programs as well. At the same time government of Pakistan also emphasized to enhance the role of different communities in development. When devolution policies were introduced in 1999 to 2008, the local governments were given more powers and a new community-based development scheme was introduced named Citizen Community Board (CCB) (Kurosaki, 2005). A number of Pakistani NGOs having experience in the community-based development were involved in running the CCB schemes during that period. However, a comprehensive economic research was not conducted to determine the impact and conditions underlying the success stories of these NGOs in Pakistan.

NGOs are playing a very active role in Pakistan for the overall development of the marginal and disadvantaged segments of country. These NGOs carry out their development plans by the collection of data, its proper analysis and service delivery. Some major working areas of these NGOs are education, health, women empowerment, socioeconomic development, and capacity building of civil society organizations. Moreover, these NGOs are playing an efficient role in creating awareness among different communities about their rights and responsibilities. Pakistani

NGOs are vigilant to watch the progress of overall development and had played very active role in the restoration of judiciary. Unfortunately, there are some NGOs which misuse the funds and resources available to them.

The role of NGOs has become diversified being now involved in social welfare, economic development, gender awareness and action, employment creation, skills training, peace and human rights, and informal economic sectors. Various NGOs are working particularly for the development of agriculture at farm level. These NGOs have played a massive role in Pakistan to uplift rural development. Some of the NGOs working for rural development in Pakistan are available online at http://www.ngos.org.pk/ruraldevelopment/rural_development_ngos.htm

A long list of NGOs available on the above cited website clearly shows the broad spectrum of these NGOs working for rural development in Pakistan. In the past these NGOs have participated in different development efforts and also complemented programs undertaken by the government. Some of their major contributions are improvement in human capital, development, and technical assistance. In addition, development of small farmers, monitoring, research, evaluation, advocacy for the poor, peace building, and enlightenment campaigns are major concerns of NGOs. However, NGOs also face some challenges that impede their working efficiency (Ngeh, 2013).

Resource poor people in developing countries hardly manage their resources over time. Credit provision enables them to manage their resources and initiatives more efficiently to build more income sources, assets, or economic security. However, traditional or conventional finance institutions occasionally lend down-market to serve the needs of the resource poor or low income and women headed families. Often these inhabitants are denied access to the capital/credit at the need of the time. Therefore, unaffordable terms of loan are not major problems but lack of access to credit itself appears to be a major constraint (Ghate et al., 1992; Mustafa, 1992). Providing better solution community banks and grassroots savings credit group globally have depicted that the micro-enterprise loans can be profitable for borrowers and for the other lenders in making appropriate poverty reduction strategies (Aziz, 2000).

It is difficult to estimate the exact number of NGOs working in Pakistan. Only a rough estimation is possible. The total number of these NGOs was suggested between 8000 and 16,000 in a publication of UNDP in 2001. While some other studies suggest that the number of both registered and nonregistered NGOs in Pakistan could be anywhere between 25,000 and 35,000. NGOs in Pakistan can be divided into several broad categories on the basis of their involvement in the following functions: advocacy and lobbying; policy issues and debates; emergency, relief and rehabilitation organizations; those involved in development projects and programs including service delivery organizations and community-based organizations (CBOs). NGOs support different activities in rural areas, in the field of agriculture, health, education and access to water among people. In case of agriculture, these organizations engage in research and guiding the rural dwellers to adopt the best practices in order to improve their agricultural output. For example, many NGOs around the world have developed programs to fight food insecurity in various parts of the world. For example, the Hunger Project and Heifer Uganda offers tangible help to the people of Uganda through training. Through their program they help people to live sustainably. They also give aid to people to boost them in kick-starting agricultural practices. There is also food security and positive social change as people embrace certain values like self-reliance and are not necessarily seeking aid from the government.

In the light of the situation described above it can be said that these NGOs have a significant role in rural development. Therefore, it is recommended that the government should embark on aggressive development of the rural and marginal areas of the country. Similarly, NGOs at the various levels should partner and work in collaboration with the state. Besides the introduction of strict rules and regulations concerning the behavior of some nonstate actors within a community, these NGOs must be provided protection, alternative financing, positive incentives, and better government help to expand their working areas and level of services for the poor. When power is shared between the government and the informal institutions of civil communities, the democracy and the rule of law can be strengthened.

32.7.2 Rural Youth

The youth is the future of any nation, and energizes it. There are 116.52 million people below the age of 29 years, a majority of whom—rural youths, mainly reside in rural areas (GOP, 2015). They are highly energetic, enthusiastic, and show potential for development. It is a dire need to mobilize this rural youth in the direction of country growth and economic prosperity (Idrees et al., 2008) Previously, farmers did not like to educate their offsprings, especially female ones. But modernization has also modified the minds of farmers, now a majority of them are sending their kids to schools for a better future for their next generation. The literacy rate in rural areas is 49% while 78% of the 5–16-year old were enrolled in rural Pakistan in 2014, which shows an accelerating trend in education towards youth mobilization (GOP, 2015).

Rural youth have improved awareness due to higher literacy. If we peep into past (1950s) there were only one or two literate persons in the whole village and rest of the people were required to approach them for reading their personnel letters. Now every farm family has literate persons not only they are well-equipped with education and information and communication tools (ICT) tools but they are better aware of society, urban customs, and opportunity of education and job than the past (Ceobanu and Boncu, 2014).

Rural youth has an opportunity of access to information sources. Majority of the farmers have TVs, radios, and mobile phones. They get information about production technologies for agricultural crops and raising of livestock and poultry (Sattar, 2007). Although rural youth is equipped with the above mentioned interventions, mainly they are utilizing their mobile phone extensively. The bias of the youth towards a cellular phone is expanding in light of the fact that it advantages them for communicating, learning, amusement, social connectivity, and perusing (Campbell, 2005; Nehra et al., 2012). The mobile phone is a basic technology for updating, retrieving, communicating with others. Rural youth can be mobilized through their phones by special updates from specific institutes (Lemoine and Ramsey, 2010).

The government has taken several initiatives to mobilize rural youth in which a number of capacity-building programs were oriented under the Umbrella of the Prime minister Youth Program; different packages were introduced for youth development such as the Laptop Scheme, Skill Development Program, Youth Training Scheme, Youth Business Loan Scheme, Interest Free Loan Scheme, and Fee Reimbursement Scheme. Rural youth are getting benefits out of these schemes and credits have been supplied to unemployed youth to start a small business. However, in all these schemes common rural youths are getting minimal benefit out of them. Monitoring and evaluation of these schemes may be ensured to approve transparency and to benefit its real stakeholder. Youth in rural areas have lots of energy, time, and potential to change and reshape their living places. They may be better involved to deal with the venture of rural development. Their interest and skill in ICT may be utilized further to enhance the knowledge base of the rural social system. Organizing them on the pattern of 4-H and FFA (Future Farmers of America) may provide us with better villages and better rural socio-economic system.

32.7.3 Gender and Development

The term "Gender" refers to both biological sexes, that is, male or female (FAO, 1997; Chegg, 2016). "Development" is a process which improves the quality of lives of human beings with three imperative perspectives *viz.* (*i*) raising peoples' living status including income, health, education, and food through relevant growth processes; (*ii*) focusing on self-esteem of human beings, (*iii*) increasing peoples' freedom of choice of goods and services (Lorenzo, 2011). Development is a continuous process with multifaceted dimensions that entails gender as a key in this phenomenon. Of all the keys of sustainable development, that is, physical, natural, social, financial and human, the last one has the utmost significance (G-DAE, 2003). In human capital gender role demands a crystal clear analysis that encompass rural development. There are six different schools of thought in this context, that is, (*i*) the welfare approach; (*ii*) women in development (WID); (*iii*) women and development (WAD); (*iv*) gender and development (GAD); (*v*) the effectiveness approach (EA); and (*vi*) mainstream gender equality (MGE). The GAD approach it is not just focused on the sexual differences among men and women but it also emphasized social, reproductive, and economic roles that are linked to gender inequalities. The WID approach focuses on assimilation of women into the workforce and increases production to raise standards of living. The most recent one, presented in 1995, is gender mainstreaming which ensures that all gender issues are dealt with and integrated in all social, politica,l and institutional stages. It was decided that inclusion of both genders in every development program/stage is the only way to succeed and progress nationally and globally (The Sociologist, 2013; Allen and Dino (n.d).

Pakistan has many social taboos, especially in rural areas where women were considered a marginal sect. It was time women were hated, burnt, slaved, encroached, cursed, victimized and deprived (Imran, 2005; Marshall and Sabhlok, 2009). But after the women's protection bill in 2006 and its later amendments, legal position of women became stronger which encourages peace and equality in society. According to it women kidnapping, abducting or forcing women into marriage is a crime and punished with imprisonment for life (Hassan, 2016). In rural areas, forced marriages were common; however, after this bill, such incidences are remarkably fewer. This is an addition in women's empowerment; now women are considered to take decisions according to their own will (Jamal, 2006).

Women of Pakistan were confined to four walls and Pakistan had lowest female labor force participation rate in the world (Sofa and Doss, 2011). It is due to the reason that farm women's work was not paid labor (Labour Force Survey, 2015). This can be addressed with gender balancing policies and strategies. On the other side, women's active participation in the economy is also increasing through a more professional approach. Trends of women's work has changed from the kitchen shelf to the office seat (Grunenfelder, 2013). In rural areas, with the passage of time there has been a shift from illiteracy to literacy. Now, in the era of education and information, women are well equipped with knowledge and awareness. Their literacy level, however, as defined in 1998 by "One who can read newspaper and write a simple letter, in any language," was only 20% of rural women (UNESCO, 2003), while in 2015 this has been increased to 49%, showing a positive trend towards an increase in education (GOP, 2015). ICTs tools are common even in rural areas where they stay connected with their near and dear ones in furlong areas. ICTs infrastructure has been expanded and continues to grow (Malik, 2016). They get updated information regarding their personnel and professional issues. They also have a political voice at all levels from town committee to parliament. They can make their decisions of education, profession, and marriage. Due to active role as breadwinners they have improved in controlling income (Sofa and Doss, 2011). Their role in the economic development is getting better.

Farmwomen provide the lion's share of agricultural development. Their roles vary considerably ranging from domestic to farm operations (Gandahi, 2016). They have a number of activities to be performed on-farm (Khan, 2007). Such activities are essential to the well-being of rural households (Sofa and Doss, 2011). The involvement of rural women in local, regional, and national politics is also increasing.

Political power has been increased among women as reserved women seats in the assemblies were increased from 2% to 20% after President Mushraffs' regime. Women reserved seats are fixed at 33% of the Union, Tehsil, District Councils, Provincial and National assembly and senate. In the 13th National Assembly (2008–2013), there were 76 women legislators, of which 16 were elected against general seats (NCSW, 2013). Women's participation was 19.9% overall in the legislative seats election in 2013. Women's political empowerment trend has also influenced grass root candidacy and voters trend, and society is absorbing the reality of gender balance in politics (Bari, 2009). In the past, women were not permitted to cast a vote, however this pattern has been changed and women can now take their decisions regarding votes. There is a great drift in Pakistani politics through involving women in politics. This trend can be a driver in modifying male dominancy into acceptability of women's status (Saed, 2013). This societal change is becoming a great opportunity for rural development, as rural women are now part of the whole development process.

The media has been playing a marvelous role in rural women awareness rising. It served a role of masterminding attitude changes. Rural women mostly like to watch TV dramas and news programs. They were updated about agricultural innovations, cooking recipes, fashion, and societal trends. Now the media has also bridged the rural–urban gap (CTA, 2009). They also get information about education and jobs at their home. Women are also empowered in the sense of decision-making. They can take better decisions about choice of a profession and can play a positive role in the rural economy. It is a positive for gender empowerment in rural areas (Media, 2010). Rural women are getting benefits in the form of awareness regarding sanitation, health, education, and they are facilitated through schools, hospitals, gynecological centers, etc.

The government of Pakistan has taken many initiatives along this line for the welfare of women. A number of career opportunities have been created exclusively for women. Multiple strategies have been designed and implemented by the government of Punjab to facilitate working women, such as the establishment of PDCF Day Care Centers (Govt. of The Punjab, 2013). In the past, this 50% portion of the rural population (females) was confined to homes and remained busy in domestic chores, but now there is better awareness and empowerment for them. This opportunity can be utilized by involving them in the holistic development process.

32.7.4 Cottage Industry

The small scale and medium scale industries play an instrumental role in enhancing rural development in developing countries. The encouragement of rural industries leads to an uplift of the political, economic, and social situation and is responsible for creating an economic balance between rural and urban areas. Due to the development of a cottage industry in rural areas people get employment, which reduces rural–urban migration of workers. The livelihood of villagers mainly depends on agricultural development but in Pakistan the yields of major crops have been stagnating for the last few years. Moreover due to transfer of agriculture land from generation to generation, most of people are cultivating small areas of land, which is not enough to fulfill their economic and social needs. This situation is compelling the rural people to search for new venues for income.

There is need to establish linkages of agriculture with the cottage industry in rural areas which is a condition for the achievement of sustainable growth of the agriculture sector. Support of native small and medium scale entrepreneurship at larger level with real participation of all stakeholders can be helpful for value addition of locally produced agricultural commodities. The manufacturing of agro-based products for international, national, and local market can change the economic face of rural areas particularly where small farmers are being engaged in agriculture. The identification potential production, injection of capital, introduction of affordable machinery/equipment, capacity building opportunities, and viable marketing activities can enhance the economic situation of rural areas. In most of the rural areas the basic infrastructure facilities are available, for example, roads, housing, electricity, etc. Therefore, there is great potential in our rural areas to produce a diversified range of products with support from the government and participation of local people.

32.7.5 The Third Wave

The futurologists (specialists forecasting future social systems) identify different time waves in the history of mankind. There was a Zero Wave called Pre-Agriculture Wave (PAW). During this wave human beings lived like animals in jungles. They depended upon natural plants and animals for food, tree leaves as their clothes, and caves as shelter. As time passed the needs of humans increased. Man learned to dig soil and sow seed. This gave birth to the first wave called agriculture. Scientific knowledge and continuous research increased man's abilities to get more foods per unit land. The economies of advanced societies and countries depended upon agriculture. Food items were sold for profit. The changing societies changed the whole wave. More and more production of agricultural products needed value addition. Resultantly, industrial wave emerged. This was called the Second Wave. There was industrial revolution all over the word. Some countries became more industrialized where as some others remained less industrialized. Overindustrialization created problems such as environmental degradation, ozone layer depletion, and climate change. A need was felt to deal with human problems scientifically. This lead to the creation of schools, colleges, and universities which served as centers of knowledge creation and dissemination. More and more knowledge was generated. Knowledge became the economy. It generated the third wave called the Information Wave. It created knowledge lead and knowledge-based economies. Information was source of income. In 1987, people experienced the third wave in the United States, but in Pakistan we are talking about it in 2016. In this wave we have the facilities to use the information highway—Internet. Our rural dwellers have access to Internet. Even the poorest one has a mobile phone with internet facility. Almost everyone is worldwide connected. Facebook is used by a majority of the rural youth. What's App, Line, Linked In, Twitter, and other social media are used by many. The cell phone alone has replaced many gadgets/tools such as radio, television, camera, answering machine, tape recorder, camcorder, torch, note book, activity planner, calendar, calculator, call recorder, gallery, and e-mail, Microsoft office, Skype, balance sheet, financial management book, and many more. FM radios are become increasingly popular in rural areas of Pakistan. Cable TV channels, Android Apps, and social media are getting more and more popular. Worldwide video calls through Facebook, Messenger, Google plus, Skype for business and other sources have turned the whole world into a chat room. Organizations, agencies, and people who develop such facilities earn millions and trillions of dollars. The OER (Open Online Resources) are available. The OER Commons can be visited at https://www.oercommons.or. It is an extensive digital library. Edutopia (http://www.edutopia.org/open-educational-resources-guide) is a type of OER. There are various such resources available to everyone in the word. Rural people, through their educated youths, can learn through these resources. Information and communication technology has facilitated human beings worldwide to interact with each other, even removing language barriers. Google can easily translate the text of one language to the text of another language. Now people don't have to memorize knowledge, but they have to use knowledge in decision-making. Therefore knowledge is available at their doorstep. This facility was not available to the kings of the past. This is an opportunity which must be utilized efficiently and effectively.

Many developed countries have crossed the third wave and they are now in the 4th wave called the Automation Wave. Most of their work is being done automatically through robots. They now don't have to depend upon human beings for performing certain sophisticated tasks, now being done by machines. They are headed towards the 5th wave called Planetary Wave. In Pakistan, we are still struggling hard to sustain in the first wave, that is, Agriculture.

32.8 BARRIERS/CHALLENGES

The law and order situation in rural areas is worse than that in urban areas. Some rich landlords and big influential criminal minded people still badly opress the poor small farmers and landless artisans. The victims are poor on one side and illiterate on the other. They have limited resources to fight against their opponents. They are forced to use the avoidance or the compromising modes of conflict resolution. In other situations, they have to face vulnerabilities, litigations, threats, violence, or in some cases severe financial damage.

The gigantic challenge faced by the farming community, which still remains true, is low return on their commodities because of higher costs of production.

32.8.1 Conflicts in the Rural Scenario

In the rural scenario, there are various impairments in the way of development. Conflicts are also obstacles that have directly and indirectly handicapped the development process. There are peculiar spheres of conflicts. Conflicts emerge from various prongs like irrigation water, land distribution, caste system, land, tenants, etc. (Shahbaz et al., 2014). Some types of conflicts of rural dwellers in Pakistan are discussed here.

32.8.1.1 Land Conflicts

Distribution of land among the descendants is a prominent root cause of disputes and conflicts among the rural people. Although the system of land distribution is available, however, the conflicts arise when people's rights appear to be snubbed. The illegal encroachment of other person's land causes disputes. The situation triggers another blind alley of litigation. This situation can last from generation after generation (Shahbaz et al., 2014).

32.8.1.2 Water Conflicts

Water conflicts are very frequent between India and Pakistan. Furthermore, interprovincial water conflicts are also critical (Ranjan 2012). These conflicts induce a negative and drastic impact on rural development. Moreover, the importance of canal water is clear because it acts as a lifeline for the agricultural sector. Getting a deep insight in water related issues, farmers are facing conflicts linked with distribution of irrigation water. In this regard, farmers at the head of watercourses have the edge and get more water compared to farmers at the tail end. The unequal advantage creates a situation of conflict. The privileged farmers at head usually do not bother about the farmers at the tail. As water is critical for crop production (Shahbaz et al., 2014), the farmers have indulged in water theft, which eventually deprives the other farmers of their share.

Farmers' disputes often appear at the time of field irrigation. Another form of conflict exists between the farming community and the irrigation department where farmers are in a situation of need of water and water is stopped due to "Warabanidi." "Warabanidi" system is not justified and stirs up the normal situation, particularly for small farmers. Conflicts arise when the department gives a lion's share of water to the powerful farmers and the small farmers are deprived of their rightful share (ibid).

32.8.1.3 Landlord and Tenant

The exploitation of the tenants by the landlords is also an impediment in the way of rural development. In a situation where the tenant's hard work is exploited by a landlord, discomfort then arises as well as agony and conflicts. Mostly absentee landlords are getting their benefits without any extra effort and on the other side the tenants have to live hand to mouth despite of their hard work.

32.8.1.4 Middleman in Marketing

Farmers are exploited at different steps/points of marketing,. Whenever farmers (mostly small farmers) do not have their own transportation to get access to the market transportation costs become a controversial point and farmers have to rely upon transport owners (carriers).

When farmers have output (crops) and they want to sell it at a reasonable price, the intervention of middlemen appears as a necessary evil. This holds true particularly in case of perishable commodities. The nature of this conflict is complicated. Ultimately, farmers have to be dependent of the middlemen regarding prices.

Apparently, it does not seem crucial, however, the gravity of the issue should not be ignored. When the price of produce is not at the desired level rather at nominal level, the grievances start on the part of farmers. Starting from the very first buyer of the produce until the final consumer there is a chain and everyone is getting his due share; only the farmer misses out. Mostly, there is a huge disparity between rates at which the farmers sell their produce and when the commodities reach the consumer. A pertinent research by Iqbal et al. (2009) revealed and confirmed this situation.

32.8.1.5 Inputs Fertilizers, Seed, Pesticides, etc.

Farmers' intentions are obviously to get agricultural inputs at low rates, while inputs providers are more inclined towards higher rates. The conflict is constantly deteriorating the situation. Seed is vital for a successful crop. In this context, farmers are conscious to get pure and authenticated seed. However, when the private/public seed agencies provide adulterated/impure/substandard seed to the farmers then a scathing situation appears in which credibility of the seed agencies vanishes and conflict arises.

The application of artificial fertilizers to soil is essential for nutrients. The black marketing and adulteration in fertilizers pose another problem for the farmer, that is, spending but getting nothing. In this situation, controversies appear.

Adulterated pesticides also create hurdles in the way of production. The agony, irritation, and frustration emerge when the farmers are faced with losses and futile efforts. Research by Iqbal et al. (2009) indicated that high prices of pesticides and adulteration are impediments for farmers.

32.8.1.6 Farmers and Government

There are more expectations on the part of farmers, (like support price of agricultural commodities, etc.) with the government. When their expectations collapse, then conflicts arise. When the policies do not seem in consonance with the farmers' benefits, reservations arise. When the price of inputs jumps up then criticism are the government. Taxations in the agricultural sector also create a situation of conflicts. Government procurement policies influence the farmers and when there is dissonance of interests conflicts occur. There are also controversies over the "Abyana" and other taxes.

32.8.1.7 Old and Young Farmers

There are also conflicts due to generational gaps between the old and young generation about various issues like decision-making, etc. The old people adhered to orthodox and traditional perceptions and are resistant to change while the youth are more eager to accept change and transitions. The conflicts may appear even between fathers and sons while making decision about various issues. Like in case of education, there is also a disparity between these two categories. Mostly the old people want to involve the young in the same profession as their own. Mostly artisan's children deviate from the artisans' profession and want to adopt other avenues for earnings.

32.8.1.8 Livestock Farmers

Tariq et al. (2008) pointed out various factors in the context of milk marketing. They indicated middlemen exploitation, price issues, seasonal fluctuations along with other dilemmas. When the outbreak of epidemics and other fatal diseases occur, livestock farmers find it difficult to tackle the situation. In this situation, they usually expect assistance from the public sector services from the government. When the diseases of livestock are not controlled then conflicts appear between the livestock farmers and the other stakeholders (service providers). There is also perception among livestock farmers that only the elite and the powerful livestock farmers are provided with facilities while small livestock farmers are deprived.

32.8.1.9 Farmers/Credit Banks

The farmers (getting loans) and the banks may have conflicts regarding loan disbursements and recovery. The farmers' prominent conflicts are over interest rates. The conflicts are more severe when farmers are unable to return loan due to crop losses. Moreover, when loans are misused then there is trouble for the banks to settle the issues. In this regard Riaz et al. (2012a,b) emphasized for ensuring the utilization of the credit for the same purpose for which it was obtained.

32.8.1.10 Small Poultry Farmers vs Control Shed Farmers

The conflicts of interest are present over poultry farmers' control of sheds. There are situations in which control shed owners try to get the monopoly with respect to the poultry market, which creates trouble for small poultry farmers.

32.8.1.11 Farmers and Agro-Based Industries

Farmers (raw material provider) to the agro-based industries face predicaments in getting payments for their produce and clashes arise. Selling produce seems logical on the part of the farmers. On the contrary, low prices as well as unnecessary delays in payments by the agro-based industry (like sugarcane) have become a major quandary (Nazir et al., 2013). Another dilemma is present when the agricultural commodity is not properly weighted and the farmers remain under paid. The conflicts are noticeable in terms of price and weight controversies between sugarcane farmers and sugar industrialists

32.8.1.12 Lack of Coordination among Nation Building Departments Impacts on Rural People

Lack of coordination and conflicts among nation building departments (such as education, agriculture, public health etc.) are having a drastic impacts on rural communities. Diversion of interests, biases, funds for exploitation, and so forth are promoting conflicts among the nation building departments. On this account, they do not coordinate with one another and the development of the rural areas suffer.

32.8.1.13 Urbanization /Industrialization

There are various schools of thought present in rural scenarios regarding urbanization. One school of thought is in favor of keeping the rural picture intact without the amalgamation of urbanization. However, the other school of thought is against this notion. The people belonging to second category are of the view that rural areas should be provided with the amenities available in urban areas. They are of the view that rural areas should look like urban areas.

There are people who are advocating industrialization for the betterment of rural people through generating employment opportunities. Moreover, the consumption of agricultural raw materials in the agro-based industries opens new avenues for selling produce. On the contrary, those people who are against industrialization are of the opinion that increasing industries in he rural areas would affect rural landscapes and pollute the environment. Converting fertile lands into buildings and industries is not a judicious decision.

32.8.1.14 Ego/Caste System/Ethnicity, etc.

Egoism is also a part of the rural picture. People consider themselves superior to one another and think only about themselves. They are indulged in fighting with other people. Nasir et al. (2015) pointed out that caste system (Biradarism) is important in the perspective of rural conflicts. Moreover, political interests are also linked with it. The caste system, ethnicity, and racism, and so on are the root cause of various conflicts. Political structures of the rural areas also create an environment for division and can lead to clashes in rural areas.

Multicultural rural scenarios are more vulnerable to conflicts. Some traditions, norms, and values create dissonances and disharmony. In multilingual environments, conflicts also spread due to biases. In rural communities, even intra and interfamily clashes are present.

Also, religions can be the source of conflicts. Religion can also be a source of conflict people belong to one religious sects try to get power and dominance by suppressing the other sect.

32.8.2 Dependence/Belief System

Rural people are much more adherent to their belief systems. They have very strong linkage with their traditions, norms, and values. Anything contradictory to their beliefs, values, traditions, and so forth faces resistance. Fatalistic behaviors are also a big hurdle. "Destiny rules over man" pushes a practical person towards the impractical. The habit of associating the loss with hard luck and bad omens causes frustration. Farooq and Kayani (2012), through their research, affirmed that a substantial proportion of people believed in various superstitions, meaningful dreams, and supernatural aspects such as black magic, taweez, and ghosts. They also reflected that people are adherent to these types of events on account of their personal experiences or observations.

Among the rural community there is also a misconception regarding the role of rural women. Although male and females both contribute towards rural development, gender related conflicts are present in the rural picture. Mostly, there is gender inequality in terms of rights and duties. Male dominance is a prevalent factor which undermines rural society. Females are also equal contributors for rural development. They work shoulder to shoulder with males but their work is not recognized. Their work in agriculture demands due respect and recognition. It is also a social dilemma that females are mostly deprived of their right of decision-making for their family, which create a conflicts. Butt et al. (2010) pointed out that cultural norms, male dominance, and traditional belief systems are impediments in the context of social milieu for women. Similar views were expressed by Arshad et al. (2010). Moreover, misinterpretation of religious teachings and cultural norms has an influence with regard to women's decision-making power.

32.8.3 Poverty—A Barrier in Rural Development

According to the Pakistan Economic Survey 2015–2016 the total population of Pakistan is 195.4 out of which about 117.48 million represent the rural population. Most of the people in rural areas are very poor. They do not have enough money to invest. About half of them are food insecure. Poverty is a barrier to development and must be eradicated. The government is trying its best through various safety nets such as Benazir Income Support Programme, Workers Welfare Fund (WWF), Zakat, Pakistan Bait-ul-Mal (PBM), and Employees Old Age Benefit Institution (EOBI). However, more sincere efforts are needed to uproot poverty.

32.8.4 Illiteracy and Lower Level of Education

Presently. The overall literacy rate in Pakistan is estimated at 60%, whereas the rural literacy rate is only 51%. It means that about half of the people in rural areas are unable to read and write. Illiteracy creates many problems. It is an obstacle in seeking and provision of justice. Although Pakistan is rich in resources, illiteracy is causing a lot of problems and crises in the country. Education teaches one to make sound scientific decisions but illiteracy is ignorance. It is a big hurdle in getting the required facts for informed decision-making.

32.9 WAY FORWARD

The following policy guidelines are recommended based on the discussion above.

1. The law and order situation in rural areas of Pakistan needs to be strictly controlled. A system of collaborative policing by including local elected institutions (such as Union Councils) can be established initially at the pilot scale and based on the lessons learned it can be scaled up in wider area.
2. Most of the rural development plans had to face failure due to the political clashes and nonsustainable democracy. The sustainability of all new policies and plans should be an important component of government planning activities.
3. Plans should be designed to utilize available opportunities such as positive aspects of nongovernment organizations; encouraging rural youth in 4-H and FFA (Future Farmers of America) type activities; getting participation of rural women in development; enhancing the rural cottage industry; and utilizing ICT.
4. The barriers to rural development should be overcome by initiating campaigns against local conflicts through counseling the individuals and communities. Frequent dialogues and debates may be organized in schools, colleges, and universities to give the belief system a scientific bases in addition to religion. Quick and promising (self-help based) poverty reduction strategies, tested worldwide, should be introduced rather than making people

beggars and cheaters through so-called safety nets. Education up to 12th grade should be made compulsory and free of cost in real terms rather than making it a political rhetoric.

5. The use of ICTs should be increased in all spheres of life. As for as possible, all decisions concerning rural issues should be carried out through computerized databases.
6. The concepts of cooperative work, teamwork, positive competition, incentives, social mobilization, community mobilization, self-help, local planning, resource management, and local government need to be promoted.
7. NGOs and community-based organizations should be provided protection, alternative financing, positive incentives, and better government to help expand their working areas and level of services for the rural poor. There should be a strong monitoring and evaluation system to throw corrupt people in jail.
8. The rural development academies and agricultural universities should be made responsible to create new but responsible rural leadership through conducting on-campus and off-campus trainings, seminars, workshops, and distance learning strategies.
9. Rural handicrafts, industrial components, and infrastructures may be developed with joint ventures of local people and local governments.
10. The subjects such as agriculture, home economics, and technical education need to be made available as optional subjects in all rural secondary and higher secondary schools.
11. Agricultural extension services may be made holistic by providing advisory and outreach services in crops, livestock, fisheries, poultry, environment, health, disaster management, youth mainstreaming, household food security and home economics through community service centers at Union Council level. It may be privatized through local service providers with heavy handed monitoring by the government. This center may be linked with cyber network, android apps, helplines, and social media. Their quarterly evaluations by anonymous third parties, such as agricultural universities, must be reported to the local and provincial governments.
12. The annual best village competitions/awards at each union council, tehsil, district, province, and finally at country level should be planned and implemented. The criteria may be developed and notified in advance.

REFERENCES

Abbas, M., Lodhi, T.E., Aujla, K.M. and Saadullah, S. 2009. Agricultural extension programs in Punjab, Pakistan. *Pakistan Journal of Life and Social Sciences* 7(1): 1–10.

Afzal, M. 2009. Population growth and economic development in Pakistan. *The Open Demography Journal*, 2: 1–7.

Ahmad, I., Shah, S.A.H. and Zahid, M.S. 2004. Why the green revolution was short run phenomena in the development process of Pakistan: A lesson for future. *Journal of Rural Development & Administration* 35 (1–4): 89–108.

Ali, M. and Byerlee, D. 2002. Productivity growth and resource degradation in Pakistan's Punjab: A decomposition analysis. *Economic Development and Cultural Change* 50(4): 839–863.

Allen, E. and Dino, F. (*n.d.*) General Introduction to Theories of Gender & Sex. *Introductory Guide to Critical Theory. Date of last update, which you can find on the home page*. Purdue U. Date you accessed the site. http://www.purdue.edu/guidetotheory/genderandsex/modules/introduction.html

Anwar, T. 1996. Structural adjustment and poverty: The case of Pakistan. *The Pakistan Development Review* 35 (4 Part II): 911–926.

Arif, G.M. and Farooq, S. 2012a. Rural Poverty Dynamics in Pakistan: Evidence from Three Waves of the Panel Survey. *Poverty and Social Dynamics Paper Series PSDPS-2*, Pakistan Institute of Development Economics, Islamabad, Pakistan.

Arif, G.M. and Farooq, S. 2012b. *Poverty Reduction in Pakistan: Learning from the Experience of China*. PIDE monograph series. Pakistan Institute of Development Economics (PIDE), Islamabad, Pakistan.

Arshad, S., Ashfaq, M., Saghir, A., Ashraf, M., Lodhi, M.A., Tabassum, H. and Ali, A. 2010. Gender and decision making process in livestock management. *Sarhad Journal of Agriculture* 26: 693–696.

Asghar, N., Awan, A. and Ur Rehmn, H. 2012. Government spending, economic growth and rural poverty in Pakistan. *Pakistan Journal of Social Sciences*, 32(2): 469–483.
Azizi, L.S. 1999. *An Analysis of Social Action Program and Education of Women in Pakistan*. Virginia Polytechnic Institute and State University, Virginia, USA.
Aziz, S. 2000. Eradicating Rural Poverty-I. *The DAWN*. 06 April. Lahore.
Banks, N. and Hulme, D. 2012. The role of NGOs and civil society in development and poverty reduction. *BWPI Working Paper 171*. Brooks World Poverty Institute (BWPI). University of Manchester, London, UK.
Bari, F. 2009 *Role and Performance Assessment of Pakistani Women Parliamentarians 2002–2007*. Friedrich Ebert Stiftung (FES) and Pattan Development Organization, Encore Islamabad.
Bennett, J. 1998. Development alternatives: NGO-Government partnership in Pakistan. *Working Paper Series # 30*. Sustainable Development Policy Institute (SDPI), Islamabad, Pakistan.
Bhutta, Z.A. 2001. Structural adjustments and their impact on health and society: A perspective from Pakistan. *International Journal of Epidemiology*, 30(4): 712–716.
Bhutto, A.W. and Bazmi, A.A. 2007. Sustainable agriculture and eradication of rural poverty in Pakistan. *Natural Resources Forum*, 31: 253–262.
Butt, T.M., Hassan, Z.Y., Mehmood, K. and Muhammad, S. 2010. Role of rural women in agricultural development and their constraints. *Journal of Agriculture and Social Sciences* 6: 53–56.
Campbell, M. 2005 *The Impact of Mobile Phone on Young People's Social Life*. Queensland, Australia.
Ceobanu, C. and Boncu, S. 2014. The challenges of the mobile technology in the young adult education. *Procedia-Social and Behavioral Sciences*, 142:647–652.
Chaudhry, M.G., Chaudhry G.M. and Qasim M.A. 1996. Growth of output and productivity in Pakistan's agriculture: Trends, sources and policy implications. *The Pakistan Development Review*, 35 (4 Part II): 527–536.
Cheema, M.S. 1980. Strategies of rural development in Pakistan. *Paper No. 136*. Directorate of Water Management, Government of the Punjab, Pakistan.
Chegg 2016 www.Chegg.com
CTA 2009 The role of media in the Agricultural and Rural Development of ACP Countries. *Compilation Document. CTA Annual seminars*. CTA, Wageningen, The Netherlands.
Davidson, A.P. and Ahmad, M. 2003. *Privatization and the Crisis of Agricultural Extension: The Case of Pakistan*. A shgate Publishing Limited, England.
Douglas, M. 2013. The Saemaul Undong: Korea's rural development miracle in historical perspective. *ARI Working Paper No. 197*. Asia Research Institute, National University of Singapore.
European Commission Pakistan. 2007. Pakistan-European Community. *Country strategy paper 2007–2013*. eeas.europa.eu/pakistan/csp/07_13_en. (accessed August 15, 2013).
FAO. 1997 What is Gender? http://www.fao.org/docrep/007/y5608e/y5608e01.htm#TopOfPage
Farooq, A. and Kayani, A.K. 2012. Prevalence of superstitions and other supernaturals in rural Punjab: A sociological perspective South Asian studies. *A Research Journal of South Asian Studies*. 27: 335–344.
Gandahi, R. 2016 Role of rural women in agriculture, Pakistan Observer, Mar 27, 2016 http://pakobserver.net/role-of-rural-women-in-agriculture/
G-DAE. 2003 *Working Paper No. 03-07: Five Kinds of Capital: Useful Concepts for Sustainable Development*. http://ase.tufts.edu/gdae.
Ghafoor, A. 1987. *Innovative Methods and Approaches Used in Social Development Planning in Pakistan. Plan 3*. Academy of Educational Planning and Management, Ministry of Education, Islamabad, Pakistan.
Ghate, P., Sen, B., Bose, S. and Srinivasan, T.N. 1992. *Informal Finance: Some findings from Asia*. Asian Development Bank and Oxford University Press, Manila.
Gill, Z.A., Mustafa, K. and Jehangir, W.A. 1999. Rural development in the 21st century: Some issues. *The Pakistan Development Review*, 38 (4 Part II): 1177–1190.
GOP. 2015. *Economic Survey of Pakistan*. Economic Advisor's Wing, Finance Division, Islamabad, Pakistan.
Government of Pakistan. 2005. *Poverty Reduction Strategy Paper Third Quarter Progress Report for the Year 2004–2005*. PRSP Secretariat – Finance Division, Pakistan.
Government of the Punjab. 2013. PDCF Day Care Centre, Women Development Department, Government of the Punjab. http://wdd.punjab.gov.pk/ (accessed June 15, 2016).
Grigorian, H. 2007. Impact of gender mainstreaming in rural development and Millennium Development Goals (MDGs). *UNDP Gender Mainstreaming Annual Conference*, Islamabad, Pakistan.
Grunenfelder, J. 2013 Discourses of gender identities and gender roles in Pakistan: Women and non-domestic work in political representations, *Women's Studies International Forum*, 40:68–77. doi:10.1016/j.wsif.2013.05.007

Haider, M. 2011. All five-year plans of Pakistan were failures. *International: The News*. https://www.thenews.com.pk/archive/print/282677-%E2%80%98all--five-year-plans-of-pakistan-were-failures%E2%80%99 (accessed September 13, 2013).

Hamid, N. 2008. Rethinking Pakistan's development strategy. *The Lahore Journal of Economics (13)*: 47–62.

Hassan, D. 2016. Womens' protection bill, A case of mens' insecurity. *Daily Dawn* May 12, 2016.

Hong, S.M. 2013. Korea's Experience on Human Resources Role in the Community Development. *Paper prepared for International Scientific Conference on "Human Capital in Kazakhstan: Status and Growth Prospects"*, February 22, 2013, Astana, Kazakhstan.

Hussain, A. 1982. Pakistan: Land Reforms Reconsidered. Group 83 Seminar *"Contradictions of Land Reforms in Pakistan"* [online]. Available at: http://www.akmalhussain.net/Publish%20Work/SouthAsia/PakistanLandReformsReconsidered.pdf (accessed 03, 2018).

Hussain, I. 2004. *Economy of Pakistan: Past, Present and Future*. Woodrow Wilson Center, Washington DC. http://www.sbp.org.pk/about/speech/2004/eco_of_pk(past_present_future).pdf (accessed August, 2013).

Idrees, M., Ali, T., Ahmad, M., Mahmood, Z., and Nasir, M. 2008. Self-perceived level of rural youth regarding social, emotional and intellectual characteristics in NWFP, Pakistan. *Sarhad Journal of Agriculture*, 24(1): 169.

IFAD. 2011. Rural poverty report. https://www.ifad.org/documents/10180/c47f2607-3fb9-4736-8e6a-a7ccf3dc7c5b (accessed 03, 2018).

Imran, R. 2005. Legal injustices: The Zina Hudood Ordinance of Pakistan and its implications for women. *Journal of International Women's Studies*, 7 (2): 78–100.

Intizar, H. 2004. Approaches to alleviating poverty in rural Pakistan. In Jehangir, W.A., I. Hussain, (Eds.). *Poverty reduction through improved agricultural water management. Proceedings of the Workshop on Pro-poor Intervention Strategies in Irrigated Agriculture in Asia, Islamabad, Pakistan, 23–24 April 2003*. Lahore, Pakistan: International Water Management Institute (IWMI). pp. 23–30.

Iqbal, A., Ashraf, I., Muhammad, S. and Chaudhry, K.M. 2009. Identification and prioritization of production, protection and marketing problems faced by the rice growers. *Pakistan Journal of Agricultural Sciences*, 46:290–293.

Islam, N. 1996. Growth, poverty and human development: Pakistan. *Occasional paper 31*. United Nation's Development Programme (UNDP). http://hdr.undp.org/en/reports/ (accessed August 10, 2013).

Israr, M.N., Ahmad, S.N., Shaukat, M.M., Shafi, A.K. and Ahmad, I. 2009. Village organizations activities for rural development in North West Pakistan: A case study of two union councils of district Shangla. *Sarhad Journal of Agriculture*, 25: 641–648.

Jaffery, S.M., Younas, and Khattak A. 1995. Income inequality and poverty in Pakistan. *Pakistan Economic and Social Review*, 33(1&2): 37–58.

Jamal, A. 2006. Gender, citizenship, and the nation-state in Pakistan: Willful daughters or free citizens? *Signs*, 31 (2) 283–304.

Kalim, R. and Salahuddin, T. 2011. Micro Financing of NGOs and Government: Collaborative impact on poverty eradication. *Information Management and Business Review*, 2 (2): 81–91.

Kemal, A.R. 1994. Structural adjustment, employment, income distribution and poverty. *The Pakistan Development Review*, 33(4): 901–911.

Khan, A. 2007. Women and Paid Work in Pakistan: Pathways of Women's Empowerment. *Collective for Social Science Research*. Karachi. Pakistan. http://www.researchcollective.org/Documents/Women_Paid_Work.pdf (accessed 03, 2018).

Khan, A.R. and Khan, A.N. 2001. An overview of rural development programmes and strategies in Pakistan. *Journal of Rural Development and Administration*, 33: 22–29.

Khan, M.M., Zhang, J. Hashmi, M.S. and Hashmi, M.S. 2011. Land distribution, technological changes and productivity in Pakistan's agriculture: Some explanations and policy options. *International Journal of Economics and Management Studies*, 1(1): 51–74.

KREI. 2010. *Agriculture in Korea*. Korea Rural Economic Institute, Seoul, Republic of Korea.

Kurosaki, T. 2005. Determinants of Collective Action under Devolution Initiatives: The case of citizen community boards in Pakistan. *Pakistan Development Review*. 44(3): 253–270.

Labour Force Survey. 2015. Available at: http://www.statpak.gov.pk/fbs/content/labour-force-survey-2010-11 (accessed September 20, 2011).

Latif, A., Nazar, M.S., Mehmood, T., Shaikh, F.M. and Shah, A.A. 2011. Sustainability of microcredit system in Pakistan and its impact on poverty alleviation. *Journal of Sustainable Development*, 4(4): 160–165.

Lemoine, M. and D. Ramsey. 2010. "Digtal Youth" ICT use by young people in Rural Southwestern Manitoba, Brandon University. *Geographical Essays*, 4: 17–24.

Lodhi, E.A., Luqman, M., and Khan, G.A. 2006. Perceived effectiveness of public sector extension under decentralized agricultural extension system in the Punjab, Pakistan. *International Journal of Agriculture and Biology* 2: 195–200.

Lorenzo, G.B. 2011. Development and Development Paradigms A (Reasoned) Review of Prevailing Visions, A policy paper of Food and Agriculture Organization, Rome. Retrieved from www.fao.org/easypol

Luqman, M. 2004. A study into the effectiveness of public sector agricultural extension after decentralization in Ditrict Muzaffargarh. *MSc (Hons.) thesis*, Department of Agri. Extension, University of Agriculture, Faisalabad, Pakistan.

Mahmood, T., Rehman, H. and Rauf, S.A. 2008. Evaluation of macroeconomic policies of Pakistan. *Journal of Political Studies*, 14:57–75.

Malik, F. 2016. 40 million smart phones in Pakistan by the end of 2016, IBEX. http://www.ibexmag.com/featured/40-million-smart-phones-in-pakistan-by-the-end-of-2016/

Mallah, M.U. 1997. Extension programs in Pakistan. In *Extension Methods*. Memon, R. A., E. Bashir (Eds.). National Book Foundation, Islamabad, Pakistan. PP: 35–60.

Marshall, G.A. and Sabhlok, A. 2009. 'Not for the sake of work': Politico-religious women's spatial negotiations in Turkey and India. *Women's Studies International Forum*, 32 (6), 406–413.

Media, W. 2010. Role of Media in Agricultural and Rural Development, *Presented at the IFA workshop on Last Mile Delivery*, 10th Feb. 2010, New Delhi, India, 5 pp. http://r4d.dfid.gov.uk/PDF/Outputs/MediaBroad/The_role_of_the_media_in_agricultural_development.pdf

Mustafa, K. 1992. The Institution of Cooperation, Credit and the Process of Development in the Indian and Pakistan Punjabs, *PhD dissertation*, University of Glasgow, U.K.

Naeem, M.R. 2005. Self-perceived role of rural youth in agricultural and rural development in district Faisalabad. *MSc (Hons.) thesis*, Department of Agri. Extension, University of Agriculture, Faisalabad, Pakistan.

Nasir, A., Chaudhry, A.G., Khan, S.E. and Hadi, S.A. 2015. Biradarism and rural conflict as a determinant of political behavior: A case study of rural Punjab. *Science International(Lahore)*, 27:703–705.

Nazir, A., Jariko, G.A., and Junejo, M.A. 2013. Factors affecting sugarcane production in Pakistan. *Pakistan Journal of Commerce and Social Sciences*, 7: 128–140.

NCSW. 2013. *Institutional Strengthening of NCSW Support to Implementation of GRAPs: Gender Review of Political Framework for Women Political Participation*, National Commission on the Status of Women, Islamabad.

Nehra, R., Kate, N., Grover, S., Khehra, N., and Basu, D. 2012. Does the excessive use of mobile phone in young adults reflect an emerging behavirol addiction? *Journal of Postgraduate Medicine Education and Research*, 46(4):177–182.

Ngeh, D.B. 2013. Non-Governmental Organizations (NGOS) and rural development in Nigeria. *Mediterranean Journal of Social Sciences*. 4(5):107.

Omer, M. and Sarah J. 2008. Pro Poor Growth in Pakistan: An Assessment of the 1970s, 1980s, 1990s and 2000s. *Published in: South Asia Economic Journal*, 9(1):51–68.

Orden, D., Salam, A., Dewina, R., Nazli, H. and Minot, N. 2006. The Impact of Global Cotton Markets on Rural Poverty in Pakistan. *Background Paper 8*, Pakistan Poverty Assessment Update, ADB Islamabad Resident Mission, Islamabad, Pakistan.

Park, Jin-Hwan. 1998. *The Sameaul Movement: Korea Approach to Rural Development in 1970s*. Korea Rural Economic Institute KREI, Seoul.

Ranjan, A. 2012. Inter-provincial Water Sharing Conflicts in Pakistan. *Pakistaniaat: A Journal of Pakistan Studies*, 4: 102–122.

Reed, P.E. 2010. Is Saemaul Undong a model for developing countries today? Paper presented at *International Symposium in Commemoration of the 40th Anniversary of Saemaul Undong*. Korea Saemaul Undong Center, September 30, 2010.

Rehman, H. and Ismail, M. 2012. Study on the role of non-governmental organizations in imparting primary education in Pakistan. *International Journal of Contemporary Research in Business*, 4(1): 751–769.

Riaz, A., Khan, G.A. and Ahmad, M. 2012a. Utilization of agriculture credit by the farming community of zarai tariqiati bank limited (ZTBL) for agriculture development. *Pakistan Journal of Agricultural Sciences* 49: 557–560.

Riaz, A., Muhammad, S., Ashraf, I. and Zafar, M.I. 2012b. Role of Punjab rural support program in improving economic conditions of rural women through micro financing. *Pakistan Journal of Agricultural Sciences* 49(2): 211–216.

Saed, M.D. 2013. Women Role in Development of Pakistani Rural Communities, Pakistan Hotline.http://www.pakistanhotline.com/2013/05/women-role-in-development-of-pakistani.html

Sattar, K. 2007. A Sustainable model of use of ICTs in Rural Pakistan. *International Journal of Education and Development Using ICT* 3(2): 116–124.

Shahbaz, B., Luqman, M. and Cho, Gyoung-Rae. 2014. Analysis of rural development timeline in Korea and Pakistan: What lessons Pakistan can learn? *Korean Journal of International Agriculture* 26(3): 197–209.

Sobhan, R. 1968. *Basic Democracies Work Program and Rural Development in East Pakistan*. Bureau of Economic Research, University of Dacca, Bangladesh.

The Sociologist. 2013. The WID, WAD, GAD Approach on Gender Development https://cn2collins.wordpress.com/2013/03/19/the-wid-wad-gad-approach-on-gender-development/ (Accessed June, 14, 2016).

Sofa and Doss, C. 2011. The role of women in agriculture, ESA Working Paper No. 11-02, *Agricultural Development Economics Division The Food and Agriculture Organization of the United Nations*. Assessed from http://www.fao.org/

Tahir, S. and Ali, S.S. 1999. *Growth with equity: policy lessons from the experiences of Pakistan*. In Growth with Equity. ESCAP, Bangkok, UN.

Tariq, M., Mustafa, M., Iqbal, A. and Nawaz, H. 2008. Milk marketing and value chain constraints. *Pakistan Journal of Agricultural Sciences*, 45:195–200.

UNESCO. 2003. *Literacy Trends in Pakistan*, UNESCO-Islamabad, cPakistan unesco.org.pk/education/life/nfer_library/Reports/4-39.pdf.

United Nations. 1978. *Industrialization and Rural Development*. United Nations Industrial Development Organization (UNIDO). United Nations, New York. ID/WG.257/23.

Waheed, S. 2009. Does rural microcredit improve well-being of borrowers in the Punjab (Pakistan)? *Pakistan Economic and Social Review*, 47(1): 31–47.

World Bank. 2007. Distortions to agricultural incentives in Pakistan. *Agricultural Distortions Working Paper 33*. www.worldbank.org (Accessed August 10, 2013).

33 Outreach and Social Mobilization

Challenges and Opportunities

Munir Ahmad, Babar Shahbaz, and Mahmood Ahmad Randhawa

CONTENTS

33.1 INTRODUCTION AND CONTEXT

How are agricultural extension and outreach services supporting agricultural development and social mobilization in Pakistan's rural areas? What is the contribution of different mega extension/outreach programs in agricultural and community development? And how can agricultural universities in general and the University of Agriculture Faisalabad (UAF) in particular catalyze agricultural growth in the country? These questions are the major thrusts for writing this chapter. This chapter give an overview on how extension systems evolved in Pakistan and what is the current extension set-up, and how extension services, coupled with agricultural universities, contribute to the development of the agriculture sector and mobilization of rural communities.

The agriculture sector is the foundation of Pakistan's economy and many economists and development practitioners predict that our country's economy will likely be based on agriculture for the foreseeable future. Being the largest employer of the country's labor force, the importance of the agriculture sector cannot be underestimated. However there has been stagnant growth in agriculture over the last many years and in 2015–16, there has been negative growth of 0.19% against 2.5% growth during the last financial year (Government of Pakistan, 2016). It has been recognized globally that an effective system of agricultural extension and outreach acts as the driving force for agricultural development. A well-organized outreach system can mobilize farmers in particular and rural masses in general for overall agricultural and rural development. On one hand, farmers perceive agricultural extension as a service offered by the government or private sector to help increase their knowledge, resource efficiency, farm productivity, and contribution to better living standards of their household. On the other hand, policy makers and legislators understand extension as an institutional component of agriculture to increase agricultural production and to alleviate rural poverty and ensure food security. At the same time economists and practitioners view agricultural outreach and extension as an instrument to ensure economic growth through the development of human capital and social mobilization (Hagmann et al., 1999). It has been generally argued that extension workers can play a central role in facilitating social mobilization processes in rural areas even beyond agriculture (ibid).

Realizing the importance of effective extension and outreach systems, several extension programs and models have been implemented in Pakistan since independence, with a few success stories and more failures. Most of the extension program provided insignificant results due to several reasons like bureaucratic hurdles, lack of training and competence, political instability, etc. (Abbas et al., 2009). Nevertheless, outreach and extension continue to be considered a critical force for overall agricultural development, and without an effective extension system we cannot meet the challenges of rapidly changing technologies of the current century.

Social mobilization and capacity building are particularly vital to meet these challenges. Development of human capital of farming communities through training and nonformal education so that people can help themselves is the building block of a competent agricultural extension and outreach system of any developing country. The role of agricultural extension in social mobilization and community development is well recognized (Christoplos, 2010). Some researchers have emphasized that the extension staff must work with researchers and educators to provide need-based solutions to farmers and to pave the way for self-help initiatives among rural communities.

Agricultural universities throughout the world have a history of catalyzing and in many cases leading the outreach and extension system of the state. This is particularly true for the United States where one of major mandates of Land Grant Universities was transfer of technological innovations to stakeholders. University of Agriculture, Faisalabad (UAF)—upgraded from Punjab Agriculture College and Research Institute in 1961—was also established to function as a land grant university. UAF has a track record in agricultural research, innovations, education, and outreach. In addition to its own outreach and community development initiatives, UAF remained a contributor to impart trained and skilled manpower and guidance to various several mega programs such as the Farm Guide Movement, Village Agricultural Industrial Development Program, Training and Visit (T&V) program, etc. UAF has also given birth to new agricultural universities in the Punjab where most of the faculty members in newly established agricultural universities are alumni of UAF.

In this context this chapter presents an analysis of the agricultural extension and outreach system with particular focus on social mobilization of rural masses. This chapter provides an analysis of several extension programs of Pakistan as given in Section 33.2, followed by the challenges and constraints faced by social mobilization (Section 33.3). The role of UAF in outreach and community development is underlined in Section 33.4. The final section presents possible ways forward and some policy suggestions.

33.2 EVOLUTION OF EXTENSION/OUTREACH SYSTEM IN PAKISTAN

This section gives an overview on the evolution of public and private extension systems in Pakistan.

33.2.1 Public Extension System

Public institutions related to agriculture were introduced in most parts of Punjab during the colonial era (the British rule) during the creation of the huge canal network during the earlier twentieth century. Most of the current agricultural institutions in Pakistan are a legacy of the British raj (Gill and Mushtaq 1998). A massive network of main canals, distributary/minor canals, link canals, and watercourses was constructed by the British rulers to irrigate the land. It was realized that farmers should be educated regarding newly introduced irrigated agriculture. Accordingly, a set-up of agricultural institutions were established across India and particularly in Punjab. During 1906, agricultural education began in Punjab province with the establishment of the Punjab Agricultural College and Research Institute in Lyallpur (now Faisalabad). Demonstration plots were also set up to establish a link between agricultural colleges and rural communities.

The public extension service in the country has always been an obligation of provincial governments, and provinces are responsible for the provision of agricultural extension services. Many researchers have claimed that extension services in Pakistan are rather traditional and top-down in nature. After separation of teaching, research, and extension, the linkages of extension with agricultural academic institutions and research are nominal (Qamar, 2012).

Several extension and outreach programs have been introduced in Pakistan after independence to uplift the country's agricultural sector. Some of the noteworthy programs include:

- Lyallpur Model
- Traditional extension system
- Training and Visit (T&V) Extension Program
- Farmer Field Schools
- Decentralized extension system
- Hub Program.

33.2.1.1 Lyallpur Model

After independence, Pakistan inherited the colonial agricultural institutions. In 1961, a separate Department of Agricultural Extension was created with the upgrading of the Punjab Agricultural College and Research Institute, Lyallpur as West Pakistan Agricultural University (WPAU), Layllpur. Before upgrading as a university, teaching, research and extension were jointly managed and worked under the Punjab Agricultural College (Chaudhry, 2002).

33.2.1.2 Traditional Extension System

As discussed above, the Lyallpur model of agricultural extension was an integration of teaching, research and extension. However this model was abolished in 1961 and a traditional extension system was introduced in 1961 which had its roots in the earlier twentieth century when the canal irrigation system was introduced in the subcontinent. Transfer of technology from government to farmers was the major emphasis of this system and therefore it was as a top-down extension system with the assumption that beneficial, hands-on and relevant technical agricultural information was available, and that the proper responsibility of agriculture extension was to transfer technology to the farmers (Abbas et al., 2009). Diffusion of innovations particularly new crop varieties, chemical fertilizers, and plant protection practices were the main thrusts of this approach. This approach worked well during the "Green Revolution" era and new high yielding varieties, fertilizers, and pest control measures were introduced among large segment of the farming community (Mallah, 1997). However lack of effective linkage between research and education remained a challenge throughout the life of this system.

33.2.1.3 Training and Visit Approach (TV)

Recognizing the shortcomings of traditional extension approaches, the TV program was introduced in 1978 to overcome the weaknesses of this approach. The idea of the TV system was based on liaisons between extension workers, researchers, and farmers. Transfer of agricultural technology to bridge the gap between the modern research farms and farmers' fields was the major purpose of this system. Farmers were recognized as extension clients under this system and the extension agents were responsible to focus their efforts on the contact farmers within their jurisdiction. The contact farmers were about 10% of the total farmers within the jurisdiction of extension worker (Abbas et al., 2009). The basic assumption behind this approach was that the contact farmers would act as opinion leaders and other (noncontact) farmers would follow these leaders in adoption of agricultural innovations. A fixed fortnightly schedule was given to each extension worker and within 15 days he had to meet the contact farmers (8 visits is 2 weeks) as well as get training and perform office work. This system was appreciated by many researchers because it brought a paradigm shift in the traditional extension system of country and efficiency of extension workers was significantly improved. However critics of T&V designated this system as too rigorous that put heavy burden on the extension workers. Similar extension message was repeated for many weeks and many contact farmers were selected improperly (Abbas et al., 2009; Shahbaz and Ata, 2014). This system is also considered as top-down in nature and farmers' participation in the planning and execution process was negligible. The emphasis was on delivery of message rather than to motivate farmers for the adoption of technology. This system remained in place until 1999 when the Punjab government made changes and implemented modified versions of the TV approach.

33.2.1.4 Farmers Field School

The Farmers Field Schools (FFS) is a participatory group learning approach and its concept is based on the idea of organizing regular training of farmers; this methodology is reported to be particularly successful in integrated pest management (IPM). The FFS system is adopted by the Punjab agriculture extension departments through Fruit and Vegetable Development Project (FVDP) in 12 districts. FVDP has established FFS in citrus, mango, and vegetable growing regions. There are four components of FFS are Master Trainer, Facilitator, Member Farmer, and Demonstration Plot. This approach should be adopted at wider scales, but is being used at a limited level because evidence has proved that FFS approach is more effective than other approaches as its philosophy is based on learning by doing.

33.2.1.5 Agricultural Hub Program

Currently a modified version of TV programs are being operated in the province. In "Hub program" a progressive farmer (labeled as hub farmer) is selected by the field assistant and AO at Union Council (UC) level (usually one hub farmers is selected from each village of the UC) and the farm of the hub farmer is designated as a demonstration plot for other farmers of area. The hub farmer is endorsed by the Deputy District Officer Agriculture (DDOA) for registration with the seed corporation for the multiplication of seeds for use in his village. The extension workers regularly visit the farm of this farmer on a fixed day (at least once a week) and conduct meetings with the hub farmers and other farmers to disseminate approved agricultural innovations and technologies. DOA and DDOA visit the demonstration plots randomly for monitoring and evaluation purposes.

The main difference between hub program and TV system is that earlier the outreach activities were controlled by the DG (Ext & AR), but now extension activities are decentralized and accomplished at the districts by DOA. In TV system, the extension field staff has to follow a rigid fortnightly schedule but in the current set-up crop specific trainings are conducted during different stages of production of major crops. Another difference is that now agricultural officers (AOs) are the frontline extension workers but in TV approach the front line extension workers were field assistants.

33.2.1.6 Plant Clinics

This is a recent initiative by the Punjab Department of Agriculture (Extension wing). In this method, local plant clinic are organized in order to ensure plant wise health and the subject experts (called plant doctors) suggests treatment to the crops based on the queries by farmers.

33.2.1.7 Decentralized Extension System

Decentralized extension system was introduced in 2001, by government led by General Pervez Musharraf who announced the Devolution of Power Plan in Pakistan, which was intended to decentralize the administration at local level and give power to local representatives (NRB, 2005). This plan brought institutional changes in most of the public departments including agricultural extension and outreach (Saeed et al., 2006). The 18th amendment of the constitution of Pakistan brought in 2010 further devolved some functions of Federal Ministry of Food, Agriculture and Livestock to the provinces.

The constitution of Pakistan* authorizes the provinces to manage the issues related to agricultural sector within their respective provinces. There is range of provincial organization in all provinces. For instance the Department of Agriculture in Punjab consists of many directorates and sections

- Director General Agriculture (Extension and Adaptive Research)
- Director General Agriculture (Field)
- Director General Agriculture (Research)
- Director General Agriculture (Pest Warning and QCP)
- Director General Agriculture (Water Management)
- Directors of Agriculture (each for crop reporting service, Information, floriculture, marketing and economics)
- Planning and Evaluation cell
- Special secretary (Agriculture Marketing)
- WTO cell

Focus of this section is Agricultural extension service, therefore extension wing of the Punjab Agriculture Department is being discussed here in more detail (Figures 33.1 and 33.2).

Director General Agriculture (Extension & Adaptive Research) Punjab keeps links with the local governments at district level. The extension wing is responsible for extending outreach services in the major crops and fruit.

The District Coordination Officer (DCO) heads the district bureaucracy and he/she reports to the Nazim† of respective district. The Executive District Officer Agriculture (EDOA) works under DCE and coordinates agricultural activities with the line departments in districts. The District Officer for Agriculture (DOA) is responsible for planning and execution of outreach and extension activities in the particular district. In Sindh, however, a Director works at the division and Deputy Director at districts (Shahbaz and Ata, 2014).

The next tier of the administrative set-up after districts in Punjab is Tehsil and here Deputy District Officers for Agriculture (DDOA) look after agricultural extension activities; however in Sindh, the Assistant Director operates at the taluka (equivalent to tehsil).

Markaz comes after the tehsils and the Agricultural Officer (AO), is based at markaz‡. The AOs are frontline extension workers and are responsible for the operation of extension and outreach activities at markaz level and they are in direct contact with farmers. Minimum qualification of AO is BSc (honours) degree in agricultural science.

---

* The functions of livestock and agriculture were transferred from Federal to Provincial government through 18th Amendment in the constitution passed in 2011.

† District Nazim (Mayor). Currently EDOs are directly under the control of DCOs because there are no district nazims.

‡ Markaz (center); a tehsil may contains 3-5 markaz, depending on the population.

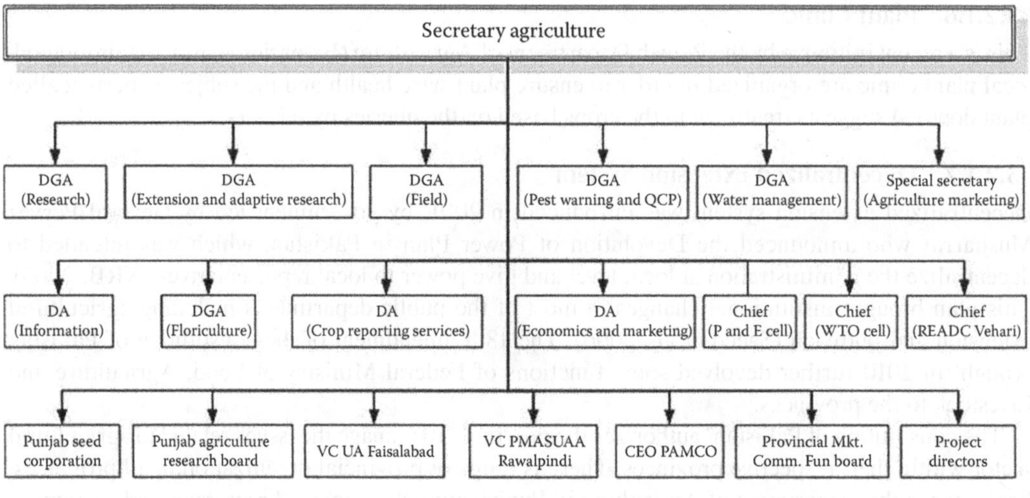

FIGURE 33.1 Organogram of Punjab Agriculture Department.

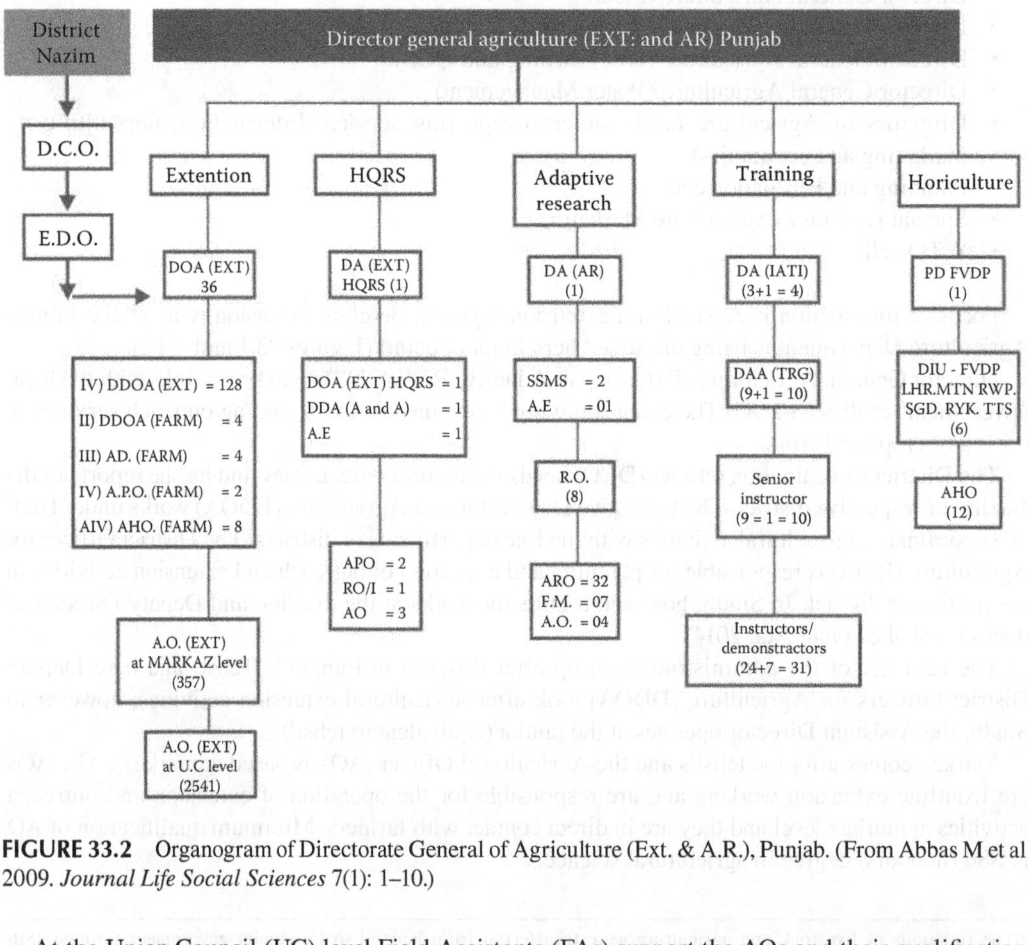

FIGURE 33.2 Organogram of Directorate General of Agriculture (Ext. & A.R.), Punjab. (From Abbas M et al 2009. *Journal Life Social Sciences* 7(1): 1–10.)

At the Union Council (UC) level Field Assistants (FAs) support the AOs and their qualification is a Diploma in Agricultural Sciences (DAS). Under each F.A. there are 2–3 laborers/fieldworkers called Beldars who are not technical persons.

33.2.2 Outreach by Private and NGO Sector

The private sector emerged in the late 1980s when the National Commission on Agriculture highlighted that the shift from subsistence to profitable agriculture would only be possible with the involvement of the private sector (Government of Pakistan, 1988). As per recommendations of the Commission the international fertilizer and pesticide supply agencies started to take part in outreach activities along with selling agricultural inputs (Riaz, 2010). Participation of the private sector in extension has had major impacts in Pakistan's agriculture. However, some authors have shown concern that the interest of private companies are more on 'marketing of their products' rather than facilitating small farmers (Siraj, 2011). Nevertheless, most of the international and national enterprises engaged with livestock and agriculture related products are also providing outreach services to their clients (Table 33.1). Riaz (2010) categorized the following areas where these companies are providing extension services in Pakistan:

- Sugarcane: by sugar factories in different sugarcane growing zones
- Plant protection: by multinational and national pesticide companies
- Tobacco: by cigarette manufacturing companies
- Seed: by seed companies
- Plant/crop nutrition: by fertilizer companies
- Maize: corn processing companies and seed companies
- Dairy: by national and multinational milk companies

Some Nongovernmental Organizations (NGOs) are also rendering extension services to farmers. For instance the World Wide Fund for Nature (WWF-Pakistan) works on different projects in Pakistan. The Pakistan Sustainable Cotton Initiative (PSCI) is one of its leading projects with the objectives of promoting best practices for cotton production, efficient water management, and reducing the use of agrochemicals. Lok-Sanjh Foundation, Sungi Development Foundations, and SDPI are some of the prominent national NGOs working on agricultural issues including outreach and social mobilization; however, their activities depend on the funding received from international donors (Shahbaz and Ata, 2014). The Agha Khan Rural Support Program (AKRSP) is the pioneer RSP in the country and its strategy is based on a participatory approach and provide training of the farmers of Northern Areas in production technology of vegetables, fruits, and value addition and marketing. The approach of the AKRSP has been regarded as exemplary in Pakistan as far as social mobilization is concerned.

33.3 SOCIAL MOBILIZATION IN PAKISTAN: CHALLENGE FOR EFFECTIVE OUTREACH

Social mobilization is an essential strategy to develop the community's capacity for undertaking collective action. A majority of Pakistan's farmers are small landholders, and mobilizing these farmers is a continuous challenge. As discussed in the previous section, Pakistan has a long history of implementing several extension approaches by the government. Additionally, private companies and NGOs are also working for outreach and extension activities. However, most of these programs have not succeeded (Luqman et al., 2011; Shahbaz and Ata, 2014). Lack of active participation by smallholder farmers in outreach programs is one of the factors behind limited success of outreach programs, and has been emphasized by international development agencies and practitioners. Social mobilization is regarded as key to the success of any outreach strategy and a central component of any sustainable rural development program. Establishment of self-help groups among rural communities is a desired philosophy of extension programs, but in Pakistan most of the outreach approaches are designated as top-down and nonparticipatory in nature. Nevertheless, there have been some attempts in mobilizing rural communities for agricultural and rural development but these initiatives remained futile. Social institutions, sociocultural norms, family, education, etc., affect social mobilization in rural masses (WDR, 2001).

TABLE 33.1
Some Private Companies and their Outreach Activities

Company	Type	Product	Outreach and Extension Methodology
Ali Akbar Group	National	Pesticides, micronutrients, seeds	Operates of two larges franchises (Target Zarai markaz and Apna Zarai Markaz). Printed material, group meetings and demonstration plots
Allah Din Group (Singh)	National	Agro-Chemicals, seed	Mostly web-based information system
Auriga Group	National	Micro fertilizers, hybrid seeds (rice)	Advisory services (On-field), website
Fatima Group	National	Fertilizer and Sugar Mill	Printed material, website, on-field outreach
Fauji Fertilizer Company (FFC) Ltd.	National	Fertilizer	Farm advisory services through five Farm Advisory Centers and 14 Regional Agricultural Services Officers. Printed literature on crops, vegetables and fruit
Jaffer Group of Companies	National	Plant nutrition and protection, irrigation	Printed material and farmers' meetings. Website
Four Brothers Group	National	Seeds, chemicals	Network of nationwide franchise (Tarzan Markaz). Model farms and advisory services
Pioneer Pakistan Seed	Joint venture with Du-Pont USA	Hybrid seeds	Advisory services through field officers. Demonstration plots, field days, training events, festivals, etc.
Monsanto	Multi-national	Seeds, Agro-chemicals	Demonstration plots, group and individual meetings printed material/crop guides, website
Syngenta Pakistan	Multi-national	Seeds, pesticides	Franchise network (Naya Savera). Pesticide trialsat demonstration sites, group meetings, printed literature, visits, field days, online advisory, etc.
Nestle Pakistan	Multi-national	Milk, juices, butter, yogurt	Technical advisory services, trainings, mobile phone
Bayer Crop Science	Multi-national	Agro-chemicals, Plant protection	Printed material, individual and group meetings.
Engro Corporation Ltd.	National	Milk, ice cream, fertilizer	Advisory services to dairy farmers
Pakistan Tobacco Company	National	Tobacco	Mostly in Khyber Pakhtunkhwa province: Advisory services in tobacco production technology. Group meetings and trainings

Source: Websites of companies, Shahbaz B, Ata S 2014. Agricultural extension services in Pakistan: Challenges, constraints and ways forward. Background Paper No. 2014/1. ACIAR project, Enabling agricultural policies for benefiting smallholders in dairy, citrus and mango industries of Pakistan—Project No. ADP/2010/091, Victoria University Sydney, Australia.

Collective action required commitment and equal participation of each member of the group. Size of group, heterogeneity/homogeneity, and local institutions do have a profound bearing on collective action and cooperation within the group (Poteete and Ostrom, 2004). Five forms of heterogeneity have been specified by the previous researcher (Velded, 2000, cited in Poteete and Ostrom, 2004). These are:

1. Political segregation
2. Heterogeneity in endowments

3. Wealth and entitlements
4. Heterogeneity in culture
5. Varied economic interests

Punjab province has a very diverse and heterogeneous culture and socioeconomic set-up. Each district has unique socioeconomic characteristic and heterogeneity is found in even most of the villages. For instance Channa (2015) in her doctoral thesis analyzed collective action and trend towards education in Punjab and Sindh provinces in the perspective of heterogeneity of land holding and caste with a focus on the fragmentary nature of kinship groups and social capital. Her analysis highlighted the salience of caste power heterogeneity in envisaging the level of collective activity for the provision of education. The results also revealed the importance of recognizing that collective action is often embedded in a system of village power relation, patronage, and politics. In this milieu, promoting collective action and social mobilization for agricultural development is a continuous challenge in Pakistan in general and in Punjab province in particular.

Another dimension of social mobilization in rural Pakistan can be traced from the emergence of Islam through Sufi and saints. According to Hussain (2006) the realization of the social action in the rural areas of Pakistan has been greatly influenced by the Sufi saints which is evident in the folklores and their philosophical traditions. For instance, during seventeenth century, Sufi Shah Inayat mobilized the farmers of Sindh and challenged the authority of feudal lords. As a matter of fact, Islam spread to masses in the subcontinent through the work of preachers and Sufi saints who gave the message of respect for all segments of society and work together for common cause of humanity and wellbeing of poor people. It means that despite heterogeneity in culture, language, and social setup there is a potential in this part of the world to mobilize rural people towards collective action and possibly agricultural development. Shami (2010) conducted case studies in rural Punjab, and proved that increased connectivity for example through roads significantly increased the farmers' capacity to be involved in collective action, even when there is heterogeneity in land holding.

Another challenge for social mobilization, in the context of agricultural and rural development, is the lessening economic dependence of smallholder farmers on farming. The evidence shows that in most rural areas of Pakistan, farming households have adopted multiple livelihood activities such as daily wage labor, migration, petty business, and so forth, and farming is being done for subsistence. In this scenario where small farmers do not see farming as the major contributor to their family income, they may show less willingness to cooperate for a common cause. Therefore transforming farming from subsistence to a business-oriented activity is a continuous challenge for extension/outreach services.

Many research studies have reported challenges in extension services in Pakistan. Some of the challenges and constraints in current outreach and extension services are listed below (for details see Shahbaz and Ata, 2014):

1. Small and fragmented landholdings
2. Targeting the small farmers—reaching last mile
3. Lack of effective linkages and coordination with line departments
4. Decentralization
5. Involvement of extension personnel in irrelevant activities
6. Climate change and natural disasters
7. Globalization and trade liberalization
8. Livelihood diversification (nonfarm income opportunities) in rural areas
9. Dearth of future farmers
10. Lack of female extension workers
11. Dearth of need based outreach

12. Lack of effective agricultural extension policy
13. In-service training of extension officers
14. Lack of coordination between extension, input supply, and markets

33.4 THE ROLE OF UAF

The University of Agriculture Faisalabad (UAF) is by far the largest agricultural university, employing more than 600 faculty staff distributed over six faculties: The Faculty of Agriculture, the Faculty of Social Sciences, the Faculty of Agricultural Engineering and Technology, the Faculty of Animal Husbandry, the Faculty of Veterinary Science, the Faculty of Basic Sciences, and Faculty of Food Sciences. The university has about 50 departments and institutes which are currently engaged in teaching, research, and development activities. It is stepping forward towards gender equality in higher education.

The UAF has a track record of outreach and community service activities. The university has the mission that its researchers and scientists will play their role to mobilize the farming community and to make sure that farmer should adopt the latest techniques of production and value addition. Some of the salient features of outreach activities are discussed in the following sections.

33.4.1 Agricultural Exhibitions and Traditional Events

The UAF regularly organizes spring festival and Rabi Kisan Mela (farmers' festivals) which brings many events including a book fair, milk competition, cultural show, agricultural exhibition, Kissan convention, greyhound race, buffalo beauty show, fancy birds show, nutrition festival, and many more. The main purpose of these festivals is to create a hub for interaction among farmers, researchers, policy makers, industrialists, entrepreneurs, and students that help enhance agricultural productivity. These festivals and exhibitions also provide opportunities to the citizen of Faisalabad to visit the university and get firsthand knowledge about the different achievements of the university (Figures 33.3 through 33.5).

33.4.2 Demonstrations

Field demonstrations are considered an effective tool of agricultural extensions for motivation farmers for adoption of agricultural innovations and modern farming practices. Demonstrations are regular

FIGURE 33.3 Goat show at UAF.

FIGURE 33.4 Distribution of chickens to the farmers by vice-chancellor.

FIGURE 33.5 Mango show in Islamabad.

features of UAF and such demonstrations are organized not only on campus and subcampuses but at the farmers' fields (Figures 33.6 and 33.7).

33.4.3 Conferences, Public Seminars, Workshops, etc.

One of the regular features of the university is public lectures, conferences, seminars, etc. Renowned scholars from Pakistan and foreign countries are invited to give lectures on current trends in the agricultural sector and social issues. Faculty members, students, and the general public participate in such events (Figure 33.8).

33.4.4 Saturday Outreach and Field Days

Saturday is working day at the UAF and the purpose is to conduct outreach and extension activities by the faculty and the students. The U.S.-Pakistan Centre for Advanced Studies in Agriculture

FIGURE 33.6 Raised bed planting demonstration.

FIGURE 33.7 Demonstration of bio-gas plant in a farmers' meeting.

FIGURE 33.8 Some resource persons during seminar on CPEC at IBMS.

FIGURE 33.9 Field day at UAF community outreach center in Samundari.

and Food Security (USPCAS-AFS) has established the Community Outreach Centers in different villages to enhance the direct interaction of UAF faculty and the farming community. These centers are equipped with ICT tools and also links the rural communities with agricultural experts of UAF (Figures 33.9 through 33.11).

33.4.5 Entrepreneurship

University of Agriculture, Faisalabad has established a business incubation center (BIC) to interpret and disseminate research findings as new products and technologies. The business incubators help in transferring technology from universities and create linkages of scientists and experts with the business community. It is a system for bringing scientifically advanced products to the market. Business incubators plan the innovative programs to support the effective development of business-oriented companies through a series of business support services and resources, developed and coordinated by incubator management. A company registered with BIC is provided with space, and access to laboratories, technical experts, library, Internet, and utilities at nominal rates. Since its inception, the BIC of UAF has formed working relationships with various startups and poorly performing companies. Relevant agreements have been signed with UM Enterprises, Karachi for

FIGURE 33.10 Awareness event about dengue fever in a village.

FIGURE 33.11 Filed day/training workshop for female farmers.

Technology Transfer, and consultancy agreements with Green Revolution, Board of Intermediate and Secondary Education (BISE) Faisalabad, JK Agricultural Farms Agriculture Department, the Government of the Punjab (Supply Chain Improvement Project), Allied Food Engineering (Pvt.) Ltd, Punjab Skill Development Fund (PSDF), Agri Support Fund (ASF), and an Agriculture Land Use Agreement with Monsanto Pakistan (Pvt.) Ltd. So far BIC has incubated 27 companies. Moreover, the university faculty in collaboration with BIC are providing consultancy services to 22 national/international companies and have developed linkages with 15 industries.

33.4.6 Community College/Regional Centers

A Community College was established in 2008 at the outskirt of Faisalabad city with the core objective to demonstrate quality education emphasizing integrated approaches to produce trained manpower in agriculture and allied disciplines. The UAF Community College has revived intermediate (Preagriculture) programs to provide capable nurseries for admission at undergraduate level in various disciplines of agriculture and its allied subjects. This program is aimed at ensuring a bright future for the young generation through amalgamation of basic sciences with agriculture for a new thinking, novel approaches, and innovative technical skills to accelerate the growth of agriculture sectors in Pakistan. So, the main incentive behind the reinitiation of this program is to provide education and learning facilities to young rural inhabitants of Punjab districts. Moreover, the UAF community college is actively engaged in imparting skills to the young in rural areas to enable them to start their own businesses.

The UAF has subcampuses at Toba Tek Singh, Bureala, and Depalpur. The university has history of establishing other universities from its subcampuses as its three constituent colleges were upgraded to universities. These are the University of Veterinary and Animal Sciences, Lahore, Pir Meher Ali Shah Arid Agri; the University Rawalpindi and Ghazi University in D.G. Khan. In the future, the subcampuses of UAF are also expected to be upgraded in to independent universities.

33.4.7 Call Centers/Cyber Extension

The UAF also provide information and outreach services to the farmers in appropriate application of site-specific solutions through call centers and the web portals www.fertilizeruaf.pk and www.kissandost.pk. Cyber extension is now widely recognized as an efficient and effective tool for agricultural and rural development. Keeping this fact in view, the UAF initiated the project

FIGURE 33.12 Information portal Zarai Baithak.

"Technology transfer through cyber extension" helping communities to help themselves. A website (www.zaraibaithak.com) was developed and community centers/village information centers were established to guide farmers about posting their queries to the website. Relevant experts then reply to these queries quickly (Figure 33.12).

33.4.8 FM Radio

The UAF has its own broadcasting station (FM 100.4), which broadcast a variety of programs related to the agriculture sector. This broadcasting station has been working since August 2012, with a coverage area of almost 25 km radius, and at the same time, FM 100.4 can also be heard online throughout the world at www.uaf.edu.pk/radio.aspx. FM 100.4 not only highlights and broadcasts the live internal academic activities such as international workshops/conferences/seminars; but it also plays a vital role through broadcasting social, cultural health, Islamic, and sports programs as well. Most of the work done is by students of the university except for the expert opinions and specialized talks; which are presented by faculty members. An important segment of UAF-FM 100.4 is the news program, broadcast daily for 10 minutes on an hourly basis. Live broadcasting is another feature of the radio. Live events include workshops, conferences, festivals, sports commentaries, and weather updates, and so on. Children's hour and health time are also popular shows among the listeners.

33.4.9 Skill Development

Continuing education is a skill oriented educational program which features participants knowledgeable in certain areas of research, knowledge, and technique. Providing Skills to the youth is considered as an important activity in most countries. The UAF's skill development program is being managed by the Department of Continuing Education and the program is highly flexible in terms of the participant's age, qualifications, fee, duration of courses, and area of interest. It is known for enhancing the productivity of individuals, profitability, and expansion of national development. The most important aspect of skill development is that it opens doors of employability, cottage industry and small business. The main aim of the skill development program is to engage the rural youth on various aspects of rural development. Another important aspect of developing the skills of rural people is that it provides frontline practitioners in the different agricultural and industrial areas with access to accredited professional qualifications, without removing them from their valuable community work. The UAF through its continuing education program has gone into collaborations with skill development agencies. We hope that such collaborations will grow to assist the overall economy of the countries involved in addition to individual economy of the trainees and skilled manpower especially in rural areas.

33.4.10 Technologies Developed

Some of the notable technologies developed and disseminated by the UAF are given below:

New mango varieties: Ten new indigenous mango varieties having good yield potential and disease resistance are identified by UAF scientist. These varieties have potential to compete other mango varieties in global market. Varieties include MLT-239, MLT-240, MLT-248, MLT-250, MLT-251, RYK-265, RYK-644, RYK 426MLT-369 MLT-658.

Solar distillation system: Solar distillation system with 40% efficiency is developed by Department of Farm Machinery and Power, UAF. It is used to extract essential oils by using solar energy. It can work on both hydro and steam distillation systems.

Solar tunnel dryer and roasting system: Tunnel dryers are used to for drying of different agricultural products like apples, chilies, medicinal plants, and so on by using solar radiation. It is easy to operate and has a size of 10 m × 1.32 m (length×Width). Temperature is controlled by DC fans (PV operated). Solar continuous roasting system is intended for the roasting of groundnuts, corn, coffee, grams, oats, and so on by utilizing solar energy. It has a Scheffler solar reflector and has a roaster capacity of 20 kg per hour.

Solar autoclave, dehydrator and solar oven: Solar autoclave is developed for sterilization (steam) of surgical instrument through solar energy. The solar dehydrator is developed at the Department of Physics, UAF. Solar radiations, allowed by a translucent glazing, are converted into heat when they collide with the dark interior surface of dehydrators. Most foods are dehydrated at 130 ^{0}F; however, meat is dehydrated at higher temperatures (155 ^{0}F). solar ovens are also developed at the Department of Physics, UAF. It also uses solar energy to cook foods like vegetables, beans, rice, chicken, etc.

Bio-energy: The UAF has developed a model biogas plant which is able to provide sufficient energy to operate a tube well

Improved agricultural machinery: Improved boom sprayers have reduced four sprays as compared to conventional sprayer because its Swivel nozzles spray both the inside and outside surfaces of cotton leave, hence ensuring 100% insect mortality. It saves 10% expenditures in spray and also increase the crop yield zone disk tiller drill, which save substantial amount of diesel, labor, and water. It is developed by the agricultural engineers of the university. About 14% increase in yield is observed due to the usage of a zone disk tiller drill. It can be used for dry rice seeding, maize, and cotton planting. The redesigned thresher is 300 kg lighter than the conventional wheat thresher. It has low diesel rates, low operational costs, and high economic returns as compared to conventional wheat threshers. There is negligible grain loss with *bhoosa* and its cleaning efficiency is 99%. Multicrop reaper is also redesigned by the agricultural engineers which is suitable for rice, wheat, sorghum, and Brassica.

Mott grass production technology: Underfeeding is one of the most important causes of low livestock productivity in Pakistan. Mott grass is a multicut fodder having great potential to provoke underfeeding of livestock. Molasses is recommended for making good quality Mott grass silage which ensures regular and uniform supply of nutrients.

Value addition in floriculture: A new strain of *Rosa centifolia* developed at UAF is capable of producing 500–700 flowers/plant/year.

Water conservation technologies: Multicrop bed planters save significant amounts of water for wheat, cotton, maize, and rice production with a yield increase. A ground water recharge system is developed bythe Department of Irrigation and Drainage in order to recharge groundwater (aquifer). Its recovery efficiency is 70–80%. Perforated pipe irrigation systems are also developed by the Faculty of Agricultural Engineering. It is an easy to manage system in which water is channeled via pipe from tube wells to the fields. It has low operational and installation cost. Its conveyance efficiency is 100% and application efficiency is 70–80%.

New chicken breed: A new breed of egg laying chicken, UniGold, has been developed at UAF through the funding of the Punjab Agriculture Research Board (PARB) for rural poultry farming. This breed has good performance in central and southern Punjab under low to medium input system. It has the potential for producing 210 eggs of approximate size of 50 gm.

Sisal fiber decorticating machine: Sisal fiber is recognized as the best substitute of jute fiber and has a great capacity to generate revenue for Pakistan. Sisal plants can grow on low fertile and uneven soils. Sisal fiber decorticator has been designed and developed in the Department of Fiber and Textile Technology, UAF. Decorticating action is more effective due to specially designed decorticating cylinder and knives. The fibers come out clean and are almost free of green matter.

Oilseed: UAF-11 is a short duration, high yielding variety of "sarson" developed at Department of Plant Breeding and Genetics. It is capable of producing a yield of 40 mounds per acre in 100–120 days. This variety is suitable for rice, sugarcane, cotton, maize, potato, and tobacco cropping zones.

Mastitis vaccine: In order to control mastitis, UAF has introduced a mastitis vaccine. Mastitis vaccines are now commercially available at cheap rates (Rs. 60 per dose and Rs. 130 per piece respectively) as a result of an agreement between UAF and UM Enterprises, Karachi.

Value addition of food products: The following value added food products are produced at National Institute of Food sciences and technology

- Omega-3 enriched eggs
- Edible coatings
- Rice bran
- Healthier meat
- Fortified bakery products
- Organic mango pickle

ICTs, mobile apps: The UAF is a pioneer institution of Pakistan in promoting the use of ICTs for outreach and extension activities. The university has developed various information portals (www.zaraibaithak.com, www.fartilizeruaf.pk, www.kissandost.pk) for the extension workers and farming community. Different mobile apps have also been developed, keeping in view the increasing use of smartphones in country.

33.5 THE WAY FORWARD AND PLAN OF ACTION

This chapter has highlighted several issues relating to the outreach/extension and social mobilization in Pakistan. The following measures are suggested to boost-up outreach and social mobilization activities in Pakistan:

Enhance the use of ICTs: Pakistan is far behind in the use of ICTs for agricultural and rural development as compared to most of the developed and developing countries. Use of ICTs particularly information portals and android based mobile applications needs to be promoted. Government departments and universities have to work together by sharing their experiences in this regard. At the village level e-Baithak or Zarai-Baithak should be introduced for guidance of the farmers.

Better linkages between universities, private sector, and public extension services: This chapter has shown that the UAF has developed an effective model of outreach and community development through several integrated activities. However public extension services, the private sector, and NGOs and universities are largely working in isolation from each other. A policy intervention is needed to integrate the work of different stakeholders for improving the overall effectiveness of outreach and social mobilization activities.

Gender and youth mainstreaming: Women share equal burden of activities in agriculture and livestock related tasks. However there are very few attempts in bringing them in the loop of overall outreach activities carried out by the public and private sector. Similarly the rural youth are not much interested in farming and they tend to migrate to the urban areas. There is need to design specific outreach and social mobilization programs for female and young farmers. The universities, public, and private sector departments should collaborate to undertake such initiatives.

Mobilizing small farmers: Evidence shows that public extension services are top-down and nonparticipatory in nature. Similarly, smallholder farmers are generally excluded from the extension system. It is suggested that the goals of outreach services be redefined to fully engage small farmers in extension services. Farmers' groups and networks may be established to integrate with the extension system. The extension workers should be provided intensive trainings on social mobilization.

Establishing farm service centers: Farm service centers at *Markaz* level should be established by the government to facilitate agricultural inputs and machinery for farmers. The private sector and NGOs can also be included in this initiative. These service centers should be ICT enabled and given proper legal cover.

Agricultural extension policy: It is a dilemma that despite the increasing recognition of extension and outreach regarding rural development, Pakistan doesn't have an agricultural extension policy. It is required that a research approved demand-driven agricultural extension policy may be formed.

Promoting the culture of collective action: The culture of collective action is diminishing in rural areas of Pakistan in general and those of Punjab province in particular. Sociocultural heterogeneity is one of the reasons, however, traditions and history show that there is good potential collective action in our rural areas if proper social mobilization campaigns are initiated. Small-scale infrastructure development programs, with the participation and ownership of local communities, can be a step in this regard.

REFERENCES

Abbas M, Lodhi TE, Aujla KM, Saadullah S. 2009. Agricultural Extension Programs in Punjab, Pak. *Journal Life Social Sciences* 7(1): 1–10.

Channa AA. 2015. Four essays on education, caste and collective action in rural Pakistan. PhD dissertation, Department of International Development London School of Economics London.

Chaudhry KM. 2002. *Community Infrastructure Services Program (CISP): HRD Manual.* Department of Local Government and Rural Development, Govt. of AJK, Muzafarabad.

Christoplos I. 2010. *Mobilizing the Potential of Rural and Agricultural Extension.* FAO, Office of Knowledge Exchange, Research and Extension. Rome, Italy.

Gill MA, Mushtaq K. 1998. *Managing Irrigation for Environmentally Sustainable Agriculture in Pakistan.* Pakistan National Programme, IIWI, Lahore, Pakistan.

Government of Pakistan. 1988. *Report of National Commission on Agriculture.* Ministry of Food and Agriculture, Islamabad, Pakistan.

Government of Pakistan. 2016. *Economic Survey of Pakistan 2015–16.* Ministry of Finance and Economic Affairs, Islamabad.

Hagmann J, Chuma E, Murwira K, Connolly M. 1999. Putting process into practice: Operationalising participatory extension. Agren Network Paper No. 94. ODI, London.

Hussain A. 2006. Participatory development praxis: A case study from Pakistan's Punjab. *The Pakistan Development Review,* 45(4 Part II): 1361–1372.

Luqman M, Shahbaz B, Ali T, Iftikhar M. 2011. Critical analysis of rural development initiatives in Pakistan: Implications for sustainable development. *Spanish Journal of Rural Development* 4(1): 67–74.

Mallah MU. 1997. Extension programs in Pakistan. In Memon RA, Bashir (ed.) *Extension Methods*. National Book Foundation, Islamabad, Pakistan.

NRB. 2005. *Devolution of Power Plan*. National Reconstruction Bureau, Islamabad Pakistan.

Poteete AR, Ostrom E. 2004. Heterogeneity, group size and collective action: The role of institutions in forest management. *Development and Change* 35(3): 435–461.

Qamar MK. 2012. *Modernizing National Agricultural Extension Systems: A Practical Guide for Policy-Makers of Developing Countries. Research, Extension and Training Division. Sustainable Development Department*. Food and Agriculture Organization (FAO) of the United Nations, Rome, Italy.

Riaz M. 2010. The role of the private sector in agricultural extension in Pakistan. *Rural Development News*, 1/2010; pp 15–22.

Saeed R, Abbas M, Sheikh AD, Mahmood K. 2006. Impact of devolution on agricultural extension system in the central Punjab: perceptions of agricultural extension workers. *Pakistan Journal of Life and Social Sciences* 4(1–2): 20–26.

Shahbaz B, Ata S. 2014. Agricultural extension services in Pakistan: challenges, constraints and ways forward. Background Paper No. 2014/1. ACIAR project, Enabling agricultural policies for benefiting smallholders in dairy, citrus and mango industries of Pakistan—Project No. ADP/2010/091, Victoria University Sydney, Australia.

Shami M. 2010. *Collective Action, Clientelism and Connectivity*. Institute of Food and Resource Economics, University of Copenhagen, Copenhagen, Denmark. (FOI Working Paper; No. 2010/14).

Siraj M. 2011. *A Model for ICT-Based Services for Agriculture Extension in Pakistan*. CABI International. USAID. Pakistan's Food and Agriculture Systems, Rawalpindi, Pakistan.

Velded T. 2000. Village politics: Heterogeneity, leadership and collective action. *Journal of Development Studies* 36(5): 105–134.

WDR. 2001. *Removing social barriers and building social Institutions*. World Development Report 2000–2001.

34 Fiscal Policy in Agriculture

Tahira Sadaf, Muhammad Ashfaq, Rakhshanda Kousar, and Qaisar Abbas

CONTENTS

34.1 BRIEF BACKGROUND

Pakistan's economy is largely based on the agricultural sector, which contributes about 20% of GDP and 42% of employment; the sector has forward and backward linkages with other sectors of the economy and plays a vital role in socioeconomic development of the country. It provides food, nutrition material, foreign trade and exchange, market for industrial outputs. The performance of the crop sector, particularly, the major crops has far reaching implications for the economy of Pakistan, particularly in meeting the food security and nutrition, and in reducing poverty of masses. However, the agricultural sector could act as an engine of growth only during the 1960s and 1970s, the era of the "Green Revolution," after that it could no longer perform the role consistently. Despite the fact that agriculture is an important sector, Pakistan still has to rely largely on imports of some key agricultural products such as cotton, vegetable oil, pulses, tea, and dairy products. Pakistan has imported USD6.0 billion of food and agricultural products only during 2014. The situation has been worsened by the ever-increasing population. The government can control markets' price policy mechanisms by utilizing various tools like stabilization of policies and income distribution; policies of taxes and subsidies on commodities; macro price policies like wage rates, interest rates, and land rent rate; and macroeconomic policy tools of fiscal and monetary policies.

The purpose of this chapter is to study the agricultural price policies of Pakistan, and make regional and international comparisons of such policies with other economies in the world in order to analyze subsidies, taxation, and investment in the agricultural sector. Different sections are formed to address different aspects related to the themes of the chapter. The first section is about domestic agricultural price policies of Pakistan, the second section presents the regional comparison of agricultural price policies, the third sections reveals international comparisons of agricultural price policies, the following section addresses the issues related to taxation and subsidies in the agricultural sector, and the final section is about investment in agriculture. Each section has three crosscutting themes, historical background, current scenario, and the way forward.

34.2 AGRICULTURAL PRICE POLICIES IN PAKISTAN

Pakistan's major crops are wheat, rice, sugarcane, cotton and maize, which contribute almost 24% of the value added in the country's agriculture with almost 5% contribution in GDP. This section presents a detailed account of discussion on Pakistan's agricultural price policies on wheat, rice, sugarcane, and cotton. First subsection is about the historical background of such policies for the above mentioned crops and some minor crops (potato and onion) too. The next subsection presents the prevailing status of agricultural price policies in Pakistan and the final subsection is about way forward for Pakistan regarding price policies in agriculture based on description presented in the first two subsections.

34.2.1 Historical Background of Agricultural Price Policies in Pakistan

Pakistan since her independence in 1947 inherited the policies of government intervention in the crop markets through altering price mechanism, particularly, in case of food grains. Such interventions were intended to support the urban consumer by providing cheap food besides benefiting urban manufacturers by providing cheap raw materials to their industries, notably to the textile industry. The main features of the government's trade and pricing policies during this period were low prices and monopoly procurement of food grain, heavy export duties on cotton, restrictions on trade of different products at district and province, and even at international level and an overvalued exchange rate, which resulted in encouraging the industrial import and discouraged the agricultural export. Such policy interventions assumed that farmers would be able to continue producing the required quantities of food and raw materials, mainly cotton for the industrial sector. Such policies were also expected to be beneficial enough for all other stakeholders. However, all ended up in a failure with huge transfers of resources from agriculture to industry, without having an overall positive impact on the economy. The country has ultimately faced food shortages and crop production below its potential level, particularly of exportable crops. Above all, it adversely affected the crop producers and could not fulfill the dream of industrialization-without-agricultural progress. Consequently, the government has to change her policy tools, in 1965 when the third Five-Year Plan was presented. It was recognized that industrial development of the nation can only be possible with sustaining agriculture at first. The governments' focus thus shifted to benefit producers in the form of price stabilization mechanism through price support (Salam, 2012). History is evident that this transfer from industry-favoring to agriculture-driven policy resulted in the progress of the agricultural sector and the economy as a whole. This success was further supported by the genetic breakthroughs in rice and wheat varieties due to the "Green Revolution." Initially, incentives to the agricultural producers were provided by subsidizing inputs (tractors, tube wells, and improved seeds) rather than just raising output prices. Output support prices were incentive to those small and medium farmers who were reluctant to adopt new input technologies at low prices, first, because of affordability, and second because of the risk associated with adoption of the new technology and subsidized inputs. Wheat and rice procurement prices were reviewed periodically in order to evaluate the dynamics of domestic supply and demand (Salam, 2012). Government focus remained on devaluation the rupee during the 1960s, which was ultimately changed as it could not increase prices and failed to benefit the agricultural sector. The situation was worsened by application of export duties and government monopolies. The input subsidy policy alone had limited impact. Farmers were not motivated to generate the marketable surplus due to apparently inadequate incentives (Bokhari 2015). The process of price support and input subsidies continued until the late 1970s. At this stage, it was recognized that a proper balance between input costs and output prices was needed to induce agricultural producers to increase agricultural production.

Until the mid-1960s, agricultural price policy, including recommendation of support prices to various agricultural commodities, has been the responsibility of the Pakistan Planning Commission. A major institutional change was the establishment of the Agricultural Prices Commission (APCom) in March 1981. Additionally, the World Banks' Structural Adjustment Program was launched to bring inputs and outputs prices closer to world prices. Subsidies on fertilizers was phased out and the rupee was devaluated and floating exchange rates were followed during 80s and 90s.

APCom used to act as an autonomous agency to advise the government on issuing support price for various agricultural commodities, as well as issuing price to agricultural factors of productions including fertilizers, pesticides, and quality seeds. Every year, the government announces the support price for each major agricultural commodity. The intention is to encourage production and protect farmers, particularly small farmers, from cost increases and to raise productivity of various crops through adopting advanced technology; to avoid wasting inputs and ensure using inputs rationally; to measure the impact of these policies on various sectors of the economy and on overall well-being; to achieve other national objectives such as food self-sufficiency (mainly through wheat), increased earnings of foreign exchange through exports of commodities like cotton and rice, and reduced dependence on agricultural imports. The price support policy for major commodities remains an important instrument of determining resource allocation and production levels. The fixed floor prices of agricultural commodities are maintained by the support of official procurement in years of low market prices.

Initially, some other government bodies remained involved in price policy formulation besides APCom, for instance, the Ministry of Agriculture, the Ministry of Commerce, the Ministry of Industries, the Federal Cabinet and its Economic Coordination Committee, and the Tobacco Board under the Ministry of Commerce. In this chain, APCom had to advise the procurement/support prices for commodities such as seet cotton, paddy, sugarcane, wheat, sugar beet, and oilseeds etc.; where cotton lint and rice prices were the responsibility of the Ministry of Commerce and those of edible oils of the Ministry of Industries. Finalization of price policy had to be done by the Federal Cabinet and its Economic Coordination Committee upon advice of the APCom. The case of support price of tobacco had to be dealt with by the Tobacco Board under the Ministry of Commerce and the final decision had to be taken by the Economic Coordination Committee as it used to happen in case of other commodities.

Currently, APCom operates with a new name, "Agricultural Policy Institute." In determining agricultural prices, API considers domestic and international market conditions, productivity trends, production targets fixed by the government, export or import parity prices,* cost of production, the prices of competing crops, the profitability in fertilizer use, and the likely impact of proposed prices on other sectors of the economy. Every year before the sowing season, and working on the recommendations of the Agricultural Policy Institute, the government reviews the prices of

* For traded commodities (imported or exported), the difference between international quoted prices and the actual import/export prices is the import or export parity price. For example rice prices are set in relation to international price trends, so the domestic prices do not exceed the export price.

crops and fixes a new price for wheat, cotton, sugarcane, and rice. The price policy for different commodities varied every year, depending upon the circumstances of the individual crop. During the 2000s, government spending has declined considerably on agriculture, which means the sector is underfunded. High oil prices, energy costs, and climatic changes are some other reasons for volatility in agricultural prices in Pakistan.

34.2.1.1 Price Policies of Various Crops

As already discussed in the previous section, crops being covered in the Output Price Policy of Pakistan are wheat, rice, sugarcane, cotton, potato, and onion. History of the policy dates back to the early 1960s, as discussed in the preceding section. Main purpose of the policy was to expand production. Tables 34.1 and 34.2 present the historical record of prices announced for various crops along with their varieties.

Among important crops, wheat has consistently enjoyed the privilege of support price since 1960 (Tables 34.1 and 34.2), the support price ranged from Rs. 14–1300 per 40 kilograms over the time period of 1960–1961 to 2015–2016. Sugarcane stood second in this regard, where support price was about Rs. 3 in 1960–1961, and has gradually increased to 172 in 2015–2016. In case of rice, the support price was withdrawn in 2010–2011. Similarly, in case of cotton, the support price announcement were stopped in 1997–1998 for both cotton lint and cotton seed except one variety of cotton seed (B-557 F-149), for which the current minimum price announced is Rs. 3000, initially it was Rs. 145. The government has been announcing support price for some minor crops like potato and onion, but after year 1999–2000, it was closed.

34.2.1.1.1 Wheat Price Policy

Wheat is Pakistan's major staple food with 78% of total food-grain production. It has been feeding the ever-increasing population of Pakistan, which is reportedly above 195 million. It contributes almost 10% of value addition in the agricultural sector and 2% of GDP of Pakistan. It is grown on almost 40% of the total cropped area of the country with yields of 2752 kg/ha.

Per capita availability of wheat varies over time period due to reliance on imports, which is why the domestic availability remains different from production of the crop. Table 34.3 reveals information on per capita availability of wheat for the period 2006–2007 to 2013–2014.

The government of Pakistan has long been intervening in the wheat crop sector because of its significance as a basic food crop of the country. Such interventions had two major purposes: first, to protect consumers by keeping the domestic price below the international price and secondly, to protect producers by controlling price fluctuations and providing them support in the form of support price. In order to meet the first objective, the government sets an import subsidy and restricts private sector trading in the international market. The policy of such a manipulation to reduce prices costs the economy, where the government has to pay a handsome amount in the form of procurement (see Table 34.4).

As evident from Table 34.4 Pakistan Agricultural Services and Storage Corporation (PASSCO) is responsible for procuring wheat at the federal level. PASSCO, through its procurement centers all throughout the country, purchases wheat from farmers at the prices set by the government right after the crop harvest. At provincial level, a similar role has to be played by the food departments. Farmers are privileged enough to sell their produce in a market, or at PASSCO's procurement centers, or to the food departments. The provincial food departments are bound to purchase any volume of wheat delivered to them at procurement price. The procurement price remains fixed during the whole year and across all centers as well. Provincial food departments sell a major chunk of wheat to private flourmills at a price known as issue price, which is also fixed. This policy controls the price of wheat at the wholesale level as well as reduces the price of flour to consumers, hence benefiting both producers and consumers.

The federal government decides the quantity of import of wheat. The level of wheat-import depends on the level of public sector stocks, the international wheat market, the capacity of ports

TABLE 34.1
Procurement/Support Price of Agricultural Commodities (PKR per 40 kgs)

		Rice		Paddy			Sugarcane			
Year	Wheat	Basmati 385	IRRi-6 (F.A.Q)	Basmati	Basmati Super/2000	IRRI-6	NWFP	Punjab	Sindh	Blochistan[c]
1960–1961	14.47	25.72	–	–	–	–	1.81	2.7	2.7	–
1961–1962	14.47	26.79	–	–	–	–	1.81	2.7	2.7	–
1962–1963	14.47	27.86	–	–	–	–	2.41	2.41	2.41	–
1963–1964	14.47	30	–	–	–	–	2.14	2.14	2.14	–
1964–1965	14.47	30	–	–	–	–	2.41	2.41	2.41	–
1965–1966	14.47	30	–	–	–	–	2.41	2.41	2.41	–
1966–1967	14.47	33.22	–	–	–	–	2.14	2.14	2.14	–
1967–1968	18.22	40.72	–	–	–	–	2.41	2.59	2.74	–
1968–1969	18.22	37.50	–	–	–	–	2.7	2.95	3.11	–
1969–1970	18.22	34.30	–	–	–	–	2.95	2.95	3.11	–
1970–1971	18.22	40.72	22.40	–	–	–	2.41	2.95	3.11	–
1971–1972	18.22	49.29	22.40	–	–	–	2.41	2.7	2.84	–
1972–1973	21.43	66.45	22.40	–	–	–	4.29	4.55	4.72	–
1973–1974	26.79	69.45	28.94	–	–	–	4.29	4.55	4.72	–
1974–1975	39.65	108.8	42.87	–	–	–	5.37	5.63	5.79	–
1975–1976	39.65	108.8	42.87	47.8	–	26.79	5.89	5.16	6.32	–
1976–1977	39.65	117.89	57.87	55.73	–	32.15	5.89	6.16	6.32	–
1977–1978	39.65	117.89	49.3	59.48	–	32.15	5.89	6.16	6.32	–
1978–1979	48.23	137	52.51	64.3	–	32.15	5.89	6.16	6.32	–
1979–1980	50	150	52.57	64.3	–	32.15	7.23	7.5	7.66	–
1980–1981	58	154	63	75	–	38.58	9.38	9.65	9.81	–
1981–1982	58	160	72.5	85	–	45	9.38	9.65	9.81	–
1982–1983	64	160	80	88	–	49	9.38	9.65	9.81	–

(Continued)

TABLE 34.1 (*Continued*)
Procurement/Support Price of Agricultural Commodities (PKR per 40 kgs)

		Rice		Paddy			Sugarcane			
Year	Wheat	Basmati 385	IRRi-6 (FAQ)	Basmati	Basmati Super/2000	IRRI-6	NWFP	Punjab	Sindh	Blochistan[c]
1983–1984	64	175	83	90	–	51	9.38	9.65	9.81	–
1984–1985	70	175	83	90	–	51	9.38	9.65	9.81	–
1985–1986	80	204	86	93	–	53	11.52	11.79	9.81	–
1986–1987	80	250	86.5	102	–	53	11.52	11.79	11.95	–
1987–1988	80	258	86.5	130	–	55	11.52	11.79	11.95	–
1988–1989	82	276	100	135	–	60	12.32	12.59	12.86	–
1989–1990	85	283	113	143	–	66	13.5	13.75	14	–
1990–1991	96	308	127	150	–	73	15.25	15.25	15.75	–
1991–1992	112	340	140	155	–	78	16.75	16.75	17.75	17
1992–1993	124	360	150	175	–	85	17.5	17.5	17.5	17.5
1993–1994	130	389	157	185	–	90	18	18	18.25	18.25
1994–1995	160	419	170	210	–	102.6	20.5	20.5	20.75	20.75
1995–1996	160	461	183	222	–	112	21.5	21.5	21.75	21.75
1996–1997	173	461	210	255	–	128	24	24	24.5	24.5
1997–1998	240	–	251	310	–	153	35	35	36	36
1998–1999	240	–	–	330	–	175	35	35	36	36
1999–2000	300	–	–	350	–	185	35	35	36	36
2000–2001	300	–	–	385	460	205	35	35	36	36
2001–2002	300	–	–	385	460	205	42	42	43	43
2002–2003	300	–	–	385	460	205	42	42[b]	43[b]	43
2003–2004	350	–	–	400[a]	485	215[a]	42	40	41	
2004–2005	400	–	–	415[a]	510	230[a]	42	40	43	43

(*Continued*)

TABLE 34.1 (*Continued*)
Procurement/Support Price of Agricultural Commodities (PKR per 40 kgs)

		Rice		Paddy			Sugarcane			
Year	Wheat	Basmati 385	IRRi-6 (FAQ)	Basmati	Basmati Super/2000	IRRI-6	NWFP	Punjab	Sindh	Blochistan[c]
2005–2006	415	–	–	460	560	300	45	45	60	–
2006–2007	425	–	–	–	–	306	65	65	67	–
2007–2008	625	–	–		–	–	65**	65**	63	–
2008–2009	950	2500	1400	1250	1500	700	80	80	81	–
2009–2010	950	0	0	1000	1250	600	100	100	102	–
2010–2011	1050	–	–	–	–	–	125	125	125	–
2011–2012	1200	–	–	–	–	–	150	150	154	–
2012–2013	1300	–	–	–	–	–	170	170	172	–
2013–2014	1300	–	–	–	–	–	170	170	172	–
2014–2015	1300	–	0	0	0	0	180	180	182	–
2015–2016	1300	0		0	0	0	180	180	172	–

Source: Government of Pakistan. 2016. *Pakistan Economic Survey 2015–2016*. Economic Advisor's Wing, Finance Division, Ministry of Finance, Islamabad, Pakistan; Ministry of Food, Agriculture and Livestock (APCom).

Note: FAQ: Fair Average Quality.

[a] Indicative price.

[b] The Federal Government did not fix any support price of sugarcane for 2002–2003 crop. Nevertheless, Government of Punjab in November, 2002 and Sindh in December 2002 fixed the minimum purchase prices of sugarcane, at factory gate as Rs. 40 and 43 per 40 kg respectively.

[c] Support price fixed by government since 1991–1992.

** Sugarcane prices are fixed by respective provincial governments.

TABLE 34.2
Procurement/Support Price of Agricultural Commodities (PKR per 40 kgs)

	Cotton Lint				SEED Cotton					
Year	Desi	AC-134,NT	B-557 149-F	Sarmasti Qallan dri DELTAPINE MS 39–40	Desi	AC-134 NT	B-557 F-149	Sarmast Qallan dri Delta pine MS 39–40	Potato	Onion
1976–1977		434.04	482.26	525	128.6	133.96	144.68	155.4	26.8	19.3
1977–1978	359	359	451.18	451	141.46	147.89	159.68	171.47	26.8	19.3
1978–1979	369.94	389	424.39	459.97	143.61	147.89	159.68	171.47	26.8	19.3
1979–1980	375	410.46	445.83	481.19	143.61	147.89	159.68	171.47	26.8	19.3
1980–1981	409.38	442.61	475.8	509	156	160	171	182	26.8	19.3
1981–1982	419	449	473	515	166	170	178	192	26.8	19.3
1982–1983	426	471.6	473	515	168	175	183	197	40.5	25
1983–1984	426	476.8	496	538	169.5	178	186	200	40.5	30
1984–1985	426	476.8	500	542.27	169.5	181	189	203	40.5	30
1985–1986	426	480	500	542.27	173.5	185	193	207	42	32.5
1986–1987	431	483	500	542.27	173.5	185	193	207	44.5	34.5
1987–1988	463	515	504	546	173.5	185	193	207	44.5	36.5
1988–1989	550	615	507	549	176.5	188	196	210	50	40
1989–1990	662	685	539	581	191.5	203	211	225	55	42
1990–1991	695	–	645	690	220	235	245	260	55	51.5
1991–1992	726	–	715	745	225	270	280	290	65	60
1992–1993	795	–	770	800	275	–	300	310	67	65
1993–1994	795	–	801	831	290	–	315[a]	325	77	78
1994–1995	–	–	986	1055	340	–	400[a]	423	84	78
1995–1996	–	–	986	1055	340	–	400[a]	423	84	85
1996–1997	–	–	–	–	440	–	500[a]	540	115	100
1997–1998	–	–	–	–	440	–	500[a]	540	145	112
1998–1999	–	–	–	–	–	–	825[a]	–	145	140

(Continued)

TABLE 34.2 (*Continued*)
Procurement/Support Price of Agricultural Commodities (PKR per 40 kgs)

	Cotton Lint				SEED Cotton					
Year	Desi	AC-134,NT	B-557 149-F	Sarmasti Qallan dri DELTAPINE MS 39–40	Desi	AC-134 NT	B-557 F-149	Sarmast Qallan dri Delta pine MS 39–40	Potato	Onion
1999–2000	–	–	–	–	–	–	725[a]	–	145	–
2000–2001	–	–	–	–	–	–	725[a]	–	–	–
2001–2002	–	–	–	–	–	–	780[a]	–	–	–
2002–2003	–	–	–	–	–	–	800	–	–	–
2003–2004	–	–	–	–	–	–	850[b]	–	–	–
2004–2005	–	–	–	–	–	–	925	–	–	–
2005–2006	–	–	–	–	–	–	975	–	–	–
2006–2007	–	–	–	–	–	–	1025	–	–	–
2007–2008	–	–	–	–	–	–	1025	–	–	–
2008–2009	–	–	–	–	–	–	1465	–	–	–
2009–2010	–	–	–	–	–	–	–	–	–	–
2010–2011	–	–	–	–	–	–	–	–	–	–
2011–2012	–	–	–	–	–	–	–	–	–	–
2012–2013	–	–	–	–	–	–	–	–	–	–
2013–2014	–	–	–	–	–	–	–	–	–	–
2014–2015	–	–	–	–	–	–	3000	–	–	–
2015–2016	–	–	–	–	–	–	3000	–	–	–

[a] Niab-79, CIM-107.
[b] Support price of seed cotton (phutti) for the base grade 3 with staple length 1–1/32 and micronaire range of 3.8–4.9 fixed.

TABLE 34.3
Availability of Wheat (May–April) (000 Tonnes)

Year	2006–2007	2007–2008	2008–2009	2009–2010	2010–2011	2011–2012	2012–2013	2013–2014
Production	21,277	23,295	20,959	24,033	23,311	25,214	23,474	25,979
Deduction for seed, feed and wastage at the rate of 10%	2128	2330	2096	2403	2331	2521	2347	2423
Import	0	1708	2685	–	175	–	200	600
Export	1027	311	143	10	1781	409	52	650
Net Availability	18,122	22,363	21,405	21,620	15,374	22,284	20,948	21,758
Population (million)**	159.84	164.7	168.2	171.73	175.31	178.91	182	188.02
Per capita availability (kgs/annum)	113.38	135.8	127.3	125.89	110.51	124.56	116.9	116.92

Source: Government of Pakistan. 2014. *Agricultural Statistics of Pakistan 2013–2014*. Pakistan Bureau of Statistics, Islamabad.

** Economic Survey of Pakistan (various issues).

TABLE 34.4
Procurement of Wheat (000 Tonnes)

Procurement Year	Punjab	Sindh	KPK	Balochistan	PASSCO	Total
1994–1995	2003	535	–	22	1180	3740
1995–1996	1891	408	9	53	1087	3448
1996–1997	1668	262	–	2	793	2725
1997–1998	2528	401	4	–	1051	3984
1998–1999	2786	573	–	–	711	4070
1999–1900	6336	630	–	44	1572	8582
2000–2001	2514	429	–	162	976	4081
2001–2002	2844	255	–	9	937	4045
2002–2003	2409	320	–	0	785	3514
2003–2004	2453	179	–	1	823	3456
2004–2005	2438	504	–	–	997	3939
2005–2006	2563	709	–	–	1242	4514
2006–2007	2569	566	–	18	1269	4422
2007–2008	2557	506	–	–	854	3917
2008–2009	5782	1216	90	–	2143	9231
2009–2010	3722	1497	300	68	1127	6714
2010–2011	3191	1395	187	108	1315.2	6196.2
2011–2012	2781	1152	317	107	1434.9	5791.9
2012–2013	3680	1056	24	50	1138.6	7910
2013–2014	3743	1204	71	90	1011.3	5948.6
2014–2015	–	–	–	–	–	6131
2015–2016	–	–	–	–	–	7050

Source: Government of Pakistan. 2014. *Agricultural Statistics of Pakistan 2013–2014*. Pakistan Bureau of Statistics, Islamabad; Government of Pakistan. 2016. *Pakistan Economic Survey 2015–2016*. Economic Advisor's Wing, Finance Division, Ministry of Finance, Islamabad, Pakistan.

in handling the stock, the potential domestic procurement, and foreign exchange reserves. The provincial food departments buy wheat import from the federal government at release price, who pays the transportation costs within the country besides buying at the CIF (cost, insurance, and freight) import price. Thus, the government has to pay amounts greater than it receives from selling to the provincial food departments at the release price. In order to finance this margin, a huge amount has to be allocated in the federal budget allocation (Faruqee et al., 1997).

Pakistan's wheat price policy has passed through a number of changes since her independence (Dorosh and Salam, 2008). From 1947 to the early 1980s, government market interventions remained at a peak. Ration shops with fixed prices remained functional but with some generous leakages and malpractices. In the late 1980s, government focus diverted to trade liberalization, where the private sector had started importing wheat, but subsequently the sector was disallowed to do so. Pakistan got her bumper harvest during fiscal year 1999–1900, which resulted in procurement of 8.6 million tons due to large increases in supply. As a result the wheat stock increased in the country, and the government had to subsidize wheat exports and again private sector exports. The government started encouraging incentives for private investment in storage and public investments in laboratories for grain testing. Restrictions were imposed on interprovincial and interdistrict transport of wheat during the early 2000s. The government issued tenders for imports during 2004, nonetheless, she rejected some shipments due to low quality. Procurement prices reached Rs. 400/40 kilograms, gradually rising over time during 2004–2005 (see Tables 34.1 and 34.2). During the same period the government lifted restrictions on transport of wheat in order to recover production and liberalize markets. Moreover, the private sector was encouraged to commercially import wheat. The procurement price kept rising from Rs. 625/40 kilograms in 2007–2008 to Rs. 1050/40 kilograms in 2011–2012 as it was a good harvest year (25.214 million tons) (see Tables 34.1 through 34.3). Pakistan had some other years of good harvest, including 2012–2013, 2013–2014, and 2014–2015 with respective productions of 24.211 million tons, 25,979 million tons, and 25,048 million tons. It is provisionally 25.482 million tons during 2015–2016. Along with rise in production, the procurement price kept rising and it is Rs. 1300/40 kilograms (for all mentioned figures see Tables 34.1 through 34.3).

34.2.1.1.2 Cotton Price Policy

Cotton being the most important nonfood cash crop and a source of raw material for Pakistan's most important industry, that is, the textile industry. It is an important source of foreign exchange earnings. It yields 3.5 million tons of cotton seeds, which contribute to 74% of edible oil production (GOP, 2016). Importance of cotton crop is evident by the freefall in the agricultural sector, which led to overall downfall in the growth rate during the current fiscal year, as cotton production declined 27% as compared to the last fiscal year. Overall, it has remained a major contributor to the agricultural growth and hence the economic growth of the country.

The cotton price policy is difficult to administer due to wide fluctuations in market prices internationally. The government has to consider proper incentives for cotton growers. For instance, if cotton lint prices increase internationally, it has to deal with pressures from cotton ginners. Tables 34.1 and 34.2 provides historical support prices since 1960s for both seed cotton and lint. The support price announcements were stopped during year 1997–1998 for both cotton lint and cotton seed except one variety of cotton seed (B-557 F-149), for which the current minimum price announced is Rs. 3000; initially it was Rs. 145.

Support prices had to be set annually on the recommendations of the APCom. The commission used to consider world prices and costs of production in deciding the support price for cotton. Support prices had to remain the same at all locations and throughout the year. Nonetheless, some required adjustments had to be made according to quality. Support prices for lint are effective at ginning. The domestic price of raw cotton was about 40%–45% of the border prices during 1960–1980. Since then, due to the decline in international prices, the local cotton prices have been approximately at the same level as the border prices at the open market exchange rate.

During normal practice, farmers used to sell seed cotton to ginners, from whom the Cotton Export Corporation (CEC) had bought lint. The CEC allowed domestic spinners to buy at market prices during harvesting season before it could enter the market. It then started purchasing cotton at market prices. As the primary implementing body of the government's cotton policy, the CEC had to balance various objectives, which included the need for foreign exchange and the desirability of keeping domestic cotton prices low (Ender, 1990). Until the late 1980s, the CEC had enjoyed monopoly over cotton exports. Since then, the private sector had been allowed to export cotton. Additionally, an export duty was imposed to earn profits arising out of increasing world prices while keeping domestic prices low. Currently, the private sector handles the export of cotton. And the imposition of export duties ensures that raw cotton is being made available to the local textile industry.

Recently, seed cotton/phutti trade has been fully liberalized, where local markets are linked with international markets. However, if local market prices are depressed in some years then the Trading Corporation of Pakistan (TCP) was established to procure lint cotton to safeguard the interest of cotton growers. The objective of TCP's procurement is to stabilize the price of seed cotton and to ensure fair returns to the growers.

34.2.1.1.3 Sugarcane Price Policy

Sugarcane has a significant importance for sugar and sugar related industries of the country. It contributes 0.6% in overall GDP. It is grown over an area of 1.1 million hectares. The government announces the support price for sugarcane annually, on the basis of the average costs of production. Sugarcane prices are normally fixed by the respective provincial governments. Reasons behind such policy of fixing high prices for sugarcane are generating revenue for the government from duties on domestic and imported sugar, and meeting the objective of attaining self-sufficient domestic sugar production. Unfortunately, such policy procedure overlooks that it is a perennial crop with a life cycle of 3 years. It also ignores the high water requirements of sugarcane and its dislocation effect on two important crops during both *rabbi* and *kharif* season in one year (such as wheat and cotton). Additionally, it provides comparatively more earning to the sugarcane farmers than other farmers. As for the *ratoon* crop, farmers can save some of the costs of production, like in case of labor, preparation of land, and cost associated with seed purchase and it's sowing.

The Pakistani government has been announcing support price for sugarcane since the 1960s (see Tables 34.1 and 34.2). However, it has announced the first National Sugar Policy in 2009–2010. The main objective of the policy is to ensure a plentiful supply of sugar at competitive prices to consumers while simultaneously protecting the interest of growers. It was also emphasized in the National Sugar Policy that there is need to educate the growers about new technologies, in order to conserve water and enhance productivity per acre. The policy further emphasizes that the introduction of alternative crops, such as sugar beet, should be encouraged to produce sugar.

34.2.1.1.4 Rice Price Policy

Rice being a crucial crop for the economy of Pakistan occupies an important place in the export trade of the country besides playing a role of second major food crop after wheat. It occupies almost 12% of Pakistan's cropped area. It contributes about 21% to total food-grain production. It is cultivated over an area of 2.8 million hectares with a production of 7 million tones.

Since the end of 1970s and start of 1980s, rice price policy has undergone many modifications. (1) The traditional procurement policy was replaced by voluntary policy of procurement. (2) The interdistrict ban of transportation was discarded, just like for other agricultural crops. (3) The Rice Export Corporation of Pakistan (RECP), founded in 1974, started procurement on its responsibility. (4) The ban on export of rice in the public sector was dissolved in 2000 and the RECP was merged with the Trading Corporation of Pakistan. (5) The government discontinued price support in order to allow market forces to operate in production and export of rice.

Due to the monopoly of the RECP, the domestic price of basmati rice has remained stagnant, between 36% and 55% of the international price, during the period 1960–1987. Since 1987, the private sector has been allowed to export the superior-quality-rice in packages of 2.5–10 kg. Like basmati-rice, the domestic price of IRRI-rice was about 40%–50% of the international prices during the period 1967–1980. However, the decline in world prices of this rice during the late 1980s ultimately balanced out to the domestic IRRI price. In some years, due to below domestic-international prices, the IRRI farmer's continued performing just because of the higher officially guaranteed price. Mostly, the government interventions in rice production and marketing were reduced in the mid-1990s. However, the government is still a key player in pursuing research and extension activities in order to encourage the expansion of hybrid rice cultivation, and promote a more efficient input utilization. Minimum producer prices are also announced every year, but mainly for indicative purposes. During some years (see Tables 34.1 and 34.2), a huge drop in domestic market prices below the target levels prompted the government purchasing paddy through the PASSCO and TCP. However, procured quantities were very minimal, as the government remained reluctant to intervene directly into the rice market. Instead, it opted for giving indirect assistance to the sector by promoting exports through government-to-government deals. In the years 1996–1997, due to the heavy losses, the government abolished RECP and replaced it with the Rice Exporters Association of Pakistan (REAP) for procuring and exporting rice. The government also made a Quality Review Committee (QRC) to certify the quality of rice before exporting. The rice trading is now being done through the private sector. Agricultural output prices are now determined by market forces, allowing market signals to be transmitted to farmers without distortion.

34.2.2 Current Scenario of Agricultural Price Policies

This section provides a glimpse of the current scenario of support price for the crops, which are at present enjoying floor price. Tables 34.1 and 34.2 provides the support price announced during the current fiscal year and the last two fiscal years, for wheat, sugarcane, and the only variety of cotton seed.

As already discussed in the preceding section, the Pakistan government encourages wheat production through intervention in the wheat market using support price policy mechanisms. Wheat is the only crop for which it sets floor purchase price, calling it the Guaranteed Minimum Price, or simply the support price. This has been helpful in setting up a stable investment atmosphere that lets farmers plan with confidence for earning returns and adopt improved technologies to enhance productivity. Still there are loopholes in the system, evident by the substantial yield gaps between leading farmers getting 6.0 tonnes per hectare, and the average farmer getting 2.6 tonnes (FAO, 2011). During last 3 years, support prices for wheat have remained Rs. 1300 per 40 kilograms.

Sugarcane prices are fixed by the respective provincial governments. During the current fiscal year the federal government announced fixed rate of Rs. 180 per 40 kg and subsequently, except in Sindh other provinces have also notified the same rate of Rs. 180 per 40 kg as support price for the sugarcane season 2015–2016. The government also subsidized export of sugar. Most recently the federal government has released Rs. 1.625 billion for payment of sugar export subsidy.* Support price for seed cotton is Rs. 3000 per 40 kilograms, such high rate has been given to farmers in response to decline in production of cotton crop during last 2 years (Table 34.5).

Besides support price, farmers are given incentives through some other measures as well. During the current fiscal year the government has announced the Prime Minister's Agriculture Package of Rs. 341 billion in order to support small farmers through direct cash support and provision of soft agriculture loans. The rice and cotton producers, growing crop on 12.5 acres would be given cash supports of Rs. 5000 per acre. The government will bear the cost of Rs. 20 billion each for rice and cotton crops.

* http://par.com.pk/

TABLE 34.5
Procurement/Support Price of Agricultural Commodities (PKR per 40 kgs)

		Year		
Province	**Crop**	**2013–2014**	**2014–2015**	**2015–2016**
	Wheat	1300	1300	1300
	Cotton Seed (B-557 F-149)	–	3000	3000
NWFP	Sugarcane	170	180	180
Punjab		170	180	180
Sindh		172	182	172
Blochistan		–	–	–

Source: Government of Pakistan. 2016. *Pakistan Economic Survey 2015–2016*. Economic Advisor's Wing, Finance Division, Ministry of Finance, Islamabad, Pakistan.

34.2.3 Way Forward

The government has good intentions of liberalizing the agricultural sector, and to increase wheat production, wheat prices have been fixed equal to international market prices. Cotton farmers are already enjoying almost double the announced procurement prices, and rice prices (for both Basmati and IRRI) have also been boosted. Sugarcane growers are also enjoying high prices despite the fact that sugarcane is highly resource intensive crop, and competes with cotton for inputs. In conclusion, due to the support price system, prices of agricultural commodities in Pakistan have been more in par with the international market prices.

Despite enjoying a thorough price support over decades, there are loopholes in the wheat crop system, evidenced by the substantial yield gaps between leading wheat farmers getting 6.0 tonnes per hectare and the average farmer getting 2.6 tonnes (FAO, 2011). Therefore, besides relying on support price mechanism, there is a dire need to increase adoption of new technologies along with using proven agronomic improvements (quality seed, timely sowing, efficient input use, adoption of balanced fertilization, and efficient water use, etc.). Productivity can also be enhanced using agricultural conservation approaches, such as using green manuring, developing rice-wheat cropping systems, reducing soil tillage, and using appropriate herbicides.

In the case of sugarcane, there is need to educate the growers about new technology in order to conserve water and enhance productivity per acre. Farmers should be educated on use of alternative crops, such as sugar beet, to encourage sugar production.

34.3 REGIONAL COMPARISON OF AGRICULTURAL PRICE POLICIES

Agricultural price policy is a mechanism of government to influence the prices of agricultural outputs and inputs. Output prices refer to support or procurement prices of crops, which provide incentive to producer to produce the desired quantity of products. Input prices include subsidies on inputs to stabilize costs of production. The basic objective of input price policy is to enhance new technology adoption and investment through production incentives. Input and output prices are interlinked as cost of production plays an important role in fixing the prices of outputs.

Agricultural price policy is an integral part of macroeconomic policy, which has influence upon resource allocation, income distribution, industrial production, and trade. Overall, the objectives of agricultural price policies vary from country to country and the stages of development within each country. Thus, agricultural prices policies play a very critical role in achieving growth in the agriculture sector. It is an important instrument to protect producers, achieving food security, and enhancing farm income. In this section, the price policies of the Indian economy will be discussed.

In this section, regional comparison of agricultural price policies with the Indian economy will be presented.

34.3.1 Indian Agricultural Price Policies: Minimum Support Price

The Minimum Support Price (MSP) is the market intervention of the Indian government in order to protect the farming community against the shortfall of farm prices. The MSPs acts as floor prices which ensure that prices of agricultural commodities will not fall below this fixed price even in case of higher supply.

34.3.1.1 History and Current Scenario

It was initiated in 1966–1967 for wheat crop in response to the green revolution and enhanced productivity in order to safeguard farmers from declining prices. After that, it was extended to many crops. In the 1970s, two kinds of administered prices were announced that were Minimum Support Prices (MSP) and Procurement Prices. Procurement prices are the prices of the cereals of both Rabbi and Kharif seasons where the public agencies such as FCI are responsible for the procurement of grains. These prices are announced just after the starting of harvest period. Generally, the procurement price is held higher than the MSP and stands lower than the market price. These both types of prices for paddy were continued with some disparity up to the years of 1973 and 1974. While for wheat these were stopped in 1969 and revived for one year in 1974–1975.

In 1980, Agriculture Prices Commission (APC) was started its functions which also dealt with the terms of trade between agricultural and nonagricultural sector, in addition to fixing support price mechanism. This commission was renamed as Commission for Agricultural Costs and Prices (CACP) in 1985. The basic objective of it was to fix Minimum Support Prices (MSP) with a view to raise productivity, adopt modern technology and safeguarding producers against the shortfall of prices.

The Minimum Support Prices (MSPs) for almost 25 major and minor crops, fixed by the Department of Agriculture and Cooperation on the recommendations of the CACP since 2008–2009 to 2016–2017 are as under in the Table 34.6.

34.3.1.1.1 Cost of Production of Different Crops in India

In deciding MSP, the government of India takes into account the cost of production of these crops, in addition to other factors. The history of the cost of production of different crops is given in Table 34.7.

As the Indian government gives substantial amounts of subsidies to farmers, the cost of production of different crops in India is almost 10%–15% lower than in Pakistan.

34.3.1.1.2 Input Subsidies of Indian Agriculture

In Indian economy, input subsidies in agriculture are the important part of the agriculture policy and cover a large share in budget. It is an attempt of government to keep prices of inputs low for farmers, which leads to high productivity with minimum cost of production. The historical record of input subsidies is given in the Table 34.8 (Figure 34.1).

34.4 INTERNATIONAL COMPARISON OF AGRICULTURAL PRICE POLICIES

In this section price policies of developed nations will be presented in order to make comparison with the South-Asia region, particularly Pakistan.

34.4.1 Agricultural Price Policy of United States of America

34.4.1.1 History and Current Scenario

The agricultural prices policy of the United States consists of federal U.S. farm bills. Up to the 1920s, agricultural prices policy of the United States was focused on small farms and provision of inputs

TABLE 34.6
Historical Record of Minimum Support Prices of Different Crops of India (INR)

S. No.	Commodity	Variety	2008–2009	2009–2010	2010–2011	2011–2012	2012–2013	2013–2014	2014–2015	2015–2016	2016–2017
1	Paddy	Common	850	950	1000	1080	1250	1310	1360	1410	1470
		Grade A	880	980	1030	1110	1280				
2	Jowar	Hybrid	840	840	880	980	1500	1500	1530	1570	1625
3	Bajra		840	840	880	980	1175	1250	1250	1275	1330
4	Maize		840	840	880	980	1175	1310	1310	1325	1365
5	Ragi		915	915	965	1050	1500	1500	1550	1650	1725
6	Arhar		2000	2300	3000	3200	3850	4300	4350	4425	4625
7	Moong		2520	2760	3170	3500	4400	4500	4600	4650	4800
8	Urad		2520	2520	2900	3300	4300	4300	4350	4425	4575
9	Cotton	Medium staple	2500	2500	2500	2800	3600	3700	3750	3800	3860
	Cotton	Long staple	3000	3000	3000	3300	3900	4000	4050	4100	4160
10	Groundnuts		2100	2100	2300	2700	3700	4000		4030	4120
11	Sunflower		2215	2215	2350	2800	3700	3700	3750	3800	3850
12	Soyabean		1350	1350	1400	1650	2200	2500	2500	2600	2675
13	Sesamum		2750	2850	2900	3400	4200	4500	4600	4700	4800
14	Niger seed		2405	2405	2450	2900	3500	3500	3600	3650	3725
15	Wheat		1080	1100	1120	1285	1350	1400	1450	1525	
16	Barley		680	750	780	980	980	1100	1150	1225	
17	Gram		1730	1760	2100	2800	3000	3100	3175	3425	
18	Masur		1870	1870	2250	2800	2900	2950	3075	3325	
19	Mustard		1830	1830	1850	2500	3000	3050	3100	3350	
20	Safflower		1650	1680	1800	2500	2800	3000	3050	3300	
21	Toria		1735	1735	1780	2425	2970	3020	3020	3290	
22	Copra	Milling	3660	4450	4450	4775	5100	5250	5250	5550	5950
23	Coconut		988	1200	1200	1200	1400	1425	1425	1500	1600
24	Jute		1250	1375	1575	1575	2200	2300		2700	3200
25	Sugarcane		81.18	129.84	139.12	139.12	170	210	220	230	280

Source: Department of Agriculture, Cooperation and Farmers Welfare, Ministry of Agriculture and Farmers Welfare, Government of India.

TABLE 34.7
Historical Record of Cost of Production of Different Crops in India (INR)

S. No.	Commodity	2009–2010	2010–2011	2011–2012	2012–2013	2013–2014
1	Paddy	645	742	888	1152	1234
2	Jowar	804	965	1141	1612	1648
3	Bajra	658	768	840	1059	1003
4	Maize	738	790	921	1070	1112
5	Ragi	861	1107	1271	1884	1687
6	Arhar	2197	2422	2702	4167	3958
7	Moong	2705	3109	3373	4699	4759
8	Urad	2257	2490	2799	4334	4112
9	Cotton	2111	2129	2528	2772	3533
10	Groundnuts	1879	2100	2633	3714	3398
11	Sunflower	1915	2257	2795	3698	3679
12	Soyabean	1200	1298	1560	2343	2216
13	Sesamum	3035	2847	3393	4186	4134
14	Niger seed	2368	2264	2945	4555	3628
15	Wheat	701	826	927	1066	1109
16	Barley	608	677	734	862	1035
17	Gram	1641	1902	2121	2328	2865
18	Masur	1626	2191	2592	3162	2760
19	Mustard	1276	1520	1786	1987	2368
20	Safflower	1884	2038	3322	3338	3501
21	Jute	1193	1301	1496	1808	2160
22	Sugarcane	81	90	102	148	185

Source: Government of India. 2016. Department of Agriculture, Cooperation and Farmers Welfare, Ministry of Agriculture and Farmers Welfare.

to the total agriculture sector. In 1933, the Agriculture Adjustment Act (AAA) was launched due to the Great Depression, which regulated the crop production by destroying crops, artificially reducing supplies, and providing subsidies to farmers to limit their production of crops.

The mechanism of providing price support to farmers dates back to post-World War I, where it was highly required given the financial crises, infrastructural damage and limited international

TABLE 34.8
Historical Record of Subsidies in India (INR)

Subsidy Head	2009–2010	2010–2011	2011–2012	2012–2013	2013–2014	2014–2015	2015–2016
Fertilizer	61,264	62,301	70,013	65,613	71,280	72,970	73,000
Petroleum	14,951	38,371	68,484	96,880	83,998	63,427	30,000
Major subsidies	134,658	164,516	211,319	247,493	247,596	251,397	NA
Total subsidies	141,351	173,420	217,941	257,079	NA	260,658	241,856
Major subsidies as % of GDP	2.08	2.11	2.35	2.45	2.18	1.95	1.5
Total subsidies as % of GDP	2.18	2.22	2.42	2.54	NA	2.02	1.7

Source: Government of India. 2016. Department of Agriculture, Cooperation and Farmers Welfare, Ministry of Agriculture and Farmers Welfare.

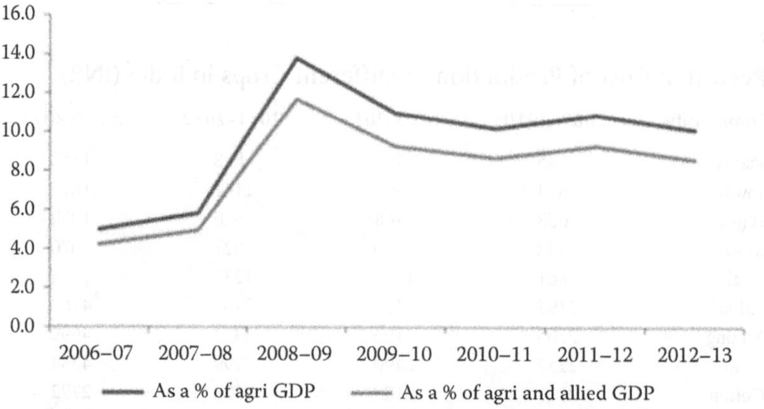

FIGURE 34.1 Fertilizer subsidy as percentage of agriculture/agriculture and allied GDP. (From Government of Pakistan. 2016. *Pakistan Economic Survey 2015–2016*. Economic Advisor's Wing, Finance Division, Ministry of Finance, Islamabad, Pakistan.)

trade relations after that war. In 1922, the Capper–Volstead Act was formulated to regulate the livestock sector and farmer's cooperatives against antitrust lawsuits. In reaction of falling agriculture commodity prices, financial crises, and the Great Depression (1929–1933), the United States initiated three bills: the Grain Future Act, 1922; the Agricultural Marketing Act, 1929; and the Agricultural Adjustment Act, 1933, that was the first comprehensive agricultural and food policy legislation.

In 2002, a farm bill also known as "Farm Security and Rural Investment Act 2002" was launched. In this bill, almost USD16.5 billion were allocated for agricultural subsidies. These subsidies had significant impacts in enhancing the productivity of cotton, oilseed, and some other crops.

Another U.S. farm bill known as the Agricultural Act of 2014 was implemented in 2014. In this act, USD956 billion for the coming ten years were procured. Direct Payments (DP) were introduced in this bill which will end almost 20 years of fixed annual payments. These payments were calculated on the basis of yield, area, and production history and paid to all farmers. In this act, Price Loss Coverage (PLC) and Agriculture Risk Coverage (ARC) were launched. U.S. farm income experienced a golden period during 2011–2014, driven largely by strong commodity prices and agricultural exports (United States Department of Agriculture 2016).

34.4.2 Agricultural Price Policy of the European Union

34.4.2.1 History and Current Scenario

In 2000, Common Agricultural Policy (CAP) was divided into two pillars: Production support and rural development. Within the "Production Support," support prices were provided on different agricultural commodities including cereal, milk, beef, etc. and direct payment were made to the farmers.

In June 2003, subsidies reforms from particular crops were decoupled. Here, the "single farm payments" were subject to "cross-compliance" conditions relating to environmental, food safety, and animal welfare standards.

In 2006, one of the important reforms "Sugar regime reform (2005–2006)" was the provision of subsidy for sugar that was not included in earlier reforms. The E.U. has decided to reduce the guaranteed price of sugar by 36% from 2006 to 2010. This is the first serious reform of sugar under the CAP in 40 years.

In 2010, the E.U. spent €57 billion on agricultural development, of which €39 billion was spent on direct subsidies. Agricultural and fisheries subsidies form over 40% of the E.U.'s budget.

TABLE 34.9
Historical Record of the Development of CAP

Period						
1960s	1970s	1992 Reforms	Agenda 2000	CAP, 2003	CAP, 2008	CAP after 2013
Price support Productivity enhancement Market stabilization	Over production Exploding expenditures International friction Supply controls	Price cut and compulsory payments Surplus reduction Income stabilization	Deep amendments in reforms Rural development	Market orientation Decoupling Cross Compliance Consumer concern Environment enlargement	Reinforcing 2003 reforms Dairy quotas	Greening Targeting Redistribution End of production constraints Food Chain R&D

Source: European Commission. 2016. Agriculture and rural Development—The common agriculture Policy after 2013. Available at: http://ec.europa.eu/agriculture/cap-post-2013/index_en.htm.

There were major reforms in June, 2013 which mainly focused on "viable food production," "sustainable management of natural resources," and "balanced development of rural areas throughout the E.U." The historical record of the development of CAP is given in Table 34.9.

34.4.2.1.1 Policy Instruments of CAP

- *Domestic price support*: Domestic support price was a key element of CAP but it was replaced with Direct Payments (DPs). The budget allocated to DP was nearly 70% of the total CAP budget. Still the floor prices play important roles in fixing the prices of major grain crops, sugar, beef, dairy products but its level of involvement now is lower than before the reforms.
- *Subsidies*: Subsidies are also support domestic prices to protect agricultural production. Other subsidies like subsidies for surplus storage and consumer subsidy also influence domestic price support. However, storage subsides were cut up to 50% in 2003.
- *Direct payments (DP)*: As support price is an importance factor of enhanced productivity and farm income, the payments made directly to producers tends to enhance more income support. In the 2003 reforms, payments to the farmers were made on the basis of average payments made in 2000–2002, instead of production records. Similarly in the livestock sector, payments were made on per animal basis and during the period 2000–2002. In addition to these payments, other minor payments were also made.

34.4.2.1.2 Benefits from the CAP

The 15 old E.U. countries are taking more benefits than the newer members of the E.U. For instance, France is at the top in term of benefits, which comprises almost 17% of CAP payments, followed by Spain (13%), Germany (12%), Italy (10.6%), and the UK (7%).

Inequality exists in CAP subsidies and benefits. For example, for individual farms the amount of average subsidy per annum is about €12,200. While the payments on the basis of acreage range from €527 (Greece) to €89 (Latvia). Likewise, giant landowners and businessmen took huge percentages of the benefits as compared to smaller farmers who have only access to traditional technology and local markets. For instance 80% of farm aid was received by almost 25% big landowners (Table 34.10).

TABLE 34.10
Summary of Major Reforms under CAP

1992	Introduction of Direct payments and set-aside
1995	Inclusion of Rural Development element
2002	Subsidy ceiling fixed until 2013
2003	Decoupling subsidies from production levels
2006	Reforms on sugar subsidies
2008	CAP' health check, milk quotas, and land set-aside
2013	Inclusion of "viable food production," "sustainable management of natural resources," and "balanced development of rural areas throughout the EU"

Source: The Washington Post. 2004; European Commission. 2016. Agriculture and rural Development—The common agriculture Policy after 2013. Available at: http://ec.europa.eu/agriculture/cap-post-2013/index_en.htm.

34.5 SUBSIDIES AND TAXATION IN AGRICULTURAL SECTOR

34.5.1 Subsidy in Agricultural Sector

34.5.1.1 Historical Background

Subsidy plays vital role in order to modernize the existing technology or to implement new technology. In a country like Pakistan, developmental subsidies particularly on fertilizers and tube wells were found to be very effective. For instance, from 1964 to 1974, use of fertilizers increased by 500% and continued to increase for the next 12 years. Same as fertilizers, the number of tube wells all over the country also increased and made water supply efficient, particularly for during the Rabi season when water supply in the rivers fluctuated. The government provided subsidies in two different forms, first form is paid on agricultural products and second is given on inputs. These products sold at lower prices than the actual prices, which were higher. Also, the government pays subsidies to the producer on input prices used in the production of agriculture products like fertilizers, tube wells, other farm machinery, pesticides, and seeds. The consumer subsidies are paid from current accounts of the government's budget while producer subsidies are paid from development budget, which is why they are also known as development subsidies. The subsidy on irrigation prices are known as implicit subsidies as their costs could not be recovered fully.

In Pakistan, subsidy on agricultural inputs were provided after the 1950s. As discussed above, first fertilizers were subsidized to enhance their use. The process of subsidizing agriculture inputs were continued during the 1960s and found that at the end of the 1960s most of the inputs including seed, insecticide, fertilizer, irrigation water, and installation of tube wells, tractors and their spare parts were subsidized. But during the 1970s subsidies started to be cut down and resulted into higher input prices. Further, these higher prices were followed by recessions all over the world. These were times of oil and credit crises, excessive devaluations of Pakistani currencies, and war with India. Although in the 1970s subsidies existed on most of agricultural inputs, due to the pressures of the World Bank and the IMF, the government removed them in the 1980s. As a result, subsidies were stopped for pesticides, tractors, tube wells and seeds and in 1985 and 1990 subsidies were removed from nitrogenous, potash, and phosphoric fertilizers consecutively (Chaudhry and Sahibzada, 1995). From 1965 to 1970, agriculture subsidy was 6.2% of the total developmental budget, 4.2% was for fertilizers and drainage purposes and about 34.2% was for irrigation purposes, which made up 44.6% of the total developmental budget. Its share continued to fall and remained 3.2% (agriculture subsidy) of the total developmental budget (GOP, 1988). It was suggested that the government has to provide subsidy on fertilizers and fertilizers must be available in time. It was also suggested to introduce new and effective technologies at subsidized prices to enhance wheat productivity. It was mentioned that removing subsidies for agricultural inputs would affect the production, consumption, and prices of agricultural products. Particularly in those countries which are mostly dependent on the agriculture sector and where their food security was based on subsidized agricultural inputs or outputs.

34.5.1.2 Current Scenario

During 2012, the government provided a subsidy of USD365 million on fertilizers, USD193 million for irrigation, and USD342 million for electricity aggregated as USD897 million on all these products that was 0.4% of the country's GDP. It was found that India provided 5 times more subsidy on agriculture than Pakistan (Dawn News, 2015) (Table 34.11).

As mentioned in Table 34.11, most of the subsidies are on fertilizers and credit was removed in the 1990s. Electricity subsidy was USD58.1 million during 1986–1988, which was reduced up to USD10.4 million. But it continued to increase and reached USD26.7 million during 1998–1999. In 2001, a tremendous increase was seen and electricity subsidies reached USD161.55 million. In 1987, the subsidy on urea was phased out, and the subsidy on DAP (diammonium phosphate) was deregulated in 1993 (Hussain and Sampath, 1996).

Removing the fertilizer subsidy in the context of stagnant or declining yields in Pakistan is an important policy issue. Fertilizer recommendations to farmers are generally made in the form of standard packages specified by research stations. Well-developed procedures, which consider fertilizer response under local soil conditions, are yet to be established in different agro-ecological zones of the country. For example, research experiments conducted in the rice-wheat zone of the Punjab show no response to phosphatic fertilizer use in rice. In the absence of proper extension advice and soil testing facilities, farmers apply fertilizer on the basis of their experience and, more importantly, on the availability of cash at the time of application. Provision of cheap fertilizer may encourage waste through overuse, particularly in a situation where the majority of farmers are not familiar with fertilizer recommendations. It has been widely argued that if fertilizer subsidies were withdrawn then farmers would reduce its use. This situation would lead to an overall slowing down or even a decline in agricultural output growth (Chaudhry and Sahibzada, 1995). However, arguments in favor of subsidies are weak as government activities in procurement, distribution, and marketing of fertilizer are inefficient and not responsive to the needs of farmers in different

TABLE 34.11
Agriculture Subsidies Provided on Fertilizers, Electricity, and Credit (PKR billion)

	Fertilizer Subsidy			
Subsidized Product	**Imported Urea**	**Other P&K Fertilizer**	**Total**	**Electricity Subsidy**
2004–2005	1.85	–	1.85	
2005–2006	4.54	–	4.54	
2006–2007	2.05	13.7	15.75	
2007–2008	2.74	17.4	20.14	
2008–2009	17.23	26.50	43.73	6.972
2009–2010	12.87	0.50	13.37	7.101
2010–2011	8.41	0	8.41	–
2011–2012	9.55	–	9.55	–
2012–2013	10.50	–	10.50	4.870
2013–2014	4.53	–	4.53	3.0

Source: National Fertilizer Development Centre. 2014.

regions. A World Bank Report argues that timely availability of this crucial input is more important than providing fertilizer at subsidized rates. Phosphate fertilizers, imported by the public sector, are rarely delivered on time and usually in insufficient quantities, which results in an imbalance between nitrogen and phosphate use. In support of benefits of liberalizing the fertilizer market, the report noted the example of widespread pesticide use on cotton when the market for pesticides was liberalized. In recent years, the government of Pakistan has subsidized fertilizer manufacturing by adopting a dual gas price policy. One rate is set for the fuel stock rate applicable for normal use of gas in the industry, and the other is set for the feedstock applicable for the gas used in manufacturing of fertilizer. The production subsidy on fertilizer was taken as the price difference between the fuel stock price and the feedstock price (per mmbtu) multiplied by the amount of fuel stock gas used in fertilizer manufacturing. The total value of production subsidy provided on fertilizer during 2011–2012 was estimated to Rs. 56.3 billion. The government has also provided distributional subsidies on imported urea and DAP worth Rs. 6.39 billion.

34.5.1.2.1 Seed Subsidy

Availability of improved seed was perhaps the most important component of the green revolution. On-farm research estimated that use of old varieties of seed can depress crop yields by 15%. The production of HYV is a function of research while a subsidy on seeds is simply a means of introducing farmers to use good quality seed. The total subsidy on seeds has varied, ranging from 3 million rupees to 44 million rupees in the 1970s and 1980s (Qureshi 1987); subsequently this subsidy on seed was gradually phased out. Seed production and distribution for the major crops (wheat, cotton, paddy) is handled by public sector departments. In recent years, private seed companies have started producing seed for cotton, wheat, rice, and for minor crops (sunflower and maize). The seed amendment bill, 2015 approved by the government in 2015, and national seed policy is in the process of finalization (GOP, 2015).

34.5.1.2.2 Subsidy on Irrigation Water and Electricity

Since the supply of surface water is limited, the installation of tube wells has played a vital role in augmenting water supplies at the farm-level. The government has encouraged their installation by giving a subsidy on their costs of installation. In the 1990s, about 27% of the total diesel tube wells installed have been subsidized (Mahmood and Walters, 1990). It has been argued that the supply of canal irrigation water at cheap rates is a concealed subsidy to agriculture. Government expenditures on irrigation have been widely debated by donor agencies. However, Chaudhry and Sahibzada (1995) argue that the benefits of these expenditures are not necessarily passed on to farmers. These expenditures include labor payments to irrigation departments and increased expenditures in maintaining public tube wells (Wolf, 1986). Similarly, as water is the critical scarce resource, officials of the irrigation department have an incentive to charge farmers illegally and these charges do not appear in their budgets (Wade, 1982; Wolf, 1986; Ilyas, 1994).

The government has been subsidizing electricity to agricultural tube wells through cheap and fixed rates. The amount of subsidy on electricity was highest in the late 1980s and early 1990s, and now the government is gradually decreasing the amount of subsidy on provision of electricity to agriculture. Solar tube wells would be provided on markup free loans to the farmers who own up to 12.5 acre of land. The mark up of 7 years would be paid by the federal government with a cost of Rs. 14.5 billion. This would ensure a saving of Rs. 1600 and Rs. 500 per day for the farmers running tube wells on diesel for 5 hours daily and on petrol for same duration, respectively. The electricity price for running tube wells at peak hours has been fixed at Rs. 10.35 per unit and Rs. 8.85 at off-peak hours. The sales tax on these bills amounting Rs. 7 billion would be borne by the federal government (GOP, 2015).

34.5.1.2.3 Agricultural Credit

In Pakistan, institutional credit is provided at subsidized rates. The government provides credit to farmers through the Agricultural Development Bank of Pakistan (ADBP), commercial banks, and Federal Bank of Co-operatives (FBC). The subsidized credit program is mainly limited to tractors, tube wells, and other investments such as land and watercourse improvements. The credit disbursement for the purchase of seasonal inputs is generally limited and not available to the majority of farmers (particularly small farmers). The size of the subsidy to institutional credit increased with the increase of credit supplied. The World Bank Report (1994) noted that a lack of dynamism and poor enforcement of the credit system are common. Loan recovery is a slow process and borrowers eventually default. Collateral requirements hinder the small farmers' access to formal credit sources. The Task Force Report on Agriculture (1993) proposed the use of crops as collateral as a solution to the lack of credit. This, however, does not seem practical as variability in crop prices increases the risk to banks, which do not have the technical capability to assess these risks. Unavailability of, or difficult procedures in obtaining bank credit force farmers to meet their short-term credit needs from informal sources, which account for 70%–80% of total agricultural credit.

Recently, SBP have taken a number of policy and regulatory initiatives to improve the access to financial services, particularly for small farmers. Some of the major initiatives are as

- *Credit Guarantee Scheme for Small and Marginalized Farmers*: Under this scheme the government has allocated Rs. 1.0 billion and will share 50% credit risk of banks. For implementation of the scheme, SBP has assigned lending targets to banks and around 200,000 farmers would benefit annually from this scheme.
- *Framework for Warehouse Receipt Financing*: In order to develop commodities' physical trade and marketing system, SBP issued a draft framework for Warehouse Receipt Financing. The framework facilitates banks in development of specialized products for providing financing to farmers, traders, processors, and other players in the value chain. SBP in collaboration with banks, MFBs, warehouse operators, and collateral management company has launched two pilot projects (Sindh and Punjab) to test the feasibility of warehouse receipt financing in the country.
- *Guidelines for Value Chain Financing*: For developing linkages between banks and small farmers, through cross guarantee by the input suppliers and traders/processor, SBP has issued financing guidelines for Value Chain Contract Farming. These guidelines are aimed at facilitating banks in the development of specialized products and also help small farmers in getting quality inputs, marketing of agriculture produce and timely payments by the traders/processors. SBP recently rolled out a project by assigning disbursement targets to bank for financing to selected value chains in the country during 2015–2016 (GOP, 2015).

Consumer subsidy provided to poorer people to raise their income when means of production did not redistribute significantly. The subsidy (as discussed earlier) provided on selected items like wheat flour, edible oil, and sugar. In 1973, when the Pakistani currency devalued, these subsides were increased remarkably and most of the benefit was taken by urban consumers as they were the major consumer of the subsidized products. In past, the major part of consumer subsidy was given on imported wheat (Table 34.12).

In Table 34.12, it was shown that a major part of the benefits of subsidy on wheat were taken by foreign producers and consumer in urban areas of the country. From 1966 to 1976, local producers came to know that government concern was not to encourage the producer in case of providing subsidy on wheat. In 2006, the government provided a subsidy of Rs. 1.40 per kg and in 7 years the

TABLE 34.12
Direct Consumer Subsidies on Agricultural Products (PKR million)

Years	Wheat	Edible Oil	Sugar	Total
1969–1970	80	0	0	80
1970–1971	101	0	0	101
1971–1972	95	0	0	95
1972–1973	939	0	0	939
1973–1974	1905	269	0	2174
1974–1975	2243	443	0	2686
1975–1976	1553	0	0	1553
1976–1977	1683	0	0	1683
1977–1978	1571	0	25	1596
1978–1979	2384	577	11	2972
1980–1981	2353	884	20	3257
1981–1982	1229	603	72	1904
1982–1983	1299	1	4	1304
1983–1984	1121	0	38	1159
1984–1985	1267	1485	15	2767
1985–1986	2878	2251	10	5139
1986–1987	3919	0	0	6396
1987–1988	4064	0	0	4064
2008–2009	20,000	1500	6300	27,800
2009–2010	25,500	1000	4000	30,500
2010–2011	15,833	0	4000	19,833
2011–2012	8388	0	4000	12,388
2012–2013	5148	0	4000	9148
2013–2014	10,098	0	4000	14,098

Source: Ministry of Food, Agriculture and Cooperatives for data up to 1978–1979; and Economic Surveys of Pakistan, for subsequent years.

government provided subsidies of Rs. 1.13, Rs. 2.15, Rs. 7.76, Rs. 4.28, Rs. 3.50, Rs. 1.08, and Rs. 4.18 per kg of wheat flour (Malik, 2015).

In case of sugar, no subsidy was provided until 1977, but an imposed Rs. 2150 per ton excise duty existed on it to make the country self-sufficient in the production of sugar. But after 1977, the government started to provide subsidy on sugar and continued until 1985. In 2013, to encourage export of sugar the Economic Coordination Committee of the Cabinet (ECC) permitted freight subsidies and reduced excise duties by the Federal Board of Revenue (FBR). Under this initiative Rs. 1.75 per kg were given to enhance export of sugar up to 1.2 million ton. Excise duties were reduced from 8% to 0.5% (Ur Rehman, 2013).

Edible oil was subsidized in 1974 and the government paid a subsidy of Rs. 269 million. It was increased in next year to Rs. 443 million but removed for next 3 years. From 1979 to 1985, the government successively provided subsidy except in 1983. In next 2 years it was removed again. In 2007, the government of Pakistan provided Rs. 6 per kg of edible oil in utility stores, but the Pakistan Vanaspati Manufacturers Association (PVMA) was against it because of limited number of utility stores in the country (Dawn News, 2007). A subsidy of Rs. 366 crore was provided on the import of edible oil in 2011–2012 that rose up to Rs. 615 crore in 2012–2013 (Union Budget, 2012–2013). The farmers cultivating rice and cotton at up to 12.5 acres of land would be given cash supports of

Rs. 5000 per acre. Further, custom duty, sales tax and withholding tax on the agriculture machinery have been reduced from 45% to 9% (GOP, 2015).

34.5.1.2.4 Other Subsidies

The government supports farmers in various other aspects, such as the provision of heavy agricultural machinery to develop wastelands so that they become fit for cultivation, and introductory subsidies on new inputs (e.g., gypsum, which reduces the soil sodicity problems). The expenditure on these subsidies is very small as compared to other subsidies and most farmers remain unaware of them.

34.5.2 Taxation and Income Transfer in Agricultural Sector

34.5.2.1 Historical Background

Although the agricultural sector in Pakistan has been subject to many direct and indirect taxes, until recently it has not come under the national taxation system. The major forms of direct taxation are land revenue and ushr (an Islamic levy). The indirect taxes include taxes on specific commodities such as export duties on cotton and rice, and cesses (a tax) on sugarcane and cotton crops. These indirect taxes are the major source of the government's farm sector revenue as compared to direct taxation on farmers.

In the beginning, West Pakistan (currently Bangladesh) was the only province which introduced agricultural income tax at provincial level. After that in 1948, it was imposed in NWFP (now KPK), then in 1951, it was imposed in Punjab and in Sindh it was imposed in 1965. Although since 1947, each taxation commission and tax imposing committee recommended to impose a central agricultural income tax but it could not be imposed due to 1935 Act. So in 1989–1990 it was implemented in Pakistan. In 1860, land revenue tax contributed to around 45% in total tax revenue. But in 1948, it was only 9% of the total tax revenue. As in 1991, it remained only 0.3% and 0.5% if aggregated with Usher (Azhar, 1991). In 1970–1971 gross implicit taxes were Rs. 0.5 billion, and in 1989–1990 they rose to Rs. 28 billion. The same picture could be seen corresponding to implicit taxes net of subsidies provided on fertilizers, pesticides, installation of tube wells, and seeds. The net taxes were Rs. 0.9 billion in 1972–1973, which increased to Rs. 25.9 billion during 1989–1990 (Chaudhry and Kayani, 1991).

Several studies in Pakistan have been conducted to calculate the overall tax burden on agriculture and the intersectoral income transfer caused by government intervention. Qureshi (1987) estimated the magnitude of income transfers for 1972–1986 and found that most taxation from agriculture was concealed and that there were declining trends in direct taxation. They also concluded that there were increasing trends in resource flows into agriculture, which suggests a more intensive taxation of agriculture. The magnitude of net resource transfers from agriculture was 156 million rupees during the 1980s. The government's pricing policies are mainly responsible for this resource transfer. It has been estimated that the resource transfer out of agriculture has decreased gradually over time and now is in the range of 6%–8% of agricultural GDP.

It was observed that Pakistan's tax system was inelastic and mostly depended on import taxes and indirect taxes. The agriculture sector, which is the major sector of the country, was exempted from direct income taxes. In this sector, only provincial governments could impose land tax but not the federal government. In 1996, only direct tax was the land tax imposed by provincial governments and accounted for Rs. 1.3 billion, which is only 0.4% of aggregated tax revenue. Although the federal government levied a tax on wealth of agricultural land, revenue collected from this tax was too low and formed only one tenth of total land revenue. In 1996–1997, the government removed exemptions on land taxes and increased rates of assessment of land that raised revenue from rupees 30 million to rupees 110 million in 1996–1997. Although in the 1990s, due to the pressure of the World Bank and the IMF, provinces imposed agricultural income taxes; however, these were not defined clearly and were a form of land tax.

34.5.2.2 Current Scenario

In 2000, these laws were amended. In Punjab, landowners of more than 12.5 acres of land have to pay land tax and above Rs. 80,000, an agriculture income progressive tax rate was imposed by the Punjab Agricultural Income Tax Act in 2001. In spite of increased agriculture prices, taxes from agriculture taxes contributed only 1% from 2001 to 2008 in agriculture tax revenue (PILDAT, 2011).

A recent study has estimated potential tax revenue from the agriculture sector under four different modes. The first three modes are a pure land tax, a combination of land and income tax, and a pure income tax. The fourth mode shows the tax that farmers would have paid if agricultural incomes were taxed at rates comparable with incomes in nonagricultural sectors. The estimates for the FY2014 shows that the tax yields under the four modes would have been Rs. 2 billion, Rs. 15 billion, Rs. 114 billion, and Rs. 54 billion, respectively (Table 34.13).

34.5.2.2.1 Indirect Taxes

34.5.2.2.1.1 Fertilizer Taxation The problem of low productivity in the agriculture sector is caused by many hurdles like degradation of soils (alkalinity, fertility depletion, erosion, and soil salinity)

TABLE 34.13
Implicit Taxes and Subsidies in Agriculture from 1970–1971 to 1989–1990

Years	Implicit Taxes	Input Subsidies	Taxes Net of Subsidies	Value Added by Agri. At Current Factor Cost	Net Taxes as % of Value Added by Agri.	Overall Tax Rate
1970–1971	521.85			16,236		12.7
1971–1972	1280.81			17,934		13.56
1972–1973	1236.14	345	891.14	21,907	4.97	12.37
1973–1974	4532.99	351	4181.99	28,084	14.89	13.47
1974–1975	5028.75	454	4574.75	33,533	13.64	12.80
1975–1976	3844.38	1012	2832.38	38,338	7.39	13.28
1976–1977	3801.38	914	2887.36	43,968	6.57	13.28
1977–1978	2283.11	1160	1123.11	50,567	2.22	14.05
1978–1979	3403.97	1983	1420.97	54,147	2.62	14.66
1979–1980	5458.17	2723	2735.17	62,164	4.40	15.25
1980–1981	8433.92	2479	5954.92	76,399	7.80	15.66
1981–1982	9893.53	1826	8067.53	92,216	8.75	14.96
1982–1983	6427.30	1980	4447.30	99,380	4.48	14.67
1983–1984	9665.40	1690	7975.40	104,550	7.63	15.69
1984–1985	6465.54	1501	4964.54	121,293	4.09	14.87
1985–1986	4917.37	2424	2493.36	128,801	1.94	15.27
1986–1987	10,561.42	1142	9419.42	135,308	6.96	16.11
1987–1988	12,909.96	2190	10,719.96	156,375	6.86	16.32
1988–1989	13,751.80	1400	12,351.80	185,498	6.66	17.46
1989–1990	28,036.31	2100	25,936.41	205,980	12.59	17.38

Source: Chaudhry, M. G. and N. N. Kayani. 1991. *The Pakistan Development Review*, 30(3): 225–242.

water resource depletion management problems of the irrigational system, land distribution in small patches, and old farming methods. Due to these reasons the use of inputs, especially fertilizer, in the agriculture sector of Pakistan is inadequate and insufficient. The provision of pesticides and hybrid seeds are limited to some specific areas of the country. The fertilizers are the most important input which enhance the production of agriculture.

The report of NFDC (1999) stated that balanced use of fertilizers in wheat crop increases its production by 77%, rice and sugarcane by 100%, and cotton production by 400%. During the past few decades—particularly for last five, fertilizer usage increased in Pakistan. The government of Pakistan realized its importance as a major input in 1952 and then introduced it during 1954 and sold about 72,000 tons in that year. The main objective was to introduce and encourage the usage via trial of fertilizers and revelation on the farmer's land and also reducing prices by giving subsidies. With the fixation of fertilizer prices its usage increased with passage of time. During 1968–1969 the demand for fertilizers enhanced up to twenty times higher than of 1953–1954. As the usage of fertilizers increased tremendously, the focus of agriculturists was changed towards a more balance use of fertilizers, which are more compatible with the crops. The continuous increase in use of fertilizers, subsidy burden was increased too much on the government so it was planned to abolish the subsidy under structural adjustment program (SAP) and reforms in the economy. In this perspective in 1986 the whole subsidy was abolished on nitrogenous fertilizers and in 1995 it was removed from DAP and in 1997 removed from NPK. In 2001, the government of Pakistan decided to impose a tax of 15% on all fertilizers that increased its prices. After that the price of urea reached to Rs. 3500 and DAP price was reached up to Rs. 700 per 50 kg bag. The main objective of fertilizer usage is to enhance the land efficiency and productivity of agriculture sector. This increase in efficiency will result into sustainable and higher growth in agriculture sector that fulfill the need of growing population for their food security and the growth of the economy as well.

With the increase in subsidy burden the government took initiative to reduce or totally removed subsidy on fertilizer under SAP as mentioned above. In 1986, 1995, and 1995 the subsidy was removed from nitrogenous, phosphate, and potassium fertilizers in respective years. The control on import of fertilizer was lifted and the government blocked importing and then private sector dominates in the market. Although the subsidy was removed partially in 1986 and fully in 1993 and due to increase in the prices a lot of private investors attracted towards it. The price of urea bag of 50 kg was Rs. 290 in 1995, which increased up to Rs. 305 in 1996. Two fertilizers companies included Dawood Hercules and Fauji Fertilizer increased the price of urea to Rs. 330 per bag in May 1996. Due to the lower occurrence of fixed charges the major fertilizer market player known as Engro chemical did not raise their prices. The reason was they had to bear low expansion cost because of buying a second hand plant and added additional capacity of ammonia; in reality the expansion in future will also be cheaper for them. The cost of urea in black was Rs. 340 and for imported variety it was Rs. 371. In 1994, the price of DAP was increased up to Rs. 560 per 50 kg bag from Rs. 410 which was to match the prices in international markets. In 2001, the government imposed a 15% GST on all types of fertilizers that increased prices sharply. In this context in 2004, the DAP prices reached up to Rs. 1000 from Rs. 670 per bag of 50 kg and the price of urea reached up to Rs. 450 from Rs. 363 per bag of 50 kg in 2002. Presently, the price of urea and DAP has increased by Rs. 730 and Rs. 3150 per bag of 50 kg, respectively, which resulted into a reduction of agricultural output. There are several reasons found for this increase in the prices of fertilizers, including the gap between demand and supply, higher oil prices, and energy crisis in the country.

34.5.2.2.1.2 Current Situation The government of Pakistan planned to provide Rs. 341 billion to provide less expensive fertilizers, loans, water, and seeds because of bad weather conditions and reduced agricultural product prices. A subsidy of 20 billion rupees provided on urea resulted into a fall of Rs. 500 fall in the prices of DAP. A concession of Rs. 15 billion were provided in taxes and duties to promote agriculture, which has been continued in 2016–2017. The price of urea was reduced to Rs. 250 and on 1st July 2016 it was reduced to Rs. 1400 per bag. This subsidy of Rs. 36 billion will be provided by the federal government. The price of DAP will be reduced from Rs. 2800 to Rs. 2500 after a subsidy of Rs. 10 billion. The government aimed to enhance the agriculture credit from Rs. 600 billion in 2016 to Rs. 700 billion in 2017 and also aimed to reduce the markup rates by 2%. The custom duties on dairy, poultry, and livestock sector decreased from 5% to 2%. The custom duties on fish feed and shrimp feed will be exempted, and were 10% before exemption. The sales tax of 7% should be abolished in 2017.

34.5.2.3 Way Forward

It has often been observed that supplies of inputs—improved seeds, fertilizers, pesticides, diesel, and so on become scarce at critical stages of the crop cycle, and farmers have to search of these intangible inputs.

There is no justification whatsoever for subsidizing inputs if adequate supplies are not available in the regular markets and the farmers have to buy the "subsidized" inputs on the black market.

When farmers are willing to pay the price, all efforts need to be directed at removing supply bottlenecks and improving market intelligence and infrastructure. Measures need to be taken that will ensure competition in the input markets, rather than wasting resources to subsidize the inputs. However, the provision of input subsidies and the policy of having a support price for important crops also has its limitations.

The agriculture sector contributes 21% to GDP of Pakistan and had an estimated annual income of Rs. 389 trillion and tax are not collected from this sector. While industrial sector contribution in GDP is also same, it contributes a 60% share to tax collections.

Why is the agriculture sector is not taxed? The possible answer of this is the reluctance of the Pakistani political landholding class do not want to see their incomes taxed. If Government applies agricultural tax, according to some experts, it will help generating at least Rs. 20 billion. While some other estimates claim this amount to be Rs. 40 to 60 billion. And if the aggressive taxes are applied on agriculture then these figures can reach Rs. 250 billion. All the above estimates are by the government's finance experts (KCCI, 2015).

There are two types of provincial taxes: The land tax and the income tax form agriculture sources. The provincial governements collected just single tax from farmers, which is the higher of the two. There are also problems in the law enforcement and monitoring system of revenue collection, which provides loopholes for tax evasion. The World Bank says that tax arbitrage chances given by the agriculture sector makes money laundering easy for drug lords and criminals. Lack of centralized records and obsolete methods of tax calculation and collection allow many loopholes to be created

TABLE 34.14
Current Collection Trends of Agriculture Income Tax (PKR Million)

Province	2010–2011	2011–2012	2012–2013	2013–2014[a]
Punjab	717.21	762.44	827.34	830.00
Sindh	210.16	122.81	406.46	426.56
KPK	17.53	20.08	22[a]	22
Baluchistan	–	0	0	0.5
Total	**944.90**	**905.33**	**1255.80**	**1279.06**
As % of Pakistan's Total Tax Revenue	0.06	0.04	0.06	0.05
As % of Total Provincial Tax Revenue	1.46	0.84	0.83	0.59

Source: KCCI Research; Ministry of Finance; Budget Statements of Provinces.
[a] Figures are Budget Estimates; N-A.

for tax evasion and avoidance. To avoid income taxes, taxable income from nonagricultural sectors is often declared as income earned from agriculture (Table 34.14).

- Punjab is the only province that has made efforts (if any) to impose higher tax on agriculture, although a lot remains to be desired.
- Higher commodity support prices have ensured that farmers gets a fair reward for their endeavors and has resulted in substantial flow of wealth to the rural economy from the urban areas.
- To create a level playing field, every source of income must be indiscriminately taxed on even grounds, whether it is from agriculture or nonagriculture.
- 51% of the country's population lives below the poverty line (USD2 per day); on the contrary, 40% of the national income is shared by the 20% highest earners. A majority of these high earners are landlords who pay minimal taxes by reporting their incomes as agriculture related.
- The state needs to urge provinces to create a centralized database, which could enable the tax collector to identify the actual land holdings and hence the tax payable on the total holding within the province.
- One measure that can be taken is the imposition of a flat tax rate on all agricultural revenue, say 1% initially. This could yield about Rs. 39 bn on an income of Rs. 3.89 trillion (Tables 34.15 through 34.18).
- FBR can be tasked to help the provincial governments with fixing the present provincial taxation setup and to construct their capability to enforce the new measures.
- The payment of support prices to farmers in a crop year can be tied to actual payment of taxes and filing of returns by the farmer in the previous fiscal year.
- Agricultural income attracts tax rates of 15%–35%. Additionally, an indirect tax of 1% is applicable on the sale of agricultural products sold as raw materials.

TABLE 34.15
Progressive Tax Rates Levied by Punjab Agricultural Income Tax Act

Income Level	Current Tax (Punjab & Sindh)
If total income is less than PKR 80,000	No tax
If total income does not exceed PKR 100,000	5%—of total income
If total income is more than PKR 100,000 but does not exceed PKR 200,000	PKR 5000 + 7.5% on over PKR 100,000
If income is more than PKR 200,000 but does not exceed PKR 300,000	PKR 12,500 + 10% on over PKR 200,000
If total income is more than PKR 300,000	PKR 22,500 + 15% on over PKR 300,000

Source: Farmers Association Pakistan. 2011. (Available at: http://www.pakkissan.com).

TABLE 34.16
Proposed Agriculture Income Tax Rates

Level of Revenue	Proposed Tax (Punjab & Sindh)
If total income is less than PKR 500,000	No Tax
If total income is more than PKR 500,000	1% of total income

Source: KCCI Research. 2013.

TABLE 34.17
Current and Proposed Land Revenue Rates for Punjab

Land Ownership	Current Tax (Punjab)	Proposed Tax (Pakistan)
Up to 12.5 acres	No tax	No Tax
12.5–25 acres	PKR 100/acre	PKR 750/acre
26–50 acres	PKR 250/acre	lPKR 1250/acre
50 acres or more/(Matured Orchards) irrigated	PKR 300/acre	lPKR 1500/acre
Mature Orchards Non-Irrigated	PKR 150/acre	PKR 750/acre

Source: PILDAT report, Taxing agriculture in Pakistan. 2011.

TABLE 34.18
Current and Proposed Land Revenue Rates for Sindh

Land Ownership	Current Tax (Sindh)	Proposed Tax (Pakistan)
Up to 4 acres (Irrigated)	No Tax	No Tax
Above 4 acres (Irrigated)	PKR 200/acre	lPKR 1000/acre
Up to 8 acres (Un-irrigated)	No Tax	No Tax
Above 8 acres (Un-irrigated)	PKR 100/acre	PKR 500/acre
Mature Orchards Banana/Betel Leaf (Irrigated)	PKR 700/acre	PKR 3500/acre
Mature Orchards Banana/Betel Leaf (Un-irrigated)	PKR 350/acre	PKR 1700/acre

- Imposition of an agricultural income tax on large and medium farmers would significantly increase revenue collection while it would also undoubtedly reduce poverty in the country.
- Enhanced agriculture taxation would make the provinces more self-sufficient with their revenue requirements and reduce the burden on the federal government.

34.6 INVESTMENT IN AGRICULTURAL SECTOR

34.6.1 Public and Private Investment in Agricultural Sector

34.6.1.1 History and Current Scenario

In the beginning of the 1980s and 1990s, accelerating external trade and exchange rate regimes in Pakistan were liberalized, input markets in the country were privatized, output markets were opened to the private sector, subsidies on inputs were removed, and public corporations who were serving these sectors were abolished. The subsidy on agriculture inputs and outputs were removed, which included large amounts for wheat that were continued until 2001. Through public corporations the public sector was dominated in both markets including input as well as product market. In this era, although markets were not fully penetrated by the private sector it nevertheless still played a vital role in marketing agriculture products and inputs including wheat. While the role of public sector corporations in some agriculture input markets could not be neglected it was still dominant in some inputs market, especially the seeds and the fertilizer markets. In this regard, it was scheduled to privatize the Sindh Seed Corporation in 2002. To promote the investment in private sector it was reported that the private investments in agriculture sector were discouraged through uncertain government interventions in input and output markets and availability of limited transparent and appropriate laws which are consistent and based on justice. Private sector plays and important role in introducing and creating new interventions through interactions between farmers and processors, examples in this regard include introduction of modern technology in dairy, introduction of off season crops in Punjab and enhanced export of horticultural crops. In the irrigational sector, privatizing shallow and deep tube wells played a vital role in controlling water logging through lowering the level of the water table.

Installment of tubewells with local participation may help solving the equity related issues and bring prosperity in the area. It is not certain if the private sector shows the ability to handle it efficiently; if not, the public sector should reintervene. However, the Ten Year Perspective Plan suggests to set up new public sector tube wells. The regulatory environment for private investment is still lacking in significant areas. The long hindrance in endorsement of biosafety strategy means that Pakistan is at least 3 years behind in competing with countries in the commercialization of AL-transgenic insect-resistant cotton, which has offered a cost benefit of 20%–25% in countries as diverse as Mexico, China, and the United States, and will soon be introduced in India.

The Punjab government has a task to expand a science-based, energetic, and globally connected agriculture sector that can not only meet the food security challenges, but also be competetive in domestic and international markets. According to FAO of the United States, GDP growth in agriculture has been shown to be at least twice as successful in reducing poverty as originating growth in other sectors.

The FAO estimates that an extra investment of USD83 billion will be desirable annually to close the space between what low and middle income countries have invested every year and what is desirable by 2050. Increased investment in agriculture may result in improved food security and reduction of poverty. Investing in agriculture is essential for reducing starvation and promoting sustainable agricultural production. Those areas where agricultural capital per worker have stagnated are epicenters of poverty and starvation today.

Private investment plays an important role in development of a country. Though, investing in small-scale manufacturers, typically women producers is especially important. This is because the 500 million small farms in developing economies support about two billion people. Yet it is these very same small-scale producer who are the most food insecure—due to lack of access to the markets, finance, land, technologies, and infrastructure enjoyed by large farms. Such investment, when delivered inclusively and sustainably, can be catalytic in providing improvement, job formation, and in accelerating broad economic growth.

Unfortunately, not all private investments in the agriculture sector have positives; present policy business and environments practices often support investments which worsen poverty. However, when in the correct policy environment, private investments can be catalytic to comprehensive economic growth, environmental sustainability, and long-term poverty reduction. Private investment should complement public sector investment. Positive agricultural investment can give benefits to the investors, small-scale producers, communities, and governments. Governments have to give priority to investments in key public goods, such as capacity building, infrastructure, and systems of research, to help small-scale producers in sustaining their livelihoods. Private investment in agriculture is essential and can't be substituted by the public sector. More than three forth of industrial activities depend on agriculture for raw material.

Figure 34.2 shows the trend in share of private and public investment in the agricultural sector. Lack of good governance, provision of subsidy on formal type of agricultural credit in the past, and the dependence on land as the only form of guarantee for the agricultural credit led to the politically strong landowners assuming this credit even when it was destined for poor farmers. Inadequate development of the formal credit market in rural areas of Pakistan has been a main restriction to improve investment by the private sector. One of the main reasons is the lack of access for landless and poor farmers and the short accessibility of agricultural credit for the development of value added and agro-processing components of agriculture outputs for export.

Agricultural sector has kept suffering due to inadequate credit credit schemes, inefficient fiscal policy and lack of advanced availability of advanced technology, which resulted in low than potential

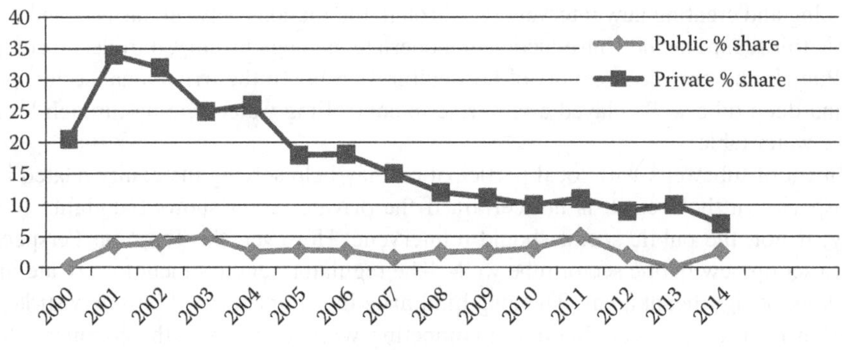

FIGURE 34.2 Trend in share of private and public investment in agriculture sector. (From Various Issues of Economic Surveys of Pakistan.)

private investment in the sector. The trend of drop in public sector investment in agriculture resulted in drop in overall investment in the sector resulting in lack of appropriate infrastructure. (Figure 34.2). This low level of investment by public sector also represented a negative indication for the private sector. A large part of this shift in investment is also because of the enormous and instant profits to be earned in the urban areas, especially in real estate and industry, which were accompanied by poor governance and government policies. The true ability of agro-processing and the strength of the underdeveloped and onward connections of the agriculture sector are only currently being completely recognized in the policy loop in Pakistan following the opening up and liberalization of agriculture markets.

The main reason of sluggish economic growth and imperfect investment of private sector in agriculture of Pakistan are the conventional technology, low quantity and quality of agriculture yield, partial accessibility to domestic and foreign markets, problems of availability of credit, non availability of the trained and skilled labor force, non availability new apparatuses and machinery related to agriculture sector as well as inadequate capability of infrastructures and resources.

Investment in agriculture sector is permitted in the actions of soil improvement, hilly areas and deserts for Agriculture purposes and Farming, recovery of water Front Areas, vegetables, fruits, Crops, Flowers, Farming and Processing of Crops. All financial institutions and banks will earmark split credit share for commercial agriculture farming. Furthermore, particular fiscal benefits including exclusion from sales tax and custom duties on imports of agricultural equipment have been awarded to support investment in the agricultural sector. Under this existing package, state land can be bought or leased for 50 years by open auction, expandable for an additional 49 years. There is a lot to be done to close the huge gap among developing nation investment needs and capital accessibility—a serious objective in meeting the food, energy, feed, fiber, and other agricultural wants of the world's quickly rising population. While private investors are obviously ready to invest in developing nations, extra efforts to maintain and continue these investments are required.

REFERENCES

Azhar, B. A. 1991. Taxation of agricultural income: A holistic view. *The Pakistan Developmental Review*, 30(4): 1065–1072.

Bokhari, A. 2015. Support price or input subsidies. Dawn news, Available online at: http://www.dawn.com/news/1179740 Accessed on: 19/06/2016.

Chaudhry, M. G. and N. N. Kayani. 1991. Implicit taxation of Pakistan's agriculture: An analysis of the commodity and input prices. *The Pakistan Development Review*, 30(3): 225–242.

Chaudhry, M. G. and S. A. Sahibzada. 1995. Agriculture input subsidies in Pakistan: Nature and impact. *The Pakistan Development Review*, 34(4):711–722.

Dawn News. 2007. Ghee, oil price subsidy to benefit only few. Online available at: http://www.dawn.com/news/235919/ghee-oil-price-subsidy-to-benefit-only-a-few Accessed on: June 18, 2016.

Dawn News. 2015. Farmers oppose duty-free imports from India. Available online at: https://www.dawn.com/news/1156350 Accessed on February 14, 2018.

Dorosh, P. and A. Salam. 2008. Wheat markets and price stabilisation in Pakistan: An analysis of policy options. *The Pakistan Development Review, Pakistan Institute of Development Economics*, 47(1): 71–87.

Ender, G. 1990. *Government Intervention in Pakistan's Cotton Sector.* United States Department of Agriculture-Economic Research Service, Agriculture and Trade Division, Washington, DC.

European Commission. 2016. Agriculture and rural Development—The common agriculture Policy after 2013. Available at: http://ec.europa.eu/agriculture/cap-post-2013/index_en.htm

FAO. 2011. *Pakistan and FAO Achievements and Success Stories.* FAO, Rome, Italy.

Faruqee, R., J. R. Coleman and T. Scott. 1997. Managing price risk in the Pakistan wheat market. *World Bank Economic Review*, 11(2): 263–292.

Government of India. 2016. Department of Agriculture, Cooperation and Farmers Welfare, Ministry of Agriculture and Farmers Welfare.

Government of Pakistan. 1988. *Government of Pakistan, Report of National Agricultural Commission.* Ministry of Food and Agriculture, Islamabad.

Government of Pakistan. 2014. *Agricultural Statistics of Pakistan 2013–2014.* Pakistan Bureau of Statistics, Islamabad.

Government of Pakistan. 2015. *Pakistan Economic Survey 2014–2015.* Economic Advisor's Wing, Finance Division, Ministry of Finance, Islamabad, Pakistan.

Government of Pakistan. 2016. *Pakistan Economic Survey 2015–2016.* Economic Advisor's Wing, Finance Division, Ministry of Finance, Islamabad, Pakistan.

Hussain, I. and R. K. Sampath. 1996. Supply response of wheat in Pakistan, Working Paper. Department of Agricultural and Resource Economics. Colorado State University, Fort Collins.

Ilyas, M. 1994. *Water Markets Plan to Harm Small Farmers. The Daily Dawn.* (Economic and Business Review). April 9–15.

KCCI. 2015. Taxation where does Pakistan Stand? To tax or not to tax agriculture, Research and Development Cell, Karachi Chamber of Commerce & Industry, Pakistan.

Mahmood, A. and F. Walter. 1990. *Pakistan Agriculture; A Description of Pakistan's Agricultural Economy.* pp: 68–76.

Malik, S. J. 2015. Agriculture policy in Pakistan- what it is and what it should be. Online Available on: http://www.pide.org.pk/pdf/Seminar/AgriculturePolicyPakistan.pdf Accessed on: 18/06/2016.

NFDC. 1999. National Fertilizer Development Center "Pakistan Fertilizer Statistics" Islamabad Pakistan.

PILDAT. 2011. Taxing the agriculture income in Pakistan. Pakistan Institute of Legislative Development and Transparency, Briefing paper no 42.

Qureshi, S. K. 1987. Agricultural Pricing and Taxation in Pakistan – Some Policy Issues. Pakistan Institute of Development Economics (PIDE), Islamabad.

Salam, A. 2012. *Review of Input and Output Policies for Cereal Production in Pakistan.* Pakistan Strategy Support Program. Discussion Paper 01223, International Food Policy Research Institute (IFPRI), USA.

Union Budget, Vol. I. 2012/2013. Agriculture sector report by Maliha Quddus. Annex II.

United States Department of Agriculture. 2016. History of price-support agriculture and adjustment programs, 1933–1984. Available at: http://www.ers.usda.gov/topics/farm-economy/farm-commodity-policy.aspx

Ur Rehman, M. S. 2013. *Global Agricultural Information Network Report on Sugar.* GAIN Report.

Wade, R. 1982. The system of administrative and political corruption: Canal irrigation in South India, *Journal of Development Studies* 18(3): 287–328.

Wolf, F. M. 1986. *Meta-Analysis: Quantitative Methods for Research Synthesis.* Sage, London.

35 Agricultural Credit and Cooperation

Khalid Mushtaq and M. Khalid Bashir

CONTENTS

35.1 INTRODUCTION AND BACKGROUND

Credit is an important input for sustainable agricultural growth. It has been recognized that narrow access to formal sources of credit can lead to "exploitation of poor farmers at the hand of informal sources of credit, to a slowdown in the adoption of modern farming techniques and inputs, resulting in slow development of this chief sector of our economy" (GOP 2012). Absence of well-defined agriculture credit policy is quoted as one of the reasons for low agricultural growth (ADB 2008). Out of total farm households of 8.3 million, around 89% comprises small farmers who for their farming needs and mainly rely on informal sources of credit (GOP 2010). A majority of small farmers and rural entrepreneurs remain financially excluded because of the reluctance of mainstream financial institutions towards them.

Lack of awareness/literacy, low incomes, asset/collateral deficiency, social exclusion on the demand side, and distance from branch, cumbersome documentation/procedures, inappropriate products, and cost of operations from the supply side are major reasons for exclusion. Together, these factors not only entangle the poor farmers in procedural hassles but also result in a higher transaction cost, making the easily accessible source of informal credit more attractive.

The onset of commercialization, especially after the adoption of green revolution technologies, implied increased demands for credit in agriculture. Diversification within agriculture in favor of high-value commodities has further boosted the demand for credit. Small farmers, with limited financial resources to invest, are the real target group for loans advanced by credit institutions. Due to the increased role of the nonfarm sector in the rural economy, the need to cater to credit requirements of this subsector has also been a motivating force for the reorientation of the rural credit system in Pakistan (ADB 2008).

35.2 SUPPLY AND DEMAND OF CREDIT (GAP ANALYSIS)

There is wide gap between what agriculture contributes to the national economy and what it gets in terms of lending from formal sources. Estimated demand for agriculture credit stands at PKR 985 billion whereas the supply stood at PKR 389 billion (only 39% of total demand)* during the year 2014–2015. Over the past 5 years, the growth in demand is about 14.6% per annum; whereas the supply has grown by only 8.6%, creating a supply-demand gap that is the market for informal sources (Figure 35.1).

While analyzing the sector-wise agriculture disbursement in depth, out of the total disbursement of Rs. 326.0 billion, the farm and nonfarm sector has received Rs. 170.0 and Rs. 156.0 billion, respectively during July–March 2015. However, the disbursement to farm sectors has declined from 52.1% to 48.8% while that of the nonfarm sector is gradually increasing from 47.9% to 51.2%—as

* Int. Conference on Innovative Agricultural Financing, 28–29 April, 2015.

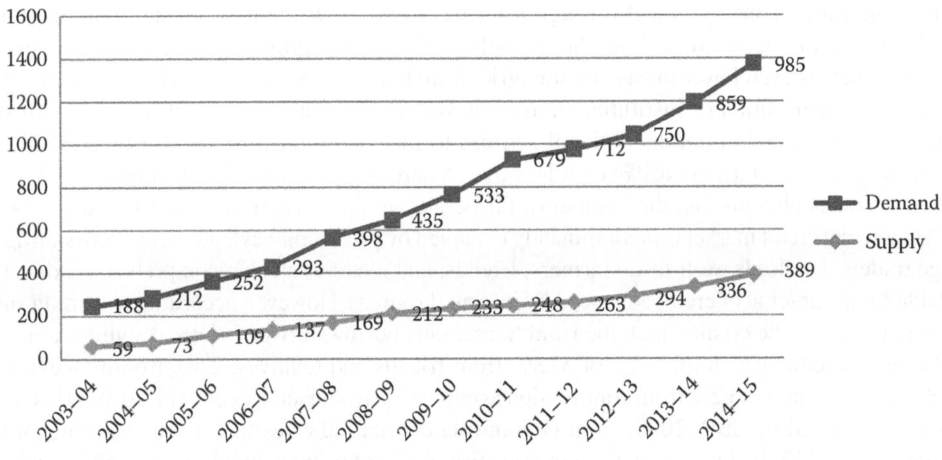

FIGURE 35.1 Demand and supply of agriculture credit (Billion Rs.).

compared with the corresponding period last year. The continued increase in nonfarm lending may be attributed to SBP's farmer's friendly credit initiatives.

Within farm sector allocations of Rs. 188.1 billion, Rs. 99.5 billion or 25.8% were allocated to subsistence holding, Rs. 41.2 billion or 10.7% to economic holding, while Rs. 47.4 billion or 12.3% to above economic holding category. However, under nonfarm sector allocation of Rs. 197.4 billion, Rs. 127.7 billion or 33.1% were allocated to large farms while Rs. 69.7 billion or 18.1% to small farm category. The comparison of farm and nonfarm sector share is given in Table 35.1.

35.3 INFORMAL SOURCES OF CREDIT

Low distribution costs, high mark-ups, and the timely availability of credit are the main characteristics of the informal credit market. Coercive loan procedure and borrower's social contacts with informal lenders and the lack of interest on the part of formal institutions for agriculture credit have made the poor to depend on the informal market. This is also leads to charging exploitative interests ranging from 50 to 100% per annum.

Access to institutional credit varies across different farm households, that is, owners, owner-cum-tenants and tenants. Data on formal and informal sources of credit show that large demand for

TABLE 35.1
Credit Disbursement to Farm and Nonfarm Sectors (Rs. Billion)

Sector		2014–2015		2015–2016	
		Disbursement	% Share in Total	Disbursement	% Share in Total
A	Farm Credit	170.0	52.1	188.1	48.8
1	Subsistence Holding	94.1	28.9	99.5	25.8
2	Economic Holding	41.0	12.6	41.2	10.7
3	Above Economic Holding	34.9	10.7	47.4	12.3
B	Nonfarm Credit	156.0	47.9	197.4	51.2
1	Small Farms	53.9	16.5	69.7	18.1
2	Large Farms	102.0	31.3	127.7	33.1
Total (A+B)		326.0	100.0	385.5	100.0

Source: GOP 2016. *Pakistan Economic Survey*, Ministry of Finance, Govt. of Pakistan, Islamabad.

credit in the rural economy is met through informal sources. Of the total outstanding debt of Rs. 723 Million within agricultural households, only 39% is being provided by institutional sources. This percentage is even lower in case of nonagriculture households and livestock holders. As can be expected, the distribution of institutional credit is skewed towards larger land holders, with small farmers largely accessing noninstitutional sources to meet their farming needs (Table 35.2). As a majority of Pakistani farmers (61%) own less than 5 acres of land and 33% own between 5 and 25 acres, it becomes obvious that the number of farmers accessing formal finance is pitifully low.

The informal credit market is predominately occupied by private moneylenders, commission agents, village traders, landlords, well-to-do farmers, friends, and relatives etc. No comprehensive data set is available for the amount of credit advanced by informal sources. However, according to a credit survey conducted in 1973, the credit which the rural households borrowed from informal sources amounted to 89% of all credits which make up of 57.8% from friends and relatives; 2.4% from moneylenders; 8.1% from landowners; 16.2% from commission agents; 0.7% from factories etc. (PIDE 1984). Recently a survey conducted by SBP (2008a,b) in Gujranwala district indicates that a very large majority of farmers, that is, 73% had taken loans from Aarthies, 63% from input suppliers, and 59% from both Aarthies and input suppliers. Further about 76% of the farmers who had taken loans from banks were also taking loans/credit from Aarthies/input suppliers. This is despite the fact that Aarthies charge higher than market rates on the inputs supplied to farmer and that most of the farmers who took credit from Aarthi are under obligation to sell the produce to Aarthi, generally at a price lower than market. Convenience and timely availability of credit without any documentary requirements are the main attributes of the informal sources of credit. Similarly, in Sukkur district, surveys indicates that most of the farmers are relying heavily on informal sources, that is, 24% have taken loans from friends, 17% from moneylenders, and 13.3% have taken loans from Aarthies and inputs suppliers; while only 5% of the respondent farmers have access to bank loans in the area (SBP 2008a,b).

SBP's Committee on Rural Finance (CRF) strongly advocates for linkages between the Aarthi's and the commercial banks and cited a disconnect between the two as "highly damaging" (SBP 2001). The agriculture credit market, which is deemed risky and unprofitable by commercial banks; however, Aarthi has embedded itself in this market and makes substantial profits. The interest rate per annum charged by Aarthi ranges between 60% and 80% (PMN 2013).

TABLE 35.2

Outstanding Debt of Households from Institutional and Noninstitutional Sources

Type of Households	Institutional (PKR Million)	Noninstitutional (PKR Million)	Total (PKR Million)	Share of Institutional (%)	Share of Noninstitutional (%)
All Households: Total	299.9	546.0	845.9	35	65
NonAgricultural households	21.1	101.9	123.0	17	83
Agricultural households	278.9	444.1	723.0	39	61
Livestock households	12.1	69.3	81.4	15	85
Farm households: Total	266.7	374.8	641.5	42	58
Under 5 acres	29.1	152.6	181.7	16	84
5–12.5 acres	75.9	106.3	182.2	42	58
12.5–25 acres	70.1	53.4	123.5	57	43
Above 25 acres	91.6	62.4	154.0	59	41

Source: ADB 2008. *Pakistan: National Agriculture Sector Strategy*. Asian Development Bank.

In agriculture, the largest source of informal credit is Aarthi. The farmer who are considered being risky and less credit worthy by banks but Aarthi has a strong binding relationship with them. Aarthi usually relies on social collateral and lends unsecured loans to farmers that they know or on a reference basis. Right borrower selection, accurate credit needs assessment, and finally having control of the farmer's cash flows by binding the farmer to sell the produce through them are the tools used by Aarthi to manage its credit risks. If crops fail, Aarthi extends the loans for the next crop to allow farmers to keep on producing cash flows.

Aarthi can act as a "intermediary" between banks and farmers in provision of banking and value added services such as access to the latest farming practices, use of modern farm implements, and productive input use, etc. Further, it would help manage the bank's risk by referring the genuine clients, correctly assessing their credit needs, ensuring proper utilization, facilitating, and managing cash flows through the crop's sale proceeds.

It is interesting to note that despite considerable expansion in formal lending, the informal sources are still the major suppliers of credit. Any attempt to drive them out, or at least to compete with them, particularly by lowering interest rates, have not been successful. This suggests that policy makers needs to think not only in terms of interest rates but also in terms of other credit traits of informal sources such as simple procedures, adequacy, accessibility, and timeliness. As Viqar and Amjad (1984) argue, "the very fact that a large number of farmers prefer to borrow from informal sources in spite of high cost shows that the demand for agricultural credit is highly service elastic, a significant factor which is overlooked."

35.4 FORMAL SOURCES OF CREDIT

Formal sources of agricultural credit includes specialized banks like the Zarai Taraqiati Bank Ltd. (ZTBL) and the Punjab Provincial Cooperative Bank Ltd. (PPCBL). Others are five major commercial banks, 14 domestic private banks, 10 microfinance banks, 4 Rural Support Programs (RSPs), NGOs, Microfinance Institutions (MFIs), Islamic financial institutions (Figure 35.2).

35.4.1 Zarai Taraqiati Bank Limited (ZTBL)

ZTBL was incorporated as a public limited company in 2002, through repeal of the ADB ordinance of 1961. ZTBL advances short, medium, and long-term loans to farmers for a very broad range of

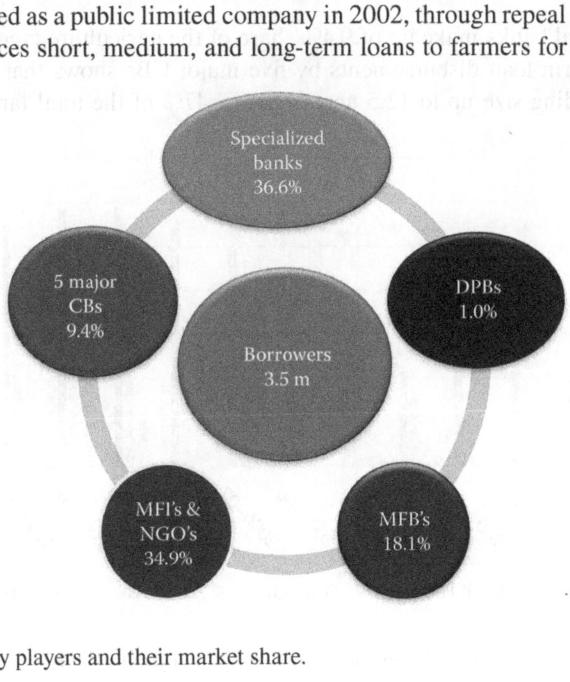

FIGURE 35.2 Industry players and their market share.

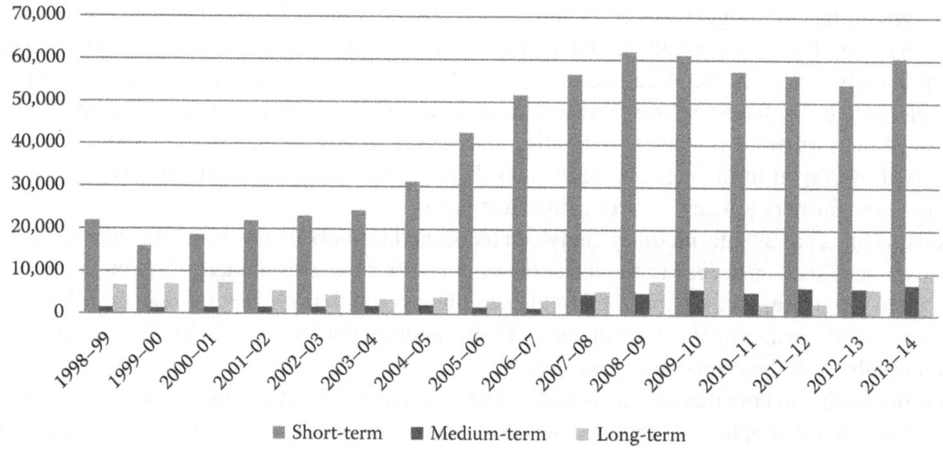

FIGURE 35.3 Term-wise position of agricultural credit advanced by Zarai Tarqiati Bank Limited (Million Rs.).

farm and off-farm activities, and to agro-based and farm-related processing units. Most of the bank's advances are of short-term duration. In 2013–2014; 78% of the amounts advanced were for short periods, 9% for medium periods, and the rest for the long term (Figure 35.3).

As regards the use of loan, the largest amount was advanced for purchase of fertilizers, seed, and pesticides. In 2013–2014, 75% of the total credit advanced was for the purchase of these three seasonal inputs; this was followed by credit for tractors (9%-Figure 35.4). Other purposes are dairy farming, poultry farming, fisheries, etc.

An analysis of loans advanced according to the size of the land holdings shows that most of the loans are advanced to owners owning up to 12.50 acres. They received 72% of the total credit in 2013–2014; followed by owners of 12.50 to 50 acres (25%); owners of 50 to 100 acres (2%); and those owning over 100 acres (0.45%). Tenants received a very negligible amount of credit.

35.4.2 Commercial Banks (CBs)

Five major commercial banks make up of 9.4% share of the agriculture credit market in Pakistan. Land holding-wise farm loan disbursements by five major CBs shows that most of the loans are advanced to land holding size up to 12.5 acres, that is, 47% of the total farm loan in 2013–2014;

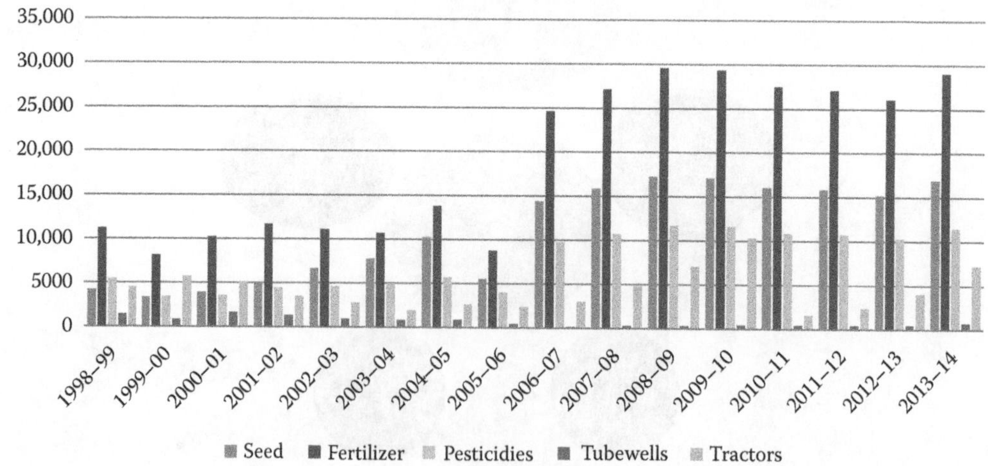

FIGURE 35.4 Item-wise break-up of Loans disbursed by Zarai Tarqiati Bank (Million Rs.).

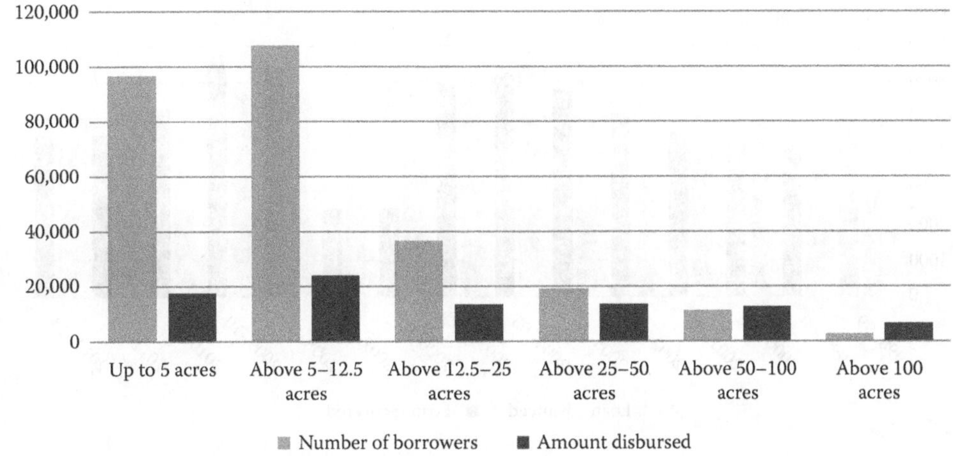

FIGURE 35.5 Land holding-wise farm loan disbursements in 2013–2014 by 5 major CBs (Million Rs.).

followed by 15% each between land holding size of 12.5 to 25 acres and 25 to 50 acres; 14% between 50 to 100 acres; and over 100 acres (8%) (Figure 35.5).

35.4.3 Domestic Private Banks (DPBs)

Fourteen DPBs make up of 1.0% share of the agriculture credit market in Pakistan. Land holding wise farm loan disbursements by DPBs shows that most of the loans are advanced to land holding size above 100 acres, that is, 29% of the total farm loan in 2013–2014; followed by 20% between land holding size of 50 to 100 acres; 19% between 25 to 50 acres; and 12.5 to 25 acres (15%) (Figure 35.6). Marginal farmers make up only 10% of the total credit disbursement.

35.4.4 Cooperative Credit Institutions

Before independence, main credit portfolios of cooperative banks consisted of commercial finance and very negligible financing was given to cooperative societies. In 1976, with the enactment of the Cooperative Banking Ordinance, the "Federal Bank for Cooperatives" (FBC) was established to provide financing to provincial cooperative banks for onward lending to cooperative societies. Subsequently, provincial cooperative banks were amalgamated to provide agricultural credit at the

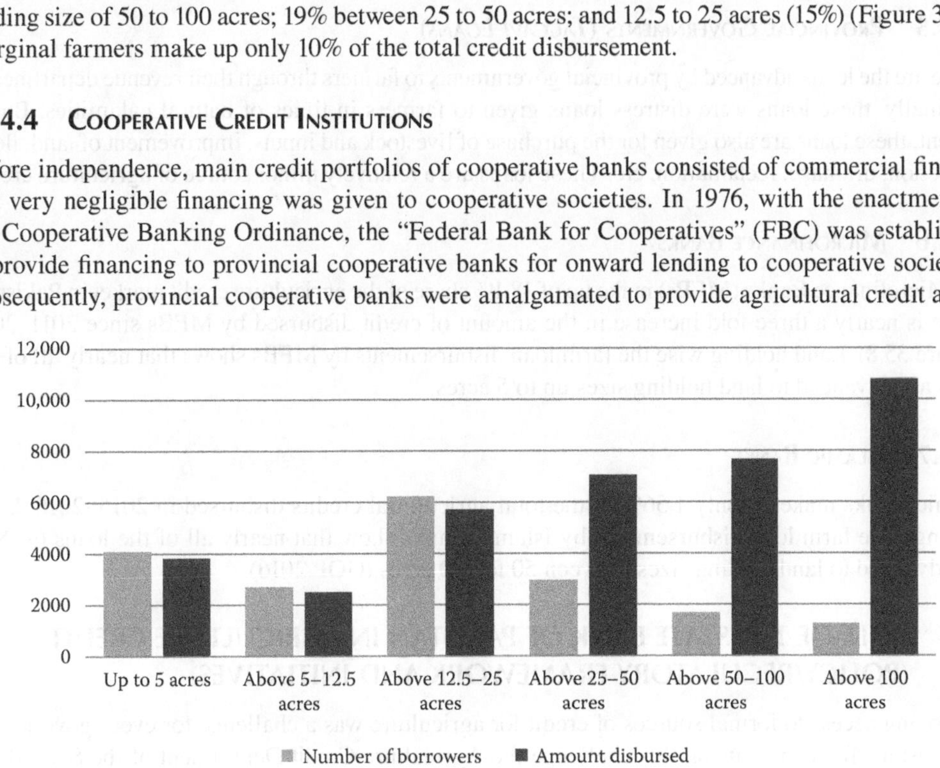

FIGURE 35.6 Land holding wise farm loan disbursements in 2013–2014 by DPBs (Million Rs.).

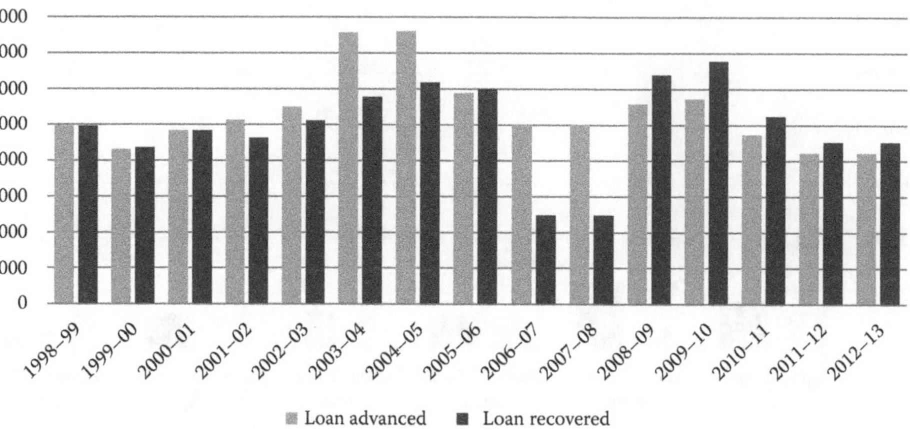

FIGURE 35.7 Loans advanced to farmers by agricultural co-operative societies (Million Rs.).

grassroots level and to encourage a cooperative structure of society in the country. However, the system failed to work properly due to subsequent default of provincial cooperative banks and fake cooperative societies. The FBC was liquidated in 2001. All provincial cooperative banks except PPCBL were liquidated in 2004. SBP provides financing to PPCBL under the guarantee of the Punjab government. PPCBL is currently the only cooperative bank scheduled with SBP and has an access to SBP credit lines for agricultural financing. The interesting thing in advances by cooperative societies is that the loan recovery rate is nearly 100% except during the years 2006–2008 when the recovery rate was nearly 50% (Figure 35.7).

35.4.5 Provincial Governments (*Taccavi Loans*)

These are the loans advanced by provincial governments to farmers through their revenue departments. Originally, these loans were distress loans given to farmers in times of natural calamities. But at present, these loans are also given for the purchase of livestock and inputs, improvement of land, flood protection, drainage, reclamation, etc. These loans are a relatively minor source of agriculture credit.

35.4.6 Microfinance Banks

Ten Microfinance Banks (MFBs) make up of 18.1% share of the agriculture credit market in Pakistan. There is nearly a three-fold increase in the amount of credit disbursed by MFBs since 2011–2012 (Figure 35.8). Land holding wise the farm loan disbursements by MFBs shows that nearly all of the loans are advanced to land holding sizes up to 5 acres.

35.4.7 Islamic Banks

Islamic Banks make up only 1.56% of the total agricultural credits disbursed in 2015–2016. Land holding wise farm loan disbursements by Islamic Banks show that nearly all of the loans (99.8%) are advanced to land holding sizes between 50 to 100 acres (GOP 2016).

35.5 ROLE OF THE STATE BANK OF PAKISTAN IN AGRICULTURE CREDIT (POLICY/REGULATORY FRAMEWORK AND INITIATIVES)

Improving access to formal sources of credit for agriculture was a challenge for every government in Pakistan. To keep focus on this challenge the Agriculture Credit Department of the State Bank of Pakistan's was created in 1953. Up till the 1990s, state owned financial institutions like the Agricultural Development Finance Corporation and the Agricultural Bank, which were later merged

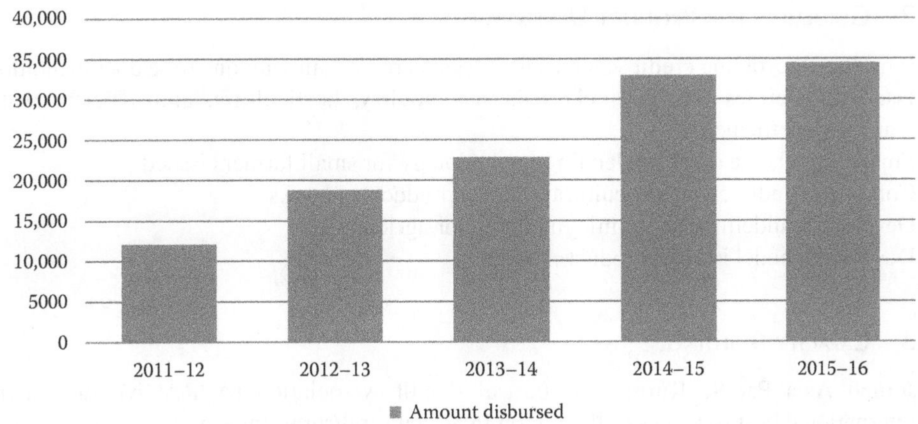

FIGURE 35.8 Credit disbursed by microfinance banks (Million Rs.).

to form the Agricultural Development Bank of Pakistan (ADBP), were used to direct credit to the agriculture sector. The Federal Bank of Cooperatives was established in 1976 to extend loans through cooperative societies to farmers. During the 1970s, mandatory agriculture credit targets were also issued to commercial banks. Since the financial reform process began, such schemes have been closed and instead SBP only provides indicative targets to the banks for agriculture lending. There are four pillars of the SBP regulatory framework for agriculture credit (SBP 2014a,b,c,d):

35.5.1 Enabling Environment

- Done away with mandatory agricultural credit regime from July 2005
- All banks are lending to agriculture on a market-based system
- Introduction of 14 domestic private banks besides 5 major banks, ZTBL & PPCBL
- Simplification and standardization of loan documents
- Separate Prudential Regulations for agriculture financing
- With the help of experts, SBP gives per acre indicative credit limits for crop production loans; (Table 35.3)
- Permanent Facility of One Window Operation (OWO) for agricultural financing
- Developed Crop loan Insurance Scheme (CLIS)

TABLE 35.3
Per Acre Indicative Agricultural Credit Limits for Crops

Crop	Amount (Rs.)
Wheat	29,000
Cotton	39,000
Rice/paddy	34,000
Sugar Cane	53,000
Potato	51,000
Tomato	37,000
Onion	34,000

Source: SBP 2008a. *Agricultural Survey of Gujranwala: Exploring District's Rural Economy*. Banking Services Corporation, Development Finance Support Department, SBP-BCS Gujranwala, State Bank of Pakistan, Karachi, Pakistan.

35.5.2 Guidelines for Product Development

- Introduced revolving credit schemes for 3-years credit limit with one time documentation;
- Issued guidelines for livestock, fisheries, poultry, horticulture, and efficient water management financing
- Financing scheme on group lending methodology for small farmers issued
- Compiled handbook on agricultural finance products of banks
- Developed guidelines on islamic financing for agriculture
- Developed model islamic product—Salam

35.5.3 Capacity Building

- Joined Asia Pacific Rural Agricultural Credit Association (APRACA) for sharing international best practices of the region in agricultural/rural finance
- Farmers' Financial Literacy and Awareness Program on Agricultural Financing (FFLP) at district level for the local farmers and bankers
- Policy adequacy seminars and internship program for agricultural universities and colleges
- Agricultural credit documentation and risk management workshops for CAD and risk management departments of banks
- Recently launched 4-days batch training programs at regional level for the agricultural field officers (AFO) of banks
- Arranged study visit/training programs for SBP officers and senior executive of banks with BAAC, Thailand, and Rabobank Netherland

35.5.4 Linkages with Stakeholders

- Agricultural Credit Advisory Committee (ACAC) of stakeholders is in place to deliberate on agricultural finance issues
- Established focus groups at SBP-BSC offices to deliberate, identify, and resolve the issues at regional level
- To facilitate timely disbursement of agricultural loans, SBP in collaboration with provincial governments launched permanent one window operation
- Arrangements of farmer's gatherings at grassroots level in collaboration with banks and other stakeholders

35.6 INTERNATIONAL BEST PRACTICES

Many examples of agribusiness financing and rural credit are available but the diverse economic, social, and geographical conditions make them very difficult to be replicated in our part of the world. However, some examples from Indonesia, Bangladesh, and Thailand are useful and can be replicated with modifications in the model (SBP 2009).

Bank Rakyat Indonesia:

- Specialized rural areas financial services with incentives for repayment
- Focus on savings mobilization—sustainability and profitability

Grameen Bank, Bangladesh:

- Collateral free lending to poor owned by the Poor
- Attractive rates on deposits—sustainability

BAAC (Thailand) Model:

- Collateral free group financing for rural development with a focus on farmers' education, coaching and training.
- Incentives for regular borrowers

Land Bank of the Philippines:

- Rural Financing to small borrowers with simplified lending procedures, zero tolerance for loan defaults and avoiding behest loans.
- Emphasis on portfolio diversification and capability-building assistance programs for farmers.

ACLEDA Bank, Cambodia:

- Easy access to group loan for small businesses with emphasis on training and education of the clients

35.7 INNOVATIVE LENDING METHODOLOGY AND DELIVERY CHANNELS

35.7.1 Value Chain Contract Farmer Financing

Value Chain Contract Farmer Financing is a binding arrangement between banks and agricultural value chain actors including producers, processors, aggregators, traders through which a farmer or group of farmers ensures supply of agricultural products to individual firms. VC contract farmer financing scheme makes financing available to farmers on processor's guarantee and in return buyers/processors may get assurance of getting required quantity and quality of agricultural produce (SBP, 2014a,b,c,d). The objectives of introducing this scheme are:

- To make small farmers who lack collateral eligible for bank finance.
- To mitigate banks risk by introducing alternative delivery channels.
- To make short-term financing available to processors, aggregators during procurement season.
- To provide mutually beneficial sustainable financial services for value chain actors by reducing cost of doing business and postdisbursement monitoring.

The guidelines may benefit in terms of:

- Capitalization on buyers and sellers relationship by financial intermediation and guarantee mechanism.
- Ensured accessibility of farmers to guaranteed markets.
- Assured supply and reliability in procurement for buyers.
- Improved planning and economies of scale will lower input costs.
- Productivity improvement due to improved technical support and extension services.

The guidelines provide a mechanism for auto liquidation of a farmer's liability. In this mechanism, on receipt of produce from grower, the lead firm will issue a payment warrant/check to settle the outstanding loans of borrowers (Figure 35.9).

35.7.2 Financing to Small Farmers

About 3.5 million borrowers are being financed by the formal sector against total farmer's population of 8.2 million. About 89% of the country's farmers is comprised of small farmers who mainly rely on informal sources for their credit requirements. The main reason of the financial exclusion of these

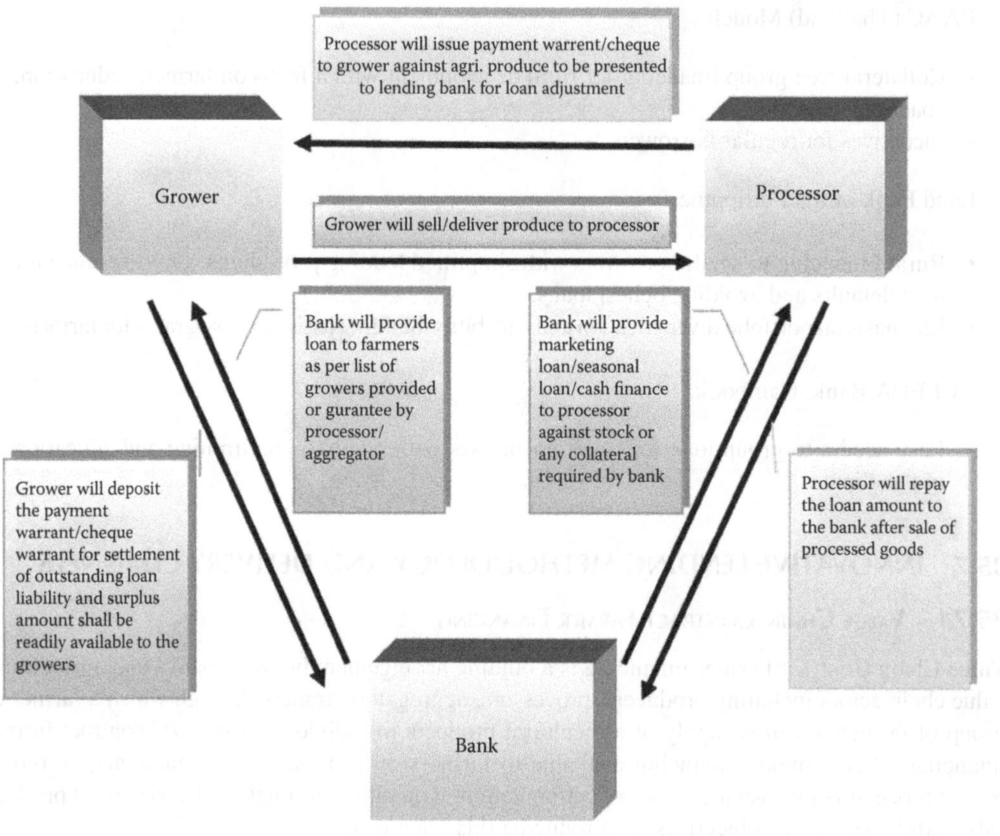

FIGURE 35.9 Value chain contract farmer financing mechanism.

small farmers is that they lack proper collateral. Therefore, SBP has launched a financing scheme for small farmers. Salient features of the scheme are (SBP 2015):

- Equally strong for micro, rural, and agriculture financing.
- Lending through group-based methodology.
- Each farmer in the group is guarantor of another farmer of the group.
- Formation of group is 5–15 farmers with similar type of financial needs.
- Blood relations and spouse are prohibited in the same group.
- Registration of SFG (small farmer group) with the bank.
- Maximum loan limit per farmer is Rs 200,000.

35.7.3 Warehouse Receipt Financing

Warehouse receipt financing is a form of secured lending to owners (farmers, traders, processors) of nonperishable commodities like grains, agricultural commodities etc. Warehouse receipts are a proven way to make use of the stored goods as collateral for loans from banks (SBP 2014a,b,c,d) (Figure 35.10 and 35.11).

Warehouse Receipt Financing—Pakistan Scenario

- SBP Task Force has developed a framework for warehousing, grading and testing, collateral management, post harvest financing, warehouse receipt system, etc. in collaboration with stakeholders to develop infrastructure of agricultural commodity physical trade & marketing.

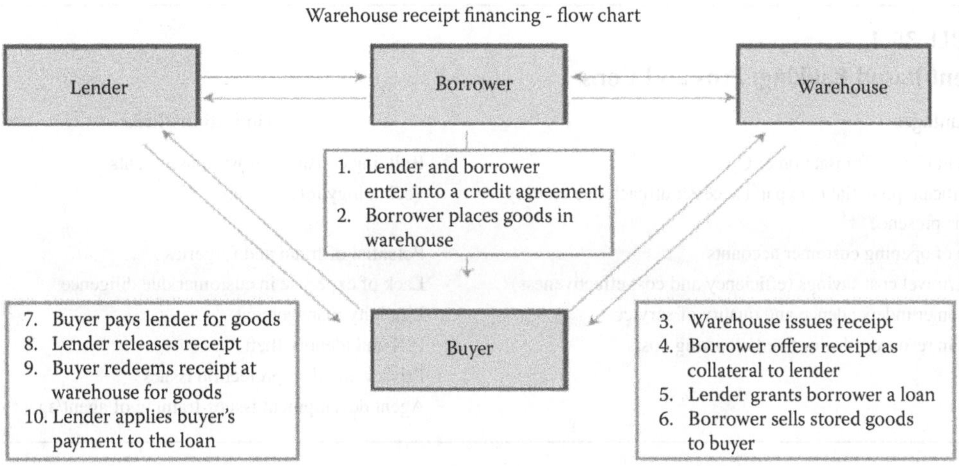

FIGURE 35.10 Warehouse receipt financing—process.

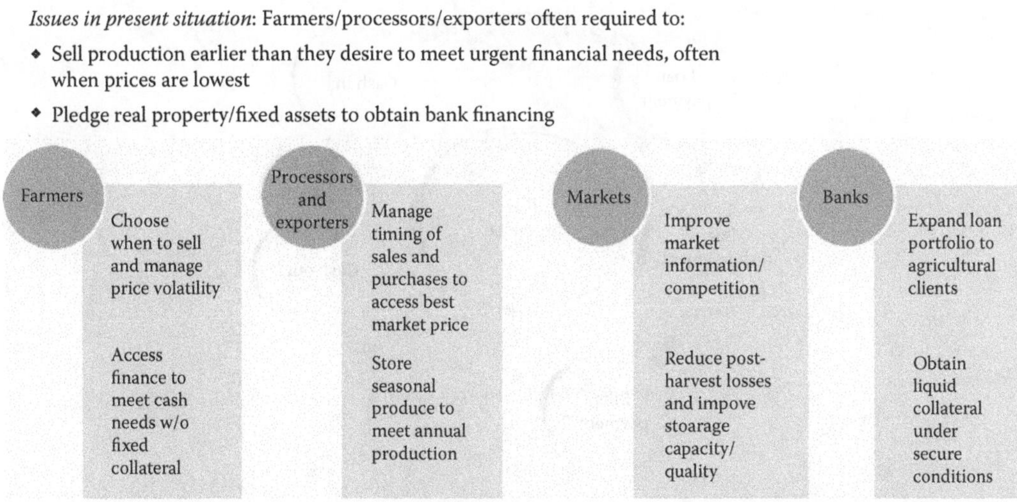

FIGURE 35.11 Benefits of warehouse financing.

- A comprehensive study has been conducted through international consultants and based on the recommendation of the study Pakistan Collateral Management Company is being formed under the Pakistan Mercantile Exchange.
- A roundtable on WHR has been conducted to sensitize the banks and other stakeholders on the initiative and best practices to facilitate them in development of specialized products, procedures and systems.
- This would ultimately evolve the fair and transparent price mechanism, food security, proper storage and post harvest financing system in Pakistan.

35.7.4 Branchless Banking-Agents Network (Easy Paisa, Omni, Mobicash, Timepay)

Branchless Banking (BB) is a distribution channel used for delivering financial services outside bank's conventional branches through use of technology and alternative delivery channels (ADCs) like retail agents, mobile network operators, cellular phones, super stores, chain stores, Pakistan Post, and so forth at cost effective, efficient, and convenient manner (SBP 2013) (Table 35.4) (Figure 35.12).

TABLE 35.4
Agent-Based Banking: Pros and Cons

Advantages	Limitations/Risks
Cuts in Cost—Transaction & Ops	Building consumer trust to use agents
Significant potential to expand access/outreach due country wide presence	Technology related risks
Ease of opening customer accounts	Potential of fraud and forgeries
Time/travel cost savings (efficiency and cost effectiveness)	Lack of expertise in customer due diligence
Customer independence and quality of service	Liquidity management
Cuts in resources-lower fixed/recurring costs	PIN and identity theft
	Privacy and data protection issues
	Agent development issues-training of agent/agent staff

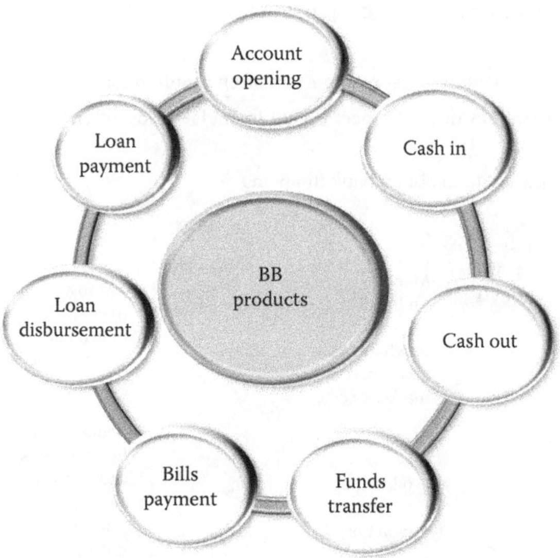

FIGURE 35.12 Branchless banking services.

35.8 WAY FORWARD

Following are the future plans adopted by SBP for the promotion/improving the access of agricultural finance in Pakistan (GOP 2016).

- Developing of Framework for Crop Insurance, where farmers will be provided insurance coverage for their major crops in case of any natural calamity.
- Development of framework for Livestock Insurance Scheme.
- Legal and regulatory framework/Guidelines for warehouse receipt financing.
- Credit Guarantee Scheme for Small Farmers.
- Model Sharia compliant products for agricultural financing for Islamic Banks.
- Agricultural Finance Awareness Program, workshops, Policy Adequacy Seminars etc. for banks, farming community and other stakeholders.
- Enhancement of agricultural credit by encouraging lending to livestock and other non farm sector activities.

35.9 COOPERATION

35.9.1 Introduction to Cooperatives

Cooperatives are the entrepreneurial form of cooperation. They are based on the principles of self help and are considered as economic enterprises. They play a significant role in improving the socio-economic conditions of their members and local communities. The locally owned and operated cooperative enterprises are in operation since the 1800s. They serve as catalysts for social organization and cohesion and represent a model of economic enterprise that gives high priority to democratic and human values. With increased events of unstable financial systems, food insecurity, global inequality, climate change, and environmental degradation the model based on cooperation is essential.

International Cooperative Alliance (1996) defined cooperative as "*an autonomous association of individuals voluntarily united to meet their common economic, social and cultural needs through a jointly-owned and democratically controlled enterprise.*"

Cooperatives can be defined as the businesses that are owned and democratically controlled by their members based on the basis of mutual cooperation for greater benefits of individuals as well as society. Such businesses operate on the principle of "one member, one vote", that is, each member has the same voting power. This differentiates them from routine businesses (investor owned) where voting rights are based on share ownership. The profit is disbursed based on the use of the cooperative services, unlike businesses where it is proportionate to their share ownership*.

35.9.2 Why Cooperate?

Cooperation can be in the form of economic, social and political. Social sciences give important to motivate cooperation among people within different group settings, that is, small groups, organizations, communities or societies, etc. It has been observed by the management scientists that organizations prosper when their members actively participate and work for its. Similarly, the law researchers have concluded that crime and community disorder issues can be easily tackled with an active involvement of community. Likewise, political scientists have recognized the importance of public involvement in building strong societies, which helps in the process of policy making, for example, stakeholder groups for policy discussion and designing policies.

In terms of cooperatives, they are business entities where economic cooperation is required. Such cooperation happens when markets fail to provide required quality goods or services at affordable prices. Such cooperation empower people enhance their economic opportunities through self-help which eventually improve their life quality.

35.10 HISTORY OF COOPERATIVES

The cooperative movement has its roots in the United Kingdom during the 1800s. There was a decline in living standard during 1800 to 1834. The wages were reduced by 3 to 4 shillings (a former British coin equal to twelve pence). There was an increase in unemployment that led to insufficient food. Due to poor economic conditions, people were forced to live in cottages where sanitary situation was worst. This increased the frequency of diseases. Due to all these conditions, women started working at relatively large farms. Child labor was also increased to the highest level. This was the time of exploitation and extreme poverty.

The prevailing capitalistic free market approach that claimed that through the free market mechanism, the situation will improve. As labor, poverty and starvation are strongly linked with unemployment, which was considered as a market commodity. An alternate school of thought emerged who are considered as the founders of the cooperative philosophy. This philosophy became the base

---

* http://www.nic.coop

for the cooperative movement. There were two significant personalities behind this philosophy: Robert Owen, an industrialist and Dr. William King, a medical practitioner.

35.10.1 Robert Owen

During the early 1800s, Robert Owen who was an industrialist and earned through cotton. He experienced poverty in his early life. He was the very first to establish a cooperative community in Scotland and then in the United States. These communities were failed, but Robert Owen identified the philosophical values of cooperation, which could become the base of organizing economic activity coupled with kindness, liberality, mutual respect, and cooperation.

35.10.2 Dr. William King

Dr. William King was a British medical practitioner from Brighton. Dr. King is known as an early supporter of the cooperative movement. He founded a cooperative store in Brighton in the early 1800s. He also started a paper "The Co-operator" in 1928 to promote these cooperative ideas. This paper acquired a wide circulation in a short time and influenced people regarding the cooperative movement. The paper was only published for about 2 years, but it greatly served to educate and unify scattered groups. Dr. King's writings provided the cooperative movement the practical guidance that was missing in its philosophical base. The most impacting statement of Dr. King (repeated in every issue of the paper) was "*knowledge and union are power. Power, directed by knowledge is happiness. Happiness is the end of creation.*"*

35.10.3 The Rochdale Pioneers

The Rochdale Pioneers Society was formed in 1844 in Rochdale in the United Kingdom. It was a society of 28 members. Around half were weavers. This society was one of the earliest cooperatives who paid a patronage dividend. It was a consumer cooperative that sold basic necessities (butter, candles, soap, flour, and blankets) of life to its members. The society worked on simple principle of supplying quality goods at cheaper prices along with returning profit to its members. They learned from the experiences of past failed attempts of cooperatives and the philosophy of Robert Owen and Dr. William King and formed their own principles.

35.11 COOPERATIVES AROUND THE WORLD

By the 1870s, the cooperative movement accumulated over £300,000 through its wholesale and insurance societies†. According to the global report on cooperatives and employment (2014), at least 250 million people are involved in cooperatives worldwide. Out of which, more than 26 million are the employees (direct employees and worker members). About 224 million people are engaged in different forms of producer cooperatives.

In the United Kingdom, retail cooperatives have a total turnover of over £7.7 billion‡. In France, there are about 21,000 cooperatives, which provide over 1 million jobs (3.5% of the active working population). In Kenya, about 250,000 people are employed or are involved with cooperatives. More than 60% of their population's livelihoods are derived from cooperatives. In Colombia, about 140,000 people are employed directly in cooperatives (3.65% of all jobs in the country). In Indonesia, cooperatives provide employment to around 300,000 individuals. In the United States, there are about 30,000 cooperatives which provide employment to more than 2 million people.

* Collis R. 2010. *The New Encyclopaedia of Brighton.* (based on the original by Tim Carder) (1st ed.). Brighton: Brighton and Hove Libraries.

† http://www.cds.coop/coop_movement/new-to-co-ops/the-rochdale-pioneers

‡ http://www.cds.coop/coop_movement/new-to-co-ops/the-rochdale-pioneers

Cooperatives have played and are playing significant role in the economies of many countries. For example, consumer cooperatives held 36.4% of the retail market in Denmark in 2007; agricultural cooperatives in Japan had a turnover of more than USD 90 billion with 91% of the farmers as their members and consumer cooperatives had a share of 5.4% in food markets; in Mauritius about 50% of the sugarcane farmers were part of cooperative setup that boosted agricultural sector; in terms of social sector, cooperatives invested USD 26 million in schools of The Ivory Coast; in New Zealand, 3% of the GDP is contributed by the cooperatives*. Cooperatives account for about 50% of the Kenya's GDP. In Rwanda, cooperatives share in the economy accounts for 8% of the GDP†.

35.12 COOPERATIVES IN PAKISTAN

35.12.1 Prepartition

The cooperative movement was started in the subcontinent in the early 1900s to solve the issue of rural indebtedness. The Cooperative Credit Societies Act was enacted in 1904 and provided many relaxations including income tax exemptions, custom duties relaxations/abandonments, registration fees reductions/withdrawals, prioritized cooperatives over ordinary creditors, free government audit, etc. Later in 1912, the Cooperative Societies Act was introduced that recognized noncredit cooperatives. Central and provincial cooperative banks were also formed as a result of this act. In 1919, the Reforms Act was passed that made cooperatives as a provincial subject. Furthermore, Bombay/Sindh Cooperative Societies Act was enacted in 1925.

There were about 1200 societies in India by 1914–1915. Total membership of these cooperatives reached 0.55 million with a working capital of 154.8 million Rs. The number of societies increased to 171,000 with a membership of 9.16 million with a working capital of 1640 million Rs in 25 States of British India. Out of these cooperative societies about 86% were agricultural cooperatives and 14% were nonagricultural cooperatives. The percentage of membership of agricultural cooperative was 62% and nonagricultural cooperatives was 38%. The percentage of the working capital was lower in agricultural cooperatives (38%) than nonagricultural cooperatives (62%).

In the smaller provinces of Coorg and Cochin the number of cooperatives exceeded the number of villages. There were one cooperative for 13 villages in Assam while the ratio was 1 cooperative to 6 villages in Orissa, Sind, Bihar and Mysore. The cooperative adoption was much better in in Madras, Bengal, and Panjab. In Madras, the ratio was one cooperative to every 2.3 villages while in Bengal the ration was one cooperative to 1.9 villages and in Punjab it was one cooperative to 1.3 villages. Bombay, Panjab, and Madras led in cooperative planning that covered all aspects of the cooperative Movement.

On the success of the cooperative movement in India, it is concluded that it could not make a remarkable progress in terms of money and energy invested in it. But, it has sufficiently improved the living of its members through breaking the moneylender's monopoly and brought interest rates down. Considerable amount of credit was available for the needy at lower interest rates. It has promoted the agricultural sector through allied industries.

35.12.2 Postpartition

After independence, Pakistan retained the Bombay/Sindh Cooperative Societies Acts of 1925. This Act is currently in operation with amendments. Furthermore, Cooperative Farming Act was enacted for agricultural cooperatives in 1976 with a Multi Unit Cooperative Societies Act of 1942, for cooperatives covering more than one province‡. However, the provinces made necessary amendments

* http://ica.coop/en/facts-and-figures

† http://www.ipsnews.net/2012/11/cooperatives-as-business-models-of-the-future/

‡ http://www.fao.org/docrep/007/ad713e/AD713E03.htm

TABLE 35.5
Cooperative Acts/Ordinances in Punjab over Time

Sr. No.	Acts/Ordinances
1	The Cooperative Societies Act, 1912.
2	The Cooperative Societies Act, 1925 Amended in 2006.
3	The Cooperative Societies Rules, 1927
4	The Multi-Unit Cooperative Societies Act, 1942
5	The Cooperative Societies (Repayment of Loan) Ordinance, 1960
6	The Cooperative Societies and Cooperative Banks (Repayment of Loans) Ordinance, 1966
7	The Cooperative Board (Dissolution) Act, 1966
8	The Cooperative Societies (Reforms) Order, 1972
9	Sind Cooperative Societies Reforms Rules, 1973
10	The Cooperative-Farming Act, 1976
11	The Interest Free Loan Utilization and Recovery Rules, 1977
12	The Establishment of the Federal Bank for Cooperative and Regulation of Cooperative Bank Act, 1977
13	The Federal Bank for Cooperative and Regulation of Cooperative Banking (Accounts) Rules, 1977
14	The Board of Directors of the Federal Bank for Cooperative and Regulation of Cooperative Banking (Meetings) Rules, 1977
15	The Cooperative Societies (Reform) Order 9 of 1980

http://cooperatives.punjab.gov.pk/system/files/ANNUAL_REPORT_2010_-2011.pdf

to the above mentioned Acts according to their needs and requirements. Table 35.5 shows the Acts/ordinances which were/are operative in Punjab province.

Currently, there are more than 33,000 registered cooperatives in Punjab with a membership of about 1.7 million. There total capital share of these societies is Rs. 1824 million with a working capital of Rs. 26547 million[7].

35.13 TYPES OF AGRICULTURAL COOPERATIVES

Agricultural cooperatives can be divided into three broad categories, that is, by function, by financial management, and by legal structure.

35.13.1 By Functions

Agricultural cooperatives can be divided into various types depending upon the functions they perform:

- *Marketing*—The cooperatives which are involved in the marketing of agricultural products are often known as marketing cooperatives. They either sell products directly to the local marketing system or to the processors.
- *Bargaining*—The bargaining cooperatives are similar to marketing cooperatives but differ in their agreements. These cooperatives usually arrange agreements relating to price and sales terms in a particular season or till the time agreed upon for renegotiation. Both marketing and bargaining cooperatives are also known as producers' cooperatives.
- *Purchasing*—Such cooperatives are also known as consumer cooperatives. They serve as wholesalers of inputs and other supplies.
- *Services*—The cooperatives which provide its members with different services are known as services cooperatives. They may further be categorized as irrigation, crop insurance, agricultural credit, human and animal health care, housing and animal shelters, electrical, and telephone, etc.

35.13.2 By Financial Management

Cooperatives can also be categorized into two categories in terms of their financial structure:

- *Stock Cooperatives*—The cooperatives whose stockholders' equity is represented by their shares of common or preferred stock. Such cooperatives restrict their members from transferring stocks and withdrawing from the cooperative.
- *Nonstock Cooperatives*—The cooperatives where ownership is represented by membership certificates. They are also known as centralized cooperatives. Such cooperatives have certain benefits such as price consistencies, quick response to changes, competent manager hiring process, more management control on purchases and marketing, expense duplication is reduced, etc.

35.13.3 By Legal Structure

Cooperatives can also be categorized into two categories in terms of their legal structure:

- *Incorporated Cooperatives*—The cooperatives which are organized under special cooperative laws.
- *Unincorporated Cooperatives*—The cooperatives which are not governed by any law. However, they follow the methods used by incorporated cooperatives such as:
  - For business purposes, the members are liable for debts and obligations, individually
  - In case of action required, the cooperative cannot take action in its name, however it may take action through its members
  - The cooperative cannot hold any property in its name

35.14 PRINCIPLES OF AGRICULTURAL COOPERATIVES

35.14.1 Cooperative Theory

The cooperative theory revolves around the famous principles of Rochdale Pioneers. These principles are considered as the basics of cooperatives around the world. These principles* are:

1. *Voluntary and open membership*—The cooperatives must have an open and voluntary membership without any discrimination of gender, social, racial, political or religious. To practice this in letter and spirit, motivations and rewards system is an integral part of this principle. Motivations can be of financial, social and voluntary type.
2. *Democratic member control*—The cooperatives must have democratic member control. Members have equal voting rights, that is, one member—one vote.
3. *Member economic participation*—Members must contribute the capital of their cooperative in a fair and impartial manner. This capital shall be the common property of cooperative. Members receive a limited compensation for this capital. Profits on the capital are to be distributed among the members.
4. *Autonomy and independence*—Cooperatives must be autonomous and independent. In case if a cooperative enters in a business deal with other business organization, the cooperative will be governed by its members and remain independent regardless the amount contributed by the business organization.
5. *Education, training, and information*—The cooperative must provide education and training to its members. Furthermore, it must provide information to the public about its nature of cooperation.

---

* http://cultivate.coop/wiki/Rochdale_Principles

TABLE 35.6
Difference between Business Organizations and Cooperative

	Sole Proprietorship	Partnership	Corporation	Cooperative
Purpose	Profit for proprietor	Profit for partners	Profit for shareholders on investment of time or money	Benefits for the members in terms of services, cost saving, and/or profits
Ownership	By sole proprietor	By partners	By shareholders	By members
Control	By sole proprietor	By partners, according to time and money	Based on the number of voting shares help per shareholder, directly or by proxy	One member-one vote. Proxy voting limited
Distribution of surplus profits	Sole proprietor	To partners according to their time and money	To shareholders according to their investment	To members in proportion to use of services
Liability	Unlimited liability of proprietor	Unlimited liability of partners	Shareholders liability limited to their shares	Members limited to shares. Directors can be liable

Source: Harris and Joynt 1997.
http://www2.gov.bc.ca/assets/gov/farming-natural-resources-and-industry/agriculture-and-seafood/farm-management/farm-business-management/agricultural_cooperatives.pdf

6. *Cooperation among cooperatives*—A cooperative is an autonomous organization, but it must work together to enhance communication across other cooperatives to strengthen cooperative movement.
7. *Concern for community*—A cooperative must be a responsible partner towards the community. Its decisions must benefit the community. Cooperatives are different from other business organizations. Table 35.6 explains this difference.

35.14.2 Objectives of Cooperatives

Following are the objectives* of cooperation:

1. Elimination of middlemen
2. Raising economic status of the poor
3. Removal of the ills of capitalism
4. Raising moral standards of its members
5. Increasing the prosperity of the whole community
6. Abolition of social inequalities
7. Political and religious neutrality
8. Development of corporate life

35.14.3 Cooperative Governance and Control

Cooperative management decisions are expected to improve the quality and quantity of service to the members. The cooperatives are governed by their members through a democratic process.

* http://www.universityofcalicut.info/SDE/co_operative_theory.pdf

Members elect their board of governors/directors through one member one vote approach. The board of governors then chooses its representative, the CEO.

- Membership responsibilities
  As a member of a cooperative society, there are some responsibilities of members which start when a person becomes a member of a cooperative/when the cooperative comes into being. These responsibilities remain throughout the membership duration or till the life span of the cooperative. These responsibilities include (Kirkman 1993):
  - Understanding the cooperative
  - Adopt legal papers
  - Select and evaluate representatives/board of governors/directors
  - Provide necessary capital
  - Use and support cooperative
  - Participate in affairs
  - Provide information
  - Help obtain new members

- Board of governors/directors responsibilities
  Like any business corporation, cooperatives appoint/select/elect boards of directors/governors. The duties of cooperative board are similar to their counterparts in other organizations. However, there are various distinctive responsibilities of cooperative boards. The general responsibilities are:
  - Representing members
  - Establishing cooperative policies
  - Hiring and supervising management
  - Overseeing acquisition and preservation of cooperative assets
  - Preserving the cooperative character of the organization
  - Assessing the cooperative's performance
  - Informing members

35.15 PROBLEMS WITH TRADITIONAL COOPERATIVES

Researchers have focused on the problems in the traditional cooperative setup that create disadvantages for members of cooperatives members. Five major problems (Ortmann and King 2007)* have been identified in the literature:

1. The free rider's problem appears when property rights are un-tradable, insecure and/or unassigned.
2. Horizon problem appears when members' residual claim on the income generated by an asset is shorter than the life of the asset.
3. Portfolio problem occurs because members invest in the cooperative in proportion to their use and the equity shares, in cooperatives, are not freely purchased or sold. Therefore, members are unable to diversify their individual investments according to their personal wealth.
4. Control problem arises due to the differences in ownership and control. The cooperatives are owned by their members and are governed by the board of governors/directors.
5. Influence cost problems arise as a cooperative is involved different activities. There may be diverse objectives of its members, which may result in costly activities. The cost may include both direct and indirect costs (poor decisions of resource allocation, etc.).

* http://ageconsearch.umn.edu/bitstream/10129/1/46010040.pdf

Besides these inherited issues, the newly formed cooperatives may face issues (Harris and Joynt 1997) like:

1. Lack of clearly identified goals
2. Inadequate planning
3. Failure to use experienced consultants and advisors
4. Lack of member leadership
5. Lack of member commitment
6. Lack of competent management
7. Failure to identify and minimize risk
8. Poor assumptions
9. Lack of financing
10. Inadequate communication

35.16 SUCCESS STORIES

35.16.1 Dairy Cooperatives in Pakistan (Idara-e-Kissan)

The member farmers of the cooperative had 29% extra income compared with nonmembers. Furthermore, the members had more buffaloes than nonmembers. They were also in a better position to feed their animals with quality feed. Similarly the members had better access to vaccination and artificial insemination facilities (Riaz 2008).

35.16.2 Producers Cooperatives in Central Punjab, Pakistan

The efficiency of member farmers of producers cooperatives in Central Punjab, Pakistan was high compared to the nonmembers. The level of quality input use was higher by the member farmers as compared to the nonmember farmers. As a result, their benefit cost ratio was 38% higher than nonmembers. Despite success these cooperatives could not be sustained for a longer time period due to conflicts between members (Sabir et al. 2012).

35.16.3 Savings and Credit Cooperatives in Swaziland

The savings and credit cooperatives contributed towards food production in Swaziland. The members of these cooperatives had more capital to buy quality inputs than nonmembers. As a result, their average yields of maize, potatoes, and beans were 2.6, 2.7, and 2.2 times higher than those of nonmembers, respectively (Mavimbela et al. 2010).

35.16.4 Agricultural Cooperatives in Ethiopia

Agricultural cooperatives in Ethiopia effectively provided support services to their member farmers. As a result technical efficiency of the member farmers was enhanced due to ease in access to inputs and extension linkages (Abate et al. 2013).

35.16.5 Farmers' Cooperatives in Nigeria

The efficiency of farmers' cooperatives were significantly corelated with income, cooperative experience, family size, types of agricultural activities, and leadership skills. To enhance the efficiency of such cooperatives, basic infrastructure facilities should be provided by government (Adefila and Joshua 2014; Innocent and Adefila 2014).

35.17 FUTURE OF COOPERATIVES

Traditional cooperatives will have to find ways to adapt to changing functional needs of the time, markets, and economies. Keeping in view the pressures, population growth, environmental degradation, water scarcity, etc., many of the cooperatives may end up in a failure or will change their structure fundamentally, As in the case of business corporations. By keeping up the pace of adaptation to the changes in the world economic scenario, the cooperatives will have a distinct competitive advantage over business organizations. They will surely reap business opportunities. Examples include, Hansmann's Cooperative Franchising, Korea's Samuel Undong Model, Empowering Youth through Cooperatives, Cooperative Business Models, Diversity in Business Activities of Cooperatives, Chains of Cooperative Convenience Stores, etc.

REFERENCES

Abate GT, Francesconi GN, Getnet K. 2013. Impact of agricultural cooperatives on smallholders' technical efficiency: evidence from Ethiopia, Euricse Working Paper number 50 13.

ADB. 2008. *Pakistan: National Agriculture Sector Strategy.* Asian Development Bank.

Adefila JO, Joshua M. 2014. Roles of Farmers' Cooperatives in Agricultural Development in Sabuwa Local Government Area of Katsina State, Nigeria. *J Econ & Sus Dev*, 5(12): 80–87.

GOP. 2010. *Agriculture Census*, Pakistan Bureau of Statistics, Islamabad.

GOP. 2012. *Pakistan Economic Survey*, Ministry of Finance, Govt. of Pakistan, Islamabad.

GOP. 2016. *Pakistan Economic Survey*, Ministry of Finance, Govt. of Pakistan, Islamabad.

Harris A, Joynt H. 1997. *Agricultural Cooperatives: A Start-Up Guide*, Canada/BC Farm business Management Program, British Columbia, Canada.

Innocent Y, Adefila JO. 2014. Farmers' cooperatives and agricultural development in Kwali Area Council Federal Capital Territory Abuja, Nigeria. *Int J Hum & Soc Sci*, 4(7): 161–170.

Krikman CH. 1993. *Cooperative Member Responsibilities and Control*, Agricultural Cooperative Service, U.S. Dept. of Agriculture, Washington, DC, USA.

Mavimbela P, Masuku MB, Belete A. 2010. Contribution of savings and credit cooperatives to food crop production in Swaziland: A case study of smallholder farmers. *Afr J Agr Res*, 5(21): 2868–2874.

Ortmann GF, King RP. 2007. *Agricultural Cooperatives I: History, Theory and Problems.* Agrekon, 46(1): 40–68.

PIDE. 1984. Pakistan Rural Credit Survey: Analytical Report, Vol. 1, Pakistan Institute of Development Economics, Islamabad.

PMN. 2013. Who is the Aarthi: Understanding the Commission Agent's role in the Agriculture Supply Chain, a joint project report of Pakistan Microfinance Network and National Institute of Banking and Finance, Islamabad.

Riaz K. 2008. A case study of milk processing: The Idara-e-Kissan Cooperative. *L J Eco*, 13(1): 87–128.

Sabir HM, Tahir SH, Arshad S, Nasir SB. 2012. Future of cooperative farming in Pakistan. *J Bio, Agr & Health*, 2(6): 42–48.

SBP. 2001. Report of SBP Committee on Rural Finance, State Bank of Pakistan, Karachi, Pakistan.

SBP. 2008a. *Agricultural Survey of Gujranwala: Exploring District's Rural Economy.* Banking Services Corporation, Development Finance Support Department, SBP-BCS Gujranwala, State Bank of Pakistan, Karachi, Pakistan.

SBP. 2008b. *Agricultural Survey of Sukkur: Exploring the District's Rural Economy.* Banking Services Corporation, Development Finance Support Department, State Bank of Pakistan, Karachi, Pakistan.

SBP. 2009. *Handbook on Best Practices in Agricultural/Rural Finance.* Agricultural Credit & Microfinance Department, State Bank of Pakistan, Karachi, Pakistan.

SBP. 2013. *Rural Bank Franchising Model in Pakistan.* Agricultural Credit & Microfinance Department, State Bank of Pakistan, Karachi, Pakistan.

SBP. 2014a. *Indicative Credit Limits and Eligibility Items for Agricultural* Financing. Agricultural Credit & Microfinance Department, State Bank of Pakistan, Karachi, Pakistan.

SBP. 2014b. *Prudential Regulations for Agricultural* Financing. Agricultural Credit & Microfinance Department, State Bank of Pakistan, Karachi, Pakistan.

SBP. 2014c. *Guidelines for Value Chain Contract Farmer Financing.* Agricultural Credit & Microfinance Department, State Bank of Pakistan, Karachi, Pakistan.

SBP. 2014d. *Framework for Warehouse Receipt Financing in Pakistan*. Agricultural Credit & Microfinance Department, State Bank of Pakistan, Karachi, Pakistan.

SBP. 2015. *Financing Scheme for Small Farmers*. Agricultural Credit & Microfinance Department, State Bank of Pakistan, Karachi, Pakistan.

Viqar A, Amjad R. 1984. *The Management of Pakistan's Economy, 1947–82*, Oxford University Press, Karachi, pp. 159–60.

36 Agricultural Marketing

Abdul Ghafoor and Hammad Badar

CONTENTS

36.1 INTRODUCTION

The agriculture sector in Pakistan is considered a major contributor to economic growth and development. This sector not only meets food demands of the population but also provides raw material to the industry besides providing surplus for exports. Often, growers have to bear with reduced prices for their produce due to poor infrastructure and postharvest practices. Moreover, frequent food surpluses and shortages have highlighted the need to modernize the marketing system.

36.2 BASIC CONCEPTS

36.2.1 Market

The market can be defined and understood in different ways, but commonly it is recognized as a place or area where buyers and sellers gather and interact for buying and selling goods and services. In a more solemn language of management sciences, market is referred as an exchange process of goods and services accompanied by price making mechanisms. Agricultural markets perform the central role of assembling agricultural commodities from scattered and distant production areas and distributing them to consumers and other stakeholders in urban and periurban areas.

36.2.2 Marketing

The term marketing may be defined in many ways. According to the American Marketing Association, marketing can be defined as "performance of business activities that direct the flow of goods and services from producer to consumer, so as to reach the consumer at the time, place and in the form he desires and at a price he is willing to pay." As such, marketing performs all those tasks, which ensure that products are available for consumption:

- At the convenient place,
- In the form desired by consumers,
- In the required quantity and quality,
- At the desired time, and
- At fair prices.

According to another definition, "marketing consists of identifying customer needs and satisfying such needs in a profitable manner." According to this definition marketing is a customer-oriented and profit-driven process based on long-term and mutually beneficial relationships between sellers and customers. Hence, the scope of marketing cannot be confined to just producing products and then making efforts to sell it. Rather, the focus is on producing only those products that can be sold at remunerative prices. This implies that consumer needs and wants should guide what to produce, how to produce and how, where, and when to sell. Firms need to analyze markets first for determining consumer needs and then develop products or services. Only in this way can they satisfy the needs of consumers and generate a profit.

Agricultural products differ from industrial products due to their perishable nature and special requirements during various farm and marketing operations. However, this does not imply that agricultural marketing is something entirely different from marketing of industrial and other products. It is simply application of principles of marketing in the agriculture sector. Agricultural marketing encompasses activities which may include picking/harvesting, drying, cleaning, sorting, grading, processing, packaging, labeling, transporting, storage, promotion, and sale of agricultural products. All these activities contribute in adding value to agricultural products as they flow from producers to consumers. While some of these activities are performed on farm, the other are carried out off-farm by market intermediaries such as traders and agroprocessors.

Successful performance of agricultural marketing activities cannot take place without adequate and timely availability of information, particularly relating to consumer requirements. Since customer orientation and profit is the key focus of marketing, all those involved in marketing chains must have an in-depth understanding of consumer requirements in terms of desired product attributes so that these can be met at minimum marketing costs. Given this, agricultural marketing can be comprehensively defined as including all business activities designed to plan, price, promote, and distribute want satisfying goods and services to household consumers and industrial users.

36.2.3 Marketing Channel

It is the unique feature of agricultural production that it is concentrated in specific geoclimatic areas, which shows its biological nature. For example, mangoes in Pakistan are produced in South Punjab, grapes in Balochistan, rice in Central Punjab, citrus in Sargodha and Bhalwal, etc. However, consumption of these commodities is spread across the country. Therefore, these commodities need to be transferred from farmers to consumers through various middlemen operating in the marketing system. Through the way an agricultural commodity flows from producers to consumers is called marketing channel. A commodity may move towards consumers through different channels. Marketing channel is like a pipeline which directs the flow of goods and services from producers to consumers.

36.2.4 Marketing Margins

When agricultural commodities move from one marketing functionary to next in the marketing chain, some functional utilities are added. For example, time utility is added through storage and form utility through processing etc. However, this incurs cost as well, which increase price, and a price differential between two stages in the marketing channel is created. This price differential is termed marketing margin. The marketing margin consists of two parts, that is, profit and explicit cost. Marketing margin may also be defined as the price of marketing functions performed by various middlemen operating in the marketing chain.

36.2.5 Middleman

Middlemen are the individuals/businesses who specialize in the performance of various marketing functions such as exchange, physical, and facilitative functions (Mohy-ud-Din and Badar, 2011). Production and consumption places of agricultural commodities are generally distantly located. Consequently, direct exchange between producers and consumers is impossible. This gives rise to the role of middlemen in the marketing system who relate producers with consumers. They add utility in the marketing system by performing various functions such as buying and selling, transportation, storage, grading, and packing of agricultural products.

36.3 EVOLUTION OF AGRICULTURAL MARKETING IN PAKISTAN: PAST POLICIES

After independence of Pakistan, the government laid a major emphasis on food production and did not pay attention to marketing of agricultural commodities. Therefore, the pace of development of the agricultural marketing system remained slow. The first 5-year plan (1955–1960) identified many weaknesses in agricultural marketing system. Amongst others, inefficiencies in the methods of buying and selling, assembling and processing, transportation of farm produce, and utilization of market resources were the key factors affecting development of agricultural marketing system in the country. In order to cope with various challenges, some policy measures (proper implementation of grades and standards for agricultural commodities, grant of agricultural loans by Agricultural Development Bank of Pakistan to farmers for the purchase of fertilizers, High Yielding Variety Seeds (HYV), farm implements and pesticides, training of stakeholders in postharvest management, and grant of subsidy for construction of cold storages in different production areas etc.) were undertaken to ensure reasonable prices to the stakeholders and improve efficiency of the agricultural marketing system in the country. Although the government focused its attention in strengthening agricultural marketing systems by bringing improvements in marketing infrastructure and postharvest management, noteworthy progress was not achieved in the implementation of the proposed measures.

The second (1960–1965) and third (1965–1970) 5-year plans underlined the need of removing various weaknesses (such as malpractices of middlemen, inefficient handling of produce during marketing, inadequate storage space, improper processing, noncompliance to standardization and grading, inadequate supply of packing material etc.) in the agricultural marketing system. The government adopted some measures to enhance efficiency of agricultural marketing on the recommendations given in the second and third 5-year plans. Some of the measures undertaken by the government for the rehabilitation of agricultural marketing in the country included promulgation of the Weights and Measures Act in major areas of Pakistan, implementation of grades and standards for major exportable agricultural commodities and upgradation and improvement of existing markets, and development of new regulated agricultural markets under the provisions of the Agricultural Produce Markets Act of 1939.

The plans identified hoarding, price manipulations by middlemen and insufficient supply of agricultural loans as some of the major impediments in the development of agricultural marketing system. High priority was assigned to overcome various obstacles in the system. Incentives were provided through grant of subsidies and by fixing floor prices of agricultural products and by strengthening the institutional framework for agricultural markets. Along with these measures, the government approved the Agricultural Produce Marketing Regulation Bill to envisage the newly emerged regulated markets. Despite these measures, lack of cold storages/warehouses for perishable commodities, improper grading and standardization, poor infrastructure (farm to market roads) continued to be the major obstacles in the smooth working of the agricultural marketing system in the country.

The policy measures of the government in the 1970s' remained primarily focused on the evolution and implementation of support price mechanisms and making necessary arrangements for the export of agricultural products keeping in view the trends and requirements of international markets. Support price programs for food grains (wheat) were successfully implemented. Adequate machinery for procurement operations was provided and storage capacity for agricultural commodities—especially for food grains—was enhanced. In order to improve terms of trade, the government increased export quotas and reduced import tariffs. The policy was further expanded to secure sanitary safeguards against the import of pesticides through provisions of the Pesticides Act.

The fifth 5-year plan (1978–1983) emphasized the need for strengthening market infrastructure. Based upon recommendations outlined in the plan, the government undertook measures to develop market infrastructure and ensure timely availability of agricultural inputs (chemical fertilizers, pesticides, and farm machinery) at reasonable prices and supplement imports when necessary. The Pakistan Agricultural Storage and Services Corporation (PASSCO) was established in 1973 to ensure better returns to producers, as well as reasonable prices to consumers. In addition, improved marketing and processing technology was adopted through collaboration between local entrepreneurs and reputed firms in the international markets. Marketing institutions (market committees, trading/export houses, commodity stabilization funds, support price cover, grading/quality standards through various institutional mechanisms) were established.

Some progress was witnessed in improving infrastructural facilities (such as grain silos, warehouses, cold storages, product quality testing laboratories, grading and primary processing plants, transportation systems for handling and speedy clearance of perishable agricultural commodities, etc.) Airfreight arrangements for promoting export of perishable products were improved. The development of a food processing industry (through availability of cheap packing material and chemicals supported with other incentives such as packaging, tax concessions, import of machinery for modernization, etc.) remained an important component of government policy. In short, agricultural marketing systems received boost as a result of the establishment of new processing plants, better procurement measures, and improved transportation and distribution systems. Notwithstanding achievements, little progress was made in the construction of new storage facilities for major food grains and for perishable commodities, and in improving management skills of the stockholders/market functionaries.

The last decade of the century was characterized with the government's focus on promoting and diversifying exports and ensuring price stability. This era witnessed the emergence of the WTO and globalization of international trade. These developments changed the dynamics and requirements of agriculture. Pakistan, like many other developing countries, had to undertake structural and institutional reforms to cope with the changing environment. As a result, the government assigned a priority to establish and develop various institutions for boosting exports of high value crops. The Pakistan Horticulture Development and Export Board was created to develop the horticulture sector and boost exports of various horticultural commodities (e.g., mango, kinnow, apples, dates, etc.) and cope with the emerging challenges encountered in international trade due to implementation of various WTO agreements (e.g., AoA, SPS, TRIPS, etc.). Export targets for various agricultural crops were not achieved due to poor compliance to the requirements of various developed countries.

The Medium Term Development Framework (MTDF) (2005–2010) advocated the policy of privatization, deregulation, and market orientation for the domestic economy. The MTDF emphasized development, expansion, and diversification of market base. Efforts were made to encourage public and private investments through tax reforms and financial liberalization in order to improve efficiency of the agricultural marketing system. Vision 2025 of Pakistan also envisages to promote a food secure Pakistan and strengthening infrastructure facilities. The government of Punjab has taken the initiative to promote rural urban linkages by renovating or establishing roads in remote and village areas so that farmers can be connected with markets in urban areas.

Although measures have been undertaken to improve the performance of agricultural marketing systems, insufficient availability of institutional credit, especially marketing credit, high postharvest losses due to poor adoption of postharvest management practices, poor implementation and adoption of standards and grades, and lesser compliance to global requirements of various WTO agreements continued to affect the working of the agricultural marketing system in the country. Establishment of new regulated markets, provisions of agricultural inputs to stakeholders at reasonable prices, compliance to grades, and standards by stakeholders are however priority areas of the government policy. The existing marketing infrastructure is incapable to cope with the emerging challenges of globalization and to meet the growing demands of food grains, fruits, vegetables, livestocks, poultry, and their products both in the domestic as well as international markets. The recent food crisis in Pakistan has highlighted various inadequacies and inefficiencies in the agricultural marketing system. There is a strong need for chalking out a comprehensive policy for marketing infrastructure and postharvest management to face the emerging challenges to agriculture and meet growing and diversified food needs of the population.

36.4 LEGAL AND INSTITUTIONAL ARRANGEMENTS FOR AGRICULTURAL MARKETING IN PAKISTAN

Agriculture is a provincial subject and provincial governments are mainly responsible for enacting legal and institutional arrangements in Pakistan. However, both the federal and provincial governments frame policies for the development of the agricultural marketing system and for various institutions supporting the system at federal and provincial levels. The legal framework for the agricultural marketing system in Pakistan has developed over time. The Agriculture Produce Markets Act, 1939 was promulgated on the recommendations of the Royal Commission on Agriculture, constituted by the British India government during 1927 with the objective to regulate agribusiness and to do away the evils and problems inherent in the agricultural marketing system. The Market Committees were established under the provisions of the above act, which were assigned the noble pursuit of safeguarding interests of growers. After independence, the same act was implemented in Pakistan.

The Act of 1939 was replaced by the Punjab Agricultural Produce Markets Ordinance, 1978. The rules to regulate the working of wholesale markets were, however, framed in 1979. All agricultural marketing activities (especially the working of wholesale markets) in the province of Punjab are legally controlled under this ordinance. It may be mentioned that 244 agricultural markets in Punjab exist for handling food grains, fruits, and vegetables. Besides these markets, there are 81 feeder markets to feed the produce to main agriculture produce markets. Grading and quality certification in the domestic markets is legally controlled under the Punjab Agricultural Produce Grading and Marking Act of 1972 (Government of Punjab, 2006). Other related legal documents in Punjab include the Punjab Pure Food (Amendment) Ordinance 2015 to look after the quality of food products and Punjab Local Government Act 2013 to deal with local issues.

The West Pakistan Municipal Committee (Cattle Market) Rules of 1969 is the only livestock market legislation that has empowered the local authorities to establish, maintain, and administer livestock markets. The Agricultural Produce Markets Ordinance of 1978 includes livestock in the list of commodities, but its coverage under this act is not properly executed. The Punjab Livestock Breeding Act 2014 has been announced by the government of Punjab to regulate livestock breeding services and to build modern cattle markets. In this connection, livestock/cattle markets are in a transition stage from traditional marketing methods to modern and value-oriented cattle markets to offer farmers and traders a secure and healthy environment. The Cattle Market Management Company (CMMC) has been structured in 2014 to establish, manage, reorganize, operate, and upgrade the existing markets in the major metropolitan cities. This initiative may eliminate and discourage extortion, corruption, role of middleman, and commission mafias from the cattle markets.

In other provinces of Pakistan, different legal documents are available which look after the affairs of agricultural marketing in respective provinces. These include The Baluchistan Agricultural Produce Market Act 1991, The NWFP Agricultural and Livestock Produce Market Act 2007, and The Sindh Wholesale Agricultural Produce Market (Development and Regulation) Act 2010.

The government established various institutions for streamlining working of agricultural marketing system in the country. Institutions supporting agricultural marketing systems in Pakistan operate both at federal and provincial levels. At federal level, the Agricultural Policy Institute (API), previously known as the Agricultural Prices Commission (APCom), chalks out domestic and international sectoral/commodity-specific policies and examines and evaluates production, processing, storage, and marketing costs of agricultural and livestock commodities. The main task of the Agricultural Policy Institute is to formulate agricultural policies including those on agricultural prices, marketing, and export competitiveness of agricultural Products.

The mandate for ensuring quality and grading assurance in the domestic market is entrusted with the Department of Agricultural and Livestock Products Marketing and Grading (DALPMG), working under the Ministry of National Food Security and Research. The DALPMG advises the federal government on all matters relating to setting/implementation of grades and standards of agricultural and livestock products for both the domestic and export markets. It also provides guidance to the provinces on various aspects of marketing (especially grading and standardization of agricultural commodities) and coordinates provincial activities at the national level.

The Pakistan Horticulture Development and Export Board, renamed Pakistan Horticulture Development and Export Company (PHDEC) was established by the federal government under the Ministry of Commerce to tackle domestic and export marketing issues in the field of horticulture in Pakistan. The PHDEC is managed by a Board of Management from the public and private sectors. The PHDEC was established for tapping enormous potential of Pakistan's horticulture products in the international high-value markets. Since a single ministry or institution with the responsibility of developing horticultural value chain at various levels was lacking, it was considered essential to establish the PHDEB. The Board has played pivotal role in boosting exports of various horticultural products (mango, kinnow, apples, dates, etc.) and important vegetables in export markets of many developed countries (PHDEB, 2005).

The Agribusiness Development and Diversification Project was established by the Government of Pakistan to promote and develop various agribusinesses and to add value to various agricultural products, diversify the cropping pattern, and impart training to stakeholders on skill development, management, and entrepreneurship. The project works in collaboration with the Ministry of Food and Agriculture (MINFA) of the government of Pakistan and the Asian Development Bank. The main objective of this project is to use the agribusiness sector to support economic growth and employment generation. This is being achieved by making the agriculture sector more competitive and dynamic. To make this sector more active agribusiness projects aim to help solve the constraints that obstruct development of agriculture sector. The project also helps the stakeholders in discovering and making use of domestic and export opportunities. An Agribusiness Support Fund has been created and is an integral component of Agribusiness Development and Diversification Project.

Agribusiness Support Fund (ASF) is a "not-for-profit company" established by Ministry of Food and Agriculture (MINFA) with the support of the Asian Development Bank (ADB). ASF provides nonreturnable matching grant for business development to individuals and firms engaged in various agribusinesses. The purpose is to enable agribusiness entrepreneurs to employ modern techniques and practices, develop business skills, know-how, expertise, understand market requirements, and improve their productivity, creditworthiness, profitability, and competitiveness.

Given the importance of the livestock sector in the economy, the government of Pakistan has also established Livestock and Dairy Development Board (LDDB). The LDDB is organized as a private sector led not-for-profit company which mainly operates in the area of livestock and dairy development in Pakistan. Major initiatives undertaken by the LDDB include milk collection and dairy development programs, livestock and meat production, food security and productivity enhancement of small farmers.

The provincial governments have also entrusted the task of developing agricultural marketing system to various organizations/institutions. In Punjab for instance, agricultural marketing is managed by the Directorate of Economics and Marketing, working under the department of agriculture. Similar arrangement are provided for other provinces where a system of market committees are working to ensure agricultural marketing functions (Aftab, 2007). These market committees are formed to protect the interests of all stakeholders working in the marketing of agricultural produce. These committees have also assumed other functions like looking after the affairs of special markets like "*Ramzan Bazars*." At the same time these committees are meant for controlling quality and prices of agricultural produce in wholesale markets. The Food Department is another organization that procures many agricultural commodities (especially staple food stuff, mainly wheat) from farmers at prices announced by the federal government.

Punjab Agriculture and Meat Company (PAMCO) has also been established by the Government of Punjab under the Public-Private initiative to energize agribusiness and marketing in Punjab. The major areas of interest of this organization include the development of integrated cold storage chains, enhancing and improving processing and marketing of agricultural commodities. The PAMCO works in the field of fruits and vegetables, poultry, livestock, dairy, fisheries, floriculture, and facilitates the stakeholders in seeking certificates (e.g., Euro GAP, Global GAP, etc.) for the export of agricultural commodities. The Punjab Institute of Agricultural Marketing (PIAM) was established in 2006 with the mandate to impart training to stakeholders and suggesting policy recommendations for improving agricultural marketing systems in Punjab.

36.5 ROLE OF PUBLIC AND PRIVATE SECTORS IN AGRICULTURAL MARKETING

Pakistan's agricultural marketing is characterized with numerous market players who perform different functions in transferring farm produce to consumer. It has been observed that an agricultural commodity changes hands seven to eight times before reaching the ultimate consumer. Functions performed by various market functionaries (especially the middlemen in the market chain) remain one of the most controversial issues in Pakistan's agricultural economy. It is argued that middlemen exploit marginal farmers and hamper their legitimate share. Infrastructure plays a vital role in facilitating and ensuring smooth functioning of the agricultural marketing system. Wholesale markets for instance, act as a cardinal link between producers and consumers and are operated by public and private sectors. Most of the wholesale markets in the country are not adequately developed and lack basic infrastructure (efficient logistics, storage and other marketing facilities), putting farmers at a disadvantageous situation while selling their produce. Currently in Punjab, there are 152 wholesale grain markets, 95 fruits and vegetable markets, 81 feeder markets, and 11 markets working in the private sector.

There is limited storage capacity (6%–7% of total agricultural production) in the public domain and that too is limited to few commodities. Existing cold storage facilities are unevenly distributed among the province. Punjab dominates with 512 units and Sindh has 25 units, KPK 16 units while Baluchistan has only 2 units. Most of these facilities are not compartmentalized, which causes transfer of odors among various commodities placed in cold stores. Another unfortunate fact about existing status is limited processing (3% of fruits, vegetables, and milk) in the country. There are 121 known pack houses in the country for horticulture crops. The Department of Agricultural and Livestock Marketing and Grading (DALPMG) has made grades and standards for 42 agricultural commodities but there still exists a need for framing grades for other commodities besides updating the existing ones. Postharvest losses are huge which amount to 35%–40% of total fruit and vegetable produced in the country.

Functions performed by middlemen in the wholesale markets of Pakistan are widely debated. It is generally argued that middlemen exploit marginal farmers and deny them their legitimate share. This allegation may not be ignored as many commission agents, bypassing the provisions of the Agricultural Produce Market Acts, have been found charging higher commission rates than prescribed. Preharvest contractors dominate the marketing system of fruits in Pakistan. They are often allegedly labeled to overutilize their power. However, despite all these allegations, the importance and role performed by middlemen cannot be underestimated.

The performance of agricultural marketing system is generally judged by market margin approaches which show the relative share of different stakeholders involved in the supply chain of agricultural commodities. It has been observed that the share of farmers in consumer rupee is relatively low in case of perishable commodities as compared to nonperishables. In the case of fruits, preharvest contractors and retailers get more profit as compared to other stakeholders (Table 36.1). Vegetables and other agricultural commodities are no exception to the observation mentioned above. The share of different stakeholders in the marketing of fruits and vegetables is presented in Tables 36.1 and 36.2.

TABLE 36.1
Marketing Margins and Producer's Share in Consumer Rupee for Various Fruits in Punjab (Percent)

Market Agency	Mango	Citrus	Date (Fresh)	Guava
Producer	20	20	37	15
Preharvest contractor	39	26	20	33
Commission agent	6	2	6	5
Wholesaler	9	8	13	5
Retailer	26	44	24	42

Source: Khushk, A. M. and A. D. Sheikh. 2004. *Structure, Conduct and Performance of the Marketing Systems Margins and Seasonal Price Variations of Selected Fruits and Vegetables in Pakistan*. PARC, Islamabad.

TABLE 36.2
Marketing Margins and Producer's Share in Consumer Rupee for Various Vegetables in Punjab (Percent)

	Potato			Onion		Tomato			Peas	Carrot	Brinjal
	Lodhi	UCL	Kokab & Smith	Kokab		Lodhi	UCL	SIddique	Lodhi	UCL	UCL
Market Agency				A	B						
Grower	56.0	62.1	63.7	49.1	55.0	57.0	55.5	54.9	25.0	56.9	60.6
Commission agent	–	8.5	11.3	1.5	1.7	7.8	3.4	–	9.0	6.9	–
Wholesaler (Pharia)	–	11.5	2.1	21.0	14.8	–	10.0	16.4	–	12.8	12.4
Retailer	–	17.9	22.9	23.4	28.5	–	26.7	25.3	–	21.3	20.1
Marketing margin	44.0	37.9	36.3	50.9	45.0	43.0	44.5	45.1	75.0	43.1	39.4

Source: Chaudhary, M. and A. Ahmad. 2000. *Food Grain Losses at Farm Level in Pakistan. Vol. 1. Department of Agricultural Marketing.* University of Agriculture, Faisalabad.
Note: A: winter onion and B: Stored winter onion.

Pakistan is a developing country and over time agriculture has proved its central importance in uplifting and supporting the economy of the country but its real potential still needs to be realized. After independence, various governments took several measures to improve the agriculture sector. However, the focus of these measures was on productivity enhancement and agricultural marketing remained a neglected area because marketing infrastructure development and postharvest management of agricultural commodities did not receive adequate attention of policy makers.

Agricultural marketing infrastructure plays a key role in facilitating and ensuring the smooth functioning of the agricultural marketing system. An efficient logistic system is a prerequisite for optimal performance of the marketing system. For example, infrequent, expensive, and poor quality transport services will put farmers at a disadvantage in selling their crops because an expensive service will result in low farm gate prices. Similarly, poor quality roads, coupled with poor storage cause enormous losses of agricultural products, Perishables products such as milk, fresh vegetables, and fruits in particular incur major losses because of their shorter shelf life. As such, all weather roads play a crucial role in enhancing market surplus for many agricultural products.

Food security and safety have emerged as prominent important concerns, particularly in the scenario of recent trade liberalization reforms under WTO. Compliance to national and international standards has become one of the most important requirements for achieving food safety and security. At present, the situation is not satisfactory in Pakistan as many processors and manufacturers lack the requisite capacity and know-how to ensure food safety. Businesses lack adequate awareness of modern hygiene and production management practices and requirements and are unable to obtain various food safety certifications such as Hazard Analysis and Critical Control Points (HACCP). As a result, they fail to capture opportunities available in high value markets of Europe, USA and other developed countries.

The food crisis in the past and occasional surpluses and shortages of agricultural commodities underline the need of taking measures by the government to improve the working of agricultural marketing system. Some of the major challenges which need urgent attention of the government include poor farm to market roads, price volatility, inadequate storage capacity, poor value addition, inefficiencies in factor and product markets, poor implementation of grades, and standards.

36.6 COMMODITY MARKETING SYSTEM

Agricultural commodities in Pakistan mainly comprises of agricultural crops, which includes food grains (wheat, rice, and maize), cash crops (cotton and sugarcane), fruits and vegetables, poultry and its products, livestock and its products, fisheries and forestry products. The marketing system and major issue in the marketing of these commodities are discussed in the following sections.

36.6.1 Food Grains

The major food grains crops in Pakistan include wheat, rice, and maize. These food grains are grown in various parts of the country. The marketing system of these three crops and the related challenges are explained as below.

36.6.1.1 Wheat

Wheat is the major food grain crop and staple food in Pakistan. From food security perspective, wheat is accorded high priority in agricultural policies. The government announces and implements support price for the wheat. As a result, the country is now self-sufficient in wheat with an average annual production of around 24–25 million tonnes. Most of the production is consumed locally.

A majority of farmers in Pakistan grow wheat and those who have a marketable surplus sell it to village dealers at market prices or at government-announced support price to procurement centers established by various government departments such as PASSCO and provincial food departments (Kurosaki, 1996). Village dealers sell the assembled wheat to commission agents in the wholesale markets who then sell it to processors and consumers. Government departments supply procured wheat to flour mills at fixed release price who, after milling, supply wheat flour to consumers through various distribution networks.

In the last few years, wheat production in Pakistan has increased considerably because of government support price policy and current production is enough to meet domestic requirements. Despite this, several inefficiencies exist in the wheat production and marketing system. The most prominent is the lack of efficient and high quality modern storage facilities. The public storage facilities are unable to store the procured wheat. Often, the wheat placed in opens spaces is lost and incur losses to public exchequer. Private storage facilities are also inadequate to compensate the public sector storage constraint. In addition there are ma y flaws in the govt. procurement system of wheat. Farmers often complain about nonavailability of bardana and exploitation by the market agents in terms of low prices and late payments.

36.6.1.2 Rice

Rice is second-most important food grain and staple food in Pakistan. It also contributes to foreign exchange earnings of the country. Pakistan Basmati rice is known for its aroma and taste. The marketing system of rice is by and large the same as that for wheat. However, the government does not extend support prices to rice. Rice growers in Pakistan sell their produce to village dealers or commission agents. Some farmers also sell their rice to agents of rice millers in their area. Village dealers and commission agents supply the rice assembled from their areas of operation to rice mills where the rice undergoes processing. Rice is then passed on to consumers through various market intermediaries such as wholesalers and retailers.

Pakistani rice industry is facing difficulties in reaching to its full potential. One major reason is the low price which may be attributed to lack of grading, branding and proper packing and noncompliance to SPS measures. Absence of accredited labs for testing and certifications hampers export of rice from Pakistan to many developed countries. Other challenges include nonavailability of certified seeds, use of unskilled labor in rice transplantation from nursery, poor postharvest techniques, higher input prices, lack of modern marketing skills.

36.6.1.3 Maize

Maize is the third most important food grain and cereal crop in Pakistan. It is grown all over Pakistan but Punjab and KPK dominate in its production. Maize was considered a subsistence crop in the past mainly grown for personal consumption, seed, livestock, etc. Now it is widely grown for commercial purposes and it is used as basic raw material for several products such as starch, sweeteners, corn oil, glucose, custard powder, and gluten. It is also used by the poultry industry for manufacturing of feed.

The last few years have witnessed some significant increases in the area under cultivation and production of maize but recently this crop has faced many problems, mainly low prices as there are few buyers of this crop. With the passage of time this crop reached to its saturation point, supply surpassing the demand, and as a result, prices dropped in the markets.

36.6.2 Cash Crops

Cotton and sugarcane are two cash crops of Pakistan and growers cultivate these crops in the hope of generating income for their families. In case of suitable conditions and good price, growers generate sizeable cash from these crops. How these crops are marketed and what type of challenges affect the marketing efficiency of these two crops is explained below.

36.6.2.1 Cotton

Cotton is a major cash crop in Pakistan and plays a key role in agricultural growth. It is the basic raw material for the textile industry, which generates more than half of Pakistan's export earnings. Growers mostly sell their marketable surplus of cotton to village dealers or commission agents who then supply it to ginning factories. A few growers, generally large landholders, sell their cotton directly to ginning factories, which are located across the cotton production zone. Cotton in the ginning process is separated into cotton lint and cotton seed. Cotton lint then passes through stages such spinning, weaving, dyeing, finishing, and making of garments and made-ups. Cotton and textile industries produce products such as readymade garments, towels, bed-wears, hosiery and knitwear, tents, canvases, and cotton bags. Cotton seed is used for extracting oil and is also used as livestock feed.

Cotton production has considerably increased over the years and average production is now more than 10 million bales. Still, there is a lot of potential for further increase. However, several factors hinder the realization of this potential. Unavailability of good and uniform quality seed cotton is the major problem. The available seed is often adulterated with foreign matter and water. Unsuitable picking methods coupled with inappropriate ginning techniques severely affect the lint quality, which lowers the commercial value of Pakistani cotton. Absence of scientifically developed and well-recognized standardization systems is another problem. Mixed grades and contamination lowers the cotton quality. Fluctuations in domestic and international prices of cotton often result in welfare losses to industry stakeholders. Other problems include the energy crisis, unskilled labor, low value addition, and lack of modernization of machines. Another major issue is the unplanned imports of seed cotton from neighboring countries and exports as well, which seriously destroy the interests of local manufacturers and traders.

36.6.2.2 Sugarcane

Sugarcane is the second most important cash crop and generates income and employment to the farming community of the country. It is the main source of raw material for the sugar industry. Other products obtained from sugarcane include molasses, bagasse, and press mud. Sugarcane production has considerably increased overtime because of sizeable expansion of the sugar industry, which now comprises of 86 sugar mills across the country. Current sugar production stands more than 65 million tonnes in the country.

The marketing of sugarcane is widely believed to be inefficient because of several problems. Growers complain of lower prices of their cane and claim that sugar mills use various tactics to exploit them. Late start of the crushing season, underweighting at purchase centers of sugar mills, inadequate and delayed payments, undue deductions are some of these tactics. On the other hand, sugar mills have high cost of production and import machinery as major reasons for many marketing ills. Other problems include stagnant cane yields, lower rate of sucrose recovery, underutilization of capacity, inadequate infrastructure, inadequate credit facility, lack of research and development, and vague government policies and political interventions.

36.6.3 Fruit and Vegetables

In Pakistan, a wide range of fruits and vegetables are grown and consumed. Major fruits include mango, citrus, apple, bananas, dates, grapes, guava, peach, and apricot. In vegetables, tomatoes, onions, potatoes, chilies, carrot, cauliflower, cabbage, lettuce, radish brinjal, cucumber, beans gourds, okra, and peas are included. Pakistan also exports fruits and vegetables and in 2014–2015 they contributed Rs. 67,864.3 million to foreign exchange earnings. In fruits, citrus, mangoes, and dates and in vegetables, potatoes, chilies, peas, and tomatoes are the major commodities exported to Middle Eastern countries, Saudi Arabia, Afghanistan, the Russian Federation, and Sri Lanka. Fruit exports to high-value markets of developed countries such as the United States and the E.U. are small because of stringent quality requirements.

The marketing channel of fruits in Pakistan is quite long and involves several intermediaries such as preharvest contractors, commission agents, wholesalers, and retailers. At orchard level, preharvest sale method is common. Under a contract in this method, growers handover their orchards to preharvest contractors who then assume the responsibility of fruit harvesting, packing and transportation to commission agents who have generally advanced them money. Some growers, however, themselves perform harvesting, packing, transportation, and selling operations. In wholesale markets, commission agents sell fruit brought either by preharvest contractor or growers to wholesalers through auction method. Wholesalers further sell fruits in smaller lots to retailers for onward transfer to consumers.

Vegetables are also marketed in a similar way. However, preharvest contractors are not very common and famers mostly sell their vegetables after harvesting through commission agents in the wholesale markets. Some large and periurban vegetable growers may sell directly to retailers. From wholesale markets, vegetables reach consumers through retailers.

Major problems which constrain Pakistan's ability to realize its full export potential of fruits and vegetables include low productivity, outdated pre and postharvest management techniques, poor packing, absence of modern infrastructures, and lack of modern marketing knowledge. In Pakistan, fruit losses are high. Of these, nearly 25% are postharvest losses, which may be attributed to improper transportation and packing, lack of cooling facilities, absence of quality controls and improper hygiene and sanitation facilities. There is need to reduce pre and postharvest losses by upgrading existing practices. In case of vegetables, the major issues include lack of quality seed of high yielding varieties, poor plant densities, narrow choice of cultivars, high incidence of pests and diseases, enormous weed infestations, and low use of fertilizers.

While other countries export a major part of their domestic production, Pakistan's fruit exports are not more than 10%–15% of its total production. Even the price of wheat is exported is also very low compared to average international prices. Shorter shelf life of our fruits is another major constraint. Presence of pesticides residues and some pests, particularly fruit flies, are major reasons for rejection of Pakistani fruits and vegetables shipments in developed countries where quarantine requirements are stringent. Pakistani fruits fail to capture European markets because of presence of excessive seeds, which consumers do not prefer. There is need to direct efforts on developing seedless varieties, particularly of kinnow and oranges, to capture the high value E.U. markets.

In the last few years, the government has started paying attention to horticultural industries. Some efforts have been made for infrastructural improvements. However, there is need to establish proper modern cool chain systems so fruits may not incur quantitative and qualitative losses across the value chain. Supermarket chains are also gradually increasing their presence. There is need to facilitate their expansion because they have the resources and capabilities to develop value chains through their large resource base, expertise, and cool chain systems. Our growers are mostly uneducated and are ignorant of how quality fruits are produced by improving these practices. Both the public and private sector need to educate and train growers for producing export quality fruits and vegetables.

36.6.4 Poultry

The poultry industry has emerged as a prominent and vibrant industry in Pakistan because of 8%–10% annual growth in the recent past. It is considered one of the organized agroindustries with a current investment of more than Rs. 200 billion. In 2014–2015, poultry industry contributed 1.3% to GDP and 6.3% and 11.2% to agriculture and livestock value added. The industry directly and indirectly provides employment to 1.5 million people. The contribution of poultry meat in total meat production is 28% (Government of Pakistan, 2015).

In the poultry marketing chain, village dealers or collectors of poultry birds collect live birds from the farmers and sell them in wholesale markets. Commission agents or wholesalers in these markets then sell live birds to retailers and processors who convert live birds into meat and other products for the final consumers.

The poultry sector is performing much better than agricultural subsectors. It has contributed a lot in fulfilling meat requirement of consumers and currently it contributes 28% to total meat production. Poultry meat has provided a relatively cheap alternative to consumers. However, there is still a lot of room for improvement. Transportation systems are inappropriate for birds, which lead to losses of poultry birds and eggs. Slaughtering and storage facilities are unhygienic and need upgrading.

36.6.5 Livestock and Dairy

Livestock is the key sector of Pakistan's economy because it contributes more than half of the agricultural value addition. A majority of the farming community supplement their incomes with livestock. The livestock sector is composed of cattle, sheep, goats, and other animals. Although numerous by-products are obtained from livestock, meat and milk are the major products.

36.6.5.1 Meat

Meat is the major livestock product and is of three types, lamb, beef, and poultry. While poultry meat is obtained from poultry birds, sheep, goats, cows, and buffaloes are used for lamb and beef. Several intermediaries are involved in the marketing of livestock. Beoparies (local traders) travel across villages and purchase small and large animals. Some of them sell them locally to butchers or supply them to commission agents in the wholesale markets in town and urban areas. From these wholesale markets, butchers and processors buy these animals and, after processing, sell their meat and products to consumers.

Despite its sizeable contribution, the livestock sector is not well developed and organized. In Punjab province, the government has taken initiative of developing model cattle markets and their contribution has yet to materialize. Existing livestock markets lacks requisite infrastructure for efficient trading of animals. Unhygienic, filthy, and exploitative conditions hinder the scope of expansion of these markets. Animals brought for sale often lack physical and medical fitness and as such there are no measures are in place to ensure quality of animals either for meat or milk. There is no grading system available for meat animals and their carcass. Normally, butchers sell their meat according to their wish and prices do not depict quality. Recent price hikes in meat are one of the current issues which show the inefficiency of this sector. Price stability in livestock production is essential for avoiding seasonal fluctuations.

36.6.5.2 Milk and Dairy Products

Milk and dairy products are a major component of livestock. Pakistan stands among the leading milk countries of the world (FAOSTAT, 2012). Milk is mainly obtained from buffaloes and cows, however, sheep and goats also contribute a small share in total production. Milk is consumed both as fresh and in processed forms. Although the milk processing industry is gradually expanding in the country with the entry of several national and multinationals, less than 5% of total milk production is processed. Several dairy products such as yogurt, butter, creams, are also consumed in Pakistan.

In the milk marketing chain, milk collectors assemble milk from small farmers and supplies it to wholesalers or processors. The village milk collector has limited resources to fulfill quality requirements and therefore often adds ice and some other materials to keep milk fresh, which may deteriorate its quality. The village dealers sell milk to wholesalers operating in cities and towns. They have their own shops and sometimes have dairy processing plants at small levels. They collect milk from various sources and supply it to retailers or processors. They sometimes separate cream and other products from milk and sell the rest. Processors change the basic form of milk into UHT milk, powdered, condensed, or skimmed milk products. These processing companies have their own distribution network. The major issue in limited expansion of processing milk industry in Pakistan is high cost of packaging material and the limited segment of consumers who like the taste of these products.

The dairy industry in Pakistan is more established as compared to any other livestock related activity. The private sector is more organized, advanced, and equipped with a better marketing system for milk in Pakistan. However, small-scale milk collection and processing activities are greatly hampered due to nonavailability of refrigerated tanks and cool reefer vehicles. Packaging cost in the processing industry is also high and needs to be controlled. For this, the government may come ahead to organize the private sector for investments in this area.

Postharvest handling of milk and related products are very poor. Due to nonavailability of proper storage and transportation system, almost 95% of total milk does not enter the marketing chain. Postharvest losses are enormous in this case. Developing a cool chain network for milk and related products may be a good option to realize the full potential of this sector and also to enhance consumer welfare by ensuring food safety and security.

36.6.6 Fisheries

Fisheries are important sources of livelihood for the coastal inhabitants. Apart from marine fisheries, inland fisheries are also important sources of animal protein. Although share of fisheries in the GDP is small, its contribution to the national income through export earnings is substantial. The major buyers of seafood from Pakistan are China, the UAE, Thailand, Korea, Malaysia, Indonesia, Hong Kong, and some Middle Eastern countries.

Many market intermediaries are involved in the marketing chain both in the case of marine and inland fish. Fish supplied by producers, passes through different channels before it reaches the ultimate consumers. There are four intermediaries involved in the flow of fresh water fish, these are contractors, commission agents, retailers, and processors cum retailers. Direct marketing, as practiced by contractors, is possible only when there is direct contact between producers and consumers. In most cases, producers sell their catch through intermediaries, particularly when consumer markets are distant from production areas. The common practice of channeling the catch is through commission agents because of the producer's desire to concentrate on production. The retailers purchase supplies from commission agents or contractors. Some sell fish to urban consumers and have permanent shops in the urban markets, generally close to the commission shops.

36.6.7 Forestry Products

The most important utility of forest plants is timber, which is mainly used for construction and furniture purposes. Pakistan has a severe shortage of timber and fuel wood resources. Instead of the desired level of 25% for sustainable economic development, forest resources cover only 4.4% of the total land area in the country, which is insufficient to meet the domestic demand for timber and fuel wood.

There are three main sources of wood, that is, state-owned forests, communal forests, and privately owned forests. Total wood consumption as timber and fuel wood in the country is around 47.73 m^3. Out of this, per capita consumption of fuel wood is approximately 0.205 m^3 per annum. Whereas, fuel wood consumption in brick industry, tobacco curing units, charcoal kilns, etc. are around 3.3% of total fuel wood consumption. Construction industry consumes 20% of the total timber consumption. Fuel wood is basically consumed in three main sectors, that is, domestic, rural, and commercial industrial sector (Zaman and Ahmad, 2011).

Marketing of timber and fuel wood is very complex. All territorial forests have their own sale depots that supply timber and fuel wood. The forest department maintains a record of forest area, growing stocks, harvesting, and auction. Two main public timber markets, that is, Forest Developing Corporation (FDC) and Azad Kashmir Logging and Saw-Mills Corporation (AKLASC) are directly involved in the sale of a major portion of the timber. Whereas, in the private sector, two markets, that is, Dargai in Khyber Pakhtunkhwa and Chilas in Gilgit Baltistan supply the timber-to-timber merchants throughout the country. The furniture industry purchases the timber directly from these timber merchants or wholesalers in their regions.

Marketing of wood is different in rural and urban areas. In rural areas, main sources are farmlands and wild lands. While in urban centers, wholesalers and retailers supply the main source. Increasing demand of timber is usually met through imports and illegal cutting of forest trees. Afghanistan, Malaysia, Singapore, China, Sweden, the United States, France, Italy, Spain are the main countries which supply wood and wood products to Pakistan.

Nontimber Forest Products (NTFPs) are materials that also come from the forests. In Pakistan, important NTFPs are morels, honey, fruits, nuts, condiments, mazri palm, silk cocoon, gums, vegetable tanning, etc. About 34% of local people have indigenous knowledge about collection, processing, drying, packing, and marketing of NTFPs, and are mainly dependent for income generation (Latif and Shinwari, 2005). Some of the NTFPs such as chalghoza, morels, walnuts, and so on are exported.

36.7 EMERGING CHALLENGES

Thegriculture sector in Paksitan is beset with infrastructural and postharvest related problems. Existing agricultural marketing infrastructure is neither adequate nor capable to meet current needs of the country. Some problems in the domain of agricultural marketing are summarized in the following sections.

36.7.1 Lack of Proper and Modern Wholesale Markets

Wholesale markets act as a cardinal link in the marketing chain of agricultural commodities. These markets absorb the bulk of the marketed surplus and are main source of supplies to retailers in the big cities and their surroundings. Wholesale markets are increasingly playing an important role not only as major centers of price formation where coordination between production and marketing takes place, but also as an important place for introducing innovations in the marketing practices. Most of the agricultural produce pass through wholesale markets for onward distribution in the consumption areas and to meet export demand.

At present there are reported over 700 fruit and vegetable wholesale markets in Pakistan. The province of Punjab occupies the largest share followed by Sindh, KPK, and Balochistan respectively. In Balochistan there are two central wholesale markets for fruits and vegetables, one at Quetta and the second at Dera Murad Jamali.

Many wholesale markets were built years ago and are unable to cope efficiently with increased transactions. Serious traffic congestion, insufficient space for efficient movement of products in and out, inadequate storage and improper management are some of the major factors for increased marketing costs and physical losses of farm products. Hygene conditions, particularly in case of fruits, vegetables, and livestock are quite dismal.

Although market committees have been constituted under the provincial statutory laws and are responsible for the smooth administration, operations, management, and development of these markets in respective provinces, their activities are much influenced by political interests. Most market committees are unable to discharge obligations. The sole concern of market committees is to regulate markets. Unfortunately, enforcement of regulations is mostly defective and is to the disadvantage of the entire marketing system.

36.7.2 Lack of Farm to Market Roads and Poor Transportation Facilities

Poor farm-to-market roads are a common features of agricultural marketing systems. These roads are often unusable during rainy months and in some cases during the chilly winter. Current length and status of farm-to-market roads is not satisfactory. The present length of farm-to-market roads (60,000 km) is crucially less than potential requirements (Government of Pakistan, 2009). There is an immediate need to increase these for improving farmer's access to markets. The establishment of all-weather roads, particularly in the remote rural areas should be assigned a priority in the upcoming policy reforms.

High freight is charged by transporters due to poor condition of roads, which ultimately increases marketing costs, largely shared by the consumers and farmers. Nonexistence of good roads limit the use of economical modes of transportation (e.g., trucks). As such, farmers and traders have to rely upon relatively less efficient mode of transportation (e.g., carloads, small vans, etc.). Poor condition of farm-to-market roads is also stumbling block in introducing innovations and new technology some time (replacement of wooden crates with fiberboard boxes in spite of their positive impact on net returns to farmers).

36.7.3 Inadequate Storage Facilities

Inappropriate storage facilities, both in the public and private sector, register the highest losses during handling operations. The perishable farm produce (fruits and vegetables), due to their specific nature and characteristics, require variable storage conditions. In most cases, produce (especially the perishable products) is stored in shallow pits covered with farm wastes without ventilation, without proper sanitation and preventive measures for insect and disease control. These conditions usually exist in on-farm storage houses.

This becomes more important in the light of the recent food crisis (for commodities like wheat, sugar, milk, maize, and meat). As stated earlier, public storage facilities are not sufficient for maintaining a stable supply of agricultural commodities to stakeholders. The investment by the private sector is nominal and can be enhanced by offering special incentives (subsidy in the construction of storages). Additional storage facilities can be established at farm level and also at market levels to avoid shortages (and handle surpluses) of foodstuffs (Table 36.3).

36.7.4 Lack of Modern Cool Chain Infrastructure

Modern cool chain infrastructure is a prerequisite for an efficient agricultural marketing system which is lacking in Pakistan. This is an important contributing in postharvest losses and quality deterioration, which results in price stabilization and loss of foreign exchange earnings. Even the existing cold storage facilities are unevenly distributed among various provinces. These facilities lack blast freezers that bring down temperatures of the produce to a level which can be maintained within the cold store. As a result, the produce is taken directly into the cold store where it loses heat and deteriorates the temperature of the commodities already present in the store. In addition to cold storage facilities, storage mechanisms and structure for other agricultural commodities also need to be evaluated.

TABLE 36.3
Government Storage Capacity ("000" Tonnes)

Agency	2006	2009	2012
1. WHEAT	**4339**	**4339**	**4339**
Provincial	3780	3780	3780
Punjab	2483	2483	2483
Sindh	709	709	709
KPK	365	365	365
Baluchistan	223	223	223
Federal	559	559	559
Food Directorate	–	–	–
AK&NA.	64	64	64
Def. Division	54	54	54
PASSCO	**441**	**441**	**441**
2. RICE	**826**	**826**	**826**
3. Cotton (In 000 Tonnes)	**77**	**77**	**77**
Total Capacity	**5242**	**5242**	**5242**

Source: Government of Pakistan. 2013. *Agricultural Statistics of Pakistan 2012–2013*. Ministry of National Food Security and Research, Islamabad. Pakistan. Agricultural Statistics of Pakistan 2012–2013, Ministry of National Food Security and Research, Islamabad.

36.7.5 Lack of Postharvest Technology and Management

Postharvest losses still remain one of the most pressing problems, particularly for perishables. Despite advances in research, enormous quantitative and qualitative losses still occur. The extent of loss depends on how the commodity is handled from farm to the market. Studies reveal that postharvest losses are greater than production losses. These losses are not due to a single contributory factor but associated with different factors in postharvest operations. Lack of farmers awareness about scientific handling of farm produce, especially the perishables, aggravates this situation. High postharvest losses, if avoided, can contribute in marketed surplus thus increasing returns to farmers and adding to supply, bridging gaps for any shortages.

36.7.6 Processing and Value Addition

Processing of agricultural commodities is performed to add value and prolong life. This is another good option to make the existing supply of agricultural commodities more sustainable. An unfortunate fact about existing status is that only a nominal amount of total production is processed (3% of fruits, vegetables, and milk) in the country. Some fruits are processed into products like jams, jellies, squashes, juices, and pulp. Even many vegetables are processed by extracting moisture/water to prolong their shelf life (e.g., dry vegetables, cutlets, and essence, etc.).

There exists an enormous potential of adding value to various agricultural commodities especially perishables in the country, which can be exploited by inculcating entrepreneurial skills among stakeholders by offering special incentives by the government to the agribusiness entrepreneurs. In addition, the role and working of food processing firms (sugar and flour mills) need to be reviewed and regulated to avoid a food crisis.

36.7.7 Poor Physical Handling of Perishable Products

Typical farm products change hands from four to ten times. Initial handling is done in the field during harvest where the product is subject to various handling operations such as picking, piling, sorting, and packaging. During this process a significant loss of produce occurs. Careless loading and unloading of perishable farm produce also causes heavy losses. As such, while analyzing marketing costs, a significant part of total marketing costs is comprised of produce handling cost.

36.7.8 Inappropriate Packing and Packaging

The types of containers used for transporting and storing products (e.g., fruits and vegetables) vary from place to place. The most popular containers for fruit packing are wooden crates. Irrespective of the structure and properties of the farm products, a common practice is to use whatever container is available. As a result, produce is pressed hard in the crates or carried in oversized containers causing huge losses. Packaging in prescribed containers (corrugated card board boxes) is an international trade norm/international requirement. Currently, the private sector enjoys an exclusive monopoly in the packaging material industry in Pakistan. There is a strong need for offering special incentives to new entrants in this industry.

36.7.9 Nonimplementation of Grades and Standards

By its very nature agricultural produce is characterized by variation in its quality. The specifications for classifying various fruits and vegetables vary and depend upon the nature of the product and requirements of the marketing system. The Department of Agricultural and Livestock Products Marketing and Grading (DAPLMG) and Pakistan Standards and Quality Control Authority (PSQCA) are entrusted with the task of setting of grades and standards and their enforcement. The Pakistan Horticultural Development and Export Company (PHDEC) has recently taken up the responsibility of setting grades and standards for various horticultural products. Although grades and standards for the exportable fruits, such as, mango, apple, and kinnow have been established but not enforced in true letter and spirit. Not only the existing grading system covers few fruits but their enforcement is also poor.

36.7.10 Lack of Agricultural Marketing Information System (AMIS)

Availability of accurate and timely marketing information plays an important role in facilitating the process of transactions. In addition, this information helps in negotiating and establishing prices for the stakeholders. Farmers are handicapped by the lack of reliable information on prices and market conditions. Many farmers take the price dictated by traders or their informal financiers. Even the traders who operate in rural areas are not well informed about the prevailing prices in wholesale markets. Even if the information is available, it is either too late or inaccurate. Information on daily prices and market arrivals are vital for farmers and village traders in planning shipment of their produce and in negotiating prices.

36.8 WAY FORWARD FOR IMPROVING AGRICULTURAL MARKETING SYSTEM

The importance of wholesale markets in the agricultural marketing system of Pakistan needs no emphasis. These markets confront many problems in the sphere of their operations, management, and control. There is strong need to address various problems hindering the proper functioning and development of these markets. As stated most of the wholesale markets in the country were established long ago to cater to needs of the population of that time, over years these markets have lost their utility due to their location and many other inherent problems (size, design, etc.). As such, there is need to establish new model markets fully equipped with requisite facilities. In this context, the role of market committees needs to be redefined and various implementations of clauses of the Agricultural Produce Markets Acts should be ensured. The private sector successfully operates the wholesale markets for poultry and its products. This model may be carefully reviewed and adopted for other commodities with the involvement of the private sector. In addition, marine fisheries markets are required to be upgraded at Gwadar and Karachi to realize full benefits of this sector.

Frequent fluctuations in food commodities (wheat flour, sugar, maize, rice, and milk) have highlighted weaknesses in two areas, that is, storage and market information systems. Available capacity and condition of various storage houses in the country are not satisfactory and consequently result into higher losses of the produce. An effective policy for creating new storage capacity is the need of time. As such, new storage facilities at farm and market levels should be created. Besides, measures should be adopted to establish cold storage facilities on scientific footings. The private sector should be encouraged to invest in this area by offering special incentives (e.g., zero rated imported equipment and tax exemptions). The government should further chalk out a comprehensive plan for establishing cool chain networks for perishables. The Punjab Agricultural Marketing Company (PAMCO) has already established one cold storage house at Lahore airport. More facilities on this pattern may be offered after carefully reviewing the merits and limitations of this experiment. Cool chain networks for other perishable commodities (e.g., milk, meat, fruits, and vegetables) and a plan in this respect should be clearly chalked out. This situation has underlined the need for preparing a comprehensive plan to cope with the surplus and shortage cycles manipulated by the stakeholders or created due to disequilibriums in the supply and demand. Amongst others, lack of market information available to stakeholders makes them dependent on market players who exploit them under one context or the other. An effective Agricultural Market Information System (AMIS) should be established by linking wholesale markets with the major producing areas through Internet and other electronic media.

Proper infrastructural facilities are the backbone of an efficient agricultural marketing system. Current farm to market road network is not in a good condition. As such, full coverage of rural areas of the country is not ensured. The government of Punjab's initiative to improve rural–urban linkages through farm-to-market roads is a good initiative but the need of the time is to look critically at its implementation as these roads should be built based on need assessments of the area. This programneeds further expansion as well.

The bulk of agricultural commodities in Pakistan are traded in their fresh/raw form. There is a need to change the mindset of stakeholders. New agribusinesses should be promoted for adding value to various commodities through processing at the farmers door steps. In this context, the Punjab Agriculture and Meat Company (PAMCO) and PHDEC may be assigned the task to accomplish this objective.

Presently, the status of human capital in the agriculture sector is not satisfactory. Farmers and other stakeholders in the supply chain are not fully equipped with technical know-how and skills required for performing various marketing functions efficiently. New agricultural marketing and postharvest management institutions should be established with the mandate to impart technical training to the stakeholders on various aspects of agricultural marketing and postharvest management. The role of the Technical and Vocational Training Authority (TEVTA) needs to be redefined in this regard. Special training programs in the area of agricultural marketing and postharvest management should be entrusted to TEVTA with a supervisory role assigned to the agriculture universities of the country in this regard.

In addition, the Department of Agricultural Extension in all provinces should be reorganized and a new mandate assigned keeping in view the emerging challenges in the field of agriculture and international trade of agricultural commodities.

Commodity boards were established for various agricultural commodities in the past but were dissolved due to mismanagement and malfunctioning. Commodity boards for major agricultural commodities may be reestablished with the participation and involvement of the private sector. Keeping in mind that in the past weaknesses in the conduct and operations of such boards may be overcome and the new institutions run on sound business footings for the welfare of the farming community and other stakeholders.

Farmer's cooperatives were established in various spheres of economic activity but experience with their working in the agriculture sector did not yield good results. Nevertheless, the idea still holds its validity in many countries having almost similar socioeconomic and cultural traits. It is suggested that farmers' cooperatives may be organized avoiding past mistakes. In this respect, models adopted by Nestlé in Pakistan and Amul in India can be a good starting point.

There is a need to analyze various impediments coming in the way of various supply chains of agricultural commodities. Research may be undertaken at various agricultural universities to develop appropriate feasible supply chain models for agricultural commodities, which will enhance interaction among stakeholders and improve their working efficiency.

There is lack of effective coordination between various institutions entrusted with the role of strengthening operations of agricultural marketing and postharvest management systems in the country. For instance, three institutions (DALPMG, PSQCA and PHDEC) have the mandate of establishing grades and standards for agricultural commodities. Policy measures should be adopted to enhance coordination between these institutions and to avoid any duplication and overlapping of tasks assigned to each organization.

The Planning Commission, Government of Pakistan seeks to establish agro processing centers, for supplying farm inputs to stakeholders and to undertake the task of marketing of their produce. It is sound proposal. 200 agro processing centers as proposed should be established across the country. The PAMCO along with the PHDEC may be assigned the task of establishing these centers in collaboration with the private sector. In this regard, pilot projects could be started at district level after reviewing the potential requirements of stakeholders in the production areas.

There is a dire need to introduce market-oriented agricultural practices. This necessitates for inculcation of entrepreneurial skills among stakeholders. Furthermore, agribusiness incubators should be established at different agribusiness clusters to provide farming community and stakeholders with required information and new business ideas. The incubators should offer necessary technical advice managerial know-how, information, and training in marketing management, advertising and sales promotions, branding and labeling, and so on to enable stakeholders to earn more profit.

There is a strong need for undertaking research on current and emerging problems in the field of agricultural marketing infrastructure and postharvest management. Agricultural universities in the country may be entrusted with this task. The private sector should be involved in research efforts undertaken at universities that are result oriented and close to reality. The private sector should also be motivated to invest in research on these areas.

Diversification of agriculture is inevitable given the emerging trends, challenges, and requirements of international trade in the context of the WTO. New agriculture ventures (floriculture, agriculture along with new avenues for value addition of agricultural and livestock products, etc.) may be identified and priorities assigned in National Plans in this regard.

The livestock sector has huge potential which could be exploited if proper investment is directed towards this sector. Policy measures should be introduced to strengthen the dairy industry in the country. The Punjab Dairy Development Company is already performing a good task by establishing Model Dairy farms in the province. There is a need to extend this initiative to other provinces as well. In this regard, import of machinery and breeds should be declared zero-rated in the upcoming Tenth Five Year Plan. There is also a need to reorganize the wholesale cattle and buffalo markets keeping in view the emerging market requirements. The hygienic conditions of various slaughterhouses need improvement. There is a need for establishing slaughterhouses on scientific footings.

A great export potential of agro-based products, particularly fruits, vegetables, and livestock products exists in Pakistan, but stringent application of international standards hampers realization of their potential. In particular, Pakistan' agricultural exports suffer due to the application of Sanitary and Phytosanitary (SPS) measures. In certain circumstances, SPS requirements are incompatible with prevailing systems of production and marketing in Pakistan because the needed resources, expertise, knowledge, and infrastructure for SPS compliance is lacking. There is a need to impart training to stakeholders and prepare them to comply with SPS measures in the production and export of agricultural products.

36.9 SOME POTENTIAL INTERVENTIONS

Irrespective of some poor performances, there are some innovative practices going on in the field of agricultural marketing, which mainly includes the following.

36.9.1 Model Cattle Markets

The livestock sector plays an important role in the economy of Pakistan. It contributes to a major part of the agriculture value added and the value of livestock is more than the combined value of major and minor crops. The major animals of Pakistan comprised of cattle, buffalo, sheep, goat, camels, horses, asses and mules. In Pakistan, livestock marketing is carried out at thousands of small rural markets called cattle markets, which operate around the larger urban areas. Most of these markets are regulated to some extent by the local authorities. Often, the market rights vested with local authorities to manage and collect fees are sold to private contractors by tender. Typically, the sale is handled through a commission agent with the price based on the trader's estimate of the likely carcass weight of the live animal and his knowledge of the wholesale meat prices.

Traditional livestock markets have been functioning under highly pathetic conditions and as a result many livestock farmers and traders hesitate to bring their animals to these markets. These markets are loosely controlled and provide few facilities for orderly marketing of livestock. Most livestock markets across the country are devoid of any infrastructure or facilities like shade, shelter, sanitation, drinking water. Mostly, no facilities are available for weighing the animals. In fact, the entire marketing of live animals is on a unit basis with weighing only at the meat wholesale and retailing stages. Besides these, security issues such as snatching of animals or cash, kidnapping of buyers, and attacks on sellers are other major problems associated with cattle markets. It was therefore a much needed desire of the people related to livestock market to establish new and innovative business models for livestock markets.

The government of Punjab, sensing the needs of stakeholders, has introduced Model Cattle Markets, which to some extent are supposed to address the livestock stakeholders' concerns. The Punjab Livestock Breeding Act 2014 was notified to regulate the livestock breeding services, improve genetic potential of breeds and build cattle markets along modern lines. Initially, these model markets are established in divisional headquarters of the province. This model has replaced the word livestock with cattle, where the term cattle includes all the animals like cows, buffaloes, horses, donkeys, camels, sheep, and goats, etc. These cattle markets are being reorganized and upgraded under the umbrella of a model private-public enterprise called the Cattle Market Management Company (CMMC). The chief minister of Punjab has already inaugurated the Sheikhupura Model Cattle Market in 2015. All the basic facilities are provided by the Lahore Division Cattle Market Management Company (LDCMMC) in this market. The Faisalabad Cattle Market Management Company (FCMMC) has also been established but this cattle market, though developed has yet to be started.

The objectives of establishing these companies include elimination of the commission mafia and abolition of outdated, exploitative systems prevailing in cattle markets. These companies are working for construction of other markets along scientific lines and introduce a simple and effective method of buying and selling. The sellers will be provided all sorts of facilities while the number of staff and officers would be kept at the minimum required level. A comprehensive strategy is being devised to bring about reforms in the system. No tax or municipality fee will be charged from those visiting the market. Other major features of these new model markets include

- No contractor
- No middlemen
- No role of TMA
- Open for business 6 days a week
- Provision of allied services

- E- Tagging of animals
- Trading of disease-free animals
- Complaint redressal system
- Data management
- Fully computerized trading transactions
- Professional management (company managed)

The government also intends to provide other necessary and free of cost services to stakeholders in these markets. These services include veterinary services, waste management services, security services, and slaughtering services. In addition, other services like fire brigades, Rescue 1122 and human dispensary will also be provided. The government will also build a small slaughterhouse and parking stand based on the model cattle market of Faisalabad.

36.9.2 Online Trading of Agricultural Commodities

The agricultural marketing system in Pakistan is facing numerous challenges, which mainly revolve around the welfare concerns of farmers and consumers. Two major issues in this context are fluctuating prices and frequent shortages/surpluses. An effort is made to address these issues from the Pakistan Mercantile Exchange Limited (PMEX) by devising an online trading model for chilies in Kunri. This initiative is regulated by the Security and Exchange Commission of Pakistan (SECP) and partnered by the Pakistan Agriculture Coalition (PAC), Agility, and SGS. PAC is involved in assembling, trading, and maintaining records whereas Agility is responsible for logistics and SGS for certification of red chilies. Other partners involved in this model include the National Bank of Pakistan (NBP), Pakistan Stock Exchange (PSX), Pak–Kuwait Investment Company (PKIC), Zarai Tarqiati Bank Limited (ZTBL), and Sindh Investment Board (SIB).

In this online trading system, they are selling chilies by grades, which are divided into A-plus, A, and B grade chilies. Through this model, they differentiate themselves from traditional agricultural marketing and claim further to provide the following services;

- Quality and certified products
- Warehousing and logistics
- Fair prices
- Helping farmers to produce quality products
- Payment to farmers in few days
- Initial screening at farm gate
- About 7% lower transactions cost to farmers compared with the traditional market
- Extension services to farmers

They are further planning to conduct the feasibility studies of wheat, rice, maize, mangoes, citrus, guava, and potatoes to extend the scope of online trading of agricultural commodities in Pakistan. Though this initiative tries to promote the participation of the private sector, there are some question marks on the scope of the private sector such as provision of required infrastructure, extension services, and quality seed. There are some issues regarding the sustainability of these ideas in light of limited access of farmers to IT services and involvement of farmers in traditional marketing system. However, overcoming limitations this can be a good initiative in solving the prevailing problems of the agricultural marketing in the country.

36.9.3 Vegetable Trading Platforms

A majority of farmers in Pakistan are small farmers having minimum resources to cultivate their lands. Their farming system is relatively small as compared to big farmers and as such they produce small marketable surpluses. As they have small amounts to sell in the market, they have low bargaining power while negotiating terms of trade and often are exploited by the market actors and forces because they have little staying power due to ever increasing needs of finance, especially for the next cropping season (Ali, 2000). As a result they have to sell their produce at relatively low prices and sometimes at further deteriorating terms of trade.

One potential solution to such problems is to motivate these small farmers to pool their resources and work jointly to enhance their bargaining power and eventually their welfare. Marketing cooperatives are the solution to such problems. But this initiative did not fetch a significant success in Pakistan though we see some good and successful case studies as well.

Sensing the ground realities and needs of small farmers, the governement of Punjab has taken an initiative to promote vegetable production and consumption in the province. This initiative worked in three stages; stage one was devoted to promote kitchen gardening of vegetables and also commercial production of vegetable by supplying certified seed. Efforts were made to promote vegetable processing and the government announced a 50% subsidy on vegetable value addition machinery. In third stage, efforts were made to connect with markets through vegetable trading platforms.

Vegetable trading platforms are an initiative taken by the government to introduce the concepts of self-help and collective action to enhance the bargaining power of small farmers. Initially, 23 districts were selected from the province for this experiment. The government has decided to give a 50% subsidy on this project where the rest of the 50% will be shared by the participants. These platforms will be managed by farmers' committee, which will be the elected body for 3 years. The committee will consist of 25 members having a president, general and finance secretaries. All these efforts are meant to make farmers self-reliant and free from exploitation of market middlemen. This initiative is in process and hopefully will contribute to empowering small farmers in the province.

REFERENCES

Aftab, S. 2007. *Retail Markets*. Ministry of Commerce, Islamabad. Pakistan.

Ali, M. 2000. *Requirements and Conditions for Perishable Products for Domestic and Export Markets: View of a Trader*. Universal Traders (Importers and exporters), Quetta, Baluchistan, Pakistan.

Chaudhary, M. and A. Ahmad. 2000. *Food Grain Losses at Farm Level in Pakistan. Vol. 1. Department of Agricultural Marketing*. University of Agriculture, Faisalabad.

FAOSTAT. 2012. Food and Agriculture Organization Online Data Base available at http://faostat.fao.org

Government of Punjab. 2006. *Agricultural Marketing System in Punjab*. Publication No. 01/2006. Directorate of Agriculture (Economic & Marketing). Government of Punjab, Lahore.

Government of Pakistan. 2009. *PC -1 Cool Chain System*. Government of Pakistan.

Government of Pakistan. 2013. *Agricultural Statistics of Pakistan 2012–2013*. Ministry of National Food Security and Research, Islamabad. Pakistan.

Government of Pakistan. 2015. *Economic Survey of Pakistan (2014–2015)*. Economic Advisor's Wing, Ministry of Finance, Islamabad.

Khushk, A. M. and A. D. Sheikh. 2004. *Structure, Conduct and Performance of the Marketing Systems Margins and Seasonal Price Variations of Selected Fruits and Vegetables in Pakistan*. PARC, Islamabad.

Kurosaki, T. 1996. Government intervention, market integration and price risk in Pakistan's Punjab. *The Pakistan Development Review*, 352(Summer): 129–144.

Latif, A. and Z. K. Shinwari. 2005. Sustainable Market Development for Non Timber Forest Products in Pakistan. Article Retrieved from www.researchgate.net

Mohy-ud-Din, Q. and H. Badar. 2011. *Marketing of Agricultural Products in Pakistan: Theory and Practice*. Higher Education Commission, Government of Pakistan.

Pakistan Horticulture Development and Export Board (PHDEB). 2005. *Mango Marketing Strategy*. Ministry of Commerce, Pakistan.

Zaman, S. B. and S. Ahmad. 2011. *Wood Supply and Demand Analysis in Pakistan: Key Issues*. Pakistan Agricultural Research Council, Islamabad, Pakistan.

37 Regional Trade
Pakistan's Perspective

Burhan Ahmad, Abdul Ghafoor, and Asif Maqbool

CONTENTS

37.1 INTRODUCTION

Different regions specialize in different commodities based upon their comparative and competitive advantage and export them which can contribute to the positive balance of trade and producea surplus balance of payments. Regions trade with each other as well as with rest of the world. Regional trade is expected to be more open and greater than the trade with rest of the world because of the smaller distances involved, low trade barriers, and smaller differences among cultures. Neighboring or geographically proximate countries should create trade blocks because these are expected to "create trade" which in turn can hoist welfare gains. Trade creation can be attributed to higher expected trade volumes because the countries are closely situated to each other. The European Union is a good example of such a trade creation. However, distortions do exist because of the regional trade policy imposing trade barriers and leads to low volume of trade, however, this is not the only factor limiting trade. South Asian Association for Regional Cooperation (SAARC)/South Asian Free Trade Area (SAFTA) is an example of low volumes of trade between the member countries. Trade liberalization policies are adopted to reduce these barriers and enhance the volume of trade.

Trade liberalization policies can be implemented in various forms, however, the simplest form is to reduce the applied tariff levels unilaterally but normally other trade partners also lower the trade barriers. Such kind of trade liberalization can take the form of multilateral agreements as laid down by the WTO in which reduction of tariffs is applicable to all the member countries. In the other forms of trade liberalization, a few countries make an agreement to reduce the trade barriers particularly the tariffs. These are termed preferential trade agreement (PTA). The salient feature of the PTA is that low levels of tariffs are applicable to the commodities of the member countries of the group (subset of nations) while higher levels of tariffs are imposed on the commodities from outside

(Panagariya 2000). The subset of nations is usually situated close to each other having geographical proximity, hence liberalization agreements are termed as regional trade agreements (RTA). Nearly all PTAs are the agreements between the nations of a certain region for reducing the trade barriers. WTO permits following three types of PTAs:

1. Custom unions and free trade agreements allowed under Article XXIV
2. Agreement between developing economies that permit partial preferential treatment
3. Agreements under the Generalized System of Preferences (GSP) permitting preferential treatment to developing economies from developed economies.

Above permission under WTO encouraged trade arrangements among the regions and closely situated countries, however, expected benefits in the form of welfare gains of member countries has been debated in the literature, for example, Panagariya (1999, 2000) and Clausing (2001). The creation of Preferential and Free Trade Areas has resulted in the application of different tariff rates for same product imported from different countries, which may not reap the true benefits of trade liberalization.

Free trade agreements and regional integration plays a vital role in the economic development and improving living standard of the people through enhancing regional trade. The E.U., the North American Free Trade Agreement (NAFTA), and the Association of South East Asian Nations (ASEAN) are the example of the success in this regard. The seven member countries of the SAARC also signed the South Asia Free Trade Agreement (SAFTA) at Islamabad (Pakistan) on 6th January 2004, which was implemented on 6th July 2006. But till to date, it could not reap its expected fruits as the intraregional trade accounts for only 5% while intraregional trade in ASEAN is about 25%. Moreover, Pakistan trade with other countries in Asia is very minimal. For example, in 2014 China imported goods worth USD 2 trillion globally in which Pakistan's share was just 0.1%. The share (0.1%) in India and Iran's global imports is similar. Two major partners in the gulf, Saudi Arabia and the UAE, receive about 0.4% of their global imports from Pakistan. In the same year, China exported about USD 2.3 trillion in which share of Pakistan was about 0.4%. Similarly, low shares exist for other mentioned countries (ITC 2016). Given the importance of regional trade and low trade shares of Pakistan with the countries in the nearby regions, the present chapter is intended to examine the bilateral trade relations of Pakistan with Afghanistan, China, India, Iran, Saudi Arabia, and the UAE and try to find out ways to increase the volume of trade with these countries.

After this introductory section of this chapter, the next section describes the strategic, economic, and political importance of Pakistan. Section 37.3 depicts the Pakistan's trade scenario while Section 37.4 onward describe the bilateral trade relations of Pakistan with Afghanistan, China, India, Iran, Saudi Arabia, and the UAE. Last section presents summary and conclusions.

37.2 STRATEGIC AND GEO-POLITICAL IMPORTANCE OF PAKISTAN

Pakistan is located in Asian region, which possesses enormous economic, political, and strategic significance. The geopolitical location of Pakistan is strategically important and enormously valuable. Following are some of the main characteristics indicating the strategic and geopolitical importance of Pakistan.

- Pakistan is situated in proximity of big and emerging powers like China and India.
- It is a gateway to Central Asian States and collectively this is emerging as a potential economic hub of the world.
- Pakistan is also emerging as transit economy and considering its importance and potential, it earns a good reputation in the world.
- Pakistan is the only Muslim country, which possesses atomic power, and at the same time it is an important member of the Muslim world.
- Pakistan has been endowed with huge natural resources, which have enhanced its importance as emerging economy.

- Considering recent changes in global politics, like increasing threat of terrorism, economic exploitation and regional politics, Pakistan has got attention of global peace seekers.
- The world especially Asian region has witnessed important economic reforms like CPEC, which not only opens opportunities for Pakistan but also alarms its competitors.
- In global logistics and trading systems seaports have got supreme importance and luckily Pakistan has two potential ports which enhance its importance in the region.

37.3 PAKISTAN'S TRADE SCENARIO

Apart from few years since its existence, Pakistan has remained a trade deficit country and this deficit has been increasing over decades. Pakistan's major exports have been concentrated in a few export items, comprised mostly of cotton manufactures, rice, leather, fish and fish preparations and sporting goods. Pakistan's exports markets are also limited to few countries. About fifty percent Pakistan's exports' are destined to seven countries, which are the United States, the UAE, China, the United Kingdom, Afghanistan, Iran, and Germany. Similar to exports and export markets, Pakistan's imports and import markets are also concentrated in few products and markets respectively. This section presents a historical view of Pakistan's trade (GOP 2016, SBP 2010).

37.3.1 Pakistan's Exports

Pakistan's major exports have been comprised of few export items, which include cotton and its manufactures, rice, leather, fish and fish preparations and sporting goods. Table 37.1 provides decade-wise share of exports of these major commodities as a share of the total value of exports from Pakistan. During the 1960s, these commodities contributed to about 39% of the total value of exports from Pakistan while this share almost doubled to 70% in the 1970s and then decreased to 62% in the 1980s. Textile manufacturers have been the largest export items ranging from 59.5% to 64.5% of total value of exports during the 1990s and 2000s.

Rice has remained Pakistan's second largest export item after cotton and cotton products and contributes (SBP 2010). The share of rice exports in the total value of exports of Pakistan increased

TABLE 37.1
Share of Exports of Various Commodities in Total Value of Exports of Pakistan (%)

Exports	1960s	1970s	1980s	1990s	2000s
Fish & fish preparations	3.02	2.58	2.57	1.88	1.20
Rice	4.80	18.12	10.04	5.71	7.81
Cotton	11.80	10.19	13.37	3.03	0.58
Leather	3.31	5.39	5.31		
Textile yarn and thread	5.06	12.37	11.83	16.38	8.07
Cotton fabrics	5.37	12.40	10.61	13.36	11.64
Sports goods	0.72	1.50	1.59	3.06	2.09
Sub-total	34.08	62.55	55.32	43.41	31.39
Other commodities	65.92	37.45	44.68	56.59	68.61
Total	100.00	100.00	100.00	100.00	100.00
Primary commodities				13.56	12.47
Textile manufactures				64.49	59.50
Other manufactures				14.14	18.44

Source: State Bank of Pakistan (SBP). 2010. *Handbook of Statistics on Pakistan Economy 2010*. http://www.sbp.org.pk/departments/stats/PakEconomy_HandBook/Chap-8.1.pdf

from 5% in the 1960s to 18% in the 1970s (SBP 2010). This can be attributed to the green revolution in the 1960s and the increase in supply of water in the early 1970s. This share steadily decreased from 10% and to 6% in the 1980s and 1990s, respectively, before recovering slightly to 8% in the 2000s (UN FAO 2012; GOP 2013). Nevertheless, it captures a large share of almost two-thirds of the value of exports of all primary commodities. Rice production accounts for almost 6% of the value added in agriculture, while contributing to 1.3% of GDP (SBP 2010; UN FAO 2012; GOP 2016).

Still Pakistan's exports are composed of cotton and its manufactures, leather, rice, chemicals and pharmaceutical products and sports goods. These five categories of exports accounted for about 70% of total exports during financial year 2016 (Jul-March) with about 60% contribution of cotton and its manufactures. Rice remained the second highest export item from Pakistan contributing to about 9% to Pakistan's total exports followed by leather which contributed about 4.6% (SBP 2010; UN FAO 2012; GOP 2016).

In addition to Pakistan's exports being concentrated on a few products, the markets to which they are exported are also concentrated indicating potential to identify and explore more markets to increase exports. The United States, the United Kingdom, Germany, France, Japan, China, Hong Kong, Kuwai,t and Saudi Arabia have been the major export markets since the 1960s. Table 37.2 exhibits shares of major export markets of Pakistan in total value of exports of Pakistan. These markets accounted for between 45% and 48% of the total value of Pakistan's exports from the 1960s through the 1980s. This share increased to 50% on average during the 1990s and the 2000s. The UAE, Italy, and the Netherlands are the other major markets. These markets, together with the UAE, Italy, and the Netherlands, amounted to 62% of the total value of Pakistan's exports. The United States alone has remained the single biggest market for the last 2 decades, capturing a share of between 17% and 23%. This share increased from 10% in the 1960s (SBP 2010). The concentration of Pakistan's exports in few markets and fluctuating behavior reflects the lack of success in identifying and exploring new markets as well as sustaining the existing markets.

Still Pakistan's exports are destined in few countries such as the United States, the United Kingdom, China, Afghanistan, and the UAE. Among these countries, the maximum contribution in Pakistan's export earnings during financial year 2016 came from The United States which was

TABLE 37.2
Export Markets Shares in Total Value of Exports of Pakistan (%)

	1960s	1970s	1980s	1990s	2000s
United States	9.74	5.64	10.75	17.02	23.28
France	3.32	2.22	3.03	3.49	2.43
Germany	3.88	5.18	2.88	5.70	4.76
United Kingdom	13.19	6.74	6.18	6.94	6.18
China	4.61	2.61	2.74	1.34	3.14
Hong Kong	4.84	8.73	3.86	7.15	4.23
Japan	6.01	9.00	9.78	6.07	1.26
Kuwait	1.17	2.02	1.17	0.53	0.54
Saudi Arabia	1.08	4.47	4.87	3.12	2.71
Sub-total	47.84	46.61	45.26	51.36	48.52
Singapore				1.22	0.55
UAE				4.19	7.92
Italy				4.01	3.26
Netherlands				3.13	2.53
Total				63.90	62.77

Source: State Bank of Pakistan (SBP). 2010. *Handbook of Statistics on Pakistan Economy 2010.* http://www.sbp.org.pk/departments/stats/PakEconomy_HandBook/Chap-8.1.pdf

about 17% while share of China export market remained about eight percent making it second largest export market. China's share in total exports has gradually increased from four percent in 2008–2009 to 10% during financial year 2016 (SBP 2010; UN FAO 2012; GOP 2016).

37.3.2 Pakistan's Imports

Pakistan's imports persistently increase especially in petroleum and petroleum related products. Other include: machinery, trucks, automobiles, computers, computer parts, civilian aircraft, defense equipments, iron, steel, aluminum wrought and worked, rubber tires and tubes, wood and cork, paper and paper board, agri and other chemicals, pharmaceutical products, food items, and other consumer items (SBP 2010; GOP 2016).

During the 1960s, 1970s, and 1980s imports shares of the broadly categorized commodities such as food and live animals, beverages and tobacco, crude materials, minerals, fuels, lubricants, animals and vegetables oils, chemicals, manufactured good, machinery and transport equipments and miscellaneous, as a share of the total imports value of Pakistan. During the 1960s, machinery and transport equipments are the dominating imports with 34.59% of the total value of imports of Pakistan. In the 1970s and 1980s, machinery and transport equipments remained dominating imports with 26.67% and 26.80% of the total value of imports of Pakistan respectively. The imports of manufacturing goods and beverages and tobacco declined from 21.83% to13.08%, and 0.35 to 0.04%, respectively in three decades. The imports of chemicals increased to 13.35% in 1970s and then decreased to 12.95% in 1980s. Throughout the three decades, the imports of crude materials, and minerals, fuels and lubricants represented the increasing trends (SBP 2010; GOP 2016).

During 1990–1994, machinery and transport equipments remained the biggest imports items expensed 31.70% of total value of imports. The imports of chemicals also mounted to 16.32% of total imports during 1990–1994, whereas during 1995–1999, the imports of machinery recorded highest shares in total imports of Pakistan. Agricultural and other chemicals' imports were found second largest import products after machinery during this span of time. During 2000–2004, the imports of Petroleum rose and got the highest imported product with 26.22% shares of the total value of imports. The shares of machinery imported were declined to 22.23% and recorded as second largest import item. During the period of 2005–2009, the imports of petroleum remain the biggest import product with 24.56% share in the total value of imports. The imports of machinery gradually declined to 20.65% and remained the second largest import item. The shares of metal imports in the total value of imports of Pakistan ascended 7.74% during 2005–2010. Currently major import items more or less same which include the food group (2.4%), machinery group (4.8%), Petroleum group (37.2%) (SBP 2010; GOP 2016).

Similar to the characteristics of exports of Pakistan, Pakistan's imports are also limited to few countries. Around 50% of Pakistan's imports came from only four countries, which include China, UAE, Saudi Arabia, and Kuwait. Other major import markets of Pakistan are Malaysia, Japan, India, and the United States. It is important to mention here that China has emerged as Pakistan's major trading partner both with respect to exports as well as imports (SBP 2010; GOP 2016).

37.4 PAK–AFGHAN TRADE

Pakistan and Afghanistan are neighboring states and are the members of the South Asian Association for Regional Cooperation (SAARC). Under the war against terrorism, the United States has declared them as major nonNATO allies. The history of relations between the two countries comprise various complexities such as issues related to the Durand Line, water and most importantly the War since the 1970s to date. However, the two states are working together to find solutions to these problems through possible defense cooperation and sharing information through sharing the intelligence, enhancing the two-way trade and uplifting the restriction of visas for diplomats from the two nations.

Despite the membership of SAARC and signatory of SAFTA, both countries have bilateral trade relations. In July 2010, both countries signed a memorandum of understanding (MoU) for

TABLE 37.3
Pakistan's Trade Balance with Afghanistan (Million USD)

Year	2005	2006	2007	2008	2009	2010	2011	2012	2013	2014	Sum 05–14	Average
Exports	1064.7	991.5	837.7	1447.6	1373.9	1684.7	2660.3	2099.3	1998.1	1879.1	16036.9	1603.7
Imports	53.2	64.9	89.5	85.5	121.2	138.4	199.5	235.1	307.6	392.2	1687.1	168.7
Trade balance	1011.5	926.6	748.2	1362.1	1252.7	1546.3	2460.8	1864.2	1690.5	1487.0	14349.8	1435.0
Trade balance (billion USD)	1.0	0.9	0.7	1.4	1.3	1.5	2.5	1.9	1.7	1.5	14.3	1.4

Source: International Trade Center (ITC) (2016). Trade Map, Market Analysis and research, ITC. http://www.trademap.org/

the Afghan-Pak Transit Trade Agreement (APTTA) and the commerce ministers of both countries signed the APTTA agreement in October 2010. Under the APTTA, Afghan trucks can be driven inside Pakistan toward the Wagah border with India and Karachi and Gwadar which are port cities. In November 2010, a joint chamber of commerce was created between the two countries to strengthen the trade relations between two countries and resolve the issues of traders. In July 2012, both countries decided to extend APTTA to Tajikistan and according to the proposed agreement Tajikistan will be facilitated Pakistan's ports at Gwadar and Karachi ports for the trade of commodities. This will be the first step for the establishment of a North-South trade corridor.

Pakistan and Afghanistan have been historic trade partners having both formal and informal trade. The informal trade can be attributed to the porous nature of the borders between them. Table 37.1 presents the data on Pakistan's exports, imports and trade balance with Afghanistan. During 2005–2014 Pakistan's exports to Afghanistan were amounted to about USD 1.6 billion while imports accounted for about USD 0.17 billion leading to a positive balance of trade of about USD 1.4 billion during the same period. Pakistan maintained a positive balance of trade throughout the above-mentioned period of recent 10 years 2005–2014. The exports of Pakistan to Afghanistan gained a pace in 2010 reflecting some effects of bilateral trade agreements mentioned above, however, fluctuations in both exports and imports can be visualized from the Table 37.3. Despite the fluctuation, an increasing trend in both exports and imports is also depicted in Table 37.4. The total formal trade volume between Pakistan and Afghanistan reached to USD 2.3 billion in 2014. Estimates indicate a total trade of about USD 5 billion including both formal and informal between Pakistan and Afghanistan (ITC 2016).

Table 37.4 presents the major export commodities of Pakistan for the afghan markets while Table 37.5 presents the major imports from Afghanistan. Major export items include mineral fuels, oils, distillation products; salt, sulphur, earth, stone, plaster, lime and cement; animal, vegetable fats and oils, cleavage products, etc; milling products, malt, starches, wheat gluten; cereals; articles of iron or steel; sugars and sugar confectionery; plastics and articles thereof; edible vegetables and certain roots and tubers; edible fruit, nuts, peel of citrus fruit, melons; and dairy products, eggs, honey and edible animal products. The major import items are comprised of cotton; iron and steel, edible fruit, nuts, peel of citrus fruit, melons; Mineral fuels, oils, distillation products, etc; edible vegetables and certain roots and tubers; salt, sulphur, earth, stone, plaster, lime and cement; wood and articles of wood, wood charcoal; raw hides and skins (other than fur skins) and leather; oil seed, fruits, grain, seed, fruit; carpets and other textile floor coverings; machinery, nuclear reactors, boilers.

37.5 PAK–CHINA TRADE

Pakistan and China relations started in 1950 and both countries possess a well-built friendship that has gone through geopolitical and economic challenges. Pakistan was one of the three countries that recognized Peoples Republic of China and both established embassies in their capitals in

TABLE 37.4
Pakistan's Exports to Afghanistan (Million USD)

Year	2005	2006	2007	2008	2009	2010	2011	2012	2013	2014
All products	1064.7	991.5	837.7	1447.6	1373.9	1684.7	2660.3	2099.3	1998.1	1879.1
Mineral fuels, oils, distillation products, etc.	253.6	278.5	248.8	467.1	394.5	662.7	776.3	5.1	25.5	108.3
Salt, sulfur, earth, stone, plaster, lime, and cement	84.3	85.1	98.9	151.5	149.4	185.3	244.5	322.1	282.7	240.2
Animal, vegetable fats and oils, cleavage products, etc.	96.8	99.4	105.5	164.5	93.9	81.7	190.4	218.6	152.5	115.1
Milling products, malt, starches, inulin, wheat gluten	101.5	120.9	94.1	4.4	3.3	38.8	228.5	227.3	196.9	195.4
Cereals	37.5	50.1	22.7	234.6	148.6	96.1	184.3	171.0	145.6	98.0
Articles of iron or steel	62.5	34.3	56.5	70.9	85.1	103.2	152.2	195.7	108.4	111.7
Sugars and sugar confectionery	38.4	28.5	10.3	72.8	18.3	18.9	28.3	75.9	202.4	250.0
Plastics and articles thereof	104.0	72.7	33.2	48.8	69.2	62.9	96.5	112.9	92.8	47.4
Edible vegetables and certain roots and tubers	7.7	10.6	5.5	8.3	43.8	60.9	165.3	136.6	143.0	95.7
Edible fruit, nuts, peel of citrus fruit, melons	11.6	15.7	6.1	9.5	41.6	71.8	105.5	112.9	108.5	101.8
Dairy products, eggs, honey, edible animal product	16.9	23.9	28.4	29.4	39.0	42.6	68.1	84.8	94.9	79.3

Source: International Trade Center (ITC) (2016). Trade Map, Market Analysis and Research, ITC. http://www.trademap.org/

1951. Keeping good friendly relations with China has remained a central part of Pakistan's policy and similar is the case with Chinese policies. Bilateral relations between the two countries have developed through neutrality policy of China to a partnership with a militarily powerful Pakistan. Diplomatic relations go back to 1950 while a military cooperation started in 1966. In 1972, formation of a strategic alliance was completed while economic cooperation started in 1979.

On the trade front, Pakistan accorded the Most Favored Nation (MFN) status to China in 1963 through a bilateral trade agreement, which provided a level playing field to Chinese products in Pakistan. Since then, both countries have closely collaborated to deepen economic and trade relations. Bilateral trade links were further boosted during the visit of Pakistan's then president to China in November 2002 when both countries agreed to sign a Preferential Trade Agreement (PTA) that provided tariff preference to a limited number of products. The process remained moving forward and resulted in signing a Free Trade Agreement (FTA) July 2007. The FTA with China envisaged the gradual liberalization of tariff on goods in various phases. The first phase was from 2007 to 2011 in which both sides agreed to abolish tariff on 30% of the products and liberalization of tariff was envisaged up to 90% of the goods in the second phase.

The tariff concessions committed by Pakistan in the first phase of the FTA predominantly contained raw materials and intermediary goods. In return, China eliminated tariff on finished goods. In 2008, an incentive package for Chinese investors was included in the FTA to facilitate Chinese FDI in China specific investment zones in Pakistan along with an agreement on trade in services leading to a framework for trade integration, market access on services and promotion of investments. The first phase negotiations on liberalization of tariffs have already been concluded in 2012 while negotiations for the second phase are going on. However, other side of the picture expects a high competition of Chinese products with domestic products and loss in government revenue.

TABLE 37.5
Pakistan' Imports from Afghanistan (Million USD)

Year	2005	2006	2007	2008	2009	2010	2011	2012	2013	2014
All products	53.2	64.9	89.5	85.5	121.2	138.4	199.5	235.1	307.6	392.2
Cotton	1.2	9.5	12.9	9.3	40.7	60.6	30.6	68.5	126.7	69.6
Iron and steel	22.5	8.2	22.4	23.9	29.6	30.4	69.3	46.8	41.8	84.5
Edible fruit, nuts, peel of citrus fruit, melons	11.1	16.6	19.5	12.8	17.0	13.3	22.9	31.9	34.1	72.6
Mineral fuels, oils, distillation products, etc.	1.2	0.2	0.0	8.7	4.1	1.9	36.8	59.6	55.7	88.5
Edible vegetables and certain roots and tubers	6.2	14.5	16.5	20.5	16.7	12.4	12.1	17.6	25.2	26.0
Salt, sulfur, earth, stone, plaster, lime, and cement	0.4	0.6	0.9	1.5	2.6	6.1	15.5	2.0	16.6	37.9
Wood and articles of wood, wood charcoal	1.8	4.7	8.4	4.2	3.3	4.0	3.4	1.5	0.6	0.4
Raw hides and skins (other than fur skins) and leather	1.1	1.6	1.8	2.0	3.0	3.1	2.5	2.9	2.7	5.0
Oil seed, oleagic fruits, grain, seed, fruit, etc.	0.8	1.0	2.3	0.6	0.5	0.5	1.1	0.5	0.2	1.5
Carpets and other textile floor coverings	0.8	1.3	1.8	0.6	0.7	1.0	1.2	0.7	0.4	0.5
Machinery, nuclear reactors, boilers, etc.	0.6	0.5	0.1	0.2	0.3	0.2	0.8	1.4	2.1	1.9

Source: International Trade Center (ITC) (2016). Trade Map, Market Analysis and Research, ITC. http://www.trademap.org/

In 2014 China imported goods worth USD 2 trillion globally in which the share of Pakistan was just 0.1%. In the same year China exported about USD 2.3 trillion in which share of Pakistan was about 0.4%. This indicates the existence of high potential to be explored to raise Pakistan's exports to China. Table 37.6 presents the data on Pakistan's exports, imports and trade balance with China. During 2005–2014 Pakistan's exports to China were amounted to about USD 2.3 billion while imports accounted for about USD 9.6 billion leading to a negative balance of trade of about USD 1.4 billion during the same period. Pakistan experienced a negative balance of trade throughout the above mentioned period of recent 10 years 2005–2014. The imports of Pakistan from China gained a pace in 2010 reflecting some effects of bilateral trade agreements mentioned above, however, fluctuations in both exports and imports can be visualized from the Table 37.6. Despite the

TABLE 37.6
Pakistan's Trade Balance with China (Million USD)

Year	2005	2006	2007	2008	2009	2010	2011	2012	2013	2014
Exports	435.7	506.6	613.8	726.7	997.9	1435.9	1679	2619.9	2652.2	2252.9
Imports	2349.4	2914.9	4164.2	4738.1	3779.8	5247.7	6470.7	6687.6	6626.3	9588.4
Trade balance	−1913.7	−2408.3	−3550.4	−4011.4	−2781.9	−3811.8	−4791.7	−4067.7	−3974.1	−7335.5
Trade balance (billion USD)	−1.9	−2.4	−3.6	−4.0	−2.8	−3.8	−4.8	−4.1	−4.0	−7.3

Source: International Trade Center (ITC) (2016). Trade Map, Market Analysis and Research, ITC. http://www.trademap.org

TABLE 37.7
Pakistan's Exports to China (Million USD)

Year	2005	2006	2007	2008	2009	2010	2011	2012	2013	2014
All products	435.7	506.6	613.8	726.7	997.9	1435.9	1679.0	2619.9	2652.2	2252.9
Cotton	271.8	358.2	376.8	382.3	701.4	910.8	1134.2	1833.6	1936.0	1525.3
Ores, slag, and ash	25.6	27.8	87.7	158.6	74.9	149.8	112.7	120.9	129.2	91.7
Cereals	0.3	0.1	0.3	0.5	0.4	1.5	11.2	256.9	144.1	137.8
Raw hides and skins (other than fur skins) and leather	29.3	31.4	38.2	43.1	34.5	47.0	47.9	62.0	57.1	55.4
Fish, crustaceans, mollusks, aquatic invertebrates	28.3	24.6	29.7	40.3	47.0	63.1	41.3	41.6	35.8	55.2
Plastics and articles thereof	2.5	5.2	12.3	14.2	19.5	40.2	47.3	35.9	43.1	33.2
Salt, sulfur, earth, stone, plaster, lime, and cement	1.4	1.9	5.0	7.5	10.9	21.0	38.2	44.2	63.5	52.3
Copper and articles thereof	3.5	7.5	6.6	11.0	21.3	25.7	30.1	41.3	36.6	34.5
Residues, wastes of food industry, animal fodder	0.0	1.3	0.0	2.1	4.7	26.5	14.6	25.9	37.8	64.7
Other made textile articles, sets, worn clothing, etc.	1.1	1.1	2.3	8.5	9.0	14.0	26.1	22.9	26.9	34.6
Lac, gums, resins, vegetable saps, and extracts	1.5	3.3	5.7	7.5	4.9	6.7	12.0	50.1	29.7	24.5

Source: International Trade Center (ITC) (2016). Trade Map, Market Analysis and Research, ITC. http://www.trademap.org/

fluctuation, an increasing trend in both exports and imports are also depicted in Table 37.6. The total formal trade volume between Pakistan and China reached to USD 11.7 billion in 2014 (ITC 2016).

Table 37.6 presents the major export commodities of Pakistan for the China markets while Table 37.7 presents the major imports from China. The selection and the order of the commodities are based on average of the data during 2005–2014. Major export items include cotton; ores, slag and ash; cereals; raw hides and skins (other than furskins) and leather; fish, crustaceans, aquatic invertebrates. The major import items are comprised of electrical, electronic equipment; machinery, nuclear reactors, boilers; organic chemicals; manmade filaments; iron and steel; fertilizers. Export of cotton including cotton yarn remained at the top with USD 1.5 billion in 2014. Cereals exports got a momentum after 2010 and reached at USD 138 million in 2014 from USD 1.5 million in 2010. Exports of residues, wastes of food industry and animal fodder also jumped from zero level to USD 65 million in 2014. Other few such examples can also be observed from the Table 37.6. The data in Tables 37.6 and 37.7 shows that differences in comparative advantage exist as Pakistan's exports are agro-based while major imports from China are from manufacturing industry. Hence, further specialization can help in boosting the exports from Pakistan. In this regard, value added products should be encouraged to be produced and exported.

37.5.1 China Pakistan Economic Corridor (CPEC)

Pakistan and China are traditional trading partners. The territories comprising the present state of Pakistan were once part of the historic silk route connecting China with the West where merchants traded in various commodities such as spices, silk and porcelain. China Pakistan Economic Corridor is also realized on these opportunities to revive the ancient silk route to facilitate China's connectivity with the West. Pak–China Economic Corridor will connect Pakistan with China and the Central Asian countries through highway connecting Kashgar to Khunjrab and Gwadar. Gwadar port in southern Pakistan is expected to serve as nerve of Chinese trade particularly with Pakistan. This can be attributed to the expected highest trade through this port particularly of oil. Currently, 60% of China's oil is transported after covering a distance of 16,000 km in about 2–3 months, which would be reduced to a great extent using Gwadar port (Table 37.8).

Pakistan and China are adding new infrastructure projects to the China-Pakistan Economic Corridor (CPEC) to connect underdeveloped areas with developed ones. Due to the rising cost of production in China, the Chinese government and its private sector want to shift some industries to Pakistan. Further Chinese want to start joint ventures in Pakistan in sectors like engineering, light engineering, automobiles, textiles, and cement. Pakistan will also setup 25 industrial zones in all the four provinces of the country. The federal government is trying to improve social sector development along with energy, infrastructure, and transport to link developed areas with less developed ones.

Regarding major infrastructure projects under CPEC, it is planned to have a complete overhauling of the main railway track from Peshawar to Karachi with an investment of USD 5 billion. It is decided to first complete the Karachi-Hyderabad and Lahore-Multan railway routes in 2.5 years while the whole track will be renovated in 5 years. This will increase the speed of the trains from current 60–80 km to 140–160 km. Now it is need of the time that long-term plan for the corridor should be developed in a comprehensive manner including utilization of spatial planning, identifying the needs of improvement in various means of communication, focusing on industrial and agricultural cooperation, which should be in the interest of both countries.

The corridor not only plans to construct and improve connectivity of transportation infrastructures, but also aims to foster bilateral cooperation in some other major areas like information and communication technology, development of industrial parks and establishment of Free Trade Areas (FTAs). Currently Pakistan is facing a severe shortage of energy problems. Present capacity of electricity generation is well short of demand, which not only disturbs domestic user but also hampers industrial activity. This problem is also posing problems, as many foreign companies are hesitant to enter Pakistan and establish their manufacturing facility in presence of such energy crisis. Under CPEC, China will invest about USD 34 billion in thermal, solar and wind power generation facilities, which may help addressing Pakistan's energy and infrastructure problems. This will lead to strengthen economic growth and development in the country with some expected mega economic benefits. Along with CPEC, Pakistan is expected

TABLE 37.8
Pakistan's Imports from China (Million USD)

Year	2005	2006	2007	2008	2009	2010	2011	2012	2013	2014
All products	2349.4	2914.9	4164.2	4738.1	3779.8	5247.7	6470.7	6687.6	6626.3	9588.4
Electrical, electronic equipment	428.9	568.1	1066.9	1392.0	989.7	1245.3	1366.2	1741.4	1756.3	2265.0
Machinery, nuclear reactors, boilers, etc.	539.9	691.5	695.6	851.6	599.3	800.1	851.1	868.6	836.8	1370.0
Organic chemicals	118.6	127.4	160.5	221.9	276.0	321.0	394.9	374.1	378.3	490.8
Manmade filaments	93.1	137.5	176.2	162.7	190.0	327.4	483.7	373.5	367.8	479.4
Iron and steel	40.6	129.6	235.1	215.7	104.3	243.2	267.9	357.7	324.8	712.7
Fertilizers	22.3	1.6	228.2	88.1	10.4	140.4	409.3	339.5	229.4	551.7
Plastics and articles thereof	65.9	85.8	100.6	106.5	98.4	135.6	202.0	203.0	232.4	335.8
Articles of iron or steel	44.3	76.5	95.6	166.8	150.3	149.7	136.2	181.0	252.4	292.9
Manmade staple fibers	6.9	20.5	91.1	96.2	86.9	186.3	303.6	177.0	170.7	310.9
Rubber and articles thereof	74.5	82.4	80.7	72.3	76.1	117.0	166.3	157.7	163.4	195.9
Vehicles other than railway, tramway	72.3	78.6	63.3	79.7	77.3	111.1	151.7	179.3	145.4	171.7

Source: International Trade Center (ITC) (2016). Trade Map, Market Analysis and Research, ITC. http://www.trademap.org/

to become an economic hub in South and Central Asia, which is an important region not only in Asia but also in the world. The Gwadar Port, once fully established and operative may be one of the most important ports in the Arabian Sea. As such CPEC is a golden opportunity, which can help Pakistan to unlock its hidden potential. In order to have successful implementation of this project, government and the people need to support the smooth execution and sustainable operation of CPEC projects.

The corridor not only plans to construct and improve connectivity of transportation infrastructures, but also aims to foster bilateral cooperation in some other major areas like information and communication technology, development of industrial parks, and establishment of Free Trade Areas (FTAs). Currently, Pakistan is facing severe shortage of energy problems. Present capacity of electricity generation is well short of demand, which not only disturbs domestic user but also hampers industrial activity. This problem is also posing problems as many foreign companies are hesitant to enter Pakistan and establish their manufacturing facility in presence of such energy crisis. Under CPEC, China will invest about USD 34 billion in thermal, solar and wind power generation facilities which may help addressing Pakistan's energy and infrastructure problems. This will lead to strengthen economic growth and development in the country with some expected mega economic benefits. Along with CPEC, Pakistan is expected to become an economic hub in South and Central Asia, which is an important region not only in Asia but also in the world. The Gwadar Port, once fully established and operative may be one of the most important ports in the Arabian Sea. As such CPEC is a golden opportunity, which can help Pakistan to unlock its hidden potential. In order to have successful implementation of this project, government and the people need to support the smooth execution and sustainable operation of CPEC projects.

37.6 PAK–INDIA TRADE

Pakistan and India are neighboring countries which came into existence after partition in 1947 and possess a central position in South Asia. Geopolitical ties can be attributed to a salient feature in the history of these countries and they had gone through wars three times since their existence. The stability of the region lies in good terms between Pakistan and India and commodity trade is a tool which may be used to stabilize relationships these between trading partners. Bilateral trade between India and Pakistan should be a matter of mutual gain. However, their trade volumes are measured by a variety of factors. Besides tariffs, other hurdles arise due to nontariff barrier, poor infrastructure resulting in costly transportation, poor trade facilitation measures like rigorous customs and procedural barriers, and strict visa regime, among others.

An historical review shows that at the time of independence, India and Pakistan were heavily dependent on each other. In fact India's share in Pakistan's global trade exports and imports accounted for 23.6% and 50.6% respectively in 1948–1949, which declined to 1.3% and 0.06%, respectively in 1975 to 1976. Pakistan's share in India's global export and import was 2.2% and 1.1% respectively in 1951–1952 which gradually reduced 0.7% and 0.13% in 2005 and 2006, respectively (GOI 2013). Considerable potential lies for greater trade between two countries. Less than 0.5% of Indian trade is accounted by Pakistan while India account for more than 3%. In 2014, India imported goods worth USD 459 billion globally in which the share of Pakistan was just 0.1%. In the same year, India exported about USD 317 billion in which share of Pakistan was about 0.7%. Informal trade and trade via third country is approximately more than 3 billion USD.

Table 37.9 presents the data on Pakistan's bilateral trade and trade balance with India for the period 2005–2014. During 2005–2014 Pakistan's exports to India amounted to about USD 0.323 billion while imports accounted for about USD 1.5 billion leading to a negative balance of trade of about USD 1.1 billion during the same period. Pakistan experienced a negative balance of trade throughout the above-mentioned period of recent 10 years 2005–2014. Both imports and exports between these countries has been fluctuating and can be visualized in Table 37.9. Despite fluctuations, an increasing trend in both exports and imports is also depicted by the Table 37.9. The total formal trade volume between Pakistan and India reached to USD 2.5 billion in 2014 (ITC 2016). Some experts estimate

TABLE 37.9
Pakistan's Bilateral Trade and Trade Balance with India (Million USD)

Year	2005	2006	2007	2008	2009	2010	2011	2012	2013	2014
Exports	337.2	326.7	291.7	354.6	235.3	275.0	272.9	348.0	402.7	392.2
Imports	576.7	1115.0	1266.2	1691.5	1080.4	1559.9	1607.3	1572.6	1874.1	2104.8
Trade balance	−239.5	−788.3	−974.5	−1336.8	−845.1	−1284.9	−1334.5	−1224.6	−1471.3	−1712.6
Trade balance billion USD	−0.2	−0.8	−1.0	−1.3	−0.8	−1.3	−1.3	−1.2	−1.5	−1.7

Source: International Trade Center (ITC) (2016). Trade Map, Market Analysis and Research, ITC. http://www.trademap.org/

that greater prospects for bilateral trade exist and it can be raised up to 40 billion USD (Zahra 2011). Increased trade will not only benefit economies of both countries, but constituencies will build for more cooperative bilateral relations and open doors to advancement on security and other core political issues.

Major imports of Pakistan from India include vegetables, artificial staple fiber, tea, chemicals, and soya bean oil cake. Detailed data on major exports and imports of Pakistan are presented in Tables 37.10 and 37.11 respectively. Agricultural imports from India accounts for 30% of total Indian exports. Dates, textiles, cement, and certain chemicals are major exports of Pakistan to India. Pakistan exported cotton worth 60 million USD to India in 2014 (GOP, 2016; ITC, 2016). Emergence of some new exports to India such as leather, medical and surgical instruments, and woven cotton

TABLE 37.10
Pakistan's Exports to India (Million USD)

Year	2005	2006	2007	2008	2009	2010	2011	2012	2013	2014
All products	337.2	326.7	291.7	354.6	235.3	275.0	272.9	348.0	402.7	392.2
Mineral fuels, oils, distillation products, etc.	151.3	162.0	94.1	137.7	22.6	26.4	9.3	3.1	21.4	7.9
Cotton	37.8	56.3	56.2	52.8	44.5	38.3	30.5	81.7	42.4	60.1
Edible fruit, nuts, peel of citrus fruit, melons	28.8	32.2	34.3	36.3	45.0	45.3	47.9	67.2	74.0	64.3
Salt, sulfur, earth, stone, plaster, lime, and cement	0.4	0.4	10.3	65.8	32.3	32.3	48.0	52.8	45.1	54.0
Organic chemicals	10.4	30.4	1.6	4.8	25.9	25.7	26.0	9.4	30.4	12.6
Raw hides and skins (other than fur skins) and leather	3.6	3.0	11.9	13.5	8.4	12.9	13.0	9.8	20.6	36.2
Copper and articles thereof	1.6	3.0	2.8	3.2	4.0	8.7	11.3	21.4	34.5	33.6
Edible vegetables and certain roots and tubers	69.6	3.3	1.9	0.1	0.1	0.1	3.4	0.9	0.5	0.0
Plastics and articles thereof	3.0	1.6	2.6	3.6	7.1	18.6	8.9	10.7	5.1	16.9
Inorganic chemicals, precious metal compound, isotopes	0.0	0.0	0.0	0.5	6.4	11.4	14.8	14.8	13.2	12.0
Oil seed, oleagic fruits, grain, seed, fruit, etc.	2.2	1.7	2.6	3.6	9.5	4.2	3.9	6.8	21.4	15.5
Wool, animal hair, horsehair yarn and fabric thereof	1.7	3.2	4.3	3.7	2.2	8.4	6.2	8.3	13.2	10.8

TABLE 37.11
Pakistan's Imports from India (Million USD)

Year	2005	2006	2007	2008	2009	2010	2011	2012	2013	2014
All products	576.7	1115.0	1266.2	1691.5	1080.4	1559.9	1607.3	1572.6	1874.1	2104.8
Organic chemicals	163.1	209.4	410.9	455.6	338.5	260.7	373.3	308.2	259.1	237.8
Cotton	32.5	73.1	281.8	448.3	139.3	338.3	305.5	190.6	408.9	381.2
Residues, wastes of food industry, animal fodder	57.3	102.2	83.8	116.8	82.0	131.8	186.3	268.3	297.4	209.9
Edible vegetables and certain roots and tubers	32.4	33.1	42.9	74.0	137.2	123.3	141.1	201.5	230.7	252.0
Plastics and articles thereof	55.9	103.1	96.5	81.8	42.9	37.5	67.5	85.8	149.5	157.1
Sugars and sugar confectionery	0.7	324.3	3.5	0.0	0.0	156.7	51.5	0.4	0.3	2.6
Tanning, dyeing extracts, tannins, derivs, pigments, etc.	20.0	23.3	30.6	37.2	35.8	42.0	41.2	45.5	57.1	89.4
Coffee, tea, mate, and spices	16.6	22.3	15.2	40.6	18.4	51.5	56.7	58.5	35.6	57.1
Miscellaneous chemical products	5.2	6.2	12.8	28.0	34.4	50.9	49.2	53.1	44.4	64.8
Rubber and articles thereof	38.0	39.9	39.4	36.7	23.1	27.3	40.3	27.8	30.7	38.8
Oil seed, oleagic fruits, grain, seed, fruit, etc.	9.3	14.2	18.0	13.9	20.9	32.2	39.5	48.9	47.5	65.3

Source: International Trade Center (ITC) (2016). Trade Map, Market Analysis and Research, ITC. http://www.trademap.org/

fabrics shows a further positive sign. Vegetable and fruit sector has the potential to generate export revenue, increase the level of national and farm income and generate new employment opportunities for the people. High potential exist for both Pakistan and India to cultivate fruit and vegetable crops for domestic as well as export market (Pasha and Imran 2012).

Punjab and the Indian Punjab (the Haryana-Delhi belt), low transportation costs, and the time advantage make trade in perishable and fresh agricultural goods quite attractive. When India falls short of onions or potatoes and Pakistan has a bumper crop or vice versa, countries, their farmers, and consumers gain from these transactions. Imports will also force Pakistani farmers, to strive to become more efficient by adopting better production, storage, and preservation techniques and reduce postharvest losses.

The above-mentioned point of view may look a bit idealistic in terms of contribution of trade that it is absolutely beneficial for everyone including producers, consumers and even government officials. This is quite a controversial argument and needs comprehensive analysis before conclusion are drawn. Agriculture is the basic pillar of economy of the two countries. The welfare concern of related stakeholders (including farmers and local communities) in this sector has got political and development priority for governments of both countries. As such, anything which may hamper growth of the agriculture sector or affect stakeholders negatively may be a point of prime importance and concern. The demand and supply gap of essential food commodities including potatoes, tomatoes, onion, garlic, and so on on either side of border may act as good instrument and opportunity to bridge gap between two nations subject to fair and merit based implementation of trade policies. The development of agriculture sector, which may be achieved on both sides of borders with mutual cooperation, may help Pakistan and India to combat social evils like unemployment, poverty, gender discrimination etc.

Trade of agricultural goods is very important for Pakistan because it is an agricultural economy and its exports mostly comprises of raw materials and agricultural goods. Pakistan is one of the world's largest producers of agricultural products including chickpeas, dates, apricot, cotton, milk, onion, sugarcane, mango, oranges, kinnow, mandarins, wheat, and rice. Rice, kinnow and citrus fruits, dates, mangoes, and apricots are major agricultural exports of Pakistan (TDAP 2013). The agricultural trade between India and Pakistan is shown in Table 37.12 below. Table 37.12 indicates that agricultural exports from India to Pakistan are relatively greater compared to imports from Pakistan. In the year 2012–2013 India's exports to Pakistan were worth USD 352.36 million, while India's imports from Pakistan wee worth USD 105.11 million.

Onion and potato are two important vegetables, particularly in Pakistan and India. One thing is common in both countries regarding these vegetables and that is the consistent demand and irregular supply. One or the other year in either country is marked with surplus or deficit of these vegetable and people's food security becomes at stake. In this scenario, if trade is practices consistently considering and handling each other concerns, then it can benefit both countries.

In 2013, onion prices skyrocketed in India due to unexpected drought in 2012 and heavy monsoon rains in 2013. The prices increased up to 500% in some parts of India. Onion being a staple food ingredient became a hot item in Indian markets and consumers were forced to cut down their consumption in response to hyper increase in its prices. In such situations, trade can be used as an effective tool to normalize relation between two countries and help bringing people even closer to each other. There is also a good opportunity for traders across the border to earn windfall profit in this seasonal hike in the prices of agricultural products.

There is significant potential for Pakistan to export mangoes to India. The reason for this being that Pakistani season peaks in July/August when India's season is almost over. Secondly, Pakistan produces more varieties of mango that is consumed fresh as compared to India where the majority varieties are used for by products such as pulping and juicing. Currently the trade of mango between two countries is almost nonexistent and the possible reason for this is high tariffs, complicated documentation and heightened SPS measures imposed by Indian side. One of most liking potential of mango trade between India and Pakistan could be that the mango season extends over 5 months, starting in mid-May in Sindh and finishing late September in Punjab, with late June to mid August being the peak production period and ending in October. The potential gains from increasing production in two countries can be capitalized through economic integration between the two countries that leads to existing export destinations, as well as emerging markets such as the People's Republic of China and Russian states for Pakistani mangoes, which has an estimated potential of one Billion dollars (TDAP, 2013).

The two countries are now progressing towards a closer economic relation with a vision to enhance prosperity in the region and to realize full potential of bilateral trade. Both governments have, in the past 2 years, taken steps to follow up. In November 2011, Pakistan took the initiative by announcing that by the end of 2012 it would apply Most Favored Nation (MFN) treatment to goods

TABLE 37.12
India–Pakistan Trade in Agricultural Products

Year	India's Exports to Pakistan (Million USD)	India's Imports from Pakistan (Million USD)
2008–2009	209.26	45.15
2009–2010	156.34	65.16
2010–2011	248.67	77.31
2011–2012	377.3	89.78
2012–2013	352.36	105.11

Source: Directorate General of Foreign Trade, Ministry of Commerce and Industry, Government of India coming from India. Both countries announced the conclusion of agreements on mutual recognition of standards, customs cooperation, and readdress of trade grievances in February 2012. In March 2012, the Pakistan Commerce Ministry replaced the relatively short "positive list" of less than 2000 items that could be imported from India with a "negative list" of 1200 prohibited items. This almost effectively freed up trade in almost 6800 product areas, which were banned previously. In September 2012, both governments announced a new visa agreement according to which provisions were designed to build an atmosphere of confidence and trust and facilitate business travel; the only way to accomplish this was economic partnership (GOP, 2013). The formation of the Pak–India Joint Business Forum is another step in reconciliation of relationships between the two states. The progressive business people from both sides of the border are meeting frequently to discuss issues hindering trade and they jointly forward recommendations to boost trade. Recent advancement like India Show in Lahore on February 15–16, 2014, formation of a sectoral task force and announcement of delivering Non-Discriminatory Market Access (NDMA) are some other right steps to build trust and confidence between the two countries.

37.7 PAK–IRAN TRADE

Iran is a neighbor country of Pakistan and is located at the west of Pakistan. Both countries can be attributed to have common historical, cultural, and economic affinities. Iran was the first country which recognized the sovereignty of Pakistan after its existence in 1947. Since then both countries are cooperating with each other at various fronts particularly to increase the trade between two countries. Aiming this, a Regional Cooperation for Development (RCD) was established in 1964 which were further strengthen Economic Cooperation Organization (ECO) and various trade agreements were signed under this umbrella. The Iran-Pakistan-India pipeline (IPI Pipeline) is currently under discussion; though India backed out from the project. The Indian government was under pressure by the United States against the IPI pipeline project, and appears to have heeded American policy after India and the United States proceeded to sign the nuclear deal. In addition, the international sanctions on Iran due to its controversial nuclear program could also become a factor in derailing the IPI pipeline project altogether. On 12 January 2001, Pakistan and Iran formed a "Pak–Iran Joint Business Council" (PIJB) body on trade disputes. The body works on to encourage the privatization in Pakistan and economic liberalization on both sides of the countries.

Table 37.13 presents the data on Pakistan's exports, imports and trade balance with Iran. During 2005–2014 Pakistan's exports to Iran were amounted to about USD 0.18 billion on an average while imports accounted for about USD 0.46 billion leading to a negative balance of trade of about USD 0.28 billion during the same period. Pakistan experienced a negative balance of trade throughout the above-mentioned period of recent 10 years 2005–2014 although fluctuating; however, it was reduced to almost half in 2014, USD 123 million, from USD 260 million mainly because of a reduction in

TABLE 37.13
Pakistan's Bilateral Trade and Trade Balance with Iran (Million USD)

Year	2005	2006	2007	2008	2009	2010	2011	2012	2013	2014
Exports	102.8	178.4	178.8	146.2	426.2	252.2	182.2	153.3	142.0	62.6
Imports	363.2	443.2	436.8	737.6	955.9	883.6	303.8	120.3	167.8	185.7
Trade balance	−260.4	−264.8	−258.0	−591.4	−529.8	−631.4	−121.6	32.9	−25.8	−123.1
Trade balance billion USD	−0.3	−0.3	−0.3	−0.6	−0.5	−0.6	−0.1	0.0	0.0	−0.1

Source: International Trade Center (ITC) (2016). Trade Map, Market Analysis and research, ITC. http://www.trademap.org/

imports. Fluctuations in both exports and imports can also be visualized from Table 37.10. However, a sharp decline can be seen after 2010 as Pakistan's exports decreased to USD 63 million in 2014 from USD 252 m in 2010, while imports reduced to USD 186 m in 2014 from USD 884 m in 2010. UN sanctions on Iran can be attributed to this decline. The total formal trade volume between Pakistan and Iran reached about USD 250 million in 2014 (ITC 2016).

A limited range of products is exported to Iran, mainly agricultural commodities, particularly rice. Major exports and imports of Pakistan are shown in Tables 37.11 and 37.12, respectively. Major exports of Pakistan to Iran include cereals; cotton; edible fruit, nuts, peel of citrus fruit, melons; meat, and edible meat offal. Major imports from Iran are composed of mineral fuels, oils, distillation products; iron and steel; organic chemicals; plastics and articles thereof; ores, slag, and ash; raw hides and skins (other than furskins) and leather; edible vegetables and certain roots and fibers. Export of cereals (mainly rice) remained highest and had been contributing more than 70% in some years such as in 2009, 2010 and 2011. Cereal exports showed a sharp decline since 2009–2010. Exports of meat and its products showed good increase in 2012 and 2013 reaching to almost USD 32 million in 2013 from a negligible level until 2010. The data in Tables 37.11 and 37.12 show that differences in comparative advantages exist as Pakistan's exports are agro-based and mainly cereals while major imports from Iran are from the oil industry or other related natural resources. Hence, exploring the Iranian markets for products other than cereals is very important to increase the exports to Iran. Pakistan's shares in the total exports and imports of Iran are 0.3% and 0.1%, respectively (Table 37.14). This low share coupled with declining trends in trade indicates a high potential of trade for Pakistan to be explored in the Iranian markets. In this regard, delegation of exporters and importers should visit Iran along with government officials, and liberalization policies should be adopted by both governments. This liberalization can also help in reducing informal trade (Table 37.15).

TABLE 37.14
Pakistan's Exports to Iran (Million USD)

Year	2005	2006	2007	2008	2009	2010	2011	2012	2013	2014
All products	102.8	178.4	178.8	146.2	426.2	252.2	182.2	153.3	142.0	62.6
Cereals	43.4	80.1	111.2	72.4	367.5	208.5	124.8	75.2	68.9	15.6
Edible fruit, nuts, peel of citrus fruit, melons	0.0	0.2	5.3	8.5	15.8	10.1	11.2	11.8	5.0	1.2
Meat and edible meat offal	0.0	0.0	0.0	0.0	0.0	0.0	1.3	23.4	31.5	5.3
Paper and paperboard, articles of pulp, paper and board	3.1	0.1	1.5	0.5	0.1	1.4	0.4	4.3	8.0	23.1
Cotton	9.2	17.4	6.2	6.0	4.9	4.1	3.8	6.7	2.4	1.2
Manmade staple fibers	23.2	16.8	9.0	10.6	4.0	0.2	0.2	0.6	0.2	0.0
Articles of leather, animal gut, harness, travel goods	0.4	21.5	12.7	0.2	0.2	0.3	0.5	0.4	0.2	0.0
Ships, boats, and other floating structures	1.6	1.3	1.6	2.7	4.4	6.0	4.3	4.2	2.4	0.7
Vegetable textile fibers, paper yarn, woven fabric	4.2	7.2	4.0	5.7	2.3	2.6	2.5	2.4	0.3	0.1
Edible vegetables and certain roots and tubers	0.0	0.9	0.6	16.8	3.7	0.0	2.1	0.2	0.0	0.0
Plastics and articles thereof	1.7	0.6	0.3	4.6	5.3	3.2	2.4	1.2	1.8	2.3
Oil seed, oleagic fruits, grain, seed, fruit, etc.	0.9	0.5	0.7	2.1	4.1	3.8	1.2	3.2	4.8	0.7
Optical, photo, technical, medical, etc., apparatus	1.0	0.7	0.7	1.4	1.4	1.3	2.5	2.2	2.3	1.9

TABLE 37.15
Pakistan's Imports from Iran (Million USD)

Year	2005	2006	2007	2008	2009	2010	2011	2012	2013	2014
All products	363.2	443.2	436.8	737.6	955.9	883.6	303.8	120.3	167.8	185.7
Mineral fuels, oils, distillation products, etc.	123.8	167.4	213.9	443.6	653.1	528.3	36.6	35.1	59.3	58.1
Iron and steel	83.2	121.5	72.6	63.3	33.4	56.1	31.1	19.4	23.2	21.1
Organic chemicals	39.2	39.5	26.8	61.4	57.2	122.5	75.6	17.5	4.7	3.8
Plastics and articles thereof	15.6	15.7	23.5	7.4	56.9	86.3	65.4	5.4	3.0	4.8
Ores, slag and ash	19.7	19.3	18.4	90.6	61.8	11.4	38.2	8.8	5.6	0.0
Raw hides and skins (other than fur skins) and leather	5.1	9.5	7.2	15.4	12.1	7.7	9.0	8.2	29.6	13.8
Edible vegetables and certain roots and tubers	13.9	7.3	2.6	6.1	8.0	2.7	4.0	5.7	5.0	15.6
Edible fruit, nuts, peel of citrus fruit, melons	11.2	11.8	13.0	5.0	5.7	6.0	4.8	3.9	2.3	7.1
Ships, boats, and other floating structures	6.1	0.0	0.0	0.0	6.3	2.6	0.4	0.0	14.3	28.0
Salt, sulfur, earth, stone, plaster, lime, and cement	5.3	5.4	2.9	7.5	3.7	2.3	5.9	3.7	6.1	9.0

Source: International Trade Center (ITC) (2016). Trade Map, Market Analysis and research, ITC. http://www.trademap.org/

37.8 PAK–SAUDI TRADE

Friendly and very close bilateral relations between Pakistan and Saudi Arabia are observed historically regarding commercial, cultural, religious, political, and strategic relations since the establishment of Pakistan in 1947. Pakistan affirms its relationship with Saudi Arabia in its current foreign policy as important, and bilateral partnership and closer bilateral ties are developing with Saudi Arabia, which is the largest country on the Arabian peninsula and host to the two holiest cities of Islam, Mecca and Medina.

In order to institutionalize bilateral economic cooperation and strengthen mutual coordination and linkages, the Pak–Saudi Joint Ministerial Commission (JMC) was established in May 1974. The JMC process is at the core of efforts to expand commercial cooperation between the two countries. Ten sessions of JMC have been held in Riyadh, most recent in April, 2014. Both sides discussed new opportunities for cooperation in various fields as Pakistan sought Saudi investments in infrastructure building, trade, banking, information technology, and telecommunications.

Saudi Arabia has remained Pakistan's one of the largest trading partner, however, still, there exist a great opportunities to raise the trade with Saudi Arabia. In 2014, Saudi Arabia imported goods worth USD 174 billion globally while its global exports amounted to about USD 347 billion. Pakistan's shares in both the global exports and imports of Saudi Arabia are 0.3% reflecting the occurrence of a highly unexploited trade potential. Table 37.16 presents the data on Pakistan's exports, imports and trade balance with Saudi Arabia. During 2005–2014 Pakistan's exports to Saudi Arabia amounted to about USD 0.4 billion while imports accounted for about USD 4.4 billion leading to a negative balance of trade of about USD 4 billion during the same period. Pakistan experienced a negative balance of trade throughout the above-mentioned period of 2005–2014. The imports of Pakistan from Saudi Arabia gained a pace in 2010 reflecting some effects of bilateral trade agreements mentioned above, however, fluctuations in both exports and imports can be visualized from the Table 37.16. Despite the fluctuation, an increasing trend in exports after 2010 is also depicted by Table 37.16. The total formal trade volume between Pakistan and Saudi Arabia reached to about USD 5 billion in 2014 (ITC 2016).

TABLE 37.16
Pakistan's Bilateral Trade and Trade Balance with Iran (Million USD)

Year	2005	2006	2007	2008	2009	2010	2011	2012	2013	2014
Exports	354.9	309.0	295.5	441.1	425.7	409.0	420.2	455.6	494.1	509.7
Imports	2650.6	3033.2	4011.8	5954.9	3500.1	3837.9	4668.3	4283.5	3847.2	4417.4
Trade balance	−2295.7	−2724.2	−3716.3	−5513.9	−3074.4	−3428.9	−4248.1	−3827.9	−3353.2	−3907.7
Trade balance billion USD	−2.3	−2.7	−3.7	−5.5	−3.1	−3.4	−4.2	−3.8	−3.4	−3.9

Source: International Trade Center (ITC) (2016). Trade Map, Market Analysis and research, ITC. http://www.trademap.org/

Table 37.17 presents the major export commodities of Pakistan entering in for the markets of Saudi Arabia while Table 37.18 presents the major imports from Saudi Arabia. The selection and the order of the commodities are based on average of the data during 2005–2014. As Saudi Arabia is an oil-based economy possessing deserts and limited agricultural commodities such as dates while Pakistan is an agro-based economy therefore major export items of Pakistan for Saudi Arabia are agricultural products, which include mainly cereals, cotton and textile products, meat, and fish. Cereals have shown an increasing trend since 2010 while all other products depicted a fluctuating

TABLE 37.17
Pakistan's Exports to Saudi Arabia (Million USD)

Year	2005	2006	2007	2008	2009	2010	2011	2012	2013	2014
All products	354.9	309.0	295.5	441.1	425.7	409.0	420.2	455.6	494.1	509.7
Cereals	43.3	39.3	46.8	136.2	126.2	137.9	96.0	99.4	125.5	133.6
Other made textile articles, sets, worn clothing, etc.	48.5	60.2	50.1	64.6	82.6	50.7	64.2	58.4	55.1	55.5
Meat and edible meat offal	9.0	10.7	11.1	17.1	23.6	38.7	42.8	54.1	58.3	67.2
Cotton	40.4	32.0	22.4	32.0	20.9	10.8	12.5	15.7	12.1	15.9
Articles of apparel, accessories, not knit or crochet	62.0	21.1	14.0	20.8	20.2	18.0	19.6	15.4	11.9	9.5
Footwear, gaiters, and the like, parts thereof	22.8	30.1	23.7	29.2	13.8	7.8	10.4	7.2	6.6	7.7
Articles of leather, animal gut, harness, travel goods	22.7	17.0	14.6	18.1	11.8	8.3	9.5	15.7	15.6	14.2
Fish, crustaceans, mollusks, aquatic invertebrates	4.0	6.4	8.7	13.0	14.3	20.2	17.1	22.7	18.9	9.3
Coffee, tea, mate and spices	2.4	3.8	5.3	8.1	10.0	13.7	16.7	20.1	19.8	21.5
Manmade staple fibers	1.6	3.6	11.7	8.8	13.2	14.0	19.0	15.9	12.0	9.0
Edible fruit, nuts, peel of citrus fruit, melons	9.4	7.4	9.0	9.8	9.7	9.5	11.2	10.5	12.5	12.4
Sugars and sugar confectionery	2.3	2.6	0.7	3.3	1.9	1.9	1.3	16.4	50.3	20.3
Articles of apparel, accessories, knit or crochet	13.0	6.3	7.3	12.1	12.2	11.9	9.8	10.3	8.2	8.7

Source: International Trade Center (ITC) (2016). Trade Map, Market Analysis and research, ITC. http://www.trademap.org/

TABLE 37.18
Pakistan's Imports from Saudi Arabia (Million USD)

Year	2005	2006	2007	2008	2009	2010	2011	2012	2013	2014
All products	2650.6	3033.2	4011.8	5954.9	3500.1	3837.9	4668.3	4283.5	3847.2	4417.4
Mineral fuels, oils, distillation products, etc.	2110.7	2554.2	3360.5	5051.0	2883.2	2953.3	3686.4	3345.7	2916.9	3211.6
Plastics and articles thereof	167.4	177.9	223.2	256.2	257.2	412.9	452.4	406.8	408.7	554.2
Organic chemicals	191.7	142.4	191.9	225.2	152.1	230.6	314.6	334.0	314.3	336.8
Fertilizers	35.3	7.1	44.1	250.9	85.0	123.9	82.1	78.4	91.2	175.6
Iron and steel	35.4	50.7	76.1	33.4	17.9	7.0	26.2	18.5	10.4	4.4
Raw hides and skins (other than fur skins) and leather	21.0	15.7	14.1	14.7	7.5	13.1	18.2	11.8	12.6	13.0
Miscellaneous chemical products	0.9	1.0	1.4	2.0	7.0	14.3	20.1	23.1	33.5	33.5
Paper and paperboard, articles of pulp, paper and board	14.0	20.6	17.3	30.9	19.0	11.2	10.5	6.7	0.6	1.7
Essential oils, perfumes, cosmetics, toiletries	10.4	12.3	16.1	16.1	11.6	7.8	3.2	3.9	3.9	7.0
Aluminum and articles thereof	12.4	8.7	22.9	3.1	2.6	3.4	3.8	4.2	3.0	4.3

Source: International Trade Center (ITC) (2016). Trade Map, Market Analysis and research, ITC. http://www.trademap.org/

trend. Oil and oil-based products have been the major exports items and sharing more than 60% of total imports of Pakistan from Saudi Arabia. In 2014 these imports amounted to USD 3.2 billion out of USD 4.4 billion. Overtime there is an increasing trend in oil-based imports from Saudi Arabia; however, many fluctuations are evident from the Table 37.18. Other major import items from Saudi Arabia include plastic products, organic chemicals, fertilizers, iron, and steel and hides and skins. Most of them are showing a uneven trend. The data in Tables 37.17 and 37.18 show that differences in comparative advantage exist as Pakistan's exports are agro-based while major imports from Saudi Arabia are comprised of oil products and product from manufacturing industry. Hence, further specialization can help in boosting the exports from Pakistan. In this regard, value-added agricultural products should be encouraged and exported. Marketing efforts can also be focused to sustain the increasing trend of exports and reduce fluctuations.

37.9 PAK–UAE TRADE

Pakistan and the UAE possess very close relations as friends as well as trade partners. These relations started since its existence in 1971 and has extended to cooperation in various fields. Pakistan was the second country after Britain, which constructed its embassy in the UAE. These time-honored relations are based on strong pillars of joint trading and collaborative understanding. After the independence from Great Britain, the UAE was lacking security and defense particularly for its major natural resourcc—oil, and decided to take external help for defense. Pakistan provided the services in strengthening the defense. The UAE is the second closest neighbor of Pakistan by sea after Oman and is an economic hub near to Pakistan. It is ranked as the seventh largest country in the world with respect to its oil reserves, which amounted to about 97.8 billion barrels while its natural gas reserves are estimated to be about 215 trillion cubic feet (US Energy Information Administration, 2017).

The UAE is a global trading hub and a major trading partner of Pakistan and both countries are keen to further boost and strengthen bilateral trade ties. In February 1977, the first meeting of the

TABLE 37.19
Pakistan's Bilateral Trade and Trade Balance with the UAE (Million USD)

Year	2005	2006	2007	2008	2009	2010	2011	2012	2013	2014
Exports	1256.8	1241.8	2114.7	2009.8	1538.6	1834.9	1921.0	2872.9	1775.1	1324.1
Imports	2480.7	3408.4	2766.9	3777.9	3349.6	5247.8	6818.8	7210.8	7751.5	7077.2
Trade balance	−1224.0	−2166.5	−652.2	−1768.1	−1811.0	−3412.9	−4897.8	−4337.9	−5976.4	−5753.1
Trade balance billion USD	−1.2	−2.2	−0.7	−1.8	−1.8	−3.4	−4.9	−4.3	−6.0	−5.8

Source: International Trade Center (ITC) (2016). Trade Map, Market Analysis and research, ITC. http://www.trademap.org/

UAE-Pakistan Joint Committee took place in Abu Dhabi. The two countries signed an agreement on avoidance of double taxation in 1999. UAE has remained Pakistan's one of the largest trading partner, however, still there exist a great potential to increase trade with the UAE. In 2014 UAE imported goods worth USD 300 billion globally, while its global exports amounted to about USD 381 billion. Pakistan's shares in the global exports and imports of UAE are 0.4% and 1.9% respectively reflecting the occurrence of high-unexploited trade potential. Table 37.19 presents the data on Pakistan's exports, imports and trade balance with UAE. During 2005–2014 Pakistan's exports to UAE were amounted to about USD 1.8 billion while imports accounted for about USD 5 billion leading to a negative balance of trade of about USD 3.2 billion during the same period. Pakistan exports to UAE reached at highest level in 2012 which amounted to USD 3 billion; however, many fluctuations can be viewed from the data. Pakistan incurred a negative balance of trade throughout the above-mentioned period of 2005–2014. The imports of Pakistan from UAE gained a pace in 2010; however, fluctuations in imports can also be visualized from the Table 37.19. The total trade volume between Pakistan and UAE reached to about USD 8 billion in 2014.

Table 37.20 presents the major export commodities of Pakistan entering in for the markets of UAE while Table 37.21 presents the major imports from UAE for the period 2005–2014. The selection and the order of the commodities are based on average of the data during 2005–2014. Mineral fuels, oils, and distillation products remained the highest trading commodities between these two countries. Major export items of Pakistan for UAE are agricultural products, which include mainly cereals, cotton, and textile products, meat, and fish. Cereals have shown quite a fluctuating trend and deceased to USD 208 million in 2014 from USD 230 million in 2005 possessing a peak of USD 385 billion in 2008. Cotton and cotton products have also been exhibiting a decreasing trend. These declining trends are not a good sign and indicating that Pakistan is losing its one of the top ten export market. Government as well as private stakeholders should give serious attention to make Pakistani products competitive in this market. Oil and oil-based products have been the major import items of Pakistan from the UAE and sharing more than 50% of total imports of Pakistan from UAE and reaching sometimes to more than 80% such as in 2014. In 2014, these imports amounted to USD 6.2 billion out of USD 7 billion. Overtime, there is an increasing trend in oil-based imports from UAE; however, many fluctuations are evident from the Table 37.21. Other major import items from UAE include plastic products, organic chemicals, fertilizers, iron and steel and hides and skins. Most of them are showing an uneven trend. The data in Tables 37.20 and 37.21 show that differences in comparative advantage exist as Pakistan's exports are agro-based while major imports from the UAE are comprises of oil products and product from manufacturing industry. Importantly, exports of major export items decreased over time indicating the Pakistan is losing its comparative and competitive advantage. Hence, cost of production should be reduced and quality of the products should be improved. Moreover, marketing efforts can also be focused to sustain/increasing the exports and value added agricultural products to encourage production and exports.

TABLE 37.20
Pakistan's Exports to UAE (Million USD)

	Value in 2005	Value in 2006	Value in 2007	Value in 2008	Value in 2009	Value in 2010	Value in 2011	Value in 2012	Value in 2013	Value in 2014
All products	1256.8	1241.8	2114.7	2009.8	1538.6	1834.9	1921.0	2872.9	1775.1	1324.1
Mineral fuels, oils, distillation products, etc.	244.2	392.6	611.1	596.0	198.1	337.5	405.3	294.2	380.1	229.5
Cereals	230.0	241.9	305.8	384.4	241.7	314.9	344.4	190.1	190.2	208.1
Other made textile articles, sets, worn clothing, etc.	102.9	85.6	77.4	117.9	128.5	85.6	83.8	103.2	94.2	95.8
Cotton	91.4	79.5	53.2	89.0	45.9	40.4	44.0	39.2	40.0	46.6
Articles of apparel, accessories, not knit or crochet	76.4	42.1	47.2	45.1	38.7	32.8	43.6	40.8	39.8	52.9
Ships, boats and other floating structures	20.6	2.8	402.4	1.2	4.4	7.3	5.6	6.1	3.4	1.1
Electrical, electronic equipment	71.2	86.5	78.5	57.0	11.1	14.1	30.3	15.1	53.4	23.7
Articles of leather, animal gut, harness, travel goods	107.5	37.2	72.0	44.3	11.9	11.6	13.4	16.8	16.9	12.4
Articles of apparel, accessories, knit or crochet	22.2	16.1	19.5	32.8	38.8	38.0	35.1	48.5	33.9	49.4
Fish, crustaceans, mollusks, aquatic invertebrates	19.8	20.2	31.5	39.5	26.5	33.2	34.4	28.7	50.8	46.6
Meat and edible meat offal	3.0	5.9	16.5	21.3	28.6	38.9	46.8	53.2	52.1	47.6
Machinery, nuclear reactors, boilers, etc.	31.0	24.8	26.9	27.7	46.5	36.5	27.1	30.7	26.3	35.9
Manmade staple fibers	3.0	9.4	35.2	22.7	33.7	39.6	33.4	34.0	33.0	29.1
Edible fruit, nuts, peel of citrus fruit, melons	16.7	13.4	17.4	19.3	17.5	23.1	25.1	26.9	48.4	50.8

Source: International Trade Center (ITC) (2016). Trade Map, Market Analysis and research, ITC. http://www.trademap.org/

37.10 SUMMARY AND CONCLUSIONS

Regional integration and free trade agreements play a vital role in the economic development and improving of living standard of people through enhancing regional trade. The EU, the NAFTA, and the ASEAN are the examples of success in this regard. The seven member countries of the SAARC also signed South Asia Free Trade Agreement (SAFTA) at Islamabad (Pakistan) on 6th January 2004, which was implemented on 6th July 2006. But till to date it could not reap its expected fruit as the intra-regional trade accounts for only 5% while intra-regional trade in ASEAN is about 25%.

TABLE 37.21
Pakistan's Imports from UAE (Million USD)

Year	2005	2006	2007	2008	2009	2010	2011	2012	2013	2014
All products	2480.7	3408.4	2766.9	3777.9	3349.6	5247.8	6818.8	7210.8	7751.5	7077.2
Mineral fuels, oils, distillation products, etc.	1320.5	2216.4	2093.1	3158.5	2589.3	4297.6	5809.3	6436.9	6588.4	6270.3
Pearls, precious stones, metals, coins, etc.	372.6	386.1	74.6	53.5	106.2	151.9	265.2	187.7	394.6	18.2
Iron and steel	82.0	77.8	107.1	90.3	116.2	102.2	87.9	87.3	114.9	127.0
Plastics and articles thereof	60.3	53.9	59.4	52.0	59.2	67.2	152.2	107.2	84.3	112.1
Machinery, nuclear reactors, boilers, etc.	68.7	71.1	67.3	74.2	55.6	88.2	82.7	84.4	117.5	98.1
Electrical, electronic equipment	119.6	136.0	75.2	73.9	50.2	97.9	84.5	49.0	36.7	70.9
Sugars and sugar confectionery	124.3	160.3	4.8	5.9	144.3	200.9	15.9	1.4	0.8	0.8
Organic chemicals	12.7	13.5	76.7	49.9	22.8	19.9	6.1	58.6	104.4	79.2
Aluminum and articles thereof	23.7	22.8	21.1	23.6	35.8	37.1	33.0	30.2	30.1	36.6
Ships, boats, and other floating structures	91.8	22.2	5.6	0.04	12.8	16.8	6.7	3.1	32.4	49.5

Source: International Trade Center (ITC) (2016). Trade Map, Market Analysis and research, ITC. http://www.trademap.org/

Moreover, Pakistan trade with other countries in Asia is very minimal. For example in 2014 China imported goods worth USD 2 trillion globally in which Pakistan's share was just 0.1%. The similar is the share (0.1%) in India and Iran's global imports. Two major partners in the gulf, Saudi Arabia and UAE, receive about 0.4% of their global imports from Pakistan. In the same year China exported about USD 2.3 trillion in which share of Pakistan was about 0.4%.

Agriculture remains a key sector of the economy contributing to about 20.9% of GDP, employing about 43.5% of the total employed labor force, and is the source of most exports. Since its independence in 1947, Pakistan has had a positive trade balance in very few years. Hence, Pakistan is a trade deficit country with a narrow range of export items and few sources of foreign exchange earnings. The major export items include raw cotton and textile manufactures, leather and related products, rice, mango, and citrus, which account for about 70% of the total export earnings. Almost half of all of Pakistan's exports are concentrated in the following major export markets which include the United States, the United Kingdom, Germany, China, the UAE, Afghanistan and France. This indicates a lot of room for diversification of export markets and commodities as well as for value addition for export purposes.

Pakistan's agriculture products' exports are continuously showing a declining trend with fluctuations because the international price of major farm products, that is, rice, wheat, sugar, and cotton are lower compared to Pakistan where cost of inputs are on the higher side due to taxation accompanies by low yields per hectare, lack of domestically produced certified seeds and high postharvest losses. The following data statistics show the examples of decline in agricultural exports. Exports of rice registered at USD 1749.7 million during July–March 2014–2015 against USD 1850.3 million during the same period of 2013–2014, showing a negative growth of 5.4%. Even Pakistan has a surplus in wheat, but its export is negligible. During the first 9 months of fiscal year 2014–2015, Pakistan exported wheat of about USD 3 million against USD 7 million in same period last year, showing a decline of 57.4%. Similarly, fish, fish preparations, oil seeds, nuts, and kernels showed negative growth.

An important reason for low and declining levels of agricultural exports is the lack of adoption of quality measures defined by the WTO's agreement of Sanitary and Phytosanitary (SPS) and the demand of our international clients. The application of the SPS is most important as it relates to food safety, and animal and plant health. This WTO agreement encourages countries to take health protection measures within their borders based on internationally established guidelines and risk assessment procedures. Exporting countries are required to follow international standards. It is necessary for farmers to use the ISO certified products for pesticides, insecticides, and seeds. We are also bound to use certified products for fulfilling the prerequisites of international standards. Due to noncompliance of SPS requirements, our exports such as rice and mangoes have been rejected by importers.

Another WTO agreement is the Trade-Related Intellectual Property Rights (TRIPS). The owners of patented products have an exclusive right to make use of this agreement. Pakistan's Basmati rice hasn't registered under the Geographical Indications (GI) of TRIPS agreement and losing its competitiveness and branding. Pakistan Intellectual Property Right Organization (PIPRO) is the relevant organization in Pakistan while the World Intellectual Property Rights Organization (WIPO) is working at world level. One of the important reasons for low competitiveness of Pakistan's agricultural exports is the support/subsidies provided by the developing and developed countries and lack of support and subsidies in Pakistan, although there are recent steps towards it such as the Kissan package, etc. This coupled with high input costs, low yields per hectare and high postharvest losses, ranging 20%–40%, make Pakistan's exports less competitive in the international market.

37.10.1 Way Forward for Improving Exports of Pakistan

Trade particularly exports can be increased by meeting the SPS requirements of the importing countries. This is one of the major trade barriers to increase exports from Pakistan and needs intervention. Knowledge provision and training of the farmers and stakeholders in meeting the SPS requirements, along with the provision of necessary infrastructure is direly needed. Moreover, various certifications required for exports should be obtained through a fair process free from any kind of corruption so that only quality products can get certifications. Branding of the product should be encouraged, and property rights under TRIPS agreement should be secured. Value added products such as mango pulp and dried mango slices should be encouraged. China is among the top 10 importers of mangoes, and mango pulp and in this regard, the CPEC would be very facilitative, enhancing trade. There is only one mango pulp plant in Multan and at least 1–2 more pulp plants, specifically for export purposes, should be established. Rahim Yar Kahan may be a good choice due to its strategic location in the mango zone. Exports can be made competitive by reducing the cost of production through investment on research and development, particularly to improve yield per hectare and reducing by at least 5%–10% postharvest losses and by reducing input costs and making efficient use of them. Public-private partnership can be helpful in achieving these prospects as well as in identifying potential markets and promoting Pakistani products through invited delegations.

REFERENCES

Clausing KA. 2001. Trade creation and trade diversion in the Canada-United States free trade agreement. *Canadian Journal of Economics*. 34 (3).

Government of India (GOI). 2013. *Union Budget and Economic Budget 2012–13*. Ministry of Finance, New Delhi, India.

Government of Pakistan (GOP). 2016. *Pakistan Economic Survey 2015–16*. Economic Advisor Wing, Ministry of Finance, Islamabad.

Government of Pakistan (GOP). 2013. *Pakistan Economic Survey 2012–13*. Economic Advisor Wing, Ministry of Finance, Islamabad.

International Trade Center (ITC). 2016. Trade Map, Market Analysis and research, ITC. http://www.trademap.org/

Panagariya A. 1999. *Preferential Trading and Welfare: the Small-Union Case Revisited*. Mimeo, U. Maryland.
Panagariya A. 2000. Preferential trade liberalization: The traditional theory and new development. *Journal of Economic Literature*. 38: 287–331.
Pasha HA and Imran M. 2012. The prospects for Indo-Pakistan Trade. *The Lahore Journal of Economics*. 17: 293–313.
State Bank of Pakistan (SBP). 2010. *Handbook of Statistics on Pakistan Economy 2010*. http://www.sbp.org.pk/departments/stats/PakEconomy_HandBook/Chap-8.1.pdf
Trade Development Authority of Pakistan (TDAP). 2013. *Normalization of Trade with India: Opportunities and Challenges for Pakistan*. Trade Development Authority of Pakistan. Government of Pakistan. http://indiapakistantrade.org/resources/Pakistan%20India%20TDAP%20Report%20Final.pdf
United Nations, Food and Agriculture Organization (UN FAO). 2012. On Line Data Base. http://www.fao.org
Zahra F. 2011. Sustaining the India-Pakistan Dialogue. Daily Times. 29th July, 2011.

38 Value Addition

Moazzam R. Khan, Aamir Shehzad,
Aysha Sameen, and Masood Sadiq Butt

CONTENTS

38.1 VALUE ADDITION OF FRUITS AND VEGETABLES

38.1.1 Introduction to Value Addition

Agriculture commodities are the biggest source of income (43.5%) of rural people and participate in 20.9% of GDP (Economic Survey of Pakistan, 2015–2016). Consumers are responsive to brand new fruits and vegetables regarding their health. In this regard, food sources having antioxidants, vitamins including vitamin A, C, and E, and minerals including calcium (Ca) and magnesium (Mg) are becoming more popular. The reduction in the risk of cardiovascular diseases, cancers of pharynx, lungs, colon and mouth, and strokes is highly associated with diets of fruits and vegetables (Riboli and Norat, 2003). Fruits and vegetables are great blessings of Almighty Allah. Value addition to fruits and vegetables is very imperative in Pakistan due to urbanization, industrial growth, globalization and socio-economic conditions. It should be done not only to satisfy processors and producers in term of profit, but also for better taste and nutrition. Value can be added to fruit and vegetable by altering their perishability, shelf life, color, and transform into different forms. The agro-ecological conditions of Pakistan are very much diverse where large varieties of fruits and vegetables can be grown. However, the losses in fruit and vegetable is very high; to prevent the loss of surplus fruit and vegetable they can be value added by processing and preservation at industrial as well as farmer level. This will enhance the processing industries development in the highly valuable area of fruits and vegetables. Additionally, this effort will increase the total production and profit of fruit and vegetable producers.

38.1.1.1 Fruit and Vegetable Production and Processing

Fruit and vegetable contain 65%–95% water and are considered as a living part of a plant. A healthy growth in the fruit and vegetable area has been observed in recent years. It plays a vital role in the economy of Pakistan by enhancing the income of people living in rural areas. Cultivation of fruit and vegetable is very laborious and time consuming, and hence generate an opportunity of employment for rural people. This sector also has a great potential to improve the agriculture growth of a country that can become the largest fruit and vegetable producer in the world. Nature blessed Pakistan with four seasons, which favors the different fruit and vegetable production. The annual production of fruit and vegetable is 6.65 million tons and 7.02 million tons respectively in Pakistan (GOP, 2010). A portion (25%–30%) of total production is wasted every year, which does not allow this sector to flourish at its peak potential. Different high temperature fruit and vegetable such as leafy vegetables, apples, tomato, oranges, plums, pears, litchis, peaches, and so on are produced in different regions of Pakistan. To avoid the wastage of fruit and vegetables, it can be processed into different products such as jams, jellies, marmalade, pickle, leather, chutney, puree, canned slices, nectar, and dehydrated slices and these products experience international reputations and have also gained importance in the European and US markets.

38.1.1.2 Supply Chain Management of Fruit and Vegetable: Issues and Challenges

The supply chain challenges for fruits and vegetables have been increased due to variations in eating habits and the increase in awareness towards food quality and safety (Reardon and Barrett, 2000). In Pakistan, the supply chain faces numerous issues and challenges which can be categorized as cold chain issues, fragmentation issues, integration issues, infrastructure issues, packaging issues, technological issues, farmers' awareness and knowledge, value addition and processing issues, financial issues, transportation issues and market demand, and information issues.

Cold chain of fruits and vegetables faces numerous issues like lack of cold chain network, insufficient capacity, and inadequate cold chain facility. Due to these issues it is very difficult for businessmen and farming communities to do their business more effectively and get more profit for their commodities. The huge numbers of intermediaries and local traders in the supply chain of fruits and vegetables has created a problem, which eats all the profit of farming communities. Integration and linkages among the farming community and other players of the fruit and vegetable supply chain is imperative to make the supply chain of fruit and vegetables more profitable and effective. However, the fruit and vegetable supply chain in Pakistan lack backward and forward integration among the stakeholders and farming community. In the fruit and vegetable sector, infrastructure is very important for the proper working of the supply chain. Adequate and proper infrastructure guides the farming community and other stakeholders of supply chain to do their business effectively and aid in delivery of commodities in wholesome condition and at a right time. Packaging is also important due to their perishability which requires suitable packaging material for proper handling. It is difficult to uphold freshness and increase shelf life without appropriate packaging. Cost of packaging material could also be considered during selection of packaging. There are many technical issues surrounding the technology, which include old machineries, obsolete techniques, inefficient technology, and advancement issues. Because of these challenges it becomes very difficult for a businessman and farming community to utilize a proper technique and technology to decrease their losses during operational activities. Farming communities in the fruit and vegetable sector have poor knowledge about the latest advancement in techniques and technologies to work efficiently and effectively. They also don't have proper information regarding the quality of seed, production technologies and postharvest management. Value addition and processing of fruit and vegetable is the best way to reduce the losses and enhance the shelf life. It also provides a great opportunity to farmers and other stakeholders to export their processed commodities to various countries. Transportation and financial issues play a very important role in fruit and vegetable supply chains. The challenges related to transportation of fruit and vegetables are very high due to lack of temperature control vehicles, high cost, and inaccessibility of good transportation modes. For efficient management of supply chain, authentic and proper information is required to farmers. Supply chain can't run in the upward direction without proper knowledge of demand and supply. The famers of fruit and vegetable in Pakistan don't have proper knowledge about the processing units, market demands, and supply, etc.

These issues can be addressed by adopting the following framework.

- Cold chain facilities should be established in an area of fruit and vegetable production, and these infrastructures can be provided by cooperative societies and private bodies
- Government agencies at farmer's level, can uphold the functions of a village level aggregator
- NGOs can solve the integration issues among the companies and farming community
- Government agencies or private bodies can take initiative to solve the infrastructure issues by providing grading equipment, installing semiprocessing units, loading and unloading equipment for value addition of fruit and vegetables
- Packaging units can be installed by private players, which can also provide opportunities for employment to peoples of that area
- Entrepreneurs in food technology, engineering, and technology should be established at rural level

- Extension services should be extended to deliver the marketing information (demand, supply) and modern technologies at farmer's level
- Govt. agencies should take steps to remove all intermediaries like retailer, wholesaler, and aggregator.

38.1.1.3 Quality Control and Food Safety in Fruit and Vegetables: Opportunities and Challenges

Fruits and vegetables are considered a highly perishable commodity that is easily deteriorated during the supply chain. Food quality and safety are important terms in the industry. In developing countries, fruit and vegetable spoilage and deterioration results in losses up to 50%. These losses, if avoided, will help the profit gain of producers and ultimately help the economy of developing countries. Fresh commodities are mostly contaminated and deteriorated by physical hazards, chemical hazards, and biological hazards, which include parasites, viruses, and bacteria.

Any foreign particle present in fruit and vegetable commodities and products that can cause injury are considered to be physical hazards. These fresh produce have a soft texture and high moisture content, which make fruit and vegetables vulnerable to injury, which can occur at any point of the supply chain. The injuries on the surface of fruit and vegetables make them susceptible to proliferation of microbes, enhance ethylene production, heat production, and fasten respiration rates. Physical hazard can enter in fresh commodity from contact with packinghouse, soil, contaminated equipment, manure, worker's habits, irrigation water, farm practices, domestic animals, fecal material from wild, improper storage, cross contamination, transportation vehicles, and distribution system (FDA, 1998; Bihn and Gravani, 2006).

Chemical hazards come from fungicides, contaminants, pesticides residues and heavy metals, and pesticide residues are considered to be highly imperative safety issues in the fruit and vegetable industry (Kader and Rolle, 2004). Recent studies indicated that 37,000 cases of cancer have been reported by the consumption of pesticides in developing countries every year (WHO, 1990). FAO (2002) reported that almost 3 million people suffered from pesticide poisoning and among them 0.2 million died every year in developing countries around the world. Tariq (2005) reported that pesticide poisoning in Pakistan should be assumed to be greater than estimated due to misdiagnosis, lack of data, and under reporting. Biological hazards may occur from microbes such as protozoans, bacteria, viruses, fungi, and helminthes. The outbreaks of biological hazards are mostly caused by Salmonella E. coli O157:H7 and investigation revealed that contamination is mostly occurring at preharvest stages. FDA (1998) issues its "Guide to Minimize Microbial Food Safety Hazards for Fresh Fruits and Vegetables." The practices in these guidelines are collectively termed as GAP (Good Agricultural Practices) which provide guidance on critical steps during production includes production, harvesting, processing, transportation, packaging, cooling and storage of fruit, and vegetables.

To overcome these safety issues following point should be considered.

- Hazards analysis critical control point (HACCP) should be adopted for identification, assessment, prevention, and control of hazard in fruit and vegetables industry
- General principle of food hygiene should be adopted to minimize the cross contamination in fresh produce
- Water used on farm should be free from chemical and biological hazards
- High degree of personal health, hygiene and sanitary facility must be established on fruit and vegetable farms
- Storage and transportation of fruit and vegetable from field to consumption should be of high quality to avoid chemical, physical, and microbial hazards
- Cooling facilities must be provided in areas of fruit and vegetable production for proper storage of fresh commodities.

38.1.2 Constraints in Value Addition

The food market in developing economies can be classified as A- B- and C-systems with different market channels and a variation in quality demand and safety. The A system is commonly composed of small-scale harvesters which distribute to a limited market (local) and a low income chain. Despite the fact that this market operates at a local level it can also be part of other market systems through middlemen. This usually builds to be part of extended chain while the added value is shared by a large number of players. The locations from production to consumption are distant and the producers have limited market information. In developing countries, despite the fact the A systems usually supply a large quantity of agricultural products, its significance is fairly low.

The B market system is the local supply chain which mostly aims at supermarkets ranging from middle to high income. The farmers in these supply chains usually operate at a small or medium level and are connected to each other in associations, cooperatives, and contracts. The amounts supplied by the B market systems are at large less than what is delivered by the A market system, however, it produces larger values yet. They also meet quality standards from domestic to, in certain cases, international safety standards for retailers to a greater extent than the A market system. The C market system is mostly focused on the export market; however, the products inappropriate for the export market are aimed to domestic markets. The C market systems are coordinated to a higher degree than the other market systems. In this system, fewer players exist and it delivers fewer products but with greater added value.

38.1.2.1 Infrastructure and Resources

The lack of affordable, reliable, and adequate infrastructure facilities touches the life of developing country's family and decreases the value of food materials. In developing markets, there are four significant constraints concerning infrastructure and resources.

- First, there is limited access to input resources
- Second, the geographic location of many producers are constrained by long distances to market location and end consumers
- Third, lack of skilled human resource and technology is a limiting factors for markets to develop for production and dissemination drives
- Moreover, there is inadequate infrastructure for information and distribution. Efficient distribution of products and information dissemination are the basic conditions for a supply chain to advance.

There are several scholars who argue that infrastructure is a certain constrain for value addition of fruits and vegetables throughout the supply chains in developing countries.

- Lack of proper infrastructure and resources for load, transport, process, and cold storage are responsible for food losses in the fresh supply chains of developing economies.
- Beyond these infrastructural problems absence of information infrastructure is a main hindrance for the option of improvement in the value addition. This problem is enormously related with the information gap between producers and consumers, which in turn results in difficulties to estimate the balance in supply and demand.

In developing countries, there are a large number of intermediaries along supply chains, which can match the immature infrastructure; however, it remains a huge cost for the chain. Local food chain systems, which denote various food systems, usually have disorganized dissemination infrastructures in developing countries. This is usually exhibited through distribution systems, which are decentralized and inflict huge transport costs for each unit. Several producers residing in the rural locations are penalized by insufficient transportation infrastructures, thus it is important in these locations to push for the development of collection centers, packaging, storage, and distribution infrastructure of agricultural products.

38.1.2.2 Low Degree of Value Addition

- Inability to manage raw material supply
- Poor financial support
- Lack of investment in supply chain
- Lack of training facilities for producer and processor
- High excise duty on packaging
- Lack of innovation
- Frequent interruption in power production/power supply
- Unequipped food analysis laboratories
- Inefficient skilled personnel and research and development activities
- Inappropriate postharvest techniques
- Value addition is hindered due to lack of operational funds
- Improper sharing of recent techniques between producer and processors
- Lack of training and demonstration activities and centers for fostering entrepreneurship
- Lack of cold chain transport facilities coupled with inadequacy of technical knowledge results in low quality value added products
- Unsatisfactory implementation of food safety & quality management system throughout food supply chain.

38.1.3 Looking Forward to a World of Value Added Foods

38.1.3.1 Supply Chain Management

Globalization and trade liberalization is inflexed by different supply chain management practices with multinational companies into the local retail sector through the establishment of "super retailers." Market structure, buyer-supplier relation, product growth, competition, marketing efficiency, innovation, price level, and producer is changed by the concentrated retail chains and the rise of highly consolidated inductries in some parts of the world. Its impact on developing economies like Malaysia is yet to be verified and the evidences is not consistent between countries. It has been observed that impacts from the rise of the consolidated retail chains in developing economies include marginalization of small market intermediaries and farmers, introduction of market innovations like new services, relating technologies, and products, and finally lower prices both to farmers and producers. The current structural problems, which are dominant in small farm sector, poses the question as to the ability of the small farmers to meet the rigid demands of the buyer, that is, large retailers. In order to improve the livelihood of rural small-hold farmers, small scale on-farm processing units must be subsidized by the government. Through contractual arrangements, the agro-industry can assist smallholders to shift from the subsistence or traditional agriculture to the production of export-orientated, high-value added products.

38.1.3.1.1 Control Atmosphere

Integration of a composite carbon dioxide absorbent scavenger material directly into a layer of the package itself or into the walls is a unique, cost effective, and safe way to remove carbon dioxide from the inside of a package. Coating of carbon dioxide absorbent is done into the layer as a blend, as the layer comes in contact with the carbon dioxide from the package. The layer possessing scavenging properties acts as multi-functional layer having barrier materials which ultimately is responsible for retarding moisture and oxygen. In the coated blend, a high percentage of scavenger absorbent filler is required for a high level of carbon dioxide absorption. It may be attained by combination of various percentages of an absorbent such as calcium oxide, calcium hydroxide, or other metal hydroxides in a polymeric matrix resin such as a low-density polyethylene. A fast and complete reaction with CO_2 is ensured by adding the moisture-retaining agent into the scavenger filler (forming the composite CO_2 scavenging filler). A twin-screw extruder is used to produce the scavenging blend. The degradation of resin is reduced by the addition of antioxidants into the blend. Compounding and absorption is also affected by the particle size of filler. Accumulation of subunits can decrease efficiency of the scavenger performance. In order to obtain the homogenous mixture of blend, the reduction of moisture content or drying of the composite scavenger fillers or polymers may be required. Surface area is the major factor for scavenging of carbon dioxide, and hence the larger the area of the coated surface, the greater the amount of the carbon dioxide absorbed by, and/or reacted with the scavenger.

38.1.3.1.2 Modified Atmosphere Packaging (MAP)

Chilling can slow down deterioration of stored foods but if the atmosphere surrounding the product is also modified to reduce oxygen concentration, the shelf life is increased considerably because of further reduction in the rate of chemical oxidation by oxygen and in the growth of aerobic microorganisms. In MAP, oxygen availability is reduced for fruit and different types of MAP could be further classified as vacuum packaging, controlled atmosphere packaging and true atmosphere packaging but its main function is to reduce the availability of oxygen. MAP was first recorded in 1927, as an extension of the shelf-life of apples by storing them in atmosphere with reduced oxygen and increased carbon dioxide concentrations. In the 1930s, it was used as modified atmosphere storage to transport fruit in the holds of ships and increasing the carbon dioxide concentration of surrounding beef carcasses transported long distances, which was shown to increase shelf-life by up to 100%.

Incorrectly designed MA packages may not be effective and even may cause adverse effects to the products, for example, shorten the shelf life. A dynamic and steady state conditions should be taken into consideration while designing the MA packaging, because if the product is exposed for a long time to unsuitable gas composition before reaching the adequate atmosphere, the package may have no benefit. The design of an MA package depends on a number of variables: the characteristics of the product, its mass, the recommend atmosphere composition, the permeability of the packaging materials to gases, and its dependence on temperature and the respiration rate of the product as affected by different gas compositions and temperatures.

The success of modified atmosphere packaging (MAP) greatly depends on the accuracy of the predictive respiration rate models. Due to the complexity of the respiration process, only empirical models have been developed. The particular variables that influence the O_2 uptake and CO_2 production should be identified and quantified for each fruit or vegetable product. Considerably, more research is needed in this area. Fresh-cut products bring more variables that may influence respiration rate, such as preparation method, cutting size, and time after cutting.

38.1.3.1.3 Edible Coating: An Innovative Value Added Tool to Mitigate Malnutrition

Nowadays, remarkable amount of drugs are being prepared from various food items for their therapeutic potential and can be used as regular medicine. Fruits and vegetables have been recognized as effective medicinal agents. Fruits and vegetables are gaining massive importance in the health of individuals and communities. Primarily, people are dependent on various plants for food as well as for sheltering themselves from physiological conditions. Developed countries have changed their daily routine diet according to their food guide pyramid and nowadays it is needed only in developing countries. These fluctuations increase the interest of consumers in foodstuffs that may mitigate the indications of diverse metabolic ailments and illness.

Fresh fruits have excellent cradle of energy, vitamins, minerals and fiber. The nutritional value of fruits is greatly depended on the quality as well as quantity of these nutrients. Among all fruits, guava, citrus and pineapple are rich source of vitamins and some of minerals. At present, there is a significant demand for fresh fruits and their processed products. As several kinds of fruits are seasonal and have limited shelf life, their processing becomes necessary to keep the quality and to ensure its availability throughout the year.

In order to curtail the challenges for processing of fruits and vegetables, quite a number of techniques have been studied aimed for value addition of fresh produce. Of the approaches, storage at low temperature, high relative humidity, and controlled/modified atmosphere packaging is deemed to be the most promising. Edible coatings can act as moisture and gas barriers, control microbial growth, preserve the color, texture, and moisture of the product, and can effectively extend the shelf life of the product.

Biofilms have received considerable attention owing to consumer demands for safe and edible grade material. It has been reported that edible coatings involving food grade emulsifying and wetting agents extend the shelf life of fruits through prevention of gas, water and solutes migration. These coatings have gained immense recognition in being environment friendly, biodegradable, nontoxic and a rather safe choice to be used in food applications.

It has been estimated that development of novel coatings with improved functionality are among the major challenges faced by the fruit processing industry. Furthermore, demand for healthier and safe foods has stemmed the trend toward edible coating. Edible coatings as oxygen and lipid barriers at low to intermediate relative humidity as the polymers can effectively make hydrogen bonds. An edible coating must have good sensory profile, acceptable color, flavor, taste, and texture with shiny look.

38.1.3.1.4 Fruit Leather

Fruit leathers are defined as the thin layers prepared from dried fruit puree with or without the addition of sugar and other preservatives. The leathery sheet like appearance is dedicated by the addition of pectin by gelatinization of matrix during dehydration that contains sugars and amino acids. Different fruits have been used for the manufacturing of fruit leathers such as apple, grapes, kiwi, guava, and mango etc. that nowadays, are common in sub continental along with North America. Contemporary, these fruit leathers are introduced as gourmet healthy foods. Fruit leathers are prevalent fruit made products that are mainly consumed in the Middle East and are very popular in Australia where children are their main consumer. Fruit leather has prolonged shelf life and consumed as snack fruit product. Usually leathers are in the form of thin leathery sheets, but it can also be processed in cubical and rectangular shapes. Processing of fruits in to leather is the best approach for conserving and utilizing of fresh fruits. Leathers are very popular, among the large number of products that can be prepared from fruits due to their ease in transportation and handling.

Processing of fruits into leather is the best approach for conserving and utilizing of fresh fruits. Ready-to-eat dried out guava produces like pieces of dried guava and pelt, are prepared with well-developed ripened guava drupes. These fruit preparations are given the name "Fruit Leather" as end product is aseptic and it has shiny texture like leather. Preparation of fruit leather at home is usually to preserve the fruit but some commercial brands also offer thin bars of guava fruit leather of varied composition. Along with minerals, vitamins and dietary fiber, energy intake of leather is very good and it also adds variety to our diet. Leather is very popular, among the large number of products that can be prepared from fruits due to their ease in transportation and handling. Countless formulations have been used to prepare fruit leather which includes diverse ingredients as honey, sugar, high fructose corn syrup, citric acid, maltodextrin, pectin, vegetable oil, lecithin, ascorbic acid, and many others.

38.1.3.1.5 Exploiting and Utilization of F&V's By-Products in Value Added Food Products to Curb Lifestyle Related Disorders

The antioxidant property of fruits and vegetables is mainly due to the presence of phytochemicals namely, flavonoids, vitamins, carotenoids, lignin, terpenoids, cumarins, saponins, and sterols. The fruit and vegetable along with its wastes are prime source of bioactive moieties including antioxidants, ascorbic acids, flavonoids, phenolic contents and pectin that are essential for the cure/prevention of many diseases hence, beneficial for human nutrition.

Phytochemicals is a term generally used for a diverse range of bioactive entities present in plants. These phytoconstituents are responsible for providing color, flavor and natural defensive system against pests however, these secondary metabolites are not essential for growth and developments however, confer some health beneficial aspects against pathogens and pests. The phytochemical profiling of these moieties avails the presence of these bioactive compounds in fruits by providing hints about the quantity and characteristics of these bioactive plant materials. These screening tests also assist as preliminary step in many research-related protocols as isolation, purification, and utilization of inherent plant-based bioactive compounds for pharmaceutics, nutraceutics, medicines, and agro-based industrial use.

Phenolic compounds present in plants are familiar as key composites responsible for preventing oxidation. Naturally occurring antioxidant phenolics can be categorized into a lipophilic cluster, tocopherols, and hydrophilic classes. They also include a number of simple phenolics, along with phenolic acids, flavonoids, flavonols, and tannins. Recently, chemists have exposed the chemical structures of different phenolics. There are still voluminous compounds that have not yet been entirely described and they are mentioned as phenolic extracts. Imran et al. (2013) concluded that high concentration of phenolics depends directly on the polarity of solids.

Antioxidants considerably avert the oxidation of a substrate if present in low concentrations. The presence of phenolic compounds, such as phenolic acids, anthocyanins and flavonoids, in addition to vitamins C and E and carotenoids impart these valuable effects to these foods. Phenolic content can be influenced by factors such as species, maturity, geographic origin, cultivation type, growth level, harvest, and storage conditions. Additionally, during the processing and storage of food, the bioactive compounds are vulnerable to oxidation reactions since a few of these compounds are unstable during thermal processing and cold storage.

Flavonoids are abundantly present in fruit juices such as citrus, water melon, and melon that have the ability to perform as antioxidants to reduce the level of oxidative oxygen species via their redox potential. Many studies have been conducted on fruit and vegetable waste to evaluate the quality and quantity of bioactive flavonoids in it, that is, polymethoxylated flavones (PMF) that have a spectrum of theraputical applications such as antidiabetic, anticancer, antiinflammatory, hypolipidemic, and anti-atherogenic properties (Ismail et al., 2016). The biomedical research also proved that by consumption of fruit and vegetable, one can attenuate the risk of gastric and stroke along with colorectal cancer.

The bioactive compounds from fruit and vegetable by-products such as peel and pulp and other fruit wastes can be evaluated as alternative to synthetic food additives, which are associated with negative effects on human health. Lipid peroxidation is a universal fact that is implicated with the promotion of rancidity in foodstuff that is associated with aging process and promotion of degenerative disorders (Khalil et al., 2017a). Numerous scientific studies have been performed to come across the functional properties of fruits as antioxidants that can be efficient in mitigating disarrays such as hypercholesterolemia, hyper glycemic index, brain dysfunction as well as cancer. Most of the food additives and supplements present in modern food and pharmaceuticals are synthetic compounds, with well-known harmful effects on human health. Many epidemiological and experimental studies certify that natural antioxidants derived from the diet can prevent the early onset of ROS (Reactive Oxygen Species) related diseases (Khalil et al., 2017b).

38.1.4 Conclusion and Future Prospects

It's almost impossible to compete with the global market nowadays by solely practicing traditional and outdated processing techniques. Pakistan is a developing country with limited land resources due to urbanization. The availability of agricultural land is decreasing proportionally daily due to continual increase in population. Therefore, there is a dire need to increase production of value added products to meet consumer demands. Purposely, appropriate funding at institutional level must be made to boost postharvest sector. Fruits and vegetables processing industry must apply latest preservation techniques to ensure quality value added products in order to increase consumer health. The economy of the country will be strengthened through value addition, quality enhancement, improved production, and greater foreign exchange earnings resulting from increased export of innovative and value added processed fruits and vegetables products.

38.2 VALUE ADDITION OF CEREALS

Cereals are considered as sources of dietary fibers, which represent almost 7%–15% of the grain. Dietary fiber has an outstanding implication as a key nutritional factor in a healthy diet. Plant foods are the only sources of dietary fiber, and among those plants cereals are considered as the best source. Dietary fiber includes polysaccharides such as pectins, gums, psyllium, and beta-glucans, which can be utilized for getting number of value, added products with allied health benefits. Although, different value added products like bread, biscuits, cakes etc. are available for masses but yet cereal processing industry is facing different challenges which need to be addressed for efficient utilization of cereals and improving overall health status of the community.

38.2.1 Barriers and Challenges in Cereal Value Addition

In Pakistan, wheat is the leading cereal crop followed by rice, whereas other cereals like maize, barley, oat, millet, rye, and sorghum are underutilized. The major challenges in cereal processing include pre and postharvest losses at different stages of the supply chain due to lack of storage and transport infrastructure, inability to manage raw material supply, lack of investment in supply chain, lack of training facilities for farmers and processors, weak regulatory system, involvement of middle men, poor technical choices, and lack of innovation. Similarly, frequent failure or interruption of power supplies, unequipped food analysis laboratories, inefficient market structures, lack of adequate manpower and lack of coordination links with academia, industry, and research organizations are also posing major problems in processing of cereals for their value addition.

38.2.2 Opportunities for Cereal Processing and Value Addition

Cereal grasses provide the grains that are staple foods for most of mankind and can be grown almost everywhere in the world. In their natural form, they are rich sources of proteins, carbohydrates, vitamins, minerals, carbohydrates, fats, and oils. However, when refined by the removal of the bran and germ, the remaining endosperm is mostly carbohydrate and lacks the majority of the other nutrients. In developed nations, cereal consumption is moderate and varied but still substantial. On the other hand, in developing countries, grain in the form of rice, wheat, millet, or maize constitutes a majority of daily sustenance. Pakistani population derives almost 60% protein from cereals and considered as cheapest and principal source of calories. Application of different technologies like, drying, aeration, storage, parboiling, pulse processing and extrusion will not only help in value addition of cereals, which are otherwise neglected or underutilized.

38.2.2.1 Seed/Grain Drying, Aeration, and Storage Technology

Sun drying is the common method for drying and is directly dependent on climatic conditions. Labor requirements and climatic conditions lead to poor operational performances. Being a fast and effective method of drying, artificial drying methods are more often used. To maintain seed quality mechanical drying methods are employed in which air is forced within the seed. Controlled temperature, air flux, and time guarantee the process efficiency. Different factors have to be considered while choosing any drying method, which includes drying time, seed volume and energy requirement, end use of seed, and also technical knowledge and purchasing power of producers.

Due to poor storage and management practices, Pakistan is facing numerous challenges. Normally, traditional approaches like godowns are used to store grains to preserve them from any foreign material and rodents and insects. Storage is the application of different techniques like fumigation, ventilation and temperature and moisture control to store seeds in heaps, bulk, bags, and store houses, also to conserve seed stocks for next year. Precautionary measures should be taken to keep away the admixture of different seed varieties. Quality is an important factor, which should be considered, and different seed grader and sieves are used for cleaning and screening of seeds. From maturity to germination time many devastating changes can occur in seed quality and seed can lose its capability. Hence, the main intention behind any storage process is to sustain the physiological quality of seed by decreasing the deterioration rate. Safe environment plays a key role to maintain a good quality storage conditions. The climatic conditions of Pakistan vary in different parts of the country, which favors all types of biological and environmental hazards. New techniques are employed to minimize seed spoilage. Formation of modern seed cleaning and drying equipment, warehouses and buildings, store rooms and silos, outside piles or bulks, and so on can be used for hygienic and batter quality grains. In Pakistan, traditional methods are still used for seed storage due to their simple application, but on a commercial scale, godowns, tower silos, and flat-type concrete storage buildings are easy to operate. Also, these storage facilities have different ventilation equipment to control moisture level and passing air out of the room to keep seed dry. The new and improved methods, on the basis of scientific principles, for seed storage will definitely lead to better quality control and reduce losses.

38.2.2.2 Rice Drying and Par-Boiling Technologies

Dying is done to lower the moisture content of grain to a level considered appropriate for storage. Delayed, incomplete, or ineffectual drying may decrease grain quality and result in losses. There are many diverse drying techniques used for drying rice involving several drying methods of different scales and complexity. These include field drying, sun drying, heated air-drying, and in store drying. Parboiling is a prehistoric way of rice processing, extensively used in underdeveloped and rice-exporting countries. Parboiled rice (PBR) is prepared by outmoded as well as current techniques. Current techniques are capital and energy demanding and are not appropriate for small-scale set-ups at the rural level. The indigenous parboiling devices range from pottery to boiler, using indirect or direct heating and single or dual steaming, which expend dissimilar energy levels. Agrowastes are the chief energy sources for local parboiling, particularly the remains of rice processing establishments. Still, sun drying is a usual practice in native parboiling techniques. The quality of PBR is influenced by the rice paddy, level of parboiling, drying parameters, moisture percentage after drying, and milling procedures (Roy et al., 2011).

38.2.2.3 Efficient Pulse Processing Technology

In the present era, consumers are demanding high quality foods that retains their freshness and nutritional level by the application of new food processing techniques. The focus on the development of new nonthermal processing technologies for cereal-based products had been increased dramatically over the past few decades owing to the demand for fresh, crispy, and natural colored-based baked products. From a safety point of view, the individuals are paying attention towards foodborne diseases, so, the usage of nonthermal treatments in baked products is recommended as irradiation and high pressure performance. In the case of irradiation, freshly baked products are treated with X-rays or gamma rays to destroy bacteria and/or molds before packaging. Hence, food irradiation is sufficient against a variety of pathogens that can easily grow on breads, cakes, and other baked products even at refrigerated temperatures due to lower shelf life. By the usage of irradiation in Pakistan, we can extend the shelf life of food by reducing the content of spoilage organisms after processing. Irradiation can also inactivate the enzymes that cause or assist the food spoilage.

Likewise, high-pressure processing (HPP) doesn't affect the nutritional value as well as color, flavor and textural properties of foodstuff. Hence, HPP is a possible solution for the production of safe, fresh foods with improved organoleptic properties, extended shelf life and enhanced functional properties. Therefore, the successful application of HPP in Pakistan can make it possible to commercialize high quality frozen baked products as well as frozen dough which can sponsor huge financial aid to baking industries. Although, the application of latest technologies, techniques, and methods can ensure that food is perfectly fresh, safe, cost effective, and nutritionally sound.

38.2.2.4 Extrusion Technology in Cereals

Application of extrusion technology in cereals can improve their protein functionality. Snack products are predominately extruded and prepared from starch or rice flour containing very low levels of protein hence, reduced biological value and less concentration of essential amino acids. Recently, a great number of scientists have attempted the fortification process for these extruded foods with amino acids along with addition of various cereals. The lysine rich cereals have higher content of proteins as well as amino acids so they can be used for fortification. Legumes, in addition to cereals, such as soy, red kidney, and corn can be used for the preparation of higher nutrient content products. Consumers also have concerns about the nutritional quality as well as functional properties of fortified products. However, extrusion on the other hand can have positive impacts on protein functionality, product texture and shelf life which can be effectively applied for producing cereal-based extruded products like wheat, rice, oat and corn flakes, modified starches, and other breakfast cereals.

38.2.3 Suggestions/Solutions for Value Addition

Different processing techniques can have positive impacts on cereal value addition:

38.2.3.1 Improving Protein Quality of Cereals through Blending

The pervasiveness of protein energy malnutrition is mounting in the developing world, particularly in Pakistan due to scarcity and the consumer's dependence on plant sources to fulfill their energy necessities. The food divergence is one means to abolish the protein energy malnutrition. Amongst all foods, pulses grasp worth for their exploitation in cereal-based foodstuff to progress the protein quality. Cereals are staple foods for human nutrition and their combination into various foodstuffs is of great financial prominence. However, wheat contains less quantity of protein and also lacking in particular amino acids causing the problem of malnutrition (De-Frias et al., 2010). Due to inadequate quantity of food proteins, legumes are used as nutritional supplements in wheat flour to curtail protein malnutrition.

Legumes are vigorous carriers of dietary protein for large segment of the world's inhabitants. Legumes are rich in protein and complex hydrocarbons in addition to the presence of considerable amounts of bioactive components and minerals. From the nutritional point of view, legumes are of specific curiosity for the reason that they cover high quantity of protein (18%–32%) (Boye et al., 2010). For these reasons, legumes are a perfect supplement to cereals in vegetarian diets with augmented attention and concentration as functional ingredient.

Similarly, chickpea flours can be an outstanding option for upgrading the nutritional significance of *chapatti* and bread. In chickpea, the chief restrictive amino acids are methionine and cystine, tailed by valine and tryptophan. Experimental work has shown that chickpea flour might be efficaciously combined into products at up to 20%, to formulate foodstuffs improved in terms of color, taste, texture, and overall acceptability. Also, chickpea can also be combined in biscuits up to a level of 50%. Several other studies have also concluded that mungbean and chickpeas are ideal candidates for refining the protein contents of cereal-based products (Pasha et al., 2011; Hefnawy et al., 2012). Wheat flour can also be blended with flours of other underutilized cereals (maize, barley, sorghum, millet, oat and rye) in order to lessen the pressure on wheat; additionally, the same areas can be utilized for the cultivation of other crops.

38.2.3.2 Rice/Corn Bran and Oils

Waste from the cereal processing industry like wheat, rice and corn, bran, corn cobs, shorts, and other byproducts can also be effectively utilized for producing value added baked products. Bran is a multifaceted structure comprised of pericarp, nucellus, seed coat, and aleurone. It is the outer covering of rice endosperm. When rice bran is attained as a by-product in the course of milling process, it covers some portion of rice endosperm, rice germ, and aleurone layer, which are ionic carriers of carbohydrates, proteins, vitamins and trace minerals. Owing to the occurrence of elevated levels of phytonutrients such as tocotrienol, oryzanol, and phytosterols, rice bran holds exceptional properties that render its appropriateness for the fabrication of value added products in neutraceuticals and pharmaceutical industry. Both corn and rice brans can be used for the preparation of cookies. After the oil has been extracted from rice and corn brans, the remainder is deoiled rice/corn bran, and is also beneficial to humans health due to a low fat content. Keeping in view the nutritional significance of defatted rice and corn bran, these can be utilized in preparing ready-to-eat breakfast cereals improved with defatted rice/corn bran. Al-Okbi et al. (2014) studied the manufacture of corn flakes and tortillas chips, by supplementation of gelatinized corn flour with rice bran from 10% to 30% and determined that the extreme breakdown viscosity and color quality was disturbed and values of sensory parameters were reduced while protein fraction was augmented depending on the intensity of rice bran. Oil from any single source has not been found to be suitable for all purposes, as oils from different sources generally differ in their composition and demands the exploitation of new sources of oils (Chatha et al., 2011).

At present, almost 3.9 million tons of rice paddy is produced in Pakistan from which about 240 thousand tons of rice bran can be obtained, which can produce about 33,000 tons of edible oil, worth Rs. one billion. It is a huge import substitution commodity. Therefore, it must be exploited as a substitute source of edible oil. Gamma-oryzanol, tocopherol, tocotrienol, squalene, and other phtyosterols in rice bran oil enjoy high antioxidant property to scavenge free radicals (Chatha et al., 2011).

38.2.3.3 Utilizing Agro-Industrial Waste

Every year, the fruit processing industry is destroying a substantial quantity of bioactive material which could perform a strong role to avert and medicate numerous illnesses. Dietary fibers obtained from fruits and vegetables have amounts of soluble dietary fiber that highlights the importance of dietary fiber as a functional ingredient. Dietary fiber amalgamation lengthens shelf life, amends the physical and structural attributes of product such as texture, water, and oil holding capacity, viscosity, and sensory assets. The biowaste of many fruits (citrus, mango, apple, and pomegranate etc.) can be dried to powder form and used in different bakery products, which can have positive effects on compositional, antioxidant, and the sensory properties of various cereal products. Plentiful research work exists highlighting the use of fruit and vegetable waste in bakery products like the use of apple and orange pomace for supplementation in the dietary fiber matter of cookies. Apple pomace, having functional ingredients, is considered a plentiful carrier of polyphenols and a good antioxidant and antiproliferative. In the same way, grape pomace, is a good source of phenolic acids, catechins, anthocyanins, and flavanoids in addition to dietary fiber (Aslam et al., 2014). This area needs special attention for improving not only the health status of individuals but also utilizing agro-industrial waste in an effective way.

38.2.3.4 Gluten Free Cereal Products

The gluten free term is associated with the foods containing less than 20 ppm of gluten, however, the foods claimed as very low gluten are used for foods that have been especially processed. The key aspect of a gluten free diet is the absenteeism of gluten which may lead to different nutritional significance including deficits and imbalances. The nutritional sufficiency of gluten free diet is chiefly important in children that need highest nutrient requirement and energy for growth, physical activity as well as development. So, maximum attention is needed towards nutritional status of gluten free food products available on the market. Amongst other alternatives, the consumption of different pseudo-cereals like quinoa, as buckwheat, amaranth, and other minor cereals can exhibit a strong alternative, which can be used as a gluten free product, as they are excellent sources of carbohydrates, dietary fiber, protein, polyunsaturated fatty acids, and vitamins (Dyner et al., 2008). If the consumption of gluten free and pseudo-cereals is successfully implemented, consumers will have more awareness on the availability of locally prepared gluten free foods which can help in increasing the consumption of these foods, resulting in an economical and balanced diet along with a wide range of varieties.

Along with quinoa, pseudo-cereals such as buckwheat, amaranth and other minor cereals can exhibit a strong alternative and can be used as a gluten free product, as they are excellent source of carbohydrates, dietary fiber, proteins, polyunsaturated fatty acids, and vitamins. Therefore, diversified food grains can meet unique and specific requirements of different communities.

38.2.3.5 Innovative Processed Foods

Baked products have been recognized as best vehicles for value addition, although there are positive effects on the physicochemical properties of baked products of value added products along with health benefits (Tuncel et al., 2014). Baking is a complex process and results in many physical and biochemical effects including structure formation, taste development, color formation, and synthesis of health promoting and health impairing constituents (Haase et al., 2012). In baking processes, the prime ingredients are wheat flour followed by fat, eggs, sugar, water, and salt along with minor ingredients such as baking powder, emulsifiers, preservatives (optional) with milk that are mixed together until a homogenous dough is formed having the proper distribution of all ingredients. Although quantity of all these elements describe about the nature and consistency of baked product in addition to effect of heating and time temperature relationship (Mamat et al., 2010). So, the production of baked products through incorporating fiber sources (fruit peels, wheat, rice, and corn brans) will not only produce a variety of products but also increase nutritional profiles by improving the overall health status of the individuals.

At the same time the following measures can ensure cereal value addition in more effective way:

- Provision of infrastructure and facilities for drying, transportation, efficient processing, and storage of cereals and their products
- Application of innovative technologies in cereal processing
- Establishing grain milling and processing training institutes
- Provision of small and easy credit facilities by banks and financial institutions
- Establishing Food Parks and Technology Transfer Centers at district level for capacity building of farmers and grain processors
- Encourage direct marketing of products by the farmers, thus avoiding middle men
- Revision of Pakistani regulatory standards and marketing strategies for improving overall quality of cereal products
- Focus should be given on brand building by creating awareness among consumers
- Developing strong linkages between industry, academia, and research organizations for providing solutions to their problems with coordinated efforts

38.3 VALUE ADDITION OF MILK IN PAKISTAN

38.3.1 Introduction

Pakistan has improved its rank in terms of milk production in the world during past few decades. The dairy industry of Pakistan contains small as well as commercial farms but is mainly dominated by small and dispersed farmlands. Milk is collected from rural areas through local milk collectors and dairy industries. The milk supply chain is very poor in Pakistan due to lack of cooling facilities, so large amounts of postharvest occur. To prevent losses, milk is consumed near the site of production or used to make products with longer shelf lives. Different dairy industries are involved in production of value added products at formal level but Nestlé is the biggest player. Mostly yoghurt, cheese, butter, UHT, and milk powder are produced. In spite of hight milk production, domestic demand cannot be fulfilled and large amounts of powdered milk are imported. Main obstacles in the development of the dairy sector are low genetic potential of current dairy animals, dispersed farmlands, poor supply chain infrastructure, insufficient research in the dairy sector, unskilled persons and the lack of laws and regulations to address quality of milk and milk products. The current situation can be improved by improving the situation of small farmers, developing corporate dairy farming, improvement of milk supply chain mainly by forming dairy hubs, integrated on farm processing, carrying out more research in the dairy sector and by the implementation of laws and regulations.

38.3.2 Current Situation of Value Addition

38.3.2.1 Milk Production and Consumption

Pakistan is amongst top five milk-producing countries with 52.632 million tons (MT) annual production as described in the National Economic Survey of Pakistan (2014–2015). Cows, buffaloes, goats, camels, and sheeps are mostly used for milk production. In 2014–2015 gross milk production from cows was 18.706 MT, buffaloes 32.180 MT, camels 0.862 MT, goats 0.845 MT, and sheeps 0.038 MT. In rural areas, people mostly keep 1–4 animals for their own needs and the excess milk is converted into products with longer shelf life as desi ghee. Urban or periurban milk producers keep animals on small or large scale (6–50 or more) near cities, which reduces the cost of transportation and also the chances of milk spoilage. They sale milk at higher prices by producing milk at low cost. In commercial farming, 30 or more animals are kept and better management practices are used for rearing. Mostly animals are produced by artificial insemination techniques so they have better genetic traits. Globalization and urbanization although has shifted trend towards commercial milk production but still 43% households keep 1–2 animals while 27%–28% has 3–4 animals (Burki et al., 2004). A study conducted by Planning and Development Division of Pakistan (2006) indicated that out of total milk produced share of smallholder farmers is 80%. They are geographically dispersed and present at far off places of the country so 60% milk produced by them is consumed at the site of production while the remainder is marketed by dodhis to urban areas. According to the Economic Survey of Pakistan (2014–2015), gross milk consumption in 2014–2015 was 42.454 MT, to which cow milk contributed 14.965 MT, buffalo milk 25.744 MT, camel milk 0.862 MT, goat milk 0.845 MT, and sheep milk 0.038MT. Out of total milk produced 71% is consumed in rural and 29% in urban areas.

38.3.2.2 Supply System of Raw Milk

In Pakistan, about 95% of the total milk produced is collected via informal channels, particularly by traditional milk collection agents called dodhis. These dodhis can be classified into three categories on basis of their operation as small or *katcha dodhis* that collect up to 100 liters of milk each day, medium sized or *pucca dodhis* that collect about 400–800 liters milk/day and large dodhis or contractor that collect milk from medium sized dodhis and distribute about 40–70 mounds/day (Ali et al., 2011). Marketable surplus fluid milk is collected by kacha dodhis from several small producers and transported either to the milk collection centers, milk shops, or sold directly to consumers. They have poor transportation facilities (bicycle or motorcycle) to transport milk while pacca dodhi have better transportation facilities (horse driven carts or vans) to collect milk in large quantities from remote areas. They supply to milk shops or collection centers usually after decreaming (Sarwar et al., 2002). As a whole, dodhis sell 80% milk to contractors, 10% to collection agents of processing plants, and 5% to bakers or confectioners while contractors sell 90% of milk to shops, processing plants, and large scale bakers, and confectioners (Ali et al., 2011).

Punjab is home to one of the largest milk supply chains in Asia. It has different milk collection units as Nestlé, Haleeb, Halla, Engro foods, HFL, Prime, and Nurpur. In 2011, Nestlé Pakistan completed 23 years of milk collection from rural areas of Punjab, while other milk processing plants have also made significant contributions in milk collection over the last 2 decades. The supply system of raw milk in Pakistan can be divided into three categories based on its function. First is self-collection of milk from farmers by dairy plants just like milk collection model of Nestlé; second is milk collection by third party on behalf of processing units like milk collection system of Nirala, Haleeb and Noon dairies; while the third one is farmer cooperatives as Halla (Idare-e-Kisan). Commercial dairy farms are although evenly distributed in Punjab but surplus amount of milk in milk supply chain is mainly collected from southern and central districts of Punjab due to low population density. Large milk processing units prefer direct procurement to ensure high quality end products, but dispersed farmlands make it difficult and mostly processors have to rely on local milk collectors. Industries have established quality criteria to ensure raw milk quality and most dodhis now use cooling facilities for milk transportation to fulfill these standards (Burki and Khan, 2011).

38.3.2.3 Milk Processing

Milk is used to produce different value added products apart from direct consumption. Out of the total milk produced only 3%–4% is processed through formal channels, while 96%–97% by processed by the informal sector. At informal level milk is processed to produce lassi, khoya, dahi, desi ghee, and butter. Production of value added products by the informal sector involves traditional processing methods so these products do not fulfill hygiene and quality standards. At formal level milk is processed into pasteurized, UHT, powdered milk, condensed milk, yogurt, butter, cheese, and ice cream. From the total milk available to plants, 50% is processed into UHT, 40% into skim milk powder (SMP), and 10% into pasteurized milk and other products (Younas, 2013). The general public usually prefer local products due to their low price but a study conducted by Unilever Pakistan Ltd. Indicates increasing trend (at the rate of 20%) for consumption of high quality processed milk with the increasing awareness in public about health and safety issues.

According to a prefeasibility study of dairy processing plants, conducted by Planning and Development Division of Pakistan (2006), milk processing was started in Pakistan in 1960 with the installation of the first milk sterilization plants. Twenty-three different types of milk pasteurization and sterilization plants were established by mid-1970s. Huge loans were sanctioned by the Agriculture Development Bank for import of these processing plants into the country to establish a milk processing industry along modern lines. Due to poor milk collection system, unskilled labor, less shelf life of final product, and after establishment of UHT plant in 1977, all of these plants had to shut down except of Idara-e-Kisan and Army Dairy plants. Punjab has a unique characteristic of having more than 20 private milk processing industries that compete in terms of milk collection as well as production. It has global giants as Nestlé, Halla, and Haleeb. Other milk processing plants include Nirala, Dairy bell, Noon, Dairy crest, Army dairies, Engro foods, HFL, Millac, Vita, Nurpur, Nirala, Military dairy farms, Karachi dairies, and Premier. Most milk processing plants are present in Lahore or its nearby areas, so milk comes here from all over Punjab and that's why it's called a dairy hub. Total processing capacity of all these plants is 5.3 million liters but average daily processing is 2.7520 million liters, which is about half of their capacity.

Nestlö is biggest player among all processing units with 1.040 million liters average daily production that collects 1040 tons milk daily from 140,000 farmers of 3500 villages (Burki and Khan, 2011). In urban areas, as in Lahore, Nirala and Gourmet process dairy products under their own brand names and use their sales points for distribution and sale of these products. Some shopkeepers and selling units purchase these products to put them on their own shelves for sale to the consumers. Halla and Adams sell their products to large number of local shops and stores by providing them freezers to improve marketing (PDDC, 2006). According to Small and Medium Enterprise Development Authority (SMEDA) market share of loose milk from milk men is 90%, of UHT milk 4.598%, pasteurized milk 3.76%, open milk sold at shops 0.98%, direct to home 0.2% and remaining is for other products. Mostly milk is produced away from the site of consumption, so transportion to remotest areas as Gilgit and Gawadar is impossible due to its perishability until it is UHT treated.

38.3.2.4 Export and Import of Value Added Dairy Products

The dairy sector of Pakistan, although very important for the national economy, contributes significantly to the GDP. Despite continuous growth and large potential it still is facing many constraints such as dispersed farmlands, low average milk per unit, supply chain issues, and less value addition due to which the export potential of the dairy sector has not been completely exploited. More than 90% of products are exported to Afghanistan, which has no legislation for import of dairy products and faces problems in domestic milk supplies due to deterioration of national herds as a result of the decades' long war. Data analysis indicates that Pakistan exported its dairy products worth USD 64.734 million to Afghanistan, Singapore, Oman, Tunisia, Saudi Arabia, the UAE, the United States, and Iran in 2011. Fresh milk was exported to Afghanistan and Tunisia, milk and cream powder to Afghanistan and the United Kingdom, buttermilk, curdled milk and cream to Afghanistan and the United States, cheese and curd to the United States and the UAE, and milk powder to Afghanistan, the United States, Hong Kong, and South Africa (TRTA, 2013).

Pakistan's domestic milk and milk products supplies are not sufficient to meet the local demands, so many dairy products are imported from other countries to fill the gap between supply and demand. In 2011, powdered milk of USD 89 million was imported from the United States, New Zealand, Turkey, France, Singapore, and Denmark. Cream powder was imported from France, Malaysia, New Zealand, and Ireland for USD 6.9 million in the same year while curd and cheese of USD 3.9 million was imported from Australia, New Zealand, Saudi Arabia, and Denmark. Milk cream of USD 1.36 million was imported from Malaysia, the UAE and the Netherland and butter was imported from Denmark, Italy, and the United States of USD 0.25 million in 2011 (TRTA, 2013).

High establishing and sustainability costs of powdered milk plants, and its greater demand in urban areas and metropolis, where access of local fresh milk suppliers is not possible, has made local milk production unable to meet the needs. Another reason for so much import was the attractive prices of powdered milk in past few years due to reductions of global milk prices by greater production. Due to these reasons, in a few years only demand for powdered milk increased from USD 24.39 million (2009) to USD 117 million (2014). According to "The Nation" (2016), Vice Chancellor of the University of Veterinary and Animal Sciences (UVAS) Prof. Talat Naseer Pasha stated, referring to the United Nations database, that Pakistan has imported 35 million kg, 22 million kg, and 34 million kg of powdered milk in 2012, 2013, and 2014; a worth of USD 102.1 million, 70.8 million, and 117 million, respectively. At the same time, import of skimmed milk powder by the Pakistan was 19.5 million kg in 2012 of worth USD 13.4 million, 18.3million kg in 2013 of worth USD 15 million, and 20.2 million kg in 2014 of worth USD 16.9 million. This unchecked import of milk powder not only hampered the growth of the dairy sector but also affected the livelihood of 40–50 million people involved in milk production in Pakistan.

38.3.2.5 Legislations Addressing Value Addition of Dairy Products

Legislative framework that addresses food processing in Pakistan includes the Pure Food Rules (1965) and the Pure Food Ordinance (1960). Some of important aspects of these laws are following:

Milk containers should be properly labelled (Section 18) improper enameling and coating of containers is illegal (Section 19), milk from diseased animal is illegal (Section 20), persons with contagious diseases cannot handle milk at any stage (Section 21), pasteurization and sterilization parameters have been considered (Section 22) and equipments that require approval from government to use in milk processing plants have been specified (Section 23). Pure Food Ordinance prevents milk adulteration, sale of unsafe food and emphasizes on laboratory testing of food for safety evaluation. Sections 272 and 273 of the Pakistan panel code deals with penalties of milk adulteration. Pakistan standards and quality control authority (PSQCA), established in 1996, is responsible for controlling the quality of all foods including dairy products. Different public sector companies formed to regulate dairy products includes Pakistan Dairy Development Company (PDDC), 2006 and Livestock and Dairy Development board (LDDB) 2007. Objectives of PDDC are to improve dairy research facilities, cold chain system, promotion of healthy pasteurized milk, developing commercial dairy farms, improving breed management, facilitating credit financing to dairy farmers, and linking rural farmers with the markets at the national level while LDDB aims to improve extension services for dairy farmers, enhancing research capabilities in priority areas of the dairy sector and enhancing national, provincial, and donor linkages. At provincial level, under the Punjab Food Safety and Standard Act (2011), the Punjab government has established the Punjab Food Authority which is contributing an important part in controlling milk and milk products quality.

38.3.2.6 Comparison with Other Countries (Developed and Developing Countries)

The Pakistan dairy sector should learn from major producer of dairy products of world, such as New Zealand and the United States. New Zealand contributes only 2% in world milk production but 30%–40% in export of dairy products. Postharvest losses are minimum due to mechanized milking and on-farm cold storage of milk so 100% milk is delivered to processors and 95% is used to produce dairy products mainly whole milk powder, butter and to some extent cheese and skim milk powder. The United States is among the largest producers of milk with an annual production of 94.5 MT in (2015), which is expected to grow up to 96.3 MT in (2016) (USDA, 2015). In the United States, cows are milked by machines, in milking parlors, and milk is stored on-farm in cooling tanks before being sent to formal processors. About 99% milk is delivered to processors for value addition (Hemme and Otte, 2010). The United States is among the top five exporters of butter, cheese, and SMP. By 2024, the United States will account for one third of SMP export.

India is world's largest milk producer with an annual production of 109 MT in 2014 (Rasheed, 2014) and a 15% share in global milk production. The country had no share in export and import of dairy products before (2000) but now is contributing in world trade after implementation of dairy flood programs. Although it had a 0.4% and 0.3% share in global imports and exports respectively, the situation is continuously being improved. The dairy sector of India showed a spectacular growth from 1971–1996 and this period is called operation flood area. During this period, milk production increased from 21 MT to 69 MT. The dairy sector development of India is also due to cooperative movements in dairying. In India, Amul has become a household name. It is operated by Gujrat Cooperative Milk Marketing Federation (GCMMF), jointly owned by 3.1 million milk producer families, of which 70% are landless workers and small marginal farmers. GCMMF consists of more than 15 district cooperatives and 5700 village dairy cooperative societies. Brand got unqualified success in 2009–2010 with daily milk procurement greater than 4.9 million Liters and annual revenue of Rs. 8000 crore. The Amul model consists of a three tier cooperative structure having Dairy Cooperative Society at village level attached at District level to a milk union and further federated at state level into a milk federation. This structure facilitates milk collection from villages, procurement and processing at district milk union and marketing of milk and milk products at state milk federation. This approach has eliminated internal competition and also improved the economy. GCMMF has largest milk handling capacity in Asia and largest cold chain network. It consists of 48 sales offices, 5000 wholesale distributors, 700,000 retail outlets, and exports milk products to 41 countries. Indian's National dairy development board has exported GCMMF model to other countries after its success in India (Faheem-ul-Islam and Qureshi, 2012).

38.3.3 Flaws in Existing Situation

38.3.3.1 Low Yield of Milk

In Pakistan, out of 52.632 MT of milk, 32.580 MT are from buffaloes. According to LDDC, buffaloes have the ability to convert poor quality roughage into milk but their genetic potential for milk production is lower (1800–2400) than cross bred cows (2000–2700/305 days). Milk yield from dairy animals is lower in Pakistan than in Germany, the United States, and New Zealand. Pakistan produces 35 billion liters milk from 5 million animals/annum while the United States produces 94.5 billion liters from 3.4 million animals it means we have 1.6 million animals more than the United States but produce 60 billion liters less milk annually. Similarly milk yields in Germany are 5 times greater then in Pakistan and in New Zealand is 3 times more. On average, a dairy animal in Pakistan yields 6–8 times less milk than animals of the developed world (Khurshedi, 2012). However, in India milk from one animal is just equal to 60% of the milk produced by one animal of Pakistan (Burki et al., 2004).

38.3.3.2 Poor Infrastructure of Milk Supply Chain

In Pakistan, milk-marketing infrastructures are insufficient to ensure product quality. Proper milk transportation requires provision of cold chain so that milk quality can be maintained during transportation. Milk is a perishable commodity and its transportation without refrigeration facilities causes postharvest losses. Due to lack of cold chain, pasteurized milk has failed in the Pakistani market, while UHT and powdered milk are successful. According to ADB report, about 20% milk is lost due to insufficient cold chain facilities (Iqbal et al., 2008).

38.3.3.3 Milk Adulteration

Milk supply and demand gap has forced man to adopt many ugly practices. Oxytocin is injected into milking animals to increase production , which leaves hormone residues. Many adulterants are also added in milk as water, ice, urea, detergents, formalin, starch, singhara, SNF (Solid not fat), and vegetable fat that cause many health problems (Kishor and Thakur, 2015). That's how we changed this valuable gift of God into maligned one. According to an annual report of PFA (2013–2014), 1106 milk samples were checked by raiding at entry points and 525 samples were found to be adulterated; 38,833 liters of milk were discarded. A total of 4406 milk shops were visited by PFA and 1319 samples were found to be adulterated and amount of milk discarded was 25,594 liters. During these visits, milk was checked for urea, formalin, detergent, starch, and specific gravity. Due to nonconformity with the standards, 1064 milk shops were sealed dealing with milk business and this number is highest among all other premises sealed dealing directly with food commodities. This is the situation in only one city of Pakistan; the same can be seen all over in Pakistan.

38.3.3.4 Low Degree of Value Addition

Seasonal availability of milk negatively affects the processing capacity of dairy sector. Milk production falls to 55% of peak production during mid-June while demand increases 60%, as compared to December (PDDC, 2006), particularly due to more consumption of yoghurt, ice creams, and other refreshing dairy products during this time, so some processors have to shut down their processing units due to unavailability of milk. Another problem faced by milk processors is payment of the highest taxes and investments although they receive only 3%–4% milk for processing. Milk production by small and dispersed farmlands increases on-site milk consumption and consequently reduces value addition at a formal level.

38.3.3.5 Lack of Awareness and Training of Different Stakeholders

"Awareness is a major hurdle in smooth running of different initiatives undertaken by public and government departments to uplift dairy development" was said by 90% of the stakeholders of the public and private sector during a survey. It's not only a problem for farmer communities, but also a major issue for workers of industries and staff of government organizations. In Sindh and Punjab provinces, different programs initiated by government for livestock and dairy development proved to be successful when awareness was provided by using the media (TRTA, 2013).

Farming communities have little knowhow about nutrition, disease management, effects of poor sanitary conditions during milking, and proper milk handling methods, which not just reduces the yield of milk but also its quality. General experience indicates that just giving balanced feed to dairy animals can improve the milk yield up to 30%/animal. Quality of nutrition in animal feed is most important during lactation and improper feeding reduces milk production per lactation. Artificial insemination for breeding of animals is not introduced in many areas of Pakistan and in others farmers are unwilling to use this technique due to lack of knowledge about its benefits. Only commercial farms can get advantage of this technique (Burki et al., 2004). Untrained veterinary persons make proper disease management unlikely that results in poor health of animals and ultimately milk quality and yield while there are strict regulations for the import of dairy products made from milk of animals suffering from contagious diseases as foot and mouth disease. Traditional milking methods increase microbial load at initial points of milk production. Similarly during processing only a few industries fulfill the standard conditions for hygiene and sanitation.

Another barrier in import of value added products is lack of traceability data, mainly due to dominance of uneducated persons in farming communities. At industry level membrane technology and other advance techniques for value addition are still inapplicable due to lack of awareness. To assist farming, communities employed work forces of the provincial government who also lacks the required knowledge. In absence of a proper evaluation and monitoring mechanism, appropriate results cannot be obtained.

38.3.3.6 Inadequate Research and Development

It's surprising to have a plethora of research describing multiple dimensions of the crop sector of Pakistan but an absence of meaningful economic data regarding the dairy sector. No serious efforts has been made regarding uplifting the potential of dairy sector in creating influence on rural economy ranging from increase of indigenous milk production, its efficient distribution, scope of dairying and its role in poverty alleviation (Burki et al., 2004). Research regarding the role of cooperatives in milk collection, its cold storage, and marketing is also deficient. The possibility of pasteurization and sterilization of milk and its processing costs is not known for milk transfer at distant areas without potential losses. There is lack of efficient methods for production of powdered milk. Potential of advance membrane techniques should be identified to utilize them in a better way in industry. Microbial culture and enzymes are mostly exported from other countries for fermented products due to absence of industry for enzyme synthesis in Pakistan that increases the cost of production of value added products.

As far as transfer of knowledge, from research to the concerned stakeholders, proper extension services are not present so there is poor utilization of developed methods; for example, experts of animal husbandry have knowledge about proper feed of animals but such education is not provided to the farming community. First there is limited knowledge in the RD sector and second there is underutilization of the present knowledge due to the gap between research and communication.

38.3.3.7 Standardization, Compliance, and Traceability Issues

Milk is among seven top foods to be adulterated and its rate is very high in Pakistan due to the absence of legislation (Awan et al., 2014). Unsanitary conditions, improper milking, and milk handling results in heavy microbial load measured in terms of Total Plate Count (TPC)/mL. Figure for TPC reaches up to millions/mL depending upon conditions making milk produced by informal sector highly unfit for human consumption (Iqbal et al., 2008). Consumers will be subjected to exploitation in the absence of well-placed and appropriate standardization and monitoring regimes. Therefore it is a challenge for Pakistan's dairy industry not only to produce safe milk for human consumption but also produce exportable surpluses which can fulfill international standards.

38.3.4 Way Foreword

38.3.4.1 Corporate Dairy Farming

For the dairy sector, development and a quantum increase in milk production which focuses on small holders will not be sufficient. A two-tier approach will be required to improve the conditions of smallholder farmers and to develop large corporate farms. Corporate dairy farm are legal entities registered with the Securities Exchange Commission of Pakistan (SECP). These farms are large and can be operated by using advanced technologies. According to Afzal (2008), the government has approved a Corporate Agriculture policy that had the following significant features.

- Public, private, local and foreign companies can invest in corporate dairy farming
- Except registration from Board of Investment no government sanction is required to undertake corporate farming
- Liberal credit will be available
- Prospective investor will determine the size of corporate farm
- Land for corporate farm can be purchased or leased for 50 years from government that will be further extendable for 49 years
- Permission for 100% foreign equity has been given
- Agriculture income tax will be applicable
- Dividends will be exempted from tax
- Appropriate labor laws will be developed for this sector due to special circumstances of Agriculture sector
- Import of Agriculture machinery and equipments will be exempted from sales tax and zero rated custom duty will be charged
- On transfer of land duty will be exempted.

Different corporate groups have started to think about investment on livestock policy due to dwindling profits in textile and other industries and greater demands of dairy products. Al-Tahur dairy farm, JK dairies, and Sapphire dairies are the initial leader in this field, while many others are thinking about this (Afzal, 2008). Concept of making corporate farms regarding dairy value chain by involving existing and new stakeholders will help to facilitate healthy competition and in meeting required standards. By this process production of producers will be improved at one end and on the other end consumers will get access to variety of quality products. This process along with other government initiatives would facilitate in formation of strong dairy industry in long run that would involve not only informal dairy producers but also integrate value added industry. Objective of getting self-reliance in dairy production and to produce exportable surplus will also be possible. Corporate farming cannot be opted in Pakistan unless better feed and animal stocks are available and credit providers are willing to provide credit to the area. In fact, research should be conducted in the dairy sector before we can understand it fully, and then a comprehensive set of measures to facilitate its development will be required.

38.3.4.2 Improvement in Cold Chain through Dairy Hub Concept

Dairy hub is a concept which has basis in the dairy development programs. It was initiated by Tetra Pak, Pakistan in 2009 and implemented through large milk processors such as Nestlé and Engro Foods. In 2009, the first dairy hub was established for Engro foods in Kassowal, District Sahiwal and second dairy hub was established in Mian Channu in the same year for Nestle. This community development program aims to organize farmers into cooperatives with focus on provision of livestock development services. It provides consultancy, training, and veterinary services to herds of rural farmers of 20 villages situated in 15–20 km radius that makes one dairy hub. Provision of on farm consultancy services and mechanized milk machines by concerned field service officer of the area is also included in this program. Implementation of Dairy Hub concept will organize dairy sector of Pakistan through coordinated and active participation of dairy processors and public sector with a sense of ownership. The purpose of this program is to implement the "one herd-one farm" concept within a particular region to focus on efficient milk collection and chilling mechanism. This process not only improves the quality and quantity of milk but also enables the documentation of the rural economy's cattle, their genetics, milk production, and medical treatment. It will also increase the traceability and application of Pakistan Quality Standards and envisage a substantial growth potential for UHT milk. Areas where dairy hubs have been established have shown improvement in milk supplied by 70%–80%, increase in milk by 3000 Liters/day and in average daily yield by over 20%/animal, which decreased the milk cost by 2%. (Mumtaz et al., 2011).

38.3.4.3 Improvement in Value Addition

Integration of the value chain in the dairy sector will facilitate provision of opportunities to different stakeholders and improvement in value addition of dairy products. Having higher milk yield and inherited edge of milk production than regional competitors will offer great opportunities to produce high yield of end products. Current scenario of dairy sector does not promise much to meet existing and anticipated domestic demands unless certain major reforms are introduced. Involvement of private and government sectors can anticipate improvement of the current situation of value addition. For this purpose complete replacement of low end or obsolete technology is required along with mobilization of financial resources by agriculture credit schemes to facilitate new entrants in dairy business. Knowledge of health and animal safety standards, modern farming and processing technologies and avenues of export businesses should also be worked out. Training of different stakeholders is also required to get the desired quality and to fulfill standards. Some ways to facilitate milk transformation into value added products are pasteurization/UHT (yogurt, cheese, butter milk and sour cream) and homogenization (skimmed milk, flavored milk, fortified milk and whipped cream). Other advanced methods include use of membrane technologies.

38.3.4.4 Integrated on Farm Processing

On-farm processing and marketing of milk and milk products is a segment of dairy industry that has recently received significant consideration by farm families throughout the world (Iqbal et al., 2008). It may prove a good solution to the problems of cold chain facilities, adulteration and transportation. Moreover, it will also help to create job opportunities for rural families in order to raise their living standards and farmer customer relationship along with reduction of selling price of milk. Examples of successful collaborations between farm and industry can be seen in other countries such as Almarai in Saudi Arabia, and milk processing plant at the McCarty farms in Kansas, the United States. Almarai is largest dairy farming and processing unit in Middle- East. In 1990, company restructured its five decentralized industries in to one large industry and 10 dispersed farms in to four farms in Al-Kharj in the central region. With 60% share of total dairy products for Saudi market, the company has gained a household name in Saudi Arabia (Pakissan, 2016). According to Business Wire (2012), in the United States, the Dannon Company (Nestlé) has established a milk condensing plant at McCarty farm in Kansas which will reduce shipping cost of milk and reuse water removed during condensing. It is only plant in the United States that will condense milk before processing it into yoghurt.

38.3.4.5 Strict Implementation of Laws and Regulations

Each country has standards to ensure that any product coming into the market is safe for human consumption and will not a cause harm to the health of animals, plants, and human beings. The World Trade Organization (WTO) agreement on Sanitary and Phytosanitary Measures (SPS) has provisions for member countries to apply standards on imported products. Therefore, in the country of import all products are subjected to quality standards. Pakistan's dairy industry is weak in terms of compliance to laws and regulations, due to the dominance of dispersed and small farms. Fulfillment of international standards is required for the dairy sector to produce internationally compliant products. Pakistan has potential to export dairy products to other countries if quality issues are addressed.

To check bacterial counts and milk and milk product adulterations, international accreditation of food testing laboratories (fulfillment of criteria of ISO/IEC 17025–2005) should be undertaken and food inspection services should be promoted. Current food inspection laboratories working under the provincial government and Pakistan Standards and Quality Control Authority (PSQCA) should be upgraded to improve their capacity. Laws should be made to deal with milk adulteration along with implementation of penalties in cases of violations. The Pure Food Ordinance should be amended to include legislations for all dairy items. For production of value added products compliance with international standards as HACCP should be made compulsory. Quality of all processed dairy products should be regulated by making a mandatory requirement for ingredient labeling, and they should be open for testing. Development, implementation, and enforcement of laws to control the quality of value added products may improve the current situation along with production of exportable surplus.

38.3.4.6 Restrictions on Import of Powdered Milk

Large amounts of powdered milk have been imported in Pakistan in the previous few years. Dumping of milk powder is due to low duty regime on its import (20pc from SAARC countries and 25pc from rest of the world). Other countries, such as India impose 68pc duty on import of powdered milk and Turkey 180 pc to protect their farmers and to become self-sufficient (The Nation, 2016). It's worth mentioning here that due to influx of whey milk powder (WMP) and skimmed milk powder (SMP), Pakistan's dairy sector is standing at the verge of destruction. Use of SMP and WMP in the dairy processing industry has deprived the local farmers from getting the right prices for their commodities resulting in a discouraging wave for the farming sector. To control import of SMP and WMP, heavy duties (100%–150%) should be imposed on its import to uplift of the dairy and raising the morale of our dairy farmers. We should strive to become self-sufficient in milk production instead of becoming a dumping ground for foreign exports over the globe.

38.3.4.7 Looking for Opportunities in Trade Market

Market analysis indicates that to penetrate in EU market is not easy for Pakistan as its import is subjected to quality evaluation by US custom and Border protection services and Ministry of Agriculture. Products to be exported from Pakistan should meet requirements of FDA, animal and plant health inspection services and should have a certificate approved by FDA before entering the markets of EU. Moreover some products are tested in EU countries, which further enhances export costs. Trade with Arabian Gulf countries is relatively easy as they require certificates of origin and health and safety accreditations for import of dairy products, but there are no requirements for traceability issues. Flexibility in food safety standards can facilitate trade in near future. However, Pakistan might not be able to immediately penetrate the Gulf region due to very strict trade regulations by intra Gulf Cooperation Council (GCC). Trade policy (2009–2012) proposed by Ministry of Commerce proposed that Pakistan being among top milk producing country can become a future exporter of milk products to countries like South East Asia, Middle East and China have increasing demand for food due to increasing population, subjected to implementation of internationally adopted standards in milk supply chain.

Trade with Arabian Gulf countries can be possible if the current hygiene and quality situation is improved. For this purpose, the federal government of Pakistan can utilize its Trade Development Authority to conduct awareness seminars in different parts of country with the collaboration of dairy traders, local chambers of commerce, exporter associations, and PDDB to create awareness among farmers regarding production of quality surplus by using appropriate milking methods. It may reduce wastage and improve hygiene standards. Creation of awareness among farmers is also important to adopt appropriate supply chains in order to reduce their exploitation by local milkmen. Workers of value added industries should also be educated on quality matters so that Pakistan can export in regulated and enforced international trade markets.

38.3.5 Conclusion and Future Prospects

Pakistan's dairy sector has much capacity for production of value added products due to high yield of milk. Nevertheless, the lack of better facilities, advanced technologies, improper management, absence of government policies and regulations are creating barriers in the development of a dairy industry. Private sectors as well as government bodies should come forward to improve dairy sector by making investment that may produce high return in near future if once import of dairy products started with other countries. For this purpose the quality of products needs to be optimized by using corporate farms, dairy hubs ,and integrated on farm processing that will allow the use of mechanical methods of milking and chilling facilities to reduce milk adulteration and contamination. Some private players such as Nestlé, Engro foods, and others have come in the field by forming their commercial farms and others are thinking to do so, which may prove to be a positive initiative for future developments.

REFERENCES

Afzal, M. 2008. Corporate dairy farming in Pakistan-Is there a future? *Pak. J. Agric. Sci.* 45:250–253.

Ali, M.R., T. Mahmood and Umm-e-Zia. 2011. *Dairy Development in Pakistan.* Food and Agriculture Organization (FAO) of the United Nations, Rome, Italy.

Al-Okbi, S.Y., A.M.S. Hussein, I.M. Hamed, D.A. Mohamed and A.M. Helal. 2014. Chemical, rheological, sensorial and functional properties of gelatinized corn- rice bran flour composite corn flakes and Tortilla chips. *J. Food Process. Preser.* 38: 83–89.

Aslam, H.K.W., M.I.U. Raheem, R. Ramzan, A. Shakeel, M. Shoaib and H.A. Sakandar. 2014. Utilization of mango waste material (peel, kernel) to enhance dietary fiber content and antioxidant properties of biscuit. *J. Glob. Innov. Agric. Soc. Sci.* 2: 76–81.

Awan, A., M. Ali, F. Iqbal, R. Iqbal and M. Naseer. 2014. A study on chemical composition and detection of chemical adulteration in tetra pack milk samples commercially available in Multan. *Pak. J. Pharm. Sci.* 27:183–186.

Bihn, E. A. and R.B. Gravani. 2006. Role of good agricultural practices in fruit and vegetable safety. In: Matthews, K.R., editor. *Microbiology of Fresh Produce.* ASM Press, Washington DC.

Boye, J., F. Zare and A. Pletch. 2010. Pulse proteins: Processing, characterization, functional properties and applications in food and feed. *Food Res. Int.* 43: 414–431.

Burki, A.A. and M.A. Khan. 2011. Formal participation in milk supply chain and technical inefficiency of smallholder dairy farms in Pakistan. *Pak. Dev. Rev.* 50: 63–81.

Burki, A.A., M.A. Khan and F. Bari. 2004. The state of Pakistan's dairy sector: An assessment. *Pak. Dev. Rev.* 43: 149–174.

Business Wire. 2012. The Dannon Company and McCarty Family Farms Inaugurate New Operations for Improved Sustainable Development. Available at: http://www.businesswire.com/news/home/20120613006274/en/DannonCompany-McCarty-Family-Farms-Inaugurate-Operations. Assessed on: June 13, 2012.

Chatha, S.A.S., A.I. Hussain, M. Zubair and M.K. Khosa. 2011. Analytical characterization of rice (oryza sativa) bran and bran oil from different agro-ecological regions. *Pak. J. Agri. Sci.* 48: 243–249.

De-Frias, V., O. Varela, J.J. Oropeza, B. Bisiacchi and A. Alvarez. 2010. Effects of prenatal protein malnutrition on the electrical cerebral activity during development. *Neurosci. Lett.* 482: 203–207.

Dyner, L., S.R., Drago, A. Pinerro, H. Sanchez, R. Stevens and M. Rashid. 2008. Gluten-free and regular foods: A cost comparison. *Can. J. Diet. Pract. Res.* 69: 147–150.

Economic Survey. 2014–2015. *Govt. of Pakistan, Finance Division.* Economic Advisor Wing, Islamabad.

Economic Survey of Pakistan. 2015–2016. *Agriculture. Ministry of Food and Agriculture (Economic Wing).* Govt. of Pakistan, Islamabad, Pakistan.

Faheem-ul-Islam and Qureshi. 2012. National economic development role of dairy sector. Available at: http://www.uvas.edu.pk/doc/society-club/dairy-club/seminars/roadmap/session-1/national_economic.pdf.

FAO (Food and Agriculture Organization). 2002. *FAO/WHO Global Forum of Food Safety Regulators.* Food and Agriculture Organization of the United Nations, Marrakech, Morocco.

FDA. 1998. *US Food and Drug Administration.* Guide to Minimize Microbial Food Safety Hazards for Fresh Fruits and Vegetables, US Department of Health and Human Services, Food and Drug Administration and Center for Food Safety and Applied Nutrition, Washington, DC.

GOP (Government of Pakistan). 2010. *Agricultural Statistics of Pakistan 2009–2010. Ministry of Food and Agriculture (Economic Wing).* Govt. of Pakistan, Islamabad, Pakistan.

Haase, N.U., K.H. Grothe, B. Mattaus, K. Vosmann and M.G. Lindhauer. 2012. Acrylamide formation and antioxidant level in biscuits related to recipe and baking. *Food Add. Contam. A.* 29: 1230–1238.

Hefnawy, T.M.H., G.A. El-Shourbagy and M.F. Ramadan. 2012. Impact of adding chickpea (Cicer arietinum L.) flour to wheat flour on the rheological properties of toast bread. *Int. Food Res. J.* 19: 521–525.

Hemme, T. and J. Otte. 2010. *Status and Prospects for Small Holder Milk Production*. Food and Agriculture Organization (FAO) of the United Nations, Rome, Italy.

Imran, M., M.S. Butt, F.M. Anjum and J.I. Sultan. 2013. Chemical profiling of different mango peel varieties. *Pak J. Nutr.* 12: 934–942.

Iqbal, A., M.I. Mustafa, H. Nawaz and M. Tariq. 2008. Milk marketing and value chain constraints. *Pak. J. Agric. Sci.* 45(2): 195–200.

Ismail, T., S. Akhtar, M. Riaz, A. Hameed, K. Afzal and A.S. Sheikh. 2016. Oxidative and microbial stability of pomegranate peel extracts and bagasse supplemented cookies. *Journal of Food Quality*, 39(6): 658–668.

Kader, A.A. and R.S. Rolle. 2004. The role of post-harvest management in assuring the quality and safety of horticultural produce. *Rome, FAO Agric. Serv. Bull.* 152: 50–51.

Khalil, A.A., M.R. Khan, M.A. Shabbir and K.U. Rahman. 2017a. Comparison of antioxidative potential and punicalagin content of omegranate peels. *JAPS, Journal of Animal and Plant Sciences*, 27(2): 522–527.

Khalil, A.A., U. ur Rahman, M.R. Khan, A. Sahar, T. Mehmood and M. Khan. 2017b. Essential oil eugenol: Sources, extraction techniques and nutraceutical perspectives. *RSC Advances*, 7(52): 32669–32681.

Khurshedi, N. 2012. Milk Industry of Pakistan. Available at: http://www.pakistaneconomist.com/pagesearch/Search-Engine2012/S.E748.php.

Mamat, H., M.O.A. Hardan and S.E. Hill. 2010. Physicochemical properties of commercial semi-sweet biscuit. *Food Chem.* 121: 1029–1038.

Kishor, K. and R. Thakur. 2015. Analysis of milk adulteration using MID-IR spectroscopy. *IJRITCC*. 3(10): 5890–5895.

Mumtaz, M.K., M.A. Hemani, N. Hameed and S., Gulzar. 2011. *Dairy Hub: A Community Dairy Development Programme*. International Growth Centre, London School of Economic and Political Science, Houghton Street, London.

Pakissan. 2016. Livestock: World Best Dairy Farming. Available at: http://www.pakissan.com/english/allabout/livestockfisheries/themes/classic/world.best.dairy.farming.shtml.

Pasha, I., S. Rashid, F.M. Anjum, M.T. Sultan, M.M.N. Qayyum and F. Saeed. 2011. Quality evaluation of wheat-mungbean flour blends and their utilization in baked products. *Pak. J. Nutr.* 10: 388–392.

PDDC. 2006. *The White Revolution-Doodh Dariya*. Pakistan Dairy Development Company, State Cement Corporation Building, Township Kot Lakhpat, Lahore, Pakistan.

Pre-feasibility study for dairy processing plants. 2006. Prepared by Mascon Associates (Pvt. Limited). *Commissioned by Employment and Research Section, Planning and Development Division, Islamabad, Pakistan*. Punjab Food Authority, Annual Report. 2013–2014.

Rasheed, D.M. 2014. Livestock development for Socio-economic uplift of rural Pakistan. Available at: http://hhrd.pk/lddc2014/wp-content/uploads/2014/12/Dr.-Rasheed-141216-presentation-in-LDDC-2014.pdf.

Reardon, T. and C.B. Barrett. 2000. Agro-industrialization, globalization and international development: An overview of issues, patters, and determinants. *Agric. Econ.* 23(3): 195–205.

Riboli, E. and T. Norat. 2003. Epidemiologic evidence of the protective effect of fruit and vegetables on cancer risk. *Am. J. Clin. Nutr.* 78: 559–569.

Roy, P., T. Orikasa, H. Okadome, N. Nakamura and T. Shiina. 2011. Processing conditions, rice properties, health and environment. *Int. J. Environ. Res. Public Health* 8: 1957–1976.

Sarwar, M., M.A. Khan, Mahr-Un-Nisa and Z. Iqbal. 2002. Dairy industry in Pakistan: A scenario. *Int. J. Agric. Biol.* 4: 420–428.

Tariq, M.I. 2005. Leachinganddegradation of cotton pesticides on different soil series of cotton growing areas of Punjab, Pakistan in Lysimeters. *PhD Thesis*, Univ. Punjab, Lahore, Pakistan.

The Nation. 2016. Available at: http://nation.com.pk/lahore/25-Apr-2016/import-of-dry-milk-harming-local-farmers-interests-uvas-vc. Assessed on April 25, 2016.

TRTA. 2013. Enhancing Dairy Sector Export Competitiveness in Pakistan. Study conducted by European Union funded Trade Related Technical Assistance (TRTA-II) programme.

Tuncel, N.B., N. Yilmaz, H. Kocabiyik and A. Uygur. 2014. The effect of infrared stabilized rice bran substitution on physicochemical and sensory properties of pan bread: Part I. *J. Cereal Sci.* 59: 155–161.

USDA. 2015. *Dairy: World Markets and Trends*. Foreign Agricultural services, United State Department of Agriculture.

WHO (World Health Organization). 1990. *Public Health Impact of Pesticides Used in Agriculture*. World Health Organization, Geneva, Switzerland.

Younas, M. 2013. The Dairy Value Chain: A promoter of development and employment in Pakistan. ICDD, working paper no. 9. University of Kassel, Germany. Available at: https://www.unikassel.de/einrichtungen/fileadmin/datas/einrichtungen/icdd/Publications/ICDD_WP9_Younas_06.pdf.

Index

B

C

E

G

H

I

J

K

N

O

P

T

U

V

W